41st European Photovoltaic Solar Energy Conference and Exhibition (EU PVSEC 2024)

Vienna, Austria
23-27 September 2024

Volume 1 of 3

Editors:

G. C. Eder   **R. Kenny**
J. Bergmiller   **J. De Gregorio**

ISBN: 979-8-3313-1538-2

Printed from e-media with permission by:

Curran Associates, Inc.
57 Morehouse Lane
Red Hook, NY 12571

Some format issues inherent in the e-media version may also appear in this print version.

Copyright© (2024) by WIP – Renewable Energies
All rights reserved.

Printed with permission by Curran Associates, Inc. (2025)

For permission requests, please contact WIP – Renewable Energies
at the address below.

WIP – Renewable Energies
Sylvensteinstr. 2
81369 Munchen
Germany

Phone: +49 89 72012735
Fax: +49 89 72012791

wip@wip-munich.de

Additional copies of this publication are available from:

Curran Associates, Inc.
57 Morehouse Lane
Red Hook, NY 12571 USA
Phone: 845-758-0400
Fax: 845-758-2633
Email: curran@proceedings.com
Web: www.proceedings.com

TABLE OF CONTENTS OF EU PVSEC 2024 PROCEEDINGS PAPERS

EU PVSEC 2024 Committees

Subject Index

Foreword

Oral SESSION 1AO.4 Silicon Material for Solar Cells: Growth, Stability and Reuse

1AO.4.3 Impact of High-Temperature Processing Steps on the Long-Term Stability in n-Type FZ Silicon   1

Melanie Mehler[1], Nicolas Weinert[1], Nicole Aßmann[2], Axel Herguth[1], Giso Hahn[1], Fabian Geml[1]
[1] University of Konstanz, Konstanz, Germany; [2] University of Oslo, Oslo, Norway

1AO.4.5 LeTID in Industrial Ga-doped Cz-Si with Melt Recharging   7

Joshua Kamphues[1], Axel Herguth[1], Juri Miech[1], Xueqi Bai[2], Yichun Wang[2], Giso Hahn[1], Fabian Geml[1]
[1] University of Konstanz, Constance, Germany; [2] LONGI Green Energy Technology, Xi'An, China

Oral SESSION 1AO.5 Processes for Highly Efficient Si Solar Cells

1AO.5.2 Wet-Chemically Grown Interfacial Oxide for Passivating Contacts Fabricated with an Industrial Inline Processing System   13

Byungsul Min[1], Philipp Noack[2], Bianca Wattenberg[2], Torsten Dippell[2], Henning Schulte-Huxel[1], Robby Peibst[1], Rolf Brendel[1]
[1] ISFH, Emmerthal, Germany; [2] Singulus Technologies, Kahl am Main, Germany

1AO.5.4 Local p+ Poly-Si Passivating Contacts Realized by Direct FlexTrail Printing of Boron Ink and Selective Alkaline Etching for High Efficiency TOPCon based Solar Cells   19

Berkay Uygun[1], Sven Kluska[2], Jana Isabelle Polzin[2], Jörg Schube[2], Mike Jahn[2], Katrin Krieg[2], Raşit Turan[1], Hisham Nasser[1]
[1] ODTÜ-GÜNAM, Ankara, Türkiye; [2] Fraunhofer ISE, Freiburg, Germany

1AO.5.5 Phosphorus- and Boron-doped Poly-Si/Siox Passivating Contacts via Inkjet Printing   25

Jiali Wang[1], Thein N. Truong[1], Jinlei Ren[2], Marie Adier[2], Laura Creon[2], Paula Peres[2], Rene Chemnitzer[2], Pierre-Yves Corre[2], Zhuofeng Li[1], Hieu T. Nguyen[1], Josua Stuckelberger[3], Daniel Macdonald[1], AnYao Liu[1], Sieu Pheng Phang[1]
[1] ANU, Canberra, Australia; [2] CAMECA, Gennevilliers, France; [3] ANU, Canberra, China

Oral SESSION 1AO.6 Highly Efficient Si Solar Cells

1AO.6.5 >24% Efficient Tunnel Back Contacted polyZEBRA Solar Cells   30

Jonathan Linke[1], Christoph Peter[1], Jan Hoß[1], Vaibhav Kuruganti[1], Saman Sharbaf Kalaghichi[1], Valentin Mihailetchi[1], Jan Lossen[1], Florian Buchholz[1]
[1] ISC Konstanz, Konstanz, Germany

1AO.6.6 Optimized Ga-Doped Cz Wafers for POLO IBC Solar Cells with High   35
Efficiency and Minimal LeTID Degradation

Thorsten Dullweber[1], Verena Mertens[1], Michael Winter[1], Sabrina Schimanke[1], M. Ripke[1], Silke Dorn[1], Yevgeniya Larionova[1], Gerrit Lange[1], Karsten Bothe[1], Jan Schmidt[1], Rolf Brendel[1], Arne K. Dahle[2], Özlem Coskun[3], Nesrin Töre Sen[3]
[1] ISFH, Emmerthal, Germany; [2] NorSun, Årdalstangen, Norway; [3] Kalyon PV, Ankara, Türkiye

Oral SESSION 1BO.1 Silicon Bottom Cells for Tandem Photovoltaics | Dielectric Layer Related Defect Characterisation

1BO.1.1 Towards TOPCon Based Bottom Cells: Current Challenges and Perspectives   40

Mario Hanser[1], Henning Nagel[1], Johannes Gry[1], Jana Polzin[1], Armin Richter[1], Jan Benick[1], Martin Bivour[1], Martin Hermle[1], Stefan Glunz[1]
[1] Fraunhofer ISE, Freiburg, Germany

1BO.1.3 Review on In-Free Recombination Junction Approaches for Two-Terminal   50
Silicon/Perovskite Tandem Solar Cells

Pia Vasquez[1], Amanda Merino Leiva[1], Perrine Carroy[1], Batiste Marteau[1], Thibaut Desrues[1], Nathalie Nguyen[1], Muriel Matheron[1], Sofia Chozas[2], Federico Ventosinos[2], Henk J. Bolink[2], Delfina Muñoz[1]
[1] CEA-INES, Le Bourget-du-Lac, France; [2] University of Valencia, Valencia, Spain

Oral SESSION 1BO.2 Advanced Silicon Solar Cell Characterisation in Laboratory and Production

1BO.2.1 Identification of Performance-Relevant Optically Detected Defects by   53
Correlative Data Analysis in Solar Cell Production

Manuel Meusel[1], A. Starke[1], Marko Turek[1]
[1] Fraunhofer CSP, Halle (Saale), Germany

1BO.2.3 Integrated Inline Characterisation Techniques for Improved Silicon   57
Heterojunction Solar Cell Production

Christian Diestel[1], Saravana Kumar[1], Alexandra Wörnhör[1], Daniel Burkhardt[1], Nico Wöhrle[1], Sebastian Pingel[1], Matthias Demant[1], Jonas Haunschild[1], Stefan Rein[1]
[1] Fraunhofer ISE, Freiburg, Germany

1BO.2.5 Expert Knowledge, AI, and Simulation: Integrative Approaches for Quality   63
Assurance in Solar Cell Manufacturing

Matthias Demant[1], Alexandra Woernhoer[1], Philipp Kunze[1], Wilkin Woehler[1], Julian Behrendt[1], Leslie Lydia Kurumundayil[1], Johannes Greulich[1], Andreas Fell[1], Stefan Rein[1]
[1] Fraunhofer ISE, Freiburg, Germany

Oral SESSION 1BO.3 Optimised Processes for the Manufacturing of TOPCon Solar Cells

1BO.3.1 Exploring the Impact and Challenges of Using Emerging Wafer Sizes in PV Manufacturing 76

Julian Reichle[1], Hardik Gohil[1], Mehul Raval[1], Avinash Kumar[1], Wolfgang Jooß[1], Peter Fath[1]
[1] RCT Solutions, Konstanz, Germany

1BO.3.2 A Horizontal Double-Sided Copper Metallization Technology Designed for Solar Cell Mass-Production 86

Lu Wang[1], Yusen Qin[1], Meilin Peng[2], Meixian Huang[2], ZhiPeng Liu[2], Yibo Lu[1], Guohua Zhou[1], Jingjia Ji[1]
[1] Jiangsu Xianghuan Technology, Wuxi, China; [2] Jiangnan University, Wuxi, China

1BO.3.6 Comprehensive Optimization of Glass Stencil Printing, Demonstrating Ultrafine Metal Fingers Below 10 μm 89

Tadeo Schweigstill[1], Niko Mielich[1], Aaron Vogt[2], Malte Schulz-Ruhtenberg[2], Jonas D. Huyeng[1], Florian Clement[1]
[1] Fraunhofer ISE, Freiburg, Germany; [2] LPKF Laser & Electronics, Garbsen, Germany

Oral SESSION 1BO.4 New Concepts for the Manufacturing of IBC and HJT Solar Cells

1BO.4.3 IBC4EU: First Results of Industrialization of Low Cost, High Efficiency IBC Technology 101

Florian Buchholz[1], Daniel Tune[1], Tobias Messmer[1], Jonathan Linke[1], Manjunath Prasad[1], Valentin Mihailetchi[1], Juras Ulbikas[2], Arne Dahle[3], Martijn Meereboer[4], Francesca Fabris[5], Erik Eikelboom[5], Tom Borgers[6], Rik Van Dyck[6], Filip Duerinckx[7], Hariharsudan Sivaramakrishnan[6], Samuel Harrison[8], Josco Kester[9], Nicolas Guillevin[10], Jan Kroon[9], Verena Mertens[11], Thorsten Dullweber[11], Ofer Shochet[12], Isaac Rosen[12], Ingo Röver[13], Wolfram Palitzsch[13], Yasmin Zaror[14], Johnnes Stierstorfer[14], Aurimas Radzevicius[15], Povilas Lukinskas[15], Julius Denafas[15], Tuomas Vanhanen[16], Tuukka Savisalo[16], Maximilian Pospischil[17], Marian Breitenbücher[17], Özlem Coşkun[18], Melodie de l'Epine[19], Philippe Macé[19]
[1] ISC Konstanz, Konstanz, Germany; [2] ProTechnologies, Vilnius, Lithuania; [3] Norsun, Årdalstangen, Norway; [4] Energyra, Westknollendam, The Netherlands; [5] Futurasun, Citadella, Italy; [6] imec, Genk, Belgium; [7] imec, Leuven, Belgium; [8] CEA INES, Le Bourget-du-Lac, France; [9] TNO, Amsterdam, The Netherlands; [10] TNO, Petten, The Netherlands; [11] ISFH, Hamelin, Germany; [12] Copprint, Jerusalem, Israel; [13] LuxChemTech, Freiberg, Germany; [14] WIP Renewable Energies, Munich, Germany; [15] UAB Valoe Cells, Vilnius,

Lithuania; [16] Valoe Cells, Mikkeli, Finland; [17] Highline, Freiburg, Germany; [18] Kalyon PV, Ankara, Türkiye; [19] Becquerel Institute, Brussels, Belgium

1BO.4.4 Self-Aligned Phase Separation for IBC Cells Using PVD Polysilicon   106

Erik Hoffmann[1], Geoffrey Gregory[1], Massimo Centazzo[1], Muhammad Khan[1], Nabeel Khan[1], Verena Mertens[2], Philip Jäger[2], Sarah Spätlich[2], Ulrike Baumann[2], Thorsten Dullweber[2]
[1] EnPV, Karlsruhe, Germany; [2] ISFH, Hamelin, Germany

1BO.4.5 Gas Phase, Selective Etching of Poly-Silicon for Layer Patterning   110

Laurent Clochard[1], Mingzhe Yu[2], Ruy Sebastian Bonilla[2], Paul Tierney[3], James Wright[3], Fiacre Rougieux[4], Yalun Cai[4]
[1] Nines Photovoltaics, Dublin, Ireland; [2] University of Oxford, Oxford, United Kingdom; [3] TUD, Dublin, Ireland; [4] UNSW, Sydney, Australia

1BO.4.6 Investigation of Ag-Reduction on Silicon Heterojunction Solar Cells with Different Approaches   113

Yu Wu[1], Eric J. Kossen[1], Astrid Gutjahr[1], M. Bruggeman[2], L.J. (Bart) Geerligs[1]
[1] TNO, Petten, The Netherlands; [2] TNO, Delft, The Netherlands

Visual SESSION 1BV.5 Silicon Material: Growth, Defects and Recycling | Manufacturing of Solar Cells and Related Tools & Processes

1BV.5.5 Investigation of Oxygen and Carbon Impurities in Mono-Silicon Wafers During Rapid Thermal Annealing   122

Nurhayat Yıldırım[1], Sertaç Eroğlu[2], Merve Çorak[1]
[1] Kalyon PV, Ankara, Türkiye; [2] Eskişehir Osmangazi University, Eskişehir, Türkiye

1BV.5.7 Increasing the Productivity of the Czochralski Process Applying Machine Learning   125

Frank Mosel[1], Lukas Kulhavy[2], Dorra Baccar[2]
[1] PVA Crystal Growing Sytems, Wettenberg, Germany; [2] THM, Friedberg, Germany

1BV.5.8 Thermal Deactivation of Boron-Oxygen Defects in Compensated n-Type Silicon   132

Rune Søndenå[1], Per-Anders Hansen[1], Bent Thomassen[1], Øyvind Mjøs[2], Tyke Naas[2]
[1] IFE, Kjeller, Norway; [2] REC Solar, Kristiansand, Norway

1BV.5.15 Highest Throughput Laser Processing for Thin Plated Contacts   135

Eduardo Alvarez-Brito[1], René Haberstroh[1], Georg Hoppe[1], Keming Du[2], Florian Roessler[3], Andreas A. Brand[1], Sven Kluska[1], Fabian Meyer[1], Jale Schneider[1], Jan Nekarda[1]
[1] Fraunhofer ISE, Freiburg, Germany; [2] EdgeWave, Würselen, Germany; [3] Moewe Optical Solutions, Mittweida, Germany

1BV.5.16 Enhancement of Photocurrent Generation in Amorphous Silicon Heterojunction (SHJ) Solar Cells through the Integration of Plasmonic Nanoparticles   139

Brahim Aïssa[1], Alessandro Sinopoli[1]
[1] QEERI, Doha, Qatar

1BV.5.21 Impact of Optimization for Mass Production PERC Solar Cell with Efficiency above 23% 144

Cheng-Wen Kuo[1], Ta-Ming Kuan[1], Yung-Chih Li[1], Chun-Wei Lee[1], Wei-Lo Chueh[1], Li-Guo Wu[1], Shih-Chieh Lin[1], Cheng-Yeh Yu[1]
[1] TSEC, Hsinchu, Taiwan

1BV.5.24 The Impact of Conductive Paste Composition on the LECO Process for TOPCon Solar Cells 148

Chun-Ping Lin[1], Chih-Jeng Huang[1], Han-Chen Chang[1], Sung-Yu Chen[1], Bang-Hao Wu[2], Cheng-Liang Cheng[2], Ying-Yuan Huang[3]
[1] ITRI, Tainan, Taiwan; [2] TeraSolar Energy Material, Miaoli, Taiwan; [3] National Cheng Kung University, Tainan, Taiwan

1BV.5.28 Realistic Estimation of Industrial TOPCon Cell Efficiency 149

Mehul Raval[1], Pirmin Preis[2], Lejo Joseph Koduvelikulathu[2], Gourab Das[1], Wolfgang Jooß[1]
[1] RCT Solutions, Konstanz, Germany; [2] ISC Konstanz, Konstanz, Germany

1BV.5.30 Optimizing the Mechanical Adhesion Properties of Plated Contacts of i-TOPCon Solar Cells 155

Christian Schmiga[1], Abdelaziz Boudellioua[1], René Haberstroh[1], Jonas Eckert[1], Sven Kluska[1], Florian Clement[1]
[1] Fraunhofer ISE, Freiburg, Germany

1BV.5.31 Addressing Edge Recombination Losses in Shingle Cells by Holistic Optimization of the Process Sequence 158

Alexander Göbel[1], Elmar Lohmüller[1], Dirk Wagenmann[1], Norbert Kohn[1], Marc Hofmann[1], Jonas D. Huyeng[1], Ralf Preu[1]
[1] Fraunhofer ISE, Freiburg, Germany

1BV.5.34 Characterization of TiOx as Electron Selective Contact Using Low-Temperature Oxidation Process via High-Pressure Sputtering 163

Franciso José Pérez Zenteno[1], Sebastian Duarte[1], Rafael Benítez-Fernandez[1], G. Godoy-Perez[1], Ignacio Torres[2], Rocío Barrio[2], Lars Rebohle[3], D. Caudevilla[1], Sari Algaidy[4], Rodgar García-Hernansanz[1], J. Olea[1], D. Pastor[1], Alvaro Del Prado[1], Eric García-Hemme[1], E. San Andrés[1]
[1] Complutense University of Madrid, Madrid, Spain; [2] CIEMAT, Madrid, Spain; [3] HZDR, Dresden, Germany; [4] Polytechnical University of Madrid, Madrid, Spain

Visual SESSION 1CV.2 Processing & Characterisation of Crystalline Si based Solar Cells | Silicon Bottom Cells for Tandem Photovoltaics | Advances in Silicon Solar Cells Characterisation and Simulation

1CV.2.2 Approaches for Reducing Metallization-Induced Losses and Cost in Industrial TOPCon Solar Cells 164

Sebastian Mack[1], Daniel Ourinson[1], Marius Messmer[1], Christopher Tessmann[2], Katrin Krieg[1], René Haberstroh[1], Sven Kluska[1], Jonas Huyeng[1], Johannes Greulich[1], Andreas Wolf[1], Florian Clement[1]
[1] Fraunhofer ISE, Freiburg, Germany; [2] Fraunhofer IAF, Freiburg, Germany

1CV.2.5 Unveiling the Synergy of Nanowires and PEDOT:PSS for Silicon Solar Cell Fabrication and Leading to Mechanical Flexibility 169

Deepak Sharma[1], Ruchi Kumari Sharma[2], Arman Ahnood[1], Sanjay Kumar Srivastava[2]
[1] *RMIT University, Melbourne, Australia;* [2] *AcSIR, Ghaziabad, India*

1CV.2.7 Polysilicon Passivation - Tunneling Oxide Routes and Annealing Conditions Effect on Passivation 172

Per-Anders Hansen[1], Junjie Zhu[1], Rune Søndenå[1]
[1] *IFE, Kjeller, Norway*

1CV.2.10 Selective p+ Poly-Si Fingers for TOPCon Front Contact Passivation 175

Jan Hoß[1], Saman Sharbaf Kalaghichi[1], Mertcan Comak[1], Pirmin Preis[1], Jan Lossen[1], Jonathan Linke[1], Lejo Koduvelikulathu[1], Florian Buchholz[1]
[1] *ISC Konstanz, Konstanz, Germany*

1CV.2.16 Review and Highlights of More Than 30 Years Research on Ever Improving Technology for PERC Solar Cells at Fraunhofer ISE 176

Elmar Lohmüller[1], Sabrina Lohmüller[1], Pierre Saint-Cast[1], Johannes Greulich[1], Stefan Glunz[1], Ralf Preu[1]
[1] *Fraunhofer ISE, Freiburg, Germany*

1CV.2.17 Investigating Interfacial Phenomena in Copper-Covered, n-Type Polysilicon-Based Contacts by Electron Microscopy 182

Reyu Sakakibara[1], Agata Lachowicz[2], Julien Hurni[1], Christophe Allebé[2], Bertrand Paviet-Salomon[2], Franz-Josef Haug[1], Christophe Ballif[1], Aïcha Hessler-Wyser[1], Audrey Morisset[2]
[1] *EPFL, Neuchâtel, Switzerland;* [2] *CSEM, Neuchâtel, Switzerland*

1CV.2.19 Robustness of Electrical Quality of Ion Implanted Black Silicon Emitters: Comparison between different Ion Implantation Service Providers 189

Olga Morozova[1], Kexun Chen[1], Behrad Radfar[1], Ulrich Kentsch[2], Luke Antwis[3], Hele Savin[1], Ville Vähänissi[1]
[1] *Aalto University, Espoo, Finland;* [2] *HZDR, Dresden, Germany;* [3] *University of Surrey, Guildford, United Kingdom*

1CV.2.22 Excellent Passivation of Silicon Surfaces by HfO2 Layers Deposited using Scalable Spatial Atomic Layer Deposition (SALD) 193

Jan Schmidt[1], Michael Winter[1], Floor Souren[2], Jons Bolding[2], Hindrik de Vries[2]
[1] *ISFH, Emmerthal, Germany;* [2] *SALD, Eindhoven, The Netherlands*

1CV.2.26 Simulation of Topcon/Perc Hybrid Bottom Structure for Perovskite/Silicon Tandem Solar Cells Using Quokka3 197

Eni Muka[1], Raşit Turan[1], Hisham Nasser[1]
[1] *ODTÜ GÜNAM, Ankara, Türkiye*

1CV.2.35 A Comprehensive Analysis of the Series Resistance for Different Interdigitated Back Contact Solar Cell Geometries 201

Telmo Isasi[1], Yeray Mateos[1], Janire Pampin[1], Vanesa Fano[1], Nekane Azkona[1], Eneko Ortega[1], Juan Carlos Jimeno[1], Eneko Cereceda[1], Alona Otaegi[1]
[1] *UPV/EHU, Bilbao, Spain*

1CV.2.37 Accuracy of Hysteresis Correction for Silicon Heterojunction Solar Cells – A   205
Simulation Study

Jonas Kern[1], Hannes Wagner-Mohnsen[2], Johannes Heitmann[1], Matthias Müller[1]
[1] *Freiberg University of Mining and Technology, Freiberg, Germany;* [2] *WAVELABS Solar Metrology Systems, Leipzig, Germany*

1CV.2.38 Contactless Carrier Lifetime Characterization of Silicon Heterojunction   209
Structures at Elevated Temperatures

Gergely Havasi[1], David Krisztián[1], Zs. Gombás[2], Zoltan Adam[2], Ferenc Korsós[1]
[1] *Semilab, Budapest, Hungary;* [2] *EcoSolifer Heterojunction, Budapest, Hungary*

1CV.2.39 Bias Light Intensity Effect on EQE Analysis for PERC Solar Cell   213

Hatice Duman[1], Özlem Coskun[1], Güven Korkmaz[1]
[1] *Kalyon PV, Ankara, Türkiye*

1CV.2.42 Improved Accuracy of Photoluminescence Images for Quality Control in   214
Solar Cell Production

Robin Wienberg[1], Jonas Haunschild[1], Saravana Kumar[1], Jurriaan Schmitz[2], Stefan Rein[1]
[1] *Fraunhofer ISE, Freiburg, Germany;* [2] *University of Twente, Enschede, The Netherlands*

1CV.2.44 Simulation and Design Optimization of Interdigitated Back Contact Silicon   219
Solar Cells with Dopant-Free Asymmetric Hetero-Contacts

You-An Li[1], Chun-Ping Lin[2], Ying-Yuan Huang[2]
[1] *NYCU, Tainan, Taiwan;* [2] *NCKU, Tainan, Taiwan*

1CV.2.45 Numerical Modeling and Design Optimization of Industrial Tunnel Oxide   222
Passivated Contact Solar Cells with Selective Passivated Contacts on the
Front

Yi-Ping Lin[1], Chun-Ping Lin[2], Jin-Cheng Chen[1], Han-Chen Chang[3], Ying-Yuan Huang[2]
[1] *NYCU, Tainan, Taiwan;* [2] *NCKU, Tainan, Taiwan;* [3] *ITRI, Tainan, Taiwan*

1CV.2.47 Modeling and Experimental Validation of Solar Cell Performance across   225
Varied Temperatures and Irradiance

Selin Cansu Gölboylu[1], Hatice Duman[1], Melisa Demir[1], Meriç Çalışkan Arslan[1]
[1] *Kalyon PV, Ankara, Türkiye*

Plenary SESSION 1EP.1 Sustainability

1EP.1.2 Copper as Cost-Effective Alternative to Silver for Si Solar Cell Metallization   226
– Status and Outlook

Florian Clement[1], Andreas Lorenz[1], Jonas Bartsch[1], Andreas Brand[1], Jonas D. Huyeng[1], Roman Keding[1], Sven Kluska[1], F. Maarouf[1], Jan Nekarda[1], Daniel Ourinson[1], Sebastian Pingel[1], J. Schube[1], Ralf Preu[1]
[1] *Fraunhofer ISE, Freiburg, Germany*

Oral SESSION 2AO.3 III-V Solar Cells & Space PV

2AO.3.2 Thermal Modeling of Triple-Junction Solar Cells Fan Out Wafer Level 250
Packaging for Concentrated Photovoltaic

Konan Kouame[1], Abdul Rehman[1], Médérick Marcotte[1], Mylana Ney[1], Artur Turala[1], Corentin Jouanneau[1], Mohamed Najah[1], Serge Ecoffey[1], David Danovitch[2], Gwenaelle Hamon[1]
[1] *University of Sherbrooke, Sherbrooke, Canada;* [2] *Université de Sherbrooke, Sherbrooke, Canada*

2AO.3.3 Overview for Tandem Solar Cell R&D Activities in Japan 254

Masafumi Yamaguchi[1], Tatsuya Takamoto[2], Kyotaro Nakamura[1], Ryo Ozaki[1], Hiroyuki Juso[2], Nobuaki Kojima[1], Yoshio Ohshita[1]
[1] *Toyota Technological Institute, Nagoya, Japan;* [2] *Sharp Corporation, Nara, Japan*

2AO.3.5 Space Applications for a Variety of Solar Cell Technologies 257

Stephen Taylor[1]
[1] *European Space Agency, Noordwijk, The Netherlands*

Oral SESSION 2BO.10 New Modelling and Characterisation - Material Properties

2BO.10.1 In-depth Characterization Methodology for the Assessment of Passivation 260
Impact in Halide Perovskite Solar Cells

Jonathan Parion[1], Santhosh Ramesh[1], Sownder Subramaniam[1], Henk Vrielinck[2], Filip Duerinckx[1], Hariharsudan Sivaramakrishnan Radhakrisnan[1], Jef Poortmans[1], Johan Lauwaert[3], Bart Vermang[1]
[1] *imec, Genk, Belgium;* [2] *University of Gent, Ghent, Belgium;* [3] *University of Gent, Zwijnaarde, Belgium*

Oral SESSION 2BO.8 Novel PV Material and Conversion Concepts

2BO.8.1 Pathways for Silicon Solar Cells with Molecular Singlet Fission 263

Phoebe Pearce[1], Nicholas Ekins-Daukes[1]
[1] *UNSW, Sydney, Australia*

2BO.8.2 Control of Hot Carrier Thermalization Rates in Nanowires for Advanced- 273
Concept Photovoltaic Solar Cells

Hamidreza Esmaielpour[1], Nabi Isaev[1], Imam Makhfudz[2], Markus Döblinger[3], Jonathan Finley[1], Gregor Koblmüller[1]
[1] *TUM, Munich, Germany;* [2] *Aix-Marseille University, Marseille, France;* [3] *LMU, Munich, Germany*

2BO.8.5 Design and Prototyping of Spectrum-Split-Type Concentrating Photovoltaic- 277
Thermoelectric Hybrid Power Generator

Kenji Kamide[1], Ryoji Funahashi[1], Tomoyuki Urata[1], Yoko Matsumura[1], Jun Sakuma[2], Hidefumi Akiyama[2], Katsuto Tanahashi[1]
[1] *AIST, Tsukuba, Japan;* [2] *University of Tokyo, Kashiwa, Japan*

Visual SESSION 2BV.1 Advances in Novel Materials, Devices and Concepts | New Modelling and Characterisation Techniques

2BV.1.1 Development of an Interdigitated Back-Contacted Solar Cell Architecture as a Platform to Assess Emerging Absorbers and New Selective Contacts   285

Juan de Dios Castillo[1], Gerard Masmitjà[1], Pau Estarlich[1], Pablo Ortega[1], Cristobal Voz[1], Arnau Torrens[1], Oriol Segura[1], Edgardo Saucedo[1], Massoud Karimipour[2], Sonia Ruiz[2], Mónica Lira-Cantu[2], Joaquim Puigdollers[1]
[1] UPC, Barcelona, Spain; [2] ICN2, Barcelona, Spain

2BV.1.2 Annealed Phosphorus-Doped Amorphous Silicon as Electron Selective Contact for Crystalline Germanium Thermophotovoltaic Cells   289

Gerard Rivera[1], Mansur Gamel[1], Gema López[1], Moisés Garín[2], Isidro Martín[1]
[1] UPC, Barcelona, Spain; [2] University of Vic, Vic, Spain

2BV.1.10 Sensitization of Crystalline Silicon with Organic Dye Molecules   294

Lukáš Gdula[1], Branislav Dzurňák[1], Tom Markvart[2]
[1] Czech Technical University in Prague, Prague, Czech Republic; [2] University of Southampton, Southampton, United Kingdom

2BV.1.11 Self-Organized Films of Carbazole Derivatives on Structured Silicon Substrates for Photovoltaic Applications   295

Sergii Mamykin[1], Daria Kuznetsova[1], Nina Roshchina[1], Petro Smertenko[1], Saulius Grigalevicius[2], Gintare Krucaite[2], Raminta Beresneviciute[2], Simona Sutkuviene[3]
[1] V. Lashkaryov Institute of Semiconductor Physics NAS Ukraine, Kyiv, Ukraine; [2] Kaunas University of Technology, Kaunas, Lithuania; [3] Lithuanian University of Health Sciences, Kaunas, Lithuania

2BV.1.12 Placement Angles for Luminescent Solar Concentrators: Simulating and Experimenting with Bifacial Photovoltaic Mosaic Devices   299

Xitong Zhu[1], Frits Reijners[1], Michael Debije[1], Angèle H.M.E Reinders[1]
[1] Eindhoven University of Technology, Eindhoven, The Netherlands

2BV.1.20 Low Emissive Molybdenum-Doped ITO for High Vacuum Photovoltaic-Thermal Application   302

Daniela De Luca[1], Umar Farooq[1], Paolo Strazzullo[1], Eliana Gaudino[1], Antonio Caldarelli[1], Anna Krammer[2], Andreas Schüler[2], Marilena Musto[1], Emiliano Di Gennaro[1], Roberto Russo[1]
[1] University of Naples Federico II, Naples, Italy; [2] EPFL, Lausanne, Switzerland

2BV.1.27 Optimization of a Planar Perovskite Solar Cell Layer Thicknesses: Optical and Electrical Effects   303

Aleksi Kamppinen[1], Kati Miettunen[1]
[1] University of Turku, Turku, Finland

2BV.1.28 Photoluminescence Imaging of Perovskite Solar Cells in Full Sunlight   308

Zhiwen Zheng[1], Felix Gayot[1], Juergen W. Weber[1], Yan Zhu[1], Ziv Hameiri[1]
[1] UNSW, Sydney, Australia

2BV.1.29 Analysis of Color Alteration as a Novel Degradation Assessment Method for Perovskite Solar Cells   309

Rustem Nizamov[1], Aapo Poskela[1], Mahboubeh Hadadian[1], Maryam Esmaeilzadeh[1], Mikael Nyberg[1], Kati Miettunen[1]

[1] University of Turku, Turku, Finland

2BV.1.30 Statistical Model of Outdoor Perovskite Performance 315

Petra Manshanden[1], Martin Späth[1], Mark Jansen[1], Valerio Zardetto[2], Arantxa Aguirre[3], Valerie Depauw[3], Mina Heydarian[4], Juliane Borchert[4]

[1] TNO, Petten, The Netherlands; [2] TNO, Petten, The Netherlands; [3] imec, Genk, Belgium; [4] Fraunhofer ISE, Freiburg, Germany

2BV.1.32 Characterization and Degradation of Perovskite Mini-Modules 323

Rita Ebner[1], Ankit Mittal[1], Gusztav Ujvari[1], Maria Hadjipanayi[2], Vasiliki Paraskeva[2], George E. Georghiou[2], Afshin Hadipour[3], Aranzazu Aguirre[4], Tom Aernouts[4], Thommaso Fontanot[5], Sabrina Pechmann[5], Silke Christiansen[5]

[1] AIT, Vienna, Austria; [2] University of Cyprus, Nicosia, Cyprus; [3] Kuwait University, Kuwait, Kuwait; [4] imec, Genk, Belgium; [5] IKTS, Forchheim, Germany

2BV.1.34 Subcell-Resolved Electroluminescence Imaging of Monolithic Perovskite-Silicon Tandem Solar Cell for High Throughput Characterization 327

Ivanol Jaurece Djeukeu[1], Jonas Horn[1], Michael Meixner[1], Enno Wagner[2], Stefan W. Glunz[3], Klaus Ramspeck[1]

[1] halm elektronik, Frankfurt, Germany; [2] Frankfurt University of Applied Sciences, Frankfurt, Germany; [3] Fraunhofer ISE, Freiburg, Germany

2BV.1.35 A Case Study of Certainly I-V Measurement of the Perovskite Solar Cell under Dim Light Intensity for Solar/ Indoor Lighting Application 328

Yean-San Long[1], Min-An Tsai[1], Hsin-Hsin Hsieh[1], Fan-Hsuan Yeh[2]

[1] ITRI, Hsinchu, Taiwan; [2] Taipei First Girls High School, Taipei, Taiwan

2BV.1.39 Perovskite Solar Cell Light-Soaking and Relaxation Modelling for Improved Energy Yield Predictions in Indoor Environments 329

Matija Pirc[1], Špela Tomšič[1], Marko Jošt[1], Marko Topič[1]

[1] University of Ljubljana, Ljubljana, Slovenia

2BV.1.41 Modelling the Effects of Tandem Module Circuit Configurations 330

M. Ignacia Devoto[1], Daniel Tune[1], Ahmer A.B. Baloch[2], Karl Wienands[1], Rüdiger Farneda[1], Bhaskar Parida[2], Omar Albadwawi[2], Vivian Alberts[2], Andreas Halm[1]

[1] ISC Konstanz, Konstanz, Germany; [2] DEWA Research & Development Center, Dubai, United Arab Emirates

Visual SESSION 2BV.2 Compound and Organic Semiconductors

2BV.2.5 III-V Thin Films Growth by RP-CVD: Towards a Reduction of Industrialization Costs 334

Lise Watrin[1], François Silva[2], Cyril Jadaud[2], Pavel Bulkin[2], Jean-Charles Vanel[2], Kassiogé Dembélé[2], Erik V. Johnson[2], Karim Ouaras[2], Pere Roca i Cabarrocas[1]

[1] IPVF, Palaiseau, France; [2] LPICM, Palaiseau, France

2BV.2.9 Modeling and Measurement of Lumped Series Resistance with Varying Illumination and Current Condition of Low-Bandgap Solar Cells 335

Shipei Zhang[1], Xiawa Wang[1]
[1] Duke Kunshan University, Kunshan, China

2BV.2.11 Color Implementation of Cu(In,Ga)Se2 Thin-film Solar Cells with Multilayered Conductive Optical Filters — 338

Yong-Duck Chung[1], Dae-Hyung Cho[1], Rina Kim[1], Woo-Jung Lee[1], Tae-Ha Hwang[1], Soyoung Lim[1], Donghyeop Shin[2], Kihwan Kim[2], Mangu Kang[1]
[1] ETRI, Daejeon, South Korea; [2] KIER, Daejeon, South Korea

2BV.2.17 Flexible Thin-Film CZTS Solar Cell based on an Electroplated Metallic Precursor Deposited on a Molybdenum/Glass Coated Stainless Steel Foil — 342

Io Mizushima[1], Peter Torben Tang[1], Christoph Kammerlander[2], Andreas Zimmermann[2]
[1] IPU, Virum, Denmark; [2] Sunplugged, Affenhausen, Austria

2BV.2.23 Manufacturing, Characterisation and Stability Tests of Printed Organic Photovoltaic Devices for Indoor Applications — 345

Ignacio Ballesteros Garcia[1], A. Khodr[2], Donia Fredj[2], Carmen M Ruiz Herrero[1], Hasan Alkhatib[2], O. Margeat[1], Sadok Ben Dkhil[2], Judikaël Le Rouzo[1], Jörg Ackermann[1]
[1] CNRS, Marseille, France; [2] Dracula Technologies, Valence, France

2BV.2.30 Fabrication of Highly Efficient CdSeTe/CdTe Thin Film Solar Cells with Emitter-Less Cell Structure — 346

Yanbo Cai[1], Hongxu Jiang[1], Kai Yi[1], Fei Liu[1], Guangwei Wang[1], Deliang Wang[1]
[1] University of Science and Technology of China, Hefei, China

Oral SESSION 2CO.2 Triple Junctions and Advanced Concepts in Perovskite-based Tandems

2CO.2.5 Characterisation of Degradation Pathways of 3-Terminal Perovskite-Silicon Tandems After Outdoor Monitoring — 349

Miha Kikelj[1], Laurie-Lou Senaud[2], Florent Sahli[2], Benjamin Lipovšek[1], Marko Topič[1], Christophe Ballif[2], Quentin Jeangros[2], Bertrand Paviet-Salomon[2]
[1] University of Ljubljana, Ljubljana, Slovenia; [2] CSEM, Neuchâtel, Switzerland

Oral SESSION 2CO.3 New Modelling and Characterisation - Device Performance

2CO.3.2 Understanding Ion-Related Performance Losses in Perovskite-Based Solar Cells by Capacitance Measurements and Simulation — 363

Christoph Messmer[1], Jonathan Parion[2], Cristian V. Meza[2], Santhosh Ramesh[2], Martin Bivour[3], Maryamsadat Heydarian[3], Jonas Schön[1], Hariharsudan S. Radhakrishnan[2], Martin C. Schubert[3], Stefan W. Glunz[1]
[1] University of Freiburg, Freiburg, Germany; [2] imec, Genk, Belgium; [3] Fraunhofer ISE, Freiburg, Germany

2CO.3.4 Analysis and Modelling of Recovery and Degradation Mechanisms in 379
Perovskite Solar Cells

Guillem Álvarez-Pérez[1], Arthur Julien[1], Karim Medjoubi[1], Jean Baptiste Puel[1], Jean François Guillemoles[1]
[1] *IPVF, Palaiseau, France*

2CO.3.5 Developments in Thermophotovoltaics (TPV) 384

Esther López Estrada[1], Alejandro Datas[1]
[1] *UPM, Madrid, Spain*

Visual SESSION 2CV.3 Perovskite-based Multijunctions | Perovskite Photovoltaics

2CV.3.4 Monolithic Series-Interconnected Two-Terminal Perovskite-CIGSe Tandem 393
Solar Cells: Voltage-Matched or Current-Matched?

Nicolas Otto[1], Christof Schultz[1], Guillermo Farias-Basulto[2], Rutger Schlatmann[1], Eva Unger[2], Bert Stegemann[1]
[1] *HTW Berlin, Berlin, Germany; [2] HZB, Berlin, Germany*

2CV.3.11 Optimisation of MA-free Lead-Tin Perovskite Absorber and Interfaces in All 397
Perovskite Tandem Solar Cells

Jules Allegre[1], Polyxeni Tsoulka[1], Noëlla Lemaitre[1], Baptiste Berenguier[2], Mathieu Frégnaux[3], Muriel Bouttemy[3], Philip Schulz[2], Solenn Berson[1], Kilian Alcocer[4]
[1] *CEA-INES, Le-Bourget-du-Lac, France; [2] IPVF, Palaiseau, France; [3] ILV, Versailles, France; [4] CEA, Grenoble, France*

2CV.3.17 Potential Induced Degradation Free Perovskite-Silicon Tandem Solar Cells 398

Kristijan Brecl[1], Matevž Bokalič[1], Gašper Matič[1], Marko Topič[1], Lisa Champault[2], Quentin Jeangros[2]
[1] *University of Ljubljana, Ljubljana, Slovenia; [2] CSEM, Neuchâtel, Switzerland*

2CV.3.18 Modeling of Metastability Behavior in Perovskite-based Solar Cells for 399
Accurate Energy Yield Estimation in Realistic Operating Conditions

Špela Tomšič[1], Marko Remec[2], Florian Scheler[2], Mark Khenkin[2], Carolin Ulbrich[2], Rutger Schlatmann[2], Steve Albrecht[2], Marko Jošt[1], Benjamin Lipovšek[1], Marko Topič[1]
[1] *University of Ljubljana, Ljubljana, Slovenia; [2] HZB, Berlin, Germany*

2CV.3.24 Microstructural Analysis on the Conformity of Chemical Vapour Deposition 404
(CVD) Perovskite Thin-Films on Silicon for Tandem PV Devices

Angela Chen[1], Emma Holder[1], Adrian Element[1], Yong Li[1], Kenrick F. Anderson[1], Tim W. Jones[1], Benjamin C. Duck[1], Noel W. Duffy[1], Gregory J. Wilson[1]
[1] *CSIRO Energy, Newcastle, Australia*

2CV.3.25 Controlling the Film Properties of SnO2 in Perovskite Solar Cells using 408
Scalable Spatial Atomic Layer Deposition

Hindrik W. de Vries[1], Floor M. M. Souren[1], S. R. Ratnasingham[2], Mehrdad Najafi[2]
[1] *SALD, Eindhoven, The Netherlands; [2] TNO, Petten, The Netherlands*

2CV.3.33 Beyond the Lab-Scale: Perovskite Photovoltaic Fabrication and Industrial   411
Assessment with Automated Slot-Die Coater

Maurizio Stefanelli[1], Simon Ternes[1], Luigi Vesce[1], Marco Balucani[2], Aldo Di Carlo[1]
[1] *University of Rome Tor Vergata, Rome, Italy;* [2] *RISE Technology, San Martino di Lupari, Italy*

2CV.3.34 Reasoning the Change in Device Parameters with Deposition Power of NiOx   414
for Low-Dimensional Perovskite Solar Cells

Bhumika Sharma[1], Vani Pawar[1], Sushobhan Avasthi[1]
[1] *Indian Institute of Science, Bengaluru, India*

2CV.3.36 Analysis of Reverse-Bias Stability of FAPbBr3 Semi-Transparent Perovskite   417
Solar Cells

Noah Tormena[1], Alessandro Caria[1], Matteo Buffolo[1], Carlo De Santi[1], Nicola Trivellin[1], Andrea Cester[1], Gaudenzio Meneghesso[1], Enrico Zanoni[1], Fabio Matteocci[2], Aldo Di Carlo[2], Matteo Meneghini[1]
[1] *University of Padova, Padova, Italy;* [2] *University of Rome, Rome, Italy*

2CV.3.43 Enhancing Efficiency and Stability of CsPbI3 Perovskite Quantum Dots   418
Through Co2+-Doping

Pouriya Naziri[1], Naeimeh Sadat Peighambardoust[1], Umut Aydemir[1]
[1] *Koc University, Istanbul, Türkiye*

2CV.3.44 Standardized Test Routines for the Assessment of Potential Induced   422
Degradation of Perovskite Solar Cells

Beyza Durusoy[1], David Adner[2], Christian Hagendorf[3], Konrad Wojciechowski[4], Samy Almosni[4], Marko Turek[5]
[1] *METU, Ankara, Türkiye;* [2] *Martin-Luther-University, Halle, Germany;* [3] *Anhalt University of Applied Sciences, Köthen, Germany;* [4] *Saule Technologies, Wroclaw, Poland;* [5] *Fraunhofer CSP, Halle, Germany*

2CV.3.45 Evaluation of Perovskite Devices Under Real and Extreme Operating   427
Conditions - A Fundamental Step Toward Practical Applications

Marília Braga[1], Lucas Augusto Zanicoski Sergio[1], Anelise Medeiros Pires[1], Ricardo Rüther[1]
[1] *UFSC, Florianópolis, Brazil*

2CV.3.46 Enhancing Measurement Protocols for Perovskite Photovoltaic Devices: 428
Insights from the VIPERLAB Project

Eugenia Zugasti[1], Ankit Mittal[2], Lucia V. Mercaldo[3], Javier Diaz[1], Giuseppe Nasti[3], Asier Murillo Marrero[1], Natalia Maticiuc[4], Ana Belén Cueli[1], Stephan Abermann[2], Paola Delli Veneri[3], Stephane Cros[5]
[1] *CENER, Sarriguren, Spain;* [2] *AIT, Vienna, Austria;* [3] *ENEA, Portici, Italy;* [4] *HZB, Berlin, Germany;* [5] *CEA-INES, Le Bourget du Lac, France*

2CV.3.51 Solvent Engineering Driven Morphology Control of Perovskite under Air   429
Ambient Device Fabrication

Nitin Kumar Bansal[1], Shivam Porwal[2], Trilok Singh[1]
[1] *IIT Delhi, New Delhi, India;* [2] *IIT Kharagpur, Kharagpur, India*

2CV.3.55 Roll-to-Roll Printed SnO₂ for Flexible N-I-P Perovskite PV   430

Thomas M. Kraft[1], Ville Holappa[1], Riikka Suhonen[1]
[1] *VTT Technical Research Centre of Finland, Oulu, Finland*

2CV.3.61 Micro Inverted Pyramid Formation in Titanium Dioxide Layer by Pulsed 432
Laser Irradation to Improve Electron Transport in MAPBI3-based
Photovoltaic Devices

*Luis Ocaña[1], Carlos Montes[1], Benjamín González-Díaz[2], Sara González-
Pérez[2], Elena Llarena[1]*
*[1] ITER, Granadilla de Abona, Spain; [2] University of La Laguna, San Cristóbal de La Laguna,
Spain*

2CV.3.63 Demonstration of Industrially Scalable Chemical Vapour Deposition (CVD) 437
Process for Production of High-Efficiency Perovskite Photovoltaics

*Emma Holder[1], Adrian Element[1], Yong Li[1], Faiazul Haque[1], Kenrick F.
Anderson[1], Tim W. Jones[1], Benjamin C. Duck[1], Noel W. Duffy[1], Gregory J.
Wilson[1]*
[1] CSIRO Energy, Newcastle, Australia

2CV.3.64 Photoluminescence and Lifetime Stability of Pentacene and Oxide 441
Perovskites Nanoparticles Films on Nanotextured Silicon Substrate

*Rémi Ndioukane[1], Diouma Kobor[1], Sergio de Armas Rillo[2], Fernando Lahoz
Zamarro[2]*
*[1] University Assane Seck of Ziguinchor, Ziguinchor, Senegal; [2] University of La Laguna,
Santa Cruz de Tenerife, Spain*

2CV.3.69 Compositional Engineering of Double-cation Single-halide Perovskite for 445
Efficient Solar Cell Fabrication under Air Ambient Conditions

Mrittika Paul[1], Binita Boro[1], Amreesh Chandra[1], Trilok Singh[2]
[1] IIT Kharagpur, Kharagpur, India; [2] IIT Delhi, New Delhi, India

2CV.3.74 Interface Engineering for Perovskite Solar Cells Using Polymer-Based 446
Antisolvent Technique

Lingeswaran Arunagiri[1], Feng Wang[2], Feng Gao[2]
[1] Linköping University, Linkoping, Sweden; [2] Linköping University, Linköping, Sweden

Oral SESSION 2DO.18 Late News: Developments in High Efficiency Tandem Cells

2DO.18.3 Perovskite Record Setting Silicon Tandem Modules: Customer Expect Lower 447
LCOE

Christopher Case[1]
[1] Oxford PV, Oxford, United Kingdom

Oral SESSION 2DO.6 Towards Improved Understanding of Perovskite Solar Cell Device Physics

2DO.6.1 Bright Insights: Exploring Perovskite Formation Mechanisms with Combined 466
Spectral Reflectance and Photoluminescence In-Situ Data

*Nasim Rezaei-Hartmann[1], Thorsten Brand[1], Adrian Adrian[1], Claudine Groß[1],
M. Leyden[2], Enno Malguth[1], Aleksandra Miaskiewicz[2], Marcel Roß[2], Viktor
Škorjanc[2], Lars Korte[2], Steve Albrecht[2], Christian Camus[1]*
[1] LayTec, Berlin, Germany; [2] HZB, Berlin, Germany

2DO.6.5 Enhancing Crystallinity of Perovskite Materials through Rapid Microwave Annealing — 481

Syed Nazmus Sakib[1], David N. R. Payne[1], Shujuan Huang[1], Binesh P. Veettil[1]
[1] Macquarie University, Sydney, Australia

Oral SESSION 2DO.8 Scalability of Perovskite Solar Modules

2DO.8.4 Fully Printed Perovskite Solar Cells and Modules — 486

Luigi Vesce[1], Karthikeyan Pandurangan[2], Maurizio Stefanelli[1], Elena Iannibelli[1], Hafez Nikbakht[1], Maria Laura Parisi[2], Adalgisa Sinicropi[2], Aldo Di Carlo[1]
[1] University of Rome Tor Vergata, Rome, Italy; [2] University of Siena, Siena, Italy

Oral SESSION 2DO.9 Lifetime and Reliability of Perovskite Devices

2DO.9.2 TÜV Rheinland Specification on the I-V Characterization of Perovskite-Based PV Modules — 490

Giorgio Bardizza[1], Qi Gao[2], Wenhao Xu[2], Yating Zhang[2], Christos Monokroussos[2], Werner Herrmann[3]
[1] TUV Rheinland, Milan, Italy; [2] TUV Rheinland, Shanghai, China; [3] TÜV Rheinland, Milan, Italy

2DO.9.5 One-Year Outdoor Testing of 4T Perovskite/Si PV Modules — 494

Matthew Norton[1], Vasiliki Paraskeva[1], Maria Hadjipanayi[1], Elias Peratikos[1], Aranzazu Aguirre[2], Anurag Krishna[2], Santhosh Ramesh[2], Tom Aernouts[2], George E. Georghiou[1]
[1] University of Cyprus, Nicosia, Cyprus; [2] Hasselt University/Imo-Imomec, Genk, Belgium

Visual SESSION 3AV.1 PV Module Design and Manufacturing | BoS Components, Operation and Aging

3AV.1.4 Challenges for Solder Interconnection pushed by High-Efficiency Solar Cell Developments — 499

Benjamin Grübel[1], Angela De Rose[1], Achim Kraft[1]
[1] Fraunhofer ISE, Freiburg, Germany

3AV.1.5 Optimizing Sustainability: Balancing Antimony Content for Enhanced Optical Properties and Environmental Impact in Solar Glass — 505

Anika Glaubitz[1], Sven Grüttner[1], Selim Yagci[1], Oliver Pfeiffer[1], Ulf Blieske[1]
[1] University of Applied Sciences Cologne, Cologne, Germany

3AV.1.6 Photovoltaic Modules Comprising III-V Cells Encapsulated in Composite Material — 511

Francisco J. Cano[1], Werther Cambarau[1], Naiara Yurrita[1], Jon Aizpurua[1], Juan M. Hernández[1], Gorka Imbuluzqueta[1], Eduardo Román Medina[1], Oihana Zubillaga[1]
[1] TECNALIA, San Sebastián, Spain

3AV.1.11 Reliability of Aluminum-Copper Contact in PV Modules 514

Tobias Messmer[1], Dominik Rudolph[1], Gernot Emanuel[2], Andreas Nägele[2], Andreas Halm[1]
[1] *ISC Konstanz, Konstanz, Germany;* [2] *Fraunhofer ISE, Freiburg, Germany*

3AV.1.12 Lightweight Photovoltaic Modules Technologies: Reliability Evaluation and Market Opportunity 519

Julien Dupuis[1], Christine Abdel Nour[1], J.V. Oliveira Santos[1], Paul Lefillastre[2]
[1] *EDF R&D, Moret-Loing-et-Orvanne, France;* [2] *EDF Renewables, Nanterre, France*

3AV.1.15 MgO/SiO$_x$ Adds Heat Dissipation Function to Crystalline Silicone Solar Cell Modules 523

Eiko Shimokata[1], Yasushi Sobajima[1], Keisuke Ohdaira[2], Atsushi Masuda[3]
[1] *Gifu University, Gifu, Japan;* [2] *JAIST, Nomi, Japan;* [3] *Niigata University, Niigata, Japan*

3AV.1.16 Investigation of Temperature Homogeneity during Infrared Soldering of Silicon Solar Cells using the Finite Element Method 527

Daniel Christopher Joseph[1], Angela De Rose[1], Dirk Eberlein[1], Onur Parlayan[1], Benjamin Grübel[1], Andreas J. Beinert[1], Holger Neuhaus[1]
[1] *Fraunhofer ISE, Freiburg, Germany*

3AV.1.17 Impact of Textured Surfaces and Cleaning on Solar Panel Glass Transmittance 530

Aapo Poskela[1], Julianna Varjopuro[1], Tommi Jokikyyny[1], Aleksi Kamppinen[1], Heikki Palonen[1], Kati Miettunen[1]
[1] *University of Turku, Turku, Finland*

3AV.1.19 Ultra-Thin Flexible Glass as Environmental Shield for CIGS Photovoltaic Modules 534

Nikolina Pervan[1], Sonja Feldbacher[1], Martina Harnisch[2], Tuuli Tettenborn[2], Andreas Zimmermann[2], Gernot Oreski[1]
[1] *PCCL, Leoben, Austria;* [2] *Sunplugged, Wildermieming, Austria*

3AV.1.20 Process Development and Material Evaluation of Photovoltaic Aluminum Facade Element for BIPV Application 535

Ringo Koepge[1], Matthias Pander[1], Stephan Großer[1], Bengt Jaeckel[1]
[1] *Fraunhofer CSP, Halle (Saale), Germany*

3AV.1.22 Material Properties Requirements for Frame Sealants and Junction Box Adhesives 539

Guy Beaucarne[1], Emmanuel Jadot[1], Dominique Culot[1], Rono Cao[2], Kayla Kenney[3], Suraj Ahuja[4], Valérie Hayez[1]
[1] *Dow Silicones Belgium, Seneffe, Belgium;* [2] *Dow (Shanghai), Shanghai, China;* [3] *Dow Silicones , Auburn, United States of America;* [4] *Dow Chemical International, Mumbai, India*

3AV.1.23 Solder Pastes in Shingled Modules 545

Karl Wienands[1], Ignacia Devoto[1], Nils Kopp[2], Carina Hallensleben[2], Rihoko Kizukuri[2], Matthias Helbig[1], Enita Kurtovic[1], Andreas Halm[1], Daniel Tune[1]
[1] *ISC Konstanz, Konstanz, Germany;* [2] *TAMURA-ELSOLD, Ilsenburg, Germany*

3AV.1.25 TiO$_2$/SiO$_x$ Surface Coating on Crystalline Silicon-Based-Solar Cell Module to Provide Anti-Soiling Functionality 549

Koshiro Iwaki[1], Yasushi Sobajima[1], Keisuke Ohdaira[2], Atsushi Masuda[3]
[1] Gifu University, Gifu, Japan; [2] JAIST, Nomi, Japan; [3] Niigata University, Niigata, Japan

3AV.1.26 Performance Analysis of Different Shading-Resistant PV Module Designs under Different Partial Shading Scenarios — 552

Andreas Maixner[1], Tales Siquera[1], Matthias Pander[2], Jens Froebel[2], Bengt Jaeckel[2], Hamed Hanifi[1]
[1] AESOLAR, Koenigsbrunn, Germany; [2] Fraunhofer CSP, Halle, Germany

3AV.1.33 Optimal Design for Flexible Solar Panels Attached Around Cylindrical Poles — 559

Hiroki Sugimoto[1]
[1] PXP Corporation, Sagamihara, Japan

3AV.1.35 Design and Implementation of a CSI Photovoltaic Microinverter Prototype with High Frequency Switching — 562

Francisco Guzman[1], Patricio Valdivia-Lefort[2], Antonio Sanchez[1], Rodrigo Barraza[1]
[1] Federico Santa Maria Technical University, Santiago, Chile; [2] Universidad de Santiago de Chile, Santiago, Chile

3AV.1.37 PV Microinverters: Balcony Power Plants, Latest Efficiency Rankings, Yield Calculation for Overpowered Mini PV Systems — 563

Stefan Krauter[1], Jörg Bendfeld[1]
[1] Paderborn University, Paderborn, Germany

3AV.1.38 Aging Behavior of Polymeric Materials used in Inverter Casings — 570

Eric Helfer[1], Petra Christöfl[1], Julia Petro[1], Margit Lang[1], Volker Reisecker[2], L. Heupl[3], A. Weiermair[3], Gernot Oreski[1]
[1] PCCL, Leoben, Austria; [2] Transfercenter für Kunststofftechnik, Wels, Austria; [3] Fronius International, Thalheim, Austria

3AV.1.39 Performance of Arc Fault Circuit Interrupters in Photovoltaic Inverters Connected to Long DC Cables — 571

Donat Hess[1], David Joss[1], Christof Bucher[1]
[1] BUAS, Burgdorf, Switzerland

3AV.1.40 Design of the Substring MPP Tracker — 578

Patrick Mader[1], Sascha Eckerter[1], Rainer Merz[1]
[1] Karlsruhe University of Applied Sciences, Karlsruhe, Germany

3AV.1.41 Testing of Electronic Interface for Diagnostic Functions of Photovoltaic Systems — 583

Edoardo Celi[1], Alessandro Minuto[1], Stefano Rizzi[1], Gianluca Timò[1]
[1] RSE, Piacenza, Italy

Visual SESSION 3AV.2 PV Modules Reliability: Components, Failure Mechanisms, Testing & Modelling

3AV.2.1 Evaluation of Degradation and Impact of Climatic Conditions on PV Modules Exposed to Extreme High UV Solar Radiation — 584

Patricio Valdivia-Lefort[1], Valentina Navarro[2], Rodrigo Barraza[2]
[1] Universidad de Santiago de Chile, Santiago, Chile; [2] Federico Santa Maria Technical University, Santiago, Chile

3AV.2.2 PV Module Brush Abrasion Testing 585

Gerhard Mathiak[1], Nithin Sha[1], Afra Seentakath[1], Prashanth Gabbadi[1], Yogesh Kumar[1], Mark Mirza[2]
[1] *DEWA R&D Center, Dubai, United Arab Emirates;* [2] *Fraunhofer ISC, Würzburg, Germany*

3AV.2.3 Numerical Simulation for Comparison of PV Module Designs based on 591
Outdoor Data in Desert Climates

Matthias Pander[1], Bengt Jaeckel[1], Klemens Ilse[1], Amir A. Abdallah[2]
[1] *Fraunhofer CSP, Halle (Saale), Germany;* [2] *QEERI, Doha, Qatar*

3AV.2.5 Impact of Modern Cell Photovoltaic Geometries on Power and Energy Loss 596
due to Cell Cracks

Ahmad Hashem[1], SL. Mortazavifar[1], Ralph Gottschalg[1]
[1] *Anhalt University of Applied Sciences, Köthen, Germany*

3AV.2.7 Performance Evaluation of the Custom-Made Small PV Modules after 600
Exposure to Saudi Arabia's Climatic Conditions over 10 Long Years

Amir Al-Ahmed[1], Amjad Ali[1], Mohammed A. Alghamdi[2], Osama Asker[2], Ridha Ben Mansour[1], Firoz Khan[1], Atif S. Alzahrani[1]
[1] *KFUPM, Dhahran, Saudi Arabia;* [2] *Gulf Renewable Lab, Dammam, Saudi Arabia*

3AV.2.8 Analyzing the Effect of Damp Heat Test on Various PV Module 601
Technologies, a Comparative Study

Ahmad Alheloo[1], Ali Almheiri[1], Baloji Adothu[1], Gerhard Mathiak[1], Vivian Alberts[1]
[1] *DEWA Research & Development, Dubai, United Arab Emirates*

3AV.2.9 Comparative Degradation Analysis of Emerging PV Module Technologies 604
Undergoing Thermal Cycling

Ali Almheiri[1], Ahmad Alheloo[1], Baloji Adothu[1], Gerhard Mathiak[1], Vivian Alberts[1]
[1] *DEWA Research & Development, Dubai, United Arab Emirates*

3AV.2.10 Assessment of Critical Laminate Temperature Increase by Fast IR-based 607
Analysis of Hot Spots on Solar Cells

Stephan Grosser[1], Matthias Schak[1], Stefan Eiternick[1], Bengt Jaeckel[1], Marko Turek[1]
[1] *Fraunhofer CSP, Halle (Saale), Germany*

3AV.2.11 Correlational Study on the Impact of Harsh Environment Stress Factors on 610
the Ageing Effects of Several Encapsulation Materials for PV Modules

Tudor Timofte[1], Maria Ignacia Devoto Acevedo [1], Joachim Glatz-Reichenbach[1], Valentina Arias Reyes[2], Andreas Halm[1]
[1] *ISC Konstanz, Konstanz, Germany;* [2] *Federico Santa María Technical University, Valparaiso, Chile*

3AV.2.12 Model Calibration of Photovoltaic Modules Photodegradation in High- 618
Radiation Environments Using UV Accelerated Exposure Testing

Patricio Valdivia[1], Valentina Arias Reyes[2], Rodrigo Barraza[3], Iván González Echeverria[2]
[1] *Universidad de Santiago de Chile, Santiago, Chile;* [2] *Federico Santa Maria Technical University, Santiago, Chile;* [3] *Universidad Adolfo Ibañéz, Santiago, Chile*

3AV.2.13 Electrical Characterization of Fresh and Degraded Photovoltaic Backsheets   619
Based on Temperature and Humidity-Dependent DC Conductivity

Anagha E R[1], Shrikrishna V Kulkarni[1], Narendra Shiradkar[1]
[1] *IIT Bombay, Mumbai, India*

3AV.2.14 Investigation of PV Module Degradation in Fixed Structure and Single-Axis   623
Tracker in Hot Desert Climate

*Baloji Adothu[1], Aafra Seentakath Puthiyapurayil[1], Shahzada Pamir Aly[1],
Gerhard Mathiak[1], Vivian Alberts[1]*
[1] *DEWA R&D Center, Dubai, United Arab Emirates*

3AV.2.17 Tackling the Fire Safety in Glass Free PV Modules   626

*Nikolina Pervan[1], Sonja Feldbacher[1], Umang Desai[2], Antonin Faes[2],
Christophe Ballif[2], Gernot Oreski[1]*
[1] *PCCL, Leoben, Austria;* [2] *EPFL, Neuchâtel, Switzerland*

3AV.2.20 Analysis and Material Modeling of Mechanical Property Degradation for   627
Simulation of Weather Exposed Polymers

*Julia Petro[1], Volker Reisecker[2], Eric Helfer[1], Gernot Oreski[1], Thomas
Antretter[3], Margit Lang[1]*
[1] *PCCL, Leoben, Austria;* [2] *TCKT, Wels, Austria;* [3] *University of Leoben, Leoben, Austria*

3AV.2.21 Reliability Investigation of Structural Colour Interlayers for Coloured PV   632
Modules

*Markus Babin[1], Roberto Boccardi[1], Aliihsan Bagci[1], Nanna Lysgaard
Andersen[1], Peter Behrensdorff Poulsen[1], Sune Thorsteinsson[1], Karlis
Petersons[2], Leif Yde[2], Jan F. Stensborg[2], Catarina G. Ferreira[3], Joel D. Cox[3],
Irina Vyalih[4], Jani Lamminaho[3], Morten Madsen[4]*
[1] *DTU, Roskilde, Denmark;* [2] *Stensborg, Roskilde, Denmark;* [3] *SDU, Odense, Denmark;* [4]
SDU, Sønderborg, Denmark

3AV.2.22 Diagnosing Potential Induced Degradation in Crystalline Silicon Photovoltaic   636
Modules

*Aysha Mahmood[1], Rodrigo del Prado Santamaria[1], Thøger Kari[1], Peter B.
Poulsen[1], Sergiu V. Spataru[1]*
[1] *DTU, Roskilde, Denmark*

3AV.2.25 On-Site Evaluation of Oxygen-Plasma Treated Glass Surfaces for Anti-   645
Soiling Properties

Brahim Aïssa[1], Ayman Samara[2]
[1] *QEERI, Doha, Qatar;* [2] *HBKU, Doha, Qatar*

3AV.2.27 Performance, Abrasion Resistivity and Anti-Soiling Testing of Innovative,   651
Nanostructured Anti-Reflection Coatings under Controlled and Standardized
Conditions

*Charlotte Pfau[1], Guido Willers[1], Christos Allagiannis[2], Ioannis Arampatzis[2],
Marko Turek[1]*
[1] *Fraunhofer CSP, Halle (Saale), Germany;* [2] *Nanophos, Lavrio, Greece*

3AV.2.30 Development of Encapsulant-Less Crystalline Silicon Photovoltaic Modules   656
and Their Durability Against Potential-Induced Degradation

Keisuke Ohdaira[1], Shuntaro Shimpo[1], Huynh Thi Cam Tu[1]
[1] *JAIST, Ishikawa, Japan*

3AV.2.35 Evaluation of the Impact of the UV Excitation Intensity on the Ultraviolet Fluorescence Measurement System for Photovoltaics   660

Zonghan Jiang[1], Carlos Meza[1], Hugo Sanchez[1], Ralph Gottschalg[1]
[1] Anhalt University of Applied Sciences, Koethen, Germany

3AV.2.36 How to Mount PV Modules: the Effect of Different Clamping Configuration on Mechanical Stresses in PV Modules   665

Pascal Romer[1], Andreas J. Beinert[1], Charlotte Hasselblatt[1], Cornelius Herr[1]
[1] Fraunhofer ISE, Freiburg, Germany

3AV.2.40 FMEA Based Degradation Rate Evaluation to Study Impact of Different Failure Modes as Function of Mission Profiles   666

Bengt Jaeckel[1], Baloji Adothu[2], Vivian Alberts[3], Matthias Pander[1]
[1] Fraunhofer CSP, Halle (Saale), Germany; [2] DEWA, Dubai, United Arab Emirates; [3] DEWA, Dubai, United Arab Emirates

3AV.2.42 Numerical Simulation of the Bypass Diode Failure Resistance and those Power Consumption in a Photovoltaic Solar Module with Failed Bypass Diode   676

Ibuki Kitamura[1], Toshiyuki Hamada[1], Ikuo Nanno[2], Norio Ishikura[3], Masayuki Fujii[4], Shinichiro Oke[5]
[1] Osaka Electro-Communication University, Osaka, Japan; [2] Yamaguchi Gakugei University, Yamaguchi, Japan; [3] Yonago College, Tottori, Japan; [4] Oshima College, Yamaguchi, Japan; [5] Tsuyama College, Okayama, Japan

3AV.2.45 Impact of the Material Combination on the Barrier Properties and their Stability in the Course of Accelerated Weathering   680

Daniel Schüsler[1], Patrick Wessel[1], Michael Wendt[1], Anton Mordvinkin[1]
[1] Fraunhofer CSP, Halle, Germany

3AV.2.46 Investigation of Thermo-Mechanical Behavior of Encapsulation Materials used in Solar Panel Production   683

Umran Dilmac[1], Merve Çorak[2], Meric Caliskan Arslan[1], Yildirim Aydogdu[3]
[1] Kalyon PV, Ankara, Türkiye; [2] Kalyon PV, Ankara, Türkiye; [3] Gazi University, Ankara, Türkiye

3AV.2.49 UV Exposure of Glass/Glass Coupons with Edge Seal and Different Encapsulants   684

Chiara Barretta[1], Lisa Meinhart[1], Andreas Brandstätter[2], Dieter Geier[3], Roland Einhaus[3], Abdulkerim Gok[4], Gernot Oreski[1]
[1] PCCL, Leoben, Austria; [2] Lenzing Plastics, Lenzing, Austria; [3] ZSW, Stuttgart, Germany; [4] Gebze Technical University, Gebze, Türkiye

3AV.2.50 Material Screening for the Development of a Photovoltaic Module Using Biodegradable Materials from Renewable Raw Materials   685

Matthias Pander[1], Ringo Koepge[1], Bengt Jaeckel[1], Anton Mordvinkin[1]
[1] Fraunhofer CSP, Halle (Saale), Germany

3AV.2.51 Failure Mode Analysis of Austria's First Road-Integrated Photovoltaic System   686

Alexander Erber[1], Bernhard Grasel[1]
[1] University of Applied Sciences Vienna, Vienna, Austria

3AV.2.52 Effects of Encapsulant-Backsheet Combinations on Durability of Optical Properties   687

Jishnu Ramachandran Nair[1], Daniel Schuesler[1], Michael Wendt[1], Ralph Gottschalg[1], Anton Mordvinkin[1]
[1] *Fraunhofer CSP, Halle, Germany*

3AV.2.53 PID Outdoor Measurements, a New Test Setup   690

Jörg Kirchhof[1]
[1] *Fraunhofer IEE, Kassel, Germany*

3AV.2.54 Coatings or Tapes? Imaging Methods to Show the Successful Repair of Backsheet Cracks   694

Raffael Schifferegger[1], Yuliya Voronko[1], Anika Gassner[1], Gabriele C. Eder[1], Eric Tilly[2]
[1] *OFI, Vienna, Austria;* [2] *ENcome Energy Performance, Klagenfurt, Austria*

3AV.2.57 Effect of Weight Percent Graphene on Barrier Properties of Ethelyne Vinyl Acetate (EVA) for Improved Photovoltaic Module Packaging Reliability   695

Emeka H. Amalu[1], Oluwagbemiga A. Fabunmi[1], David J. Hughes[1], Yongxin Pang[1], Michael Short[1]
[1] *Teesside University, Middlesbrough, United Kingdom*

3AV.2.58 Tests beyond Standards on Bifcacial PV Modules with Transparent Backsheets   702

Alessandro Anderlini[1], Angelika Beinert[2], Ingrid Hädrich[2], Luigi D'arco[1]
[1] *Coveme, Gorizia, Italy;* [2] *Fraunhofer ISE, Freiburg, Germany*

Visual SESSION 3AV.3 PV Modules Performance: Testing, Modelling Techniques and Outdoor Performance

3AV.3.1 Enhanced Performance of PV Modules using Hierarchically Structured Glass in Different Climatic Conditions   703

Cristina Leyre Pinto[1], Jaione Bengoechea[1]
[1] *CENER, Sarriguren-Navarra, Spain*

3AV.3.9 A Data-Driven Calibration of the FEM Temperature Model with Wind Direction Input   709

Anastasios Kladas[1], Bert Herteleer[1], Jan Cappelle[1]
[1] *KU Leuven, Leuven, Belgium*

3AV.3.13 Areal Cell Temperature Monitoring Using Array of In-Laminate Integrated Sensors for Partial Shading Detection   712

Seyed Mojtaba Sadati Faramarzi[1], Georgi H. Yordanov[1], Arvid van der Heide[2], Jan Genoe[1], Jef Poortmans[1]
[1] *KU Leuven, Leuven, Belgium;* [2] *imec, Leuven, Belgium*

3AV.3.14 Maximum Power Output Predicting Algorithm of Solar Modules Based on Artificial Intelligence Technology   716

Ju-Hee Kim[1], Joonyoung Jeon[1], Yong Hyun Kim[1]
[1] *KOPTI, Gwangju, South Korea*

3AV.3.16 A Parametric Approach for Estimation of PV Short-Circuit Current   717

Sergiu Mihai Hategan[1], Marius Paulescu[1]
[1] *West University of Timişoara, Timişoara, Romania*

3AV.3.22 Comparison of Changes in the Parameters of Five PV Module Types after one Year in the Swiss Jura Mountains 721

Donat Hess[1], Fabio Panduri[1], Matthias Burri[1], Christof Bucher[1], Mauro Caccivio[2], Gabi Friesen[2]
[1] *BFH, Burgdorf, Switzerland;* [2] *SUSPI, Mendrisio, Switzerland*

3AV.3.25 Accurate Energy Performance Model for Bifacial PV Modules 738

Kristijan Brecl[1], Matevž Bokalič[1], Marko Topič[1], Antonin Faes[2]
[1] *University of Ljubljana, Ljubljana, Slovenia;* [2] *CSEM, Neuchâtel, Switzerland*

3AV.3.28 Uncertainty Assessment in the Measurement of Solar Cells under Standard Test Conditions 739

Yating Zhang[1], Wenhao Xu[1], Qi Gao[1], Giorgio Bardizza[2], Werner Herrmann[2], Christos Monokroussos[1]
[1] *TÜV Rheinland, Shanghai, China;* [2] *TÜV Rheinland Energy, Cologne, Germany*

3AV.3.29 Stabilization of Field-Aged Crystalline PV Modules Before STC Power Determination 743

Soha Essbai[1], Marcus Rennhofer[1], Ankit Mittal[1], Gusztáv Újvári[1], Thomas Weber[2], Brian Azzopardi[3]
[1] *AIT, Vienna, Austria;* [2] *PI Berlin, Berlin, Germany;* [3] *FIR Malta, Valetta, Malta*

3AV.3.30 AC/DC Electroluminescence. The War of the Currents 746

Mario Martínez[1], Sergio Suarez[1], Daniel Villoslada[1], Jose Manuel Rivas[1], Sofía Rodríguez-Conde[1]
[1] *Enertis Applus, Madrid, Spain*

3AV.3.31 Evaluation of the Contact Quality in Silicon Solar Cells and Modules Using LBIC Phase Mapping 751

Majid Salari[1], Jonas Buddgård[2], Markus Rinio[1]
[1] *Karlstad University, Karlstad, Sweden;* [2] *StickySolarPower, Sollentuna, Sweden*

3AV.3.32 Nomenclature and Description of EL Observations: Cell Cracks and Other Findings 755

Bengt Jaeckel[1], Matthias Pander[1], Paul Schenk[1], Aswin Linsenmeyer[2], Jochen Kirch[3]
[1] *Fraunhofer CSP, Halle (Saale), Germany;* [2] *Sunset Energietechnik, Adelsdorf, Germany;* [3] *Ing.-Büro Jochen Kirch, Leeder, Germany*

3AV.3.38 Daylight Electroluminescence Inspection of PV Panels On-site vs. Traditional EL Inspection with Silicon Cameras 758

Luis Alberto Carpintero[1], Diego Gónzalez-Francés[2], Kabir Paul Sulca[2], Cristian Terrados[2], Carmelo de Castro[2], Victor Alonso[2], Míguel Ángel Gónzalez Rebollo[2], Oscar Mártinez[2]
[1] *Cobra Instalaciones y Servicios, Madrid, Spain;* [2] *University of Valladolid, Valladolid, Spain*

3AV.3.39 Photovoltaic Module Array Luminescence Image Preprocessing: Heuristic Algorithms for Perspective Correction and Cell Segmentation 761

Brendan Wright[1], Ali Shakiba[1], Rama Sharma[1], Ziv Hameiri[1]
[1] *UNSW, Sydney, Australia*

3AV.3.42 Comparing Measured PV Module Power to Nameplate Values 765

Frank Weinrich[1], Stefan Riechelmann[1], Laura Stenzig[1], Stefan Winter[1]
[1] *PTB, Braunschweig, Germany*

3AV.3.44 Finding the Cell to Module Performance Values for Industrial TOPCon and HJT Technologies 768

Sraisth[1], Hardik Gohil[1], Mehul Raval[1], Wolfgang Jooss[1]
[1] *RCT Solutions, Konstanz, Germany*

3AV.3.45 The Impact of Module Degradation on the Economics of PV Projects 772

Harry Apostoleris[1], Baloji Adothu[1], Bengt Jaeckel[2], Gerhard Mathiak[1], Sgouris Sgouridis[1]
[1] *DEWA R&D, Dubai, United Arab Emirates;* [2] *Fraunhofer CSP, Halle (Saale), Germany*

3AV.3.50 Improved Sampling of IV Measurements 773

Maximilian Schönau[1], Elisabeth Schönau[2], Darwin Daume[3], Markus Panhuysen[1], Achim Schulze[4], Bernd Hüttl[3], Dieter Landes[3]
[1] *smartblue, Munich, Germany;* [2] *Catholic University Eichstätt-Ingolstadt, Eichstätt, Germany;* [3] *Coburg University of Applied Sciences, Coburg, Germany;* [4] *Rosenheim University of Applied Sciences, Rosenheim, Germany*

3AV.3.52 Enhancing Production Forecasting of Grid-Connected PV Strings Operating under Semi-Arid Climate Conditions 776

Khadija El Ainaoui[1], Mhammed Zaimi[1], Imane Flouchi[2], Said Elhamaoui[2], Yasmine El Mrabet[2], Abdellatif Ghennioui[2], El Mahdi Assaid[1]
[1] *University of Chouaib Doukkali, El Jadida, Morocco;* [2] *Green Energy Park, Ben Guerir, Morocco*

3AV.3.53 Evaluation of the Glare Function and Description of Key Measurement Procedures 780

Wolfgang Nemitz[1], Roman Trattnig[1], Jakob Zehndorfer[2], Markus Babin[3], Lukas Plessing[4]
[1] *Joanneum Research, Weiz, Austria;* [2] *Zehndorfer Engineering, Klagenfurt, Austria;* [3] *DTU, Roskilde, Denmark;* [4] *TPPV, Vienna, Austria*

Oral SESSION 3BO.11 Reliability of PV Modules: The Impact of Solar Cell Technology

3BO.11.1 Reliability of Commercial TOPCon PV Modules – An Extensive Comparative Study 786

Paul Gebhardt[1], Jochen Markert[1], Ulli Kräling[1], Esther Fokuhl[1], Ingrid Haedrich[1], Daniel Philipp[1]
[1] *Fraunhofer ISE, Freiburg, Germany*

3BO.11.3 Study and Mitigation of Moisture-Induced Degradation in Silicon Heterojunction Solar Modules 792

Lucie Pirot-Berson[1], Romain Couderc[1], Romain Bodeux[2], Frédéric Jay[1], Julien Dupuis[3]
[1] *CEA, Le Bourget-du-Lac, France;* [2] *IPVF, Palaiseau, France;* [3] *EDF, Moret Loing et Orvanne, France*

3BO.11.4 Investigation of Potential-induced Degradation and Recovery in Perovskite Minimodules 800

Junchuan Zhang[1], Haodong Wu[1], Yi Zhang[2], Fangfang Cao[1], Zhiheng Qiu[1], Minghui Li[1], Xiting Lang[1], Yongjie Jiang[1], Yangyang Gou[1], Xirui Liu[1], Abdullah M. Asiri[3], Paul J. Dyson[2], Mohammad Khaja Nazeeruddin[2], Jichun Ye[1], Chuanxiao Xiao[1]
[1] *CAS, Ningbo, China;* [2] *EPFL, Lausanne, Switzerland;* [3] *KAU, Jeddah, Saudi Arabia*

Oral SESSION 3BO.12 Reliability of PV Modules: The Impact of Polymers

3BO.12.4 Recent Developments in PV Module Backsheets - What Do We Really Know about Them?   809

Gernot Oreski[1], Chiara Barretta[1], Karl-Anders Weiß[2]
[1] *PCCL, Leoben, Austria;* [2] *Fraunhofer ISE, Freiburg, Germany*

Oral SESSION 3BO.14 Failure Modes and Degradation Mechanisms in PV Modules

3BO.14.2 Analyses of Glass Quality and its Influence on Mechanical Stability of Large Area PV Modules   812

Jochen Markert[1], Aditya Girish Belawadi[1], Pascal Romer[1], Frank Ensslen[1], Enzo Job[1], Ingrid Hädrich[1], Daniel Philipp[1], Tobias Rist[2]
[1] *Fraunhofer ISE, Freiburg, Germany;* [2] *Fraunhofer IWM, Freiburg, Germany*

3BO.14.3 Polarization-Type Potential-Induced Degradation in Bifacial PERC Modules in the Field   826

Peter Hacke[1], Cecile Molto[2], Dylan J.Colvin[2], Ryan Smith[3], Farrukh Ibne Mahmood[4], Fang Li[4], Jaewon Oh[5], Govindasamy Tamizhmani[4], Hubert Seigneur[2], Christopher DiRubio[6], Matthew Gardeski[6]
[1] *NREL, Golden, United States of America;* [2] *FSEC Energy Research Center University of Central Florida, Cocoa, United States of America;* [3] *Pordis, Austin, United States of America;* [4] *ASU, Mesa, United States of America;* [5] *University of North Carolina, Charlotte, United States of America;* [6] *First Solar, Tempe, United States of America*

3BO.14.5 LeTID in Real Life: The Relevance and Importance of Accelerated Tests and Treatments   834

Alison Ciesla[1], Arastoo Teymouri[1], Petra Manshanden[2], Alvin Mo[1], Astrid Gutjahr[2], Moonyong Kim[1], Li Wang[1], Catherine Chan[1], Ran Chen[1], Gianluca Coletti[1], Jakob Jan Dijksterhuis[3], Bas Van Aken[2]
[1] *UNSW, Sydney, Australia;* [2] *TNO, Petten, The Netherlands;* [3] *Elsun, Roden, The Netherlands*

3BO.14.6 Towards Establishing Criteria for Electrical Safety in Second-Use Photovoltaic (PV) Modules   846

Tadanori Tanahashi[1], Takashi Oozeki[1]
[1] *AIST, Koriyama, Japan*

Oral SESSION 3BO.15 Reliability of PV Modules: Testing and Modelling Approaches

3BO.15.2 Material Selection and Novel Reliability Testing for Floating Photovoltaic   854
Modules

Nikoleta Kyranaki[1], Arvid van der Heide[1], Hamed Javanbakht Lomeri[1], Ismail Kaaya[1], Sara Bouguerra[1], Jens D. Moschner[2], Arnaud Morlier[1], Michaël Daenen[1]
[1] Hasselt University, Genk, Belgium; [2] KU Leuven, Leuven, Belgium

3BO.15.3 Outdoor Accelerated Ageing Test Using Additional Thermal and   862
Thermomechanical Stresses

Ebrar Özkalay[1], Gabi Friesen[1], Alessandro Virtuani[2], Mauro Caccivio[1], Christophe Ballif[3]
[1] SUPSI, Mendrisio, Switzerland; [2] CSEM, Neuchâtel, Switzerland; [3] EPFL, Neuchâtel, Switzerland

3BO.15.4 Development of PV Module Hot Desert Test Cycle Protocol Extended Failure   878
Modes and Effective Analysis

Baloji Adothu[1], Jim Joseph John[1], Gerhard Mathiak[1], Vivian Alberts[1], Bengt Jäckel[2], Ralph Gottschalg[2], Narendra S Shiradkar[3], Amir A. Abdallah[4], Juan Lopez Garcia[4], Michael Salvador[5], Bram Hoex[6], Hussein A Kazem[7], Muhammad Ashraful Alam[8]
[1] DEWA, Dubai, United Arab Emirates; [2] Fraunhofer CSP, Halle, Germany; [3] IIT Bombay, Mumbai, India; [4] QEERI, Doha, Qatar; [5] KAUST, Thuwal, Saudi Arabia; [6] UNSW, Sydney, Australia; [7] Sohar University, Sohar, Oman; [8] Purdue University, West Lafayette, United States of America

3BO.15.6 Solar Cell Crack Image Generation for Power Loss Prediction   886

Norman Jost[1], Emma Cooper[1], Benjamin G. Pierce[1], Brandon Byford[1], Ojas Singh[1], Jennifer L. Braid[1]
[1] Sandia National Laboratories, Albuquerque, United States of America

Oral SESSION 3CO.10 Materials and Processes for PV Modules

3CO.10.1 Benchmarking of Encapsulant Materials for c-Si/Perovskite Tandem Modules   892

Petra Christöfl[1], Chiara Barretta[1], Marcel Kühne[2], Frans Opden Buijsch[3], Sem Sals[3], Quentin Jeangros[3], Bernd Stannowski[4], Gernot Oreski[1]
[1] PCCL, Leoben, Austria; [2] Hanwha Q CELLS, Thalheim, Germany; [3] The Compound Company, Geleen, The Netherlands; [4] HZB, Berlin, Germany

3CO.10.2 Reliability Studies of PV Minimodules Using an Ethylene – Butyl Acrylate   898
(EBA) Based Encapsulant and High Efficiency n-Type PV Cells

Ignacio Fidalgo[1], Inmaculada Campoy Felipe[2], Andreas Halm[3]
[1] Polaris Open Innovation, Oviedo, Spain; [2] Repsol Química, Madrid, Spain; [3] ISC Konstanz, Constance, Germany

3CO.10.4 Reducing Process Time of PV Module Lamination by Using Double-Side   911
Heating System

Sraisth[1], Djamel Eddine Mansour[2], Aksel Kaan Öz[2], Paul Gebhardt[2], Daniel Klaus[3], Christine Wellens[2]
[1] RCT Solutions, Constance, Germany; [2] Fraunhofer ISE, Freiburg, Germany; [3] Robert Buerkle, Freudenstadt, Germany

Oral SESSION 3CO.11 Emerging Interconnection Technologies

3CO.11.2 Design Roadmap to Modules with 24 % Efficiency 920

Max Mittag[1], Christian Reichel[1], Alexander Protti[1], Dirk Holger Neuhaus[1]
[1] Fraunhofer ISE, Freiburg, Germany

3CO.11.3 Effect of Lowering Curing Temperature of Electrically Conductive Adhesives 924
on Ribbon Connected Solar Cells

Veronika Nikitina[1], Tim Riehle[1], Leonhard Böck[1], Torsten Rößler[1]
[1] Fraunhofer ISE, Freiburg, Germany

3CO.11.5 To Bypass or Not to Bypass: Integrating and Evaluating Parallel Connections 929
and Bypasses in c-Si PV Laminates

*Tom Borgers[1], Jonathan Govaerts[1], Hamed Javanbakht Lomeri[1], Apostolos
Bakovasilis[1], Rik Van Dyck[1], Bart Reekmans[1], Hariharsudan
Sivaramkrishnan Radhakrishnan[1], Jef Poortmans[1], Manuel Van den Storme[2],
Guy Van den Storme[2]*
[1] imec, Genk, Belgium; [2] VdSWeaving, Oudenaarde, Belgium

Oral SESSION 3DO.12 Low Environmental Impact Module Design and Technologies

3DO.12.1 Steps Towards a 100% Renewable Material Solar Module: Evaluating 933
Material Substitutions for Encapsulation and Interconnection

Ringo Koepge[1], Matthias Pander[1], Anton Mordvinkin[1], Stephan Großer[1]
[1] Fraunhofer CSP, Halle (Saale), Germany

3DO.12.2 New Encapsulant for PV Modules Designed for Recycle: A Lab Scale 940
Prototype

*Margot Landa[1], Alexis Brastel[2], Eeva Mofakhami[1], Timea Bejat[1], Pierre
Piluso[2]*
[1] CEA-INES, Le Bourget-du-Lac, France; [2] CEA Liten, Grenoble, France

3DO.12.3 Laser-Assisted Delamination for Si Modules Recycling 943

*Remi Aninat[1], Maarten van der Vleuten[1], Johan Bosman[1], Henri Fledderus[2],
Anne Biezemans[1], João Gomes[1], Veronique Gevaerts[1], Ando Kuypers[1],
Mirjam Theelen[1]*
[1] TNO, Eindhoven, The Netherlands; [2] TNO, Eindhoven, The Netherlands

3DO.12.4 Innovative Design-for-Recycle for Critical Material-Free Interconnection of 953
PV Modules

Antoine Perelman[1], Vincent Barth[1], Fabien Mandorlo[2], Eszter Voroshazi[1]
[1] CEA-INES, Le Bourget-du-Lac, France; [2] INSA, Lyon, France

3DO.12.5 Bifacial Lightweight Solution without Glass 960

*Alicia Buceta[1], Ana Belén Cueli[1], Miguel Aguirre[1], Ana Linares[2], Elena
Llarena[2], Silvia Cal[2], Jaione Bengoechea[1]*
[1] CENER, Sarriguren, Spain; [2] ITER, Granadilla de Abona, Spain

3DO.12.6 Development of Novel Frontsheets with Protective Coatings to Increase the 966
Durability and Reliability of Glass-free Lightweight PV Modules

Yuliya Voronko[1], Gabriele C. Eder[1], Elisabeth Reiser[2], Markus Babin[3], Gernot Oreski[4]
[1] OFI, Vienna, Austria; [2] KANSAI HELIOS, Vienna, Austria; [3] DTU Electro, Roskilde, Denmark; [4] PCCL, Leoben, Austria

Oral SESSION 3DO.16 PV Module Assessment and Classification

3DO.16.3 Quantitative Description of the Quality of Daylight Electroluminescense (dEL) Images Against Dark Room EL Images   974

Kabir Paul Sulca[1], Carmelo de Castro[1], Diego González-Francés[1], Cristian Terrados[1], Julián Anaya[1], Victor Alonso[1], Miguel Angel González[1], Oscar Mártinez[1]
[1] University of Valladolid, Valladolid, Spain

3DO.16.4 Photovoltaic Cell Defect Classification from Luminescence Images: Embedding and Clustering with Unsupervised Machine Learning   980

Brendan Wright[1], Rama Sharma[1], Ziv Hameiri[1]
[1] UNSW, Sydney, Australia

3DO.16.5 Daylight Photoluminescence of Silicon Solar Panels in Operation by Electrical Modulation   983

Cristian Terrados[1], Diego González-Francés[1], Kabir Paul Sulca[1], C. de Castro[1], Miguel Ángel González[1], Oscar Martínez[1]
[1] University of Valladolid, Valladolid, Spain

Oral SESSION 3DO.17 Outdoor Performance and Energy Yield Estimation

3DO.17.1 PV Module Degradation in Hot Deserts: Laboratory and Outdoor Data Analysis   988

Gerhard Mathiak[1], Shahzada Pamir Aly[1], Kaushal Chapaneri[1], Baloji Adothu[1], Jim Joseph John[1]
[1] DEWA R&D Center, Dubai, United Arab Emirates

3DO.17.2 Incidence Angle Effect: Results of an Interlaboratory Comparison of Measurements on Commercial-Size Modules   995

Mauro Pravettoni[1], Min Hsian Saw[1], Giorgio Bardizza[2], Giovanni Bellenda[3], Romain Couderc[4], Gabi Friesen[3], Werner Herrmann[2], Shin Woei Leow[5], Stefan Riechelmann[6], Flavio Valoti[3], Arvid van der Heide[7], Frank Weinrich[6], Stefan Winter[6]
[1] TII, Abu Dhabi, United Arab Emirates; [2] TÜV-Rheinland, Cologne, Germany; [3] SUPSI, Mendrisio, Switzerland; [4] CEA, Le Bourget-du-lac, France; [5] SERIS, Singapore, Singapore; [6] PTB, Braunschweig, Germany; [7] imec, Genk, Belgium

3DO.17.3 Climate Specific Energy Rating (CSER) Analysis of Outdoor PV Field Data   999

Ismael Medina[1], Teodora S. Lyubenova[1], Ewan Dunlop[1]
[1] European Commission JRC, Ispra, Italy

3DO.17.4 Module Parameters Extraction for Assessing Photovoltaic Energy Yield: A Comparative Approach   1005

Ahmad Hashem[1], Hugo Sanchez[2], Frank Xu[3], SL. Mortazavifar[1], Christos Monokroussos[3], Ralph Gottschalg[4]
[1] *Anhalt University of Applied Sciences, Koethen, Germany;* [2] *Anhalt University of Applied SciencesUniversity of Applied Sciences, Köthen, Germany;* [3] *TÜV Rheinland, Shanghai, China;* [4] *Anhalt University of Applied Sciences, Köthen, Germany*

3DO.17.5 Performance and Degradation Evaluation of C-Si Modules Under Different Open-Rack and Residential Mounting Configurations   1009

Gabi Friesen[1], Ebrar Özkalay[1], Mauro Caccivio[1]
[1] *SUPSI ISAAC, Mendrisio, Switzerland*

Oral SESSION 3DO.19 Modelling Techniques for PV Modules

3DO.19.1 An Accurate Data-Driven Physical Model for Bifacial PV Power Estimation   1020

Ali Sohani[1], Marco Pierro[2], David Moser[2], Cristina Cornaro[1]
[1] *University of Rome Tor Vergata, Rome, Italy;* [2] *Eurac Research, Bolzano, Italy*

3DO.19.2 Comparative Analysis of Temperature Estimation Models in Bifacial Photovoltaic Modules   1029

Aline Kirsten Vidal de Oliveira[1], Marília Braga[1], Isadora Maciel Queiroz[1], Helena Naspolini[1], Ricardo Rüther[1]
[1] *UFSC, Florianópolis, Brazil*

3DO.19.5 Apparent Intensity Dependence of Shunts in PV Modules - Revision of the Shunt Parameterization in the De Soto Model and PVsyst   1033

Nils-Peter Harder[1], José Cano Garcia[1]
[1] *TotalEnergies, Palaiseau, France*

Oral SESSION 3DO.20 Shading and Soiling on PV Modules

3DO.20.2 Correlating Field Experimentation and Image Analysis for the Assessment of Induced Losses from Thin Object Shading on Photovoltaic Sources   1040

Matthew Axisa[1], Luciano Mule'Stagno[1], Marija Demicoli[1]
[1] *University of Malta, Marsaxlokk, Malta*

3DO.20.3 Laboratory Intercomparison on a Shading Resistance Classification of PV Modules   1046

Stefan Riechelmann[1], Hendrik Sträter[1], Laura Stenzig[1], Giorgio Bardizza[2], Werner Herrmann[2], Ebrar Özkalay[3], Gabi Friesen[3], Özcan Bazkir[4], Alexandra Schmid[5], Stefan Winter[1]
[1] *PTB, Braunschweig, Germany;* [2] *TÜV Rheinland, Cologne, Germany;* [3] *SUPSI, Manno, Switzerland;* [4] *TÜBITAK, Ankara, Türkiye;* [5] *Fraunhofer ISE, Freiburg, Germany*

3DO.20.4 Exploring Dust Particle Properties and PV Soiling Mapping: a Case Study in the Arid Landscape of a Desert Environment   1050

Brahim Aissa[1], Atef Zekri[2], Mosab I. A. Kareem Subeh[1]
[1] *QEERI, Doha, Qatar;* [2] *HBKU, Doha, Qatar*

3DO.20.6 PV Module Cleaning under Hot Desert Conditions   1055

Gerhard Mathiak[1], Afra Seentakath[1], Nithin Sha[1], Shashank Suvarn[1], Prashanth Gabbadi[1], Arumugham Muthusamy[1], Nabeel Ibrahim[1], Kaushal Chapaneri[1]
[1] *DEWA R&D Center, Dubai, United Arab Emirates*

Oral SESSION 3EO.1 In Field Characterisation of PV Modules | BoS Components in Operation

3EO.1.2 From Fab to Field - Quality Control with a Mobile PV Laboratory — 1065

Magnus Herz[1], Hamza Maaroufi[1], Giorgio Bardizza[2]
[1] *TÜV Rheinland, Cologne, Germany;* [2] *TÜV Rheinland, Milan, Italy*

3EO.1.3 Reduction of Uncertainty of Outdoor PV Module Characterization: Test Field Experiences — 1072

Mariella Rivera[1], Christian Reise[1]
[1] *Fraunhofer ISE, Freiburg, Germany*

3EO.1.5 MPP Tracking Losses of Module Level Power Electronics at Partial Module Shading — 1079

Franz P. Baumgartner[1], Markus Klenk[1], Adrian Widler[1], Linus Baumann[1]
[1] *ZHAW, Winterthur, Switzerland*

3EO.1.6 Improvement of Tracking Algorithms using Machine Learning — 1085

Sarra Ben Brahim[1], Kai Saegebarth[1], Martin Dennenmoser[2], Alsayed Algergawy[3]
[1] *BayWa r.e., Munich, Germany;* [2] *BayWa r.e., Freiburg, Germany;* [3] *University of Passau, Passau, Germany*

Oral SESSION 4AO.7 Advanced O&M Strategies and Methods

4AO.7.1 Best Practice Guidelines for the Use of PV System KPIs — 1089

Sascha Lindig[1], Magnus Herz[2], Julián Ascencio-Vásquez[3], Marios Theristis[4], Bert Herteleer[5], Julien Deckx[6], Kevin Anderson[7], Karel De Brabandere[6], Erik Stensrud Marstein[8]
[1] *UNIVERS, Munich, Germany;* [2] *TÜV Rheinland, Cologne, Germany;* [3] *UNIVERS, Redwood, United States of America;* [4] *Sandia, Albuquerque, United States of America;* [5] *KU Leuven, Leuven, Belgium;* [6] *3E, Brussels, Belgium;* [7] *NREL, Golden, United States of America;* [8] *IFE, Lillestrøm, Norway*

4AO.7.2 Hybrid Decision Support System: a Framework for Data-Driven Troubleshooting and Reporting — 1095

Sandra Gallmetzer[1], Mousa Sondoqah[1], Pablo Sebastian Enriquez Paez[2], Atse Louwen[1], David Moser[1]
[1] *EURAC Research, Bolzano, Italy;* [2] *BayWa r.e., Rome, Italy*

4AO.7.3 Identifying Distinct Performance Patterns in Utility-Scale Photovoltaic Plants Using an Unsupervised Machine Learning Model — 1104

Ali Shakiba[1], Brendan Wright[1], Ziv Hameiri[1]
[1] *UNSW, Sydney, Australia*

4AO.7.5 Enhancing Fault Diagnosis in Photovoltaic Plants: a Comprehensive Approach to Simultaneous Failures   1108

Giosué Maugeri[1], Salvatore Guastella[1], Andrea Rossetti[1]
[1] *RSE, Milan, Italy*

4AO.7.6 Design and Application of Intelligent Scalable Automatic Fault Detector for Commercial Photovoltaic Systems   1118

Mücahid Candan[1], David Melgar[1], Christian Schill[1], Mete Çubukçu[2], Eduardo Sarquis Filho[3], Björn Müller[3], Duarte Kazacos[4]
[1] *Fraunhofer ISE, Freiburg, Germany;* [2] *Solar Energy Institute of Ege University, Bornova, Türkiye;* [3] *Enmova, Freiburg, Germany;* [4] *Mondas, Freiburg, Germany*

Oral SESSION 4AO.8 PV Plant Performance, Analysis, Monitoring and Fault Detection in Inverters

4AO.8.1 Uncertainty-Aware Estimation of Inverter Field Efficiency Using Bayesian Neural Networks in Solar Photovoltaic Plants   1124

Gerardo Guerra[1], Pau Mercade-Ruiz[1], Gaetana Anamiati[1], Lars Landberg[2]
[1] *GreenPowerMonitor a DNV Company, Barcelona, Spain;* [2] *DNV Denmark, Hellerup, Denmark*

4AO.8.2 Analysis of Fault Detection and Defect Categorization in Photovoltaic Inverters for Enhanced Reliability and Efficiency in Large-scale Solar Energy Systems   1132

Stephanie Malik[1], David Daßler[1], Dharm Patel[1], Carola Klute[1], Robert Klengel[1], Andreas Dietrich[2], Kai Kaufmann[3], Carsten Hennig[4], Danny Wehnert[5], Matthias Ebert[1], Leonard Kraft[5]
[1] *Fraunhofer IMWS, Halle (Saale), Germany;* [2] *DiSUN, Werder (Havel), Germany;* [3] *DENKweit, Halle (Saale), Germany;* [4] *saferay holding, Berlin, Germany;* [5] *Leipziger Energiegesellschaft, Leipzig, Germany*

4AO.8.3 Anomaly Detection in Similarly Behaving Solar Inverters   1141

Pau Mercade Ruiz[1], Gerardo Guerra[1], Gaetana Anamiati[1], Lars Landberg[2]
[1] *GreenPowerMonitor, Barcelona, Spain;* [2] *DNV Denmark, Copenhagen, Denmark*

4AO.8.6 Towards Higher Efficiency: Data Analysis and Optimization of PV String Wiring in a Long-Running Solar Power Plant   1146

Žiga Miklič[1], Janez Krč[1], Marko Topič[1]
[1] *University of Ljubljana, Ljubljana, Slovenia*

Oral SESSION 4AO.9 The Impact of Soiling on PV Systems

4AO.9.3 Qatar Dust Atlas Project: Deployment of a National Field Soiling and Environmental Parameters Monitoring Network   1156

Brahim Aissa[1], Mohamed Abdelrahim[2], Mosab Subeh[1], Amir A. Abdallah[1], Benjamin W. Figgis[1], Juan Lopez-Garcia[1], Veronica Bermudez Benito[1]
[1] *QEERI, Doha, Qatar;* [2] *Bin Omran Trading & Telecommunications, Doha, Qatar*

4AO.9.4 Quality Assurance from Laboratory to Field: Novel Test Solutions for Soiling-Prone PV Systems   1160

Ioannis (John) Tsanakas[1], Rodrigo Moretón[2], Eric Pilat[1], Jorge Solórzano[3], Kévin Garcia[4]
[1] *CEA - INES, Le Bourget-du-Lac, France;* [2] *QPV, Madrid, Spain;* [3] *Entec Solar, Madrid, Spain;* [4] *CNR, Lyon, France*

4AO.9.6 Degradation Root-Cause Numerical Analysis of Around 100 PV Modules   1168
Installed in Hot and Arid Desert Environment

Shahzada Pamir Aly[1], Kaushal Chapaneri [1], Baloji Adothu[1], Jim Joseph John[1], Gerhard Mathiak[1], Vivian Alberts[1]
[1] *DEWA R&D, Dubai, United Arab Emirates*

Oral SESSION 4BO.16 Technology, Performance and Economics of PV in/on Buildings

4BO.16.1 A Systematic Approach for the Integration of BIPV Planning into the   1172
Construction Planning Process

Frank Ensslen[1], Mona Mühlich[1], Jan-Bleicke Eggers[1], Tilmann E. Kuhn[1], Bruno Bueno[1]
[1] *Fraunhofer ISE, Freiburg, Germany*

4BO.16.2 Semitransparent Bifacial PV Windows with Integrated Blinds: Experimental   1178
and Modelling Results

Simona Villa[1], Martin Hurtado Ellmann[1], Roland Valckenborg[1]
[1] *TNO, Eindhoven, The Netherlands*

4BO.16.3 Cost-Effective Energy Transition: Rooftop PV in European Union Buildings   1183

Carmen Maduta[1], Delia D'Agostino[1], Sofia Tsemekidi-Tzeiranaki[2], Luca Castellazzi[1]
[1] *European Commission JRC, Ispra, Italy;* [2] *NRB, Herstal, Belgium*

4BO.16.5 Dynamic BIPV Shading Systems: Performance Analysis for High TRL   1187
Validation and Market Transfer

Tian Shen Liang[1], Paolo Corti[1], Pierluigi Bonomo[1], Francesco Frontini[1]
[1] *SUPSI, Mendrisio, Switzerland*

4BO.16.6 Study on Improvement of Power Generation for a Window by Solar   1194
Radiation Reflected from the Low-E Coating of a Semi-Transparent
Photovoltaic Module that is Equally Arranged Linear Double-Sided Solar
Cells

Kazuhiko Umeda[1], Nobusato Kobayashi[1], Akira Yamaguchi[1], Akihiko Nakajima[2], Kengo Maeda[2], Akihiro Kuraoka[2], Naoki Kadota[2]
[1] *TAISEI, Tokyo, Japan;* [2] *KANEKA, Tokyo, Japan*

Oral SESSION 4BO.17 Characterisation, Reliability and Safety of PV in/on Buildings

4BO.17.3 Experimental Investigation of the Temperature Distribution in a BIPV Facade   1199

Nanna Lysgaard Andersen[1], Markus Babin[1], Sune Thorsteinsson[1]
[1] *DTU Electro, Roskilde, Denmark*

Oral SESSION 4BO.6 Performance and Degradation of PV Systems

4BO.6.4 Trend-Based Predictive Maintenance and Fault Detection Analytics for Photovoltaic Power Plants 1209

Demetris Marangis[1], Andreas Livera[1], George Makrides[1], George E. Georghiou[1]
[1] University of Cyprus, Nicosia, Cyprus

4BO.6.6 PV Module Operating Temperature: Reliable Extraction of Model Parameters from Dynamic Field Data 1214

Anton Driesse[1], Jesus Polo[2]
[1] PV Performance Labs, Freiburg, Germany; [2] CIEMAT, Madrid, Spain

Oral SESSION 4BO.7 Data Driven Field Inspection based on Imaging

4BO.7.1 From Pixels to Insights: A Software Prototype for AI-Driven Complete Diagnostics of PV Plants 1220

John (Ioannis) A. Tsanakas[1], Murielle Stepec[1], Philippe Marechal[1], Duy-Long Ha[1]
[1] CEA - INES, Le Bourget-du-Lac, France

4BO.7.3 Redefining Failure Detection in PV Systems: A Comparative Study of GPT-4o and ResNet's Computer Vision in Aerial Infrared Imagery Analysis 1225

Sandra Gallmetzer[1], Lukas Koester[1], Evelyn Turri[1], Mousa Sondoqah[1], Atse Louwen[1], David Moser[1]
[1] EURAC, Bolzano, Italy

4BO.7.4 Evaluation of Field Measurements on Hail Damage to Photovoltaic Modules 1234

Evelyn Bamberger[1], Alexandre Voirol[1]
[1] OST, Rapperswil, Switzerland

4BO.7.5 Evaluation of Daylight Filters for Electroluminescence Imaging Inspections of c-Si PV Modules 1241

Gisele Alves dos Reis Benatto[1], Thøger Kari[1], Rodrigo Del Prado Santamaria[1], Aysha Mahmood[1], Liviu Stoicescu[2], Sergiu V. Spataru[1]
[1] DTU, Roskilde, Denmark; [2] Solarzentrum Stuttgart, Stuttgart, Germany

Visual SESSION 4BV.3 Operation, Performance and Maintenance of PV Systems

4BV.3.1 In-Situ Maintenance-Free Measurement of Soiling-Induced Power Losses in PV Arrays 1245

Michael Gostein[1], Damien Cosme[2], Quentin Berthet-Rayne[2], Julien Chapon[3], Lluvia Ochoa[3], William Stueve[1], Dhanup Somasekharan Pillai[4], Brahim Aïssa[4], Benjamin W. Figgis[4], Juan Lopez-Garcia[4], Veronica Bermudez Benito[4]

[1] Atonometrics, Austin, United States of America; [2] TotalEnergies, Doha, Qatar; [3] TotalEnergies, Paris, France; [4] QEERI, Doha, Qatar

4BV.3.3 Improving Performance Ratio Calculations through Optimizing Front POA Irradiance Sensor Positioning — 1249

Marc A. N. Korevaar[1], Damon Nitzel[1], Shuo Wang[2], Nate Solofra[3]
[1] OTT Hydromet, Delft, The Netherlands; [2] TUAS, Turku, Finland; [3] Merit Controls, Somerville, United States of America

4BV.3.6 A Method for Detecting PV Module's Degradation due to Increased Local Resistance in Power Plant — 1253

Tohru Kohno[1], Jun Tsunoda[1]
[1] Hitachi, Tokyo, Japan

4BV.3.8 Dependence of Series Resistance on Ideality Factor and Shunt Resistance in Online Photovoltaic Module Parametric Identification — 1258

Heidi Kalliojärvi[1], Kari Lappalainen[1]
[1] Tampere University, Tampere, Finland

4BV.3.9 Predictive Maintenance and Anomaly Detection Analytics for Utility-Scale Photovoltaic Plants — 1264

Jesus Montes-Romero[1], Demeteris Marangis[2], Andreas Livera[2], George Makrides[2], Juergen Sutterlueti[3], Steve Ransome[4], George E. Georghiou[2], Nino Heinzle[3]
[1] University of Jaen, Jaen, Spain; [2] University of Cyprus, Nicosia, Cyprus; [3] Gantner Instruments, Schruns, Austria; [4] Steve Ransome Consulting, Kingston upon Thames, United Kingdom

4BV.3.12 Safety Analysis of PV Systems for Soundproof Tunnel Based on Voltage and Current Mismatch — 1269

Juhee Jang[1], Chongmin Kim[1], Sujeong Oh[1]
[1] Korea Electrical Safety, Wanju, South Korea

4BV.3.14 Improved Modelling of PV Systems with Snow Soiling for Optimized Local Energy Sharing — 1273

Ida Fuchs[1], Ole-Morten Midtgård[1]
[1] NTNU, Trondheim, Norway

4BV.3.15 Ensuring Photovoltaic Module Integrity through Electroluminescence Imaging and Machine Learning Solutions — 1279

Daniel J. Castillo Patton[1], Lucas Viani[1], Fernando García[2], Vicente Parra[1], Sofía Rodríguez-Conde[1], Jesús Cuaresma[1]
[1] Enertis Applus+, Madrid, Spain; [2] UC3M, Madrid, Spain

4BV.3.16 RACONT2050 - Reliability and Comparison of New PV Technologies — 1286

Domenico Chianese[1], Mauro Caccivio[1], Gabi Friesen[1]
[1] SUPSI, Mendrisio, Switzerland

4BV.3.18 Comparative Analysis of String IV Measurement Methods for Fault Detection in Photovoltaic Systems — 1287

Martin Bartholomäus[1], Peter Behrensdorff Poulsen[1], Sergiu Viorel Spataru[1]
[1] Technical University of Denmark, Roskilde, Denmark

4BV.3.22 AI-SafePV: An AI-Based Fault Detection Software Package to Provide Safety in Photovoltaic Arrays — 1293

Aref Eskandari[1], Jafar Milimonfared[2], Amir Nedaei[2], P. Parvin[2], M. Braga[3], Mohammadreza Aghaei[4]
[1] Iran University of Science and Technology, Tehran, Iran; [2] Amirkabir University of Technology, Tehran, Iran; [3] UFSC, Florianópolis, Brazil; [4] NTNU, Ålesund, Norway

4BV.3.23 DetectivePV: A Detection Package for Electrical Faults in Photovoltaic   1297
Arrays based on Machine Learning

Aref Eskandari[1], Jafar Milimonfared[2], Amir Nedaei[2], P. Parvin[2], M. Braga[3], Mohammadreza Aghaei[4]
[1] Iran University of Science and Technology, Tehran, Iran; [2] Amirkabir University of Technology, Tehran, Iran; [3] UFSC, Florianópolis, Brazil; [4] NTNU, Ålesund, Norway

4BV.3.24 Wet Leakage and Insulation Test on String Level Through IEC 61215   1301

Mario Martínez[1], Sergio Suarez[1], Jose Cantisano[1], Jonathan Vilela[1], Jose Maria Alvarez[1], Jose Manuel Rivas[1], Sofia Rodríguez-Conde[1]
[1] Enertis Applus, Madrid, Spain

4BV.3.25 TALOS: Robotics and Artificial Intelligence Living Labs Improving   1304
Operations in PV Scenarios

Nicolas Congouleris[1], Athanasios T. Balafoutis[1], Lisandro Puglisi[2], João Formiga[3], Daniel Albuquerque[3], Bruno Barrionuevo[1]
[1] CERTH, Thermi, Greece; [2] EDP Renewables, Madrid, Spain; [3] EDP NEW, Lisbon, Portugal

4BV.3.26 Harmonising Multi-Sites Measurement of Photovoltaic Systems: 1305
Comprehensive Framework for Real-Life Test Conditions in a Maltese
Environment

Brian Bartolo[1], Brian Azzopardi[1], Alexandre Mignonac[2], Marcus Rennhofer[3], Bernhard Kubicek[3], Rita Ebner[3], Carlos Meza[4], Melodie de l'Epine[5], Eugenia Zugasti[6], Steve Zerafa[7], Kenneth Scerri[8]
[1] The Foundation for Innovation and Research, Birkirkara, Malta; [2] CEA, Cadarache, France; [3] AIT, Vienna, Austria; [4] Anhalt University of Applied Sciences, Anhalt, Germany; [5] Becquerel Institute, Brussels, Belgium; [6] CENER, Pamplona, Spain; [7] PIXAM, Msida, Malta; [8] The University of Malta, Msida, Malta

4BV.3.27 Mediterranean Climate Impact on Photovoltaic Systems: Insights from Malta   1309
and Implications for Future European Integration

Brian Bartolo[1], Brian Azzopardi[1], Alexandre Mignonac[2], Marcus Rennhofer[3], Bernhard Kubicek[3], Rita Ebner[3], Carlos Meza[4], Melodie de l'Epine[5], Eugenia Zugasti[6], Steve Zerafa[7], Kenneth Scerri[8]
[1] The Foundation for Innovation and Research, Birkirkara, Malta; [2] CEA, Cadarache, France; [3] AIT, Vienna, Austria; [4] Anhalt University of Applied Sciences, Anhalt, Germany; [5] BI, Brussels, Belgium; [6] CENER, Pamplona, Spain; [7] PIXAM, Msida, Malta; [8] UoM University of Malta, Msida, Malta

4BV.3.28 Comparison of Physical, Machine Learning and Hybrid Models of   1313
Monofacial and Bifacial PV Systems

Jonas Petzschmann[1], Dirk Stellbogen[1], Manuel Heim[1]
[1] ZSW, Stuttgart, Germany

4BV.3.29 Quantitative Shade Detection for PV Systems Based on Clearsky Data   1314

Achim Schulze[1], Markus Panhuysen[2], Darwin Daume[3], Maximilian Schönau[2]
[1] Rosenheim Technical University of Applied Sciences, Rosenheim, Germany; [2] Smartblue, Munich, Germany; [3] Coburg University of Applied Sciences, Coburg, Germany

4BV.3.30 Assessing Electroluminescence Image Quality with Machine-Learning and 1317
Grey-Level Co-Occurrence Matrix Texture Descriptors

Thøger Kari[1], Aysha Mahmood[1], Rodrigo del Prado Santamaria[1], Gisele Alves dos Reis Benatto[1], Peter Behrensdorff Poulsen[1], Sergiu V. Spataru[1]
[1] DTU, Roskilde, Denmark

4BV.3.31 Forecasting the Lifetime of Photovoltaic Modules through Coupling a 1322
Physics-Based Degradation Model with 3D Heat Transfer Simulations

Timofey Golubev[1]
[1] ThermoAnalytics, Calumet, United States of America

4BV.3.32 Development of a Model to Ensure the Safety of PV Systems Using FMEA 1328

Sujeong Oh[1], Chongmin Kim[1], Juhee Jang[1]
[1] KESCO, Wanju County, South Korea

4BV.3.35 Long-Term Monitoring of Degradation and Defect in High-Voltage Strings 1331
through Dark I-V Measurements

Samuele Chiesa[1], Gian Carlo Dozio[1], Domenico Chianese[2]
[1] SUPSI-ISEA, Lugano, Switzerland; [2] SUPSI-ISAAC, Lugano, Switzerland

4BV.3.36 Machine Learning Techniques for the Assesment of Open Circuit Voltage 1335
Losses in Photovoltaic Systems

Sandra Riaño[1], Jose Domingo Santos[1], Miguel Esteras[1], Amaia Abanda[1], Javier del Ser[1]
[1] TECNALIA, Derio, Spain

4BV.3.39 Real-Time Monitoring and Diagnostic of Rooftop Monofacial PV System 1341
Validated with Thermography

Amr Osama[1], Giuseppe Marco Tina[1], Antonio Gagliano[1], Gabino Jiménez-Castillo[2], Francisco Jose Muñoz-Rodriguez[2]
[1] University of Catania, Catania, Italy; [2] University of Jaén, Jaén, Spain

4BV.3.40 Single Image Geospatial Referencing 1347

Evgenii Sovetkin[1], Andreas Gerber[1], Bernhard Kubicek[2], Bart E. Pieters[1]
[1] Forschungszentrum Jülich, Jülich, Germany; [2] AIT, Vienna, Austria

4BV.3.41 Outdoor Exposure Study on the Performance of Nine Different Types of 1348
Industrial PV Modules under 35° and under 90° Tilt

Carolin Ulbrich[1], Niklas Albinius[1], Luka Wernke[1], Björn Rau[1], Rutger Schlatmann[1]
[1] HZB, Berlin, Germany

4BV.3.43 Photovoltaic Output Power Modeling: a Hybrid Approach 1354

Leticia de Oliveira Santos[1], Francisco Alexandre Andrade Souza[2], Tarek AlSkaif[3], Paulo C. M. Carvalho[1]
[1] UFC, Fortaleza, Brazil; [2] imec-NL, Wageningen, The Netherlands; [3] Wageningen University, Wageningen, The Netherlands

4BV.3.45 Estimation of Annual Power Loss of a Solar PV System due to Rise in the 1357
Cell Temperature: A Case Study for Indian Climate

Shubham Kumar[1], P. M. V. Subbarao[1]
[1] IIT Delhi, New Delhi, India

4BV.3.46 Snow Losses for Different PV Module Designs: Modelling and Validation in 1361
Southern Finland

Shuo Wang[1], Hugo E. Huerta[1], Sami Jouttijärvi[2], Aleksi Heinonen[1], Juha A. Karhu[3], Anders V. Lindfors[3], Kati Miettunen[2], Samuli Ranta[1]
[1] Turku University of Applied Sciences, Turku, Finland; [2] University of Turku, Turku, Finland; [3] Finnish Meteorological Institute, Helsinki, Finland

4BV.3.52 Defect Quantification System Through Aerial Inspections   1365

Mario Martínez[1], Sergio Suarez[1], Daniel Jason[1], Daniel Villoslada[1], Jose Rivas[1], Sofia Rodríguez-Conde[1]
[1] Enertis Solar, Madrid, Spain

4BV.3.53 Shaping European Collaboration on Photovoltaics: A Collaborative Platform   1369
for Simulation and Monitoring (COPLASIMON)

Simone Vitale[1], Jonathan Leloux[2], Hervè Colin[3], Eric Pilat[3], Stéphane Mollier[3], Basem Idlbi[4], Rodrigo Moretón[5], Oscar Anchorena[5], Christophe Salperwyck[6], David Melgar[7], Christian Schill[7]
[1] LuciSun, Sart Dames Avelines, Belgium; [2] LuciSun, Sart-Dames-Avelines, Belgium; [3] CEA, Le Bourget-du-Lac, France; [4] Ulm University of Applied Sciences, Ulm, Germany; [5] Qualifying Photovoltaics, Madrid, Spain; [6] MyLight150, Auvergne-Rhône-Alpes, France; [7] Fraunhofer ISE, Freiburg, Germany

Visual SESSION 4BV.4 Photovoltaic in/on Buildings

4BV.4.1 Performance of Vertically Mounted Bifacial Photovoltaics on High-Rise   1375
Buildings in the Nordic Conditions

Bergpob Viriyaroj[1], Sami Jouttijärvi[2], Matti Jänkälä[1], Kati Miettunen[2]
[1] Aalto University, Espoo, Finland; [2] University of Turku, Turku, Finland

4BV.4.3 Reducing the Angular Colour Dependence of Building Integrated   1381
Photovoltaic Modules Based on Optical Interference Coatings

Chang Chuan You[1], Ørnulf Nordseth[1], Arne Røyset [2], Tore Kolås[2]
[1] Institute for Energy Technology, Kjeller, Norway; [2] SINTEF Industry, Trondheim, Norway

4BV.4.6 Design and Optimization of Structural Colored Interlayers for Building-   1385
Integrated Photovoltaic Applications

Catarina G. Ferreira[1], Irina Vyalih[1], Jani Lamminaho[1], Markus Babin[2], Nanna Lysgaard Andersen[2], Peter Behrensdorff Poulsen[2], Sune Thorsteinsson[2], Karlis Petersons[3], Joel D. Cox[1], Morten Madsen[1]
[1] University of Southern Denmark, Odense, Denmark; [2] DTU, Roskilde, Denmark; [3] Stensborg, Roskilde, Denmark

4BV.4.8 Comparative Analysis of Individual and Collective PV Integration Strategies   1390
for a Residential Neighborhood

Qiuxian Li[1], Natasa Vulic[2], Hanmin Cai[2], Philipp Heer[2]
[1] KU Leuven, Ghent, Belgium; [2] Urban Energy Systems Laboratory, Empa, Duebendorf, Switzerland

4BV.4.10 Modelling Framework for Optimizing Hybrid Photovoltaic-Thermal Systems   1391
in Combination with Seasonal Heat Storage

Zain Ul Abdin[1], Aron van Rossum[1], David Martinez Aguilera[1], D. N. Kanawala[1], Olindo Isabella[1], Rudi Santbergen[1]
[1] TU Delft, Delft, The Netherlands

4BV.4.11 Performance Assessment of Novel Solar Energy Systems for Aged Neighbourhoods and Buildings in Dutch Cities   1392

Edward Otoo[1], Guang Hu[1], Roel C. G. M. Loonen [1], Angèle H.M. E. Reinders [1]

[1] *Eindhoven University of Technology, Eindhoven, The Netherlands*

4BV.4.12 Steel Framing/Structure as a Solution to Support BIPV Competitiveness   1396

Simon Boddaert[1], Jean-Pierre Reyal[2], Michel Dernis[3], Philippe Alamy[4]
[1] *CSTB, Sophia Antipolis, France;* [2] *Semperstyl, Eragny Sur Oise, France;* [3] *Atrium Data, Paris, France;* [4] *EnerBim, Donneville, France*

4BV.4.13 Advanced PV and Thermal Modeling for a Feasible and Efficient BAPV-T System Design and Evaluation   1400

Iñaki Cornago[1], Mikel Ezquer[1], Patxi Sorbet[1], Alicia Kalms[1], Gonzalo Diarce[2], Olatz Irulegi[3], Fritz Zaversky[1]
[1] *CENER, Sarriguren, Spain;* [2] *UPV/EHU, Bilbao, Spain;* [3] *UPV/EHU, San Sebastian, Spain*

4BV.4.14 PV on Green Roofs. Two Years of Comparative Measurement Data from Various System Concepts, Supplemented by Simulation Results and General Considerations   1401

Markus Klenk[1], Roger Glarner[1], Selina Pfyffer[1], Hartmut Nussbaumer[1], Stephan Brenneisen[1], Andreas Dreisiebner[2]
[1] *ZHAW, Winterthur, Switzerland;* [2] *A777 Gartengestaltung, Seuzach, Switzerland*

4BV.4.20 PV Façades > 30 m - Fire Prevention Guidelines on High-Rise Buildings   1409

Urs Muntwyler[1], Eva Schüpbach[1]
[1] *Dr. Schuepbach & Muntwyler, Bern, Switzerland*

4BV.4.22 Semi-Transparent CIGS Thin-Film PV Modules   1412

Peter Borowski[1], Thomas Schutt[2], Julian Röder[1], Maik Schubert[2], Martin Hillmann[2], Kristian Herath[2], Subarna Sapkota[2], Volker Speer[2], Marko Stölzel[1], Rene Reichel[2], Thomas Dalibor[1]
[1] *AVANCIS, Munich, Germany;* [2] *AVANCIS, Torgau, Germany*

4BV.4.23 Assessing Photovoltaic-Thermal System Performance across Diverse Climates: an Economic and Environmental Comparative Analysis   1417

Zain Ul-Abdin[1], Olindo Isabella[1], Rudi Santbergen[1]
[1] *TU Delft, Delft, The Netherlands*

4BV.4.26 A Strategic Approach to Enable Large-Scale Photovoltaic Energy Systems Deployment in Urban Areas   1418

Joyce Arthllan Oliveira de Sousa[1], Martin Thebault[1], Lamia Berrah[1]
[1] *USMB, Annecy, France*

4BV.4.27 Performance Assessment of Colorful BIPV Facade in Norway   1430

Junjie Zhu[1], Jørgen Young[2]
[1] *Institute for Energy Technology, Kjeller, Norway;* [2] *Isola Solar, Larvik, Norway*

4BV.4.28 Implementing Strain Relief for Improved Reliability of BIPV Modules Built on Aluminum Façade Elements   1433

Wiebke Wirtz[1], Kevin Meyer[1], Susanne Blankemeyer[1], Thomas Daschinger[1], Henning Schulte-Huxel[1]
[1] *ISFH, Emmerthal, Germany*

4BV.4.29 CONIPHER BIPV Facades: Design and Performance Prediction   1434

Ya-Brigitte Assoa[1], Philippe Thony[1], Emmanuel Schmitt[2], Olivier Bizzini[3], Stephane Gelibert[3], Vincent Bressy[4], Olivier Wiss[1], Alexandre Plissonnier[1], Zeina Hamam[1]
[1] *CEA, Le Bourget-du-Lac, France;* [2] *Vicat, L'Isle-d'Abeau, France;* [3] *Araymond, Grenoble, France;* [4] *Workspaces-architecture, Grenoble, France*

4BV.4.30 The Potential of Plug&Play PV in Switzerland   1435

Jan Remund[1], Anne-Kathrin Weber[1], Lukas Meyer[1], David Joss[2], Christof Bucher[2], Theo Zwahlen[2]
[1] *Meteotest, Bern, Switzerland;* [2] *BFH, Burgdorf, Switzerland*

4BV.4.32 Integration of Transparent Photovoltaic Panels into Buildings   1439

Nilşah Özar[1], Müjde Altın[1]
[1] *Dokuz Eylül University, Izmir, Türkiye*

4BV.4.33 Integrating FIDES Reliability Prediction into Building-Integrated Photovoltaic Systems   1443

Fereshteh Poormohammadi[1], Martijn Deckers[2], Johan Driesen[1]
[1] *KU Leuven, Leuven, Belgium;* [2] *Energy Ville, Genk, Belgium*

Oral SESSION 4CO.8 Solar Resource Assesment

4CO.8.4 Fast Horizon Algorithm – Case of Integrated PV   1444

Evgenii Sovetkin[1], Andreas Gerber[1], Bart E. Pieters[1]
[1] *Forschungszentrum Jülich, Jülich, Germany*

4CO.8.5 Global Patterns of Solar Resource Short-Term Variability Based on Solargis Time Series Data   1451

Juraj Betak[1], Martin Opatovsky[1], Konstantin Rosina[1], Marcel Suri[1]
[1] *Solargis, Bratislava, Slovakia*

Oral SESSION 4CO.9 Solar Forecasting

4CO.9.1 Can Deep Learning Replace Cloud Motion Vectors?   1456

Nils Straub[1], Steffen Karalus[1], Wiebke Herzberg[1], Elke Lorenz[1]
[1] *Fraunhofer ISE, Freiburg, Germany*

4CO.9.2 Skill-Driven Model Training for Solar Forecasting with Sky Images   1462

Amar Meddahi[1], Arttu Tuomiranta[1], Sebastien Guillon[1]
[1] *TotalEnergies, Palaiseau, France*

4CO.9.3 Ramp Rate Metric Suitable for Solar Forecasting and Nowcasting   1466

Bijan Nouri[1], Yann Fabel[1], Niklas Blum[1], Dominik Schnaus[2], Luis F. Zarzalejo[3], Andreas Kazantzidis[4], Stefan Wilbert[1]
[1] *DLR, Almería, Spain;* [2] *TUM, Munich, Germany;* [3] *CIEMAT , Madrid, Spain;* [4] *University of Patras, Patras, Greece*

4CO.9.4 Fog and Snow Detection to Improve Regional Photovoltaic Power Prediction   1476

Elke Lorenz[1], Steffen Karalus[1], Wiebke Herzberg[1], Tobias Zech[1], Babak Jahani[2], Eva Pauli[3], Jan Cermák[3], Tjade Appel[4], Merle Vespermann[4], Heidrun Misfeld[4], Jan Kühnert[4]

[1] Fraunhofer ISE, Freiburg, Germany; [2] SRON, Leiden, The Netherlands; [3] KIT, Karlsruhe, Germany; [4] energy & meteo systems, Oldenburg, Germany

4CO.9.5 Photovoltaic Power Plants as Efficient Cloud Motion Detectors   1480

Magnus Moe Nygård[1], Erling Ween Eriksen[1], Heine Nygard Riise[1]
[1] IFE, Kjeller, Norway

4CO.9.6 How Connected Cars can Improve Solar Forecasting - Expanding the Scale of   1491
Local Sensor Networks

Tobias Veihelmann[1], Maximilian Lübke[1], Norman Franchi[1]
[1] Friedrich-Alexander-University, Erlangen, Germany

Plenary SESSION 4CP.1 PV Everywhere

4CP.1.1 Dynamic Agrivoltaics: An Agronomical Tool to Protect Crops from Climate   1496
Change - Feedback from 15 Years of Research

Damien Fumey[1], Sophie Bellacicco[1], Gerardo Lopez-Velasco[1], Jérôme Chopard[1], Severine Persello[1], Perrine Juillion[1], Vincent Hitte[1], Yassin Elamri[1], Isaac A. Ramos-Fuentes[1], Jean Garcin[2], Benoît Valle[2], Francis Sourd[2]
[1] Sun'Agri, Paris, France; [2] Sun'R, Paris, France

Plenary SESSION 4CP.2 Performance and Reliability | Thin Films and Tandems

4CP.2.3 Performance of Partial Shaded PV Generators Operated by Optimized Power   1501
Electronics an IEA PVPS T13 Activity

Franz P. Baumgartner[1], Sara Golroodbari[2], Christof Bucher[3], Matthew Berwind[4], Felipe Valencia[5], Ulrike Jahn[6]
[1] ZHAW, Winterthur, Switzerland; [2] University Utrecht, Utrecht, The Netherlands; [3] Bern University, Bern, Switzerland; [4] Fraunhofer ISE, Freiburg, Germany; [5] ATAMOSTEC, Atacama, Chile; [6] Fraunhofer CSP, Halle (Saale), Germany

Visual SESSION 4CV.1 Solar Resource and Forecasting

4CV.1.4 Hindcasting Solar Irradiance by Machine Learning using Photovoltaic Data   1509

Maximilian Schönau[1], Darwin Daume[2], Markus Panhuysen[1], Tristan Kreller[2], Joseph Jachmann[2], Achim Schulze[3], Bernd Hüttl[2], Dieter Landes[2]
[1] Smartblue, Munich, Germany; [2] Coburg University of Applied Sciences, Coburg, Germany; [3] Rosenheim Technical University of Applied Sciences, Rosenheim, Germany

4CV.1.5 Climate Clustering for Photovoltaic Interest   1514

Anastasios Kladas[1], Karel Lagast[1], Bert Herteleer[1], Jan Cappelle[1]
[1] KU Leuven, Leuven, Belgium

4CV.1.6 Advancing Solar Resource Data: the Validation Journey of 3E's   1516
Satellite-Based Irradiation Data

Philippe Malcorps[1], Gofran Chowdhury[1]
[1] 3E, Brussels, Belgium

4CV.1.8 Resource-Efficient PV Energy Yield Nowcasting with Sky Images: a Hybrid Global Annealing Schedule   1519

Markos Kousounadis-Knousen[1], Apostolos Bakovasilis[2], Francky Catthoor[3], Pavlos Georgilakis[1]
[1] *NTUA, Athens, Greece;* [2] *imo-imomec, Genk, Belgium;* [3] *imec, Leuven, Belgium*

4CV.1.9 Variability of Solar Radiation in the Context of a Flat Region Highly Loaded with Aerosols   1524

Dunia A. Bachour[1], Daniel Perez-Astudillo[1]
[1] *QEERI, Doha, Qatar*

4CV.1.10 Towards Climate-Neutral Energy: Assessing Equations for Optimization of Photovoltaic Production Estimates   1528

Mahesh Sutariya[1], Luiz Fonseca[2], Raphael Abrahão[2], Haresh Vaidya[1]
[1] *University of Applied Sciences, Feuchtwangen, Germany;* [2] *Federal University of Paraíba, João Pessoa, Brazil*

4CV.1.14 Irradiance Transposition and Reflections in BIPV Installations   1534

Stefan Grünsteidl[1], Peter Borowski[1], Thomas Dalibor[1]
[1] *Avancis, Munich, Germany*

4CV.1.21 Irradiance Modeling for Integrated PV with OpenStreetMap   1543

Michael Gordon[1], Evgenii Sovetkin[1], Bart E. Pieters[1], Andreas Gerber[1]
[1] *Forschungszentrum Jülich, Jülich, Germany*

4CV.1.23 Availability of Solar Energy on Vehicle Roofs in German Road Network; Validation of Surface Structure Data for Shadow Loss Modelling   1544

Christian Braun[1], Alexander Kleinhans[1], Christian Schill[1], Elke Lorenz[1], Felix Basler[1], Martin Kaiser[1], Nicolas Holland[1]
[1] *Fraunhofer ISE, Freiburg, Germany*

4CV.1.25 Physics Informed Graph Neural Networks for Multi-Site Solar Forecasting   1548

Jelena Simeunovic[1], Baptiste Schubnel[1], Pierre-Jean Alet[1], Pascal Frossard[2], Rafael E. Carrillo[1]
[1] *CSEM, Neuchâtel, Switzerland;* [2] *EPFL, Lausanne, Switzerland*

4CV.1.27 Dimensionality Reduction of Environmental Data for Long-Term PV Performance Analysis Using Graph Based Methods   1552

Srijani Mukherjee[1], Laurent Vuillon[2], Denys Dutykh[3], Ioannis Tsanakas[1]
[1] *CEA, Le Bourget-du-Lac, France;* [2] *CNRS, Chambéry, France;* [3] *Khalifa University, Abu Dhabi, United Arab Emirates*

4CV.1.28 Statistical Methods for Monitoring Pyranometer Drift in Solar Radiation Operational Data   1556

Lucas T. Silva[1], Rodrigo S. Queiroz[1], Nathianne M. Andrade[1], Danielle B. Cavalcante[1]
[1] *Delfos Energy, Barcelona, Spain*

4CV.1.29 Performance Evaluation of Utility-Scale Solar PV Projects in the State of Gujarat, India   1559

Saurabh Motiwala[1], Sudarshan Kumar[1], Ashish Kumar Sharma[2], Ishan Purohit[3]
[1] *IIT Bombay, Mumbai, India;* [2] *University of Petroleum and Energy Studies, Dehradun, India;* [3] *International Finance Corporation, New Delhi, India*

Oral SESSION 4DO.1 PV System Design and Optimisation

4DO.1.1   Enhancing Bifacial Gain: Addressing Tracker Installation Challenges for   1565
Optimized Performance

*Ismail Kaaya[1], David Moser[2], Richard de Jong[1], Olivier Dupon[1], Arnaud
Morlier[1]*
[1] Imo-Imomec, Genk, Belgium; [2] Eurac Research, Bolzano, Italy

4DO.1.2   Assessing the Performance, Reliability, Economic and Environmental Impact   1576
of PV Systems Installation Parameters in Harsh Climates: Case Study Iraq

*Mohammed Adnan Hameed[1], Ismail Kaaya[2], Richard de Jong[2], Roland
Scheer[3], Ralph Gottschalg[1]*
*[1] Fraunhofer CSP, Halle (Saale), Germany; [2] Imec, Genk, Belgium; [3] MLU, Halle (Saale),
Germany*

4DO.1.3   A Techno-Economic Comparison Analysis for Optimal PV Revamping   1584
Strategies

Elina Bosch[1], Philippe Macé[1], Caroline Plaza[2], Gaëtan Masson[1]
[1] Becquerel Institute, Brussels, Belgium; [2] Becquerel Institute, Lyon, France

4DO.1.6   Innovative Setups for Photovoltaic Solar Trackers to Really Boost the   1590
Electricity Generation per Square Meter of Occupied Surfaces

Rosario Carbone[1], Cosimo Borrello[1], Ferdinando Gioia[1]
[1] University "Mediterranea" of Reggio Calabria, Reggio Calabria, Italy

Oral SESSION 4DO.2 The Integrated Agrivoltaic Performance: Approaches, Modelling, Experiences

4DO.2.1   Europe's Agrivoltaic Future: Design of Four Innovative Demonstrators   1596
through Advanced Modeling in the SYMBIOSYST Project

*S Prithivi Rajan[1], Jesus Robledo[1], Jonathan Leloux[1], Christian A.
Gueymard[1], Angelo Pignatelli[2], Giovanni Borz[3], David Moser[3], Ismail
Kaaya[4], Shu-Ngwa Asaa[4], Alexandros Katsikogiannis[5], Martin Thalheimer[6],
Walter Guerra[6], Marcel Macarulla[7], Irma Roig[7], Gil Gorchs[7], Niels Groen[8],
James MacDonald[9], Giuseppe Demofonti[10], Cinja Seick[11], Giacomo Bosco[12]*
*[1] LuciSun, Brussels, Belgium; [2] EF Solare, Milano, Italy; [3] EURAC, Bolzano, Italy; [4] Imec,
Leuven, Belgium; [5] TU Delft, Delft, The Netherlands; [6] Laimburg, Laimburg, Italy; [7] UPC,
Barcelona, Spain; [8] KUBO, South Holland, The Netherlands; [9] Engie-Lab, Barcelona, Spain;
[10] Convert, Roma, Italy; [11] Aleo, Prenzlau, Germany; [12] Physee, Delft, The Netherlands*

Oral SESSION 4DO.3 The Integrated Agrivoltaic Performance: Different Climatic Conditions, Crops and Technologies

4DO.3.3   A Computational Comparison and Validation Between Ray Tracing   1606
Techniques Under Special Light-Sharing Trade off Scenarios in Photovoltaics

Hugo Sánchez Ortiz[1], Roxane Bruhwyler[2], Sebastian Dittmann[1], Nicolas De Cook[2], Carlos Meza[1], Frederic Lebeau[2], Ralph Gottschalg[1]
[1] Hochschule Anhalt, Koethen, Germany; [2] Liege University, Gembloux, Belgium

4DO.3.4 Automatic Agrivoltaic Site Selection: a User-Friendly Interface powered by AHP Multicriteria Decision-Making   1610

Andressa de Sousa Cardoso[1], Alfonso López Ruiz[1], María Isabel Ramos Galán[1], Juan Manuel Jurado[1], Francisco Ramón Feito Higueruela [1]
[1] University of Jaén, Jaén, Spain

Oral SESSION 4DO.4 Vehicle Integrated PV

4DO.4.1 SolarMoves: The Impact on Grid Electricity Demand of VIPV   1616

Anna J. Carr[1], Ashish Binani[1], Akshay Bhoraskar[2], Oscar van de Water[2], Michiel Zult[2], René van Gijlswijk[2], Lenneke Slooff-Hoek[1]
[1] TNO, Petten, The Netherlands; [2] TNO, Den Haag, The Netherlands

4DO.4.3 Simulation and Concept Evaluation of Extendable Lightweight Photovoltaic Modules for Vehicle Integration under Wind Loads   1620

Cornelius Herr[1], Marc Andre Schüler[1], Felix Basler[1], Christopher Daniel Joseph[1], Andreas Beinert[1], Pascal Romer[1], Martin Heinrich[1]
[1] Fraunhofer ISE, Freiburg, Germany

4DO.4.6 VIPV: Urban Shading Effect to Solar Irradiation Estimation Method Using GIS: Case Study in Fukushima, Japan   1626

Pawita Bunme[1], Hidenori Mizuno[1], Takumi Takashima[1], Takashi Oozeki[1]
[1] AIST, Fukushima, Japan

Oral SESSION 4DO.5 Floating, Integrated and Hybrid PV

4DO.5.2 Exploiting the Full Performance Potential of (Offshore) Floating Photovoltaics through Thermal Approaches: an Overview of Options   1634

Oscar Delbeke[1], Jens D. Moschner[1], Johan Driesen[1]
[1] KU Leuven/EnergyVille, Leuven, Belgium

4DO.5.5 Performance of Zigzag Photovoltaics Noise Barrier near a Belgian Highway   1638

Sara Bouguerra[1], Richard de Jong[1], Philip Le[1], Fabio Di Giusto[1], Fallon Colberts[2], Ismail Kaaya[1], Nikoleta Kyranaki[1], Marta Casasola Paesa[1], Elke Deckers[1], Arnaud Morlier[1], Michaël Daenen[1]
[1] IMO-IMOMEC, Diepenbeek, Belgium; [2] Zuyd University, Heerlen, The Netherlands

4DO.5.6 Hybrid (Tandem?) Implementation: Solar Spectrum Splitting PV/CSP for Thermal and Electrical Energy Harvesting   1647

Jonathan Govaerts[1], Bart Reekmans[1], Patrick Choulat[1], Filip Duerinckx[1], Loic Tous[1], Bin Luo[1], Tom Borgers[1], Hariharsudan Sivaramakrishnan Radhakrishnan[1], Jef Poortmans[1], Hannes Laget[2], Qizheng Dou[3], Francis Costa[3], Lieven Stalmans[3], Ravi Kishore[4], Youri Meuret[4], Georgi H. Yordanov[4], Jens Moschner[4], Tatjana Vavilkin[5], Stefan Dewallef[5]
[1] imec, Genk, Belgium; [2] Azteq, Genk, Belgium; [3] Borealis, Beringen, Belgium; [4] KULeuven, Leuven, Belgium; [5] Soltech, Genk, Belgium

Visual SESSION 4DV.1 Dual Use (Floating PV, Agrivoltaics, VIPV) and other Innovative PV Applications

4DV.1.3 Bifacial Panels for Agrivoltaics and Crop Influence: Expected Benefits — 1651

Miguel-Ángel Muñoz-García[1], María Beatriz Nieto[2], Guillermo Pedro Moreda[1], Carmen Alonso-García[2], Luís Fialho[3], Fátima Baptista[3]
[1] UPM, Madrid, Spain; [2] CIEMAT, Madrid, Spain; [3] University of Évora, Évora, Portugal

4DV.1.4 Analysis of the Use of Bifacial Solar Panels in Vertical Placement and their Temporal Coupling in Agrivoltaic Irrigation — 1654

Guillermo-Pedro Moreda[1], Raúl Sánchez-Calvo[1], Luis Juana[1], Delia Rodríguez-Lucas[2], Miguel-Ángel Muñoz-García[1]
[1] UPM, Madrid, Spain; [2] Harvard University, Cambridge, United States of America

4DV.1.6 Design and Methodology for an Agrovoltaic Pilot Project in the Alentejo Region — 1658

Helena Oliveira[1], Lisa Bunge[1], José A. Silva[1], Luís Fialho[1], Paulo Infante[1], Pedro Horta[1]
[1] University of Évora, Évora, Portugal

4DV.1.8 Growing Greener. First Step on the Journey to Maximize Agri-Voltaic Potential. The SYMBIOSYST Project: Monitoring System and Platform — 1663

Giovanni Borz[1], Enrico Dalla Maria[1], David Moser[1], Maitheli Nikam[2], Gofran Chowdhury[2], Alba Perez[3], David Caballero[3], Niels Groen[4], Jennifer Porter[5]
[1] Eurac Research, Bolzano, Italy; [2] 3E, Brussels, Belgium; [3] Universitat Politecnica de Catalunya, Barcelona, Spain; [4] KUBO Greenhouse Projects, Monster, The Netherlands; [5] Above Surveying, Colchester, United Kingdom

4DV.1.9 Assessing the Agrivoltaic Potential in Hot Desert Climates — 1664

Juan Lopez-Garcia[1], Sachin Jain[1], Daniel Perez-Astudillo[1], Dunia Bachour[1], Dhanup Pillai[1], Veronica Bermudez-Benito[1]
[1] HBKU, Doha, Qatar

4DV.1.10 AgriPV in Norway: Evaluating the Initial Performance and Lessons Learned — 1669

Steve Völler[1], Marisa Di Sabatino[1], Richard J. Randle-Boggis[2], Gaute Stokkan[2]
[1] NTNU, Trondheim, Norway; [2] SINTEF Industry, Trondheim, Norway

4DV.1.12 Dual-Use Potential of Agrivoltaics in Portugal – a Case Study in Baixo Alentejo — 1675

Cláudia Fernandes[1], Jose Almeida Silva[2], Jeremias dos Santos[2], Lisa Bunge[2], André Soeiro[3], Luís Fialho[2], Pedro Horta[2], Daniel Albuquerque[1], Filipe Serra[1], Diogo Cordeiro[3]
[1] EDP NEW, Sacavém, Portugal; [2] University of Évora, Évora, Portugal; [3] EDP Generation, Lisbon, Portugal

4DV.1.16 IEA HEV TCP PVPS Task 17: VIPV Business Plan - the Long Way to the Mass Market — 1679

Urs Muntwyler[1], Eva Schüpbach[1]
[1] Dr. Schüpbach & Muntwyler, Bern, Switzerland

4DV.1.17 Cost-Competitiveness Analysis of Infrastructure Integrated PV 1682

André Penas[1], Elina Bosch[1], Philippe Macé[1], Gaëtan Masson[1], Caroline Plaza[2], Jose Maria Vega de Seoane[3]
[1] *Becquerel Institute, Brussels, Belgium;* [2] *Becquerel Institute, Lyon, France;* [3] *Becquerel Institute, San Sebastián, Spain*

4DV.1.22 Sierra Brava Floating Photovoltaic Plant: Real Data vs Simulation Software 1687

Dorivaldo Duarte[1], Luis Fialho[1], Sara Pereira[1], José Silva[1], Manuel Collares-Pereira[1], Pedro Horta[1], Maria Cebria[2], Nerea Vidal[2]
[1] *University of Évora, Évora, Portugal;* [2] *ACCIONA Energía, Madrid, Spain*

4DV.1.23 Numerical Model for Wave Motions and Loads of Multibody Floating Photovoltaic Structures 1691

Antonio Mikulić[1], Ivan Catipovic[1], Neven Alujević[1], Inno Gatin[2]
[1] *University of Zagreb, Zagreb, Croatia;* [2] *Cloud Towing Tank, Zagreb, Croatia*

4DV.1.24 Port of Sines Energy Transition: Photovoltaic Solutions Addressing R[4] Concept 1696

Joana Correia[1], Luís Fialho[1], José Silva[1], Pedro Horta[1]
[1] *University of Évora, Évora, Portugal*

4DV.1.27 Accelerate Product Development for PV in Alpine Installations 1704

Anika Gassner[1], Ebrar Özkalay[2], Gabriele C. Eder[1], Gabi Friesen[2], Markus Feichtner[3], Mauro Caccivio[2], Friedrich Bleicher[4]
[1] *OFI, Vienna, Austria;* [2] *SUPSI PVLab, Mendrisio, Switzerland;* [3] *Sonnenkraft Energy, St. Veit a.d. Glan, Austria;* [4] *TU Wien, Vienna, Austria*

4DV.1.29 Back Irradiance Measurements and Influence of the Ground Coverage on the Production of a Bifacial Agrivoltaics System 1705

Diogo Vicente[1], Dmitri Boutov[1], João M. Serra[1]
[1] *University of Lisbon, Lisbon, Portugal*

4DV.1.33 Assessing the Energy Yield and Irradiation Distribution in Fixed and Tracking Agrivoltaic Orchards 1709

Shu-Ngwa Asa'a[1], Ismail Kaaya[1], Olivier Dupon[1], Richard de Jong[1], Arvid van der Heide[1], Arnaud Morlier[1], Hariharsudan Sivaramakrishnan Radhakrishnan[1], Jef Poortmans[2], Michael Daenen[1]
[1] *Hasselt University/Imo-Imomec, Genk, Belgium;* [2] *Imo-Imomec, Genk, Belgium*

4DV.1.34 Economic Attractiveness of Agrivoltaics in Different Regulation Statuses – Case Study 1715

Carolina Plaza[1], Julien Van Overstraeten[2], André Penas[2], Elina Bosch[2], Melodie de l'Epine[1], Philippe Macé[2], Gaëtan Masson[2]
[1] *Becquerel Institute, Lyon, France;* [2] *Becquerel Institute, Brussels, Belgium*

4DV.1.38 Optimizing Land Productivity with Customized Tracking Algorithms for Single-Axis Trackers in Agrivoltaic Systems 1719

Gaurang Chhapia[1], Djaber Berrian[1], Johannes Linder[1]
[1] *Belectric, Kolitzheim, Germany*

4DV.1.39 Potential and Techno-Economic Feasibility Assessment of Utility-Scale Floating Solar Photovoltaics (FSPV) in India 1724

Saurabh Motiwala[1], Sudarshan Kumar[1], Ashish Kumar Sharma[2], Ishan Purohit[3]

[1] IIT Bombay, Mumbai, India; [2] University of Petroleum and Energy Studies, Dehradun, India; [3] International Finance Corporation, New Delhi, India

Visual SESSION 4DV.4 PV System Engineering | Control and Systems for Power Systems with Renewables Integration

4DV.4.1 Assessing Glare Hindrance Three Ways in Fixed Tilt PV Systems　　1729

Ashish Binani[1], Antonius R. Burgers[1], Kay Cesar[1], Bas Van Aken[1]
[1] TNO, Petten, The Netherlands

4DV.4.3 Complementary Guide for the Electrical Design of Grid-Connected PV　　1733
Systems

Bruno Gaiddon[1], Marielle Perrin[1], Elika Saidi-Chalopin[2], Salomé Durand[3], David Gréau[4], Dimitri Gagnaire[5], Mathieu Mansouri[6], François Saugues[7], Olivier Verdeil[8], Gérard Moine[8]
[1] Hespul, Lyon, France; [2] Consuel, Paris, France; [3] SER, Paris, France; [4] Enerplan, La Ciotat, France; [5] CEATECH-INES, Le Bourget-du-Lac, France; [6] CRER, La Crèche, France; [7] Stäubli, Hésingue, France; [8] Solarcoop, Mornant, France

4DV.4.5 Increasing the Proportion of Winter Electricity through Design Optimisation　　1736
of Photovoltaic Roof Systems

Hartmut Nussbaumer[1], Roger Hiltebrand[1], Selina Pfyffer[1], Andreas Dreisiebener[2], Markus Klenk[1]
[1] ZHAW, Winterthur, Switzerland; [2] A777 Gartengestaltung, Seuzach, Switzerland

4DV.4.7 Implementation of a Sub-Hourly Clipping Correction in PVsyst　　1740

Michele Oliosi[1], Bruno Wittmer[1], André Mermoud[1], Agnes Bridel-Bertomeu[1], Robin Vincent[1]
[1] PVsyst, Satigny, Switzerland

4DV.4.10 Highest Energy Yields per Area for PV Systems on Flat Roofs　　1747

Hartmut Nussbaumer[1], Roger Hiltebrand[1], Selina Pfyffer[1], Lona Tulinski[1], Janis Preisig[1], Markus Klenk[1]
[1] ZHAW, Winterthur, Switzerland

4DV.4.12 Impacts of Measures to Achieve Dispatchability on the Cost of PV-BESS　　1752
Power Plants

Alex Renan Arrifano Manito[1], Pedro Torres[1], Marcelo Pinho Almeida[1], Gilberto Figueiredo[2], José Cesar Almeida[3], Roberto Zilles[1]
[1] USP, Sao Paulo, Brazil; [2] Fluminense Federal University, Niterói, Brazil; [3] Mackenzie Presbyterian University, Sao Paulo, Brazil

4DV.4.13 Analysis of Irradiation Differences on Substring Level of Modules in Solar　　1758
Parks

Sascha Eckerter[1], Krisztián Kerekes[2], Patrick Mader[2], Rainer Merz[2]
[1] University of Applied Science Karlsruhe, Ettlingen, Germany; [2] University of Applied Science Karlsruhe, Karlsruhe, Germany

4DV.4.14 Using Standard PV Mounting Structures with Spaced Modules in Agrivoltaic　　1763
Applications

Alex Renan Arrifano Manito[1], Marcelo Pinho Almeida [1], Bruno Jacomel Vieira[1], Maria Cristina Fedrizzi[1], Roberto Zilles[1]
[1] USP, São Paulo, Brazil

4DV.4.15 Optimization Analysis for the Best Sizing and Operation of Photovoltaic   1769
Generators in Distributed Electricity Systems

Jacopo Baldacci[1], Ciro Lanzetta[1], Antonio Piazzi[1], Nabi Taheri[2], Mauro Tucci[2]
[1] *i-EM, Livorno, Italy;* [2] *University of Pisa, Pisa, Italy*

4DV.4.19 Optimising Solar Asset Performance through Smart Module Installation using   1770
Above's Digital Twin Technology

Imke Meyer[1], Chisanupong Thawanyavitchajit[2], Inaki Perez[3], Will Hitchcock[4], Henrique Balchada[4], Jennifer Porter[4]
[1] *Mott MacDonald, Brighton, United Kingdom;* [2] *Mott MacDonald, Bangkok, Thailand;* [3] *Mott MacDonald, Madrid, Spain;* [4] *Above Surveying, Colchester, United Kingdom*

4DV.4.20 Experimental Comparison of Solar Absorption Characteristics Using   1775
Different Colors

Sedong Kim[1]
[1] *KITECH, Chungcheongnam-do, South Korea*

4DV.4.21 Solar Roof Potential Analysis Case Study: Test Area in South of Germany   1776

Sabrina Krähmer[1], Basem Idlbi[1], Kaouther Belkilani[1], Dietmar Graeber[1]
[1] *Ulm University of Applied Sciences, Ulm, Germany*

Oral SESSION 4EO.2 Planning of PV Systems | Digital PV

4EO.2.1 BIPV and PV in a Multidisciplinary Building Information Modelling (BIM)   1782
Planning and Asset Management System

Astrid Schneider[1], Karin Stieldorf[1], Christian Schranz[1], Harald Urban[1], Alfred Waschl[2], Markus Feichtner[3], Fedele Rende[4], Andreas Aiello[5], Martin Hauer[6], Kurt Battisti[7], Markus Dörn[7], Jaqueline Scherret[7], Martin Treberspurg[8], Christoph Treberspurg[8]
[1] *TU Wien, Vienna, Austria;* [2] *buildingSMART, Vienna, Austria;* [3] *Sonnnenkraft Energy, Veith, Austria;* [4] *ACCA Software, Bagnoli, Italy;* [5] *ACCA Software, Vienna, Austria;* [6] *Bartenbach, Aldrans, Austria;* [7] *A-Null Development, Vienna, Austria;* [8] *Treberspurg and Partner, Vienna, Austria*

4EO.2.2 Assessing Yield Disparities: Anticipated Versus Optimal Rooftop Solar   1790
Photovoltaic Systems and Implications for Prosumer Viability

Dominik Keiner[1], Dmitrii Bogdanov[1], Stefan Krauter[2], Christian Breyer[1]
[1] *LUT University, Lappeenranta, Finland;* [2] *Paderborn University, Paderborn, Germany*

4EO.2.3 Energy Yields and Wind Loads of Alternative PV Designs for Roofs in   1799
Snowy Climates

Maria Svedjeholm[1], Josefin Lampa[1], Anna Malou Petersson[1], Arvid Olofsson[1], Robin Andersson[2], Ehsan Fooladgar[1], Pirjo Estola[3], Mattias Lindh[1]
[1] *RISE, Umeå, Sweden;* [2] *Luleå Technical University, Luleå, Sweden;* [3] *Luleå Energi, Luleå, Sweden*

4EO.2.4 Digital Twin of Photovoltaic Power Plants Considering Spatio-Temporal   1806
Characteristics

Faruk Ugranlı[1], Eşref Deniz[2], Engin Karatepe[3]
[1] Izmir Bakircay University, Izmir, Türkiye; [2] Entegro Enerji Sistemleri, Izmir, Türkiye; [3] Ege University, Izmir, Türkiye

4EO.2.5 Fully Privacy Preserving Net-load Prediction with Federated Learning and Homomorphic Encryption   1812

Grazia Barchi[1], Mousa Sondoqah[1], Atse Louwen[1], David Moser[1]
[1] EURAC, Bolzano, Italy

Oral SESSION 5CO.4 PV Module Recycling

5CO.4.2 Comparative Analysis of Layer Thickness Measurement Methods for Photovoltaic Modules: a Comprehensive Study   1825

Lukas Neumaier[1], Martin De Biasio[1], Gabriele C. Eder[2], Anika Gassner[2]
[1] Silicon Austria Labs, Villach, Austria; [2] OFI, Vienna, Austria

5CO.4.4 Characterization of the Output-Fractions from Different Mechanical PV-Recycling Approaches   1828

Anika Gassner[1], Gabriele C. Eder[1], Ferozan Azizi[2], Sonja Feldbacher[3], Friedrich Bleicher[4]
[1] OFI, Vienna, Austria; [2] MUL, Leoben, Austria; [3] PCCL, Leoben, Austria; [4] TU Vienna, Vienna, Austria

Oral SESSION 5CO.5 End-of-Life PV Modules & Ecology

5CO.5.2 Comparative Analysis of Recycled Content in Metals used for Photovoltaic Applications   1835

Martina Goverts[1], Simona Villa[2], Mirjam Theelen[2]
[1] Eindhoven University of Technology, Eindhoven, The Netherlands; [2] TNO Energy and Materials Transition, Eindhoven, The Netherlands

5CO.5.3 PV Module ID: Data Driven Results to Enable PV Circularity and Address Toxicity Concerns   1841

Taylor L. Curtis[1], Ashley Gaulding[1], Ligia Smith[1]
[1] NREL, Golden, United States of America

5CO.5.4 Standardisation Activities on the Reuse of PV Modules in IEC TC82   1849

Arvid van der Heide[1], Serge Noels[2], Jan Clyncke[2], Rich Strömberg[3]
[1] imec/imo-imomec, Genk, Belgium; [2] PV CYCLE, Brussels, Belgium; [3] University of Alaska Fairbanks, Fairbanks, United States of America

Oral SESSION 5CO.6 Life Cycle Assessment of PV

5CO.6.1 Sustainability Improvement of C-Si PV Manufacturing through Technology Choices   1853

Moritz Fath[1], Mehul Raval[1], Wolfgang Jooss[1], Peter Fath[1]
[1] RCT Solutions, Constance, Germany

5CO.6.3 A Simplified Model to Assess the Greenhouse Gas Emissions of
Perovskite/Silicon Tandem Modules 1859

Lu Wang[1], Paula Perez-Lopez[1], Raphaël Jolivet[1], Mathilde Marchand[1], Lars Oberbeck[2]
[1] *PSL University, Sophia Antipolis, France;* [2] *TotalEnergies, Paris Saclay, Norway*

5CO.6.4 Carbon Footprint vs Reliability of Solar Photovoltaic Modules: A New
Dilemma? 1862

Alessandro Virtuani[1], Alexis Barrou[1], Bertrand Paviet-Salomon[1], Gianluca Cattaneo[1], Matthieu Despeisse[1], Christophe Ballif[1]
[1] *CSEM, Neuchâtel, Switzerland*

5CO.6.5 Are BIPV Contributing to Environmental Sustainability? An Environmental
LCA Analysis of Innovative BIPV Solutions 1873

Cristina Polacchi[1], Atse Louwen[1], Mirjam Theelen[2], David Moser[1]
[1] *Eurac Research, Bolzano, Italy;* [2] *TNO partner in Solliance, Eindhoven, The Netherlands*

5CO.6.6 The Influence of Climate Specific Degradation on the Greenhouse Gas
Emissions of PV Electricity 1885

Karl-Anders Weiß[1], Sina Herceg[1], Marie Fischer[1], Ismail Kaaya[2], Julian Ascencio-Vásquez [3], Liselotte Schebek[4]
[1] *Fraunhofer ISE, Freiburg, Germany;* [2] *EnergyVille, Genk, Belgium;* [3] *Envision Digital, Redwood, United States of America;* [4] *Technical University of Darmstadt, Darmstadt, Germany*

Plenary SESSION 5CP.1 PV Everywhere

5CP.1.2 Where Agriculture meets Energy: Assessing EU's Agrivoltaic Potential 1894

Anatoli Chatzipanagi[1], Georgia Kakoulaki[1], Nigel Taylor[1], Robert Kenny[1], Sandor Szabó[1], Ana Martinez Fernandez[1], Arnulf Jaeger-Waldau[1]
[1] *European Commission JRC, Ispra, Italy*

Oral SESSION 5DO.10 Manufacturing PV in Europe | Social Aspects of PV

5DO.10.1 Would an Increase in PV Modules Prices Impact the European PV Market? 1903

Johan Lindahl[1], Gaëtan Masson[2], Elina Bosch[2], Amelia Oller Westerberg[1]
[1] *Becquerel Sweden, Stockholm, Sweden;* [2] *Becquerel Institute, Brussels, Belgium*

Oral SESSION 5DO.11 Value and Competitiveness of PV in the Growing Market

5DO.11.1 A Snapshot of Global PV Market - 2023 1906

Gaëtan Masson[1], Melodie de l'Epine[2], Arnulf Jäger Waldau[3], Izumi Kaizuka[4], Amelia Oller Westerberg[5], Jose Donoso[6]
[1] *IEA PVPS Task 1, Brussels, Belgium;* [2] *IEA PVPS Task 1, Lyon, France;* [3] *European Commission JRC, Ispra, Italy;* [4] *RTS Corporation, Tokyo, Japan;* [5] *Becquerel Institute, Knivsta, Sweden;* [6] *UNEF, Madrid, Spain*

5DO.11.2 Driving the Quest for Reliable and Bankable PV in Europe - Status and Targets in 2030   1909

Ulrike Jahn[1], David Moser[2], Delfina Muñoz[3], Paula Sánchez-Friera[4]
[1] *Fraunhofer CSP, Munich, Germany;* [2] *EURAC, Bolzano, Italy;* [3] *CEA, Le Bourget du Lac, France;* [4] *Solkeys, Gijón, Spain*

5DO.11.3 Is the Value of (BI)PV Increasing or Decreasing Over Time?   1917

Wouter L. Schram[1], Elham Shirazi[1]
[1] *University of Twente, Enschede, The Netherlands*

5DO.11.4 The Role of Flexible Demand in Reducing the Utility-Scale PV Integration Costs: an Italian Case-Study   1921

Elisa Veronese[1], Giampaolo Manzolini[1], Grazia Barchi[1], David Moser[1]
[1] *EURAC Research, Bolzano, Italy*

5DO.11.6 Cost Analysis for a Small-Scale Hybrid, Hydrogen-Based PV Energy System   1932

Marius C. Möller[1], Stefan Krauter[1]
[1] *University of Paderborn, Paderborn, Germany*

Oral SESSION 5DO.14 Energy System Integration with Storage

5DO.14.2 Effects of the Operating Point on PV Systems Equipped with Energy Storage   1937

Kari Lappalainen[1]
[1] *Tampere University, Tampere, Finland*

Oral SESSION 5DO.15 Resilience and Security of Supply

5DO.15.1 Extraction of PV Yield Data from Smart Meter Data Disaggregation   1943

Bas van der Ploeg[1], Wilfried van Sark[1]
[1] *Utrecht University, Utrecht, The Netherlands*

5DO.15.2 Development of an Architecture for Power Interchange by Linking Photovoltaic and Electrification Vehicles   1953

Jun Tsunoda[1], Tohru Kohno[1], Issei Suemitsu[1], Kengo Kumano[1]
[1] *Hitachi, Tokyo, Japan*

5DO.15.3 Possibilities of PV Maximization for Achieving Positive Energy Districts with Respect to Building Density   1957

Helmut Bruckner[1], Maarten Verkou[2], Simon Schneider[3], Miro Zeman[2], Zain Ul Abdin[4], Rudi Santbergen[4], Olindo Isabella[4]
[1] *Sonnenplatz Grossschoenau, Grossschoenau, Austria;* [2] *PV Works, Delft, The Netherlands;* [3] *UAS Technikum Wien, Vienna, Austria;* [4] *TU Delft, Delft, The Netherlands*

5DO.15.5 Reliability Analysis of Coupled PV-Electrolyser Systems – Evaluation of Onsite Factors   1961

Stefan Niederhofer[1], Marcus Rennhofer[1], Rene Hofmann[2]
[1] *AIT, Vienna, Austria;* [2] *TU Wien, Vienna, Austria*

5DO.15.6 Grid Supporting Power Plants with 100% Energy from Wind and PV   1966

Gerhard Mütter[1], Andreas Hensel[2], Jan Winkelmann[3]
[1] *Gerhard Mütter, Waldneukirchen, Austria;* [2] *Fraunhofer ISE, Freiburg, Germany;* [3]
VENSYS Elektrotechnik, Diepholz, Germany

Visual SESSION 5DV.2 Energy System Integration; Resilience and Security of Supply; Solar Fuels, Storage | PV Sustainability

5DV.2.1 Techno-Economic Analysis of Residential PV-Battery Energy System in Nordics 1969

Lauri Karttunen[1], Sami Jouttijärvi[1], Johannes Niskanen[1], Jerzy J. Jasielec[1], Hugo Huerta[2], Samuli Ranta[2], Kati Miettunen[1]
[1] *University of Turku, Turku, Finland;* [2] *Turku University of Applied Sciences, Turku, Finland*

5DV.2.2 On the Statistics of Photovoltaics in Europe 1976

Wilfried van Sark[1], Anton Driesse[2]
[1] *Utrecht University, Utrecht, The Netherlands;* [2] *PV Performance Labs, Freiburg, Germany*

5DV.2.4 Sizing of Energy Storage Systems for Different Levels of PV and Wind Power in Combined PV-Wind Power Plants 1977

Micke Talvi[1], Kari Lappalainen[1]
[1] *Tampere University, Tampere, Finland*

5DV.2.6 Quantitative Evaluation Method for Regional Variations in Electricity Supply-Demand Balance Fluctuation by Weather Forecast Error 1982

Issei Suemitsu[1], Tohru Kohno[1], Jun Tsunoda[1], Kengo Kumano[1]
[1] *Hitachi, Tokyo, Japan*

5DV.2.8 From Predictions to Profit of a Hybrid Prosumer Pilot: a Forecast-based Robust Battery Dispatch 1988

Mojtaba Eliassi[1], Anouk Hut[1], Gofran Chowdhury[1]
[1] *3E Belgium, Brussels, Belgium*

5DV.2.9 Hybrid Energy Storage Systems Design Tool 1989

Ana Foles[1], Luís Fava[1], Luís Fialho[1], Pedro Matos[2], José Silva[1], Pedro Horta[1]
[1] *University of Évora, Évora, Portugal;* [2] *Capwatt Services, Maia, Portugal*

5DV.2.10 Solar PV and Battery Microgrid for Electric Cooking - Case Study Eco Moyo Education Centre in Kenya 1995

Audun Bangsund[1], Stian Rummelhoff[1], Ida Fuchs[1]
[1] *NTNU, Trondheim, Norway*

5DV.2.11 Integrating Bifacial PV Power Forecasting into Energy Management Systems at High Latitudes 2001

Hugo E. Huerta[1], Shuo Wang[1], Samuli Ranta[1]
[1] *Turku UAS, Turku, Finland*

5DV.2.12 Optimization of Vanadium Redox Flow Battery Performance for Solar PV Integrated Electric Vehicle Charging Station 2005

Ankur Bhattacharjee[1]
[1] *BITS Pilani, Hyderabad, India*

5DV.2.14 Challenges and Lessons in Residential Energy Storage Projects 2009

Amanda Mendes Ferreira Gomes[1], Aline Kirsten Vidal de Oliveira[1], Marília Braga[1], Ricardo Rüther[1]
[1] UFSC, Florianopolis, Brazil

5DV.2.17 Load Shifting in Energy Communities by Providing User-Centered Recommendations – Forecast, Optimization and Potential   2015

Lukas Gaisberger[1], Georgios Chasparis[2], Wolfgang Traunmüller[3]
[1] University of Applied Sciences Upper Austria, Wels, Austria; [2] Software Competence Center Hagenberg, Hagenberg, Austria; [3] BLUE SKY Wetteranalysen, Attnang, Austria

5DV.2.18 Fast Oscillations Damping Control for PV-BESS Power Plants   2021

Alex Renan Arrifano Manito[1], Pedro Torres[1], Marcelo Pinho Almeida[1], Gilberto Figueiredo[2], José Cesar Almeida[3], Roberto Zilles[1]
[1] USP, São Paulo, Brazil; [2] Fluminense Federal University, Niterói, Brazil; [3] Mackenzie Presbyterian University, São Paulo, Brazil

5DV.2.19 Optimal Use of Batteries on PV Systems for Solving Problems Caused by Predictable Partial Shadings   2027

Rosario Carbone[1], Cosimo Borrello[1], Ferdinando Gioia[1]
[1] University "Mediterranea" of Reggio Calabria, Reggio Calabria, Italy

5DV.2.20 Techno-Economic Assessment of Pumped Storage Hydro Power in Hybrid Operation with Floating Photovoltaic and Battery Energy Storage   2033

Andreas Patha[1], Sebastian Steinlechner[1], Johannes Kathan[1], Antonia Golab[2], Johann Auer[2]
[1] AIT, Vienna, Austria; [2] Technical University of Vienna, Vienna, Austria

5DV.2.22 Open Architecture for Battery Interfaces: Opportunities for Technological Advancements and Community Benefits   2040

Anna Ponomarenko[1], Konstantin Rozanov[2], Claudia Gutierrez Collave[2], Saif Al-Bajjali[2]
[1] Lauder Business School, Vienna, Austria; [2] CF Energy, Vienna, Austria

5DV.2.27 Guideline on Life Cycle Assessment of Agrivoltaic Systems   2044

Maria Anna Cusenza[1], Andrea Danelli[1], Pierpaolo Girardi[1]
[1] RSE, Milan, Italy

5DV.2.31 Intermediate Environmental Assessment   2045

Rene Peche[1], Karsten Wambach[1]
[1] Bifa Environmental Institute, Augsburg, Germany

5DV.2.37 Holistic Assessment of Scenarios for Future PV Deployment Considering Circular Economy in the EU Using PV ICE   2046

Fabian Spera[1], Andreas Schwarz[1], Robin Graeber[1], Oliver Pfeiffer[1], Ulf Blieske[1]
[1] Cologne University of Applied Sciences, Cologne, Germany

5DV.2.38 Development and Testing of a Thermomechanical Procedure to Assess the Disassembly Potential of a Photovoltaic Module   2052

Asier Murillo[1], Cristina Pinto[1], Alicia Buceta[1], Eugenia Zugasti[1], Antonio Urbina[2], Jaione Bengoechea[1]
[1] CENER, Sarriguren, Spain; [2] UPNS, Pamplona, Spain

5DV.2.40 Which is the Most Environmentally Friendly PV Technology: c-Si Solar Cell or Perovskite Silicon Tandem Solar Cell?   2056

Elisabetta Brivio[1], Andrea Danelli[1], Maria Anna Cusenza[1], Sofia Spagnolo[1], Pierpaolo Girardi[1]
[1] *RSE, Milan, Italy*

5DV.2.41 Considering the Environmental Consequences of the Evolution of the Risk of Extreme Natural Events on a PV Installation: a Morphological Analysis-based Prospective Method Applied to Life Cycle Assessment   2057

Alejandra Cue Gonzalez[1], Eric Rigaud[1], Paula Perez-Lopez[1], Philippe Blanc[1]
[1] *PSL University, Sophia Antipolis, France*

5DV.2.42 Riding the Wave: Opportunities and Constraints to Reuse and Resale of Photovoltaic PV Modules in South Africa   2062

Nicole M. Crozier[1], Jacqueline L. Crozier McCleland[2], Ernest E. van Dyk[2], Catherina Schenck[1], Palisa G. Ntsala[2]
[1] *University of the Western Cape, Cape Town, South Africa;* [2] *Nelson Mandela University, Nelson Mandela Bay, South Africa*

5DV.2.47 Circularity in the PV Industry Analysis of Environmental Impacts for Reused PV Panels   2068

Alejandra Galarza[1], Pierre-Philippe Grand[1], Nicolas Vandamme[1], Anaïs Gouabault[2], Juan Alzate[2], Nicolas Defrenne[2], Marie Lacombe[2], Lars Oberbeck[1]
[1] *IPVF, Palaiseau, France;* [2] *SOREN, Paris, France*

5DV.2.51 High Vacuum Flat Plate Hybrid Photovoltaic-Thermal Collectors: Economic and Environmental Comparison over Stand-Alone Devices   2072

Annalisa Di Napoli[1], Paolo Strazzullo[1], Roberto Russo[2], Marilena Musto[1]
[1] *University of Naples Federico II, Naples, Italy;* [2] *National Research Council of Italy, Naples, Italy*

5DV.2.53 Separation of EoL PV Modules Using Liquid-Based Methods to Achieve Better Recycling Quality   2077

Sonja Feldbacher[1], Daniel Schwabl[2], Ferozan Azizi[3], Gabriele Eder[4], Anika Gassner[4], Thomas Nigl[3], Gernot Oreski[1]
[1] *PCCL, Leoben, Austria;* [2] *Circulyzer, Leoben, Austria;* [3] *University of Leoben, Leoben, Austria;* [4] *OFI, Vienna, Austria*

Visual SESSION 5DV.3 PV Diversification Upstream and Downstream - from Industry to Applications | Costs, Economics, Finance and Markets | The Revolution of PV

5DV.3.6 The Role and Impact of Rooftop PV in the Norwegian Energy System under Different Energy Transition Pathways   2078

Stine Fleischer Myhre[1], Eva Rosenberg[1], Heine Nygard Riise[1]
[1] *IFE, Kjeller, Norway*

5DV.3.7 Towards a Common Strategy for Agri-PV in Europe - the Italian Perspective   2084

Celeste Mellone[1], Alessandra Scognamiglio[2], Giancarlo Ghidesi[3], Giulia Guidetti[4], Fabio Salis[5]
[1] *Green Horse Advisory, Rome, Italy;* [2] *ENEA, Rome, Italy;* [3] *RemTec, Rome, Italy;* [4] *Green Horse Legal Advisory, Milan, Italy;* [5] *Iberdrola Renovables, Roma, Italy*

5DV.3.9 The Impacts of Large-Scale Implementation of Solar Power in the Nordic Power Market   2087

Dilshika Heenatigala Kankanamge[1], Jaakko Jääskeläinen[1], Sanna Syri[1]
[1] *Aalto University, Espoo, Finland*

5DV.3.11 Technical and Economic Analysis of the Implementation of Battery Energy Storage Systems (BESS) for Nodes in the National Electric System (SEN) with High Concentration of Solar Energy   2092

Fernando Flores Lizana[1], Patricio Valdivia-Lefort[2]
[1] *Federico Santa Maria Technical University, Santiago, Chile;* [2] *Universidad de Santiago de Chile, Santiago, Chile*

5DV.3.13 The Role of Coupling the Heating, Cooling and Power Sectors to Achieve 100% Renewable Heating and Cooling in Europe   N/A

Olgu Birgi[1], Dominik Rutz[1], Rainer Janssen[1]
[1] *WIP Renewable Energies, Munich, Germany*

5DV.3.21 Fabrication Planning of Module Manufacturing Plants – Analysis of Site Parameters and Modelling Tools   2095

Max Mittag[1], Hannah Hoffman[1], Christian Reichel[1], Dirk Holger Neuhaus[1]
[1] *Fraunhofer ISE, Freiburg, Germany*

5DV.3.26 Techno-Economic and Life-Cycle Assessments of Recycling Pathways for Perovskite on Silicon Tandem Modules   2099

Lian Duan[1], Alejandra Galarza[1], George Wong[1], Lars Oberbeck[2]
[1] *IPVF, Palaiseau, France;* [2] *TotalEnergies OneTech, Paris La Défense, France*

5DV.3.27 Sensitivity of Electricity Price in the Finnish Market Conditions with Increasing Solar Energy Production   2102

Sami Jouttijärvi[1], Lauri Karttunen[1], Seela Tervo[2], Hugo Huerta[3], Samuli Ranta[3], Sanna Syri[2], Kati Miettunen[1]
[1] *University of Turku, Turku, Finland;* [2] *Aalto University, Espoo, Finland;* [3] *TUAS, Turku, Finland*

5DV.3.30 Photovoltaic Systems and Data Centers in Africa: a Bottom-Up Analysis   2108

Marco Pittalis[1], Georgia Kakoulaki[1], Iolanda Saviuc[1]
[1] *European Commission JRC, Ispra, Italy*

5DV.3.33 The Benefits of a Hybrid Wind-PV Power Plant at Competitive Wholesale Electricity Market – Case Finland   2109

Simeon Seppälä[1], Sanna Syri[1], Iraj Moradpoor[1]
[1] *Aalto University, Helsinki, Finland*

5DV.3.34 Profitability of Utility-Scale Photovoltaic Systems in Finland   2119

Seela Tervo[1], Sami Jouttijärvi[2], Kati Miettunen[2], Sanna Syri[1]
[1] *Aalto University, Espoo, Finland;* [2] *University of Turku, Turku, Finland*

5DV.3.35 Enhancing Energy Generation of Bifacial Photovoltaic Systems with Permeable Albedo Enhancement Composite   2124

Filippos V. Farmakis[1], Alexandros I. Droudakis[2], George I. Tzinoglou[2]
[1] *Democritus University of Thrace, Xanthi, Greece;* [2] *THRACE NG, Xanthi, Greece*

5DV.3.40 Energy Communities-Challenge and an Opportunities for Energy Decentralization and Efficiency. A Comparison of PV based Case-Studies with Different Control Strategies   2129

Domenico Vito[1], Martina Bosone[2], Barbara Pirelli[3]
[1] *Metabolism of Cities Living Lab, San Diego, United States of America;* [2] *Università degli Studi di Napoli Federico II, Naples, Italy;* [3] *Foro di Taranto, Taranto, Italy*

5DV.3.42 Hands-On Training in Photovoltaic Reliability Assessment: A Multinational Educational Approach under the PROMISE Project   2134

Carlos Meza[1], Brian Azzopardi[2], Bernhard Kubicek[3], Aritz Legarrea Oyarzun[4], Ana Gracia-Amillo[4], Melodie de L'Epine[5], Steve Zerafa[6], Austeja Mockeviciute-Azzopardi[2], Carmel Azzopardi[2], Brian Bartolo[2]
[1] *Anhalt University of Applied Sciences, Koethen, Germany;* [2] *The Foundation for Innovation and Research, Valletta, Malta;* [3] *AIT, Vienna, Austria;* [4] *CENER, Sarriguren, Spain;* [5] *ICARES Consulting, Brussels, Belgium;* [6] *PIXAM, Valletta, Malta*

5DV.3.43 Challenges of Energy Communities at Universities – A Virtual Approach   2138

Matevž Bokalič[1], Matej Guštin[1], Marko Topič[1], Ana Belen Cristóbal[2], Marta Victoria[3], Afonso Cavaco[4], Luis Fialho[4], Alexander Gerber[5]
[1] *University of Ljubljana, Ljubljana, Slovenia;* [2] *UPM, Madrid, Spain;* [3] *Aarhus University, Aarhus, Denmark;* [4] *University of Évora, Évora, Portugal;* [5] *inscico, Kleve, Germany*

5DV.3.45 Developing Communication Formats for a Positive Energy Transition Focusing on Photovoltaic – A Delphi Design Sprint Approach   2139

Eva-Maria Grommes[1], Sofia Scroppo[2], Stefanie Könen[1], Laura Züll[1], Anne Karrenbrock[1], Anne-Maren Feldhof[1], Ulf Blieske[1], Thorsten Schneiders[1], Valérie Varney[1], Laura Popplow[1]
[1] *University of Applied Sciences Cologne, Cologne, Germany;* [2] *University of Applied Science Cologne, Cologne, Germany*

5DV.3.46 TRANSIT: Empowering Sustainable Energy Futures through Innovative Education and Grid-Integrated Roadmap Development   2145

Brian Azzopardi[1], Daniel Busuttil[2], Araceli Hernandez Bayo[3], Ali Ehsan[4], Eduardo Maritinez Cesenia[4]
[1] *The Foundation for Innovation and Research, Birkirkara, Malta;* [2] *MCAST, Paola, Malta;* [3] *Madrid Polytechnic University, Madrid, Spain;* [4] *The University of Manchester, Manchester, United Kingdom*

5DV.3.48 Coincidence of Photovoltaic Electric Generation During Heat Waves: An Example Analysis for Northern Italy   2149

Danny S. Parker[1], Karthik Panchabikesan[1], Delia D'Agostino[2], Dru B. Crawley[3], Linda K. Lawrie[4]
[1] *Florida Solar Energy Center, Cocoa, United States of America;* [2] *European Commission JRC, Ispra, Italy;* [3] *Bentley Systems, Ismaning, Germany;* [4] *DHL Consulting, Pagosa Springs, United States of America*

Oral SESSION 5EO.3 Challenges and Opportunities along the PV Value Chain

5EO.3.2 Comparative Global PV Manufacturing Cost and Sustainable Pricing Assessment: China, Southeast Asia, India, USA, and Europe   2152

Sebastian Nold[1], Baljeet Singh Goraya[1], Ralf Preu[1], Jochen Rentsch[1], Julian Reichle[2], Wolfgang Jooß[2], Peter Fath[2], Michael Woodhouse[3]
[1] *Fraunhofer ISE, Freiburg, Germany;* [2] *RCT Solutions, Konstanz, Germany;* [3] *NREL, Golden, United States of America*

5EO.3.3 Assessing the Potential of Agrivoltaic Systems in Korea through Geospatial   2160
Analysis and Multi-Criteria Scenarios

ChangYeol Yun[1], Changki Kim[1], Jinyoung Kim[1], Sangmin Jo[2], Yongil Kim[3]
[1] *Korea Institute of Energy Research, Daejeon, South Korea;* [2] *Korea Energy Economics Institute, Ulsan, South Korea;* [3] *Seoul National University, Seoul, South Korea*

5EO.3.4 Integration of Photovoltaic Systems in the Austrian Power Plant Portfolio – a   2169
Geospatial Data Analysis

Stefan Übermasser[1], Fabian Leimgruber[1], Bernhard Kubicek[2]
[1] *AIT, Vienna, Austria;* [2] *AIT , Vienna, Austria*

5EO.3.5 Distributed Photovoltaics Provides Key Benefits for a Highly Renewable   2183
European Energy System

Parisa Rahdan[1], Elisabeth Zeyen[2], Cristobal Gallego-Castillo[3], Marta Victoria[1]
[1] *Aarhus University, Aarhus, Denmark;* [2] *TU Berlin, Berlin, Germany;* [3] *Technical University of Madrid, Madrid, Spain*

5EO.3.6 Identifying the Ecological Implications of the Repowering of Photovoltaic   2192
Systems

Karl-Anders Weiß[1], Sina Herceg[1], Marie Fischer[1], Liselotte Schebek[2]
[1] *Fraunhofer ISE, Freiburg, Germany;* [2] *Technical University of Darmstadt, Darmstadt, Germany*

Plenary SESSION 5EP.1 Sustainability

5EP.1.3 Towards Reuse-ready PV: a Perspective on Recent Advances, Practices and   2200
Future Challenges

Ioannis (John) Tsanakas[1], Gernot Oreski[2], Gabriele Eder[3], Anika Gassner[3], Arvid van der Heide[4], Daniela Ariolli[5], Guillermo Oviedo Hernandez[5], David Moser[6], Karsten Wambach[7]
[1] *CEA, Le Bourget-du-Lac, France;* [2] *PCCL, Leoben, Austria;* [3] *OFI, Vienna, Austria;* [4] *imo-imomec, Genk, Belgium;* [5] *BayWa r.e., Milan, Italy;* [6] *Eurac Research, Bolzano, Italy;* [7] *Wambach-Consulting, Petersdorf, Germany*

5EP.1.4 Enhancing Citizens' Participation in PV Deployment   2205

Silvia Caneva[1], Duygu Celik[1], Chiara Busto[2], Chiara Candelise[3], Alessia Cornella[4], Letizia Bua[5], Edouard Breniaux[6], Nouha Gazbour[7], Ivan Gordon[8], Wander Jager[9], Rudolf Kapeller[10], Gökhan Kirkil[11], Paola Mazzucchelli[12], Osbel Almora Rodríguez[13], Marcello Passaro[14], Alessandro Sciullo[15], Sebastien Lizin[16], Alessandro Martulli[16], Atse Louwen[4], Hanna Dittmar[17], Thomas Garabetian[17], Rania Fki[1], Johannes Stierstorfer[1], Melanie Kern[1]
[1] *WIP Renewable Energies, Munich, Germany;* [2] *Eni, Novara, Italy;* [3] *Bocconi University, Milan, Italy;* [4] *Eurac Research, Bolzano, Italy;* [5] *Eni, Milan, Italy;* [6] *Carnot Institute Chimie Balard Cirimat, Toulouse, France;* [7] *CEA, Le Bourget-du-Lac, France;* [8] *imec, Genk, Belgium;* [9] *University College Groningen, Groningen, The Netherlands;* [10] *Johannes Kepler University, Linz, Austria;* [11] *Kadir Has University, Istanbul, Türkiye;* [12] *CIRCE, Zaragoza, Spain;* [13] *URV, Tarragona, Spain;* [14] *Sunzest Solar, Rotterdam, The Netherlands;* [15] *University of Turin, Turin, Italy;* [16] *Hasselt University, Hasselt, Belgium;* [17] *SolarPower Europe, Brussels, Belgium*

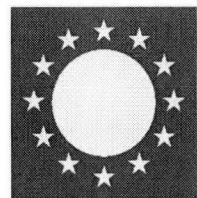

41st European Photovoltaic Solar Energy Conference and Exhibition

Proceedings of the International Conference
23 September – 27 September 2024

Edited by:

GABRIELE C. EDER
OFI
Austria

J. BERGMILLER
WIP Renewable Energies
Germany

R. KENNY
European Commission
Joint Research Centre
Italy

J. DE GREGORIO
WIP Renewable Energies
Germany

Photos at:

Coordination of the Technical Programme:

European Commission Joint Research Centre
Via E. Fermi 1
21020 Ispra (VA)
Italy

Institutional Support:

European Commission

Institutional PV Industry Cooperation:

SolarPower Europe

ESMC – European Solar Manufacturing Council

Supporting Organisations:

ASOM – Aliance for Solar Mobility

CDER - Renewable Energy Development Center

ETIP PV – European Technology & Innovation Platform PV

FOTOPLAT – Plataforma Tecnológica Española Fotovoltaica

GÜNDER – Turkish Solar Energy Society

INES – National Solar Energy Institute

NUS /SERIS – National University of Singapore / Solar Energy Research Institute of Singapore

ODTÜ-GÜNAM – Center for Solar Energy Research and Applications

PV THIN - Thin-Film PV Technology

Supporting Associations:

ARE – Aliance for Rural Electrification

EERA – European Energy Research Aliance

EREF – European Renewable Energies Federation

EUREC – The Association of European Renewable Energy Research Centres

UNEF – Unión Española Fotovoltaica

VDMA Photovoltaic Equipment

Local Support:

Klima und Energiefonds

Meeting Destination Vienna

Österreichs Energie

Photovoltaic Austria

Stadt Wien

TPPV - Austrian Technology Platform PhotoVoltaic

TU Wien – Technical University of Vienna

Wirtschaftsagentur Wien

EU PVSEC 2024 realised by:

WIP Renewable Energies
Sylvensteinstr. 2, 81369 München, Germany
Tel: +49 89 720 12 735, Fax: +49 89 720 12 791
Email: pv.conference@wip-munich.de
www.eupvsec.org
www.wip-munich.de

Proceedings produced and published by:

WIP Renewable Energies
Sylvensteinstr. 2, 81369 München, Germany
Tel: +49 89 720 12 735, Fax: +49 89 720 12 791
Email: pv.conference@wip-munich.de
www.eupvsec.org
www.wip-munich.de

Legal notice

Neither the European Commission, the Organiser or the Publisher nor any person acting on their behalf is responsible for the use which might be made of the following information.

© 2024 WIP Renewable Energies

All rights reserved. No part of this publication may be reproduced in any form or by any electronic or mechanical means, including photocopying, recording or by any information storage and retrieval system without permission in writing from the copyright holder and the publisher.

Despite due diligence no liability for accuracy and completeness of the information and material offered in this document can be assumed by WIP Renewable Energies.

ISBN 3-936338-90-6
ISSN 2196-0992

41st EUROPEAN PHOTOVOLTAIC SOLAR ENERGY CONFERENCE AND EXHIBITION
23 SEPTEMBER – 27 SEPTEMBER 2024

EU PVSEC 2024 COMMITTEES

INTERNATIONAL SCIENTIFIC ADVISORY COMMITTEE (ISAC)

Chair
P. Szymanski, European Commission Joint Research Centre, Director of Energy, Transport and Climate, Petten, The Netherlands

Committee Members
P. Frankl, Head of the Renewable Energy Division, International Energy Agency, France
M. Getsiou, European Commission, DG RTD, Brussels, Belgium
S.W. Glunz, Head of Division Photovoltaics - Research, Fraunhofer ISE, Freiburg, Germany
R. Kenny, European Commission Joint Research Centre, Directorate for Energy and Transport and Climate, Ispra, Italy
P. Malbranche, General Director, CEA INES, France
S. Nowak, Managing Director of NET Nowak Energy & Technology, St. Ursen, Switzerland
R. Schlatmann, Chairman of ETIP PV, Head of the Solar Energy Division at Helmholtz-Zentrum Berlin, Germany
W.C. Sinke, University of Amsterdam, The Netherlands
M. Topič, Head of Laboratory of Photovoltaics and Optoelectronics of the University of Ljubljana, Slovenia
P. Verlinden, Director at Amrock, Visiting Professor at Sun Yat-Sen University, Guangzhou, China
J. Bergmiller, Managing Director Events & Knowledge Transfer, WIP Renewable Energies, Munich, Germany
J. de Gregorio, Head of Unit, Scientific Services and Cooperation, WIP Renewable Energies, Munich, Germany

CONFERENCE EXECUTIVE COMMITTEE

Conference General Chair
G.C. Eder, OFI, Vienna, Austria

Technical Programme Chair
R. Kenny, European Commission Joint Research Centre, Directorate for Energy and Transport and Climate, Ispra, Italy

Committee Members
W.C. Sinke, University of Amsterdam, The Netherlands
S. Nowak, Managing Director of NET Nowak Energy & Technology, St. Ursen, Switzerland
M. Topič, Head of Laboratory of Photovoltaics and Optoelectronics of the University of Ljubljana, Slovenia
V. Bermudez Benito, Senior Research Director of the Energy Center at QEERI, Qatar, United Arab Emirates
E. Voroshazi, Head of PV Module Process Laboratory, CEA, Le Bourget-Du-Lac France
H. Ossenbrink, Band Gap, Germany
J. Bergmiller, Managing Director Events & Knowledge Transfer, WIP Renewable Energies, Munich, Germany
J. de Gregorio, Head of Unit, Scientific Services and Cooperation, WIP Renewable Energies, Munich, Germany

2024 SCIENTIFIC COMMITTEE

Programme Technical Chair
R. Kenny, European Commission, Joint Research Centre, Italy

Topic Chairs

Topic 1: Silicon Materials and Cells
D. Muñoz, CEA, France

Topic 2: Thin Films and New Concepts
M. Edoff, Uppsala University, Sweden

Topic 3: Photovoltaic Modules and BoS Components
G. Bardizza, TÜV Rheinland Solar, Germany

Topic 4: PV Systems Engineering, Integrated/Applied PV
F. Baumgartner, Zurich University of Applied Sciences, Switzerland

Topic 5: PV in the Energy Transition
M. Victoria, Aarhus University, Switzerland

Topic Organisers and Paper Review Experts

Topic 1: Silicon Materials and Cells
T. Liu, GCL, China
F. Schindler, Fraunhofer ISE, Germany
P. Roca i Cabarrocas, CNRS-LPICM, France
K. Ding, Forschungszentrum Jülich, Germany
G. Hahn, University of Konstanz, Germany
S. W. Glunz, Fraunhofer ISE, Germany
D. Muñoz, CEA / INES, France
A. W. Weeber, TNO Energy Transition, The Netherlands
K. Bothe, ISFH, Germany
M. Topic, University of Ljubljana, Slovenia
P. Fath, RCT-Solutions, Germany
S. Peters, Hanwha Q CELLS, Germany

M. P. Bellmann, STIFTELSEN SINTEF, Norway
A. Ciesla, UNSW Australia, Australia
C. Fischer, Wacker Chemie, Germany
C. Hagendorf, Fraunhofer CSP, Germany
X. Yu, Zhejiang University, China
R. Brendel, ISFH, Germany
T. Dullweber, ISFH, Germany
M. Ernst, ANU, Australia
W. Favre, CEA / INES, France
J. Horzel, Fraunhofer ISE, Germany
J. Meier, Meier Technologies, Switzerland
F. Menchini, ENEA, Italy
A. Morisset, CSEM, Switzerland
R. C. G. Naber, Hanwha Q CELLS, The Netherlands
W. Nemeth, NREL, USA
J. Rentsch, Fraunhofer ISE, Germany
J. Schmidt, ISFH, Germany
R. Turan, METU, Türkiye

M. Wright, University of Oxford, UK
J. Zhao, CSEM, Switzerland
E. Bruhat, HOLOSOLIS, France
P. Delli Veneri, ENEA, Italy
S. Dubois, CEA/ INES, France
M. Hermle, Fraunhofer ISE, Germany
T. Matsui, AIST, Japan
Y. Ohshita, Toyota Technological Institute, Japan
R. Preu, Fraunhofer ISE, Germany
B. Terheiden, University of Konstanz, Germany
A. Augusto, Arizona State University, USA
H. Duman, KalyonPV, Türkiye
F. Ferrazza, ENI, Italy
A. Otaegi, UPV/EHU, Spain
M. C. Schubert, Fraunhofer ISE, Germany
D. L. Bätzner, Meyer Burger Research AG, Switzerland
D. Brunner, RENA Technologies
A. Danel, CEA / INES, France
C. Gerardi, ENEL Green Power, Italy
H. J. Nonnenmacher, Meyer Burger, Germany
P. J. Verlinden, AMROCK, Australia
Q. Wang, Zhejiang Jinko Solar, China
W. Wu, Tongwei Solar (Meishan), China
W. Zhang, Gonda Electronic Technology, China

Topic 2: Thin Films and New Concepts
N. Beaumont, Oxford PV, United Kingdom
J. C. Goldschmidt, University of Marburg, Germany
I. Gordon, imec, Belgium
T. Aernouts, imec, Belgium
N. Kyranaki, Hasselt University, Belgium
S. Veenstra, TNO Energy Transition, The Netherlands
M. Edoff, Uppsala University, Sweden
G. Siefer, Fraunhofer ISE, Germany
S. Taylor, ESA, The Netherlands
A. N. Tiwari, EMPA, Switzerland
A. Martí Vega, UPM, Spain
J. J. Poortmans, imec, Belgium
T. Magorian-Friedlmeier, ZSW, Germany
I. Ramiro, UPM, Spain

S. Albrecht, HZB, Germany
S. Berson, CEA / INES, France
A. Bruno, Nanyang Technological University, Singapore
P. Carroy, CEA / INES, France
C. Case, Oxford PV, UK
G. Coletti, TNO Energy Transition, The Netherlands
S. De Wolf, KAUST, Saudi Arabia
K. Makita, AIST, Japan
U. W. Paetzold, KIT Karlsruher Institut für Technologie, Germany
P. Schulze, Fraunhofer ISE, Germany
H. Sivaramakrishnan Radhakrishnan, imec, Belgium
L. Wang, Technology Innovation Institute, Abu Dhabi, UAE
Y. Smirnov, Caelux, USA
B. Stannowski, HZB, Berlin
A. Aguirre, imec, Belgium
M. J. Brites, LNEG, Portugal
S. Cros, CEA / INES, France
C. Fell, CSIRO, Australia
S. Hayase, The University of Electro-Communications, Japan
S. Huang, Macquarie University, Australia
M. Khenkin, HZB, Germany
D. Lan, Zhejiang University, China
C. F. Lin, National Taiwan University, Taiwan
P. Manshanden, TNO Energy Transition, The Netherlands
B. Mihaylov, European Commission JRC, Italy
M.S.H. Norton, University of Cyprus
P. Pistor, Universidad Pablo de Olavide de Sevilla, Spain

M. Saliba, University of Stuttgart, Germany
W. Tress, Zurich University of Applied Sciences, Switzerland
L. Vesce, University of Rome "Tor Vergata", Italy
Z. Zonglong, City University of Hong Kong
G. Brammertz, imec, Belgium
R. Campesato, CESI, Italy
T. Dalibor, Avancis, Germany
D. Dimova-Malinovska, Bulgarian Academy of Sciences, Bulgaria
G. Flamand, imec, Belgium
B. Friedel, PTB, Germany
S. Ishizuka, AIST, Japan
V. Khorenko, Azur Space Solar Power, Germany
R. R. King, Arizona State University, USA
S. Paetel, ZSW, Germany
A. Redinger, University of Luxembourg, Luxembourg
A. Romeo, University of Verona, Italy
V. Sittinger, Fraunhofer IST, Germany
M. Theelen, TNO/Solliance, The Netherlands
G. Timò, RSE, Italy
J. P. Connolly, GeePs-CentraleSupelec, France
J. P. Kleider, CNRS/GeePs, France
I. Konovalov, University of Applied Sciences Jena, Germany
H. Meddeb, DLR, Germany
Y. Okada, University of Tokyo, Japan
F. Roca, ENEA, Italy
M. Rusu, HZB, Germany
E. Saucedo, UPC, Spain
C. Becker, HZB, Germany
D. Kuciauskas, NREL, USA
M. Ochoa, University of Cantabria, Spain
P. Pearce, UNSW Australia, Australia
T. Tayagaki, AIST, Japan
S. Wasmer, WAVELABS Solar Metrology Systems, Germany

Topic 3: Photovoltaic Modules and BoS Components
V. Bermudez Benito, QEERI, Qatar
E. Voroshazi, CEA / INES, France
T. Barnes, NREL, USA
R. Gottschalg, Fraunhofer CSP, Germany
I. Tsanakas, CEA / INES, France
G. Bardizza, TÜV Rheinland Energy, Germany
G. Friesen, SUPSI, Switzerland
C. Buerhop-Lutz, HI ERN, Germany
J. Moschner, KU Leuven, Belgium

V. Barth, CEA / INES, France
A. Faes, CSEM, Switzerland
S. Feldbacher, PCCL, Austria
A. Halm, ISC Konstanz, Germany
H. Hanifi, AESOLAR, Germany
P. Kratzert, Holyvolt, Germany
A. Lennon, Sundrive Solar, Australia
M. Mittag, Fraunhofer ISE, Germany
M. Á. Muñoz-García, UPM, Spain
H. Nagel, Fraunhofer ISE, Germany
S. M. Pietralunga, CNR – ISMAC, Italy
T. Timofte, ISC Konstanz, Germany
E. Warren, NREL, USA
C. Barretta, PCCL, Austria
G. Beaucarne, Dow Silicones, Belgium
T. Bejat, CEA, France
C. Camus, LayTec, Germany
B. Figgis, QEERI, Qatar
E. Fokuhl, Fraunhofer ISE, Germany
U. Jahn, Fraunhofer CSP, Germany
E. Krassowski, CE Cell Engineering, Germany

S. Lindig, Eurac Research, Italy
A. Morlier, imec, Belgium
G. Oreski, PCCL, Austria
M. Pander, Fraunhofer CSP, Germany
T. Sample, European Commission JRC, Italy
M. Bokalic, University of Ljubljana, Slovenia
S. Bordihn, ISFH, Germany
A. J. Carr, TNO Energy Transition, The Netherlands
M. Despeisse, CSEM, Switzerland
J. Govaerts, imec, Belgium
W. Herrmann, TÜV Rheinland Solar, Germany
Y.S. Long, ITRI, Taiwan
J. Lopez-Garcia, QEERI, Qatar
T. Lyubenova, European Commission JRC, Italy
J. Moereke, Avancis, Germany
C. Monokroussos, TÜV Rheinland (Shanghai), China
M. Paverttoni, SERIS, Singapore
R. Sinton, Sinton Instruments, USA
C. Ulbrich, HZB, Germany
S. Bouguerra, Hasselt University, Belgium
S. Krauter, University of Paderborn, Germany
N. M. Pearsall, Northumbria University, UK

Topic 4: PV Systems Engineering, Integrated/Applied PV
A. M. Gracia Amillo, Fraunhofer ISE, Germany
W. M. van Sark, Utrecht University, The Netherlands
F. P. Baumgartner, Zurich University of Applied Sciences, Switzerland
K. Lappalainen, Tampere University, Finland
J. M. Almeida Serra, University of Lisbon - FCiencias.ID, Portugal
D. Moser, Eurac Research, Italy
G. Mütter, Alternative Energy Solutions, Austria
G. Eder, OFI, Austria
F. Frontini, SUPSI, Switzerland
I. Antón Hernández, UPM, Spain
T. Reindl, NUS, Singapore
A. Scognamiglio, ENEA, Italy
R. M. E. Valckenborg, TNO Energy Transition, The Netherlands
G. Adinolfi, ENEA, Italy

A. Alcañiz, Delft University of Technology, The Netherlands
P. Blanc, MINES ParisTech, France
V. Lara-Fanego, Solargis, Slovakia
A. Louwen, Eurac Research, Italy
A. Martinez Fernandez, European Commission JRC, Italy
T. Oozeki, AIST, Japan
C. Protogeropoulos, EEPS, Greece
J. Remund, Meteotest, Switzerland
M. Sengupta, NREL, USA
M. Zehner, Rosenheim Technical University of Applied Sciences, Germany
C. Alonso-Tristán, UBU, Spain
D. Berrian, OPES Solar Mobility, Germany
M. Carbone, ENEL Green Power, Italy
M. Dennenmoser, BayWa re, Germany
C. W. Hansen, Sandia National Laboratories, USA
A. Neubert, DNV, Germany
M. Oliosi, PVsyst, Switzerland
M. Bolen, SB Energy, USA
D. Daßler, Fraunhofer CSP, Germany
A. Driesse, PV Performance Labs, Germany
R. Einhaus, ZSW, Germany
P. Hacke, NREL, USA
A. Heimsath, Fraunhofer ISE, Germany
J. Lin, PV Guider, Taiwan
O. Mayer, Bayern Innovativ, Germany
A. Migan-Dubois, GeePs, France

A. Nobre, SERIS, Singapore
M. Ogaard, Institute for Energy Technology, Norway
M. Rinio, University of Karlstad, Sweden
J. S. Stein, Sandia National Laboratories, USA
D. Stellbogen, ZSW, Germany
M. Theristis, Sandia National Laboratories, USA
A. Virtuani, CSEM, Switzerland
P. Alonso Gomez, BayWa re, Germany
Y. B. Assoa, CEA, France
P. Bonomo, SUPSI, Switzerland
D. D'Agostino, European Commission JRC, Italy
V. D'Ambrosio, University of Naples Federico II, Italy
T. Del Caño, Onyx Solar Energy, Spain
M. La Rosa, Glass to Power, Italy
E. Román Medina, Fundación Tecnalia, Spain
L. H. Slooff-Hoek, TNO Energy Transition, The Netherlands
S. Villa, TNO Energy Transition, The Netherlands
X. Zhihao, AIST, Japan
K. Araki, University of Miyazaki, Japan
A. Chatzipanagi, Joint Research Center, Italy
N. Cherradi, Desert Technologies, Saudi Arabia
F. Colucci, ENEA, Italy
M. Dörenkämper, TNO Energy Transition, The Netherlands
M. Heinrich, Fraunhofer ISE, Germany
J. Leloux, LuciSun, Belgium
B. Newman, Lightyear, The Netherlands
K. Nishioka, University of Miyazaki, Japan
A. Reinders, Eindhoven University of Technology, The Netherlands
R. Schlatmann, HZB, Germany
E. Shirazi, University of Twente, The Netherlands
T. Tanahashi, AIST, Japan
G. Barchi, European Commission JRC, Italy
R. Bründlinger, AIT, Austria
V. Efthymiou, University of Cyprus, Cyprus

Topic 5: PV in the Energy Transition
R. Pestana, R&D NESTER, Portugal
J. Stierstorfer, WIP Renewable Energies, Germany
M. Victoria, Aarhus University, Denmark
M. de Wild-Scholten, SmartGreenScans, The Netherlands
K. Wambach, Wambach-Consulting, Germany
M. Getsiou, European Commission DG RTD, Belgium
P. Malbranche, International Solar Aliance, India
S. Nowak, NET Nowak Energy & Technology, Switzerland
C. Breyer, Lappeenranta University of Technology, Finland
I. Kaizuka, RTS Corporation, Japan
G. Masson, Becquerel Institute, Belgium
S. Caneva, WIP Renewable Energies, Germany
N. Della-Valle, European Commission JRC, Italy

M. C. Brito, University of Lisbon - Faculty of Science, Portugal
F. Carigiet, Zurich University of Applied Sciences, Switzerland
J. S. da Costa Fernandes, University of Applied Sciences Offenburg, Germany
B. Gaiddon, HESPUL, France
T. Merdzhanova, FZ Jülich, Germany
A. C. Neves Foles, University of Évora, Portugal
F. Z. Ouchani, Green Energy Park, Morocco
M. Rennhofer, AIT, Austria
I. Weiss, independent consultant, Germany
M. Aleman, Becquerel Institute, Belgium
A. Anctil, Michigan State University, USA
S. Arancón, Plug and Play, Spain
S. Capaccioli, ETA - Florence Renewable Energies, Italy
V. Fthenakis, Columbia University, USA
G. Heath, NREL, USA

K. Komoto, Mizuho Research & Technologies, Japan
W. Palitzsch, LuxChemtech, Germany
S. Ovaitt, NREL, USA
S. Guastella, RSE, Italy
H. Ossenbrink, Band Gap, Germany
D. Polverini, European Commission DG GROWTH, Belgium
M. Ray, IIT Kharagpur, India
N. Taylor, European Commission JRC, Italy
R. Blanchard, Loughborough University, United Kingdom
S. De Iuliis, ENEA, Italy
T. Haarberg, BNW-Energy AS, Germany
A. Nayfeh, Khalifa University, United Arab Emirates
E. Vartiainen, Fortum Growth, Finland
C. Candelise, Bocconi University, Italy
A.B. Cristobal, UPM, Spaon
G. Ruggieri, Insubria University, Italy
S. Tay En Rong, National University of Singapore, Singapore

Awards Coordinators

Student Awards Coordinator
A.H.M. Smets, Delft University of Technology, The Netherlands

Student Awards Committee
J. M. Almeida Serra, University of Lisbon - FCiencias.ID, Portugal
F. Baumgartner, ZHAW, Switzerland
V. Bermnudez Benito, QEERI, Qatar
G. Eder, OFI, Austria
I. Kaizuka, RTS, Japan
R. Kenny, European Commission JRC, Italy
A. Marti Vega, UPM, Spain
D. Munoz, CEA/ INES, France
J. Poortmans, imec, Belgium
P. Roca i Cabarrocas, CNRS-LPICM, France
R. Schlatmann, HZB, Germany
G. Siefer, Fraunhofer ISE, Germany
W. C. Sinke, University of Amsterdam, The Netherlands
A. Tiwari, EMPA, Switzerland
M. Topic, University of Ljubljana, Slovenia
M. Victoria, Aarhus University, Denmark

Poster Awards Coordinator
P. Malbranche, Solar Action, France

Poster Awards Committee
J. M. Almeida Serra, University of Lisbon - FCiencias.ID, Portugal
I. Anton, UPM, Spain
F. Baumgartner, ZHAW, Switzerland
S. Caneva, WIP Renewable Energies, Germany
P. Delli Veneri, ENEA, Italy
G. Eder, OFI, Austria
G. Friesen, SUPSI, Switzerland
A. M. Gracia Amillo, CENER, Spain
G. Hahn, University of Konstanz, Germany
G. Heath, NREL, USA
G. Kakoulaki, European Commssion JRC, Italy
R. Kenny, European Commssion JRC, Italy
H. Ossenbrink, Band Gap, Germany
P. Roca i Cabarrocas, CNRS-LPICM, France
F. Schindler, Fraunhofer ISE, Germany
I. Tsanakas, Cea/ INES, France
R. Valckenborg, TNO Energy Transition, The Netherlands

Highlights Committee
G.C. Eder, OFI, Vienna, Austria
R. Kenny, European Commission Joint Research Centre, Directorate for Energy and Transport and Climate, Ispra, Italy
W.C. Sinke, University of Amsterdam, The Netherlands
S. Nowak, Managing Director of NET Nowak Energy & Technology, St. Ursen, Switzerland
M. Topič, Head of Laboratory of Photovoltaics and Optoelectronics of the University of Ljubljana, Slovenia
V. Bermudez Benito, Senior Research Director of the Energy Center at QEERI, Qatar, United Arab Emirates
E. Voroshazi, Head of PV Module Process Laboratory, CEA, Le Bourget-Du-Lac France
H. Ossenbrink, Band Gap, Germany
J. Bergmiller, Managing Director Events & Knowledge Transfer, WIP Renewable Energies, Munich, Germany
J. de Gregorio, Head of Unit, Scientific Services and Cooperation, WIP Renewable Energies, Munich, Germany
S. Leanza, Marketing and PR Manager, WIP Renewable Energies, Munich, Germany
D. Muñoz, CEA, France
M. Edoff, Uppsala University, Sweden
G. Bardizza, TÜV Rheinland Solar, Germany
F. Baumgartner, Zurich University of Applied Sciences, Switzerland
M. Victoria, Aarhus University, Switzerland

SUBJECT INDEX

Silicon Materials and Cells

Sessions 1AP.1, 1EP.1, 1AO.4, 1AO.5, 1AO.6, 1BO.1, 1BO.2, 1BO.3, 1BO.4, 1BV.5, 1CV.2

Thin Films and New Concepts

Sessions 2CP.3, 2AO.1, 2AO.2, 2AO.3, 2BO.8, 2BO.9, 2BO.10, 2CO.1, 2CO.2, 2CO.3, 2DO.6, 2DO.7, 2DO.8, 2DO.9, 2DO.18, 2BV.1, 2BV.2, 2CV.3

Photovoltaic Modules and BoS Components

Sessions 3CP.2, 3BO.11, 3BO.12, 3BO.14, 3BO.15, 3CO.10, 3CO.11, 3DO.12, 3DO.16, 3DO.17, 3DO.19, 3DO.20, 3EO.1, 3AV.1, 3AV.2, 3AV.3

PV Systems Engineering, Integrated/Applied PV

Sessions 4CP.1, 4CP.2, 4AO.7, 4AO.8, 4AO.9, 4BO.5, 4BO.6, 4BO.7, 4BO.16, 4BO.17, 4CO.8, 4CO.9, 4DO.1, 4DO.2, 4DO.3, 4DO.4, 4DO.5, 4EO.2, 4BV.3, 4BV.4, 4CV.1, 4DV.1, 4DV.4

PV in the Energy Transition

Sessions 5CP.1, 5EP.1, 5CO.4, 5CO.5, 5CO.6, 5DO.10, 5DO.11, 5DO.14, 5DO.15, 5EO.3, 5DV.2, 5DV.3

Topic Code	Session Type	Day Codes
1 Silicon Materials and Cells	P = Plenary Session	A = Monday, 23 September 2024
2 Thin Films and New Concepts	O = Oral Session	B = Tuesday, 24 September 2024
3 Photovoltaic Modules and BoS Components	V = Visual Session	C = Wednesday, 25 September 2024
4 PV Systems Engineering, Integrated/Applied PV		D = Thursday, 26 September 2024
5 PV in the Energy Transition		E = Friday, 27 September 2024

e.g. 1AO.4 ⇒ 1= Silicon Materials and Cells, A=Monday, O=Oral session, 4=Session 4

FOREWORD

The European Photovoltaic Solar Energy Conference and Exhibition (EU PVSEC) is the World's leading and most renowned forum for PV research and development and the biggest conference on PV solar energy. In 2024, celebrating its 41st edition, the EU PVSEC was the essential meeting and exchanging point for global PV experts from research, development, and industry.

The EU PVSEC 2024 which took place from 23 – 27 September 2024 in Vienna, Austria, was a resounding success providing diverse and state of the art research findings. Combining Conference and Exhibition, this year, more than 1800 participants from 60 countries attended this highly inspiring platform delivering over 1100 presentations. It provided a vital forum for information exchange in the field of PV solar energy research, innovations and applications. In the exhibition area 71 companies from all parts of the world welcomed visitors and presented their products and services.

Conference Highlights

The EU PVSEC covered a broad range of topics with an extensive programme that offers an opportunity for workers from across the entire field of photovoltaics to share their findings, as well as an opportunity for multidisciplinary learning. The programme was arranged into 5 topics as follows:
- Silicon Materials and Cells;
- Thin Films and New Concepts;
- Photovoltaic Modules and Balance of System Components;
- PV Systems Engineering, Integrated/Applied PV;
- PV in the Energy Transition.

Communicating the key messages from the conference, not only to participants, but also to other researchers and key stakeholders, policy makers and the general public was an important added value. We thank the Highlights Committee, composed of selected members of the Scientific Committee, for providing a comprehensive summary of the findings and state of the art research that were delivered during this year´s event. Some key highlights are listed below, while further details may be found in the dedicated highlights presentation in the annex of these proceedings.

Latest Solar Innovations in Materials, Cells, Modules, Systems and Machine Learning:

Silicon solar cells remain the backbone of the PV market, continuing to deliver efficiency improvements. A major focus is on reducing the use of critical raw materials (CRM), like silver, while simplifying manufacturing processes to lower costs. Advances (and even the first shipments) in perovskite-silicon tandem cells and triple-junction devices have also shown great potential, pushing efficiency further and progressing in stability and reliability. Presentations also covered perovskite-CIGS tandems, all-perovskite tandems, Chalcogenides, and space applications of silicon cells due to growing interest from low earth orbit satellite constellations.

The design, manufacturing, and reliability of PV modules continue to be pivotal areas of focus. Innovations in module production, such as advanced encapsulants and interconnection methods, are driving improvements in performance. However, emerging technologies like perovskite materials introduce new challenges regarding characterization and standards.

The PV systems and applications topic is the largest at the conference, underlining that PV is absolutely mainstream. Operations and Maintenance contributions are growing year by year and increasingly incorporating non-intrusive techniques, such as daylight photoluminescence imaging. Machine learning is increasingly utilized for performance monitoring of PV systems, transforming how systems are managed and optimized.

Sustainability and Circular Economy:

Sustainability was at the forefront of discussions, with sustainable manufacturing practices and new module designs focusing on easier end-of-life recycling. Ongoing research into recycling techniques and eco-friendly materials is a key step toward a fully circular PV economy.

Grid Integration and Storage:

As PV capacity increases globally and we transit a multi-Terawatt PV era, grid integration, flexibility and storage are becoming essential. Discussions focused on new solutions for handling high PV penetration rates, particularly through smart grid management, hybrid plants, storage solutions, sector coupling, and innovative electricity market designs.

New Applications and Integration:

The conference showcased emerging applications like Agrivoltaics and Floating PV, both showing strong potential in diverse (even extreme) environments. In Building-Integrated PV (BIPV), research focuses on custom design elements, including colored modules, to enhance architectural integration.

EU PVSEC 2024 Proceedings

The EU PVSEC 2024 Proceedings contain the full papers of most of these highlights and more, providing a comprehensive overview of the PV solar sector, its current status and future prospects in science, research, innovation, development and deployment on almost 3,900 pages. Selection for inclusion in the conference was made by the Scientific Committee's paper review experts and topic organisers (see the listing on pages 010002-001-004), to whom we express our sincere gratitude for their comprehensive review work and overall contribution to the success of the conference. In addition to the 306 submitted papers, the proceedings include 110 presentations (slides) shown during the plenary and oral presentations as well as 153 poster files of the visual presentations. In total this amounts to 569 publications.

The proceedings of the EU PVSEC 2024 strengthen the commitment to providing quick and open access to high quality scientific results. This is a powerful source for targeted and quick information search and retrieval, enabling you to search by topic, keywords, paper title, DOI, author, or organization. We are sure that these new features will help to simplify the use and exploit the full potential of this extensive fountain of knowledge.

The Conference Proceedings are published as downloadable files and are also fully accessible online. A DOI code (Digital Object Identifier) has been assigned to each paper. This ensures unequivocal and permanent identification and full citability. The EU PVSEC 2024 papers can be viewed and downloaded in a full free open access from the EU PVSEC's Proceedings website https://userarea.eupvsec.org/proceedings.

We are confident that these Proceedings will play an important role in providing a comprehensive overview of the current actors and activities in the global PV sector and that they will disseminate information on the state-of-the-art of technologies and applications. This can generate further research, add momentum to innovation and promote interest in PV worldwide.

We would like to cordially thank all authors and participants of the EU PVSEC 2024 for their contributions and look forward to welcoming you in Bilbao, Spain from 22 – 26 September 2025 at the EU PVSEC 2025, the 42nd European Photovoltaic Solar Energy Conference and Exhibition

The Editors

NETWORKING | Meet and share

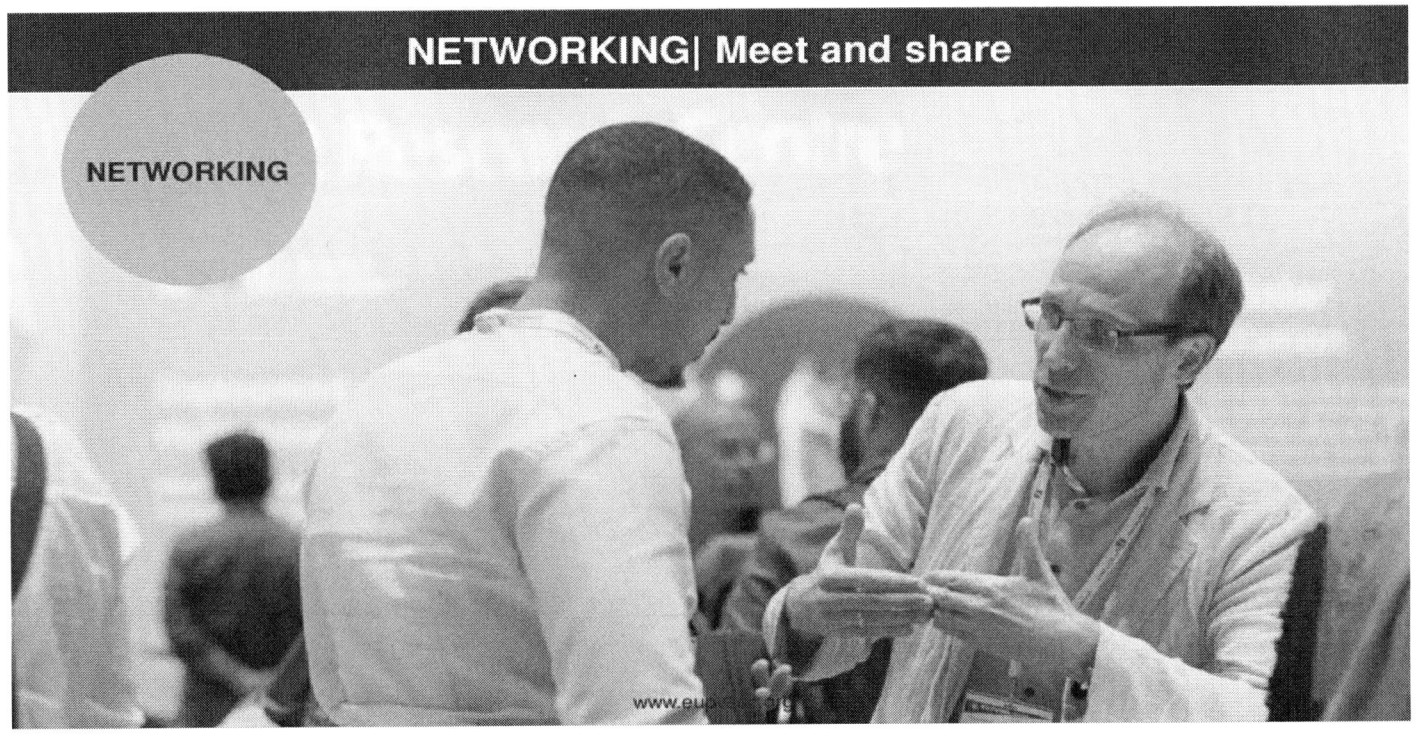

FACTS & FIGURES | Presentations

EU PVSEC Programme -
**Distribution of
Presentations per Type**

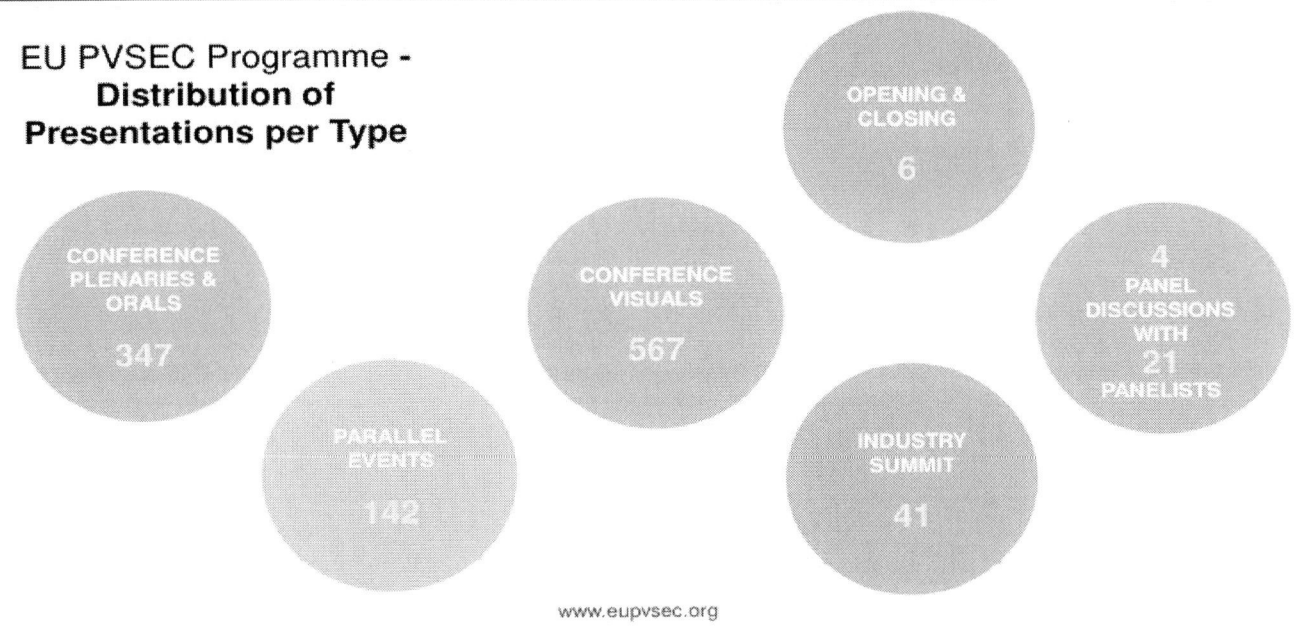

www.eupvsec.org

FACTS & FIGURES | Presentations

EU PVSEC Scientific Conference Programme - Distribution of Presentations per Topic

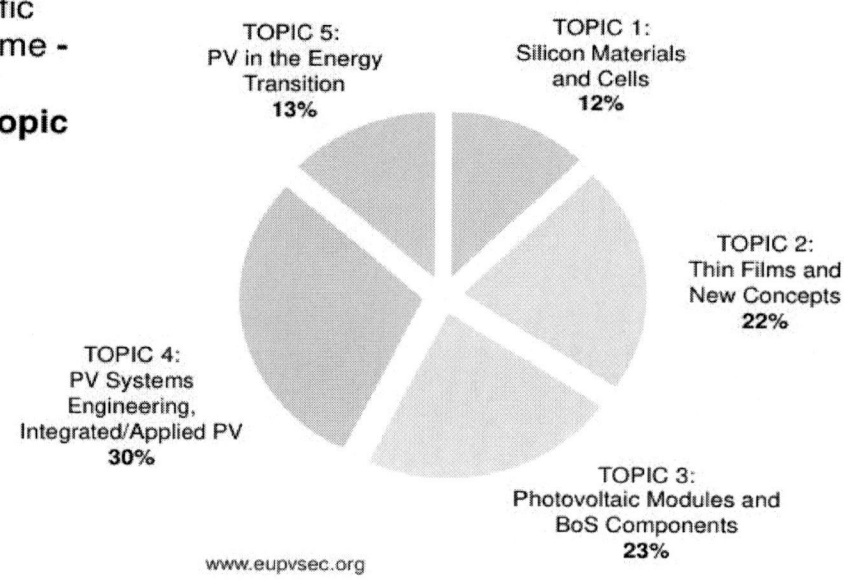

TOPIC 5: PV in the Energy Transition **13%**

TOPIC 1: Silicon Materials and Cells **12%**

TOPIC 2: Thin Films and New Concepts **22%**

TOPIC 4: PV Systems Engineering, Integrated/Applied PV **30%**

TOPIC 3: Photovoltaic Modules and BoS Components **23%**

www.eupvsec.org

FACTS & FIGURES | Participants

EU PVSEC Programme – **Participants by Countries – Top 10**

No	Country	Participants
1	Germany	456
2	Austria	144
3	Italy	109
4	France	104
5	South Korea	99
6	The Netherlands	85
7	Switzerland	83
8	Spain	76
9	Belgium	66
10	Japan	64

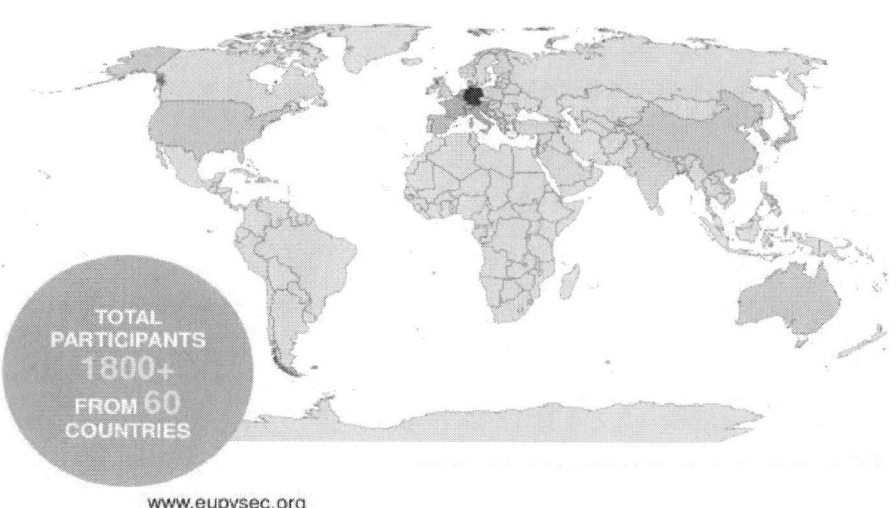

TOTAL PARTICIPANTS 1800+ FROM 60 COUNTRIES

www.eupvsec.org

OPENING | Monday, 23 Sept. 2024

Plenary Session "Manufacturing"

Moderated Panel Discussion "Solar Everywhere: Addressing the Challenges and Potential of Global PV Expansion"

OPENING | Monday, 23 Sept. 2024

Becquerel Prize Ceremony

Welcome Message, Opening Addresses & Keynote Speech

www.eupvsec.org

PANEL DISCUSSIONS | Four sessions

OPENING PANEL DISCUSSION
Solar Everywhere:
Addressing the Challenges and Potential of Global PV Expansion

BO.13
Stable Perovskite Tandems: Hype or Hope?

CO.7
Social Acceptance of Ubiquitous-PV: The Era of
Integrated Photovoltaics - Interactive Workshop

DO.13
Renewables 24/7: The challenges of integrating
renewables into the future Electricity System

www.eupvsec.org

VISUAL SESSIONS | Poster Awards

POSTER
AWARDS

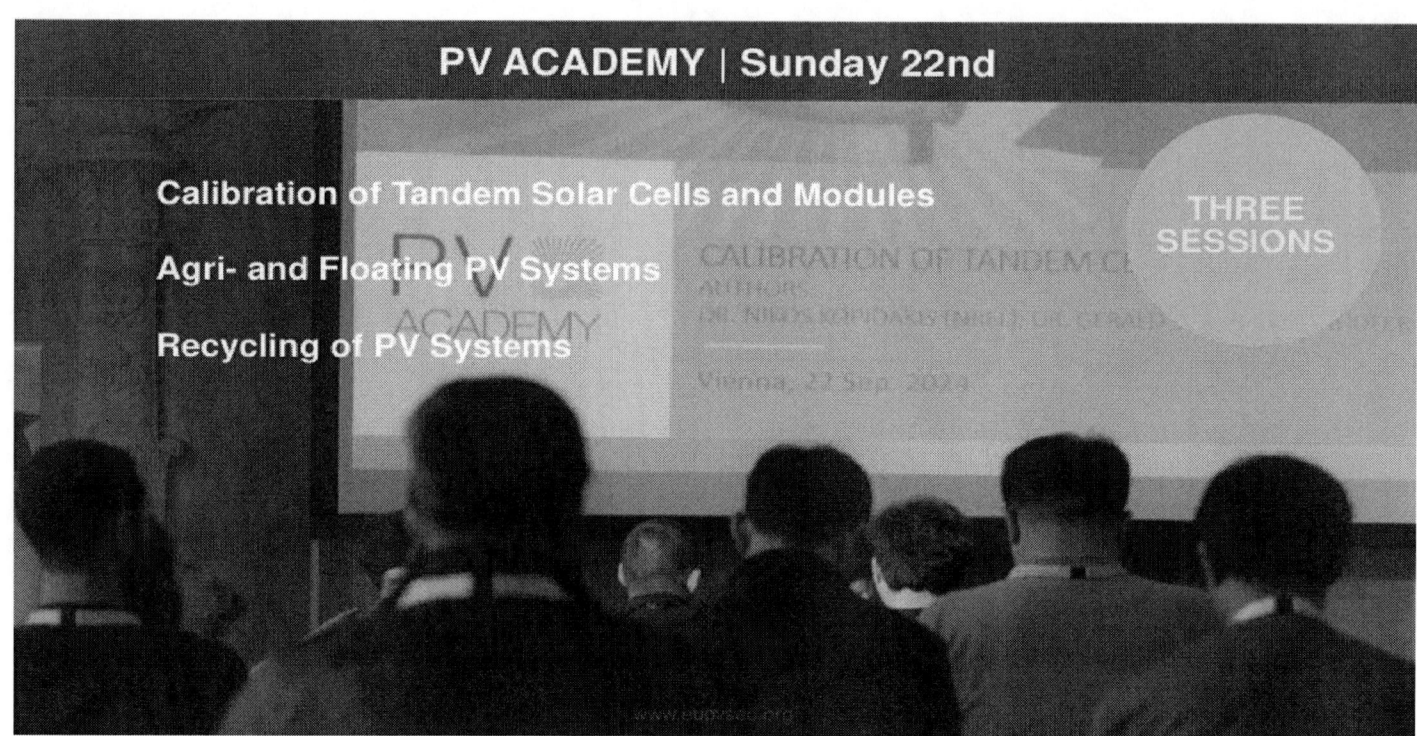

CONFERENCE | Topic 1

TOPIC 1: SILICON MATERIALS AND CELLS

Silicon solar cells continue to demonstrate efficiency improvements due to continued research. Nevertheless a focus of research is now on sliver replacement and CRM reduction (including Si). Process simplification also furthers the above aims, while at the same time reducing manufacturing costs. Silicon bottom cells are the basic building blocks of perovskite silicon tandem solar cells and much research is ongoing, including recombination junction assessment and optimization, for this application. There is an increased interest in silicon cells for space applications due to the deployment of low earth orbit communications satellite constellations.

www.eupvsec.org

CONFERENCE | Topic 2

TOPIC 2: THIN FILMS AND NEW CONCEPTS

Chalcogenides (Kesterite, CIGS) continue to demonstrate increased efficiencies. Perovskites and tandems with perovskites are an increasing area of research. Good progress on tandem and triple junction devices with perovskite-CIGS tandems (19.9% module), all-perovskite tandems (19.7% flexible tandem module), perovskite-Si tandems, perovskite-perovskite-Si (3V open circuit voltage) and all-perovskite triple junction tandems. Outdoor testing of perovskite devices are really important to detect failure mechanisms.

www.eupvsec.org

CONFERENCE | Topic 3

TOPIC 3: PHOTOVOLTAIC MODULES AND BOS COMPONENTS

The design, manufacturing, and measurement and reliability of photovoltaic modules continues to be one of the larger areas of interest. Module manufacturing is improved through new encapsulants and interconnection techniques. Modules are increasingly being designed to facilitate recycling at their end of life an important sustainability objective. New technologies, such as perovskite materials are introducing new challenges in characterisation and standands.

www.eupvsec.org

CONFERENCE | Topic 4

TOPIC 4: PV SYSTEMS ENGINEERING, INTEGRATED / APPLIED PV

The field of systems and applications of PV is the largest at the conference, demonstrating the increasing maturity of the sector. PV systems O&M is increasingly employing non-intrusive methods such as daylight photoluminescence imaging. Machine learning is being used to monitor the performance. The wide range of applications of PV is rapidly developing, such as for agrivoltaics, with numerous studies looking at system performance and also impact on crop yield – which can also be positive for the case of arid climates. Floating PV, for example on reservoirs or onshore, is being field tested and showing great potential. In BIPV, a lot of work is looking into improving the integrating custom elements, such as coloured modules.

www.eupvsec.org

CONFERENCE | Topic 5

**TOPIC 5:
PV IN THE
ENERGY
TRANSITION**

An important objective is the achievement of a fully circular economy, and recycling techniques and several schemes for module recycling are studied. The integration of PV into energy systems is a major theme. Novel approaches to grid integration, storage and new electricity market designs were presented. The social dimension, is increasingly being studied to understand the needs of society and also to evaluate the social acceptance of PV.

www.eupvsec.org

PLENARIES | Thematic

Manufacturing

PV Everywhere

Performance and Reliability | Thin Films and Tandems

Sustainability

FOUR SESSIONS

www.eupvsec.org

PARALLEL EVENTS | Varied programme

PARALLEL EVENTS

Some examples

Women in PV: A Platform for Dialogue and Networking

The Landscape of Agrivoltaics: In Between People and Technology

Unveiling the Future of Solar Energy with Perovskite PV

Roundtable-BIPV Opportunities in the Indian Marketplace

www.eupvsec.org

INDUSTRY SUMMIT | Debate

INDUSTRY SUMMIT

Industry Stakeholders

European PV Manufacturing in stormy times – will we see the tide turning any time soon?

Global Perspectives on Solar PV Manufacturing: Time-to-market, Costs and Technology Transfer in USA, India, and Europe

Next era of European PV Manufacturing

Company Presentations & Round Tables

www.eupvsec.org

SESSIONS | In detail

The following sections provide further details of the presented works, session by session, based on the valuable feedback provided by the Session Chairs.

The sessions are ordered by topic, and include oral and visual session highlights. The session code and title are included for easy reference.

www.eupvsec.org

TOPIC 1 | Session by session highlights

1AP.1	Manufacturing	The presentation of Mr. Zhang showing the development of solar cell sizes and thickness and what were the drivers for this development was really good. Also he showed the measures/activities taken to improve the cell production pricess in respect to economics and ecology
1AO.4	Silicon Material for Solar Cells: Growth, Stability and Reuse	1. Joshua Kamphues showed us that we can use Fe contamination as a proxy metric to assess the reproducibility in resistance and impurity concentration of recharged Ga Cz Si, including a correlation with LeTID. A creative method for monitoring industry critical Ga-doped melt recharged wafers. 2. Guo Li presented results that show H as an important mitigator or healing agent that could enable improved stability Si for space applications, where H could heal the high radiation damage and improve the competitiveness of Si as a technology for space PV. The session also included many new insights on thermal defects in a range of silicon wafers and PV structures, as well as considerations for recycling silicon based on dopant type.
1AO.5	Processes for Highly Efficient Si Solar Cells	- Advances in leaner processes and passivation improvement for TOPCon and HJT solar cells - Critical research in reducing silver consumption in HJT
1AO.6	Highly Efficient Si Solar Cells	TU Delft showed an IBC-tunnel SHJ solar cell with a MoOx/(n) a-Si based tunnel contact at the rear side and copper-plated metallization with a conversion efficiency of 23.1%. Fraunhofer ISE showed TOPCon solar cells fabricated on epitaxially grown Si wafers. They could demonstrate 24.4% and 24.7% conversion efficiency on n-type and p-type epi wafers, respectively (i.e., as good as on reference FZ silicon wafers). ISC Konstanz showed a poly-Zebra IBC solar cell with 24.12% efficiency and demonstrated it holds potential for 25% based on experimental-based simulations. ISFH showed POLO IBC solar cells with 23.9% efficiency.
1BO.1	Silicon Bottom Cells for Tandem Photovoltaics | Dielectric Layer Related Defect Characterisation	Silicon bottom cells are the fundament of perovskite silicon tandem solar cells. TOPCon and HJT bottom cells have been presented in the session. Both technological progress (thinner wafers, improved voltage) and a better in-depth understanding were discussed in the excellent presentations. A special focus was set on different approaches for the interface to the perovskite top cell as tunnel junctions or recombination layers.
1BO.2	Advanced Silicon Solar Cell Characterisation in Laboratory and Production	The papers in this session highlighted ways in which large data sets from production lines can be collected and analyzed to understand the impact of specific variations in production parameters on final cell performance. In this regard, there was no single highlight contribution, but each presentation provided pieces of the puzzle to substantially improve solar cell manufacturing with big data analysis.
1BO.3	Optimised Processes for the Manufacturing of TOPCon Solar Cells	- Julian Reichle; RCT Solutions: review of emerging wafer sizes industry - Lu Wang ; Jiangsu Xianghuan Technology: 1GW Cu platting tool for solar cell manufacturing - Damian Brunner ; RENA : 37%USD Cost saving for alkaline batch edge isation tool compared to acidic solutions - Eric Schneiderlochner ; Von Ardenne : Full PVD route for TOPCon including SiN, poly-Si and Silicon tunnel oxide - Xutao Wang ; UNSW; laser assisted firing (LECO/JISIM) 0.6% abs efficiency gain compared to selective emitter - Tadeo Schweigstill; FISE: Printing down to 8um with glass stencil
1BO.4	New Concepts for the Manufacturing of IBC and HJT Solar Cells	2 important European projects (BUSSARD and IBC4EU) were presented with relevant technical data from the material to the module, including cells, equipment and recycling tasks ; and several process flow for simple and efficient IBC technologies were presented. This made the session really interesting with alternative and promising results paving the way to new industrial processes. Solutions for more simple and sustainable process have been presented, including metallisation with reduced silver consumption and selective etching for patterning process.
1BV.5	Silicon Material: Growth, Defects and Recycling | Manufacturing of Solar Cells and Related Tools & Processes	This session showed progress on the application of copper metallisation for high efficiency cell concepts like HJT and TOPCon with cell efficincies comparabel to Ag references.
1CV.2	Processing & Characterisation of Crystalline Si based Solar Cells | Silicon Bottom Cells for Tandem Photovoltaics | Advances in Silicon Solar Cells Characterisation and Simulation	A very rich session with many novelties going from processing to characterization and modelling. Award to the impactful poster on simple method to reduce Cu diffusion.

www.eupvsec.org

TOPIC 2 | Session by session highlights (1)

2CP.3	Performance and Reliability	Thin Films and Tandems	CP.2.1 Increase in size and reduction of glass thickness to 2mm reduces strength and increases probability of failure; Longi announced to go back to 3.2mm front glass thickness. CP.2.2 Watch out manufacturers, Daylight PL has become even more powerful. 9000 modules at once, 50 at once when resolution on cell level is needed. CP.2.3 Power optimizers don't deliver 30% perfomance increase of roof systems under partial shading conditions as advertized. CP.3.4 Outdoor testing of encapsulated perovskite modules (statistics) is crucial for better understanding of their performance and stability. CP.3.5 triumph project succesfully worked on triple junction and achieved 26% in the endeavor to make tandems/triple junctions great again. CP.3.6 Comeback of CIGS? New record efficiency of ~24% reported, in 4 terminal tandem with Perovskite: 29,9%
2AO.1	Inorganic and Organic Compound Solar Cells and Tandems	In session 2AO.1 about Inorganic and Organic Compound Solar Cells and Tandems, several novel contributions were presented. Two major highlights were on some new devices: Both IREC and UPC showed recent progress in kesterite development. The latter showed a kesterite device with Li doping and Ag alloying reaching 14.1%. A second highlight was on the rear surface passivation of CuIn[S,Se]2 by patterned Al2O3, presented by the University of British Columbia.
2AO.2	Solar Cells based on CIGS and its Alloys	Upsala University presented record efficiency CIGS solar cells alloyed with Ag and reaching 23,6% after a ligth soaking step. She gave an in-depth analysis on the evolution of the structural, chemical and electronics properties of the devices. Another highlight was from ZSW on transparent CIGS semitransparent cells for tandem devices. Results focused on the need to control the ITO/CGS interface by adding a thin Mo layer to prevent GaOx formation. Alternative TCO layers are also proposed to achieve back contacts combining both high transparency and good electrical behaviour.
2AO.3	III-V Solar Cells & Space PV	It seems silicon came back to space again, due to marked volume demand a lower cost. All space related presentations were talking (iii-v also) about Silicon.
2BO.10	New Modelling and Characterisation - Material Properties	Spatially resolved determination of implied open-circuit voltage and thickness of perovskites and perovskites silicon tandem solar cells.
2BO.8	Novel PV Material and Conversion Concepts	* two talks could be highlighted for this session, namely 2BO8.1 and 2BO8.3, however the 2BO8.3 makes one step further in view of practical implementation of the contribution 2BO8.1. * it is practically applicable on an IBC solar cell and could potentially lead to 34% efficiency using 1 junction only * There might be doubt about additional effects due to band bending related to MoOx and heavy-doping induced BGN, but at least it sketches a way to practical implementation.
2BO.9	Advanced Materials for PV Devices	In this session, we saw a trend towards trying to improve the performance and manufacturability of emerging materials; in particular, we saw Sarallah Hamtaei from Imo-imomec (Hasselt University/imec) present results showing PV-grade WSe2 deposition, a transition metal dichalcogenide, on much larger areas (up to 6 inch wafers) using ALD than previously achieved using mechanical exfoliation.
2CO.1	Processing, Characterisation, and Modelling of Perovskite/Silicon Tandems	2CO.1.1: Oussama Er-raji from Fraunhofer ISE, student finalist, showed how incorporating the urea-treated perovskite absorber in a fully textured tandem cell passivation were improved and series resistance were reduced achieving over 31% stabilized PCE.
2CO.2	Triple Junctions and Advanced Concepts in Perovskite-based Tandems	Perovskite-CIGS mini-module with scalable processes with 19.9% and 24.5% on solar cell level. 3V Perovskite-Perovskite-Si Triple Junction Device. All-Perovskite Flexible Tandem Module with 19.7 efficiency.

www.eupvsec.org

TOPIC 2 | Session by session highlights (2)

2CO.3	New Modelling and Characterisation - Device Performance	This session covered various aspects of modelling and characterization of perovskite solar cells solo or in tandems, as well as a bonus talk on thermophotovoltaics. We heard about the fundamentals of cell selective techniques to determine loss mechanisms, along with correlations between capacitance-based measurements and simulations identifying e.g. effects of ion migration and mobility. Outdoor testing of bifacial perovskite-silicon tandems were also investigated with a special focus on spectral and temperature effects. Modelling using machine learning techniques help to identify recovery mechanisms in perovskite solar cells. Finally we were educated about thermophotovoltaic systems and how to measure them.
2DO.18	Late News: Developments in High Efficiency Tandem Cells	HZB: Two main innovations in all-PVK tandems. The first is piperazinium iodide treatment to displace the contact between WBG PVK and C60. This results in better QFLS and Delta_Voc > 100 mV. The second innovation is a new SAM-based bilayer HTL for NBG PVK. This enables good transport while allowing crystallization of the Tin-based PVK absorber. Deploying these innovations, an all-PVK tandem with PCE ~ 30% and an all-PVK triple junction with Voc ~ 3 V are demonstrated. Jinko: Certified 33.24% 2T tandem based on PVK/TOPCon combination (certified at SIMIT, China). Key innovations are: high temperature ITO as Rl; double side textured TOPCon (boron diffusion at the rear side), one step coating process for the top cell with buried interface in the absorber as well as novel additive to quench hysteretic behavior and new surface treatment. Oxford PV: Focus of the company in on high PCE to push down the LCOE. Main market for them is not necessarily rooftop but rather power plants. Their module is sturdy anyway against partial shading and passed IEC 61215 (draft certification). Datasheet value is 545 Wp with 15 year warranty. Right now, full area module PCE is 25% (certified at Fraunhofer). Polytechnic University of Turin: Preparation of 3T PVK/SHJ in 3T hybrid bipolar junction configuration (PNP in terms of front/back contacted PVK onto front/back contacted SHJ). Technical challenges have been overcome to extract the common terminal in the middle of the 3T HBT.
2DO.6	Towards Improved Understanding of Perovskite Solar Cell Device Physics	2DO.6.5: Enhancing crystallinity of perovskite materials through rapid microwave annealing. The innovative modification of perovskite annealing, and the implications for upscalability were impressive.
2DO.7	Process Innovations for Perovskite Devices	Stronger focus on a combination of industrialization (scale up, higher throughput) and stability of perovskite solar cells and modules. The relatively low fracture energy in PSM compared to cSi solar modules can be largely compensated by a laser scribe process. In this approach the POE bonds to the substrate and handles the mechanical load.
2DO.8	Scalability of Perovskite Solar Modules	The session impressively demonstrated how the perovskite technology is maturing and moving towards industrialization. UNSW showed in 2DO.8.1 a new metrology approaches for mass production whilst the other talks demonstrated the increasing number of high-level pilot lines for supporting industrial ramp-up. Scaling is happening right now!
2DO.9	Lifetime and Reliability of Perovskite Devices	Extended dataset on small and large (28x28 cm2) perovskite modules outdoors in 4 locations up to 3 years [2DO.9.1] as well as 20 months-long testing of 4T perovskite/Si tandems in Cyprus [2DO.9.5].
2BV.1	Advances in Novel Materials, Devices and Concepts	New Modelling and Characterisation Techniques	The session showcased a wide variety of new concepts and characterization techniques in different stages of development. It was very interesting to learn about the new materials, device designs and methods for predicting the performance of new device technologies.
2BV.2	Compound and Organic Semiconductors	The poster session 2BV2 included thin film CIGS, kesterite and CdTe solar cells as well as III-V materials and solar cells for space. For the CIGS solar cells, the poster by Arivazhagan Valluvar Oli et al from University of Luxembourg with a 15.7 % efficient high bandgap Cu(In,Ga)S2 device was nominated to the poster award. Both the results and the presentation were excellent. Kesterite solar cells with Cd-free buffer layers were presented by IREC, Spain (Yudania Sanchez et al) and show good promise for the future of this material. Novel architectures with bottom-up grown CISe micro-cells had potential for high efficiency concentrator cells with selective growth, presented by (Lucassen et al from University of Duisburg-Essen, Germany). The jury was also impressed by the poster presented by Jennifer Teixeira from INL (Portugal) on modeling of passivation of Ultrathin CIGSe solar cells. For the III-V, remote-plasma process gave good results for growing high quality thin films. The poster was presented by IPVF, France (Lise Watrin et al).
2CV.3	Perovskite-based Multijunctions	Perovskite Photovoltaics	Highlight of the session was a all vacuum-based approach to form perovskite layers on planar substrates and textured silicon via an evaporation + CVD sequential deposition. Their innovative approach shows a process sequence by scalable techniques, including for the organic precursor deposition, which typically is a critical step. (posters 2CV3.63 and 2CV3.24 by CSIRO, Australia)

www.eupvsec.org

TOPIC 3 | Session by session highlights (1)

3BO.11	Reliability of PV Modules: The Impact of Solar Cell Technology	Our session covered different cell technologies from TOPCon over HJT to Perovskites and poked at various degradation mechanisms. A common theme of all presentation was the importance to have a well designed Voc in order to keep degradation limited.
3BO.12	Reliability of PV Modules: The Impact of Polymers	The importance of studying the reliability studies of the backsheet materials that are mostly used today in production, namely PET with fluoropolymer coating was highlighted. First results of combined stress on this type of backsheets were presented (NIST). The session was focused on investigating the reliability and long-term performance of new encapsulating materials and backsheets. Initial results on the reliability of fluoro-coated backsheets were presented, which is very critical for the field but still unsufficiently studied.
3BO.14	Failure Modes and Degradation Mechanisms in PV Modules	The talks highlighted a variety of microstructure failure modes for new cell technologies and provided insights into mitigation strategies.
3BO.15	Reliability of PV Modules: Testing and Modelling Approaches	An innovative approach for reliability testing of PV modules, tailored for hot desert climates. Authors from four different continents presented results from outdoor testing on degradation rates and failure modes and effect analysis (FMEA). Slide #6 as a nice example-reference.
3CO.10	Materials and Processes for PV Modules	Very interesting session with material topics covering: > overview encapsulation materials with recommended selection for next generation cells (e.g. for perovskite based tandem cells), > presentation of alternative encapsulation material, > new developments for multilayer ARC for solar glasses, some of which could be also implemented on other substrates like sensitive cells (e.g. customized multilayer ARC applied on perovskites to improve their stability by reducing their exposure to certain wavelength) And module manufacturing processes topics covering: > review advantages of double side heated laminators > proposal to avoid thickness decrease (and mechanical stress) at glass-glass laminates with membrane laminators (partial cross-linking in edge region).
3CO.11	Emerging Interconnection Technologies	3 CO.11.3 - necessary step towards lowT interconnection - hopefully relevant study, need to developed towards industrially relevant cycle times. 3 CO.11.5 - nice to see new concepts being featured in presentations; very interesting for niche applications where shading is hard to avoid.
3DO.12	Low Environmental Impact Module Design and Technologies	3DO.12.1 Steps Towards a 100% Renewable Material Solar Module: Evaluating Material Substitutions for Encapsulation and Interconnection, Ringo Koepge: Full PV-module produced with recycled backsheet and biodegradable encapsulant, recycled frame. 3DO.12.2 New Encapsulant for PV Modules Designed for Recycle: A Lab Scale Prototype, Timea Bejat: New innovative silicone based vitrimer developed for recycling purposes with easy lamination and delamination approach, it promote repair and reuse in the future. 3DO.12.3 Laser-Assisted Ablation for Silicon Solar Panels Recycling, Remi Aninat: Solution with NIR laser for ablation and separation of silicon and encapsulation interface, fabrication and decomposition are both show for samples in a clean process. 3DO.12.4 Innovative Design-for-Recycle for Critical Material-Free Interconnection, Antoine Perelman: Assembly of a module with a liquid instead of polymeric encapsulation with led to full decomposition of the stack with recovery of the material, it also showed satisfying stability under TCT 350. 3DO.12.5 Bifacial Lightweight Solution without Glass, Alicia Buceta: Light weight module with reinforced composites and UV protection film was tested for hail impact, damp heat and UV, it showed positive results for hail test. 3DO.12.6 Development of Novel Protective Coatings to Increase the Durability and Reliability of Glass-Free Lightweight PV Modules, Yuliya Voronkon: Comparison of UV-curable coatings developed for PET-foil protection with good results of accelerated aged samples in characterisation of optical transmission
3DO.16	PV Module Assessment and Classification	Noted interest in also monitoring potential Pb leaking from Pk-Si tandem, reported by Atse Louwen (EURAC research) from Nexus project. Brendan Wright (UNSW) reported on significant improvements in luminescence imaging (together with contributions), in particular also in machine learning methods to implement this automatically on a large scale.

www.eupvsec.org

TOPIC 3 | Session by session highlights (2)

3DO.17	Outdoor Performance and Energy Yield Estimation	Both presentations on degradation rates highlight the need to analyse more technologies for longer timeframes, the importance of bill of materials and the re-definition of main stressors in extreme weather conditions. The other 2 presentations, on energy rating standards highlighted that the scientific community is still working on their improvement, by validating them through outdoor performance data and round robins between laboratories to validate the parameters extraction. The results are promising and will be implemented in new versions.
3DO.19	Modelling Techniques for PV Modules	Blind PV Performance modelling comparison of commercial software (3DO.19.6). The paper reveal critical shortcomings of currently used modules and provided valuable feedback for the PV community which will help improve the accuracy of energy yield simulation.
3DO.20	Shading and Soiling on PV Modules	1) SHADING: A methodology for shading resistance classifictation of PV modules have been presented by Richelmann (PTB) 3DO.20.3. The methalody has been proved to be simple to carry out and reproducible across different labs. A proposal for a new standard about shading resistance has been suggested. 2) SOILING: Different type of dust has been observed and analyzed in desert environment. Correlation with Isc losses have been presented. (3DO.20.5). PV cleaning has been shown to be important in desert environment. Standardize evaluation of cleaning procedures has been proposed (3DO.20.6).
3EO.1	In Field Characterisation of PV Modules | BoS Components in Operation	The session covered various topics of outdoor module performance testing and optimization. Visual inspection was shown to be a useful screening tool for low performing modules that might require further testing. Different encapsulant compositions degrade very differently. A new UVF and NIRA technique allows for a more informative inspection. Variations between similar BOM modules were also shown. A mobile module testing lab including a A+A+A+ simulator with an EL capability was also shown that can test a large amount of modules. It has been applied both to initial pre-installation inspections and to already fielded modules. Lower performance than Pnom was demonstrated in initial tests that correlates to manufacturer and, seemingly, not to the specific technology of the module. Reduction of uncertainty components arising from irradiance and temperature variations in fielded modules was also shown. A couple module types were used as test cases, fielded in different locations and different mounting configurations. Combined expanded (k=2) uncertainty of Pmax was found to range between 3.3 and 5.7%, which could be improved further. A power electronics device was demonstrated and fully integrated to a module package to produce an AC module. This included connectivity between the modules. The structure avoids multiple connections by integrating everything to the module package itself in a larger "junction-box" that has been shown to be durable and operate safely. Typical commercial DC power optimizers were tested to evaluate the algorithm for finding maximum power under partial shading. It was determined that the devices actually do not find maximum power but they rather settle on a low voltage local maximum of power. This seems reasonable for avoiding bypass diode activation and possible hotspot problems, however it does not conform with the specifications provided by the manufacturer. A 1-axis tracker algorithm optimization effort was shown that aims to find optimized tracker angles for diffuse conditions. A neural network was trained and initially validated using lab data. The next step is for further application and validation on an outdoor system.
3AV.1	PV Module Design and Manufacturing | BoS Components, Operation and Aging	Shingling with soldering technology: process optimisation, reliability testing and perspectives for further optimisation. Detailed encapsulant formulation, coloring optimisation through design of experiments.
3AV.2	PV Modules Reliability: Components, Failure Mechanisms, Testing & Modelling	We are witnessing an increasing interest and research activity on failures assessment, testing and modelling approaches related to glass, as critical component of current and future PV module technologies.
3AV.3	PV Modules Performance: Testing, Modelling Techniques and Outdoor Performance	Characterization remains a topic of ongoing research, with the need to establish clear measurement methods, in particular for bifacial measurements. There is also concern about underperforming new technologies, both in performance and reliability.

www.eupvsec.org

TOPIC 4 | Session by session highlights (1)

4CP.1	PV Everywhere	With 15 years research data on over 30 implemented projects in Southern France, Sun'Agri demonstrated that dynamic agrivoltaics proves to be very effective for improving the crops performance, supporting a better response of plants to mitigated extreme climatic effects. Shading offered from PV modules is a clear advantage for the crops, and helps in saving the crops in case of adverse seasons, reducing the water demand around 30%-40%. JRC through an extensive and detailed analysis of the potential for agrivoltaics in Europe demonstrated that the land consumption is not the issue. Countries like Italy, Spain, Germany and France demonstrate the highest potential for agrivoltaics in Europe. Still important challenges need to be faced: Regulatory framework, Social acceptance; impact on agricultural yield. Christophe Ballif gave a very comprehensive overview of the 'history of colored PV'. His talk included all the different techniques to apply color, the major breakthroughs, efficiency ranges, and even euro/m2 best estimates. Colored BIPV is entering a new era in which the reliability of the color becomes important. That requires a new color measurement methodology and maybe even a new chapter 'color reliability/stability' in the IEC BIPV-standard (although that was not explicitly mentioned by Christophe). Minne de Jong gave a nice comprehensive overview of the 'rise of Floating PV'. Without running into confidentiality issues, he showed quite open all the challenges in the field of FPV. He summarized all important issues in the Wave Category (WC) 1 systems. And he gave a first glimpse of what aspects will be very important research (and development) topics for the more challenging WC2,3 or even WC4 (off-shore) FPV: 'Birds have a preference of soiling the more horizontal systems'.
4AO.7	Advanced O&M Strategies and Methods	Several Data driven Models were presented for PV System failure prediction Even with Integrated Human in the Loop Maschine learning Concept, Deeplearning fault detection based on Artificial and convolutional Network And diagnostics Hybrid Modell. The last one could Report on 89% accuracy fault detection for each measurement.
4AO.8	PV Plant Performance, Analysis, Monitoring and Fault Detection in Inverters	There is an increased need for PV performance diagnostics on inverter level and common failure syntax. For example: 8.2 by Stephanie and Malik on failure diagnostics and curtailment and 8.5 by Bernhard Kubicek a common syntax and georeferencing.
4AO.9	The Impact of Soiling on PV Systems	We see more models addressing soiling estimation in large regions; assumptions related to rain cleaning were revisited and updated and an interesting paper with soiling chambers to generate soiling more quickly.
4BO.16	Technology, Performance and Economics of PV in/on Buildings	Transparent Pv for windows and IGU; Shading effect and energy management; Policy at EU level; Architect as a new stakeholder; Agri-BIPV.
4BO.17	Characterisation, Reliability and Safety of PV in/on Buildings	The session presented progress in bipv characterizations and reliability assessment: new approaches and experiments to assess the impact of color and air gaps in BiPv facade together with the use of lightweight substrate have been presented. Fire is a concern for researchers to find the proper material but in the Netherlands out of 10000 fire occurred in buildings only 152 had Pv system and 0,005% of total fire events. PVROOF IN EU27 has a potential of 2,4TWh/y based on a new open access model developed by JRC.
4BO.5	Design and Coloring Techniques of PV for Buildings	This session explored innovative approaches to integrating photovoltaic (PV) systems into buildings, focusing on colored modules. Experts discussed techniques to enhance architectural aesthetics while maintaining efficiency and durability, with key topics including methods to assess the visual and functional impact of these solutions. Marie Courtant from EPFL presented a significant contribution as a finalist of students award—a novel colorimetry tool designed for building-integrated PV (BIPV) applications, addressing current market limitations. The research's relevance extends to an alternative methodology, based on a measurement tool, offering further insights into optimizing the visual appeal and functionality of PV modules for greater acceptance in architectural design.
4BO.6	Performance and Degradation of PV Systems	Analysis of five PV systems with more than 30 years operation in Switzerland showed that degradation rates were lowest at higher elevations (due to lower temperatures?) and in general lower than degradation rates from comparable studies. Author: Hugo Quest, 35 Swiss Solar Solutions.
4BO.7	Data Driven Field Inspection based on Imaging	The section had a wide variation of imaging techniques and specialties within the field, what enriched largely the quality of the session. Relatively new imaging techniques for field inspections such as electro- and photoluminescence, and ultraviolet fluorescence are upscaled and fault detection better understood. More highly accurate imaging inspection techniques are available to be performed during daylight, what maximizes the trust in the available data for PV plant diagnostics. Finally, automatic AI tools for imaging analysis and fault classification are shown to make utility power plant diagnostics possible.
4CO.8	Solar Resource Assesment	Absolute Highlight was the talk 4CO.8.1 'The Fourth Edition of the Best Practices Handbook for Solar Resource Data: An Introduction' - given by Jan Remund (Meteotest, Berne, Switzerland). The Solar Resources Handbook is truly exceptional, offering essential insights into all aspects of energy meteorology while also documenting the current state of research. It is a collaborative work authored by many globally renowned experts, freely accessible to the public, and is expected to be published soon after the conference.

www.eupvsec.org

TOPIC 4 | Session by session highlights (2)

4CO.9	Solar Forecasting	Machine learning based method to do satellite based solar forecasts using Unit architecture show significant improvements from current state of the art. Cloud index based prediction better than cloud motion vector based method. GHI forecasts also demonstrate similar improvements. Hybrid model provides better forecast for longer lead time. Presentation on optimal data usage to reduce the number of training samples. Shows that skills driven sampling strategy can result in 30% of data being eliminated with no impact on data driven sky imager based forecasts. Deep learning model vs video processing method with regression (generative models) for solar ramp forecasting shows that the generative models are much better at forecasting ramps as they do not smooth the forecasts. XGBoost using infrared data from satelite can improve fog detection with a Probability of detection of 80%. Connected cars with data from light sensors can improve solar forecasts.
4DO.1	PV System Design and Optimisation	The session presented impressive work on optimizing the design plan of PV systems and PV plants often based on elaborated simulation techniques. The studies addressed not just optimal power performance but total delivered energy and LCOE taking in account degradation and soiling besides other impacts. An especially interesting presentation was given by Dr. Hesan Ziar of Delft University of Technology on a software tool for the allocation of irradiance sensors in PV plants. Starting from the purpose of the irradiance monitoring, it determines the minimum number of irradiance sensors and their optimal placement based on the plant layout and the topography of the site and its surroundings. (4.DO.1.5 A Matlab-Based Software Solution for Irradiance Sensor Allocation in European Solar PV Farms).
4DO.2	The Integrated Agrivoltaic Performance: Approaches, Modelling, Experiences	Some of the highlights and key learnings from the session were: * Very detailed modelling of agrivoltaics systems is possible, which include both the PV generator (and its yield) and the plants based on a 3-dimensional crop model. This allows to predict the expected photosynthesis active radiation (PAR) for different plants and the crop yield. * Important to distinguish between "food for humans" and "food for animals" (= food for animal husbandry). Crop used for "food for humans" tend to be more suitable for Agrivoltaics and hence should be addressed first. Experiments showed different optimisations for crops in arid conditions, where some plants (e.g. beans) grow better under PV and even under 50% less irrigation needs. * New use case for agrivoltaics: Agromining = phytoextraction of metals from hyper-accumulating plants, which were found to also work well under PV systems. * Sophisticated modelling platform and method is presented for optimizing solar tracker position boosting synergies for crop growth in agrivoltaics systems which support societal acceptance of these projects—validation in Germany follows soon. * Detailed irradiance validation is presented in greenhouses fitted with 30% and 50% cover ratio of PV panels in a checkered patterns in Spain. First results show expected crop yield reductions, large variance in sugar content and that number of fruits increase towards end of crop cycle.
4DO.3	The Integrated Agrivoltaic Performance: Different Climatic Conditions, Crops and Technologies	4DO.3.2 Reducing wind speeds in vertical PV farms is shown to have a generally positive effect on the agricultural yield, offsetting the reduction of light in Nordic countries as shown by Erlend Honningdalsnes from IFE, Norway. 4DO.3.5 Alex Katsikogiannis on modelling of 3D apple trees in orchard, shading by PV panels brings the light conditions to near optimal for photosynthesis. 4DO.3.6 Innovative product for PV in greenhouses by Jacques Levrat, CSEM, already at TRL8 and 1 hectare installation. Growth of tomatoes shows no reduction, even in Netherlands conditions, but insects are less attracted by the plants as they appear less green.
4DO.4	Vehicle Integrated PV	New record 33.7 % efficiency of a module based on triple junction III-V on Silicon (InGaP/GaAs/Si) with high potential for VIPV. Toyota Prius equipped with a III-V 860 Wp module demonstrated a driving range up to 26.6 km/day without heating/air conditioning and 17.7 km/day with air conditioning and 100% energy self-sufficient ratio of for driving scenarios of 15km/day.
4EO.2	Planning of PV Systems	Digital PV	Very diverse session, addressing key aspects of PV integration into the digital energy system. The diversity of issues and problem solving methodologies presented in this session is a sign that the PV community reaches a good level of maturity and understanding of their place and impact on the renewable energy transition. - Energy yields and yield disparities for energy transition modelling - Effects of large scale extreme weather events on a national power supply system with significant shares of PV - Solutions to address privacy issues with PV generation data.

www.eupvsec.org

TOPIC 4 | Session by session highlights (3)

4BV.3	Operation, Performance and Maintenance of PV Systems	Posters in session 4BV.3 highlighted many innovations that are improving the efficiency and scalability of PV O&M and reliability monitoring such as AI-powered analyses, drone-based monitoring, robotic automation, as well as many new monitoring methods that can extract new information from existing data streams. The most promising results combine the needs from industry with the creative innovation from research institutes and universities and are able to scale to fleets of systems.
4BV.4	Photovoltaic in/on Buildings	How to optimize color in BIPV is a challenge for many cases. New techniques are proposed. A new texture is proposed to improve the angle dipendence of BIPV. Optimization of community PV has been simulated considering different surfaces, costs and co2 emission, Still to early to integrate batteries.
4CV.1	Solar Resource and Forecasting	More and more machine learning methods are used, compared and validated for forecasts, data assimilation, updated irradiance databases and all-sky imaging. Also physics-based and physics-informed models are employed, which enhances understanding of the various results, especially regarding to forecasting.
4DV.1	Dual Use (Floating PV, Agrivoltaics, VIPV) and other Innovative PV Applications	The PV community has a better understanding of the limitations and specific requirements of integrated PV applications such as agrivoltaics, floating PV, and low-energy PV. Several posters addressed design conditions, performance, and economics from the point of view of the end application and not just the PV part, which is a (good) sign that these industries are becoming more mature.
4DV.4	PV System Engineering	Control and Systems for Power Systems with Renewables Integration	Battery Systems sizing in combination with PV Grid Integration to optimise the electricity price; Assessing Potential gain of bifacial modules in different Applications using albedo enhancing material in arid climate and alpine locations.

www.eupvsec.org

TOPIC 5 | Session by session highlights (1)

5CO.4	PV Module Recycling	The number of attendess: 140-150 demonstrates the growing interest in the topic. New methods to measure the layer thickness of different PV module components were discussed. Innovative approaches to separate the polymers were presented. High value recycling of perowskites is demonstrated. Cost reduction models for the collection of PV models were presented. Innovative approach for generating energy in the form of heat and hydrogen from waste to create new products.
5CO.5	End-of-Life PV Modules & Ecology	We can see that there is a growing interest in sustainability for PV but some challenges remain. Presentations on PV sustainability on multiple perspectives: Test recycling solutions for novel PV technologies (PK) Develop metrics to evaluate circularity Identify challenges on lack of transparency on PV module material composition Ensure safe and transparent labeling for re-used modules with soon to be published IEC technical report Design of bifacial modules to improve ecosystems services Monitor microenvironnement for development of nature positive PV through bio and physical perspective models.
5CO.6	Life Cycle Assessment of PV	Developments in the field of LCA have made very good progress. - Low carbon solar panel for technology choice Good news from the PhD student Lu Wang - she introduced a methology to simplify LCA's by focus on key parameters. CI should not reduce for reliability. Fraunhofer ISE delivered new results (climate specific degradation) can be used for optimized PV deployment strategies.
5DO.10	Manufacturing PV in Europe	Social Aspects of PV	Crowdfunding as a way to commit people to public engagement of RES as done at the University of Aarhus in Denmark as presented by Marta Victoria.

www.eupvsec.org

TOPIC 5 | Session by session highlights (2)

SDO.11	Value and Competitiveness of PV in the Growing Market	SDO.11.1 G. Masson: Task1 IEA PVPS gave an overview of Global PV Market 2023 Global PV installed capacity reached 1.6TW. China is the largest supplyer and consumer of PV modules. Supply and demand gap causes lower prices stimulating the growth of the market. SDO.11.2 U. Jahn give a presentation titled "Driving the Quest for Reliable and Bankable PV in Europe - Status and Targets in 2030" Key items of Strategic Research and Innovation Agenda (SRIA) related to reliability and bankability were introduced. Enhancement of lifetime, reliability, and sustainability of PV technology are focused. Predictive mentenance, field inspection, lifetime modelling are picked up. SDO.11.3 Dr. E Shiraz is The Value of (BI)PV increasing or Decreasing over Time? East and west facade PV value is promising option in comparison with South faced PV. Decline of the valu-factor is slower. SDO11.4 The Role of Flexible Demand in Reducing Utility-Scale PV Integration - Costs: An Italian Case Study, Elisa Veronese, EURAC Important topic SDO.11.5 Wholesale Electricity Market Prices Forecast Considering Building Conditions Using Price Sensitivity J. Hirota JPEX, Japnese Electricity market forecast SDO.11.6 Cost Analysis for Small scale hybrid Hydorgen based PV Energy System - Marius Möller, Univ. Paderborn, showed that depending on the further cost development of all the relevant system components and different locations in Europa, such systems can be in the upcoming years become more cost effective the electricity from the grid. Moreover, costs might not be the only reason for deciding for such systems, independency from public grids might be another factor, which makes this analysis very valuable in order to discuss the further needs and the financing of public electricity grids.
SDO.14	Energy System Integration with Storage	PV modules indirectly connected to electrolyzers improve the H2 conversion efficiency, relative to direct PV-electrolyzer connection. When using batteries to balance solar PV participating in the grid: if the settlement time (flexibility market) is shorter, the battery degrades faster. Power-heat-Power storage using thermo-photovoltaic for discharging can be used, together with Lithium-ion, to supply electricity and heat to buildings. Sandia is considering combining PV, battery, CSP and thermal energy storage to supply 100% renewable energy to facilities with loads more than 50 MW. Thermal energy storage is key for the extreme days in the year.
SDO.15	Resilience and Security of Supply	Grid supporting powerplants on 100% energy from PV and wind (Gerhard Mütter) - overcapacity (with reference to the nominal grid capacity)) of wind and PV at one grid connection point combined with a battery (4h - 8h) result in very high RE share in the grid with low lossees due to curtailment.
SDV.2	Energy System Integration; Resilience and Security of Supply; Solar Fuels, Storage	PV Sustainability	Storage for PV now includes not only Lithium-Ion batteries but also redox-flow batteries. Energy communities with many PV installations are getting better in forecasting PV generation and demand and giving information to members of the communities regarding when to consume or not to electricity to maximize self-consumption.
SDV.3	PV Diversification Upstream and Downstream - from Industry to Applications	Costs, Economics, Finance and Markets	The Revolution of PV	The session highlighted the importance of supply chain economics, engaging communities, and training programmes for the industry.

www.eupvsec.org

SESSIONS | In detail

Acknowledgements

The Scientific Committee is thanked for its work in creating the Technical Programme, and the authors for presenting their work.

The highlights committee and session chairs are thanked for its input into the creation of these conference highlights.

Special thanks are also due to the staff of WIP who have supported the organization of the entire event.

www.eupvsec.org

LIST OF EXHIBITORS
(in alphabetical order)

Afore New Energy Technology (Shanghai) Co., Ltd. .... China
Austrian Institute of Technology .... Austria
Austrian Pavilion... Austria
Austrian Technologyplatform Photovoltaics .... Austria
Autarq... Switzerland
Avalon .... Belgium
Biosphere Solar.. The Netherlands
BIPV.world.. The Netherlands
Bottero.. Italy
Cosmotaics... Austria
Coveme .... Italy
ECM Greentech.. France
Ecoprogetti .... Italy
EcoSys Abatement LLC .... USA
EDF ENR PWT.. France
EKO Instruments .... The Netherlands
Energy3000 solar .... Austria
Energyra... The Netherlands
Enerlution .... The Netherlands
EnGreen .... Italy
Ertex Solartechnik .... Austria
Eternal Sun... The Netherlands
European Commission - Joint Research Centre .... Italy
European Solar Manufacturing Council... Belgium
exateq .... Germany
Fluxim... Switzerland
FUTURASUN .... Italy
Habemax... Austria
halm elektronik .... Germany
HUIZHOU LYCN ELECTRONICS CO., LTD...................................... China
IEA-PVPS.. Germany
ISC Konstanz .... Germany
Lunovon .... Germany
METU Center for Solar Energy Research and Applications .... Türkiye
my-PV .... Austria
Nagase Chemtex America... USA
NEO .... Austria
Netherlands Enterprise Agency (RVO).. The Netherlands
Niederhuber & Partner Rechtsanwälte .... Austria
NOARK Electric Europe .... Czech Republic
OTT HydroMet (Kipp & Zonen).. The Netherlands
Perovskite Network .... Belgium

pv magazine group	Germany
PVsyst	Switzerland
RCT Solutions	Germany
RENAC POWER TECHNOLOGY CO.,LTD.	China
Risen Energy Co.,Ltd	China
SALD	The Netherlands
Semilab	Hungary
Singulus Technologies	Germany
Sinton Instruments	USA
SolarNL	The Netherlands
SolarPower Europe	Belgium
Sonnenkraft	Austria
Sonnnig	Austria
Sunplugged	Austria
Suzhou Maxwell Technologies	China
Terramo	Belgium
ThermoAnalytics	USA
Thrace Group	Greece
TW (Tongwei) Solar	China
TT VISION	Malaysia
University of Ljubljana	Slovenia
Vienna Business Agency	Austria
VON ARDENNE	Germany
WAVELABS Solar Metrology Systems	Germany
Weidmueller Interface	Germany
Welser Profile	Germany
WIP Renewable Energies	Germany
Zentrum für Sonnenenergie- und Wasserstoff-Forschung Baden-Württemberg	Germany
ZHEJIANG HECHUAN	China

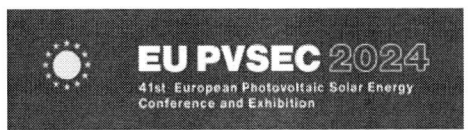

We thank the EU PVSEC 2024 Sponsors

GOLD SPONSOR

SILVER SPONSORS

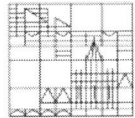

Impact of High-Temperature Processing Steps on the Long-Term-Stability in n-type FZ Silicon

**Melanie Mehler[1], Nicolas Weinert[1], Nicole Aßmann[2],
Axel Herguth[1], Giso Hahn[1], Fabian Geml[1]**

[1]University of Konstanz, [2]University of Oslo

EUPVSEC 2024, Vienna, September 23 2024

Motivation

- Significant progress in the development of solar cell concepts using n-type Si wafers [1]

- LeTID can also occur in n-type mc-, Cz- and FZ-Si material [2-4]

- Solar cell concepts involve high-temperature steps during cell processing

- Investigation of high-temperature processing steps
    - Tabula Rasa
    - P-gettering
    - Fast Firing

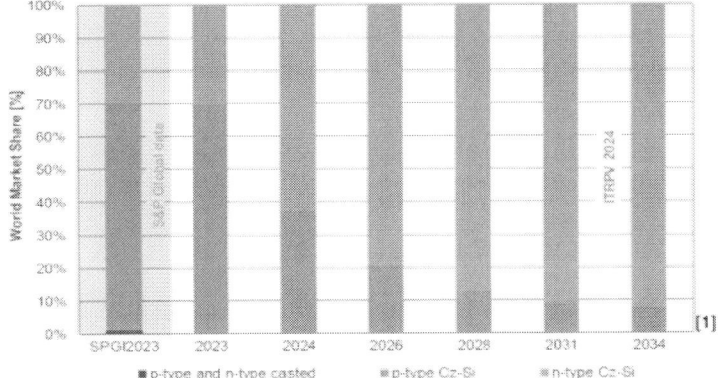

[1] International Technology Roadmap for Photovoltaic (2024)
[2] D. Chen et al., Solar Energy Materials and Solar Cells, 185, 174-182 (2018)
[3] C. Vargas et al., IEEE Journal of Photovoltaics, 9(2), 355-363 (2018)
[4] M. Mehler et al., Solar Energy Materials and Solar Cells, 278 (2024)

Effect of Tabula Rasa

Effect of high temperature
- Dissolution of O-precipitates/agglomerates
- Formation of vacancies and Si self-interstitials (Frenkel pairs)
- More interstitials than vacancies reach the surface and annihilate there
- → vacancy-rich bulk

TR in O_2 atmosphere
- Growing of SiO_2 layer
- Excess of interstitials injected in Si bulk
- → vacancy-lean bulk

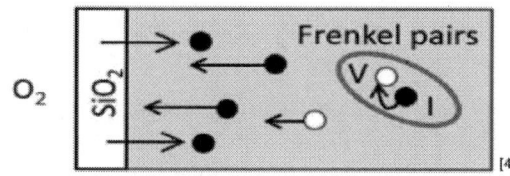

[4] A. Meyer et al. *IEEE Journal of Photovoltaics*, 10(6), 1557-1565 (2020)

Sample Preparation and Characterisation

Thermal anneal (tabula rasa (TR))
- 1050°C in O_2 atmosphere for 30 min

Cryo-FTIR measurement
- Performed at 5 K

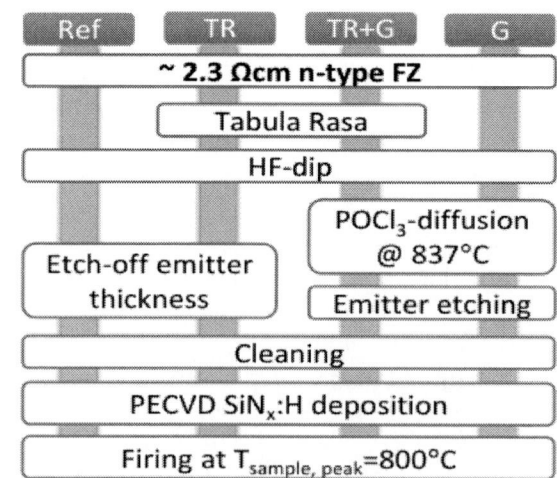

FT-IR Measurements: O_i

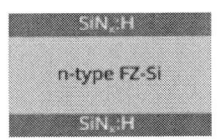

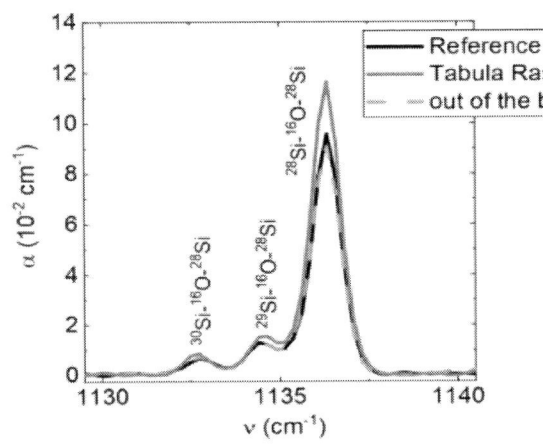

- Minimal differences in $[O_i]$ between reference and non-processed sample
  → firing step has no impact on $[O_i]$

- Sample with TR step exhibits higher $[O_i]$
  → dissolution of O-precipitates/agglomerates

	Reference	Out of the box	Tabula Rasa
O_i [10^{14} cm^{-3}]	10.9±0.2	10.5±0.2	13.6±0.2

FT-IR Measurements: VH_4

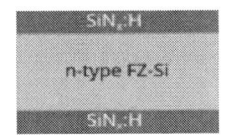

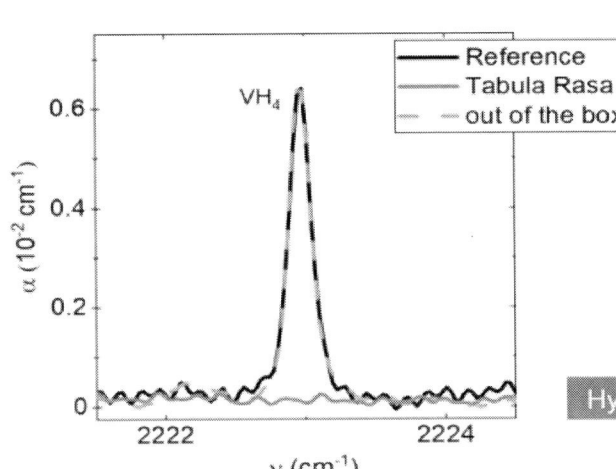

- Minimal differences in VH_4 between reference and non-processed sample

- No local vibrational mode of VH_4 detected for sample with TR step

Hypothesis: TR step reduces the number of vacancies

Sample Preparation and Characterisation

Thermal anneal (tabula rasa (TR))
- 1050°C in O_2 atmosphere for 30 min

Cryo-FTIR measurement
- Performed at 5 K

Lifetime measurements (PCD)
for long-term-stability testing
- τ_{eff} evaluated at $\Delta n = 0.1 \cdot n_0$
- "Defect" density $\Delta N_{\text{leq}} = \frac{1}{\tau_{\text{eff}}(t)} - \frac{1}{\tau_{\text{eff}}(0)}$
- Change in saturation current density Δj_0 [5]

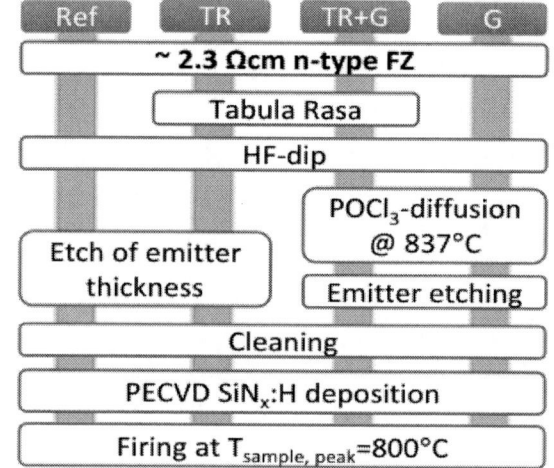

[5] A. Herguth & J. Kamphues, *IEEE Journal of Photovoltaics*, 13(5) (2023)

Lifetime Measurements under Illumination

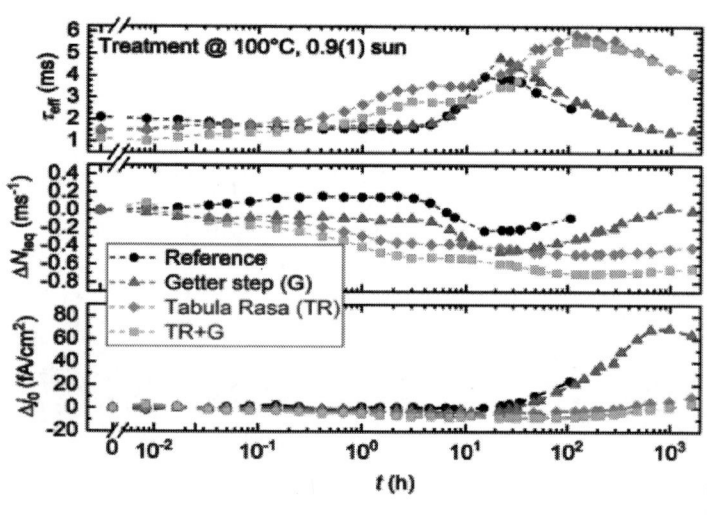

- Slight degradation and strong regeneration for the reference sample

- Gettering shows no significant impact on degradation and regeneration behavior

- Samples with TR step show no degradation, but a two-step regeneration

TR step changes degradation and regeneration

Lifetime Measurements in Darkness

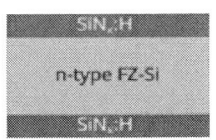

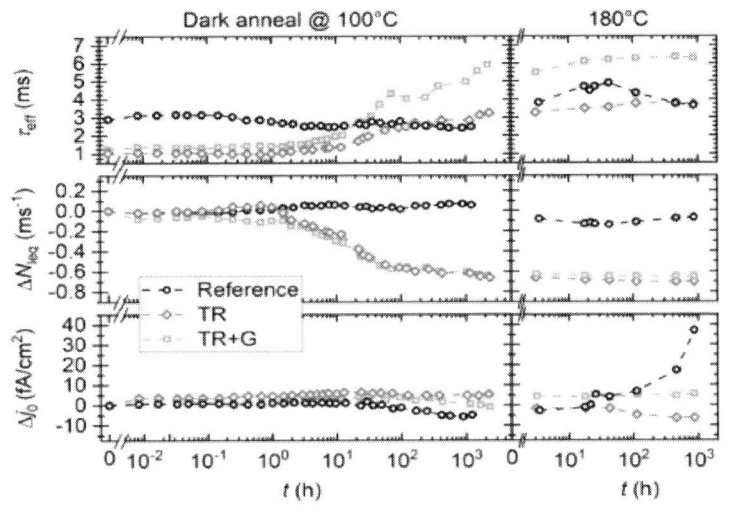

Samples with TR step show no bulk-related degradation, but regeneration with slower kinetics

Effect appears to be carrier-induced

Increase in τ_{eff} could be caused by deactivation of bulk defects activated during firing

Deactivation of defects does not appear to be caused by vacancies

---

Conclusions

- Gettering shows no influence on the long-term stability behavior

- Samples with TR step show only bulk-related regeneration under LeTID conditions

- Same behavior could also be observed in darkness, but with slower kinetics, so the effect seems to be charge carrier-induced

- Cryo-FTIR measurements:
    - Samples with a TR step show higher O_i concentration
    - No local vibrational mode of VH_4 could be detected for samples with TR step

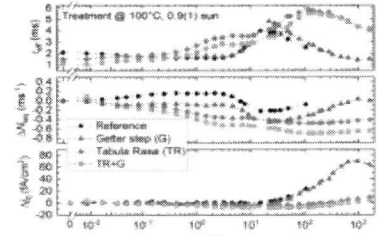

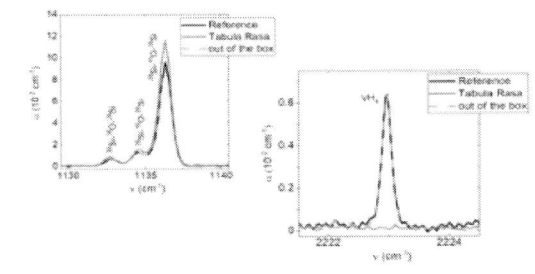

 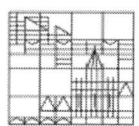

Thank you
for your attention!

Fabian Geml
Department of Physics
University of Konstanz

Tel.: +49 (0) 75 31/88 - 4995
fabian.geml@uni-konstanz.de

LeTID in Industrial Ga-Doped Cz-Si with Melt Recharging

J. Kamphues[1], A. Herguth[1], J. Miech[1], X. Bai[2], Y. Wang[2], G. Hahn[1], F. Geml[1]
[1]University of Konstanz, [2]R&D Center-Wafer B.U, LONGI Green Energy Technology Co., Ltd.

41st EUPVSEC, Vienna, September 23, 2024

Motivation

- Ga-doped material: no BO-defects, but LeTID

- Degradation phenomena like LeTID impact solar cell efficiency

- Factors influencing LeTID:
  - Hydrogen (passivation layer and firing) [1,2,3]
  - Impurity levels [4]

- Cz-ingot growth for Ga dopant incorporation is different compared to B
  → melt recharging for cost-efficient Ga-doping [5]

[1] Eberle et al., Phys. Status Solidi RRL 10(12), 861-865 (2016)
[2] Sharma et al., Sol. RRL 2, 1800070 (2018)
[3] Varshney et al., IEEE J. Photovolt. 11(1), 65-72 (2021)
[4] Wagner et al., Sol. Energy Mater. Sol. Cells 187, 176-188 (2018)
[5] Mosel et al., Proc. 32nd EUPVSEC 2016, 1064-1068

Cz-Si:Ga Growth by Melt Recharging

- Segregation coefficient $\kappa = \dfrac{C_{S0}}{C_{L0}}$
    B: $\kappa = 0.8$
    Ga: $\kappa = 0.008$
    $\rightarrow$ pulling shorter ingots to lower variation in dopant conc.

- Scheil equation: $C_S(f) = C_{S0} \cdot (1-f)^{\kappa-1}$

- Calculate solidified fraction: $f = 1 - (C_S/C_{S0})^{\frac{1}{\kappa-1}}$

- For Ga: $\kappa \ll 1$ and demand that $C_S(f) = 2 \cdot C_{S0}$
    $\rightarrow$ only ~50% of the melt can be used

- Recharging of poly-Si allows to pull multiple Ga-doped ingots subsequently

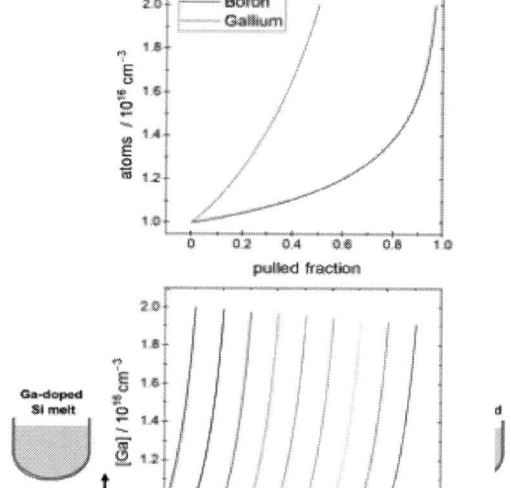

after: J. Friei

3   23.09.2024   LeTID in Industrial Ga-Doped Cz-Si with Melt Recharging   Joshua Kamphues   Universität Konstanz

---

Cz-Si:Ga Growth by Melt Recharging

- Impurity levels increase in the melt due to low segregation coefficient ($\kappa \ll 1$, e.g. Fe: $\kappa \approx 6.4 \cdot 10^{-6}$) [6]
    $\rightarrow$ rising impurity concentrations for subsequently pulled ingots

- Under the assumption that:
    * Same fraction is crystallized for each pull
    * Contamination by recharged Si
    $\rightarrow$ accumulation in the melt is linear in $n \cdot f$
    $\rightarrow$ independent of κ

> Fe$_i$ can be used as a tracker for any species with $\kappa \ll 1$

[6] Hopkins et al., J. Cryst. Growth 42, 493-498 (1977)

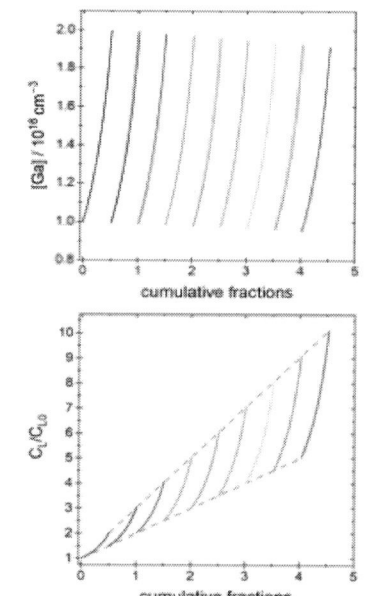

4   23.09.2024   LeTID in Industrial Ga-Doped Cz-Si with Melt Recharging   Joshua Kamphues   Universität Konstanz

Experimental Concept

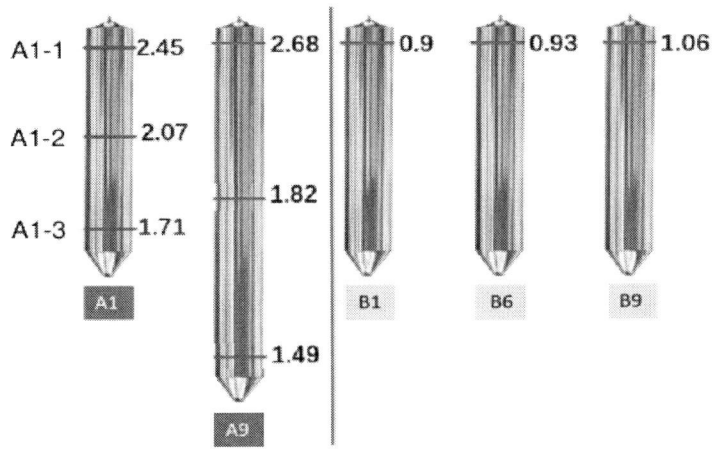

- A1, A9: First and last ingot of one recharged Cz-Si growth process

- B1, B6, B9 from comparable growth process with lower resistivities

- Impurity level is expected to rise with each ingot pulled from melt while base resistance is very similar

- Ingot A9 deliberately pulled longer to increase impurities towards the tail

---

Experimental Concept

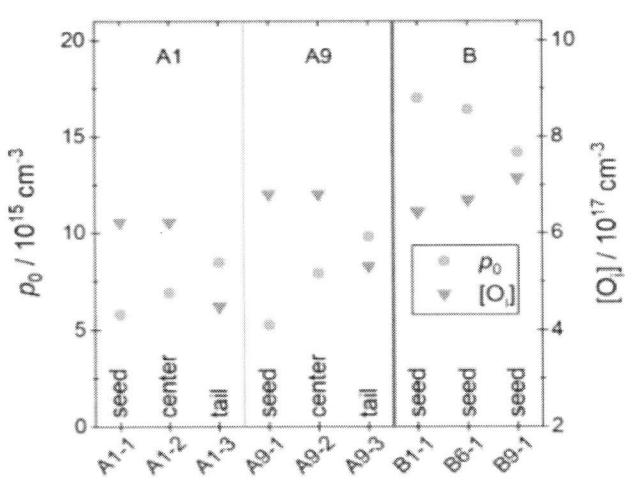

- A1, A9: First and last ingot of one recharged Cz-Si growth process

- B1, B6, B9 from comparable growth process with lower resistivities

- Impurity level is expected to rise with each ingot pulled from melt while base resistance is very similar

- Ingot A9 deliberately pulled longer to increase impurities towards the tail

Process Flow

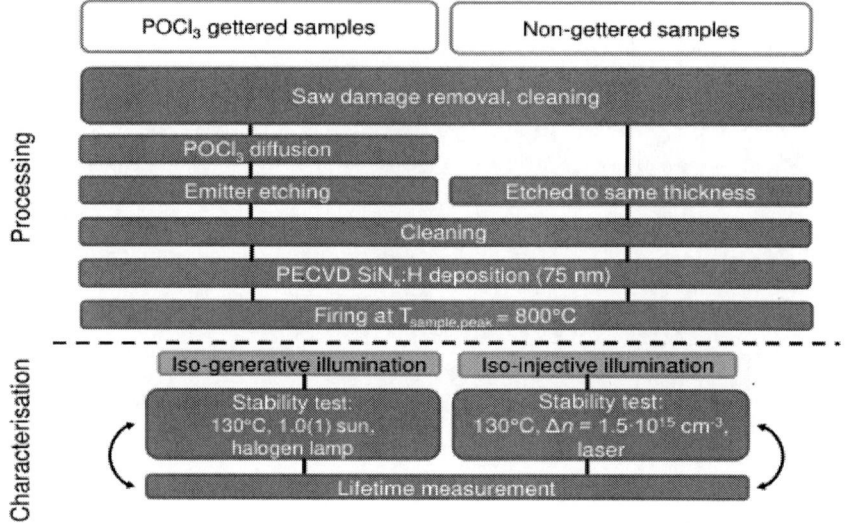

Iron Concentrations for Gettered and Ungettered Samples

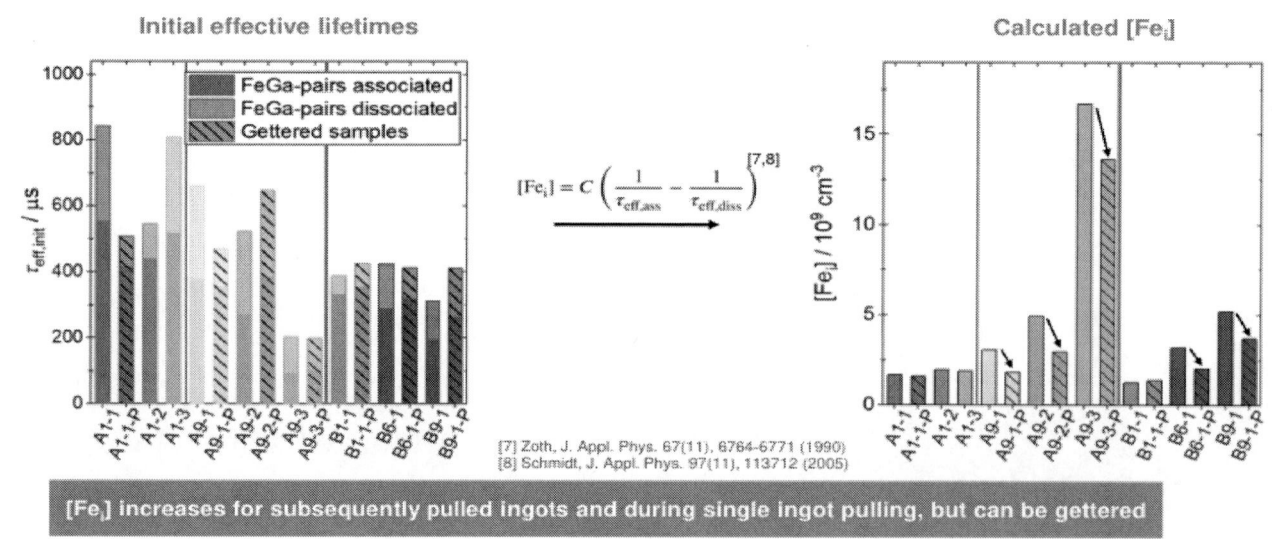

$$[Fe_i] = C \left(\frac{1}{\tau_{eff,ass}} - \frac{1}{\tau_{eff,diss}} \right)^{[7,8]}$$

[7] Zoth, J. Appl. Phys. 67(11), 6764-6771 (1990)
[8] Schmidt, J. Appl. Phys. 97(11), 113712 (2005)

[Fe_i] increases for subsequently pulled ingots and during single ingot pulling, but can be gettered

Long-Term Stability of Gettered and Ungettered Samples

$$\Delta N_{leq} = \frac{1}{\tau_{eff,t}} - \frac{1}{\tau_{eff,t_0}} \quad [9]$$

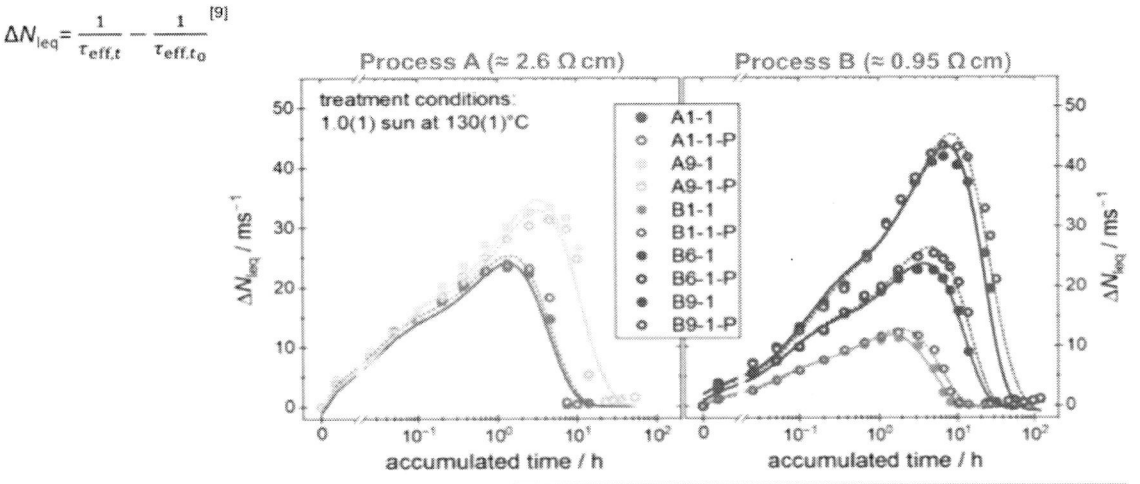

$\Delta N_{leq,max}$ increases for subsequently pulled ingots and does not change for gettered samples

[9] Herguth, IEEE J. of Photovolt 9(5), 1182-1194 (2019)

Long-Term Stability Along Single Ingots and Iso-Injective Treatment

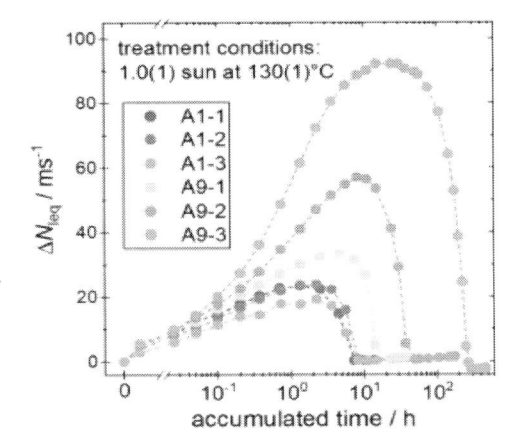 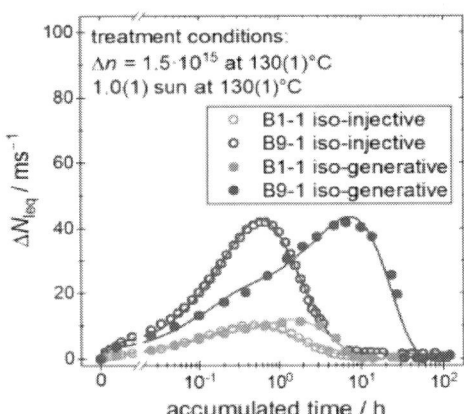

[Fe$_i$] can be used as tracker → contaminated samples show increased LeTID extent

Increase observed under iso-injective treatment, too (same LeTID kinetics for iso-injective conditions)

Conclusions

Fe_i as tracker validates impurity accumulation for $\kappa \ll 1$

Gettering lowers $[Fe_i]$, but LeTID extent is unchanged

Extent of LeTID related to impurity concentration

Iso-injective treatment shows same LeTID extent as iso-generative LeTID kinetics not dependent on impurity concentration

*Paper submitted to Sol. Energy Mater. Sol. Cells

Thank You
For Your Attention!

Joshua Kamphues, M.Sc.

University of Konstanz
Department of Physics / Photovoltaics Division

Tel.: +49 (0) 75 31/88 3732
joshua.kamphues@uni-konstanz.de

Part of this work was financially supported by the German Federal Ministry for Economic Affairs and Climate Action (FZK 03EE1148D). The content is the responsibility of the authors.

Supported by:

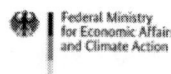

Federal Ministry
for Economic Affairs
and Climate Action

on the basis of a decision
by the German Bundestag

41th EU PVSEC
Vienna, Austria, 23th September 2024, 1AO.5.2

Wet-Chemically Grown Interfacial Oxide for Passivating Contacts Fabricated with an Industrial Inline Processing System

B. Min[1], P. Noack[2], B. Wattenberg[2], T. Dippell[2],
H. Schulte-Huxel[1], R. Peibst[1] and R. Brendel[1,3]

[1]Institute for Solar Energy Research Hamelin (ISFH), Germany

[2]SINGULUS Technologies AG, Germany

[3]Dep. Solar Energy, Inst. Solid-State Physics, Leibniz University Hannover, Germany

---

Motivation

- Passivating polysilicon on oxide (POLO) contacts increased significantly the efficiency in mass production

- 1–2 nm thin interfacial oxide layer + highly doped n+-type poly-Si

  mostly in the production:
  in situ grown oxide in LPCVD or PECVD

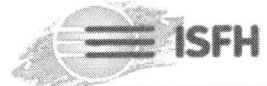

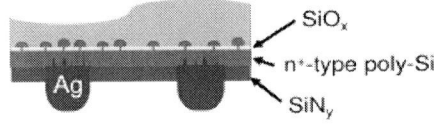

- SiO_x
- n+-type poly-Si
- SiN_y

Motivation

- Passivating polysilicon on oxide (POLO) contacts increased significantly the efficiency in mass production

- 1–2 nm thin interfacial oxide layer + highly doped n+-type poly-Si

  mostly in the production:
  in situ grown oxide in LPCVD or PECVD

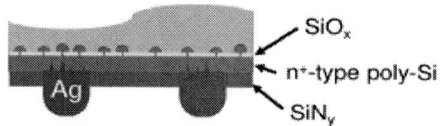

- Wet-chemically grown interfacial oxide → lower process cost and fabrication energy consumption

- Process type for wet-chemical interfacial oxide: *batch* vs. *inline*

  + well established in R&D[1,2]
  − less suitable for production if the previous step uses an inline tool

  − no publications, possibly more challenging i.e. due to larger surface-to-volume ratio
  + Easy to implement by extending the existing inline tool

[1] S. Bordihn et al., Energy Procedia, vol. 8, pp.654-659 (2011)
[2] A. Moldovan et al., IEEE 42nd PVSC (2015)

POLO Back Junction (BJ) and its advantages

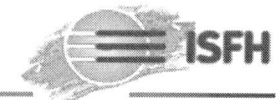

POLO BJ[1]

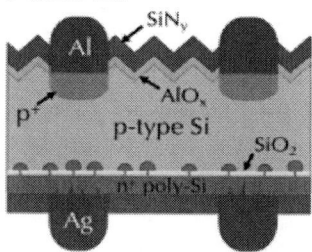

POLO IBC

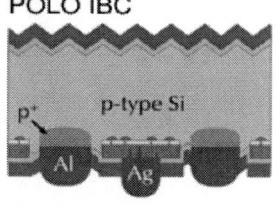

- n+-type passivating poly-Si on oxide (POLO) rear contacts

- Very similar process flow to PERC
  → **Minor modification of proven PERC lines**

- **Leaner process flow** compared to TOPCon
  (less process steps, no boron diffusion)

- Up to **50% less Ag consumption** compared to TOPCon & HJT

- Best efficiency so far[2]: 24.2% with a V_{oc} of 725 mV

- Further innovations:
  - **Upgrade to POLO IBC** (Dullweber et al. 1AO.6.6, Monday, 5:00 pm)
  - **Ag-free metallization**[3]

[1] R. Brendel et al., 35th EUPVSEC (2018)
[2] B. Min et al., Prog Photovolt Res Appl., 1-9 (2024)
[3] B. Min et al., 40th EUPVSEC (2023)

Advantages of the inline wet-chemical processing system

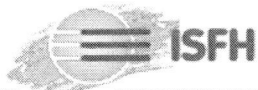

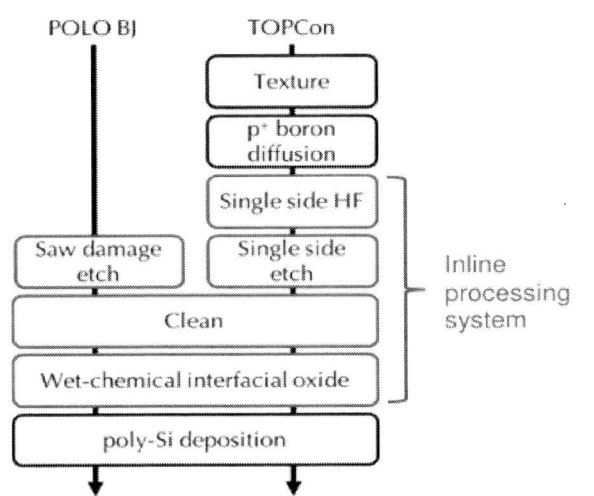

- Extend the existing process (SDE for POLO BJ & edge isolation for TOPCon) with an additional ozone-containing process bath

- Prevents an uncontrolled native oxide growth

- No contamination of the wafer surface due to handling between two different process systems

---

Passivation quality after hydrogenation

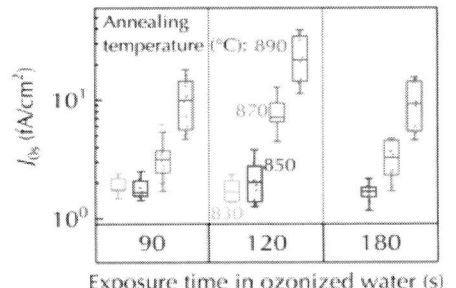

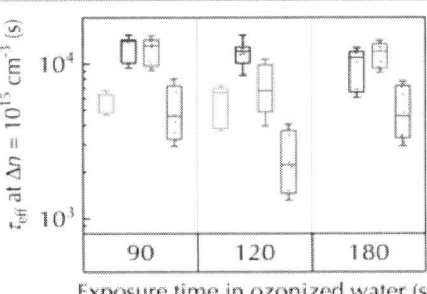

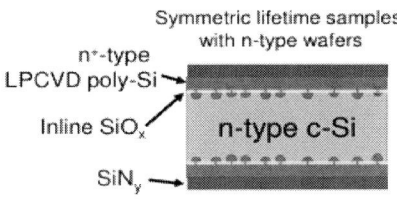

- J_0 on the same level for 830 °C and 850 °C, but τ_{eff} values higher for 850 °C

- Possibly due to more efficient gettering of defects in the bulk by stronger in-diffusion of P

- Lowest J_0 values with 850 °C: 1.4, 1.3 and 1.2 fA/cm² for 90, 120, and 180 s

- For solar cell batch → exposure time of 90 s for higher throughput in mass production

Interfacial oxide from batch vs. inline process system

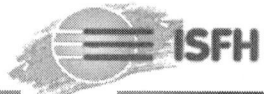

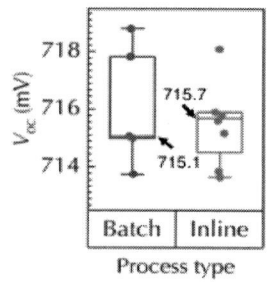 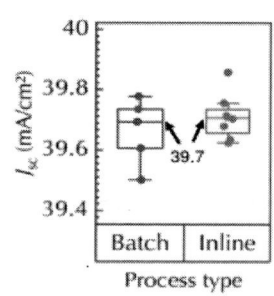 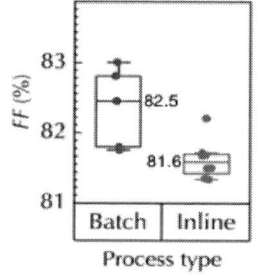 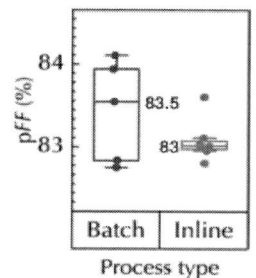 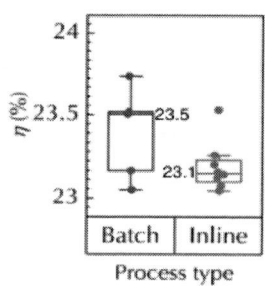

- Reference: POLO BJ with a wet-chemical interfacial oxide from a batch processing system

- V_{oc} and J_{sc} on the same level for both process types

- Difference in *FF* and p*FF* caused by different poly-Si and wafer material qualities, since the results are from two different cell batches

B. Min et al., 1AO.5.2, 41st EU PVSEC, 23rd September 2024

Excellent passivation quality with interfacial oxide from inline process system

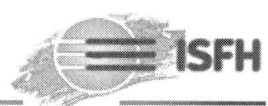

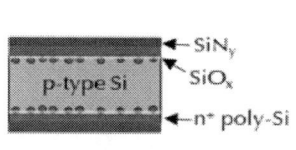

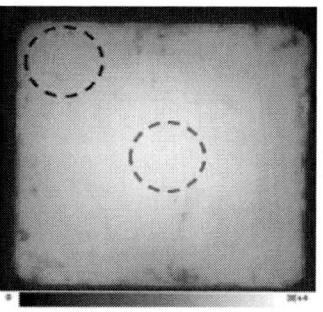

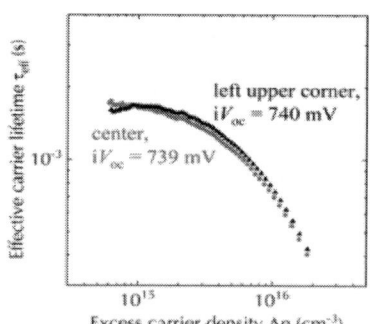

- Homogenous bright PL image under illumination intensities equivalent to 1 sun → homogenous deposition of interfacial oxide and the poly-Si

- iV_{oc} up to 740 mV with a Ga-doped p-type wafer

- Difference in measured iV_{oc} between wafer center and corner → only 1 mV

B. Min et al., 1AO.5.2, 41st EU PVSEC, 23rd September 2024

Economic impact

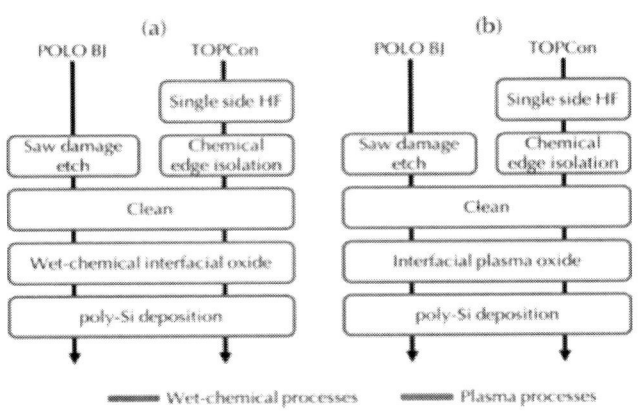

- CAPEX, OPEX and footprint for M10 wafer size

- Only process steps depicted in the figure
  → all other steps remain unaffected by changing the interfacial oxide formation step

- Reference: interfacial plasma oxide from the inline PECVD system GENERIS from SINGULUS

- Wafer throughput of 7000 wph for each system

- Interfacial plasma oxide and the poly-Si deposited in separated process chamber

B. Min et al., 1AO 5.2., 41st EU PVSEC, 23rd September 2024

Saving for applying interfacial oxide from an inline processing system

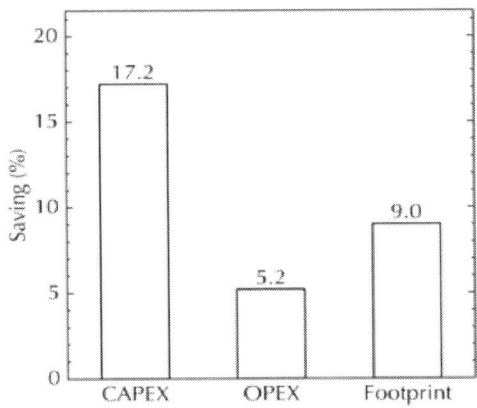

- Representative of both POLO BJ and TOPCon

- Wet-chemical interfacial oxide avoids additional process chamber in the inline PECVD system

  - CAPEX saving up to 17.2%

  - Footprint saving up to 9%

- OPEX can be saved yearly up to 5.2%

B. Min et al., 1AO.5.2., 41st EU PVSEC, 23rd September 2024

Summary

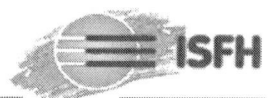

- Sucessful demonstration of passivating contacts with an **interfacial oxide from an inline processing system**

- Excellent passivation quality with a J_0 of 1.4 fA/cm² after the hydrogenation step achieved

- **Homogenous deposition of interfacial oxide** and the poly-Si demonstrated

- **Very short exposure time of 90 s** in ozonized water

- 23.6%-efficient POLO BJ solar cell with a V_{oc} of 719 mV on M2-size Ga-doped p-type c-Si wafers as independently confirmed by ISFH CalTeC

- Wet-chemical interfacial oxide from an inline tool instead of a interfacial plasma oxide **saves 17.2% in CAPEX, 5.2% p.a. in OPEX and 9% in the footprint**

- For more detailed information:
  B. Min et al., IEEE JOURNAL OF PHOTOVOLTAICS, VOL. 14, NO. 2, MARCH (2024)

Thank you for your attention!

The authors thank to **M. Pollmann, B. Gehring, T. Brendemühl, T. Neubert** and **D. Sylla** for processing solar cells, **L. Nasebandt** for characterizations, and **P. Jäger** for his support by ECV measurements (all ISFH).

This work was financially supported by the state of Lower Saxony and the German Federal Ministry for Economic Affairs and Climate Action (BMWK) under grant number 03EE1012A (NanoPERC).

For more detailed information:
B. Min et al., IEEE JOURNAL OF PHOTOVOLTAICS, VOL. 14, NO. 2, MARCH (2024)
https://doi.org/10.1109/JPHOTOV.2024.3352836

Supported by:

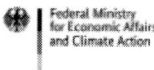

Federal Ministry
for Economic Affairs
and Climate Action

on the basis of a decision
by the German Bundestag

41st European Photovoltaic Solar Energy Conference and Exhibition

This presentation was selected by the Sc. Committee of the EU PVSEC 2024 for submission of a full paper to one of the EU PVSEC's collaborating peer-reviewed journals.

LOCAL P+ POLY-SI PASSIVATING CONTACTS REALIZED BY DIRECT FLEXTRAIL PRINTING OF BORON INK AND SELECTIVE ALKALINE ETCHING FOR HIGH EFFICIENCY TOPCON BASED SOLAR CELLS

Berkay Uygun[a,b], Sven Kluska[c], Jana-Isabelle Polzin[c], Jörg Schube[c], Mike Jahn[c], Katrin Krieg[c], Raşit Turan[a,b,d], Hisham Nasser[a]

[a]ODTU-GUNAM, Middle East Technical University, 06800 Ankara, Turkey
[b]Micro and Nanotechnology Graduate Program of Natural and Applied Sciences, Middle East Technical University (METU), 06800 Ankara, Turkey
[c]Fraunhofer Institute for Solar Energy Systems ISE, Heidenhofstr. 2, 79110 Freiburg, Germany
[d]Department of Physics, Middle East Technical University (METU), 06800 Ankara, Turkey

ABSTRACT: In this study, we present the development of localized p+ SiOx/poly-Si structures, doped with boron through the direct application of boron-ink using FlexTrail printing, combined with wet chemical etching. This approach facilitates the formation of mask-free local TOPCon structures, optimized for high-efficiency tunnel oxide passivated contact (TOPCon) solar cells. We provide a detailed analysis of the factors influencing etch-back selectivity, with particular emphasis on the 1 wt% diluted HF pre-treatment, which is a crucial preparatory step before KOH wet chemical etching. This process optimally removes the native oxide from the intrinsic poly-Si, while preserving the boron silicate glass (BSG) layer on the local p+ poly-Si, thus enabling the selective removal of intrinsic poly-Si to achieve the desired p+ TOPCon structure. For both planar and textured surfaces, line widths between 86.0 - 100.0 µm and 24.0 - 40.0 µm, respectively, were achieved. The test structures incorporating a line grid of local TOPCon features exhibited a maximum open-circuit voltage (iV_{OC}) of 720 mV and the lowest saturation current density (J_{0SE}) of the local p+ SiOx/poly-Si was in the range of 90-120 fA/cm². In future work, the developed local p+ poly-Si structures will be integrated into high-efficiency TOPCon solar cells, positioning the p+ poly-Si beneath the metal contacts to further enhance cell performance.
Keywords: Local TOPCon, Boron Ink, FlexTrail Printing, Selective TOPCon, Etch-Back of Poly-Si

1 INTRODUCTION

In 2021, at the Fraunhofer Institute for Solar Energy Systems (ISE), a photoconversion efficiency of 26.0% was achieved on a laboratory scale using the tunnel oxide passivated contact solar cell (TOPCon) rear emitter (TOPCoRE) structure [1]. Similarly, significant results have been obtained through the application of poly-Si passivating contacts in the interdigitated back contact (IBC) cell configuration, achieving a conversion efficiency of 26.1% on laboratory-scale solar cells [2]. Additionally, passivating contact-based solar cells have demonstrated notable efficiencies in industrial-scale applications. Jolywood reported an impressive 26.7% efficiency on M10-size solar cells in 2023 [3], and Jinko Solar achieved a 26.4% efficiency in 2022 [4].

Enhancing the efficiency of TOPCon remains a critical challenge, particularly in addressing recombination losses under metal contacts. One approach involves integrating the TOPCon layer on the front surface to improve carrier selectivity. However, the parasitic absorption caused by the highly doped poly-Si layer necessitates limiting the use of the TOPCon layer beneath metal contacts. Several methods have been proposed to create localized TOPCon structures under metal contacts, such as the masked diffusion of intrinsic poly-Si followed by a wet chemical etch-back or overcompensation [5], masked etching techniques (including inkjet, laser oxidation, and dielectric methods) on doped poly-Si [6], [7], [8], [9], [10], [11], and the masked deposition of doped TOPCon layers [12]. Additionally, techniques such as 3D printing and secondary LPCVD/phosphorus diffusion [13] have been explored. Masking can be achieved through the use of oxides or solid materials, applied via laser oxidation or inkjet printing. These methods effectively protect the desired poly-Si layer during wet chemical etching, while allowing the removal of undesired sections.

In this study, we employ an innovative printing method, FlexTrail [14], combined with a commercially available sol-gel boron-ink [15], [16], to create p+ poly-Si structures. FlexTrail offers enhanced stability during the printing process for a wide range of fluids and is characterized by its high tolerance to variations in fluid properties, such as viscosity, which contrasts with other methods like inkjet printing or dispensing. The process is simplified by room-temperature printing of boron ink at high speeds (350 mm/s) while maintaining consistently well-controlled finger widths in the local p+ TOPCon structures. At the core of the FlexTrail printer is a thin, flexible glass capillary that holds and dispenses the fluid.

During the wet chemical etching of poly-Si, alkaline solutions such as ammonia hydroxide (AH) [17], tetramethylammonium hydroxide (TMAH) [18], [19], and potassium hydroxide (KOH) [20], [21], [22], [23]—the latter of which is utilized in this investigation—have demonstrated high efficacy. The evolution of hydrogen gas during the etching process plays a critical role in its effectiveness and uniformity. The presence of hydrogen bubbles on the poly-Si surface, formed by hydrogen atoms adhering to the surface, can hinder the etching process [24]. To mitigate this, additives and etchants are incorporated into the solution to facilitate the removal of hydrogen by-products during the etching process [25]. The selective etching of poly-Si regions is made possible by the differential etch rates of doped and intrinsic poly-Si in alkaline solutions [26]. Research indicates that n-type silicon etches more rapidly than p-type silicon in these alkaline environments, which further contrasts with the behavior of intrinsic silicon [26]. This differential increases proportionally with the level of dopant carriers in the material.

Despite multiple attempts to integrate the local TOPCon layer into industrial solar cell fabrication, the complete integration of this structure into commercial-scale TOPCon solar cells remains elusive. This study aims to address this issue by producing a locally diffused boron surface field using a p+ TOPCon layer, thereby enhancing carrier selectivity and reducing recombination beneath the metal contacts in TOPCon solar cells positioned on the front side. The formation of the local p+ TOPCon layer is achieved through wet chemical etching and direct printing of boron ink on intrinsic a-Si using FlexTrail, a mask-free structuring approach. Furthermore, the factors affecting etch-back selectivity between doped and undoped poly-Si during wet chemical etching are systematically analyzed, including the temperature, concentration, and composition of the alkaline solutions, as well as the pre-treatment procedures prior to KOH etching.

2 EXPERIMENTAL

To fabricate the local p+ TOPCon structures, two different test groups were prepared: the etch-back test structures (G1 group) and the lifetime structures (G2 group). Both groups were formed on textured, M2-sized (156.75 x 156.75 mm²) n-type Czochralski silicon wafers with a resistivity of 1.5 ohm.cm. The procedures followed for the formation of these two groups are summarized in Figure 1.

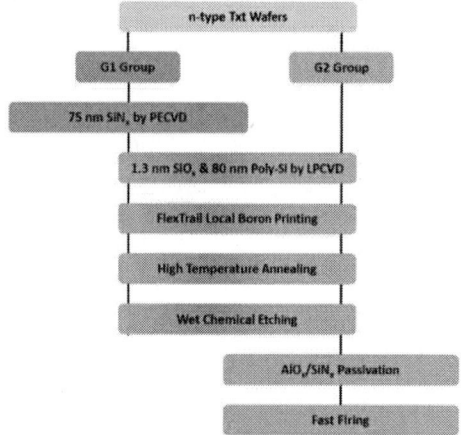

Figure 1: Process flow chart for formation of G1 (etch-back test design) and G2 (life-time) groups.

For the fabrication of the local p+ TOPCon structures, the wafers in the G1 group first received a 70 nm SiN_x interfacial layer, deposited through plasma-enhanced chemical vapor deposition (PECVD). Subsequently, all samples, including both G1 and G2 groups, were coated with a 1.3 nm SiO_x layer, followed by the deposition of an 80 nm poly-Si layer in a low-pressure chemical vapor deposition (LPCVD) system. This poly-Si deposition resulted in a color change in the G1 wafers, facilitating etch-back tracking in later stages.

Following the deposition, local FlexTrail boron ink printing was conducted to optimize printing parameters, first for the G1 group and then for the G2 group. The printing speed was set up to 350 mm/s, and the poly-Si feature size was adjusted by altering the printing parameters. After printing, the organic residues were evaporated using a hot plate, and the samples were annealed at 300°C for 2 minutes under a fume hood. Boron doping activation occurred during high-temperature annealing, conducted at 925°C for durations of 10 and 30

minutes, as well as at 950°C for 30 and 60 minutes. The width of the FlexTrail printed poly-Si fingers was monitored using an optical microscope.

Several different KOH etching recipes were tested using alkaline KOH solutions. Prior to immersion in the KOH solution, some samples received pre-treatment in 1 wt% diluted HF for varying times, while others did not undergo HF treatment. Etching was performed at KOH concentrations of 3 and 7 wt% and at temperatures of 40°C and 60°C. The impact of a chemical edge isolation additive on etch-back selectivity was also tested. The etching time was monitored visually based on the etching rate.

The G1 group and the G2 lifetime test group followed similar processes, with the only difference being that interfacial layer of 1.3 nm of SiO_x TOPCon layers were deposited on G2 samples through LPCVD instead of using the SiN_x interface layer. The etch-back and FlexTrail printing results obtained from the G1 group were applied to the G2 group using the most optimized conditions. AlO_x and SiN_x passivation were then applied to the G2 group. For the G2 lifetime samples, which had a total area of 35 × 35 mm², nine local FlexTrail poly-Si structures were printed, each with varying line widths of 100–120 µm, 60–70 µm, and 30–40 µm.

Figure 2 shows the resulting structures for the G1 and G2 groups along with the test design structures, while Figure 3 illustrates the design of the lifetime samples. The spatial distribution of the implied open-circuit voltage (iV_{OC}) and saturation current densities (J_{0SE}) of the selective p+ TOPCon lines was determined using photoluminescence (PL) images at 1 sun illumination, calibrated with a modulated PL measurement [27]. PL measurements for the G2 group samples were conducted using the Fraunhofer ISE Modulum tool.

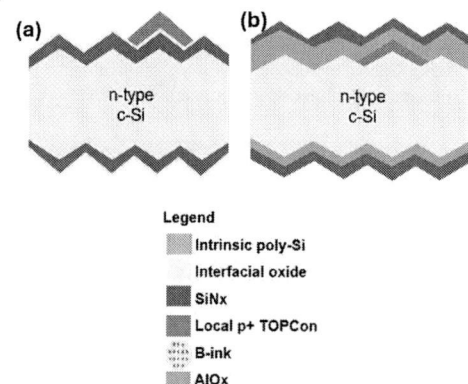

Figure 2: The schematic representation of G1 and G2 group samples on (a) and (b), respectively.

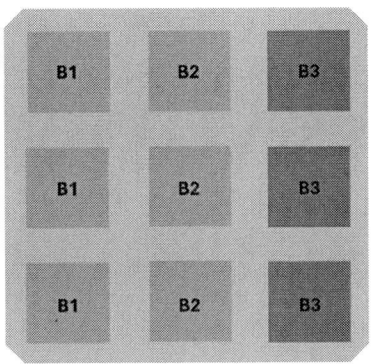

Figure 3: The test design groups of life-time test group (G2). The linewidths are in the orders of B1 > B2 > B3.

3 RESULTS AND DISCUSSION

The poly-Si line widths printed using the FlexTrail method on both textured (TXT) and saw-damaged etched (SDE) surfaces are depicted in Figure 4. Maximum line widths of 100.5 μm and 86.0 μm, and minimum widths of 24.0 μm and 40.0 μm were achieved on SDE and TXT surfaces, respectively. These values can be adjusted by modifying printing parameters, such as printing speed and pressure. However, considering the alignment requirements for subsequent process steps, the fine-line potential of the FlexTrail printing technique was not fully leveraged for this specific application. For further details regarding the relationship between specific FlexTrail printing parameters and the resulting line geometries, the reader is kindly referred to the relevant literature [14].

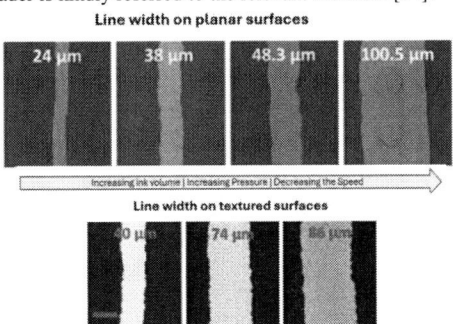

Figure 4: The microscope images showing the widths of local poly-Si lines on SDE and TXT on top and bottom, respectively.

Figure 5 presents the images after KOH etch-back, highlighting the effects of different KOH treatment conditions on etch selectivity. All samples were treated using the KOH solution and settings that offered the highest selectivity. While it is difficult to draw definitive conclusions about KOH etch-back patterns and overall performance, it is evident that increasing both temperature and concentration accelerates the etch-back rate. However, selectivity diminishes when the temperature reaches 60°C and the concentration is 7 wt%. The optimal KOH etching conditions are achieved at 40°C with a concentration of 3 wt%.

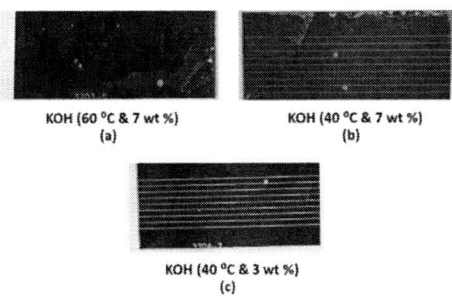

Figure 5: Images of the G1 samples after different KOH etch-back process. 60 °C and 7 wt %, 40 °C and 7 wt %, 40 °C and 3 wt % used in (a), (b) and (c), respectively. Note that in all KOH solutions additives were used.

The pre-treatment with HF dipping prior to the application of the alkaline solution is considered crucial for effective etching. In the absence of HF pre-treatment, alkaline solutions failed to etch the poly-Si, as illustrated in Figure 6a. Conversely, excessive duration of HF pre-treatment led to a loss of selectivity, as shown in Figure 6b. During the KOH etch-back, both the printed p+ poly lines and the intrinsic poly-Si layers were removed from the wafers simultaneously, emphasizing the importance of timing in HF applications. A brief HF dip of 25 seconds was employed, which successfully facilitated selectivity between the doped and undoped poly-Si structures during the subsequent KOH etch-back, as demonstrated in Figure 6c.

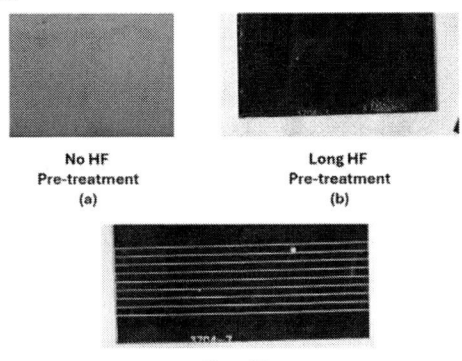

Figure 6: Images of G1 samples following KOH etch-back at 40°C demonstrate the correlation between HF pre-treatment and etch-back. None of the poly-Si was etched in a) and all in b) if no and lengthy HF treatments (110 seconds) were applied. In c), selectivity was maintained throughout a short (25-second) HF drop.

High-temperature annealing activates the boron doping process, leading to the formation of boron silicate glass (BSG) on top of the FlexTrail printed TOPCon lines, as well as a native oxide layer on the intrinsic poly-Si. In scenarios where HF pre-treatment is omitted, the poly-Si layer remains intact after KOH etching, since both the native oxide and BSG layer act as etch stops during the etching process. Similarly, prolonged HF pre-treatment results in the removal of all oxides, causing a significant reduction in selectivity during KOH etch-back. To achieve optimal selectivity, it is essential to etch the native oxide while preserving the BSG layer on the printed lines prior to KOH etching. In this approach, the BSG functions as an etch barrier, allowing KOH to effectively etch away the intrinsic poly-Si layer.

Figure 7 presents a photoluminescence (PL) image of the G2 life-time group following fast firing, including the implied open-circuit voltage (iV_{OC}) and saturation current density (J_{0SE}) values. The FlexTrail local poly-Si in the B3 group achieved an average iV_{OC} of 720 mV, whereas the reference passivated region, devoid of intrinsic or doped poly-Si, exhibited a higher value of 724 mV. Notably, thicker poly-Si widths corresponded to lower iV_{OC} results, with maximum values recorded at 691 mV in the B1 group. These iV_{OC} results align with the J_{0SE} calculations, as the FlexTrail local TOPCon demonstrated the lowest local saturation current density, calculated at J_{0SE} = 90-120 fA/cm².

To evaluate the passivation improvement between intrinsic and doped local TOPCon structures, one G2 sample underwent an inadequate etch-back process, leaving a substantial amount of intrinsic poly-Si on the wafer surface. Consequently, the intrinsic J_{0SE} values ranged from 2000 to 2500 fA/cm². This indicates that proper etch-back significantly enhances passivation. The tendency for lower iV_{OC} values with increasing line thickness may be attributed to the relatively low doping concentration within the crystalline silicon (c-Si) resulting from the printed doped lines, as the ink volume in thicker lines exceeds that of thinner lines, thereby hindering adequate doping injection. To produce thicker lines, adjustments were made to the printing speed and/or pressure during FlexTrail printing, transitioning from the B3 to the B1 group, where the FlexTrail-printed lines were at their narrowest. Consequently, greater amounts of dopants could be administered from the capillaries in the B1 group.

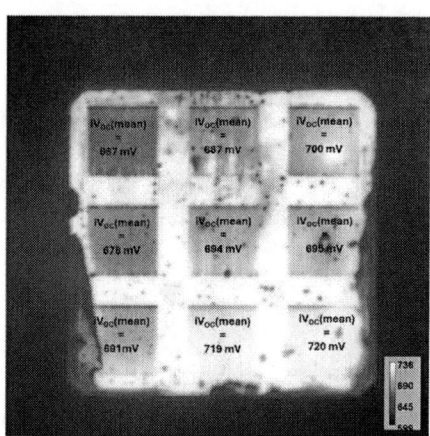

Figure 7: PL images of textured M2 n-type wafers treated with local p+ TOPCon and annealed to 925 °C for 30 minutes. J_{0SE} and iV_{OC} values are shown at the top and bottom, respectively.

4 SUMMARY

This investigation produced a maximum iV_{OC} of ~720 mV and a minimum J_{0SE} of ~90-118 fA/cm². The factors influencing the etch-back selectivity by KOH solution are investigated, and it is discovered that pre-treatment of HF prior to KOH treatment is critical for this purpose. The optimal condition for selectivity is to place the BSG layer on top of the FlexTrail printed poly-Si layer and then etch the native oxide on the intrinsic poly-Si following high temperature annealing. Poly-Si finger widths of 24 - 105 µm and 40 - 86 µm were achieved on flat and textured surfaces, respectively. The width may be readily modified throughout a large range by adjusting printing process parameters. In further work, feature sizes can be significantly reduced. In the near future, the created local p+ poly-Si will be incorporated into a high efficiency TOPCon solar cell, with the p+ poly-Si inserted under the metal contact.

5 ACKNOWLEDGEMENTS

The authors are grateful to Sven Kluska, Jana-Isabelle Polzin, Jörg Schube, Mike Jahn, Katrin Krieg, Raşit Turan, and Hisham Nasser for their crucial collaboration and processes (ODTÜ-GÜNAM and Fraunhofer-ISE). TÜBİTAK provided financial assistance for this work through projects SUPERTOP and CETP WamTech (number 20AG002).

REFERENCES

[1] A. Richter *et al.*, "Design rules for high-efficiency both-sides-contacted silicon solar cells with balanced charge carrier transport and recombination losses," *Nature Energy 2021 6:4*, vol. 6, no. 4, pp. 429–438, Apr. 2021, doi: 10.1038/s41560-021-00805-w.

[2] F. Haase *et al.*, "Laser contact openings for local poly-Si-metal contacts enabling 26.1%-efficient

POLO-IBC solar cells," *Solar Energy Materials and Solar Cells*, vol. 186, pp. 184–193, Nov. 2018, doi: 10.1016/J.SOLMAT.2018.06.020.

[3] E. Bellini, "Jolywood claims 26.7% efficiency for n-type TOPCon solar cell – pv magazine International." Accessed: Jun. 21, 2024. [Online]. Available: https://www.pv-magazine.com/2023/04/12/jolywood-claims-26-7-efficiency-for-n-type-topcon-solar-cell/

[4] Ltd. JinkoSolar Holding Co., "JinkoSolar's High-efficiency N-Type Monocrystalline Silicon Solar Cell Sets Our New Record with Maximum Conversion Efficiency of 26.4%." Accessed: Jun. 21, 2024. [Online]. Available: https://www.prnewswire.com/news-releases/jinkosolars-high-efficiency-n-type-monocrystalline-silicon-solar-cell-sets-our-new-record-with-maximum-conversion-efficiency-of-26-4-301700102.html

[5] F. Haase *et al.*, "Interdigitated back contact solar cells with polycrystalline silicon on oxide passivating contacts for both polarities," *Jpn J Appl Phys*, vol. 56, no. 8, p. 08MB15, Aug. 2017, doi: 10.7567/JJAP.56.08MB15/XML.

[6] S. Schäfer *et al.*, "Role of oxygen in the UV-ps laser triggered amorphization of poly-Si for Si solar cells with local passivated contacts," *J Appl Phys*, vol. 129, no. 13, Apr. 2021, doi: 10.1063/5.0045829/157428.

[7] J. Linke, F. Buchholz, C. Peter, J. Hoß, J. Lossen, and R. Kopecek, "The role of masking layers during metallization of poly-Si/SiOxcontacts," *AIP Conf Proc*, vol. 2826, no. 1, Jun. 2023, doi: 10.1063/5.0141842/2900402.

[8] S. Singh *et al.*, "Large area co-plated bifacial n-PERT cells with polysilicon passivating contacts on both sides," *Progress in Photovoltaics: Research and Applications*, vol. 30, no. 8, pp. 899–909, Aug. 2022, doi: 10.1002/pip.3548.

[9] B. Yu *et al.*, "Selective tunnel oxide passivated contact on the emitter of large-size n-type TOPCon bifacial solar cells," *J Alloys Compd*, vol. 870, p. 159679, Jul. 2021, doi: 10.1016/J.JALLCOM.2021.159679.

[10] M. Recaman Payo *et al.*, "LPCVD polysilicon-based passivating contacts for plated bifacial n-type PERT solar cells," in *35th European Photovoltaic Photovoltaic Solar Energy Conference and Exhibition - EU PVSEC*, 2018. Accessed: Jun. 21, 2024. [Online]. Available: https://imec-publications.be/handle/20.500.12860/31644

[11] R. Peibst *et al.*, "For none, one, or two polarities—How do POLO junctions fit best into industrial Si solar cells?," *Progress in Photovoltaics: Research and Applications*, vol. 28, no. 6, pp. 503–516, Jun. 2020, doi: 10.1002/PIP.3201.

[12] S. Dasgupta *et al.*, "Novel Process for Screen-Printed Selective Area Front Polysilicon Contacts for TOPCon Cells Using Laser Oxidation," *IEEE J Photovolt*, vol. 12, no. 6, pp. 1282–1288, Nov. 2022, doi: 10.1109/JPHOTOV.2022.3196822.

[13] Q. Wang *et al.*, "High-efficiency n-TOPCon bifacial solar cells with selective poly-Si based passivating contacts," *Solar Energy Materials and Solar Cells*, vol. 259, p. 112458, Aug. 2023, doi: 10.1016/J.SOLMAT.2023.112458.

[14] J. Schube *et al.*, "FlexTrail Printing as Direct Metallization with Low Silver Consumption for Silicon Heterojunction Solar Cells: Evaluation of Solar Cell and Module Performance," *Energy Technology*, vol. 10, no. 12, Dec. 2022, doi: 10.1002/ente.202200702.

[15] Z. Kiaee *et al.*, "INKJET-PRINTING OF PHOSPHORUS AND BORON DOPANT SOURCES FOR TUNNEL OXIDE PASSIVATING CONTACTS," in *36th European Photovoltaic Solar Energy Conference and Exhibition*, 2019, pp. 187–191.

[16] Z. Kiaee *et al.*, "Inkjet printing of phosphorus dopant sources for doping poly-silicon in solar cells with passivating contacts," *Solar Energy Materials and Solar Cells*, vol. 222, Apr. 2021, doi: 10.1016/j.solmat.2020.110926.

[17] D. Shin, K. Kim, Y. Ahn, and T. Kim, "Study on wet etching of dummy polysilicon in narrow pattern gap using alkaline solution," *Mater Sci Semicond Process*, vol. 143, Jun. 2022, doi: 10.1016/j.mssp.2022.106561.

[18] H. Takahashi *et al.*, "Wet Etching Behavior of Poly-Si in TMAH Solution," *Solid State Phenomena*, vol. 195, pp. 42–45, 2013, doi: 10.4028/WWW.SCIENTIFIC.NET/SSP.195.42.

[19] T. Park and S. Lim, "Reaction Kinetics of Poly-Si Etching in TMAH Solution," *Solid State Phenomena*, vol. 314, pp. 60–65, 2021, doi: 10.4028/WWW.SCIENTIFIC.NET/SSP.314.60.

[20] N. Ohtani, T. Katayama, H. Yamamoto, and H. Koyama, "Detection of Gate Oxide Defects Using Electrochemical Wet Etching in KOH:H20 Solution," *Conference Proceedings from the International Symposium for Testing and Failure Analysis*, vol. 1997-October, pp. 279–283, Sep. 1997, doi: 10.31399/ASM.CP.ISTFA1997P0279/8643/DETECTION-OF-GATE-OXIDE-DEFECTS-USING.

[21] B. Yang and M. Lee, "Fabrication of honeycomb texture on poly-Si by laser interference and chemical etching," *Appl Surf Sci*, vol. 284, pp. 565–568, Nov. 2013, doi: 10.1016/J.APSUSC.2013.07.134.

[22] J. W. Faust and E. D. Palik, "Study of the Orientation Dependent Etching and Initial Anodization of Si in Aqueous KOH ," *J Electrochem Soc*, vol. 130, no. 6, pp. 1413–1420, Jun. 1983, doi: 10.1149/1.2119964/XML.

[23] K. Krieg, S. Mack, J. Vollmer, T. Dannenberg, D. Brunner, and M. Zimmer, "Wet chemical poly-Si(n) wrap-around removal for TOPCon solar cells," *EPJ Photovoltaics*, vol. 15, p. 9, 2024, doi: 10.1051/EPJPV/2024002.

[24] K. Singh, S. K. Gupta, A. Azam, and J. Akhtar, "A wet-etch method with improved yield for realizing polysilicon resistors in batch fabrication of MEMS pressure sensor," *Sensor Review*, vol. 29, no. 3, pp. 260–265, 2009, doi: 10.1108/02602280910967675/FULL/PDF.

[25] K. H. Jun, B. J. Kim, and J. S. Kim, "Effect of additives on the anisotropic etching of silicon by using a TMAH based solution," *Electronic Materials Letters*, vol. 11, no. 5, pp. 871–880,

Sep. 2015, doi: 10.1007/S13391-015-4499-X/METRICS.

[26] P. K. Singh, R. Kumar, M. Lal, S. N. Singh, and B. K. Das, "Effectiveness of anisotropic etching of silicon in aqueous alkaline solutions," *Solar Energy Materials and Solar Cells*, vol. 70, no. 1, pp. 103–113, Dec. 2001, doi: 10.1016/S0927-0248(00)00414-1.

[27] M. Drießen *et al.*, "Simultaneous Boron Emitter Diffusion and Annealing of Tunnel Oxide Passivated Contacts Via Rapid Vapor-Phase Direct Doping," *IEEE J Photovolt*, vol. 12, no. 5, pp. 1142–1148, Sep. 2022, doi: 10.1109/JPHOTOV.2022.3190772.

41st European Photovoltaic Solar Energy Conference and Exhibition

PHOSPHORUS- AND BORON-DOPED POLY-SI/SIOₓ PASSIVATING CONTACTS VIA INKJET PRINTING

Jiali Wang[1*,] Thien Truong[1], Jinlei Ren[2], Marie Adier[2], Laura Creon[2], Paula Peres[2],
Rene Chemnitzer[2], Pierre-Yves Corre[2], Zhuofeng Li[1], Hieu T. Nguyen[1], Josua Stuckelberger[1],
Daniel Macdonald[1], Anyao Liu[1], Sieu Pheng Phang[1]
[1]School of Engineering, The Australian National University, Australia
[2]CAMECA, Gennevilliers, France
Phone: +61 0481271158; Email: Jiali.Wang1@anu.edu.au

ABSTRACT: In this work, we utilize the advantages of inkjet printing technology to achieve maskless localized doping of both n- and p-type poly-Si/SiOₓ passivating contacts, offering flexibility and simplicity in the fabrication process for advanced structures, such as interdigitated back contact (IBC) solar cells. By using commercially available spin-on solutions, full-area printed samples show promising passivation quality and electrical contacts, yielding an iV_{oc} of 729 mV with ρ_c of 5.4 mΩ·cm² for n-type poly-Si/SiOₓ, and iV_{oc} of 718 mV with ρ_c of 6.1 mΩ·cm² for p-type poly-Si/SiOₓ after hydrogenation. Optical microscopy shows that localized n- and p-type dopant lines with minimum feature sizes down to ~ 60 µm can be achieved. We demonstrated the simultaneous formation of localized phosphorus- and boron-doped poly-Si/SiOₓ passivating contacts in a single annealing step. Micro-photoluminescence (µPL) mappings confirmed the improved surface passivation of the locally doped region. Besides, high-resolution dynamic secondary ion mass spectrometry (D-SIMS) measurements revealed elevated dopant concentrations ([B] $\approx$ 1×10¹⁹ atom/cm³, [P] $\approx$ 1×10²⁰ atom/cm³) at the doped regions. Additionally, D-SIMS results indicated the presence of unintended doping, likely due to volatile dopant species entering the gas phase during the high-temperature annealing.
Keywords: Inkjet-printing, Passivating contacts, TOPCon, Doping

1 INTRODUCTION

The integration of poly-Si/SiOₓ passivating contacts to silicon-based solar cells has yielded remarkable research results with world-record efficiencies above 26 % [1-3] and is experiencing rapid commercialization. Currently, solar cell manufacturers are evaluating economically viable fabrication routes for advanced solar cells structures [4]. In recent years, inkjet printing technology has garnered growing attentions in silicon solar cell fabrication due to its abilities to pattern materials without masking [5-10]. In 2021, Kiaee *et al.* pioneered the application of inkjet printing technology to poly-Si/SiOₓ passivating contacts utilizing a phosphorus dopant source, yielding an iV_{oc} of 733 mV and an implied fill factor (iFF) of 86.4 % [11]. In 2022, they advanced their research by showcasing TOPCon cells featuring inkjet-printed phosphorus-doped regions (~300 µm wide) underneath the metal contacts. This notable development yielded a conversion efficiency of 22.0 % [12]. Nevertheless, it was suggested that the outgassing of dopants from the printed layers led to reduced shunt resistance in the solar cells.

In our work, we utilized inkjet printing technology to demonstrate both n- and p-type poly-Si/SiOₓ passivating contacts using commercially available phosphorus- and boron-containing solutions. Localized doped n- and p-type poly-Si/SiOₓ passivating contacts, with minimal dimensions as small as ~60 µm on mechanically polished surfaces, were successfully fabricated. Comprehensive characterizations of the fabricated localized poly-Si/SiOₓ passivating contacts were conducted to assess the effects of unintended doping spreads to the unprinted region. Subsequentially, we realized the formation of alternating localized doped n- and p-type poly-Si/SiOₓ passivating contacts within a single annealing step at 950 °C. These results pave the way for fabricating high-efficiency IBC solar cells with enhanced flexibility and simplicity.

2 EXPERIMENTAL DETAILS

2.1 Full area control sample

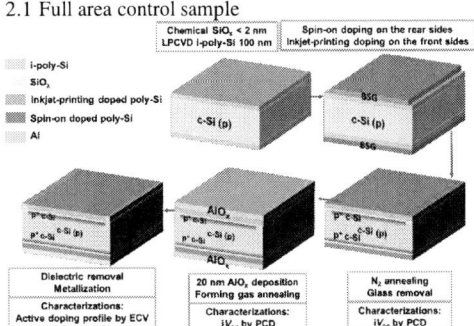

Figure 1: Process flow for full area printed boron-doped poly-Si/SiOₓ passivating contacts.

Figure 1 outlines the experimental process and sample structure for boron-doped samples. The process flow for phosphorous-doped samples is similar and is detailed in our published paper [13]. The substrates were 175 µm gallium-doped Czochralski-grown (Cz) silicon wafers with a resistivity of 1 Ω·cm. The SiOₓ (< 2 nm) and i-poly-Si (100 nm) layers were grown by hot nitric acid bath and low-pressure chemical vapour deposition (LPCVD) respectively. After RCA cleaning, the samples were pre-annealed in a quartz tube furnace at 1050 °C for 1 hour in N₂. Then the rear sides of the samples were spin-coated with the boron-containing spin-on-glass (SOG) solution (B-1500, [B] = 7.2×10²¹ cm⁻³, Desert Silicon), while the front sides were printed with a 4 cm × 4 cm solid SOG layer by inkjet printing (FUJIFILM Material Printer DMP-2850). Thereafter, the samples were annealed at various temperatures for 1 hour in N₂. In the next step, the samples were hydrogenated by depositing AlOₓ:H layers followed by a forming gas annealing at 425 °C for 30 minutes.

The iV_{oc} values were measured using a Sinton WCT-120 lifetime tester based on photoconductance decay method (PCD). The contact resistivities with thermally evaporated Al contacts were measured by the Cox and

Strack method after a sintering process at 250 °C for 5 minutes in forming gas.

2.2 Localized poly-Si/SiOₓ passivating contacts

Additional samples with localized poly-Si/SiOₓ passivating contacts were fabricated to examine the localized doping performance. The substrates were mechanically polished boron- or phosphorus-doped Cz wafer for a clear visualization. The SiOₓ and i-poly-Si growth were the same as the full area control samples. The front surfaces of the substrates were spin-on coated with a non-doped glass. After a curing process at 600 °C, the non-doped glass layers were transferred to ~ 200 nm thick SiOₓ layers. Afterwards, the rear surfaces were spin-on coated with phosphorus-containing solution followed by a 950 °C annealing for 1 hour to form rear surface passivation.

We prepared control samples by using photolithography to open areas of 60 μm × 5 mm on the SiOₓ layer. The unprinted regions were protected by the thick SiOₓ layer to prevent doping during the subsequent annealing step. In contrast, for the test samples, the SiOx layer was completely removed using HF before printing the dopant lines. By comparing test samples with the control samples, we can determine whether the doping is confined to the printed patterns. Localized boron- and phosphorous-doped poly-Si/SiOₓ passivating contacts on test samples and control samples were characterized after a 1-hour 950 °C N₂ annealing by optical microscopy, μPL, and D-SIMS.

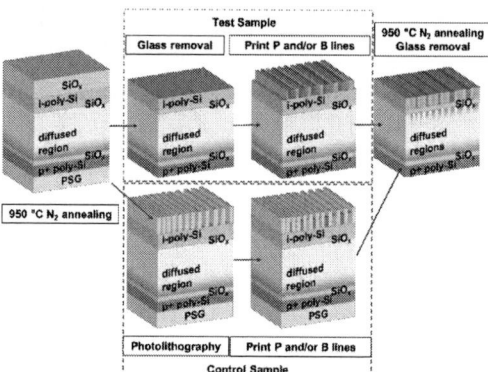

Figure 2: Process flow for localized poly-Si/SiOₓ passivating contacts.

3 RESULTS AND DISCUSSION

3.1 Passivation quality of full area printed sample

As shown in Figure 3 (a), the iV_{oc} value for phosphorus-doped sample improves as the annealing temperature increases from 900 °C to 975 °C, while further increasing the annealing temperature to 1000 °C leads to a sharp decline in the iV_{oc} value. The hydrogenation step enhanced the iV_{oc} values across the range of the applied annealing temperatures. The optimal annealing temperature for phosphorus-doped sample is 950 °C, reaching an iV_{oc} value of 729 mV after hydrogenation. For boron-doped samples, the iV_{oc} values exhibit above 700 mV until an annealing temperature of 975 °C is reached, as shown in Figure 3 (b). Further enhancement in iV_{oc} is observed across all annealing temperatures after hydrogenation, with a peak iV_{oc} of 718 mV achieved at an annealing temperature of 950 °C.

According to Figure 4, the contact resistivity values for the fabricated samples remain below 10 mΩ·cm² across all annealing temperatures for both phosphorus- and boron-doped samples, which are suitable for high efficiency solar cell fabrication. At the optimal annealing temperature, the contact resistivities are below 5.4 mΩ·cm² for the full area phosphorus-doped samples and 6.1 mΩ·cm² for the full area boron-doped samples, respectively.

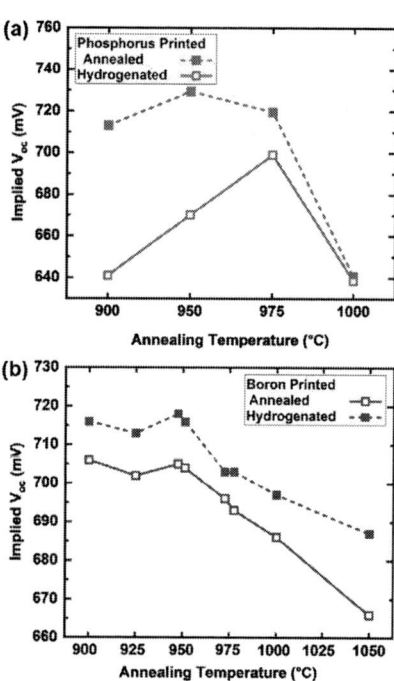

Figure 3: iV_{oc} values for inkjet-printed (a) phosphorus-doped and (b) boron-doped samples after annealing and hydrogenation.

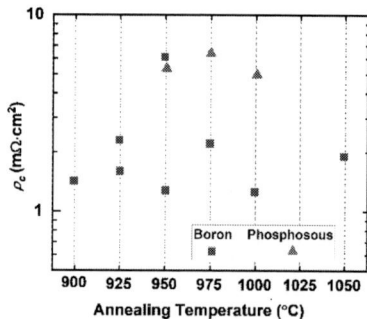

Figure 4: Contact resistivities for full-area phosphorus-doped and boron-doped samples.

3.2 Localized poly-Si/SiOₓ passivating contacts

Optical microscope images in Figure 5 illustrate the well-defined inkjet-printed boron and phosphorus lines on polished intrinsic-poly-Si/SiOₓ/c-Si substrates. With a printing drop spacing (center-to-center distance between adjacent droplets) of 40 μm, single-droplet boron and phosphorus lines with a minimal line width of ~ 60 μm could be attained. By increasing the number of droplets or decreasing the drop spacing, thicker dopant lines can be produced, depending on the application requirements. The line morphology varies depending on the liquid properties. For instance, the single-droplet boron line exhibits a scalloped shape, while the phosphorus lines have straight and uniform edges due to a smaller viscosity of the applied solution. Reducing the drop spacing improves the morphology of the boron line, as demonstrated by the smooth, straight lines printed with a 20 μm drop spacing.

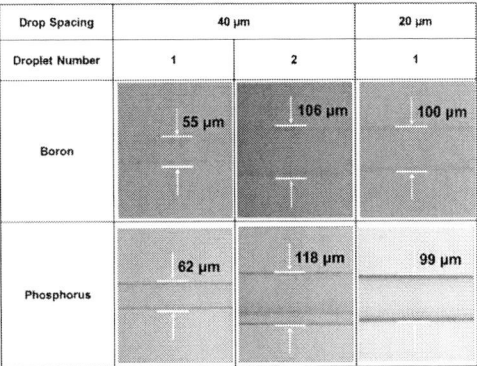

Drop Spacing	40 μm		20 μm
Droplet Number	1	2	1
Boron	55 μm	106 μm	100 μm
Phosphorus	62 μm	118 μm	99 μm

Figure 5: Microscope images of inkjet-printed boron and phosphorus lines on mechanically polished poly-Si/SiO$_x$/c-Si substrate with different printing drop spacing and droplet numbers.

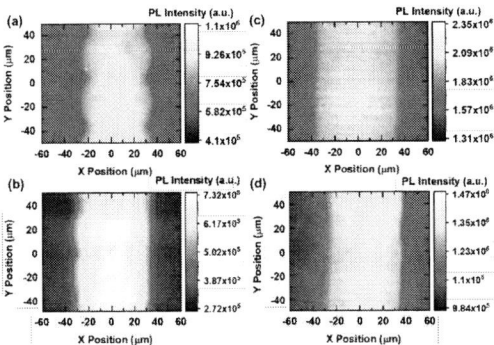

Figure 6: Room temperatue μPL maps for single-droplet boron lines on (a) test sample and (b) control sample as defined in Figure 2 above, and single-droplet phosphorus lines on (c) test sample and (d) control sample. The lines were printed with a drop spacing of 40 μm on mechanically polished substrate.

We firstly used an non-destructive μPL mapping to characterize the passivation quality of the localized poly-Si/SiO$_x$ passivating contact. Figure 6 presents the μPL maps for boron and phorphous lines after 950 °C anealing, measured at room temperature using the Horiba Lab RAM μPL system. The maps distinctly exhibit the band-to-band PL emission intensity, with the doped region emitting a stronger PL intensity due to the improved surface passivation for single-droplet boron and phosphorus lines, with and without the SiO$_x$ mask at unprinted regions as shown in Figure 2. A slightly higer PL emission at the unprinted region for the test samples (Figure 6 (a) and (c)) are observed when compared to the control samples (Figure 6 (b) and (d)) where the unprinted regions were protected by the SiO$_x$ mask during high temperature annealing. This suggests that there might be unintended doping on the test samples when the unprinted regions were fully exposed during the high temperature processes.

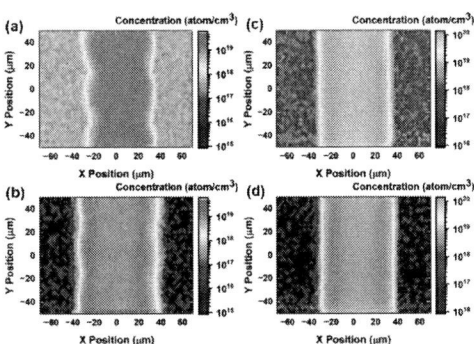

Figure 7: SIMS concentration maps for single-droplet boron lines on (a) test sample and (b) control sample, and single-droplet phosphorus lines on (c) test sample and (d) control sample. The lines were printed with a drop spacing of 40 μm on mechanically polished substrate.

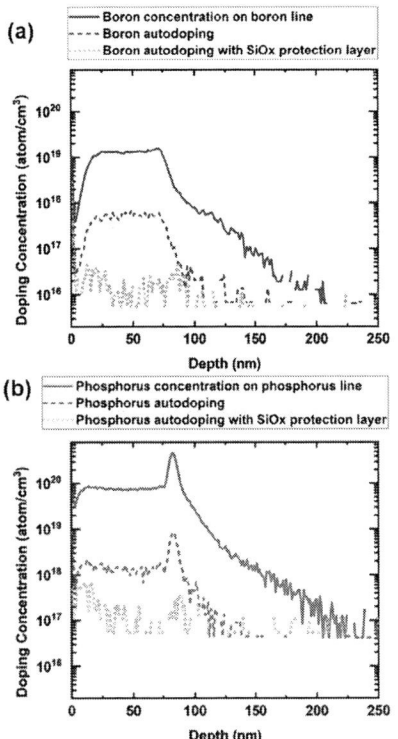

Figure 8: SIMS depth concentration profiles at printed lines, unprinted region without SiO$_x$ mask and unprinted region covered by SiO$_x$ mask for samples with (a) boron lines and (b) phosphorus lines.

To further confirm and quantify the unintended doping effects, D-SIMS mappings were performed on the same samples using a CAMECA IMS 7f-Auto system. Figure 7 summarizes the total dopant atom concentration maps for boron and phosphorus on test samples and control samples. Figure 8 displays the depth concentration profiles within the dopant lines and in the unprinted regions, with and without the SiO$_x$ mask. According to the maps and depth profiles, the doping concentration within the dopant lines is about [B] $\approx 1 \times 10^{19}$ atom/cm^3, [P] $\approx 1 \times 10^{20}$ atom/cm^3. Meanwhile, unintended doping of [B] $\approx 5 \times 10^{17}$

atom/cm^3 and [P] $\approx 2\times10^{18}$ atom/cm^3 were detected at the unpritned region on the test samples, as shown in Figure 7 (a), (c) and dashed profiles in Figure 8.

Apparently, the unintended doping effects are negligible on the control samples ([B] $\approx 5\times10^{15}$ atom/cm^3, [P] $\approx 5\times10^{16}$ atom/cm^3) and the dopant concentrations are close to the substrate doping level, as indicated by Figure 7 (b), (d) and the dotted lines in Figure 8. Although unintended doping issues were identified, the heavily doped regions still align with the printed pattern, showing a sharp drop in doping concentration at the line edges. This suggests that the unintended doping is likely caused by excess dopant species entering the gas phase during the high-temperature process and affecting the unprinted regions, rather than lateral diffusion of the dopants that would broaden the doped lines.

The results clearly highlight a limitation of the inkjet-printed doping method if applied as it is (i.e. the test sample structure in Figure 2), which could lead to shunting in solar cell devices due to unintended doping, as also observed by Kiaee *et al.*[12]. As the effect is likely caused by unintended dopants in the gas phase, applying a full-area capping layer over the printed pattern could help contain the outgassing of volatile dopant species and prevent unintended doping. We are currently working on this approach.

3.3 Single annealing for localized n-type and p-type poly-Si/SiO$_x$ passivating contacts

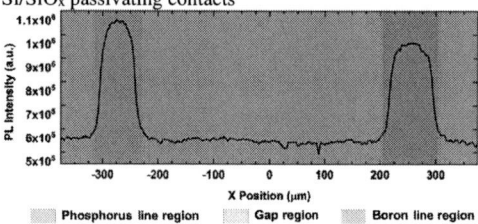

Figure 9: Room temperature μPL line scan of inkjet-printed phosphorous (left) and boron (boron) lines on a mechanically polished substrate after 950 °C annealing.

Sample with alternating boron- and phosphorous-doped lines on poly-Si/SiO$_x$/c-Si mechanically polished substrates was fabricated and characterized to explore the capability of forming poly-Si/SiO$_x$ passivating contacts of both polarities within a single annealing step. The μPL intensity profile in Figure 9 clearly illustrates improved surface passivation from both the phosphorous (left) and boron (right) lines after 950 °C annealing. The unprinted region between the lines emitted a consistently low and flat background PL signal. However, based on the studies in section 3.2, unintended doping issues are likely present in this sample as well. Despite these issues, we have demonstrated that inkjet-printing technology is capable of forming localized n-type and p-type poly-Si/SiO$_x$ passivating contacts in a single annealing step. Further optimizations, such as incorporating a full-area capping layer to prevent the unintended doping, will be explored to enhance the reliability of the inkjet printing processes.

4 CONCLUSIONS

We demonstrated that inkjet printing technology can be successfully utilized as an alternative doping method for both p-type and n-type poly-Si/SiO$_x$ passivating contacts.

Full-area printed samples show promising passivation quality and electrical contacts, yielding an iV_{oc} of 729 mV with ρ_c of 5.4 mΩ·cm^2 for n-type poly-Si/SiO$_x$, and iV_{oc} of 718 mV with ρ_c of 6.1 mΩ·cm^2 for p-type poly-Si/SiO$_x$ after hydrogenation. Analysis of localized n- and p-type poly-Si passivating contacts reveal improved passivation quality at the locally phosphorus and boron printed regions. A more quantitative assessment of the passivation quality of the printed lines is underway. High resolution D-SIMS measurements identified and quantified the unintended doping at unprinted region close to the printed dopant lines. We also demonstrated the formation of alternating inkjet-printed n-type and p-type lines with a single N$_2$ annealing step. By fine-tuning the inkjet printing parameters, it is possible to achieve various widths for localized n-type and p-type poly-Si/SiO$_x$ passivating contacts, providing greater flexibility in fabricating high-efficiency solar cells. Ongoing optimizations are focused on addressing unintended doping issues to make this process more reliable for producing high-efficiency solar cells with localized poly-Si/SiO$_x$ passivating contacts.

5 REFERENCE

[1] F. Haase *et al.*, "Laser contact openings for local poly-Si-metal contacts enabling 26.1%-efficient POLO-IBC solar cells," *Solar Energy Materials and Solar Cells,* vol. 186, pp. 184-193, 2018.

[2] C. Hollemann, F. Haase, S. Schäfer, J. Krügener, R. Brendel, and R. Peibst, "26.1%-efficient POLO-IBC cells: Quantification of electrical and optical loss mechanisms," *Progress in Photovoltaics: Research and Applications,* vol. 27, no. 11, pp. 950-958, 2019.

[3] A. Richter *et al.*, "Design rules for high-efficiency both-sides-contacted silicon solar cells with balanced charge carrier transport and recombination losses," *Nature Energy,* vol. 6, no. 4, pp. 429-438, 2021/04/01 2021, doi: 10.1038/s41560-021-00805-w.

[4] B. Kafle, B. S. Goraya, S. Mack, F. Feldmann, S. Nold, and J. Rentsch, "TOPCon – Technology options for cost efficient industrial manufacturing," *Solar Energy Materials and Solar Cells,* vol. 227, p. 111100, 2021/08/01 2021, doi:https://doi.org/10.1016/j.solmat.2021.111100.

[5] S. K. Karunakaran *et al.*, "Recent progress in inkjet-printed solar cells," *Journal of Materials Chemistry A,* vol. 7, no. 23, pp. 13873-13902, 2019.

[6] O. Khaselev *et al.*, "Novel inkjet inks for complete off contact fabrication of silicon solar cell," in *Proceedings of the 23rd European Photovoltaic Solar Energy Conference*, 2008.

[7] T. Masuda, M. Nakayama, K. Saito, H. Katayama, and A. Terakawa, "Inkjet Printing of Liquid Silicon," (in English), *Macromolecular rapid communications.,* vol. 41, no. 23, pp. e2000362-n/a, 2020, doi: 10.1002/marc.202000362.

[8] A. J. G. S.-P. Su, H.-T. Chen, P.-S. Huang, A.W. Lu, Z. Ding, W. Fan, H.M. Huang, R. Leung, "Progress of Mask-Less Selective Emitter Solar Cells by Inkjet Doping Technology," in *24th European Photovoltaic Solar Energy Conference*, Hamburg, Germany, 21-25 September 2009 2009.

[9] K. Ryu *et al.*, "High efficiency n-type silicon solar cell with a novel inkjet-printed boron emitter," 2011, no. Conference Proceedings: IEEE, pp. 001131-001133, doi: 10.1109/PVSC.2011.6186152.

[10] D. Stüwe, D. Mager, D. Biro, and J. G. Korvink, "Inkjet technology for crystalline silicon photovoltaics," *Advanced Materials,* vol. 27, no. 4, pp. 599-626, 2015.

[11] Z. Kiaee *et al.,* "Inkjet printing of phosphorus dopant sources for doping poly-silicon in solar cells with passivating contacts," *Solar Energy Materials and Solar Cells,* vol. 222, p. 110926, 2021.

[12] Z. Kiaee *et al.,* "TOPCon Silicon Solar Cells With Selectively Doped PECVD Layers Realized by Inkjet-Printing of Phosphorus Dopant Sources," (in English), *IEEE journal of photovoltaics,* vol. 12, no. 1, pp. 31-37, 2022, doi: 10.1109/JPHOTOV.2021.3129073.

[13] J. Wang *et al.,* "Development of Phosphorus-Doped Nanoscale Poly-Si Passivating Contacts via Inkjet Printing for Application in Silicon Solar Cells," *ACS Applied Nano Materials,* vol. 6, no. 1, pp. 140-147, 2023/01/13 2023, doi: 10.1021/acsanm.2c04148.

41st European Photovoltaic Solar Energy Conference and Exhibition

This presentation was selected by the Sc. Committee of the EU PVSEC 2024 for submission of a full paper to one of the EU PVSEC's collaborating peer-reviewed journals.

>24% EFFICIENT TUNNEL BACK CONTACTED POLYZEBRA SOLAR CELLS

Jonathan Linke, Christoph Peter, Jan Hoß, Vaibhav Kuruganti, Saman Sharbaf Kalaghichi, Valentin Mihailetchi,
Jan Lossen, Florian Buchholz
International Solar Energy Research Center Konstanz e.V.
Rudolf-Diesel-Str. 15, D-78467 Konstanz, Germany
Corresponding author: Jonathan Linke, +49 7531 36 183 362, jonathan.linke@isc-konstanz.de

ABSTRACT: The polyZEBRA tunnel back contact technology is a low-cost solar cell concept, which utilizes industrially established equipment only. In particular, the patterning of both polarities on the rear side is done with commercially available lasers. A recent cell batch achieved a solar cell efficiency mean value of 24.0% with a certified champion efficiency of 24.12%. The next steps for efficiency improvements of the polyZEBRA technology are explored in an experiment-based simulation. Actual measured values from improved test structures are used as input parameters for the iterative simulation study. This includes reduced recombination currents at the surface of the gap and front side region as well as in the (p) poly-Si/SiO$_x$ emitter region, the increase of the emitter coverage ratio and the reduction of the recombination at the metal contacts. Under the assumptions that these improvements will soon be implemented in the process flow, a short-term efficiency potential of >25% was estimated.

Keywords: polyZEBRA, TBC, poly-IBC, TOPCon, Simulation

1 INTRODUCTION

Back-contacted solar cells featuring both polarities poly-Si/SiO$_x$ based passivating contacts, also known as tunnel back contact (TBC), have been commercially available since many years and also addressed in scientific literature [1,2]. However, the complexity of the fabrication process and application of non-standard methods lead to high production costs. In contrast, the polyZEBRA TBC solar cell concept relies on industrially established equipment only and thus aims at lower cost production. The key technologies distinguishing it from industrially realized TBC processes are the patterning of the rear side with commercially available lasers and the metallization with standard screen-printing for both polarities in one printing step. Though the principle applicability of the process has been demonstrated before [3,4], the achieved solar cell efficiency still has to be improved in order to be realized in a future production line. Details of the recent progress in the past year is discussed in [5].

In this work, the current status of the polyZEBRA technology development is reported before the short-term efficiency improvement potential is explored. This is done via experiment-based simulations, in which actual measurement values from improved test structure are used as input parameters. In this sense, the simulation results reflect the efficiency potential for the short-term improvement scenario, in which all the optimization results on test structures will be implemented into the process flow.

2 METHODS

2.1 Fabrication of solar cells

PolyZEBRA solar cells were produced according to the process flow shown in Fig. 1, which is an adjusted version of the previous published process [3]. The base material were Cz-grown n-type wafers in M6 format with a thickness of 150 µm. On the rear side, the emitter and base regions feature poly-Si/SiO$_x$ contacts and are separated by a p-type diffused gap region. The interfacial tunnel oxides were thermally grown and capped with low-pressure chemical vapor deposited (LPCVD) (n) poly-Si (Fig. 1, step 2) and (i) poly-Si (Fig. 1, step 6),

respectively. A first BBr$_3$-diffusion was applied for the background p-type doping of the initially undoped (i) poly-Si and the resulting boron silicate glass (BSG) was removed in hydrofluoric acid (HF) (Fig. 1, step 7). The thermal budget of the second BBr$_3$-diffusion for p-type doping of the gap and front side region also activates the passivation of the poly-Si/SiO$_x$ contacts on both polarities (Fig. 1, step 10). Details regarding the laser-based patterning steps (Fig. 1, step 4, 8) can be found in [6,7]. Plasma-enhanced chemical vapor deposited (PECVD) stacks of AlO$_x$ capped with a SiN$_x$ layer were used for passivation of the surfaces. The metal contacts were screen-printed with Ag-containing paste for both fingers and busbars (BB) and a pitch of the metallization pattern of 800 µm and 6BB. In principle, also the low-cost Ag/Cu screen-printing approach would be applicable [8].

The finished solar cells were characterized by IV measurements using a Halm IV flasher. Not metallized cell precursors and test wafers were fired and the implied open-circuit voltage (iV$_{oc}$) as well as the surface recombination J$_{o,pass}$ was measured by quasi steady-state photoconductance decay (QSSPC) using a standard Sinton lifetime tester.

2.2 Experimental-based simulations

The cell parameters were simulated using Quokka3 [9]. The most important input parameters are based on measurement data of test structures, which reflect the process flow and parameters used for the fabrication of the solar cells (Sec. 2.1). They are listed in Table 1. The bulk lifetime (τ_{bulk}) was adjusted in order to match the simulated open-circuit voltage (V$_{oc}$) to the experimentally determined iV$_{oc}$ with the metal recombination turned off. Subsequently, the metal recombination of the (n) poly-Si/SiO$_x$ was fixed to J$_{0,met,(n) poly-Si}$ = 50 fA/cm² while the metal recombination of the (p) poly-Si/SiO$_x$ was adjusted in order to match the simulated V$_{oc}$ to the experimentally determined V$_{oc}$. This resulted in a value of J$_{0,met,(p) poly-Si}$ = 500 fA/cm².

41st European Photovoltaic Solar Energy Conference and Exhibition

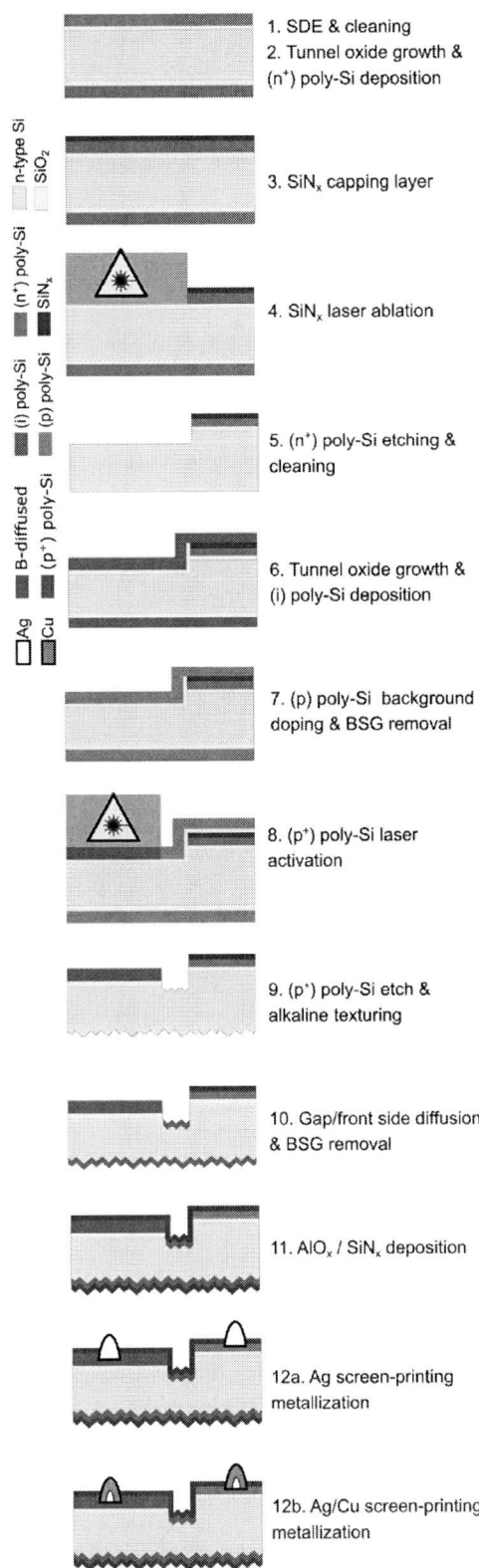

1. SDE & cleaning

2. Tunnel oxide growth & (n+) poly-Si deposition

3. SiNx capping layer

4. SiNx laser ablation

5. (n+) poly-Si etching & cleaning

6. Tunnel oxide growth & (i) poly-Si deposition

7. (p) poly-Si background doping & BSG removal

8. (p+) poly-Si laser activation

9. (p+) poly-Si etch & alkaline texturing

10. Gap/front side diffusion & BSG removal

11. AlOx / SiNx deposition

12a. Ag screen-printing metallization

12b. Ag/Cu screen-printing metallization

n-type Si
SiO$_2$
(n+) poly-Si
SiN$_x$
(i) poly-Si
(p) poly-Si
B-diffused
(p+) poly-Si
Ag
Cu

Figure 1: polyZEBRA solar cell fabrication process flow based on [3].

Table 1: Most important simulation input parameters based on measurement data of test structures, which represent the process flow and parameters used for the fabrication of the solar cells.

	Input parameter	Value
General properties	Wafer size	M6
	#BB	6BB
	τ_{bulk}	3 ms
Rear pattern	Cell pitch	800 μm
	W_{base}	290 μm
	W_{emi}	360 μm
	W_{gap}	75 μm
Surface recombination	$J_{0,pass,(n)\ poly-Si}$	1 fA/cm²
	$J_{0,pass,(p)\ poly-Si}$	10 fA/cm²
	$J_{0,pass,gap/front}$	13 fA/cm²
Metal recombination	$J_{0,met,(n)\ poly-Si}$	50 fA/cm²
	$J_{0,met,(p)\ poly-Si}$	500 fA/cm²
Sheet resistance	$R_{sheet,(n)\ poly-Si}$	55 Ω/sq
	$R_{sheet,(p)\ poly-Si}$	175 Ω/sq
	$R_{sheet,gap/front}$	490 Ω/sq
Resistivity	$\rho_{c,(n)\ poly-Si}$	0.9 mΩcm²
	$\rho_{c,(p)\ poly-Si}$	2.8 mΩcm²

3 SOLAR CELL RESULTS

The measured solar cell parameters are summarized in Table 2, showing the mean values and standard errors of nine cells. Figure 2 shows the results of the certified IV measurement of the champion cell at CalTech, ISFH Hamelin, Germany. The iV_{oc} of not metallized cell precursors was measured on three wafers and yielded a value of (718.0 ± 0.2) mV.

Table 2: Cell parameters mean values and standard errors of nine cells as measured by IV.

η (%)	V_{oc} (mV)	J_{sc} (mA/cm²)	FF (%)
24.0 ± 0.1	711 ± 0.5	41.7 ± 0.1	81.1 ± 0.2

The iV_{oc}-to-V_{oc} loss during metallization is a measure for the recombination at the metal contacts $J_{o,met}$. Ideally, a passivating contact yields no metal recombination at all, while in practice 1-2 mV is accepted. However, the observed iV_{oc}-to-V_{oc} loss mean value of ≈7 mV is a clear indication for strong recombination at the metal contacts, which should be reduced to yield even higher efficiencies. From photoluminescence images of test structures it is obvious, that this recombination takes place in the emitter regions of the (p) poly-Si/SiO$_x$ contacts. Approaches for a reduction are an optimized AlO$_x$/SiN$_y$ stack [4], the insertion of a blocking layer for the metal paste [10] or the application of laser-enhanced contact optimization (LECO) [11]. Apart from the reduction of the metal recombination, an enhancement of the surface passivation will boost the iV_{oc} of not metallized cell precursors and in turn also the final cells V_{oc}.

The mean value of the IV measured pseudo fill factor (pFF) is $(82.7 \pm 0.1)\%$. The pFF-to-FF loss of ≈1.6%$_{abs}$ is a measure for the resistive losses in the cell. Since it is already challenging to maintain this value during a reduction of the $J_{0,met}$, a further FF improvement is mostly expected from a higher pFF. A higher pFF in turn will inherently be achieved by an improvement of the iV_{oc} and V_{oc}.

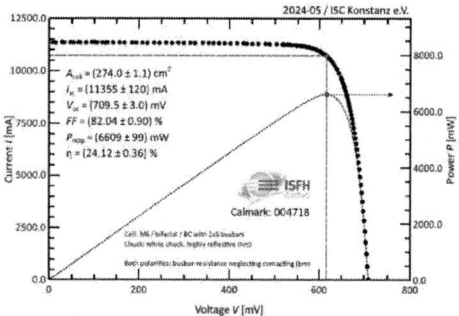

Figure 2: Certified IV measurement of the champion polyZEBRA solar cell at ISFH CalTec.

4 EXPERIMENT-BASED SIMULATIONS

Simulations have been conducted using a Quokka3 model representing the polyZEBRA solar cell structure. The simulation is used to explore the cell efficiency potential using actual measurement values from test structures as input parameters. To do so, improved input values are replaced in the simulation step-by-step so that the improvements add up gradually. All improvements are summarized in Table 3. The corresponding simulation output cell parameters for all improvements are summarized in Table 4.

Table 3: Gradual improvements and corresponding value changes of the simulation input parameters based on measurements on test structures.

Improvement	Parameter	Reference→Improved value
Gap/front side passivation	$J_{0,pass,gap/front}$	$(13 \rightarrow 5.1)$ fA/cm²
(p) poly-Si passivation	$J_{0,pass,(p) poly-Si}$	$(10 \rightarrow 1.9)$ fA/cm²
Rear pattern	W_{base} W_{emi} W_{gap}	$(290 \rightarrow 200)$ µm $(360 \rightarrow 540)$ µm $(75 \rightarrow 30)$ µm
Metal recombination	$J_{0,met,(n) poly-Si}$ $J_{0,met,(p) poly-Si}$	$(50 \rightarrow 31)$ fA/cm² $(500 \rightarrow 39)$ fA/cm²

4.1 Simulation baseline

The baseline of the simulation was adjusted to the certified cell parameters following the procedure described in the methods section 2.2. The corresponding output cell parameters are shown in Table 4 and match with the certified cell parameters within their uncertainties (Fig. 2). Thus, the simulation is assumed to properly describe the experimental cells. In the following sections, the gradual improvements are discussed one by one.

Table 4: Simulation output cell parameters based on the improvements and values in Table 3.

Improvement	η (%)	V_{oc} (mV)	J_{sc} (mA/cm²)	FF (%)
Baseline	24.14	709.0	41.33	82.38
Gap/front side passivation	24.33	715.5	41.33	82.27
(p) poly-Si passivation	24.42	718.2	41.33	82.28
Rear pattern	24.57	718.6	41.54	82.30
Metal recombination	25.07	730.7	41.55	82.57

4.2 Gap/front side passivation

The gap and front side region covers roughly 60% of the total cell's surface and thus an excellent surface passivation in this region is mandatory. This is ensured by a BBr₃-diffusion, which creates a very shallow doping profile and low surface concentration. The corresponding sheet resistance is as high as 5,000 Ω/sq. On symmetrically textured and diffused test wafers an average over the whole wafer of $J_{0,pass,gap/front} = (5.1 \pm 0.2)$ fA/cm² was achieved (Fig. 3). Insertion of this value in the simulation yields and improvement of the V_{oc} of 6.5 mV. At the same time, the FF reduces by 0.11%abs due to the high sheet resistance in the gap region. In sum, the efficiency increases by 0.19%abs.

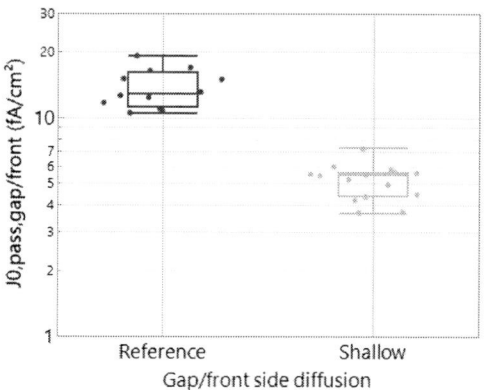

Figure 3: Surface recombination improvement of the gap/front side region for a shallow BBr₃-diffusion with low surface concentration.

4.3 (p) poly-Si/SiOₓ passivation

The process parameters of the fabricated solar cells yielded an emitter surface passivation of $J_{0,pass,(p) poly-Si} = 10$ fA/cm². An improvement was achieved on symmetric test structures featuring (p) poly-Si/SiOₓ contacts on both sides by a post-deposition treatment of the interfacial tunnel oxide, lowering the recombination to $J_{0,pass,(p) poly-Si} = (1.9 \pm 0.4)$ fA/cm² (Fig. 4). Insertion of this value in the simulation yields an improvement of the V_{oc} of 2.7 mV and in turn a gain of 0.09%abs in efficiency. These values are lower than for the gap/front side region's improvement, as the area fraction of the emitter with respect to the total area of the solar cell is only ≈23% compared to ≈60% for the gap/front side region.

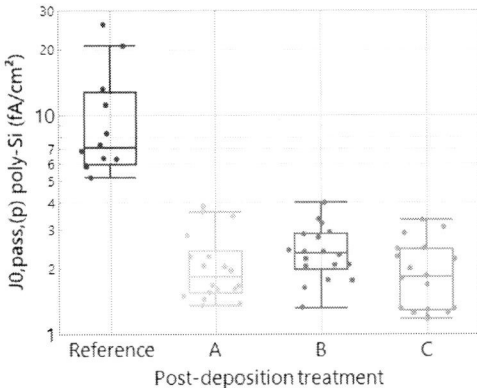

Figure 4: Surface recombination improvement of the (p) poly-Si/SiO$_x$ contact region for a post-deposition treatment of the interfacial tunnel oxide.

4.4 Rear pattern

The rear pattern was previously dictated by the accuracy of the laser system used to structure the emitter and base regions (Fig. 1, step 4, 8). After a technical upgrade, an improved fiducial alignment allows to precisely align the second laser step to the first one. This enables to reduce the widths of the gap and base region in order to increase the width of the emitter and as such the emitter coverage ratio from 45% to 68%. This value is close to a simulated optimum of ≈70% [12] and so a gain in solar cell efficiency is expected. Indeed, an increase of the simulated short-circuit current density of 0.21 mA/cm² is observed.

4.5 Metal recombination

Following the procedure to adjust the simulation to the certified cell parameters as described in the methods section 2.2, it was necessary to increase the recombination at the metal contacts in the emitter region to $J_{0,met,(p) poly-Si}$ = 500 fA/cm². This value is very large considering that an ideal passivating contact should show vanishing metal recombination.

Fig. 5 shows PL images of wafers featuring alternating squared regions of (p)/(n) poly-Si/SiO$_x$ contacts and varying metal fractions. On the wafer fired at the reference conditions (Fig. 5, left), differences in brightness are observable on the (p) poly-Si/SiO$_x$ squares. The darker the square is, the larger is the metal fraction, which is a clear evidence for significant recombination at the metal contacts. In contrast, considering colder firing conditions (Fig. 5, right), a similar brightness is observed for all (p) poly-Si/SiO$_x$ squares. This is a clear evidence for vanishing recombination at the metal contacts.

Colder firing conditions are able to reduce the metal recombination, but typically at the same time increase the contact resistivity. This can be mitigated by the LECO technique, which is able to form a proper contact between semiconductor and metal after low temperature firing [11]. However, investigations on this approach are ongoing and so far the resulting FF has not yet reached the level of the reference process without LECO treatment.

Nevertheless, assuming that the FF issue will be solved soon, insertion of the $J_{0,met}$ values for both polarities measured on the wafers fired at colder conditions yields a significant improvement of the V$_{oc}$ of 12.1 mV. Such a large voltage gain naturally also increases the FF, here

with a gain of 0.27%$_{abs}$. Altogether the simulated cell efficiency is increased by 0.50%$_{abs}$ to 25.07%.

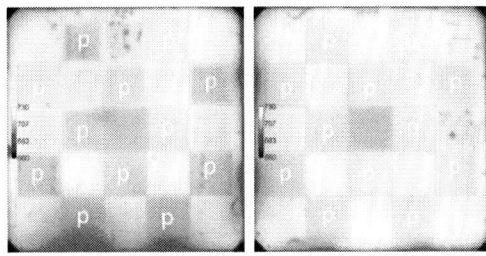

Figure 5: PL images converted to iV$_{oc}$ maps of test structures with alternating (p)/(n) poly-Si squares with different metal fractions under reference firing conditions (left) and colder firing conditions (right). The (p) poly-Si squares marked with a "p".

4.6 Simulation summary

Fig. 6 shows a summary of the simulated improvements of the solar cell parameters that can be expected if the improved measurement values on test structures will be integrated in the polyZEBRA fabrication process. The largest gains in V$_{oc}$ are expected by a reduction of $J_{0,met}$, in particular in the emitter region, and by an improvement of the surface passivation of the gap/front side region. Increasing the emitter coverage ratio of the rear pattern is expected to increase the J$_{sc}$. The improvement of the surface passivation of the (p) poly-Si/SiO$_x$ contact is expected to have a minor influence on the V$_{oc}$, though this optimization is mandatory for high efficiency solar cells.

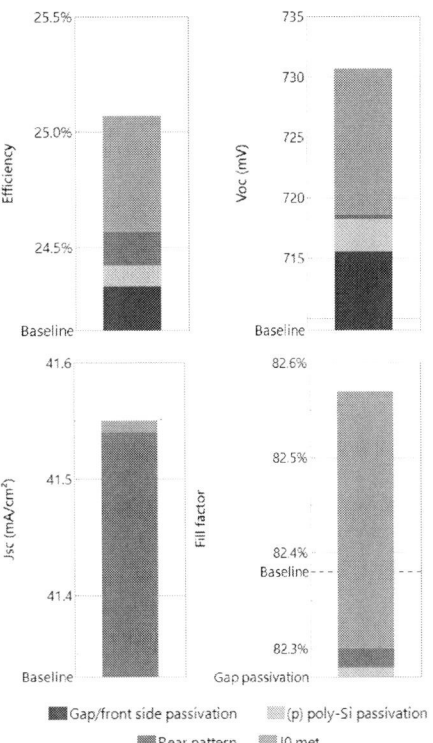

Figure 6: Summary of the gradual improvements in solar cell parameters from the experiment-based simulations.

5 CONCLUSIONS

The current development status of the polyZEBRA TBC technology is a mean solar cell efficiency of 24.0% with a certified champion efficiency of 24.12%. Experiment-based simulations have been done in order to determine the efficiency potential of >25% for the short-term scenario, in which the recent improvements achieved on test structures will be implemented in the process flow.

6 ACKNOWLEDGEMENTS

This work was supported by the project IBC4EU and has received funding from the European Union's Horizon Europe research and innovation programme under grant agreement No.101084259.

7 REFERENCES

[1] F. Haase *et al.*, "Laser contact openings for local poly-Si-metal contacts enabling 26.1%-efficient POLO-IBC solar cells", *Solar Energy Materials and Solar Cells* 18 (2018), 184–193, 10.1016/j.solmat.2018.06.020.

[2] G. Yang *et al.*, "High efficiency black IBC c-Si solar cells with poly-Si as carrier-selective passivating contacts", *Solar Energy Materials and Solar Cells* 186 (2018), 9–13, 10.1016/j.solmat.2018.06.019.

[3] J. Linke *et al.*, "Fully Passivating Contact IBC Solar Cells Using Laser Processing", *Proc. of 8th World Conference on Photovoltaic Energy Conversion WCPEC*, Milan (2022), 102–106, 10.4229/WCPEC-82022-1CV.2.11.

[4] J. Linke *et al.*, "Progress in Development of polyZEBRA IBC Solar Cells", *Proc. of 40th European Photovoltaic Solar Energy Conference and Exhibition EUPVSEC*, Lissabon (2023), 020046 10.4229/EUPVSEC2023/1CV.3.3.

[5] J. Linke *et al.*, EPJ Photovoltaics (2024), submitted

[6] F. Buchholz *et al.*, "Local Passivating Contacts from Laser Doped p+ Polysilicon", *Proc. of 38th European Photovoltaic Solar Energy Conference and Exhibition EUPVSEC*, Online (2021) 10.4229/EUPVSEC20212021-2BO.11.3.

[7] S. Sharbaf Kalaghichi *et al.*, "Laser Activation for Highly Boron-Doped Passivated Contacts", *Solar* 3 (2023), 362–381, 10.3390/solar3030021.

[8] N. Chen *et al.*, "Thermal Stable High-Efficiency Copper Screen Printed Back Contact Solar Cells", *Solar RRL*, 7 (2022), 2200874, 10.1002/solr.202200874.

[9] A. Fell, *Quokka3, v2.4.6.* (2024), Online available: www.quokka3.com.

[10] R. Glatthaar *et al.*, "Silver Metallization with Controlled Etch Stop Using SiO$_x$ Layers in Passivating Contacts for Improved Silicon Solar Cell Performance", *Solar RRL* 7 (2023), 2300491, 10.1002/solr.202300491.

[11] A. Mette *et al.*, "Q.ANTUM NEO with LECO exceeding 25.5 % cell Efficiency", *Solar Energy Materials and Solar Cells* 277 (2024), 113110, 10.1016/j.solmat.2024.113110.

[12] P. Procel *et al.*, "Numerical Simulations of IBC Solar Cells Based on Poly-Si Carrier-Selective Passivating Contacts", *IEEE Journal of Photovoltaics* 9 (2019), 374–384, 10.1109/JPHOTOV.2019.2892527.

41st European Photovoltaic Solar Energy Conference and Exhibition

This presentation was selected by the Sc. Committee of the EU PVSEC 2024 for submission of a full paper to one of the EU PVSEC's collaborating peer-reviewed journals.

OPTIMIZED GA-DOPED CZ WAFERS FOR POLO IBC SOLAR CELLS WITH HIGH EFFICIENCY AND MINIMAL LETID DEGRADATION

Thorsten Dullweber*[1], Verena Mertens[1], Michael Winter[1], Sabrina Schimanke[1], M. Ripke[1], Silke Dorn[1], Yevgeniya Larionova[1], Gerrit Lange[1], Karsten Bothe[1], Jan Schmidt[1,2], Rolf Brendel[1,2], Arne K. Dahle[3], Özlem Coskun[4], Nesrin Töre Sen[4]

[1]Institute for Solar Energy Research Hamelin (ISFH), Am Ohrberg 1, 31860 Emmerthal, Germany

[2]Institute of Solid-State Physics, Leibniz Universität Hannover, Appelstrasse 2, 30167 Hannover, Germany

[3]NorSun AS, Tangenveien 3A, 6885 Årdalstangen, Norway

[4]Kalyon PV, Başkent OSB Şadi Türk Bulvarı No: 23 Malıköy, Ankara, Turkey

*corresponding author: Tel.: +49 (0)5151-642, dullweber@isfh.de

ABSTRACT: The PV industry is using 0.4 to 0.8 Ωcm low-resistivity Ga-doped Cz-Si wafers with carrier lifetimes < 700 µs for mass production of PERC and bifacial PERC+ cells. Our Quokka3 simulations predict, that a Ga-doped Cz wafer resistivity around 1.5 Ωcm will increase V_{oc} and J_{sc} and the POLO IBC solar cell efficiency from the presently best measured efficiency of 23.9% up to a simulated efficiency of 24.3%, because the charge carrier density in POLO IBC solar cells depends less on the doping concentration due to their much higher V_{oc}. In this paper, we investigate Ga-doped Cz wafers with higher resistivities ranging from 0.7 to 4.8 Ωcm from Cz-Si ingots supplied by NorSun and Kalyon PV. POLO IBC lifetime precursors processed at ISFH without metal contacts show, that the effective carrier lifetime linearly increases with higher wafer resistivity up to 3 ms at 4.0 Ωcm. The measured implied V_{oc} of up to 740 mV and the implied FF of up to 86% correspond to a calculated implied efficiency of 26.7% for a wide wafer resistivity range. The measured IV parameters of POLO IBC solar cells processed with different Ga-doped Si wafer resistivities agree qualitatively with our Quokka 3 simulations. During LeTID degradation conditions at 0.5 sun, 80°C, the POLO IBC lifetime samples show a moderate degradation in implied V_{oc} and implied FF and reveal, that the LeTID-related wafer bulk defect density is independent of the Ga doping concentration. The POLO IBC solar cells show a maximum LeTID degradation of the efficiency of 4%$_{rel}$ which fully recovers during extended testing. The LeTID degradation could be further minimized by adjusting the firing temperature profile. In addition, we present our near-term POLO IBC efficiency roadmap towards 25% by optimizing the POLO IBC screen-printed Ag and Al metallization.

1 INTRODUCTION

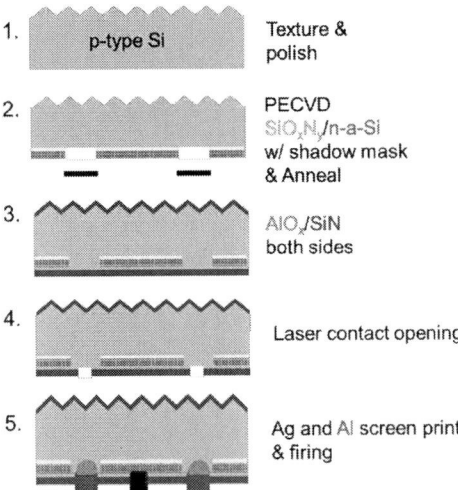

Figure 1: Schematic drawing of the lean POLO IBC process flow applying the local PECVD SiOxNy/n-a-Si deposition through a glass shadow mask.

The POLO IBC solar cell design by ISFH replaces the POCl3 emitter of PERC and bifacial PERC+ solar cells by a passivating poly-Si on oxide (POLO) contact thereby increasing the V_{oc} potential towards 733 mV and the efficiency potential to 25.5% [1]. At the same time the POLO IBC design is reusing many processing steps from PERC+ technology like Ga-doped p-type silicon wafers,

AlOx/SiN passivation, and Aluminum finger BSF contacts [1, 2]. For the POLO IBC solar cells, ISFH applies a lean processing sequence as shown in Fig. 1 using industrial processing tools for all process steps with local in-situ deposition of PECVD SiOxNy/n-a-Si through a glass shadow mask [2,3]. The resulting POLO IBC cells exhibit a best conversion efficiency of 23.8% using 0.8 Ωcm Ga wafers [3]. ISFH is developing the POLO IBC solar cell towards pilot production readiness in collaboration with Kalyon PV within the EU-funded R&D project IBC4EU. The Chinese cell manufacturers Tongwei and LONGi are evaluating very similar IBC cell designs named p-TBC [4] and HPBC [5] for mass production achieving efficiencies above 25%.

In this paper, we evaluate the optimum Ga-doped Cz-Si wafer properties to achieve high POLO IBC solar cell conversion efficiencies with minimal Light and elevated Temperature Induced Degradation (LeTID). For mass production of PERC and PERC+ solar cells, the PV industry is applying Ga-doped wafers with very high Ga doping concentration and correspondingly low wafer resistivity of 0.4 to 0.8 Ωcm and low carrier lifetimes below 700 µs. The low wafer resistivity minimizes resistive carrier transport losses to the ca. 1 mm spaced local Al base contact fingers whereas the recombination is dominated by the PERC+ cells front side limiting the iV_{oc} to around 710 mV. In POLO IBC solar cells, the n-poly-Si contact increases the iV_{oc} up to 745 mV [3] thereby increasing the minority carrier density beyond the doping level which may allow to use higher resistivity Ga-doped Cz wafers. This may be favorable since the low carrier lifetime of the low-resistivity Ga wafers limits the V_{oc} and J_{sc} and hence the efficiency of POLO IBC cells. Hence, we investigate Ga-doped Cz wafers with higher resistivity range of 0.7 Ωcm to 4.8 Ωcm produced by Norsun and Kalyon for POLO IBC solar cells processed at ISFH and report the resulting Ga-doped Cz wafer and POLO IBC solar cell properties.

2 CARRIER LIFETIME PROPERTIES OF POLO IBC PRECURSORS WITH OPTIMIZED GA-DOPED CZ WAFERS

In order to estimate the optimum Ga-doped Cz wafer resistivity range for POLO IBC solar cells, we simulate the POLO IBC IV parameters in dependence of the Ga wafer resistivity with Quokka 3. We model the bulk carrier lifetime in dependence of the wafer resistivity by assuming boron-oxygen-type SRH recombination after regeneration [6] since a recent study demonstrated a good agreement of this model with measured Ga-doped Cz wafer lifetimes [7]. The Quokka 3 simulation results are shown in green in Figure 3 and reveal, that the low carrier lifetimes due to high Auger recombination of wafer resistivities < 0.8 Ωcm strongly reduce V_{oc} and J_{sc} of POLO IBC cells and hence their efficiency to well below 24.0%. The optimum wafer resistivity is around 1.5 Ωcm and increases the efficiency up to 24.3%. For higher wafer resistivities beyond 2 Ωcm, the simulation predicts a decrease of the conversion efficiency below 24.0% since the increased series resistance decreases the FF values to below 80%. The Quokka 3 cell input parameters [8] are derived from our present POLO IBC cells with 23.8% efficiency using 0.8 Ωcm wafers and they also nicely reproduce the IV parameters of a new best POLO IBC solar cell with 23.9% efficiency shown as yellow triangles in Fig. 3. Based on the simulation results in Fig. 3, we estimate that Ga-doped Cz wafer resistivities between 0.8 Ωcm and 3.5 Ωcm are suitable for our present POLO IBC solar cells with conversion efficiencies around 24%.

Norsun and Kalyon took the effort to grow Ga-doped Cz ingots in their Cz ingot and wafer production lines according to the simulated optimum resistivity range. The resulting Ga-doped wafers exhibit a measured resistivity range of 0.7 Ωcm to 4.8 Ωcm due to the Ga doping concentration variation across the ingots from seed to tail. The Ga-doped wafers were processed at ISFH with the POLO IBC process sequence in Fig. 1 up to process step 3 and a subsequent firing step into carrier lifetime samples, which correspond to POLO IBC solar cells without metal contacts. Since the in-situ PECVD SiO_xN_y process in Fig. 1 (step 2), was temporarily not available due to an issue with the graphite boat, we used a wet chemical ozone oxide instead which was grown at the end of step 1 in Fig. 1. The lifetime samples were measured at ISFH by QSSPC to determine the effective carrier lifetime τ_{eff} at an excess carrier density $\Delta n = 1\times10^{15}$ cm^{-3}, the one-sun implied V_{oc} (iV_{oc}), and the implied FF (iFF).

Figure 2 a) shows that the POLO IBC lifetime precursors exhibit effective carrier lifetimes τ_{eff} up to 0.8 ms at 1.0 Ωcm and 3 ms at 4.0 Ωcm similar to the values reported in [7]. The lifetime values correspond to iV_{oc} values around 740 mV over the whole resistivity range with a best value of 745 mV at 2 Ωcm as shown in Fig. 2 b). The low flyer values are mainly caused by blistering of the AlO_x/SiN passivation layer on the textured front surface and partly on the rear surface which is subject to further improvement. The iFF derived from the QSSPC measurements in Fig. 2 c) exhibits best values of 86%. We conclude that the Ga-doped Cz wafers from Norsun and Kalyon exhibit very similar properties. The slightly lower

iFF value of the Kalyon wafers is likely due to variations in the surface passivation, not due to different wafer bulk properties.

To assess the efficiency potential of the Ga-doped Cz wafers, based on the results in Fig. 2 we calculate the so-called implied efficiency iη as follows

$$i\eta = iV_{oc} \times iFF \times J_{sc} / P_{light}$$
$$= 740 \text{ mV} \times 86\% \times 42 \text{ mA/cm}^2 / 100 \text{ mW/cm}^2$$
$$= 26.7\%$$

The values for iV_{oc} and iFF are derived from Fig. 2 as their upper limit. The J_{sc} of 42 mA/cm^2 is assumed as a typical value for high-efficiency IBC solar cells. The calculated implied efficiency iη = 26.7% demonstrates the excellent quality of the Ga-doped Cz wafers allowing very high solar cell efficiencies. In addition, the high iV_{oc} and iη values demonstrate the very good surface passivation quality of the AlO_x/SiN and SiO_x/n-poly-Si layer stacks as well as the good quality of the Ozone-based wet cleans prior to the layer stack depositions, processed with industrial PECVD and Wet-Batch tools. The calculated implied efficiency iη = 26.7% is, however, not applicable to the final POLO IBC solar cells, since the metallization induced recombination losses and series resistance losses limit its potential to 25.5%, as published in [1].

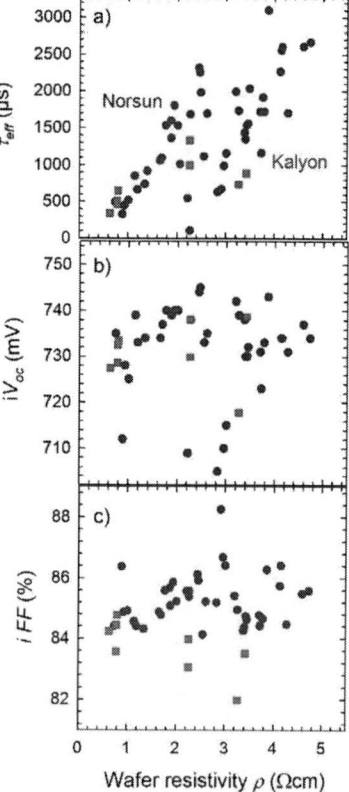

Figure 2: a) Effective carrier lifetime τ_{eff} (at $\Delta n = 10^{15}$ cm^{-3}), b) implied V_{oc} (iV_{oc}) and c) implied FF (iFF) of POLO IBC lifetime precursors processed at ISFH with Ga-doped Cz wafers from Norsun and Kalyon exhibiting a targeted wafer resistivity range between 0.7 and 4.8 Ωcm.

Tongwei recently reported strong light soaking effects of their p-TBC cells attributing it to FeGa defect pairs with Fe concentrations up to 1.3×10^{11} cm^{-3} [4]. In contrast, our POLO IBC precursors exhibit an iV$_{oc}$ increase of only 3 mV after light soaking, from which we calculate a relatively low Fe concentration of $(1.5 \pm 0.5) \times 10^{10}$ cm^{-3} for several Ga wafers with different resistivities. This low Fe value could be due to a low Fe concentration in the as-grown Ga-doped wafers as well as due to good Fe gettering capabilities of the PECVD n-poly Si. Also, the minimized alkaline chemistry usage during the POLO IBC cell manufacturing process flow [2,3] helps to maintain low Fe concentrations in the wafer.

3 POLO IBC SOLAR CELL RESUTS WITH OPTIMIZED GA-DOPED CZ WAFERS

Subsequently, full POLO IBC solar cells with screen-printed and fired Ag and Al metal contacts were processed at ISFH according to the process flow in Fig. 1 using Ga-doped Cz wafers with different resistivities from Norsun and Kalyon. Since the in-situ PECVD SiO$_x$N$_y$ process in Fig. 1 (step 2), was temporarily not available due to an issue with the graphite boat, we used a wet chemical ozone oxide instead which was deposited at the end of step 1 in Fig. 1. The POLO IBC solar cell busbar design includes several contact pads per busbar. For measuring the current-voltage (I-V) characteristic under illumination, a measurement chuck built by ISFH contacts these pads with contact pins to extract the current. Dedicated sense pins measure the voltage at 1/5 of the distance of 2 adjacent current pins resulting in a busbar resistance neglecting (brn) contacting scheme. The solar cells are vacuum-sucked to the black chuck and the temperature is controlled to 25°C.

The measured IV parameters efficiency η, open circuit voltage V$_{oc}$, short circuit current density J$_{sc}$, and fill factor FF are shown in Figure 3 in blue for Cz wafers from Norsun and in red for Cz wafers from Kalyon. Both wafer types obtain best POLO IBC solar cell efficiencies up to 23.4% for wafer resistivities between 0.8 and 1.8 Ωcm and exhibit very similar trends in all IV parameters. In particular, as predicted by the Quokka 3 simulations shown in green, with increasing Ga-doped Cz wafer resistivity the J$_{sc}$ increases by up to 1 mA/cm^2 whereas the FF declines from 82% to 78%. Both trends result in an optimum POLO IBC efficiency for wafer resistivities between 0.8 and 1.8 Ωcm similar as predicted by the simulation. For comparison, the yellow triangles in Fig. 3 show the IV parameters of a new best POLO IBC solar cell from a recent different batch with a conversion efficiency of 23.9%, which matches the simulated IV results very well for 0.8 Ωcm.

Compared to the Quokka 3 simulations and the best POLO IBC cell, the efficiency level of the POLO IBC solar cells processed in this study with the wafers from Norsun and Kalyon is about 0,5% to 1%abs. lower mainly due to about 1 mA/cm^2 lower J$_{sc}$ values. The root cause is not yet fully understood and still has to be further analyzed. A possible root cause is, that this batch suffered from partial blistering of the AlO$_x$/SiN layers, which can result in additional "ghost" Al and Ag metal contacts leading to increased carrier recombination. Nevertheless, the experimental results in Fig. 3 confirm that Ga-doped Cz wafers from a wide resistivity range between 0.8 and 3.0 Ωcm are suitable for POLO IBC solar cells with an optimum efficiency around (1.3 ± 0.5) Ωcm.

Based on a recent detailed test structure analysis of POLO IBC solar cells including a Quokka 3 simulated Synergistic Efficiency Gain Analysis (SEGA) published in [8], we target to increase the best POLO IBC cell efficiency from presently 23.9% up to 25.0% in the next 12 months in particular by reducing the Ag to n-poly contact resistance (+ 0.5%), by optimizing the local Al-BSF contact geometry (+ 0.3%), and by minimizing resistive losses with an optimized busbar and pad design (+ 0.3%).

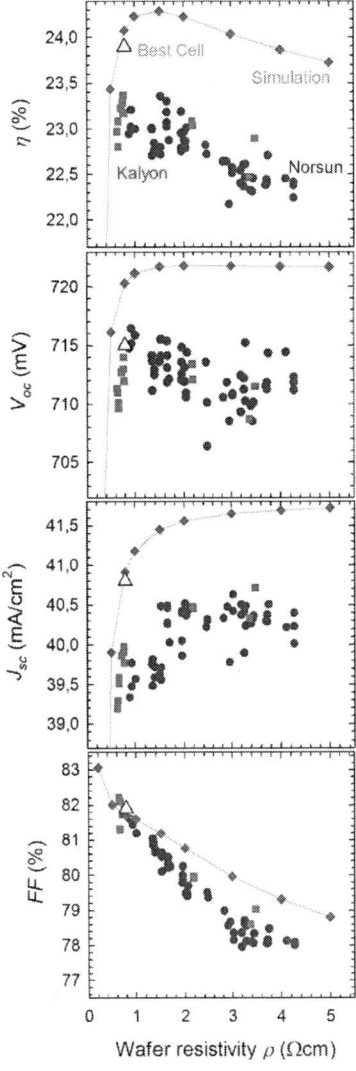

Figure 3: IV parameters of POLO IBC solar cells processed at ISFH with Ga-doped Cz wafers from Norsun (blue circles) and Kalyon (red squares) exhibiting a targeted wafer resistivity range between 0.7 and 4.8 Ωcm. The yellow triangles show the IV parameters of the best POLO IBC cell with 23.9% efficiency processed at ISFH in a previous batch. The green line refers to POLO IBC IV parameters simulated with Quokka3 assuming a lifetime vs. resistivity parametrization published for B-doped Cz wafers after regeneration [6].

4 LETID INVESTIGATIONS

4.1 LeTID investigation of POLO IBC lifetime test structures

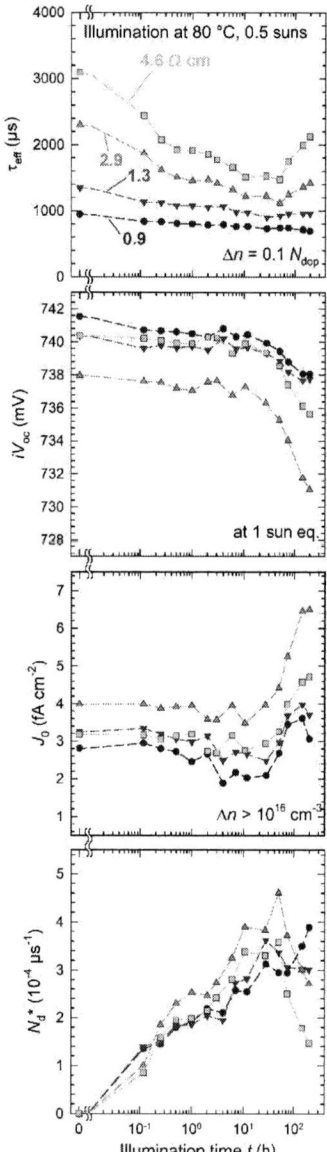

Figure 4: LeTID investigations of four POLO IBC lifetime precursors of Fig. 2 using Ga-doped Cz wafers from Norsun with resistivities 0.9, 1.3, 2.9, and 4.6 Ωcm

To investigate the carrier lifetime stability of the Ga-doped Cz-Si wafers in LeTID tests (light and elevated temperature induced degradation), we select four POLO IBC lifetime test wafers of Fig. 2 with Norsun Cz wafers of 0.9, 1.3, 2.9, and 4.6 Ωcm resistivity. We illuminate these wafers at 0.5 suns and 80°C for up to 300 h. In-between at defined intervals, we measure their injection dependent carrier lifetime by QSSPC. Figure 4 shows the measured effective carrier lifetime τ_{eff} at an excess carrier

density $\Delta n = 0.1\ N_{dop}$, the iV_{oc}, and the average saturation current density J_0 of the front and rear surface passivation of the POLO IBC lifetime sample according to Fig. 1 (step 3.). With increasing illumination time, the effective carrier lifetime τ_{eff} decreases by up to 30% for all four POLO IBC lifetime test structures which causes a slight decrease of their iFF values by about 1%abs. (not shown in Fig. 4). The iV_{oc} is stable around 740 mV until 10 h of degradation time before it starts to degrade by 4 to 7 mV. The average J_0 surface value is stable around 3 fA/cm² (see Fig. 4) until 10 h of degradation time before it starts to increase up to 6 fA/cm² explaining most of the iV_{oc} degradation. The root cause of the surface passivation degradation will be further investigated in the future.

From the QSSPC measurements we calculate the effective defect concentration $N_d^*(t)=1/\tau_d(t)-1/\tau_0$ from the effective lifetime with τ_0 being the initial lifetime before degradation and $\tau_d(t)$ being the degrading lifetime. As shown in Fig. 4, the four different Ga wafer resistivities exhibit very similar effective defect concentrations N_d^* for all illumination times. Hence, we conclude that the Ga atoms and the majority carrier density are not involved in the defect formation caused by LeTID. This finding confirms a previous result [9] now with better statistics and for the first time in the Ga-doped Cz-Si wafer resistivity range of 0.9 to 4.6 Ωcm.

4.2 LeTID investigation of POLO IBC solar cells

Finally, the LeTID stability of POLO IBC solar cells is studied by selecting one POLO IBC solar cell with a Ga-doped Cz wafer from Norsun with a resistivity of 2.0 Ωcm and one POLO IBC solar cell with a Ga-doped Cz wafer from Kalyon with a resistivity of 2.2 Ωcm. We illuminate these solar cells at 0.5 suns and 80°C for up to 1000 h. In-between at defined intervals, we measure their IV parameters as shown in Fig. 5. The initial conversion efficiency of 23.3% and 23.4 % degrades to 22.4% and 22.9% after 60 h of illumination followed by a recovery to 23.4% and 23.5% after 1000 h illumination. Whereas the V_{oc} is stable around 716 mV, the degradation and recovery trends in J_{sc} and pFF / FF explain the efficiency variations. The POLO IBC solar cell LeTID trends in Fig. 5 are in good qualitative agreement with the POLO IBC lifetime test structure LeTID trends in Fig. 4 with respect to iV_{oc} vs V_{oc} and iFF vs. pFF in particular when considering, that the POLO IBC solar cell operates at lower excess carrier densities compared to the lifetime test structure which may accelerate the LeTID defect kinetics for the solar cell.

In summary, the POLO IBC lifetime and solar cell LeTID results in Figs. 4 and 5 show that independent of the Ga-doped Cz wafer resistivity the bulk lifetime slightly degrades during LeTID exposure causing an efficiency drop of up to 4% relative followed by a full recovery. It is difficult to judge how this result forecasts long term stability in the field, since there is a LeTID test norm only for modules [10], not for cells, and since there is no pass or fail criteria defined for this module LeTID test norm.

If there should be a need to further improve the LeTID stability of POLO IBC cells in future module outdoor applications, one option is to optimize the firing temperature profile. It has been recently published for Ga-doped Cz wafers [11,12], that the LeTID degradation can be strongly reduced when decreasing the cool down ramp rate of the firing profile from 70 K/s to 50 K/s. The firing profile used in this paper for the POLO IBC solar cell manufacturing still applies a cool down ramp rate of 70

K/s, which could be reduced to 50 K/s in the future.

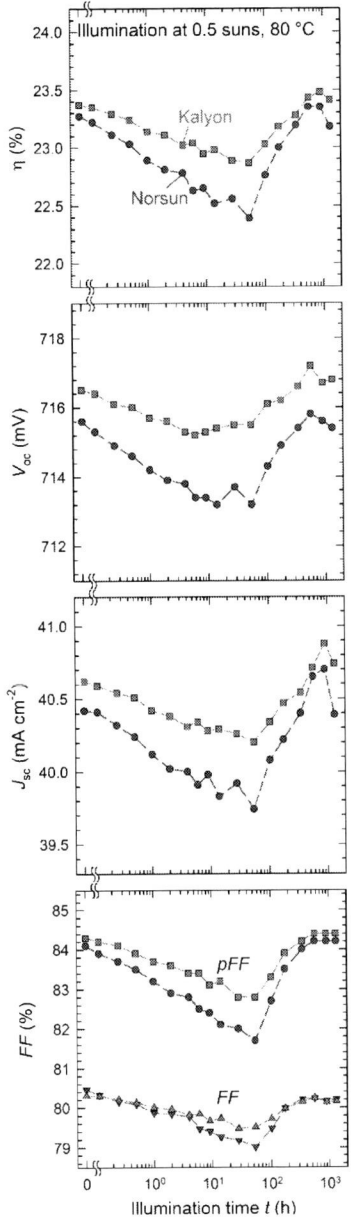

Figure 5: LeTID investigations of two POLO IBC solar cells using a Ga-doped Cz-Si wafer from Norsun with a resistivity of 2.0 Ωcm and a Ga-doped Cz wafer from Kalyon with a resistivity of 2.2 Ωcm.

5   CONCLUSIONS

We applied Ga-doped Cz wafer resistivities ranging from 0.7 to 4.8 Ωcm to optimize POLO IBC solar cells. The POLO IBC lifetime test structures without metal contacts revealed, that the effective carrier lifetime linearly increases from 0.8 ms at 1.0 Ωcm up to 3 ms at 4.0 Ωcm. From the lifetime measurements, we determined an iV_{oc} of up to 740 mV and an iFF of up to 86% and

calculated an implied efficiency of up to 26.7% for a wide wafer resistivity range demonstrating the excellent quality of the Ga-doped Cz wafers and of the POLO IBC surface passivation. The measured IV parameters of POLO IBC solar cells processed with different Ga wafer resistivities agree qualitatively with our Quokka 3 simulations and confirm an optimum resistivity range between 0.8 and 1.8 Ωcm. However, compared to the best measured POLO IBC cell efficiency of 23.9% and the simulated increase towards 24.3% with higher wafer resistivity, the POLO IBC solar cells processed with the higher resistivity Ga-doped Cz wafers exhibited a best efficiency of only 23.4% due to significantly lower J_{sc} values. A possible root cause could be, that in these cell batches the AlO_x/SiN passivation partly blistered which could alloy additional Al ghost contacts in the Al finger area thereby increasing carrier recombination, which is subject to further analysis and optimization.

During LeTID degradation conditions at 0.5 sun, 80°C, the lifetime samples showed a moderate degradation in implied V_{oc} of up to 7 mV and implied FF up to $1\%_{abs.}$ and revealed, that the LeTID-related wafer bulk defect density is independent of the Ga doping concentration. This finding confirmed data published in a recent publication, now with better statistics and for the first time over a broad resistivity range of 0.9 to 4.6 Ωcm. The POLO IBC solar cells showed a maximum LeTID degradation of the efficiency of $4\%_{rel}$ with full recovery during the extended testing of up to 1000 h. The LeTID degradation could be further minimized by reducing the firing temperature cool down rate from the present 70 K/s to 50 K/s. We target to improve the POLO IBC efficiency towards 25.0% near-term by optimizing the POLO IBC metallization, in particular by reducing the Ag to n-poly contact resistance, by optimizing the Al-BSF contact geometry and the busbar and pad design.

References

[1] C.N. Kruse et al., Scientific Reports, 11 (2021) 996

[2] T. Dullweber et al., Proc. 8th World Conf. Photovolt. Energy Conv. (2022), p. 35 - 39

[3] V. Mertens et al., Proc. 40th Europ. Photovolt. Solar Energy Conf. (2023), p. 020015

[4] X. Meng, presented at the 11th BC Workshop (2023), https://www.backcontact-workshop.com/pdf/2023/8_Tongwei-Solar.pdf

[5] P. Shen, presented at the 10th BC Workshop (2022), https://www.backcontact-workshop.com/pdf/2022-2/10_Longi.pdf

[6] D. C. Walter et al., Progr. Photovolt. Res. Appl. 24 (2016)

[7] N. E. Grant et al., Sol. RRL 5 (2021) 2000754 920.

[8] V. Mertens et al., Sol. RRL 8 (2024) 2300919

[9] F. Maischner et al, Solar Energy Mat. & Solar Cells 260 (2023) 112451

[10] Modul test norm IEC TS 63342

[11] F. Maischner et al., Solar Energy Mat. & Solar Cells, 262 (2023) 112529

[12] M. Winter et al., IEEE JPV 13 (6), p. 849-857 (2023)

Towards TOPCon based bottom cells:
Current challenges and Perspectives

Mario Hanser, Henning Nagel, Johannes Gry, Jana Polzin, Armin Richter, Jan Benick, Martin Bivour, Martin Hermle, Stefan Glunz
Tuesday, 24. September 2024

EUPVSEC 2024

Single Junction Silicon Solar Cells with Passivating Contacts
Motivation

Heterojunction

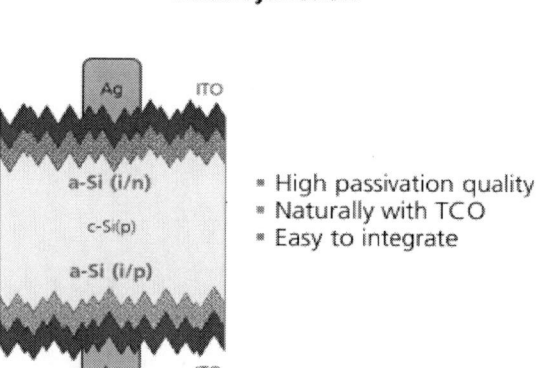

- High passivation quality
- Naturally with TCO
- Easy to integrate

iTOPCon

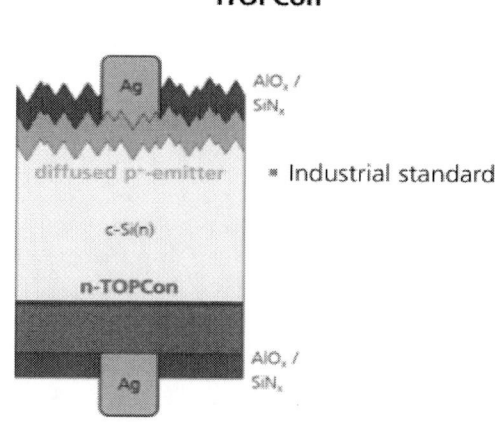

- Industrial standard

Bottom Cell Concepts

Overview of TOPCon based Bottom Cell Concepts

iTOPCon[-1]

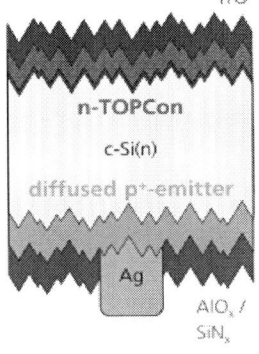

TOPCon[2]

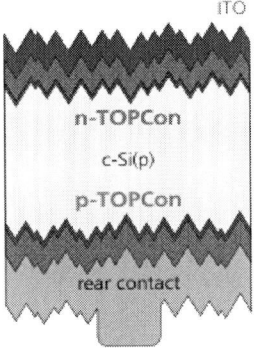

TOPCon[3]

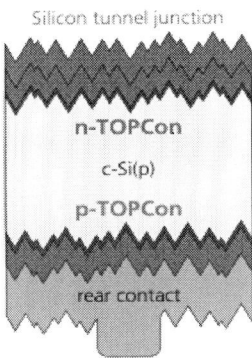

From iTOPCon to iTOPCon[-1]

iTOPCon Pilot Line

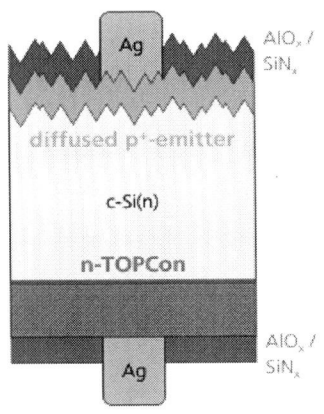

Inline single side etching tool

Diffusion furnace

Screen printing line

Firing furnace

Small lab-scale cells
2.5 x 2.5 cm²

From iTOPCon to iTOPCon[-1]
Adaptions Towards a Lab-Scale Bottom Cell

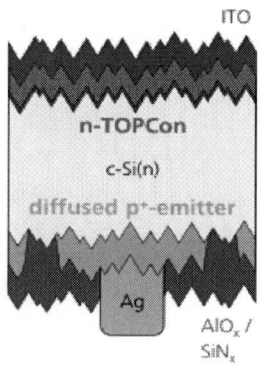

- Inverted iTOPCon (iTOPCon[-1])

- Thin n-TOPCon on textured surface

- **Structuring** of the emitter [1]

- Removable, fire stable front side **hydrogenation**

- ITO **sputtering** and curing

- Laser separation of small cells

[1] Andreas Fell et al., SiliconPV 2024, Understanding and Minimizing Perimeter Losses of Perovskite-Silicon Tandem Solar Cells

---

iTOPCon[-1]
Laser Structuring for Small Lab-Scale Cells

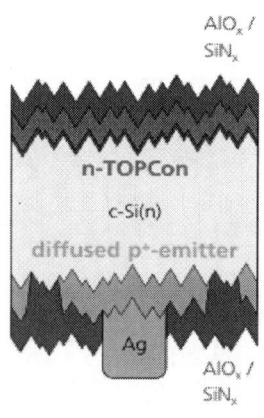

- Laser structuring of emitter
- AlO_x/SiN_x passivation of the rear side [1], [2]
- High passivation quality in lasered area

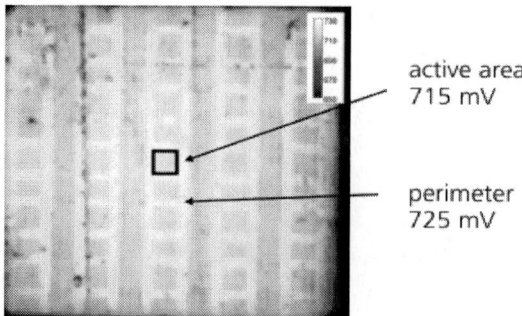

active area
715 mV

perimeter
725 mV

[1] Andreas Fell et al., SiliconPV (2024), Understanding and Minimizing Perimeter Losses of Perovskite-Silicon Tandem Solar Cells
[2] Yiliang Wu et al., Advanced Energy Materials (2022), DOI: 10.1002/aenm.202200821

iTOPCon[-1]
Hydrogenation

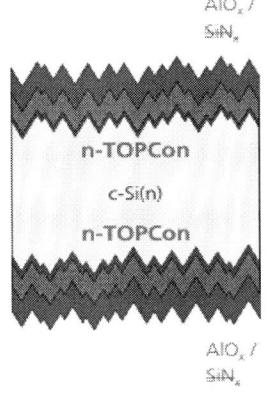

- Fire stable hydrogenation layer
- Removable after firing process
- 10 nm AlO_x fullfills the requirements

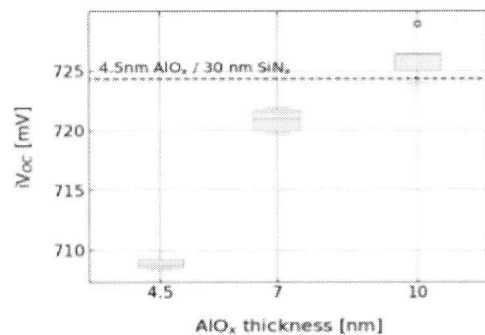

---

iTOPCon[-1]
Finished Bottom Cell

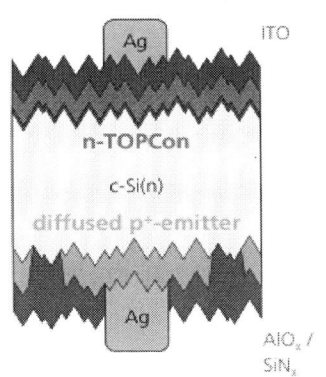

- ITO sputtering
- 300 °C curing anneal for 5 min
- iV_{OC} around 710 mV
- pFF of 82.4 %

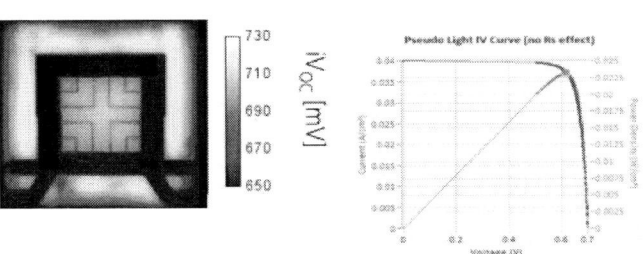

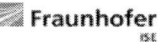

iTOPCon^{-1}
Screen Printing Particles and Cleaning

Front side

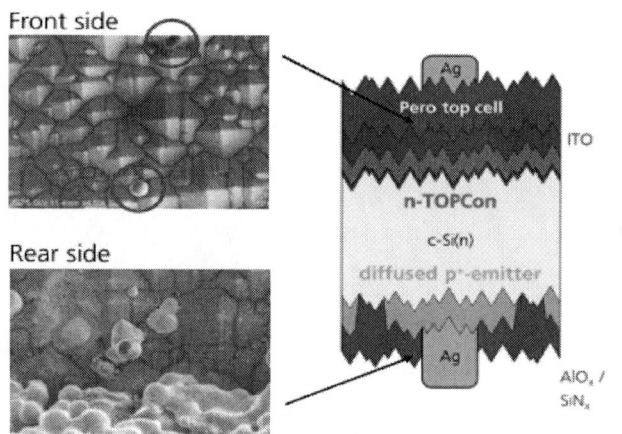

Rear side

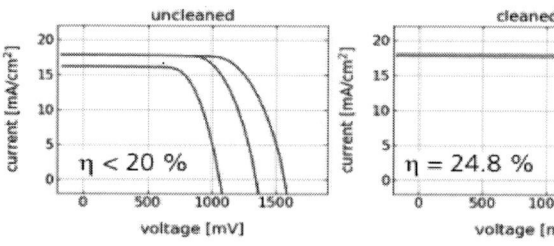

* Particles from screen printing process on active cell area
* Cleaning of the front side with conc. HNO_3
* Tandem cell efficiencies up to 24.8 %

---

iTOPCon^{-1}
Conclusion

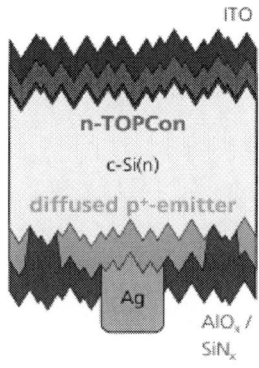

* 150 µm thin M2 wafer
* Established pilot line adaptions for iTOPCon^{-1} bottom cells
  * Laser structured emitter
  * Fire stable, removeable front side hydrogenation
  * ITO sputtering and curing

* Screen printing leads to particles
* Cleaning possible
* Problem not solved for pilot line

* Limited efficiency as processes such as e.g. LECO (laser-enhanced contact optimization) are not possible in its current form

From iTOPCon[-1] to TOPCon[2]
Cell Design

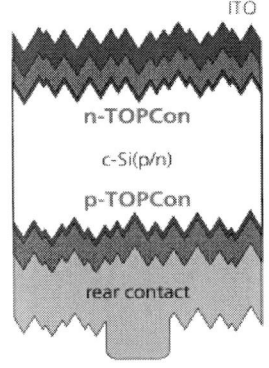

* Replacement of diffused emitter with p-TOPCon

* Full area passivating contacts on both sides

* Leaner process chain

TOPCon[2]
iV$_{OC}$ Measurements

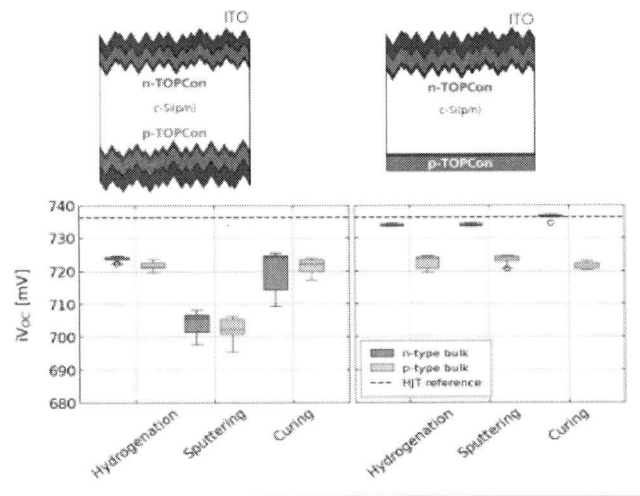

Experimental
* 1 Ωcm p- vs. n-type FZ material
* Textured vs. planar rear side
* 50 nm in-situ doped PECVD TOPCon

Results
* High passivation quality
* Sputter damage can be cured with 300 °C hotplate anneal for 5 min
* No influence on contact resistivity observed

TOPCon²
Tandem Cell Results

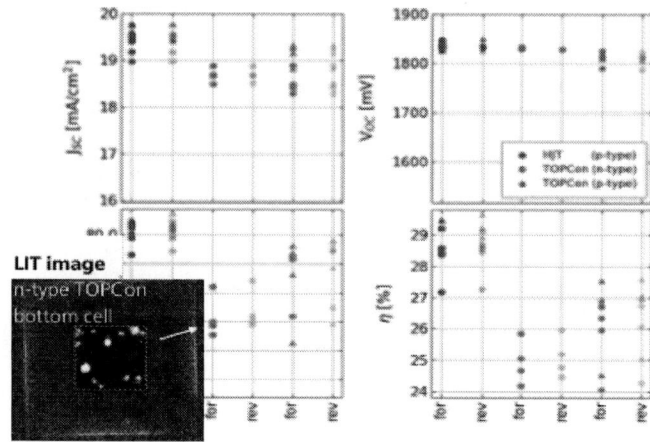

LIT image
n-type TOPCon bottom cell

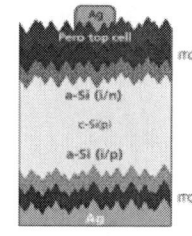

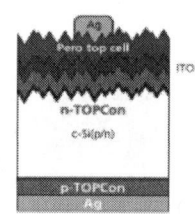

- Tandem cell results
- Little V_{OC} loss 10 – 20 mV
- J_{SC} loss of 0.5 mA
- FF loss due to shunts in TOPCon emitter
- Best results with 27.5 %

TOPCon²
Conclusion

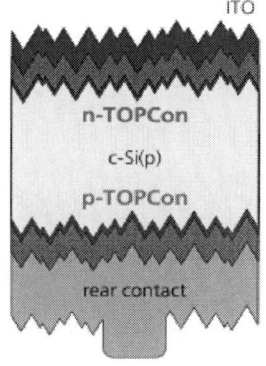

- Full area passivating contacts

- Efficiencies up to 27.5%

- Clear roadmap for further improvements

- Metallization concept under research

TOPCon³
Cell Design

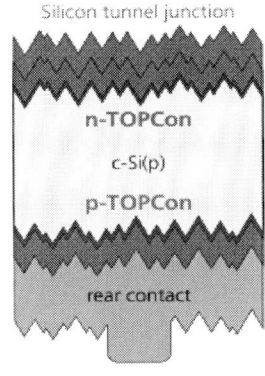

* Indium free interconnection

* Silicon process technology

* Optical transparent

* Good electrical properties [3]

* Open Challenge:
  Hole transport layers (HTL) are developed for TCO

[3] Mario Hanser et al., SiliconPV 2024, Development of Silicon Tunnel Junctions for Perovskite/Silicon Tandem Devices

---

TOPCon³
HTL Choice

Self assembling monolayers (SAM) [4]
Me-4PACz

* Polar
* Requires hydrophilic surface

TaTm

* Unpolar
* No additional surface requirements

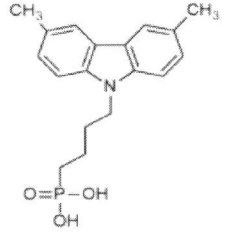

[4] Zheng, Jingming et al., Nature Energy (2023), DOI: 10.1038/s41560-023-01382-w

TOPCon³
Contact Resistivity Screening of Surface Treatments
—

- Experiments on simple ohmic test structures

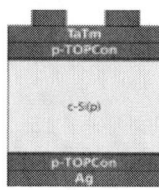

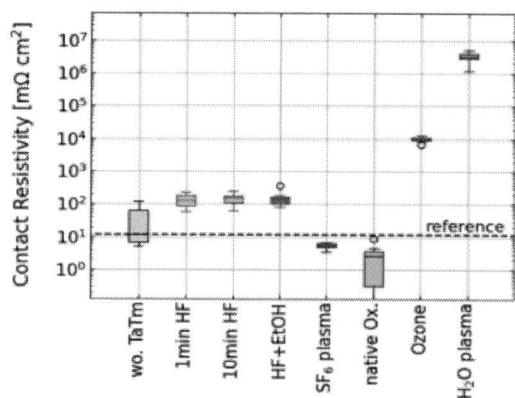

- Screening of large parameter space
- Structures with high contact resistivity can be excluded

17

©Fraunhofer ISE

Fraunhofer ISE

TOPCon Based Bottom Cells
Conclusion, Comparison and Open Challenges
—

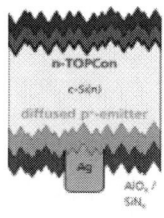

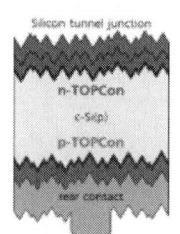

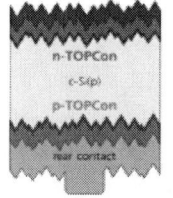

- **iTOPCon[-1]**
  - Close to mainstream silicon solar cell
  - Successful integration in pilot line
  - Solutions for laser structuring, hydrogenation and sputter damage
  - Screen printing particles could be cleaned for single solar cells

- **TOPCon²**
  - Full area passivating contacts
  - Successful implementation in tandem solar cell with 27.5 %
  - Clear roadmap for further improvements

- **TOPCon³**
  - Working silicon tunnel junction
  - Challenging tandem integration

18

©Fraunhofer ISE

Fraunhofer ISE

Thank you for your attention.

Get in contact:
Mario Hanser
mario.hanser@ise.fraunhofer.de

Bundesministerium
für Wirtschaft
und Klimaschutz

This work was funded by the German Federal Ministry for Economic Affairs and Climate Action (BMWK) and industry partners as part of the projects Pero-Si-SCALE (03EE1191), RIESEN (03EE1132A) and EPoBoC (03EE1192)

REVIEW ON IN-FREE RECOMBINATION JUNCTION APPROACHES FOR TWO-TERMINAL SILICON/PEROVSKITE TANDEM SOLAR CELLS

P. Vasquez[1,3], A. Merino[1], P. Carroy[1], B. Marteau[1], T. Desrues[1], N. Nguyen[1], M. Matheron[1], S. Chozas[2], F. Ventosinos[2], H. J. Bolink[2], D. Muñoz[1]

[1]Univ Grenoble Alpes, CEA, LITEN, DTS, INES, F-38000, Grenoble
[2]Instituto de Ciencia Molecular, Universidad de Valencia, Calle Catedratico Jose Beltran 2, 46980 Paterna, Spain
[3]LMGP, 123, Grenoble, 38000, France
Corresponding author: Delfina.munoz@cea.fr

ABSTRACT: This contribution reviews the development of In-free recombination junction (RJ) approaches for two-terminal (2T) silicon/perovskite (PK) tandem solar cells, motivated by the increasing demand for scarce raw materials like indium. While the utilization of In-based transparent conductive oxide (TCO) thin films as RJ remains extensively used, in particular in devices reporting efficiencies greater than 30%, alternative In-free methods, including interlayer-free and nano-/poly-crystalline silicon tunnel junctions, have demonstrated their potential but are much less reported in literature. Four In-free RJ architectures are presented here, utilizing different techniques or materials and with a wide diversity in terms of integration in tandem devices: PIN and NIP configurations are reported, as well as two types of bottom cells (silicon heterojunction -SHJ-) and TOPCon-like) and different top cell processing (either solution- or vacuum-processed). A low-temperature tunnel RJ based on heavily doped nc-Si:H layers deposited by PECVD SHJ bottom cells was teste in PIN configuration tandem devices, either on a planar bottom cell and solution-processed top cell, achieving an efficiency of over 12%, or an a fully textured bottom cell with vacuum-processed top cell reaching efficiency of 16%. On planar TOPCon bottom cells, the deposition of a nc-Si:H(n+) layer by PECVD on the poly-Si:H(p+) allowed the formation of tunnel RJ in tandem NIP configuration resulting in an efficiency close to 17%. Second, a direct RJ structure allowing the recombination at the interface between the poly-Si:H(p+) and the SnO2 based electron transport layer of the top cell reaches an efficiency above 19%. Finally, a direct RJ architecture employing a NiOx layer directly sputtered on the poly-Si:H(n+) contact of a planar TOPCon bottom cell lead to a tandem efficiency above 20% in tandem PIN configuration.
Keywords: Silicon/perovskite tandem solar cells, Recombination junction, Nanocrystalline silicon thin films.

1 AIM AND APPROACH

During the last decade, great progress in high efficiency silicon-based solar cells –both in single junction and monolithically integrated (2T) silicon / perovskite (PK) tandem solar cells has been made, thanks to the efforts of many research groups in laboratories and industrial manufacturers worldwide. Only nine years since the first demonstration of a 2T silicon heterojunction (SHJ) / PK tandem which achieved an efficiency of 13.7% in an active area of 1 cm² [1], current efficiencies have been pushed out up to 33.7% (KAUST) in laboratory-scale demonstrations and 33.9% in commercial substrates (LONGi). On the single junction side, efficiencies are rapidly approaching the practical limits in SHJ and TOPCon technologies [2]. Additionally, the tremendous rise in PV generation (mainly Si) and the projected increase in SHJ, TOPCon and PK/Si tandem solar cells production by 2030 make these technologies strong candidates for a fully decarbonated energy mix.

Nevertheless, the promising rise on the solar cells market will further stress the already pressed scarce raw materials production, particularly indium. The subject of In-suppression in commercial solar cells has been widely discussed [3,4], but in the case of 2T silicon / perovskite tandems, addition of ITO thin films at the recombination junction (RJ) for achieving efficiencies > 30% is still the trend. However, alternative In-free RJ approaches such as the interlayer-free (or direct junction) [5], and nano- and poly-crystalline silicon tunnel junctions [6-8] with interesting efficiencies, approaching 29% (0.1 cm²) have been also demonstrated.

In order to further study the potential and challenges of an In-free approach for recombination junction for PK/Si, here we present advances done at CEA on over 8

cm² devices both in PK/SHJ and PK/TOPCon tandem cells.

2 METHODS

The devices discussed in this work are the following:

1. Low temperature nc-Si:H based tunnel RJ in a PIN configuration tandem device: The highly doped nc-Si:H tunnel layers forming the tunnel junction were deposited by PECVD on a SHJ bottom cell based on double-side polished wafers, functionalized at their surface with an organic, carbazole-based self-assembled monolayer (SAM) and covered with a FACsPbIBr PK absorber deposited by spin-coating. An evaporated C60/BCP stack formed the electron transport layer (ETL). Sputtered ITO was used as top electrode and the Ag grid was evaporated on top of it. The tandem device achieved > 12% efficiency on an active surface of 8.49 cm² (Fig. 1a).

2. Low temperature nc-Si:H based tunnel RJ in a PIN configuration tandem device fabricated with a SHJ bottom cell based on a fully textured commercial wafer and a fully vacuum-processed the top cell. The fabrication of the wide bandgap FACsPbIBr perovskite was carried out via a three-source vacuum deposition process using FAI, CsI and a pre-synthesized mixed Br/I-Pb "alloy" as precursors. For this tandem device, a PbBr2:PbI2 ratio of 1:8 was implemented, leading to a bandgap of around 1.68 eV. The hole transport layer (TaTm-CS9/TaTm), electron transport layer (C60/BCP) and Ag grid were deposited via vacuum deposition, and the ITO top electrode was deposited via soft pulsed laser deposition (PLD). An efficiency over 16% was achieved on a 1 cm² tandem device for this

configuration compared to our reference process at 17% (Fig. 1b).

3a. High temperature nc-Si :H based tunnel RJ in an NIP configuration tandem device fabricated on TOPCon-like bottom cell based on a double-side polished wafer: a highly-doped nc-Si:H(n+) layer was deposited by PECVD directly on the poly-Si:H(p+) contact of the bottom cell, thus forming the tunnel junction. The top cell ETL was formed by spin-coated SnO2 nanoparticles that were then covered by the FACsPbIBr PK absorber. The device yielded an efficiency close to 17% over an active surface of 8.49 cm² (Fig. 3a).

3b. High temperature direct RJ in an NIP configuration tandem device fabricated on TOPCon-like bottom cell: the SnO2-based ETL as described above was directly spin-coated on the poly-Si:H(p+) contact of the bottom cell, followed by the subsequent steps of the top cell fabrication already described. An efficiency above 19% was reached on an 8.49 cm² active area in this case (Fig. 3a).

4. High temperature direct RJ in an PIN configuration tandem device fabricated on TOPCon-like bottom cell based on a double-side polished wafer: a NiOx layer was directly sputtered on the poly-Si:H(n+) contact of the bottom cell and further functionalized with an evaporated SAM layer. The FACsPbIBr absorber was then spin-coated on top of it and the device was finalized with the same front stack as described in step 1. The tandem device reached an efficiency above 20% on an active surface of 8.49 cm² (Fig. 3b).

It is worth to mention that all devices are non-optimized in terms of metallization schemes and without any double antireflection coating, so shadowing effects are close to 15%, especially for large area devices.

3 PRELIMINARY RESULTS

As shown in Figure 1a, the TRJ alone does not perform a good hole contact with the PK absorber, in contrast, addition of the SAM interlayer results in an increment of 10.0% in efficiency. The s-shape observed around Voc (~1800 Mv), responsible of a FF loss and a difference of 7.5% efficiency with respect to the ITO RJ reference, can be attributed to uneven coverage of the SAM on the TRJ surface, given that its processing by spin coating is optimized for ITO.

The reduction in the S-shape in Figure 1a could be explained by a better passivation effect of the SAM as shown by photoluminiscence and transmission electron microscopy (TEM) measurements. Fig. 2 shows TEM measurements performed one month after deposition. The degradation of the perovskite, evident by the white spots, on the PK/TJ device (Fig. 2a) is spread throughout the whole layer, as compared to the PK/SAM/TJ device (Fig. 2b), where the perovskite degradation is restricted to the surface.

The same TRJ, deposited on a textured SHJ substrate (1 cm²) with evaporated top cell components shows only 1% efficiency loss with respect to an ITO RJ, as shown in Fig. 1b.

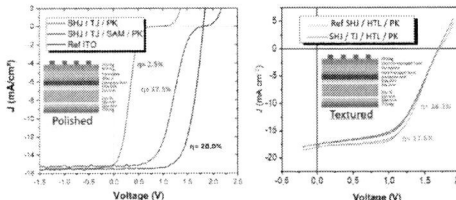

Figure 1: JV characteristics of the (n+/p+)nc-Si:H tunnel junction featuring the RJ of a SHJ/PK tandem. a) Polished SHJ bottom cell with the perovskite directly deposited on the TRJ (red), with a SAM interlayer between the TRJ and the PK (blue) and the reference cell with an ITO RJ. b) Textured SHJ bottom cell with nc-Si:H based TRJ (green) or with a reference RJ (yellow) and vacuum-processed top cell.

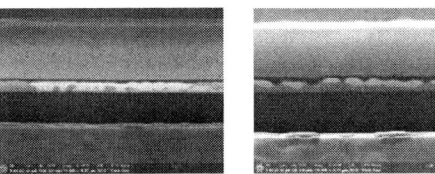

Figure 2: TEM measurements of a) the PK/TJ and b) PK/SAM/TJ devices performed one month after deposition.

TopCON-like bottom cells using poly-Si/SiOx passivating contacts on both sides can be used for PIN and NIP configurations by simply flipping the devices. Results for NIP tandem cells are shown in Fig. 3a, where the best efficiency is obtained without any interface layer between both subcells ("interlayer free"). For PIN tandem devices (Fig. 3b), almost the same performances are obtained with ITO or NiOx interface layers. These results show that In-free interface structures can lead to efficient devices both in NIP and PIN configurations, using TopCON-like bottom cell.

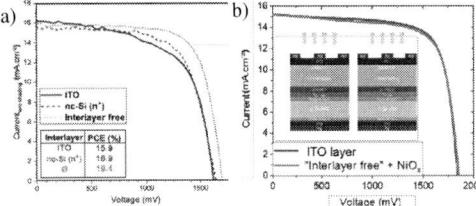

Figure 3: a) JV characteristics of NIP tandem devices including high temperature TOPCon-like bottom cells with a poly-Si:H(p+)/nc-si:H(n+) based TRJ (red), interlayer-free RJ (yellow) and ITO RJ as reference (blue). b) JV characteristics of PIN tandem devices including high temperature TOPCon-like bottom cells with a poly-Si:H(n+)/NiOx interface structure compared to a standard ITO recombination layer.

4 CONCLUSIONS

Efforts on developing recombination junction for both low and high temperature silicon solar cells have been made in both literature and experiments presented in this work, showing potential for an In-free approach as a recombination junction in PK/Si tandem devices.

It is important to note that all indium free recombination junction solar cells developed by CEA are over 8 cm² devices.

Results for nanocrystalline silicon tunnel junction show the need for optimized top subcell and deposition processes for this new indium free recombination junction.

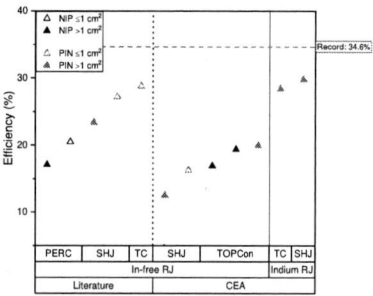

Figure 3: Summary of results for PK/Si tandem solar cells from literature and advances done in CEA

4 REFERENCES

[1] M. Smith, A. Miller, Proceedings 17th European Photovoltaic Solar Energy Conference, Vol. I (2002) 903.

[1] Mailoa, J. P. et al. A 2-terminal perovskite/silicon multijunction solar cell enabled by a silicon tunnel junction. Appl. Phys. Lett. 106, (2015).

[2] Best Research-Cell Efficiency Chart. https://www.nrel.gov/pv/cell-efficiency.html.

[3] Reduction in Indium Usage for Silicon Heterojunction Solar Cells in a Short-Term Industrial Perspective - Jay - 2023 - Solar RRL - Wiley Online Library. https://onlinelibrary.wiley.com/doi/full/10.1002/solr. 202200598.

[4] Gageot, T. et al. Feasibility test of drastic indium cut down in SHJ solar cells and modules using ultra-thin ITO layers. Solar Energy Materials and Solar Cells 261, 112512 (2023).

[5] Zheng, J. et al. Large area efficient interface layer free monolithic perovskite/homo-junction-silicon tandem solar cell with over 20% efficiency. Energy & Environmental Science 11, 2432–2443 (2018).

[6] Li, Y. et al. Nanocrystalline silicon-oxygen based tunneling recombination junctions in perovskite/silicon heterojunction tandem solar cells. Solar Energy Materials and Solar Cells 262, 112539 (2023).

[7] Mazzarella, L., Morales-Vilches, A. B., Korte, L., Schlatmann, R. & Stannowski, B. Versatility of Nanocrystalline Silicon Films: from Thin-Film to Perovskite/c-Si Tandem Solar Cell Applications. Coatings 10, 759 (2020).

[8] Zheng, J. et al. Polycrystalline silicon tunnelling recombination layers for high-efficiency perovskite/tunnel oxide passivating contact tandem solar cells. Nat Energy 8, 1250–1261 (2023).

41st European Photovoltaic Solar Energy Conference and Exhibition

IDENTIFICATION OF PERFORMANCE-RELEVANT OPTICALLY DETECTED DEFECTS BY CORRELATIVE DATA ANALYSIS IN SOLAR CELL PRODUCTION

M. Meusel, A. Starke and M. Turek
Fraunhofer Center for Silicon Photovoltaics CSP
Fraunhofer Institute for Microstructure of Materials and Systems IMWS
Otto-Eißfeldt-Straße 12, D-06120 Halle, Germany
marko.turek@csp.fraunhofer.de

ABSTRACT: The quality control of a solar cell production relies on a large variety of sensors, cameras, and measurements. Typically, each individual measurement is analyzed by its own criteria. For example, an optical inspection yields the number of optically detectable artefacts, such as finger interruptions. On the other hand, performance parameters are available for each cell. Thus, linking these different data sources in a quantitative manner is a major objective in quality control. However, the impact of a single defect found in optical inspection on a single cell's performance can be much smaller than other process-induced effects. Therefore, we evaluate three different statistical approaches to identify whether defects observed by optical inspection quantitatively relate to the performance of larger cell sets. This analysis results in a separation of relevant optically detected defects from defects without impact on the cell's performance. From a data analysis point of view, this is challenging as the optical data set is rather sparse, e.g. only few cells show many optically detected defects, while the performance data set is described by a smooth distribution of values around some average value. The presented methods are rather universal and can thus be transferred to establish quantitative relations between any two characterization methods in a production line.

Keywords: solar cell characterization, optical inspection, correlative production data analysis

1 INTRODUCTION

Current-voltage measurements under standard test conditions (STC I-V-curve) are a central characterization step in a solar cell production as this data is the basis for a cell pricing. On the other hand, it is also an important means for process and quality control during production. Several innovations regarding the performance measurement have been proposed during the last years, like contactless measurements [1], spectrally resolved inline measurements [2], or advanced analysis methods for more complex cell models [3]. Using suitable physical models, valuable information about the solar cell's quality can be extracted and possible issues during production identified.

Additionally, there are many further characterization approaches, cameras, and sensors along the production line. One example is the optical print inspection after the metallization print. Typically, images of inspection tools are analyzed by image analysis algorithms resulting in a set of quantitative parameters like the type, size or number of optically detectable defects such as finger-thinnings or finger-interruptions.

The correlative data analysis in a production is of significant relevance for an advanced data-driven quality control. [4, 5, 6] In particular, the combination of different sensors, cameras, or measurements into one data set promises to yield much more detailed information on the cell quality and the involved production or measurement processes. [7, 8, 9] While most scientific works focus on a physics-based investigation of one measurement type, e.g. the I-V-data analysis using equivalent circuit models or the analysis of EL-images using material-related models for the luminescence signal, we demonstrate the usefulness of combining several data sets in one approach. In particular, we focus on how it can be decided which optically detected defects are relevant for the cell's performance. This correlation is crucial when the production process is to be optimized as it allows to focus on the relevant defects. Furthermore, it allows for an early detection of production drifts by not only considering the changes in distributions of one set of values, like the performance

values, but also time-dependent drifts in the correlation between data sets from various data sources.

Drawing conclusions about the relevance of a certain optically detected defect requires algorithmic approaches that quantitatively relate the optical inspection with the performance measurement. Only in this way, one can decide whether a certain optically detected defect is severe enough to impact a solar cells performance and which of the I-V-parameters is affected. However, depending on the type of defect, a typical distribution of the optically detected defects is such that most cells show no defect at all or only a very low number or size of defects. The number of defects thus resembles, for example, a highly unsymmetric Poisson distribution which is strongly peaked around zero. On the other hand, the distribution of I-V-parameters is typically more Gaussian-like distributed, e.g. the values are spread with a certain width around the mean values. This means that a simple statistical correlation analysis between these two quantities also includes a large number of cells showing no or few optically detectable defects. In such a case, a direct analysis of, for example, linear correlation coefficients might not be sufficient. Furthermore, the quantitative impact of a single optically detected defect might be very small compared to the overall spread in the solar cell's performance parameters. This implies that the impact of these defects cannot be determined by comparing a single cell without this particular type of defect with another single cell showing this defect. It is – from a statistical point of view – not unlikely that a cell without defects shows a lower performance than another cell with defects due to the intrinsic process variations leading to the overall spread in performance parameters.

Therefore, suitable statistical approaches to identify relevant defects found in optical inspection are required. The aim of this work is to evaluate three different data analysis approaches than can be applied: a) analysis of the correlation matrix, i.e. linear correlation coefficients, between optical inspection parameters and I-V-parameters, b) machine-learning based detection of I-V-inliers and I-V-outliers using the I-V-data set and comparison of the statistical occurrence of the defects within these two sub-groups, and c) separation of the data set into two sub-groups with and without optical defects and comparison of the statistical distribution of I-V-parameters for these two sub-groups.

We apply these three methods to both production data of about 60000 cells and a better controllable simulation data set of equivalent size. Using the simulation approach, we can establish a consistent description of the production data which includes a quantification of the statistical impact of a single defect on the performance, in particular the fill factor, of the solar cells.

Furthermore, the advantages and dis-advantages of each approach are discussed. Finally, we argue that the developed methodology can be directly applied to any other set of quality control parameters which substantiates the relevance of our approach.

2 EXPERIMENTAL APPROACH AND DATA ANALYSIS METHODS

The production data set we used to develop and test our approaches includes I-V-data together with data of an optical inspection of about 60.000 cells. The I-V-data includes the cell's efficiency η, its short circuit current I_{sc}, open circuit voltage V_{oc}, and fill-factor FF. As an example, the asymmetric Gauss-like distribution of the efficiency (shifted and scaled by the mean value) is presented in Figure 1. The optical data includes the number of finger interrupts and finger-thinnings per cell and several other parameters. While some cells exhibit these optically detected defects a large set of cells does not exhibit these defects. For our approach, we use one single parameter which combines all types of detected finger interrupts in one number. The distribution of this defect parameter is shown in Figure 2 and follows a Poisson-like, i.e. exponentially decaying, shape.

The objective is to provide a tool set of algorithms that allow to decide whether a defect observed in the optical inspection is statistically severe enough to result in a cell's performance loss. We have developed and implemented algorithms for three different approaches that aim at the identification of performance-relevant optically detected defects:
a) "Correlation": Determination of the matrix of linear correlation coefficients between number of optically detected defects and solar cell performance parameters,
b) "I-V-outlier": Identification of outliers (about 1%) in the I-V-parameter data space using the isolation-forest-algorithm and comparison of the distributions of number of defects for the sub-goups of "I-V-outliers" vs. "I-V-inliers",
c) "Distribution": Separation of the data set into two sub-groups with and without defects and subsequent comparison of the distribution of solar cell parameters, e.g. the fill-factor.

Furthermore, we have implemented an extended two-diode-model where the optical defects can be included for each individual solar cell with a varying quantitative impact on the series resistance. Using this model, we have calculated I-V-curves and -parameters for several data sets containing large number of solar cells.

Eventually, we aim at the application of all three data analysis methods to a production data set of 60.000 cells. To first qualify the three methods, we generated several simulation data sets of equal size with random distributions of the two-diode-model parameters. These

simulation data sets are created in such a way that their resulting I-V-parameter distributions for I_{sc}, V_{oc} and FF, respectively, are very similar to the production data set, see Figure 1 and Figure 2. Each of these simulation data sets contains one subset where the optical defects are present but do not have any impact on the series resistance (subset "no impact") and one subset where they do have an impact on the series resistance (subset "with impact"). In all cases, these two subsets are designed such that the resulting distributions for the I_{sc}, V_{oc} and FF values are very similar to each other. This can be achieved by "switching on" the impact of the defects on the series resistance and simultaneously modifying the random choice of two-diode-model parameters. The key objective is then to assess whether applying the three data analysis methods on the various simulation data sets can distinguish the cases of defects without impact (subset "no impact") from cases of defects with impact (subset "with impact") on the FF.

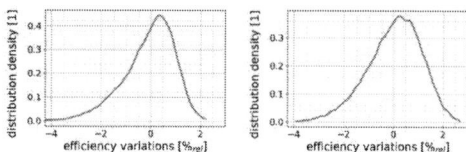

Figure 1: Asymmetric Gauss-like distribution of cell efficiencies shifted and scaled relative to mean value (left) for a production data set and (right) for the simulated data of the implemented model.

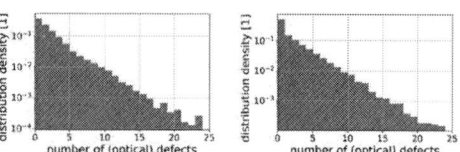

Figure 2: Exponentially decaying distribution of number of optically detected defects (left) for a production data set and (right) for the simulated data of the implemented model.

Each of the three data analysis methods is then first applied to several versions of the simulated data and then to the production data set.

3 RESULTS

3.1 Evaluation using simulation data

First, we test and evaluate the three methods "correlation", "I-V-outlier", and "distribution" on the simulation data sets. As a first analysis, we calculate the correlation matrix between the parameter describing the optical defect and the I-V-parameters efficiency, short circuit current I_{sc} and fill-factor FF. In Figure 3, the results are shown for the fill-factor and efficiency vs. number of optical defects. The left diagram shows the case where the defects do not have an impact on performance and the right diagram represents the case "with impact". One can observe that – as expected – a zero correlation coefficient is found between number of defects and fill-factor in the case "no impact". In the "with impact" case, a non-zero correlation coefficient is found. However, this coefficient is rather small, and the scatter plots are also rather inconclusive. This is in particular the case, if the individual impact of the defect in the model is chosen to be small. Also, the I-V-parameters are Gauss-like distributed while the optical defects show an exponential decaying Poisson-like distribution which puts much emphasis on the large number of cells with no or only a few defects.

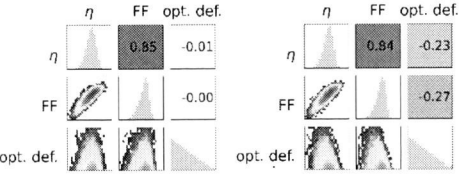

Figure 3: Correlation matrix between the optical defect under consideration and the relevant I-V-parameters efficiency and fill-factor FF for a simulation data set: (left) case "no impact" and (right) case "with impact".

In the second analysis ("I-V-outliers"), we identify inliers and outliers of the entire I-V-data set based on the three-dimensional data space made of current I_{sc}, voltage V_{oc}, and fill-factor FF. To identify the outliers, an isolation forest algorithm has been applied such that 1% of all data points have been labelled as outliers. Then, the distribution of the occurrence of optical defects for each of these two sub-groups – "I-V-inliers" and "I-V-outliers" – has been calculated. In the case "no impact", see upper diagram in Figure 4, the two sub-groups of inliers and outliers exhibit a very similar distribution of the number of defects. severe enough to affect the cell's performance.

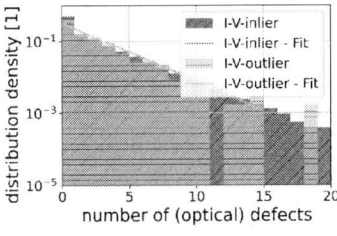

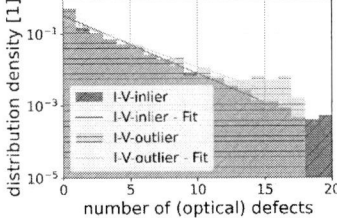

Figure 4: Exponential decaying distribution of defects for the sub-groups of "I-V-inliers" and "I-V-outliers": (top) case "no impact" and (bottom) case "with impact".

This relies on the fact that in this case, the distribution of I-V-parameters that leads to the outliers is simulated to be unrelated to the defects. For the other case "with impact", on can observe that the I-V-outliers typically contain more cells with many defects, i.e. the exponential decay of the distribution is less pronounced. This can be seen in the semi-logarithmic diagram in Figure 4 (bottom) where the slopes of the distributions are indicated by the green (I-V-inliers) and yellow (I-V-outliers) lines, respectively. This is consistent with the following interpretation: If the optical defect had a statistical impact

on the I-V-parameters, one would have expected that the I-V-outliers show a qualitatively different distribution of defects than the I-V-inliers. In particular, one would expect that the I-V-outliers are characterized by a larger number of optically detected defects if these defects are

The third method ("distribution") separates the data set into a sub-group with optically detected defects and another sub-group without any defects. Then, the fill-factor distribution among all cells of these two sub-groups is analyzed. The distributions related to the entire data set and to the two sub-groups are shown in Figure 5. In the "no impact" case, where the defects are unrelated to the I-V-parameters, one finds rather identical FF distributions expect for statistical fluctuations. In the other case "with impact", the FF-distribution of the sub-group without defects is slightly shifted towards higher values while the FF-distribution of the sub-group with defects is slightly shifted towards lower values.

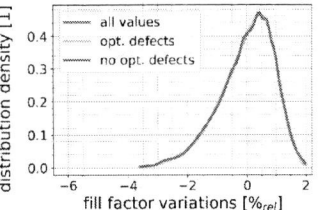

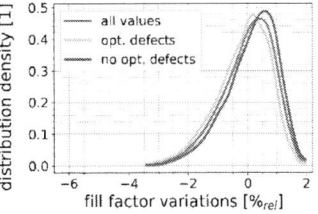

Figure 5: Distribution of fill-factor values for the entire data set (green line), the sub-group that is free of defects (red line) and the sub-group that contains defects (yellow): (top) case "no impact" and (right) case "with impact".

3.2 Application to production data

In a second step, we apply these three data analysis approaches to the production data set to check for any of the described signs that would indicate an impact of the optical defects on the I-V-parameters, in particular on the fill-factor. Here, the correlation matrix exhibits some non-zero linear correlation coefficients, see Figure 8. However, these coefficients are extremely small and the scatter plots could not be interpreted as clearly showing some correlation.

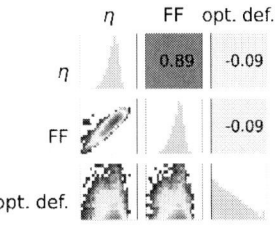

Figure 6: Application of the method "correlation" to the

production data set resulting in the correlation matrix

The "I-V-outlier" approach, see Figure 7, results in a slightly different slope for the exponential decaying distribution of defects for the "I-V-outlier" sub-group compared to the "I-V-inlier" sub-group as indicated by the fitted straight lines in the semi-logarithmic diagrams.

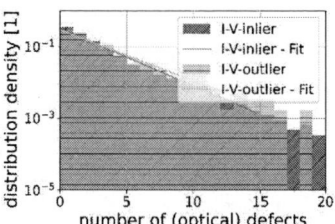

Figure 7: Application of the methods "I-V-outlier" to the production data set: distribution of optical defects for "I-V-outlier" and "I-V-inlier" sub-groups.

Third, the FF-distributions of the sub-groups "with defects" and "without defects", see Figure 8, show some minor deviations however a clear shift cannot be identified.

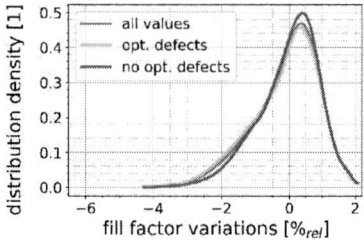

Figure 8: Application of the method "distribution" to the production data set: distribution of fill-factor for sub-groups "with defects" and "without defects" compared to the overall distribution.

Thus, comparing these three approaches, the method "I-V-outliers" appears to be most sensitive to a small impact of the optically detected defects on the fill-factor. To quantify this impact, i.e. to give an upper limit, we setup a simulation data set by successively reducing the impact of the defects on the series resistance in the model until we achieved similar linear correlation coefficients and similar "I-V-outlier" distributions as for the production data set. Here, we found that if a single defect contributes to the series resistance of about 0.7% then the statistical results for the production data set are well described by the simulation data. Hence, we conclude that in the investigated case study of the production data, the optical defects contribute very weakly with around 0.7% increase in series resistance thus slightly reducing the average fill-factor for a large set of solar cells.

4   SUMMARY AND CONCLUSIONS

In summary, we present a data-driven approach that allows to analyze and correlate production data coming from different data sources. In particular, we investigate how optically detectable defects with impact on the cell's

performance can be distinguished from defects without impact on I-V-parameters. Typically, the impact of a single optically detected defect on a specific cell is rather small. Also, if a larger cell batch is considered, the number of defects per cell is rather low with the number of defect-free cells being rather large. This is reflected in a Poisson-like distribution for the number of defects. On the other side, the I-V-parameters such as the fill factor typically resemble an asymmetric Gauss-like distribution. For these two reasons, the analysis of correlation matrices is inconclusive due to very small correlation coefficients. A second approach relies on the detection of outliers using the entire I-V-data set. Then, the distribution of the number of defects can be compared for the "I-V-inlier" sub-group and "I-V-outlier" sub-group. Here, we find that this distribution is rather sensitive to an impact of the defects in the I-V-parameters. It can be quantified by fitting the linear slope of the distribution in a semi-logarithmic representation. Finally, the impact of the defects on a large set of cells, i.e. an entire production run, can be assessed by analyzing the I-V-parameter distributions for the defect-free sub-group compared to the sub-group showing some defects.

If a simulation model is implemented to resemble the statistical properties of a production data set, one can quantify the impact of defects or corresponding mitigation measures in different scenarios. For example, the economic impact of exchanging the screen in the screen printing can be quantified by analyzing the changes output of the respective production line.

The investigate data analysis approaches are not restricted to quantitatively linking Poisson-like distributed imaging data with Gaussian-like distributed performance data. Thus, they can be applied to any two or more data sources in production lines which yield quantitative data to identify performance relevant quality control parameters.

References:

[1] J.M. Greulich et. al., Sol Mat 248, p111931 (2022)
[2] M. Turek et. al., Sol Mat 194, p142 (2019)
[3] M. Turek, Jour. Appl. Phys. 115, p144503 (2014)
[4] J.A. Harding et. al., Jour. Manufact. Science and Engineering 128, p969 (2006)
[5] H. Wagner-Mohnson et. al., IEEE Jour. Photovolt. 10, p 1441 (2020)
[6] B. Klöter et. al., IEEE 50th Photovoltaic Specialists Conference (PVSC), 1-3 (2023)
[7] M. Alt et.al., IEEE 7th WCPEC, pp. 3298 (2018)
[8] M. Turek et. al., Sol. Ener. Mat. & Sol. Cells 260, p112483 (2023)
[9] H. Wagner-Mohnsen and P.P. Altermatt, 2021 Intern. Conf. on Numer. Simul. of Optoelectronic Dev. +NUSOD (2021)

41st European Photovoltaic Solar Energy Conference and Exhibition

INTEGRATED INLINE CHARACTERISATION TECHNIQUES
FOR IMPROVED SILICON HETEROJUNCTION SOLAR CELL PRODUCTION

Christian Diestel[1], Saravana Kumar[1], Alexandra Wörnhör[1], Daniel Burkhardt[1], Nico Wöhrle[1], Sebastian Pingel[1],
Matthias Demant[1], Jonas Haunschild[1], Stefan Rein[1]
[1]Fraunhofer Institute for Solar Energy Systems ISE, Freiburg, Germany
christian.diestel [at] ise.fraunhofer.de, +49 761 4588 5052
Heidenhofstr. 2, 79110 Freiburg, Germany

ABSTRACT: Silicon heterojunction (SHJ) technology is gaining market share in photovoltaics due to its lean process sequence, high efficiency potential and low CO_2 footprint. Margins for improvement in performance, yield and cost are decreasing, while at the same time the benefit of even small gains is growing with production volume. It is therefore worthwhile to implement detailed monitoring of process parameters as well as inline characterisation of the cell precursors during production. In this contribution, we present an overview of inline characterisation techniques that are relevant to SHJ cell production and critically discuss their benefits and weaknesses with the help of some showcase examples. We find that inline characterisation is useful at multiple stages of production. Measurements between processing steps allow early detection of defective wafers, preventing unnecessary processing. They also permit evaluation of the performance of individual manufacturing processes. As one prominent example, we demonstrate the generation of highly-resolved thickness maps of the amorphous silicon and transparent conducting oxide layers using reflection spectroscopy combined with multispectral imaging, as well as physical and machine learning models. This method provides information at an unprecedented level of detail about the layer deposition processes that are at the heart of SHJ technology.
Keywords: inline characterisation, silicon heterojunction, SHJ, HJT, silicon solar cells

1 INTRODUCTION

Inline characterisation during the production of silicon heterojunction (SHJ) solar cells offers great advantages for the manufacturer. Testing the finished cells under operating conditions is essential in order to discard defective ones and sort the rest for tiered pricing and optimal module integration. However, characterisation at earlier stages in the production line is beneficial, too. Inspecting the incoming raw wafers allows defective wafers to be rejected, thus saving unnecessary production costs. Cost analysis simulations with SCost [1] predict that if a theoretical portion of 1% of wafers is correctly identified as defective and extracted before metallisation, 20 €ct/kWp are saved. If a critical defect is detected even before deposition of the transparent conductive oxide (TCO), which contains expensive indium, a further 5 €ct/kWp is saved. For an annual production of 1GWp, this equates to savings of 250 k€.

Inline characterisation can further assist in process monitoring and fine-tuned prediction of final cell performance. In this work, we show as examples how an inhomogeneity in the chemical composition of the texturing bath can be detected by spectral reflectometry even before it becomes critical for final cell efficiency. The predictive strength of carrier lifetime measurements after double-sided amorphous silicon deposition, on the other hand, is demonstrated by a good correlation with the final cells' open circuit voltage.

By combining different tools with sophisticated analysis software and machine learning, a detailed picture of the critical attributes of SHJ precursors can be formed. As a prime example, we present an overview over our developed methods for high-resolution thickness mapping of the amorphous silicon (a-Si) and TCO layers based on reflection spectroscopy and multispectral imaging, as well as optical modelling and machine learning. By associating these data with the positions of the wafers in the deposition trays, we obtain a detailed map of the deposition inhomogeneity within the coating machine.

While literature is available on individual characterisation methods, there is little that gives a comprehensive overview of which methods are relevant to the SHJ production line and how they complement each other. Based on our experience in inline inspection along the whole cell manufacturing process and our close collaboration with both in-house SHJ production and industry partners, we evaluated a number of dedicated tools and methods ranging from established standards to novel approaches. Our findings can aid (i) SHJ cell manufacturers in deciding which tools could best improve their production, and (ii) metrology suppliers in identifying wafer quality parameters that are worth measuring and where there is potential for further improvement of inline characterisation technology. The paper aims at bridging the gap between highly technical research papers on the topic and superficial reviews, targeting readers within the PV community but who are not necessarily familiar with either SHJ technology, cell manufacturing or inline characterisation.

2 SIMPLIFIED DESCRIPTION OF THE MANU-FACTURING PROCESS FOR SHJ SOLAR CELLS

Figure 1 shows the structure and processing sequence of a silicon heterojunction (SHJ) cell. The as-cut wafers

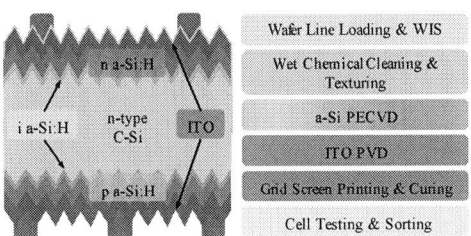

Figure 1: Schematic structure and processing sequence of a SHJ solar cell

57

are textured on both sides in a series of wet-chemical baths. To passivate the surfaces and to form the junction for carrier separation, differently doped amorphous silicon (a-Si) layers are then deposited on both sides of the wafers in a sequence of plasma-enhanced chemical vapor depositions (PECVD). As a final step in frontend processing, indium-doped tin oxide (ITO) is deposited on both sides via physical vapor deposition (PVD) and forms transparent conductive (TCO) layers, which encapsulate the sensitive a-Si layers and facilitate lateral carrier transport to the contact grid.

In backend processing, the front and back side grids are screen-printed. The wafers are then cured to solidify the paste and allow it to bond to the underlying surface. Finally, the full cells may optionally be laser-cut into half cells or even narrower shingle cells.

3 OVERVIEW OF INLINE CHARACTERISATION METHODS

3.1 The finished cell

The most important inline characterisation step is measuring the electrical performance of the finished cells under standard operating conditions (25°C, 1000 W/m², AM1.5G spectrum). Based on the conversion efficiency η or current at the maximum power point I_{MPP}, the cells are binned into quality classes for module integration and sorted out in case of severe defects.

To provide deeper insight into the potential origin of reduced cell efficiencies, spatially- resolved cell inspection techniques have to be added. For instance, electroluminescence (EL) and photoluminescence (PL) imaging not only provide detailed maps of local carrier lifetime; they can also be compared to separate the effects of electrical and optical injection from measured local excess carrier density. The line resistance of the grid fingers can be isolated via grid-resistance neglecting contacting. For bifacial cells, rear illumination and / or using chucks of different reflectivity can emulate the influence of different types of module integration.

3.2 The as-cut wafer

Some unnecessary processing can already be spared, however, by testing the as-cut wafers before they enter the production line to sort out wafers with damage from prior transport and handling. Such incoming wafer inspection may also include measurements associated with the outgoing quality control of the wafer fabrication line. These highly resolved data can help to quantify the influence of the raw material on final cell performance. The simulation in Figure 2 shows the predicted impact of bulk resistivity and lifetime on the final efficiency of SHJ cells. While resistivity can be measured on the as-cut wafer, obtaining accurate bulk lifetime measurements requires passivation of the surfaces. This is already achieved after PECVD of the a-Si layers, though, which is comparatively early in the production sequence. As such, wafers with low lifetime can be sorted out before the deposition of expensive ITO and silver-containing metallisation. This is unique to the SHJ architecture – others, such as PERC or TOPCon, require a late-stage diffusion step that precludes prior knowledge of bulk carrier lifetime. With the impact of the electrical properties shown by the simulation, taking the opportunity to measure them at such early stages provides further cost saving potential compared to other technologies.

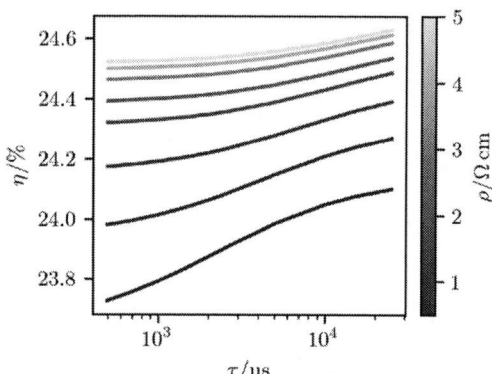

Figure 2: Simulation of efficiency dependence on bulk resistivity and bulk lifetime.

Apart from analysing the bulk material itself, it is equally important to determine the specific mechanical properties of the wafers. High-resolution imaging can be used to identify broken wafers and those whose shape and size are not within the required margins. PL and infrared transmission (IR) imaging can localize microcracks that cause charge carrier recombination, reduced carrier transport and potential breakage of the wafer, even at a later stage during production or handling. These methods also identify other localized defects, although with modern high-lifetime monocrystalline silicon, surface recombination is the dominant cause of contrast in PL images of non-passivated wafers.

Wafer thickness and the extent of saw marks can be measured via laser triangulation or capacitive distance sensors. Wafers that are too thin can result in too low absorption or breakage, while thick wafers may exhibit reduced charge extraction.

3.3 Characterisation after texturing

The quality of the individual processing steps can be assessed by incorporating characterisation in between the steps and ideally evaluating anomalies in terms of their impact on final cell efficiency. The first stage to do this would be after the wet chemical cleaning and texturing steps. Here, care must be taken to reduce the wafers' time spent exposed to the atmosphere, in order to minimize

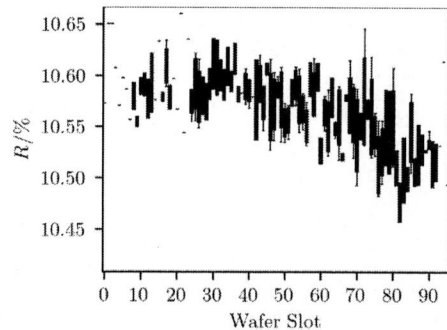

Figure 3: Reflection intensity at 600 nm after texturing. Each box-and-whiskers shows the reflectance distribution of all the wafers in that slot in the carrier. As the carrier lies horizontally in the texturing bath, the slot numbers also correspond to positions across the bath.

oxidation of the bare silicon. Hence, integration of the characterisation equipment into the existing automated production line is critical.

Figure 3 shows a gradient in mean reflection at 600nm against the position of wafers in the wet chemical baths. These wafers were a subset of the same batch of precursors mentioned in Section 3.1. No deliberate variation of texturing conditions was intended. Nevertheless, this gradient indicates slightly inhomogeneous etching of the surface texture across the length of the bath. In fact, OPAL2 simulations show that an increase in reflection intensity can be a direct result of increased planar fraction (Figure 4, blue).

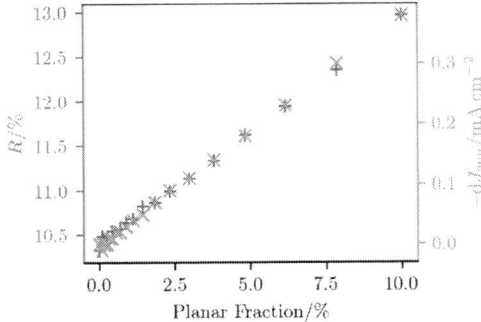

Figure 4: Simulated impact of the planar fraction on the reflection intensity at 600 nm (blue) and the resulting loss in photogeneration current (orange).

Furthermore, this is predicted to have a proportional detrimental effect on the photogeneration current of the finished solar cell (Figure 4, orange). However, comparing the scales on the Y axes of Figure 4 with that of Figure 3, it is clear that no significant loss of performance is to be expected from the small variation in the constitution of the chemical bath observed in this production. This means, on the other hand, that reflection measurements after texturing are very sensitive towards texture quality and can serve as an early warning system against irregularities in the etching process, well before they significantly affect final cell performance.

This measurement has to be done directly after texturing, however, as after a-Si and TCO deposition the reflection intensity is dominated by the properties of the applied layers and no longer shows a clear trend with surface texture.

3.4 Characterisation after layer deposition

As mentioned in the introduction, intermediate characterisation after TCO coating, i.e. before metallisation, has high potential for cost savings. Furthermore, at this stage, many of the physical properties of the finished cell are already fully defined and measurable without the influence of the grid. Meanwhile, the quality of the surface texture as well as the a-Si and TCO layers can still be evaluated.

Due to the robustness of the precursors after TCO coating compared to earlier stages, handling and exposure to the environment can be tolerated at this point, making it an ideal place for extensive characterisation in our study. In fact, during our recent production of 1500 SHJ cells mentioned above, the wafers were transported back and forth between cities for characterisation between ITO deposition and metallisation. They were stacked with separating paper and packed in plastic at atmospheric pressure. Their median performance showed no significant reduction (see Figure 5). Merely the number of outliers

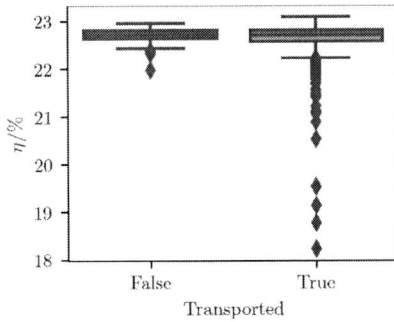

Figure 5: Effect of transport to an off-site frontend wafer inspection system (FWIS) after ITO coating.

increased, which can be attributed to the outer wafers in each package, which experienced increased abrasion.

As surface passivation quality is fully established after TCO, measurements of the effective minority carrier lifetime τ_{eff} are meaningful in terms of cell performance. Figure 6 shows the τ_{eff} values measured by means of the quasi-steady-state photoconductance (QSSPC) technique and plotted against the open-circuit voltage (V_{OC}) for the same batch of SHJ cells as mentioned above. While V_{OC} is recorded on the finished cells, lifetime is measured directly after ITO deposition. As expected, a correlation between

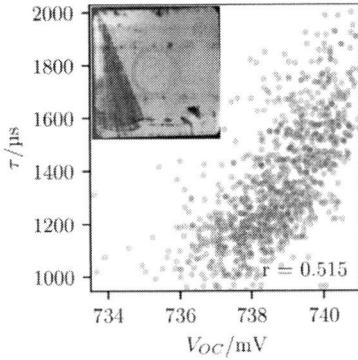

Figure 6: Minority carrier lifetime after PVD vs. open circuit voltage V_{oc} of the finished cell. Inset shows a PL image of a wafer with damage outside the detection area of the QSSPC lifetime tool (orange circle).

τ_{eff} and V_{OC} is visible. However, lateral lifetime inhomogeneities, as seen in the inset PL image, are generally not detected within the central integration area (orange circle) of the QSSPC lifetime tool although they do affect cell performance – this explains the strong scattering observed in Figure 6. To account for these lateral inhomogeneities, PL images have to be recorded and evaluated with machine learning methods or ELBA analysis [2].

4 LAYER THICKNESS MAPS

The layers deposited onto the silicon wafer via PECVD and PVD are the core features of the heterojunction architecture. The 5-20 nm thick amorphous silicon (a-Si) layers form the carrier-selective layers that enable charge separation after photogeneration, which is the basic functionality of a solar cell [3]. They also passivate the bare silicon surface, reducing recombination. The encapsulating 20-70 nm thick layer of transparent conductive oxide, usually indium-doped tin oxide (ITO), provides lateral transport to the grid, while at the same time allowing light to pass through it as well as protecting the delicate a-Si layers underneath.

Achieving good coverage of these materials with optimal and uniform layer thicknesses is essential for the performance of the solar cell. Too thin an a-Si layer creates not enough of a step in the relevant band, leading to poor charge selectivity. Too thick a layer prevents light from reaching the absorption region; in fact, the parasitic absorption in the a-Si layers reduces the short-circuit current density of the solar cell by 0.16 mAcm^{-2} per nanometre thickness [4]. With ITO, a balance has to be struck between optical transmission and lateral conductivity when choosing the right thickness [5]. Furthermore, the rear-side wafer edge is masked during ITO deposition, in order to prevent wrap-around and consequent short-circuiting of the opposite terminals. An optimal width of the exclusion region must be found to effectively prevent shunts while maximizing the active area of the wafer.

To fine-tune the processing parameters that determine the quality of these critical layers, accurate spatially-resolved thickness maps would be ideal. A precise, non-destructive method of determining thin-film thicknesses is (spectral) ellipsometry. The technique requires precise sample positioning and integration times on the order of minutes for clear results, thus only allowing measurements at individual spots, rather than maps, and not at inline-compatible speeds to date.

4.1 Using a physical model to obtain layer thickness from reflection spectra

An alternative approach uses the thickness dependence of reflection spectra of the layers [6]. Figure 7, for example, shows the variation of the reflection spectrum for different a-Si thicknesses. We have developed a method for determining the thickness based on the reflection spectrum [7]. In this method, an optical model of the layer stack is created, and the Fresnel equations are used to set up a series of transfer matrices, describing the optical transmission and reflection at each interface. With this transfer matrix method (TMM), a simulated reflection spectrum can be generated based on the optical constants and the thicknesses of the constituent layers. The optical constants can be determined once for each given material using ellipsometry, and the layer thickness can be used as a variable to fit the simulated spectrum to a measured one. Such reflection measurements can be carried out with an inline spectrometer, which yields fast and well-calibrated data but is restricted to provide data only along traces [8].

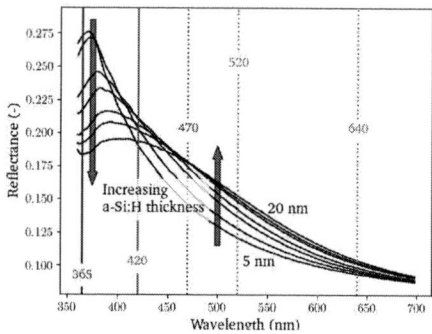

Figure 7: Dependence of the reflection spectrum on a-Si layer thickness. The discrete wavelengths used in multispectral imaging are indicated by the vertical lines.

4.2 Thickness maps from multispectral images

We have further determined that only a few reflection measurements at critical points in the spectrum are sufficient to obtain an accurate fit. For an a-Si layer, a selection of such discrete wavelengths is shown in Figure 7 by the coloured vertical lines. Particularly in the UV (at 365nm) and in the green (520nm), strong variations in reflectance occur with changing layer thickness [6]. For ITO layers, the variation in the red, green and blue channels in the visible spectrum is sufficient.

This enables us to use multispectral imaging to obtain highly resolved reflection data in the wavelengths required for thickness determination. The method works by recording multiple images in quick succession, each illuminated by narrow-band LEDs of different peak wavelengths. The pixel values are calibrated to reflection intensities using a photospectrometer, and all unique combinations of the different wavelength intensities are fitted to thickness values, creating a lookup table. With this table, a detailed thickness map can be computed in minutes.

Figure 8 shows a thickness map of the a-Si layer stack generated by this method [6]. For demonstration purposes, wafer chips were placed onto the sample during the coating process to prevent a-Si deposition in certain patches. They were removed at different times during the

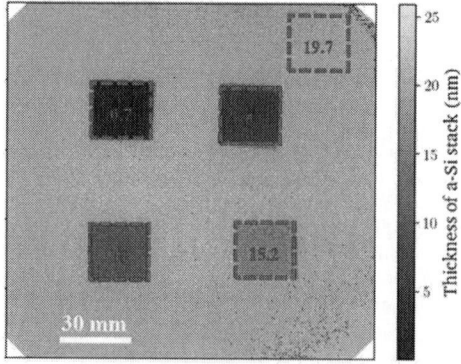

Figure 8: Thickness map of the a-Si stack on a SHJ precursor, determined via inline photospectrometry and multispectral imaging. The dark patches were covered by wafer chips for different durations during PECVD, resulting in different amounts of a-Si deposition in those areas.

process, leading to different coating thicknesses in the four patches. The thickness variation is clearly visible in the image. The top-right patch shows an overlap of two misaligned, partially coated squares. This is due to the covering chip having moved during deposition. The sharpness of the transition between the patches and the surrounding coated area is limited not by the spatial resolution of the measurement but by the vapor creeping in between the sample surface and the covering chips during deposition. Comparisons with ellipsometry data show excellent accuracy of the method [6].

4.3 Fast and robust prediction of thickness maps with machine learning

Taking a few minutes to calculate thickness maps from multispectral images is still not fast enough for real-time inline processing. Furthermore, systematic errors such as those due to inhomogeneous illumination cannot easily be eliminated by some fixed calibration routine, as their magnitude and position with respect to the wafer depend on its positioning within the tool: there is significant coupling between the wafer and the illumination dome, as light bounces between the two multiple times before hitting the camera.

We addressed these issues by developing a machine learning algorithm for fast prediction of layer thicknesses. Our current algorithm has been developed for ITO; a corresponding one for a-Si is in development. The algorithm is based on a physics-informed convolutional neural network, which is trained on synthetic image data with both random and systematic errors added to them. This approach allows a large volume of training data to be used and results in a robust algorithm that is suitable for real-time inline use. It is not only able to create accurate thickness maps of the deposited ITO layer within one second per wafer, but it also accurately and precisely detects the wafer edge and characterises the ITO exclusion region [9].

Such an edge exclusion evaluation is shown in Figure 9, where a detailed map of the ITO thickness gradient is shown for each of the four edges. From this analysis, we can clearly see for this particular example wafer that there are some sections along the top and right edges that are

only partially masked. These could lead to shunts. The fine spatial resolution not only reveals unevenness in the mask edge, which causes local fluctuations of the exclusion width, but also a misalignment of the mask, which is visible in this example by the left and bottom edges having a wider average exclusion region than the top and right edges [9].

We used our machine learning algorithm to generate high-resolution thickness maps of the ITO layer on ~900 wafers of the aforementioned experiment. Associating these maps to the positions of the wafers in the PVD chamber can be used qualitatively to monitor the uniformity of the deposition within the chamber. Figure 10 shows the mean thickness map for each position on the deposition tray. Averaging over ~30 maps per position, random inhomogeneities and effects of other processing steps are cancelled out, and localised deposition density is clearly visible.

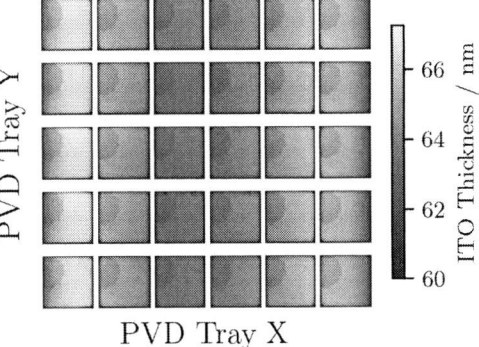

Figure 10: ITO thickness maps vs. position in the deposition tray.

This example shows a clear variation along the X axis. The highest thickness is deposited in two stripes close to the left and right edges of the tray, with minima in the middle of the tray and at the outermost edges. Note that the polygonal shape visible in the top-left corner of each image is a lens artifact of our setup; the problem has since been addressed.

5 OUTLOOK

Modern SHJ cell production has reached levels of reproducibility such that the spread of efficiencies is low, see Figure 11. It is no longer enough to record one statistical datapoint for each wafer with each measurement tool. We have seen for the evaluation of a-Si and TCO layer thicknesses that spatial information is crucial to further improve the characterisation of anomalies. The only significant difference between the left-hand-side and the right-hand-side PL images inset in Figure 11 is the local horizontal dark areas, likely due to damage of an underlying layer. Yet they correspond to wafers on opposite ends of the η distribution, so it is important to identify such defects (and many more subtle ones) automatically.

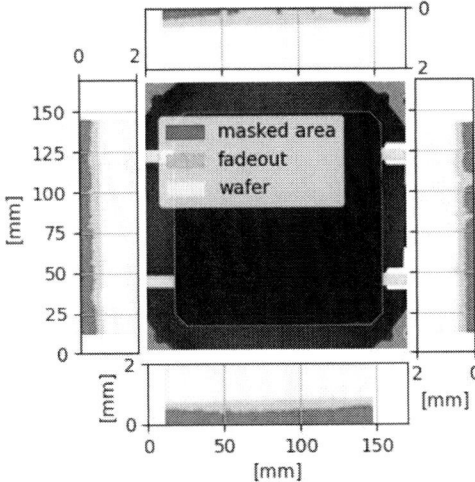

Figure 9: Edge exclusion mapping using multispectral imaging and machine learning.

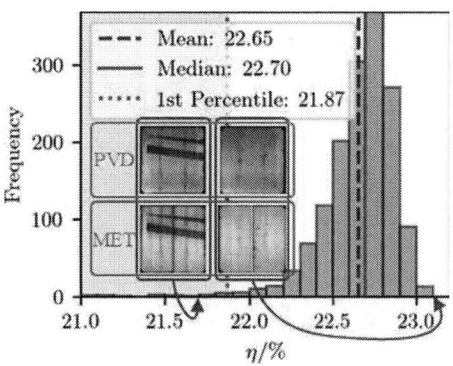

Figure 11: Efficiency distribution for our recent industrial production of SHJ cells. Inset: PL images of a wafer with $\eta = 21.70\,\%$ (left) and one with $\eta = 23.06\,\%$. Top images were recorded after ITO PVD, bottom ones on the finished cells (after metallisation).

Fortunately, we can use machine learning for such tasks, which we have already applied to the layer thickness mapping (Section 4.3), and which we are currently developing further for other characterisation techniques. And such methods have great impact when the defects are visible early in the production sequence. In the example in Figure 11, the top row of images was recorded after PVD of the ITO layer, i.e. before metallisation.

It will require such early measurements of spatialy-resolved precursor parameters combined with sensitive yet robust machine learning algorithms in order to further reduce processing costs by sorting out such wafers, as discussed in the Introduction. The development of such methods is ongoing.

6 CONCLUSION

In conclusion, there are numerous inline characterisation techniques that provide valuable quality assurance and thus enable efficient process development in silicon heterojunction solar cell production. Precision of these techniques is high enough that they are able to resolve the small differences remaining in the quality of cells in modern manufacturing. They can be combined to produce detailed quantitative maps of wafer and cell parameters, and they can help to predict final cell performance early in the production line as well as to monitor and improve individual processes.

7 ACKNOWLEDGEMENTS

This work was funded by the German Federal Ministry for Economic Affairs and Climate Action within the project SALSA under contract number 03EE1096A.

8 REFERENCES

[1] S. Nold, "Techno-ökonomische Bewertung neuer Produktionstechnologien entlang der Photovoltaik-Wertschöpfungskette," 2019.

[2] O. Breitenstein, F. Frühauf, J. Bauer, F. Schindler, and B. Michl, "Local Solar Cell Efficiency Analysis Performed by Injection-dependent PL Imaging (ELBA) and Voltage-dependent Lock-in Thermography (Local I-V)," *Energy Procedia*, vol. 92, pp. 10–15, https://www.sciencedirect.com/science/article/pii/S1876610216304301, 2016.

[3] P. Procel, H. Xu, A. Saez, C. Ruiz-Tobon, L. Mazzarella, Y. Zhao, C. Han, G. Yang, M. Zeman, and O. Isabella, "The role of heterointerfaces and subgap energy states on transport mechanisms in silicon heterojunction solar cells," *Progress in Photovoltaics*, vol. 28, no. 9, pp. 935–945, 2020.

[4] Z. C. Holman, A. Descoeudres, L. Barraud, F. Z. Fernandez, J. P. Seif, S. de Wolf, and C. Ballif, "Current Losses at the Front of Silicon Heterojunction Solar Cells," *IEEE J. Photovoltaics*, vol. 2, no. 1, pp. 7–15, 2012.

[5] Z. C. Holman, M. Filipič, A. Descoeudres, S. de Wolf, F. Smole, M. Topič, and C. Ballif, "Infrared light management in high-efficiency silicon heterojunction and rear-passivated solar cells," *Journal of Applied Physics*, vol. 113, no. 1, 2013.

[6] S. Kumar, C. Diestel, S. Al-Hajjawi, J. Schmitz, M. Hemsendorf, J. Haunschild, S. J. Rupitsch, and Rein S., "Inline Mapping of Amorphous Silicon Layer Thickness of Heterojunction Precursors using Multispectral Imaging," *TIB Open Publishing*, 2024.

[7] S. Kumar, H. Vahlman, S. Al Hajjawi, C. Diestel, J. Haunschild, S. J. Rupitsch, and S. Rein, "Inline Characterization of Ultrathin Amorphous Silicon Stacks in Silicon Heterojunction Solar Cell Precursors With Differential Reflectance Spectroscopy," *IEEE J. Photovoltaics*, vol. 13, no. 5, pp. 711–715, 2023.

[8] S. Al-Hajjawi, H. Vahlman, J. Haunschild, S. Kumar, H. Schremmer, L. Jablonka, and S. Rein, "Spectroscopic Inline Characterisation of Partially Processed Epi Wafers after Porosification," in *EU-PVSEC*.

[9] Alexandra Wörnhör, Saravana Kumar, Daniel Burkhardt, Jonas Schönauer, Sebastian Pingel, Ioan Voicu Vulcanean, Anamaria Steinmetz, Stefan Rein, and Matthias Demant, "Physics-Informed Machine Learning for TCO-Layer Thickness Prediction and Process Analysis from Multi-Spectral Images," in *SiliconPV 2024*.

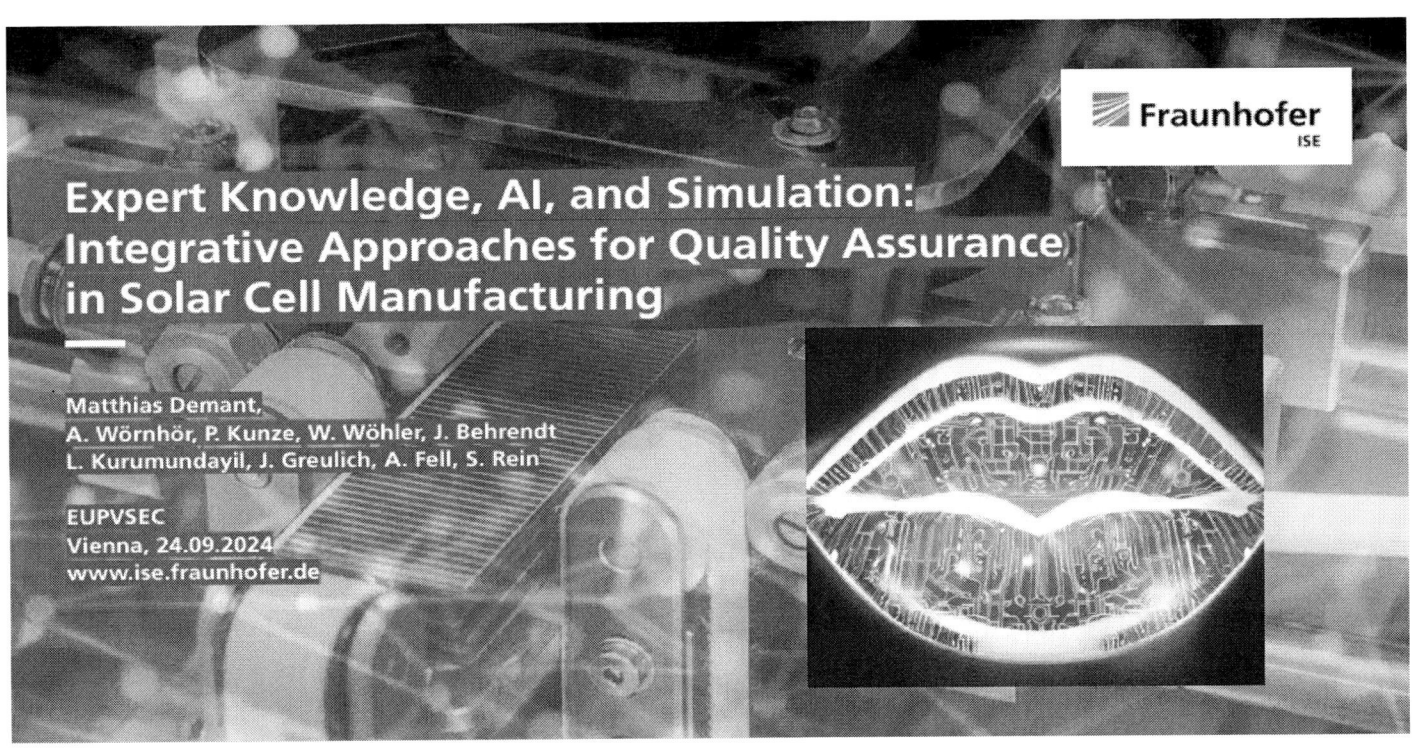

Motivation

„Data avalanche" - Challenges

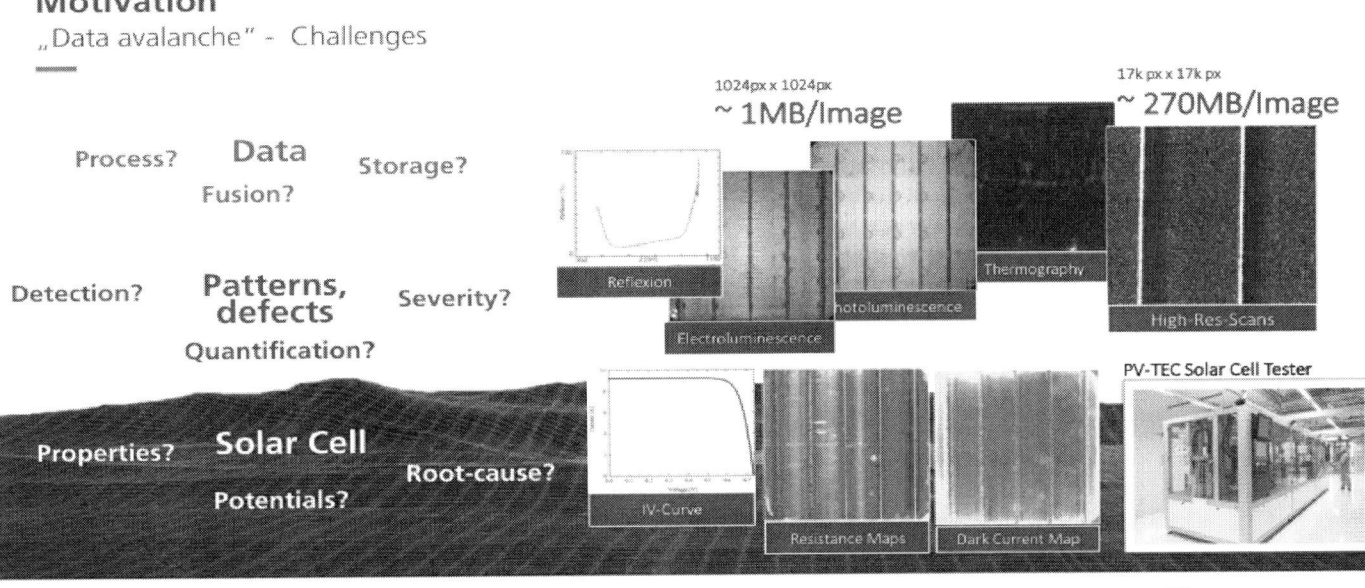

Overview

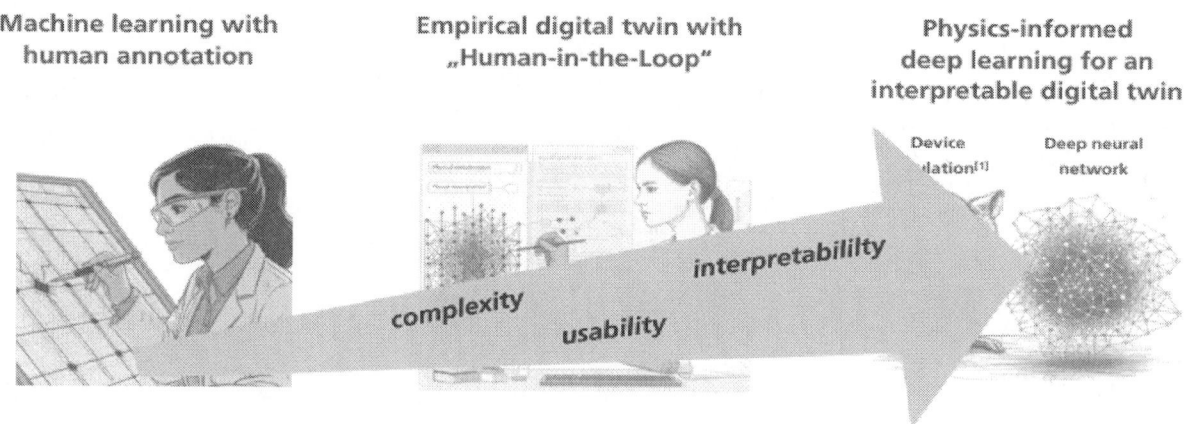

Machine learning with human annotation

Empirical digital twin with „Human-in-the-Loop"

Physics-informed deep learning for an interpretable digital twin

Device simulation[1]

Deep neural network

complexity   usability   interpretability

Images: OpenAI. (2023). *DALL-E 3* [AI-Model]. https://openai.com/dall-e-3

Overview

Machine learning with human annotation

Empirical digital twin with „Human-in-the-Loop"

Physics-informed deep learning for an interpretable digital twin

Device simulation[1]

Deep neural network

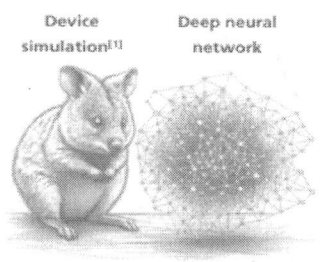

Images: OpenAI. (2023). *DALL-E 3* [AI-Model]. https://openai.com/dall-e-3

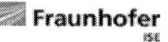

Deep Learning
Elements

Model Input
* Luminescence Image

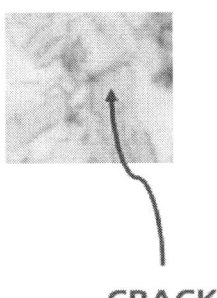

Model Output
Defect segmentation

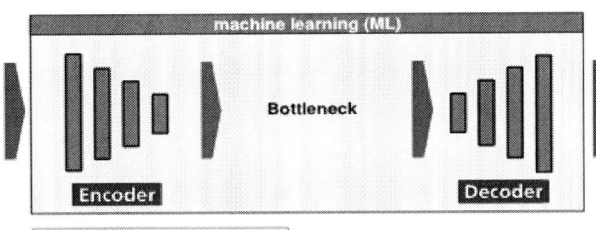

CRACK

Annotation

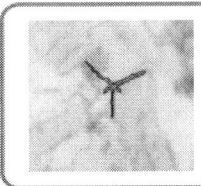 

Operator 1

Learning Based on Human Annotations
„State-of-the-Art"

Model Input
* Luminescence Image

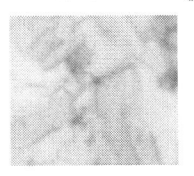

Model Output
Defect segmentation

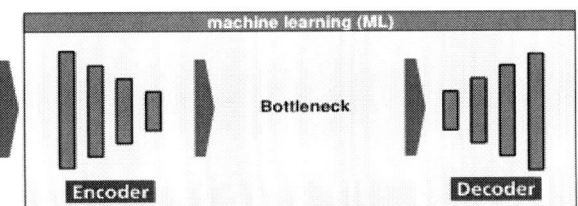

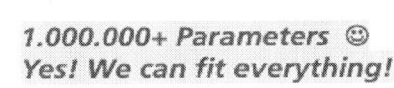

Data scientist

1.000.000+ Parameters ☺
Yes! We can fit everything! [3]

[3]A. Zhang et al., *ICLR.* 2017

Learning Based on Human Annotations

„State-of-the-Art"

Model Input
* Luminescence Image

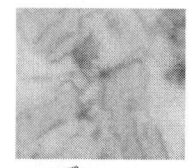

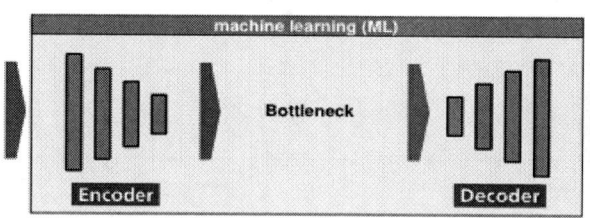

Model Output
Defect segmentation

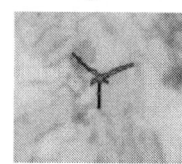

Data scientist
➢ Cumbersome

Production
➢ Operation

Device
➢ Severity

Equipment
➢ Transferability

Operator 2
➢ Reproducability[1]

Operator 1
➢ Complexity

[1] JM Greulich, IEEE JPV, 2020

Overview

Machine learning with human annotation

Empirical digital twin with „Human-in-the-Loop"

Physics-informed deep learning for an interpretable digital twin

Device simulation[1]

Deep neural network

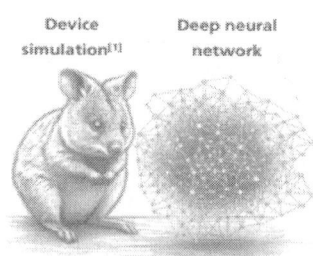

Images: OpenAI. (2023). *DALL-E 3* [AI-Model]. https://openai.com/dall-e-3

Empirical Digital Twin[1,2,3]

Representation learning

subsecond

Model Input
- High-dimensional data
- Inline measurements

machine learning (ML)

Encoder — Sensor fusion & compression

value — „empirical" digital twin

Decoder — Quality prediction

Model Output for Optimization

1. Global IV [1,2]
Efficiency
Fill factor

2. C-DCR [1,3]
j_0 image
R_s image

Electroluminescence
Photoluminescence
Thermography
Reflexion
High-Res-Scans

[1] Patent granted (FHG-ISE, EP3627568B1)
[2] P. Kunze et al., Solar RRL 2021
[3] P. Kunze et al., IEEE PVSC, 2024

© Fraunhofer ISE

Empirical Digital Twin[1,2,3]

Representation learning

subsecond   ✓ Fast processing   ✓ Quick update

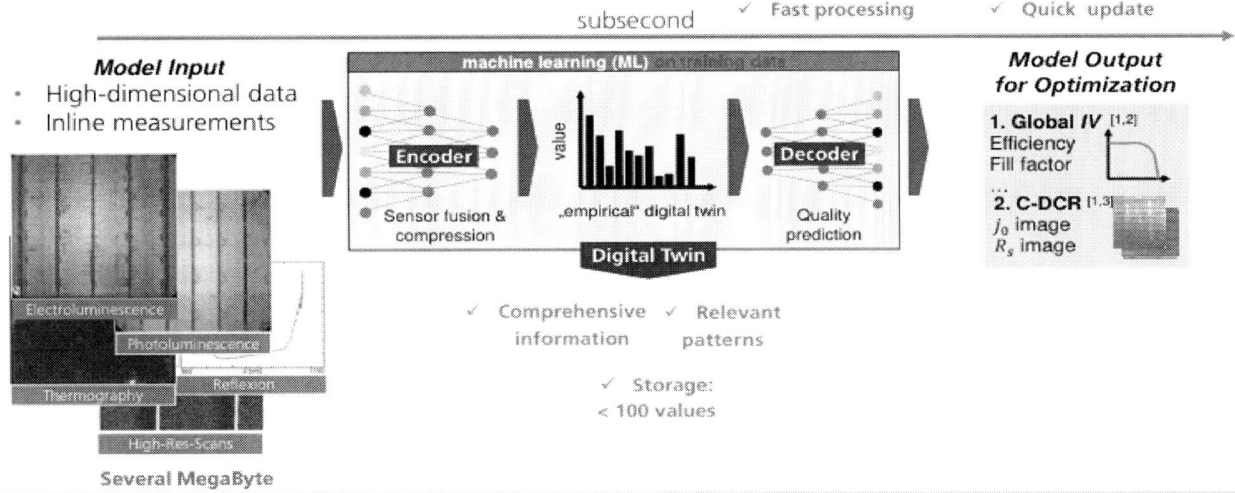

Model Input
- High-dimensional data
- Inline measurements

machine learning (ML)

Encoder — Sensor fusion & compression

value — „empirical" digital twin

Decoder — Quality prediction

Digital Twin

✓ Comprehensive information   ✓ Relevant patterns

✓ Storage: < 100 values

Electroluminescence
Photoluminescence
Thermography
Reflexion
High-Res-Scans

Several MegaByte

Model Output for Optimization

1. Global IV [1,2]
Efficiency
Fill factor

2. C-DCR [1,3]
j_0 image
R_s image

[1] Patent granted (FHG-ISE, EP3627568B1)
[2] P. Kunze et al., Solar RRL 2021
[3] P. Kunze et al., IEEE PVSC, 2024

© Fraunhofer ISE

Empirical Digital Twin[1,2,3]

Representation learning

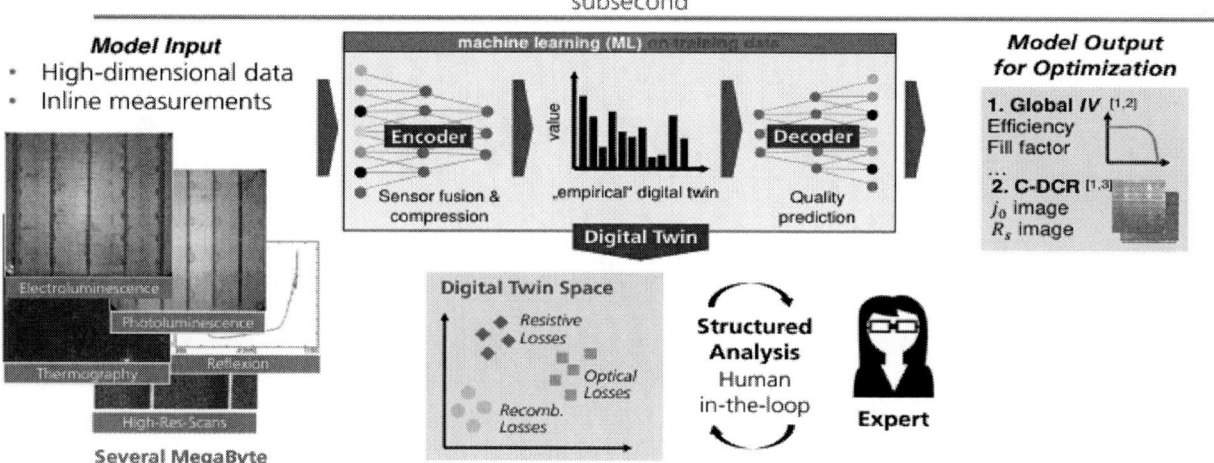

subsecond

Model Input
- High-dimensional data
- Inline measurements

Electroluminescence
Photoluminescence
Thermography
Reflexion
High-Res-Scans

Several MegaByte

machine learning (ML) on training data

Encoder — Sensor fusion & compression
value — "empirical" digital twin
Decoder — Quality prediction

Digital Twin

Digital Twin Space
- Resistive Losses
- Optical Losses
- Recomb. Losses

Structured Analysis
Human in-the-loop

Expert

Model Output for Optimization
1. **Global IV** [1,2]
Efficiency
Fill factor
...
2. **C-DCR** [1,3]
j_0 image
R_s image

©Fraunhofer ISE
[1] Patent granted (FHG-ISE, EP3627568B1)
[2] P. Kunze et al., Solar RRL 2021
[3] P. Kunze et al., IEEE PVSC, 2024

Empirical Digital Twin

Physics-guided structured analysis

Visualization in 2D
with embedding technique

Compute Digital Twins

value — Sample 1
value — Sample 2
value — Sample 3

Structured Analysis
Human in-the-loop

Expert

$Embedding_2$ (a.u.)

$Embedding_1$ (a.u.)

η (%)

©Fraunhofer ISE
[1] Patent granted (FHG-ISE, EP3627568B1)
[2] P. Kunze et al., Solar RRL 2021
[3] P. Kunze et al., IEEE PVSC, 2024

Empirical Digital Twin

Physics-guided structured analysis

Results

* Cluster of High-Quality Cells identified

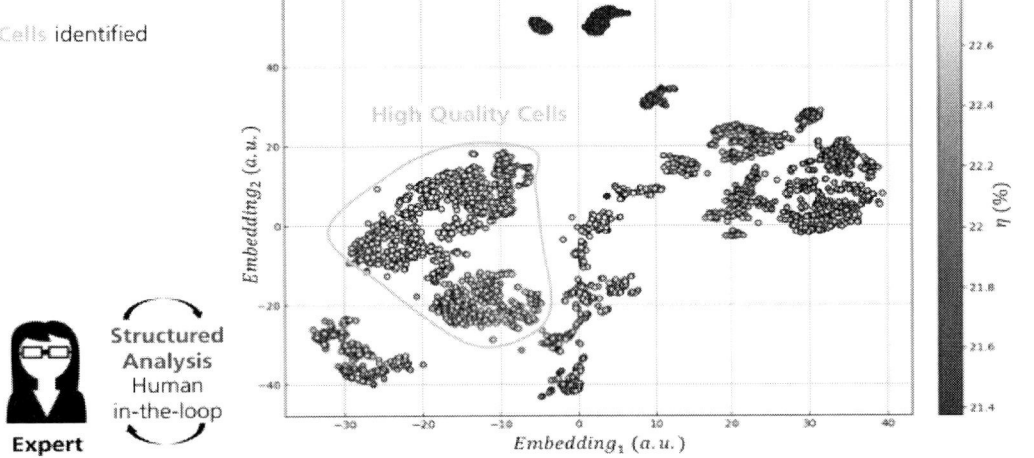

Expert — Structured Analysis Human in-the-loop

Empirical Digital Twin

Physics-guided structured analysis

Results

* Cluster of High-Quality Cells identified
* **Resistive losses:** High R_s, sub-groups via
  * Sheet resistance high
  * Grid resistance high
  * Finger geometry problem
  * Others

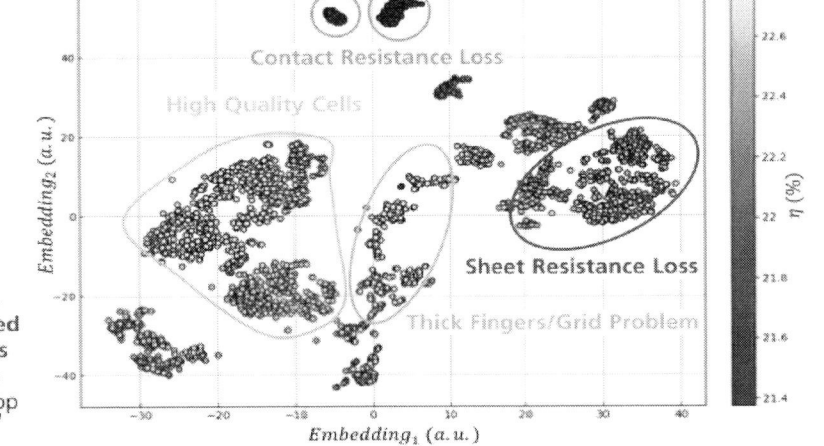

Expert — Structured Analysis Human in-the-loop

Empirical Digital Twin

Physics-guided structured analysis

Results

- Cluster of High-Quality Cells identified
- **Resistive losses:** High R_s, sub-groups via
  - Sheet resistance high
  - Grid resistance high
  - Finger geometry problem
  - Others
- **Optical losses:** J_{sc} low, V_{oc} ok, sub-groups via
  - $1 - EQE - R$ high ⇨ Parasitic absorption
  - $R * AM1.5g$ high ⇨ Reflectance

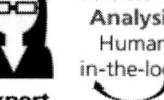

Structured Analysis Human in-the-loop

Expert

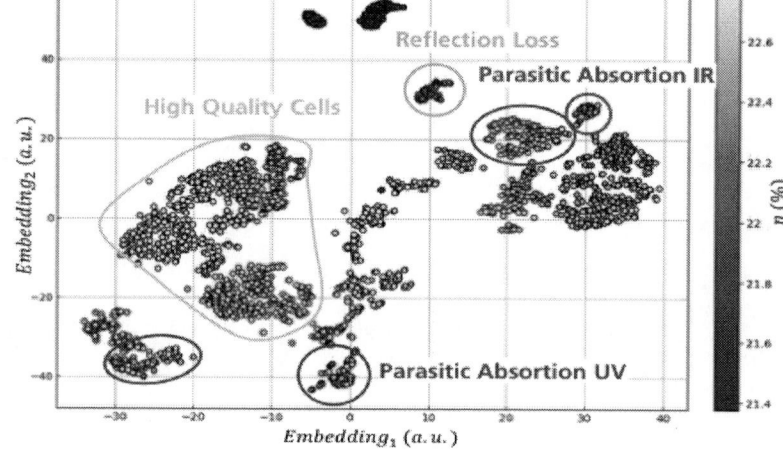

Empirical Digital Twin

Physics-guided structured analysis

Results

- Cluster of High-Quality Cells identified
- **Resistive losses:** High R_s, sub-groups via
  - Sheet resistance high
  - Grid resistance high
  - Finger geometry problem
  - Others
- **Optical losses:** J_{sc} low, V_{oc} ok, sub-groups via
  - $1 - EQE - R$ high ⇨ Parasitic absorption
  - $R * AM1.5g$ high ⇨ Reflectance
- **Recombination losses:**
  j_0 high, V_{oc} low

Structured Analysis Human in-the-loop

Expert

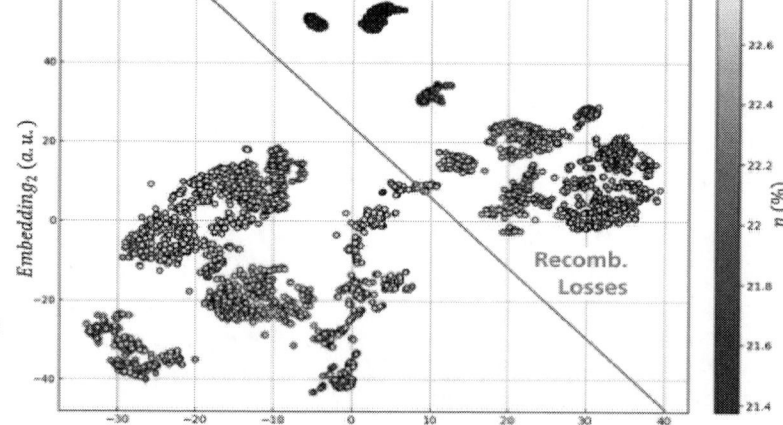

Empirical Digital Twin

Physics-guided structured analysis

Model benefit

- **Identify clusters with same properties**
- **Classify new data inline** (X=?)

✓ Efficient clustering

✓ Quality sorting

✓ Expert guided loss analysis

„Loss-analysis" requires
screening of reverence data!

Physical digital twin possible ?

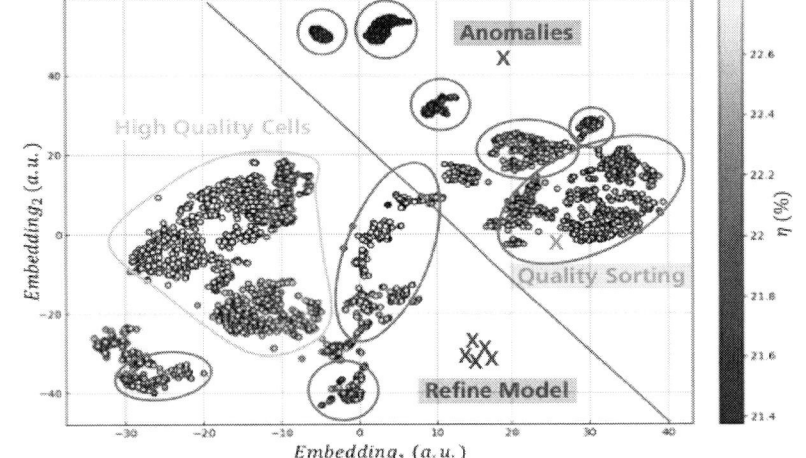

Overview

Machine learning with
human annotation

Empirical digital twin with
„Human-in-the-Loop"

Physics-informed
deep learning for an
interpretable digital twin

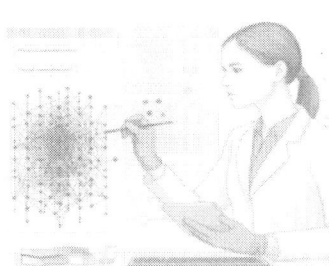

Device simulation[1]   **Deep neural network**

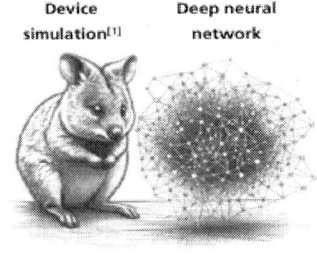

[1]Quokka3, Solar Cell Simulation Software, https://www.quokka3.com
Images: OpenAI. (2023). *DALL-E 3* [AI-Model]. https://openai.com/dall-e-3

Physics-informed Deep Learning

R&D Physicist

—

days!

Model Input
- Offline measurements
- few cells, plus separately processed test structures

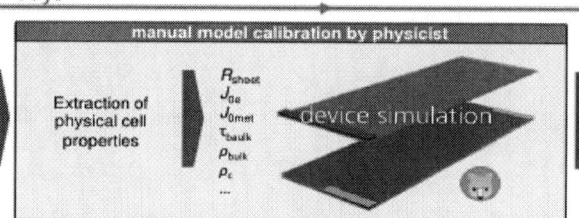

Model Output
- Cell performance (= IV curve)
- Many more measurable quantities (e.g. PL, EQE)

Application
- Detailed loss analysis and performance potential analysis
- Design optimization (e.g. #fingers)

Physics-informed Deep Learning

KISS: join forces for high speed AND high interpretability

—

subsecond

Model Input
- High-dimensional data
- Inline measurements

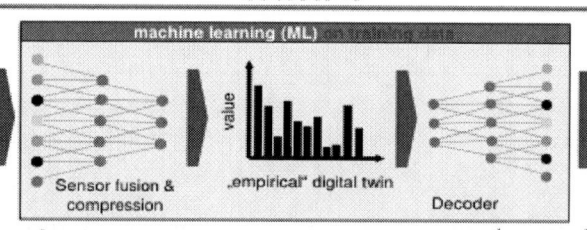

Model Output
- cell performance (= IV / two-diode parameters)
- More quality parameters (e.g. j_0, R_s images)

days!   subsecond

- **Offline** measurements
- few cells, plus separately processed test structures

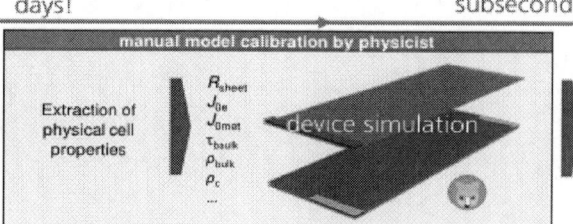

Model Output
- Cell performance (= IV curve)
- Many more measurable quantities (e.g. PL, EQE)

Physics-informed Deep Learning

KISS: join forces for high speed AND high interpretability

subsecond!

Model Input
- High-dimensional data
- Inline measurements

Electroluminescence
Photoluminescence
Thermography
Reflexion
IV-Curve
High-Res-Scans

ML + physical model (=theory guided data analysis)

R_{sheet}
J_{0b}
J_{0met}
τ_{bulk}
ρ_{bulk}
ρ_c
...

Sensor fusion & compression

Model Output for Optimization
- cell performance
  (= IV / two-diode parameters)
- More parameters (e.g. EQE)

Model Application
- Loss and performance-potential analysis for design optimization

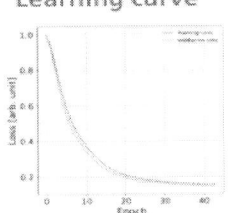

R_{sheet}

τ_{bulk}

Model Output
Physical DT
- reveals hidden solar cell properties
- to identify common issues

Experimental

Optimization and generalizability

Optimization

☑ Collect experimental data of PERC cells
☑ Setup metamodel (fast & derivable)
☑ Train hybrid model on experimental data
☑ Check convergence

✓ Smooth model which works for test data = not overfitted

Learning curve

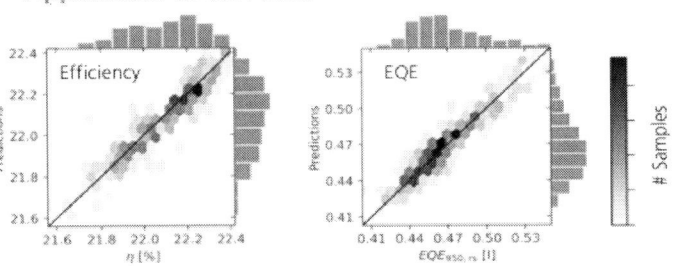

Application to test data

Experimental
Plausability

Optimization

☑ Collect experimental data of PERC cells
☑ Setup metamodel (fast & derivable)
☑ Train hybrid model on experimental data
☑ Check convergence

Physical digital twin

☑ Check, if latent parameters in value range

Physical digital twin

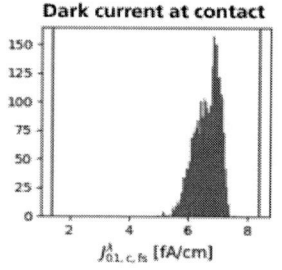
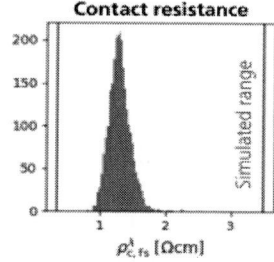
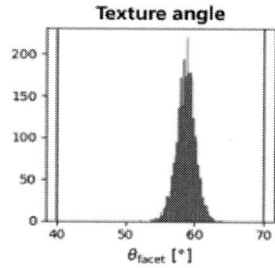
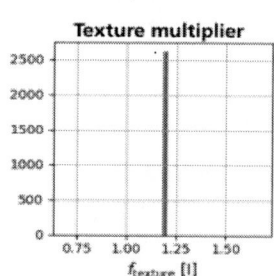

Experimental
Validity

Optimization

☑ Collect experimental data of PERC cells
☑ Setup metamodel (fast & derivable)
☑ Train hybrid model on experimental data
☑ Check convergence

Physical digital twin

☑ Check, if latent parameters in value range
☑ Compare to reference data => rough estimates
➔ Avoid ambiguities …
 ♪ by reducing the number of latent parameter
 ♪ by adding known constraints as loss functions
 ♪ by adding optimization parameters
➔ Identify losses and potentials

Emitter sheet resistance

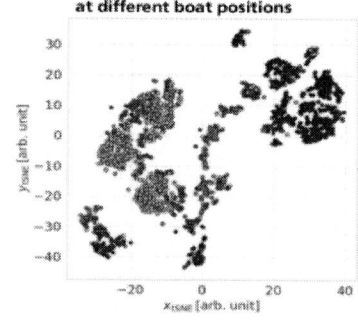
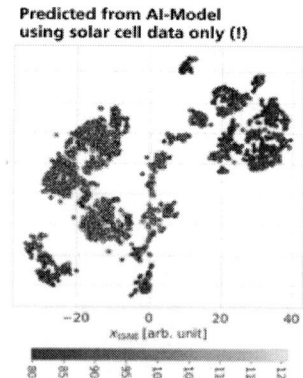

Conclusion

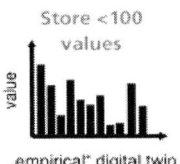
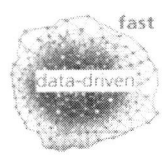

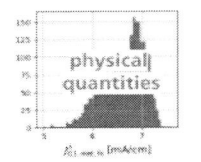
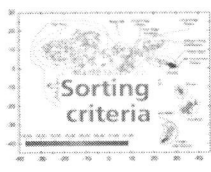

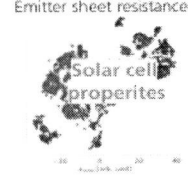
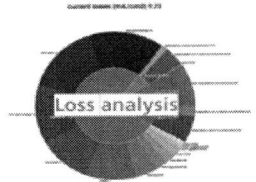

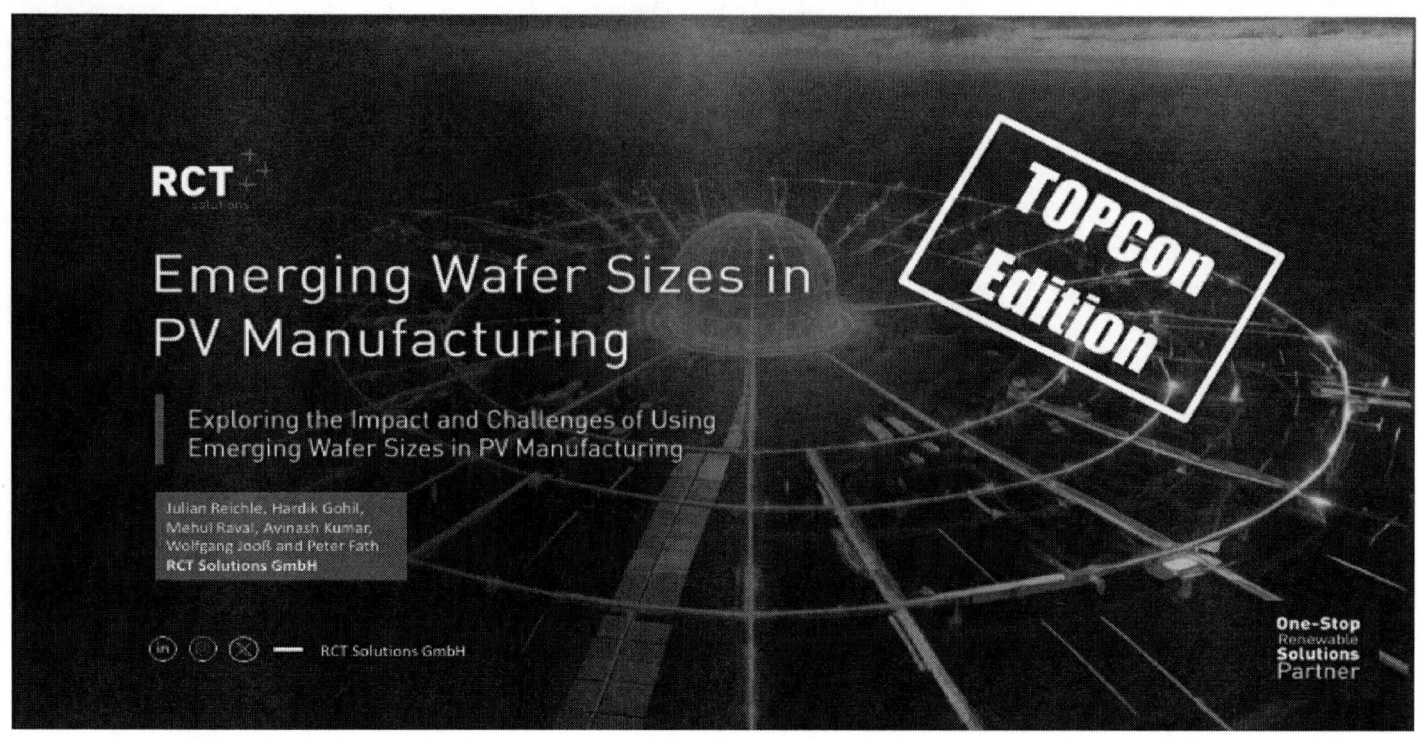

Latest history of wafer sizes used in industry

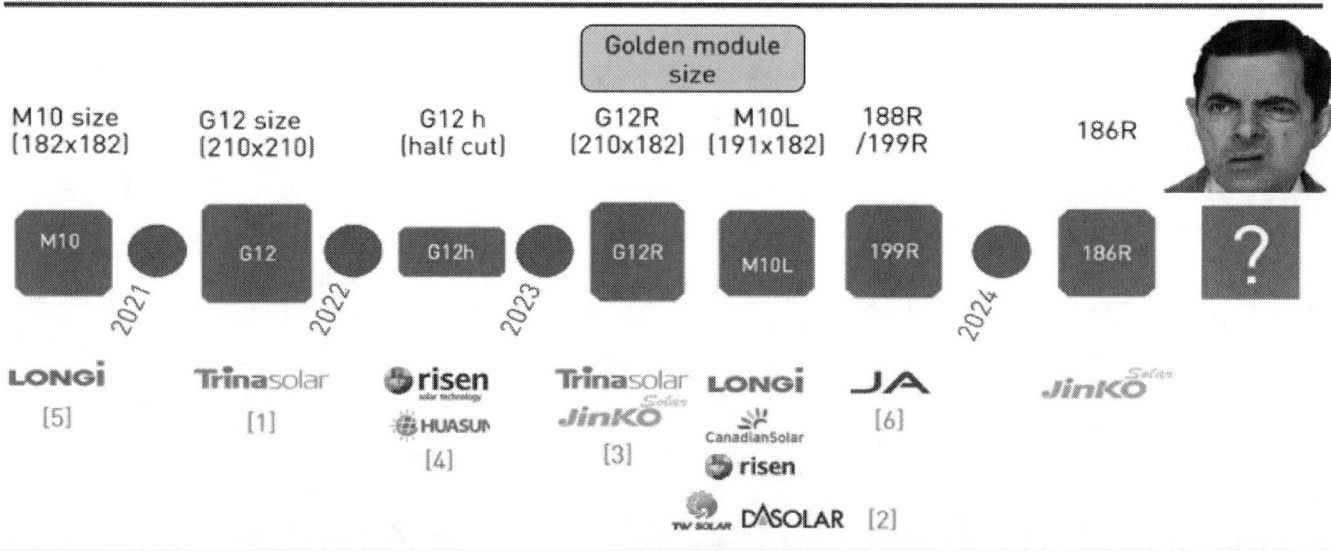

Content – wafer sizes for TOPCon modules

1. Market overview for n-type modules
   a) Present data on the **current market share of various wafer sizes** and types in 2023.
   b) Introduce the **top 5 manufacturers** and their **adoption of wafer/cell sizes**, with a focus on **n-type** modules.
2. Challenges in **manufacturing** for **different wafer sizes** from ingot to cell:
3. Factors affecting **module design**
4. Customer value analysis
   1. Metholodgy
   2. Results
5. **Conclusion:** Impact of wafer sizes on TOPCon modules

What is the status in wafer type & size in 2023?

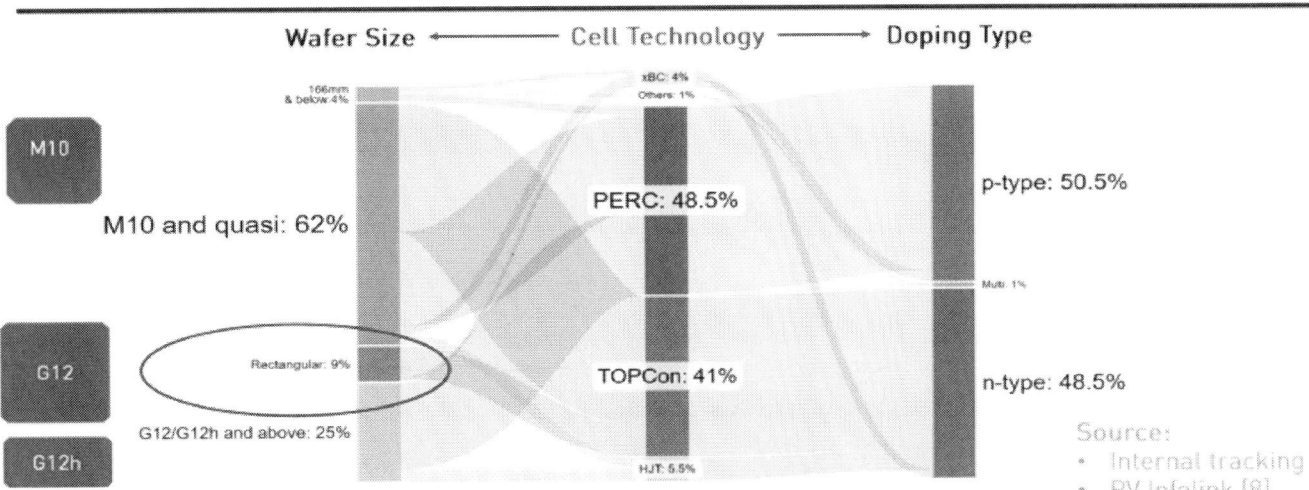

Source:
- Internal tracking
- PV Infolink [8]
- CPIA [9]

→ Biggest streams are M10-PERC, G12-PERC and M10 TOPCon wafers

Wafer sizes share in 2023

Wafer Size	M10	M11-L	199-R	G12-R	G12h	G12 (210)
Dimensions	182 x 182*	182.2 x 191.6	182 x 199	182 x 210	210 x 105	210 x 210
Introduced	2020: LONGi [5]	2023 : LONGi, Risen, Canadian Solar, Tongwei, DAS, Astronergy [2]	2023: JA Solar [6]	2023: Trina, Jinko [3]	2022: Huasun / Risen [4]	2021: Trina [1]
Share 2023: ITRPV [7] Infolink [8] CPIA [9] RCT	64% 68% 68% 63%		2% Missing 10% 9%		Missing Missing Missing 5%	20% 27% 20% 20%
Used for Technology:	TOPCon / PERC / xBC	TOPCon / xBC	TOPCon / xBC	TOPCon / xBC	HJT	PERC / HJT

- M10 wafer size dominated the market with a share of ca. 63%–68% in 2023.
- G12 wafer size held a market share of about 20%-29% in the same year.
- Large G12 wafers are primarily used for PERC and HJT technologies.
- Half-cut wafers have become increasingly popular for HJT solar cells (5%).

Tracking TOP 5 module manufacturers wafer/cells for n-type module variants (August 2024)

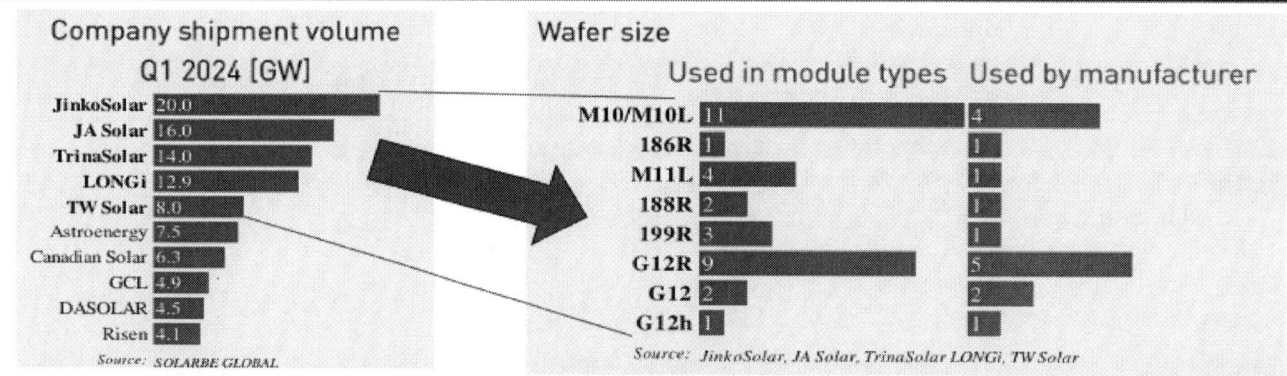

Source: SOLARBE GLOBAL.

Source: JinkoSolar, JA Solar, TrinaSolar LONGi, TW Solar

- The top five manufacturers using **8 distinct cell sizes**
- **M10** and **G12R** are the most used formats, appearing 11 and 9 times in products
- **G12R is adopted by all** the TOP 5 manufacturer
- **2-4 different wafer sizes** per manufacturer

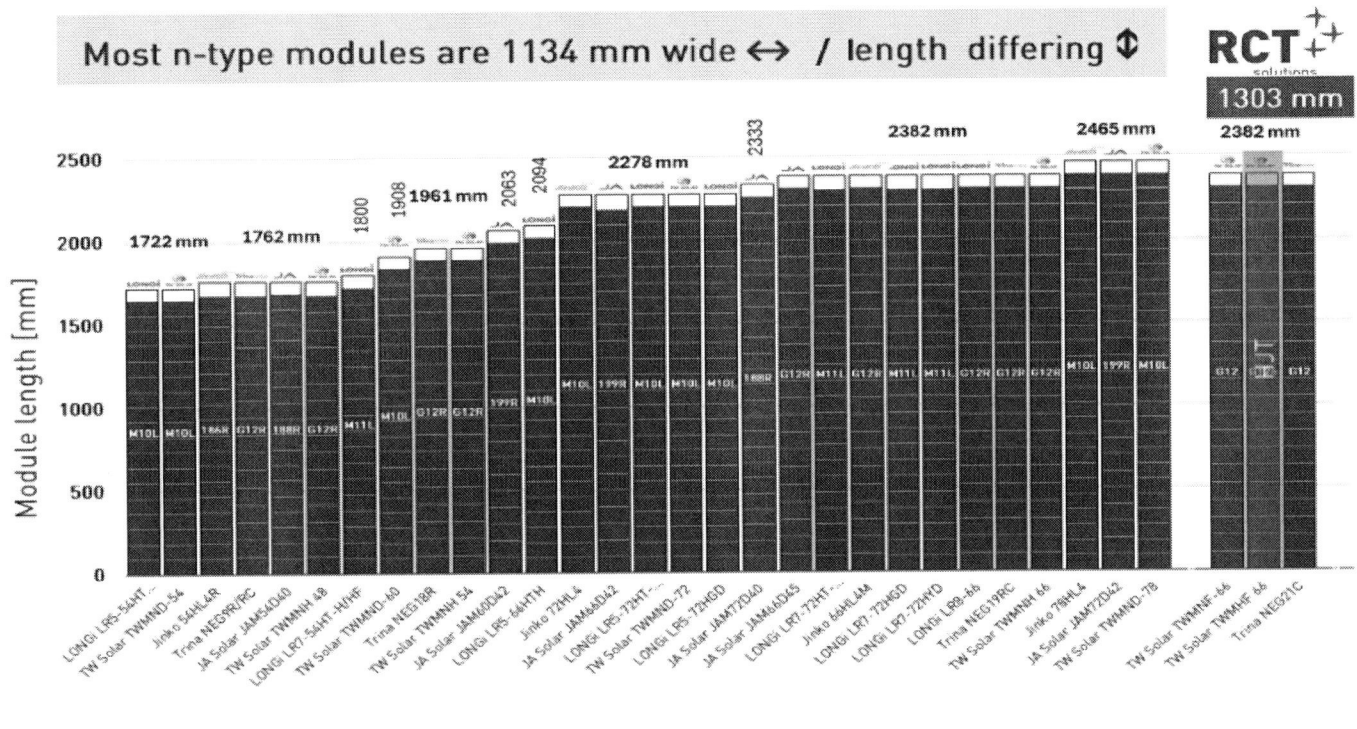

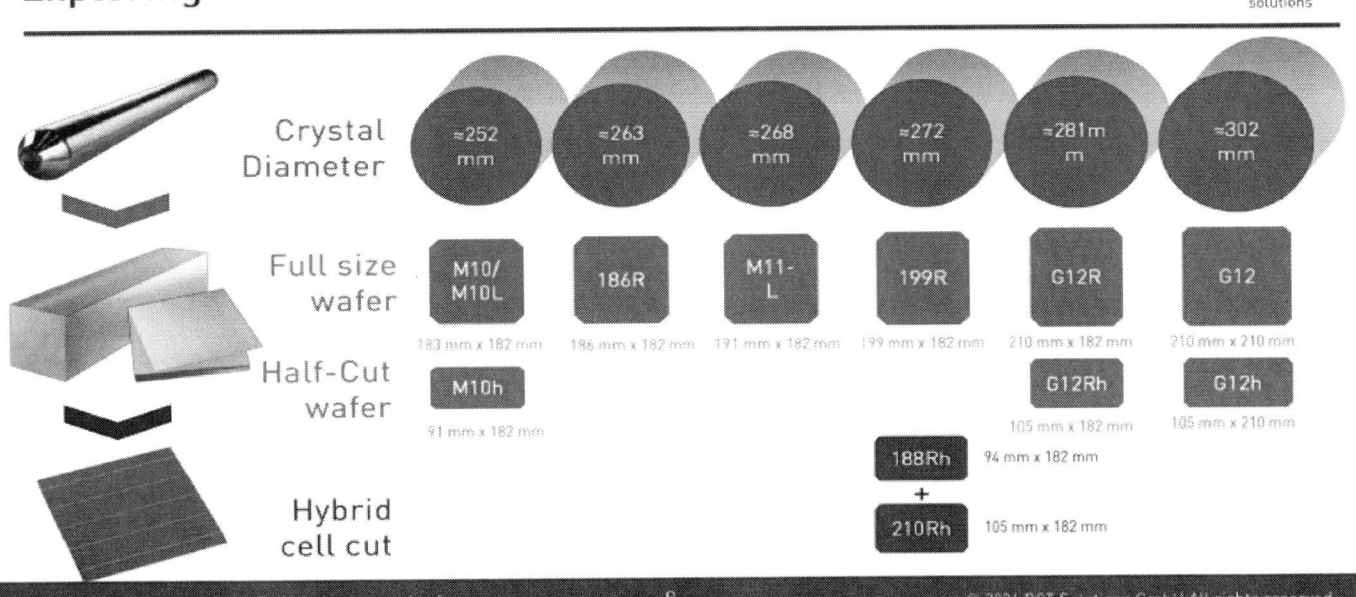

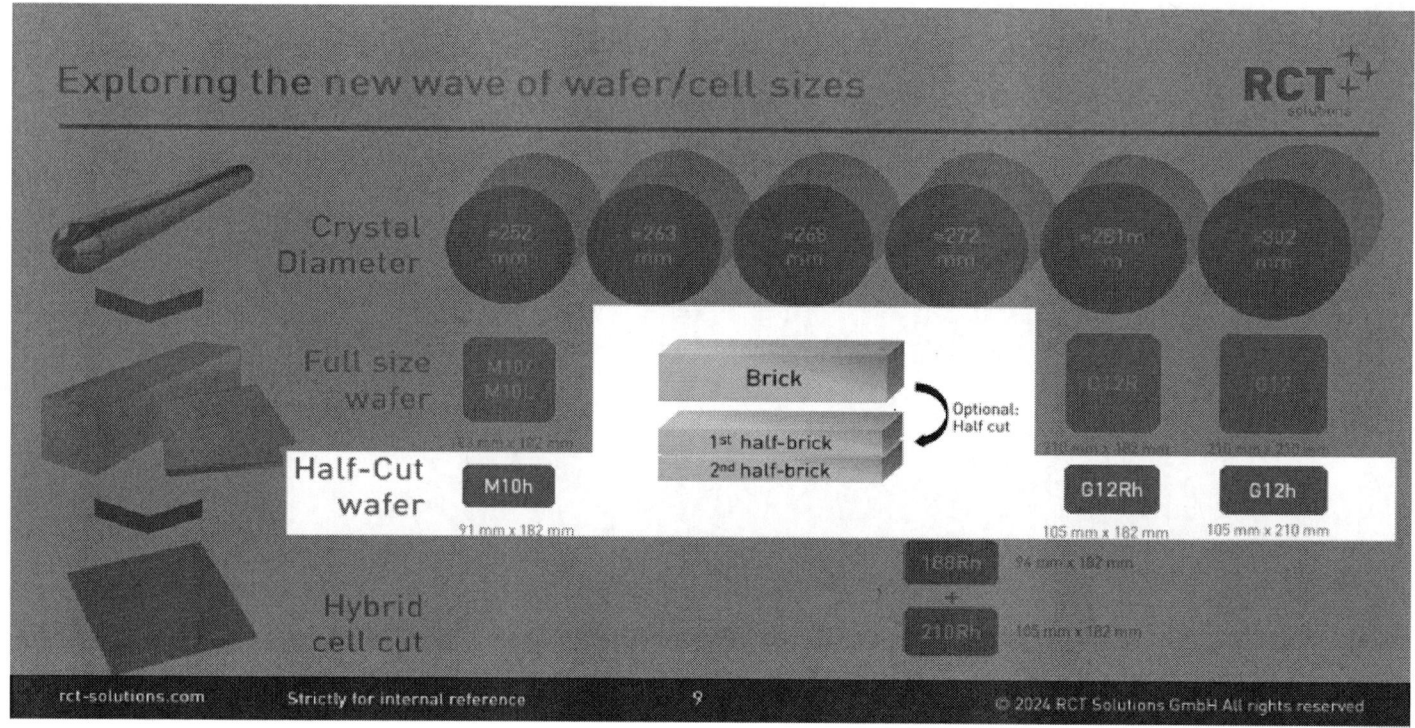

Solar wafer specifications – considered dimensions

- Most common wafer size M10 (182 x 182 mm) with 63% in 2023
- Wafer size and thickness are selected based on the cell and module technology selected
- Exact sizes can differ according to each manufacturer's preference

Wafer size	M10	M11-L	199-R	G12-R	G12h	G12 (210)	Tolerance
Introduced	2019/2020	2023			2022	2021	
Share 2023:	63%	9%			5%	20%	
Used for Cell Technology:	All	TOPCon/xBC	TOPCon/xBC	TOPCon/xBC	HJT	PERC/HJT	
Width (A1) [mm]	182.2	182.2	182.2	182.2	105.0	210.1	± 0.25 m
Height (A2) [mm]	182.2	191.6	199.0	210.0	210.1	210.1	± 0.25 m
Diagonal (B) [mm]	247.0	262.5	267	272.0	233.0	295.0	± 0.35 m
Area [mm^2]	≈ 330.8	≈ 349.0	≈ 362.5	≈ 382.2	≈ 220.4	≈ 440.9	N/A
Thickness (T)	110 - 150	120 - 140	120 - 140	110 - 130	100 - 130	130 - 150	± 8 - 10

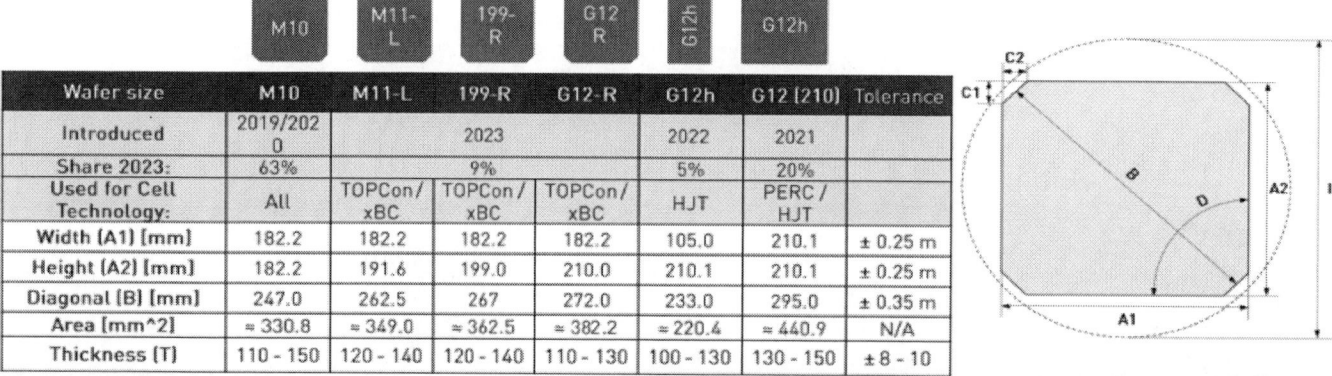

Challenges in the ingot and wafer factory

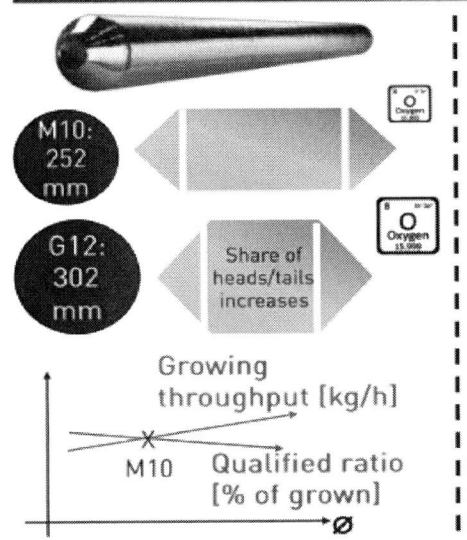

M10:
252
mm

G12:
302
mm

Share of
heads/tails
increases

Growing
throughput [kg/h]

× M10
Qualified ratio
[% of grown]

∅

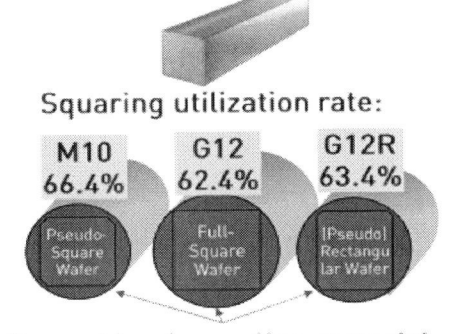

Squaring utilization rate:

M10 66.4%	G12 62.4%	G12R 63.4%
Pseudo-Square Wafer	Full-Square Wafer	(Pseudo) Rectangular Wafer

Downside of recycling material
1. Energy and material were wasted for growing this part of the ingot
2. During recycling for reuse, losses occur

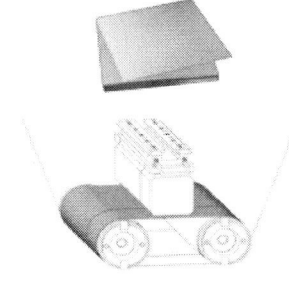

Slicing:
- Set-up time share to cutting time reduced if brick side length increases
- Feed rate increases, with shorter

Challenges in cell manufacturing

Thermal tube processes, diffusion
- Bigger wafers needs to be spaced wider in boats
- Quartz tubes optimized for rather M10 or G12 diameter
- Impact on space, electricity and gas consumption

Wet chemical batch process:
- Due toe capillary effects, bigger wafers needs to be spaced wider in cassettes

Single line inline processes:
+ Firing and printing line benefits from bigger wafers

Factors to consider to define a module

Module size/power:

Width of module usually fixed

- 1134/1303 mm due to inline manufacturing of glass

Height of the module:

- Resolves in **weight, mounting differences** and **shipping differences**
- Together with the package density and cell technology → module voltage and power is determined

→ **Strategical portfolio decision required**

Somehow the wafer sizes have adapted ←

Supply chain availability:

- Do I get the right wafer / cell sizes?
- Do I get the right module glasses?
- Do I get the right sized equipment?
- Is there a competing market for all these Items?

| Module efficiency [%] | \multicolumn{10}{c|}{Module heights used by TOP 5 Manufacturers} | |
	1722 ×1134	1762 ×1134	1800 ×1134	1961 ×1134	2063 ×1134	2094 ×1134	2278 ×1134	2333 ×1134	2382 ×1134	2465 ×1134	
182	22.9%	22.0%	21.5%	21.9%	20.8%	22.6%	22.9%	22.1%	21.7%	22.7%	M10
183	22.6%	22.1%	21.6%	22.0%	21.0%	22.7%	22.8%	22.2%	21.8%		
184	20.2%	22.2%	21.7%	22.2%	21.1%	20.8%	21.0%	22.4%	21.9%	21.2%	M10L
185	20.3%	22.3%	21.8%	22.3%	21.2%	20.9%	21.1%	22.5%	22.0%	21.3%	
186	20.4%	22.4%	22.0%	22.4%	21.3%	21.0%	21.2%	22.6%	22.1%	21.4%	
187	20.5%	22.6%	22.1%	22.5%	21.4%	21.1%	21.3%	22.7%	22.3%	21.5%	
188	20.6%	22.7%	22.2%	22.6%	21.5%	21.2%	21.4%	22.8%	22.4%	21.6%	188R
189	20.7%	20.9%	22.3%	22.8%	21.6%	21.3%	21.6%	22.0%	22.5%	21.7%	
190	20.8%	20.4%	22.4%	20.6%	21.8%	21.4%	21.7%	21.2%	22.6%	21.8%	
191	21.0%	20.5%	22.6%	20.7%	21.9%	21.5%	21.8%	21.3%	22.7%	22.0%	M11L
192	21.1%	20.6%	22.7%	20.6%	22.0%	21.7%	21.9%	21.4%	22.8%	22.1%	
193	21.2%	20.8%	21.9%	20.9%	22.1%	21.8%	22.0%	21.5%	21.1%	22.2%	
194	21.3%	20.8%	20.4%	21.0%	22.2%	21.9%	22.1%	21.6%	21.2%	22.3%	
195	21.4%	20.9%	20.5%	21.1%	22.3%	22.0%	22.2%	21.7%	21.3%	22.4%	
196	21.5%	21.0%	20.6%	21.2%	22.4%	22.1%	22.4%	21.8%	21.4%	22.5%	
197	21.6%	21.1%	20.7%	21.4%	22.6%	22.2%	22.5%	21.9%	21.5%	22.7%	
198	21.7%	21.2%	20.8%	21.5%	22.7%	22.3%	22.6%	22.0%	21.6%	22.8%	
199	21.8%	21.3%	20.9%	21.6%	22.8%	22.4%	22.7%	22.2%	21.7%	22.9%	199R
200	21.9%	21.4%	21.0%	21.7%	20.6%	22.6%	22.8%	22.3%	21.8%	21.1%	
201	22.1%	21.6%	21.1%	21.8%	20.7%	20.9%	20.9%	22.4%	21.9%	21.2%	
202	22.2%	21.7%	21.2%	21.9%	20.8%	22.8%	20.9%	22.5%	22.0%	21.3%	
203	22.3%	21.8%	21.3%	22.0%	20.9%	20.6%	21.0%	22.6%	22.1%	21.4%	
204	22.4%	21.9%	21.4%	22.1%	21.0%	20.7%	21.2%	22.7%	22.3%	21.5%	
205	22.5%	21.9%	21.5%	22.2%	21.1%	20.8%	21.3%	22.8%	22.4%	21.6%	
206	22.6%	22.1%	21.6%	22.3%	21.2%	20.9%	21.4%	20.9%	22.5%	21.7%	
207	19.9%	22.2%	21.7%	22.4%	21.3%	21.0%	21.5%	21.0%	22.6%	21.8%	
208	20.0%	22.4%	21.8%	22.5%	21.4%	21.1%	21.6%	21.1%	22.7%	21.9%	
209	20.1%	22.4%	21.9%	22.7%	21.5%	21.2%	21.7%	21.2%	22.8%	22.0%	
210	20.2%	22.5%	22.0%	22.8%	21.6%	21.3%	21.8%	21.3%	22.9%	22.1%	G12R

Factors after shipping and other BOS cost

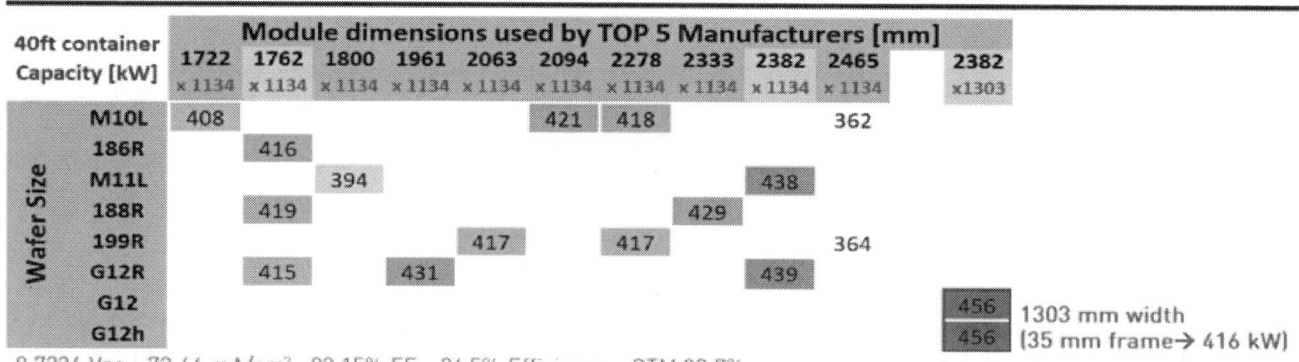

0.7236 Voc – 72.44 mA/cm² – 83.15% FF – 24.5% Efficiency – CTM 98.7%
All modules considered with 30 mm frame

- The right module voltage is key, to optimize the usage of the Inverter
- Higher module efficiency results in more effective use of area
- Longer modules with longer strings show higher voltage, resulting in less ohmic losses

Assessing customer value variations due to wafer size differences.

The **Customer Lifetime Value** (CLV) - estimate the total net profit a business can expect from a customer over the duration of their relationship:

$$CLV = \sum \left(\frac{Revenue\ per\ period - Cost\ per\ period}{(1 + Discount\ rate)^{period}} \right)$$

Simplified approach used:

$$Customer\ Value\ =\ Total\ Customer\ Benefits\ -\ Total\ Customer\ Costs$$

Sufficient if **no time-dependent factors required** for the comparison, e.g.:

* Rental costs for land
* Differences in degradation
* Performance under high-temperature
* Performance under low light

Assessing customer value variations due to wafer size differences.

Value chain step	Differences by wafer Sizes investigated	Potential positive impact of change	Total customer cost
Polysilicon	Differences in polysilicon consumption	OPEX reduction	$\Delta\ Customer\ cost$
N-type ingot	Differences in qualified throughput	Reduced OPEX + CAPEX	$=\Delta \dfrac{Module\ cost}{power\ output}$
N-Type wafer	Differences in qualified throughput	Reduced OPEX + CAPEX	
TOPCon cell	Higher power output per cell	Enhanced power output	
TOPCon module	Changes in packaging density	Enhanced power output	
		Module is sold	
Value chain step	Differences by wafer sizes investigated	Impact of change	Total customer benefits
Distribution	Container package density	Reduced transportation costs	$\Delta\ Customer\ benefits$
Module installation	Change in module efficiency	Reduction of area-related BOS cost	$=\Delta \dfrac{BOS\ cost}{power\ output}$
Inverter installation	Optimized module voltage	Reduction of inverter cost	
		Module is operating	
Power generation	Change in module efficiency	Reduction of area-related running cost	Customer benefits differences

Assessing customer value variations due to wafer size differences.

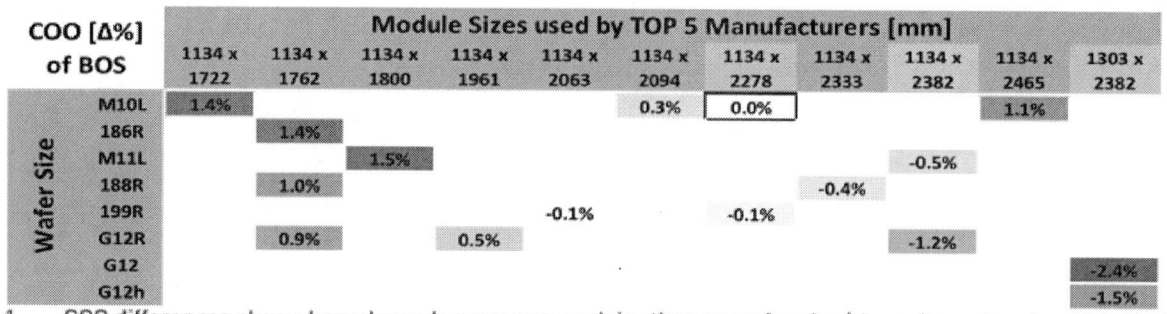

COO [Δ%] of BOS	Module Sizes used by TOP 5 Manufacturers [mm]										
	1134 x 1722	1134 x 1762	1134 x 1800	1134 x 1961	1134 x 2063	1134 x 2094	1134 x 2278	1134 x 2333	1134 x 2382	1134 x 2465	1303 x 2382
M10L	1.4%					0.3%	0.0%			1.1%	
186R		1.4%									
M11L			1.5%						-0.5%		
188R		1.0%						-0.4%			
199R					-0.1%		-0.1%				
G12R		0.9%		0.5%					-1.2%		
G12										-2.4%	
G12h											-1.5%

(Row label: Wafer Size)

1. COO differences shown here depends more one module- than on wafer size (→ equipment and module design more efficient for larger modules, known fact)
2. Rectangular wafers reduce the cost especially in larger scale modules, which is the most important market for the producers
3. Only slight improvement for going to bigger + rectangular in the smaller sized modules
4. Going for G12 and larger would be beneficial, but the wafers are currently not fitting into the standard glass sizes
5. Producing half-wafers will start benefiting, if the wafer thickness can be reduced in advance (here similar with 130 μm)! and the cell production equipment is adapted

---

Results of different wafer sizes for TOPCon modules

Recommendation:

- As the cost difference is small, we recommend to follow the market and take a solution with a diverse Supply chain.

- We highly recommend emerging markets to achieve standardization.

  → This will reduce conversion cost and productivity loss

  due to production line changeover caused by different products

- Still, we need to investigate each individual case:
  - Which steps op the value chain we are considering (Ingot, Wafer, Cell or/and Module)
  - Where we source the feedstock and materials?
  - Who will be our final customer?
  - What will be the cell technology?

References

1) Y. Chen, "Technology Evolution of the PV Industry," in 30th Annual NREL Silicon Workshop, 2023.
2) A. Bhambhani, "9 Leading Integrated Solar Module Manufacturing Companies Reach Consensus Over Rectangular Silicon Wafer Modules," 10 July 2023. [Online]. Available: https://taiyangnews.info/standardized-module-size-gets-unanimous-votes/.
3) A. Bhambhani, "6 Leading Solar PV Manufacturers Reach Consensus For Using Rectangular Silicon Wafers," IEA, 21 August 2023. [Online]. Available: https://taiyangnews.info/technology/silicon-wafer-size-standardization-update/.
4) Risen Energy, "Risen Energy," 05 11 2022. [Online]. Available: https://en.risenenergy.com/news/case_cont?nid=319.
5) LONGi, "Analysis of the development of high-power modules and the logic behind the increase in size of photovoltaic silicon wafers," 07 07 2020. [Online]. Available: https://www.longi.com/en/news/6887/. [Accessed 02 02 2023].
6) JA Solar, "JA Solar Releases its Next Generation Module DeepBlue 4.0 Pro," 26 05 2023. [Online]. Available: https://www.jasolar.com/index.php?m=content&c=index&a=show&catid=399&id=123. [Accessed 2022 02 01].
7) VDMA, "International Technology Roadmap for Photovoltaic – 2023 Results," March 2024. [Online].
8) PV Infolink Consulting, "Technology Market Repot Q1 2024," 2024.
9) CPIA China PV Industry Development Roadmap 2023 Published March 2024 [Online]

RCT Solutions GmbH
Line-Eid-Strasse 1
D-78467 Konstanz, Germany

Regd. HRB 708952.
Executive Board: Dr. Peter Fath

Get in touch with us

Phone +49 7531 58470 12
info@rct-solutions.com
julian.reichle@rct-solutions.com

rct-solutions.com

Thank you

41st European Photovoltaic Solar Energy Conference and Exhibition

This presentation was selected by the Sc. Committee of the EU PVSEC 2024 for submission of a full paper to one of the EU PVSEC's collaborating peer-reviewed journals.

A HORIZONTAL DOUBLE-SIDED COPPER METALLIZATION TECHNOLOGY DESIGNED FOR SOLAR CELL MASS-PRODUCTION

Lu Wang[1], Yusen Qin[1], Meilin Peng[2], Meixian Huang[2], ZhiPeng Liu[2], Yibo Lu[1], Guohua Zhou[1], Jingjia Ji[1]
1. Jiangsu Xianghuan Technology Co., Ltd 2.Jiangnan University

348977155@qq.com

ABSTRACT: In the international photovoltaic market, the "carbon footprint" label has received great attention, and low-carbon footprint products are widely popular. Compared with metallic silver, copper has the characteristics of low carbon emissions and low cost, which can effectively reduce the carbon footprint. Based on this, this article reports a horizontal double-sided Copper metallization technology. This technology can not only metallize the front and back sides of various types of silicon solar cells at the same time, but also has fast speed, good uniformity, and simple process, making it suitable for the industrial mass production of solar cells. This article describes the structure and manufacturing process of TOPCon solar cells patterned with ultrashort pulse laser and metallized using this novel horizontal double-sided copper metallization technology. In this batch, average efficiency reached above 26%.
Keywords: HDPLATE; TOPCon; Ni/Cu metallization

1 INTRODUCTION

With the increasing development of science and technology, environmental awareness has gradually attracted people's attention. In order to better face the international photovoltaic market, reducing the "carbon footprint" of products has become a top priority. In upstream photovoltaic processes, copper is far superior to silver in terms of carbon emissions due to its large reserves and ease of mining and purification. The overall cost of copper plating metallization is much less than screen printing, HJT screen printing cost 0.14 CNY per watt, TOPCon screen printing cost 0.08 CNY per watt, copper plating cost can be only 0.03 CNY per watt. Therefore, the copper crystal electrode technology that replaces the traditional screen-printed silver electrode has more advantages in terms of cost and carbon footprint [1].

In this paper, a horizontal double-sided plating (HDPLATE) metallization device is designed, which can simultaneously plate both sides of the solar cells. Therefore, the HDPLATE has high plating efficiency and simple process, which is suitable for industrial mass production. In this study, the HDPLATE was used on TOPCon solar cells. It is found that the uniformity of plating is excellent and damage rate of the thin solar cells is quite low. Therefore, the industrialization of low-cost and high-efficiency Cu metallization solar cells can be realized by using this HDPLATE technology.

2 SOLAR CELLS STRUCTURE AND EQUIPMENT

2.1 Solar cells structure
In this paper, a TOPCon solar cell featuring double-sided Ni/Cu metallization was employed, and its 3D structure is depicted in Figure 1.

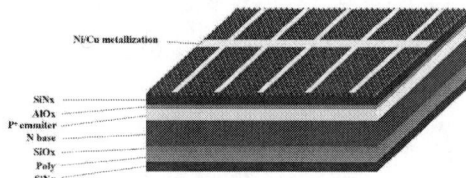

Figure 1: Diagram of TOPCon solar cell featuring double-sided Ni/Cu metallization

The structure used N-type CZ silicon wafers with a side length of 210 mm, a thickness of 135 μm and a resistivity of 1 Ω·cm. The front diffusion formed a P+ emitter with a concentration of 1.2e19 cm-3 and a thickness of 1.1 μm.

The front surface design includes texturing, passivation and Ni/Cu metallization. The front surface used 3 nm AlOx layer and a passivation of 80 nm SiNx with a refractive index of 1.93. The dielectric layer opening was about 10 μm. The thickness of Cu metallization exceeded 5 μm. The back surface used 2 nm ultra-thin tunneling oxide layer, approximately 85 nm phosphorus-doped polysilicon with the back sheet resistance about 50 ohm/sq. and a passivation of 82 nm SiNx with a refractive index of 2.18. The opening of the dielectric layer on the back was about 16 μm, and Ni/Cu is plated by light induced plating.

2.2 laser patterning process
Since the plating process cannot be directly performed on the SiNx structure on the surface of the crystalline silicon solar cells, the SiNx layers in the plating area need to be opened before plating [2]. In this study, an ultra-short pulse laser with high energy density was used to remove the SiNx on the front and back surfaces of the solar cells. This process is the graphical patterning processing before the Ni/Cu metallization process. Figure 3 (a) and (b) present the microscope image and the scanning electron microscopy (SEM) image after laser ablation process respectively . It can be seen that the width of the laser ablation is about 10 μm, which meets the requirement of the thinner fingers.

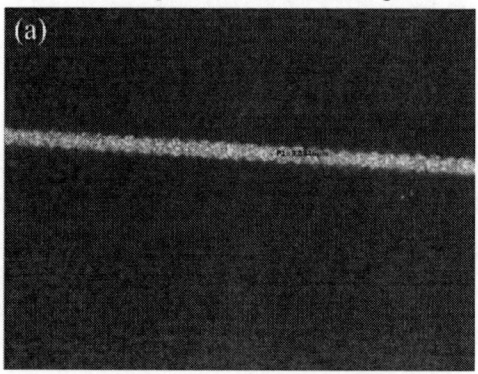

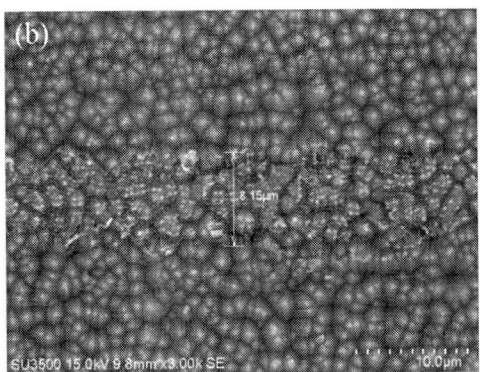

Figure 2: The solar cell surface images after laser ablation process (a) the microscope image; (b) SEM image

Moreover, Figure 3 (b) shows that the SiNx layer on the surface of the solar cell is perfectly removed by the ultrafast laser, while the original pyramid structures can still be retained, indicating that the laser ablation process has little damage to the silicon substrate, which will be beneficial to the subsequent metallization process.

2.3 Metallization equipment

The horizontal double-sided plating metallization equipment is designed as shown in Figure 3. Compared with other plating methods, the HDPLATE has the advantages of low fragmentation rate and excellent conductive effect, which is of great significance for the stable and efficient plating without damaging the the solar cells.

Figure 3: Horizontal double-sided plating metallization equipment

In addition, the HDPLATE can regulate the plating conditions of the two surfaces of the cells separately. For instance, it allows simultaneous plating on one surface while light-induced plating is performed on the other surface. Moreover, the plating rates of the two surfaces can be controlled separately. The HDPLATE has no restrictions on solar cells types and plating metal materials.

The recyclable water system is designed to recycle the plating solution, so that the plating solution can be continuously recycled without replacement. In addition, the design of the recyclable water system is also in line with the concept of green production. The waste water treatment of this system can meet the national level standard and realize the recycling of Cu chemicals.

3 RESULT AND DISCUSSION

3.1 Tensile test

To ensure the fingers quality of the solar cell meets standards, a tensile test with a force exceeding 0.8 N is implemented. Plating and screen printing processes are employed to fabricate 20 solar cells, each subjected to the tensile test. The relevant test results are shown in Table I.

Table I: Comparison of tensile test results

(N)	Plating (N)	Screen printing (N)
Maximum	3.9	3.4
Minimum	1.2	1.25
Average	2.5	2.466
Qualification rate	100%	100%

It can be seen from the table I that the tensile force of plating is larger than the specified 0.8 N, achieving a 100% qualification rate, signifying the fingers produced by plating is qualified.

3.2 Ni/Cu contacts

Figure 4 (a) shows the SEM image of the fingers. It can be observed that after Ni/Cu metallization, the width is only 17.6 μm. The aspect ratio of the fingers refers to the ratio of the height to the width of the finger [3], which is usually expressed as h/w. In Figure 4 (b), the aspect ratio of the display finger is 0.4.

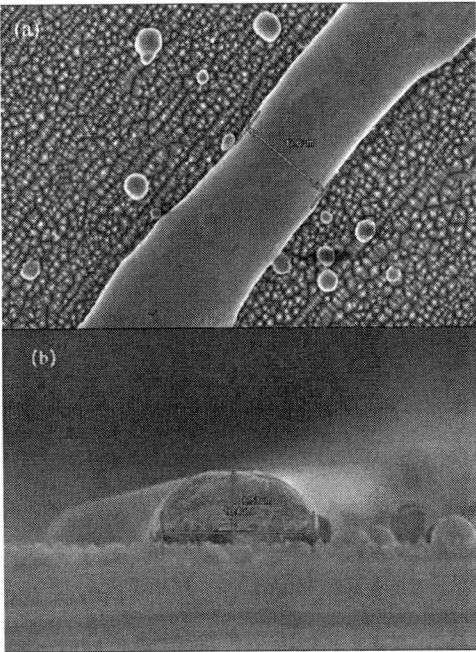

Figure 4: SEM images (a) the finger width; (b) the aspect ratio of the finger

3.3 High-temperature repair

In this study, a horizontal double-sided Ni/Cu metallization process is used, and assisted by high-temperature repair to improve the electrical performance of solar cells. Finally, a photoelectric conversion efficiency of up to 25.819 % is successfully achieved, as shown in Figure 5. Group A undergoes double-sided plating with assisted high-temperature repair, while Group B undergoes double-sided plating without assisted high-temperature repair.Through the observation of the above diagram, it can be found that the efficiency distribution of solar cells after high temperature repair is more concentrated, and the overall efficiency is 0.422 % higher than that without high temperature repair.

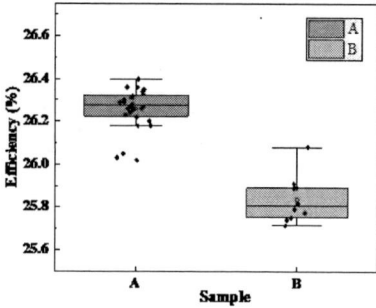

Figure 5: Efficiency distribution graphs for two conditions (A: double-sided plating with assisted high-temperature repair; B: double-sided plating without assisted high-temperature repair) double-sided plating metallization equipment

3.4 Electrical performance

In this paper, laser ablation followed by Ni and Cu metallization based on two-sided silicon nitride passivated silicon wafers results in excellent electrical performance of solar cells distribution, showing in Figure 6. The narrow distribution of electrical performance showed in Figure 6 indicate the excellent effect of horizontal double-sided Ni/Cu plating.

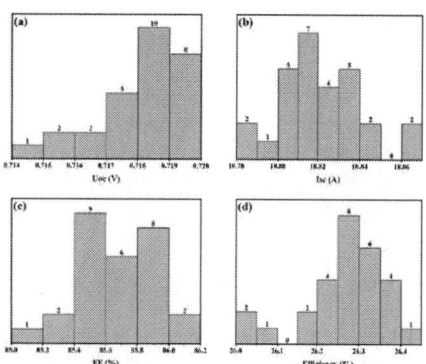

Figure 6: Electrical performance distribution of Ni/Cu metallization

In this study, an ultra-short pulse laser with high energy density was used to laser ablation, which can perfectly remove the surface SiNx layer without damaging the underlying silicon, which is conducive to forming good and reliable contact between metal and silicon substrate during subsequent Ni/Cu metallization, with low contact resistance and effectively improving current transmission efficiency. In addition, the conductivity of copper is better than that of silver paste. Therefore, the contact resistance and grid resistance of the Ni/Cu metallization are lower than that of the screen-printed Ag/Al metallization. Low contact and grid resistance can reduce the series resistance of the solar cell, thereby increasing FF, which can be observed in Figure 6(c).

4   CONCLUSION

In the photovoltaic industry, it is important to obtain reliable contacts while reducing the cost of metallization in the manufacturing of solar cells. In the future, replacing Ag paste contacts by plated Cu contacts is the development trend of metallization technology of solar cells. The introduction of horizontal double-sided Copper metallization technology provides an effective way for Ni/Cu metallization and improving the performance of solar cells.

Overall, this study has achieved substantial advancements in solar cell performance through the utilization of HDPLATE technology patterned by an ultra-short pulse laser. This offers robust practical support for enhancing the photoelectric conversion efficiency of solar cells. At the same time, HDPLATE enabling high throughput is more suitable for large-scale production compared with traditional single-side electroplating. Moreover, the overall cost of electroplating is lower than screen printing, and the efficiency of plated cell is higher than that of screen printed cell. The water reuse system is increased, and the discharge meets the environmental protection requirements.

5   REFERENCES

[1] Frédéric L, Faustine L, et al, Minerals Engineering, 15 (2021),170.
[2] Wenham, Alison, Chong, et al. Solar Energy Materials and Solar Cells: An International Journal Devoted to Photovoltaic, Photothermal, and Photochemical Solar Energy Conversion, 169 (2017) 151-158.
[3] Wu S , Zhang J, et al. Solar Energy Materials and Solar Cells: An International Journal Devoted to Photovoltaic, Photothermal, and Photochemical Solar Energy Conversion, (2023) 259.

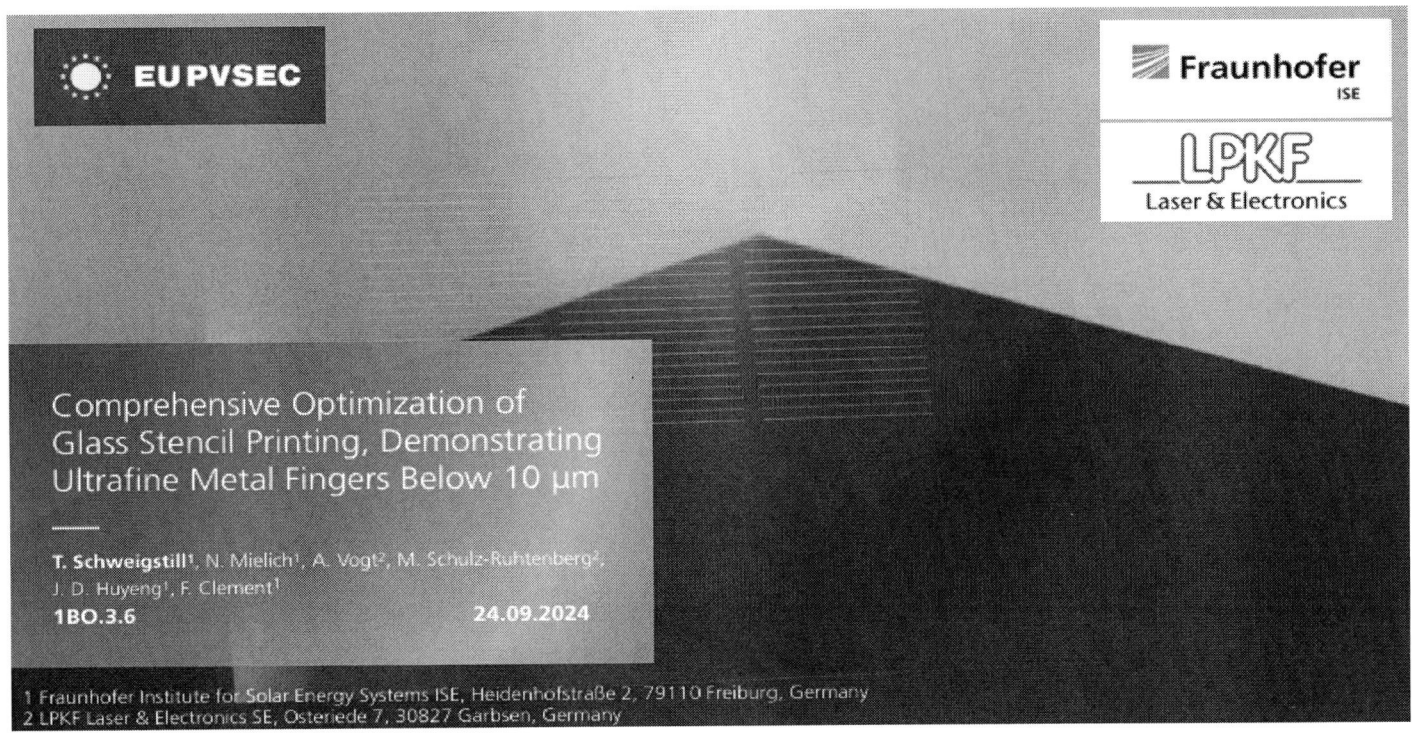

Motivation

Overview front side metallization development

- Impressive finger width reduction over the last decades

- Screen printing as the main metallization process [1]
  - Severe challenges for very narrow openings [2]

- Trend will continue [3]
  - Towards photolitho. structured contacts

- Process technology window glass stencil printing: 5 µm to 20 µm

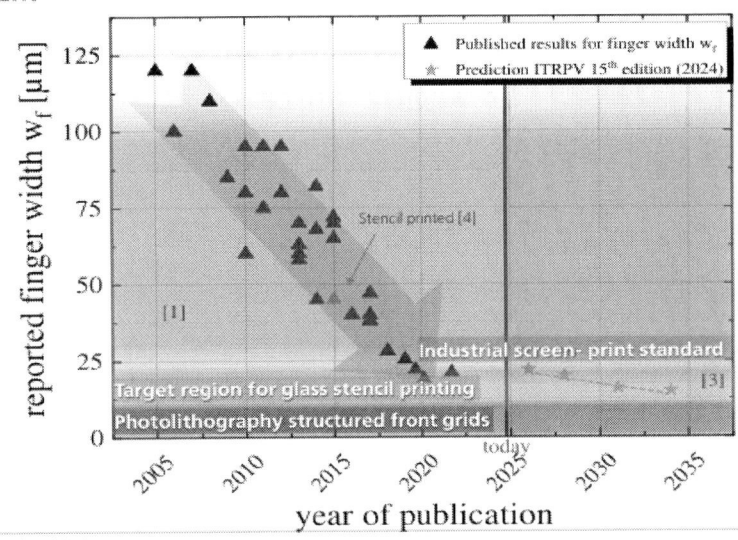

[1] Wenzel, T. et al. (2022). Metallization of silicon solar cells. *SSRN*, 244, 111804.
[2] Tepner, S. et al. (2021). Model for screen utility in solar cell metallization. *Scientific Reports*, 11(1), 4352.
[3] ITRPV 15th edition, March 2024, VDMA
[4] Hannebauer, H. et al. (2015). Optimized stencil print for low Ag paste consumption. *Energy Procedia*, 67, 108-115.

Motivation
Screen printing

Schematic of the screen-printing process

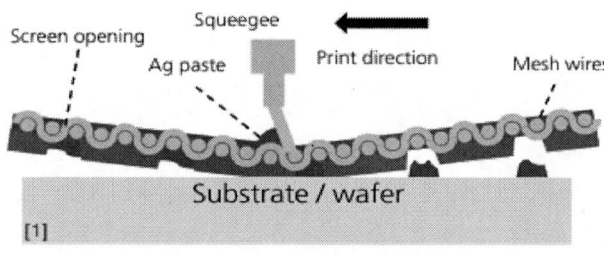

Simulations of mesh angle on relative opening size in a screen

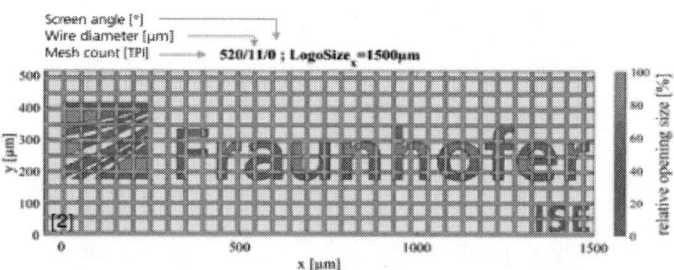

[1] Tepner, S., & Lorenz, A. (2023). Printing technologies for silicon solar cell metallization: A comprehensive review. *Progress in Photovoltaics: Research and Applications*, 31(6), 557-590.
[2] Singler, M. et al. (2024). Prediction of screen-printed electrodes with fine-lines and arbitrary structures. *Energy Technology*, in press. https://doi.org/10.1002/ente.202401346

Motivation
Stencil printing

- Mesh wires crossing the openings are disadvantageous

- Leads to low relative opening size

- Reduced paste transfer

- Possible solutions:
  - Knotless screens → reduced durability / layout flexibility
  - Thinner wires → durability issues [2]

[1] Tepner, S., Ney, L., Singler, M., Preu, R., Pospischil, M., & Clement, F. (2021). A model for screen utility to predict the future of printed solar cell metallization. *Scientific reports*, 11(1), 4352.
[2] Singler, M. et al. (2024). Prediction of screen-printed electrodes with fine-lines and arbitrary structures. *Energy Technology*, in press. https://doi.org/10.1002/ente.202401346

Motivation
Stencil printing

- Stencil printing as great alternative to screen-printing

- Additional benefits stencil printing [1] :
  - No mesh marks
  - Increased lifetime
  - Reduced distortion in print image

- Conventional stencils made from stainless steel or Ni

- Stencil printing was used for solar cell metallization [2]

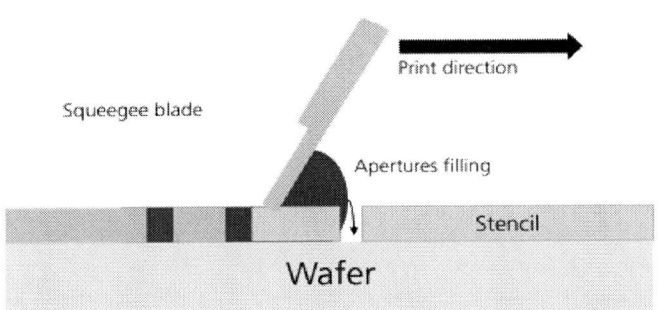

[1] Tepner, S., & Lorenz, A. (2023). Printing technologies for silicon solar cell metallization: A comprehensive review. Progress in Photovoltaics: Research and Applications, 31(6), 557-590.
[2] Hannebauer, H., Schimanke, S., Falcon, T., Altermatt, P. P., & Dullweber, T. (2015). Optimized stencil print for low Ag paste consumption and high conversion efficiencies. Energy Procedia, 67, 108-115.

Motivation
Stencil printing

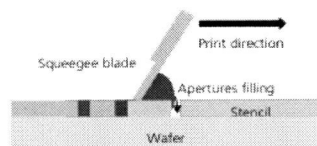

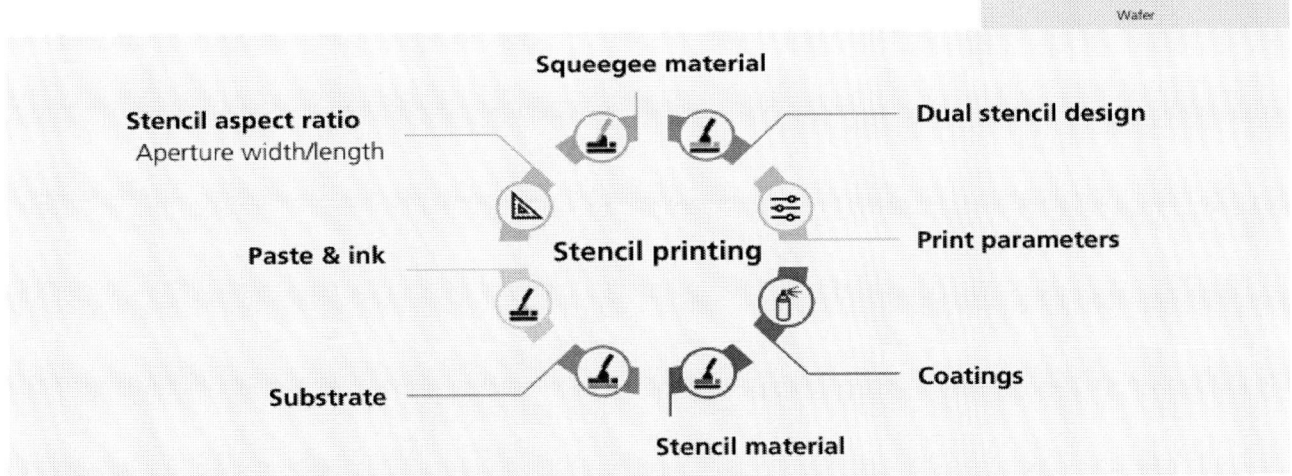

Motivation
Conventional stencil printing

- Paste adhesion to stencils walls reduces paste transfer efficiency

- In solder paste stencil printing:
  - Definition of stencil area ratio
  - Target area ratio > 0.66 [1,2]

- Narrow aperture openings requires extremely thin stencils

- Conventional stencil printing assumes quasi rectangular aperture channel geometries

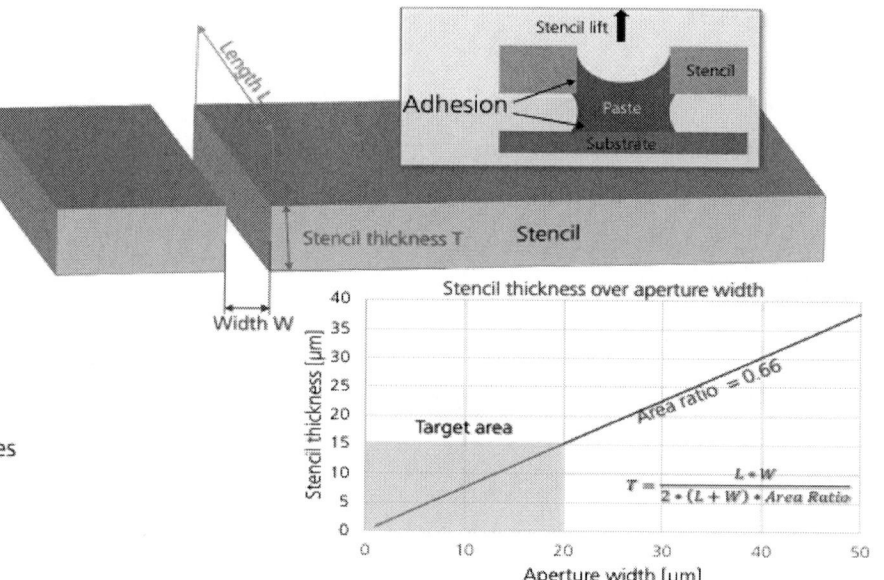

[1] IPC Association Connecting Electronics Industries. (2020). IPC-7522: Guidelines for Stencil Design. IPC.
[2] Durairaj, R., Nguty, T. A., & Ekere, N. N. (2001). Critical factors affecting paste flow during the stencil printing of solder paste. Soldering & Surface Mount Technology, 13(2), 30-34.

Motivation
Glas stencils

- To overcome thickness constrains: Decouple aperture channel length from stencil thickness

- CFD simulations to determine suitable stencil cross section geometry [1]

- Conventional stencil manufacturing not enable a precise geometrical aperture channel adaptation

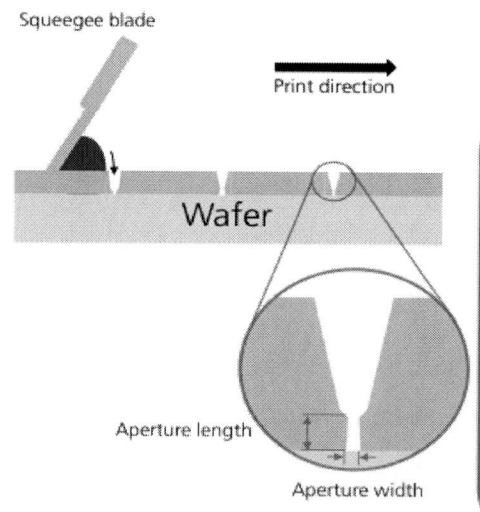

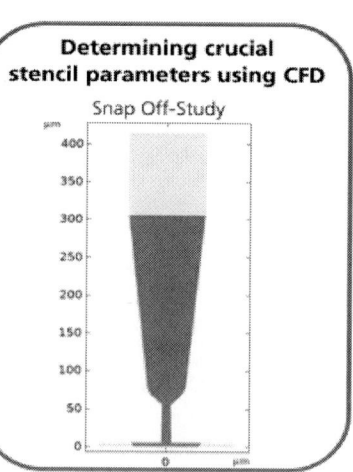

[1] Lorenz, A., Tepner, S., Gensowski, K., Wenzel, T., Karimi, K., Vogt, A., ... & Clement, F. (2022). Project" Innomet"-Evaluation of Innovative Glass-Based Printing Forms for Solar Cell Metallization.

Laser Induced Deep Etching (LIDE®)
Process overview

- LIDE process is used for various applications, focus on microelectronic packaging [1,2]

- Enables clean drilling or cutting of glass with precision

- Supports high aspect ratio openings in glass foils
  - Aspect ratio typically 1:10; achievable up to 1:100

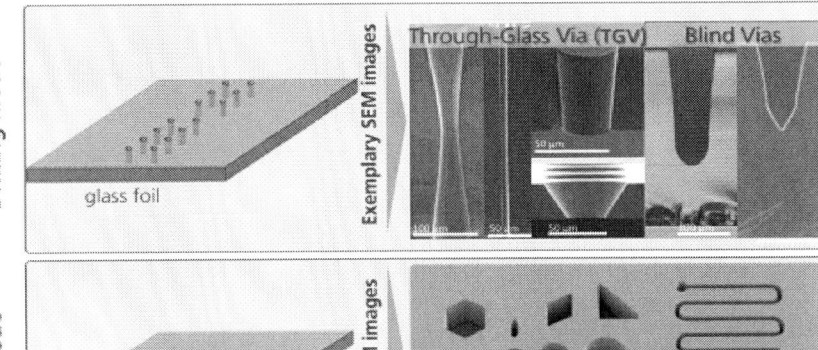

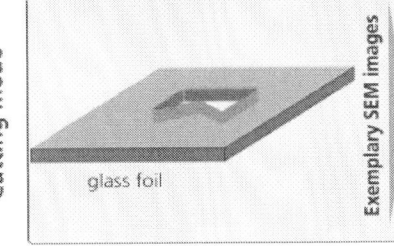

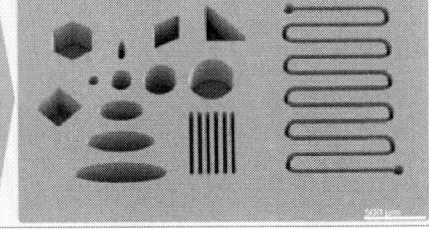

[1] Santos, R., Delrue, J.-P., Ambrosius, N., Ostholt, R., & Schmidt, S. (2020). Glass substrate processing. 2020 IEEE ECTC, 1922–1927. https://doi.org/10.1109/ECTC32862.2020.00300
[2] Ostholt, R., Ambrosius, N., Dunker, D., & Delrue, J. P. (2017). Via formation in glass. 2017 MiNaPAD Conference, 17-18.

Laser Induced Deep Etching (LIDE®)
Stencil manufacturing

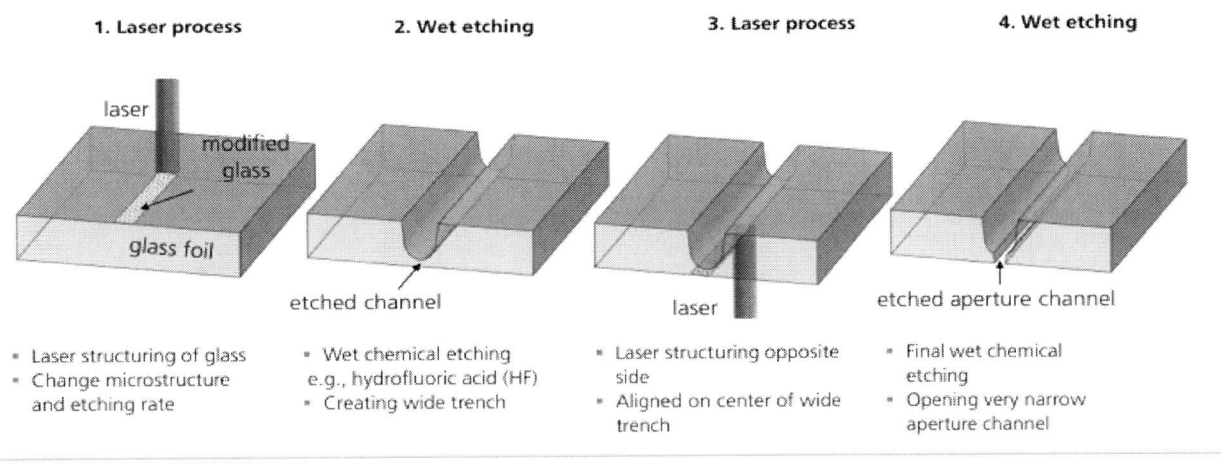

1. Laser process
- Laser structuring of glass
- Change microstructure and etching rate

2. Wet etching
- Wet chemical etching e.g., hydrofluoric acid (HF)
- Creating wide trench

3. Laser process
- Laser structuring opposite side
- Aligned on center of wide trench

4. Wet etching
- Final wet chemical etching
- Opening very narrow aperture channel

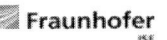

Laser Induced Deep Etching (LIDE®)
Stencil manufacturing

- Geometrical modification by laser and etching parameter adaptation

- Print channel modification
  - Aperture length
  - Aperture width

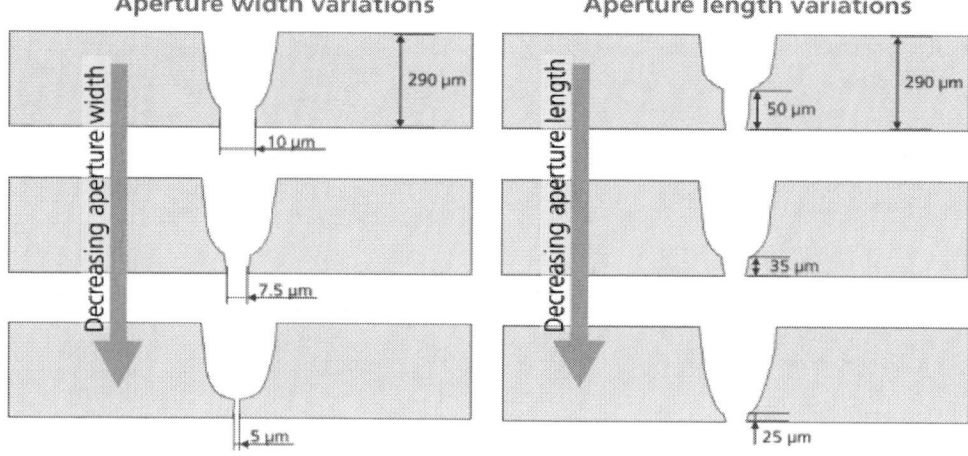

Laser Induced Deep Etching (LIDE®)
Stencil manufacturing

- Four-step process to achieve desired print channel cross section geometry

- Stencil test pattern on
  - 10 cm x 10 cm glasses
  - 69 test finger segments
  - Segment length 20 mm

- Light microscope analysis shows reliable process

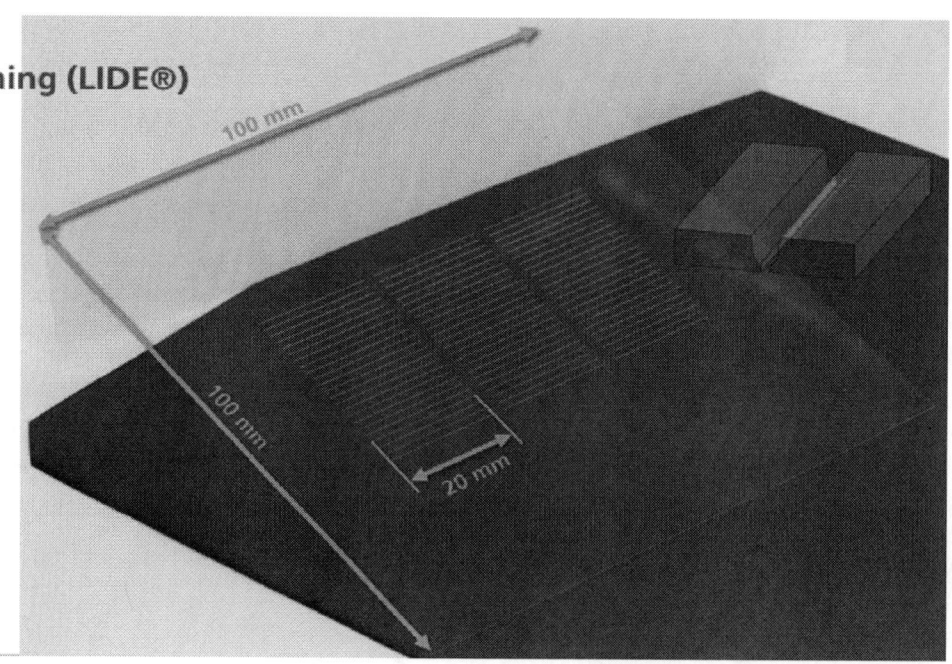

Laser Induced Deep Etching (LIDE®)
Stencil characterization

- Light microscope analysis shows reliable process

- Three nominal aperture widths
  - 5 μm
  - 7.5 μm
  - 10 μm

- Three nominal aperture lengths
  - 25 μm
  - 35 μm
  - 50 μm

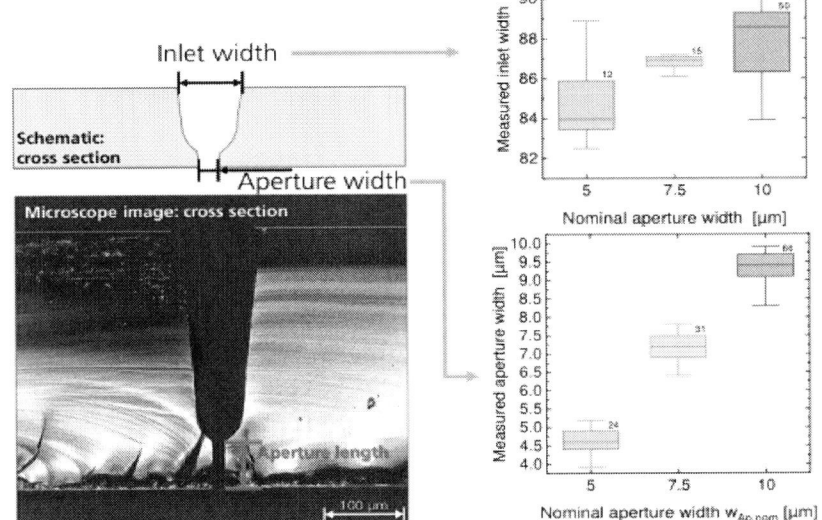

Glass stencil printing
Printing setup and conditions

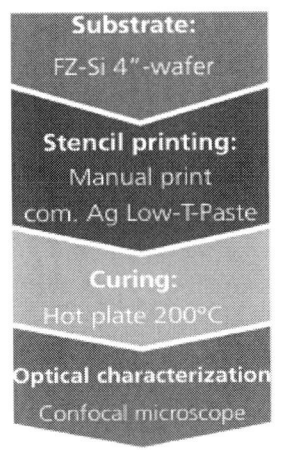

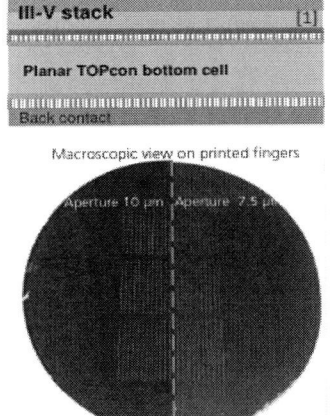

[1] Schygulla, P. et al. Two-terminal III–V//Si triple-junction solar cell with power conversion efficiency of 35.9 % at AM1.5g. Prog. Photovolt. Res. Appl. 30, 869–879.

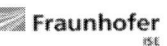

Glass stencil printing
Results
___

- Despite manual print good paste transfer

- Increased errors for 7.5 μm opening

- Print defects observed:
  - Bleeding
  - Smearing
  - Necking / Interruptions

- 5 μm opening with current paste not printable

Macroscopic view: printed fingers   **Microscopic view: printed fingers**

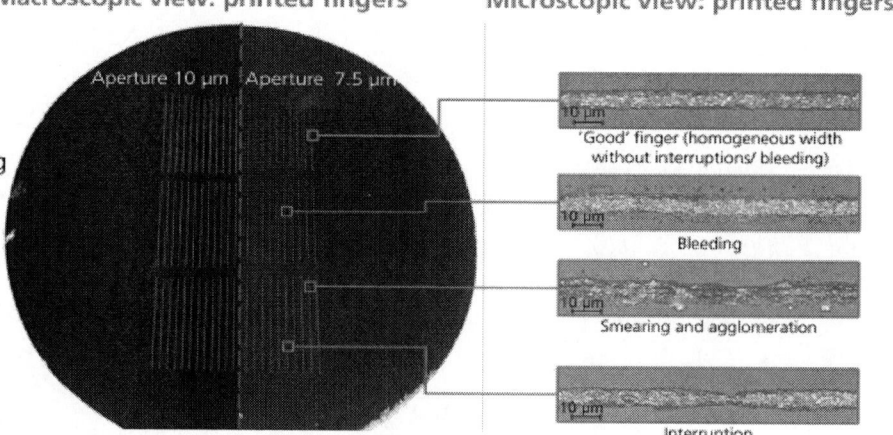

'Good' finger (homogeneous width without interruptions/ bleeding)

Bleeding

Smearing and agglomeration

Interruption

Glass stencil printing
Results
___

- Best avg. finger width 8 μm

- Low finger height leads to small aspect ratio

Macroscopic view on printed fingers

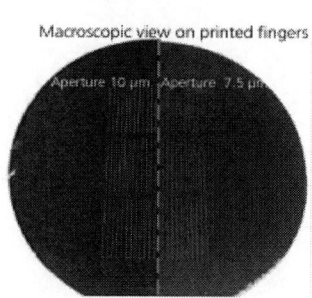

Comparison

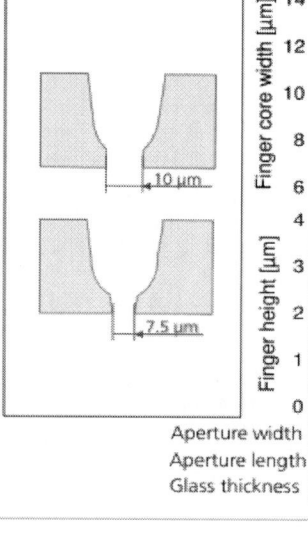

Aperture width variations

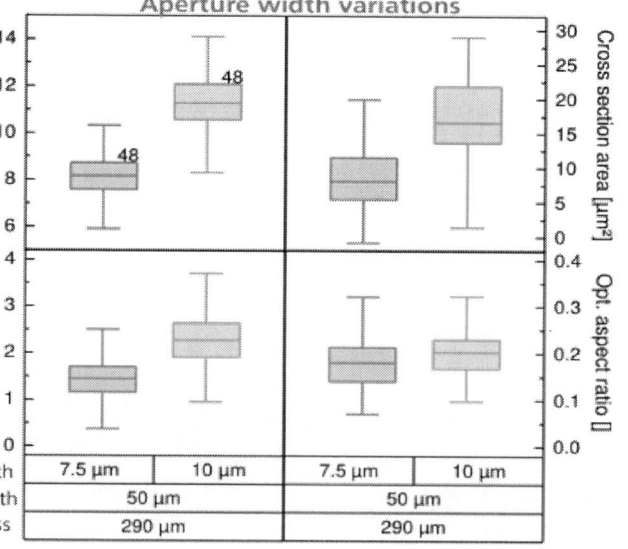

Aperture width	7.5 μm	10 μm	7.5 μm	10 μm
Aperture length	50 μm		50 μm	
Glass thickness	290 μm		290 μm	

Glass stencil printing
Results

- Low finger height leads to small aspect ratio

- Reduction of aperture channel length increases aspect ratio

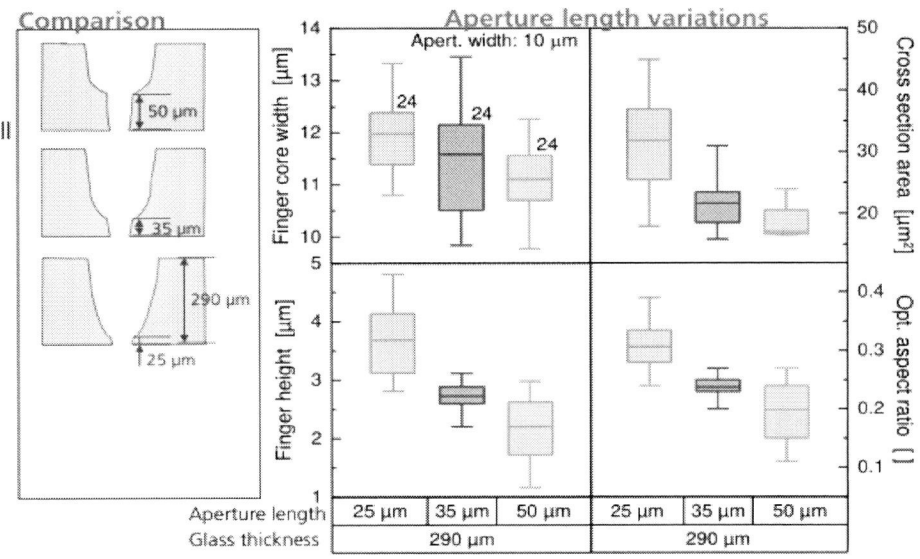

Glass stencil printing
Results

- Low finger height leads to small aspect ratio

- Reduction of aperture channel length increases aspect ratio

- Reduction of glass stencil increases aspect ratio further

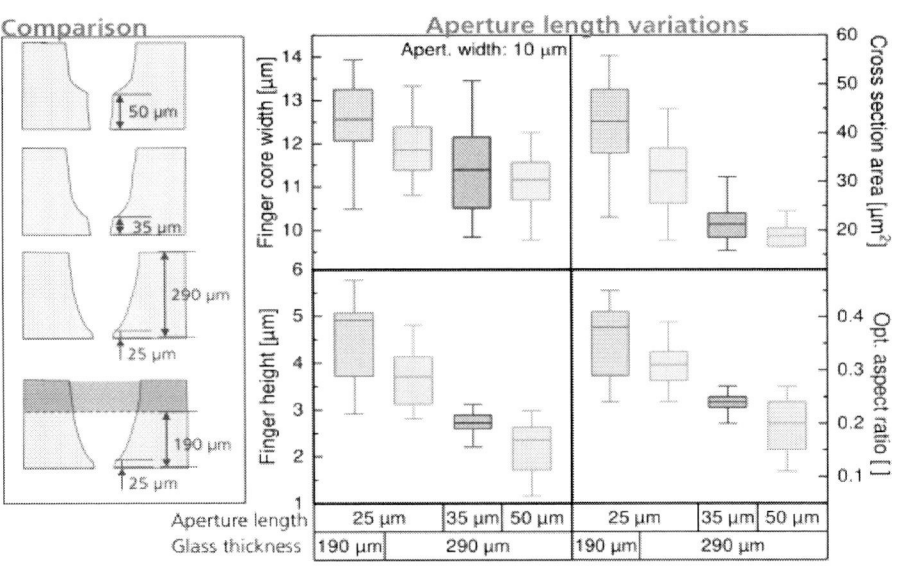

Glass stencil printing
Results

- Low finger height leads to small aspect ratio

- Reduction of aperture channel length increases aspect ratio

- Reduction of glass stencil increases aspect ratio further

- Paste properties with significant impact on print results

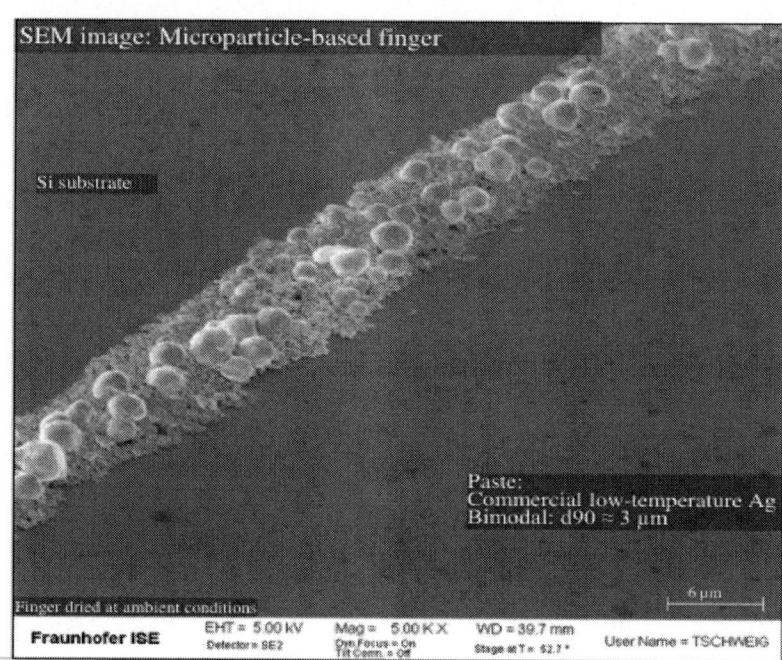

Glass stencil printing
Results

- Low finger height leads to small aspect ratio

- Reduction of aperture channel length increases aspect ratio

- Reduction of glass stencil increases aspect ratio further

- Paste properties with significant impact on print results

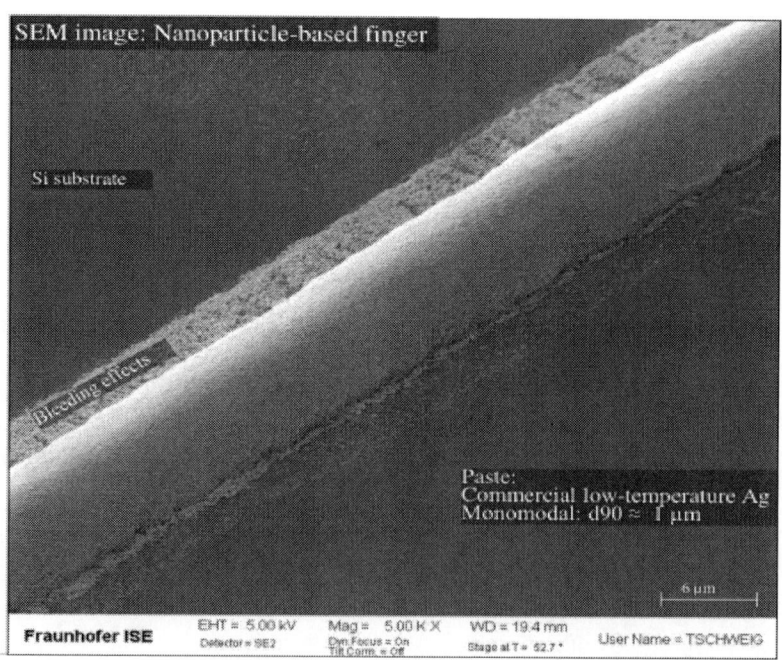

Glass stencil printing
Results

- Low finger height leads to small aspect ratio

- Reduction of aperture channel length increases aspect ratio

- Reduction of glass stencil increases aspect ratio further

- Paste properties with significant impact on print results

- Nanoparticle based paste enables aspect ratios > 0.5

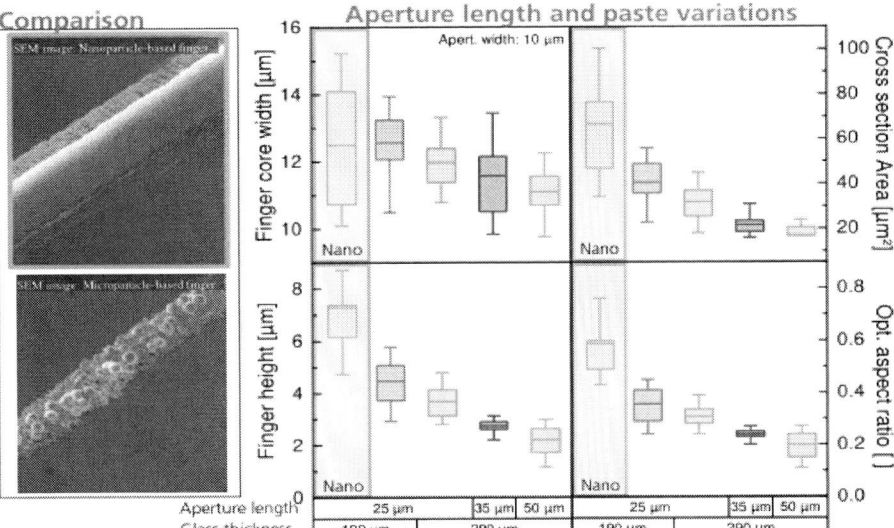

Conclusion
In a nutshell…

- Successful proof of principle: Glass stencil printing

- Decouple aperture channel length from mask thickness using the LIDE process

- Finger width below 10 μm are printable

- Best results require nanoparticle-based paste

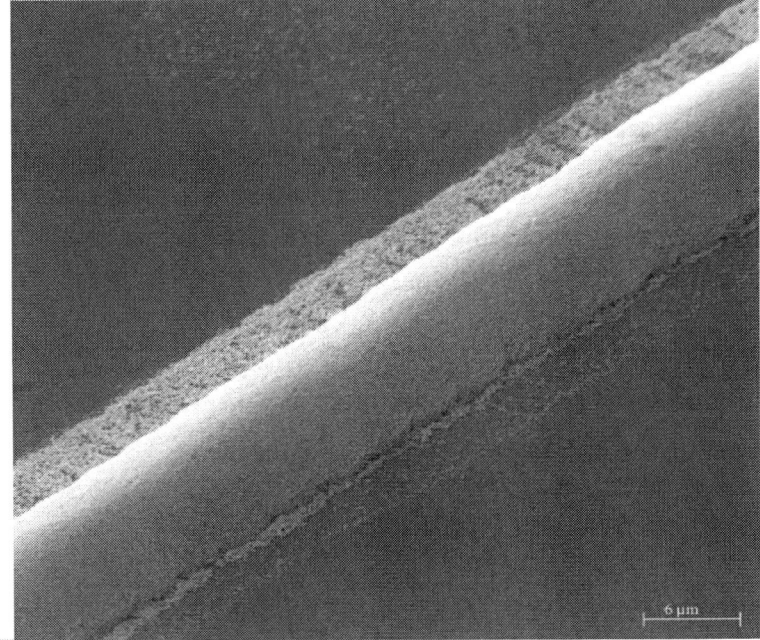

Conclusion
In a nutshell…

- Successful proof of principle: Glass stencil printing

- Decouple aperture channel length from mask thickness using the LIDE process

- Finger width below 10 µm are printable

- Best results require nanoparticle-based paste

Next Steps
- Proof of concept on automated setup
  - Optimization of print parameter

- Further stencil improvements: coatings, dual layers

Outlook:
Dual stencil design

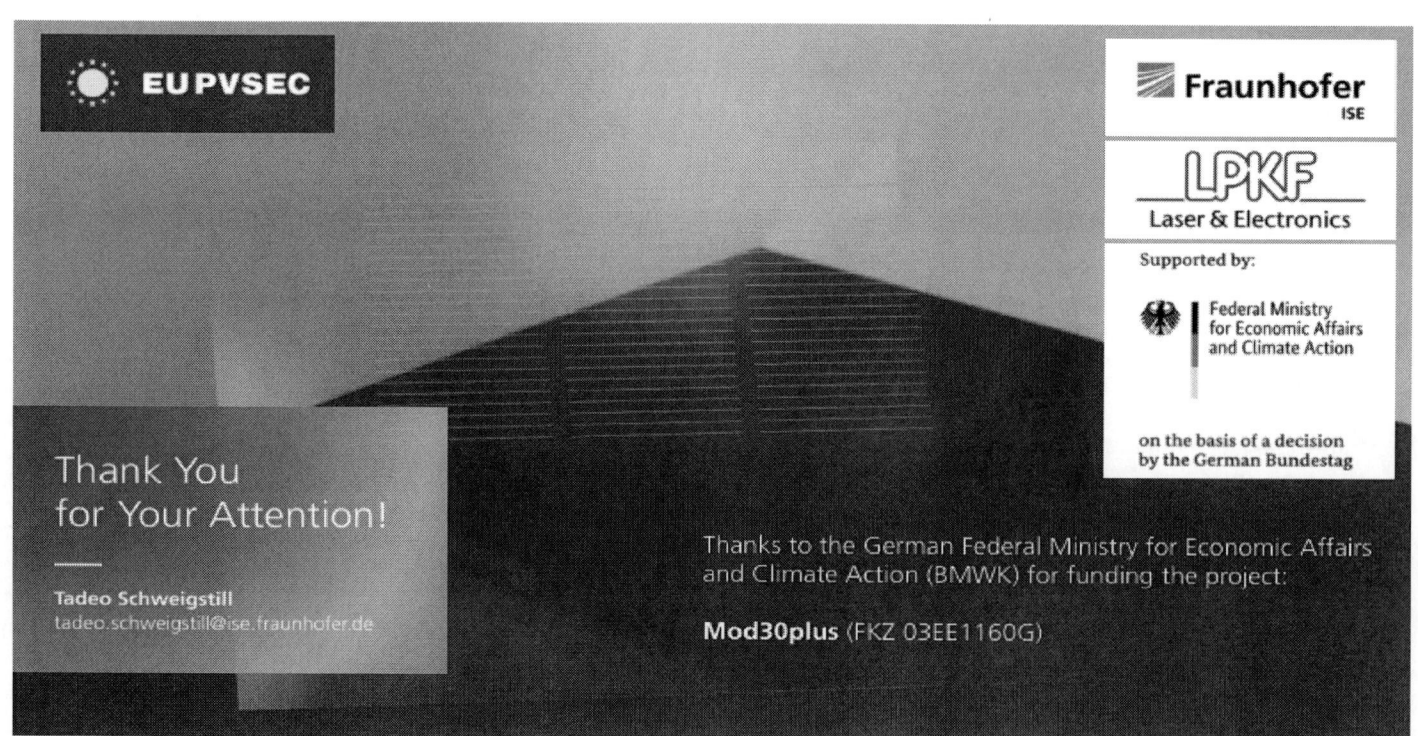

IBC4EU: FIRST RESULTS OF INDUSTRIALIZATION OF LOW-COST, HIGH-EFFICIENCY IBC TECHNOLOGY

Florian Buchholz[1], Daniel Tune[1], Tobias Messmer[1], Jonathan Linke[1], Manjunath Prasad[1], Valentin Mihailetchi[1], Juras Ulbikas[2], Arne Dahle[3], Martijn Meereboer[4], Francesca Fabris[5], Erik Eikelboom[5], Tom Borgers[6], Rik Van Dyck[6], Filip Duerinckx [6], Hariharsudan Sivaramakrishnan[6], Samuel Harrison[7], Josco Kester[8], Nicolas Guillevin[8], Jan Kroon[8], Verena Mertens[9], Thorsten Dullweber[9], Ofer Shochet[10], Isaac Rosen[10], Ingo Röver[11], Wolfram Palitzsch[11], Yasmin Zaror[12], Johannes Stierstorfer[12], Aurimas Radzevicius[13], Povilas Lukinskas[13], Julius Denafas[13], Tuomas Vanhanen[14], Tuukka Savisalo[14], Maximilian Pospischil[15], Marian Breitenbücher[15], Özlem Coşkun[16], Melodie de l'Epine[17], Philippe Macé[17]

Corresponding author florian.buchholz@isc-konstanz.de
1 ISC Konstanz e.V., Rudolf-Diesel-Straße 15, 78467 Konstanz, Germany
2 ProTech, Vismaliukų g. 34, Vilnius LT 10243, Lithuania
3 NorSun, Karenslyst Allé 9C, 0278 Oslo, Norway
4 Energyra, Handelsweg 45, 1525 RG Westknollendam, The Netherlands
5 FuturaSun, Riva del Pasubio, 14, 35013 Cittadella PD, Italy
6 imec, Energyville, Thor Park 8320, 3600 Genk, Belgium
7 CEA INES, 50 avenue Lac Léman Bâtiment Lynx 3. Technopôle Savoie Technolac 73375 Le Bourget-du-Lac, France
8 TNO, Westerduinweg 3, 1755 LE Petten, The Netherlands
9 ISFH, Am Ohrberg 1, 31860 Emmerthal
10 Copprint, 8 Hamarpe street, Jerusalem, 9777408, Israel
11 LuxChemTech, Alfred-Lange-Straße 18, 09599 Freiberg, Germany
12 WIP – Renewable Energies, Sylvensteinstraße 2, 81369 München, Germany
13 UAB Valoe Cells, Mokslininkų str. 6A, LT-08412 Vilnius, Lithuania
14 Valoe Cells, Insinöörinkatu 8, 50150 Mikkeli, Finland
15 Highline Technology GmBH, Tullastraße 87B, 79108 Freiburg, Germany
16 Kalyon PV, Malıköy, Şaditürk Blv., 06909 Malıköy Başkent Osb/Sincan/Ankara, Turkey
17 Becquerel Institute, Rue de Praetere 2, 1000 Brussels, Belgium

ABSTRACT: This paper introduces the Horizon Europe project IBC4EU with the goal to establish a European value chain based on innovative passivated contact back contact solar cells and modules. The two key solar cell technologies – POLO-IBC and polyZEBRA – are introduced. We present simulation studies and experiments on the suitability of bulk resistance ranges for use in the two solar cell concepts indicating best solar cell efficiencies for similarly low resistance values as standard TOPCon solar cells. The higher the minority carrier lifetime of the given material, the smaller this impact becomes. The p-type based solar cell (POLO-IBC) shows to require a more confined range that is slightly higher in sheet resistance than the standard PERC requirements. Reliability data on the cells (LeTID) and modules (DH, TC) proves the excellent long-term stability of the n-typed back contact cells and modules. In addition, an innovative way to interconnect the IBC cells based on printed conductive patterns printed on glass is introduced as well as a novel recycling route for state-of-the-art back contact solar cells.

Keywords: IBC, back-contact, solar cell, solar module, recycling

1 INTRODUCTION

The IBC4EU project, launched in November 2022, is a European Union initiative aimed at developing cost-competitive and sustainable solar technology, focusing on interdigitated back contact (IBC) photovoltaic (PV) cells. The goal is to create a fully integrated European production line for solar modules, from ingots and wafers to cells and final modules, and to scale production to gigawatt levels by 2030.

The IBC4EU project is funded under the Horizon Europe Program and coordinated by the International Solar Energy Research Center Konstanz (ISC Konstanz). The project brings together 21 partners from various countries, including research institutions, private companies, and renewable energy firms. An overview can be found in Figure 1. The project focuses on further advancing the efficiency and reducing the environmental

impact of bifacial IBC technology, which already offers high efficiencies and greater sustainability by reducing material usage in general (such as silicon) and especially the use of scarce materials (relative to PV demand) like silver.

Key technological advances made within the project are the polyZEBRA and POLO IBC cell designs, which aim to outperform other solar technologies such as PERC and heterojunction (HJT) in terms of cost and efficiency [1]. The project also incorporates Industry 4.0 solutions for predictive maintenance and quality control.

Running until October 2025, IBC4EU aims to pave the way for large-scale, competitive solar production in Europe, supporting the EU's renewable energy goals.

This paper summarizes a number of results obtained within the project. It focuses on the impact of silicon material properties on the solar cell performance, degradation studies on cells, namely light- and elevated temperature-induced degradation (LeTID), and on modules, namely damp heat (DH) and thermal cycling (TC) tests. In addition, a novel recycling route for solar cells and modules is shown.

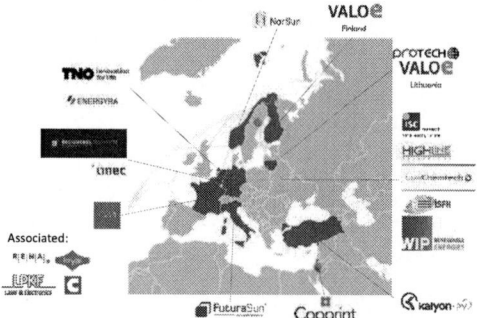

Figure 1: Overview of the IBC4EU partners.

The structures of the two solar cell concepts that are optimized by the partners are shown in Figure 2. While the POLO IBC is based on processes known from PERC solar cell technology [2,3], polyZEBRA draws from the TOPCon technology. Silver saving potentials, e.g. by moving to Cu-based screen printing [4,5], are investigated using its predecessor, the ZEBRA IBC technology, with which polyZEBRA shares its metallization approach [6].

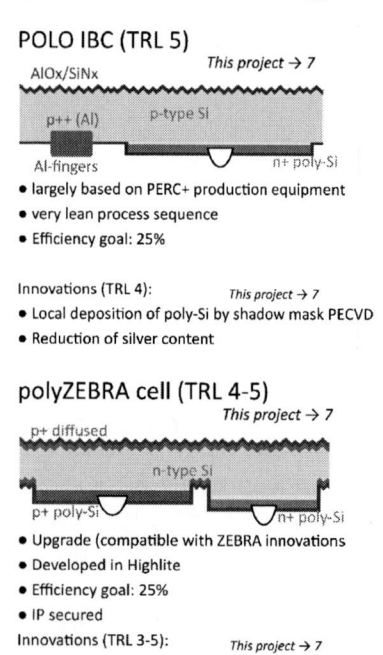

Figure 2: Cross section sketches of the two solar cell concepts with the TRL levels at the beginning of the projects and the key innovations targeted with the final TRL goals. The polyZEBRA cell was developed in the H2020 project HighLite.

While POLO IBC can be piloted in existing PERC lines, polyZEBRA shares its majority of processes with TOPCon standard equipment and may therefore be easily put into pilot production in existing or new TOPCon production lines [7].

2 EXPERIMENTAL SECTION

2.1 POLO IBC solar cells

The POLO IBC solar cells were fabricated using the Institut für Solarenergie Forschung Hameln (ISFH) pilot line [6]. The wafers were textured and single side polished, then an SiO_xN_y, n-a-Si, tunnel oxide stack is deposited through shadow masks to create the local emitter regions. After annealing, AlO_x and SiN_x passivation layer stacks are deposited on both sides. The AlO_x/SiN_x layer is locally opened by laser ablation on the base region. Al paste lines are printed to contact the base and Ag paste is printed to contact the n-poly-Si regions. For the purpose of this paper, p-type wafers with different base resistivity were provided by Norsun and Kalyon.

2.2 polyZEBRA solar cells

The process to produce polyZEBRA solar cells is described in [7], it features laser ablation and local laser activation for structuring the n-poly-Si and p-poly-Si regions, respectively. The a-Si layers were deposited by LPCVD, a process that will be transferred to PECVD within the scope of the project.

2.3 ZEBRA modules

In IBC4EU, different methods are investigated to produce the bifacial modules targeted by the project. At ISC Konstanz, Zebra modules are produced using the industry-dominant conventional stringing approach based on solder-coated flat copper ribbons. This is realized on an industrial stringer tool using an adapted process for the back contact cells. The strings are then bussed and laminated into bifacial modules using conventional processes and a variety of encapsulants and backsheets are investigated to maximize performance, bifaciality and durability.

TNO has developed a module interconnection technology consisting of conductive tracks directly deposited on the rear-side glass. The electrical connection of the Zebra solar cells to the conductive tracks is realized via low temperature solder paste during the lamination process. This module interconnection technology enables the manufacturing of bifacial back-contact PV modules while benefiting from the performance, process and design flexibility advantages of TNO's conductive foil-based interconnection technology already used in the industry.

2.4 Recycling

LuxChemTech received polyZEBRA solar cells and Cu screen-printed solar cell scrap from the ISC Konstanz pilot line. Classical routes were explored to recycle the solar cells, however, these were unsuccessful due to the given material mix. A new approach was thus developed, and this is presented below.

3 RESULTS AND DISCUSSION

3.1 POLO IBC results

While industrial PERC+ cells use Ga-doped Cz wafers with a base resistivity between 0.3 Ωcm and 1.0 Ωcm, simulations (Quokka3) indicated that the optimum resistivity for POLO IBC cells lies in the range of wafer resistivities between 1 Ωcm and 2 Ωcm, as indicated in Figure 3. The best POLO IBC cell achieved to date is indicated in the graph and has a 23.9% conversion efficiency [9].

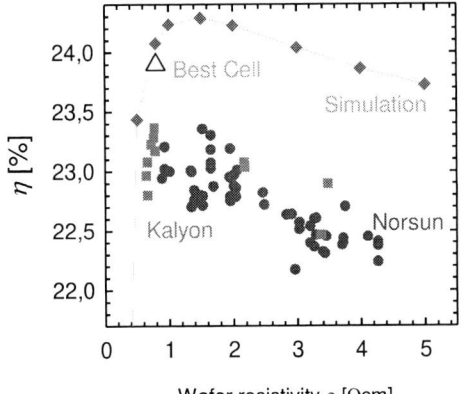

Figure 3: POlO IBC solar cell results using different Ga-doped Cz-Si wafers from the project partners Norsun and Kalyon PV. In green: Quokka 3 simulation result using input parameters from [3].

From the graph, the trendline across the wafer resistivity follows nicely the prediction by the simulations. The slightly lower experimental efficiencies are caused by a lower J_{sc} values which occured in this run, probably due to slight blistering of the AlO$_x$/SiN$_y$ passivation which is subject to further analysis and improvement. The precursor lifetime values (not shown) indicate excellent bulk and surface passivation quality. To increase the POLO IBC solar cell efficiency to 25.0%, further optimization of the Ag and Al metallization and the use of 1.5 Ωcm Ga wafers is planned for the coming months.

As p-type wafers tend to degrade under illumination at elevated temperature, LeTID tests were performed. These showed a maximum LeTID of POLO IBC cells < 4%rel., which fully recovers after extended testing. From lifetime data, we conclude that the LeTID-related defect concentration is independent of the Ga doping concentration [3].

3.2 PolyZEBRA results

Quokka 3 simulations of polyZEBRA solar cells were conducted varying bulk resistivity from 1 to 20 Ωcm and bulk lifetimes ranging from 1 to 10 ms.

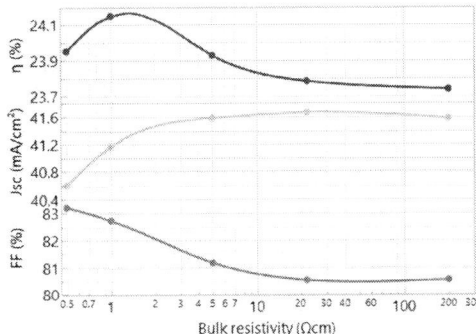

Figure 4: Quokka3 simulation results for the efficiency plotted against base resistivity for different bulk lifetime levels.

The results of the simulations indicate that, in particular for lower lifetime values, lower bulk resistance values are favorable – even lower than 1 Ωcm.

For the experimental study, wafers were provided by Norsun with 1.8 Ωcm, 5.16 Ωcm. and 200 Ωcm, and material from our reference supplier with a resistivity of 22 Ωcm. Lifetime values that were obtained by different structures (passivation by SiN$_x$ only, passivation by n-poly-Si (annealed at 925°C) and SiN$_x$) indicate excellent effective lifetime values in the range from 4 ms to 11 ms (1 Ωcm to 200 Ωcm). However, when processing solar cells from the given material, no clear trend observed, as shown in Figure 5. However, the 5.16 Ωcm material performed substantially worse. This was understood to be due to minor faults in the lifetime samples, as PL imaging revealed substantial oxygen-related stacking faults (OSFs). It is presumed that the high processing temperature of the boron diffusion triggered the defects to be activated.

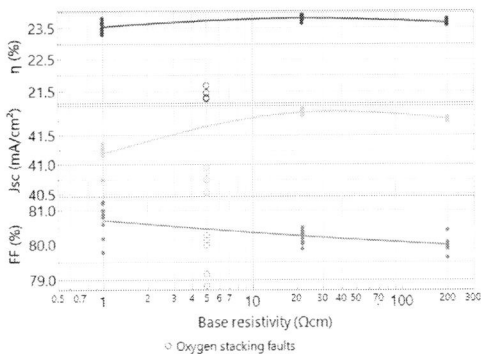

Figure 5: Solar cell efficiency, short current density and fill factor obtained for the different base material resistivities. Open symbols mark cells affected by oxygen stacking faults.

From the cell data we see a similar trend as predicted by the Quokka 3 simulations: For lower base resistivity the FF increases, but at the same time the current decreases. Contradicting to the simulation, the measured cell efficiency is highest for the 22 Ωcm material. This is because it was provided by a different wafer supplier and showed higher bulk lifetime on test wafers. Nevertheless, the differences are small and so the project material shows good quality.

Last, solar cells produced from the different materials were subjected to LeTID testing (80°C, 1 sun illumination). The results of the cells' V_{oc} is plotted in Figure 6. For this purpose, solar cells were cut to 5 by 5 cm pieces, which explains the lower V_{oc} as compared to the solar cells. The lower graph indicates measurements of the reference cell that was not subjected to the light and temperature treatment and stored in the dark. The V_{oc} increases also for the references over time, which indicates a systematic change in the IV measurement setup. Overall, no degradation is observed within the time frame investigated.

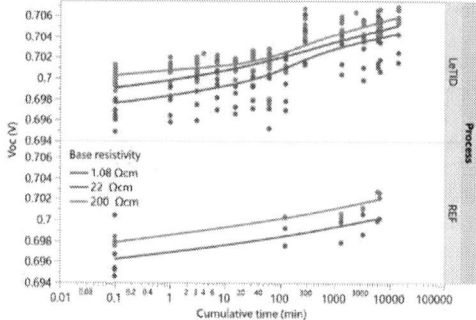

Figure 6: Top-row: Open circuit voltage of solar cells after different durations of light and elevated temperature treatment. Bottom-row: Voltages of non-degraded cells.

3.3 ZEBRA module results

ZEBRA solar cells (with identical metallization scheme as the polyZEBRA cells) were interconnected to produce bifacial modules using an industrial tabber stringer at ISC Konstanz. Bifaciality factors of the mini modules in glass-glass configuration as shown in Figure 7 were measured to be 82.4% with ZEBRA cells. However, due to the poly-Si layers on the rear of the passivating contact polyZEBRA IBC cells, the bifacialty is reduced to 69.3% which is in the range of standard PERC+ modules [10].

Figure 7: Bifacial mini module in glass-glass configuration (left: front, right: rear).

TC and DH tests results of the modules are shown in Figure 8. The degradation of less than 2% after 3,000 h of damp heat testing and 600 thermal cycles prove excellent long-term performance (3xIEC).

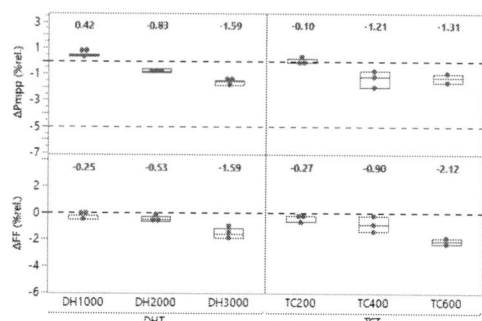

Figure 8: TC and DH test results of bifacial glass-glass modules.

In an alternative approach, Zebra solar cells were interconnected using TNO's bifacial interconnection technology with conductive tracks directly deposited on the rear-side glass. As shown in Figure 9, fully functional glass-glass bifacial module prototypes were made using silver and copper-based interconnection tracks. Excellent initial performance was obtained on the silver-based interconnection modules with FF above 79% corresponding to a relative cell-to-module FF loss of around 1.5%. Bifaciality factors above 83% were reached, similar to the bifaciality of the solar cells.

Due to higher resistivity of the metal tracks, suboptimal FF values (around 74%) were measured on the copper-based interconnection modules. High resistive paths in the copper electrical circuit were identified using dark lock-in thermography measurements. Based on these measurements, the interconnection design is currently being optimized to reduce resistive losses and improve FF.

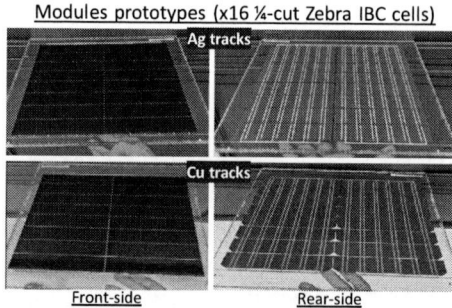

Figure 9: TNO's bifacial mini-module prototypes in glass-glass configuration.

A silver-based interconnection module is currently under thermal cycling reliability tests. As shown in Figure 10, after a degradation of 2.5% in the first 100 cycles, the power of the bifacial module was stable for the following 200 cycles. Despite a slightly larger initial degradation compared to the monofacial reference module based on conductive foil interconnection technology, the performance loss of the bifacial glass-glass module is well below the 5% mark.

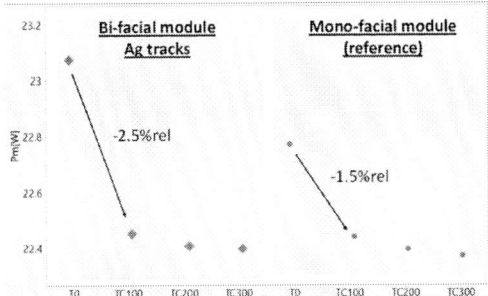

Figure 10: Module power changes of TNO's bifacial and monofacial modules as a function of number of thermal cycles.

Additional reliability testing will also be done on the copper-based interconnection modules, and upscaling of the technology to full-size modules is currently on-going.

3.3 Recycling results of IBC cells

Initial tests showed that the fact that insulation layers are employed on the rear side of the cells and different materials such as silver and copper are increasingly used, classical wet chemical separation routes cannot be used for recycling. In addition, ever thinner solar wafers make the reuse of silicon in elemental form less and less attractive. In the below presented route (Figure 11), the wafers are milled and the resulting particles dissolved in sodium hydroxide (NaOH). Due to the size of the particles, the reaction in which hydrogen and dissolved sodium silicate is produced is exothermic. With sedimentation or filtration, metals can be recovered.

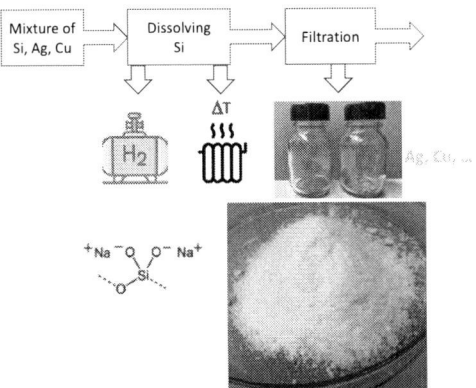

Figure 11: New recycling route for modern solar cell scrap.

After some cleaning steps, solid waterglass is obtained, a raw material for manifold applications from cosmetics to the chemical industry.

4 CONCLUSION

Experimental results show the factors that need to be taken into account when sourcing wafers for p- and n-type based IBC solar cells. For the p-type based POLO IBC cells, a higher base resistance range is required compared to PERC. For the polyZEBRA n-type cells, it was demonstrated that the performance hardly depends on the base resistivity as long as the minority carrier lifetime is sufficiently high and the oxygen stacking faults are sufficiently few.

We could demonstrate the excellent long-term durability of IBC cells and modules indicating the high TRL of the product already achieved - both solar cell concepts and their module interconnection approaches show excellent performance even after extended aging tests. Furthermore, a new recycling approach for modern IBC solar cells has been introduced.

5 ACKNOWLEDGEMENTS

This project has received funding from the European Union's Horizon Europe research and innovation programme under grant agreement No.101084259

6 REFERENCES

[1] Kopecek, Radovan, et al. Interdigitated back contact technology as final evolution for industrial crystalline single-junction silicon solar cell. In: Solar. MDPI, 2022. S. 1-14.

[2] Mertens, V., et al. Local PECVD SiOxNy/n-poly-Si deposition through a shadow mask for POLO IBC solar cells. In: Proc. 38th Eur. Photovolt. Sol. Energy Conf. Exhib. 2021. S. 135-139.

[3] T. Dullweber et al., Optimized Ga-doped Cz wafers for POLO IBC cells, EUPVSEC (2024), submitted

[4] Chen, N., et al. Thermal Stable High-Efficiency Copper Screen Printed Back Contact Solar Cells. *Solar RRL*, 2023, 7. Jg., Nr. 2, S. 2200874.

[5] Chen, N., et al. Screen printed copper paste for metallization of IBC solar cells. In: AIP Conference Proceedings. AIP Publishing, 2023.

[6] Kopecek, R., et al. ZEBRA technology: low cost bifacial IBC solar cells in mass production with efficiency exceeding 23.5%. In: *2020 47th IEEE Photovoltaic Specialists Conference (PVSC)*. IEEE, 2020. S. 1008-1012.

[8] Linke, J., et al. Fully passivating contact IBC solar cells using laser processing. In: Proceedings of the 8th World Conference on Photovoltaic Energy Conversion, Milan, Italy. 2022. S. 26-30.

[9] Min, B., et al. 24.2% efficient POLO back junction solar cell with an AlOx/SiNy dielectric stack from an industrial-scale direct plasma-enhanced chemical vapor deposition system. *Progress in Photovoltaics: Research and Applications*.

[10] Dullweber, T., et al. Present status and future perspectives of bifacial PERC+ solar cells and modules. *Japanese Journal of Applied Physics*, 2018, 57. Jg., Nr. 8S3, S. 08RA01.

41st European Photovoltaic Solar Energy Conference and Exhibition

This presentation was selected by the Sc. Committee of the EU PVSEC 2024 for submission of a full paper to one of the EU PVSEC's collaborating peer-reviewed journals.

SELF-ALIGNED PHASE SEPARATION FOR IBC CELLS USING PVD POLYSILICON

Erik Hoffmann*[1], Geoffrey Gregory[1], Massimo Centazzo[1], Muhammad Khan[1], Nabeel Khan[1], Verena Mertens[2], Philip Jäger[2], Sarah Spätlich[2], Ulrike Baumann[2], Thorsten Dullweber[2]

[1]EnPV GmbH, Durlacher Allee 93, 76131 Karlsruhe, Germany
[2]Institute for Solar Energy Research Hamelin (ISFH), Am Ohrberg 1, 31860 Emmerthal, Germany

*Corresponding author: Email: e.hoffmann@enpv.de, phone: +49 151 54 75 08 18

ABSTRACT: We introduce an innovative IBC solar cell process leveraging the directional deposition nature of doped polycrystalline silicon (poly-Si) through physical vapor deposition (PVD). This method enables the self-alignment of passivated contacts, effectively separating the polarities. The self-aligned back contact (SABC) cell incorporates n-type and p-type passivated contacts, achieved through interfacial oxide (SiO_X) and doped n- and p-type poly-Si layers respectively, arranged in an interdigitated design on the back side. The insulation between the p-type and n-type poly-Si layers requires only a single structuring process of the firstly deposited p-type poly-Si. The subsequent blanket deposition of n-type poly-Si by PVD remains insulated from the p-type poly-Si layer due to the self-alignment properties inherent in our structuring and deposition processes. The SABC target process sequence can be implemented into existing TOPCon manufacturing lines requiring only two additional processing tools.
Keywords: IBC, TOPCon, passivating contacts, self-alignment

1 INTRODUCTION

The SABC solar cell[1] implements an interdigitated back contact (IBC) design, featuring passivating poly-Si on silicon oxide (poly-Si/SiO_X) contacts of both polarities on the rear side on n-type silicon wafers. The passivating contacts of opposite polarity are not insulated by a dedicated trench, but by an undercut in the trench wall itself. The trench is needed anyway to pattern the first poly-Si and create space for the second poly-Si to form the contact to the base. When etching the trench into the p-type poly-Si, the poly-Si layer is intentionally under-etched. We then utilize the directional nature of PVD of the second poly-Si to prevent its deposition underneath the under-etched first poly-Si layer, thereby achieving an insulation of the opposite doped poly-Si layers without the need for additional structuring or masking. The n-type poly-Si, located at the bottom of the trench, has already demonstrated its excellent passivating properties yielding $iV_{OC} > 735$ mV [1,2].

The proposed process flow for manufacturing the SABC solar cell builds upon the standard TOPCon process. The transition from TOPCon to SABC requires minimal adjustments, involving only four additional processing steps, two of which require new tools: laser structuring of the first poly-Si layer and for the PVD deposition of the second poly-Si layer. The other two steps can be implemented on existing tools of the TOPCon process: PECVD deposition of an etch mask and wet chemical cleaning. Beyond these additions, the SABC process also requires the deposition of in-situ doped p-type poly-Si in a LPCVD tool formerly used for n-type or intrinsic poly-Si deposition.

In this contribution we present the SABC cell concept and evaluate a possible process sequence. Furthermore, our optimized p-type poly-Si/SiO_X layer stack deposited by LPCVD demonstrates an excellent $J_0 = 2.3$ fA/cm², similar to previously reported values [3], and we show SEM images of the under-etched trench. We also provide proof of concept with a first solar cell yielding an efficiency of $\eta = 20.2\%$.

---
[1] Patent pending.

2 EXPERIMENTAL

2.1 SABC solar cell concept and process flow

Figure 1 a) illustrates the SABC solar cell design and figure 1 b) outlines a possible processing sequence. Unlike other IBC concepts with two TOPCon passivating contacts, often referred to as TBC [5], this approach requires no dedicated trench to insulate the poly-Si layers of opposite polarity. Instead, the undercut of the trench wall separates the two polarities, where the first p-type poly Si layer (emitter) lies atop the trench surface, while the second n-type poly-Si is located at the bottom of the

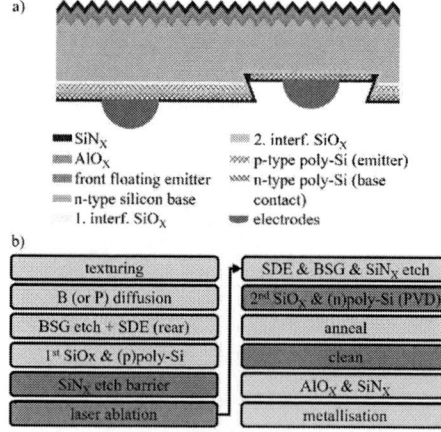

Figure 1: a) Sketch and b) process sequence of a self-aligned back contact (SABC) cell. The IBC cell with passivating poly-Si/SiO_X contacts features an insulation of the poly-Si layers of opposite polarity by separating the layers only across the undercut along a trench wall. Compared to a conventional TOPCon process, this process only requires two additional processing tools (orange) and another two fabrications steps (blue) which can be implemented on existing tools.

trench. Underneath the undercut, which realised by wet chemical etching of the trench, no poly-Si is deposited during the PVD process. The n-type poly-Si also covers the p-type poly-Si emitter forming a low resistive tunnelling contact, as both poly-Si layers are highly doped with dopant concentrations exceeding $C = 10^{20}$ cm^{-3}. The n-type poly-Si layer covers the full surface and enables the use of the same metallization scheme for both polarities.

Figure 1b) illustrates one possible process sequence and highlights the additionally required fabrication steps to a standard TOPCon process in orange and blue. The orange highlighted steps require two additional tools, while for the blue coloured steps an existing PECVD tool and wet bench can be used respectively. Furthermore, the n-poly-Si in a TOPCon process is deposited either in-situ doped or intrinsic in the LPCVD tool. This tool requires retrofitting to deposit p-type silicon, while the $POCl_3$ diffusion furnace (in the case of ex-situ doped n-poly) is used for SABC's annealing step. Other slightly modified sequences may be advantageous, depending on the TOPCon process of the base line.

In accordance with TOPCon, the SABC target process starts with texturing and boron diffusion on both front and rear. A single side BSG etch followed by alkaline etching of the boron emitter on the rear prepares this surface for the poly silicon deposition. Instead of the boron diffusion forming the front floating emitter (FFE), a phosphorus diffusion forming a front surface field (FSF) or an un-diffused surface could be used for the SABC process flow as well. Advantageously, a phosphorus-diffused surface typically yields better surface passivation than a boron-diffused surface [6]. After depositing interfacial oxide and p-type poly-Si by LPCVD on both the front and rear side, the rear side receives a PECVD-deposited SiNx etch barrier. Laser ablation locally removes the SiNx layer and exposes the poly-Si in the typical IBC pattern. Alkaline wet chemical etching strips the poly-Si on the front side up to the BSG layer and on the rear etches the p-type poly-Si in the exposed areas. The etching on the rear not only removes the poly-Si layer, but also etches the silicon base vertically and lateral, creating an undercut of the surface. Even after removal of the barrier layer, the poly-Si and trench edge will shadow a significant part of the trench wall. After etch barrier (rear) and BSG (front) removal and growing a second interfacial oxide, the subsequently deposited n-type poly-Si by PVD will not cover the full surface on the rear. With the highly directional deposition technique, the doped silicon mainly deposits on vertically exposed surfaces, i.e. surfaces not being shadowed by the undercut. Hence, at the bottom of the trench n-type poly-Si forms a passivating contact but there will also be an n-type poly-Si on top of the p-type poly-Si forming a tunnel junction to the emitter. The shadowed part of the trench wall insulates both passivating contacts, without the need for additional processing steps like structuring, local etching or masking. The SABC process finishes with high temperature annealing, crystallising the poly-Si layers and activating the dopants, standard passivation with AlOx and SiNx and metallisation by screen printing. Since the n-type poly-Si covers the surfaces of both passivating contacts, the same screen-printing paste may be applied in a single printing step for both polarities.

2.2 Sample preparation

The passivation quality of the p-type poly-Si/SiOx layer was investigated and optimized at ISFH on symmetrically processed n-type wafers, passivated with a PECVD SiNx layer and fast firing. Three different interfacial oxides were annealed at different temperatures to find the optimum process window in terms of sheet resistance and surface passivation.

Scanning electron microscopy (SEM) images inspect the n-type poly-Si deposited at the trench cross section and show separation of the layer at the under-cut. The sample preparation includes laser ablation of an etch barrier on top of a p-type poly-Si, wet chemical etching of the poly-Si and up to 10 μm of the bulk, and subsequent etching the barrier layer. Then, to enhance the contrast between n-poly silicon and bulk wafer, an 80 nm dielectric layer by PECVD is deposited followed by the PVD n-type poly-Si layer. Laser scribing enables cleaving in a 90 ° angle to the trench direction.

For the SABC cell processing at ISFH, we are currently using a slightly different process flow than depicted in figure 1b): Utilizing BSG as an etch mask for alkaline etching requires additives that are not yet available in the standard wet bench. Hence, the cells feature a non-diffused front surface and single side acidic polish on the rear to remove the texture. Additionally, we apply ISFH's standard p-type poly-Si LPCVD deposition process yielding the excellent results in figure 2, which includes a dedicated annealing of the p-type poly-Si layer. Presently, in the SABC cell processing the p-poly layer receives a second anneal required for n-type poly-Si crystallisation, leading to an iV_{OC} drop below the target value, as too much boron diffuses into the silicon bulk, increasing J_{0e} of the emitter, which is subject to future improvements.

3   RESULTS AND DISCUSSION

3.1 p-type poly-Si surface passivation

Figure 2 a) shows sheet resistance R_{sh} and figure 2 b) J_{0e} of a 200 nm thin p-type poly-Si emitter deposited by LPCVD on three different oxides and annealed in an inert atmosphere at temperatures between 820 °C and 900 °C. All oxides achieved their lowest R_{sh} at the maximum investigated annealing temperature T = 900 °C. However, the J_0 yields its lowest value $J_0 = 2.3$ fA/cm² at 880 °C with a median of $J_0 = 5.1$ fA/cm². This corresponds to maximum $iV_{OC} = 739$ mV and, in combination with the low $R_{sh} \approx 154$ Ω/sq, promises a high efficiency potential when applied to a SABC solar cell. The same test run included wafers annealed with an oxygen atmosphere at T = 860 °C (not shown). The resulting oxide layer can serve as an etch barrier for the subsequent wet chemical etching of the poly-Si. For this particular group the median $J_0 = 6.7$ fA/cm² stands only marginally higher than the samples subjected to the inert annealing. The sheet resistance $R_{sh} \approx 303$ Ω/sq is notably higher, as the surface of the poly-Si oxidizes and consumes a substantial amount of the boron dopants. This oxidized p-type poly-Si still forms the emitter of the solar cell described below, but the oxidation is targeted to be skipped in the future, as we expect too much in-diffusion of boron into the bulk, if a second annealing step is applied for crystallization of the n-type poly-Si.

3.2 N-type poly-Si insulation

Figure 3 shows a SEM image of the cross section of one side of an anisotropic etched trench with a dielectric and n-type poly-Si layer. The p-type poly-Si is etched slower than the lowly doped n-type c-Si wafer resulting in

41st European Photovoltaic Solar Energy Conference and Exhibition

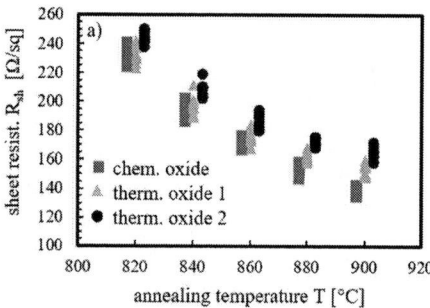

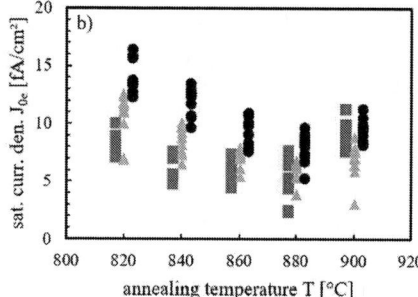

Figure 2: a) Sheet resistance R_{sh} and b) saturation current density J_{0e} for p-type poly-Si deposited by LPCVD at ISFH. All oxides were annealed at the same temperature and are only displayed next to each other for better visualisation. Annealing at $T = 880°C$ yields the lowest $J_{0e} = 2.3$ fA/cm² and sufficiently low $R_{sh} = 154$ Ω/sq.

approximately 0.5 µm poly-Si overhang on the wafer surface. Additionally, the anisotropic etch removes the Si faster into the <111> direction resulting in an additional undercut with a 54° angle. The SiNx layer, deposited be PECVD to enhance the contrast between n-type poly-Si and c-Si wafer, wraps itself around all exposed surfaces, but does appear to be thinner underneath the trench. The n-type poly-Si however is mostly deposited on the vertically exposed and non-shadowed surfaces. Underneath the undercut its thickness starts to decrease

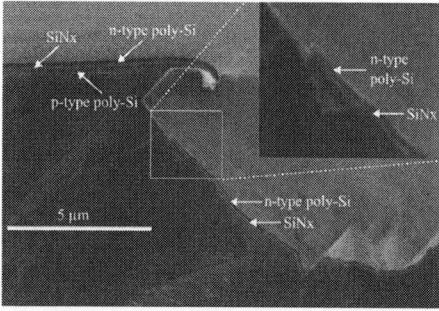

Figure 3: SEM image showing the gap in the n-type poly-Si layer generated by the undercut of the p-type poly-Si at the trench edge. An additional SiNx dielectric layer, deposited before the n-type poly-Si sputtering enhances the contrast between poly-Si and bulk, but will not be included in a solar cell device.

until the n-type poly-Si layer fully disappears before the undercut reaches its maximum lateral depth. On the underside of the undercut we cannot find any n-poly silicon with SEM. With this etched geometry we achieve an approximately 2-3 µm gap between the n-type poly-Si at the bottom of the trench and the n-type poly-Si on top of the p-type poly-Si forming the emitter. Tailoring the etch depth or using isotropic etching allows varying the gap width.

3.3 Proof-of-concept SABC solar cell

Figure 4 a) shows the IV-curve and b) a Lock-In-Thermography (LIT) image of a first SABC cell with an efficiency of 20.2 %. The open circuit voltage V_{OC} yields 708 mV and indicates a reasonable surface passivation, considering that the p-type poly-Si emitter is annealed twice. However, both short circuit current density J_{SC} and fill factor FF are rather low and limit the cell's efficiency. A possible reason for the low J_{SC} could be a process-related low wafer bulk lifetime, which would also affect the FF, which is subject to further analysis and improvement. The series resistance $R_S = 0.72$ Ωcm² is not the primarily factor limiting the FF with pFF-FF = 3.3 %. The shunt resistance $R_{sh} = 3.3$ kΩcm², extracted from the dark IV-curve, still needs improvement but shows that the gap across the trench edge insulates the two poly-Si layers quite well.

Figure 4 b) shows a thermal map of the cell by Lock-In-Thermography (LIT) under reverse bias. Defects, such as shunts because of incomplete insulation of oppositely doped surfaces, generate heat due to increased current flow. In a typical IBC cell with interdigitated positive and negative contacts, the oppositely doped surfaces come into close contact across most of the back side of the cell and therefore may form shunts. In this cell, the LIT shows mostly local shunts at, or close to, the busbars interconnecting the positive electrodes, while most p-type/n-type poly-Si at the indigitated electrodes appear well insulated. Underneath all busbars interdigital pattern is interrupted and the poly-Si is doped according to the busbar polarity. Two possible causes of shunts at the emitter busbars are currently under investigation: the positive busbars are piercing through the emitter after fast firing of the paste and contact the base, or local shunts across the trench wall only in the busbar area. Given that a non-fire-through paste is used to form the positive busbar, the first option seems unlikely. Instead, we suspect that the laser patterning parameters near the emitter busbar are suboptimal, leading to the formation of shunts across the trench in this region. An improved laser pattern at the emitter busbar should solve these local shunts in the next cell run.

4   CONCLUSIONS

In this contribution, we present the SABC back contact cell concept and show first results on the passivation quality of the p-type poly-Si emitter, the self-aligned insulation and a proof-of-concept solar cell. The solar cell only shows local shunting mostly at the positive busbars contacting the emitter. Across most part of the cell, e.g. between the positive and negative electrodes, the poly-Si layers of opposite polarity are well insulated. Solving the cause of the local shunts and implementing a single annealing step of the emitter should increase the efficiency significantly.

108

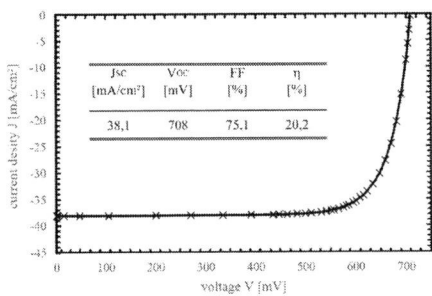

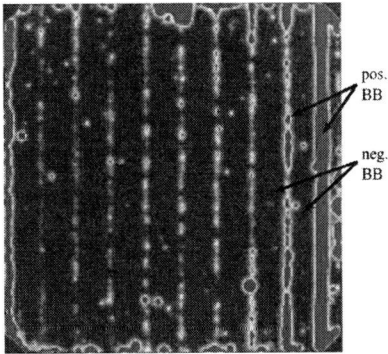

Figure 4: a) IV curve of the best SABC cell so far and b) a lock in thermography image thereof. The cells efficiency is mostly limited by FF and J_{SC} with a V_{OC} = 708 mV showing good surface passivation. The LIT image shows shunts mostly located at the emitter busbars, proving a good insulation between the n-type and p-type poly-Si layer across most of the cell.

5   ACKNOWLEDGEMENTS

We would like to thank the team at ISFH for processing the samples and their support of this work.

6   REFERENCES

[1] Schneiderlöchner, E.; Dietsch, T.; Hoss, J.; Linke, J.; Polzin, J.; Mack, S.; Nagel, H.; Linß, V.; "i-TOPCon Solar Cells Prepared by High Throughput Magnetron Sputtering of In-Situ Doped n-Type Amorphous Silicon Layers." in *Proceedings of 50th IEEE Photovoltaic Specialists Conference IEEE-PVSC*, San Juan, Puerto Rico, 2023

[2] Gregory, G. et all. in *Proceedings of 52nd IEEE Photovoltaic Specialists Conference IEEE-PVSC*, Seattle, USA, 2024

[3] Mack, S.; Schube, J.; Fellmeth, T.; Feldmann, F.; Lenes, M.; Luchies, J.-M. Metallisation of Boron-Doped Polysilicon Layers by Screen Printed Silver Pastes. Phys. Status Solidi-Rapid Res. Lett. 2017, 11, 1700334.

[4] Fell, A, *Quokka3 Software*, Version 2.4.5

[5] Kopecek, R.; Buchholz, F.; Mihailetchi, V.D.; Libal, J.; Lossen, J.; Chen, N.; Chu, H.; Peter, C.; Timofte, T.; Halm, A.; et al. Interdigitated Back Contact Technology as Final Evolution for Industrial Crystalline Single-Junction Silicon Solar Cell. *Solar* **2023**, *3*, 1-14

[6] Schmidt, J.; Peibst, R.; Brendel, R: Surface Passivation of Crystalline Silicon Solar Cells: Present and Future, *Sol. Energy Mater. Sol. Cells* **2018**, 187, 39–54

41st European Photovoltaic Solar Energy Conference and Exhibition

GAS PHASE, SELECTIVE ETCHING OF POLY-SILICON FOR LAYER PATTERNING

Author(s): Laurent Clochard (1), Mingzhe Yu & Ruy Sebastian Bonilla (2) ,
Paul Tierney & James Wright (3) , Fiacre Rougieux, & Yalun Cai (4)

1) Nines Photovoltaics,, Dublin, Ireland l.clochard@nines-pv.com
2) University of Oxford, Electronic and Interface Materials Laboratory, UK
3) Technical University Dublin (TUD), Ireland
4) UNSW, Australia

ABSTRACT: This paper will present patterning process developments using a single-side, gas-phase etching process based on the thermal reaction of poly-Silicon and a spontaneous etching Halogen gas, that results in a high etching selectivity between layers, and a high etching rate. This work was carried out in the context of the development of solar cell architectures beyond PERC and TOPCON, where more sophisticated etching steps are required in order to accurately pattern poly-silicon layers across the wafer surface.

1 Background

Tunnel oxide passivated contacts (TOPCON) are now commonplace in industrial manufacturing to increase solar cell efficiencies beyond PERC cells performance. The rear contacts are the most straightforward to passivate, as a full surface layer stack of an ultra-thin oxide of a few nanometers, topped with a layer of doped amorphous silicon is relatively easy to realize and deliver good efficiency improvements. However, when applying the same method to the front contact, the full area poly-Si leads to high parasitic absorption of ~ 0.5 mA/cm² per 10 nm poly- Si, and surface passivation degradation for poly-Si layers <10 nm [1]. Therefore, applying a full area poly-Si to the front side result in a net loss. If the front poly-Si layer was only applied under the metal contact, the efficiency gain is calculated to be up to +0.5% absolute [2] from a Voc increase of about 10mV; there is therefore an interest in developing a suitable patterning process to selectively etch the front solar cell. Similarly, back contacted cell architectures require patterning of mostly poly-silicon layers of various doping.

2 Aim and approach

In this work we investigate the patterning of poly-Si using a single sided gas phase etching process to remove the poly-Si from the unwanted areas. This gaseous etching process has been used successfully for texturing applications of multi-crystalline wafer [3], and for the poly-Si wrap-around removal produced by commonly used industrial deposition tools such as LPCVD and PECVD tools [4]. Although not yet used in production, its fundamental characteristics makes it suitable for high throughout/low cost processing.

The validation of the suitability of the process can be split in a number of building blocks: first the ideal etch process needs to demonstrate a good selectivity between layers, i.e. the layers used to protect the useful areas versus the layers to be etched. Second, the process window needs to allow for a good etch stop control [REF], in order to minimize the impact on the underlayers. Finally, the resulting etched surface needs to show a good passivation, deposition of the passivating layers[REF]. In this paper we will mostly focus on the first block, and look at the various ways the selectivity of the process can be leveraged to achieve our patterning goals.

3 Results

For this set of experiments, p-type Cz wafers of M0 and M2 sizes were used to characterize the etch rate of various deposited layers. The wafers kept in a shipping wafer box, stacked on each other requires a short pre-clean to remove organic contamination and achieve a more uniform results. They are typically pre-cleaned in a 5%HF solution for 1 minutes, followed by 5minutes D.I. rinsing, and 10minutes drying in 95%Nitrogen at 80degC, before being process through an atmospheric pressure(AP) gas-phase Etching reactor (ADE) using a halogen gas as etchant, in this case molecular Fluorine (F2). The reactor temperature is set below 300 DegC and does not require a plasma source. The etch rate is directly affected by temperature, gas concentration. The overall gas flow setting affects the efficiency of the process, i.e. utilization rate of the gas. The wafers are weighted before and after the process, with a 1mg balance accuracy. The average etched thickness across the wafer surface can be calculated from the mg measured value, the wafer dimensions, and the theoretical density of the etch materials.

3.1 Etch selectivity

The layers were deposited on flat surfaces after saw damage removal (no texture).
The etch rate is the speed of the etching process for a material [nm/min]. The etch selectivity is the ratio of etch rates between materials.

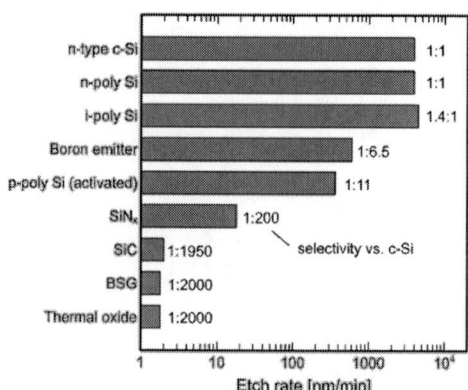

Fig 1. ADE Etch rate for various layers using 20%F2/N2 at 220degC

It was found that the etch rate is highly dependant on the nature of the layer, and therefore overall, the etch is very selective. Dielectric layer such as oxides and nitride are very slow to etch. PSG and BSG glass have similarly very slow etch rate when compared to silicon. It was also found that Boron doped silicon can slow down the etch rate considerably, as it is the case for heavily doped p-poly silicon layers.

Considering the overall result and the nature of layer, we devised a considered a number of ways to implement the selectivity results into a patterning scheme.

3.2 PATTERNING METHODS & OPTIONS
The graph below shows a summary of the various options we have been considering.

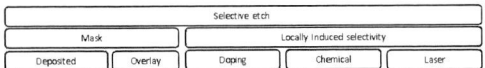

Fig 2. Options for patterning using selective atmospheric gas-phase etching

The methods described can be divided in two main strands. The first one entails the use of a mask either deposited or overlayed. The second strand are methods that are designed to locally induce selectivity.

3.3 Mask
As a baseline, our first experiment used the traditional masking method used in semi-conductor manufacturing.
Here we selected a thermal SiOx, with a very high selectivity of 2000:1 as per table above, to create the mask. We used 100 p-type, 4'' polished wafers of semi-conductor grade. An oxide thickness of circa. 120nm was grown onto the wafer top surface by dry oxidation at 1100 °C. A photoresist (S1818 G2) was then applied and soft baked at 115 °C for 60 seconds. Then a POLOS uPrinter was used to expose a pattern, that was subsequently developed in Microsposit 351 developer, then hard baked at 135 °C for 180 sec. The wafer was then etched in 10:1 BOE for 120 sec to pattern the oxide. The photoresist was then stripped in Microposit SVC175 PR stripper.
The wafer was then ready to be processed our selective gas-phase etch, at a temperature of 193 degC and gas concentration of 30%. Finally the SiO2 mask was fully removed by dipping the wafer in HF. The sample was characterised at the various stages of the process by optical microscope imaging.

The pattern contains lines and dots of micrometer sizes. For the first pattern, the SiO2 mask had lines of 7um width with a gap of 9.7um.

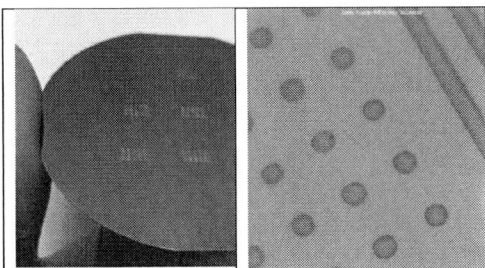

Fig 3. Pictures of the patterned wafer prior to gas-phase etching

After the silicon etch using gas-phase F2, the lines engraved in the silicon had very similar dimension to the mask itself.

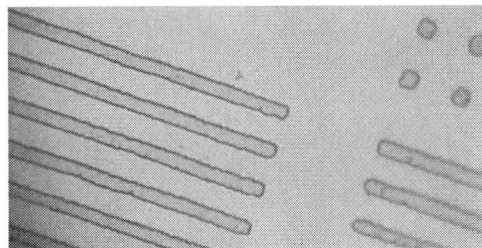

Fig 4. Pictures of the patterned wafer after gas-phase etching and removal of the remaining SiO2 mask.

	Stage	Line width	Line spacing
Mask	Resist Design	4um	14um
	Photoresist measured	7um	9.7um
	SiO2	7um	9.7um
Result	F2 etch	7um	9.7um

Table 1. Parameters of the first mask design /experiment

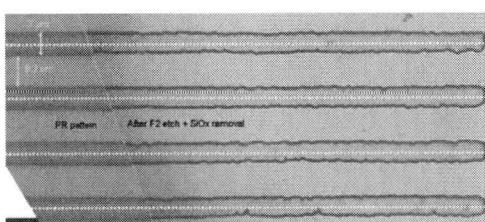

Fig 5. Photo montage of the patterned wafer before (left side) and after (right side) gas-phase etching for the first experiment.

A second sample was prepared with smaller features (lines of 3um width), and over-etched (about 2.2 times the time of the previous experiment).

	Stage	Line width	Line spacing
Mask	Resist Design	3um	21um
	Photoresist measured	3.4um	20.1um
	SiO2	3.8um	19.9um
Result	F2 etch	13.3um	10.3um

Table 2. Parameters of the second mask design/experiment

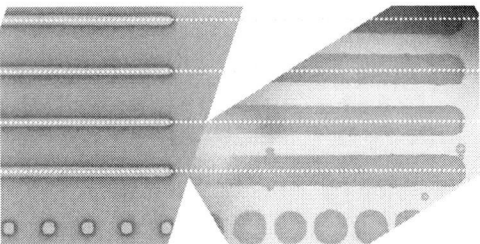

Fig 6. Photo montage of the patterned wafer before (left side) and after (right side) gas-phase etching for the second experiment.

In this second experiment, we can see that despite the high selectivity of the mask used, the extended etching as resulted in enlargement of the feature beyond the SiO2 mask boundaries. This is most likely due to the silicon being etched under the SiO2 in a first stage, eventually lifting the oxide above, allowing the continuation of the line enlargement.

3.4 Overlay

Another masking method that can be used with this gas-phase etch process is the use of a overlay; instead of being deposited on the wafer, a mask made of low etching material such as quartz or glass can be laid on top of the wafer and processed as such. This type of shadow mask can be quite sophisticated and provide very fine apertures in the micrometer range. To demonstrate the process, a proof of concept experiment was carried out using a simple disk made of quartz, positioned on top of a poly-silicon wafer, before processing using the atmospheric gas-phase etching process.

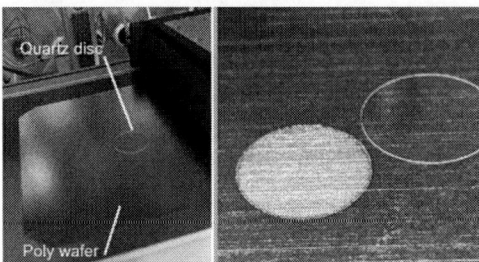

Fig 7.Left: wafer entering the gas-phase reactor with quartz disc on top. Right: after the etch, zoom on the area that was protected by the quartz disc.

In contrast to the previous masking method, this method is very simple.

3.5 Locally induced selectivity

Another pathway we have developed is to change locally the properties of the selectivity surface by using a number of local application method such as changing the local doping concentration, applying chemicals or lasering the surface. Each of these local surface conditioning methods affect the etch rate, and can either catalyze the etching reaction, or slow the reaction down.

Changing locally the doping:
Referring back to the selectivity graph, p doped poly-silicon layers typically etch 11 times slower than un-doped poly layers. There are a number of methods available to locally change the doping of the layer, such as flex trail, APCVD, laser or ink jet printing. Once the doping level of the desired area, the sample can be etched using the very selective atmospheric gas phase etch to remove the lowly dopped area, and leave the more heavily dopped part, hence creating a patterned surface.

Applying chemical locally to catalyze the etch rate:
Here we referred to a method described in patent [6] where the use of a chemical to pre-condition the polysilicon surface can catalyze the etch process locally. Similarly to the previous method, if very fine lines or patterns are required, inkjet printing can be used to deliver precisely the chemicals to the surface.

Laser
Here the authors found that applying a laser to the poly-

silicon surface will enhance the gas-phase etch rate locally. Lines of various width (25um, 100um and 1mm) were lasered on a M2 c-Si wafer that had a stack of 15nm SiOx and 300nm of a-Si:H deposited by PECVD. The intensity of the laser was also varied between "low", "medium" and "high" settings.

After processing the wafer through atmospheric gas-phase etch using halogen gas, the lasered areas etched preferentially.

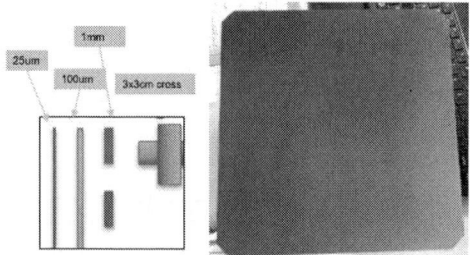

Fig 8. Lasered Wafer prior to gas-phase etching

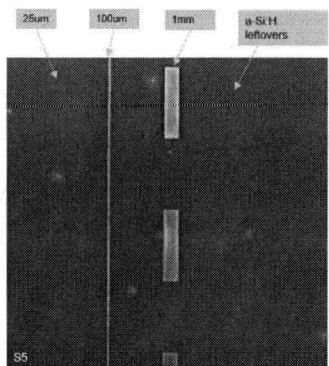

Fig 9. Zoom on the wafer after gas-phase etching, showing preferential etching of the lasered areas.

4. Conclusion

The high selectivity of the atmospheric gas-phase etch using Fluorine Halogen gas enables a number of method to pattern silicon. In particular, it provides convenient processes to pattern poly-silicon and amorphous silicon for applications in solar cell devices such as TOPCON or back contacted cells.

4. References

[1] Feldmann F., et al., 43rd IEEE PVSC, (2016)
[2] J. Stuckelberger, et al, EUPVSEC (2020) Industrial Solar Cells Featuring Carrier Selective Front Contacts
[3] B. Kafle, T. Freund, S. Werner, J. Schon, A. Lorenz, A. Wolf, L. Clochard, E. Duffy, P. Saint-Cast, M. Hofmann, J. Rentsch, IEEE J. Photovoltaics. 2017 7, 136, DOI: 10.1109/JPHOTOV.2016.2626921.
[4] Kafle B., Mack S., Teßmann C., Bashardoust S., Clochard L., Duffy E., Wolf A., Hofmann M. and Rentsch J. 2022 Sol. RRL 6 2100481
[5] L. Clochard, D. Young, M. Yu and R. S. Bonilla, SILICON PV (2024), Session 9: TOPCON Solar Cells, TIB Open publishing
[6] IE20200018A3

INVESTIGATION OF AG-REDUCTION ON SILICON HETEROJUNCTION SOLAR CELLS WITH DIFFERENT APPROACHES

Yu Wu, TNO Solar Energy
1BO4.6 EUPVSEC, Sep-24-2024

OUTLINE

❑ Background

❑ Performance evaluation of HJT cell with-low Ag metallization

❑ Investigation of fine line printing on low Ag metallization

❑ Summary & outlook

Background

THE TREND OF AG REDUCTION

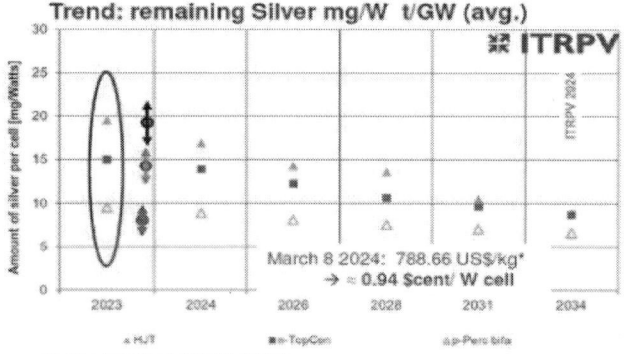

M. Fischer et al, ITRPV. 2024

Challenge!!!

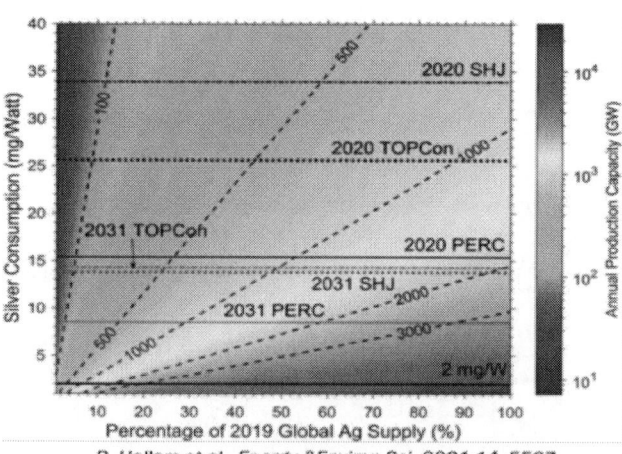

B. Hallam et al, Energy &Environ Sci. 2021 14, 5587

"to allow 3 TW production capacity of PV using only 20% of the global supply at 2019, regardless of technology, silver consumption needs to be below **2 mg/W**",

CURRENT ROUTES FOR AG REDUCTION

❖ Ag fine line print & mBB/0BB
❖ Ag coated Cu

- o In mass production
- o Complies to screen printing technology
- o *Ag consumption still exists*

❖ Cu plating

- o Towards to mass production
- o Fully replace Ag contact
- o *Extra investment on equipment and large footprint*
- o *Complex process*

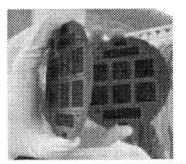

PV magazine March-2 2022

❖ Cu printing

- o Fully replaced Ag contact
- o Comply to screen printing technology
- o *Risk of oxidation*

MOTIVATION

❖ To reduce Ag consumption in the HJT cell process with minimal penalty on cell performance.

❖ Adhere to screen printing metallization process

❖ Investigate different alternatives including Ag-coated Cu (Ag/Cu) and pure Cu

Performance evaluation of HJT cell with low Ag metallization

SCENARIOS TEST FOR AG REDUCTION

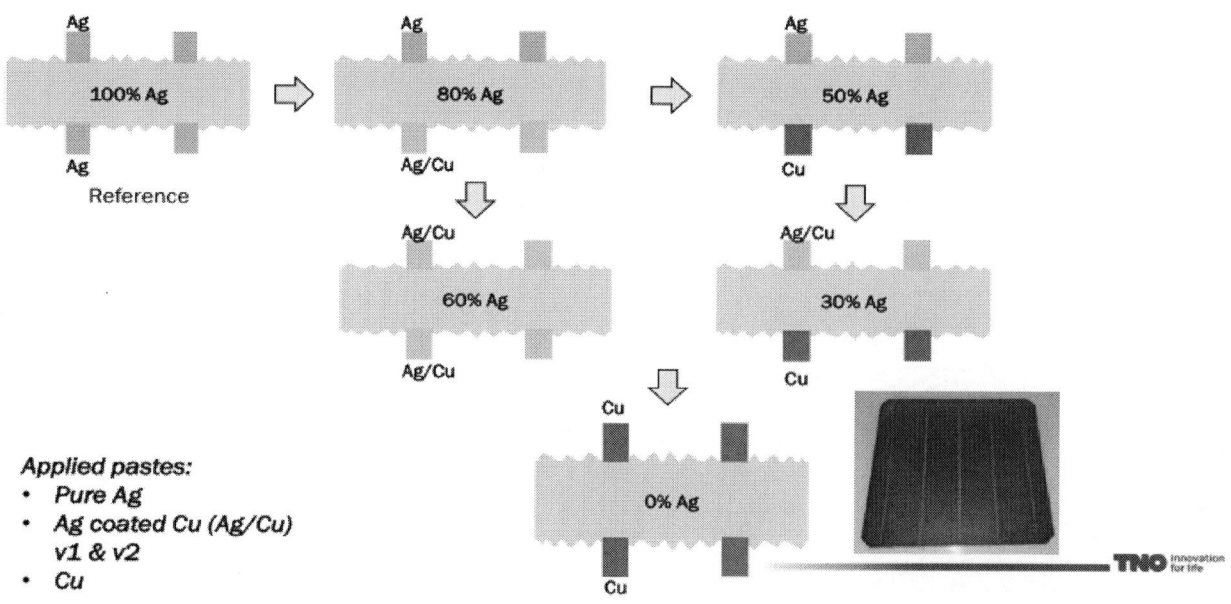

Applied pastes:
- Pure Ag
- Ag coated Cu (Ag/Cu) v1 & v2
- Cu

POWER LOSS OF HJT CELL WITH LOW AG METALLIZATION

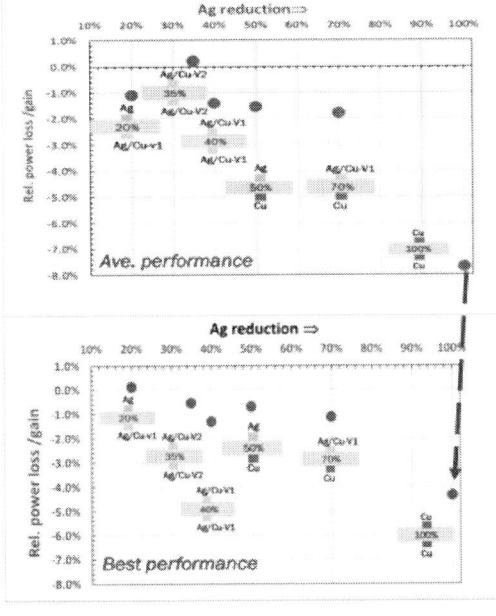

- Cu and low Ag pastes from commercial partners have been applied in this test. < 2% rel. power loss has been achieved up to 70% Ag reduction.

- HJT metallized with only Cu contacts shows the highest power loss

- For fully-Cu metallized cells, best cell has ~50% less power loss compared to reference than average, mostly because Cu metallization process execution was not completely stable yet.

POWER LOSS OF HJT CELL WITH LOW AG METALLIZATION

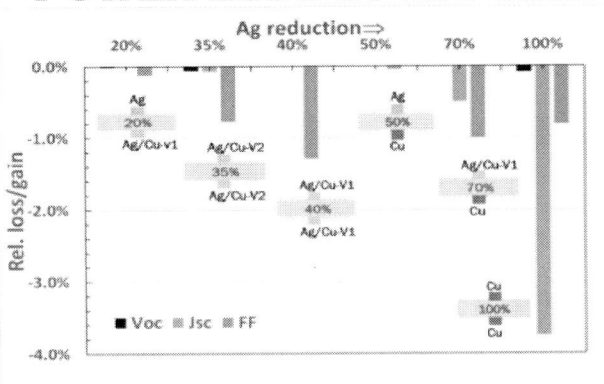
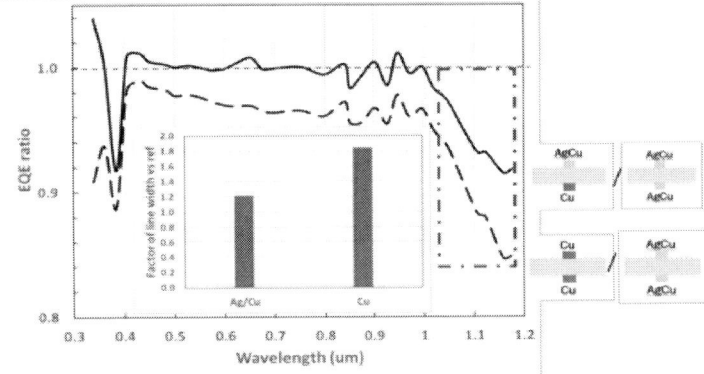

- V_{oc} variation is very small, no indication of surface passivation degradation.

- FF loss increases with using less Ag however most samples show <~1% FF loss.

- A significant J_{sc} loss of fully Cu metallized cell is due to optical shading. EQE also shows slight optical change of rear stack probably related to Cu metallization.

Investigation of fine line printing on low Ag metallization

FINE LINE PRINTING WITH DIFFERENT METAL PASTES

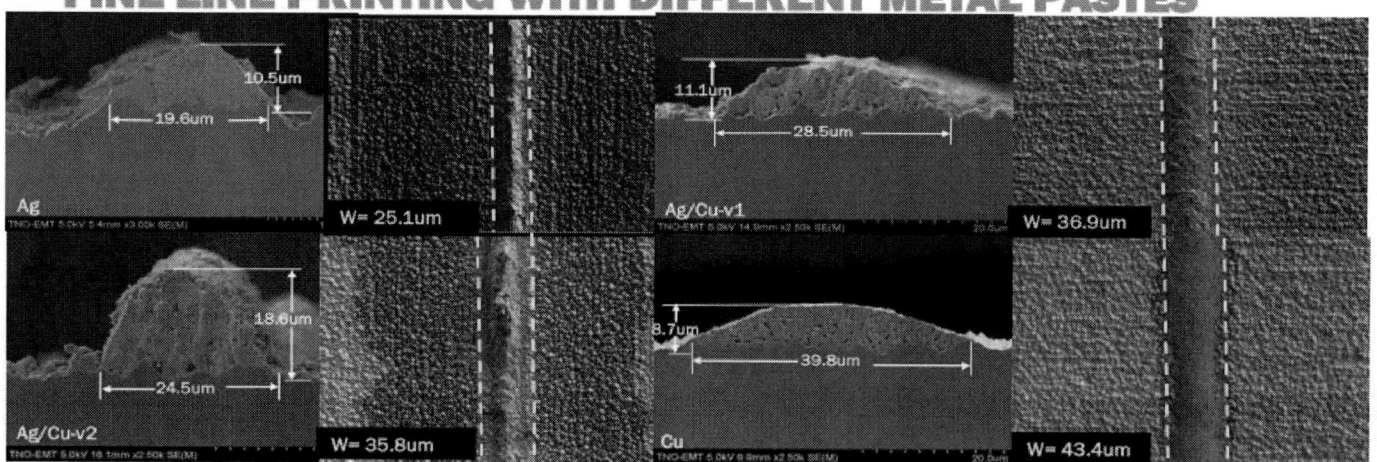

- Narrow finger width has been achieved on pure Ag.

- AgCu-v2 shows similar line width but better aspect ratio than AgCu-v1.

- Cu printing shows even large finger width with the lowest aspect ratio.

RESISTANCE (R_{LINE} & R_{CON})

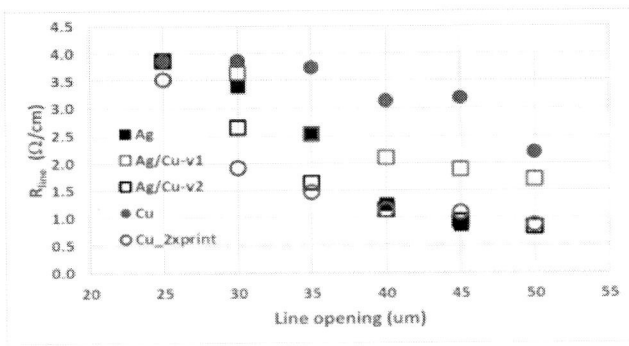 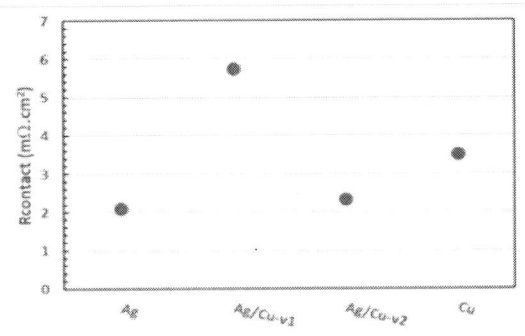

- With double print on Cu contact, R_{line} can be dramatically reduced compared to single printed Cu and it is even comparable to pure Ag contact.

- R_{con} of all metal contacts is <6mΩ.cm^2 showing good ohmic contact at interface.

MODELING: POWER LOSS OF LOW AG METALLIZATION WITH FINE LINE PRINT & MBB

Estimation of power loss for H-pattern grid:
- o The minimum line width achieved in test for different metallization is applied.
- o mBB & fineline metal grids are optimized theoretically.

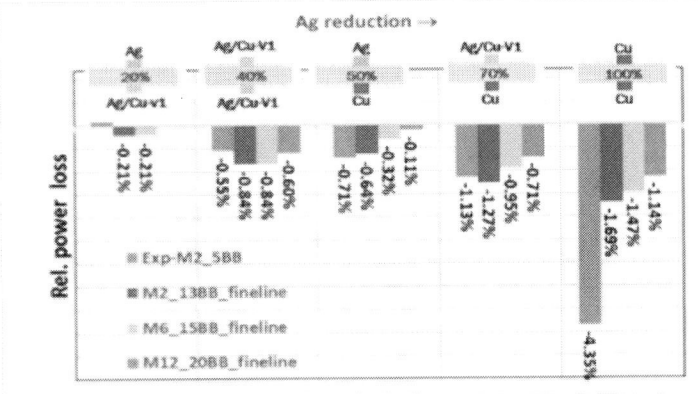

- ➤ In general, for applying very fine lines and low metallization, mBB or 0BB is beneficial to avoid significant power loss

- ➤ With mBB, line resistance of Cu contact with single print is sufficient to achieve low power loss

- ➤ We find that this solution works just as well for Cu as for Ag or Ag/Cu

SUMMARY & OUTLOOK

❖ Ag reduction in the HJT cell process with minimal penalty on cell performance could be approached with up to 70% Ag reduction metallization by screen printing.

❖ Cu metallized HJT cell has been successfully demonstrated showing a great opportunity to realize sustainable metallization process without any Ag consumption. A stable process is required and under investigation.

❖ Implementation of mBB and fineline contact design for low Ag and Cu contacts shall be beneficial for large Ag reduction with lese power loss.

❖ Interconnection, encapsulation and reliability of HJT module with low Ag content metallization will be investigated in the near future.

ACKNOWLEDGEMENT

E. J. Kossen

A. Gutjahr

M. Bruggeman

L. J. Geerligs

Fundings:

This work was supported by SolarNL, a national research, innovation and industrial development program funded by the Netherlands National Growth Fund

This work was supported by TKI-Energie from the Toeslag voor Topconsortia voor Kennis en Innovatie (TKI's) of the ministry of Economic Affairs and Climate, project SusCon.

Dutch industrial partner

41st European Photovoltaic Solar Energy Conference and Exhibition

INVESTIGATION OF OXYGEN AND CARBON IMPURITIES IN MONO-SILICON WAFERS DURING RAPID THERMAL ANNEALING

Nurhayat Yıldırım[*,1,2], Sertaç Eroğlu[2], Merve Çorak[1]

[1] KalyonPV Research and Development Center, Kalyon Güneş Teknolojileri Üretim A.Ş., 06909, Ankara, Turkey
[2] Eskişehir Osmangazi University, Faculty of Science, Department of Physics, TR-26480 Eskişehir, Turkey
*email: n.yildirim@kalyonpv.com

ABSTRACT: Silicon-based materials are essential in many industries but face challenges due to impurities like oxygen, carbon, and metals, affecting their performance. These impurities originate during growth processes from raw materials, leading to defects that lower their efficiency in applications like solar cells. Research highlights the impact of oxygen and carbon on crystal structures and carrier lifetimes. Especially, boron-oxygen complexes further influencing performance in boron-doped silicon. Impurity effects in silicon-based materials are prominent during Rapid Thermal Annealing (RTA) process. This study investigates the structural impact of oxygen and carbon contamination on silicon ingot, during RTA with different temperature ranges. Samples extracted from the silicon ingot were subjected to Fourier Transform Infrared Spectroscopy (FTIR) and X-ray diffraction (XRD) methods to analyse the change in oxygen and carbon bonding behaviour in the structure. The aim is to provide insights into the alterations induced by annealing processes and their implications for the silicon-based materials in the context of solar cell fabrication.
Keywords: Czochralski process, oxygen impurities, carbon impurities, rapid thermal annealing (RTA).

1 INTRODUCTION

Silicon-based materials play a vital role in various technological and global industries; However, they face ongoing difficulties because of impurities like oxygen, carbon, and metals. These impurities have a notable impact on the performance of silicon-based materials, despite their widespread utilization. The introduction of impurities in the structure occurs during the growth processing and mainly originates from the raw materials used in silicon wafer production. As a result, a comprehensive examination and improvement studies are essential to address the issue of overall quality and efficiency of silicon-based materials. Recent research has illuminated the adverse effects of oxygen and carbon contamination in silicon ingots and wafers, for instance oxygen impurities in silicon have been linked to defects and decreased carrier lifetimes, impacting the efficiency of solar cells [1]. Similarly, carbon impurities can contribute to the formation of crystal defects, affecting both the electronic and structural properties of silicon [2].

The study of boron-oxygen complexes (B-O complexes) has gained prominence in the pursuit of improving silicon solar cells. Boron-doped p-type silicon ingots and wafers can form B-O complexes during the process, influencing both the immediate crystal structure and module efficiency in subsequent solar cell development stages [3]. Understanding and characterizing the behaviour of these complexes has become essential for optimizing solar cell performance. According to recent studies, the method that is used for this research, Rapid Thermal Annealing (RTA), may provide valuable insight into the mechanism behind oxygen and carbon's induction on silicon wafers, offering a perspective for future material enhancement [4,5].

2 EXPERIMENTAL

This research employed boron-doped monocrystalline silicon ingots, which were cultivated using the Czochralski (Cz) method. Samples were extracted from the central region of the ingot approximately 2 mm thick and subjected to a lapping-polishing process to achieve a more uniform surface. Subsequently, they were laser-cut into perfect squares measuring 1cm x 1cm on each side. The annealing process was uniformly applied to all samples for a duration of 5 minutes at 300, 400, 500 and 600 ºC temperatures by Rapid Thermal Annealing (RTA) system (RTA-CT, Createc, Berlin, Germany) XRD (APD 2000 PRO, GNR Optica, Mliano, Italy) and FT-IR (Bruker-Vertex 80, Bruker, Billerica, Massachusetts) analyses were done each sample.

3 RESULTS

Based on FT-IR results, it is observed that all samples exhibit specific characteristic vibrational modes at each temperature value. Specifically, important modes include: the O=C=O stretching vibrations at 2334 and 2359 cm⁻¹, C-C stretching vibrations at 1303 and 1448 cm⁻¹, Si-O-Si stretching vibrations at 1107 cm⁻¹, out-of-plane O-H bending vibrations at 962 cm⁻¹, Si-O stretching vibrations at 887 cm⁻¹, C-C stretching vibrations at 817 and 738 cm⁻¹, C-H bending vibrations at 689 cm⁻¹, and Si-Si stretching vibrations at 610 cm⁻¹ (Figure 1).

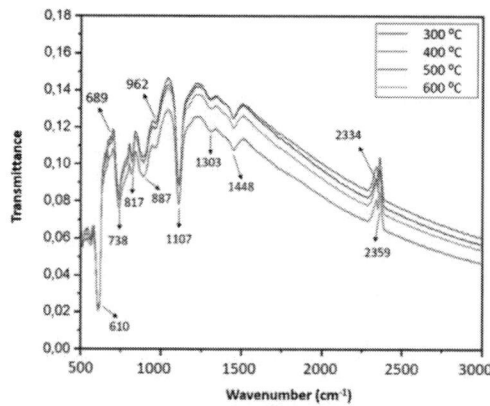

Figure 1. FT-IR results of 5 min annealed mono-Si wafers with different temparetures in RTA.

Upon a detailed examination of the data, it has been concluded that, in addition to characteristic vibrations, increase in the applied annealing temperature exhibits peaks at different points for certain bond structures. Moreover, in some occurrences, peaks not observed at low temperatures are observed at high temperature values. This observation suggests that different bond structures within the material are affected due to temperature variations. These changes in bond structures might be due to the increased mobility of oxygen and carbon atoms under high-temperature conditions, leading to a reconfiguration of the crystal lattice. This reconfiguration can impact the material's overall performance, particularly in semiconductor applications.

The peak value for the C=O=C stretching vibration at 2334 cm⁻¹ is not observed at 300°C. On the other hand, the peak, which begins to become apparent at 400 and 500°C, is more clearly observed at 600°C (Figure 2).

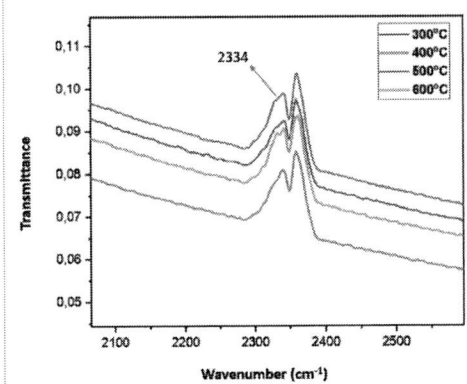

Figure 2. FT-IR peak at 2334 cm⁻¹ which is not seen at 300 ⁰C.

The C-H bending vibration, nearly absent at 300 and 400°C, becomes distinctly observable at 500 and 600°C, particularly at 663 cm⁻¹ (Figure 3).

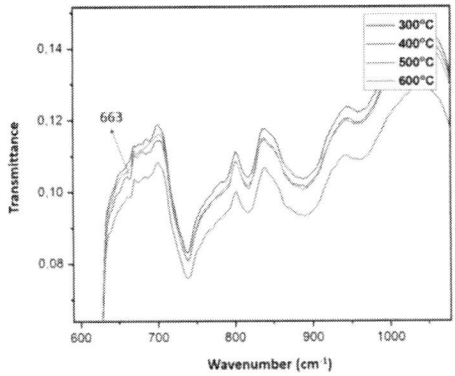

Figure 3. FT-IR peak at 663 cm⁻¹ which is not seen at 300 ⁰C.

In the XRD analysis, peaks around $2\theta=69.13°$ at 300°C, $2\theta=69.09°$ at 400°C, $2\theta=69.08°$ at 500°C, and $2\theta=69.09°$ at 600°C were observed on the (400) plane. It is observed that the crystal structure of the produced material is aligned in the desired plane. The accurate comment of the formation of oxygen and carbon

precipitates, can be observed from the vibrations structures at high annealing temperatures. However, the XRD results will be further analyzed with additional sample measurements for a more detailed understanding (Figure 4).

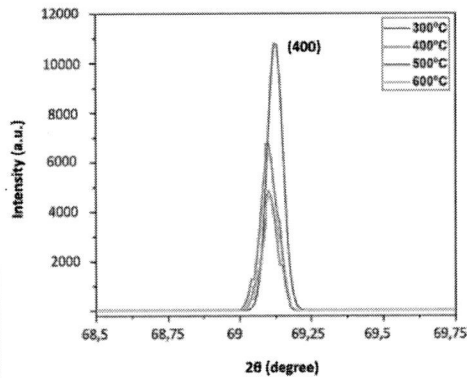

Figure 4. XRD results of 5 min annealed mono-Si wafers with different temperatures in RTA.

In general, the crystal structure is deformed in the same way as the lattice deforms with temperature. Accordingly, a sinusoidal curve was obtained. In addition, the minimum value was seen at 400°C, while the maximum value was seen at 500°C. The lower FWHM value indicates that crystallization degree is better at 400°C (Figure 5).

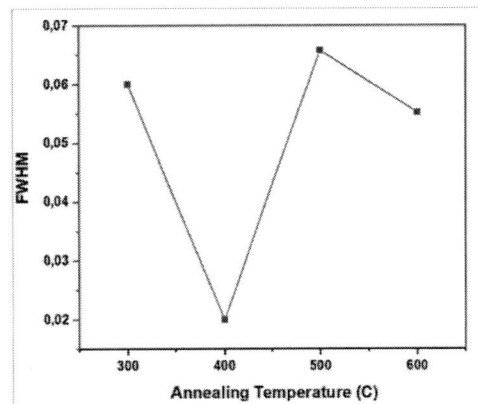

Figure 5. FWHM results of mono-Si wafers according to annealing temperatures.

5   CONCLUSIONS

Based on the investigation at the bulk level, we studied temperature annealing affects on oxygen and carbon impuruties' behaviour in boron-doped mono-Si samples. The behavior of oxygen and carbon bonds within the crystal structure becomes more apparent at high annealing temperatures. Examining the changes in the bond structure due to inreased applied temperature and varying application time will be necessary for comparison in the future.

6   REFERENCES

[1] Green, M. A., & Keevers, M. J. (1995). Optical

properties of intrinsic silicon at 300 K. Progress in Photovoltaics: Research and Applications, 3(3), 189-192.

[2] Abelson, J. R., & Vook, J. R. (1983). Carbon impurities and the lifetime of charge carriers in silicon. Journal of Applied Physics, 54(3), 1255-1259.

[3] Zhang, Z., Que, D., Xu, Y., & Huang, H. (2019). Influence of boron–oxygen complexes on the performance of p-type crystalline silicon solar cells. Journal of Materials Science: Materials in Electronics, 30(3), 2531-2537.

[4] Que, D., Zhang, Z., Xu, Y., & Huang, H. (2019). Impact of rapid thermal annealing on oxygen precipitation in silicon wafers. Journal of Applied Physics, 126(2), 025701.

[5] Liao, W., Wei, J., Zhang, L., & Cai, L. (2017). Rapid thermal annealing induced modification of carbon concentration and electrical properties in silicon wafers. Applied Surface Science, 422, 382-386.

5 ACKNOWLEDGEMENTS

The authors thank to Kalyon Solar Technologies Factory for providing samples and Gazi University Faculty of Applied Sciences Department of Photonics for their support in the measurement of XRD and FTIR analyses.

INCREASING THE PRODUCTIVITY OF THE CZOCHRALSKI PROCESS APPLYING MACHINE LEARNING

F. Mosel[1], L. Kulhavy[2], D. Baccar[2]

[1] PVA Crystal Growing Sytems GmbH, Im Westpark 10-12, 35435 Wettenberg, Germany
[2] Technische Hochschule Mittelhessen, Wilhelm-Leuschner-Strasse 13, 61169 Friedberg
e-mail: frank.mosel@pvatepla.com
phone: +49 64168690-125, fax: +49 64168690-822

ABSTRACT: Monocrystalline silicon for the production of solar cells has to be provided cost-effectively in large quantities. The Czochralski (Cz) technique is currently the only manufacturing method that can ensure the availability of high-quality material in sufficient quantities. Even if the crystal growth technology is highly developed, increasing productivity while reducing crystallization costs remains a major challenge. The outstanding advantage of Cz technology is the possibility of the visual control of the growing crystal. This means that disturbances during the crystal growth process that may result in a loss of monocrystallinity referred to as structure loss can be identified in an early stage and appropriate countermeasures can be initiated. Visual monitoring of the entire growth process has to be automated to such an extent that a failure is automatically detected and an alarm is triggered. A reliable visual assessment must still be given by an experienced operator for the further course of action. In this article, we report on an ongoing project in which we are testing different possibilities to ensure an effective automated monitoring of the Cz process. In a continuation of the project, we also want to apply methods of machine learning in order to find any meaningful correlations between process parameters and crystal growth disorders.
Keywords: Czochralski, crystal growth, structure loss, machine learning

1 INTRODUCTION

Rapidly advancing climate change has highlighted the need for rapid transformation in energy production. Fossil fuels as energy sources have a global impact on CO_2 emissions and has to be replaced by renewable energy sources. In addition to the expansion of wind energy, photovoltaic solar cell technology based on monocrystalline silicon is currently the most cost-effective energy supplier. However, China has become the leader in module production on the global market. In the area of ingot production and wafering, China owns practically a monopoly position. This situation similarly affects the engineering construction required for photovoltaics, particularly the cost-effective production of competitive crystal-growing systems based on the Czochralski-technology (Cz-puller).
The geopolitical situation has triggered resilience efforts across Europe to regain a certain degree of independence from the Chinese market, especially in the photovoltaics sector. In addition to innovative construction of crystallization systems, crystal growing technology has to be optimized also from the perspective of cost pressure and maximum productivity. Unproductive process times and downtimes of the crystal pulling systems have to be minimized. The main trigger for unproductive process times is the loss of structure, which means a phase transition from the monocrystalline to the polycrystalline state of the growing crystal. Polycrystalline crystal material can no longer be used for further processing and means a loss of yield that has to be kept as low as possible. If there is a loss of structure during the crystal growth, the following options are available: The growth process can be continued, the crystal can be removed or the crystal can be melted back and a new crystallization process can be started. In any case, timely detection of structural loss is absolutely necessary. Since it is practically no longer possible for an operator to visually observe the crystallization process on an industrial scale, an automatic warning message in the event of a structure loss is indispensable.
In addition to the detection of a loss of structure, its avoidance has the highest priority. In the Cz-process, with

ever larger diameters and simultaneously increasing pulling speeds of the growing crystal, the thermo-mechanical stresses increase, which ultimately lead to an increasing probability of structure loss. The connection between the thermal stresses that occur in industrial Cz growth arrangements and the formation of dislocations has not been fully understood yet. The causes that initiate a loss of structure are diverse and still the subject of investigations. The distribution and level of the thermal stresses in the growing crystal, process parameters, the quality of the starting materials and consumables as well as process conditions that are difficult to quantify have to be taken into account. Hendawi's diagram provides an overview of some of the possible triggers for the occurrence of a structure loss.

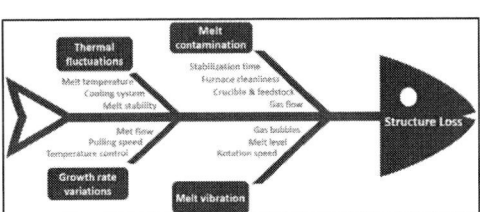

Figure 1: Fishbone diagram illustrating the main root causes of structure loss during the Cz growth of silicon [1]

2 APPROACH

In this paper we present an ongoing project in which a reliable method for automatic structure loss detection is being developed. The monocrystallinity of a growing crystal is accompanied by the formation of growth lines, which appear as vertical linear structures on the crystal surface in the growth direction. In the case of <100> crystal orientation, a monocrystal exhibits four growth lines. As long as the growing crystal shows continuous growth lines, it has a monocrystalline structure. The disappearance of these growth lines during the growth process is an indication that dislocations have formed that have multiplied due to the thermo-mechanical stresses in the growing crystal and ultimately triggered polycrystalline growth, i.e. a loss of structure. By means of the process camera, which is used for diameter control, the presence of these growths lines, which are rotating with the crystal rotation rate, can be monitored. Two different possibilities for the detection of growth lines were taken into consideration. The growth line can be detected using a grayscale evaluation of the monochrome process image in the area of the bright meniscus region, which reflects the crystallization front of the growing crystal as shown in fig.2. Fig.3 shows the corresponding grey values along the red line sketched in Fig.2. This method of image evaluation is very complex because the position of the melt meniscus constantly changes due to diameter fluctuations and fluctuations of the melt level. Complicating matters are pendulum movements of the crystal and varying lighting conditions during the process. In addition, the growth lines may have different patterns depending on the temperature conditions at the phase boundary [2]. This type of image evaluation seems to be automatable only with great effort. We are therefore currently investigating various methods of convolutional neural networks (CNNs) to develop a reliable and robust method for the detection of the presence or absence of the growth lines.

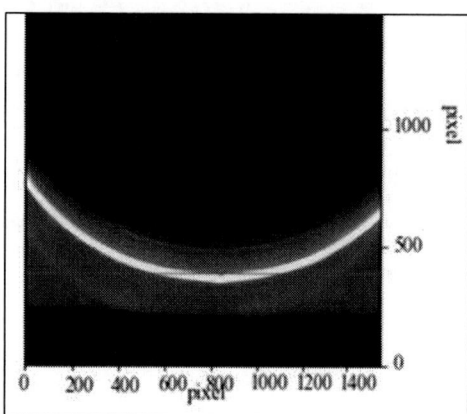

Figure 2: Detail of a process image showing the growth line entering the bright meniscus region and the red line of grey level evaluation.

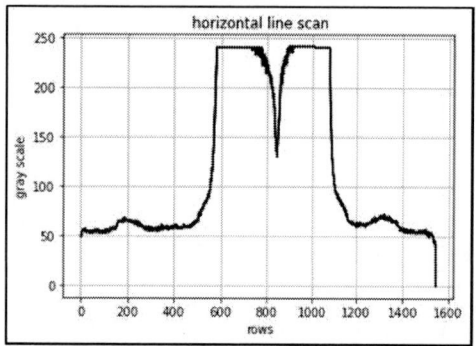

Figure 3: Distribution of the grey values along the red line in fig.2

3 EXPERIMENTAL

3.1 Crystal growth

The crystal growing process sequence can be divided into individual process steps (fig.4), which are already mostly automated. Nevertheless, observation of critical process phases, particularly the dipping phase and the subsequent start of the necking, are often still carried out manually by experienced operators.

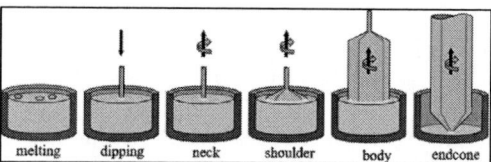

Figure 4: Process phases of a crystal growth run

The melting phase is followed by a certain period of time for the homogenization and stabilization of the melt volume. During the subsequent entire growth process from neck to endcone the diameter of the growing crystal is controlled by means of the process camera and monitored by visual observation. For our investigations, the images of the process camera were recorded from crystal growth processes executed in two different Cz-growers under completely different process conditions. In one of the two growers (SC24) 8-inch crystals were grown and in the other one (SC32) 12-inch crystals were grown under magnetic field conditions.

3.2 Data preparation

The difference in 8-inch and 12-inch crystal growth processes was taken into account in the data preparation as well as in the architecture of the convolutional neural networks. This means that in the first step the images were processed separately according to the two different Cz-growers. The process camera of both Cz-growers is an industrial monochrome camera with a resolution of 2064 x 1545 pixels. Since the CNNs (Convolutional Neural Networks) involve supervised learning, the images had to be classified into the two categories "mono" and structure loss "sl". The classification itself, i.e. deciding based on the appearance of the growth lines in the recorded process images whether there is a structure loss or not, proved to be very challenging and not always clear.

Figure 5: Abrupt end of a growth line, indicated by the red line

The growth lines often have an indifferent appearance. A sudden rupture of growth lines as shown in fig.5 is clear, but is only one possibility among many that indicate a loss of structure. Fig.6 shows different characteristics of a growth line on the surface of a crystal. In the upper area up to the green line the growth line is undisturbed, in the middle area up to the red line the growth line shows an indifferent appearance, which suggests the presence of dislocations. In the lower part of the green line the crystal exhibits clearly a polycrystalline structure. Due to these uncertainties in the classification, the process images had to be reclassified or to be classified differently after the first training sessions, in order to remove uncertainties of the image selection. In the context of this report, only the body phase of the growing crystal was evaluated, since no structure losses occurred in the shoulder phase in our crystal growth experiments.

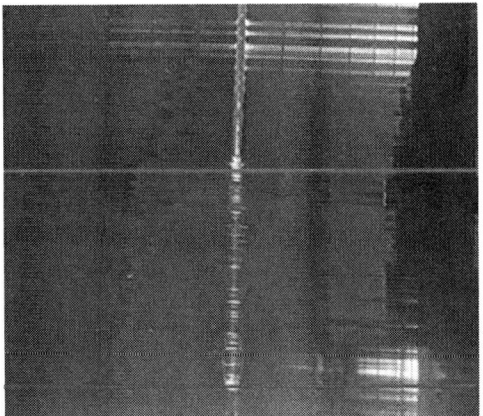

Figure 6: Different appearances of the growth line on the crystal surface

3.3 Convolutional Neural Network (CNN) Setup

Fig.7 shows the architecture of a CNN in a simplified representation. In several layers, the essential features of the classified images are extracted through several training sequences (Feature Extraction) and fed into a one-dimensional vector for classification. The result is given by the output vector ("mono" or "sl"). The training runs are repeated until the error between input and output is minimal, for details see [3].

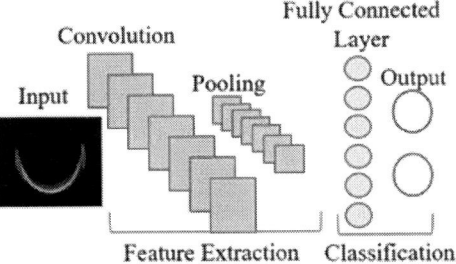

Figure 7: Architecture of a Convolutional Neural Network

For the project presented in this contribution, D. Baccar has developed an extremely efficient software tool that allows the user to create and use convolutional neural networks without writing any explicit program code [4]. The developer has the option of designing the networks from scratch or using the transfer learning strategy [5]. Various transfer networks are available in this software

tool, which can be selected via a pull-down menu (Fig.8). Well-known basic networks are, for example, VGG16 and VGG19, which have a linear network architecture and the different ResNets, which use so-called residual blocks in their structures. The number at the end of the network name represents the number of convolutional layers used. Basic networks with a few layers are interesting for applications with short inference times. During transfer learning, the transfer networks are adapted to the specific data of the new classification task. This can be done by training the entire transfer network or just specific layers. This type of model creation saves time and resources. The software tool presented here is based on the Pytorch framework, the use of a GPU is recommended. All models in our project were trained on an Intel Core i9/2.6 GHz with GeForce RTX 4070 GPU, 32GB RAM.

3.4 Model training and evaluation of CNN

Various transfer networks with different hyperparameters were tested. Fig.8 and fig.9 show two screenshots of the software tool with the applied parameters for the example of ResNet 18. The selected hyperparameters proved to be suitable for several different transfer networks. The corresponding performance metrices of a training run of 6695 images are given in fig.10 and fig.12.

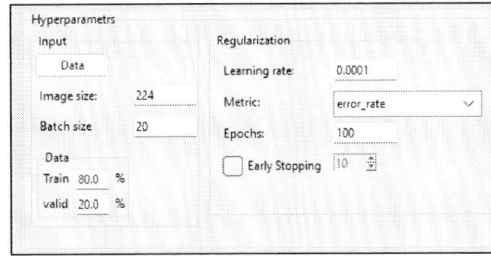

Figure 8: Screenshot of the user interface

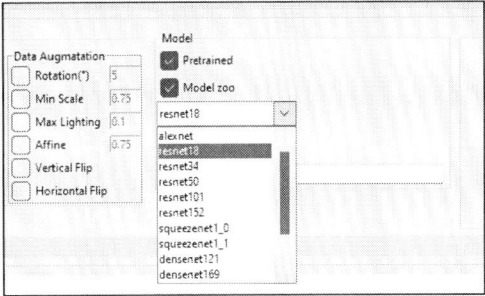

Figure 9: Screenshot of the user interface showing a small selection of available transfer nets

During training, the train loss and validation loss are recorded versus the epochs to show the progress of training performance (fig.11). The best model is continuously updated and saved. The corresponding data with a time stamp are recorded. All relevant training data as well as the model are stored and can be reloaded for optimization purposes. The training ends after the last epoch is completed and shows the confusion matrix of the validation data (fig.12). The rows of the matrix show the labeled values, the columns the calculated values. Using the confusion matrix, the key figures "accuracy" and "precision" of a classification can be deduced.

41st European Photovoltaic Solar Energy Conference and Exhibition

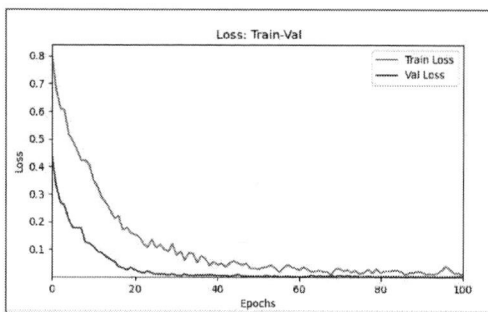

$$accuracy = \frac{T_{mono}+T_{SL}}{T_{mono}+T_{SL}+F_{mono}+F_{SL}}$$

$$precision = \frac{T_{mono}}{T_{mono}+F_{mono}}$$

Figure 10: Confusion matrix and key figures of evaluation (T: true values, F: false values)

In our models, mainly based on transfer nets, the validation curves consistently show better values than the training curves, which may be due to the simplicity of the classification. A performance of 100% of the validation could be achieved for all applied transfer models.

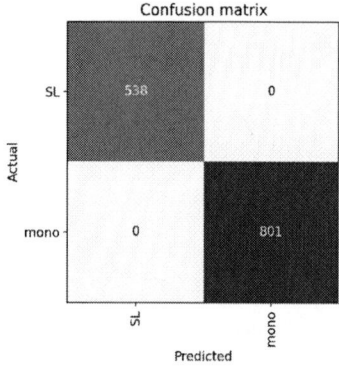

Figure 11: Train-loss and validation-loss versus training runs (epochs)

Figure 12: Confusion matrix of a completed training

However, the inference of our CNN-models showed a completely different picture for all models. 1000 test images, solely used for the inference, consisted of equal parts of mono-classified and SL-classified images (500 images each part). In both the training data and the test data, the absence of a growth line in the bright meniscus region was classified as a structure loss ("sl"). During monocrystalline growth, the slightly darker growth line extends into the bright meniscus ring-shaped region, as shown in fig.13 (red circle), which is no longer observed in the event of a structure loss, as shown in fig14 (yellow circle).

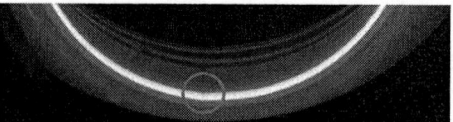

Figure 13: Process image classified "mono" with growth line visible in the meniscus region

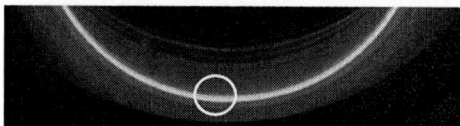

Figure 14: Process image classified "sl" with growth line outside the meniscus region

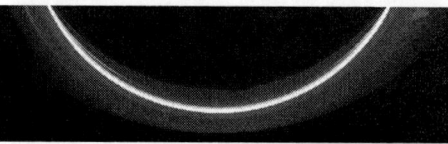

Figure 15: Process image classified "sl" with absence of growth line

Further stricter image classification, i.e. without the visibility of a growth line (fig.15), revealed that all images were classified as mono as soon as even a small fraction of a growth line was visible in the test images. Only images in which a growth line, i.e. a vertical structure, was not visible, provided the required inference result "sl". Fig.16 shows feature maps of ResNet18 without vertical structure (left), with partly vertical (middle) and completely vertical structure (right).

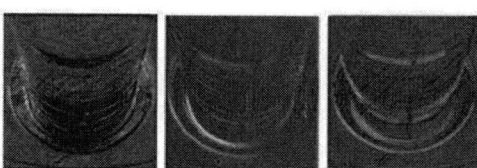

Figure 16: Feature maps of ResNet 18.

Corresponding histograms of the inference results are shown in fig.17 and fig.18.

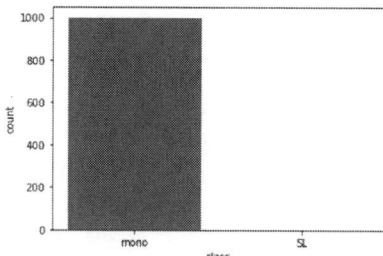

Figure 17: Histogram of inference executed on test-images, before strict classification (with partly vertical structure in the test images)

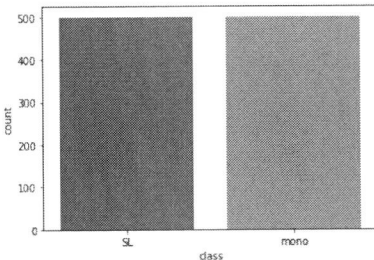

Figure 18: Histogram of inference executed on test-images after strict classification (without vertical structure in the test images)

The test results show that an early detection of a structure loss by means of a standard CNN is difficult to realize. Therefore, a region-based convolutional network (R-CNN) could be used in the area of the bright meniscus ring. Based on Zhang et al., a Faster R-CNN architecture was set up and trained [6].

3.5 Region based Convolutional Neural Network (Faster R-CNN) Setup
For the detection of the growth lines near the crystallization front, the use of a Faster R-CNN architecture is recommended as a model with a fast inference time for real-time detection. The principle of the architecture is shown in fig.19.

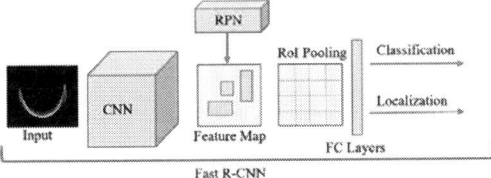

Figure 19: Architecture of the Faster R-CNN network

The Faster R-CNN architecture consists of two networks, Fast R-CNN and RPN (Region Proposal Network). The Fast R-CNN module creates a feature map from the input image using a pretrained CNN model. The RPN module is a fast network that recognizes objects and their location (bounding box) in the feature map of the CNN. The kind of the objects are unknown. These object proposals are fed into the Fast R-CNN network as RoIs (Region of Interest). A RoI pooling layer performs a max-pooling operation on the RoIs of nonuniform shape to obtain small fixed-size feature maps. The output contains two fully connected layers, a classification layer for the class probability ("mono" and "sl") and a bounding box regression for localization of the object, which is unimportant for our purposes. During training, a combined error of classification and localization is calculated and compared with the classified input until the error is minimal. The big advantage of the Faster R-CNN architecture for our application is the ability to limit an area in the image for training the network. In order to detect structural loss early, we limited the object search area by setting the ground truth boxes in a small area over the crystallization front as sketched (red rectangles) in fig.20 and fig.21.

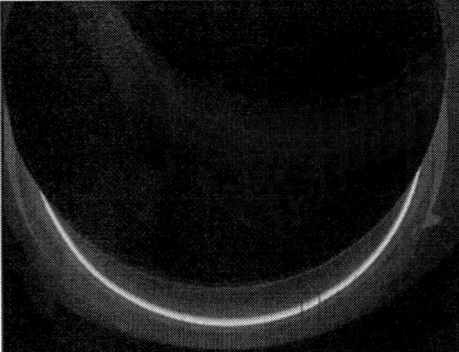

Figure 20: Bounding box for detection of the growth line in process image

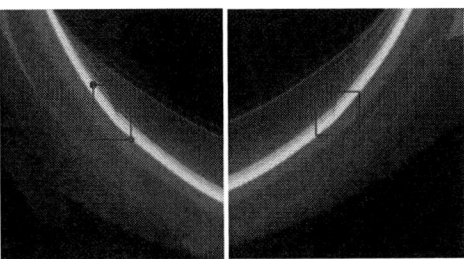

Figure 21: Detail enlargements of process images

3.6 Model training and evaluation of Faster R-CNN

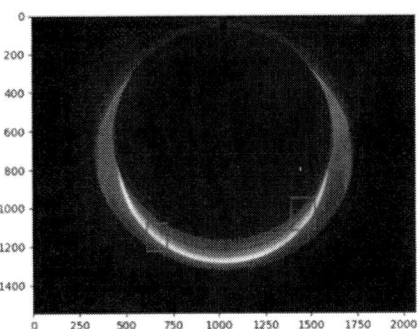

Figure 22: bounding box localization with growth line on a test image

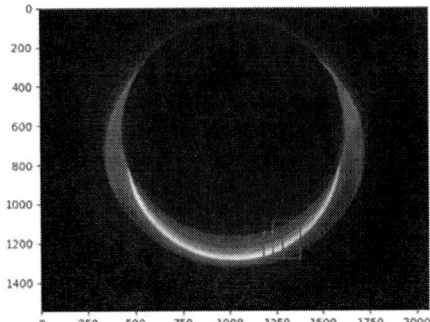

Figure 23: Overlapping bounding boxes on a growth line on a test image

The classified process images with the original size of 2064 x 1545 pixels together with the corresponding coordinates of the ground truth boxes are loaded as input into the transfer network ResNet50. The hyperparameters used are a batch size of 4 images and 100 epochs with a learning rate of 0.0001. The data split was 90% for training and 10% for inference. Also in this Faster R-CNN architecture, the training loop is run through until the deviation from the output combined from classification and localization to the input is minimal. Fig.22 and fig.23 show two examples. In both cases, the growth lines were located in the light meniscus area and were also correctly classified as mono. The amount of images, especially in the case of structure loss, was not sufficient to present a meaningful model. However, the results so far are promising.

3.7 Analysis of process parameters from the perspective of structure loss

The connection between dislocation formation and thermal stresses in the growing crystal is established by a critical shear stress CRSS-model (Critical Resolved Shear Stress) [7]. The temperature dependence of the critical shear stresses in the silicon glide system is shown in fig.24 for various authors [8].

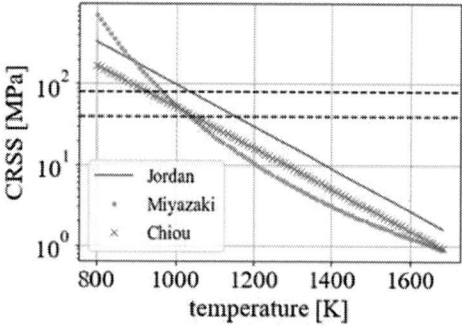

Figure 24: CRSS in the glide system for silicon [8]

The published values are in the range of 1-2 MPa at the solidification temperature. The thermal stresses that occur in industrial Cz-crystal growth configurations are in the range of 40 and 80 MPa (dashed lines) [9]. This means that monocrystalline silicon crystals can be grown successfully in the region of this supercritical thermal stress field (fig. 24). This discrepancy can be explained by a qualitative model of Muiznieks et al [10]. It is assumed that the crystal is in a metastable state concerning the critical thermal stress during growth and requires some additional energy contribution to leave this metastable state. It is assumed that only the local level of thermal stress is crucial for dislocation formation and not the mean thermal stress distribution in the crystal volume. This additional energy contribution can be delivered by temperature fluctuations, local melt-back effects or other growth disturbances in the area of solidification. This additional energy contribution decreases as the level of stress in the crystal increases. Since the root-causes that trigger a structure loss can be diverse and in some cases have not yet been clarified, they cannot be classified. Therefore, an unsupervised learning method has to be used in machine learning in order to find correlations to process parameters. AutoEncoder seems to be a suitable neural network to identify process anomalies. AutoEncoders are trained exclusively with process data representing perfect process conditions. The functionality is outlined in fig.25.

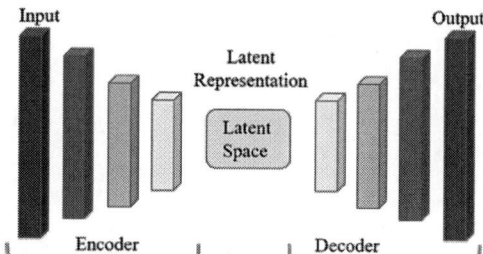

Figure 25: Diagram of AutoEncoder network

The input data are compressed into the latent space in the neural network of the AutoEncoder. All information from the undisturbed process are available in reduced form in the latent space. Decoding the latent space trains the network to reconstruct the input data as best as possible. The aim of this procedure is to learn undisturbed process data. During inference, i.e. applying the trained model to undisturbed process data, the AutoEncoder will reconstruct the process data as best as possible. If the process data are disturbed or show anomalies, the reconstruction will reveal anomalies.

3.8 Model training and evaluation of AutoEncoder

To analyse the process variables that can trigger a loss of structure, among other things, we had access to the process data of 103 crystal growth runs, all of which took place in a Cz puller under process conditions that were as similar or at least comparable as possible. A total of 150 process variables were recorded every 5 seconds. The structure losses were recorded in time by operators. Of the 103 processes, 71 processes were finished completely monocrystalline, 5 processes were completed with a loss of structure, which resulted in a complete loss of the material, in the other processes the crystals with a structure loss were still usable to a large extent. Of the 71 processes that were completed as mono crystals, 32 processes were carried out without any failure, i.e. remelting after structure loss. These process data are therefore predestined for training processes for perfect process conditions. In the first step of our model building, the input data were converted into matrices, with 93000 time steps and 12 process parameters (out of 150). The output data had the same format. The batch size was one process each, the training duration was 100 epochs with a learning rate of 0.0001. The network which was programmed from scratch consisted of 5 hidden layers, two for the contracting and two for the expanding path, one layer for the latent space. Fig.26 and fig.27 show two example results of an inference shown as a rectangular matrix. The time steps over the reconstruction errors of 12 process parameters are shown in a heatmap representation. Fig27. shows anomalities in the process parameters 3 and 4 that lie in the same range of time. If a loss of structure is recorded at a slightly later point in time, this could be an indication of a correlation of a structure loss and process parameters.

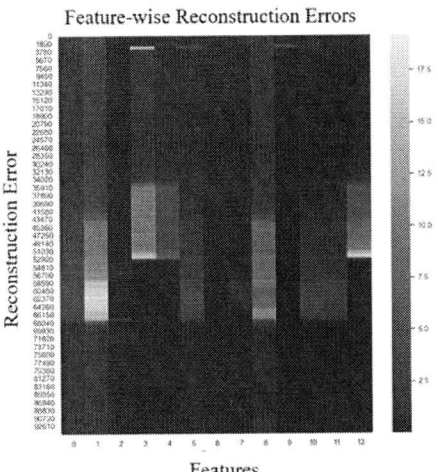

Figure 26: Heat map of the reconstruction error (time scale) versus 12 process parameters (out of 150)

The examples shown here are intended to demonstrate a general possibility of examining correlations in process parameters through the use of machine learning.

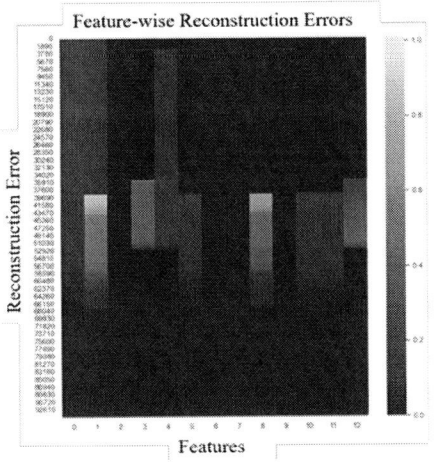

Figure 27: Heat map of the reconstruction error (time scale) versus 12 process parameters (out of 150)

At this point, it should be noted that we are just at the very beginning of analyzing the process parameters by means of machine learning.

4   SUMMARY

In this report, we present the status of our ongoing project work on structure loss of silicon monocrystals grown by the Cz-technology. The project is split into two parts. In one part, an automatable method is to be developed, that bases on machine learning methods, detects a structure loss or other anomalies of the growing crystal at an early stage. The results achieved so far with a region-based convolutional neural network are very promising. However, the model creation is not yet been complete, which is also due to the fact that only a limited number of suitable process images are available. Alternative network architectures need to be investigated.

In the second part of our project, various machine learning methods will also be examined in order to find correlations to the process conditions in the recorded process data. The root causes that led to the correlations found should be clarified in order to contribute to process optimization.

5   ACKNOWLEDGEMENTS

This work was supported by the German Federal Ministry for Economy Affairs and Climate Action under contract number 03EE1126B

6   REFERENCES
[1] R. Hendawi, Marisa Di Sabatino, Journal of Crystal Growth **629** (2024), 127564
[2] L. Stockmeier et al., Journal of Crystal Growth **512** (2019), 26-31
[3] Ian Goodfellow, Yoshua Bengio, Aaron Courville, Deep Learning, Adaptive Computation and Machine Learning Series, Cambridge, Mass.: MIT press Ltd, (2017)
[4] D. Baccar, Technische Hochschule Mittelhessen, dorra.baccar@m.thm.de
[5] Karl Weiss et al.,Journal of Big Data 3.1 (2016)
[6] Jun Zhang et al., Appl. Sci. 2020, 10, 7799; doi:10.3390/app10217799
[7] A. S. Jordan, R. Caruso, A. R. Von Neida, Bell. Syst. Tech. Journ. **59** (1980) 593
[8] G. Raming, Fortschritt-Berichte VDI Reihe 9 Nr. 330, VDI-Verlag Düsseldorf, 2001, S.133
[9] W. von Ammon, E. Dornberger, P.O. Hansson, Journal of Crystal Growth **198/199** (1999) 390
[10] A. Muiznieks et al., Journal of Crystal Growth **230** (2001), 305-313

41st European Photovoltaic Solar Energy Conference and Exhibition

THERMAL DEACTIVATION OF BORON-OXYGEN DEFECTS IN COMPENSATED N-TYPE SILICON

Rune Søndenå[1,a)], Per-Anders Hansen[1], Bent Thomassen[1], Øyvind Mjøs[2], and Tyke Naas[2]
[1]Institute for Energy Technology, Norway
[2]REC Solar Norway AS, Norway
[a)]Rune.Sondena@ife.no

ABSTRACT: There are at least two distinct pathways for deactivating the defect complexes responsible for the detrimental light induced degradation in silicon wafers containing both boron and oxygen. One method widely used in the photovoltaic industry is regeneration of the fully formed recombination active defects through simultaneous carrier injection and annealing. This regeneration is a post-processing step that also mitigates light- and elevated temperature-induced degradation often occurring in hydrogenated solar cells. A second approach capable of reducing the extent of the boron-oxygen related degradation involves thermal deactivation of the defect complexes in a pre-processing step. This work shows that thermal deactivation will reduce the scope of the boron-oxygen related degradation in compensated n-type silicon wafers, i.e. wafers containing a minority concentration of boron in addition to the phosphorus donors. Thermal deactivation is a pre-processing step that can be combined with the low temperature silicon heterojunction solar cell production process.
Keywords: Crystalline silicon, degradation,

1 INTRODUCTION

The minority charge carrier lifetime (here: lifetime) in Czochralski (Cz) silicon wafers containing boron dopants may degrade by as much as 90% of the initial value under carrier injection due to boron-oxygen (BO) related light induced degradation (LID)[1], [2], [3]. It has been shown that the BO defect responsible for the degradation can be permanently regenerated by carrier injection, either by illumination or an applied bias voltage, at an elevated temperature[4], [5]. Figure 1 shows a proposed four-state model describing the BO-LID. Anti-degradation treatments through the degradation and regeneration path are quite common in the PV industry, even for n-type solar cells, as they also mitigate the hydrogen related light- and elevated-temperature induced degradation (LeTID). The concentration of BO defects, N_{BO}, and, hence, the extent of the degradation is also dependent on the thermal history of the wafers. Phosphorus in-diffusion and oxidation steps can reduce the N_{BO} [6], [7], [8]. Fast firing processes will also reduce the N_{BO} [9], [10], [11], [12]. A dissociation of the defect precursor, with a corresponding reduction of defect complexes available for degradation has been hypothesized[13].

Thermal deactivation

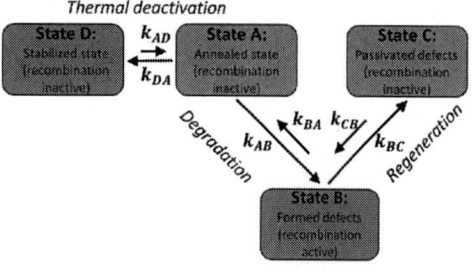

Figure 1. Proposed four-state-model for the boron-oxygen defect system (adapted from [13] and [14]).

Conventional n-type silicon wafers, produced with polysilicon from the Siemens process, is not affected by BO-related degradation due to the absence of boron. However, in metallurgically purified silicon feedstock traces of boron may be compensated for with an increased amount of the phosphorus dopant, hence the name compensated silicon. Such phosphorus doped n-type silicon with boron present as a minority dopant is also susceptible to BO-LID upon illumination[15], [16], [17], [18].

Silicon heterojunction (SHJ) solar cells can reach very high conversion efficiencies. LONGi Solar has recently developed a crystalline silicon heterojunction back contact (HBC) solar cell with an efficiency exceeding 27.3%[19]. SHJ cells require a high-lifetime wafer as a starting point, as the low-temperature processing of a SHJ cell does not involve any defect engineering. By defect engineering we mean processes that mitigate grown in defects, e.g. tabula rasa, phosphorus diffusion gettering, or hydrogenation. PERC and TOPCon solar cells typically include both emitter-in-diffusions and fast firing of the metal contacts, thus, implementing a certain degree of defect engineering in their production lines. In lieu of such high temperature steps, SHJ cells can benefit notably from pre-fabrication treatment for increasing the lifetime in the wafer. Such a pre-annealing step mimicking gettering is indeed becoming increasingly adopted in the SHJ process sequence, easing the tight requirements for wafers as well as increasing the performance[20].

In this work we will study the thermal deactivation effects of a pre-fabrication process on the BO-related LID in compensated n-type silicon wafers produced by metallurgically refined silicon sawdust, a.k.a. kerf, from wafer slicing ("E2M").

2 EXPERIMENTAL DETAILS
Cz-Si wafers have been produced with 75% E2M kerf-

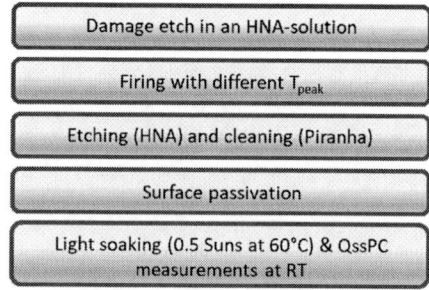

Figure 2. The process flow of the Cz-wafers.

based feedstock and 25% Siemens polysilicon. Phosphorus dopants are added to reach the desired resistivity. The resulting n-type feedstock contains about 0.135 ppmw boron. Neighbouring wafers with a resistivity of about 1 Ω-cm has been processes as shown in Figure 2. After an initial damage etch in a HNA solution (Hydrofluoric acid:Nitric acid:Acetic acid − 2:10:5) the wafers were fired in a belt furnace at different peak temperatures (T_{peak}), shown in Figure 3. A second HNA-etch and cleaning in a Piranha solution (Sulphuric acid: Hydrogen peroxide – 4:1) was performed before the wafers were surface passivated using PECVD deposition. Passivation was achieved with a stack of amorphous silicon capped with silicon nitride (a-Si:H/a-SiN$_x$:H) deposited at 230°C. A surface recombination velocity of less than 5 cm/s is expected [21]. The degradation kinetics is evaluated under approximately 0.5 Suns LED illumination at 60°C. The wafers were manually moved from the illuminated hotplate to the QssPC (Sinton WCT-120TS) for lifetime measurements. Lifetimes have been extracted at room temperature (approx. 25°C) for an injection level of $0.1 \times n_0$, where n_0 is the equilibrium electron concentration.

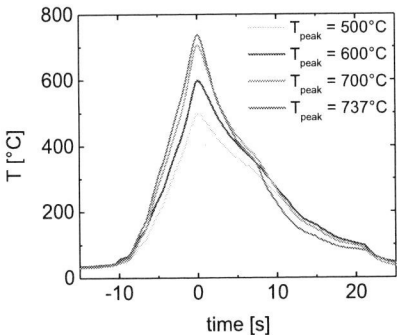

Figure 3. Fast firing temperature profiles measured using a datapaq and a thermocouple on a wafer.

As the BO defect complex responsible for light induced degradation has yet to be observed directly its capture cross sections as well as its concentration in a wafer is difficult to determine. A normalized defect density (NDD) is therefore commonly used as a measure of the concentration. $NDD(t)$ is defined as

$$NDD(t) = \frac{1}{\tau_{effective}(t)} - \frac{1}{\tau_0}$$

Where τ_0 and $\tau_{effective}(t)$ are the initial lifetime and the effective lifetime measured at a time t, respectively.

3 RESULTS AND DISCUSSION

The lifetime evaluation under illumination at 60°C of wafers fired with different temperature profiles as well as an unfired wafer are displayed in Figure 4a. As expected from wafers containing boron dopants as well as oxygen the lifetime in the wafers degrade considerably upon illumination. Despite few measurements in the initial stage, the measurements also seem to indicate a fast and a slow degradation characteristic for BO-LID. We interpret this degradation as BO-related defects transitioning from the annealed state A to the degraded state B in the four-state-model sketched in Figure 1. Figure 4 shows us that the initial lifetimes can vary quite a bit depending on the thermal history. An initial lifetime (τ_0) of around 600 µs was measured in the unfired wafer, whereas a slightly higher starting lifetime was measured in the wafer fired with a peak temperature of 500°C. A T_{peak} at 600°C, however resulted in a τ_0 exceeding 1.8 ms. Even higher firing temperatures led to a decrease in the τ_0.

Figure 4b shows the normalized defect densities calculated from the lifetime data in Figure 4a. The evolution over time of $NDD(t)$ indicates the concentration

of defect complexes in the recombination active state B (Fig. 1). Figure 4b shows that the largest concentration of defects is generated in the non-fired wafer upon illumination. Fast firing with a T_{peak} at 500°C will reduce the defect concentration somewhat. Even lower defect concentrations after illumination can be observed with increasing peak temperatures, levelling off at around 700°C. This shows that the concentration of BO-related defect centres available for activation can be considerably reduced using thermal deactivation.

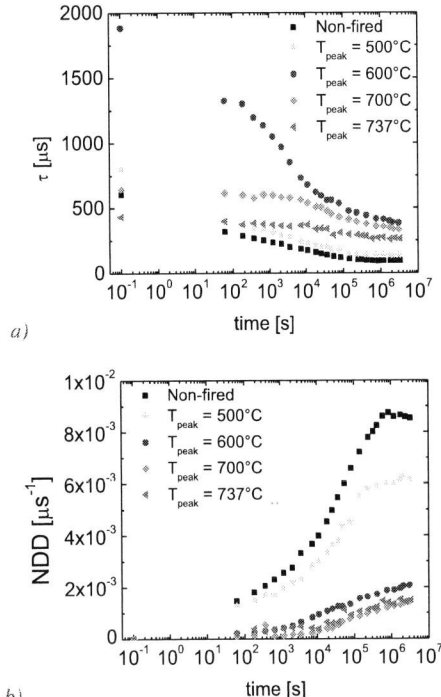

a)

b)

Figure 4. The lifetime development (a) and the corresponding NDD (b) as a function of illumination time at 60°C for wafers with different thermal history.

During the fast firing processes the wafer surfaces have been bare. There is no dielectric film present during firing that could be a potential source for hydrogen in-diffusion. Thus, the findings support the hypothesis that defect precursors are dissociated during the fast-firing processes[13]. The regeneration process has been closely linked to the presence of hydrogen in the silicon wafers. Thus, thermal deactivation of BO-defects seems to be a distinct mechanism unrelated to the regeneration process, as no hydrogen was introduced into these wafers.

The wafer processing shown in Figure 2 includes a surface passivation step performed by PECVD deposition. This process is performed after the thermal deactivation of the BO-defects in the wafers, proving that the deactivation effect remains even after processing at 230°C total time exceeding 20 minutes. Thus, thermal deactivation by a fast-firing process is a viable option for pre-processing of SHJ solar cells based on compensated n-type wafers.

4 SUMMARY

The lifetime evolution upon illumination at 60°C in compensated n-type Cz-Si sister wafers subjected to different thermal deactivation processes has been studied. By quantifying the degradation using normalized defect densities, we find that fast-firing processes with a peak temperature above 500°C will reduce the scope of the BO-degradation. Decreasing degradation is observed with increasing T_{peak} up to about 700°C. In this work the fast-firing process is performed on bare wafers, showing that the mechanism involved is not related to hydrogen.

5 ACKNOWLEDGMENTS

Funding for this work was provided by the Norwegian Research Council through the EnergiX programme (IPN REFORM, #296309) and the Research Center for Sustainable Solar Cell Technology (FME SUSOLTECH, #257639). The data that support the findings of this study are available from the corresponding author upon reasonable request.

[1]   T. U. Nærland, H. Haug, H. Angelskår, R. Søndenå, E. S. Marstein, and L. Arnberg, *IEEE J. Photovolt.*, 3 (2013) 1265–1270.

[2]   T. Niewelt, J. Schön, W. Warta, S. W. Glunz, and M. C. Schubert, *IEEE J. Photovolt.*, 7 (2017) 383–398.

[3]   K. Bothe and J. Schmidt, *Appl. Phys. Lett.*, 87 (2005) 262108.

[4]   B. Hallam, A. Herguth, P. Hamer, N. Nampalli, S. Wilking, M. Abbott, S. Wenham, and G. Hahn, *Appl. Sci.*, 8 (2018) 10.

[5]   A. Herguth, G. Schubert, M. Kaes, and G. Hahn, *Prog. Photovolt.*, 16 (2008) 135–140.

[6]   S. Glunz, S. Rein, W. Warta, J. Knobloch, and W. Wettling, in 2nd WCPEC (1998) Vienna, Austria.

[7]   S. Glunz, S. Rein, J. Lee, and W. Warta, *J. Appl. Phys.*, 90 (2001) 2397–2404.

[8]   K. Bothe, J. Schmidt, and R. Hezel, in 29th IEEE PVSC (2002). New Orleans, Louisiana.

[9]   J. Youn Lee, S. Peters, S. Rein, and S. W. Glunz, *Prog. Photovolt. Res. Appl.*, 9 (2001) 417–424.

[10]   D. Lin, Z. Hu, Q. He, D. Yang, L. Song, and X. Yu, *Sol. Energy Mater. Sol. Cells*, 226 (2021) 111085.

[11]   S. Wilking, A. Herguth, and G. Hahn, *Energy Procedia*, 38 (2013) 642–648.

[12]   D. C. Walter and J. Schmidt, *Sol. Energy Mater. Sol. Cells*, 158 (2016) 91–97.

[13]   N. Nampalli, H. Li, M. Kim, B. V. Stefani, S. R. Wenham, B. J. Hallam, and M. D. Abbott, *Sol. Energy Mater. Sol. Cells*, 173 (2017) 12–17.

[14]   A. Herguth and G. Hahn, *J. Appl. Phys.*, 108 (2010) 114509.

[15]   C. Sun, D. Chen, F. Rougieux, R. Basnet, B. Hallam, and D. Macdonald, *Sol. Energy Mater. Sol. Cells*, 195 (2019) 174–181.

[16]   T. Schutz-Kuchly, J. Veirman, S. Dubois, and D. R. Heslinga, *Appl. Phys. Lett.*, 96 (2010) 93505.

[17]   B. Lim, F. Rougieux, D. Macdonald, K. Bothe, and J. Schmidt, *J. Appl. Phys.*, 108 (2010) 103722.

[18]   V. V. Voronkov, R. Falster, K. Bothe, B. Lim, and J. Schmidt, *J. Appl. Phys.*, 110 (2011) 63515.

[19]   M. A. Green, E. D. Dunlop, M. Yoshita, N. Kopidakis, K. Bothe, G. Siefer, D. Hinken, M. Rauer, J. Hohl-Ebinger, and X. Hao, *Prog. Photovolt. Res. Appl.*, 32 (2024) 425–441.

[20]   S. K. Chunduri and M. Schmela, "Heterojunction Solar Technology Report 2023" Accessed: Feb. 14, 2024.   [Online].   Available: https://taiyangnews.info/heterojunction-solar-technology-report-2023/

[21]   H. Haug, R. Søndenå, M. S. Wiig, and E. S. Marstein, *Energy Procedia*, 124 (2017) 47–52.

HIGHEST THROUGHPUT LASER PROCESSING FOR THIN PLATED CONTACTS

E. Alvarez-Brito[1], R. Haberstroh[1], G. Hoppe[1], K. Du[2], F. Roessler[3], A. A. Brand[1], S. Kluska[1], F. Meyer[1], J. Schneider[1], J. Nekarda[1]

[1] Fraunhofer Institute for Solar Energy Systems ISE, Heidenhofstraße 2, 79110 Freiburg, Germany
[2] EdgeWave GmbH, Carlo-Schmid-Straße 19, 52146 Würselen, Germany
[3] Moewe Optical Solutions GmbH, Leipziger Strasse 27, 09648 Mittweida, Germany

The market for solar cells and modules is growing rapidly, but it also poses significant challenges to existing production technologies, including laser processing tools. In this paper, we present our approach for the next generation of laser processing and demonstrate its capabilities for the demanding application of laser contact opening. Our concept is based on on-the-fly processing employing an ultrafast polygon scanner combined with a high repetition rate laser and high-end lens. The on-the-fly process consists of a linear axis transport system synchronized with the scanner, allowing for operation within a quasi 1D scanning field, which enhances the homogeneity of the laser contact openings and the number of elements processed in a time period. The scanning system could achieve a potential throughput of 31,500 M10 (182 mm x 182 mm) wafers on a single lane, with small structures (< 10 µm) at record speeds.
Keywords: ultrafast laser processing, laser contact opening, copper electroplating

1 AIM AND MOTIVATION

The demand for solar cells and modules is increasing quickly, but this growth presents substantial challenges for current manufacturing technologies. The desired wafer format has evolved from M2 to M10 and even M12+, necessitating adapted machine concepts [1]. Another major challenge is the need to reduce the amount of silver used in solar cell production, as it could otherwise exceed global resources [2]. Simultaneously, more efficient yet complex cell architectures, such as IBC or TOPCon, which require large-area laser structuring, must be supported for automated high-throughput production. In summary, all major trends either depend on or significantly benefit from rapid laser processing.

Particularly, laser contact opening (LCO) has been a crucial step for several cell concepts. LCO for electroplating is gaining even more attention, as it enables metallization with copper, providing a cost-effective alternative to silver contacts [3]. Here, the size, homogeneity, and quality of the laser structure significantly influence the contact width, the contact and series resistance, as well as recombination and adhesion properties. The main objective of this work is the development of a novel machine concept that would allow the creation of small structures over a large area at high speed for laser contact opening for metallization with

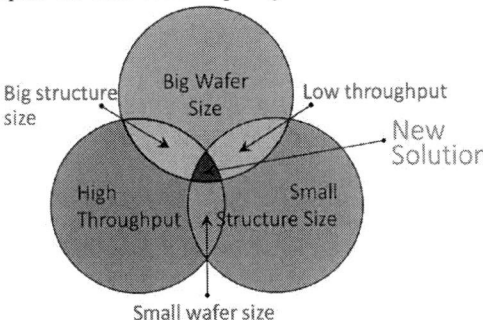

copper electroplating.
Figure 1: Traditional trade-off between throughput, structure size and wafer size.

Larger wafer format, small structure size and high system throughput are conflicting factors with conventional laser processing technologies. Figure 1 illustrates these requirements. For example, a large beam

diameter and small focal length of the focusing optics result in small structures on the workpiece plane. However, a large beam diameter means that the scanner mirrors must be correspondingly large, which in turn can be electromechanically accelerated more slowly than smaller ones, thus reducing throughput [4]. Small structures can be achieved within small field of views as well. By stitching them together, a large workpiece can be addressed with multiple small working fields, but again at the expense of reduced throughput or greatly increased system complexity, multiplexing and cost. With a conventional laser system setup, this leads to ratios that make it appear impossible to apply small structures to large formats at high throughput.

This situation requires a rethinking of laser system technology to find the new approach to reach the goals as shown in Figure 1.

2 SCIENTIFIC INNOVATION AND APPROACH

2.1 Scanning System and Laser Source

For the application of LCO for copper electroplating metallization, the laser must operate with ultrashort pulses in UV wavelengths to ensure short penetration depths of a few nanometers on the emitter side of the cells. Small structure size below 10 µm on big wafer formats is essential since results from Kluska et al. [3] have already shown an absolute increase in total conversion efficiency of 0.65% abs. of the solar cell when the opening size is reduced from 14 µm to 5.5 µm.

Our development is based on the implementation of a UV polygon mirror scanner (MOEWE Optical Solution GmbH) and its combination and synchronization with a pulse on-demand high repetition rate UV laser system (EdgeWave GmbH), high-end optics, and axis system.

The scanner consists of a multifaceted mirror rotating at high speed, which allows to reach beam deflection speeds up to 1 km/s at a focal length of 560 mm, which is much higher than conventional galvo scanning systems [5]. The polygon scanner is capable of handling large beam diameters due to its free aperture of 30 mm, which allows small focal diameters meeting the challenging structure size requirements of 10 µm or less.

To take advantage of the high scanning speed a laser with high repetition rate must be used. The described setup includes a prototype laser (EdgeWave GmbH) that currently offers sufficient pulse energy for the ablation at the highest on the market available repetition rate of

16.2 MHz at 343 nm UV wavelength with 2.4 ps ultrashort pulses.

2.2 On-the-fly processing

This method achieves focusing with reduced optical distortion by guiding the beam in a narrow area at the center of the scan field (quasi-1D) orthogonal to the transport direction, while the target is moved across the scanning area. In contrast, conventional 2D f-theta lenses exhibit significant distortion in the corners of the scan field. Additionally, the on-the-fly approach increases the number of wafers processed per hour [4].

This technique requires synchronization between the scanning and transport systems. A quadrature encoder signal couples the motion of the conveyor to the scanner, ensuring that scanning is synchronized with transport.

Figure 2a illustrates the schematic of the setup's functionality, showing how the laser beam is directed into the scanner and deflected by the polygon mirror. The output includes an f-theta lens designed to meet the stringent requirements for achieving a small focal spot size for ultrashort UV pulses across the entire field of 210 mm x 210 mm.

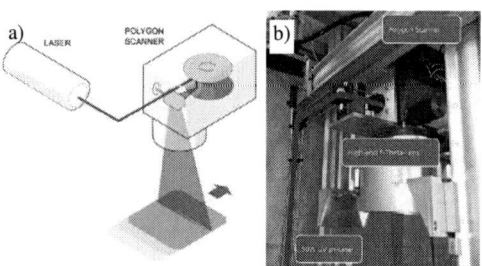

Figure 2: a) Schematic of how the polygonal mirror works b) Photo of the prototype, the laser path to the scanner is shown in purple.

The performance of the proposed on-the-fly process is tested and compared to a laser scanning process where the substrate is static and processed over the full area. In both cases, the same scanning system and laser described previously are used.

Laser ablation is performed at different intensities along the entire surface of the wafer. The ideal opening is made by a Gaussian beam, resulting in a round spot, but due to optical distortions this may not be the case. Size and roundness of the openings made in the passivation layer are evaluated over the entire field. Both processes are performed on a TOPCon silicon wafer of size M10 (182 mm x 182 mm). The resulting structures are observed and imaged by a microscope.

2.3 Metallization with electroplating

In this work, TOPCon precursors (M10 wafer area 330.23 mm², pseudo square) were metallized using electroplating, which allows for the deposition of nickel and primarily copper in the metal stack. At Fraunhofer ISE, an industrial pilot system (InCellPlate; RENA Technologies) is used, with a throughput capacity of up to 5,000 wafers per hour. The metallization process is conducted in an inline system, where one side of the solar cell is metallized during a single pass. This inline system includes four process baths: pretreatment (1.5% hydrofluoric acid), nickel electrolyte, copper electrolyte,

and silver electrolyte. For bifacial metallization, the process is repeated for the second side of the solar cell, as shown in Figure 3. [6]

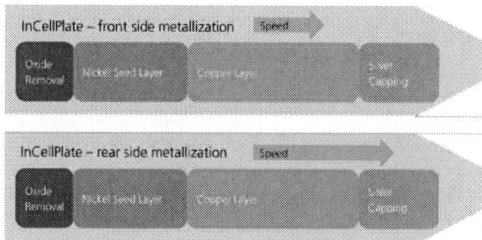

Figure 3: Schematic of the bifacial electroplating metallization sequence made at Fraunhofer ISE.

The metallization process comprises a nickel seed layer (1 μm height), a highly conductive copper layer (8 μm height), and a very thin silver layer (<0.5 μm) to prevent oxidation of the copper layer and to ensure good solderability. In total, the finger height is 10 μm and the finger width is 25 μm. The reference samples were metallized by printing silver-aluminum and silver pastes on the front and rear sides, respectively.

3 RESULTS

3.1 LCO in static process

Although the f-theta lens was specifically designed to scan large areas, up to 210 mm x 210 mm M12 wafers, a clear distortion of the structure shape can be observed at the corners of the wafer, as shown in Figure 3, while the center of the wafer has rounder spots.

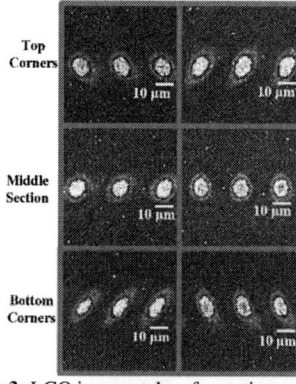

Figure 3: LCO images taken for static process in different locations over the wafer.

These spots at the extremes of the wafer are evaluated and compared to those in the middle section to verify their quality. The metric used for comparison is ellipticity, or flattening, which measures the compression of a circle to form an ellipse, as shown in Figure 4.

The results of these measurements reflect the distortion obtained far from the center of the scan field. In addition to the increase in ellipticity, the consistency of this value is decremental. The roundness and homogeneity obtained in the center are desirable results and support the argument for reducing the scan field as much as possible and lasering in the center of the field in on-the-fly mode.

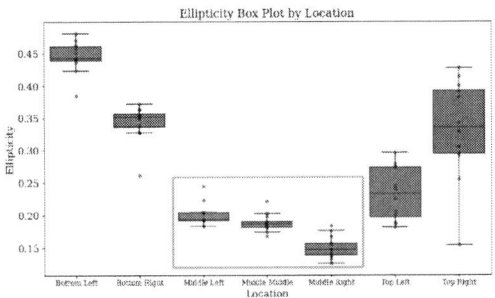

Figure 4: Evaluation of openings ellipticity over the wafer. The orange box highlights the desirable results.

3.2 LCO in on-the-fly process

To test the proposed technique, a linear axis is used. Thanks to the encoder signal, the scanner can follow the movement of the wafer and perform the desired process. By implementing this technique, the scanning field in the transport axis was reduced to 2 mm. Figure 5 shows the results for two different LCO processes in the center and corners of the wafer.

From Figure 5a it can be noted that the distortion observed in Figure 4 is not present. Both the ellipticity and the opening size are reduced and the homogeneity in the transport axis is improved.

Figure 5b shows that it is possible to create thin, continuous openings in the substrate with a width of approximately 10 μm, which is the target value for subsequent metallization steps.

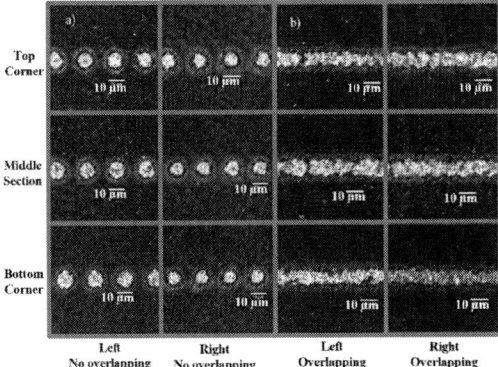

Figure 5: Collage of images taken at different positions on the wafer for two different processes a) LCO without overlap at a scan speed of 250 m/s, resulting in dots b) LCO with overlap at a scan speed of 125 m/s, resulting in continuous lines.

3.3 Laser throughput calculation

For this case study, two types of structures required for electroplating are realized. The first one is an LCO pattern of continuous lines along the wafer with a 1 mm spacing between each opening. The second pattern are bus bars that are placed every 14 mm each one consisting of 3 lines. Figure 6 shows a close-up image of these structures.

Figure 6: Close-up image of the resulting pattern. bus bars can be seen as vertical ticker lines, meanwhile the connectors are horizontal thinner lines.

To predict the production capacity and time requirements of the process, calculation tools have been developed to account for dynamic factors. The most relevant variables to consider include the laser repetition rate, which limits the scanning speed to ensure overlapping openings. This, in turn, dictates the line frequency of the polygon scanner and the optimal speed of the axis system [4]. The potential throughput is determined by calculating and extrapolating the time required to scan a given pattern. Wafer handling and positioning are not included in these calculations.

The results of the throughput calculations are shown in Table I, which compares a galvanometer-based scanning system and our approach with the polygon scanning system.

Table I: Throughput results for a system with a polygon scanner and another with galvanometer scanner.

LCO structure	Scanner type	time [s]	Throughput [wph]
Lines	Galvanometer	4.637	776
Lines + Bus bars	Galvanometer	5.579	645
Lines	Polygonal	1.241	2900
Lines + Bus bars	Polygonal	2.482	1450

The results presented in Table I show that the setup implemented by the polygon scanner has a processing capacity 2x to 4x times higher than that of the galvanometer scanner. However, it should be noted that these values are limited by the structure to be created on the substrate.

When calculating the throughput for the on-the-fly case with the polygon scanner, the laser frequency proved to be the main limitation. Although the polygon scanner can achieve processing speeds in excess of 1 km/s, 125 m/s was chosen. Otherwise, the spot-to-spot distance, or pulse pitch, would have been too big, avoiding formation of a continuous line. In this study, the maximum laser frequency at which ablation is possible was 16.2 MHz with a pulse pitch of 7.72 μm, ensuring sufficient overlap and line homogeneity. Overall, the constraint is shifted from the scanning system to the laser unit.

3.4 Electroplating and cell results

In this study, it was demonstrated for the first time that LCO produced by the high-throughput system is applicable for the metallization of large M10 (330.23 cm²) TOPCon wafers through electroplating. In Figure 7 the resulting contacts are shown for the different regions of interest over the wafer. Comparable wafers were produced using silver screen printing. The IV data were collected using an industrial measurement setup (LOANA, pv-tools) according to the categorization established by Rauer et al. [7], following the brn | grn, hrc classification. Table II presents the characteristic IV data for the best lasered, electroplated, and screen-printed solar cells. These results demonstrate that comparable efficiency can be achieved with the plated cell as with the standard screen printing process.

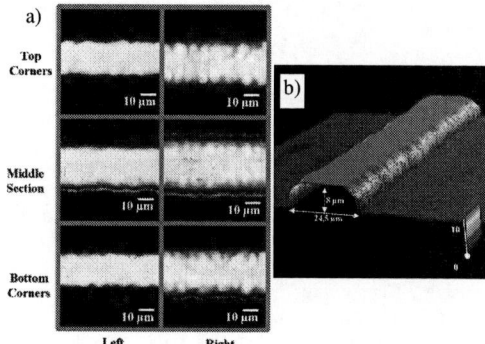

Figure 7: Result of the electroplating metallization process a) Collage of images taken at different positions on the wafer b) 3D magnification of a single contact line.

Table II: Comparison of silver screen printed (SP) and electroplated copper (plated) contacts in respective to the IV results.

	Eta [%]	Jsc [mA/cm²]	Voc [mV]	FF [%]
SP cell	22,7	40,2	707,2	80
Plated cell	22,8	40,2	697,7	84

4 CONCLUSIONS AND OUTLOOK

The presented on-the-fly laser micromachining concept enables application of laser contact openings with following achievements:

- The opening size is below 10 μm, exhibiting minimal ellipticity.
- Optical distortions are minimized in the central region of the scanner field.
- The LCO of continuous lines obtained meets the requisite standards for electroplating for M10 wafers.
- The resulting throughput is higher by a factor between 2x and 4x than that of other systems, depending on the scanned structure. However, the throughput is limited by the pattern and repetition rate of the laser, and thus, the theoretical maximum throughput is 31,500 wph, which is 40 times higher than that of the reference system.
- The new laser-contact opening method for metallizing M10 TOPCon wafers via electroplating achieves comparable efficiency to standard silver screen-printing with an equivalent grid layout.

This work has demonstrated that high throughput in a large wafer format (M10+) is achievable while maintaining a small opening size uniformly across the entire area. This approach overcomes the classical trade-off between speed, area, and structural quality, which was the primary motivation for developing the on-the-fly method and this setup.

To fully leverage the 1 km/s speed while maintaining the necessary pulse-to-pulse spacing for generating a continuous line, a laser operating at approximately 130 MHz would be needed. Even though this type of laser is unavailable yet; if it were accessible [8], the potential throughput could reach an impressive 31,500 wafers per hour. The upcoming work should include further benchmarking of this prototype with the state-of-the-art systems on M12 wafers and with different patterns.

6 REFERENCES

[1] M. Fischer, M. Woodhouse, P. Baliozian e J. Trube, International Technology Roadmap for Photovoltaic (ITRPV), Frankfurt am Main, Germany: VDMA e. V., 2023.

[2] Y. Zhang, M. Kim, L. Wang, P. Verlinden e B. Hallam, «Design considerations for multi-terawatt scale manufacturing of existing and future photovoltaic technologies: challenges and opportunities related to silver, indium and bismuth consumption,» *Energy Environ. Sci.,* vol. 14, n. 11, pp. 5587-5610, 2021.

[3] S. Kluska, R. Haberstoh, B. Grübel, G. Cimiotti, C. Schmiga, A. A. Brand, A. Nägele, B. Steinhauser, M. Kamp, M. Passig, M. Sieber, D. Brunner e S. Fox, «Enabling savings in silver consumption and poly-Si thickness by integration of plated Ni/Cu/Ag contacts for bifacial TOPCon solar cells,» *Solar Energy Materials and Solar Cells,* 2022.

[4] G. Hoppe, E. Alvarez-Brito, G. Emanuel, J. Nekarda, M. Diehl, R. Preu e F. Meyer, «Overcoming Throughput Limitations of Laser Systems in Solar Cell Manufacturing via On-The-Fly Processing Using Polygon Scanners,» *Energy Tech (Energy Technology),* 2023.

[5] S. Klötzer e A. Streek, «Polygonscantechnik für Lasermikrobearbeitung,» in *Lasertagung Mittweida,* 2015.

[6] S. Kluska, B. Grübel, G. Cimiotti, C. Schmiga, H. Berg, A. Beinert, I. Kubitza, P. Müller e T. Voss, «Plated TOPCon solar cells & modules with reliable fracture stress and soldered module interconnection,» *EPJ Photovolt. (EPJ Photovoltaics),* vol. 12, p. 10, 2021.

[7] M. Rauer, A. Krieg, A. Fell, S. Pingel, N. Wöhrle, J. M. Greulich, S. Rein, M. C. Schubert e J. Hohl-Ebinger, «Assessing current-voltage measurements of busbarless solar cells,» *Solar Energy Materials and Solar Cells,* vol. 248, 2022.

[8] «UKP Workshop,» [Online]. Available: https://www.ultrakurzpulslaser.de/de/ukp-workshop/rueckblick/6--ukp-workshop.html.

This work has been supported by the German Federal Ministry of Economic Affairs and Climate Action under the project name Miracle (No.:03EE11487A).

41st European Photovoltaic Solar Energy Conference and Exhibition

ENHANCEMENT OF PHOTOCURRENT GENERATION IN AMORPHOUS SILICON HETEROJUNCTION (SHJ) SOLAR CELLS THROUGH THE INTEGRATION OF PLASMONIC NANOPARTICLES

Brahim Aïssa[1*], Alessandro Sinopoli[1]

[1]Qatar Environment & Energy Research Institute (QEERI), Hamad Bin Khalifa University (HBKU), Education City,
P.O. Box: 34110, Qatar Foundation, Doha, Qatar
*Corresponding author: baissa@hbku.edu.qa

ABSTRACT: Gold (Au) nanoparticles were meticulously deposited onto the indium tin oxide (ITO) layer of silicon heterojunction solar cells through the pulsed laser deposition method. This innovative technique allowed for precise control over the density of the nanoparticles by adjusting the number of laser pulses applied during the deposition process. The primary focus of this study is to investigate the effects of Au nanoparticles on the photovoltaic characteristics of the solar cells, with a specific emphasis on the short-circuit current. The findings of this research indicate that a notable enhancement in the short-circuit current—approximately 5.5%—was achieved when applying six laser pulses. This optimal enhancement highlights the delicate interplay between the deposition parameters and the resultant photovoltaic performance. Additionally, both the surface density and the average size of the Au nanoparticles were identified as significant factors influencing the generated photocurrent. In essence, this research emphasizes the critical importance of nanoparticle density and size in optimizing the performance of silicon heterojunction solar cells. The insights garnered from this study not only deepen our understanding of the role of metal nanoparticles in photovoltaic applications but also pave the way for the refinement of solar cell designs. By considering these factors, researchers and engineers can make informed decisions to enhance the efficiency and output of solar cells, ultimately contributing to advancements in renewable energy technology.
Keywords: Silicon heterojunction solar cell, Plasmonic, Photocurrent.

1 INTRODUCTION

The integration of plasmonic nanostructures into solar cells presents a promising avenue for enhancing absorption within the active layer, which can ultimately lead to improved solar cell efficiency. Among the various plasmonic nanostructures, metallic nanoparticles (NPs) are particularly notable, as they can be strategically deposited either on the surface of the solar cell or at the metal-semiconductor interface, effectively incorporating them into the device architecture.

Two primary mechanisms have been proposed to elucidate how metallic nanoparticles contribute to the improvement of conversion efficiency in solar cells: near-field concentration of light and light scattering. When metallic NPs are subjected to an electromagnetic wave, they undergo collective oscillations of their conduction electrons. If the frequency of these oscillations aligns with that of the incident light, surface plasmon resonance (SPR) occurs. This phenomenon generates an electric field around the surface of the nanoparticles, with its intensity being significantly enhanced when the nanoparticles are closely spaced. This effect is referred to as near-field concentration of light. However, to capitalize on the benefits of the localized electric field generated by the nanoparticles, it is crucial for them to be situated in proximity to the p–n junction of the solar cell. This necessitates the deposition of nanoparticles directly on the emitter or on the transparent conducting oxide (TCO) layer [1-2].

The second mechanism by which nanoparticles influence photovoltaic efficiency involves light scattering due to the excitation of their surface plasmon resonance frequency (SPRF), commonly referred to as plasmonic light scattering (PLS). The interaction between incoming light and these nanoparticles can result in either absorption or scattering, depending on the size of the particles. The effectiveness of this interaction is determined by the ratio of the effective cross-sectional areas for both processes. Research has shown that smaller nanoparticles typically exhibit a higher effective cross-sectional area for absorption compared to scattering, which can lead to significant absorption losses. Conversely, larger nanoparticles are more effective at scattering incident light into the intrinsic absorbing layer of the solar cells, thereby increasing the optical path length and enhancing overall performance. The enhancement of photocurrent observed with metal nanoparticles placed on the surface of solar cells can primarily be attributed to the light scattering mechanism, positioning these nanostructures as highly efficient light harvesters [3-4].

Various metallic nanoparticles, such as copper (Cu), aluminum (Al), gold (Au), silver (Ag), and palladium (Pd), have been explored for use in plasmonic solar cells. While Cu and Al nanoparticles are cost-effective, they are prone to rapid oxidation, leading to instability in their SPR. Most research efforts have concentrated on noble metals, which exhibit strong resonances in the visible spectrum and offer the advantage of chemical inertness. Gold and silver are the most frequently utilized metals in plasmonic applications, with silver often preferred due to its lower absorption losses stemming from a higher scattering cross-section. However, silver can form an absorbing oxide layer that causes damping and increases overall absorption losses. In contrast, gold does not form such an oxide layer, but for it to exhibit preferential light scattering over absorption, gold nanoparticles must be larger than their silver counterparts. Studies have demonstrated that larger gold nanoparticles scatter more light in the forward direction compared to similarly sized silver nanoparticles, making them a compelling choice for placement on the upper surface of solar cells [5-6].

In this study, gold nanoparticles were deposited on the surface of indium tin oxide (ITO)/silicon heterojunction (SHJ) solar cells using the pulsed laser deposition technique. The density of the resulting gold nanoparticle network was meticulously controlled by varying the number of laser pulses during the deposition process. This paper explores the effect of different Au network densities on the photovoltaic characteristics of SHJ solar cells. Notably, it was found that an optimal short-circuit current enhancement of 5.5% could be achieved with the application of six laser pulses; however, this improvement came at the expense of other photovoltaic performance metrics. This observation highlights the necessity of making a clear trade-off based on the specific optoelectronic applications being targeted.

Furthermore, the performance of the SHJ solar cells was evaluated under varying temperature conditions to mimic the climate of the state of Qatar. While these results are preliminary, they indicate the potential for further development in this specific area of research, suggesting that optimizing nanoparticle deposition and density could lead to significant advancements in solar cell technology. Overall, this work contributes valuable insights into the nuanced relationship between plasmonic nanostructures and solar cell efficiency, offering a foundation for future explorations in the field [7-10].

2 METHODOLOGY

2.1 Fabrication:

Indium tin oxide (ITO) thin films were synthesized using the direct current (DC) sputtering technique, employing an ITO target composed of 90 wt.% indium oxide (In_2O_3) and 10 wt.% tin oxide (SnO_2) at room temperature. This method was chosen for its ability to produce high-quality thin films with uniform properties. The deposition process was carried out under a DC power density of 1.9 W/cm², which is critical for achieving a good balance between film quality and deposition rate. During this process, the oxygen to total flow ratio was carefully controlled, defined as:

$$r(O_2)=O_2\backslash Ar+O_2$$

Was maintained at a precise value of 1%. This specific ratio is vital for ensuring optimal film characteristics and for preventing unwanted oxidation or contamination of the ITO layer. Each ITO film was fabricated to a consistent thickness of 60 ± 5 nm, which is an important parameter for the electrical and optical performance of the solar cells.

To evaluate the electrical properties of the ITO films, several key metrics were measured, including electrical conductivity (σ), carrier concentration (N_e), and Hall mobility ($\mu Hall$). These measurements were performed using an HMS-5000 system, employing the Van der Pauw method, a reliable technique for determining the properties of thin films. The results of these assessments provide critical insights into the effectiveness of the ITO layer as a transparent conductive electrode, which is essential for the overall efficiency of the solar cells. A schematic representation of the rear junction silicon heterojunction (SHJ) solar cell can be found in Figure 1, illustrating the various components and their arrangement.

The SHJ solar cells were constructed on high-quality n-type float-zone crystalline silicon (c-Si) wafers, specifically oriented in the (100) direction. These wafers had a thickness of 180 μm and a resistivity range of 1–5 Ω cm, making them ideal substrates for solar cell applications due to their excellent electronic properties. The fabrication process commenced with random pyramidal texturing of the silicon wafers in an alkaline solution, a critical step designed to enhance light absorption by increasing the surface area and minimizing reflection. Following this texturing process, the wafers underwent a thorough wet chemical cleaning to remove any contaminants that could adversely affect subsequent layers.

After the cleaning procedure, the wafers were treated with hydrofluoric acid to eliminate any native oxide layer that may have formed on the silicon surface. This step is crucial, as the presence of an oxide layer can hinder the deposition of subsequent layers and negatively impact device performance. The prepared wafers were then subjected to the deposition of intrinsic and doped hydrogenated amorphous silicon (a-Si) layers using plasma-enhanced chemical vapor deposition (PECVD). This technique allows for the growth of high-quality a-Si films that serve as the active layers in the SHJ solar cells.

The ITO films, which serve as the front electrode of the solar cells, were deposited using a shadow mask measuring 2×2 cm². This precise technique effectively defined the size of each solar cell during the fabrication process, ensuring uniformity across multiple devices. By carefully controlling the deposition parameters and the fabrication process, this approach aims to optimize both the electrical and optical properties of the solar cells. These attributes are essential for enhancing the overall performance and efficiency of SHJ solar cells, making them competitive in the renewable energy market. Through this meticulous fabrication process, the research aims to advance the understanding of how each layer contributes to the functionality of the solar cells, ultimately leading to innovations in solar technology [11-18].

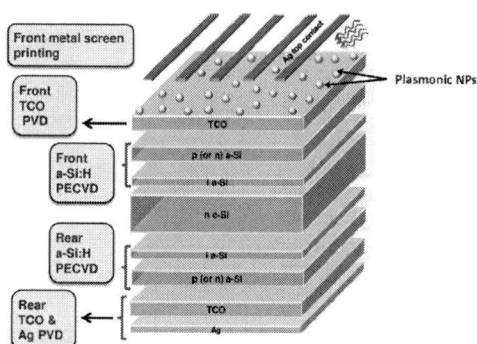

Figure 1: Schematic illustration of the fabrication process for silicon heterojunction (SHJ) solar cells, highlighting the pulsed laser deposition (PLD) technique used to deposit gold (Au) plasmonic nanoparticles onto the surface of the indium tin oxide (ITO) layer. This diagram outlines the sequential steps involved in the fabrication, showcasing the integration of Au nanoparticles aimed at enhancing the photovoltaic performance of the solar cells. The specific parameters for the PLD process, such as laser pulse frequency and deposition conditions, may also be detailed to illustrate their role in controlling nanoparticle density and distribution on the ITO surface.

2.2 Characterizations:

On the rear side of the silicon wafer, an indium tin oxide (ITO) layer was applied across all devices, serving as a transparent conductive oxide to enhance the efficiency of the solar cells. This ITO layer was immediately followed by the deposition of a silver back reflector, which was also sputtered onto the wafer. The silver back reflector plays a crucial role in improving light trapping and minimizing reflective losses, thereby contributing to the overall performance of the solar cells.

On the front side of the device, a silver front grid was meticulously screen-printed onto the surface of the five

ITO pads. This front grid is essential for electrical contact while simultaneously allowing for maximum light exposure to the active layers of the solar cell. Following the screen-printing process, the solar cells underwent a curing process for 30 minutes at a temperature of 210 °C. This curing step is vital as it ensures the adhesion of the silver grid to the ITO layer and promotes the formation of a stable electrical contact.

Once the solar cells were fully assembled, they were subjected to characterization under standard test conditions (STC). This included various temperature settings, allowing for a comprehensive analysis of the devices' performance. The current-voltage (J–V) measurements were conducted using an AAA-certified solar simulator under Air Mass 1.5 global illumination, simulating real-world sunlight conditions to evaluate the efficiency and operational characteristics of the solar cells.

To further enhance the optical properties of the solar cells, gold (Au) nanoparticles were deposited directly onto the ITO surface using the Pulsed Laser Ablation (PLD) technique. The deposition process was carefully controlled to manage the density of the deposited Au nanoparticles, which was achieved by varying the number of laser pulses directed at a pure gold target (99.999% purity). This meticulous control over the deposition parameters is critical for optimizing the plasmonic effects of the nanoparticles, which can significantly enhance the solar cell performance. A more detailed explanation of the PLD process and its parameters will be provided in the full version of this article, highlighting the innovative aspects of this technique.

Additionally, the surface morphology of the solar cells integrated with gold nanoparticles was thoroughly investigated using atomic force microscopy (AFM) in contact mode, employing a NanoScope III Quadrexed Dimension 3000. This advanced imaging technique allows for high-resolution analysis of the surface topography and structural characteristics of the solar cells, providing valuable insights into how the incorporation of gold nanoparticles influences the overall device performance. By examining the surface morphology, researchers can assess the distribution, size, and uniformity of the Au nanoparticles, which are essential factors in understanding their impact on light absorption and charge carrier dynamics within the solar cells.

3 RESULTS AND DISCUSSION

Figure 2 presents typical atomic force microscopy (AFM) images captured in contact mode, showcasing the surface morphology of the plasmonic gold (Au) nanoparticles deposited onto the indium tin oxide (ITO) surface using 1 and 6 laser pulses. These images illustrate the distribution and size of the Au nanoparticles, highlighting the effects of varying the number of laser pulses on their deposition characteristics.

In the lower panel of Figure 2, the associated short-circuit current (Jsc) measured from these n-type silicon heterojunction (n-SHJ) solar devices is displayed for different densities of the Au nanoparticle network. This data reveals the relationship between the nanoparticle density and the photovoltaic performance of the solar cells, specifically emphasizing how the increased density of Au nanoparticles influences the Jsc values. The findings illustrated in this figure provide crucial insights into the optimal conditions for maximizing the efficiency of n-SHJ devices through the strategic incorporation of plasmonic

Au nanoparticles.

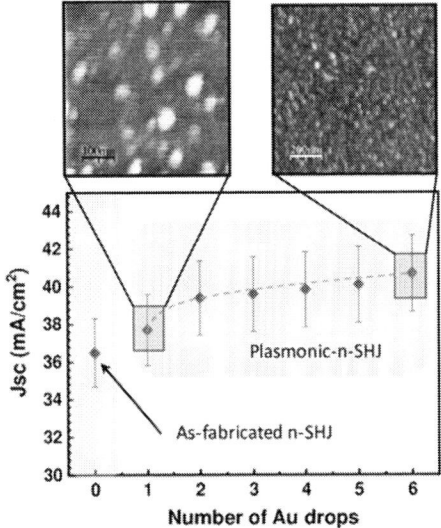

Figure 2: Top: Atomic force microscopy (AFM) images depicting the surface morphology of plasmonic gold (Au) nanoparticles deposited onto the indium tin oxide (ITO) layer using 1 and 6 laser pulses. These images provide insight into the distribution and size of the nanoparticles on the ITO surface. Bottom: Graph illustrating the short-circuit current (Jsc) of the n-type silicon heterojunction (n-SHJ) solar cells as a function of the density of the plasmonic Au nanoparticle network. The nanoparticle density is controlled by the number of laser pulses applied during the deposition process, highlighting the relationship between Au nanoparticle density and the photovoltaic performance of the solar devices.

The short-circuit current (Jsc) exhibits a clear increasing trend as the density of gold (Au) nanoparticles (NPs) rises. This correlation indicates that higher nanoparticle density contributes to enhanced light absorption within the active layer of the solar cells. Supporting this observation, the reflection and absorption spectra (not displayed here) measured in the wavelength range of 200 to 1200 nm align well with the Jsc data, showing that more light is absorbed as the density of Au nanoparticles increases.

However, it is important to note that a comparison of the reflection and absorption spectra with the photovoltaic performance of the solar cells reveals that lower reflection coefficients do not always translate to improved Jsc. The concentration of nanoparticles on the surface is a critical factor, as excessive nanoparticle density can lead to significant shadowing effects, obstructing incident solar radiation from reaching the active layer of the solar cell. This shadowing can, in turn, result in a deterioration of the photoelectric parameters, highlighting the need for careful optimization of Au nanoparticle density to balance the benefits of enhanced light absorption with the potential drawbacks of light blockage. Thus, achieving the optimal nanoparticle concentration is essential for maximizing the overall efficiency and performance of the solar cells.

41st European Photovoltaic Solar Energy Conference and Exhibition

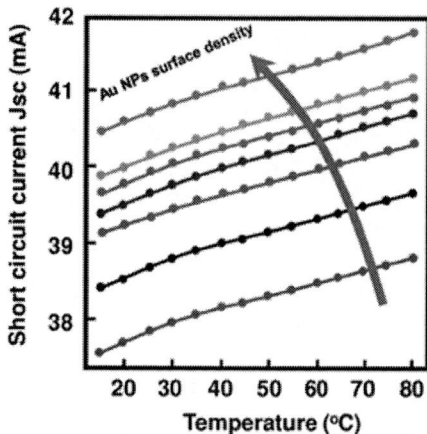

Figure 3: The short-circuit current (Jsc) of the n-type silicon heterojunction (n-SHJ) solar cells is significantly influenced by temperature variations. As the temperature increases, the behavior of charge carriers within the solar cell can change, which in turn affects the Jsc values.

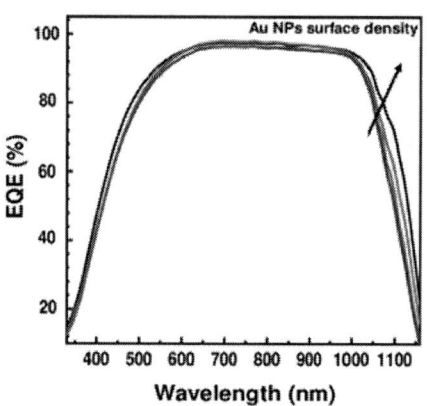

Figure 4: External Quantum Efficiency (EQE) (%) for different plasmonic Gold (Au) Network Densities.

As a matter of fact, studies have shown that the deposition of gold nanoparticles (NPs) directly onto the surface of solar cells via electrodeposition can lead to a significant decrease in the short-circuit current (Jsc). This counterintuitive result highlights the complexity of interactions between nanoparticles and the photovoltaic performance of solar cells. In contrast, employing alternative deposition techniques, such as immersion and centrifugation methods, has resulted in a clear enhancement of the photovoltaic properties of the cells. These methods appear to facilitate better integration and distribution of the NPs, which positively affects light absorption and charge carrier generation.

Among the various techniques available for nanoparticle deposition, pulsed laser deposition (PLD) stands out due to its ability to offer a higher degree of freedom in controlling the surface density or concentration of Au NPs on indium tin oxide (ITO) layers. This flexibility is crucial for optimizing the plasmonic effects that enhance the efficiency of solar cells.

Figure 3 illustrates the relationship between the short-

circuit current of n-type silicon heterojunction (n-SHJ) solar cells and temperature, alongside the corresponding external quantum efficiency (EQE) percentages for different densities of the plasmonic Au network. This data provides critical insights into how varying temperatures and nanoparticle densities impact the overall performance of the solar cells.

Determining the optimal number of nanoparticles is a complex challenge primarily due to the fact that the resonance frequency related to the effective cross-section area of scattering for a nanoparticle is often much larger than its geometric size. For instance, if a nanoparticle exhibits an effective scattering cross-section that is ten times greater than its diameter, it means that even if the surface is covered by just 10% of the area with these nanoparticles, it can still effectively absorb or dissipate the incident light, depending on whether the nanoparticles are small or large. This phenomenon underscores the nuanced interplay between nanoparticle size, concentration, and their optical properties, making it essential to fine-tune these parameters for optimal performance.

Current investigations into the influence of nanoparticle size on photoelectric characteristics are ongoing, with preliminary data aligning closely with observations made in reflectance spectra analyses. Notably, it has been found that the increase in short-circuit current for larger nanoparticles is significantly greater than that for smaller ones. For example, the deposition of 30 nm gold nanoparticles via PLD results in a modest 2.6% increase in Jsc. In contrast, utilizing larger gold nanoparticles, specifically those measuring 45 nm, leads to a more substantial increase of 5.5% in current.

It is important to highlight that the presence of these nanoparticles does not have a similarly pronounced effect on the open-circuit voltage (Voc). The deviation in Voc observed in the presence of nanoparticles remains relatively minor, approximately 3%. This phenomenon can be attributed to the fact that the deposition of gold nanoparticles on the surface of the solar cell does not significantly alter the potential barrier at the interface between the film and substrate. This understanding is vital for developing strategies to optimize solar cell performance through nanoparticle integration while minimizing adverse effects on critical electrical parameters. Overall, ongoing research will further clarify these relationships and refine approaches for leveraging plasmonic nanoparticles in solar energy applications.

4 SUMMARY AND CONCLUSIONS

In this study, we thoroughly investigated the impact of gold nanoparticles (Au NPs) deposited on indium tin oxide (ITO) layers through the pulsed laser deposition (PLD) method on the photovoltaic parameters of n-type silicon heterojunction (n-SHJ) solar cells. The findings revealed that utilizing the PLD technique for Au NP deposition significantly enhances the short-circuit current (Jsc) of the solar cells by as much as 5.5%. This improvement is primarily attributed to the increased light absorbance facilitated by the presence of these plasmonic nanoparticles, which effectively enhances the interaction of incident sunlight with the solar cell structure.

The enhanced light absorption resulting from the gold nanoparticles can be attributed to the plasmonic effects they exhibit. When these nanoparticles are excited by incoming electromagnetic radiation, they resonate at specific frequencies, leading to localized electric field enhancements. This phenomenon allows for greater penetration of light into the active layers of the solar cell, resulting in a more efficient generation of electron-hole pairs and, consequently, an increase in the short-circuit current. The systematic increase in Jsc as a function of the density of the Au NPs further emphasizes the importance of optimizing nanoparticle parameters to maximize solar cell performance.

In addition to the improvements in short-circuit current, our study also highlighted the influence of temperature on Jsc. It was observed that increasing the operational temperature of the solar cells further contributes to an enhancement in Jsc. This effect can be explained by the fact that elevated temperatures can increase the kinetic energy of charge carriers, thereby enhancing their mobility and promoting more efficient collection at the electrodes. However, this temperature dependence introduces a complexity in balancing performance with thermal stability, which is crucial for the real-world application of solar cells in various environmental conditions.

While the presence of plasmonic gold nanoparticles demonstrates a pronounced effect on the short-circuit current, we noted that the influence on short-circuit voltage (Voc) appears to be less significant. Initial assessments indicate that the incorporation of Au NPs does not lead to substantial variations in Voc, which remains relatively stable. This observation prompts further investigation into the underlying mechanisms that govern the interaction between the nanoparticles and the electrical characteristics of the solar cells.

Ongoing work aims to systematically study the plasmonic effects on additional key photovoltaic parameters, including the fill factor (FF%) and the associated power conversion efficiency (PCE). These metrics are essential for fully understanding the performance landscape of n-SHJ solar cells modified with Au NPs. The comprehensive analysis will consider several critical factors: the density of the Au NP network, the average size of the nanoparticles, and the temperature at which the solar cells operate.

The interplay between these parameters is expected to yield valuable insights into optimizing the design and fabrication of high-efficiency solar cells. By elucidating the relationships among nanoparticle characteristics, operational temperature, and solar cell performance, this research could pave the way for more advanced solar technologies that leverage plasmonic effects for enhanced energy conversion. Ultimately, the findings from this work could contribute to developing more efficient, cost-effective solar energy solutions that meet the growing demand for sustainable energy sources in the future.

References

[1] D. L. Wang, H. J. Cui, G. Su, Solar Energy, vol. 120, pp. 505–513, October 2015.

[2] A. Tamang, et al. Solar Energy Materials and Solar Cells, vol. 144, pp. 300–308, January 2016.

[3] D. Lockau, et al. Solar Energy Materials and Solar Cells, vol. 125, pp. 298–304, June 2014.

[4] K.A. Salman, Sol. Energy, vol. 147, pp. 228–231, May 2017.

[5] K. Islam, A. Alnuaimi, E. Battal, A. K. Okyay, and A. Nayfeh, Sol. Energy, vol. 103, pp. 263–268, May 2014.

[6] K. Islam, F. I. Chowdhury, A. K. Okyay, and A. Nayfeh, Sol. Energy, vol. 120, pp. 257–262, October 2015.

[7] Islam, K., Alnuaimi, A., Battal, E., Okyay, A. K., & Nayfeh, A. (2014). Solar Energy, 103, 263–268.

[8] Islam, K., Chowdhury, F. I., Okyay, A. K., & Nayfeh, A. (2015). Solar Energy, 120, 257–262.

[9] Ranjbar, S., Wörth, K., & Dey, S. (2015). Materials Letters, 139, 363–366.

[10] Cai, Z., & Zhang, G. (2018). Solar Energy Materials and Solar Cells, 183, 142–150.

[11] Zhang, J. et al. Nanotechnology, 28(22), 225402.

[12] Yadav, A., Kumar, A., & Gupta, S. (2020). Renewable Energy, 152, 1020–1031.

[13] Tame, J. R. (2013). Journal of Physics D: Applied Physics, 46(29), 293001.

[14] Kumar, S., Sharma, S., & Sinha, R. (2016). Solar Energy, 128, 181–198.

[15] Grinberg, A. (2018). Energy Reports, 4, 679–685.

[16] Chen, Y., & Wang, H. (2019). Solar Energy Materials and Solar Cells, 201, 109973.

[17] Gavi, N. M. H., Ngom, B. D., Beye, A. C., Strydom, A. M., Aissa, B., Srinivasu, V. V., and Chaker, M. Journal of Magnetism and Magnetic Materials, 324(6), 1172-1176, 2012.E

[18] El Khakani, M. A., Le Borgne, V., Aïssa, B., Rosei, F., Scilletta, C., Speiser, E., et al. Applied Physics Letters, 95(8), 2009.

[19] Habib, M. A., Barkat, M., Aissa, B., and Denidni, T. Progress In Electromagnetics Research, 88, 135-148, 2018

Impact of optimization for mass production PERC solar cell with efficiency above 23%

Cheng-Wen Kuo, Ta-Ming Kuan, Yung-Chih Li, Chun-Wei Lee, Wei-Lo Chueh, Li-Guo Wu, Shih-Chieh Lin and Cheng-Yeh Yu
TSEC Corporation, No.85, Gaungfu N. Rd., Hsin-Chu 30351, Taiwan.

ABSTRACT: In this work, we demonstrate that p-type PERC solar cells can also improve mass production efficiency and productivity based on the current PERC structure process. By changing the POCl3 parameters in the diffusion process, the productivity of the diffusion process can be effectively increased by 7.5% and the efficiency is slightly improved. For the front anti-reflective layer process, it could be found that the reducing the n value of the anti-reflective multi-layer would increase the light absorption of the solar cells, which can effectively increase the photocurrent of about 25~30mA. In the process of rear side passivation layer, the temperature of CVD process optimization can effectively enhance the passivation effect, and the photocurrent and open circuit voltage are both improved. By applying above three process-optimized parameters to the production line process, it can be found that the conversion efficiency can be increased by 0.15% to 0.2%, and the productivity can be improved as well. Finally, we took the same efficiency level of experiment cells and the reference cells for module testing. It was found that the cell to module (CTM) degradation of the Optimized group was lower than the reference group, which means the quality of solar cells to modules could be effectively improved. The related quality test of all solar cells also complied with IEC specifications.
Keywords: PERC, mass production, temperature

1 INTRODUCTION

As the push towards net-zero emissions accelerates, the energy trend is shifting towards replacing traditional fossil fuels with renewable energy sources. Research institutions are extensively studying renewable energy, focusing on both its sources and environmental impacts [1]. Among renewable energy sources, solar energy stands out as both abundant and environmentally friendly. Utilizing photovoltaics (PV) to convert solar energy into electricity has been shown to reduce CO2 emissions [2]. Photovoltaic systems operate on the principle of the photovoltaic effect, which generates a direct current (DC) voltage between two different semiconductor materials when exposed to sunlight [3]. However, thel efficiency can be compromised by various factors, including significant reflection at the interface between the solar cell and the surrounding air. This reflection occurs due to the high refractive index of the solar cell substrate, which results in a substantial portion of incident solar radiation being reflected away, thereby not contributing to carrier pair generation and reducing overall efficiency [1,4]. To enhance efficiency in the mass production of solar cells, effective solutions are needed to minimize photon loss at the cell's front surface. One approach involves depositing antireflective layers [5]. In this context, the amorphous hydrogenated form of silicon nitride (SiNx) is particularly advantageous due to its low absorption in the solar spectrum and excellent passivation properties for both mono- and multi-crystalline silicon materials. Plasma-enhanced chemical vapor deposition (PECVD) is a preferred method for depositing SiNx thin films due to its lower thermal budget and high hydrogen content [6].

Similarly, the amorphous hydrogenated form of silicon oxynitride (SiOxNy) offers comparable benefits. It can be deposited using PECVD at low temperatures and features low absorbance with an adjustable refractive index ranging from below 1.5 to as high as 3.0, depending on the precursor gas flow ratio during deposition [7]. By optimizing the optical index and thickness of SiON films, they can be effectively used in antireflection multilayer coatings for solar cells. Previous studies [8] have highlighted that the refractive index of SiNx anti-reflection coatings (ARC) significantly impacts both optical absorption and potential-induced degradation (PID) sensitivity.

In this study, our primary goal was to enhance productivity while maintaining a certain level of performance. We improved a previously developed anti-reflective coating structure composed of SiOx/SiOyNx/SiNx layers [9], which had already shown improved reflectivity. The structure composed also introduced into the rear side passivation layer and adjusted the process parameters. The results indicated that this new multi-layer structure not only enhance productivity but also increases efficiency.

2 EXPERIMENTAL

Figure 1 showed the PERC+ structure in this work. The M10 size (182.0 mm x 182.0 mm) bifacial PERC solar cells were fabricated on 155 μm thick p-type boron doped Cz-silicon wafers with a resistivity of 0.4~1.1 Ω-cm.

The process sequence in this work was shown in Figure 2. Chemical wet etching methods were used for saw damage removal and surface texturing, which include alkaline etching for mono-crystalline silicon wafers. The solution composition of alkaline etching was potassium hydroxide (KOH) : deionized water (DI) = 1 : 50 and additives. The diffusion process of this work also improves the temperature curve and its uniformity.Conventional phosphorous diffusion was employed by furnace to form the n+ emitters with a sheet resistance of 160 – 170 ohm/sq.

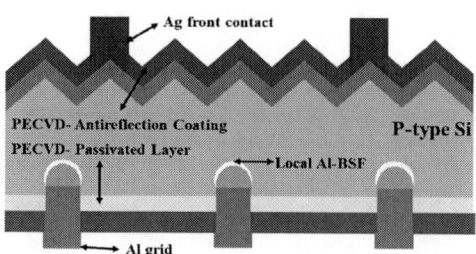

Figure 1: PERC+ structure as investigated in this work.

The phosphosilicate glass (PSG) was then removed by HF-contained solutions, formed edge isolation in the meantime. Then, the rear surface is polished which removes 3-6 μm silicon in this work, in order to reduce the roughness and its impact on the effective charge carrier lifetime [10]. In order to decrease PID effect and increase efficiency, we optimized the CVD process parameter to improve. An antireflection coating (ARC) layer of silicon nitride (SiN_x: H) and/or $SiO_x/SiO_xN_y/SiN_x$ multi ARC was deposited by plasma enhanced chemical vapor deposition (PECVD) with thickness of 80-100 nm, as shown in Fig 3. The rear side passivation was carried out on rear side also using a PECVD [11]. The Fig 3 shown that the thickness of rear side passivation thickness was about 90~110 nm. Afterward, the rear side laser opening ratio is also optimized. All PERC solar cells have the laser-ablated opening widths in the rear-side passivation stacks which were about 30 μm. The rear Al finger width is 150 μm in this work. The detail parameters were listed in Table 1. The front side silver and rear side aluminum metallization was formed by screen printing technique and followed by co-firing process, which resulted in formation of front silver grid contact and local back-surface-field. Finally, the current-voltage measurements of solar cells under AM 1.5 illumination were performed.

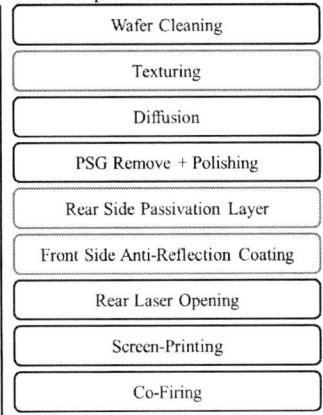

Figure 2: Process flow for the fabrication of PERC solar cell as developed in this work

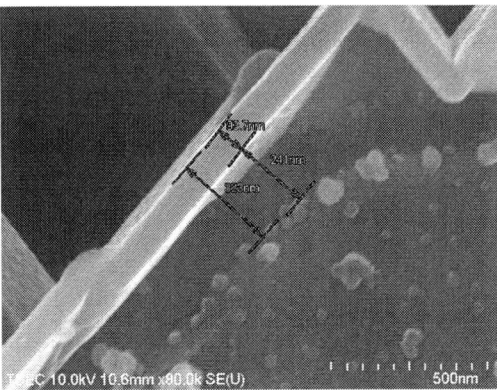

Figure 3: The scanning electron microscope (SEM) image of front side passivation layer for this work.

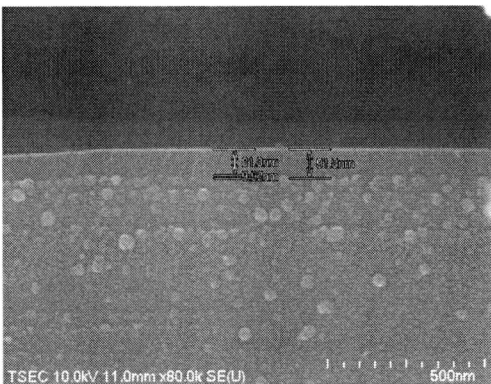

Figure 4: The scanning electron microscope (SEM) image of rear side passivation layer for this work.

3 EXPERIMENTAL RESULTS AND DISCUSSION

The auxiliary heating time of the rear side passivation layer process was adjusted for optimizing the uniformity of the passivation layer. The thickness of different time for rear side passivation layer is similar. The Table 1 shows the average electrical performance of the different auxiliary heating time PERC solar cells.

Table 1: The average electrical results of M10 different auxiliary heating time PERC solar cells.

	EFF (%)	FF (%)	V_{OC} (V)	I_{SC} (A)
RSD temp_310s (REF)	23.02	80.14	0.6884	13.779
RSD temp_190s	23.07	80.14	0.6882	13.807
RSD temp_100s	23.10	80.31	0.6874	13.814

It can be found that the RSD temp_100s PERC cell has the highest efficiency from table 1. The efficiency distribution is also the best, as shown in Figure 5. It means that in the future, the recipe can be used in the process to reduce time and increase production capacity.

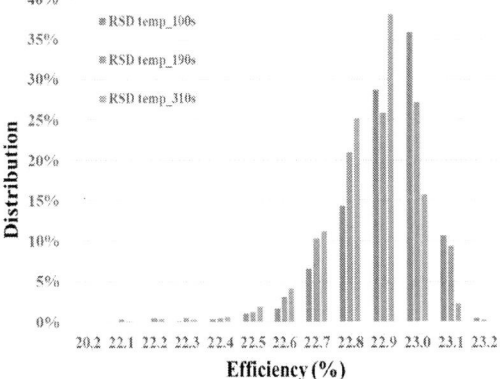

Figure 5: The efficiency distribution of different time for rear side passivation layer PERC solar cells.

We have performed an optical reflectance analysis for the different ARC layer, as shown in Figure 6. The distinct reduction of the reflectance at long wavelengths (>780 nm) is apparent for the optimized ARC layers compared with the baseline REF, the front average reflection of wavelengths (350-1100nm) about 2.44% and 2.17%, respectively.

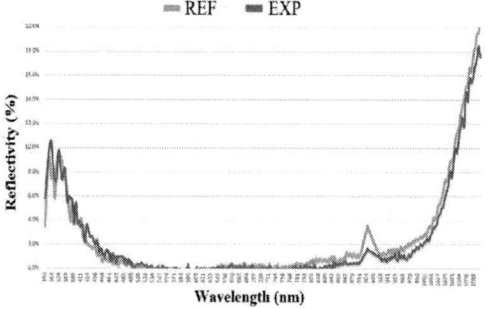

Figure 6: Measured reflectivity for the experiment ARC PERC solar cell and Ref PERC solar cell.

Table 2 shows the average electrical performance of REF and Experiment ARC PERC solar cells. The average efficiency of Experiment ARC PERC solar cell was 23.18%, the average efficiency of 23.1% was obtained for the REF PERC solar cells. It shows that the photocurrent can also be increased by about 20mA and 0.08% the absolute gain in conversion efficiency can be obtained by introducing $SiO_x/SiO_yN_x/SiN_x$ as an ARC instead of multi $SiNx$ layer. According to the measurement, the average thickness of the REF ARC PERC solar cells is 82.73nm and the average thickness of the Experiment ARC PERC solar cells is 82.74nm, but the average refractive index is 2.023 and 2.006 respectively. It can be found that the ARC thickness is similar to the process setting which means that the use of Experiment ARC layers can improve the optical characteristics

Table 2: The average electrical results of M10 REF and experment ARC PERC cells.

	EFF (%)	FF (%)	V_{OC} (V)	I_{SC} (A)
REF ARC PERC	23.1	80.14	0.6879	13.836
Experiment ARC PERC	23.18	80.22	0.6886	13.856

	Pre-PID		Post-PID		EL		Decay Rate(%) (Spec<5%)
Sample	Rsh	Pmpp	Rsh	Pmpp	Pre-PID	Post-PID	△Pmpp
REF-1	221.7	6.9953	306.3	6.9304			0.93%
REF-2	168.1	6.9966	251.2	6.9367			0.86%
REF-3	151.1	7.0157	158.8	6.9670			0.69%
Exp-1	153.7	6.9701	142.3	6.8713			1.42%
Exp-2	499.5	6.9502	734.4	6.8454			1.51%
Exp-3	1382.9	6.8900	739.3	6.8434			0.68%

Figure 7: Summary of power loss and EL images of pre-PID and post-PID of REF and optimized ARC PERC solar cells. PID test condition is 85 degree Celsius / 85% RH biased -1000V for 96 hours.

On the other hand, the ARC layer structure of PERC solar cells has an effect on PID [8], the potential-induced degradation (PID) test of the solar cell also conforms to the IEC TS 62804-1 specification and was used in this experiment. The REF and Experiment ARC PERC solar cells also passed PID test which condition was indicated to be 85 degree Celsius and 85% RH in climatic chamber apply -1000 V for 96 hours as shown in Figure 7. Observed the power loss < 5% and no EL darkened area were found after PID test for the REF and Experiment ARC PERC solar cells. Besides, the environmental test of the damp heating (DH) 1000 hours and the thermal cycle 200 cycle for REF and Experiment ARC PERC solar cells were shown in Fig. 8 and 9. The DH 1000 test shows that decay rate < 3% and TC 200 test show that the decay rate < 1%. It could be found that there is not much difference in the change of ARC when the screen printing conditions are the same from the DH 1000 and TC 200 test. It also shows that the experimental results are effective.

Finally, we summed up the improvement results and obtained the electrical data as shown in Table 3. It can be found that the performance has been improved. The same efficiency solar cell of Ref and optimization for modules, it could be observed that the CTM loss is less than 0.5%. However, it was found that the CTM degradation of the Optimized group was lower than the Ref group, which meets the relevant specifications.

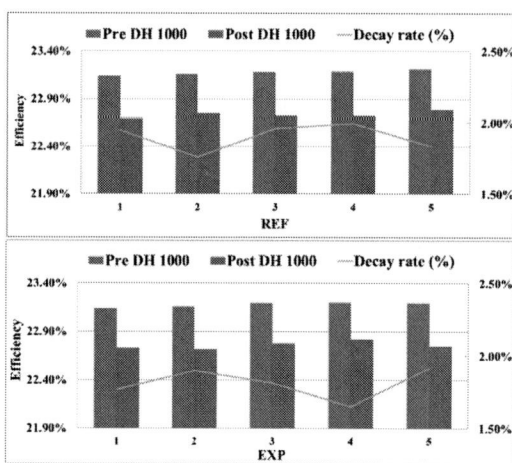

Figure 8: DH 1000 test for REF and Experiment ARC PERC

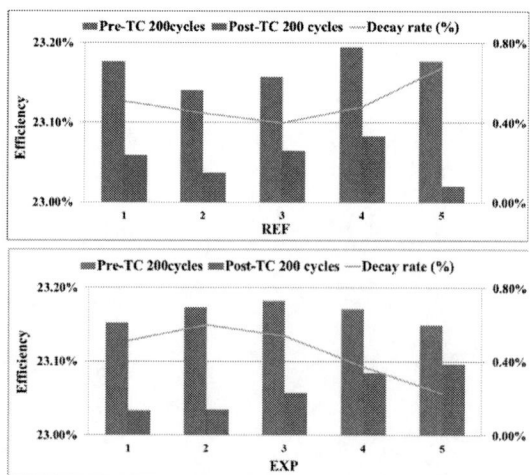

Figure 9: TC 200 test for REF and Experiment ARC PERC

Table 3: The average electrical results of REF and optimized PERC solar cells.

	EFF (%)	FF (%)	V_{OC} (V)	I_{SC} (A)
REF PERC	23.00	80.29	0.6874	13.759
Optimized PERC	23.20	80.23	0.6892	13.850

4   CONCLUSIONS

In this work, we have demonstrated that productivity can be enhanced while maintaining a consistent level of performance. By adjusting the parameters of the anti-reflection coating (ARC) layer through the PECVD process, we achieved a significant reduction in reflectivity at longer wavelengths (>780 nm) compared to the REF. This optimization led to an improvement in conversion efficiency of approximately 0.2% over the reference solar cell. Furthermore, both the REF and optimized ARC PERC solar cells passed reliable testing and met the IEC TS 62804-1 standard, with power losses in this experiment remaining below 1%.

5   REFERENCES

[1] H. J. El-Khozondar, R.J. El-Khozondar, R. Al Afif, and C. Pfeifer, "Modified Solar cells with antireflection coatings" Int. J. Thermofluids Vol. 11, pp. 100103, 2021.

[2] C.-H. Park, Y.-J. Ko, J.-H. Kim, H. Hong, "Greenhouse gas reduction Effect of solar energy systems applicable to high-rise apartment housing structures in South Korea", *Energies*, vol. 13, pp. 1, 2020.

[3] N. Khordehgah, A. Żabnieńska-Góra, and H. Jouhara, "Analytical modelling of a photovoltaics-thermal technology combined with thermal and electrical storage systems", Renew. Energy, Vol. 165, pp. 350, 2021.

[4] O.V. Semenova, V.A. Yuzova, T.N. Patrusheva, F.F. Merkushev, M.Y. Railko, and S.A. Podorozhnyak, "Antireflection and protective films for silicon solar cells", IOP Conf. Ser.: Mater. Sci. Eng., Vol. 66, pp. 012049, 2014.

[5] A. Dieye, N. Mbengue, O. A. Niasse, F. Dia, M. Diagne, and B. Ba, "Influence of Multi Anti – Reflective Layers on the Silicon Solar Cells", Journal of Materials Science and Surface Engineering, Vol. 5, pp. 729, 2017.

[6] J. Dupuis, E. Fourmond, J. Lelièvre, D. Ballutaud, M. Lemiti, "Impact of PECVD SiON stoichiometry and post-annealing on the silicon surface passivation." *Thin Solid Films*, vol.516, pp.6954, 2008.

[7] J. Dupuis, E. Fourmond, D. Ballutaud, N. Bererd, M. Lemiti, "Optical and structural properties of silicon oxynitride deposited by plasma enhanced chemical vapor deposition" *Thin Solid Films*, vol. 519, pp.1325, 2010.

[8] C. Zhou, J. Zhu, S.E. Foss, H. Haug, Ø. Nordseth, E.S. Marstein, W. Wang, "SiOyNx/SiNx Stack Anti-reflection Coating with PID-resistance for Crystalline Silicon Solar Cells," Energy Procedia, vol. 77 pp.434, 2015.

[9] C.-W. Kuo, T.-M. Kuan, Y.-C. Li, C.-W. Lee, W.-L. Chueh, L.-G. Wu, S.-C. Lin and C.-Y. Yu., "Impact of the Front Side Multi Anti-Reflection Coating Layer on PERC Solar Cells" 26th EUPVSEC, pp. 020021, 2023.

[10] E. Urrejola, R.Petres, J. Glatz-Reichenbach, K. Peter, E. Wefringhaus, H. Plagwitz, G. Schubert., "High Efficiency Industrial PERC Solar Cells with all PECVD-Based Rear Surface Passivation" 26th EUPVSEC, pp. 2233, 2011.

[11] C.-W. Kuo, T.-M. Kuan, W.-L. Chueh, L.-G. Wu, S.-C. Lin, C.-Y. Yu, "Investigation of the Rear Side Passivation Layer on Bifacial PERC Solar Cells" 38th EUPVSEC, pp. 288, 2021

41st European Photovoltaic Solar Energy Conference and Exhibition

THE IMPACT OF CONDUCTIVE PASTE COMPOSITION ON THE LECO PROCESS FOR TOPCON SOLAR CELLS

1BV.5.24

Chun-Ping Lin[1,3], Chih-Jeng Huang[1], Han-Chen Chang[1], Sung-Yu Chen[1], Bang-Hao Wu[2], Cheng-Liang Cheng[2], Ying-Yuan Huang[3]

[1]Photovoltaic Technology Division, Green Energy and Environment Research Laboratories, Industrial Technology Research Institute, Tainan, Taiwan
[2]TeraSolar Energy Materials Corp., Miaoli, Taiwan
[3]Institute of Microelectronics, Department of Electrical Engineering, National Cheng Kung University, Tainan, Taiwan
*Phone: +886 6 363 6808; *E-mail: itriB10475@itri.org.tw

LECO technology leverages laser-induced energy to generate a substantial number of electrons in the solar cell instantaneously. Under reverse bias external current, these electrons accumulate at the interface between the metal and silicon, facilitating the diffusion of silver into silicon and the formation of silver-silicon alloy. This process improves the metal contact resistance significantly. Consequently, with the employment of LECO technology, it is possible for TOPCon solar cells to achieve the desired metal contact resistance with little amount aluminum in the front electrode, thus substantially reducing J0,metal loss and further enhancing the efficiency of TOPCon solar cells. Unfortunately, so far no relevant researches have been reported that specifically address how to adjust the front silver paste in TOPCon solar cells combined with LECO technology to further enhance solar cell efficiency. Therefore, this study will focus on investigating this aspect.

Table 1：Four different conditions of conductive pastes.

Front-Ag paste	BSL	Paste A	Paste B	Paste C
Glass Frit	Glass X	Glass X	Glass X	Glass Y
Al frit	2%	0%	0.5%	0.5%

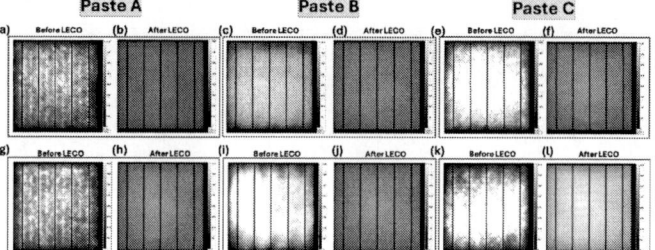

Figure 1 Spatial EL measurements of cells utilizing Paste A, B and C under different sintering temperatures before and after LECO processing. (a-f) 740°C (g-l) 720°C.

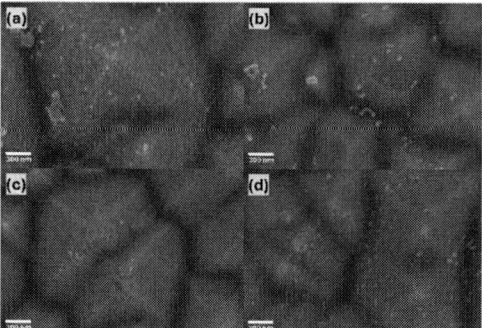

Figure 3 Top-view SEM images of samples: (a) BSL, (b) Paste A, (c) Paste B, (d) Paste C.

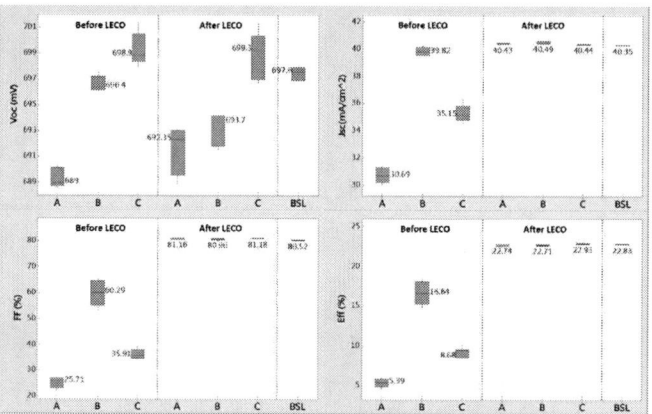

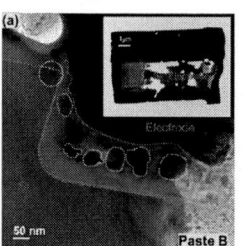 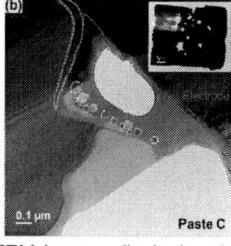

Figure 4 Cross-sectional TEM images displaying the interfaces via the employment of (a) Paste B and (b) Paste C.

Figure 2 Comparison of various cell performances based on pastes A, B, and C before and after the LECO process, with peak firing temperatures reduced by 40°C compared to that of BSL.

SUMMARY

The results demonstrate that reducing the peak firing temperature while adjusting the aluminum content and glass frit composition significantly improves the contact resistance and overall efficiency of the solar cells. It is found that the paste with a 0.5% aluminum content and modified glass frit, presents the best cell performances, achieving a Voc of 699.3 mV and an efficiency of 22.93% after LECO processing. This improvement is attributed to the formation of highly dense silver-silicon alloy particles at the contact interfaces, which improve electrical contact characteristics and reduce the series resistance, ultimately leading to boost the cell efficiency. The findings suggest that the optimization of paste compositions and sintering conditions, combined with LECO processing, can substantially improve the efficiency of TOPCon solar cells.

ACKNOWLEDGEMENTS

The financial support provided by Energy Administration, Ministry of Economic Affairs, Taiwan (R.O.C.) (Grant No. 113-S0102) is gratefully acknowledged.

41st European Photovoltaic Solar Energy Conference and Exhibition

REALISTIC ESTIMATION OF INDUSTRIAL TOPCON CELL EFFICIENCY

Mehul Raval[1]*, Pirmin Preis[2], Lejo Joseph Koduvelikulathu[2], Gourab Das[1], Wolfgang Jooß[1]
1-RCT Solutions GmbH, Line-Eid-Str. 1, Konstanz-78467, Germany
2- ISC Konstanz e.V., Rudolf-Diesel-Straße 15, Konstanz-78467, Germany
* Email: mehul.raval@rct-solutions.com

ABSTRACT: Over the past year, there has been a sharp increase in the installed manufacturing capacity and efficiency of n-type Tunnel-Oxide Passivated Contact (TOPCon) solar cells with reported efficiency and open-circuit voltage (V_{oc}) values of above 26.0% and 730 mV, respectively. Modules based on commercial TOPCon cells with Cell to Module (CTM) values of close to 100% are expected to deliver higher wattage than the achieved wattage in production. In this work, commercial 16 Busbar (BB) M10 TOPCon solar cells of labelled efficiency of 25.65% and 25.11% are characterized via Quantum Efficiency (QE) and I-V measurements under STC and Quokka3 based simulations of industrial TOPCon cells have been carried out to correlate with the QE and I-V measurements. The study reveals that the real efficiency of commercial M10 TOPCon is higher by 1.0% absolute compared with the labelled efficiencies. The main difference is attributed to the higher short-circuit current density (J_{sc}). The simulations also show that the commercial TOPCon cells should have efficiency in the range of 24.0% to 24.8% based on the variation of the base resistivity and minority carrier lifetime.

Keywords: TOPCon; CTM; IV Measurement; Quokka3; Efficiency

1. INTRODUCTION

PV installations reached over 1,600 GW, with more than 400 GW solar PV installation in 2023 [1]. More than 1,100 GW of new production capacity expansion based on high efficiency solar cell has been announced in 2023 amongst which more than 850 GW is TOPCon which shows the acceptability and dominance of TOPCon in the following years.

Surface passivation is the key parameter of any high efficiency solar cells resulting in improvement in open V_{oc} and fill factor (*FF*). Modified study reveals the new limit of V_{oc} and *FF* for passivated solar cells due to change in diode ideality factor (n) from 1 to 2/3 [2] in which recombination are mostly limited to intrinsic recombination rather than recombination at the surfaces. The new limits for the V_{oc} and FF_0 as obtained from the revised equations in [2] are illustrated in figures I and II below. It can be observed that the new FF_0 values for commercial solar cells is above 87% and can have a value of up to 89%. Similarly, the upper limit for V_{oc} is above 750 mV and can touch 760 mV for reduced wafer thickness. It is evident that lower base doping and lower wafer thickness lead to higher V_{oc} and FF_0 values.

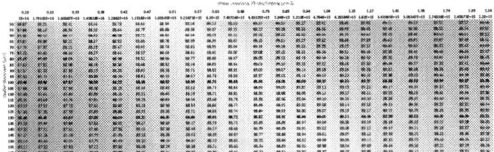

Figure I: New upper limit for V_{oc} for varying wafer thickness and doping.

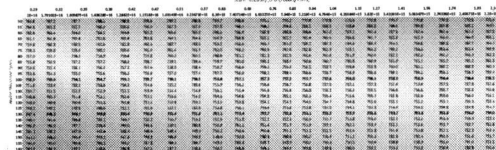

Figure II: New upper limit for FF_0 for varying wafer thickness and doping.

TOPCon is a rear passivated solar cell which has great potential to reach 28+% [3] and already 26.5%+ has been reported by some of the leading industries in their pilot production with rapid technological and production improvements [4] [5].

The rated power output of TOPCon cell-based PV modules has difficulties to attain expected powers which are expected with Cell to Module (CTM) values of ≈ 100%. Reported CTM values have a loss of more than 3% [6] which motivates to investigate the I-V parameters of commercial TOPCon solar cells. The datasheets of various manufacturers offer varying I-V characteristics for commercial TOPCon solar cells, therefore it is important to comprehend the "realistic" efficiency and I-V parameters of these cells. This is crucial because it has an immediate effect on the CTM loss and the final PV module power.

2. SIMULATION AND CHARACTERIZATION STUDIES

Quokka3 [7] based simulation has been carried out to model commercial M10 TOPCon cell. The main cell parameters considered in the simulations are outlined in Table I. The optical fitting was performed with reflectance measurements (both sides) on semi-finished blue wafers from production. In addition, the base resistivity (0.3 Ω-cm to 2.1 Ω-cm) and minority carrier lifetime of the wafers were varied to check their influence on the cell I-V parameters. The minority carrier lifetime of the n-type wafers are much higher than the typical criteria of ≥ 500 μs and value of 2,000x, 4,000x and 6,000x (times base resistivity) were used for the simulations [8]. It is to be noted that when the studies were carried out, Laser Enhanced Contact Optimization (LECO) was not considered, which is an established process in TOPCon solar cell production as of today.

Table I: Commerical TOPCon solar cell parameters considered for the simulations.

Parameter	Value	Notes
Cell thickness	123 μm	Wafer thickness of 130 μm with 5 μm loss due to texture and 2 μm due to polishing
FS Finger width	19.5 μm	Based on microscope measurements.
FS Finger pitch	1.12 mm	Based on 160 fingers on front-side
FS Contact resistivity	1.5 mΩ-cm^2	Based on current industrial status
Emitter sheet resistance	225 Ω/□	Based on current industrial status
FS J_0 in non-contact area	8 fA/cm^2	[9]
FS J_0 in contact area	200 fA/cm^2	[9]
RS Finger width	37 μm	Based on microscope measurements.
RS Finger pitch	0.873 mm	Based on 200 fingers on rear-side
RS Contact resistivity	1.0 mΩ-cm^2	Based on current industrial status
Poly-Si thickness & doping	120 nm & 2*10^{20} cm^{-3}	Based on current industrial status
RS J_0 in non-contact area	2 fA/cm^2	[10]
RS J_0 in contact area	50 fA/cm^2	[10]
External Rs (for simulation)	0.173 Ω-cm^2	Based on Grid Calculator for grid pattern [11]

Further, commercial 16 BB M10 TOPCon solar cells with labelled efficiency of 25.65% and 25.11%, were characterized via QE and I-V measurements under STC. Since the I-V test measurement set-up cannot contact all the BBs, the I-V measurements were performed with application of BB resistance correction based on previous experience of measurement of commercial TOPCon solar cells. It is common in the industry to contact cells with many BBs with reduced BBs and apply the appropriate

correction factors. The Quokka3 based simulations were then correlated with the QE and I-V measurements.

3. RESULTS AND ANALYSIS

The simulation results and the solar cell characterization are discussed in the section ahead.

3.1 Simulation results for TOPCon solar cells

The simulation results for varying base resistivity and minority carrier lifetime are discussed in section 3.1.1, while the influence of varying wafer thickness, passivated contacts on the front-side and influence of varying rear-side poly-Si thickness are discussed subsequently.

3.1.1 Impact of base resistivity & minority carrier lifetime

The trend of efficiency and I-V parameters versus base resistivity for different lifetime scenarios is shown in the figures III to VI below. The simulated efficiency is in the range of 24.0% to 24.7% and it is observed that higher base resistivity leads to higher cell efficiency. This suggests that lighter doping is better suited for TOPCon solar cells. An important point to be noted is that the dependence of efficiency on base resistivity is unlike for PERC solar cells which has a bell-shaped dependence with base resistivity.

The V_{oc} is in the range of 713 mV to 726 mV with higher base resistivity leading to better V_{oc} values. J_{sc} shows a weak dependence on the base resistivity for value of more than 0.6 Ω-cm. The *FF* increases for higher base resistivity values, while it decreases for 2x resistivity case by around 0.5% to 0.6% absolute. Increase of the *FF* with base resistivity is also unlike for the PERC cells and is attributed to the availability of entire rear-side for transport and carrier collection improving for higher lifetime substrate.

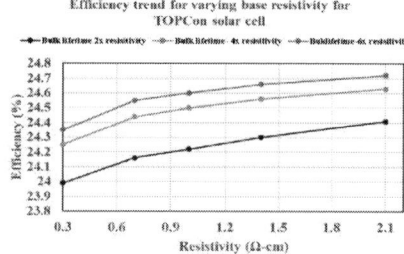

Figure III: Simulated TOPCon cell efficiency *vs* base resistivity trend for different minority carrier lifetime cases.

A summary of the I-V parameters derived from the simulations is presented in table II below. The range of J_{sc} is from 40.57 mA/cm^2 to 40.74 mA/cm^2 based on the variation of the wafer quality. Similarly, the *FF* is expected to be in the range of 82.9% to 83.6% for the commercial TOPCon solar cells.

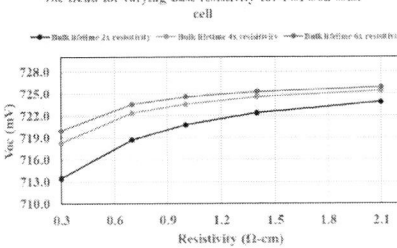

Figure IV: Simulated TOPCon cell V$_{oc}$ *vs* base resistivity trend for different minority carrier lifetime cases.

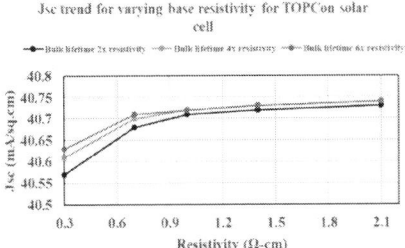

Figure V: Simulated TOPCon cell J$_{sc}$ *vs* base resistivity trend for different minority carrier lifetime cases.

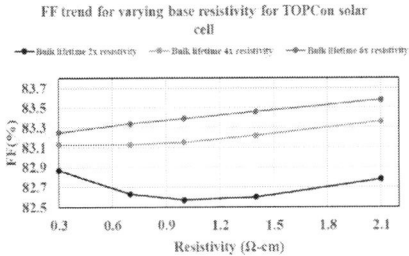

Figure VI: Simulated TOPCon cell *FF vs* base resistivity trend for different minority carrier lifetime cases.

Table II: Summary of simulated I-V parameters for industrial TOPCon cells for different base resistivities and minority carrier lifetime.

Resistivity (Ω-cm)	Bulk life time (ms)	V$_{oc}$ (mV)	J$_{sc}$ (mA/cm²)	FF (%)	η (%)
2 x Minority carrier lifetime					
0.3	0.6	713.5	40.57	82.87	23.99
0.7	1.4	718.7	40.68	82.63	24.16
1.0	2.0	720.8	40.71	82.57	24.22
1.4	2.8	722.4	40.72	82.6	24.3
2.1	4.2	723.9	40.73	82.87	24.41
4 x Minority carrier lifetime					
0.3	1.2	718.3	40.61	83.13	24.25
0.7	2.8	722.4	40.70	83.13	24.44
1.0	4.0	723.6	40.72	83.15	24.5
1.4	5.6	724.6	40.73	83.22	24.56
2.1	8.4	725.4	40.74	83.36	24.63
6 x Minority carrier lifetime					
0.3	1.8	720.0	40.63	83.25	24.35
0.7	4.2	723.6	40.71	83.34	24.55
1.0	6.0	724.6	40.72	83.39	24.6
1.4	8.4	725.3	40.73	83.46	24.66
2.1	12.6	725.9	40.74	83.58	24.72

3.1.2 Impact of wafer thickness

Decreasing the wafer thickness from 130 µm to 90 µm (for 1 Ω-cm & 4 ms case), V$_{oc}$ & *FF* increases by ≈ 2.5 mV & 0.15% abs and J$_{sc}$ decreases by 0.6 mA/cm². Over-all η decreases by 0.25% absolute from 24.5% to 24.25%. Improved optical trapping is required to increase J$_{sc}$ for reduced wafer thickness. To reduce the per W$_p$ cost of the TOPCon solar cell, it is expected that the wafer thickness will reduce from 130 µm to 120 µm by 2025. However, no wafer thickness reduction is observed in 2024, indicating that there might be production-level wafer handling aspects leading to higher yield loss, which would negate. the cost reduction by using thinner wafers.

3.1.3 Improvement with passivated front-side contacts

The comparison of simulation results without and with passivated front-side contacts for TOPCon solar cells is shown in table III below. The J$_{0met}$ for the front-side was reduced from 200 fA/cm² to 8 fA/cm² for all scenarios assuming there will be no additional parasitic absorption on the front-side due the poly-Si layer. A V$_{oc}$ gain of 4-5 mV is expected for TOPCon solar cells with passivated front-side contacts, while no major impact is expected on the J$_{sc}$ and *FF*. An absolute efficiency gain of 0.10%-0.15% is expected compared to conventional TOPCon solar cells.

Given the reduced front-side metallization coverage (2.7%), increased process complexity for introducing selective passivated contacts on the front-side and the nominal efficiency gain, it appears that introduction of TOPCon with front-side passivated contacts might not be commercially beneficial to the solar cell manufacturers.

Table III: Comparison of I-V parameters for commercial TOPCon cell with and without passivated contacts on the front-side.

Resistivity (Ω-cm)	Bulk life time (ms)	V$_{oc}$ (mV)	J$_{sc}$ (mA/cm²)	FF (%)	η (%)
4 x Minority carrier lifetime					
0.3	1.2	718.3	40.61	83.13	24.25
0.7	2.8	722.4	40.7	83.13	24.44
1.0	4.0	723.6	40.72	83.15	24.5
1.4	5.6	724.6	40.73	83.22	24.56
2.1	8.4	725.4	40.74	83.36	24.63
4 x Minority carrier lifetime (with passivated front-side contacts)					
0.3	1.2	721.9	40.61	83.03	24.34
0.7	2.8	726.4	40.70	83.04	24.55
1.0	4.0	727.8	40.72	83.08	24.62
1.4	5.6	728.8	40.73	83.17	24.69
2.1	8.4	729.7	40.74	83.34	24.77

3.1.4 Impact of reducing poly-Si thickness

Decreasing the rear-side poly-Si thickness from 120 nm to 70 nm (for 1 Ω-cm & 4 ms case with same phosphorus doping concentration), leads to an estimated gain of ≈ 0.1 mA/cm² in the J$_{sc}$, without any influence on the V$_{oc}$ and the *FF*. The reduction in the poly-Si thickness will also be related to the development of the rear-side metallization pastes to avoid firing through of the pastes and damaging the passivation at the rear-side interface. The impact on the bifaciality factor might be more significant which needs to be evaluated further.

3.2 Measurement of industrial M10 TOPCon solar cells

Industrial M10 16 BB TOPCon solar cells with labelled efficiency of 25.65% and 25.11% were characterized at ISC, Konstanz. The I-V and QE measurements are discussed ahead.

3.2.1 I-V measurement of M10 TOPCon solar cells

The corelation of the series resistance, measured FF and the efficiency with respect to bus-bar correction factor are shown in the figures VII to X below. A correction factor of 2.2 mΩ was found to be most suitable for an appropriate I-V measurement of the cells and the series resistance was in the range of 0.5-0.7 mΩ and the estimated the FF values are 83.8% and 83.0%, respectively for the 25.65% and 25.11% labelled cells, respectively. The difference in the FF is attributed to higher pFF (86.8%) for the 25.65% rated cell compared to 85.7% for the 25.11% rated cell. The difference in the solar cell efficiency is attributed to increased recombination or inferior bulk material which reduces the V_{oc} and the pFF.

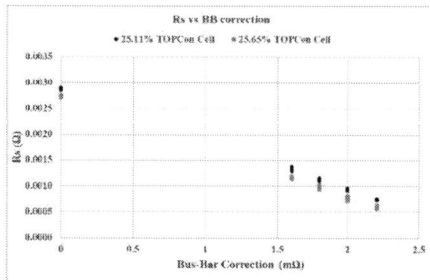

Figure VII: Series resistance vs BB correction correlation for the measured industrial M10 TOPCon cells.

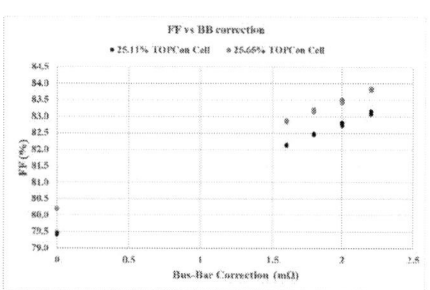

Figure VIII: Measured FF vs BB correction correlation for the measured industrial M10 TOPCon cells.

The corresponding V_{oc} values for the 25.65% and 25.11% labelled cells are $\approx$ 720 mV and 710 mV, respectively. The measured J_{sc} values for the 25.65% and 25.11% labelled cells for BB correction factor of 2.2 mΩ are 40.63-40.65 mA/cm^2 and 40.60-40.61 mA/cm^2, respectively. Based on the measurements, the estimated 'real' efficiency of the 25.65% and 25.11% TOPCon cells are 24.5% and 24.0%, respectively. An important inference from the measurements is that the 'real' efficiency of the TOPCon solar cell can be lower by up to 1.0% absolute compared to the labelled efficiency.

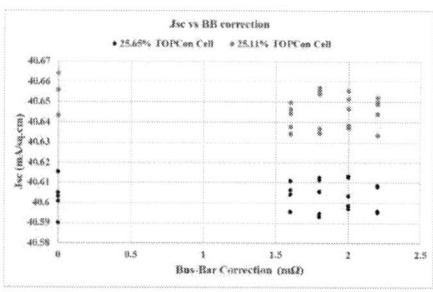

Figure IX: Measured J_{sc} vs BB correction correlation for the measured industrial M10 TOPCon cells.

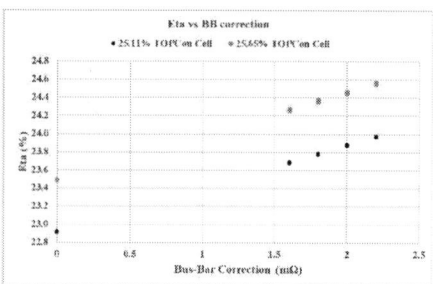

Figure X: Measured efficiency vs BB correction correlation for the measured industrial M10 TOPCon cells.

3.2.2 Quantum Efficiency measurement of M10 TOPCon solar cells

The QE and reflectance curves for the 25.65% labelled TOPCon cell is shown in Figure XI. The expected cell-level J_{sc} for the cell was calculated based on the photon flux density and the EQE response taking into account the front-side metallization shading. The J_{sc} values for both cells (25.65% & 25.11%) were very similar (40.66 mA/cm^2 and 40.62 mA/cm^2, respectively). The corresponding cell-level I_{sc} for an M10 size (182 mm x 182 mm) TOPCon solar cell are 13.42 A and 13.41 A, respectively.

Additionally, we have examined the EQE of commercially 22.5% PERC solar cells and compared with the EQE of labelled 25.65% TOPCon cell. It is evident that the response of the TOPCon solar cell is evidently poor at longer wavelengths compared to the PERC solar cell. The reduction in the long wavelength response is attributed to the thick (120 nm) poly-Si layer on the rear-side of the bulk Si wafer which acts as a potential recombination centre for the carriers (electron) while UV response is higher in TOPCon solar cells. At the time of writing the paper, the J_{sc} of an industrial 23.45% M10 PERC solar cell was known to be 40.68 mA/cm^2, which is slightly better than the J_{sc} value of the characterized TOPCon cell.

41st European Photovoltaic Solar Energy Conference and Exhibition

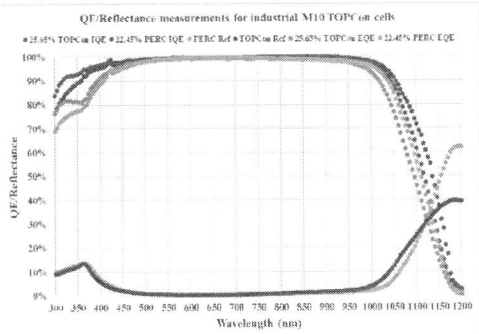

Figure XI: QE and Reflectance measurements for 25.65% labelled M10 TOPCon cell. Same curves are also shown for a 22.45% efficiency PERC solar cell for comparison.

Based on these results, the J_{sc} values for commercial TOPCon cell are expected to be similar to industrial PERC cells. For M10 wafer size of 182.2 mm x 182.2 mm, the short-circuit current (I_{sc}) of the TOPCon cells is expected to be in the range of 13.45-13.50 A.

3.3 Comparison of simulation results with commercial solar cell datasheets

A comparison of simulated I-V parameters with the actual TOPCon solar cell datasheets for some of the Tier 1 manufacturers is exhibited in the figures XII to XIV below. The x-axis is the efficiency bins mentioned in the datasheets, while the corresponding I-V parameters (from different manufacturers) are indicated in round symbols, while the simulated values are indicated in green-colored triangular symbols. The commercial solar cells have V_{oc} in the range of 692 mV to 735 mV, for labelled efficiencies in the range of 23.3%to 25.5% with a significant scattering in the V_{oc} values for a given efficiency bin. The corresponding V_{oc} range from simulations is from 715 mV to 726 mV for an efficiency range of 24.0% to 24.7%. This indicates that the labelled efficiencies might be higher based on the labelled J_{sc} and FF values in the datasheets.

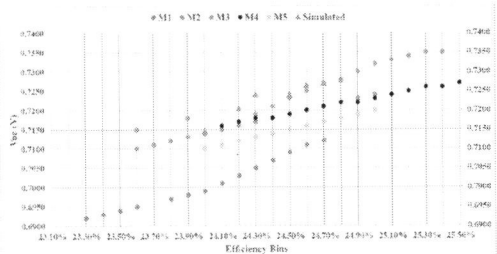

Figure XII: Comparison of simulated and commercial datasheet V_{oc} values for different efficiency bins.

Most of the datasheets have a higher J_{sc} than simulated value over the efficiency range of 24.0% to 24.7%. Only M2 and M4 are closest to the simulated values. The J_{sc} variation over the efficiency range should be flatter as observed from the simulation results, so the linear trend of

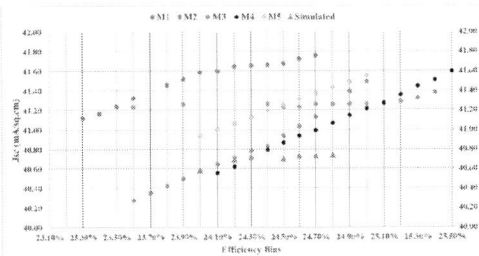

Figure XIII: Comparison of simulated and commercial datasheet J_{sc} values for different efficiency bins.

J_{sc} variation over the efficiency range doesn't seem to correlate with actual measured values. Based on the significantly higher J_{sc} values in the datasheets, the labelled TOPCon solar cell efficiencies will also be much higher.

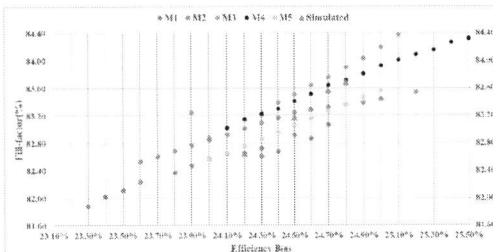

Figure XIV: Comparison of simulated and commercial datasheet FF values for different efficiency bins.

The FF for commercial cells are in the range of 82.4% up to 84% for efficiency range of 24.0% to 25.0%. The deviation from the simulated values are much less compared to the variation in the V_{oc} and J_{sc} values.

4   SUMMARY

Detailed Quokka3 based simulation of TOPCon solar cells were carried out based on industrial status of the solar cell parameters. It is observed that lighter based doping is more suitable for TOPCon solar cells with increase in V_{oc} and FF values with increasing base resistivity. Efficiency and FF don't exhibit the bell-shaped dependence curve as for PERC solar cells which is related to high-level of passivation on both sides and a fully conducting rear-side for lateral transport of carriers. Simulations over base resistivity range of 0.3 to 2.1 Ω-cm for different life-time scenarios lead to efficiency, V_{oc}, J_{sc} and FF in the range of 24.0%-24.7%, 713 mV- 726 mV, 40.57 mA/cm^2 to 40.74 mA/cm^2 and 82.9% to 83.4%, respectively. Introduction of passivated contacts on the front-side can lead to an absolute efficiency gain of 0.1%-0.15%, while reduction of the rear-side poly-Si thickness can lead to a J_{sc} gain of 0.1 mA/cm^2.

The I-V measurements indicate that the labelled TOPCon cell efficiencies were higher by 1.0% absolute, while the J_{sc} values obtained from the QE measurement are in the range of 40.62-40.66 mA/cm^2 for the industrial TOPCon solar cells. Comparison of the simulated I-V parameters with commercial datasheets indicate that the V_{oc} values are lower compared to the expected range, while the J_{sc} values are significantly higher. Based on these results, it is inferred that the CTM loss on the TOPCon modules can be artificially higher due to the rated higher J_{sc} values in the datasheets. It is recommended that the module manufacturers to have a cell-level I-V flasher for IQC to verify the efficiency of externally procured solar cells and ensure solar cells with accurate efficiencies for reliable CTM calculations.

ACKNOWLEDGMENTS

Mehul Raval would like to thank Andreas Fell for discussions related to Quokka3 based simulations of TOPCon cells.

REFERENCES

[1] "IEA_PVPS_Snapshot_2024," IEA, 2024.

[2] K. Bothe, D. Hinken and R. Brendel, "Extended FF and VOC Parameterizations for Silicon Solar Cells," *IEEE JOURNAL OF PHOTOVOLTAICS*, vol. 13, no. 6, pp. 787-792, 2023.

[3] Q. Wang et al., "High-efficiency n-TOPCon bifacial solar cells with selective poly-Si based passivating contacts," *Solar Energy Materials and Solar Cells*, vol. 259, p. 112458, 2023.

[4] A. Bhambhani, "taiyangnews," taiyangnews, January 2024. [Online]. Available: https://taiyangnews.info/technology/chinese-company-touts-new-record-topcon-efficiency.

[5] E. Bellini, "PV Magazine," PV Magazine, April 2023. [Online]. Available: https://www.pv-magazine.com/2023/04/12/jolywood-claims-26-7-efficiency-for-n-type-topcon-solar-cell/.

[6] W. Wang, "The Progress of HJT solar cell in China," in *PVSEC-34*, Shenzhen, China, 2023.

[7] A. Fell et al., "The concept of skins for silicon solar cell modeling," *Solar Energy Materials and Solar Cells*, vol. 173, pp. 128-133, 2017.

[8] J.W. Müller et al., "Recent Progress of Q Cells' High Efficiency Solar Cell Development, October, 2022, Hanwha Q Cells Cz Workshop CSP," [Online].

[9] F. Fertig et al., "Q.Antum Neo: Qcells Silicon Technology with > 25% Cell Conversion Efficiency Fabricated with Mass-Production Processes,," in *40th EUPVSEC, 2023*, Lisbon., 2023.

[10] J. Ye, "Recent development, challenges, and future strategies of TOPCon solar-cell technologies," in *PVSEC-34*, Shenzhen, November 2023.

[11] K. McIntosh et al., "Grid Calculator, PVLIGHTHOUSE," PVLIGHTHOUSE, [Online]. Available: https://www2.pvlighthouse.com.au/calculators/Grid%20calculator/Grid%20calculator.aspx.

41st European Photovoltaic Solar Energy Conference and Exhibition

OPTIMIZING THE MECHANICAL ADHESION PROPERTIES
OF PLATED CONTACTS OF I-TOPCON SOLAR CELLS

Christian Schmiga, Abdelaziz Boudellioua, Rene Haberstroh, Jonas Eckert,
Sven Kluska and Florian Clement
Fraunhofer Institute for Solar Energy Systems (ISE)
Heidenhofstraße 2, 79110 Freiburg, Germany
Phone: +49(0)761/4588-5701, E-mail: christian.schmiga@ise.fraunhofer.de

ABSTRACT: In this study, we have investigated and optimized the laser-induced modification of the silicon surface to improve the mechanical adhesion of Ni/Cu/Ag-plated contacts on the rear side of i-TOPCon Si solar cells. We discuss different approaches to create the laser contact openings (LCO) for the fingers by varying the laser parameters and the geometry of the contact structure, targeting at an enhanced micro surface roughness at the Si/Ni interface. We evaluate the mechanical adhesion via tape tests and the effect on electrical degradation due to increased recombination via photoluminescence measurements. We demonstrate that a well-defined laser parameter set allows for improved adhesion while maintaining no significant loss in implied open-circuit voltage.
Keywords: Plated Contacts, i-TOPCon Solar Cells, Mechanical Adhesion

1 INTRODUCTION

Electrochemical plating is a growing alternative solar cell metallization process that aims to replace screen-printed (SP) silver (Ag) contacts by nickel/copper/silver (Ni/Cu/Ag) plated contacts on silicon solar cells (see Figure 1). This substantially cuts costs through drastically reducing silver consumption, which is projected to be a limiting factor in terms of cost and availability in a multi-TW PV production market [1]. The process involves using an ultra-short pulse laser for LCO (Laser Contact Opening) prior to the deposition of the Ni/Cu/Ag contacts to locally open the passivation layer and create the necessary microstructure for plating. One technical challenge of this technology is the mechanical contact adhesion on non-textured silicon surfaces.

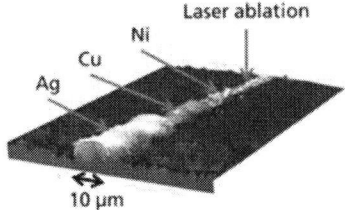

Figure 1: Cross section of electrochemically plated contact of an i-TOPCon solar cell.

Since in industrial TOPCon solar cell (see Figure 2) production the rear side is typically planarized during chemical edge isolation, the back surface contact adhesion can be challenging. We hypothesize that adjusting the laser contact opening parameters and, thereby, modifying the Si surface is a key to improving the mechanical adhesion. However, these parameters cannot be adjusted freely, as excessively laser ablation of the passivation layer to improve the mechanical contact adhesion will result in higher laser-induced damage and by that increased contact recombination. The textured surface of the front side of the cells results in a significantly better contact adhesion due to laser-induced anchor points that act as hooks for the contact fingers [2].

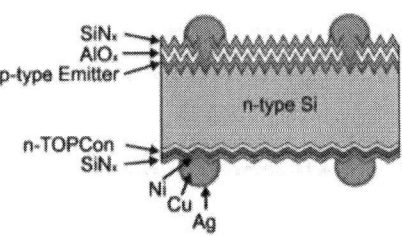

Figure 2: Cross section illustration of plated i-TOPCon solar cell structure.

2 EXPERIMENTAL

Different combinations of laser parameters, targeted at improving the surface roughness within the LCO and, thus, the adhesion of the plated contacts, were applied during the LCO process on the planarized rear sides of industrial i-TOPCon solar cell precursors, which were then characterized, plated and tested for adhesion.

A tape peel test was used to evaluate the adhesion performance of the matrix of contact fingers. The adhesion results of the different laser field structures were then compared against the previously characterized electrical properties obtained through photoluminescence (PL) measurements.

To create the necessary LCO variations for the required tests, the laser pulse energy, overlap density of laser pulses, the number of laser lines per finger, and the laser pulse passes per finger were all varied and adjusted to create the combinations seen in the 8°x°8 field matrix of finger structures described in Table 1 below. These parameters were selected as they are the most relevant for creating variations in the laser induced surface roughness. This matrix was created on M6 size i-TOPCon precursors with fields that are 18 x 18 mm and a finger-to-finger distance of 1.5 mm as seen in Figure 3.

155

Table 1: Variable LCO parameters used to create the test structures within the 8°x°8 field matrix.

Laser pulse energy	LCO3 < LCO2 < LCO1		
	low	mid	high
Pulse overlap in x and y direction	0%, 20%, 40%, 60%, 80%		
Number of laser lines per finger	1, 2, 3, 4		
Number of laser pulse passes per line	1, 2, 3, 4, 5, 6, 7, 8		

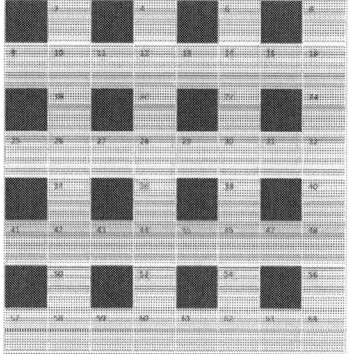

Figure 3: 8x°8 field matrix on the rear side of an M6 i-TOPCon wafer, designed for a large variation of laser parameter. This layout contains 48 fields with LCO finger structures (13 fingers per field with a distance of 1.5 mm), and 16 non-lasered reference fields.

Figure 4 shows representative examples of confocal microscope images of different LCO structures which have been investigated within the field matrix.

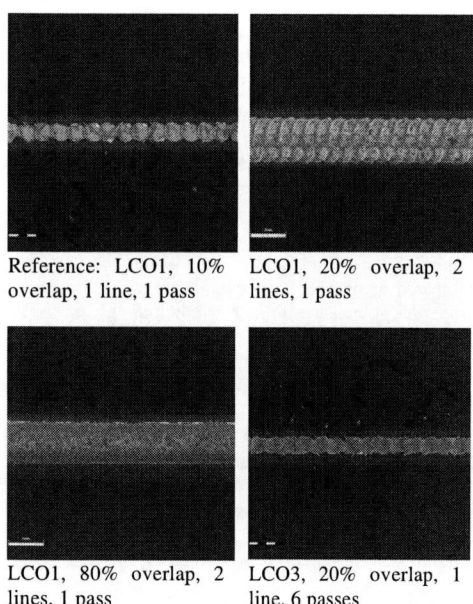

Reference: LCO1, 10% overlap, 1 line, 1 pass

LCO1, 20% overlap, 2 lines, 1 pass

LCO1, 80% overlap, 2 lines, 1 pass

LCO3, 20% overlap, 1 line, 6 passes

Figure 4: Confocal microscope images (100x, top view) of different LCO finger structures.

After the i-TOPCon precursors are laser-opened and annealed during a short high-temperature firing step, but before they are plated, microscope images are taken to document the laser contact openings. PL measurements (at 1 sun illumination) to obtain the implied open-circuit voltage V_{oc} (average of each LCO test field) as a measure of the laser-induced damage of the silicon surface are performed.

Contact adhesion was evaluated via a tape peel test. This is done through the following steps:
1. On each of the 8 columns of the field matrix, a strip of tape is placed.
2. A metal mask sheet is placed on top of the cell to ensure even adhesion of the tape and to prevent cell breakage during the peel test.
3. The tape is slowly pulled at a 90 degree angle, then the remaining contact fingers are counted and documented.

Each of the field structures of the matrix shown in Figure 3 has 13 contact fingers, and the remaining contact fingers after the peel test is the documented result, thus a higher number of remaining contacts indicates better adhesion.

RESULTS

For the test field structures, the graphs in Figure 5 and 6 show the results of what was found to be the best performing groups out of all LCO parameter variations tested. As is expected, the trendline of Figure 5 suggests that an increased laser pulse energy (LCO3 < LCO2) or number of laser pulse repetitions (1...8) yield better mechanical contact adhesion, with the downside of a drop in implied V_{oc} as seen in Figure 6. The results imply that roughening of the silicon surface by applying e.g. a lower (LCO3) laser pulse energy in combination with about 6 passes is a promising approach for LCO, as about an average > 90% of the plated contact fingers are retained with a corresponding V_{oc} drop < 5mV. Thus, we hypothesise that a laser-induced modification of the Si surface creates an advantageous micro roughness, leading to improved mechanical contact adhesion similarly to the under-cut structures reported in [2].

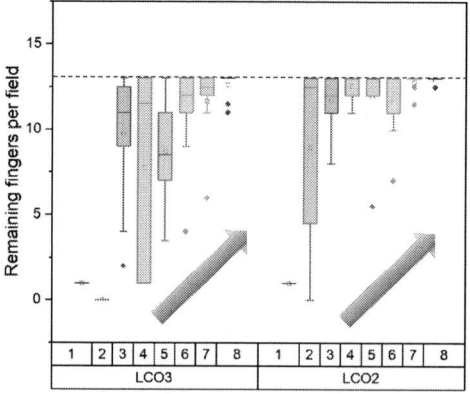

Figure 5: Number of remaining contact fingers per field (out of 13) after tape peel test for different laser pulse energies (LCO3 < LCO2) and per number of laser pulse passes at one laser line and 20% pulse overlap.

41st European Photovoltaic Solar Energy Conference and Exhibition

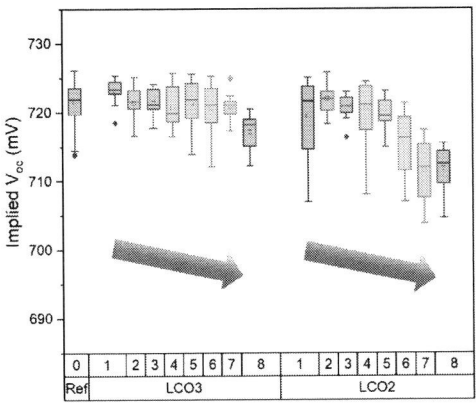

Figure 6: Implied open circuit voltage V_{oc} for different laser pulse energies (LCO3 < LCO2) and number of laser pulse passes at one laser line and 20% pulse overlap. Reference (0) without LCO.

Additionally, we found that increasing the number of laser pulse lines leads to a large drop in i-V_{oc}. This results from the combination of the enhanced laser-opened area and the increased laser-induced damage due to the additionally overlapped laser pulses. The effect of different overlap densities on the implied V_{oc} is displayed in Figure 7.

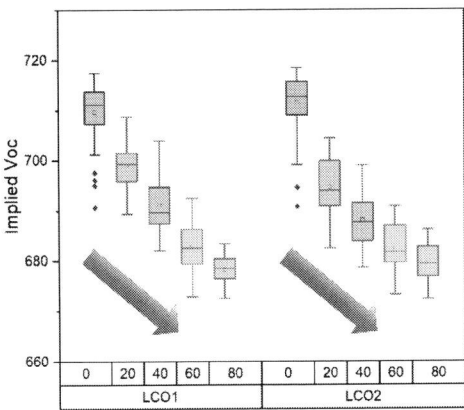

Figure 7: Implied open circuit voltage V_{oc} for different laser pulse energies (LCO1 > LCO2) and laser pulse overlap densities.

CONCLUSIONS

Generally, we have demonstrated that increasing the number of lines, laser energy, or overlap density improves the mechanical adhesion strength of the plated Ni/Cu/Ag contact fingers, but at different costs of damage to the silicon surface. It is observed that an optimal result is achieved using a lower laser pulse energy, in combination with 6 or 7 passes of repetition, as relatively good adhesion is achieved with only a small drop in implied V_{oc} of under 5 mV at a level of around 720 mV, thus showing minimal electrical degradation.

The results strongly imply that roughening of the silicon surface using a low-medium laser energy with many repetitions is a low-damage approach for LCO. Applications of such LCO parameters on full size M6 solar cells will be made in further work to confirm a successful translation of these optimal adhesion results.

ACKKNOWLEDGEMENT

This work was funded by the Federal Ministry for Economic Affairs and Climate Action of Germany under the contract number: 03EE1146B „Indianapolis".

REFERENCES

[1] Hallam, Brett, et al. "The silver learning curve for photovoltaics and projected silver demand for net-zero emissions by 2050. Progress in Photovoltaics: Research and Applications 31.6 (2023): 598-606.

[2] Büchler, A. (2019) Interface study on laser-structured plated contacts for silicon solar cells. dissertation.

[3] Grübel, B., Kluska, S., Cimiotti, G., Schmiga, C., Arya, V., Steinhauser, B., Goraya, B. S., Nold, S., Hermle, M., Kamp, M., Passig, M., Sieber, M., & Brunner, D. (2022). Plating metallization for bifacial i-TOPCon silicon solar cells. In SiliconPV 2021, The 11th International Conference on Crystalline Silicon Photovoltaics. AIP Publishing.

[4] Lacombe, R. (2005). Adhesion Measurement Methods: Theory and Practice. Taylor & Francis Group.

[5] Solar Cell Characterization - Fraunhofer ISE. (n.d.). Fraunhofer Institute for Solar Energy Systems ISE. https://www.ise.fraunhofer.de/en/business-areas/photovoltaics/iii-v-and-concentrator-photovoltaics/iii-v-epitaxy-and-solar-cells/solar-cell-characterization.html

[6] Copper Replaces Silver: TOPCon Solar Cells with Electroplated Metallization Achieve Peak Efficiency of 24 Percent - Fraunhofer ISE. (n.d.). Fraunhofer Institute for Solar Energy Systems ISE. https://www.ise.fraunhofer.de/en/press-media/news/2022/topcon-solar-cells-with-electroplated-metallization-achieve-peak-efficiency-24-percent.html

[7] Kluska, S. et al. (2022) 'Enabling savings in silver consumption and poly-si thickness by integration of plated Ni/Cu/AG contacts for bifacial TOPCon solar cells', Solar Energy Materials and Solar Cells, 246, p. 111889.

41st European Photovoltaic Solar Energy Conference and Exhibition

ADDRESSING EDGE RECOMBINATION LOSSES IN SHINGLE CELLS
BY HOLISTIC OPTIMIZATION OF THE PROCESS SEQUENCE

Alexander Göbel[1,2], Elmar Lohmüller[1], Dirk Wagenmann[1],
Norbert Kohn[1], Marc Hofmann[1], Jonas D. Huyeng[1], Ralf Preu[1,2]
[1]Fraunhofer Institute for Solar Energy Systems ISE, Heidenhofstr. 2, 79110 Freiburg, Germany
[2]University of Freiburg, Department of Sustainable Systems Engineering, INATECH, 79110 Freiburg, Germany
Corresponding author: alexander.goebel@ise.fraunhofer.de

ABSTRACT:
This work focuses on edge recombination losses of cut silicon solar cells, with special attention to shingle cells. We discuss a high-throughput method to form emitter windows by laser ablation of the *p/n*-junction along the cut line. To show the proof of concept, we fabricate symmetrical test structures that reveal a superior surface passivation in the ablation areas. Further, we present results where emitter windows are implemented into our PERC baseline sequence by a single additional structuring step. Shingle cells with emitter window show 83 %rel less losses in pseudo fill factor *pFF* after thermal laser separation (TLS) compared to cells with full area emitter. Additionally, we introduce an enhancement of the passivated edge technology (PET) by optimizing of atomic layer deposition. We prepared lifetime samples and cut solar cells. We find an increased effective lifetime of the excited carriers by a factor of 1.75. The *pFF* gain for cut solar cells is increased by 24 %rel. In total, our work provides two methods that can save up to 80 % of the power losses in cell cutting by TLS, and the methods are feasible for integration into existing industrial process chains.
Keywords: shingling, edge passivation, emitter window, laser ablation

1 INTRODUCTION AND MOTIVATION

Nowadays, cut solar cells are the standard in the silicon photovoltaics (PV) industry [1], as they reduce electrical losses, which leads to increased module efficiencies [2]. Utilizing silicon solar cells in the form of shingled interconnection presents several advantages. Notably, an expected 10 %rel increase in module power density [3, 4] and their unobtrusive appearance make them excellent for integrated PV applications [5].

Usually, the cells are cut in the end of the cell's fabrication. This leads to a significant drawback as the new cut edges lack passivation compared to all other surfaces, which causes eventually losses in the energy conversion efficiency [6–8]. Main causes for this are defects at the edges, e.g., dangling bonds or other defects [9].

To reduce the recombination losses that occur at the edges, which are preferably prepared with very little damage, the passivated edge technology (PET) was first introduced on passivated emitter and rear cells (PERC) by Fraunhofer ISE [10, 8] and later demonstrated also on tunnel-oxide passivated contact (TOPCon) [11, 12] and silicon heterojunction (SHJ) solar cells [13, 14]. The PET harnesses two effects by coating the cell's edge with, e.g., AlOx: the chemical and the field effect passivation [15]. The first one saturates open crystal bonds and reduces thereby the amount of crystal defects at the edge. The field effect passivation works by forming stationary charges on the edge, which thus reinforce an imbalance of the free charge carriers near the edge, resulting in fewer recombination pairs of charge carriers.

Dicker has shown that the surface recombination rate is highest at the *p/n*-junction [16]. Despite its small dimension, it causes still most of the edge recombination as has been also simulated by Wöhrle et al. [7]. This leads to an alternative strategy wherein the *p/n*-junction is locally removed along the subsequent separation line through laser treatment. Thus, the severance of the cut *p/n*-junction can be avoided, which is called emitter window. This concept may benefit from the cell's surface passivation that can cover the emitter's edge [17]. Additionally, the current towards the edge is limited [18].

While this approach has been tested on lab scale [19, 17], to the best of the authors' knowledge it has not yet been realized for standard industrial cells. Simulations predict that the emitter window structure is able to shield 80 % of edge recombination losses in the fill factor [20].

In recent studies, we have pursued the emitter window approach by implementing a single additional structuring step into our PERC baseline sequence, revealing proof of successful emitter removal on test structures. Furthermore, this achievement is transferred on cell level showing a significant reduction in losses by the separation step. This novel avenue of exploration holds promise for mitigating edge recombination effects in shingle solar cells. For the PET we pursued an optimization of our processes that leads to improved edge passivation properties.

2 EXPERIMENTAL

2.1 Process sequence PERC

Our process sequence for PERC solar cells is illustrated as flowchart in Fig. 1. As-cut wafers are initially textured, and the emitter is formed by a phosphorous diffusion. Following this, the emitter is locally ablated in the vicinity of the designated separation pathways. This can be done by similar laser equipment as that used for laser doped selective emitter. Therefore, the emitter window in our case refers to all those cells where new cell edges are characterized by an emitter-free area at the end due to separation. The next step is the wet chemical removal of the rear side emitter in a batch process. In this step, any damage caused by the laser ablation on the front side is removed at the same time. The next steps are the passivation of the front and back and the anti-reflective coating (ARC).

The solar cell is finalized by laser contact opening on the backside, screen printing a metallization layout, that is optimized for shingle solar cells, and the contact formation. To obtain shingle solar cells, the full host cells are cut by thermal laser separation (TLS).

Another pair of groups is not treated with the laser, but the PET is applied to the newly formed edges after cutting. The PET consists of two subsequent treatments: application of an AlOx coating by, e.g., plasma enhanced atomic layer deposition (PE-ALD) and its activation through an annealing step [12]. The cut cells are stacked and held tightly during the application of the PET.

158

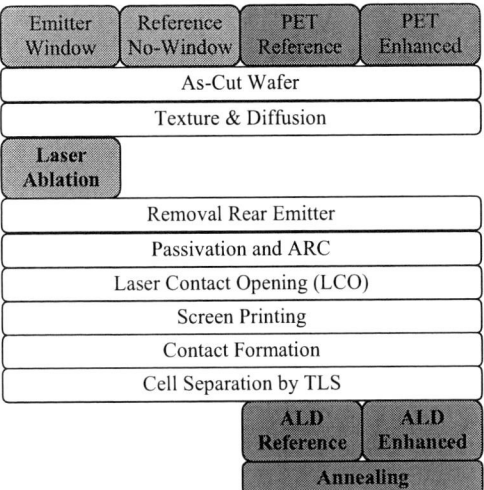

Figure 1: Flow chart of the experimental process flow for PERC solar cells. Two process strategies are evaluated: The emitter windows aim for a suppressed current flow to the edge. Whereas the PET is a very shallow edge damage singulation technology combined with a highly charged dielectric layer passivation.

2.2 Emitter windows

Before utilizing the laser for ablation of the emitter on a complete cell, the process to do so needs to be set up. We fabricate three identical test structures, i.e., n-type wafers are textured and undergo the diffusion for the n+ emitter. Subsequently, laser ablation is performed on both sides of the sample in a pattern similar to a chessboard, creating fields that represent all combinations of ablation and no ablation on the same wafer. The laser emits 532 nm light pulses of about 30 ns duration. A photoluminescence (PL) image of this sample is shown in Fig. 2. Eventually, the effective lifetimes τ_{eff} for all nine fields are measured via quasi steady state photoconductance (QSSPC) from both sides.

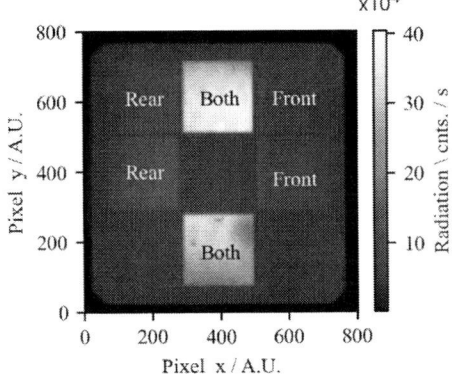

Figure 2: The photoluminescence image reveals a higher amount of radiative recombination in the emitter free fields. By this superior surface passivation, it is found that the emitter is absent, the laser damage in the silicon is etched away and the lateral conductivity is reduced. The labels "Rear", Front" and "Both" indicate which side of the symmetrical sample is treated with the laser ablation.

For the cells, we use a metallization layout that is designed for shingle cells on M2 wafers. This pattern consists of four full square (fsq) shingle cells and two pseudo square (psq) shingles. After cutting, the fsq shingles have two new edges and the psq shingles have one new edge each. For reference and evaluation of losses caused by cutting, we fabricate twin groups differing only by the emitter ablation. For the ablation, we choose stripes with a width of about 400 μm at the separation line. In the data's evaluation, we show only the fsq-shingles. The *IV*-data is gathered at an automized cell tester unit.

2.3 Passivated edge technology (PET)

For the optimization of the PET, float-zone silicon (FZ-Si) wafers are used. They suffer only little from impurities and are therefore predestinated for the characterization of the effectiveness of surface passivation layers, such as AlOx. One group of the FZ-Si wafers is coated following our reference recipe, other groups receive variations of the PE-ALD coating process, without changing the layer's thickness (i.e. cycle time). All groups are annealed equally and the effective lifetimes τ_{eff} are obtained by QSSPC before and after annealing. After evaluation of the results, the reference and the most promising variation ("enhanced") of the PET are applied to cut solar cells. The gain in the pseudo fill factor *pFF* for one sun illumination is chosen as quantity for the recovery of the cutting losses.

3 RESULTS AND DISCUSSION

3.1 Emitter windows

The samples, shown in Fig. 2, undergo the QSSPC measurement yielding the effective lifetimes τ_{eff} for each field whose data is depicted in the boxplot in Fig. 3. Overall, an increase in effective lifetime is observed for fields with laser treatment. The analysis is simplified by holding the factor for the light in coupling constant. This causes the deviation in τ_{eff} for the areas that have the ablation on one side.

In average an increase from 470 μs (no ablation) to 590 μs (one-side ablation) is observed. Notably, ablation on both sides leads to a lifetime of 960 μs on average which is approximately a doubling in τ_{eff} compared to no ablation. This confirms the successful removal of the emitter by laser ablation without damaging the cell and enables the transfer of our process to shingle solar cells.

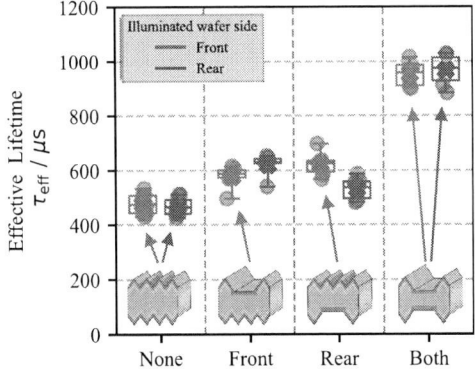

Figure 3: The effective lifetimes τ_{eff} are measured for each test field in both QSSPC configurations. The factor for the light in coupling in not changed.

With the proof of successful removement of the emitter we take complete shingle cells into focus. According to Fig. 1, two almost identical groups of shingles are made. They differ only by having the emitter ablation in the vicinity of the cut edge or being the ordinary case without the window. The light microscopic image of shingle cells with emitter window can be found in Fig. 4, and the scanning electron microscope (SEM) image is presented in Fig. 5.

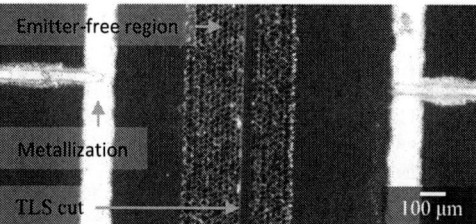

Figure 4: Light microscopic image of two shingle cells side by side with the emitter window after cutting using TLS. The ablation stripe has a width of about 400 μm.

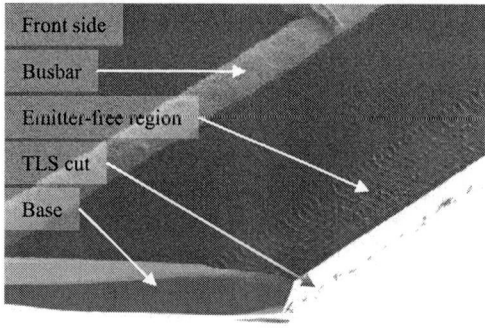

Figure 5: In the SEM image the emitter-free region, made by laser ablation and subsequential etching, is well visible.

Fig. 6 shows the pFF for both groups before and after cutting. In the reference group, 1.2 %$_{abs}$ in pFF are lost by separating the cell. The shingle cells with the emitter window lose only 0.2 %$_{abs}$ in pFF. Thus, we find that the use of emitter windows causes only 17 % of the losses in the pFF when cut by TLS compared to the reference group.

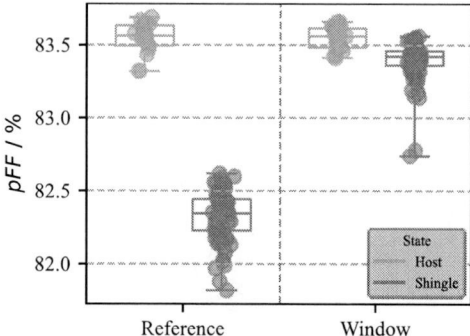

Figure 6: The host cells with the emitter window layout and the reference group start on equivalent pFF levels. Post TLS cutting, the shingles with emitter window suffer fewer losses.

An obvious drawback, which is already noticeable at the host level, is the decrease in absolute current for the emitter window group compared to the reference. As a result, we find that the short-circuit current density j_{sc} is 0.24 mA/cm² lower, which compensates for the decreased pFF loss in the energy conversion efficiency. Note that we used a quite larger emitter window with 400 μm width for this first trial. Fortunately, when going to a shingling module, only half of the j_{sc} loss come into action due to the principle of shingling.

3.2 Passivated edge technology (PET)

For the investigation of the PET, we measured the τ_{eff} of the FZ-Si samples before and after annealing. The data is illustrated in Fig. 7. In the as-deposited state, no significant difference in τ_{eff} is found for both coatings, averaging 1.5 μs. After annealing, τ_{eff} reaches 2.1 ms for the reference ALD process, while an average of 3.8 ms is achieved for the enhanced process. This result shows a 75 %$_{rel}$ improvement in effective lifetime.

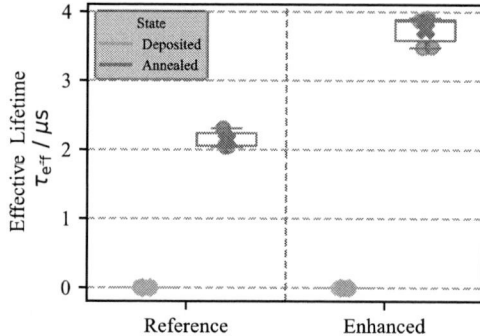

Figure 7: The graph illustrates the effective lifetimes of charge carriers on FZ-Si samples for two implementations of the PE-ALD process. Both coatings underwent the same annealing process.

We applied these encouraging findings to solar cells cut on two sides by TLS. First, we looked at the increased performance from the time just after the cut to after the PET treatment. Second, we examined the overall change in pFF from the original cell to the cut cell after passivation.

Fig. 8 displays the increase in pFF after applying the PET for both ALD recipes. The data is normalized to the average of the reference run. Our new process results in a 24 %$_{rel}$ increase in pFF recovery. In other words, the edge passivation is almost a quarter more effective than before. The relative improvement is less than the 75 %$_{rel}$ improvement seen with the FZ-Si samples. The reason for this smaller improvement is that PET can only make up for the losses caused by cutting the cell, which is not limiting the FZ-Si samples. When we include the pFF at the host level in our calculations, we find that we can recover up to 80 % of the separation losses regarding pFF.

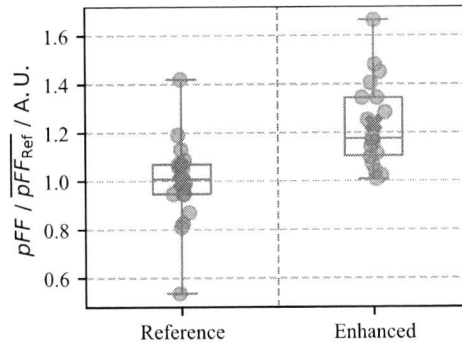

Figure 8: On equal cells that are cut on two sides by TLS the PET was done in two versions. The enhanced process sequence reveals a benefit of 24 %rel regarding *pFF*.

4 CONCLUSIONS

Our analysis of cut silicon solar cells has revealed significant options for improving cell efficiency and mitigating the effects of edge recombination.

Our experimental data show a notable increase in effective lifetime in cases of ablation, suggesting that laser ablation with subsequent wet-chemical etching can successfully remove the emitter. Furthermore, the reduction of losses in *pFF* when using the emitter window approach highlights the potential of this technique for industry application.

The effectivity of the passivated edge technology (PET) was improved, showing a 75 %rel rise of the effective charge carrier lifetime on FZ-Si samples. On separated solar cells, our new process enables 24 %rel more gain, yielding a recovery in *pFF* of up to 80 %rel.

The successful implementation of the emitter window approach and the improvement in the application of PET indicate promising avenues for future research and development in this field.

5 ACKNOWLEDGEMENTS

We thank our colleagues at Fraunhofer ISE, who supported the presented experiments, especially Alexander Krieg, Andreas Brand, Christian Harmel and Leander Kniffki.

This work was supported by the German Federal Ministry for Economic Affairs and Climate Action (BMWK) through the projects "GutenMorgen" (FkZ: 03EE1101A) and "Liebesbrief" (FkZ: 03EE1151A). Further, Stiftung Nagelschneider supports Alexander Göbel with a PhD scholarship.

6 REFERENCES

[1] ITRPV, "International Technology Roadmap for Photovoltaic (ITRPV): 2023 Results," VDMA, 2024.

[2] J. Schneider, S. Schoenfelder, S. Dietrich, and M. Turek, "Solar Module with Half Size Solar Cells," (eng), pp. 185–189, 2014.

[3] M. Mittag, T. Zech, M. Wiese, D. Bläsi, M. Ebert, and H. Wirth, "Cell-to-Module (CTM) Analysis for Photovoltaic Modules with Shingled Solar Cells," in *44th IEEE Photovoltaic Specialists Conference (PVSC)*, Washington, DC, USA, 2017, pp. 1531–1536.

[4] D. Tonini, G. Cellere, M. Bertazzo, A. Fecchio, L. Cerasti, and M. Galiazzo, "Shingling Technology For Cell Interconnection: Technological Aspects And Process Integration," *Energy Procedia*, vol. 150, pp. 36–43, 2018.

[5] B. Blasi, T. Kroyer, T. Kuhn, and O. Hohn, "The MorphoColor Concept for Colored Photovoltaic Modules," *IEEE J. Photovoltaics*, vol. 11, no. 5, pp. 1305–1311, 2021.

[6] M. Hermle, J. Dicker, W. Warta, S. W. Glunz, and G. Willeke, "Analysis of edge recombination for high-efficiency solar cells at low illumination densities," in *3rd World Conference on Photovoltaic Energy Conversion: Joint Conference of 13th PV Science & Engineering Conference, 30th IEEE PV Specialists Conference, 18th European PV Solar Energy Conference*, Osaka, Japan, 2003, pp. 1009–1012.

[7] N. Wöhrle, T. Fellmeth, E. Lohmüller, A. Fell, J. Greulich, R. Preu, T. Fellmeth, P. Baliozian, A. Fell, and R. Preu, "The SPEER solar cell – simulation study of shingled PERC technology based stripe cells," in *33rd European Photovoltaic Solar Energy Conference and Exhibition*, Amsterdam, The Netherlands, 2017.

[8] P. Baliozian, M. Al-Akash, E. Lohmuller, A. Richter, T. Fellmeth, A. Munzer, N. Wohrle, P. Saint-Cast, H. Stolzenburg, A. Spribille, and R. Preu, "Postmetallization "Passivated Edge Technology" for Separated Silicon Solar Cells," *IEEE J. Photovoltaics*, vol. 10, no. 2, pp. 390–397, 2020.

[9] W. Shockley and W. Read, "Statistics of the recombinations of holes and electrons," *Phys. Rev.*, vol. 87, no. 5, pp. 835–842, 1952.

[10] E. Lohmüller, R. Preu, P. Baliozian, T. Fellmeth, N. Wöhrle, P. Saint-Cast, F. Clement, and A. Brand, "Verfahren zum Auftrennen eines Halbleiterbauelements mit einem pn-Übergang," Germany DE 10 2018 123 485.

[11] E. Lohmüller, P. Baliozian, L. Gutmann, L. Kniffki, A. Richter, L. Wang, R. Dunbar, A. Lepert, J. D. Huyeng, and R. Preu, "TOPCon shingle solar cells: Thermal laser separation and passivated edge technology," *Progress in Photovoltaics*, 2023.

[12] E. Lohmüller, P. Baliozian, L. Gutmann, L. Kniffki, V. Beladiya, J. Geng, L. Wang, R. Dunbar, A. Lepert, M. Hofmann, A. Richter, and J. D. Huyeng, "Thermal laser separation and high-throughput layer deposition for edge passivation for TOPCon shingle solar cells," *Sol. Energy Mater. Sol. Cells*, vol. 258, p. 112419, 2023.

[13] A. Münzer, P. Baliozian, A. Steinmetz, T. Geipel, S. Pingel, A. Richter, S. Roder, E. Lohmüller, A. Spribille, and R. Preu, "Post-Separation Processing for Silicon Heterojunction Half Solar Cells With Passivated Edges," *IEEE J. Photovoltaics*, vol. 11, no. 6, pp. 1343–1349, 2021.

[14] F. Dhainaut, R. Dabadie, B. Martel, T. Desrues, M. Albaric, O. Palais, S. Dubois, and S. Harrison, "Edge passivation of shingled poly-Si/SiO x passivated contacts solar cells," *EPJ Photovolt.*, vol. 14, p. 22, 2023.

[15] B. Hoex, Heil, S. B. S., E. Langereis, van de Sanden, M. C. M., and Kessels, W. M. M., "Ultralow surface recombination of c-Si substrates

passivated by plasma-assisted atomic layer deposited Al2O3," *Appl. Phys. Lett.*, vol. 89, no. 4, p. 42112, 2006.

[16] J. Dicker, "Analyse und Simulation von hocheffizienten Silizium-Solarzellenstrukturen für industrielle Fertigungstechniken," Dissertation, Fakultät für Physik, Universität Konstanz, Konstanz, 2003.

[17] K. Ruhle, M. K. Juhl, M. D. Abbott, L. M. Reindl, and M. Kasemann, "Impact of Edge Recombination in Small-Area Solar Cells with Emitter Windows," *IEEE J. Photovoltaics*, vol. 5, no. 4, pp. 1067–1073, 2015.

[18] E. Lohmüller, R. Preu, P. Baliozian, T. Fellmeth, N. Wöhrle, P. Saint-Cast, and A. Richter, "Verfahren zum Vereinzeln eines Halbleiterbauelements mit einem pn-Übergang," Germany DE 10 2018 123 484.

[19] S. W. Glunz, J. Dicker, M. Esterle, M. Hermle, J. Isenberg, F. J. Kamerewerd, J. Knobloch, D. Kray, A. Leimenstoll, F. Lutz, D. Oßwald, R. Preu, S. Rein, E. Schäffer, C. Schetter, H. Schmidhuber, H. Schmidt, M. Steuder, C. Vorgrimler, G. Willeke, D. Osswald, and E. Schaffer, "High-efficiency silicon solar cells for low-illumination applications," in *29th IEEE Photovoltaic Specialists Conference (PVSC)*, New Orleans, LA, USA, 2002, pp. 450–453.

[20] S. Xue, G. Yang, X. Zhao, J. Wu, R. Li, B. Li, and Y. Xu, "Contactless edge for edge recombination optimization in solar cell," *Micro & Optical Tech Letters*, vol. 66, no. 7, 2024.

Characterization of TiO_x as electron selective contact using low-temperature oxidation process via high-pressure sputtering

F. Pérez-Zenteno[1], S. Duarte[1], R. Benítez-Fernandez[1], G. Godoy-Pérez[1], I. Torres[2], R. Barrio[2], L. Rebohle[3], D. Caudevilla[1], S. Algaidy[4], R. García-Hernansanz[1], J. Olea[1], D. Pastor[1], A. Del Prado[1], E. García-Hemme[1], E. San Andrés[1]

[1]Dpto. EMFTEL, Fac. de Ciencias Físicas, Universidad Complutense de Madrid. Avda. Complutense s/n, Ciudad Universitaria, 28040 Madrid, Spain
[2]Unidad de Dispositivos Energéticos y Ambientales, División de Energías Renovables. CIEMAT. Av/ Complutense 40, E-28040 Madrid, Spain
[3]Institute of Ion Beam Physics and Materials Research, Helmholtz-Zentrum Dresden Rossendorf, Bautzner Landstrasse 400, 01328 Dresden, Germany
[4]Instituto de Energía Solar E.T.S.I. Telecomunicación Universidad Politécnica de Madrid, Ciudad Universitaria28040 Madrid, Spain

francp05@ucm.es

EU PVSEC 2024

Motivation & Experimental

- *High-Pressure Sputtering* **(HPS)** TiO_x layer as electron selective contact for solar cells.

  Reduction of sputtering damage (0.5 mbar)

  $$\lambda = \frac{kT}{\sqrt{2}\pi d^2 (p)}$$

  Mean free path between collisions

- Mitigate oxidation of **Si** substrate by 2-step process fabrication

Deposition step:
Power: 45 W
Temp. RT
Dep. rate: ~0,80 nm/min

Oxidation step:
Power: 45 W
Temp. 150°C or 200°C
Time: 0,5 h – 2 h
Gasses flux: 5%/95% (O_2/Ar)

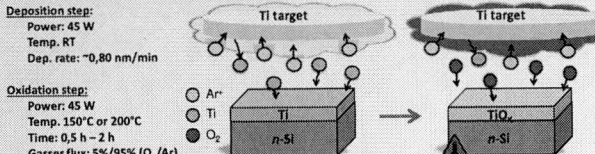

- **Characterization of TiO_x thin film**

  - Physical (FTIR, XPS, Transmittance)
  - Contact resistivity (Cox & Strack)
  - Lifetime (Photoconductance)

Material characterization

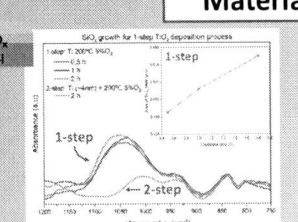
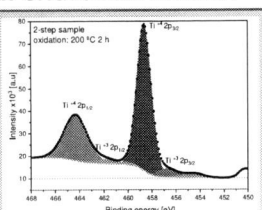

FTIR: 2-step process reduce the oxidation of the substrate

XPS: Sub-stoichiometric film, $TiO_{x=1,8}$

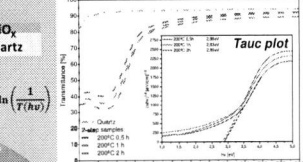
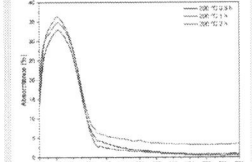

$$\alpha(h\nu) = \frac{1}{t}\ln\left(\frac{1}{T(h\nu)}\right)$$

Theoretical value: 3,2 eV [1]
Indirect bandgap: 2,8 eV

Low absorptance in the visible range

Cox&Strack

To ensure good measurements, we followed the next criteria [2]:

- More Accurate expression of spreading resistance

$$R_T - R_S = \frac{\rho_c}{\frac{1}{4}\pi d^2} + R_0 \qquad R_S \approx \frac{2\rho_w}{\pi d}\left(\frac{d}{2t} + \frac{4}{\pi} + \left(1-\frac{4}{\pi}\right)\arctan\left(\frac{2d}{3t}\right)\right)^{-1}$$

- Circular contacts d < 2 thickness (Si) ➔ d < 520 μm

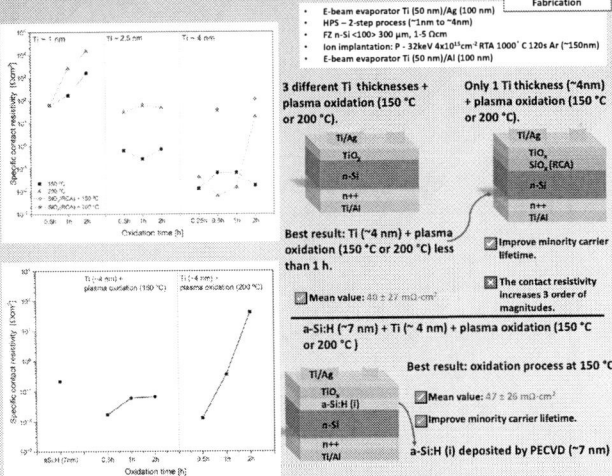

Fabrication
- E-beam evaporator Ti (50 nm)/Ag (100 nm)
- HPS – 2-step process (~1nm to ~4nm)
- FZ n-Si <100> 300 μm, 1-5 Ωcm
- Ion implantation: P - 32keV 4x10^{11}cm^-2 RTA 1000° C 120s Ar (~150nm)
- E-beam evaporator Ti (50 nm)/Al (100 nm)

3 different Ti thicknesses + plasma oxidation (150 °C or 200 °C).

Only 1 Ti thickness (~4nm) + plasma oxidation (150 °C or 200 °C).

Best result: Ti (~4 nm) + plasma oxidation (150 °C or 200 °C) less than 1 h.
☑ Improve minority carrier lifetime.
☑ Mean value: 49 ± 27 mΩ·cm²

☑ The contact resistivity increases 3 order of magnitudes.

a-Si:H (~7 nm) + Ti (~4 nm) + plasma oxidation (150 °C or 200 °C)

Best result: oxidation process at 150 °C
☑ Mean value: 47 ± 26 mΩ·cm²
☑ Improve minority carrier lifetime.

a-Si:H (i) deposited by PECVD (~7 nm)

Photoconductance

Carrier lifetime of TiO_x, TiO_x/SiO_x(RCA), and TiO_x/a-Si:H was obtained by Sinton WCT-120 equipment.

Soft annealing to increase the lifetime (Hotplate or FGA)

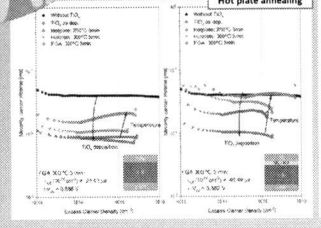
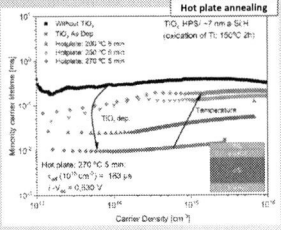

Flash-lamp annealing?
☑ **Faster** alternative to conventional annealing processes.
☑ **Reduce** thermal budget expenditure (3 ms process).
☑ **Avoid** rear contact to be heated.

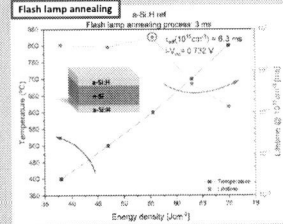
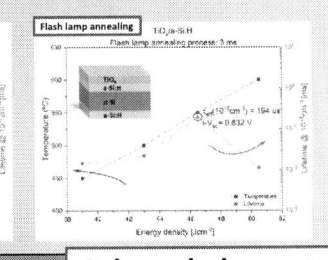

Conclusion

- The 2-step process by HPS controls the oxidation of Si substrate.

- Low contact resistivity was obtained with TiO_x (40 mΩ·cm²) and TiO_x/a-Si:H (47 mΩ·cm²)

- TiO_x/a-Si:H shows a lifetime of ~200 μs with a hotplate or flash lamp annealing (3 ms, reached temperature: 550 °C).

[1] R. Gannenberg and P. Greene, "Reactive sputter deposition of titanium dioxide", Thin Solid Films, vol. 368, no. 1-2, pp. 120–127, 2000, doi: 10.1016/S0040-6090(98)00530-4.
[2] E. van Wijngaarden, J. Yang, and S. Schreuer, "Discrepancies in contact resistivity from the Cox-Strack method: A review" Sol. Energy Mater. Sol. Cells, vol. 245, no. 2022, 2022, doi: 10.1016/j.solmat.2022.111903.

Acknowledgments

The authors would like to acknowledge the "CAI de Técnicas Físicas" of the Universidad Complutense de Madrid, as well as the Centro Nacional de Investigaciones Metalúrgicas (CENIM). This project was funded by projects PID2020-116508RB-I00, TED2021-130894B-C21 and PID2020-117498RB-I00 from the spanish AEI, F. Pérez-Zenteno acknowledges UCM for the predoctoral (call CT58/21 – CT59/21). R. Benítez-Fernandez acknowledges the research contracts under the Investigo Program (CT19/23-INVM-27) of the Ministerio de Trabajo y Economía Social.

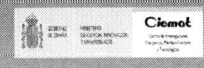

41st European Photovoltaic Solar Energy Conference and Exhibition

This presentation was selected by the Sc. Committee of the EU PVSEC 2024 for submission of a full paper to one of the EU PVSEC's collaborating peer-reviewed journals.

APPROACHES FOR REDUCING METALLIZATION-INDUCED LOSSES AND COST IN INDUSTRIAL TOPCON SOLAR CELLS

S. Mack, D. Ourinson, M. Meßmer, C. Tessmann[1], K. Krieg, R. Haberstroh, S. Kluska,
J. D. Huyeng, J. Greulich, A. Wolf, F. Clement
Fraunhofer Institute for Solar Energy Systems ISE, Heidenhofstr. 2, 79110 Freiburg, Germany
[1] now with Fraunhofer Institute for Applied Solid State Physics IAF, Tullastr. 72, 79108 Freiburg, Germany
Phone +49 761/4588-5048, sebastian.mack@ise.fraunhofer.de

ABSTRACT: Reducing carrier recombination in silicon solar cells is essential for enhancing conversion efficiency, as it influences both fill factor and open circuit voltage. Recombination occurring at metal-semiconductor interfaces is particularly significant; however, processing conditions that minimize recombination often lead to higher contact resistivities. This is overcome by the implementation of laser-enhanced contact optimization, which enables both high fill factors and elevated open circuit voltages. To maximize the gain in efficiency, front end processing has been optimized. We show an implied open circuit voltage limit of industrial cell precursors exceeding 740 mV, indicating the extremely high voltage potential of modern TOPCon solar cells. Ag based metallization pastes without Al are chosen for contacting, as they exhibit a lower metallization related recombination. Contact optimization leads to an increase in conversion efficiency by almost 24 % (absolute). Such treated samples feature a similar conversion efficiency as conventional AgAl pastes. Light induced plating of nickel, copper, and Ag on M10 cells leads to a higher conversion efficiency than screen-printed contacts, even without the additional need for contact treatment, while at the same time reducing Ag consumption by more than 90 % to only 1.1 mg/W_p, which lowers metallization related cost.

Keywords: TOPCon, metallization, screen-printing, plating

1 INTRODUCTION

The tunnel oxide passivating contact (TOPCon) structure has become widely accepted in industrial manufacturing. These solar cells are made from n-type silicon wafers and feature a tunnel oxide combined with an n-doped polysilicon layer on the rear side. The synergy between passivation from the interfacial oxide layer and a very shallow dopant profile beyond the oxide in the silicon wafer results in an exceptionally low recombination current density. Additionally, metal-induced recombination can be reduced significantly. Overall, the implementation of passivating contacts in industrial solar cells (iTOPCon) has led to conversion efficiencies surpassing 25% [1–5].

To further enhance solar cell efficiency, it is crucial to minimize other loss mechanisms in iTOPCon solar cells. Key factors include carrier recombination within the wafer itself, as well as on front and rear side of the sample, which encompasses dielectrically passivated areas and contacted regions. Front side recombination at the metal contacts $j_{0e,met}$ is influenced by the emitter dopant profile, as well as by the type of metallization technique used. In the case of flatbed screen printing technology this includes the composition of the metallization paste, the contact firing profile, and the metallization fraction, such as finger widths and pitch. For iTOPCon solar cells, metallization typically involves the use of Ag based pastes on both front and rear, which is a significant cost driver for this type of solar cell.

This paper describes approaches to address metallization-related carrier recombination in iTOPCon solar cells. First, the impact of laser-enhanced contact optimization (LECO) on cell parameters is described, on solar cells, which feature a front side metallization formed by either an AgAl paste or an Ag paste without any Al content. In a second section, screen-printing technology is compared to laser ablation of dielectrics in combination with light induced plating of a metallization stack of Ni/Cu/Ag. Here, a metallization approach based on plating strongly reduces Ag consumption and thus cost.

2 SAMPLE PREPARATION

2.1 Solar cells

The process sequence for iTOPCon solar cells, illustrated in Fig. 1, utilizes commercial n-type silicon wafers. Our lab is compatible with different wafer formats, and in the experiments described below, wafers with either M2 (156.75 mm edge length) or M10 (182 mm) size are being used. Following saw-damage removal, etching in an alkaline solution creates upright random pyramids. Gas-phase diffusion in an atmospheric tube furnace with boron tribromide (BBr_3) forms the emitter, which is then followed by inline glass etching and batch alkaline etching to eliminate the parasitic emitter doping on the rear side of the wafer, as well as on the edges. After growing an interfacial oxide layer of 1 nm to 1.5 nm thickness, a phosphorus-doped silicon layer is deposited on the rear using a chemical vapor deposition process, either low pressure (LPCVD) or plasma-enhanced (PECVD). Another inline etching step removes the wrap-around on the front side and the wafer edge, as well as the borosilicate glass (BSG) stack formed during diffusion. Following a cleaning step, thermal annealing drives phosphorus deeper into the silicon wafer and enhances the crystallinity of the deposited silicon layer. The passivation sequence includes another cleaning step, the deposition of an Al_2O_3 layer by means of atomic layer deposition, outgassing, and the deposition of SiN_x layers on both the front and rear sides via PECVD. For metallization, we employ established flatbed screen printing technology with silver-containing pastes, followed by contact firing in an inline conveyor belt furnace. Optionally, laser-enhanced contact optimization (LECO) [6, 7] is used to reduce contact resistivity ρ_c. An alternative metallization approach consists of laser ablation of front and rear side dielectrics, firing for thermal activation of passivation layers, and light induced plating of Ni, Cu, and Ag.

41st European Photovoltaic Solar Energy Conference and Exhibition

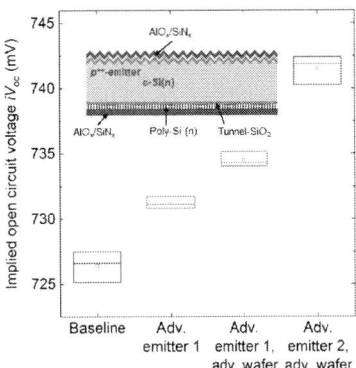

Figure 1: Process sequence for fabrication of industrial *i*TOPCon solar cells. The route with screen-printing technology is displayed on the left, whereas the right side represents the alternative with plated contacts.

2.2 Characterization

Throughout processing, samples are removed from processing and characterized. This includes measurement of the passivation quality by means of photoconductance decay in a Sinton lifetime tester WCT-120, the measurement of the emitter sheet resistance R_{sheet} by four-point-probe (4pp), and current-voltage (*I-V*) measurements of the fabricated solar cells in a lab-type cell tester. Additional transfer length method (TLM) analysis yields contact resistivity ρ_c on prepared samples.

3 RESULTS

3.1 Lifetime results

The passivation quality has been determined on asymmetric lifetime samples, *i.e.* solar cells without metallization. To determine the potential of our baseline, we process samples as depicted in Fig. 1, but leave out the processes screen printing, laser ablation, and plating. Next to our baseline emitter process with a $R_{sheet} = 160\ \Omega/sq$, as measured by 4pp on textured surfaces, we also include two other emitters with sheet resistances of $220\ \Omega/sq$ and $330\ \Omega/sq$, respectively. In addition, we also include wafers of a second source. The samples have been fired for activation of the passivation layers at a set temperature of 800°C, which is a typical temperature also for *i*TOPCon solar cells in our laboratory.

Optimizations at Fraunhofer ISE show how a low level of minority carrier recombination can be achieved, as visible from Fig. 2, with the inset showing a schematic cross section of the investigated structure. The current baseline process with $R_{sheet} = 160\ \Omega/sq$ (black) yields a median $iV_{oc} = 727\ mV$ and represents a starting point compared to current industrial standards. The implementation of an advanced emitter with $R_{sheet} = 220\ \Omega/sq$ (red) increases this limit to over 730 mV, and 734 mV with another wafer stock. Replacing advanced emitter 1 with advanced emitter 2 with $R_{sheet} = 330\Omega/sq$ (blue) increases iV_{oc} to a remarkable level of 742 mV, closing the gap towards silicon hetero-junction (SHJ) solar cells. While advanced emitter 1 is still within range of what is compatible with screen-printed contacts today, emitter 2 currently requires either the use of a selective emitter structure at the contacts, due to low

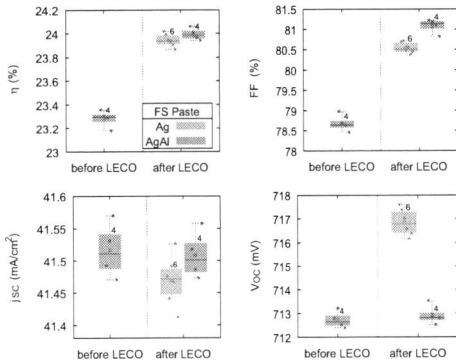

Figure 2: Implied open circuit voltage results for non-metallized *i*TOPCon precursors, measured after firing.

surface dopant concentration, which otherwise would make contacting with low contact resistivity challenging. Alternatively, a contact optimization such as LECO would be required for a low ρ_c. The demonstrated high iV_{oc} on TOPCon precursors illustrates that the reduction of metallization induced losses is key for closing the V_{oc} gap to SHJ cell technology.

3.2 Impact of front metallization paste

An experiment was conducted to compare the effects of two distinct metallization pastes for contacting the front side of the solar cell. The first paste is a pure Ag paste, whereas the second one includes additional Al. The *I-V* parameters are illustrated in Fig. 3.

Figure 3: *I-V* parameters of *i*TOPCon solar cell results on M2 wafers, indicating the impact of LECO treatment, for samples with a front side metallization by either Ag or AgAl paste determined at an industrial cell tester.

Prior to LECO treatment, the AgAl paste exhibits a median conversion efficiency $\eta = 23.3\%$, while a low fill factor $FF < 30\%$ in case of the Ag paste (not shown) indicates that the front contact formation is incomplete, which results in a conversion efficiency of 0.0%. LECO processing strongly reduces the series resistance R_s for cells of both types, leading to similar FF and a $\eta = 24\%$, with a slight advantage for the AgAl paste, while the short circuit current density j_{sc} is on the same level of 41.5 mA/cm². Thus, in case of the Ag paste, LECO treatment has to be considered as a contact forming process, which is in accordance with other work [5].

165

However, completely different behaviors are observed for FF and open circuit voltage V_{oc}. The Ag paste results in a 4 mV higher V_{oc} but a 0.6% lower FF, which is due to the absence of Al. The addition of Al leads to deeper contacts, as previously shown [8]. The emitter in this experiment features a depth of more than 1 µm. However, if metal penetrates too deep into the wafer and approaches the depth of the junction, a strongly increasing recombination current density $j_{0e,met}$ is the result [8], which needs to be prevented. On the other hand, the larger metallized area fraction due to deeper spikes [8] (i.e., the actual 3D contact area, which exceeds the 2D projection) and the Al doping from the paste itself lead to a lower series resistance R_s ($\rho_{c,AgAl} = 1$ mΩcm², compared to $\rho_{c,Ag} = 3$ mΩcm²). Thus, to exploit the higher V_{oc} in case of the Ag paste and to increase η even further, challenges have to be undertaken to decrease the final $\rho_{c,Ag}$ after LECO, either by changes to the emitter dopant profile, the use of an optimized metallization paste, or by modification to the LECO parameters.

Based on the results above and the described lower ρ_c, however, we chose to apply an AgAl paste for contacting of emitters with higher R_{sheet} and fabricated iTOPCon solar cells with the other processes as highlighted in Fig. 1. In the final I-V data, we did in fact find slightly higher V_{oc} and j_{sc} for advanced emitters 1 and 2, however, lower FF values originating again from higher R_s lead to slightly lower cell efficiencies, compared to our baseline process, even after LECO treatment. Therefore, at this point, the high V_{oc} potential previously identified could not be utilized effectively with the metallization process that we currently have at our disposal.

3.3 Screen printing vs. plating

For fabrication of M10-sized iTOPCon solar cells, LPCVD has been chosen as the technology for formation of the TOPCon stack. The I-V measurement results are depicted in Fig. 4, for both screen-printed (SP) metallization and plated contacts. Please note that here only an AgAl paste has been used for front side metallization. In addition, results are shown both directly after processing as well as after additional LECO treatment. For the solar cells with screen printed metallization, an additional firing set temperature variation (780°C to 820°C) has been performed.

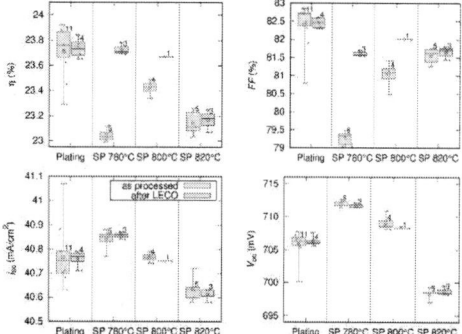

Figure 4: I-V results for iTOPCon cells on M10 wafers with either plated or screen printed metallization, measured with an industrial cell tester. In case of screen printing (SP), the experiment includes a variation in the firing set temperature (780°C to 820°C).

Directly after processing, which refers to the states "after plating", or "after contact firing", respectively, a considerably conversion efficiency up to $\eta = 23.9\%$ is found for plated contacts, while screen printed contacts allow for solar cells up to 23.5% at 800°C firing. Lower or higher firing set temperatures result in lower η, as an increasing FF for 820°C firing goes hand in hand with a decreasing V_{oc} due to increasing $j_{0,met}$, and vice versa for 780°C. This means, 800°C represents a trade-off between V_{oc} and FF. The comparison of plating and screen printing reveals a considerably higher FF and lower V_{oc} for plated contacts. The origin for this lies in the metallized area fraction, as laser ablation leads to a very efficient opening of the dielectrics on front and rear. This is especially detrimental on the front side, where typical $j_{0,met}$ values are in the range of 1000 fA/cm² [9, 10], compared to a $j_{0,met}$ in the estimated range of 10 to 50 fA/cm² on the rear side. In general, j_{sc} follows the trend visible in V_{oc}.

Looking at values after LECO, it becomes apparent that LECO has no positive effect in case of the plated contacts here, whereas screen printed metallization strongly benefits, with a significant increase of FF by 2.6% to 81.6%, which boosts efficiency to 23.8%, close to the initial value of plated metallization. The reason for this different behavior during LECO treatment is most probably linked to the actual contact area for plated contacts, where a larger Si area is in direct contact to the Ni contact material. Further work has to be underdone to investigate this in more detail.

Table 1 shows the I-V parameters of our champion M10 solar cell with screen-printed metallization. The calibrated measurement on a highly reflective chuck has been independently performed at ISFH CalTec, neglecting grid resistances on both front and rear. A conversion efficiency of 24% is found, with especially V_{oc} indicating significant room for improvement. Thus, future work needs to focus on decreasing minority carrier recombination on front and rear side, as well as minimizing metallization related recombination, while keeping FF on a high level. Please note that so far, the cell with plated metallization has not been subjected to a calibrated measurement.

Table 1: I-V data of our champion M10 iTOPCon solar cell. Independent, calibrated measurement at ISFH CalTec, measurement condition grid resistance neglected (grn) on both sides, measurement on highly reflective chuck (hrc) "grn | grn, hrc" [11].

iTOPCon	η (%)	V_{oc} (mV)	j_{sc} (mA/cm²)	FF (%)
M10	24.0	712	41.1	82.1

In terms of silver consumption and processing cost the implementation of Ni/Cu/Ag plating has the potential to significantly lower both. The fabricated solar cells feature a silver consumption of about 15 mg/W for screen printed metallization and down to 1.1 mg/W for plated metallization. This lies even below the targeted silver consumption of 2 mg/W [12] for a sustainable silver consumption in a multi TW PV production market. A direct comparison of the cost of ownership of screen printed vs. plated metallization of M10 TOPCon solar cells shown in Fig. 5 reveals a main difference in the cost structure of the two metallization approaches. The COO screen printing approach is mainly driven by process consumable costs (silver paste). The plating approach on the other hand is due to its low supply chain maturity mainly driven by equipment costs. Overall, the plating approach shows potential to significantly reduce processing cost especially in times of volatile and increasing silver prices.

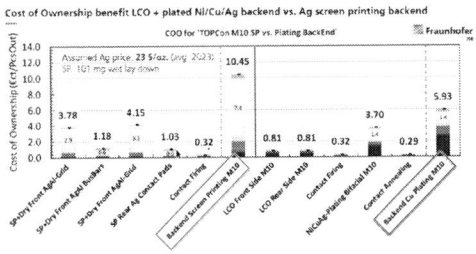

Figure 5: Cost of ownership calculations of the screen printing and plating metallization approach for M10 TOPCon solar cells.

SUMMARY

This study investigated the passivation quality and performance of iTOPCon solar cells using various emitter configurations and metallization techniques. Asymmetric lifetime samples with an emitter sheet resistance of $R_{sheet} = 330\ \Omega/sq$ achieve a remarkable $iV_{oc} = 742$ mV, demonstrating the potential to close the voltage gap to SHJ cells. The investigation of Ag and AgAl front side pastes reveals a significantly different η directly after processing, but LECO treatment changes this to an absolute minimum, with a similar η in the 24% range. Nevertheless, I-V parameter strongly differ in terms of V_{oc} and FF. Quite similarly, also screen-printed metallization using an AgAl paste achieves a similar efficiency as cells with plated Ni/Cu/Ag metallization, with plated contacts achieving higher FF, but lower V_{oc}. The development culminates in a champion M10 cell with an independently measured conversion efficiency of 24 %. While cells with plated contacts are not affected by LECO treatment, plating technology represents a promising approach to reduce metallization related cost in iTOPCon solar cells, as it reduces Ag consumption by over 90%, to a level of only 1.1 mg/W_p, which is also relevant for reasons of sustainability.

However, challenges still remain to achieve lowest contact resistivities in order to fully exploit the high iV_{oc} potential of our optimized frontend process.

Acknowledgments
The authors would like to thank our lab staff for excellent processing.

Funding
The authors acknowledge the financial support by the German Federal Ministry for Economic Affairs and Climate Action in the projects "StroKoTOP" (Fkz 03EE1178A) and "WamTec" (Fkz 03EE1193, Funding via the Clean Energy Transition Partnership (CETP)).

REFERENCES

[1] Ms. Stella Wang, *JinkoSolar's High-efficiency N-Type Monocrystalline Silicon Solar Cell Sets New Record with Maximum Conversion Efficiency of 26.89%*. [Online] Available: https://ir.jinkosolar.com/news-releases/news-release-details/jinkosolars-high-efficiency-n-type-monocrystalline-silicon-3. Accessed on: Jul. 26, 2024.

[2] E. Bellini, *Jolywood claims 26.7% efficiency for n-type TOPCon solar cell*. [Online] Available: https://www.pv-magazine.com/2023/04/12/jolywood-claims-26-7-efficiency-for-n-type-topcon-solar-cell/#:~:text=The%20China%20Academy%20of%20Metrology,rating%20for%20a%20similar%20cell. Accessed on: Jul. 26, 2024.

[3] DAH Solar, *26.5%! A New Record of Mass Production Conversion Efficiency of TOPCon PV Modules by DAH Solar*. [Online] Available: dahsolarpv.com/26-5-a-new-record-of-mass-production-conversion-efficiency-of-topcon-pv-modules-by-dah-solar_n167. Accessed on: Jul. 26, 2024.

[4] A. Mette, S. Hörnlein, F. Stenzel, R. Hönig, I. Höger, M. Schaper, K. Petter, M. Junghänel, C. Klenke, A. Weihrauch, H.-C. Ploigt, O. Kwon, A. Schönmann, O. Tobail, K. Kim, A. Schwabedissen, M. Kauert, K. Duncker, B. Faulwetter-Quandt, J. Scharf, J. Cieslak, F. Kersten, B. Lee, S. T. Kristensen, O. Schnelting, C. Baer, M. Queck, G. Zimmermann, L. Burtone, L. Niebergall, M. Schütze, S. Schulz, M. Fischer, S. Peters, F. Fertig, and J. W. Müller, "Q.ANTUM NEO with LECO Exceeding 25.5 % cell Efficiency," *Solar Energy Materials and Solar Cells* 277 (2024) 113110.

[5] Y. Fan, S. Zou, Y. Zeng, L. Dai, Z. Wang, Z. Lu, H. Sun, X. Zhou, B. Liao, and X. Su, "Investigation of the Ag–Si Contact Characteristics of Boron Emitters for n-Tunnel Oxide-Passivated Contact Solar Cells Metallized by Laser-Assisted Current Injection Treatment," *Sol. RRL* 8 (2024) 2400268.

[6] R. W. Mayberry, K. Myers, V. Chandrasekaran, A. Henning, H. Zhao, and E. Hofmüller, "Laser enhanced contact optimization (LECO) and LECO-Specific pastes – A novel technology for improved cell efficiency," in Proceedings *36th EU PVSEC*, Marseille, France, 2019.

[7] T. Fellmeth, H. Höffler, S. Mack, E. Krassowski, K. Krieg, B. Kafle, and J. Greulich, "Laser-enhanced contact optimization on i TOPCon solar cells," *Prog Photovolt Res Appl.* 30 (2022) 1393.

[8] N. Wöhrle, E. Lohmüller, J. Greulich, S. Werner, and S. Mack, "Towards understanding the characteristics of Ag–Al spiking on boron-doped silicon for solar cells," *Sol. Energy Mater. Sol. Cells* 146 (2016) 72.

[9] Q. Wang, W. Wu, Y. Li, L. Yuan, S. Yang, Y. Sun, S. Yang, Q. Zhang, Y. Cao, H. Qu, N. Yuan, and J. Ding, "Impact of boron doping on electrical performance and efficiency of n-TOPCon solar cell," *Solar Energy* 227 (2021) 273.

[10] H. Höffler, F. Schindler, A. Brand, D. Herrmann, R. Eberle, R. Post, A. Kessel, J. Greulich, and M. C. Schubert, "Review and Recent Development in Combining Photoluminescence- and Electroluminescence Imaging with Carrier Lifetime Measurements via Modulated Photoluminescence

at Variable Temperatures," in *37th European Photovoltaic Solar Energy Conference and Exhibition*, Online, 2020, pp. 264–276.

[11] M. Rauer, A. Fell, W. Wöhler, D. Hinken, C. Reichel, K. Bothe, M. C. Schubert, and J. Hohl-Ebinger, "The impact of measurement conditions on solar cell efficiency," *Sol. RRL* 8 (2024) 2300873.

[12] Brett Hallam, Moonyong Kim, Yuchao Zhang, Li Wang, Alison Lennon, Pierre Verlinden, Pietro P. Altermatt, and Pablo R. Dias, "The silver learning curve for photovoltaics and projected silver demand for net-zero emissions by 2050," *Prog Photovolt Res Appl* 31 (2023) 598.

41st European Photovoltaic Solar Energy Conference and Exhibition

UNVEILING THE SYNERGY OF NANOWIRES AND PEDOT:PSS FOR SILICON SOLAR CELL FABRICATION AND LEADING TO MECHANICAL FLEXIBILITY

Deepak Sharma[1,2,3]†, Ruchi K. Sharma[1,2], Arman Ahnood[3], Sanjay K. Srivastava[1,2]*

[1]Photovoltaic Metrology Section, Advanced Materials and Device Metrology Division,
CSIR-National Physical Laboratory, Dr. K.S. Krishnan Marg, New Delhi-110012, India
[2]Academy of Scientific and Innovative Research (AcSIR), Ghaziabad-201002, India
[3]School of Engineering, RMIT University, Melbourne, VIC-3000, Australia
E-mail: †deepaksharma24x7@gmail.com; *srivassk@nplindia.org

ABSTRACT: This study elucidates a pioneering approach to fabricating flexible crystalline silicon (c-Si) solar cells with low temperatures by leveraging the light-trapping capabilities of silicon nanowires (SiNWs) and the effective junction formation between n-type c-Si and organic p-type PEDOT:PSS. Through comprehensive experimentation, we demonstrate a low-cost fabrication method that utilizes SiNWs and PEDOT:PSS and then achieves mechanical flexibility using 50 μm thin c-Si wafers prepared by alkali etching. Followed by comparing the performance of thin-flexible HSCs to traditional 200 μm thick rigid c-Si hybrid heterojunction solar cells (HSCs). The results indicate that SiNWs, combined with PEDOT:PSS, significantly improve the efficiency of both flexible and rigid HSCs. The fabricated devices achieve power conversion efficiencies (PCE) of 8.72% for thin-flexible and 10.26% for thick SiNW-based HSCs with optimized 170 nm long SiNWs. This work contributes to the advancement of flexible photovoltaic (PV) technology, offering potential for practical applications in portable and resilient energy solutions.
Keywords: Hybrid solar cell; Silicon nanowire; PEDOT:PSS; Thin flexible solar cell.

1 INTRODUCTION

The global shift towards renewable energy solutions has intensified interest in the development of cost-effective, flexible, lightweight, and efficient solar cells. Crystalline silicon (c-Si) remains the dominant material in photovoltaic (PV) technology account to its high power conversion efficiency (PCE) and abundance [1]. However, traditional homojunction c-Si solar cells are rigid, exhibit reflection losses, and require high temperatures for doping [2]. These limitations hinder their use in flexible and portable applications. In this work, we explore the integration of silicon nanowires (SiNWs) of optimal length with an organic p-type polymer poly(3,4-ethylenedioxythiophene):poly(styrenesulfonate) (PEDOT:PSS) to form a heterojunction with n-type c-Si at low-temperature, aiming to enhance the light absorption and flexibility of c-Si solar cells. By using thin (50 μm) c-Si wafers prepared through alkali etching, we achieve significant mechanical flexibility while maintaining high device performance. Additionally, we compare the performance of these thin-flexible solar cells with thicker (200 μm) rigid cells, demonstrating the potential of this low-cost, flexible technology.

2 MATERIALS AND METHOD

2.1 Fabrication of SiNWs

The SiNW structure was created using a silver-assisted chemical etching (Ag-ACE) process [3]. We employed a 0.02 M AgNO₃ and 5 M HF aqueous solution at room temperature for various etching durations (0, 0.5, 1.5, 3 mins) in a Teflon container, as presented in Fig. 1(a). The Ag particles were eliminated by immersing the wafers in an NH₄OH and H₂O₂ solution (3:1 ratio). This process resulted in uniformly distributed SiNWs on n-type c-Si wafer's surface, enhancing light absorption by reducing surface reflection [4]. Fig. 1(b) and (c) present the digital image of planar Si and SiNW wafers, which depicts that SiNW traps more light and hence has a darker contrast. Figure 2(a) schematically presents the mechanism of SiNW fabrication via Ag-ACE. Here Ag⁺ ions get reduced to Ag by capturing electrons, and the formation of SiOₓ takes place, which is then etched by HF, leading to the formation of SiNW. A detailed discussion on SiNW formation is present in our earlier works [3-5]. Figure 3(b) depicts the FESEM image of SiNW fabricated over Si wafer via Ag-ACE process.

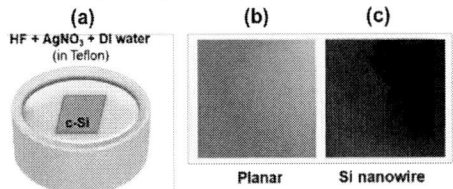

Figure 1: (a) Schematic presentation of SiNW fabrication process via Ag-ACE; (b) Digital image of planar Si and (c) SiNW textured Si.

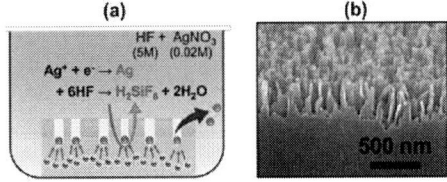

Figure 2: (a) Mechanism of SiNW fabrication via Ag-ACE; (b) FESEM image of SiNW.

2.2 Device Fabrication

The step involved for the junction formation is the deposition of PEDOT:PSS as the p-type layer. Fabrication steps are presented in Fig. 3. Commercially available n-type Si (100) wafers were cleaned and preheated at 110°C to form a hydrophilic 1.5 nm thin SiOₓ tunnel layer [6]. A mixture of PEDOT:PSS, 2-propanol and 7 wt% ethylene glycol was spin-coated onto the SiNWs to obtain a 100 nm thick layer [6]. This polymer blend forms an effective heterojunction with the n-type c-Si, allowing for low-temperature fabrication. After spin-coating, the devices were annealed at 110°C, followed by the deposition of silver (Ag) as the top electrode and indium-gallium

(In:Ga) as the back contact, completing the 'Ag/PEDOT:PSS/SiNW-Si/In:Ga' device structure and device area of 2.25 cm^2 [6]. The fabricated solar cells were named as HSC$_x$, where x represents the etching time.

Figure 3: Fabrication steps involved in hybrid-heterojunction solar cell fabrication (HSC).

The 50 µm thin-flexible Si wafers, as presented in Fig. 4(a), were prepared using alkali etching, detailed in our previous works [5,7]. Figure 4(b) presents the schematic device structure of thin-flexible HSC.

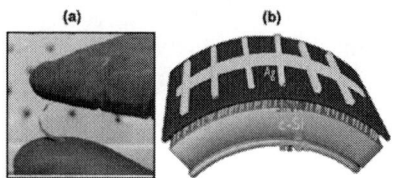

Figure 4: (a) Digital image presenting flexible Si wafer; (b) Flexible SiNW-based HSC device structure.

2.3 Characterization

A 'FEI Nova NanoSEM 200' field emission scanning electron microscope (FESEM) was utilized to analyze the surface morphology. A 'CRAIC Apollo' UV-vis-NIR microspectrophotometer was employed for reflection measurement. A 'WCT-120' Sinton's method was used to investigate how variations in SiNW length and PEDOT:PSS affect the minority carrier lifetime. PV characteristics of HSCs were measured using equipment from Bunkoukeiki, Japan (Model: CEP-25HS-50).

3 RESULTS AND DISCUSSION

3.1 Surface Morphology and Optical Property

The etching time was varied to achieve different nanowire lengths, and the length increases with the increase in etching time. The SiNW length for 0.5-, 1.5-, and 3.0-mins etching times were obtained as ~60, ~170, and ~340 nm. Figure 5(a) presents the FESEM image of HSC$_{1.5}$ where PEDOT:PSS is coated over SiNWs. Reflection measurements were conducted to compare the light absorption of the various solar cell structures, which are presented in Fig. 5(b). The SiNWs enhance light trapping and reduce reflection. The PEDOT:PSS-coated SiNW-wafers exhibited a reflection of only 8% across the 400-1100 nm wavelength range for 1.5 min etched Si, compared to 27% for planar cells, while uncoated Si has ~40% reflection. This reduction is attributed to the multiple scattering and trapping of incident light within the nanowire array, which increases the optical path length and enhances photon absorption [8,9]. The reduction in reflection significantly enhanced light trapping, particularly for the thin-flexible Si cells, compensating for the reduced thickness of the absorber layer.

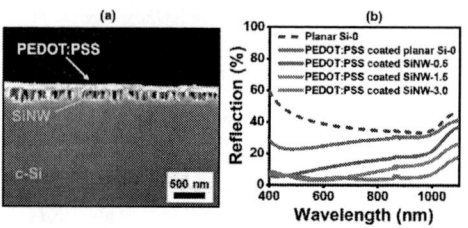

Figure 5: (a) FESEM image of SiNW-based HSC; (b) Reflection curves.

3.2 Passivation

Minority carrier lifetime (MCL) measurements were performed to evaluate recombination losses in the various solar cells and are presented in Fig. 6. The SiNW-based devices exhibited a lower carrier lifetime compared to their planar Si. The increase in length of SiNW is beneficial for anti-reflection but has a drastic impact on MCL. However, PEDOT:PSS improves the MCL to a certain extent and acts as a passivation layer [10]. Thus, the organic layer, which is a hole transport layer, also provides surface passivation, allowing for improved charge collection.

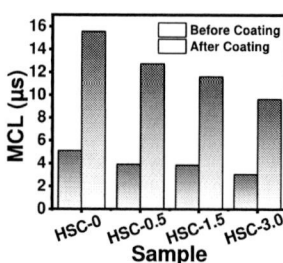

Figure 6: MCL values for before and after coating PEDOT:PSS over SiNW etched Si for different etching duration.

3.3 Device performance

The significant improvement in PCE for the SiNW-based devices is attributed to enhanced light trapping and better charge carrier collection through the organic-inorganic heterojunction formed by PEDOT:PSS and n-type Si. Figure 7 presents the illuminated J-V curve for thick HSCs with varying etching duration. We found that SiNW length plays a critical role in balancing optical and electrical performance. While longer SiNWs enhanced light trapping, they also increased recombination, leading to reduced performance. Optimal device performance with PCE of 10.26% was achieved with 170 nm SiNWs for HSC$_{1.5}$, which provided the best balance between light trapping and minimal recombination.

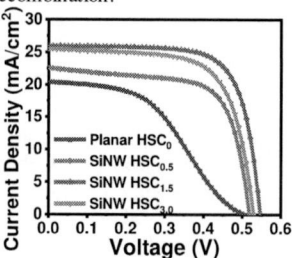

Figure 7: Illuminated J-V characteristics curve for SiNW-based HSCs with etching time variation.

3.4 Comparison of thick and thin HSC

Figure 8 presents the comparative (a) illuminated J-V (b) external quantum efficiency (EQE) curves for thick and thin-flexible HSCs. The electrical performance of the devices was evaluated by measuring the PCE, which depends on short-circuit current density (J_{sc}), open-circuit voltage (V_{oc}), and fill factor (FF). The SiNW-based devices demonstrated a significant increase in J_{sc} compared to their planar counterparts. The results showed that the J_{sc} for thick SiNW-based HSCs reached 25.93 mA/cm^2, compared to 24.05 mA/cm^2 for thin-flexible cells and only 20.40 mA/cm^2 for planar thick cells and 19.91 mA/cm^2 for planar thin HSC. The reduction in J_{sc} in the case of thin HSC is attributed to lower absorption in the higher wavelength range, which is clearly observed in the EQE curve. Table I present the PCE value of different HSCs. The PCE values for the SiNW-based devices were 8.72% for thin-flexible cells and 10.26% for thick-rigid cells. These values represent a substantial improvement over the planar cells, which exhibited PCEs of 6.59% and 4.52%, respectively. This enhancement in PCE is attributed to the improved light absorption provided by the nanowires, as well as the better charge collection enabled by the heterojunction formed with PEDOT:PSS. The performance could be further improved by effective coverage of PEDOT:PSS over SiNWs [11].

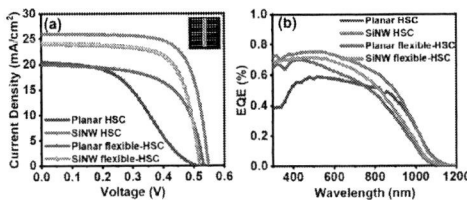

Figure 8: (a) Illuminated J-V and (b) EQE curves presenting the comparison of thick and thin-flexible Si HSCs fabricated over planar and SiNW etched Si wafers.

Table I: PCE of fabricated HSCs

HSC	PCE
Planar thick HSC	4.52%
SiNW-based thick HSC	10.26%
Planar thin-flexible HSC	6.59%
SiNW-based thin-flexible HSC	8.72%

4 CONCLUSIONS

This study presents a comprehensive approach to fabricating Si solar cells by integrating SiNW and PEDOT:PSS followed by attaining flexibility by using 50 μm Si wafers. The HSCs have a simple device structure and a cost-effective low-temperature fabrication process. The results show that SiNW-based devices achieve significantly higher PCE compared to planar devices, with a maximum PCE of 10.26% for thick SiNW-based cells and 8.72% for thin-flexible cells. The enhanced light trapping validated by reflection measurement demonstrates improvement in optical properties via multiple interactions of light by SiNW, passivation by PEDOT:PSS and mechanical flexibility of thin Si wafers, which highlights the potential of this technology for flexible PV applications. The findings from this work offer promising avenues for the development of resilient, adaptable, and efficient solar energy solutions.

5 REFERENCES

[1] C. Battaglia, A. Cuevas, S. De Wolf, Energy Environ Sci 9 (2016) 1552.

[2] D. Y. Khang, J Phys D: Appl Phys 52 (2019) 503002.

[3] S. K. Srivastava, D. Kumar, S. W. Schmitt, K. N. Sood, S. H. Christiansen and P. K. Singh, Nanotechnology 25 (2014) 175601.

[4] S. K. Srivastava, D. Kumar, P. K. Singh, M. Kar, V. Kumar and M. Husain, Sol Energy Mater Sol Cells 94 (2010) 1506.

[5] D Sharma, R. K. Sharma, A. Srivastava, Vamsi K. Komarala, A. Ahnood, P. Prathap, Sanjay K. Srivastava, Sustainable Energy Fuels 8 (2024) 2969.

[6] A. Srivastava, D. Sharma, P. Kumari, M. Dutta and S. K. Srivastava, ACS Appl Energy Mater, 4 (2021) 4181.

[7] D. Sharma, A. Srivastava, J. S. Tawale, P. Prathap and S. K. Srivastava, J Mater Chem C, 11 (2023) 13488.

[8] H. D. Omar, M. R. Hashim and M. Z. Pakhuruddin, Opt Laser Technol 136(2021) 106765.

[9] S. Chattopadhyay, Y. F. Huang, Y. J. Jen, A. Ganguly, K. H. Chen and L. C. Chen, Mater Sci Eng R, 69 (2010) 1.

[10] J. He, P. Gao, M. Liao, X. Yang, Z. Ying, S. Zhou, J. Ye and Y. Cui, ACS Nano, 9 (2015) 6522.

[11] J. Wang, H. wang, A. B. Prakoso, et al., Nanoscale 7 (2015) 4559.

41st European Photovoltaic Solar Energy Conference and Exhibition

POLYSILICON PASSIVATION - TUNNELING OXIDE ROUTES AND ANNEALING CONDITIONS EFFECT ON PASSIVATION

Per-Anders Hansen*, Junjie Zhu, Rune Søndenå
Institute for Energy Technology
Instituttveien 18, 2007 Kjeller, Norway
*Per-anders.hansen@ife.no

TOPCon type structures can result in exceptional wafer passivation, even surpassing 500 ms for high-resistivity n-type wafers. These passivation structures are also known to be very time stable. However, the processing is energy intensive. This structure involves a very thin tunnelling layer followed by a heavily doped polysilicon layer. The highest passivation levels are often obtained when growing this tunnelling layer by dry oxidation of the silicon wafer substrate at or above 600 °C. The doped polysilicon is then deposited by PECVD, followed by a high temperature annealing step. In this work, we have investigated several routes to the tunneling oxide in TOPCon passivation structures and how different annealing temperatures and annealing atmospheres (inert, hydrogen) affect the passivation. The tunneling oxides are divided into oxidation and deposition types. It was found that for our set of samples, oxidation through dry oxidation or nitric acid gave superior passivation than deposition of silicon oxide and silicon oxynitride through PEALD and PECVD. After the tunneling oxide, an approximately 70 nm thick layer of phosphorous doped a-Si was deposited at 200 °C. The wafers were then annealed at different temperatures from 300 to 900 °C.
Keywords: TOPCon, passivation, low temperature, hydrogenation

1 INTRODUCTION

The TOPCon passivation structure results in very low surface recombination, allowing wafer lifetimes even surpassing 500 ms for high-resistivity n-type wafers [1]. The basic structure is a thin tunneling layer, often based on silicon oxide or oxynitride, and a highly doped silicon layer which is generally micro- or polycrystalline. This doped layer provides high conductivity as well as surface field effect passivation. One drawback of this structure though is that it requires energy-intensive high temperature growth and annealing steps [2]. The tunneling oxide can be created either by oxidizing the silicon surface or by deposition, both which can be accomplished at moderate to low temperatures [3]. But the highest lifetimes are often achieved through dry-oxidation at or above 600 °C [1]. In addition, the doped Si layer usually either benefits or fully requires a high temperature crystallization step. Either due to being deposited in the amorphous state or benefiting from further crystallization even when it is (partly) crystalline as-deposited.

Liu *et al.* found that these structures can result in very high lifetimes when this last annealing step is either at 3-400 or 8-900 °C [4]. In that work, the highest lifetime was obtained at 850 °C annealing, in line with the general processing of TOPCon passivation. In our work, however, the best passivation was obtained at 350 °C. Combined with a 400 °C anneal in hydrogen, this lifetime became even higher. A hydrogen anneal without the initial anneal step provided similarly high lifetimes comparable to the best sample without a hydrogen anneal, This shows that the thermal budget of these structures may be reduced depending on synthesis details.

2 EXPERIMENTAL

The wafer processing is shown in Figure 1. N-type wafers (~1.7 Ωcm) was first saw-damage etched in an HNA solution followed by piranha cleaning. The tunnelling oxide was then added wither through oxidizing by gas-phase dry oxidation or aqueous HNO_3, deposition through PEALD or PECVD (SiO_xN_y), or no oxide at all. Doped a-Si(n+) was then deposited by PECVD, using SiH_4 and PH_3. The wafers were annealed in a tube furnace at various temperatures for 30 minutes. Some wafers finally got a H_2 anneal at 400 °C. For some wafers, this H_2 anneal was the only annealing step. All annealing (inert and H_2) was done in a Tempress tube furnace, while all PEALD and PECVD was done in an Oxford Instruments PlasmaLab133 direct plasma system with an excitation frequency of 13.56 MHz. Some wafers (Figure 5Figure 6) was annealed by a simulated contact firing process in a belt furnace. The temperature profile of this firing is given in [5]. Lifetime measurements were done on a Sinton WCT-120-TS, while automatic lifetime measurements were done by continuous white LED illumination at ~1 sun equivalent intensity without moving the wafer, only interrupted for lifetime measurements [6].

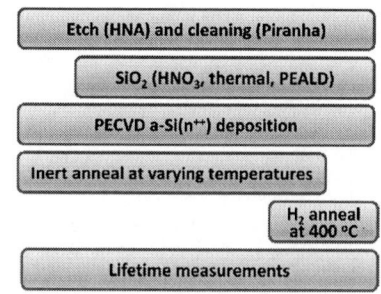

Figure 1: Processing steps for the n-type wafers in this work.

3 RESULTS AND DISCUSSION

To evaluate our TOPCon structure, passivated wafers were compared to wafers passivated by an a-Si/SiNx stack. The lifetime stabilities of both these samples during illumination at room temperature are shown in Figure 2. The TOPCon passivation is both superior in surface

172

passivation level, but also towards light-soaking stability, showing a weak lifetime increase whereas a-Si degrades somewhat during prolonged light-soaking.

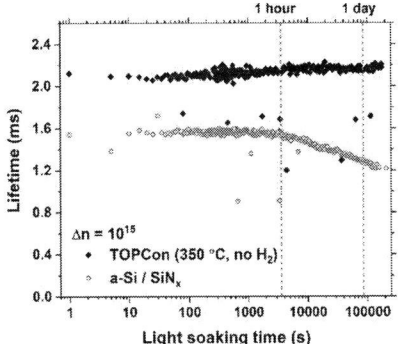

Figure 2: Comparison of lifetime stability during room temperature light soaking at approximately 1 sun white LED illumination.

Figure 3 shows the obtained lifetimes obtained after the inert anneal. Two lifetime maxima appears; a low-temperature, and a high temperature maximum at 350 °C and 800 °C, respectively. These two low/high-temperature peaks are similar to those reported in [4].

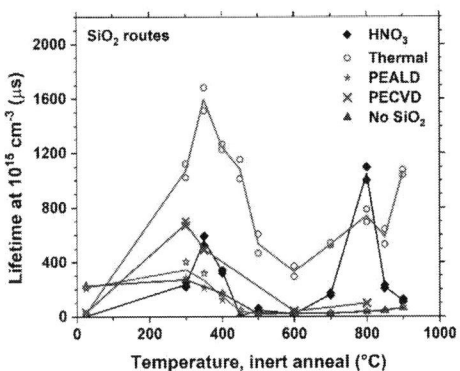

Figure 3: Resulting lifetime obtained after final inert anneal.

Hydrogenation is often incorporated in the final processing, either through a dedicated hydrogenation process or through the contact firing process. In this work, selected wafers underwent a second annealing step in a H_2 atmosphere at 400 °C for 30 minutes *after* the inert anneal. The lifetime results are shown in Figure 4. Note that the samples at x = 25 °C (i.e. room temperature) did not receive the first inert anneal, but only the H_2 anneal. It is seen that the H_2 anneal further increases the obtained lifetime from around 1.6 ms to 1.9 ms. In addition, the sample that only got the H_2 anneal show high lifetimes as well. This indicates that the TOPCon annealing step and hydrogenation step can be combined, in particular for samples like ours which results in high lifetimes also for low temperature anneals.

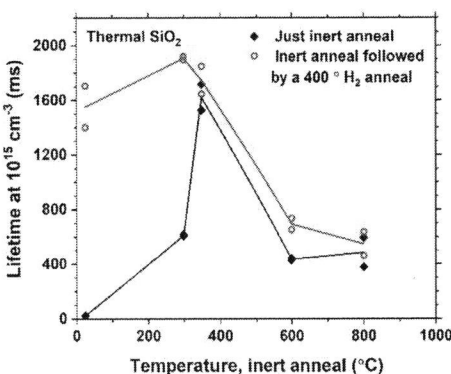

Figure 4: Resulting lifetime obtained after a 30 minute H_2 anneal at 400 °C, following the inert anneal. The x-axis shows the temperature during the inert anneal, while the H_2 anneal was always at 400 °C. The sample at x = 25 °C was not given an inert anneal, just a H_2 anneal.

We also investigated the failure of wafers annealed at medium temperatures (5-700 °C). Four wafers that got oxidized tunneling layers (HNO_3 and thermal) and two grown layers with different thicknesses of PEALD SiO_2 got a simulated contact firing in a belt furnace reaching about 700 °C. The PEALD wafer show a higher initial lifetime in the as-deposited state, but all samples are expected to perform poorly after this medium temperature anneal as seen in Figure 3. Figure 5 shows that this is what we obtain as well.

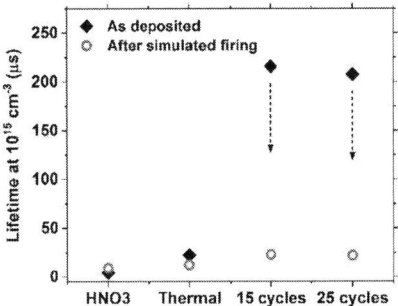

Figure 5: Lifetime before and after a simulated contact firing in a belt furnace, for oxidized (HNO_3 and thermal) and PEALD grown SiO_2.

To understand the failure mode of this sample, we compared the wafer resistivity before and after the simulated contact firing, shown in Figure 6. It is seen that the resistivity is reduced after the firing. This is likely from the in-diffusion of phosphorous from the poly-Si(n^{++}) layer into the n-type Si wafer bulk. It is clear that the oxidized tunneling layer routes results in better diffusion barrier properties than the deposited layers. The poor barrier properties of the deposited layers means that such layers can be expected to cause strong degradation during high-temperature annealing.

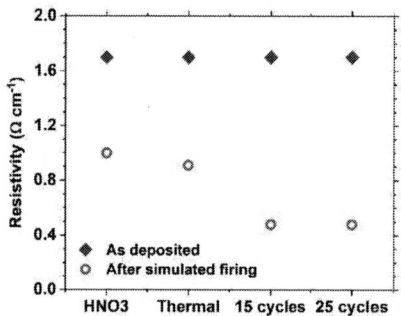

Figure 6: Wafer resisitivity before and after a simulated contact firing in a belt furnace, for oxidized (HNO₃ and thermal) and PEALD grown SiO₂.

4 CONCLUSION

In this work, we have compared both oxidizing and deposition routes for the tunneling layer as well as an additional hydrogenation step in TOPCon passivation structures for n-type Si wafers. In our results, we obtained the highest lifetimes for low-temperature anneals which was further improved to achieve 1.9 ms lifetime when combined with hydrogenation. This shows that it is possible to achieve high passivation levels in TOPCon structures also without high temperature anneals, but that this strongly depends on the processing and in particular on the type of tunneling layer.

5 REFERENCES

[1] B. Steinhauser, T. Niewelt, A. Richter, R. Eberle, and M. C. Schubert, Solar RRL **5** (11), 2100605 (2021).

[2] B. Kafle, B. S. Goraya, S. Mack, F. Feldmann, S. Nold, and J. Rentsch, Solar Energy Materials and Solar Cells **227**, 111100 (2021).

[3] H. Yousuf, M. Q. Khokhar, M. A. Zahid, M. Rabelo, S. Kim, D. P. Pham, Y. Kim, and J. Yi, Energies **15** (15), 5753 (2022).

[4] W. Liu, X. Yang, J. Kang, S. Li, L. Xu, S. Zhang, H. Xu, J. Peng, F. Xie, J.-H. Fu, K. Wang, J. Liu, A. Alzahrani, and S. De Wolf, ACS Applied Energy Materials **2** (7), 4609 (2019).

[5] R. Søndenå and M. S. Wiig, Journal of Applied Physics **125** (8) (2019).

[6] T. U. Nærland, H. Haug, H. Angelskår, R. Søndenå, E. S. Marstein, and L. Arnberg, IEEE Journal of Photovoltaics **3** (4), 1265 (2013).

41st European Photovoltaic Solar Energy Conference and Exhibition

research for a sunny future

Selective p⁺ Poly-Si Fingers for TOPCon Front Contact Passivation

J. Hoß, S. Sharbaf Kalaghichi, M. Comak, P. Preis, J. Lossen, J. Linke, L. Koduvelikulathu, F. Buchholz

ISC Konstanz e.V., Rudolf-Diesel-Str. 15, 78467 Konstanz, Germany
jan.hoss@isc-konstanz.de / +49-7531-36183-366

Advanced TOPCon solar cells with local p⁺ poly-Si front contacts

- Avoid metal-induced recombination for TOPCon cells by integrating local passivated contacts at front side → Selective Finger (SelFi) TOPCon
- Upgrade solution for existing TOPCon lines using only technologies with proven track-record in PV industry
- Based on laser process for p⁺ poly-Si dopant activation[1,2,3]

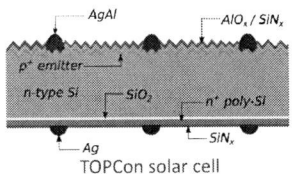

TOPCon solar cell

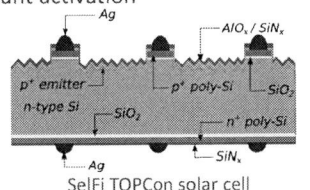

SelFi TOPCon solar cell

Process flow

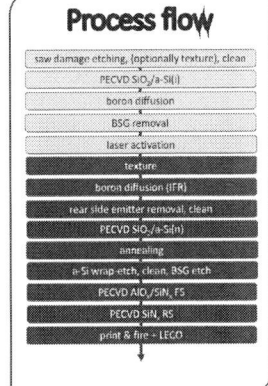

Integration of p⁺ poly-Si fingers

- Laser-treated p⁺ poly-Si fingers withstand all wet chemical steps of TOPCon baseline process with "cluster" etching[4]
- Poly-Si removed and substrate textured in non-lasered regions
- Laser fiducials for print alignment

100 µm p⁺ poly

IV results

- V_{oc} up to 719 mV → gain of 11 mV w.r.t TOPCon
- All cells edge-isolated
- Efficiency so far 0.1-0.2%^{abs} lower than reference → so far limited by optical losses and ρ_c front

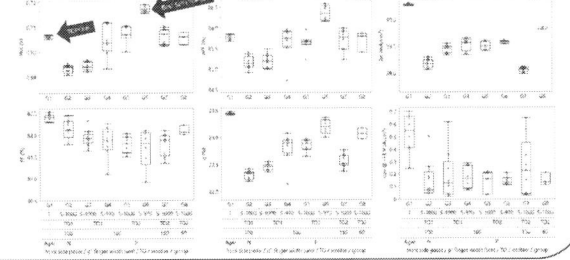

Vanishing metal-induced recombination

- TOPCon ref. loses ≈ 12 mV by metallization
- SelFi TOPCon (G6): iV_{oc}-to-V_{oc} loss < 2 mV → near-perfect contact passivation!
- Outlook: minimize optical losses by reducing p⁺ poly-Si width and printing narrower fingers

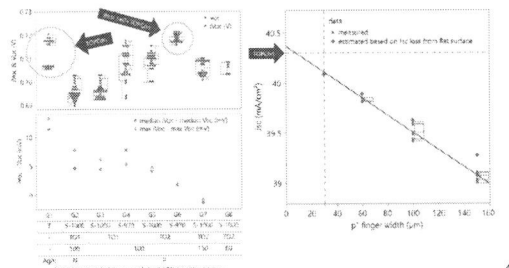

LECO effect on p⁺ poly-Si contacts

- Laser-enhanced contact optimization (LECO lab tool, CE Cell Engineering) significantly improves series resistance (R_s) and FF
- No impact on V_{oc}, pFF, J_{sc} and I_{rev}
- Efficiency gain 0.16-0.29%^{abs} for SelFi TOPCon → η up to 23.4% for SelFi cells

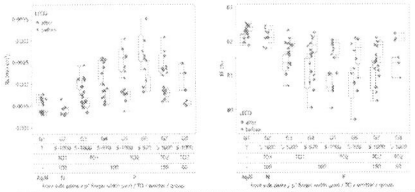

References
[1] F. Buchholz et a., EUPVSEC, 2021
[2] S. Sharbaf Kalaghichi et al, Solar 3(3), 2023
[3] S. Sharbaf Kalaghichi et al, Energies 17(6), 2024
[4] T. Dannenberg et al, Sol. RRL 7, 2023

Acknowledgements
This work was partly funded by Bundes-ministerium für Wirtschaft und Klimaschutz (BMWK), grant agreement No. 03EE1138A.

Further reading
Results published in „Advanced TOPCon solar cells with patterned p-type poly-Si fingers on the front side and vanishing metal induced recombination losses", submitted to EPJ Photovoltaics, 2024

41st European Photovoltaic Solar Energy Conference and Exhibition

This presentation was selected by the Sc. Committee of the EU PVSEC 2024 for submission of a full paper to one of the EU PVSEC's collaborating peer-reviewed journals.

REVIEW AND HIGHLIGHTS OF MORE THAN 30 YEARS RESEARCH ON EVER IMPROVING TECHNOLOGY FOR PERC SOLAR CELLS AT FRAUNHOFER ISE

E. Lohmüller[1], S. Lohmüller[1], P. Saint-Cast[1], J. Greulich[1], S. Glunz [1,2], R. Preu[1,2]

[1] Fraunhofer Institute for Solar Energy Systems ISE, Heidenhofstrasse 2, 79110 Freiburg, Germany
[2] INATECH, University of Freiburg, Emmy-Noether-Str. 2, 79110 Freiburg, Germany
Phone: +49 761 - 4588 5701; e-mail: elmar.lohmueller@ise.fraunhofer.de

ABSTRACT: The transition of the passivated emitter and rear cell (PERC) from laboratory to industrial production is one of the most important chapters in photovoltaic (PV) technology. PERC-based technologies started to dominate the PV industry in 2018. However, considering the long-time span between the first laboratory prototypes in 1984, the introduction in industrial pilot lines in 2011, and the subsequent successful implementation in mass production since 2012, this development is also an excellent lesson of the challenges that must be overcome in the development of efficient industrial PV technology. Fraunhofer ISE has advanced this path over the past decades and has provided important technological and scientific impulses. This paper gives an overview about milestones in the more than 30 years lasting research activities on PERC at Fraunhofer ISE. With a conversion efficiency progress of +0.4%$_{abs}$ per year achieved for PERC devices since 2007, Fraunhofer ISE's PV-TEC pilot-line is close to the progress reported from industrial pilot lines. We present insights into our latest process optimizations for PERC devices. Our champion power conversion efficiency of 23.4% is achieved on monofacial M2-format gallium-doped Cz-Si PERC solar cells with homogeneous emitter and front silver contacts featuring core finger widths of only 14 μm.
Keywords: PERC, milestones, efficiency improvement, screen printing, Cz-Si

1 INTRODUCTION

In 1989, the concept of passivated emitter and rear cells (PERC) was introduced by Blakers et al. [1].

It then took more than 20 years, until 2011, before PERC solar cells were introduced in industrial pilot lines by solar cell manufacturers like, e.g., Q CELLS [2–4], SolarWorld [5], SCHOTT Solar [6], or Bosch Solar [7].

Only one year later, as of 2012, volume production of monofacial PERC devices started.

With glass-glass module production in place, SolarWorld was the first to implement bifacial PERC modules in record time initiated by collaboration with Fraunhofer ISE in early 2014 [8].

Almost 30 years after its introduction, the PERC technology took over the majority of global photovoltaic (PV) solar cell production in 2018 [9].

Fraunhofer ISE has been researching this cell technology almost from the beginning and can now look back on a PERC era spanning more than 30 years. Beginning in 1989 with small area PERC solar cells fabricated in clean room environment [10], the PERC concept was transferred to industrial cell formats in the course of the opening of the Photovoltaic Technology Evaluation Center (PV-TEC) pilot-line in 2006 [11]. Since then, large-area PERC devices have been a steady companion in everyday laboratory life at the PV-TEC.

As an example, in 2011, Fraunhofer ISE could surpass the "20% efficiency barrier" for large wafer formats as one of the first and achieved 20.1% energy conversion efficiency for a 149 cm² metal wrap through boron-doped Czochralski-grown silicon (Cz-Si) PERC solar cell metallized by dispensing on the front side [12].

This paper gives a brief overview on the most important milestones regarding PERC research at Fraunhofer ISE and shares close-up insight into the latest p-type silicon PERC technology that has been developed with industrial focus. A technical cell structure for a monofacial PERC solar cell is exemplified in Figure 1.

Detailed reviews on the PERC technology can be found in the literature, e.g., in [8,13–16].

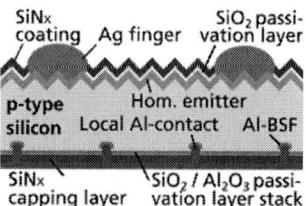

Figure 1: Schematic cross section of a monofacial PERC solar cell with homogeneous emitter.

2 PERC MILESTONES AT FRAUNHOFER ISE

2.1 Progress in energy conversion efficiency

The performance of the PERC development in the Fraunhofer ISE PV-TEC pilot line since 2007 is illustrated in the graph shown in Figure 2.

With a mean energy conversion efficiency progress of +0.4%$_{abs}$ per year (from 2007 to 2023), the PV-TEC pilot-line is close to the value reported for, e.g., the (pilot) line of Q CELLS with +0.5%$_{abs}$ per year (from 2014 to 2021) [17].

2.2 Overview on milestones

Table I gives an excerpt on important milestones regarding PERC research at and/or by Fraunhofer ISE.

PERC became industrially viable, when the front side recombination could be sufficiently reduced by, e.g., narrower screen-printed front side contacts, improved silver pastes, improved emitter doping optionally combined with selective emitters, etc. As example for those developments, the reduction of screen-printed finger width over time is reported in [8,18,19], which supported the benefit of a low recombination active rear side.

Further important roles have played bifacial modules and half cells and, above all, the cost reduction/quality improvement of the mono-crystalline silicon wafers through diamond wire sawing, crystallization improvements, and gallium doping as well as an improved understanding of LID and its mitigation.

Fraunhofer ISE was involved in almost all these developments.

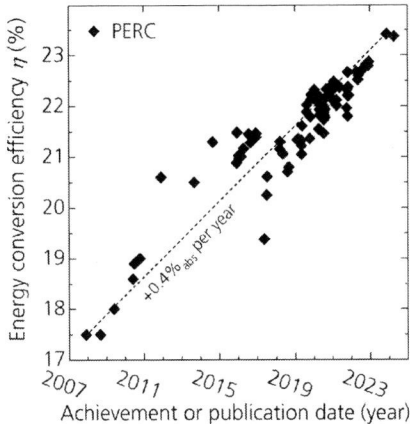

Figure 2: Development of maximum energy conversion efficiency within the PV-TEC pilot-line for PERC solar cells (partly determined at CalLab PV Cells). The efficiency progress is +0.4%$_{abs}$ per year.

Table I: Excerpt on PERC milestones at/by Fraunhofer ISE.

Year	Description
1988	Introduction of clean room-based research activities. Originally, they had been directed to work on the so-called local back surface field (LBSF) solar cell process [20].
1989	Early use of aluminum as local dopant. First solar cells with local Al-BSF yielded open-circuit voltages of 660 mV [10].
1993	A 35-step boron-LBSF (B-LBSF) process was implemented [21].
1994	Implementation of a front random pyramid texture (standard since then until today) and omitting the B-LBSF [22]. This solar cell type, denoted as RP-PERC, had to be paid by an efficiency reduction to 21.6%, but helped to omit photo-lithographical (PL) structuring steps.
1997	Optimization of this B-LBSF process led to an efficiency of 23.3% on float-zone silicon (FZ-Si) and 22.0% on high-resistivity Cz-Si [23]. With single layer and double layer ARCs, a $V_{OC} = 700$ mV and a short-circuit current density $j_{SC} = 42.0$ mA/cm² were reached, respectively.
1999	Ga-doped Cz-Si proved to be free of light-induced degradation (LID) [24].
2000	Substantial attention gained the introduction of the laser ablation of the dielectric [25] and the laser fired contact (LFC) [26,27] processes on the rear side, which form disc-like local contact structures. Both approaches substituted the complex PL-structuring process and have then later been implemented in mass production.
2001	For front SiO₂ passivation, an efficiency up to 21.3% was demonstrated [27].
2003	LFC successfully applied to fabricate high-efficiency ultra-thin solar cells with an efficiency of 20.1% for a wafer thickness of 37 μm [28].

2004 Important milestone for multicrystalline silicon solar cells: demonstration of an efficiency of 20.3% (later corrected to 20.4%) for a plasma-textured multicrystalline silicon solar cell with LFC, a world record for multicrystalline silicon cells that lasted 10 years [29].

The combination of screen-printed front contacts with laser-fired local rear contacts through a thick thermal SiO2 passivation layer was first demonstrated [30].

2006 PERC processing was transferred to the PV-TEC pilot-line, incorporating latest technologies [11].

Proof-of-concept of the FoilMet approach, in which an aluminum foil is used to realize the rear side contact [31,32].

2009 Demonstration of high-throughput inline PECVD of Al2O3 [33] being implemented in a pilot line tool by tool manufacturer Roth&Rau [34]. PECVD Al2O3 became then (one) standard for PERC rear side passivation in mass production.

Laser doping from phosphosilicate glass after POCl3 diffusion offers the possibility to fabricate a selective emitter in a single additional processing step [35]. The implementation of a selective emitter can improve the energy conversion efficiency by decoupling the requirements of the emitter for light conversion and metallization.

2010 Introduction of the thermal oxide passivated all sides (TOPAS) approach, which became the industry standard in PERC surface passivation [36,37].

2011 Surpass of the "20% efficiency barrier" for large wafer format (149 cm²) using screen printing or dispensing for the font side finger metallization on a selective emitter [12].

2014 First solar cell with LFC FoilMet metallization with an efficiency larger than 21% and fully automatic integration of the FoilMet technology into an industrial tool [38].

2016 The introduction of bifacial PERC cells and modules was generally supporting the successful transition to PERC [39].

2017 Investigation of the impact of in-situ oxidation, second deposition and further parameters on the phosphosilicate glass (PSG)/silicon dioxide (SiO₂) stack layer properties grown on the silicon surface during POCl3 diffusion [40].

Development of an ultrafast regeneration (UFR) process to prevent up to 98% of the LID effect in less than 4 seconds process-time [41].

Re-establishing of a PERC baseline after the fire in the PV-TEC laboratory in February [42].

2018 Development of the passivated edge technology (PET) to address the cutting losses in today's solar cells by a simple, high-throughput post-processing on separated solar cells [43]. This proprietary development for reduction of edge recombination was initially trialed on bifacial PERC shingle solar cells [44].

2022   PERC solar cells fabricated out of 100% recycled silicon (without the addition of commercial ultra-pure silicon) [45].

2023   Champion monofacial PERC solar cell with homogeneous emitter in M2 format (156.75 mm edge length, 210 mm diameter) achieves an energy conversion efficiency of 23.4%; see Chapter 3.

3 RECENT PERC TECHNOLOGY

The latest cell efficiency improvements at the end of 2023 have been achieved thanks to an optimized front side texture and ultra fine line front side metallization, taking advantage of the high-quality homogeneous emitter formation and passivation.

3.1 Process flow

Figure 3 shows our most recent PERC baseline process flow. Pseudo-square p-type gallium-doped Cz-Si wafers with M2 format and a base resistivity of about 0.6 Ωcm serve as starting material. After alkaline texturing, a tube furnace diffusion with phosphorus oxychloride (POCl$_3$) as liquid dopant precursor and in-situ oxidation forms the homogeneous emitter with an emitter sheet resistance $R_{sh} \approx 100$ Ω/sq.

Subsequently, the rear side phosphosilicate glass (PSG)/silicon dioxide (SiO$_2$) layer stack [40] is removed in an inline wet-chemical etching process followed by the alkaline etching of the rear emitter and the PSG etching on the front side in a batch process. A wet-chemical batch cleaning step precedes thermal oxidation, in which a SiO$_2$ layer is grown on the silicon surface.

The rear surface passivation is then formed by plasma-enhanced chemical vapor deposition (PECVD) and consists of a layer stack including aluminum oxide (AlO$_X$), silicon-oxynitride (SiO$_X$N$_Y$), and silicon nitride (SiN$_X$).

The front side is capped by a dual layer anti-reflection coating consisting of PECVD SiN$_X$ and PECVD silicon oxide (SiO$_X$), serving as anti-reflection coating and surface passivation.

An infrared laser process locally ablates the rear layer stack to form the disc-shaped laser contact openings (LCO) with a pitch of 600 µm in a periodic square arrangement.

The front and rear metallization is applied by screen printing using commercially available metal pastes. For the formation of the rear electrode, an aluminum paste is printed on the full area of the laser structured rear dielectric layer stack. The busbarless front silver grid features either (Gr1) a nominal screen opening width of $w_{f,screen} = 20$ µm with 120 fingers or (Gr2) $w_{f,screen} = 15$ µm with 156 fingers. For both groups, a 520/11-knotless mesh is used. The nominal metallization fraction is almost equal for both screens.

Finally, contact formation by inline furnace firing with temperature variation, laser-enhanced contact optimization, and current-voltage testing are performed.

3.2 Results

With the optimizations discussed at the beginning of this chapter and the process flow depicted in Figure 3, we could achieve an energy conversion efficiency $\eta = 23.4\%$ with open-circuit voltage $V_{OC} = 693$ mV, short-circuit current density $j_{SC} = 41.5$ mA/cm², and fill factor $FF = 81.5\%$ for a monofacial PERC solar cell from Gr2 (measurement at CalLab PV Cells on a black chuck: 'grn|grn, nrc'

according to [46]). A photograph of a sister to the champion cell is shown in Figure 4.

The measured current-voltage data for both groups is given in Figure 5. The PERC solar cells from Gr2 yield a mean efficiency benefit of about $\Delta\eta_{mean} = +1.0\%_{abs}$. While the values for j_{SC} and V_{OC} are very similar for both groups, the mean FF in Gr2 is increased by $\Delta FF_{mean} = +3.3\%_{abs}$. This corresponds to a mean series resistance reduction of $\Delta R_{S,mean} = -0.58$ Ωcm (not shown in Figure 5) for Gr2 with the screen with $w_{f,screen} = 15$ µm and the larger finger number.

As can be seen in Figure 6, the silver finger of Gr1 shows mesh marks, while for Gr2, the finger contour is quite straight. For the latter, a core finger width of about 14 µm is found. The benefit in R_S for Gr2 not only originates from the increased finger number, but also from the fact that the silver paste used showed an improved printability for the narrower finger openings for Gr2. As can be seen in the scanning electron microscope (SEM) image depicted in Figure 7, the silver fingers of Gr2 feature a desirable homogeneity.

The SEM image of the cross section of this finger, shown in Figure 8, illustrates the core finger width being about 14 µm with a similar value for the finger height.

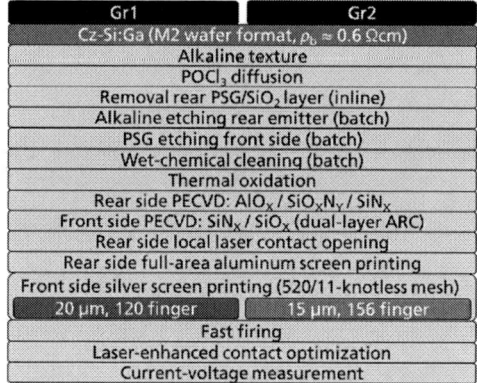

Figure 3: Schematic illustration of the latest PERC baseline process flow with homogeneous emitter at the PV-TEC pilot-line of Fraunhofer ISE as of the end of 2023.

Figure 4: Photograph of one of the PERC solar cells with dual layer anti-reflection coating from Gr2.

41st European Photovoltaic Solar Energy Conference and Exhibition

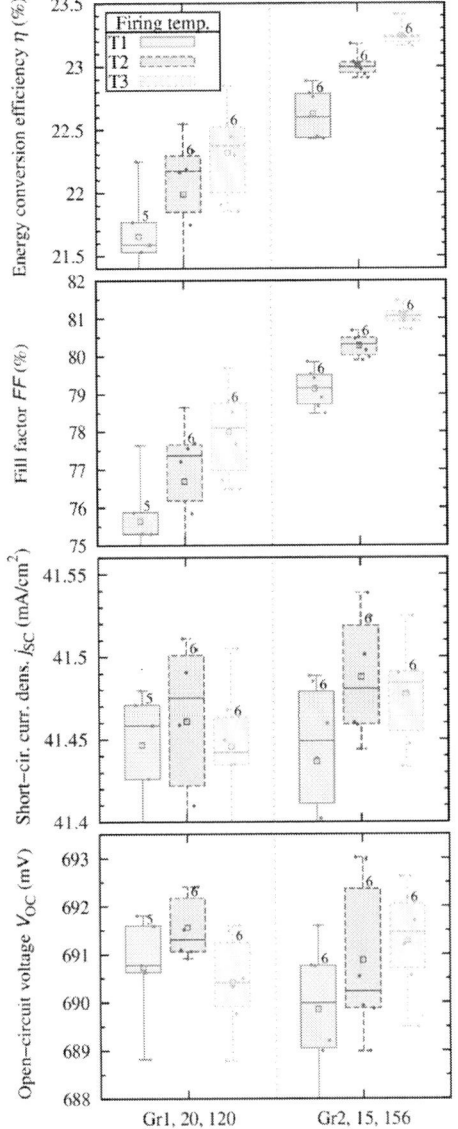

Figure 7: Scanning electron microscope image of a front silver finger from Gr2 at an angle from above.

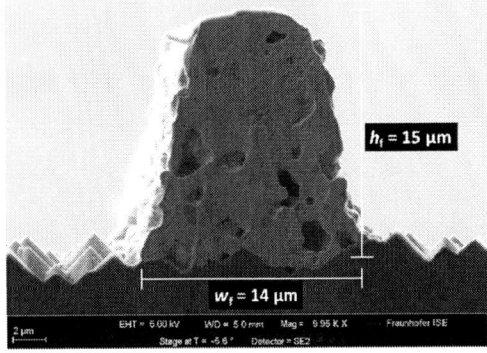

Figure 8: Scanning electron microscope image of a cross section of the front silver finger from Figure 7 (Gr2).

4 SUMMARY AND CONCLUSION

The PERC technology has been a constant topic of research since its introduction 35 years ago. After achieving several milestones, the technology was successfully transferred to mass production 12 years ago.

We review these milestones in the light of our most recent progress and share our insights gathered from more than 30 years PERC experience with the community.

PERC technology is supposed to remain an industrially interesting and important cell architecture for the foreseeable future. Until today, PERC is the most deployed PV technology ever since.

However, the largest capacity increase today occurs with the very related tunnel-oxide passivated contact (TOPCon) technology. Thereby, the TOPCon device is a hybrid of a PERC-like front side and a poly-silicon rear side.

Figure 5: Current-voltage data for the monofacial PERC solar cells fabricated according to Figure 3. The data is corrected with respect to the CalLab PV Cells measurements.

ACKNOWLEDGEMENTS

The authors would like to thank all colleagues and former colleagues at Fraunhofer ISE for their contributions to the PERC development over the last decades. For the reported fine line screen printing results thanks also go out to Asys Automatisierungssysteme GmbH, Murakami Co., Ltd., and ASADA MESH CO., LTD.

This work was funded by the German Federal Ministry for Economic Affairs and Energy within the research project "GutenMorgen" (contract no. 03EE1101A).

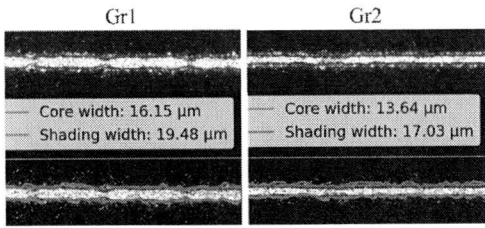

Figure 6: Confocal laser-scanning microscope images of a front silver finger for both groups from Figure 3. The widths are evaluated using the image analysis software "Dash" developed at Fraunhofer ISE [47,48].

REFERENCES

[1] A. W. Blakers, A. Wang, A. M. Milne et al., "22.8% efficient silicon solar cell", *Appl. Phys. Lett.*, vol. 55, no. 13, pp. 1363–1365, 1989.

[2] P. Engelhart, J. Wendt, A. Schulze et al., "R&D pilot line production of multi-crystalline Si solar cells exceeding cell efficiencies of 18%", *Energy Proced.*, vol. 8, pp. 313–317, 2011.

[3] P. Engelhart, D. Manger, B. Klöter et al., "Q.ANTUM – Q-Cells next generation high-power silicon solar cell & module concept", *Proc. 26th EU PVSEC*, Hamburg, Germany, 2011, pp. 821–826.

[4] A. Mohr, P. Engelhart, C. Klenke et al., "20%-efficient rear side passivated solar cells in pilot series designed for conventional module assembling", *Proc. 26th EU PVSEC*, Hamburg, Germany, 2011, pp. 2150–2153.

[5] T. Weber, G. Fischer, A. Oehlke et al., "High volume Pilot Production of High Efficiency PERC Solar Cells-Analysis Based on Device Simulation", *Energy Proced.*, vol. 38, pp. 474–481, 2013.

[6] A. Metz, D. Adler, S. Bagus et al., "Industrial high performance crystalline silicon solar cells and modules based on rear surface passivation technology", *Sol. Energy Mater. Sol. Cells*, vol. 120, pp. 417–425, 2014.

[7] T. Böscke, R. Hellriegel, T. Wüthrich ct al., "Fully screen-printed PERC cells with laser-fired contacts—An industrial cell concept with 19.5% efficiency", *Proc. 37th IEEE PVSC*, Seattle, USA, 2011, pp. 3663–3666.

[8] R. Preu, E. Lohmüller, S. Lohmüller et al., "Passivated emitter and rear cell – Devices, technology, and modelling", *Appl. Phys. Rev.*, vol. 7, no. 4, 041315-1–041315-41, 2020.

[9] ITRPV, 2020, "International Technology Roadmap for Photovoltaic (ITRPV) - Results 2019.

[10] J. Knobloch, A. Aberle, B. Voss, "Cost effective processes for silicon solar cells with high performance", *Proc. 9th EU PVSEC*, 1989, pp. 777–780.

[11] D. Biro, R. Preu, S. W. Glunz et al., "PV-TEC: Photovoltaic technology evaluation center - design and implementation of a production research unit", *Proc. 21st EU PVSEC*, Dresden, Germany, 2006, pp. 621–624.

[12] E. Lohmüller, B. Thaidigsmann, M. Pospischil et al., "20% efficient passivated large-area metal wrap through solar cells on boron-doped Cz silicon", *IEEE Electron Device Lett.*, vol. 32, no. 12, pp. 1719–1721, 2011.

[13] M. A. Green, "The passivated emitter and rear cell (PERC): From conception to mass production", *Sol. Energy Mater. Sol. Cells*, vol. 143, pp. 190–197, 2015.

[14] T. Dullweber, J. Schmidt, "Industrial silicon solar cells applying the passivated emitter and rear cell (PERC) concept—A review", *IEEE J. Photovoltaics*, vol. 6, no. 5, pp. 1366–1381, 2016.

[15] J. Liu, Y. Yao, S. Xiao et al., "Review of status developments of high-efficiency crystalline silicon solar cells", *J. Phys. D: Appl. Phys.*, vol. 51, no. 12, p. 123001, 2018.

[16] A. Blakers, "Development of the PERC solar sell", *IEEE J. Photovoltaics*, vol. 9, no. 3, pp. 629–635, 2019.

[17] A. Mette, S. Hörnlein, F. Stenzel et al., "Q.ANTUM NEO with LECO exceeding 25.5 % cell efficiency", *Sol. Energy Mater. Sol. Cells*, vol. 277, p. 113110, 2024.

[18] A. Lorenz, M. Linse, H. Frintrup et al., "Screen printed thick film metallization of silicon solar cells - recent developments and future perspectives", *Proc. 35th EU PVSEC*, Brussels, Belgium, 2018, 819 - 824.

[19] S. Tepner, A. Lorenz, "Printing technologies for silicon solar cell metallization: A comprehensive review", *Prog Photovolt Res Appl.*, vol. 31, no. 6, pp. 557–590, 2023.

[20] J. Knobloch, A. G. Aberle, W. Warta et al., "Dependence of surface recombination velocities at silicon solar cells surfaces on incident light intensity", *Proc. 8th EU PVSEC*, 1988, pp. 1165–1170.

[21] J. Knobloch, A. Noel, E. Schaffer et al., "High-efficiency solar cells from FZ, Cz and mc silicon material, Louisville, KY, USA, 1993, pp. 271–276.

[22] S. Sterk, J. Knobloch, W. Wettling, "Optimization of the rear contact pattern of high-efficiency silicon solar cells with and without local back surface field", *Prog Photovolt Res Appl.*, vol. 2, no. 1, pp. 19–26, 1994.

[23] S. W. Glunz, J. Knobloch, C. Hebling et al., "The range of high-efficiency silicon solar cells fabricated at Fraunhofer ISE", *Proc. 25th IEEE PVSC*, Anaheim, USA, 1997, pp. 231–234.

[24] S. W. Glunz, S. Rein, J. Knobloch et al., "Comparison of boron- and gallium-doped p-type Czochralski silicon for photovoltaic application", *Prog. Photovolt.: Res. Appl.*, vol. 7, no. 6, pp. 463–469, 1999.

[25] R. Preu, S. W. Glunz, S. Schäfer et al., "Laser ablation—a new low-cost approach for passivated rear contact formation in crystalline silicon solar cell technology", *Proc. 16th EU PVSEC*, Glasgow, UK, 2000, pp. 1181–1184.

[26] R. Preu, E. Schneiderloechner, S. Glunz et al., "Verfahren zur Herstellung eines Halbleiter-Metallkontaktes durch eine dielektrische Schicht", Patent application DE 10 046 170, Germany, 2000.

[27] E. Schneiderlöchner, R. Preu, R. Lüdemann et al., "Laser-fired contacts (LFC)", *Proc. 17th EU PVSEC*, Munich, Germany, 2001, pp. 1303–1306.

[28] D. Kray, H. Kampwerth, A. Leimenstoll et al., "Analysis of very thin high-efficiency solar cells, Osaka, Japan, 2003, pp. 1021–1024.

[29] O. Schultz, S. W. Glunz, G. P. Willeke, "Multicrystalline silicon solar cells exceeding 20% efficiency", *Prog. Photovolt.: Res. Appl.*, vol. 12, no. 7, pp. 553–558, 2004.

[30] E. Schneiderlöchner, G. Emanuel, G. Grupp et al., "Silicon solar cells with screen printed-front contact and dielectrically passivated, laser-fired rear electrode", *Proc. 19th EU PVSEC*, Paris, France, 2004.

[31] J. Nekarda, A. Grohe, O. Schultz-Widmann, "Verfahren zur Metallisierung von Solarzellen und dessen Verwendung", Patent application DE 10 2006 044 936, Germany, 2006.

[32] J.-F. Nekarda, A. Grohe, O. Schultz et al., "Aluminum foil as back side metallization for LFC cells", *Proc. 22nd EU PVSEC*, Milan, Italy, 2007, pp. 1499–1501.

[33] P. Saint-Cast, D. Kania, M. Hofmann et al., "Very low surface recombination velocity on p-type c-Si by high-rate plasma-deposited aluminum oxide", *Appl. Phys. Lett.*, vol. 95, no. 15, 151502-1–151502-3, 2009.

[34] H.-P. Sperlich, D. Decker, P. Saint-Cast et al., "High productive solar cell passivation on Roth&Rau MAiA® MW-PECVD inline machine - a comparison of Al_2O_3, SiO_2 and SiN_X-H process conditions and performance", *Proc. 25th EU PVSEC*, Valencia, Spain, 2010, pp. 1352–1357.

[35] U. Jäger, M. Okanovic, M. Hörteis et al., "Selective emitter by laser doping from phosphosilicate glass", *Proc. 24th EU PVSEC*, Hamburg, Germany, 2009, pp. 1740–1743.

[36] S. Mack, U. Jäger, A. Wolf et al., "Verfahren zur Herstellung einer photovoltaischen Solarzelle", Patent application DE102010024309, Germany, 2010.

[37] S. Mack, U. Jäger, G. Kästner et al., "Towards 19% efficient industrial PERC devices using simultaneous front emitter and rear surface passivation by thermal oxidation", *Proc. 35th IEEE PVSC*, Honolulu, Hawaii, USA, 2010, pp. 34–38.

[38] M. Graf, J. Nekarda, D. Eberlein et al., "Progress in laser-based foil metallization for industrial PERC solar cells", *Proc. 29th EU PVSEC*, Amsterdam, The Netherlands, 2014, pp. 532–535.

[39] K. Krauß, F. Fertig, J. Greulich et al., "biPERC silicon solar cells enabling bifacial applications for industrial solar cells with passivated rear sides", *Phys. Status Solidi A*, vol. 213, no. 1, pp. 68–71, 2016.

[40] S. Werner, S. Mourad, W. Hasan et al., "Structure and composition of phosphosilicate glass systems formed by $POCl_3$ diffusion", *Energy Proced.*, vol. 124, pp. 455–463, 2017.

[41] A. A. Brand, K. Krauß, P. Wild et al., "Ultrafast in-line capable regeneration process for preventing light induced degradation of boron-doped p-type Cz-silicon PERC solar cells", *Proc. 33rd EU PVSEC*, Amsterdam, The Netherlands, 2017, pp. 382–387.

[42] Fraunhofer ISE (press release), 2017, "Fraunhofer ISE Laborbrand", available online: https://www.ise.fraunhofer.de/de/presse-und-medien/news/2017/brand-in-labor-im-solar-info-center.html, accessed on: 09-17-2024.

[43] E. Lohmüller, R. Preu, P. Baliozian et al., "Verfahren zum Auftrennen eines Halbleiterbauelements mit einem pn-Übergang", Patent application DE 10 2018 123 485, Germany, 2018.

[44] P. Baliozian, M. Al-Akash, E. Lohmüller et al., "Postmetallization "passivated edge technology" for separated silicon solar cells", *IEEE J. Photovoltaics*, vol. 10, no. 2, pp. 390–397, 2020.

[45] Fraunhofer ISE (press release), 2022, "PERC solar cells from 100 percent recycled silicon", available online: https://www.csp.fraunhofer.de/en/press/press-releases/perc-solar-cells-from-100-percent-recycled-silicon.html, accessed on: 09-17-2024.

[46] M. Rauer, A. Fell, W. Wöhler et al., "The impact of measurement conditions on solar cell efficiency", *Sol. RRL*, vol. 8, no. 3, p. 2300873, 2023.

[47] T. Strauch, M. Demant, A. Lorenz et al., "Two Image Processing Tools to Analyse Alkaline Texture and Contact Finger Geometry in Microscope Images", *Proc. 29th EU PVSEC*, Amsterdam, The Netherlands, 2014, pp. 1132–1137.

[48] T. Wenzel, A. Lorenz, E. Lohmüller et al., "Progress with screen printed metallization of silicon solar cells - Towards 20 μm line width and 20 mg silver laydown for PERC front side contacts", *Sol. Energy Mater. Sol. Cells*, vol. 244, 111804-1–111804-9, 2022.

41st European Photovoltaic Solar Energy Conference and Exhibition

INVESTIGATING INTERFACIAL PHENOMENA IN COPPER-COVERED, N-TYPE POLYSILICON-BASED CONTACTS BY ELECTRON MICROSCOPY

Reyu Sakakibara[1], Agata Lachowicz[2], Julien Hurni[1], Christophe Allebé[2], Bertrand Paviet-Salomon[2], Franz-Josef Haug[1], Christophe Ballif[1, 2], Aïcha Hessler-Wyser[1], and Audrey Morisset[2]

[1]PV-Lab, Institute of Electrical and Micro Engineering, School of Engineering, École Polytechnique Fédérale de Lausanne, Neuchâtel, Switzerland

[2]PV-Center, Centre Suisse d'Électronique et de Microtechnique, Neuchâtel, Switzerland

Copper is a promising alternative to silver as a metallization material, but because it is a notorious contaminant for silicon, it is important to understand when and how copper can induce the degradation of photovoltaic cells. Here, we prepare intentionally copper-covered, symmetric, n-type samples with polysilicon/silicon oxide passivating contacts to study the impact of annealing at 200-500°C on passivation and contact resistivity. We image cross sections using electron microscopy to correlate changes in functional properties with interfacial phenomena. Copper diffuses into polysilicon as low as room temperature, but passivation only begins to degrade after annealing at 225°C. At 250°C, passivation is fully ruined, which corresponds copper diffusing fully through the polysilicon and forming nanometer-sized spikes at the wafer surface. We also introduce a diffusion barrier layer between the copper and polysilicon, which improves temperature stability to about 375°C. At 400°C, passivation degrades, which appears to correspond not to copper diffusion but instead to morphological changes at the barrier layer / polysilicon interface. This study may help elucidate the conditions under which copper is suitable for TOPCon metallization.

Keywords: TOPCon, Copper metallization

1 INTRODUCTION

Although most crystalline silicon (c-Si) photovoltaic cells manufactured today are metallized with silver, there is a need for an alternative material because silver is both expensive and scarce [1]. While copper is considered a promising alternative, because it is also a notorious contaminant for silicon [2, 3, 4, 5], it is important to understand when and how copper can induce the degradation of cells, especially where fabrication imperfections (dust or plating issues) may cause a diffusion barrier to be absent.

There have been several interactions documented (many of which come from a microelectronics context) that could affect copper-covered silicon layers, such as the rapid diffusion of copper into silicon, even at room temperature [6, 7, 8, 9], interdiffusion [10], potential interactions of copper and silicon oxide / silicon [11, 12], the formation of copper silicides [13, 14, 15] and precipitates [16, 17]. In particular, precipitate formation is reported to be more likely in n-Si [3] than in p-Si and to impact minority carrier lifetime [18, 19, 20]. It is common practice to use diffusion barriers in copper-containing silicon devices [21, 22].

Here, we sputter deposit copper thin films on flat, symmetric, n-type, float zone, crystalline silicon wafers with polysilicon/silicon oxide contacts (like those used at the rear of a TOPCon, or tunnel oxide passivated contact, cell, the now-mainstream silicon single-junction technology [23]) and anneal these at moderate temperatures (200-500°C) in nitrogen in order to observe how their passivation and contact resistivities change, and to study how these changes may be related to copper / polysilicon (poly-Si) / silicon oxide interactions. We use transmission electron microscopy (TEM) and elemental mapping via energy-dispersive spectroscopy (EDX) to image morphological and/or chemical changes in sample cross sections.

In addition, we insert a diffusion barrier layer between the poly-Si and copper to study its effectiveness in preventing passivation loss and its impact on contact resistivity.

2 EXPERIMENTAL METHODS

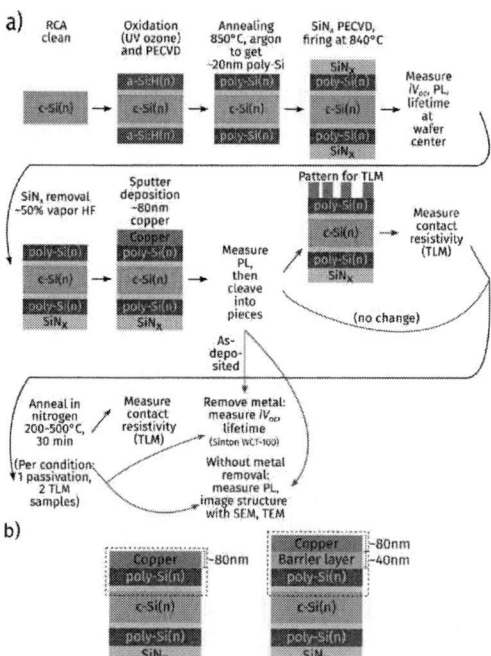

Figure 1: a) We prepared symmetric n-type TOPCon-style samples, then sputtered copper on a single side. We similarly processed the samples with a barrier layer inserted between the copper and poly-Si. Schematic not to scale. b) We prepared two stack types, one with copper directly sputtered onto poly-Si and one with a barrier layer and copper sputtered onto poly-Si. Dotted lines show the regions of the samples (interfaces) we examined with TEM.

Sample preparation (as shown in Figure 1a) began

with the fabrication of flat, symmetric wafer samples with polysilicon / silicon oxide (SiO_x) passivating contacts. After RCA (Radio Corporation of America) cleaning crystalline silicon (c-Si) wafers (4-inch, 200 μm thick, 2 Ω cm, <100>, shiny-etched, float zone), we oxidized both sides by exposing them to UV ozone for 2 min (Jelight UVO-cleaner 42), then deposited phosphorus-doped amorphous silicon (a-Si(n)) using plasma enhanced vapor deposition (PECVD) with silane, hydrogen, phosphine, and methane precursor gases in a parallel-plate reactor with a plasma frequency of 40.6 MHz (KAI-M, Unaxis). We then annealed the wafers in a quartz tube furnace (PEO 603, ATV) in argon to crystallize the amorphous silicon and activate the dopants, with a ramp rate of 10°C / min up to a maximum temperature 850°C (no hold time). This resulted in an ~18 nm thick poly-Si(n) layer. On similarly processed samples, ECV (electrochemical capacitance-voltage) measurements (not shown here) showed a shallow dopant diffusion profile (with a decay around 40 nm into the wafer) and a surface concentration of ~5 × 10^{18} cm^{-3}.

After annealing, we hydrogenated the samples by depositing a layer of hydrogen-rich silicon nitride (SiN_x:H) by PECVD and then firing for a few seconds at 840°C in a ceramic roller furnace (CAMINI, Meyer Burger). At this point, before silicon nitride removal and metal deposition, we measured the passivation through PL (photoluminescence), minority carrier lifetime, and iV_{oc} (implied open-circuit voltage) measurements (the latter two with a Sinton WCT-100) at the wafer center.

Before metal deposition, we removed silicon nitride from one side through a vapor hydrofluoric acid (HF) etch and water rinse. We prepared two types of stacks by sputtering directly onto the full area of the wafer: (1) "Cu only" with ~80 nm of copper, and (2) "CBL+Cu" with ~80 nm of copper on ~40 nm of CBL (copper barrier layer).

After metal deposition, we cleaved the wafers into pieces and separated them into two types of samples: passivation samples and TLM samples. The passivation pieces were at least 2.5 cm on each side.

We measured the passivation "as-deposited" by first removing the copper or copper and CBL by wet etch, then measuring the PL, lifetime, and iV_{oc}.

We measured the contact resistivity (ρ_c) with TLM (transfer length method or transmission line model). We prepared TLM samples by printing a hotmelt pattern of pads, wet etching the copper or copper and CBL, then removing the hotmelt. We first measured all TLM samples as-deposited (no anneal), then remeasured them after annealing.

We did all anneals in a furnace in nitrogen ambient for a 30 min duration each, at temperatures of 200-500°C. The passivation samples were wafer pieces with full-area coverage, in other words the entire surface of the pieces was fully covered by either copper only or barrier layer and copper. After annealing, we (a) measured PL maps, (b) cleaved a small piece to reserve for TEM analysis, (c) removed the metal, and (d) measured iV_{oc} and lifetime. On the other hand, for contact resistivity, as the TLM samples were already prepared post metal deposition, we simply annealed the TLM samples and then remeasured them.

We did TEM cross section imaging on a subset of samples, on the copper /(barrier layer)/ poly-Si(n) / SiO_x / c-Si(n) interfaces. We used a gallium focused ion beam (FIB, Zeiss NVision 40) to prepare lamellae of 80-150 nm thickness using the standard lift-out process. We imaged

the cross sections (FEI Talos F200S) in bright field and high resolution TEM imaging as well as HAADF STEM (high-angle annular dark field scanning transmission electron microscopy) modes. In all cases we left the metal unetched. We also did some SEM top view imaging with the Zeiss NVision 40.

Tables I and II in Section 3 show details about anneal conditions for Cu only and CBL+Cu samples, respectively, as well as indicate the subset of samples for which we did TEM cross section imaging

3 RESULTS AND DISCUSSION

Table I: Samples with copper only: at-a-glance on anneal conditions and results.

Temp.	Cu only			
	iV_{oc} (mV)	τ (ms)	ρ_c (mΩ cm²)	TEM
No anneal	730	4.5-5.3	~30	Fig. 4
200°C	731	4.5	no change	-
225°C	732	4.9	-	Fig. 4
250°C	603	25.8 μs	-	Fig. 5
300°C	569	9.5 μs	>100	Fig. 6

Table II: Samples with both copper and barrier layer: at-a-glance on anneal conditions and results.

Temp.	CBL+Cu			
	iV_{oc} (mV)	τ (ms)	ρ_c (mΩ cm²)	TEM
No anneal	729	4.1	~30	Fig. 8
200°C	732	6.3	no change	-
300°C	734	6.8	~17	Fig. 8
350°C	730	8.2	-	-
375°C	736	6.6	-	-
400°C	694	2.1	-	Fig. 9
450°C	716	2.7	-	-
500°C	642	111 μs	unclear	-

3.1 As-deposited (no anneal)

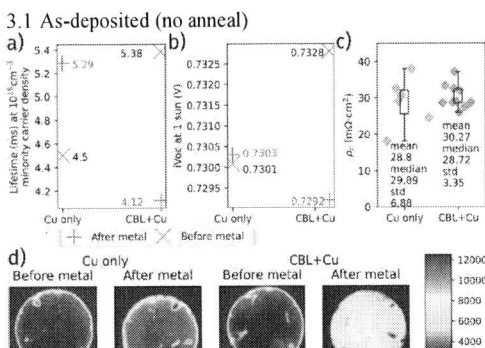

Figure 2: As-deposited results: sputtering the metal (either copper only or barrier layer (CBL) and copper) did not seem to significantly decrease the passivation. a) Lifetime, b) iV_{oc}, c) contact resistivity, d) PL. Note "before metal" indicates both before silicon nitride removal and before metal deposition, and "after metal" is with the metal on the rear side.

After depositing either copper only or barrier layer and

copper, we characterized the as-deposited passivation and contact resistivity, as shown in Figure 2 (note "before metal" indicates both before silicon nitride removal and before metal deposition, and "after metal" is with the metal on the rear side.). Passivation decreased slightly after metal deposition, as indicated by the decrease in PL signal (Figure 2d), but passivation remained reasonable overall, with iV_{oc} values consistent both before and after metal around ~730 mV (Figure 2b). The increase in lifetime for the Cu only sample after metal deposition, as shown in Figure 2b, may be because we measured different parts of the wafer: we measured the wafer center before metal deposition, and a small piece after metal deposition. In addition, the Cu only wafer PL seems higher than that of the CBL+Cu sample in Figure 2d, but this may be because the copper, on the rear side, is more reflective.

In terms of contact resistivity, the values measured for Cu only and CBL+Cu were consistent around 30 mΩ cm², which would be compatible with full-area passivating contact and metallization [24, 25, 26, 27]. This result seems to indicate that inserting the CBL between the poly-Si and copper had little adverse impact on the contact resistivity.

3.2 Anneals: copper only

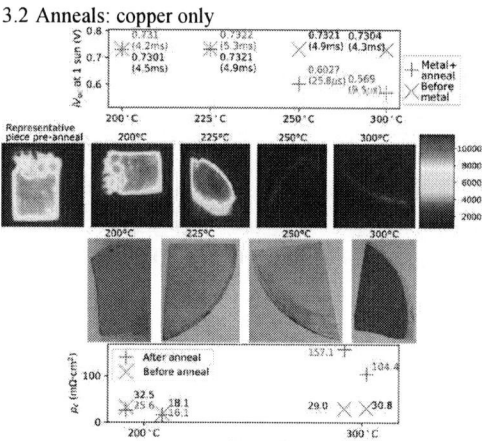

Figure 3: Copper only anneal results. The passivation began to degrade at 225°C (low PL at bottom corners of sample) and was fully ruined by 250°C and 300°C. By eye, however, the 250°C and 300°C looked different. The contact resistivity did not change at 200°C and worsened (increased) at 300°C. From top: first row lifetime / iV_{oc}, second row PL maps, third row photographs, fourth row contact resistivity.

The passivation of the copper only samples withstood annealing at 200°C but began to degrade at 225°C and had fully degraded by 250°C, as shown in Figure 3. The degradation of passivation as a function of anneal temperature is most obvious from the PL images (second row of Figure 3): at 225°C, the bottom left and bottom right corners of the piece showed low PL, and at both 250°C and 300°C there was almost no signal.

We annealed TLM samples at 200°C and 300°C and measured them (Figure 3 fourth row). At 200°C we observed very little change (though the plot seems to show a decrease, the measured raw resistivity data from which we calculated ρ_c appeared to have changed only minimally), while at 300°C the contact resistivity increased by at least an order of magnitude.

In other words, at 300°C, both metrics (passivation

and contact resistivity) worsened; we did not find a condition at which passivation worsened (decreased) but contact resistivity improved (decreased).

Also, while both the 250°C and 300°C samples had very low PL signal, they looked quite different optically (by eye), as shown in the third row. The samples annealed at the three lowest temperatures (200°C, 225°C, and 250°C) looked similar, with a stereotypical "copper-color," but the sample annealed at 300°C had darkened significantly, and only the rim, which had residual silicon nitride (the wafer rim was unetched during the vapor HF etch), remained copper-colored. This appears to be due to different phenomena occurring at the two temperatures, as we will explain in the next subsection.

3.3 TEM cross section imaging of copper only samples

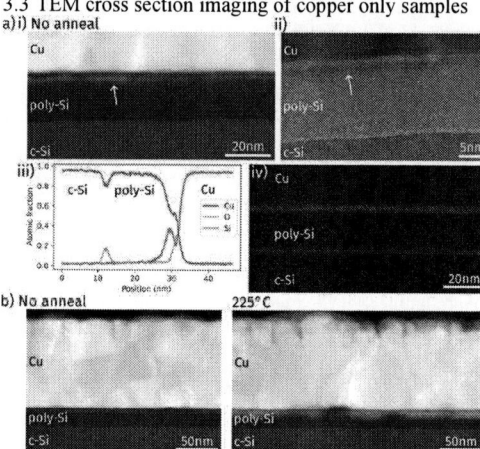

Figure 4: Copper diffused into the poly-Si as early as room temperature (no anneal) through an oxide layer on the poly-Si top surface; continued diffusion into the poly-Si at 225°C is the likely cause of a degradation in passivation. a) No anneal sample: i) an HAADF image and ii) a high resolution TEM image. White arrows indicate a line of copper (white and black respectively) in the poly-Si. iii) EDX profiling and iv) oxygen elemental map show a line of oxygen. b) The cross section HAADF images of the no anneal (room temperature) and 225°C samples are shown side by side: at 225°C, the diffusion front advanced further into the poly-Si, and the top of the copper film became less flat and more uneven.

We show a few different types of TEM cross section images with different types of contrast in Figures 4, 5, and 6, including (i) bright field and high resolution TEM images, where copper appears dark and silicon bright, (ii) HAADF (high angle annular dark field) images, with the opposite contrast where copper appears bright and silicon dark, and (iii) elemental maps based on EDX with false colors.

We observed copper diffusion into the poly-Si layer as low as room temperature (no anneal), as shown in Figure 4ai, 4aii, and left side of 4b, but this did not appear to cause a significant decrease in the passivation. In all of these cross section images a white line (HAADF, 4ai or left side of 4c) or black line (high resolution TEM, 4aii) corresponding to copper is visible in the poly-Si.

The line of copper appeared separately from the bulk copper thin film (rather than being continuous) because, it seems, this line of copper first diffused through an oxide present on the surface of the poly-Si. That oxide is clearly visible in the oxygen elemental map (4aiv). We have observed a similar parasitic oxide on poly-Si in previous experiments [28]. This oxide could potentially contribute to relatively high contact resistivity values. In addition, EDX profiling (4aiii) shows two distinct oxygen peaks (cyan color), one corresponding to the tunnel oxide (left) and the other to the parasitic oxide on the poly-Si. A copper peak (magenta) appears to the left of the parasitic oxide peak, suggesting copper diffusion into the poly-Si layer through the parasitic oxide.

The right side of Figure 4b shows a cross section of the 225°C sample in the area corresponding to low PL. Here we show side by side the no anneal and 225°C sample cross sections to demonstrate that the copper diffused further into the poly-Si at 225°C, which likely explains the reduction in PL signal.

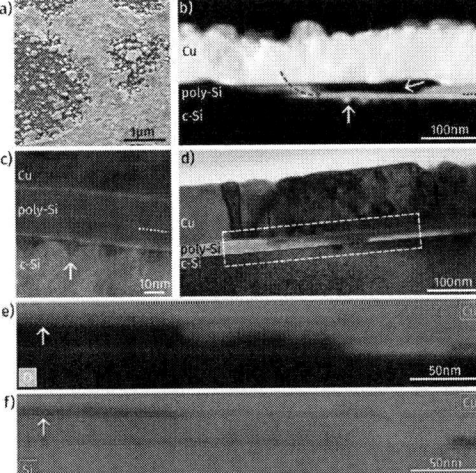

Figure 5: Several phenomena can be observed in the sample annealed at 250°C: full (sometimes lateral) diffusion through the poly-Si, spike formation, an uneven top surface. a) Top view SEM image. b) HAADF cross section image c) High resolution image. White arrows in b and c indicate spikes. Red dashed arrow shows potential lateral diffusion; yellow area indicates dark area, above the spikes, without copper signal. d) Bright field image of a different area; white box indicates region of interest whose elemental maps are shown in subfigures e) (Cu and O) and f) (Cu and Si). Arrows point to areas where copper film might have delaminated and that are oxygen-rich. Note: no spikes were visible by EDX in e and f.

We observed multiple phenomena, one of which is copper diffusion, in the 250°C sample for which the passivation is completely ruined. As might be expected, the copper diffused fully through the poly-Si layer (Figure 5b, d, e, f), though only in parts of the sample (meaning not uniformly). Some of this diffusion seems to have occurred laterally; Figure 5b shows an area where the part of the poly-Si layer above the spikes does not show copper signal (indicated by the yellow arrow). The red dashed arrow shows potential lateral diffusion. Also, the EDX signal intensity in the copper elemental maps seems to reflect the composition of the poly-Si. The elemental maps seem to show two sublayers in the poly-Si with slightly different EDX signal intensities. This corresponds to the two different sublayers in the poly-Si, one more crystalline and the other more amorphous; these results suggest a

difference in copper interaction between amorphous and polycrystalline silicon. We have added small dotted lines in Figure 5b and c, as a visual guide to help delineate these two sublayers.

In addition to copper diffusion, small spikes, ~10 nm in size, formed at the wafer surface. The spikes are visible in areas where copper reached the wafer surface. Some are indicated by white arrows in Figure 5b and c. They appear to be composed mainly of copper, given that they appear white in 5b and dark in 5c and there is copper signal in EDX elemental maps (not shown).

Also, the top surface was uneven (Figure 5a). This roughness is also visible from the top surface of the copper in the cross section image of 5b.

Finally, we have done elemental mapping on an area of the sample that shows full and partial copper diffusion, as well as what might be copper film delamination (white dotted box in Figure 5d; it is a different area from the one depicted in b). Elemental maps of Cu and O together are represented in 5e and Cu and Si together in 5f. In the maps, there are two areas to the right and left sides that have no copper signal and especially high oxygen and low silicon signals compared to poly-Si and c-Si. They appear cyan in 5e and dark red in 5f. The area on the right side is near the poly-Si / c-Si interface; it is not clear whether the high oxygen signal is at all related to copper diffusion (whether it prevents or precedes diffusion). The area on the left side is close to the poly-Si / copper interface; there, it looks almost as though the copper delaminated from the poly-Si, though this also is unclear.

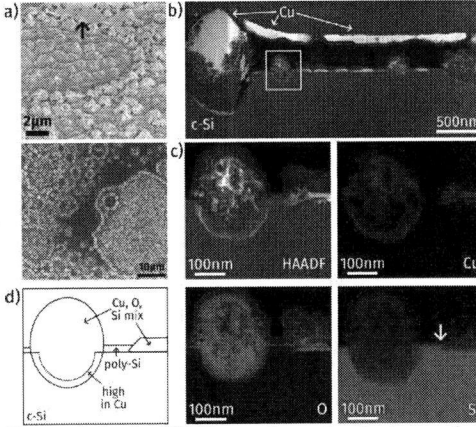

Figure 6: In the 300°C anneal sample, the top surface shows spike-shaped particles and what appear to be suspended films of copper; cross section imaging shows that discrete, continuous thin films have disappeared and particles have become embedded in the wafer surface a) Top view images, top taken at 54° tilt and bottom at 0°. b) HAADF cross section image. c) Elemental maps (Cu, Si, O) of particle indicated by white square in b). d) Schematic of particle shown in c).

After annealing at 300°C, both the top view and cross section images look significantly different from those of the sample annealed at 250°C. Although the 250°C sample also had an uneven top surface, the impression given by the 250°C top view image (Figure 5a) is of a copper film from which pieces had partially flaked off. On the other hand, the top view images at 300°C (Figure 6a, top image taken at 54° tilt) show what appear to be micron-sized, spike-shaped particles, on some of which appear to be thin films suspended, as indicated by the black arrow. The bottom image of Figure 6a, taken at 0° tilt, shows the particles over a large area.

Cross section imaging (Figure 6b) of the 300°C samples showed that the discrete, continuous thin films of copper and poly-Si had disappeared. Instead, particles of different sizes were embedded in the wafer, and a discontinuous film of copper (the white lines towards the top of 6b) suspended over the largest particles. The suspended copper film is likely similar in nature to the one we observed in top view imaging.

We did elemental mapping on one of the smaller particles indicated by a box in Figure 6b; the elemental maps of copper, silicon, and oxygen are shown in 6c. The particle seemed to be composed of an inhomogeneous mix of copper, oxygen, and silicon. In particular, the bottom of edge of the particle appeared bright and therefore high in copper content. Some of the other particles in 6b also showed a bright bottom edge high in copper (leftmost particle, for example). Also, while the edges of the spikes at 250°C were straight and appeared to follow the crystalline planes of the wafer, here the shapes of the particles seemed to not have any straight edges.

The silicon elemental map (Figure 6c bottom right) shows what could be a trace of the poly-Si layer. In this image, the white arrow indicates a step in silicon signal intensity; to the right of the arrow there is a drop in silicon intensity. The area to the left of the arrow could correspond to an area with poly-Si remaining, as shown in the schematic in Figure 6d.

The disappearance of discrete layers and formation of embedded particles likely explain the change in appearance of the annealed sample as shown in Figure 3.

Given that the anneals were done in an inert nitrogen ambient, it was unexpected to see oxygen accumulation and potentially oxygen-mediated interactions of copper and silicon. We do not consider it likely that the source of oxygen was within the sample itself. It has been reported [29, 30] that argon is a better inerting gas than nitrogen because it is denser than air, and therefore more effective at displacing air (and water vapor and oxygen) than nitrogen and more protective. So it is possible that nitrogen ambient still allowed oxygen to come into contact with the samples.

3.4 Anneals: barrier layer and copper

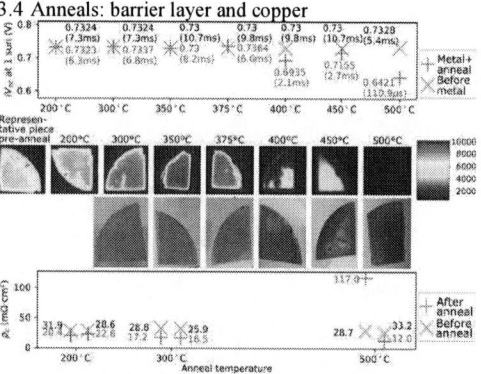

Figure 6: Anneal results for samples with copper and barrier layer. The passivation begins to degrade between 375°C and 400°C; contact resistivity improves (decreases) at 300°C. From top: first row lifetime / iV_{oc}, second row PL maps, third row photographs, fourth row contact resistivity.

The barrier layer increased the temperature stability of the stack to above 375°C (an increase of ~150°C), as shown in Figure 7. In particular, PL signal increased at 350°C and 375°C, which could potentially correspond to hydrogen diffusing from the silicon nitride on the rear side of the samples.

Also, contact resistivity improved after annealing at 300°C, from ~30 mΩ cm^2 to ~17 mΩ cm^2. After annealing at 400°C and 450°C, the passivation decreased, though inhomogeneously: in some areas of the sample the passivation significantly decreased, while in other areas there was only a partial loss, as can be seen in the PL images of Figure 6. However, the pattern of the passivation loss did not match the patterns visible by eye on the surface. At 500°C, the passivation of the piece was completely ruined, and there was a color change.

3.5 TEM cross section imaging of samples with copper and barrier layer

We did TEM cross section imaging on the three following samples: no anneal (room temperature), 300°C where the contact resistivity improved, and 400°C in an area with very low PL signal. All cross section images are shown in Figures 8 and 9 with false colors, where copper is magenta, the barrier layer "CBL" lime green, silicon red, and oxygen cyan (but oxygen is not always very visible).

The sample without annealing (Figure 8a) had discrete, continuous, and flat layers of copper, barrier layer, and silicon. The tunneling oxide SiO$_x$ layer was visible as a line with low silicon intensity.

After annealing at 300°C, the sample looked unchanged (TLM sample, Figure 8b); TEM analysis was not sufficient to explain the improvement in contact resistivity. While the top surface of the copper film looked perhaps slightly less flat compared to 8a, the rest of the layers looked undisrupted and unchanged: flat, discrete, and continuous. (While the intensity of the colors in Figure 8a is higher than that of 8b, this can be attributed to slight differences in lamellae thickness and acquisition time). Nor did we see any copper diffusion through the barrier layer and into the poly-Si layer. There was practically no signal for copper in the silicon and CBL layers.

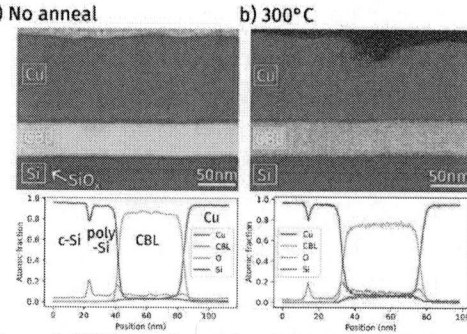

Figure 8: While we observed an improvement in contact resistivity in the samples with copper and barrier layer (CBL+Cu) after annealing at 300°C, TEM imaging did not show a difference that could explain this improvement. a) No anneal sample with HAADF image and EDX profiling b) 300°C sample with HAADF image and EDX profiling.

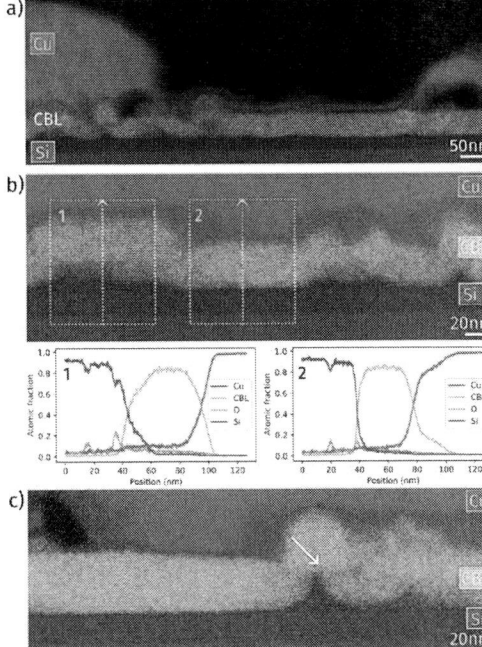

Figure 9: We observed no obvious copper diffusion in the sample annealed at 400°C but instead changes at the CBL / poly-Si interface. a) Elemental maps. b) Elemental maps with boxes indicating two areas of profiling, and their corresponding profiles. c) High magnification elemental map with white arrow indicating potential CBL delamination and silicon displacement (possibly diffusion).

When we imaged the low PL area of the 400°C sample (Figure 9), we did not observe any obvious copper diffusion but rather changes in the morphology of the poly-Si and CBL interface. One of the most prominent changes was that the copper film had "balled up," becoming thin in some places and thick in others (Figure 9a). Also, voids or empty pockets formed at the copper and CBL interface. We did not observe any "islands" of copper; where we imaged, the copper film appeared continuous, even if the thickness varied significantly.

The CBL and poly-Si films in the 400°C sample were also uneven and continuous, but unlike the copper, we did not see dramatic variations in the thickness. In fact, given that the CBL layer thickness remained roughly the same, the overall impression is that the CBL delaminated from the poly-Si, forming a groove, and some silicon diffused into the areas where CBL delaminated. Figure 9c shows a close-up of an area (see white arrow) where CBL (green) appeared to have delaminated, and in the space left behind by the CBL is a high silicon signal (red).

This behavior of silicon displacement has some similarity with the interactions of aluminum and poly-Si investigated previously [31]. This work involved fired stacks of aluminum / silicon nitride / poly-Si(p) / c-Si(p), where, after firing at 580°C, it seemed that both aluminum had diffused into the poly-Si and silicon had diffused into the original aluminum layer. Cross section images showed the original layer of aluminum full of pockets of silicon, and vice versa. Unlike this study, however, one difference with our copper samples is that neither Cu nor CBL appeared to have diffused into the poly-Si layer; all

displacement / diffusion was away from, and not towards, the poly-Si.

We also considered that it was unlikely a reaction between CBL and silicon had occurred, given that in our elemental maps and profiling (Figure 9b), we did not see any areas where there was simultaneously silicon and CBL signal. (Profile 1 does not show a sharp transition between CBL and silicon because the interface itself is not a sharp, horizontal line.)

Profiling also does not show high copper signal in the poly-Si or c-Si layers, although this does not mean that there was no copper diffusion into the silicon.

4 CONCLUSIONS

We compared the passivation and contact resistivity performance of poly-Si(n)/SiO$_x$ stacks with copper only or copper and a barrier layer as a function of anneal temperature, and used primarily TEM cross section imaging to investigate interfacial phenomena.

We have shown several interactions of sputtered copper with a poly-Si(n)/SiO$_x$ stack. The dominant mechanism for passivation degradation appears to be copper diffusion through the poly-Si(n) layer to the wafer, as per the conventional wisdom, although other phenomena include spike formation, oxygen accumulation, and the formation of particles made up of oxygen, silicon, and copper. We did not see any improvement in contact resistivity with annealing.

We have also demonstrated an improvement in both passivation stability and contact resistivity by inserting a barrier layer. Passivation stability improved by about 150°C, and the contact resistivity improved (decreased) at 300°C. The barrier layer seemed to effectively reduce copper diffusion, although TEM cross section imaging was not adequate to fully reveal the barrier layer's failure mode or the mechanism for improvement in contact resistivity, or to completely rule out copper diffusion.

5 ACKNOWLEDGMENTS

We would like to thank our PV-Lab colleagues Xavier Niquille for wet cleaning the wafers and Joël Spitznagel for sample fabrication support, as well as Dr. Lucie Navratilova and Dr. David Reyes Vasquez of the Centre Interdisciplinaire de Microscopie Électronique at EPFL for assistance with FIB lamellae preparation and TEM imaging and interpretation, respectively.

Project COMET is supported under the umbrella of SOLAR-ERA.NET Cofund 2 Additional Joint Call N° 018 by the Swiss Office of Energy (Grant contract SI/502483-01). SOLAR-ERA.NET is supported by the European Commission within the EU Framework Programme for Research and Innovation HORIZON 2020 (Cofund ERA-NET Action, Grant Agreement N° 786483).

6 REFERENCES

[1] B. Hallam, M. Kim, Y. Zhang, L. Wang, A. Lennon, P. Verlinden, P.P. Altermatt, and P.R. Dias. Prog Photovolt Res. Appl. 31 (2023) 598–606.

[2] J.R. Davis, A. Rohatgi, R.H. Hopkins, P.D. Blais, P. Rai-Choudhury, J.R. McCormick, and H.C. Mollenkopf. IEEE Trans. Electron Devices, 27 (1980)

677–687.

[3] A.A. Istratov and E.R. Weber. J. Electrochem. Soc., 149 (2002) G21.

[4] J. Bartsch, A. Mondon, K. Bayer, C. Schetter, M. Hörteis, and S.W. Glunz. J. Electrochem. Soc., 157 (2010) H942–H946.

[5] G. Coletti, P.C.P. Bronsveld, G. Hahn, W. Warta, D. Macdonald, B. Ceccaroli, K. Wambach, N. Le Quang, and J.M. Fernandez. Adv. Funct. Mater. 21 (2011) 879–890.

[6] T. Heiser and A. Mesli. Appl. Phys. A., 57 (1993) 325–328.

[7] A.A. Istratov, C. Flink, H. Hieslmair, E.R. Weber, and T. Heiser. Phys. Rev. Lett., 81 (1998) 1243.

[8] A.A. Istratov, C. Flink, H. Hieslmair, S.A. McHugo, and E.R. Weber. Materials Science and Engineering B, 72 (2000) 99–104.

[9] E.R. Weber. Technical Report NREL/SR-520-31528 2002, NREL, 2002.

[10] S.H. Corn, J.L. Falconer, and A.W. Czanderna. J. Vac. Sci. Technol. A (1988) 6, 1012–1016.

[11] H. Wendt, H. Cerva, V. Lehmann, and W. Pamler. J. Appl. Phys., 65 (1989) 2402–2405.

[12] A. Correia, D. Ballutaud, and J.-L. Maurice. Jpn. J. Appl. Phys., 33 (1994) 12217–1222.

[13] M. Setton, J. Van der Spiegel, and B. Rothman. Appl. Phys. Lett., 57 (1990) 357–359.

[14] M. Seibt, M. Griess, A.A. Istratov, Λ. Hedemann, A. Sattler, and W. Schröter. Phys. Stat. Sol. A, 166 (1998) 171.

[15] B.O. Kolbesen and H. Cerva. Phys. Stat. Sol. B, 222 (2000) 303.

[16] C. Flink, H. Feick, S.A. McHugo, A. Mohammed, W. Seifert, H. Hieslmair, T. Heiser, A.A. Istratov, and E.R. Weber. Physica B, 273–274 (1999) 437–440.

[17] V. Kolkovsky and K. Lukat. Microelectron. Reliab., 53 (2013) 1342–1345.

[18] W.B. Henley, D.A. Ramappa, and L. Jastrezbski. Appl. Phys. Lett., 74 (1999) 278–280.

[19] K. Stewart, A. Cuevas, D. Macdonald, and J. Williams. Workshop on Crystalline Silicon Solar Cell Materials and Processes (2001) 212–215.

[20] G. Gaspar, C. Modanese, H. Schøn, M. Di Sabatino, L. Arnberg, and E.J. Øvrelid. 5th International Conference on Silicon Photovoltaics, SiliconPV 2015 77 (2015) 586–591.

[21] S.-Q. Wang. MRS Bulletin, 19 (1994) 30–40.

[22] A.E. Kaloyeros and E. Eisenbraun. Annu. Rev. Mater. Sci., 30 (2000) 363–85.

[23] International Technology Roadmap for Photovoltaics (ITRPV): Results 2023, Fifteenth Edition. Technical report, VDMA, 2024.

[24] M.A. Green. Solar Cells, 7 (1982) 337–340.

[25] D.K. Schröder and D.L. Meier. IEEE Trans. Electron Devices, 31 (1984) 637–647.

[26] R. Brendel and R. Peibst. IEEE J. Photovolt., 6 (2016) 1413–1420.

[27] T.G. Allen, J. Bullock, X. Yang, A. Javey, and S. De Wolf. Nature Energy, 4 (2019) 914–928, 2019.

[28] S. Libraro, L.J. Bannenberg, T. Famprikis, D. Reyes, J. Hurni, E. Genç, C. Ballif, A. Hessler-Wyser, F.-J. Haug, and Audrey Morisset. ACS Appl. Mater. Interfaces, 16 (2024) 47931–47943.

[29] H.-J. Reinhardt and H.-R. Himmen. Inerting in the chemical industry. Technical report, Hydrocarbon Processing, 2010.

[30] Mary O'Kane. Argon vs Nitrogen for Glove Boxes.

https://www.ossila.com/pages/argon-nitrogen-glove-box.

[31] S. Libraro, M. Lehmann, J.J. Diaz Leon, C. Allebé, A. Descoeudres, A. Ingenito, C. Ballif, A. Hessler-Wyser, and F.-J. Haug. Sol. Energy Mater. Sol. Cells, 249 (2023) 112051.

41st European Photovoltaic Solar Energy Conference and Exhibition

**ROBUSTNESS OF ELECTRICAL QUALITY OF ION IMPLANTED BLACK SILICON EMITTERS:
COMPARISON BETWEEN DIFFERENT ION IMPLANTATION SERVICE PROVIDERS**

Olga Morozova[1*], Kexun Chen[1], Behrad Radfar[1], Ulrich Kentsch[2], Luke Antwis[3], Hele Savin[1], Ville Vähänissi[1]

[1]Department of Electronics and Nanoengineering, Aalto University, Tietotie 3, 02150, Espoo, Finland
[2]Institute of Ion Beam Physics and Materials Research, Helmholtz-Zentrum Dresden-Rossendorf, Bautzner Landstrasse
400, 01328 Dresden, Germany
[3]Surrey Ion Beam Centre, University of Surrey, Guildford GU2 7XH, United Kingdom

*Corresponding author, electronic mail: olga.morozova@aalto.fi, phone +358456478459
Electronic mails of other authors: chen.kexun22@gmail.com, behrad.radfar@aalto.fi, u.kentsch@hzdr.de,
luke.antwis@surrey.ac.uk, hele.savin@aalto.fi, ville.vahanissi@aalto.fi

ABSTRACT: Ion implantation provides precise control over the resulting dopant atom density, enabling high-quality
optical and electrical performance of nanostructured (black silicon, b-Si) emitters. In this work, we study how
sensitive the performance of Al_2O_3-passivated b-Si emitters is to small variations in the implantation conditions and
the equipment used to perform it. We carried out boron emitter implantations for identical nanostructured and planar
wafers at four different implantation service providers and with both beam line and parallel beam tool configurations.
We then benchmarked the results against the earlier optimised b-Si emitter process. The results show that there are
some differences in obtained sheet resistance and emitter saturation current depending on the service provider and the
used tool configuration. Finally, there are also some deviations in terms of possible bulk contamination among the
different service providers.

Keywords: ion implantation, black silicon, emitter, emitter saturation current

1 INTRODUCTION

Silicon nanostructures, also known as black silicon
(b-Si), can effectively eliminate reflectance losses in
solar cells [1]. In Passivated Emitter and Rear Cell
(PERC) and Tunnel Oxide Passivated Contact
(TOPCON) cell (or any front contact cell type), emitter
formation presents an additional challenge, as it needs to
be accomplished on the nanostructured surface. Due to
the extensive surface area of b-Si, controlling the dopant
profile through a conventional diffusion process is
difficult. Consequently, dopant diffusion on b-Si often
results in significant Auger and Shockley-Read-Hall
(SRH) recombination within the nanostructure [2-4]. This
issue has recently been addressed through a more
controlled doping method—ion implantation—which has
achieved emitter saturation currents of less than
20 fA/cm² and external quantum efficiencies (EQEs)
exceeding 100% in the UV [5]. Studies on ion-implanted
b-Si emitters have focused on optimizing implantation
parameters by testing different doses, energies, and
annealing conditions to achieve optimal emitter
performance [6]. However, there is no data on how
sensitive the performance of b-Si emitters is to minor
variations in implantation conditions and the
equipment/technology used. This study aims to
investigate the robustness of the performance of
implanted b-Si emitters against potential small variations
in implantation conditions. Specifically, it involves
comparing the performance of b-Si and planar emitters
processed using identical ion implantation parameters
across four different service providers and two different
tool configurations.

2 EXPERIMENTAL

2.1 Samples

The substrates used in this work were phosphorus-
doped 4-inch float-zone (FZ) (100) oriented Si wafers
with a resistivity of 3 Ωcm and a shiny etch surface
finish. First, the wafers were cleaned with standard RCA
solutions and HF dip, followed by dry oxidation at
1050°C for 30 minutes to remove any possible bulk
defects present in as-grown FZ material [7]. The resulting
54 nm silicon oxide film was then removed, i.e., no
screen oxide was used during the following implantation.

The next processing step was nanostructure formation
on both sides of the samples. This was achieved using
inductively coupled plasma reactive ion etching (ICP-
RIE) with SF_6/O_2 plasma for 10 minutes at a pressure of
5 mTorr and a temperature of -120°C, similar to the
method reported by Repo et al.[8].

Surface texturization was followed by emitter
formation through boron (p⁺) implantation. In earlier
studies, the implantation parameters (energy, dose and
drive-in anneal) have already been optimized for b-Si
surface [6] for beamline implantation. In this study, we
use these optimized parameters and present the earlier
data obtained during the optimization as a reference.
Thus, the implantation energy used was 10 keV, and the
dose was 3×10^{15} cm⁻². Four different ion implantation
providers were chosen for this test, namely *i*) Institute of
Ion Beam Physics and Materials Research at HZDR in
Dresden, Germany, *ii*) Institute of Microelectronics of
Barcelona (IMB-CNM) in Barcelona, Spain, *iii*) Ion
Beam Services in Peynier, France and *iv*) Surrey Ion
Beam Centre in Gilford, UK. The implantations are
labelled as A-G in randomized order. The implantations
A-E were done with conventional beamline implanters
while implantations F-G were done with parallel beam
implanters. All implantations were done at room

temperature with a 7º angle. Implantations A-D and G were done without wafer rotation, while implantations E and F were so-called quad implantations (the dose was divided by four and implanted from four different directions). These variations in samples are summarized in Table I.

Table I: Summary of the implantation samples. Please note that each implantation (A-G) was made with different tool / implantation provider.

Sample name	Implanter type	Stationary (S) /Quad (Q)	Black /Planar
A	beam line	S	yes/yes
B	beam line	S	yes/yes
C	beam line	S	yes/yes
D	beam line	S	yes/yes
E	beam line	Q	yes/no
F	parallel beam	Q	yes/no
G	parallel beam	S	yes/yes

Each provider received a set of samples, which included both nanostructured and planar samples. The planar samples underwent the same process flow as the nanostructured ones, except for the texturization step. Note that none of the planar samples experienced quad implantation. After implantation, the samples were cleaned with standard RCA solutions and HF dip, and drive-in annealing at 1050°C for 45 min was performed [6], which also resulted in a thin silicon oxide.

Emitter formation was followed by removing the thin silicon oxide and passivating the surface of the samples with a 20 nm thick Al_2O_3 film to eliminate surface recombination losses. The Al_2O_3 film was deposited using atomic layer deposition (ALD) at 200°C with trimethylaluminum (TMA) and water as precursors. Finally, the wafers were annealed in forming gas at 400°C for 30 min to activate the passivation. Previous studies have shown that these parameters result in a high-quality interface with silicon and a negative Al_2O_3 charge, combined with conformal coverage of the nanostructures, providing excellent passivation both for planar and nanotextured surfaces [8]. The cross-section of a finished b-Si sample is shown in Fig. 1.

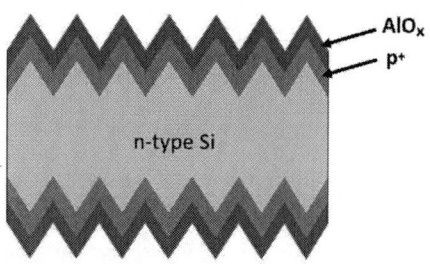

Figure 1: A schematic of an implanted and surface passivated b-Si sample.

2.2 Characterization

Sheet resistance (R_s) was measured from all samples after the drive-in and oxide removal steps using the four-point probe (4PP) method from five different locations on the wafers. The emitter saturation current, J_{0e}, was

extracted from quasi-steady state photoconductance (QSSPC) measurements at an injection level of 1.0×10^{16} cm^{-3}. In addition to emitter properties, we also characterized the effective/bulk lifetime (τ_{eff}) of the samples to see possible traces of contamination resulting from the implantation. Note that while J_{0e}, and τ_{eff} are both determined from the photoconductance measurements, the former was determined at a generalised measurement mode while the latter utilised a transient measurement mode. Finally, photoluminescence imaging (PLI) with lifetime calibration was performed to investigate the lateral uniformity of the samples.

3 RESULTS

The electrical quality of the emitters was assessed by studying the sheet resistance and emitter saturation current. For industrially relevant applications such as front junction solar cells, a sheet resistance of 70-140 Ω/□ is preferred [9]. For a good ohmic contact and minimal resistive losses, the sheet resistance cannot be too high. Low sheet resistance can be achieved by increasing doping density; however, a higher amount of dopant atoms increases Auger and SRH recombination leading to a higher emitter saturation current [5]. Therefore, there is often a tradeoff between sheet resistance and emitter saturation current.

3.1 Sheet resistance R_s

In this study, our reference value for the sheet resistance is 105 Ω/□, which was optimised earlier [6] for b-Si emitter keeping in mind the above mentioned tradeoff between contact resistance and recombination.

The measured sheet resistances of the implanted nanostructured samples prepared in this work are presented in Fig. 2a. There is some variation (97–125 Ω/□) in the sheet resistance between the samples but all of them are relatively close to the target value. Regarding beam line implantations, the sample from provider C has a lower sheet resistance as compared to all the other four. Please note that the sample C is also quite close to the reference value from the optimisation study. In addition, one can observe that samples F and G, which are implanted with parallel beam, seem to have even lower sheet resistances than sample C and the reference. Regarding stationary vs quad implantations, quad implantation results in somewhat higher sheet resistance but this effect is only minor.

The planar counterpart samples are shown in Fig. 2b. As expected, the sheet resistances are less than half of that of the nanostructured samples and thus demonstrate much higher doping concentration. Such emitters are likely to suffer from heavy recombination. The much higher doping concentration in planar references can be explained by the much smaller surface area and thus more condensed doping with the identical implantation dose. When we compare the service providers and implantation methods, in planar samples we can see the same difference than in Fig. 2a, i.e, sample C has a lower sheet resistance as compared to others implying that their implantation tool produces somewhat higher amount of active dopants to the samples. Regarding parallel beam vs. beam line, no clear impact can be observed unlike in the nanostructured samples.

41st European Photovoltaic Solar Energy Conference and Exhibition

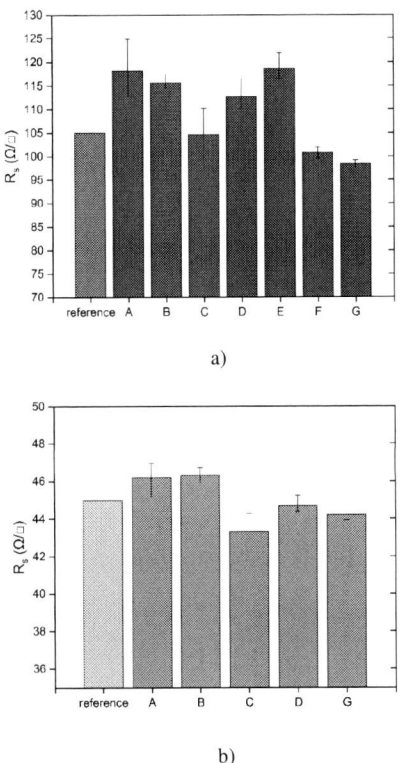

a)

b)

Figure 2: The sheet resistance of implanted a) b-Si and b) planar samples.

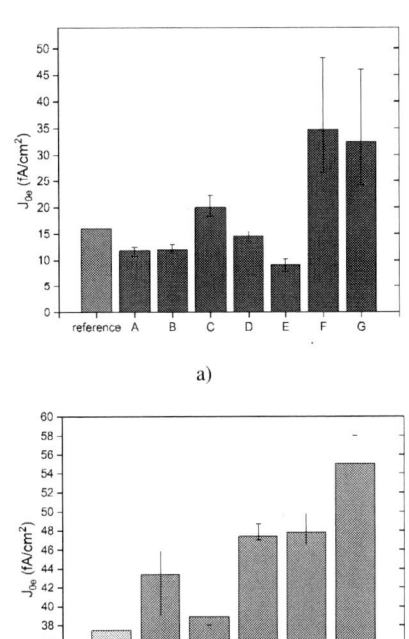

a)

b)

Figure 4: The emitter saturation current J_{0e} of a) b-Si and b) planar samples.

3.2 Emitter saturation current J_{0e}

Emitter saturation current density, J_{0e}, is a widely used recombination parameter in photovoltaics that considers all recombination processes taking place inside the junction as well as at the surface [10]. For high-efficiency solar cells, J_{0e} should be well below 50 fA/cm^2 [5]. The goal is, however, to minimize J_{0e} as much as possible. Our reference value is 16 fA/cm^2, as it was the value that was obtained previously during the b-Si emitter optimization.

Measured J_{0e} values for b-Si and corresponding planar reference samples fabricated in this study are presented in Fig. 4. All the beam line samples (A-E) perform surprisingly well, some reaching even lower values than the reference value. They show a good correlation with the sheet resistance seen in Fig. 2a, i.e., the lower doping concentration is likely to reduce emitter recombination. Sample E shows J_{0e} even below 10 fA/cm^2. The parallel beam samples (F and G) perform clearly worse with a relatively high J_{0e}. The difference to other samples is so high that the lower sheet resistance alone may not be able to explain it.

As expected, the emitter saturation current is much worse in planar counterpart samples (Fig. 4b) due to the higher doping density. None of the planar samples achieve the reference value and the correlation to sheet resistance is much less clear. More specifically, the parallel beam sample G shows really high J_{0e} even though the sheet resistance is in the same range than in the other planar samples. This supports the earlier speculation that the differences seen in J_{0e} between

parallel beam and beam line implantations in nanostructured samples cannot be explained with lower sheet resistance alone. It is worth noting that it could be possible to obtain better results also with parallel beam implantation as parameters used here were optimized based on beam line. However, that would require a separate optimization.

3.3 Effective recombination lifetime τ_{eff}

In addition to emitter quality, it is important to keep the samples clean during implantation to prevent unintentional contamination that can affect the solar cell efficiency. Effective recombination lifetime, τ_{eff}, is a common parameter to study the cleanliness of the processes. Although the samples were carefully RCA cleaned after implantation to remove any surface contamination present during implantation and/or transportation/wafer handling, it is possible that some bulk contamination occurs. Moreover, since there was no screen oxide present, the risk for contamination is increased.

Fig. 5 shows τ_{eff} as a function of injection level in all the nanostructured samples. The lifetimes are close to the 1 ms range in almost all samples at low injection level. Based on the J_{0e} results seen in Fig. 4a, the surface and emitter recombination should be minimal, and thus the large differences seen at low injection level can be explained by differences in bulk recombination. Particularly striking is sample C, which has a reduced effective lifetime of close to 100 µs. In addition, both parallel beam samples have also reduced lifetime. These are likely to be seen in the efficiency of solar cells if advanced cell architecture is used.

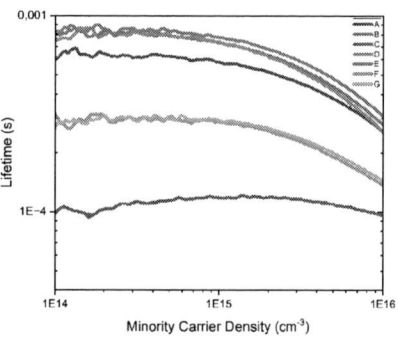

Figure 5: The effective recombination lifetime of minority carriers measured in transient mode as a function of injection level.

Finally, the lateral uniformity of each sample and sample-to-sample variation were studied with photoluminescence imaging and selected images are shown in Fig. 6. In b-Si samples, the uniformity is very high in all samples. PLI imaging confirms a notable difference in lifetime in b-Si sample C. Somewhat surprisingly, notable lateral variation was observed for planar reference samples. We noticed local spot-like defect areas lowering the lifetime in all planar samples (e.g Fig. 6 c). Additionally, in planar sample C, there was also some indication that the implantation might not be uniform at the edges of the wafer (much lower lifetime in the centre of the wafer). Unfortunately, we could not measure parallel beam samples due to PLI image tool breakage.

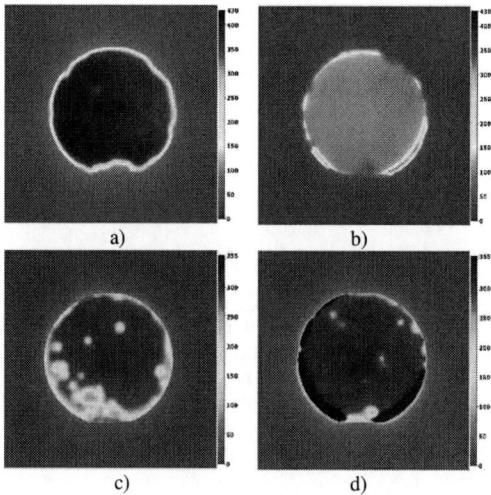

Figure 6: Lifetime calibrated PLI maps for selected wafers: a) B (b-Si), b) C (b-Si), c) D (planar), d) C (planar). The unit shown is microseconds.

4 CONCLUSIONS

The findings of this study suggest that b-Si emitters are surprisingly robust against small variations in implantation conditions. In most cases the reduced emitter saturation current correlated with higher sheet resistance. With beam line implanters, the emitter saturation current was low in all samples and even below 10 fA/cm^{-2} was reached in some samples. For parallel beam implanters, slightly worse values were observed indicating that separate parameter optimization for them would be required.

5 ACKNOWLEDGEMENTS

The authors acknowledge the provision of facilities and technical support by Micronova Nanofabrication Centre in Espoo, Finland within the OtaNano research infrastructure at Aalto University.

The work is part of the Research Council of Finland Flagship Programme, Photonics Research and Innovation (PREIN), decision number 346529.

Parts of this research were carried out at Ion Beam Center at the Helmholtz-Zentrum Dresden-Rossendorf e. V., a member of the Helmholtz Association.

6 REFERENCES

[1] P. Repo et al., Energy Procedia, 38 (2013), 866
[2] J. Oh, H.-C. Yuan, H. M. Branz, Nature Nanotechnology, 7 (2012), 743
[3] S. H. Zhong et al., Solar Energy Materials and Solar Cells, 108 (2013), 200
[4] P. Li et al., RSC Advances, 6 (2016), 104073
[5] K. X. Chen et al., IEEE Photonic Technology Letters, 33 (2021), 1415
[6] G. von Gastrow et al., Journal of Applied Physics, 121 (2017)
[7] N. E. Grant et al., Physica Status Solidi – Rapid Research Letters, 10 (2016), 443
[8] P. Repo et al., IEEE Journal of Photovoltaics, 3 (2013), 90
[9] X. Y. Zhang et al., Progress in Photovoltaics 31 (2023), 369
[10] A. Cuevas, Proceedings of the 4th International Conference on Crystalline Silicon Photovoltaics, Vol. 55 (2014), 53

41st European Photovoltaic Solar Energy Conference and Exhibition

EXCELLENT PASSIVATION OF SILICON SURFACES BY HfO₂ LAYERS
DEPOSITED USING SCALABLE SPATIAL ATOMIC LAYER DEPOSITION (SALD)

Jan Schmidt,[1,2] Michael Winter,[1] Floor Souren,[3] Jons Bolding,[3] Hindrik de Vries[3]
[1]Institute for Solar Energy Research Hamelin (ISFH), Am Ohrberg 1, 31860 Emmerthal, Germany
[2]Institute of Solid-State Physics, Leibniz University Hannover, Appelstr. 2, 30167 Hannover, Germany
[3]SALD B.V., Luchthavenweg 10, 5657 EB Eindhoven, The Netherlands

ABSTRACT: Spatial Atomic Layer Deposition (SALD) had been successfully applied in the past for the Al_2O_3 surface passivation on silicon solar cells. In contrast to conventional sequential ALD techniques, as typically used in the labs, SALD allows for high deposition rates of a few nm per second, which are compatible with industrial solar cell production. In this contribution, we apply SALD for the first time to the electronic passivation of moderately doped ($\sim 10^{16}$ cm⁻³) p-type crystalline silicon surfaces with thin layers of hafnium oxide (HfO_2). For 10 nm thick HfO_2 layers annealed at 400°C in air, an effective surface recombination velocity S_{eff} of only 4 cm/s is achieved, which is below what has been reported before using sequential ALD techniques. The one-sun implied open-circuit voltage amounts to $iV_{oc} = 727$ mV. Firing is shown to reduce the passivation quality, however, by adding a capping layer of plasma-enhanced-chemical-vapor-deposited hydrogen-rich silicon nitride (SiN_x) onto the HfO_2, the firing stability is found to improve. The presented study demonstrates that SALD-deposited HfO_2 layers and HfO_2/SiN_x stacks have the potential to evolve into an attractive surface passivation scheme for future silicon solar cells.
Keywords: Silicon, Surface passivation, Hafnium oxide, SALD

1 INTRODUCTION

Although Al_2O_3 has proven to be an outstanding dielectric layer for the surface passivation of silicon, single Al_2O_3 layers are not perfectly suitable in all application cases [1]. One reason is that Al_2O_3 easily reacts/dissolves with metal pastes used for the solar cell metallization during firing as well as in contact with chemicals such as hydrofluoric acid. Both is usually avoided by adding a SiN_x protection/capping layer [2]. Another problem, which also applies to Al_2O_3/SiN_x stacks, is related to the fundamental passivation mechanism. The predominant passivation mechanism of Al_2O_3 on silicon is based on the very large negative fixed charge density within the Al_2O_3 layer ("field-effect passivation") [3]. This passivation mechanism leads e.g. to problems on phosphorus-diffused n^+-type silicon surfaces [4] and leads to shunt-causing inversion layers on n-type silicon surfaces, a problem for the rear passivation of interdigitated back-contact (IBC) solar cells [5].

In this contribution, we examine the surface passivation performance of thin layers of hafnium oxide (HfO_2) and HfO_2/SiN_x stacks on moderately doped 1.4 Ωcm p-type silicon surfaces. In the past, atomic-layer-deposited HfO_2 layers were applied as high-κ dielectrics in microelectronics [6] and are well known to show a much better resistance against acidic etching compared to Al_2O_3 [7]. In addition, there have been several studies published in recent years demonstrating a promising potential of HfO_2 for the electronic passivation of crystalline silicon surfaces [8-12]. Most of these studies found that the fixed charge density in the HfO_2 layers deposited on silicon is smaller compared to that in Al_2O_3 layers [13]. Hence, the surface passivation mechanism of HfO_2 is assumed to be less field-effect passivation and a larger fraction of "chemical" interface passivation, i.e. a reduction of the interface state density. One might hence expect that HfO_2 provides a comparable passivation level on p- and on n-type silicon surfaces. However, the majority of studies published so far have examined n-type silicon surfaces, resulting in much lower surface recombination velocities compared to the very few results on p-type silicon wafers. In this contribution, we try to fill this gap and examine

HfO_2 passivation of p-type silicon surfaces. For the first time, we also examine the firing stability of HfO_2 layers and of HfO_2/SiN_x stacks in an industrial conveyor-belt firing furnace, as used for the metal contact formation in the production of commercial silicon solar cells. The technique applied for the HfO_2 deposition is Spatial Atomic Layer Deposition (SALD) [14] with a movable substrate table, which is a sufficiently fast deposition technique ready for the industrial scale-up.

2 EXPERIMENTAL

6-inch-diameter shiny-etched 1.4-Ωcm boron-doped float-zone (FZ) silicon wafers are used for our experiments. The 300 μm thick (100)-oriented wafers are RCA-cleaned with HF dip as last step. Subsequently, HfO_2 layers are symmetrically deposited on both wafers sides using Spatial Atomic Layer Deposition (SALD). More details on the deposition technique can be found in a recent publication [15]. In contrast to the conventional sequential ALD process, as mostly used, the two half-reactions in the SALD process are spatially separated, eliminating the need for intermediate pumping steps, which accelerates the deposition significantly.

Tetrakis(dimethylamido)hafnium (TDMAHf) is used as precursor and oxidation is performed using water vapor (H_2O). H_2O and TDMAHf inlets are separated by N_2, which acts as a gas curtain between H_2O and TDMAHf, preventing any reaction of both components before they react on the Si wafer surface. One pass of the wafer through the three zones corresponds to one ALD cycle, resulting in one monolayer of HfO_2. The thickness of ALD HfO_2 is proportional to the number of ALD cycles. We have determined a Growth per Cycle for the SALD deposition of HfO_2 of 0.12 nm/cycle. The deposition rate is scalable to several nanometers per second, depending on the hardware configuration, compared to typically much less than a nm per minute for the conventional sequential ALD processes usually used in the lab. The movable substrate table is kept at a constant temperature of 200 °C during our depositions.

Some of the 6" wafers received silicon nitride (SiN_x)

capping layers on top of the SALD-HfO$_2$ films, resulting in HfO$_2$/SiN$_x$ stacks. 90 nm thick SiN$_x$ layers were deposited at ~400°C onto the HfO$_2$ layers on both wafer sides using microwave-remote plasma-enhanced chemical vapor deposition (PECVD) using silane (SiH$_4$) and ammonia (NH$_3$) and hydrogen (H$_2$) as process gases. We apply an industrial deposition system (Meyer Burger, SiNA) [16]. The resulting SiN$_x$ capping layers have a refractive index of 2.05 (at a wavelength of 633 nm) and are very rich in hydrogen (>10 at.%).

After HfO$_2$ or HfO$_2$/SiN$_x$ depositions on both wafer sides, the 6" wafers were laser-cut into quarters to increase the number of samples. Sample annealing was performed on the HfO$_2$-passivated samples in the temperature range between 350 and 550 °C in air on a temperature-controlled hot-plate with cover. Some HfO$_2$- and all HfO$_2$/SiN$_x$-passivated samples were fired in an industrial conveyor-belt furnace (centrotherm, DO-FF-8.600-300) at a belt speed of 7.2 m/min, as typically applied for the contact firing in solar cell production. The actual temperature profiles on the wafer surface were measured using a temperature tracker (Datapaq DQ1860A) and a type-K thermocouple (Omega, KMQXL-IM050G-300) attached to the wafer surface. The reported peak temperatures are all measured temperatures, the set-peak temperatures are significantly higher.

The effective carrier lifetime τ_{eff} of the processed samples was measured as a function of the excess carrier concentration Δn at a temperature of 30 °C using a WCT-120 lifetime tester from Sinton Instruments. For effective lifetimes larger than 200 µs, the transient photoconductance decay (PCD) mode was applied, whereas for lifetimes lower than 200 µs, the quasi-steady-state photoconductance (QSSPC) mode was used. Further details on the lifetime analysis can be found in Ref. 15.

3 RESULTS AND DISCUSSION

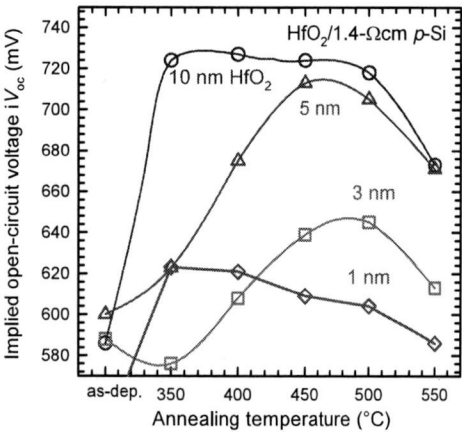

Figure 1: Measured implied open-circuit voltages iV_{oc} of 1.4-Ωcm p-type FZ silicon wafers passivated (on both sides) with HfO$_2$ layers of 10 nm (blue circles), 5 nm (red triangles), 3 nm (green squares), and 1 nm (pink diamonds) thickness as a function of hot-plate annealing temperature.

Figure 1 shows the implied open-circuit voltages iV_{oc} measured using the Sinton WCT-120 system. The silicon wafers are passivated on both sides with HfO$_2$ layers of 10

nm (blue circles), 5 nm (red triangles), 3 nm (green squares), and 1 nm (pink diamonds) thickness. Very low iV_{oc} values ≤600 mV measured directly after deposition indicate that there is no significant level of surface passivation achieved without additional annealing. Subsequent annealing on a hot-plate in air drastically increases iV_{oc} with maximum values of iV_{oc} = 727 mV for the 10 nm HfO$_2$ annealed at 400°C (for 1 min), demonstrating an excellent level of surface passivation. For the 5 nm HfO$_2$ layer, still a very good passivation with iV_{oc} = 713 mV is achieved after annealing at an increased annealing temperature of 450°C (for 8 min). However, reducing the HfO$_2$ film thickness further to 3 nm does not provide any sufficient level of surface passivation anymore, as indicated by a maximum iV_{oc} of 645 mV after annealing at 500°C. For even thinner HfO$_2$ films, the level of surface passivation further deteriorates, as indicated by a maximum iV_{oc} of only 623 mV for the 1 nm thick HfO$_2$ layer after annealing at 350°C.

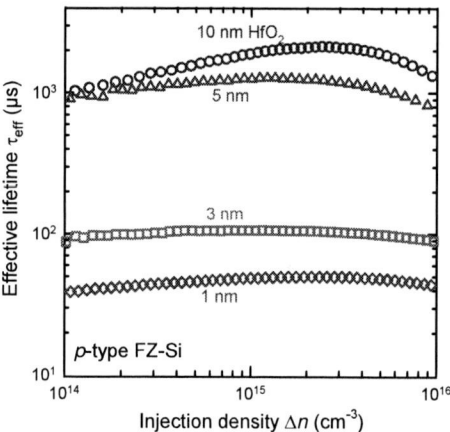

Figure 2: Injection-dependent measurements of the effective lifetime $\tau_{eff}(\Delta n)$ of the best HfO$_2$ layers of Fig. 1.

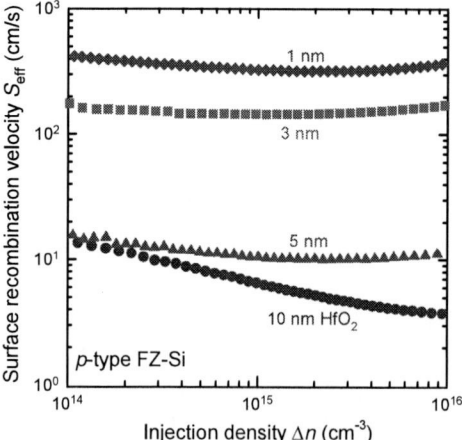

Figure 3: Injection-dependent effective surface recombination velocities $S_{eff}(\Delta n)$ deduced from the lifetimes shown in Fig. 2.

The injection-dependent lifetime measurements of the best-passivating HfO$_2$ layers of Fig. 1 are shown in Fig. 2. The maximum effective lifetime of τ_{eff} = 2.14 ms is measured in the injection range $\Delta n = 2$ to 3×10^{15} cm^{-3} on the silicon sample passivated with 10 nm HfO$_2$ and annealed at 400 °C. Note that the $\tau_{\text{eff}}(\Delta n)$ dependence is not very pronounced and that in the entire injection range relevant to silicon solar cells, effective lifetimes exceeding 1 ms are measured for the 10 nm HfO$_2$ thickness.

The corresponding injection-dependent effective surface recombination velocities $S_{\text{eff}}(\Delta n)$ are plotted in Fig. 3. Due to the decreasing intrinsic lifetime with increasing injection level [17], S_{eff} of the wafer passivated with 10 nm HfO$_2$ is continuously decreasing with increasing Δn. The decrease of τ_{eff} with increasing Δn for $\Delta n > 3 \times 10^{15}$ cm^{-3} is obviously caused be the increasing intrinsic recombination alone. For $\Delta n > 5 \times 10^{15}$ cm^{-3}, the surface recombination velocity approaches a constant value of $S_{\text{eff}} = 4$ cm/s, which is comparable to what has been reported for Al$_2$O$_3$ on p-type silicon with comparable doping concentration [14]. For HfO$_2$, this is the lowest S_{eff} reported so far for surface passivation on moderately doped p-type silicon [10]. For the 5 nm thick HfO$_2$ layer, an even weaker injection dependence of the effective lifetime and the corresponding surface recombination velocity is observed. From $\Delta n = 10^{14}$ to 10^{15} cm^{-3}, S_{eff} is slightly decreasing from 16 to 10 cm/s and for $\Delta n > 10^{15}$ cm^{-3}, S_{eff} is constant at 10 cm/s. Note that such a weak $S_{\text{eff}}(\Delta n)$ dependence is beneficial for optimal solar cell operation. For layer thicknesses below 5 nm, the measured lifetimes strongly decrease and the corresponding S_{eff} values increase. For the 3 nm thick HfO$_2$ layer, over the measured injection range, S_{eff} varies between 145 and 170 cm/s and for the 1 nm thick HfO$_2$, S_{eff} is even in the range 320 to 420 cm/s, not providing any sufficient level of surface passivation.

We conclude that a minimum thickness of 5 nm HfO$_2$ is required to allow for a high level of surface passivation and for excellent passivation, the HfO$_2$ layers should be 10 nm thick. HfO$_2$ layers thinner than 5 nm do not provide a sufficient level of passivation on p-type silicon surfaces after low-temperature annealing.

Note that recently, a decreasing S_{eff} with increasing HfO$_2$ film thickness has also been reported by Tomer et al. [18] for n-type as well as p-type silicon surfaces, where the HfO$_2$ layers were deposited via traditional sequential thermal ALD. On the other hand, in another study, Pain et al. [19] reported an increasing S_{eff} with increasing film thickness. This discrepancy might be due to the fact that they used an O$_2$ plasma (i.e. plasma-enhanced ALD) rather than a thermal oxidation by H$_2$O, as in our SALD process and also the sequential thermal ALD processes applied by Tomer et al. Obviously, depending on the deposition conditions, different thickness dependences of the HfO$_2$ passivation quality can be achieved.

Any surface passivation layer has to keep a stable level of surface passivation during the final process in a solar cell production line, the firing step, which is performed in an infrared conveyor belt furnace heating up the solar cell to a peak temperature of 650 – 800°C for a few seconds. It is well known that such high temperatures can be detrimental to some passivation layers, such as ultrathin (i.e. a few nm thick) Al$_2$O$_3$ layers [2]. For ultrathin Al$_2$O$_3$ layers, hydrogen-rich SiN$_x$ capping layers were added, which led to a significant improvement of the thermal stability of the surface passivation during firing [2].

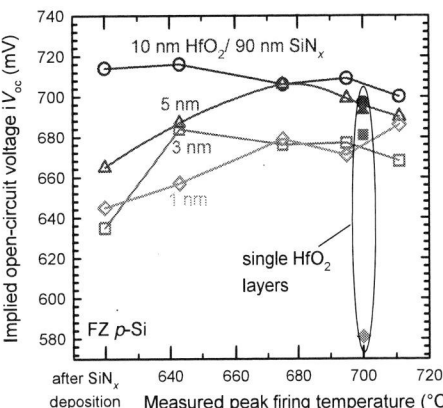

Figure 4: Measured implied open-circuit voltages iV_{oc} of 1.4-Ωcm p-type silicon wafers passivated on both sides with HfO$_2$/SiN$_x$ stacks (open symbols) directly after SiN$_x$ deposition on top of the HfO$_2$ layers and after firing at measured peak temperatures from 640 to 710°C. For comparison, iV_{oc} values measured on silicon wafers passivated with HfO$_2$ single-layers and fired at 700°C are also shown (closed symbols).

This was the motivation in the present study, to examine HfO$_2$/SiN$_x$ stacks for the first time during firing at different peak temperatures and compare their firing stability with that of HfO$_2$ single-layers. Figure 4 shows measured iV_{oc} values of p-type silicon wafers passivated on both sides with HfO$_2$/SiN$_x$ stacks (open symbols). The HfO$_2$ thickness was varied, whereas the SiN$_x$ thickness was kept constant at 90 nm. Directly after the deposition of the SiN$_x$ layers on both HfO$_2$-coated sides of each sample, the measured iV_{oc} values depend strongly on the HfO$_2$ layer thickness and $iV_{\text{oc}} > 710$ mV is only obtained for the 10 nm thick HfO$_2$ layer. The improvement compared to the status directly after HfO$_2$ deposition (see Figure 1) is probably just due to the annealing during the PECVD-SiN$_x$ deposition at ~400°C, which is less effective for the thinner HfO$_2$ layers, in agreement with the results shown in Figure 1. Firing at measured peak temperatures in the range 640 – 710°C does not show a significant impact on iV_{oc} for the 10 nm thick HfO$_2$ layer, but improves the passivation quality for all thinner HfO$_2$ layers. However, only the 5 nm thick HfO$_2$ layer with SiN$_x$ capping slightly exceeds the iV_{oc} of 700 mV (at 675°C peak firing temperature). Also shown in Figure 4 are iV_{oc} values measured on samples passivated with HfO$_2$ single-layers (without hydrogen-rich SiN$_x$ capping layers) and fired at 700°C peak temperature (filled symbols). These samples show a reduced iV_{oc} by 12 and 6 mV for the 10 nm and the 5 nm HfO$_2$, respectively, compared to the respective HfO$_2$/SiN$_x$ stacks, demonstrating a small hydrogenation effect of the hydrogen-rich SiN$_x$ capping layers. However, the impact of the SiN$_x$ capping, and probably of the hydrogenation of interface states, is much more pronounced for the ultrathin, only 1 nm thick HfO$_2$ layer, where we measure a drastic improvement by adding the SiN$_x$ capping layer. After firing at ~700°C peak temperature, the 1 nm thick HfO$_2$ single-layer provides practically no surface passsivation, as shown by a measured iV_{oc} of only 580 mV. On the other hand, when capped by the hydrogen-rich SiN$_x$, the 1 nm HfO$_2$ provides a much higher iV_{oc} of 670 nm after firing at ~700°C. These results demonstrate for the first time the benefit of hydrogen-rich capping layers concerning the firing stability of the surface passivation provided by HfO_2 films.

4 CONCLUSIONS

We have demonstrated that the surfaces of moderately doped p-type silicon wafers (as e.g. used in today's PERC solar cells), can be effectively passivated by thin HfO_2 layers. The passivation quality was found to improve with increasing HfO_2 thickness with the highest implied voltages of iV_{oc} = 727 mV measured for 10 nm HfO_2, annealed at 400°C. The HfO_2 layers were thereby deposited using the high-rate SALD technique, which is applicable to industrial solar cell production. Firing was found to deteriorate the passivation quality, the extent depending on the film thickness, but adding a hydrogen-rich SiN_x capping layer improved the firing stability and led to iV_{oc} values > 710 mV after firing.

The presented results on p-type silicon surfaces hence clearly demonstrate that HfO_2 and HfO_2/SiN_x stacks could evolve into an interesting option for the rear&front surface passivation of PERC or IBC cells as well as to the front passivation of POLO or TOPCon solar cells.

Acknowledgements
The authors thank C. Marquardt and D. Sylla for sample processing and R. Brendel for fruitful discussions (all ISFH). Funding was provided by the German State of Lower Saxony.

References

[1] J. Schmidt, R. Peibst, R. Brendel, *Sol. En. Mat. Sol. Cells* **187**, 39 (2018).

[2] J. Schmidt, B. Veith, R. Brendel, *Phys. Status Solidi RRL* **3**, 287 (2009).

[3] B. Hoex, J. Schmidt, P. Pohl, M. C. M. van de Sanden, W. Kessels, *J. Appl. Phys.* **104**, 044903 (2008).

[4] B. Hoex, M. C. M. van de Sanden, J. Schmidt, R. Brendel, W. Kessels, *Phys. Stat. Sol. RRL* **6**, 4 (2012).

[5] B. Veith, T. Ohrdes, F. Werner, R. Brendel, P. Altermatt, N.-P. Harder, J. Schmidt, *Solar Energy Materials & Solar Cells* **120**, 436 (2014).

[6] G. D. Wilk, R. M. Wallace, J. Anthony, *J. Appl. Phys.* **89**, 5243 (2001).

[7] A. Wratten, D. Walker, E. Khorani, B. F. M. Healy, N. E. Grant, J. D. Murphy, *AIP Advances* **13**, 065113 (2023).

[8] J. Wang, S. S. Mottaghian, M. F. Baroughi, *IEEE Trans. Elec. Dev.* **59**, 342 (2012).

[9] F. Lin, B. Hoex, Y. H. Koh, J. Lin, A. G. Aberle, *ECS Journal of Solid State Science and Technology* **2**, N11 (2013).

[10] J. Cui, Y. Wan, Y. Cui, Y. Chen, P. Verlinden, A. Cuevas, *Appl. Phys. Lett.* **110**, 021602 (2017).

[11] A. B. Gougam, B. Rajab, A. B. Afif, *Materials Science in Semiconductor Processing* **95**, 42 (2019).

[12] A. Wratten, S. L. Pain, D. Walker, A. B. Renz, E. Khorani, T. Niewelt, N. E. Grant, J. D. Murphy, *IEEE J. Photovolt.* **13**, 40 (2023).

[13] A. Wratten, S.L. Pain, A. Yadav, E. Khorani, T. Niewelt, L. Black, G. Bartholazzi, D. Walker, N. E. Grant, J. D. Murphy, *Solar Energy Materials and Solar Cells* **259**, 112457 (2023).

[14] F. Werner, B. Veith, V. Tiba, P. Poodt, F. Roozeboom, R. Brendel, J. Schmidt, *Appl. Phys. Lett.* **97**, 162103 (2010).

[15] J. Schmidt, M. Winter, F. Souren, J. Bolding, H. de Vries, *Phys. Status Solidi RRL* **2024**, in press

[16] J. D. Moschner, J. Henze, J. Schmidt, R. Hezel, *Prog. Photovolt: Res. Appl.* **12**, 21 (2004).

[17] T. Niewelt, B. Steinhauser, A. Richter, B. Veith-Wolf, A. Fell, B. Hammann, N. E. Grant, L. Black, J. Tan, A. Youssef, J.D. Murphy, J. Schmidt, M. C. Schubert, S. W. Glunz, *Solar Energy Materials and Solar Cells* **235**, 111467 (2022).

[18] S. Tomer, M. Devi, A. Kumar, S. Laxmi, S. Satapathy, K. K. Maurya, P. Singh, P. Pathi, Vandana, *IEEE J. Photovolt.* **13** (5), 691 (2023).

[19] S. L. Pain, E. Khorani, T. Niewelt, A. Wratten, G. J. Paez Fajardo, B. P. Winfield, R. S. Bonilla, M. Walker, L. F. J. Piper, N. E. Grant, J. D. Murphy, *Adv. Mater. Interfaces* **9**, 2201339 (2022).

SIMULATION OF TOPCON/PERC HYBRID BOTTOM STRUCTURE FOR PEROVSKITE/SILICON TANDEM SOLAR CELLS USING QUOKKA3

Eni Muka[1], Raşit Turan[1,2,3], Hisham Nasser[1]
[1]ODTÜ-GÜNAM, Middle East Technical University, Ankara, 06800, Türkiye
[2]Micro and Nanotechnology Department, Middle East Technical University, Ankara, 06800, Türkiye
[3]Department of Physics, Middle East Technical University, Ankara, 06800, Türkiye

ABSTRACT: This work emphasizes the potential of perovskite/silicon tandem solar cells for increased power conversion efficiencies. By employing crystalline silicon (c-Si) as the bottom cell, particularly with p-type PERC technology, there are cost-effective and advantageous physical properties. However, traditional phosphorus-doped emitters in PERC Si bottom cells are hindered by high surface recombination, which limits their performance. This research introduces a novel hybrid PERC/TOPCon structure that integrates a phosphorus-doped poly-Si (n+ TOPCon) layer as the front emitter to address these challenges. Numerical simulations using Quokka3 confirmed the feasibility of the design, focusing on optimizing the rear side metallization to enhance implied open-circuit voltage (V_{oc}) and fill factor (FF). A two-step process systematically varied local contact openings to examine their impact on performance metrics. Results highlighted optimal rear metallization parameters, achieving optimal metal fractions approximately 2%. This innovative approach demonstrates the effectiveness of combining TOPCon and PERC technologies for bottom cells in tandem structures, providing valuable insights into their development and optimization. The study underscores the potential of the hybrid PERC/TOPCon structure in enhancing the functionality and efficiency of perovskite/silicon tandem solar cells.
Keywords: tandem solar cells, PERC technology, TOPCon layer, numerical simulations, metallization optimization

1 INTRODUCTION

The progression toward enhanced solar efficiencies through tandem cells is remarkable, particularly with the rapid advancements in the last decade [1]. The choice of crystalline silicon (c-Si) for perovskite/silicon tandem solar cells is promising due to its advantageous physical properties and cost-effectiveness, and optimizing c-Si as the bottom cell is crucial for improving the overall functionality and efficiency of such devices [2]. Given its widespread adoption and market presence, the utilization of p-type PERC technology as the foundational concept for the bottom cell in perovskite/silicon tandem devices is reasonable. However, conventional phosphorus-doped emitters in PERC Si bottom cells can lead to high surface recombination, presenting a challenge [3]. To address this challenge, a novel solution integrates a PERC structure with a phosphorus-doped poly-Si (n+ TOPCon) layer as the front emitter, resulting in a hybrid PERC/TOPCon structure. This innovative approach shows promise in overcoming the limitations of conventional PERC-based bottom cells in perovskite/silicon tandem cells [4]. This work focuses at developing a viable PERC- based bottom cell structure with front TOPCon for perovskite/silicon tandem solar cells are presented. Using numerical simulations conducted by the Quokka3 software, the aim is to validate the feasibility and performance potential of this design. Our investigation emphasizes optimizing the rear side metallization design of the PERC/TOPCon structure, intending to enhance key performance metrics such as open-circuit voltage (V_{oc}) and fill factor (FF). By systematically varying the local contact opening (LCO) and analyzing its impact on device performance, the development of a comprehensive roadmap for achieving optimal results has been put into focus. Overall, this study aims to contribute valuable insights into developing and optimizing bottom-cell structures for perovskite/silicon tandem solar cells. This study shows that combining TOPCon and PERC technologies is effective for the bottom cell in a tandem structure.

2 METHODS

Simulations were conducted on a 2.2 x 2.2 cm² hybrid PERC/TOPCon structure, with the specific layer compositions and parameters outlined in Figure 1 and Table I, all of which were measured experimentally in our lab.

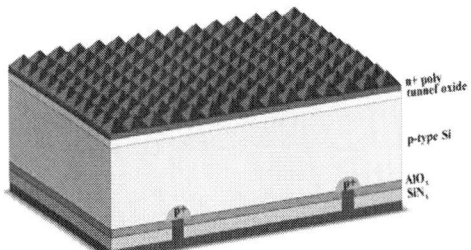

Figure 1: Simulated Hybrid TOPCon/PERC Structure

Table I: Simulation Parameters for Quokka 3

Type of Parameters		Value
Optical Parameters	Front Side Transmission	Obtained by OPAL 2
Bulk	Thickness	150 µm
	Resistivity	1 Ωcm
	Lifetime	1 ms
Front Side Emitter	J_{TOPCon}	3 fA/cm2
	R_{sheet}	60 Ω/□
Local BSF	R_{sheet}	40 Ω/□
	$J_{0,metal}$	1126.24 fA/cm2
	$J_{0,passivated}$	5 fA/cm2

While the front side has no metal, the rear side has local metal contacts, forming the back surface field (BSF) in monofacial design. The configuration for the rear metallization and contact is illustrated on Figure 2. To accurately establish optical parameters, the wavelength-dependent external front surface transmission (T_{ext}) was determined using OPAL2 [5]. The front side layers consisted of an n+ type poly-Si layer and a thin silicon oxide layer, with thicknesses of 70 nm and 1.5 nm, respectively. The choice of the TOPCon stack on the front side is guided by the need to balance its optical and passivation properties. A key consideration is that the poly-Si layer can degrade during the high-temperature processes typically required for rear-side metal contact formation. To mitigate this, a successful solution has been developed by introducing a sacrificial layer that prevents degradation [6].

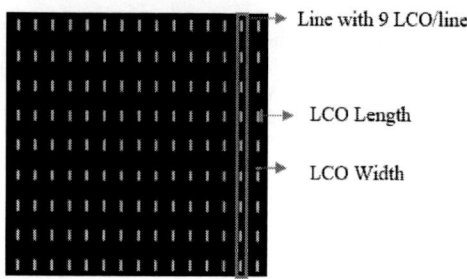

Figure 2: Rear Side Metal and Contact Configuration

The optical model used in this study is based on the standard AM 1.5G spectrum. Initially, the current-voltage (*I-V*) characteristics of the structure were calculated under 1-Sun irradiance. However, this does not represent practical conditions, as the bottom cell in a perovskite/silicon tandem configuration does not absorb the full spectrum [7]. To address this, a second set of simulations was conducted under 0.5-Sun irradiance. In both scenarios, the main goal is to assess the impact of the rear-side metal on V_{oc} and FF values. The study aims to identify the optimal metal fraction (f) based on these parameters, without considering the effects on short-circuit current and power conversion efficiency.

To optimize the rear side contact, a two-step process was employed as detailed in Table II. This approach allowed for a comprehensive exploration of the impact of key parameters on the metal fraction and facilitated the identification of optimal conditions for the hybrid PERC/TOPCon structure.

Table II: Optimization Plan

	Parameters
STEP I f = 0.6-4.1%	LCO/line = 24 LCO Length = 400 µm
	Number of lines = 10-35 LCO Width = 30-60 µm
STEP II f = 1.4-3%	Optimal values in STEP I are kept constant
	LCO/line = 17-31 LCO Length = 350-410 µm

3 RESULTS AND DISCUSSION

To comprehensively understand the impact of rear-side metal configuration on the performance parameters V_{oc} and FF, Figure 3 and Figure 4 illustrate Step I and Step II under 1 Sun illumination, while Figure 5 and Figure 6 depict the same steps under 0.5 Sun illumination. To explore the trade-off between V_{oc} and FF as metal fraction increases, their product was calculated as well.

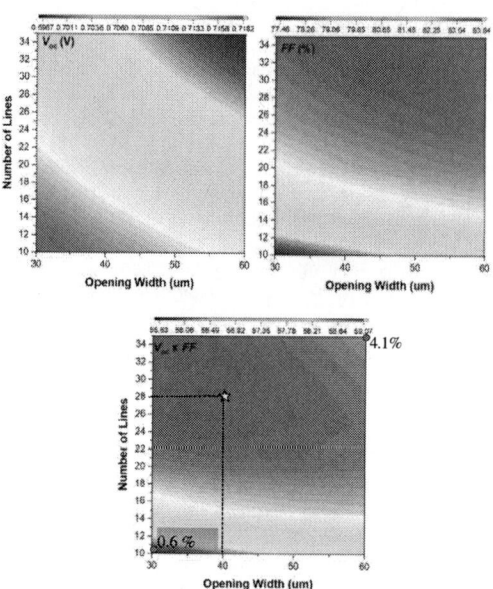

Figure 3: V_{oc}, FF and V_{oc} x FF dependency on opening width and number of lines for a fixed number of LCO/line of 24 with fixed length of 400 µm (Step I) under 1 Suns illumination

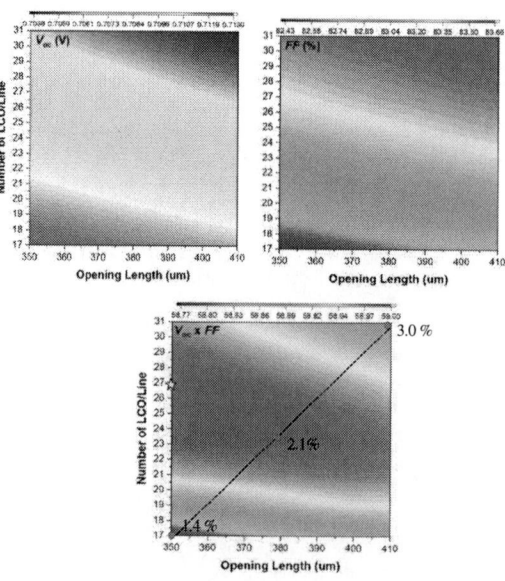

Figure 4: V_{oc}, FF and V_{oc} x FF dependency on number of LCO/line and opening length for fixed number of lines of 28 and opening width of 40 um (Step II) under 1 Suns illumination

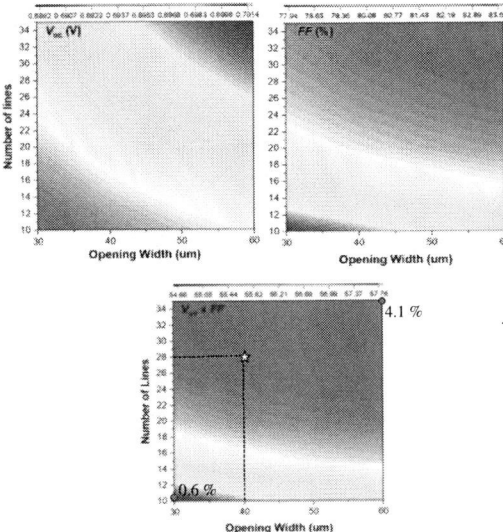

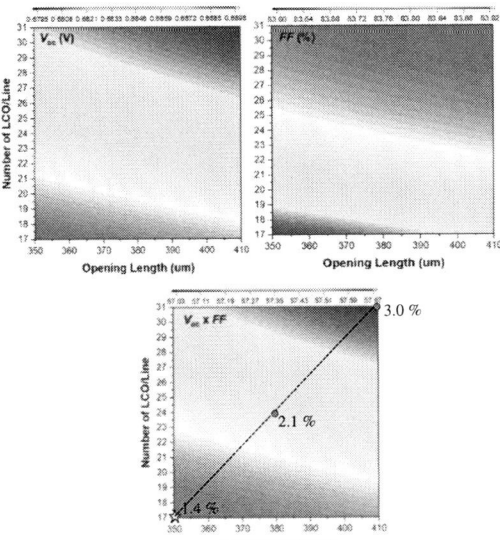

Figure 5: V_{oc}, FF and V_{oc} x FF dependency on opening width and number of lines for a fixed number of LCO/line of 24 with fixed length of 400 µm (Step I) under 0.5 Suns illumination

Figure 6: V_{oc}, FF and V_{oc} x FF dependency on number of LCO/line and opening length for fixed number of lines of 28 and opening width of 40 um (Step II) under 1 Suns illumination

Under 1 Sun illumination, the behavior of both V_{oc} and FF shows a clear pattern as the metal fraction increases. Specifically, V_{oc} experiences a steady decline, while FF increases, reflecting a typical trade-off between these two performance parameters. However, when the two steps are analyzed separately, it becomes evident that the product of V_{oc} × FF does not exhibit the same trend across both steps in relation to the metal fraction. In Step I, FF reaches a saturation point at approximately f ≈ 1%, indicating that further increases in the metal fraction beyond this value

yield diminishing improvements in FF. On the other hand, in Step II, the saturation point occurs later, at around f ≈ 2.5%, suggesting that FF can be further optimized with a higher metal fraction in this step.

The indications for these behaviors are shown in the production plots as well, where the product of both parameters is fully dominated by the FF in the first step. As the opening length and their pitch on the second step change, a higher trade-off between the series resistance loss and rear shading area which corresponds to the losses in V_{oc} and gains in FF are seen [5]. This trade-off is balanced with a metal fraction of 2.02%. This metal fraction represents the optimal choice between the V_{oc} losses and FF gain. However, the current literature indicates that the optimal metal fraction for conventional PERC typically falls within the 3-5% range. In this case, a lower optimal metal fraction is observed, likely due to the front emitter's excellent surface passivation (low J_0) compared to the diffused emitter of PERC.

For the 0.5 Sun case, Step I mirrors the trends seen under 1 Sun, with similar patterns of V_{oc} reduction and FF increase as the metal fraction increases. However, Step II exhibits a distinct shift in behavior compared to the 1 Sun case. Unlike the 1 Sun case where FF dominates, under 0.5 Sun illumination, V_{oc} becomes the key determining factor in the trade-off. This indicates that at lower illumination levels, the effects of series resistance and other electrical losses become more pronounced, making V_{oc} the critical parameter to optimize in Step II under reduced light conditions. Despite these differences, the analysis under both illumination levels reinforces the importance of fine-tuning the metal fraction to achieve the best balance between V_{oc} and FF across both optimization steps.

4 CONCLUSIONS

To conclude, the presented study underscores the significance of optimizing the rear side metal configuration in the TOPCon/PERC hybrid structure. The insights gained from this two-step process shed light on the balance V_{oc} and FF, emphasizing the importance of parameters such as opening width, number of lines, the number of openings per line and length of local opening contacts for achieving optimal photovoltaic performance.

ACKNOWLEDGEMENT: Funded by the European Union. Views and opinions expressed are however those of the author(s) only and do not necessarily reflect those of the European Union or RIA. Neither the European Union nor the granting authority can be held responsible for them. NEXUS project has received funding from the European Union's Horizon Europe research and innovation program under grant agreement No. 101075330.

REFERENCES

[1] Messmer, C., Goraya, B. S., Nold, S., Schulze, P. S. C., Sittinger, V., Schön, J., Goldschmidt, J. C., Bivour, M., Glunz, S. W., & Hermle, M. (2020). The race for the best silicon bottom cell: Efficiency and cost

evaluation of perovskite–silicon tandem solar cells. Progress in Photovoltaics: Research and Applications, 29(7), 744–759.

[2] OPAL2. (Nov.2023) https://www2.pvlighthouse.com.au/calculators/opal%202/opal%202.aspx

[3] Madani, K., Rohatgi, A., Rounsaville, B., Kang, M., Song, H., & Ok, Y. (2021). Enhanced Stability of Exposed PECVD Grown Thin n + Poly-Si/SiOx Passivating Contacts with Al2O3 Capping Layer During High Temperature Firing. IEEE Journal of Photovoltaics, 11(2), 268–272.

[4] Sadhukhan, S., Acharyya, S., Panda, T., Mandal, N. C., Bose, S., Nandi, A., Das, G., Maity, S., Chaudhuri, P., Chakraborty, S., & Saha, H. (2021). Evaluation of dominant loss mechanisms of PERC cells for optimization of rear passivating stacks. Surfaces and Interfaces, 27, 101496. https://doi.org/10.1016/j.surfin.2021.101496

[5] Tang, H., Ma, S., Lv, Y., Li, Z., & Shen, W. (2020). Optimization of rear surface roughness and metal grid design in industrial bifacial PERC solar cells. Solar Energy Materials & Solar Cells/Solar Energy Materials and Solar Cells, 216, 110712. https://doi.org/10.1016/j.solmat.2020.110712

[6] Madani K., Rohatgi A., Rounsaville B., Kang M., Song H., & Ok Y. Enhanced Stability of Exposed PECVD Grown Thin n + Poly-Si/SiOx Passivating Contacts with Al2O3 Capping Layer During High Temperature Firing. IEEE J Photovoltaics. 2021 Mar 1;11(2):268–72. doi: 10.1109/jphotov.2020.3043259.

[7] Kuang Y, Ma Y, Zhang D, Wei Q, Wang S, Yang X, et al. Enhanced optical absorption in perovskite/Si tandem solar cells with nanoholes array. Nanoscale Res Lett. 2020 Nov 12;15(1). doi: 10.1186/s11671-020-03445-3.

A COMPREHENSIVE ANALYSIS OF THE SERIES RESISTANCE FOR DIFFERENT INTERDIGITATED BACK CONTACT SOLAR CELL GEOMETRIES

Telmo Isasi, Yeray Mateos, Janire Pampin, Vanesa Fano, Nekane Azkona, Eneko Ortega, Juan Carlos Jimeno, Eneko Cereceda, and Aloña Otaegi
Technological Institute of Microelectronics, UPV/EHU, 48013, Bilbao, Spain
Torres Quevedo ingeniaria plaza 1. 48013, Bilbao, Spain

ABSTRACT: Top contact-free solar cells achieve higher efficiencies by eliminating shading losses from front metal fingers since the front metal finger is brought to the rear surface of the device. However, since the positive contact and the negative contact coexist in the rear side of the cell, the geometry of the metallization is vital. In this context, we propose a busless geometry, for which ohmic losses are deduced according to a basic analytical 1D model and focusing on the emitter series resistance and the metallic series resistance. This work helps us to understand the role of both partial components for the optimization of the ohmic losses in the interdigitated back contact solar cells.
Keywords: series resistance, interdigitated back contact solar cell, geometry, grid of dots

1 INTRODUCTION

Al-BSF technology dominated the market up to the second half of the decade of 2010 leading to the predominance of PERC technology in the beginning of the decade of 2020. Predictions for the near future stands that the mainstream technology will be marked by top contact free solar cells and specifically by interdigitated back contact solar cells, IBC, since efficiencies around 24% have been recently reported by industrially feasible screen-printed solar cells technologies [1-3]. Eliminating shading losses from front metal fingers make IBCs potentially attractive: the resistive power loss is reduced largely as the rear surface can be designed for best design of metal contact formation. In that context, analytical models are developed [4], and 2- and 3-d simulations are carried out [5, 6] to characterise IBC cells. Several geometries are also subject of analysis if ohmic losses are to be reduced [7-11].

In this work we present a simple analytic model for a first approximation of the partial components of the series resistance. We focus in the the series resistance of the emitter and on the metallization, emphasizing the relevance of the geometry and we finish the work justifying a new geometry without buses.

2 THE RELEVANCE OF THE SERIES RESISTANCE

From the equivalent electronic circuit model of a solar cell, a power loss of $I^2 \cdot R_{series}$ occurs when delivering current to a load. The series resistance is therefore critical for the efficiency of solar cells.

$$R_s = R_{s,b} + R_{s,e} + R_{s,c} + R_{s,m} \qquad (1)$$

With $R_{s,b}$ representing the ohmic losses in the base, $R_{s,e}$ the losses due to the lateral flow of charge carriers in the emitter, $R_{s,c}$ the contact resistance at the emitter-metal interface, and $R_{s,m}$ the ohmic metal losses due to the electron flow in metal electrodes.

Moreover, if the solar cell presents symmetrical geometries, it can be divided into unit cells and the series resistance, normalized by area, of each unit cell matches the total normalized series resistance.

$$\overline{R}_{s,total} \, (\Omega \cdot cm^2) = \overline{R}_{s,unit} \, (\Omega) \cdot \text{Area} \, (cm^2) \qquad (2)$$

All this entails that the calculation of the total series resistance of a solar cell with symmetric patterns is reduced to the calculation of each of the components of the series resistance for one of the unit cells, which is commonly known.

3 ANALYSIS OF PARTIAL SERIES RESISTANCE IN IBC CELLS

In this section it is determined the contribution of the partial resistances in the total value of a solar cell following (1). The equations obtained are specific to the geometry of IBCs proposed in Fig. 1, hereafter referred to as A geometry.

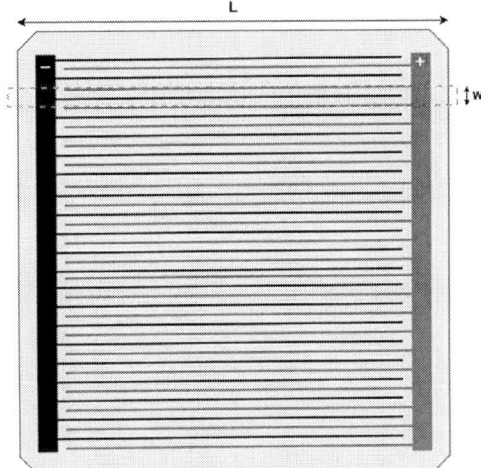

Figure 1: IBC Cell structure. A geometry

The IBC cell has a length, L, of 12,5 cm and features two busbars —one for the positive contact and one for the negative contact. Additionally, not represented in the figure, the cell has 127 interdigitated fingers designed to collect the current of the cell for each polarity. The unit cell applicable to the IBC structure is as follows in Fig. 2.

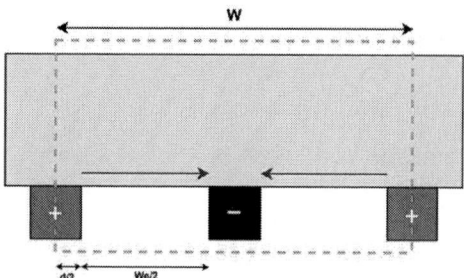

Figure 2: IBC cell's unit cell structure

Being W the width of the unit cell, d the width of a finger and w_e width of the emitter.

In a basic model, the series resistance is decomposed in the contributions of the emitter, fingers, and base.

Emitter series resistance, $R_{s,e}$

$$R_{s,e}(\Omega \cdot cm^2) = \tfrac{1}{3} \cdot R_{sheet,e}(\Omega/\square) \cdot \square_e \cdot W \cdot L \quad (3)$$

Since the geometry of the unit cell determines the the square, (4) determines the square for this specific geometry.

$$\square_e = \frac{W - 4\frac{d}{2}}{L} \quad (4)$$

Metallic fingers series resistance, $R_{s,m}$

Metallic fingers are usually characterized by their sheet resistance, $(\Omega/\square)$, consisting on the resistivity and metal height ratio:

$$R_{sh,m} = \rho_m \, (\Omega \cdot cm) \, / \, h \, (cm) \quad (5)$$

$$R_{s,m}(\Omega \cdot cm^2) = \tfrac{1}{3} \cdot R_{sheet,m}(\Omega/\square) \cdot \square_m \cdot W \cdot L \quad (6)$$

With (7) determining the square for this specific geometry.

$$\square_m = \frac{4 \cdot L}{\frac{d}{2}} \quad (7)$$

Base series resistance $R_{s,b}$

$$R_{s,b}(\Omega \cdot cm^2) = \rho_b (\Omega \cdot cm) \cdot W_b \quad (9)$$

4 OPTIMIZATION OF THE SERIES RESISTANCE

Even though the backside of A geometry cell completely metallized by the metal fingers —which is optimal since emitter loss would be infimal— the buses at the extrems of the cell jeopardize its efficiency, since an electron generated below the positive contact could recombine before reaching its negative contact [15]. An optimization of a busless geometry would be interesting to avoid electrical shadows, and in that context, we have looked for an alternative geometry that optimizes the series resistance. The proposed geometry is hereafter referred to as B geometry and it is shown in Fig. 3.

Figure 3 shows a first insight into the geometry of the cell, in which we have interdigitated the positive and the negative contacts in the center of the cell, being the metallic fingers length l. The rest of the parameters remain the same as in A geometry.

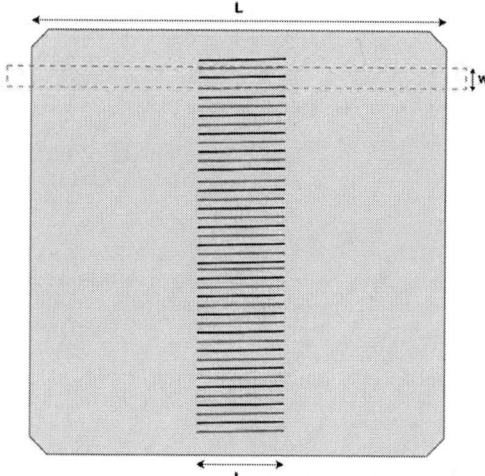

Figure 3: Proposed IBC Cell Structure. B geometry

The new equations adapted the proposed geometry are shown below.

Emitter series resistance, $R_{s,e}$

In the new design, the electron flow is different depending on the position, as can be seen in Fig. 4.

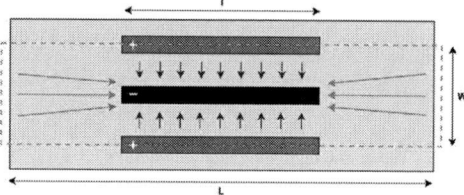

Figure 4: Electron flow in the proposed IBC Cell

The electrons between the fingers only need to travel a maximum of $w_{e1}/2$ distance to reach the fingers (blue arrows in Figure 4).

$$R_{s,e1}(\Omega \cdot cm^2) = \tfrac{1}{3} \cdot R_{sheet,e}(\Omega/\square) \cdot \square_{e1} \cdot W \cdot L \quad (9)$$

With:

$$\square_{e1} = \frac{W_{e1}}{l/2} = 2 \cdot \frac{W - 4\frac{d}{2}}{l} \quad (10)$$

Nevertheless, the electrons outside the fingers need to travel more distance (pink arrows in Figure 4). Although they actually travel some vertical distance, the horizontal distance is large enough to ignore the vertical one. Therefore, the emitter resistance for those electrons is extrapolated to the horizontal flow:

$$R_{s,e2}(\Omega \cdot cm^2) = \tfrac{1}{3} \cdot R_{sheet,e}(\Omega/\square) \cdot \square_{e2} \cdot W \cdot L \quad (11)$$

With:

$$\square_{e2} = \frac{W_{e2}}{W} = \frac{L - l}{W} \quad (12)$$

Therefore, the total $R_{s,eT}$ is the sum of the horizontal flow and vertical flow as in (13):

$$R_{s,eT}(\Omega \cdot cm^2) = R_{s,e1}(\Omega \cdot cm^2) + R_{s,e2}(\Omega \cdot cm^2) \quad (13)$$

Metallic fingers series resistance, $R_{s,m}$

The metallic fingers series resistance is the same in both cases:

$$R_{s,m}(\Omega \cdot cm^2) = \frac{1}{3} \cdot R_{sheet,m}(\Omega/\square) \cdot \square_m \cdot W \cdot L \quad (14)$$

Where

$$\square_m = \frac{4 \cdot l}{\frac{d}{2}} \quad (15)$$

5 OPTIMIZATION OF L AND l

In this section it is sought to optimise the variables L (length of the cell) and l (length of the metallic finger), with the aim of minimising the ohmic losses generated by R_e and R_m.

For this purpose, the formulas developed in section 4 have been introduced and plotted to visualise where the two curves intersect.

In the following figure, the relationship between R_e and R_m is displayed in a 3D graph, for different values of l and L.

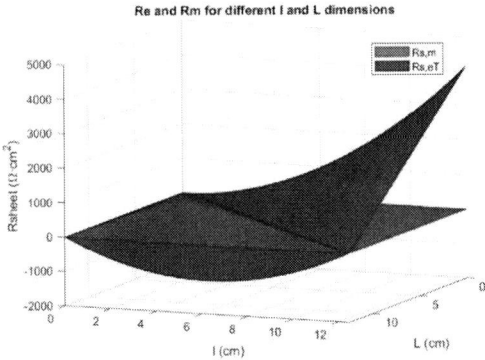

Figure 5: 3D plot of R_e (blue) and R_m (red) depending on L and l

In order to better visualise where the two equations converge, the 2D graph is shown in Fig. 6. As can be seen, the optimal value is achieved by equating L and l. Moreover, by decreasing the unit cell, the value of both resistances goes down.

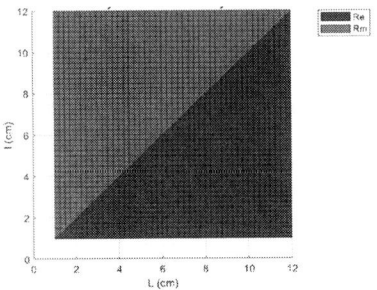

Figure 6: 2D plot of R_e (blue) and R_m (red) depending on L and l

6 NUMERICAL RESULTS

In this section specific values are applied to (9)-(15) to represent what is described in section 5. Table I shows

some estandard generic for sheet resistances and cell geometry. The l parameter, length of the metallic finger is leaving undetermined, since is the value we want to determine.

Table I: Solar cell parameters applied for numerical results

	Value	Description
$R_{sh,e}\ (\Omega/\square)$	80	emitter sheet resistance
$R_{sh,m}\ (\Omega/\square)$	$5,3 \cdot 10^{-4}$	metallic sheet resistance
W (cm)	$916 \cdot 10^{-4}$	width of unit cell
L (cm)	12,5	lenght of solar cell
d (cm)	$245 \cdot 10^{-4}$	width of metallic finger
l	variable	length of metallic finger

From parameters in Table I, (9), (11) equations are applied in order to determine the partial emitter and metal series resistances for several values of l and being L 12,5 cm. Results are shown in Table II.

Table II: Partial emitter series resistances for several lenghts of metallic finger

	$l = L/3$	$l = 2L/3$	$l = L$
$R_{s,e1}\ (\Omega \cdot cm^2)$	0,624	0,312	0,21
$R_{s,e2}\ (\Omega \cdot cm^2)$	2780	1393	0

To conclude, Table III shows the total emitter series resistance, the metallis series resistance and the sum of them, to reach an optimal series resistance value. Metallic sheet resistance has been here overestimated, for sure, and values will be significantly lower.

Table III: Optimal series resistance as the sum of the partial resistances for several finger lengths

	$l=1/3 \cdot L$	$l=2/3 \cdot L$	$l=L$
$R_{s,e}\ (\Omega \cdot cm^2)$	2780	1390	0
$R_{s,m}\ (\Omega \cdot cm^2)$	0,28	0,55	0,78
$R_s\ (\Omega \cdot cm^2)$	2780	1390	0,78

The series resistance is drastically reduced being l=L.

7 METAL DOTS ALONG THE AREA

Based on the results obtained in the previous section, it can be concluded that the emitter resistance is minimised when the length of the metal finger equals the width of the cell.

In addition, the metallic resistance can be reduced without increasing the emitter resistance, with a geometry in which metallic fingers are replaced by a grid of dots, such as the geometry shown in Fig. 7. In such case, the back area of the cell would we completele covered by dots, and the ratio l/L would recommended to be as close as possible to 1 (one). The total series resistance would be minimum in that case. The positive and the negative contacts are differentiated in the figure by showing different colors.

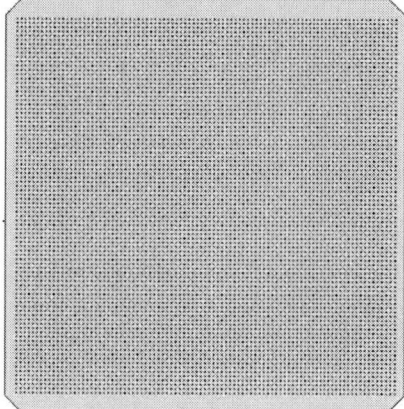

Figure 7: Proposed design to reduce series resistance

8 SUMMARY

A basic analytical model for IBC solar cells is developed in order to look for a geometry that minimises the ohmic losses. Partial terms have been given form for the emitter resistance and the metallic resistance, as a function of L, the length of the cell and l, length of the metal finger. Several ratios of l/L have been discussed, up to the point to identify l reaching almost L the geometry in which resistive losses are minimised.

In addition, instead of a model with metallic fingers, if we turn into a grid of dots simmetrically located, the unit cell is reduced and metallic losses are optimised.

ACKNOLEDGEMENTS

The Spanish Agencia Estatal de Investigación MCIN/AEI/10.13039/ 501100011033 is acknowledged for financial support through the GREASE project (PID2020-113533RB-C32)

REFERENCES

[1] Clean Energy Reviews: Most Efficient Solar Panels. 2024. Availabe online. https://www.cleanenergyreviews.info/blog/most-efficient -solar-panels (accesed on 13/09/2024).
[2] D. D. Smith et al, "Toward the practical limits of silicon solar cells", IEEE Journal of Photovoltaics, vol. 4 (2014).
[3] R. Kopecek et al. "Interdigitated back contact technology as final evolution for industrial crystalline single-junction silicon solar cell", Solar, vol. 3, 1-14. https://doi.org/10.3390/solar3010001 (2023).
[4] D. Giaffreda et al. "A Distributed Electrical Model for Interdigitated back Contact Silicon Solar Cells", Energy Procedia, vol. 55, pp. 71-76 (2014).
[5] O. Nichiporuka et al, "Optimisation of interdigitated back contacts solar cells by two-dimensional numerical simulation" Solar Energy Materials & Solar Cells 86,

pp. 517–526 (2005). https://doi.org/10.1016/j.solmat.2004.09.010
[6] G. Micard et al, "Extraction of individual components of series resistance using TCAD simulations meeting the requirements of 2- and 3-dimensional carrier flow of IBC solar cells". Energy Procedia, vol. 124, pp. 113-119. https://doi.org/10.1016/j.egypro.2017.09.325
[7] M. Hendrichs et al, "Screen-Printed Metallization Concepts for Large-Area Back-Contact Back-Junction Silicon Solar Cells", IEEE Journal of Photovoltaics, vol. 6 (2016).
[8] G.G. Untila et al, "Multi-wire metallization for solar cells: Contact resistivity of the interface between the wires and In2O3:Sn, In2O3:F, and ZnO:Al layers", Solar Energy, 142, pp. 330-339 (2017). https://doi.org/10.1016/j.solener.2016.12.049
[9] Jonas D et al, "Influence of interconnection concepts for IBC solar cell performance by simulation", AIP Conf. Proc. 1999, 020011 (2018). https://doi.org/10.1063/1.5049250
[10] D. Rudolph et al, "Screen printable, non-fire-through copper paste applied as busbar metallization for back contact solar cells", AIP Conference Proceedings 2709, 020006 (2022).
[11] Y. -W. Peng et al, "Design, Fabrication, and Characterization of n-Si IBC Solar Cells Using PERC Technology," in IEEE Journal of Photovoltaics, 10, 2, pp. 383-389 (2020). https://doi.org/10.1109/JPHOTOV.2019.2958186.
[12] A.R Burgers, J.A. Eikelboom, "Optimizing metallization patterns", in Proc. 26th IEEE Photovoltaic Spec. Conf., pp. 219-222 (1997). https://doi.org/10.1109/PVSC.1997.654068
[13] F. Recart, thesis, "Evaluación de la serigrafía como técnica de metalización para células solares eficientes" (2001).
[14] Yang Yang et al, "Analysis of series resistance of industrial crystalline silicon solar cells by numerical and analytical modelling", Proc. 28th EUPVSEC, pp. 1558-61 (2013).
[15] P. Magnone, M. Debucquoy, D. Giaffreda, N. Posthuma, and C. Fiegna, "Understanding theInfluence of Busbars in Large-Area IBC Solar Cells byDistributed SPICE Simulations", IEEE Journal of Photovoltaics, 5, 2 (2015).

41st European Photovoltaic Solar Energy Conference and Exhibition

ACCURACY OF HYSTERESIS CORRECTION FOR SILICON HETEROJUNCTION SOLAR CELLS – A SIMULATION STUDY

Jonas Kern[1], Hannes Wagner-Mohnsen[2], Johannes Heitmann[1], Matthias Müller[1]
[1] TU Bergakademie Freiberg, Institute of Applied Physics, Leipziger Str. 23, 09599 Freiberg, Germany
[2] WAVELABS Solar Metrology Systems GmbH, Spinnereistr. 7, 04179 Leipzig, Germany
Mail: Jonas.Kern@physik.tu-freiberg.de
Phone: +493731392593

ABSTRACT: In photovoltaic industry, the time used for current-voltage measurements should be as short as possible in order to achieve a high throughput. However, for short measurement times, capacitive effects and hysteresis might occur which cause deviations of the measured current-voltage data from steady state. In this work, we simulate capacitive effects and hysteresis for a silicon heterojunction solar cell using TCAD (technology computer-aided design) and compare the results with analytical approaches. A differential equation for modelling the transient current-voltage curves based on the diffusion capacitance is found to be in excellent agreement with the TCAD simulations. Furthermore, we investigate a method for hysteresis correction, which is shown to be very accurate even for short measurement times. This makes it a feasible approach for usage in photovoltaic industry.
Keywords: Simulation, TCAD, Hysteresis, Silicon, Heterojunction

1 MOTIVATION

Due to the demand for a high throughput in photovoltaic industry, current-voltage (J-V) measurements have to be performed in several tens of milliseconds [1,2]. In this case, capacitive effects can occur and lead to strong deviations of the measured transient J-V curves from the steady state J-V curve and to a pronounced hysteresis. For proper classification of the solar cells, the measured data have to be corrected. One approach for this correction is based on the diffusion capacitance and was proposed by Sinton et al. [3]. Basically, their method considers a weighted average of two transient J-V curves with the only additionally required parameter being the lumped series resistance of the solar cell. While this approach was already successfully applied in practice [4], literature still lacks a systematic study of the accuracy of this hysteresis correction for small measurement times (as far as the authors could survey).

In this work, electrical TCAD (technology computer-aided design) simulations of a silicon heterojunction solar cell are performed in order to understand the effect of different device parameters on the hysteresis and its correction accuracy. The considered device parameters are the finger distance, the wafer doping density and the Shockley-Read-Hall (SRH) recombination parameters τ_n and τ_p. Transient J-V curves are simulated with TCAD and a Python script is used to model the data and to apply the hysteresis correction. The corrected data are then compared to the steady state J-V curve simulated with TCAD.

2 APPROACH

The electrical simulations are performed with Sentaurus TCAD from Synopsys. The model of a silicon heterojunction solar cell is based on literature [5,6] and contains semiconductor ITO layers as well as trap-assisted tunneling through the amorphous silicon layers. A 1-sun correspondent constant optical generation rate is assumed in the crystalline silicon absorber which is sufficient to investigate transient behavior in highly efficient solar cells. The simulated domain is schematically shown in Figure 1.

According to Sinton et al. [3], the capacitive effects due to the diffusion capacitance of an operating solar cell can be modelled by Equation 1:

$$\frac{dJ}{dt} = \frac{1}{R_s}\left[(J_{SS} - J)\frac{kT}{q}\frac{\sqrt{N_D^2 + 4n_i^2\exp\left(\frac{qV_j}{kT}\right)}}{qwn_i^2\exp\left(\frac{qV_j}{kT}\right)} - \frac{dV}{dt}\right] \quad \text{(Eq. 1)}$$

In Equation 1, J is the current density, t is the time, R_s is the lumped series resistance, J_{SS} is the steady state current density, k is the Boltzmann constant, T is the temperature, q is the elementary charge, N_D is the wafer doping density, n_i is the effective intrinsic charge carrier density, w is the wafer thickness, V is the voltage and $V_j = V + JR_s$ is the junction voltage. Equation 1 is an ordinary differential equation of first order for the current density J and can be solved either for the forward case (voltage increases) or the reverse case (voltage decreases).

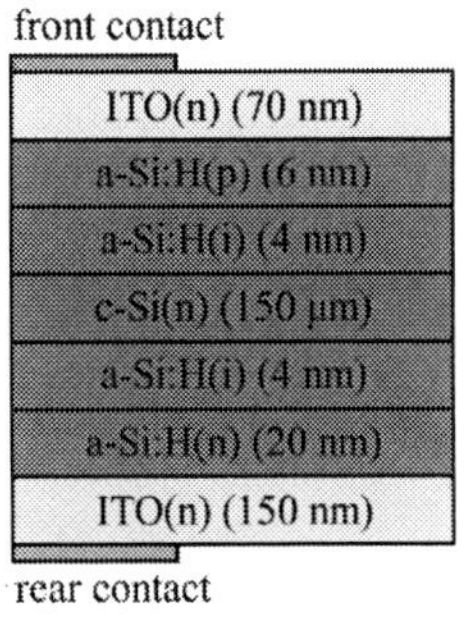

front contact

rear contact

Figure 1: Scheme of the simulated domain of a silicon heterojunction solar cell.

In addition, Sinton et al. [3] provide a formula (Equation 2) for the calculation of the steady state (SS) J-V curve from the forward (F) and the reverse (R) J-V curves:

$$J_{SS} = \frac{J_R\frac{dV_j}{dt}\big|_F - J_F\frac{dV_j}{dt}\big|_R}{\frac{dV_j}{dt}\big|_F - \frac{dV_j}{dt}\big|_R} \quad \text{(Eq. 2)}$$

205

In this work, Equations 1 and 2 are implemented in Python. With J_{SS}, R_s, w, T, N_D and n_i from the steady state TCAD simulation, Equation 1 can be applied to model the transient J-V curves which are then compared to the transient TCAD simulations. Furthermore, with J_F, J_R and R_s from the transient TCAD simulations, Equation 2 can be used to reconstruct the steady-state J-V curve which is then compared to the steady state TCAD simulation. A scheme of the workflow is schematically shown in Figure 2.

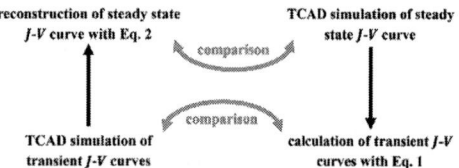

Figure 2: Scheme of the simulation and calculation workflow.

The TCAD simulations are performed for different finger distances, wafer doping densities and Shockley-Read-Hall recombination parameters τ_n and τ_p. The transient TCAD simulations consist of a forward scan from $V = 0.00$ V to $V = 0.76$ V (which is above the open-circuit voltage) and a subsequent reverse scan from $V = 0.76$ V to $V = 0.00$ V. The measurement time for the transient TCAD simulations is varied by changing the voltage scan rate. The lumped series resistance is determined at maximum power point using the triple light level method [7].

3 RESULTS

Figure 3 shows the simulated steady state J-V curve and the simulated transient J-V curves. In addition, the transient J-V curves calculated with Equation 1 are depicted. Near the maximum power point, the forward and reverse currents are lower and higher than the steady state current, respectively. In the forward case, the solar cell (i.e. a large capacitor) is getting charged with excess charge carriers during the voltage scan. On the contrary, for the reverse J-V curve, there is a discharge of the solar cell as excess charge carriers are flowing out of the device. Apart from that, Figure 3 visualizes the great match between the transient J-V curves simulated with TCAD and the J-V curves calculated with Equation 1. This means that the theory of diffusion capacitance by Sinton et al. [3] appears to be sufficient for the description of the transient behavior. These findings are in agreement with results from literature suggesting that the diffusion capacitance is the dominant contribution to the capacitance of a solar cell near its maximum power point [8] and that the diffusion capacitance of silicon heterojunction solar cells is significantly larger than for other solar cell architectures [3].

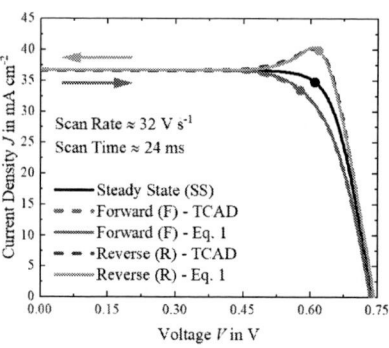

Figure 3: With TCAD simulated steady state (black) and transient (red forward, blue reverse) current-voltage curves, as well as calculated current-voltage curves (green forward, orange reverse) based on Equation 1. The maximum power points are highlighted with dots. In this example, the finger distance is 2.5 mm.

In Figure 4, the simulated and calculated power conversion efficiency is depicted in dependence of the voltage scan rate. It can be seen that the deviation of the transient data from the steady state data strongly increases with increasing scan rate. Again, the results from Equation 1 match the transient TCAD results very well. Furthermore, Figure 4 shows the error of the steady state reconstruction based on Equation 2. As visible, the error is less than 0.3%rel for scan rates up to 32 V s^{-1} (scan time of about 24 ms for one scan direction, i.e. one forward or reverse J-V curve) and increases for higher scan rates.

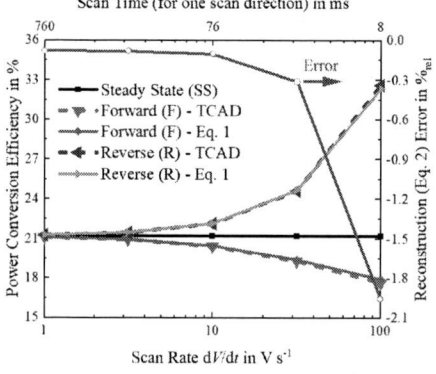

Figure 4: With TCAD simulated steady state (black, depicted just for reference) and transient (red forward, blue reverse) power conversion efficiency in dependence of the voltage scan rate, as well as calculated power conversion efficiency (green forward, orange reverse) based on Equation 1. In addition, the error of the steady state reconstruction (violet) based on Equation 2 is shown. In this example, the finger distance is 2.5 mm.

Device parameter variations for a high scan rate of 100 V s^{-1} (scan time of about 8 ms for one scan direction, i.e. one forward or reverse J-V curve) are visualized in Figure 5 for the finger distance, Figure 6 for the SRH recombination parameters τ_n and τ_p and Figure 7 for the wafer doping density. As visible, the error of the steady state reconstruction based on Equation 2 increases with increasing finger width (i.e. increasing lumped series resistance), increasing SRH recombination parameters τ_n and τ_p (i.e. increasing SRH and effective charge carrier lifetime) and increasing wafer doping density. From these results, it can be deduced that the reconstruction error increases with increasing excess charge carrier density. Apart from that, it can be seen that the reconstruction is very accurate with an error of less than 0.5%rel (0.3%rel for usual parameter ranges) for scan times down to 24 ms and less than 2.0%rel (1.5 %rel for usual parameter ranges) for scan times down to 8 ms.

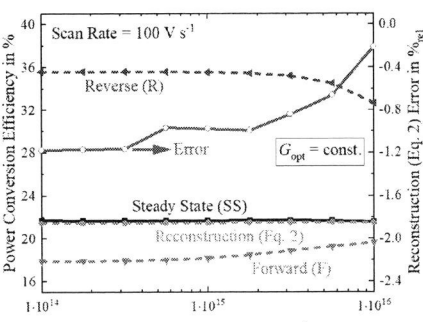

Figure 7: With TCAD simulated steady state (black) and transient (red forward, blue reverse) power conversion efficiency in dependence of the wafer doping density for a scan rate of 100 V s^{-1}. In addition, the steady state reconstruction (magenta) based on Equation 2 is shown, as well as its deviation (violet) from the steady state TCAD data.

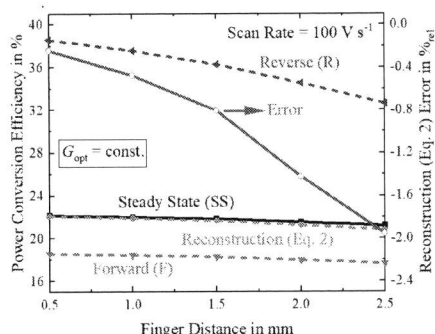

Figure 5: With TCAD simulated steady state (black) and transient (red forward, blue reverse) power conversion efficiency in dependence of the finger distance for a scan rate of 100 V s^{-1}. In addition, the steady state reconstruction (magenta) based on Equation 2 is shown, as well as its deviation (violet) from the steady state TCAD data.

4 CONCLUSIONS

By using electrical TCAD simulations, we validated two equations [3] describing the capacitive effects and their correction for current-voltage measurements of solar cells. It was found that transient current-voltage curves can be accurately modelled by a differential equation considering the transient current flow due to diffusion capacitance. Additionally, the hysteresis correction method proposed by Sinton et al. [3] was found to be very accurate even for short current-voltage measurement times of about 8 ms for one scan direction, where the deviation of the reconstructed steady state data from the real data is less than 1.5%rel. This accuracy demonstrates the feasibility of the method as an approach for hysteresis correction in industry.

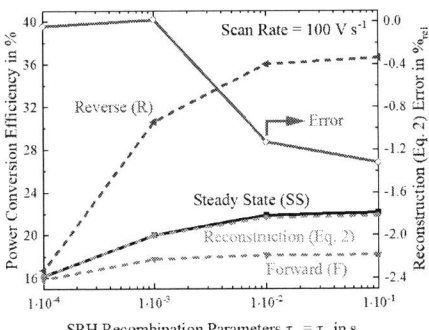

Figure 6: With TCAD simulated steady state (black) and transient (red forward, blue reverse) power conversion efficiency in dependence of the SRH recombination parameters τ_n and τ_p for a scan rate of 100 V s^{-1}. In addition, the steady state reconstruction (magenta) based on Equation 2 is shown, as well as its deviation (violet) from the steady state TCAD data.

5 REFERENCES

[1] WAVELABS Solar Metrology Systems GmbH (2022), wavelabs.de/wcpec-8-bernhard-kloeter-accurate-high-throughput-current-voltage-classification-of-high-efficiency-silicon-solar-cells (accessed on September 22, 2024).
[2] Meyer Burger Technology AG (2021), meyerburger.com/de/eroeffnung-thalheim (accessed on September 22, 2024).
[3] R.A. Sinton, H.W. Wilterdink, A.L. Blum, IEEE J. Photovolt. 7 (2017) 1591.
[4] T. Kemmer, J.M. Greulich, A. Krieg, S. Rein, Sol. Energy Mater. Sol. Cells 248 (2022) 111953.
[5] A. Fell, K.R. McIntosh, P.P. Altermatt, G.J.M. Janssen, R. Stangl, A. Ho-Baillie, H. Steinkemper, J. Greulich, M. Müller, B. Min, K.C. Fong, M. Hermle, I.G. Romijn, M.D. Abbott, IEEE J. Photovolt. 5, 1250 (2015).
[6] J. Kern, J. Heitmann, M. Müller, ACS Appl. Energy Mater. 6 (2023) 2199.
[7] K.C. Fong, K.R. McIntosh, A.W. Blakers, Prog. Photovolt.: Res. Appl. 21 (2013) 490.
[8] A. Cuevas, F. Recart, J. Appl. Phys. 98 (2005) 074507.

6   ACKNOWLEDGEMENTS

This work was funded by the German Federal Ministry for Economic Affairs and Climate Action BMWK (contract no. 03EE1175D) and by the foundation "Dr. Erich-Krüger-Stiftung".

CONTACTLESS CARRIER LIFETIME CHARACTERIZATION OF SILICON HETEROJUNCTION STRUCTURES AT ELEVATED TEMPERATURES

G. Havasi[1, 2], D. Krisztián[1, 2], Zs. Gombás[3], Z. Ádám[3], F. Korsós[1]
[1]Semilab Co. Ltd. 4/A Prielle K. str., H-1117 Budapest, Hungary
[2]Department of Physics, Budapest University of Technology and Economics
[3]EcoSolifer Heterojunction Ltd. 49-51 Csörsz. str. H-1124 Budapest, Hungary
gergely.havasi@semilab.hu

We realized the contactless transient photoconductance decay (PCD) lifetime measurement method at elevated temperatures. We utilized illumination-controlled heating to adjust the sample temperature instead of using a hotplate to simulate the conditions of realistic solar cell operation. Heating and lifetime probing laser pulses are alternating continuously to keep the samples at stable temperature. Pseudo current density – voltage (J-V) characteristics and the emitter saturation current density, J_{0e}, were recorded in the temperature range between 28°C and 62°C from silicon heterojunction (SHJ) precursor samples with and without transparent conductive oxides (TCO). For the proper evaluation, state-of-the-art intrinsic recombination, band gap narrowing and mobility models are applied. In terms of efficiency and fill factor, similar trends are acquired as in the case of conventional I-V measurements of finished solar cells at higher temperatures. Beyond the well-known efficiency reducing effect of the increased intrinsic carrier concentration at higher temperatures, we concluded, that the carrier lifetime gets larger at the maximum power point (MPP) of the cells around $1\text{-}2\cdot10^{15}$ $1/cm^3$ injection level, which results in improved efficiency compared to the expectations. We also found that the slope of the Kane-Swanson fit decreases with increasing temperatures as well, so both bulk and surface recombination properties improve. This effect explains the low temperature coefficient (< 0.3 %/°C) of the SHJ type solar cells.
Keywords: Lifetime, Characterization, Implied J-V

1 INTRODUCTION

The electrical characterization of solar cells is typically performed at room temperature, although the operational temperature of solar panels can reach 60 °C. In this study, we investigate the temperature dependent behavior of silicon heterojunction (SHJ) precursor cells by injection level dependent carrier lifetime measurement to understand the underlying physical processes behind the changes of the solar cell efficiency at an elevated temperature range.

We realized the contactless transient photoconductance decay (PCD) carrier lifetime measurement method at elevated temperatures. We utilized light-controlled heating to adjust the sample temperature instead of using a hotplate to simulate the conditions of realistic solar cell operation.

We investigated the characteristics of a set of SHJ precursor cells with and without transparent conductive oxide (TCO) layers. The investigated temperature range was between 28 °C and 62 °C.

Although temperature coefficients of various solar cells structures are known from conventional I-V testing, only few studies discuss the underlying recombination processes at elevated temperatures[1,2]. We developed a novel and unique platform to perform very accurate carrier lifetime and injection level measurements at elevated temperatures.

The illumination-controlled heating instead of a hotplate is a novel approach to mimic the on-field operation of solar panels. It enables to study the physical phenomena affecting the temperature coefficient of solar cells. Such measurements are important both for the scientific community and for the PV industry as well to further improve the performance of solar panels on-field.

2 THE MEASUREMENT SETUP

The schematic drawing of the experimental setup is depicted in Fig. 1. The PCD measurement is realized using an electrically modulated 980 nm laser and a calibrated radiofrequency eddy current sensor (RF antenna) below the wafer.

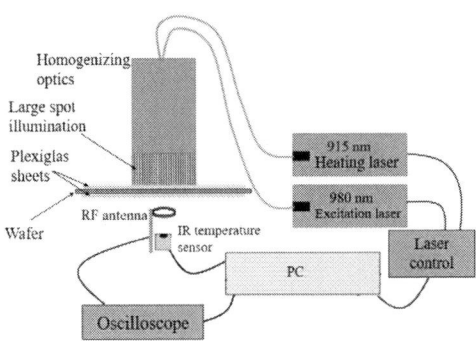

1. Figure: Schematic illustration of the measurement setup.

A high power 915 nm CW laser is responsible for the sample heating. The wafer was placed between two plexiglass sheets as "encapsulation" and therefore serving the thermal insulation of the wafer. The temperature of the wafer was measured using a calibrated in-situ infrared temperature sensor below the RF antenna. Heating and PCD probing laser pulses are alternating periodically with a period time adjusted according to the carrier lifetime of the given sample as illustrated in Fig. 2/a. The induced periodic carrier density waveforms (Fig. 2/b.) enable to obtain the PCD decay curves. The temperature of the sample was controlled by the power of the heating laser. The variation of the wafer temperature during the PCD measurement was negligible (<0.1 °C) thanks to the insulating plexiglass sheets.

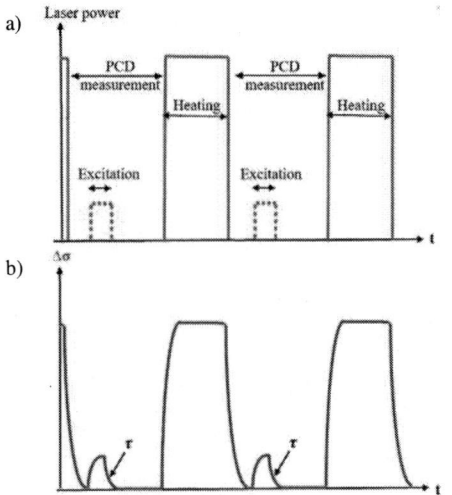

2. Figure: the PCD probing and heating laser pulses (a) and the induced excess carrier density $\Delta n(t)$ (b).

3 RESULTS

The recorded effective carrier lifetime versus injection level $\tau(\Delta n)$ curves of a typical precursor cell without TCO layer is depicted in Fig. 3. at different temperatures. The shows indicate an increasing trend of the carrier lifetime elevating the temperature as well as the Δn position at the maximum power point (MPP) marked as red dots in the graph .

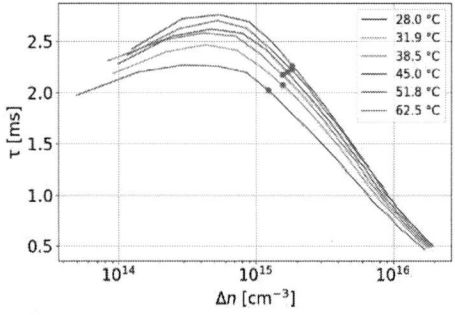

3. Figure: $\tau(\Delta n)$ curves and the MPPs according to the pseudo J-V curves.

The PCD method enables to obtain the pseudo J-V curve in a contactless way since this is governed by the dependence of the recombination rate on Δn [3,4]. The implied open-circuit voltage V_{oc} and short-circuit current J_{diode} density values can be calculated from the recorded $\Delta n(t)$ decay curve as follows:

$$V_{oc}(\Delta n) = \frac{k_B T}{q} \ln \frac{\Delta n(\Delta n + N_{dop})}{n_i^2(T)}, \quad (1)$$

$$J_{diode}(\Delta n) = -\frac{\partial \Delta n}{\partial t} qW. \quad (2)$$

The Green model [5] was applied to determine $n_i(T)$ intrinsic carrier concentration. Then the $V_{oc}(\Delta n)$, and the $J(\Delta n) = J_{sc}(1sun) - J_{diode}(\Delta n)$ functions can be interpreted as the terminal voltage and current values under different

loads at 1 sun illumination level. We used $J_{sc}(1\,sun) = 40$ mA/cm² as a typical data of SHJ cells to convert the short circuit current to suns unit. The J_{sc}-V_{oc} method characterizes the recombination losses and therefore defines an upper limit to the final cell performance at each temperature steps. The 1 sun equivalent implied J-V curves plotted in Fig. 4. different temperatures. The MPPs according to the pseudo J-V curves shift to lower voltages due to the increased n_i.

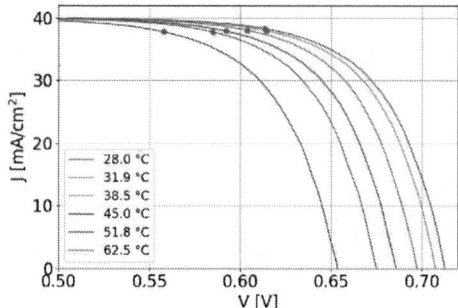

4. Figure: The calculated pseudo J-V curves and the operating points.

The driving force behind the temperature dependency of electrical parameters is the decrease of $V_{oc}(T)$ (Fig. 5.) due to the increased $n_i(T)$.

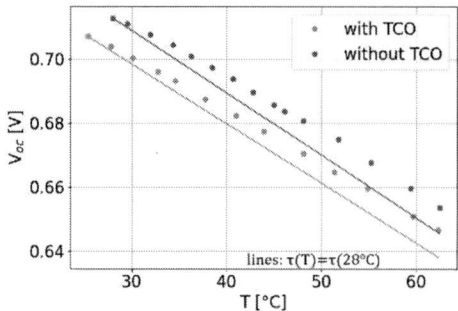

5. Figure: The decreasing $V_{oc}(T)$ with and without TCO layers. The impact of $n_i(T)$ is investigated separately using room temperature recombination properties (demonstrated by solid lines).

We also found a decreasing efficiency $\eta(T)$ and fill factor FF(T) at elevated temperatures (Fig. 6. and Fig. 7.). The average, practically constant relative reduction in efficiency was found to be -0.329 rel%/°C with TCO and -0.299 rel%/°C without TCO.

41st European Photovoltaic Solar Energy Conference and Exhibition

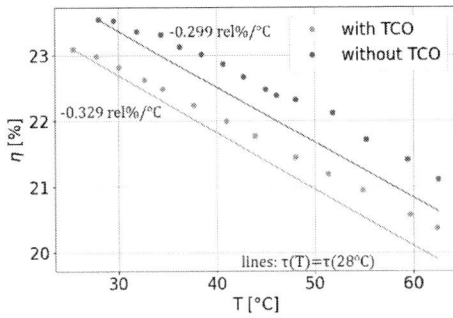

6. Figure: The decreasing $\eta(T)$ with and without TCO layers. The impact of $n_i(T)$ is investigated separately using room temperature recombination properties (demonstrated by solid lines).

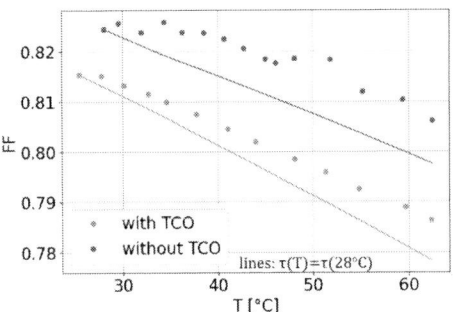

7. Figure: The *decreasing* FF(T) *with and without TCO layers. The impact of $n_i(T)$ is investigated separately using room temperature recombination properties (demonstrated by solid lines).*

These results are similar to the values from typical contact measurements on final cells indicating the reliability of our method [6]. To separate the effect of the temperature dependence of $\tau(\Delta n)$ from the inevitable impact of $n_i(T)$, the measured $V_{oc}(T)$, efficiency $\eta(T)$ and fill factor FF(T) curves are compared to $V_{oc}(T)$, $\eta(T)$ and FF(T) considering only $n_i(T)$ so using $\tau(\Delta n)$ measured at 28°C to calculate them (lines in Fig. 5-7). Our results show efficiency values significantly above this theoretical line, which means the improving recombination properties partially balancing the η reduction due to the increased n_i.

We determined the sum of emitter and back side surface recombination current $J_{0e}+J_{0b}$ using the Kane-Swanson method [7] to investigate the surface recombination processes independently. We applied Auger carrier lifetime corrections using the most recent Auger lifetime model [8], then using the slope of corrected inverse lifetime value we calculated the $J_{0e}+J_{0b}$ sum, since:

$$\frac{1}{\tau_{eff}(\Delta n)} - \frac{1}{\tau_{Auger}(\Delta n)} = \frac{1}{\tau_{SRH}} + \frac{(J_{oe}+J_{ob})(N_{dop}+\Delta n)}{qWn_i^2}. \quad (3)$$

We primarily found $J_{0e}(T) + J_{0b}(T)$ as an exponential increasing function governed by the temperature dependency of $n_i(T)$. The observable small changes are caused by the improved surface passivation. This improvement is rather obvious, if comparing the slope of the Kane-Swanson plots (Fig. 8). In this case the n_i^2 factor, which is included in the J_0 formulas, is compensated, and therefore the slope reflects the temperature dependence of

the surface recombination processes, the diffusion length and diffusivity of charge carriers in the emitter layer, which is proven to be significant.

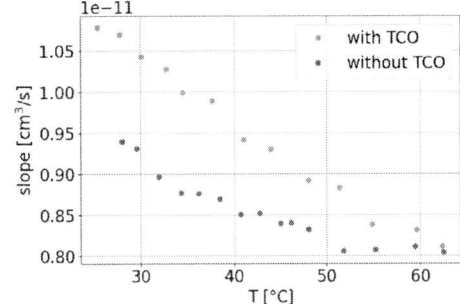

8. Figure: The slope of Kane-Swanson fit in the function of temperature.

We compared the $\eta(T)$ curves (Fig. 6.) and the slopes of the Kane-Swanson plots (Fig. 8.) for similar SHJ cells without TCO. The measured $\eta(T)$ curves are above the theoretical $n_i(T)$ dependent curves for the SHJ cells in both cases, however, this effect is smaller in the case of the TCO layer. The surface recombination is obviously stronger in the presence of the TCO layer, and the slope of the Kane-Swanson fit is changing similarly in both cases. The investigation of samples with TCO layer shows decreased lifetime and similar $\eta(T)$ dependence. However, the Kane-Swanson slope represents more sensitive temperature dependent surface recombination properties.

4 CONCLUSIONS

We realized a PCD measurement setup with alternating probing and light induced heating cycle. Stable elevated temperature can be reached between 30 °C and 60 °C. The main driving force behind the temperature dependency of implied $\eta(T)$ and $J_0(T)$ is $n_i(T)$ which has exponential temperature dependency. Our measurements show that both the bulk and surface recombination processes improved at higher temperatures. So, $\tau(\Delta n)$ increases at elevated temperatures resulting in an increase of the implied $\eta(T)$ and reduction of $J_0(T)$ respectively to the $n_i(T)$ changes. The $\tau(\Delta n,T)$ affects the temperature coefficient of the cells positively.

[1] Priyanka Singh, S.N. Singh, M. Lal, M. Husain, "Temperature dependence of I–V characteristics and performance parameters of silicon solar cell", Solar Energy Materials and Solar Cells, Volume 92, Issue 12, 2008, Pages 1611-1616, ISSN 0927-0248, https://doi.org/10.1016/j.solmat.2008.07.010.
[2] Sun, C., Zou, Y., Qin, C. *et al.* Temperature effect of photovoltaic cells: a review. *Adv Compos Hybrid Mater* **5**, 2675–2699 (2022). https://doi.org/10.1007/s42114-022-00533-z
[3] R. A. Sinton, A. Cuevas and M. Stuckings, "Quasi-steady-state photoconductance, a new method for solar cell material and device characterization", Conference Record of the Twenty Fifth IEEE Photovoltaic Specialists Conference - 1996, Washington, DC, USA, 1996, pp. 457-460, doi: 10.1109/PVSC.1996.564042.

[4] M. J. Kerr, A. Cuevas, R. A. Sinton, "Generalized analysis of quasi-steady-state and transient decay open circuit voltage measurements" Journal of Applied Physics 91, 399 (2002)

[5] M. A. Green, "Intrinsic concentration, effective densities of states, and effective mass in silicon" J. Appl. Phys. 67, 2944 (1990); doi: 10.1063/1.345414

[6] Le AHT, Dréon J, Michel JI, et al. Temperature-dependent performance of silicon heterojunction solar cells with transition-metal-oxide-based selective contacts. Prog Photovolt Res Appl. 2022; 30(8): 981-993. doi:10.1002/pip.3509

[7] D. E. Kane, R. M. Swanson, "Measurement of the emitter saturation current by a contactless photoconductivity decay method", Stanford Electronics Laboratories, McCullough 204 Stanford, 1985

[8] T. Niewelt, B. Steinhauser, A. Richter, B. Veith-Wolf, A. Fell, B. Hammann, N. E. Grant, L. Black, J. Tan, A. Youssef, J. D. Murphy, J. Schmidt, M. C. Schubert, S. W. Glunz, "Reassessment of the intrinsic bulk recombination in crystalline silicon", Solar Energy Materials and Solar Cells, 235 (2022), 111467. https://doi.org/10.1016/j.solmat.2021.111467.

41st European Photovoltaic Solar Energy Conference and Exhibition

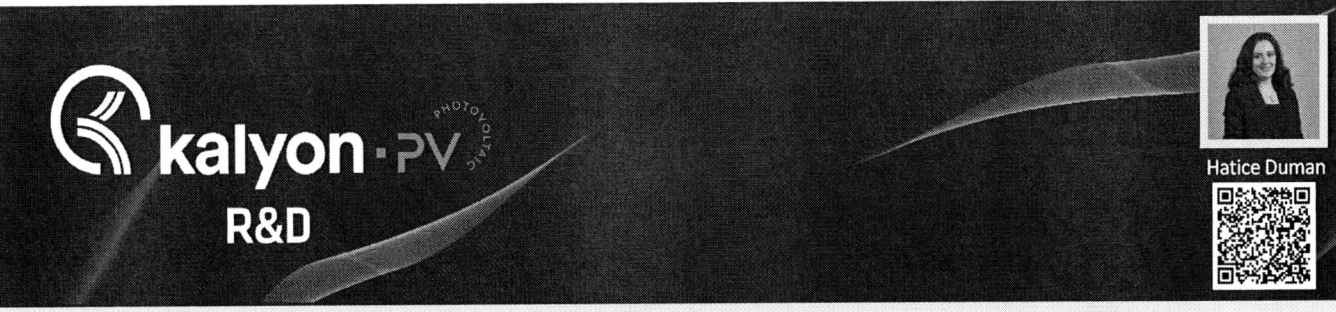

kalyon·PV
R&D

Hatice Duman

BIAS LIGHT INTENSITY EFFECT ON EQE ANALYSIS FOR PERC SOLAR CELL

*Hatice Duman, Özlem Coşkun

Kalyon PV Research and Development Center, Kalyon Güneş Teknolojileri Üretim A.Ş., 06909 Ankara, Türkiye
e-mail: hduman@kalyonpv.com

INTRODUCTION & MOTIVATION

PERC solar cells are currently dominating the PV market. Most installed PV modules have PERC-type cells, with efficiencies ranging between 21.5-23.5%. As solar energy continues to play a crucial role in the global transition to sustainable energy sources, understanding and optimizing Quantum Efficiency (QE) becomes increasingly vital. Therefore, a more precise information on the quantum efficiency of industrially produced PERC cells is crucial for reliable yield estimations. In essence, quantum efficiency quantifies the ability of a photovoltaic device to harness incident energy, shedding light on the overall conversion process. The External Quantum Efficiency (EQE) is formally defined as the ratio of collected charge carriers per incident photon (φin) and is quantified by measuring the short circuit current response when illuminated with monochromatic light.

In this study, we report on the effects of bias light intensity conditions in solar cell testing for obtaining meaningful absolute EQE values. Meanwhile, it examines the importance of bias light, its purpose, characterization, and impact on quantum measurements, the detailed aspects of solar cell characterization, explaining how non-linearity and underlying mechanisms cause deviations from the expected proportionality in EQE measurements.

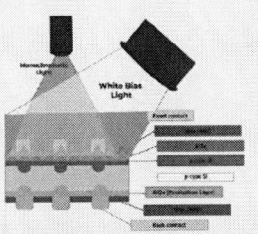

$$EQE(\lambda) = \frac{J_{SC}(\lambda)/q}{\phi_{in}(\lambda)}$$

$$EQE_{abs} = \frac{\Phi}{\int_0^{1sun} \frac{1}{EQE_{diff}} d\Phi}$$

EXPERIMENTAL APPROACH

In the scope of this work, we conducted a preliminary characterization of two efficiency groups of M10 PERC cells, with efficiencies of η =22.0% and η=22.5%, respectively. We used the LOANA tool from pv-tools to measure the External Quantum Efficiency (EQE) under various bias light conditions.

Different bias light intensities were applied to assess their impact on measurement results. Additionally, the sensitivity of cells from different efficiency groups to bias light was compared, using various bias light levels and wavelengths under different conditions.

Scenario	Efficiency Class (22.0%)				Efficiency Class (22.5%)			
Bias ramp Wavelength (λ)	1050 nm		750 nm		1050 nm		750 nm	
Different Bias Light Intensity (suns)	0,00 0,05 0,10 0,15 0,20 0,25 0,30				0,00 0,05 0,10 0,15 0,20 0,25 0,30			

RESULTS

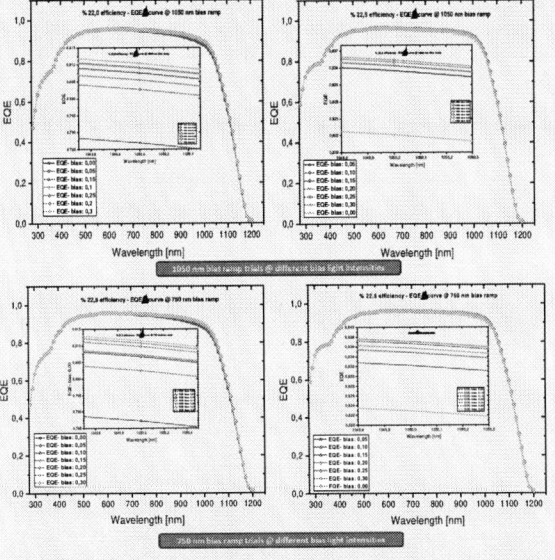

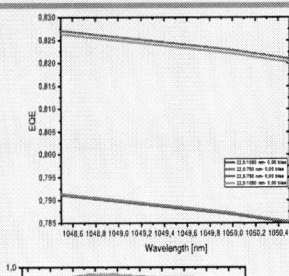

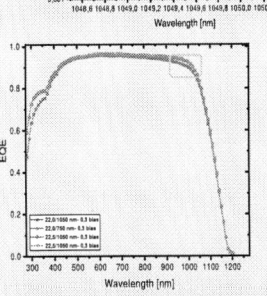

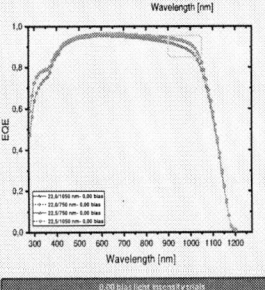

CONCLUSIONS

❖ Through this systematic analysis, we studied the varying sensitivities of cells with different efficiency to bias changes, providing valuable insights for researchers and practitioners in the field of photovoltaics. Bias light is used in quantum efficiency (QE) measurements of solar cells to simulate real-world conditions, ensuring that laboratory results are consistent with the cell's actual performance. It influences the spectral response by affecting the generation, separation, and collection of charge carriers.

❖ This work has studied bias light effect on EQE analysis. When the bias light intensity was 0.00 (i.e., without using bias light), the quantum efficiency was obtained at the lowest value under all conditions. While the effect of bias change on quantum efficiency can be observed in low-efficiency cells, our research reveals that the effect of bias light intensity change on quantum efficiency is much less in high-efficiency cells. There is no significant difference in the change of the bias ramp.

❖ Bias light can also correct spectral mismatches, making the measurements more accurate and representative of real-world conditions. This study aims to better understand the relationship between bias conditions and quantum efficiency, ultimately improving the accuracy of efficiency assessments in solar cell technologies.

❖ Higher intensity means more incident photons, leading to increased electron-hole pair generation, enhanced charge carrier generation and so higher QE. Additionally, higher bias intensity may reduce charge carrier recombination, allowing more carriers to contribute to the photocurrent and increasing QE. On the other hand, higher intensity can influence the electric field, improving charge carrier collection and contributing to higher QE.

REFERENCES

1) Fischer, B. (2003). "Loss analysis of crystalline silicon solar cells using photoconductance and quantum efficiency measurements." Cuvillier Verlag.
2) https://enlitechnology.com/blog/qe/quantum-efficiency-01/ Access Time:31.01.2024 16:32
3) LOANA-PVtools – User Manual

41st European Photovoltaic Solar Energy Conference and Exhibition

**IMPROVED ACCURACY OF PHOTOLUMINESCENCE IMAGES
FOR QUALITY CONTROL IN SOLAR CELL PRODUCTION**

[1]Robin Wienberg, [1]Jonas Haunschild, [1]Saravana Kumar, [2]Jurriaan Schmitz, [1]Stefan Rein
[1]Fraunhofer Institute for Solar Energy Systems (ISE), Heidenhofstraße 2, 79110 Freiburg, Germany
[2]MESA+ Institute for Nanotechnology, University of Twente P.O. Box 217, 7500 AE Enschede, The Netherlands

ABSTRACT: Photoluminescence (PL) imaging has become an important method for assessing solar cell quality. In practice, laser or camera inhomogeneities can cause a 10-40% PL-intensity variation across the PL-image of a sample, which obscure the valuable PL-information. Using the PL-image of a homogeneous reference sample, the system-induced inhomogeneities are quantified and corrected for. But as there is no perfect homogeneous reference sample, this reference method introduces new inhomogeneities to the corrected PL-image. In this paper, a method is presented for inline-PL-imaging that can deliver more accurate PL-images without the need for a homogeneous reference sample. In this correction method, a sample needs to be measured twice, with a 90° rotation in-between. The rotation enables the distinction between system imperfections and sample imperfections. A rigorous approach is followed to overlay the two PL-images precisely and merge them into a corrected PL-image. The result can be used as a reference to correct all other PL-images (assuming that the laser was not restarted in-between the measurement). With that procedure, we introduce an effective workaround that helps to strongly improve the representation of sample imperfections in PL-images within a production environment without the need for a homogenization of the illumination.
Keywords: quality control, photoluminescence-imaging, inline metrology, system imperfections, calibration

1 INTRODUCTION

In a solar cell, all generated electron hole pairs that do not recombine (via Shockley-Read-Hall (SRH), via band-band or via auger recombination) and do not disappear via shunts contribute to the electrical current. Under open-circuit conditions (OCC), the total recombination rate equals the generation rate. Though, the rates of the individual recombination processes depend on the doping density and the SRH lifetime (which is independent from the generation rate for a low-level injection), in addition to the generation rate. One of the basic performance parameters of a solar cell is the open-circuit voltage. It is completely determined by the temperature, the physical constants and the band-band recombination rate under OCC and increases logarithmically with the latter. Already in the as-cut state, wafer quality can be inspected using the photoluminescence (PL) effect by measuring the rate of the band-band recombination under OCC upon illumination. The goal of PL-imaging is to create a two-dimensional top view map of the sample quality.

The illumination sources of inline-PL-units used for the quality control of solar cells, often exhibit significant spatial inhomogeneities. The resulting gradient of the generation rate in the sample obscures the valuable information about the sample quality variations within a PL-image. A precise standard calibration is difficult to define as even high-quality reference samples have inhomogeneities and furthermore may lose quality over time. Using the PL-image of an inhomogeneous reference sample can lead to incorrectly assessed sample qualities [1]. This work derives a correction method of system imperfections of two inline-PL-modules for as-cut wafers and precursors to improve the accuracy of PL-images without the need for a homogeneous reference sample.

2 PL-MODULES AND THEIR IMPERFECTIONS

All experimental results were obtained with two inline-inspection systems (referred to below as WIS and PIS), which analyzed either silicon (Czochralski grown) as-cut wafers (WIS) or silicon heterojunction (SHJ) precursors (PIS). A conveyor belt transports the sample

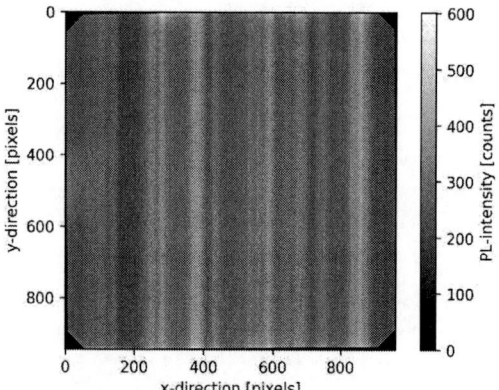

Figure 1: Uncorrected (raw) PL-image of an as-cut wafer, measured with an inline-PL setup in an automated wafer inspection system (WIS). The vertical stripe pattern is a system artefact and explained in chapter 2.

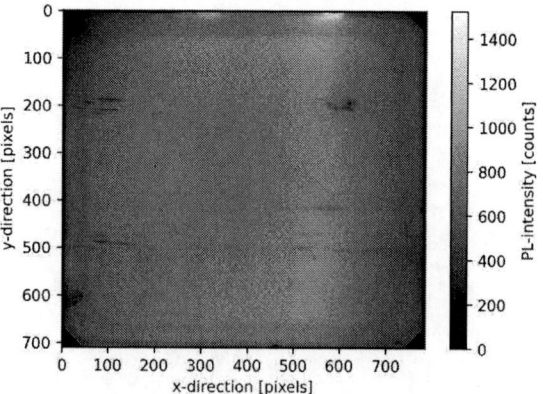

Figure 2: Raw PL-image of a silicon heterojunction (SHJ) precursor measured with an inline-PL setup in an automated precursor inspection system (PIS). The vertical stripe pattern is less dominant as only one laser-diode is used.

through the whole inspection system. In the integrated inline-PL-modules, laser-diodes line-wise illuminate the inserted samples. The absorbed light generates electron-hole pairs in the sample. The light reflected at the surface of the sample is filtered out, and a line scan camera measures the remaining PL-radiation. The transport direction of the sample on the belt occurs in y-direction in the presented PL-images. A single snapshot from the line scan camera consists of a series of 1024 pixels that form a row in the x-direction of the PL-image, which corresponds to a length of 168 mm. The y-direction is built by placing 1024 rows of PL-intensities on top of each other, which are measured for different positions of the sample in the PL-module. The laser diodes in both systems operate at 808 nm wavelength and have an intensity of 200 W in the WIS unit and of only 3 W in the PIS unit (which is sufficient in PIS due to a higher lifetime of SHJ precursors resulting in higher excess carrier densities and thus higher rates of band-band recombination). To realize powerful line-wise illumination in x-direction across the whole sample width, in WIS PL-module, seven laser diodes are mounted in a row in x-direction. In the PIS PL-module, only one laser diode is used and combined with a one-dimensional beam expander. In the y-direction, the laser intensity profile follows a Gaussian shape with a 4σ-width of (1.0 ± 0.3) mm [1].

Examples of raw PL-images are shown in Figure 1 and 2. When the samples are illuminated line by line, a peculiarity occurs which is explained in detail below, as it later (in chapter 3) plays a key role to correct the PL images without considering a sample as homogeneous: All PL-intensity values with the same x-coordinate (within the same column) of a PL-image represent locations on the sample, which were all measured at the same position in the PL-module. During a measurement, the time dependence of the illumination intensity can be neglected. This means that all points of the sample that are assigned to one column in the PL-image are irradiated by the light source with the same intensity during the measurement of their PL-intensity.

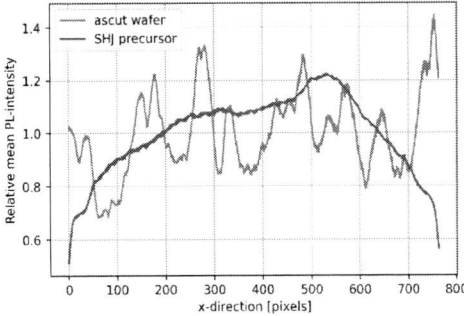

Figure 3: Raw PL-intensity profiles in x-direction for the as-cut wafer (grey, measured with PL-imaging in WIS) and the SHJ precursor (blue, measured with PL-imaging in PIS), shown in Figure 1 and 2. If only this data is given, it is unclear whether the sample is completely homogeneous or has significant inhomogeneities.

Changes in the PL-intensity along a column (y-direction) of a proper inline-PL-image represent changes in the sample quality only. On the contrary, PL intensity variations observed in x-direction (rows in the PL-image) may arise from either variations of the sample quality or from spatial inhomogeneities of the illumination intensity

(illustrated in Figure 3). For this reason, there is a one-dimensional stripe pattern in the x-direction of the raw PL-image that describes the intensity profile of the line illumination. So, the stripe pattern in inline-PL-images is an artifact and presents a common system imperfection of inline-PL-modules. For sufficiently homogeneous samples, the stripe pattern can be eliminated by homogenizing the mean (averaged in y-direction) PL-intensity of the columns of a PL-image (procedure A "mean correction"). Furthermore, the PL-intensity profile in x-direction of a sufficiently homogeneous sample can be used to correct PL-images of other samples (procedure B "mean reference correction"). However, a fundamental drawback of the mean correction (procedure A) is that any remaining PL variations in x-direction of a mean corrected PL-image cannot be interpreted as quality variations in the sample. Also, small inhomogeneities in the reference sample already lead to unprecise corrections when conducting the reference correction (procedure B), like remaining stripes in reference corrected PL-images. If the samples are not approximated as homogeneous, a different approach is required for the calibration of the PL-module. The suggested procedure is to restore the mean PL-intensity change in the x-direction caused by the imperfections of the sample and obscured by the imperfections of the system [2].

3 NEW APROACH: SUPERPOSITION

3.1 Creating a sample independent reference – elimination of material variations by superposition

The new approach presented here makes use of the invariance of the artifacts under translations along the y-axis: PL-intensity variations along a column of a PL-image exclusively come from changes in the sample quality, but variations along a row can also be artifacts. This means that if a sample is measured always in the same orientation, only the sample quality change in one direction is precisely known. If a sample is measured twice with a 90° rotation in-between, the quality changes in both directions can be determined (procedure C "correction by superposition"). This procedure allows the sample quality to be mapped independent from the laser profile in x-direction without assuming a homogeneous reference sample. The sample which is measured twice, with a 90° rotation in-between, is referred to below as "superposition sample".

To merge the information of both PL-images of the superposition sample, it is necessary to determine a transformation function. This function assigns every pixel of one PL-image to a different pixel of the second PL-image, whereby both pixels describe the same location on the sample. To define the transformation, at least four parameters are necessary: (i and ii) the shift of the x- and y-coordinates of the sample center between the two PL-images, (iii) the angle of rotation between the orientations of the sample in the two PL-images and (iv) a constant stretching factor of both PL-images in one direction. The stretching constant corrects the distorted aspect ratio of the raw PL-image, which comes from another imperfection of the system. These parameters can be inserted in a Galilean transformation to obtain the transformation function [3]. All four parameters can be calculated roughly by determining the edges of the superposition sample in both PL-images. To reduce the uncertainty of the transformation parameters, these parameters can be varied intentionally by a small factor: For every new parameter

41st European Photovoltaic Solar Energy Conference and Exhibition

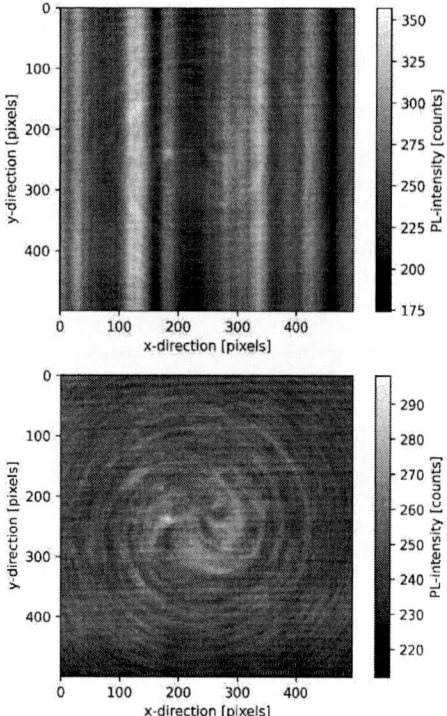

Figure 4: PL-images of an as-cut wafer: (top) raw data and (bottom) data after correction by superposition. The stripe pattern vanishes after the correction and the Czochralski-specfic ring structure and saw marks dominate.

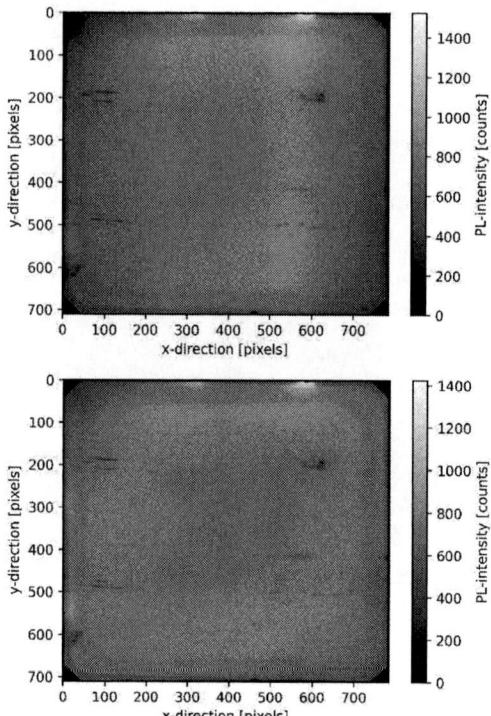

Figure 5: PL-image of an SHJ precursor: (top) raw data and (bottom) data after correction by superposition. In the corrected PL-image, one can hardly identify any stripe pattern.

set, the transformation function needs to be updated, which sorts the PL-intensities of the two PL-images in pairs (transformation pairs). The parameter set that maximizes the correlation coefficient between the transformation pairs is used for the further procedure, as the resulting deviation between two PL-intensities of a transformation pair comes only from the inhomogeneous illumination.

The next step is to calculate a subtractive image by subtracting the PL-intensities of the transformation pairs. Both PL-intensities of a transformation pair describe the same sample quality but measured at different generation rate (due to the inhomogeneities of the laser-diodes in the x-direction). By subtracting the PL-intensities of a transformation pair, the dependence on the imperfections of the sample mostly vanishes. Thus, the PL-intensities of the subtractive image, averaged along the y-direction, can be used as a reference to correct the PL-images of other samples, which were measured in the same run as the superposition sample. To apply the reference on the raw PL-images of the superposition sample, the reference is subtracted from the raw PL-intensity column by column. The resulting PL-image is normalized, so that it has the same mean PL-intensity as initially [4].

3.2 Reweighting the reference – correcting PL-images of samples with deviating material parameters

Furthermore, the reference can also be applied to PL-images of another arbitrary sample, which was measured in the same run as the reference. If the arbitrary sample has significantly different material properties, a reweighting of the reference might be necessary. This is because the

relationship between the PL-intensity and the generation rate is not proportional and also depends on the doping density and (via the excess carrier density) on the SRH lifetime of the sample. In the next part, it is summarized how to obtain the functional dependence between the intensity of the light source and the PL-intensity for a given doping density and SRH lifetime:

The approach is based on the fundamental relation (1) in an OCC describing the balance between the generation rate G, the SRH recombination rate R_{SRH}, the band-band recombination $R_{band-band}$ and the auger recombination R_{auger}.

$$G = R_{SRH} + R_{band-band} + R_{auger} \qquad (1)$$

After inserting our knowledge about these recombination processes in (1), the excess carrier density Δn needs to be calculated in dependence of the generation rate G by solving the resulting cubic equation (2) for Δn.

$$G = \frac{\Delta n}{\tau_{SRH}} + B\left(\Delta n + N + \frac{n_i^2}{N}\right)\Delta n$$
$$+ (\Delta n + N)\left(\Delta n + \frac{n_i^2}{N}\right)(C_e(\Delta n + N) \qquad (2)$$
$$+ C_h(\Delta n + n_i^2/N))$$

with the SRH lifetime τ_{SRH}, the coefficient B of the band-band recombination, the doping density N, the intrinsic charge carrier density n_i and the coefficients C_e and C_h of the auger recombination [5]. This step includes the approximation of τ_{SRH} as constant, which is allowed for low injections [6]. In a second step, the rate of the band-band recombination can be calculated by inserting

216

the dependency of Δn on G in equation (3).

$$R_{band-band} = B\left(\Delta n + N + \frac{n_i^2}{N}\right)\Delta n \qquad (3)$$

The generation rate differs only by a factor from the power of the light source as illustrated by equation (4).

$$G = \frac{P}{A}\frac{1}{hf}(1-R)a \qquad (4)$$

with the laser power per illuminated area P/A (laser intensity), the Planck constant h, the frequency f of the light source, the reflectance R at the surface and the light absorption coefficient a in the bulk of the sample [7]. The band-band recombination rate only differs by a factor from the PL-intensity. This factor can be determined by calculating the mean band-band recombination (from the power of the laser-diodes) and divide it by the mean PL-intensity.

Using equations (2) to (4), the relationship between the laser intensity and the PL-intensity can be determined. As this relationship is not proportional, the reference needs to be reweighted by calculating the PL-intensities of the reference as if it has the same properties as the other sample. The reweighted reference is applied to the arbitrary sample as if it is the superposition sample (as before by subtracting and renormalizing). For a 20% raw PL-intensity variation across the sample (like in PIS), the relationship between the raw and the reweighted PL-intensity of the reference can be approximated to be linear. The parameters of the linear function are obtained by minimizing the chi square of an expected linear relationship between the PL-intensities of the superposition sample and the other sample. This procedure allows to apply the correction by a reference on PL-images of other samples without knowing the properties of the samples.

Combining the subtractive image and the reweighting of the reference allows to correct the PL-images of all samples, which were measured in the same run as the reference.

4 RESULTS

The resulting PL-images of the correction by superposition can be seen in Figures 4 and 5. The human eye is well trained in finding patterns in images. Nevertheless, it is hard to identify any left artifacts in these corrected PL-images, which do not describe the sample. To check if PL-intensity variations along the x-direction of a corrected PL-image represent the actual sample quality variations (and not artifacts of the system or of the correction procedure), a more precise analysis is required:

A SHJ precursor with fewer defects was measured four times with rotations around the y-axis by 180° (flip over) and z-axis by 90° (usual change in orientation). By mirroring the PL-images of the flipped SHJ precursor on the y-axis, the same sample imperfections can be analyzed with a mirrored laser profile. The correlation of a PL-image with the flipped one shows the influence of the SHJ precursor imperfections compared to the influence of the system imperfections on the PL-image. The corresponding correlation coefficient was calculated for the uncorrected PL-images, for the mean corrected PL-images (with the approximation of a homogeneous SHJ precursor) and for the PL-images corrected by superposition.

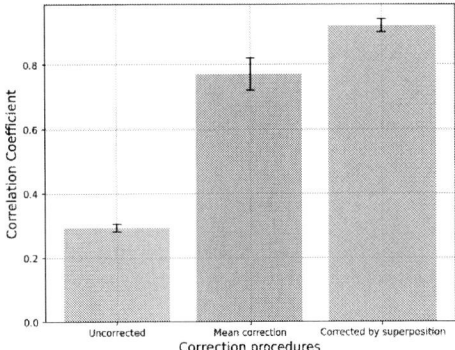

Figure 6: Correlation of the mean PL-intensities of the columns of two (corrected) PL-images of the same sample, measured with different laser profiles. 50 rows of a rather homogeneous part (with a standard deviation of 7% in the raw data) of a SHJ precursor were used to analyze the correlation.

The results are illustrated in Fig. 6. The correlation coefficient increases from 0.293 ± 0.012 (uncorrected) to 0.77 ± 0.05 (with mean correction) and 0.92 ± 0.02 (with superposition correction), which underlines the improvement in the representation of the sample with the correction by superposition. Last, the corrected PL-images of many different samples are evaluated.

To quantify the success of a correction of a PL-image with resulting PL-intensities $I(x,y)$, a stripe strength S is introduced in equation (5) which is based on the derivation of the PL intensity in x-direction.

$$S = \left\langle \left\langle \left| \frac{d}{dx} I(x,y) \right\rangle_y^2 \right|_x \right\rangle \qquad (5)$$

The PL-images of 1000 SHJ precursors were measured with PIS. A sample which looked particularly homogeneous was chosen as reference sample (it is the sample, which is already shown in Figure 2 and 5). By approximating it as homogeneous, all PL-images were reference-corrected (using the approximation of a linear (not proportional!) relationship between the PL-intensities of the reference and another sample).

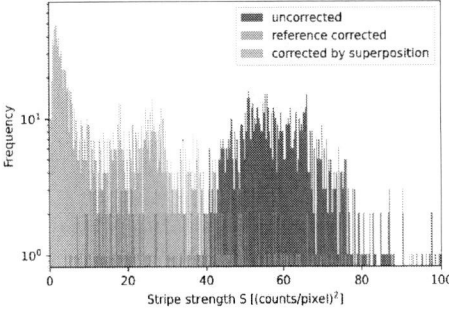

Figure 7: Histogram of S (eq. 5) of the PL-images of 1000 SHJ precursors; raw (blue), reference corrected (orange) or corrected by superposition (green).

The reference sample was also measured a second time with a 90° rotation in-between to enable the distinction between system imperfections and sample imperfections. Overlaying the two PL-images results in another reference. This new reference was used to correct (like before when applying the reference correction) the PL-images of all samples (correction by superposition).

An evaluation shows that the stripe strength of the raw PL-images is $S_{raw}=(60\pm10)$(counts/pixels)², the reference correction results in $S_{ref} = (25\pm9)$(counts/pixels)² and the correction by superposition results in a stripe strength of $S_{sp} = (5\pm4)$(counts/pixels)². So, the artifacts of the system were reduced by $\frac{S_{raw} - S_{sp}}{S_{raw}} = 0.6\pm0.2$ by the reference correction and by 0.9 ± 0.2 by the correction of superposition (see Figure 7).

5 CONCLUSIONS

If a sample is known to be homogeneous, the mean correction can be applied. In addition, this sample can be used as a reference sample to correct PL-images of other samples without the need of a superposition correction. However, if the reference sample is not homogeneous, a mean correction destroys information, and the respective reference correction obscures the information about the quality change of other samples in x-direction. In this study, a correction procedure has been developed which does not rely on the approximation of a homogeneous sample. Though, the superposition correction relies on the approximation of a time-independent laser profile in the x-direction (which is also needed for the reference correction). For the correction by superposition, a sample is measured twice with a 90° rotation in-between. The rotation makes it possible to distinguish between the sample imperfections and the system imperfections in the x-direction of a raw PL-image. For this purpose, the two resulting PL-images are merged and a reference is calculated, which describe the system imperfections. The rotation has the same function as if a small sample is measured with the given setup while being transported along the illuminated line (which needs a more complex setup). The resulting reference can be applied to the PL-images of all samples measured in the same run - even if they have different material parameters. As the laser intensity is in general not proportional to the PL-intensity, the reference should be reweighted when correcting PL-images of samples with deviating material parameters. An evaluation shows a reduction of the artifacts of the system by (60 ± 20) % by the reference correction and by (90 ± 20) % by the correction by superposition.

6 ACKNOWLEDGEMENTS

This work was funded by the German Federal Ministry for Economic Affairs and Climate Action within the project SALSA under contract number 03EE1096A.

7 REFERENCES

[1] H. Höffler et al. Comparison of line-wise pl-imaging and area-wise pl-imaging. 2017. doi: 10.1016/j.egypro.2017.09.341.

[2] H. Höffler. Lumineszenz-Imaging Anwendungen in industrieller Fertigungsumgebung von Silicium-Solarzellen. PhD thesis, Albert-Ludwigs-Universität Freiburg, 2015.

[3] V. I. Arnold. Mathematical Methods of Classical Mechanisc. Springer New York. 1989. ISBN 0-387-96890-3.

[4] R. Wienberg. Untersuchung und Korrektur von experimentell bedingten Inhomogenitäten in Photolumineszenz-Topogrammen. Bachelor's thesis, Albert-Ludwigs-Universität Freiburg, 2023.

[5] A. Smets et al. Solar Energy - The physics and engineering of photovoltaic conversion, technologies and systems. 2016. ISBN 978-1-906-86032-5.

[6] S. Rein. Lifetime Spectroscopy – A Method of Defect Characterization in Silicon for Photovoltaic Applications. Springer, Berlin Heidelberg New York, 2005. ISBN 978-3-540-25303-7.

[7] J. Haunschild. Lumineszenz Imaging - Vom Block zum Modul. Fraunhofer Verlag, 2012. ISBN 978-3-8396-0397-0.

41st European Photovoltaic Solar Energy Conference and Exhibition

SIMULATION AND DESIGN OPTIMIZATION OF INTERDIGITATED BACK CONTACT SILICON SOLAR CELLS WITH DOPANT-FREE ASYMMETRIC HETERO-CONTACTS

You-An Li[1], Chun-Ping Lin[2,3], Ying-Yuan Huang[2,4*]

[1] Institute of Photonic System, National Yang-Ming Chiao Tung University, Tainan, Taiwan
[2] Institute of Microelectronics, Department of Electrical Engineering, National Cheng Kung University, Tainan, Taiwan
[3] Photovoltaic Technology Division, Green Energy and Environment Research Laboratories, Industrial Technology
[4] Program on Semiconductor Manufacturing Technology, Academy of Innovative Semiconductor and Sustainable Manufacturing, National Cheng-Kung University, Tainan, Taiwan
*Corresponding author (yingyuanhuang@gs.ncku.edu.tw)

ABSTRACT: In this study, we employ Quokka2 simulations to optimize the design of dopant-free interdigitated back-contact (IBC) solar cells. The investigation focuses on the effects of varying the widths of the hole and electron transport layers (HTL and ETL) and the gap between them on the efficiency of solar cells with N-type and P-type crystalline silicon (c-Si) substrates. The results indicate that optimizing the gap from 150 μm to 5 μm enhances the efficiency of N-type c-Si substrate cells, increasing from 24.6% to 25.1%, assuming the 5 ms midgap SRH bulk lifetime. Similarly, the efficiency of P-type c-Si substrate cells improves from 24.6% to 25.4%. These findings indicate a potential approach for manufacturing high-efficiency dopant-free silicon-based IBC solar cells by optimizing the dimensions of the interlayers.
Keywords: Silicon-based solar cells, interdigitated back contact, dopant-free materials, semiconductor process, device simulation

1 INTRODUCTION

Owing to the elimination of metal shading on the front surface and the excellent passivation quality of hydrogenated amorphous silicon films, interdigitated back contact (IBC) silicon solar cells have emerged as one of the most promising technologies for approaching the theoretical efficiency limits of crystalline silicon (c-Si) solar cells. A distinguished example is the heterojunction back-contact (HBC) solar cell, which integrates an interdigitated back-contact structure with doped/intrinsic amorphous silicon (a-Si:H) heterojunctions, achieving an efficiency of 26.7% [1]. Nevertheless, manufacturing IBC cells typically requires multiple high-precision patterning steps, such as photolithography, to define the electron and hole transporting layers (ETL and HTL, respectively). This complexity poses a significant challenge for the mass production of IBC cells. In this study, we explore the feasibility of employing dopant-free asymmetric hetero-contacts, patterned using cost-effective, mass-producible shadow masks during the evaporation processes, and simulate the potential of this approach to achieve high-efficiency performance. Figure 1 compares the complexity of the photolithography and shadow mask processes. While photolithography offers superior precision than the shadow mask process, it comes at the cost of increased complexity and investment.

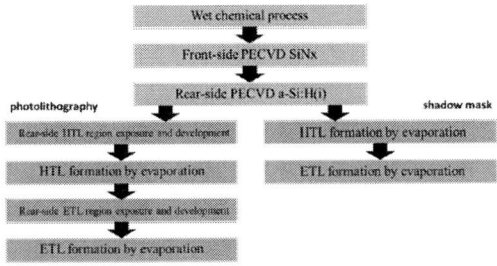

Figure 1: Comparison of the photolithography process and the shadow mask process.

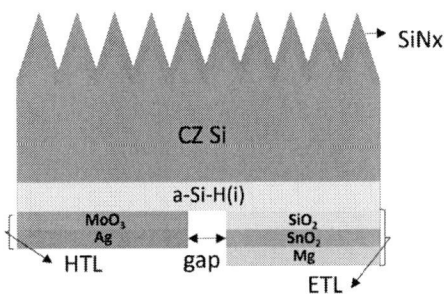

Figure 2: Schematic of dopant-free asymmetric hetero-contacts IBC solar cells

2 SIMULATION OF OPTIMAL HTL AND ETL WIDTH, WITH P- OR N-TYPE C-SI SUBSTRATES

To investigate the optimal design and performance potential of IBC silicon solar cells with dopant-free asymmetric hetero-contacts defined by shadow masks, we utilized Quokka 2 [2]. for device simulation. Materials commonly utilized for this purpose include alkali/alkaline earth ionic compounds, such as lithium fluoride (LiF)[3], tin dioxide (SnO_2) [4], and transition metal oxides like molybdenum oxide (MoO_x). We selected the a-Si:H/MoO_3/Ag [5] and a-Si:H/SiO_2/SnO_2/Mg [4] as the HTL and ETL passivation layers and electrodes due to their superior passivation quality while maintaining good contact resistivity [6]. The schematic of the dopant-free IBC solar cell is shown in Fig. 2. The key parameters used in the simulation are detailed in Table 1. To simulate the cell efficiency that standard shadow-mask processes can achieve, the initial gap width is assumed to be 150 μm wide.

Figure 3 shows the simulated IBC cell efficiencies as a function of ETL and HTL width, with n-type and p-type Si substrate, respectively. For n-type Si substrates, an optimal efficiency of 23.2% was achieved with an HTL width of 1000 μm and an ETL width of 220 μm. In contrast, for p-type Si substrates, an optimal efficiency of 23.58% was attained with an HTL width of 330 μm and an ETL width of 400 μm.

Table I: Key parameters used in Quokka2 simulations

Parameters	Values
Hole Transport layer (HTL)	a-Si:H/MoO$_3$ /Ag [5]
HTL passivation J$_0$ (fA/cm^2)	10 [5]
HTL contact resistivity ρ (mΩcm^2)	200 [5]
Electron Transport layer (ETL)	a-Si:H/SiO$_2$ /SnO$_2$/Mg [4]
ETL passivation J$_0$ (fA/cm^2)	2.4 [4]
ETL contact resistivity ρ (mΩcm^2)	35 [4]
Bulk resistivity (Ωcm)	1
Midgap SRH bulk lifetime (ms)	1
Wafer thickness (μm)	220
Front passivation J$_0$ (fA/cm^2)	7
Gap passivation J$_0$ (fA/cm^2)	4.3 [7]
J$_{gen}$ (mA/cm^2)	43

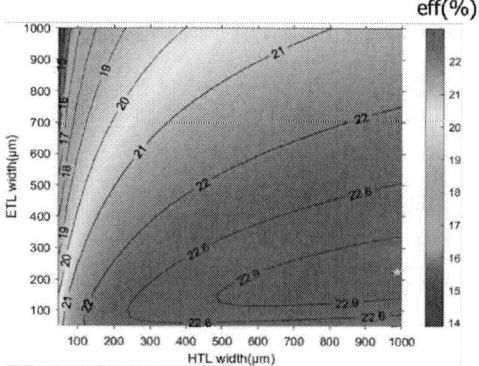

Figure 3: Simulated IBC efficiencies for N-type Si substrates with dopant-free contacts as a function of various HTL and ETL widths.

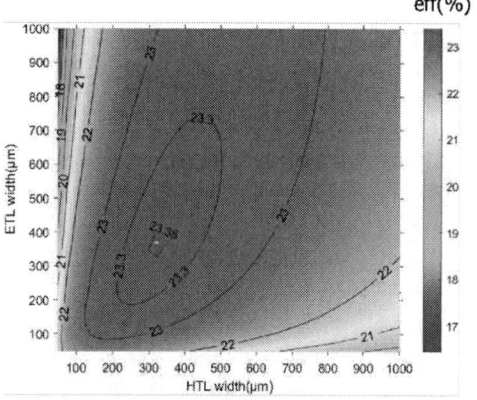

Figure 4: Simulated IBC efficiencies for P-type Si substrates with dopant-free contacts as a function of various HTL and ETL widths.

3 EFFECT OF PATTERNING ACCURACY (GAP WIDTH) AND BULK SRH LIFETIME ON CELL EFFICIENCY

In optimizing the performance of Si IBC solar cells with dopant-free asymmetric hetero-contacts, the gap width between the hole transport layer (HTL) and electron transport layer (ETL) plays a crucial role. A narrower gap facilitates more efficient current collection by minimizing the distance carriers must traverse to reach the contacts. However, reducing the gap width also elevates the risk of shunting or short circuits, particularly if the gap size falls below the accuracy limits of the patterning process. Fig. 4 presents the simulated cell efficiency as a function of the gap width with P- and N-type silicon substrates, along with 1 and 5 ms mid-gap SRH lifetime.

According to existing literature, the smallest feasible gap achievable through cost-effective shadow-mask patterning is approximately 5 μm [8]. Our simulated results, illustrated in Fig. 4, reveal that cell efficiency can be significantly enhanced from 24.6%—achieved using a conventional 150 μm gap with standard shadow masks— to 25.4% when employing a 5 μm gap with advanced shadow masks and 5 ms silicon SRH lifetime. This improvement underscores the importance of optimizing the HTL and ETL widths, as discussed in the preceding section, in conjunction with gap width adjustments. Furthermore, with similar SRH lifetime, P-type substrates achieve higher efficiency than N-type substrates. However, in practice, the n-type silicon wafers tend to be more tolerant to metal contamination and have a higher lifetime. The suitable type of silicon for higher efficiencies depends on the actual lifetime of the finished cells.

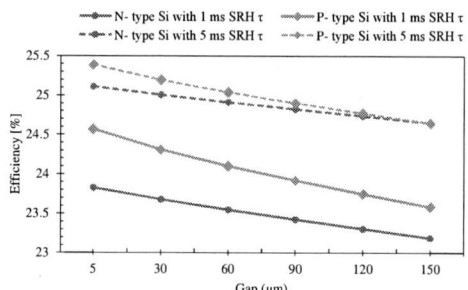

Figure 5: Simulated IBC efficiencies for N-type Si and P-type Si substrates with various SHR and the gap between ETL and HTL.

4 CONCLUSION

This study utilizes Quokka2 simulations to improve the design of IBC solar cells featuring undoped asymmetric hetero-contacts. By investigating the impact of varying the widths of the HTL and ETL layers, as well as the gaps between them, on the efficiency of both N-type and P-type substrates, the study finds that optimizing the silicon gap can enhance the efficiency of N-type c-Si substrate cells from 24.6% to 25.1% and P-type c-Si substrate cells from 24.6% to 25.4%. Furthermore, the study proposes that P-type Si substrates demonstrate higher efficiency than N-type Si substrates when they possess the same SRH lifetime under the assumption of stable manufacturing quality for monocrystalline silicon material.

5 ACKNOWLEDGEMENTS

The financial support provided by the National Science and Technology Council, Taiwan (R.O.C.) (Grant No. 113-2222-E-006-004), the Yushan Fellow Program (MOE-112-YSFEE-0005-002-P1) and National Cheng Kung University within the framework of the Higher Education Sprout Project by the Ministry of Education (MOE), Taiwan, and the Yu-Jen Scholar program, National Cheng Kung University, Taiwan, are gratefully acknowledged.

6 REFERENCES

[1] K. Yoshikawa et al., "Silicon heterojunction solar cell with interdigitated back contacts for a photoconversion efficiency over 26%," Nature Energy, Article vol. 2, no. 5, 2017, Art no. 17032.

[2] A. Fell, "A Free and Fast Three-Dimensional/Two-Dimensional Solar Cell Simulator Featuring Conductive Boundary and Quasi-Neutrality Approximations," Electron Devices, IEEE Transactions on, vol. 60, pp. 733-738, 02/01 2013.

[3] J. Bullock et al., "Lithium Fluoride Based Electron Contacts for High Efficiency n-Type Crystalline Silicon Solar Cells," Adv. Energy Mater., vol. 6, no. 14, p. 1600241, 2016/07/01 2016.

[4] M. Liu et al., "SnO2/Mg combination electron selective transport layer for Si heterojunction solar cells," Sol. Energy Mater. Sol. Cells, vol. 200, p. 109996, 2019/09/15/ 2019.

[5] J. Dréon et al., "23.5%-efficient silicon heterojunction silicon solar cell using molybdenum oxide as hole-selective contact," Nano Energy, vol. 70, p. 104495, 2020/04/01/ 2020.

[6] J. I. Michel et al., "Carrier-selective contacts using metal compounds for crystalline silicon solar cells," (in English), Prog. Photovoltaics, Review vol. 31, no. 4, pp. 380-413, Apr 2023.

[7] W. J. Wang et al., "Realization and simulation of interdigitated back contact silicon solar cells with dopant-free asymmetric hetero-contacts," (in English), Sol. Energy, Article vol. 231, pp. 203-208, Jan 2022.

[8] P. B. Agarwal et al., "Reusable silicon shadow mask with sub-5 mu m gap for low cost patterning," (in English), Sens. Actuator A-Phys., Article vol. 242, pp. 67-72, May 2016.

41st European Photovoltaic Solar Energy Conference and Exhibition

NUMERICAL MODELING AND DESIGN OPTIMIZATION OF INDUSTRIAL TUNNEL OXIDE PASSIVATED CONTACT SOLAR CELLS WITH SELECTIVE PASSIVATED CONTACTS ON THE FRONT

Yi-Ping Lin[1], Chun-Ping Lin[2,3], Jin-Cheng Chen[1], Han-Chen Chang[3], Ying-Yuan Huang[2,4]*

[1] Institute of Photonic System, National Yang Ming Chiao Tung University, Tainan, Taiwan
[2] Institute of Microelectronics, Department of Electrical Engineering, National Cheng Kung University, Tainan, Taiwan
[3] Photovoltaic Technology Division, Green Energy and Environment Research Laboratories, Industrial Technology Research Institute, Tainan, Taiwan
[4] Program on Semiconductor Manufacturing Technology, Academy of Innovative Semiconductor and Sustainable Manufacturing, National Cheng-Kung University, Tainan, Taiwan
*Corresponding author (yingyuanhuang@gs.ncku.edu.tw)

ABSTRACT: The exceptional passivation of TOPCon on the rear of Si solar cells has motivated researchers to explore its application on the front side. However, having a full-area front poly-Si passivated layer may lead to high absorption loss, subsequently reducing photo-generated current density (J_{gen}) and overall efficiency. To address this issue, the implementation of a selective poly-Si layer on the front has been proposed, aiming to increase J_{gen} and efficiency. It is generally recognized that narrower selective poly-Si widths contribute to decreased parasitic absorption losses, thus resulting in enhanced efficiency. However, identifying the optimal poly-Si width—one that strikes a balance between maximum benefits and the feasibility of industrial mass-production—necessitates further in-depth research. This study investigates the benefits of front selective (localized) TOPCon in industrial bifacial passivated contacts (biPC) solar cells, employing Sentaurus TCAD for optical ray tracing and electrical simulations to analyze the effects of varying selective poly-Si widths and thickness on the efficiency of industrial biPC solar cells. Our simulations project a potential efficiency of approximately 25.2%, guided by optimal parameters from literature and the industrial standards reported in ITRPV 2023.
Keywords: Si solar cells; TOPCon; Selective passivated contacts; Sentaurus TCAD

1 INTRODUCTION

The exceptional passivation of Tunnel Oxide Passivated Contacts (TOPCon) on the rear of Si solar cells has motivated researchers to explore its application on the front side. However, parasitic absorption by poly-Si presents a significant challenge. However, having a full-area front poly-Si passivated layer may lead to high absorption loss, subsequently reducing J_{gen} and overall efficiency. To address this issue, the implementation of a selective poly-Si layer on the front has been proposed, aiming to increase J_{gen} and efficiency. It is generally recognized that narrower selective poly-Si widths contribute to decreased parasitic absorption losses, thus resulting in enhanced efficiency. However, identifying the optimal poly-Si width—one that strikes a balance between maximum benefits and the feasibility of industrial mass-production—necessitates further in-depth research. In this research, we will utilize a industrial-compatible screen-printed metal line width of 30 μm, and use Sentaurus TCAD electrical simulation to analyze the effects of varying selective poly-Si widths and thickness on the efficiency of industrial bifacial TOPCon solar cells.

2 EXPERIMENT

2.1 Optical simulation

To explore the parasitic optical absorption loss arising from the full-area and the selective-area poly-Si layer on the front surface, we utilize the optical raytracer in Sentaurus Device[1]. In this simulation, an incident spectrum of AM1.5G is utilized, 5000 rays are specified, and the structure adheres to an upright pyramid configuration in Fig. 1. In this study, we employ an optimized pitch of 1.56 mm for the front metal grid and a practical 100 μm wide selective TOPCon positioned beneath 30 μm wide screen-printed metal grid lines in our optical simulation. This configuration leaves

approximately 4.49% of the selective TOPCon area unobscured by the metal grid lines, and thickness of poly-Si J_{gen} is 150 nm. Our optical simulation results in Fig. 2 illustrate that for full-area poly-Si on the front surface, even an ultra-thin 40 nm poly-Si can result in an absorption loss of over 3 mA/cm² in current. Conversely, selective TOPCon enables the use of much thicker poly-Si (> 100 nm) to mitigate any potential metal-induced recombination while maintaining an absorption loss of less than 0.55 mA/cm².

Figure 1: Optical simulation configuration in Sentaurus TCAD.

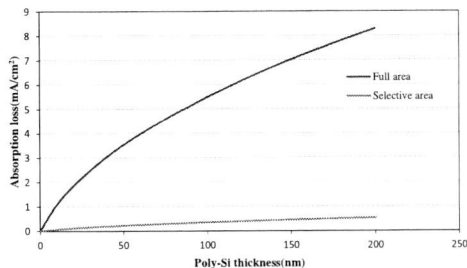

Figure 2: The Sentaurus TCAD simulated absorption loss from the front poly-Si layer as a function of poly-Si thickness. For the selective-area poly-Si, a practical 100 µm wide selective TOPCon with area coverage of 4.49% is assumed.

2.2 Electrical simulation

Sentaurus TCAD is also utilized for the electrical simulation of biPC solar cells. The schematic diagram of the solar cell design is displayed in Fig. 3. A rear-junction configuration is employed to facilitate lateral carrier transport predominantly through the silicon bulk, thereby minimizing the influence from front surface sheet resistance. Selected parameters employed in the simulation are presented in Table 1. To evaluate the benefits of selective TOPCon and ascertain its alignment and patterning precision requirements, we vary the widths of the front selective TOPCon between 40 and 200 µm. Additionally, we also consider different poly-Si

thicknesses and their corresponding J_0 values, among other parameters, in this simulation. An external resistance, which reflects the combined effects of measured contact resistivity and metal grid resistivity on the fill factor, was incorporated into the simulation. This external resistance—representing influences outside the cell structure—is calculated based on actual contact and grid line resistance measurements to simulate their impact on the fill factor accurately. Furthermore, to evaluate the quality of surface passivation, we modeled carrier selectivity on passivated contacts by considering the asymmetric electron and hole recombination velocities, which are related to the area-weighted J_0 on the front and rear surfaces.

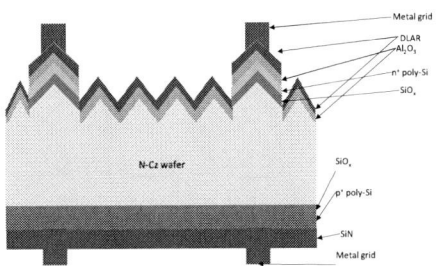

Figure 3: Schematic diagrams of biPC solar cell with selective TOPCon the front side

Table 1: Selective parameters used in Sentaurus TCAD simulations

	Parameter	Value	
Front Selective TOPCon	Selective area width [µm]	40-200	
	Front poly-Si type	N	N
	Front poly-Si thickness [nm]	150	200
	$J_{0frontpoly,pass}$ [fA/cm²]	0.4[2]	1[4]
	$J_{0frontpoly,metal}$ [fA/cm²]	24[3]	30[4]
Bulk	Bulk Type	N	
	Bulk thickness [µm]	160	
	Bulk resistivity [Ω-cm]	1	
	Lifetime [ms]	20[5]	
Rear TOPCon	Rear poly-Si type	P	
	Rear Poly thickness [nm]	250	
	Rear Poly sheet resistance[Ω/□]	200	
	$J_{0rearpoly,pass}$ [fA/cm²]	1[6]	
	$J_{0rearpoly,metal}$ [fA/cm²]	112[7]	
Front grid	Pitch [µm]	1560	
	Line width[µm]	30[8]	
	Contact resistivity [mΩ/cm²]	0.5[3]	
Rear grid	Pitch [µm]	520	
	Line width[µm]	30[8]	
	Contact resistivity [mΩ/cm²]	2[6]	
r_s	External resistance[Ω-cm²]	0.072	

223

3 RESULT AND CONCLUSION

The simulation results are presented in Fig. 4. With a selective area width of 100 μm and a grid line width of 30 μm, which means the alignment tolerance is ±30 μm, the cell efficiency reaches 24.6% with 200 nm thick poly-Si layer, and 24.8% efficiency with 150 nm poly-Si layer. Efficiencies exceeding 25% are attainable with poly-Si widths of less than 60 μm. These findings demonstrate that, through the use of processes compatible with industrial standards, the application of selective passivating contacts on the front side can effectively alleviate the trade-off between generation current J_{gen} and metal-induced recombination, achieving efficiencies greater than 24.5%.

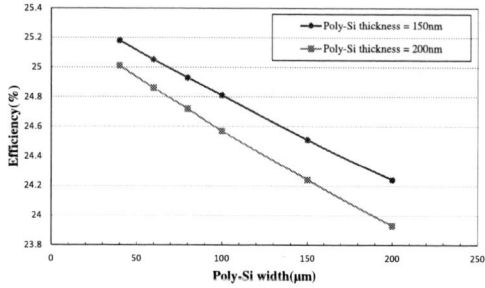

Figure 4: Sentaurus simulated cell efficiency as a function of selective area width of front selective TOPCon.

4 ACKNOWLEDGMENT

The financial support provided by Bureau of Energy, Ministry of Economic Affairs, Taiwan (R.O.C.) (Grant No. 113-S0102), the National Science and Technology Council, Taiwan (R.O.C.) (Grant No. 113-2222-E-006-004), the Yushan Fellow Program (MOE-112-YSFEE-0005-002-P1) and National Cheng Kung University within the framework of the Higher Education Sprout Project by the Ministry of Education (MOE), Taiwan, and the Yu-Jen Scholar program, National Cheng Kung University, Taiwan, are gratefully acknowledged.

5 REFERENCES

[1] Synopsys, Sentaurus™, Device User Guide, 2023.
[2] R. Peibst et al., Implementation of N+ and P+ Polo Junctions on Front and Rear side of Double-side.
[3] P. Padhamnath et al., Development of thin polysilicon layers for application in monoPoly™ cells with screen-printed and fired metallization, Solar Energy Materials and Solar Cells, vol. 207, p. 110358, 2020.
[4] Y.-Y. Huang et al., Fully screen-printed bifacial large area 22.6% N-type Si solar cell with lightly doped ion-implanted boron emitter and tunnel oxide passivated rear contact, Solar Energy Materials and Solar Cells, vol. 214, 2020, doi: 10.1016/j.solmat.2020.110585.
[5] D. D. Smith, P. Cousins, S. Westerberg, R. D. Jesus-Tabajonda, G. Aniero, and Y.-C. Shen, Toward the Practical Limits of Silicon Solar Cells, IEEE Journal of Photovoltaics, vol. 4, no. 6, pp. 1465-1469, 2014, doi: 10.1109/jphotov.2014.2350695.

[6] J. S. Sebastian Mack, Tobias Fellmeth, Frank Feldmann, Martijn Lenes, and Jan-Marc Luchies, Metallisation of Boron-Doped Polysilicon Layers by Screen Printed Silver Pastes., 2017, doi: 10.1002/pssr.201700334.
[7] W.-J. Choi, A. Jain, Y.-Y. Huang, Y.-W. Ok, and A. Rohatgi, Quantitative Understanding and Implementation of Screen Printed p+ Poly-Si/Oxide Passivated Contact to Enhance the Efficiency of p-PERC Cells presented at the IEEE, 2020.
[8] International Technology Roadmap for Photovoltaic (ITRPV)2022 Results., April 2023.

41st European Photovoltaic Solar Energy Conference and Exhibition

Modeling and Experimental Validation of Solar Cell Performance Across Varied Temperatures and Irradiance

Selin Cansu Gölboylu[a,b] , Hatice Duman[a], Melisa Demir[a] , Meriç Çalışkan Arslan[a]

[a] *Kalyon PV Research and Development Center, Kalyon Güneş Teknolojileri Üretim A.Ş., 06909 Ankara, Turkey*
[b] *Micro and Nanotechnology Program, Middle East Technical University, Ankara 06800, Turkey*
e-mail: scgolboylu@kalyonpv.com

INTRODUCTION & MOTIVATION

The performance of photovoltaic (PV) modules is strongly influenced by environmental conditions such as temperature and solar irradiance. As temperature increases, efficiency tends to decrease due to changes in electrical characteristics like open-circuit voltage and current [1]. While these effects can be assessed through I-V measurements, such testing is often time-consuming and requires expensive equipment. Consequently, modeling and simulation of PV modules have increasingly become a focus as a more efficient and cost-effective alternative.

In this study, mathematical modeling and simulation of full-size PV cells were carried out using MATLAB software to enhance the efficiency of solar cells and promote the wider adoption of solar energy. The single-diode electrical equivalent circuit model, incorporating a series resistor (R_s) and shunt resistance (R_{SH}) to account for contact resistance, was applied [2]. Experimental measurements were conducted on solar cells employing Passivated Emitter and Rear Contact (PERC) technology, produced by Kalyon PV. Key electrical parameters, including open-circuit voltage (Voc), short-circuit current (Isc), and maximum power point values (Vmp, Imp), were obtained.

The model was also adapted to simulate 108 half-cell PV modules in order to assess its broader applicability. The results demonstrated strong agreement between the experimental data and the simulated I-V characteristics for both single cells and module configurations, validating the accuracy of the model. This approach offers an efficient method to evaluate the performance of solar cells under varying conditions and can facilitate the optimization of PV module efficiency without extensive physical testing.

MODELING APPROACH & EXPERIMENTAL METHODS

Current–Voltage Relationship For a Single Solar Cell

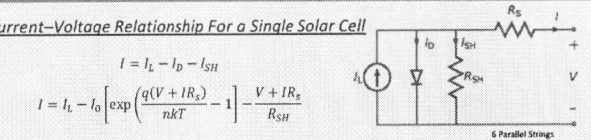

$$I = I_L - I_D - I_{SH}$$

$$I = I_L - I_0 \left[\exp\left(\frac{q(V + IR_s)}{nkT} - 1 \right) \right] - \frac{V + IR_s}{R_{SH}}$$

Current–Voltage Relationship for a 108 Half-cell PV Module

$$I = I_L - I_0 \left[\exp\left(\frac{q(V_M + I_M N_c R_s)}{N_c nkT} - 1 \right) \right] - \frac{V_M + I_M N_c R_s}{N_c R_{SH}}$$

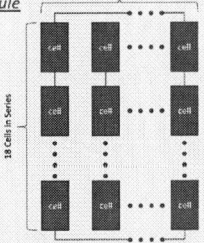

- The equations were solved by MATLAB fzero function with 500 data points. All measurements and calculations made under 1000 W/m2 solar irradiance.

- The laboratory measurement results of electrical I-V characterization tests for a single full size PERC cell obtained by LOANA - PVTools device.

- For the 108 half-cut cell PV module, Eternalsun Spire Solar Similator, was used to take measurements under different temperatures.

RESULTS

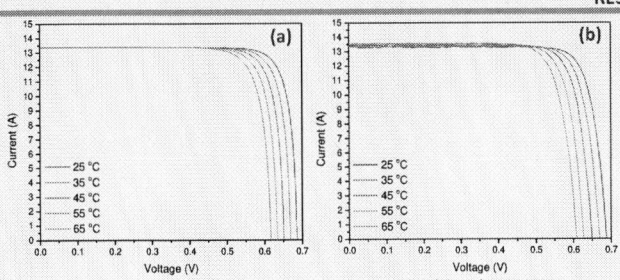

Figure 1. Modeling (a), and experimental (b) I-V results of the KalyonPV PERC solar cell at different temperatures.

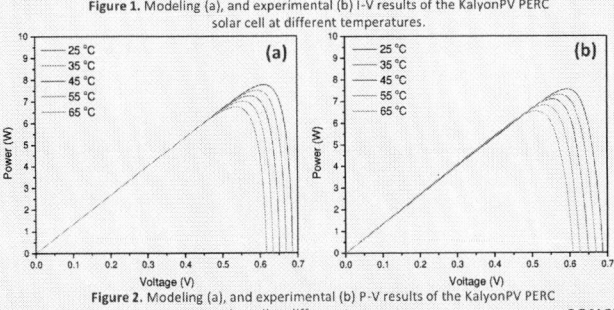

Figure 2. Modeling (a), and experimental (b) P-V results of the KalyonPV PERC solar cell at different temperatures.

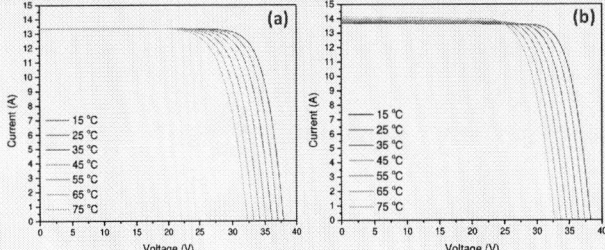

Figure 3. Modeling (a), and experimental (b) I-V results of the KalyonPV M10-108 Half-cut cell PV Module at different temperatures.

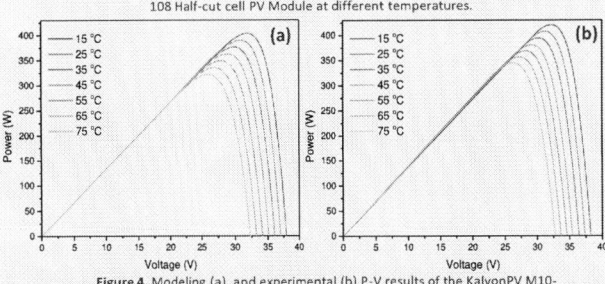

Figure 4. Modeling (a), and experimental (b) P-V results of the KalyonPV M10-108 Half-cut cell PV Module at different temperatures.

CONCLUSIONS

- Mathematical modeling to simulate the performance of full-size PV cells under varying temperature conditions was successfully demonstrated by using MATLAB.
- The electrical parameters of the solar cells obtained from both simulation and experimental data showed strong agreement, validating the accuracy of the developed simulation model.
- While there were small differences between the experimental and simulated results, these discrepancies could be attributed to factors such as measurement inaccuracies, non-ideal environmental conditions during testing, or slight variations in material properties not fully captured by the model.

- Temperature significantly impacts the performance of PV cells, as shown by the results. The modeling approach effectively captures this effect, providing valuable insights for PV optimization.
- The results emphasize the importance of incorporating temperature variations in solar cell simulations for more accurate performance prediction.
- This work provides a strong basis for further research and optimization of solar cell designs, enabling improved efficiency and reliability of PV systems under different environmental conditions.

REFERENCES

[1] Ponnusamy, L., & Desappan, D. (2014). An investigation of temperature effects on solar photovoltaic cells and modules. International Journal of Engineering, 27(11), 1713-1722.
[2] H. Tian, F. Mancilla-David, K. Ellis, E. Muljadi, and P. Jenkins, "A cell-to-module-to-array detailed model for photovoltaic panels," Solar Energy, vol. 86, no. 9, pp. 2695–2706, Sep. 2012

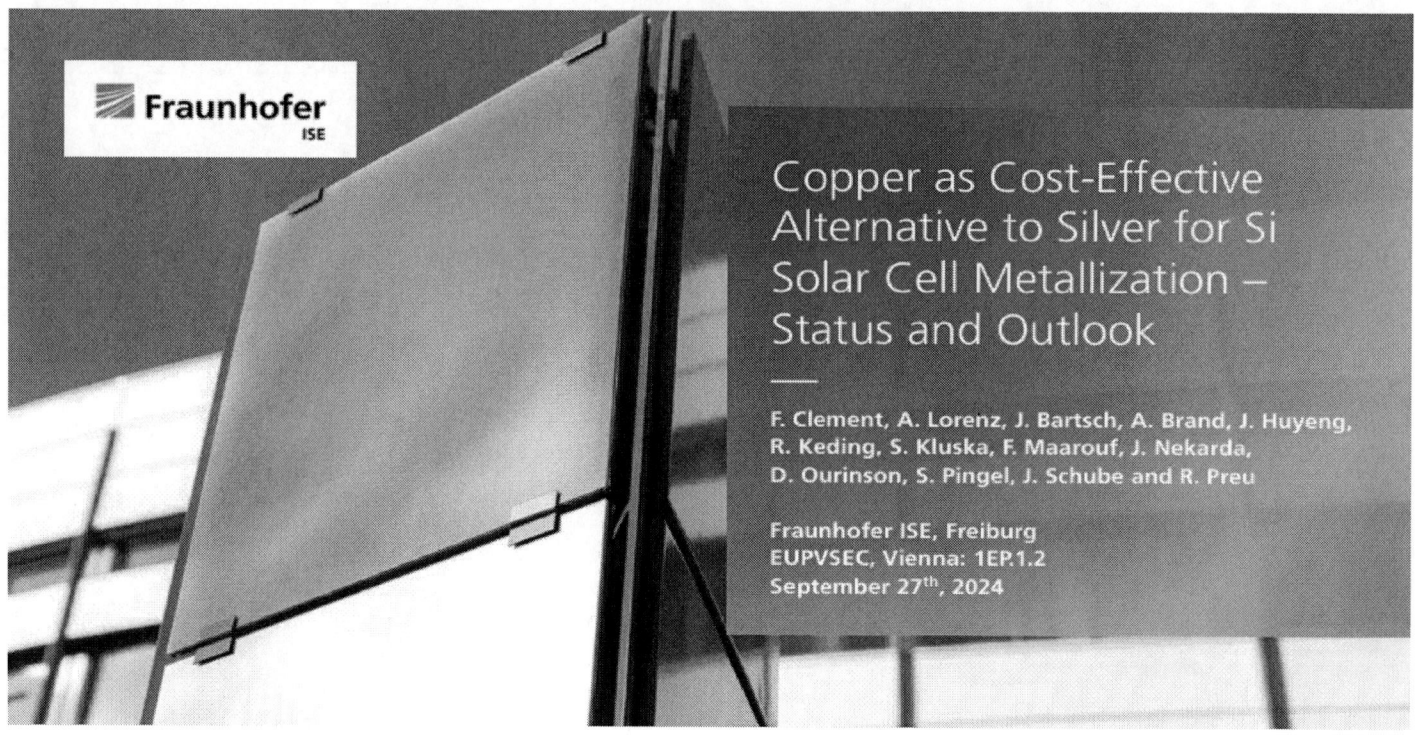

Copper as Cost-Effective Alternative to Silver for Si Solar Cell Metallization – Status and Outlook

F. Clement, A. Lorenz, J. Bartsch, A. Brand, J. Huyeng,
R. Keding, S. Kluska, F. Maarouf, J. Nekarda,
D. Ourinson, S. Pingel, J. Schube and R. Preu

Fraunhofer ISE, Freiburg
EUPVSEC, Vienna: 1EP.1.2
September 27th, 2024

Background and Motivation
Rapid growth of the PV Industry — Risk and Challenges

Rapid growth of PV:

- Installed capacity 2023: **447 GW$_p$** [1]

FIGURE 4 WORLD ANNUAL SOLAR PV MARKET SCENARIOS 2024–2028

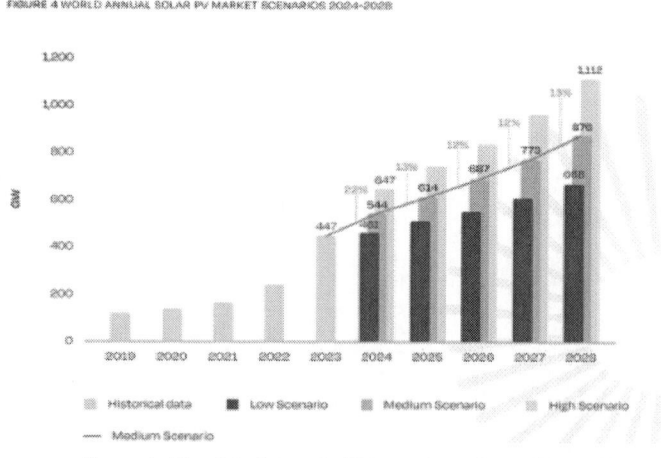

Forecast of the globally installed PV capacity until 2028. Source: [2]

[1] Solar Power Europe 2024 .
[2] https://api.solarpowereurope.org/uploads/Global_Market_Outlook_for_Solar_Power_2024_a083b6dcd5.pdf

Background and Motivation
Rapid growth of the PV Industry – Risk and Challenges

Rapid growth of PV:

- Installed capacity 2023: **447 GW$_p$** [1],

Risk and Challenge:

- Rising demand for scarce material (**Ag**, Bi, In…)

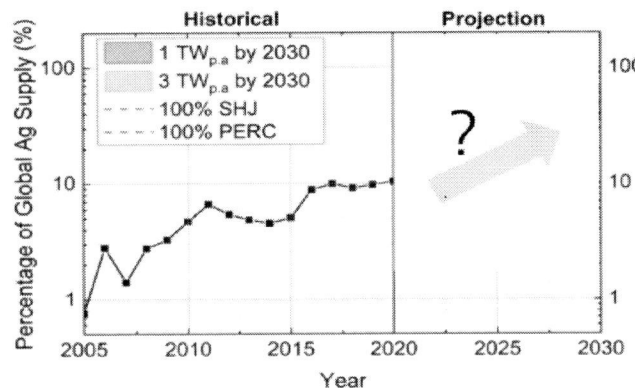

Development of global silver consumption until 2030 with a projected yearly increase of 1 or 3 TW/year. Source: [6]

3
©Fraunhofer ISE
Public

[1] Solar Power Europe 2024
[3] Verlinden P.J. J Renew Sustain Energy. 2020;12(5)   [4] Hallam B et al., Prog Photovolt: Res Appl. 2022;31(6) [5] Chang et al., Prog Photovolt: Res Appl. 2024; 1-12 [6] J. Bartsch et al., PV International 48, 2021

Background and Motivation
Rapid growth of the PV Industry – Risk and Challenges

Rapid growth of PV:

- Installed capacity 2023: **447 GW$_p$** [1],

Risk and Challenge:

- Rising demand for scarce material (**Ag**, Bi, In…)
- Most critical resource: Silver usage in metallization
- Today: 5 mg/W (PERC) – 17 mg/W (SHJ) Ag is used

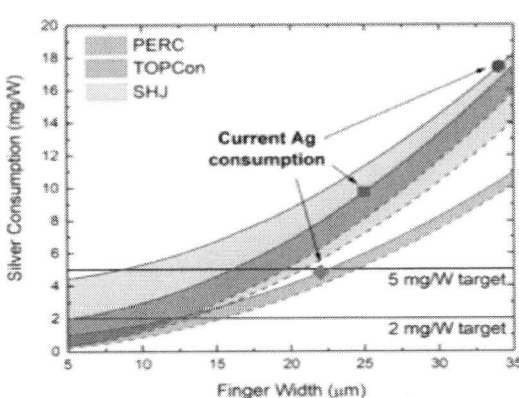

Calculated silver consumption as a function of printed width of ag fingers in industrial screen-printed PERC, TOPCon and SHJ solar cells. Solid lines show the total silver consumption in the finger, busbar and soldering tab regions, and dash lines show finger silver consumption only; Source: [5]

4
©Fraunhofer ISE
Public

[1] Solar Power Europe 2024
[3] Verlinden P.J. J Renew Sustain Energy. 2020;12(5)   [4] Hallam B et al., Prog Photovolt: Res Appl. 2022;31(6) [5] Chang et al., Prog Photovolt: Res Appl. 2024; 1-12 [6] J. Bartsch et al., PV International 48, 2021

Background and Motivation
Rapid growth of the PV Industry – Risk and Challenges

Rapid growth of PV:

- Installed capacity 2023: **447 GW$_p$** [1],

Risk and Challenge:

- Rising demand for scarce material (**Ag**, Bi, In...)
- Most critical resource: Silver usage in metallization
- Today: 5 mg/W (PERC) – 17 mg/W (SHJ) Ag is used

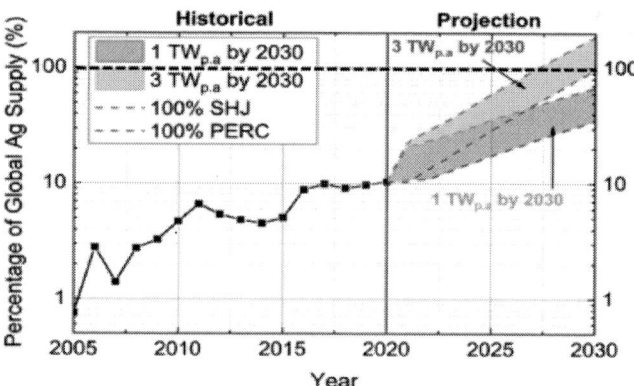

Development of global silver consumption until 2030 with a projected yearly increase of 1 or 3 TW/year. Source: [6]

[1] Solar Power Europe 2024
[3] Verlinden P.J. J Renew Sustain Energy. 2020;12(5) [4] Hallam B et al., Prog Photovolt: Res Appl. 2022;31(6) [5] Chang et al., Prog Photovolt: Res Appl. 2024; 1-12 [6] J. Bartsch et al., PV International 48, 2021

Background and Motivation
Rapid growth of the PV Industry – Risk and Challenges

Rapid growth of PV:

- Installed capacity 2023: **447 GW$_p$** [1],

Risk and Challenge:

- Rising demand for scarce material (**Ag**, Bi, In...)
- Most critical resource: Silver usage in metallization
- Today: 5 mg/W (PERC) – 17 mg/W (SHJ) Ag is used
- Sustainable TW production: reducing Ag usage to **5 or even 2 mg/W** [3-5]

➔ Silver has to be reduced or replaced!

➔ This talk: Copper as alternative

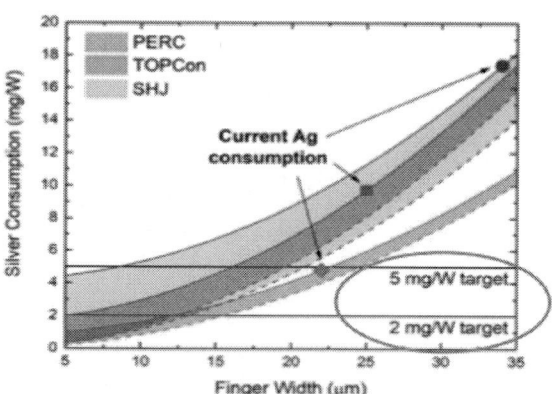

Calculated silver consumption as a function of printed width of ag fingers in industrial screen-printed PERC, TOPCon and SHJ solar cells. Solid lines show the total silver consumption in the finger, busbar and soldering tab regions, and dash lines show finger silver consumption only; Source: [5]

[1] Solar Power Europe 2024
[3] Verlinden P.J. J Renew Sustain Energy. 2020;12(5) [4] Hallam B et al., Prog Photovolt: Res Appl. 2022;31(6) [5] Chang et al., Prog Photovolt: Res Appl. 2024; 1-12 [6] J. Bartsch et al., PV International 48, 2021

Technology Roadmap
How can we reduce / replace Silver?

foto from www.tradestation.com

foto from www.degussa-goldhandel.de

Copper as Alternative for Silver for Solar Cell Metallization?

Perspectives for Solar Cell Metallization with Copper
Advantages and Challenges of Copper Metallization

Benefits:

- Resistivity comparable to Ag
- Substantial cost reduction
- More sustainable production
- Reduction of economic risks and material dependency

	Copper	Silver
Resistivity	0.018 µΩm	0.016 µΩm
Price [$/Kg] [7]	9.03	920.28
Carbon footprint [Kg CO_2 Equivalents per Kg] [8]	1.71	98.1
Abundance in earth's crust [%] [9]	0.0055	7×10^{-6}

[7] https://www.dailymetalprice.com [Sep 11, 2024]
[8] LCA data MFD & Printers (epa.gov) [accessed June 1st, 2024]
[9] Pandey, B. State-of-the-Art Report on Technology for Producing Rare Metals in India (2012).

Perspectives for Solar Cell Metallization with Copper
Advantages and Challenges of Copper Metallization

Risk and Challenge:

- Cu particularly critical for high-temperature n-type solar cell concepts (e.g. TOPCon)
  - Diffusion of Cu into Silicon, i.e. n-type Si, and formation of recombination-active precipitates (deep-level traps)

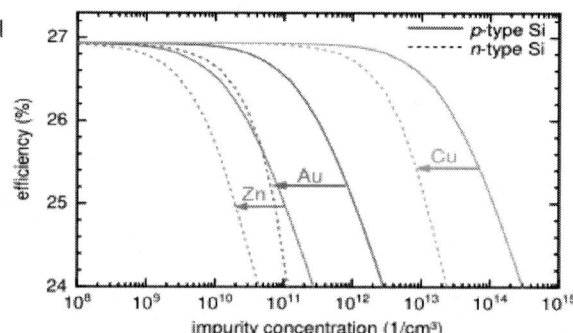

Simulated cell eff. depending on Impurities that are more detrimental for n-type Si [10]

[10] A. Richter et al. MRS Meeting 2022.

Perspectives for Solar Cell Metallization with Copper
Advantages and Challenges of Copper Metallization

Risk and Challenge:

- Cu particularly critical for high-temperature n-type solar cell concepts (e.g.TOPCon)
  - Diffusion of Cu into Silicon and formation of recombination-active precipitates (deep-level traps)
  - Oxidation of contacts (loss of conductivity)
- Long-term stability of copper contacts
  → effects on module reliability

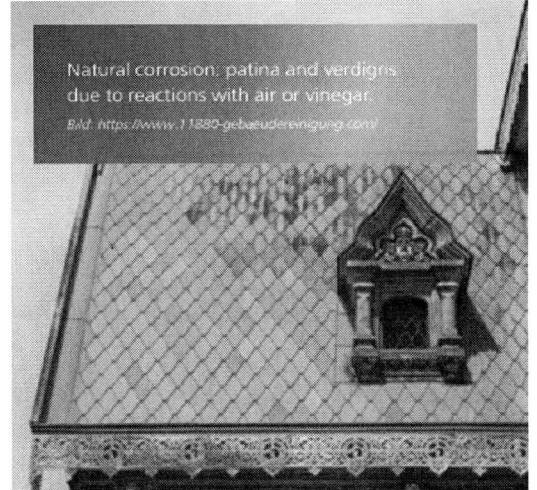

Natural corrosion: patina and verdigris due to reactions with air or vinegar.
Bild: https://www.11880-gebaeudereinigung.com/

Perspectives for Solar Cell Metallization with Copper
Advantages and Challenges of Copper Metallization

Risk and Challenge:

- Cu particularly critical for high-temperature n-type solar cell concepts (e.g.TOPCon)
  - Diffusion of Cu into Silicon and formation of recombination-active precipitates (deep-level traps) [1]
  - Oxidation of contacts (loss of conductivity)
- Long-term stability of copper contacts
  → effects on module reliability

➢ How can we replace silver with copper without performance losses?

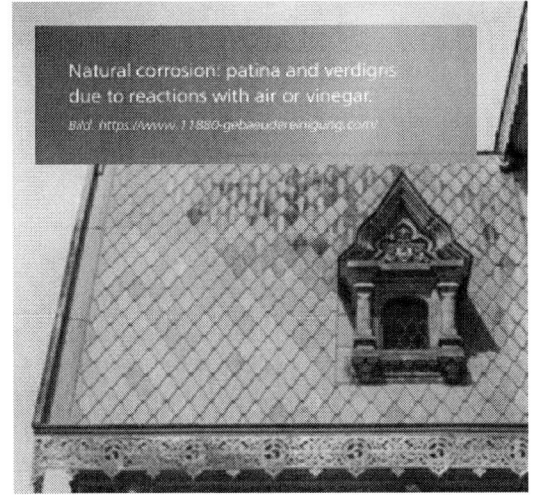

Natural corrosion: patina and verdigris due to reactions with air or vinegar.
Bild: https://www.11880-gebaeudereinigung.com/

Technology Roadmap
How can we reduce / replace Silver?

foto from www.tradestation.com

Ag reduction / replacement .....
- Fine line printing
- Screen printing
  - Ag(/Cu) pastes
  - 100% Cu pastes
- Cu plating

foto from www.degussa-goldhandel.de

Technology Roadmap
How can we reduce / replace Silver?

foto from www.tradestation.com

Ag reduction / replacement .....
- Fine line printing
- Screen printing
  - Ag(/Cu) pastes
  - 100% Cu pastes
- Cu plating

foto from www.degussa-goldhandel.de

Technology Roadmap
How can we reduce / replace Silver?

foto from www.tradestation.com

Ag reduction / replacement .....
- Fine line printing
- Screen printing
  - Ag(/Cu) pastes
  - 100% Cu pastes
- Cu plating

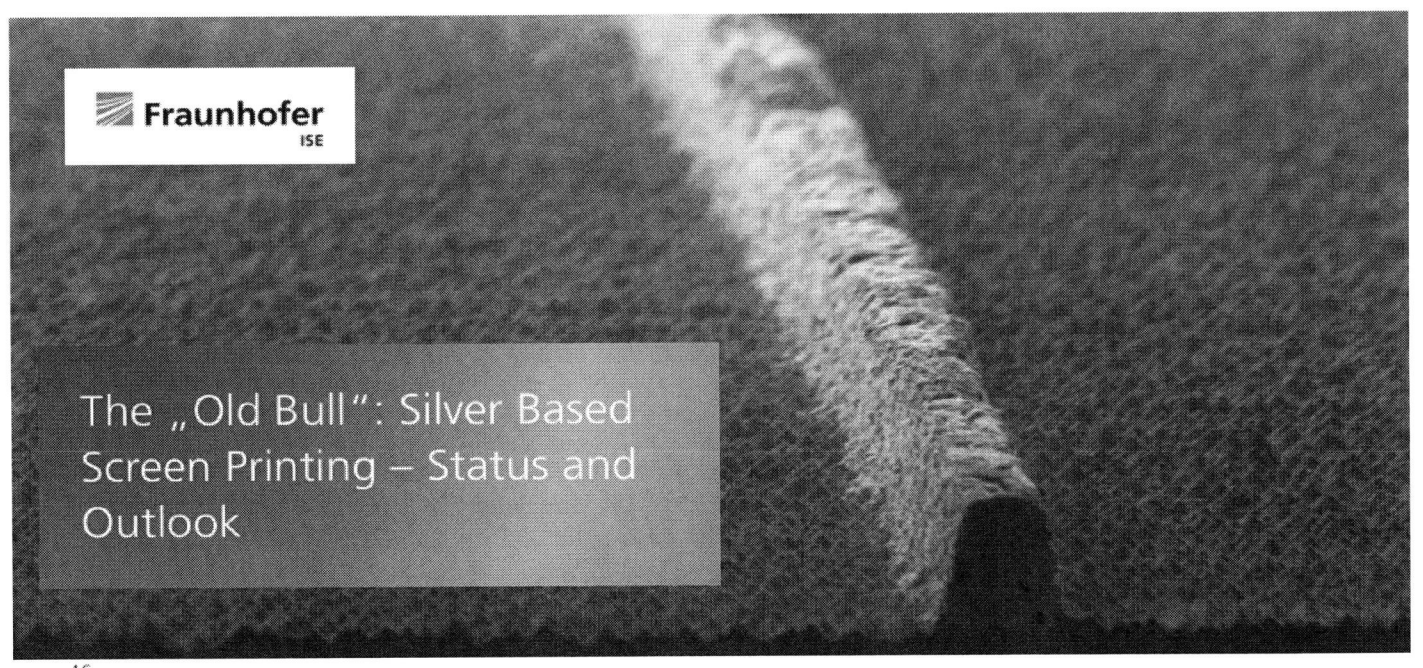

The „Old Bull": Silver Based Screen Printing – Status and Outlook

Screen Printed Metallization for Si Solar Cells
Towards Minimizing Silver Dependence

- State-of-the-art: flatbed screen printing

Screen Printed Metallization for Si Solar Cells
Towards Minimizing Silver Dependence

- State-of-the-art: flatbed screen printing
- Finger width reduction by factor 8 since 2010 [11]

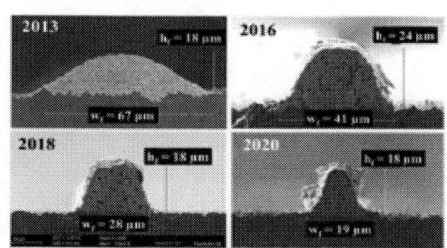

Evolution of screen-printed finger width illustrated by selected SEM images of screen printed contacts. Source: Fraunhofer ISE

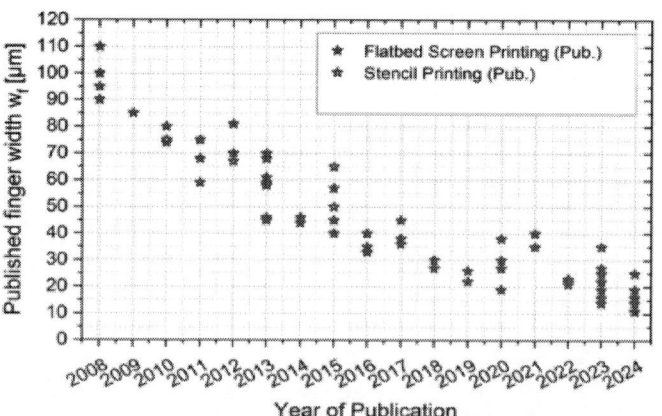

Evolution of screen-printed finger width based on published results from 2008 to 2004. Updated version, based on [11]

[11] Tepner S. & Lorenz, A.. Progress in Photovoltaics: Research and Applications 31 (2023) 557–590.

Screen Printed Metallization for Si Solar Cells
Towards Minimizing Silver Dependence

* State-of-the-art: flatbed screen printing
* Finger width reduction by factor 8 since 2010 [2]
  * "Learning rate" increases in 2013 due to technological leap from 2 or 3 to 5 or more busbars
  * Further Increase with current technological leap to half cells and multi-busbar technology?

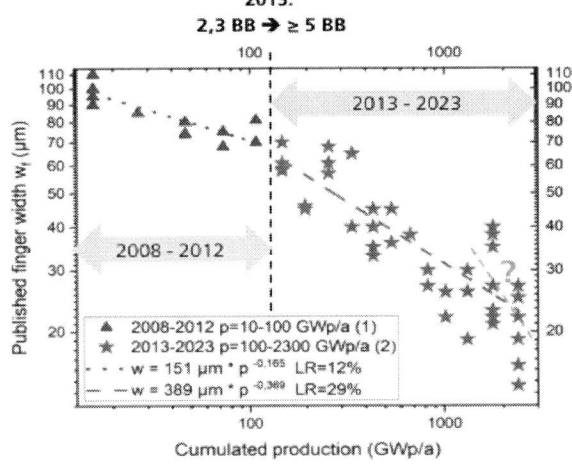

Evolution of screen-printed finger width based on published results from 2008 to 2004. Updated version, based on [11]

[11] Tepner. S. & Lorenz. A.. Progress in Photovoltaics: Research and Applications 31 (2023) 557–590.

Screen Printed Metallization for Si Solar Cells
Towards Minimizing Silver Dependence

* State-of-the-art: flatbed screen printing
* Finger width reduction by factor 8 since 2010 [2]
  * "Learning rate" increases in 2013 due to technological leap from 2 or 3 to 5 or more busbars
  * Further Increase with current technological leap to half cells and multi-busbar technology?
* Recent result:
  * PERC front side metallization w_f = 14 µm [12,13]

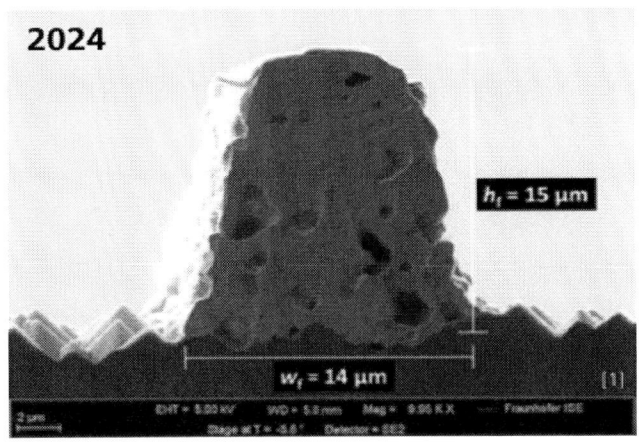

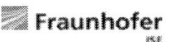

[12] Source: Fraunhofer ISE, Lohmüller et al. (submitted for publication on EUPVSEC 2024)
[13] https://www.pv-magazine.com/2024/05/01/fraunhofer-ise-develops-23-4-efficient-perc-solar-cells-with-limited-silver-amount/

Screen Printed Metallization for Si Solar Cells
Towards Minimizing Silver Dependence

- State-of-the-art: flatbed screen printing
- Finger width reduction by factor 8 since 2010 [2]
  - "Learning rate" increases in 2013 due to technological leap from 2 or 3 to 5 or more busbars
  - Further Increase with current technological leap to half cells and multi-busbar technology?
- Recent result:
  - PERC front side metallization w_f = **14 μm** [3,4]
  - SHJ front side metallization w_f = **17 μm** [14]

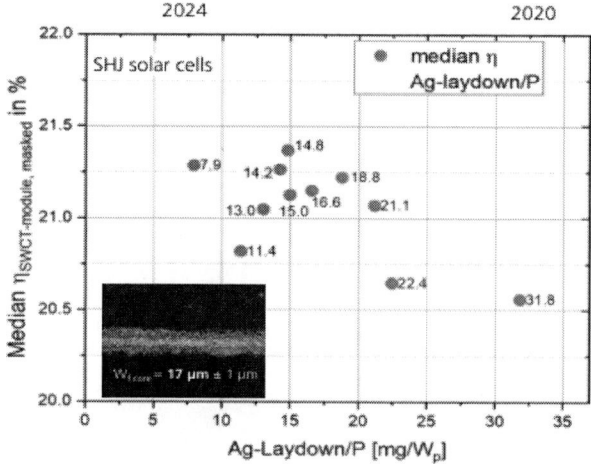

Reduction of silver consumption @Fraunhofer ISE of SHJ solar cells without busbars / pads for smart wire (SWCT) module integration

[14] S. Pingel et al., Metallization Workshop 2024 (to be published)

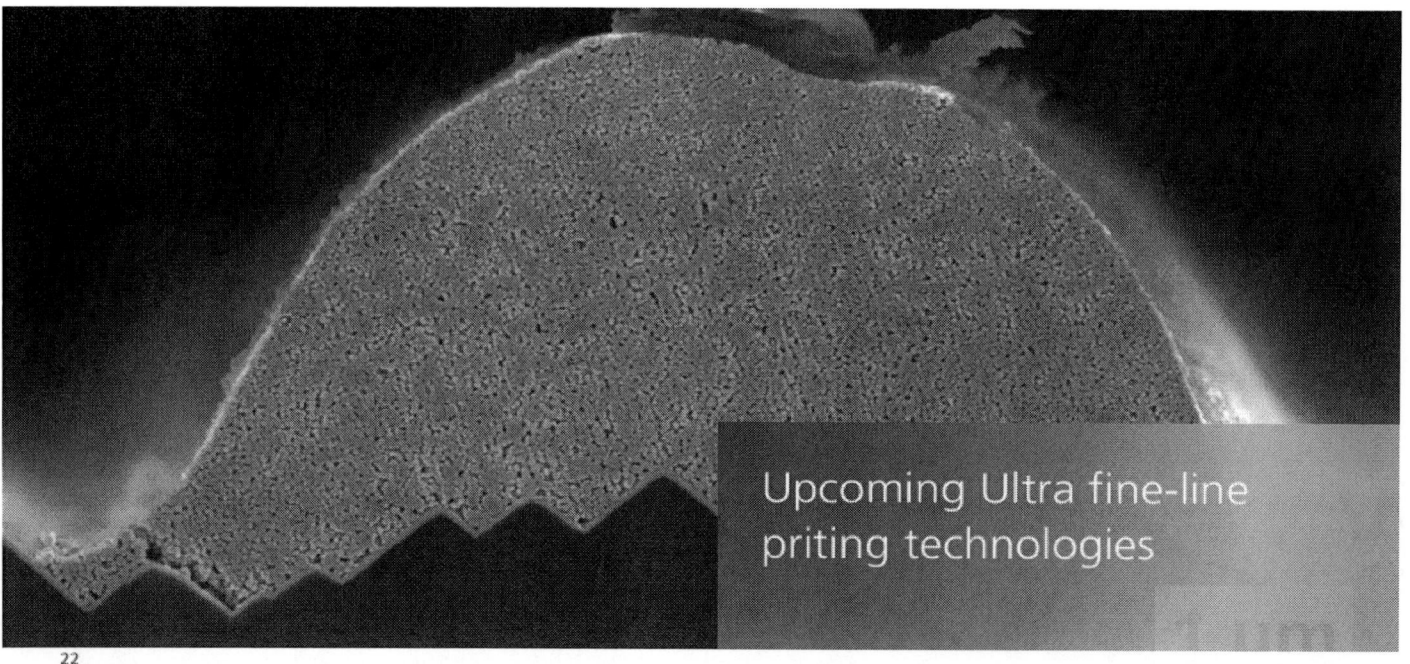

Upcoming Ultra fine-line priting technologies

Ultra Fine-Line Printing
Upcoming Technologies

- **Parallel Dispensing Technology**
  - allows more homogenous finger geometry
    - → about 20% less silver consumption demonstrated
  - Industrial printheads available
    - → HighLine Technology (Fraunhofer ISE Spin-off)

In-line Intermitted Parallel Dispensing printhead [16]

SEM image of a dispensed contact finger [15]

[15] Pospischil et al. (Met WS 2019)
[16] www.highline-technology.de

Ultra Fine-Line Printing
Upcoming Technologies

- Parallel Dispensing Technology

- **Flextrail Printing Technology**[17,18]
  - finger width down to 10µm
  - Significant silver saving potential
  - technology upscaling is ongoing

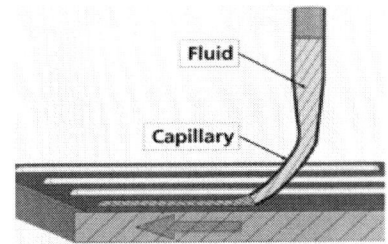

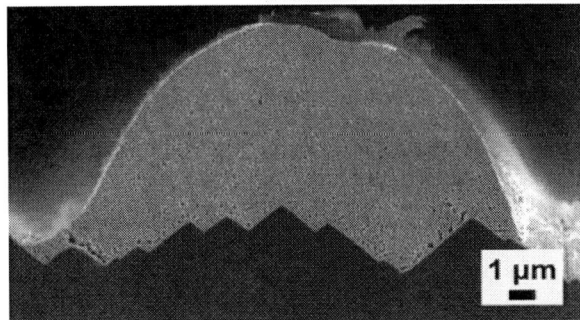

[17] J. Schube et al., Energy Technology (2022
[18] M. Jahn, EP 3 736 050 (2022)...

Ultra Fine-Line Printing
Upcoming Technologies

- Parallel dispensing technology

- Flextrail printing technology

- Further high potential ultra fine-line printing technologies:
  - Glass stencil printing [19]
  - Pattern transfer printing (PTP) [21]
  - „Lumet Metallization Technology" [20]
  - ....

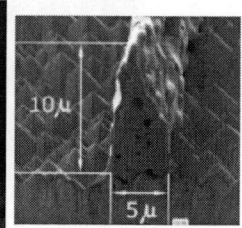

[19] T. Schweigstill this conference (1BO.3.6)   [20] https://www.lumet.com/..
[21] https://www.utilight.com/

Technology Roadmap
How can we reduce / replace Silver?

foto from www.tradestation.com

Ag reduction / replacement (First step): Fine line printing allows ultra-fine finger geometries (below 15 μm)

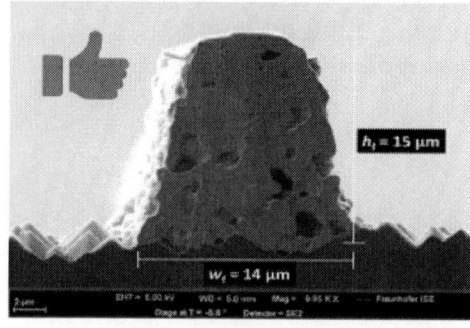

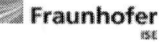

Technology Roadmap

How can we reduce / replace Silver?

foto from www.tradestation.com

Ag reduction / replacement (first step):
Fine line printing allows ultra-fine finger geometries (below 15 µm)

Ag reduction / replacement (second step):
- Ag/Cu and Cu pastes
- Cu plating

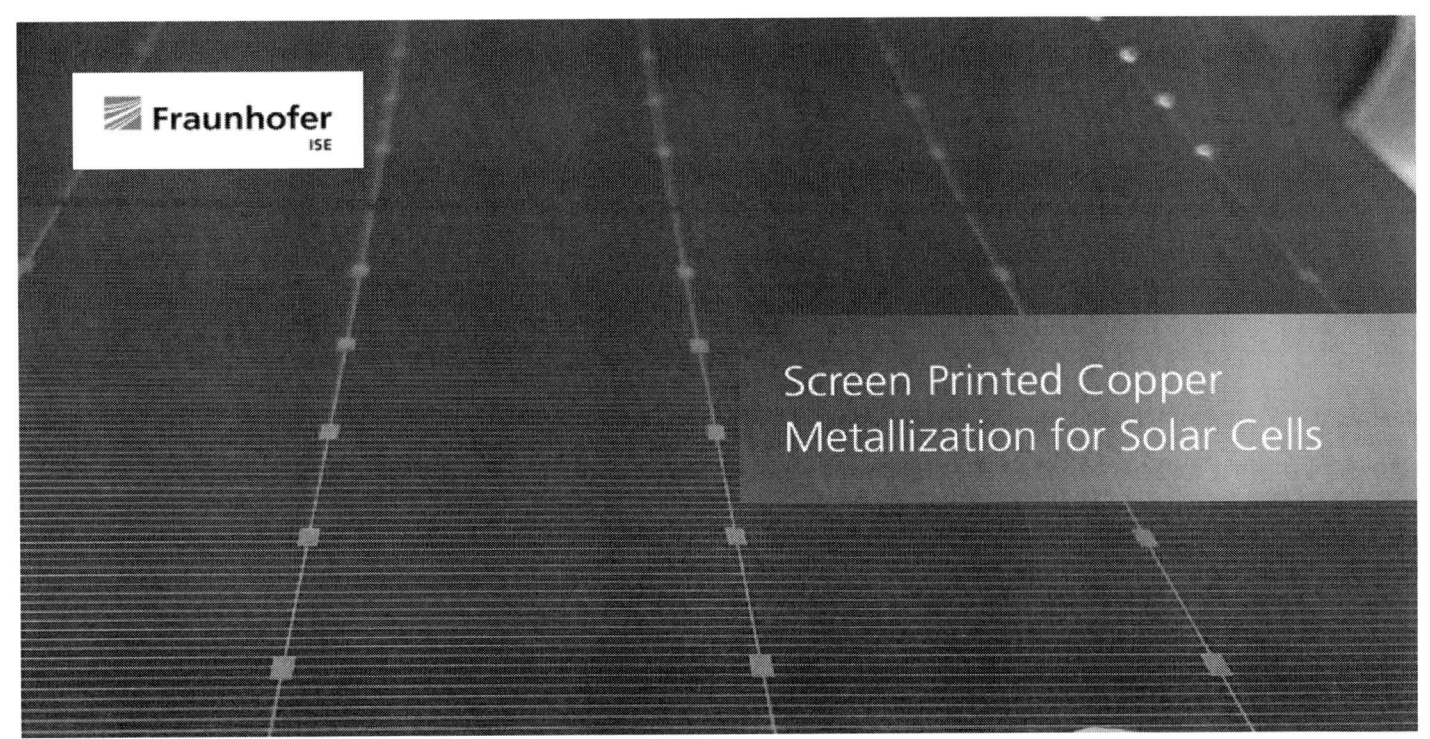

Screen Printed Copper Metallization for Solar Cells

Screen Printing of Copper Contacts

SHJ solar cells with screen printed silver-copper (AgCu) metallization

- Low temperature approach with silver-copper pastes
- Latest versions of AgCu pastes can compete with the best Ag pastes
- AgCu fingers allow to reduce Ag consumption by at least 50% for SHJ cell

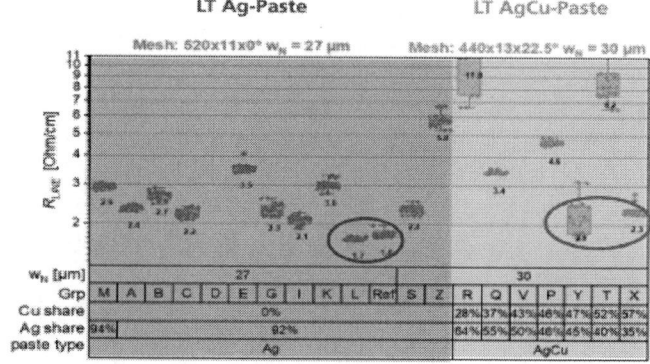

R_{LINE} data for Ag and AgCu pastes printed with varied w_H (27 µm and 30 µm are shown).

Screen Printing of Copper Contacts

SHJ solar cells with screen printed silver-copper (AgCu) metallization

- Low temperature approach with silver-copper pastes
- Latest versions of AgCu pastes can compete with the best Ag pastes
- AgCu fingers allow to reduce Ag consumption by at least 50% for SHJ cell
- Ag consumption below 5 mg/W achievable
- Significant step towards the 2 mg/W long-term goal for sustainable SHJ production

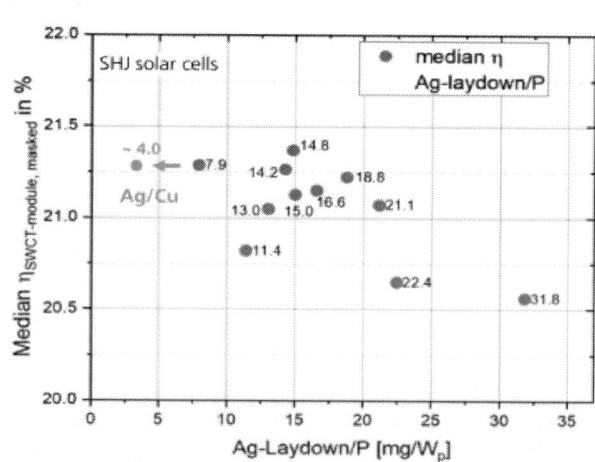

Reduction of silver consumption @Fraunhofer ISE of SHJ solar cells without busbars / pads for smart wire (SWCT) module integration

Screen Printing of Copper Contacts
SHJ solar cells with screen printed silver-copper (AgCu) metallization

* Low temperature approach with silver-copper pastes
* Latest versions of AgCu pastes can compete with the best Ag pastes
* AgCu fingers allow to reduce Ag consumption by at least 50% for SHJ cell
* Ag consumption below 5 mg/W achievable
* Significant step towards the 2 mg/W long-term goal for sustainable SHJ production

➤ Will 100% Cu pastes be the next step?

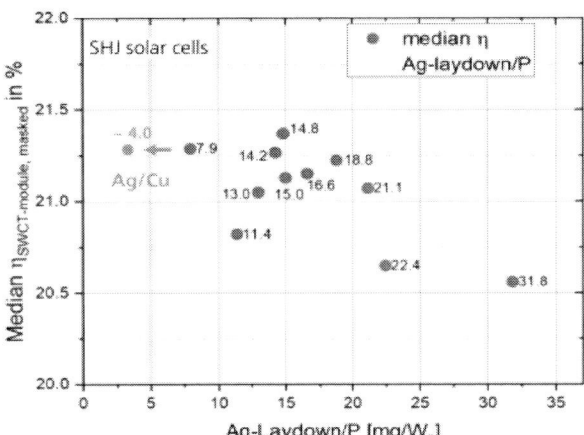

Reduction of silver consumption @Fraunhofer ISE of SHJ solar cells without busbars / pads for smart wire (SWCT) module integration

---

Screen Printing of Copper Contacts
SHJ solar cells with screen printed copper metallization

* SHJ solar cells with Ag front side and different rear side metallization
    * Reference: standard Ag paste
    * Silver Copper paste with around 50% Cu
    * Pure Copper paste 100% Cu
* Solar cell results show....
    * Ag/cu pastes allow same efficiency level
    * Cu pastes show promising results, but still need some improvement (printing geometry and line resistance)

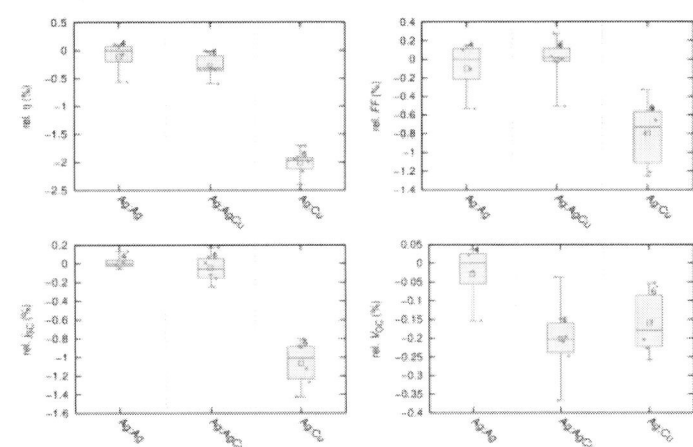

IV results for AHJ cells: relative values in comparison to both side Ag reference. The rear side metallization is varied (Ag, Ag/Cu, Cu)

Screen Printing of Copper Contacts
SHJ solar cells with screen printed copper metallization

- SHJ solar cells with Ag front side and different rear side metallization
  - Reference: standard Ag paste
  - Silver Copper paste with around 50% Cu
  - Pure Copper paste 100% Cu
- Solar cell results show....
  - Ag/cu pastes allow same efficiency level
  - Cu pastes show promising results, but still need some improvement (printing geometry and line resistance)
- ➤ **Ag/Cu pastes "ready for production"***
- ➤ **Cu pastes not "far away"***

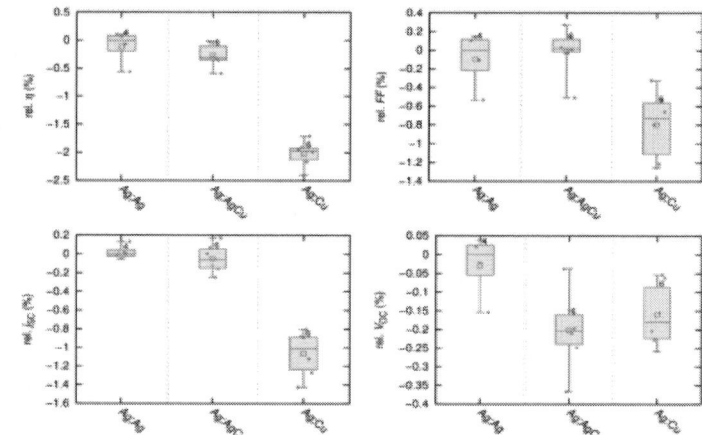

IV results for AHJ cells: relative values in comparison to both side Ag reference. The rear side metallization is varied (Ag, Ag/Cu, Cu)

* See also talks by CEA INES and TNO this conference

Screen Printing of Copper Contacts
TOPCon solar cells with screen printed low temperature copper metallization

---

Previous Work:

- Cu metallization successfully demonstrated on IBC solar cells [22]

Previous Study at ISE:

- TOPCon solar cells with rear side **Ag contacts + full area low-temp copper** conduction layer
- Similar conversion efficiency as reference group
- 75 % silver reduction on the rear side
- No detectable Cu oxidation (r_S) or Cu diffusion into Si (V_{OC})
- Results published by D. Ourinson [23]

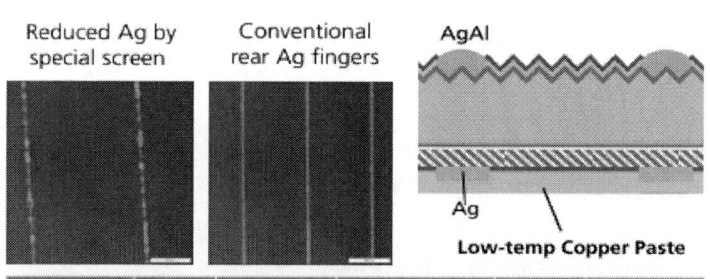

Best cell	η (%)	V_{OC} (mV)	j_{SC} (mA/cm²)	FF (%)	r_S (Ωcm²)
„LT-Cu"	23.0	701	40.8	80.2	0.6
Ag-Ref	23.0	699	40.7	80.7	0.5

[22] Chen et al., Solar RRL 7 (2023)   [23] Ourinson et al., SolMat 266 (2024)

Screen Printing of Copper Contacts
TOPCon solar cells with screen printed copper rear side metallization [24]

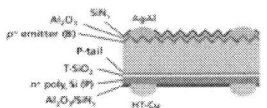

- High temperature approach with copper paste
- Fully functional TOPCon solar cells with screen printed & fired copper metallization on the rear
- Screen printed Cu contacts: w_f ~ 130µm
  → Process Optimization for fine line contacts ongoing
- Silver reduction: ~ 60% less silver
  → 2-5 mg/W silver consumption reasonable

➢ Feasibility confirmed, further optimization of Cu paste and firing process is ongoing

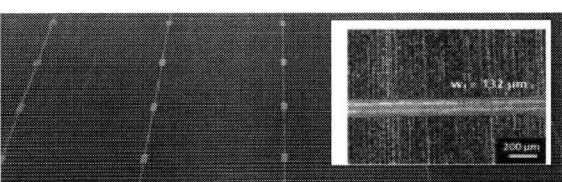

Group	η (%)	V_{OC} (mV)	j_{SC} (mA/cm²)	FF (%)
1 (Cu-RS)	21.6	679	40.2	79.3
2 (Ag-Ref)	23.5	712	40.9	80.5

Group	Ag Front [mg]	Ag Rear [mg]	Total Ag [mg]
1 (Cu-RS)	39	-	39
2 (Ag-Ref)	39	65	104

[24] Lorenz et al., IEEE PVSC 2024, Seattle (USA)

Technology Roadmap
Screen Printing: How can we reduce / replace Silver?

Silver (Ag)

foto from www.tradestation.com

Low Temperature Metallization (SHJ)

High Temperature Metallization (TOPCon)

Technology Roadmap
Screen Printing: How can we reduce / replace Silver?

foto from www.tradestation.com

Low Temperature
Metallization (SHJ)

High Temperature
Metallization (TOPCon)

- Metallization with Ag/Cu pastes ready for mass production
→ at least 50% less Ag
- Approaches with 100% Cu pastes show high potential

Technology Roadmap
Screen Printing: How can we reduce / replace Silver?

foto from www.tradestation.com

Low Temperature
Metallization (SHJ)

High Temperature
Metallization (TOPCon)

- Metallization with Ag/Cu pastes ready for mass production
→ at least 50% less Ag
- Approaches with 100% Cu pastes show high potential

- Print (Ag paste) on Print (Cu pastes) approach promising
- Approaches with 100% Cu pastes still challenging

→ Plating as alternative ?

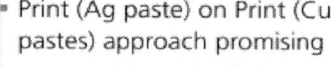

Copper Plating of High-Efficiency Solar Cells

Electroplating of Copper Contacts
TOPCon Solar Cells with Plated Copper Metallization

Objective:

- TOPCon solar cells with plated Ni/Cu metallization
- First pilot fabrication of TOPCon solar cells with format M10 (182 mm x 182 mm) at Fraunhofer ISE

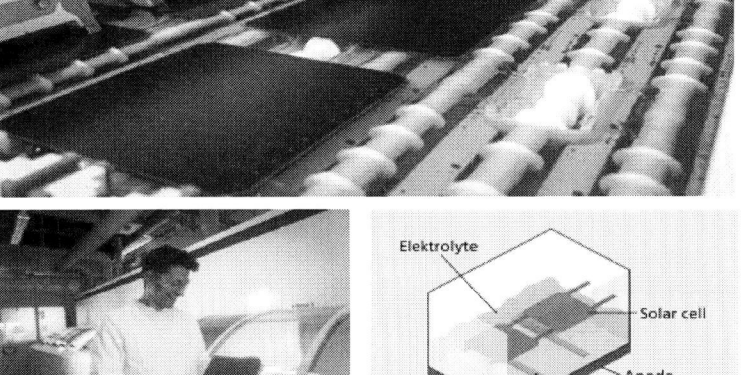

Electroplating of Copper Contacts
TOPCon Solar Cells with Plated Copper Metallization

Experimental:

- TOPCon precursors fabricated at Fraunhofer ISE, format M10 (182 mm x 182 mm)
- Industrial Cz-Si (n-type) Silicon wafers, base resistivity $\rho_{Si} = 0.3\text{-}2.1\ \Omega cm$
- Two experimental groups:
  - Group 1: Laser contact opening (LCO) + Electroplating (Ni/Cu/Ag) on front and rear side (RENA InCell Plating Device)
  - Group 2 (Reference): Screen printing (commercial AgAl and Ag paste)

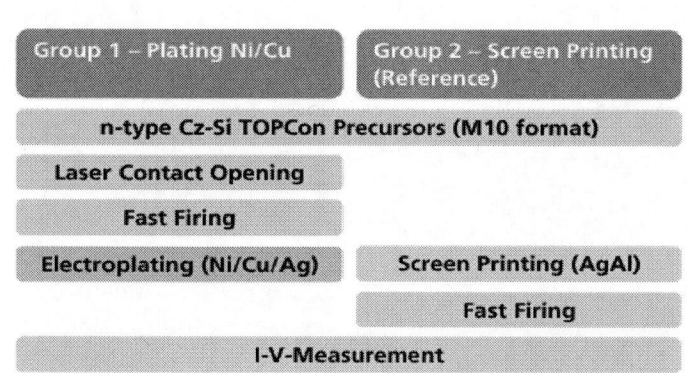

Electroplating of Copper Contacts
TOPCon Solar Cells with Plated Copper Metallization

Results:

- First TOPCon solar cells with format M10 completely fabricated at Fraunhofer ISE [25]
- Champion cell efficiencies:
  - Screen printing: η_{max} = **24.0 %**
  - Electroplating: η_{max} = **24.0 %**
- Silver capping: **~ 9 mg / cell (1 mg/W)** Ag reduction by **~93 %** to SP reference
- Perspective: Silver can be completely avoided by using Sn as capping layer

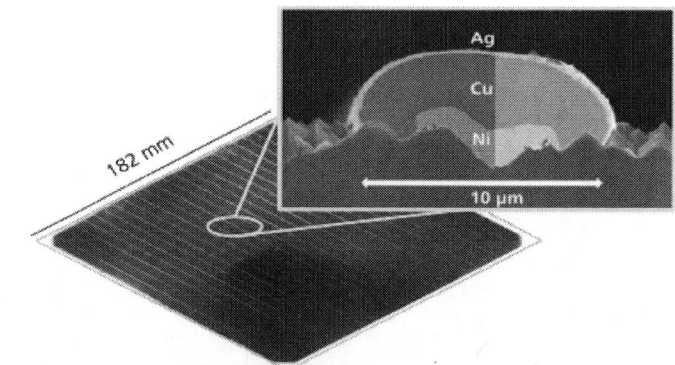

Group	η (%)	V_{OC} (mV)	j_{SC} (mA/cm²)	FF (%)
1 (PL)	24.0	708	41.0	82.7
2 (SP)	24.0	713	41.0	82.0

[25] https://www.ise.fraunhofer.de/en/press-media/news/2024/fraunhofer-ise-successfully-produces-topcon-solar-cell-with-24-percent-efficiency-in-m10-format.html

Electroplating of Copper Contacts
SHJ Solar Cells with Plated Copper Metallization

Results:

- Patented process for plated Cu contacts on SHJ

  - PVD metal masking   → organic free
  - Laser patterning   → narrow width, fast
  - Bifacial plating Cu/Sn   → **Silver free**

- Succesfull piloting on industrial SHJ solar cells M2-G12 format
- Cost of ownership benefits compared to screen printing for SHJ
- Market Introduction:
  → Fraunhofer ISE Spin-off: PV²⁺ GmbH

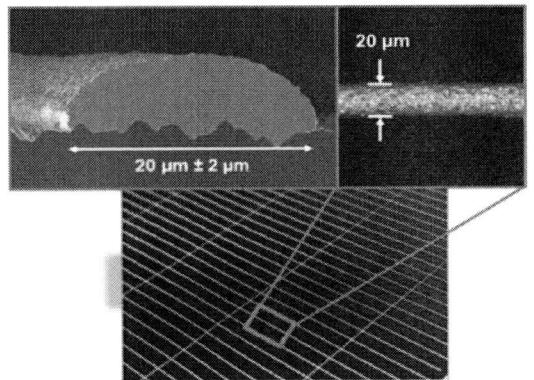

Mask and Plate Copper Contacts
Ultra-Low-Temperature Metallization Approach

Approach

- Masking using hotmelt inkjet printing [26]
- Cu electroplating directly on ITO (no seed layer)
- Mechanical adhesion improved by pre-treatment
- Process temperature well below 100 °C

Process integration

- III-V//Si tandem solar cells with $\eta = 31.6$ % [27,28]
- Electrical performance on ITO and texture
  - (2.5 ± 0.1) µΩ cm finger resistivity
  - (0.4 ± 0.2) mΩ cm² contact resistivity
- Transfer to SHJ and Pero-Si tandem (tbp.)

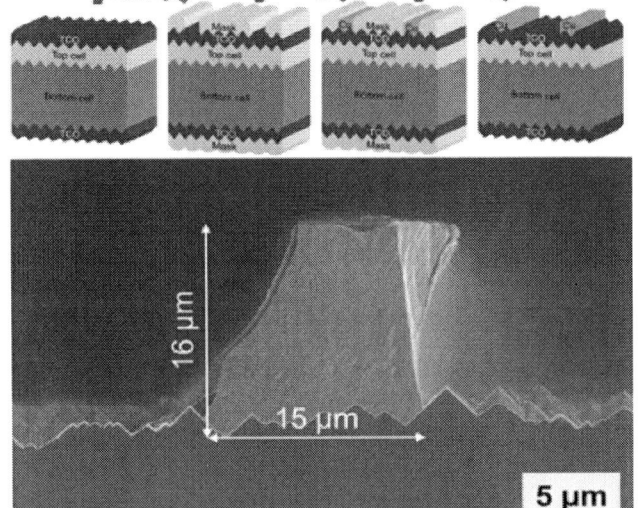

[26] J. Hermans et al. SNEC, 2015
[27] J. Schube et al., *Scientific Reports*, 2023.
[28] R. Keding et al., *EUPVSEC*, 2023.

Technology Roadmap
Plating: How can we reduce / replace Silver?

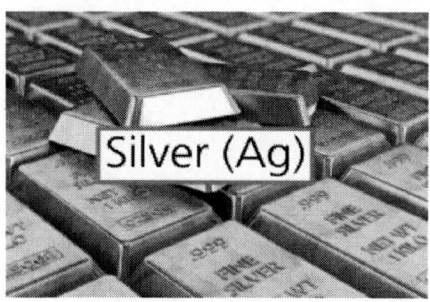

Silver (Ag)

foto from www.tradestation.com

Plating (SHJ, TOPCon)

Technology Roadmap
Plating: How can we reduce / replace Silver?

Silver (Ag)

foto from www.tradestation.com

Plating (SHJ, TOPCon)

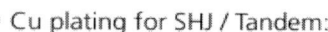

- Ni/Cu(/Ag) Plating for TOPCon:
  - almost Ag free
  - and ready for mass production

- Cu plating for SHJ / Tandem:
  - high potential demonstrated
  - process upscaling is ongoing

Copper as Cost-Effective Alternative to Silver
Summary and outlook

Summary & Outlook:

- There is a strong need to reduce / replace silver
- Different printing technologies allow ultra fine line contacts with less silver consumption
- Ag/Cu pastes replace Ag pastes for low temperature metallization step by step
- Cu pastes are a promising alternative to Ag pastes for low and high temperature metallization
- Cu plating is ready for production especially for TOPCon solar cells
- Cu will replace Ag step by step in the future
- Other metals (e.g Al) will also play a role

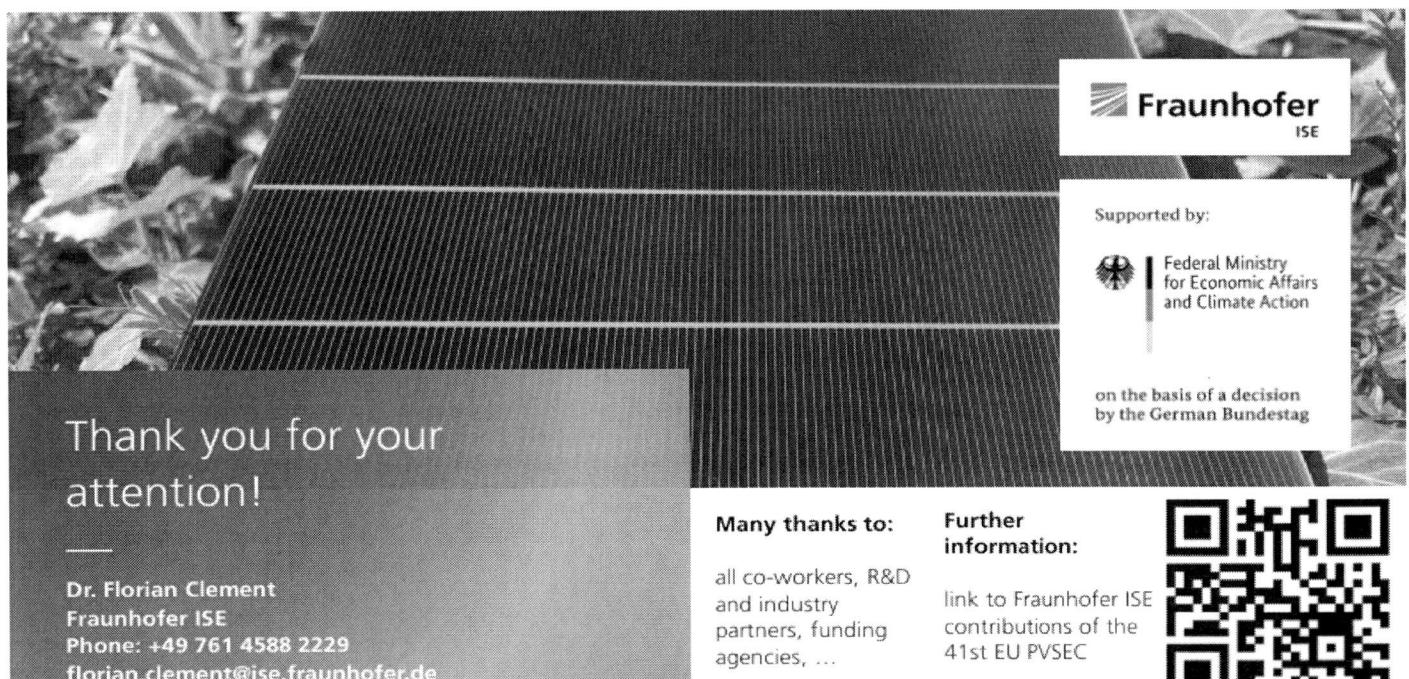

Thank you for your attention!

Dr. Florian Clement
Fraunhofer ISE
Phone: +49 761 4588 2229
florian.clement@ise.fraunhofer.de

Many thanks to:

all co-workers, R&D and industry partners, funding agencies, ...

Further information:

link to Fraunhofer ISE contributions of the 41st EU PVSEC

Supported by:

Federal Ministry for Economic Affairs and Climate Action

on the basis of a decision by the German Bundestag

41st European Photovoltaic Solar Energy Conference and Exhibition

THERMAL MODELING OF TRIPLE-JUNCTION SOLAR CELLS FAN OUT WAFER LEVEL PACKAGING FOR CONCENTRATED PHOTOVOLTAIC

Konan Kouame[1,2], Abdul Rehman[1,2], Médérick Marcotte[1,2], Mylana Ney[1,2], Artur Turala[1,2], Corentin Jouanneau[1,2], Mohamed Najah[1,2], Serge Ecoffey[1,2], David Danovitch[1,2], Gwenaelle Hamon[1,2]

[1]Laboratoire Nanotechnologies Nanosystèmes (LN2) - CNRS IRL-3463 Institut Interdisciplinaire d'Innovation Technologique (3IT), Université de Sherbrooke, 3000 Boulevard Université, Sherbrooke, J1K 0A5 Québec, Canada

[2]Institut Interdisciplinaire d'Innovation Technologique (3IT), Université de Sherbrooke, 3000 Boulevard Université, Sherbrooke, J1K 0A5, QC, Canada
konan.jean.herbert.kouame@usherbrooke.ca

ABSTRACT: We explore Fan-Out Wafer Level Packaging (FOWLP) as a novel packaging approach, for Concentrated Photovoltaic (CPV) modules. This approach involves using reconstituted wafers containing multijunction solar cells embedded in Epoxy Molding Compound (EMC), with the cells interconnected by heat dissipating, thick interconnexion metal layer. This study proposes a detailed Finite Element Modeling (FEM) analysis to assess the thermal performance of a macro-cell and a micro-cell CPV module configuration. The FEM results indicate maximum solar cell temperatures of 63°C for the macro-cells and 60°C for the micro-cells. Those numbers are comparable to cells packaged using flip-chip, 58°C, or wirebonding, 69°C [1], assembly methods (flip chip and wire bonding, respectively and), demonstrating the potential of FOWLP, and below the maximum temperature recommended for such devices (80ºC). The FEM also reveals that the FOWLP module with micro-cells exhibits more efficient heat dissipation compared to the one using macro-cells. A parametric study of heat sink thicknesses shows the viability of this approach for heat sink with thicknesses equal to or greater than 35 μm maintain the maximum solar cell temperature below 80°C, meeting literature-recommended thermal performance criteria. These results highlight the potential of FOWLP for concentrated photovoltaic applications.

Keywords: photovoltaics, concentrator photovoltaic, CPV, fan-out wafer level packaging, FOWLP, surface mount technologies, SMT, thermal simulation, finite elements, solar cell assembly.

1 INTRODUCTION

Concentration Photovoltaics (CPV) represents a solar electricity generation technology that harnesses concentrated incident sunlight onto a high-efficiency solar cell, achieving a record conversion efficiency of 47.6% with a 6-junction solar cell [2]. However, CPV remains less competitive than crystalline silicon-based photovoltaics due to high material costs and technological complexity. Most industrial CPV modules use the FLATCON technology (Fresnel Lens All Glass Tandem cell CONcentrator module) [3], [4], [4], [5], [6], [7], [8], [9]. These modules involve the positioning on carrier of macro solar cells through pick-and-place, followed by wire bonding for interconnection. However, wire bonding leads to long assembly times, especially for sub-millimeter cells (micro-CPV), resulting in low production throughput. Parallel connection methods such as Semprius transfer printing [10], fluidic self-assembly [11], and Surface Mount Technology (SMT) approaches [12], [13], [14], [15] have been developed to reduce assembly time and improve alignment accuracy. Another alternative method for manufacturing CPV modules is Fan Out Wafer Level Packaging (FOWLP). This technique aims to reduce assembly time and manufacturing costs of CPV modules through parallel interconnection. It consists of reconstructing a wafer with an epoxy molding compound (EMC), and then interconnect the solar cells using interconnexion layers [16], [17], [18]. However, to ensure the viability of FOWLP in the CPV domain, effective heat dissipation must be ensured, given the high concentration of incident light on CPV cells generating a significant amount of heat. Maintaining cells at temperatures below 100°C-120°C [19], [20], [21] is generally recommended,

but operating temperatures exceeding 80°C can significantly compromise cell performances and long-term reliability [22].

The ability of a CPV module to dissipate heat depends on the geometry and thermal properties of its materials. Finite Element Modeling (FEM) is a numerical tool used to predict temperature and optimize the design of various solar module technologies [1], [23], [24], [25], [26], [27]. In this paper, a finite element thermal simulation is proposed to assess the thermal performance of a micro and macro CPV module assembled through FOWLP and evaluate the potential of this approach for macro- and micro-CPV.

2 THERMAL PERFORMANCE EVALUATION BY FINITE ELEMENT MODELING

The Fig. 1 illustrates the principle of a CPV module where solar cells are assembled using FOWLP. This module consists of Fresnel lenses that concentrate sunlight onto solar cells contained within an Epoxy Molding Compound (EMC). The combination of solar cells and Epoxy Molding Compound (EMC) is referred to as a reconstituted wafer. The front (-) and back (+) contacts of the solar cells are interconnected by thick copper metal, which also serve as a thermal dissipator. The front and back thick Cu are respectively call interconnexion layer and heat sink. The entire reconstituted wafer and metal layers constitute the assembly.

The concentrated light on the solar cells generates both electricity and heat. This thermal flux on the solar cells is exchanged with the environment through conduction, radiation, and convection. Thermal conduction is carried

41st European Photovoltaic Solar Energy Conference and Exhibition

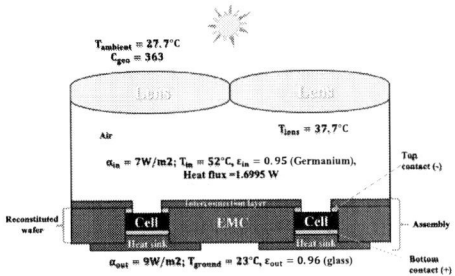

Figure 1: Cross-Sectional View of the CPV Module Model Based on Simulated Fan-Out Wafer Level Packaging using Finite Element Analysis. EMC on the figure means Epoxy Molding Compound EMC.

out by the atoms of the various components of the assembly, with the thermal conductivity of each component considered in the simulation. Two types of radiation must be considered: radiation between the front face of the assembly and the lens, as well as radiation between the back face of the assembly and the ground. The emissivity coefficients of Germanium (ε_{Ge}) and Copper (ε_{Cu}) are used to account for radiation on the front face, while the emissivity coefficients of the EMC (ε_{EMC}) and copper are used for radiation on the back face. These emissivity coefficients are intrinsic properties of each material. Two types of convection must also be considered: convection from the front face of the assembly inside the module and convection from the back face of the assembly to the outside of the module. The convection coefficients α_{in} and α_{out} are used to represent internal and external convections, respectively. The temperatures of the lens (T_{lens}), the inside of the module (T_{in}), the ambient (T_{amb}), and the ground (T_{sol}) are also applied as boundary conditions in the simulation.

The Workbench module of the ANSYS software was used to perform the finite element thermal simulation. In this study, the focus is exclusively on the thermal distribution within the assembly. The simulation is carried out for an assembly consisting of 4 triple-junction III-V/Ge solar cells, each with a thickness of 190 μm and an efficiency of 40%. However, in this simulation, the solar cell is represented by a germanium die, given that the thickness of III-V materials (10 μm) in a triple-junction solar cell is negligible compared to that of Ge (180 μm) and does not impact the simulation results. The thickness of the front interconnection layer is set at 35 μm, a chosen value due to its feasibility for this type of application, as demonstrated in the relevant literature [28]. Two simulation cases were conducted:

Case 1: For the macrocells, the surface area dimension of the solar cells is 3×3 mm^2. The EMC has a surface area of 57.14×57.14 mm^2 and a thickness of 500 μm. The surface area of the thick heat sink layer at the back of the solar cell is 41×41 mm^2.

Case 2: For the microcells, the surface dimensions of the solar cells are 0.52×0.52 mm^2. The EMC has a surface area of 9.91×9.91 mm^2 and a thickness of 500 μm. The surface area of the heat sink at the back of the solar cell is 7.11×7.11 mm^2.

For both cases, a thermal flux of 1.67E5 W/m^2 is applied to the surface of each solar cell, corresponding to a geometric concentration ratio of 363 and an electrical efficiency of the solar cell of 40%. The thermal conductivity coefficients of the assembly materials are extracted from the literature [23], [29]. The emissivity of

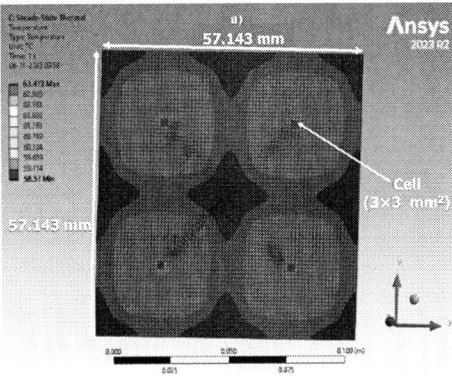

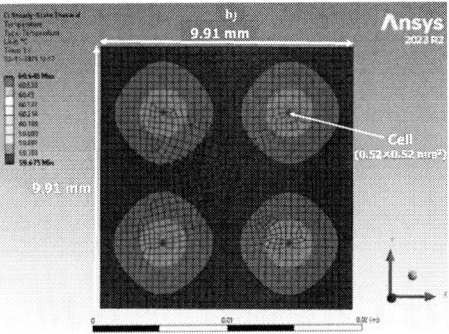

Figure 2: a) Finite Element Modeling of a macro-CPV Module Based on Fan-Out Wafer Level Packaging, displaying a Temperature of 63.5°C. a) Finite Element Modeling of a micro-CPV Module Based on Fan-Out Wafer Level Packaging, Displaying a Temperature of 60.6°C.

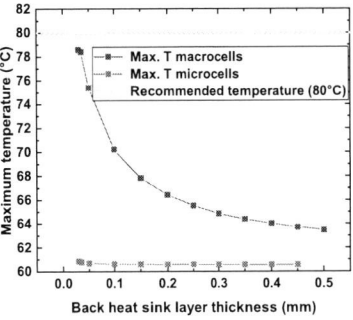

Figure 3: Graph depicting the variation of the maximum temperature of the solar cell as a function of changes in the thickness of the heat sink on the rear side of the FOWLP assembly, comparing modules utilizing macrocells and modules utilizing microcells.

the EMC, germanium, and coppers are 0.97, 0.95, and 0.88, respectively. The convection coefficients inside and outside the module are 7 W/m^2 and 9 W/m^2, respectively. These coefficients are referenced and derived from the literature [24]. The choice of reference [23], [24], [29] is motivated by the fact that these data are validated experimentally and/or commonly used for modeling in the field. The entire front surface of the assembly is covered with a thick metal layer, except for the active areas of the solar cells. The dimensions of the heat sink surfaces were set to maintain an identical coverage ratio on the back surface of the assembly in both simulation cases.

The simulation parameters mentioned earlier, such as convection coefficients, emissivity coefficients, thermal conductivities, temperatures, and heat fluxes, are identical to those used in our previous works [1].

The Workbench module of the ANSYS software was used to perform the finite element thermal simulation. Figures 2-a and b respectively present the temperature distribution within the FOWLP module using macro-cells and the one using micro-cells, obtained through finite element simulation. These results indicate that the maximum temperature of the FOWLP module employing macro-cells is 63.5 °C, while that of the FOWLP module using micro-cells is 60.6 °C. Thus, under equivalent boundary conditions and with the use of the same type of heat dissipator, the FOWLP module using micro-cells exhibits a lower temperature. The maximum temperature in the FOWLP module using micro-cells is lower due to the smaller size of the cells compared to macro-cells. Indeed, with a similar heat density for all solar cells, the total heat generation is reduced when the solar cell is smaller. Additionally, a smaller solar cell size promotes faster lateral dissipation, thereby accelerating the heat dissipation process which is precisely the effect expected from micro-cells [30]. The FEM results indicate maximum solar cell temperatures comparable to those achieved with other assembly methods (flip chip and wire bonding, respectively 69°C and 58°C [1]), demonstrating the effectiveness of FOWLP.

To minimize metallization costs while optimizing manufacturing time, a parametric study was undertaken involving the variation of the thickness of the heat sink at the back of the assembly from 0.035 mm to 0.5 mm in both cases to observe the variation in maximum temperature. The other dimensions of the assembly remain unchanged, as previously mentioned. The choice of the minimum thickness (0.035 mm) for the parametric study is because this thickness corresponds to the industry standard for electroplated heat sinks [31], [32]. Fig. 3 illustrates the variation in the maximum temperature of the solar cell as a function of thickness in both simulation cases. Temperatures range from 78.6 °C to 63.5 °C, and from 60.9 °C to 60.6 °C, respectively, for modules using macrocells and modules using microcells when the thicknesses of the heat sink at the back of the assemblies vary. The maximum temperature of the micro-cells consistently remains lower than that of the macro-cells. The maximum temperature significantly increases with the reduction of the heat sink thickness in the case of modules using macrocells, while the temperature remains nearly constant for modules using microcells. This highlights the advantage of microcells over macrocells in terms of thermal dissipation. It is observed that for a thickness equal to or greater than 0.035 mm, the maximum temperature of the solar cells remains below 80°C, in line with literature recommendations for both assemblies. These results confirm the viability of modules fabricated using FOWLP approach for concentrated photovoltaic applications.

4 CONCLUSION

Our study explores Fan Out Wafer Level Packaging (FOWLP) as an innovative assembly approach for Concentrated Photovoltaic (CPV) modules. Through Finite Element Modeling (FEM) thermal analysis, we assessed the thermal performance of two CPV module configurations, one utilizing macro-cells and the other micro-cells. The FEM results revealed maximum solar cell temperatures (63°C and 60°C for the FOWLP module with macro-cells and micro-cells, respectively) comparable to other assembly methods such as flip chip and wire bonding (69°C and 58°C, respectively [1]), thus demonstrating the viability of studied approach. The FEM modeling also indicated that the FOWLP module with micro-cells exhibits more efficient heat dissipation than the one using macro-cells. A parametric study further demonstrated that heat sink with thicknesses equal to or greater than 35 µm maintain the maximum solar cell temperature below 80°C, in accordance with literature-recommended thermal performance criteria. These results show the potential of FOWLP in the renewable energy sector, providing a competitive assembly alternative for Concentrated Photovoltaic applications. A Fan-Out Wafer Level Packaging process has been conceptualized and is currently in the development phase within our laboratory. One of the main challenge lies in the adhesion of the metal to the Epoxy Molding Compound (EMC). Detailed results of these efforts will be presented in our upcoming publications.

6 ACKNOWLEDGMENT

We acknowledge the Fonds de recherche du Québec – Nature et technologies (FRQNT) and the MiQro Innovation Collaborative Centre (C2MI) for the support in the scope of the C3P Research Chair in microelectronic assembly technologies for energy and optoelectronics. LN2 is a joint International Research Laboratory (IRL 3463) funded and co-operated in Canada by Université de Sherbrooke (UdeS) and in France by CNRS as well as ECL, INSA Lyon, and Université Grenoble Alpes (UGA).

7 REFERENCES

[1] K. Kouame et al., « New Triple-Junction Solar Cell Assembly Process for Concentrator Photovoltaic Applications », in 2023 IEEE 73rd Electronic Components and Technology Conference (ECTC), IEEE, 2023, p. 2223-2229.

[2] M. A. Green et al., « Solar cell efficiency tables (Version 61) », Progress in Photovoltaics, vol. 31, no 1, p. 3-16, janv. 2023, doi: 10.1002/pip.3646.

[3] J. S. Foresi, R. Hoffman, D. King, et P. Ponsardin, « Performance of silicone-on-glass Fresnel lenses in EMCORE's Gen 3 high-concentration concentrator photovoltaic system », in High and Low Concentrator Systems for Solar Electric Applications VII, SPIE, oct. 2012, p. 84-89. doi: 10.1117/12.929348.

[4] A. Gombert, N. Abela, T. Gerstmeier, S. Bambrook, et F. Rubio, « Annex: Soitec Power Plants », in Handbook of Concentrator Photovoltaic Technology, John Wiley & Sons, Ltd, 2016, p. 513-520. doi: 10.1002/9781118755655.ch08c.

[5] M. Steiner et al., « FLATCON® CPV module with 36.7% efficiency equipped with four-junction solar cells », Progress in Photovoltaics: Research and Applications, vol. 23, no 10, p. 1323-1329, 2015, doi: 10.1002/pip.2568.

[6] W. Belt' et al., « FLATCON~~-MODULETSE: CHNOLOGY AND CHARACTERISATION ».

[7] A. Luque et V. M. Andreev, Éd., Concentrator photovoltaics. in Springer series in optical sciences, no. v. 130. Berlin: Springer, 2007.

[8] M. Wiesenfarth, S. Gamisch, T. Dorsam, et A. W. Bett, « Challenges for thermal management and production technologies in concentrating photovoltaic (CPV) modules », in Proceedings of the 5th Electronics System-integration Technology Conference (ESTC), Helsinki, Finland: IEEE, sept. 2014, p. 1-4. doi: 10.1109/ESTC.2014.6962842.

[9] M. Wiesenfarth, I. Anton, et A. W. Bett, « Challenges in the design of concentrator photovoltaic (CPV) modules to achieve highest efficiencies », Applied Physics Reviews, vol. 5, no 4, p. 041601, déc. 2018, doi: 10.1063/1.5046752.

[10] S. Burroughs et al., « A New Approach For A Low Cost CPV Module Design Utilizing Micro-Transfer Printing Technology », présenté à 6TH INTERNATIONAL CONFERENCE ON CONCENTRATING PHOTOVOLTAIC SYSTEMS: CPV-6, Freiburg, (Germany), 2010, p. 163-166. doi: 10.1063/1.3509179.

[11] T. Nakagawa et al., « High-efficiency Thin and Compact Concentrator Photovoltaics with Micro-solar Cells Directly Attached to Lens Array », in Light, Energy and the Environment, Canberra: OSA, 2014, p. RF4B.5. doi: 10.1364/OSE.2014.RF4B.5.

[12] C. Domínguez, N. Jost, S. Askins, M. Victoria, et I. Antón, « A review of the promises and challenges of micro-concentrator photovoltaics », AIP Conference Proceedings, vol. 1881, no 1, p. 080003, sept. 2017, doi: 10.1063/1.5001441.

[13] N. Jost, T. Gu, J. Hu, C. Domínguez, et I. Antón, « Integrated Micro-Scale Concentrating Photovoltaics: A Scalable Path Toward High-Efficiency, Low-Cost Solar Power », Solar RRL, vol. 7, no 16, p. 2300363, 2023, doi: 10.1002/solr.202300363.

[14] C. Jouanneau, T. Bidaud, M. Darnon, et G. Hamon, « Compact and high efficiency micro-CPV module with high wafer utilization rate », juin 2023. Consulté le: 1 février 2024. [En ligne]. Disponible sur: https://hal.science/hal-04119376

[15] C. Jouanneau, T. Bidaud, G. Hamon, M. Darnon, D. Danovitch, et A. Turala, « Fabrication de cellules solaires submillimetriques III-V/Ge `a haut rendement et haut taux d'utilisation de wafer pour le micro-CPV ».

[16] J. H. Lau, « Fan-in Wafer-Level Packaging Versus FOWLP », in Fan-Out Wafer-Level Packaging, J. H. Lau, Éd., Singapore: Springer, 2018, p. 69-113. doi: 10.1007/978-981-10-8884-1_3.

[17] M. LaPedus, « Fan-Out Wars Begin », Semiconductor Engineering. Consulté le: 1 février 2024. [En ligne]. Disponible sur: https://semiengineering.com/fan-out-wars-begin/

[18] M. Töpper, J. Simon, et H. Reichl, « Redistribution technology for chip scale package using photosensitive BCB », 1996, Consulté le: 1 février 2024. [En ligne]. Disponible sur: https://publica.fraunhofer.de/handle/publica/189734

[19] AZURSPACE, « Concentrator Triple Junction Solar Cell (0004357-00-01_3C44_AzurDesign_3x3.pdf) ». 4 janvier 2017. [En ligne]. Disponible sur: https://www.azurspace.com/images/products/0004357-00-01_3C44_AzurDesign_3x3.pdf

[20] SPECTROLAB, « CPV Point Focus Solar Cells, C4MJ Metamorphic Fourth Generation CPV Technology ». BOEING COMPANY, 16 juillet 2018. Consulté le: 21 août 2023. [En ligne]. Disponible sur: https://www.spectrolab.com/photovoltaics/C4MJ_40_Percent_Solar_Cell.pdf

[21] G. Segev, G. Mittelman, et A. Kribus, « Equivalent circuit models for triple-junction concentrator solar cells », Solar Energy Materials and Solar Cells, vol. 98, p. 57-65, mars 2012, doi: 10.1016/j.solmat.2011.10.013.

[22] P. Espinet-González et al., « Temperature accelerated life test on commercial concentrator III-V triple-junction solar cells and reliability analysis as a function of the operating temperature: Temperature accelerated life test », Prog. Photovolt: Res. Appl., vol. 23, no 5, p. 559-569, mai 2015, doi: 10.1002/pip.2461.

[23] T.-L. Chou, Z.-H. Shih, H.-F. Hong, C.-N. Han, et K.-N. Chiang, « Thermal performance assessment and validation of high-concentration photovoltaic solar cell module », IEEE Transactions on Components, Packaging and Manufacturing Technology, vol. 2, no 4, p. 578-586, 2012.

[24] M. Wiesenfarth et al., « Technical boundaries of micro-CPV module components: How small is enough? », présenté à 17TH INTERNATIONAL CONFERENCE ON CONCENTRATOR PHOTOVOLTAIC SYSTEMS (CPV-17), Freiburg, Germany / Online, 2022, p. 030008. doi: 10.1063/5.0099878.

[25] J. F. Martínez, M. Steiner, M. Wiesenfarth, S. W. Glunz, et F. Dimroth, « Thermal analysis of passively cooled hybrid CPV module using Si cell as heat distributor », IEEE Journal of Photovoltaics, vol. 9, no 1, p. 160-166, 2018.

[26] L. Micheli, E. F. Fernández, F. Almonacid, T. K. Mallick, et G. P. Smestad, « Performance, limits and economic perspectives for passive cooling of High Concentrator Photovoltaics », Solar Energy Materials and Solar Cells, vol. 153, p. 164-178, 2016.

[27] K. J. H. Kouame et al., « Finite Element Modeling and Experimental Validation of Concentrator Photovoltaic Module Based on Surface Mount Technology ». Rochester, NY, 22 septembre 2023. doi: 10.2139/ssrn.4580254.

[28] J. Amanlim, K. Cao, Z. Li, et K. H. Tan, « 30μm Thick Cu RDL in WLCSP », IMAPSource Conference Papers, vol. 2019, no DPC, p. 195-211, janv. 2019, doi: 10.4071/2380-4491-2019-DPC-Presentation_TA2_043.

[29] N. Jiang, A. G. Ebadi, K. H. Kishore, Q. A. Yousif, et M. Salmani, « Thermomechanical Reliability Assessment of Solder Joints in a Photovoltaic Module Operated in a Hot Climate », IEEE Trans. Compon., Packag. Manufact. Technol., vol. 10, no 1, p. 160-167, janv. 2020, doi: 10.1109/TCPMT.2019.2933057.

[30] N. Jost, T. Gu, J. Hu, C. Domínguez, et I. Antón, « Integrated Micro-Scale Concentrating Photovoltaics: A Scalable Path Toward High-Efficiency, Low-Cost Solar Power », Solar RRL, vol. 7, no 16, p. 2300363, 2023, doi: 10.1002/solr.202300363.

[31] M. Lueck et al., « Fan-Out Packaging of Microdevices Assembled Using Micro-Transfer-Printing », in 2016 IEEE 66th Electronic Components and Technology Conference (ECTC), Las Vegas, NV, USA: IEEE, mai 2016, p. 37-42. doi: 10.1109/ECTC.2016.269.

[32] S. K. Kumar, S. C. Warn, et R. Lee, « LED thermal management of an automotive electronic control module with display », in 2012 IEEE 14th Electronics Packaging Technology Conference (EPTC), Singapore: IEEE, déc. 2012, p. 764-769. doi: 10.1109/EPTC.2012.6507187.

41st European Photovoltaic Solar Energy Conference and Exhibition

Overview for Tandem Solar Cell R&D Activities in Japan

Masafumi Yamaguchi[1*], Tatsuya Takamoto[2], Kyotaro Nakamura[1], Ryo Ozaki[1], Hiroyuki Juso[2], Nobuaki Kojima[1] and Yoshio Ohshita[1]

1. Toyota Technological Institute, Nagoya 468-8511, Japan
2. Sharp Corporation, Nara, 639-1186, Japan

Introduction

The development of high-performance solar cells offers a promising pathway toward achieving high power per unit cost for many applications. Because state-of-the-art efficiencies of single- junction solar cells are approaching the Shockley-Queisser limit [1], the multi-junction (MJ) solar cells [2] are very attractive for high-efficiency solar cells. Although the concept of MJ solar cells was first and most successfully implemented using III-V compound materials, there is a need to further improve their conversion efficiency and rduce cost. The Si tandem solar cells [3] have great potential of cost reduction by maining high-efficiency with an efficiency of more than 37% with 2-junction and 43% with 3-junction.

This paper reviews Japanese R&D activities for III-V compound MJ solar cells.. As the most recent R&D activities, developments of high-efficiency and low cost III-V/Si tandem solar cells and modules are presented in this paper.

2. Progress in R&D activities for III-V MJ solar cells in Japan

Based on R&D activities conducted at NTT Labs since 1982 [4], R&D activities for III-V MJ solar cells in Japan have been progressed.

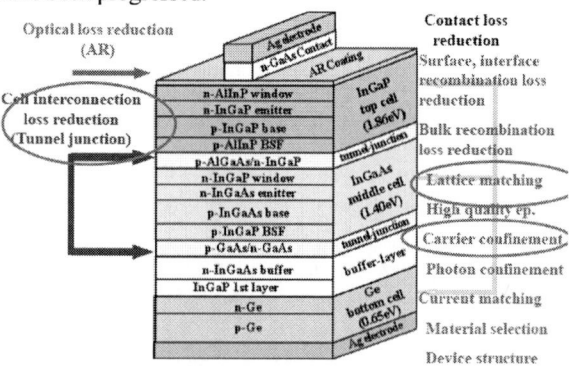

Fig. 1. Key technologies for high-efficiency MJ solar cells.

Figure 1 shows key technologies for high-efficiency MJ solar cells. The authors have contributed to the following issues marked as red circles in Fig.1 :1) Proposal of double hetero (DH) structure tunnel diode for sub-cell   interconnection, 2) proposal of lattice matched InGaAs middle cell, 3) proposal of AlInP BSF layer, and

discovery of superior radiation-resistance of InP, InGaP solar cells and materials.

As a results of such contributions, the Japanese group has contributed to developments of high-efficiency MJ and Si tandem solar cells and modules as follows [5]:
* 1st 20% efficiency AlGaAs/GaAs 2-junction solar cells in 1987.
* 1st 30% efficiency InGaP/GaAs 2-junction solar cell in 1996.
* 37.9% efficiency InGaP/GaAs/InGaAs 3-junction solar cell in 2013.
* 44.4% efficiency InGaP/GaAs/InGaAs 3-junction concentrator solar cell in 2013.
* 31.2%, 32.7% InGaP/GaAs/InGaAs 3-junction solar cell modules in 2016, 2022
* 33.7% InGaP/GaAs/Si 3-junction tandem solar cell module in 2023

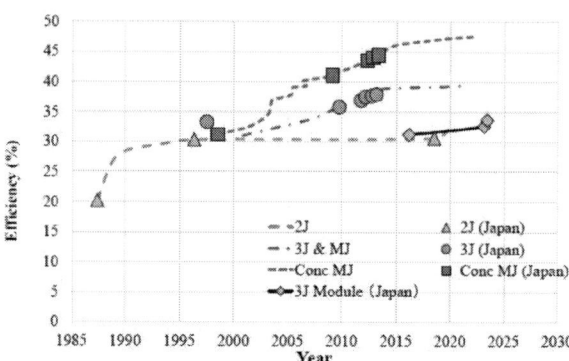

Fig. 2. Japanese contributions to development of high-efficincy III-V 2-junction, 3-junction, concentrator multi-junction solar cells and 3-junction solar cell modules including III-V/Si 3-junction tandem solar cell module.

Figure 2 shows Japanese contributions to development of high-efficincy III-V 2-junction, 3-junction, concentrator multi-junction solar cells and 3-junction solar cell modules including III-V/Si tandem solar cell module.

3. Development of high-efficiency III-V/Si tandem solar cells and modules

High-efficiency and low-cost solar cells such as Si tandem cells have great potential for PV as a vital energy source in mobility application where the installation area is limited. According to the NEDO's report [6], new broader PV market with more than 50GW in 2050 are expected to be established. Cumulative PV capacity for PV-powered vehicles will be 0.4TW in 2050.

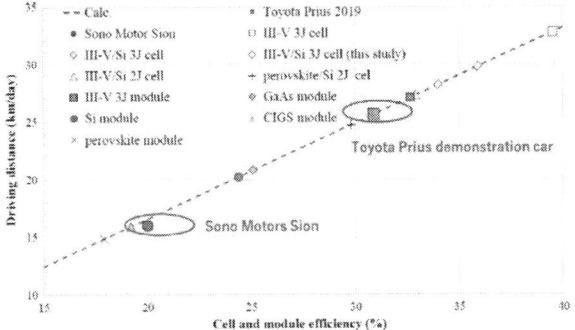

Fig. 3. Calculated results for the PV driving range of vehicles powered by current various solar cell modules and various candidate high-efficiency solar cells as a function of cell and module efficiency in comparison with the actual PV driving range calibrated of the Prius demonstration car powered by the 3-junction solar cell module and the Sono Motors Sion powered by back-contact Si solar cell module.

Figure 3 shows calculated results for the PV driving range of vehicles powered by current various solar cell modules and various candidate high-efficiency solar cells as a function of cell and module efficiency in comparison with the actual PV driving range calibrated of the Toyota Prius demonstration car [7] powered by the 3-junction solar cell module and the Sono Motors Sion [8] powered by back-contact Si solar cell module. Our analysis show that the use of high-efficincy vehivcle integrared PV (VIPV) modules with a conversion efficiency of higher than 35 % enables more than 30 km/day of PV-driving range under average irradiance of 4 kwh/m²/day without

external charging. Cost reduction of VIPV modules is also very important for accelerating the uptake of VIPV-powered electric vehicles (VIPV-EV). Thus, we are developing high-efficiency and low-cost Si tandem solar cells and modules for automobile applications.

Development of high-efficiency solar cell modules is very important for the VIPV-EV applications. Most recently, high-efficiency (33.66%) InGaP/GaAs/Si 4 terminal 3-junction solar cell module with an area of 775 cm² has been demonstrated as a result of collaboration between Sharp and Toyota Tech. Inst. under the NEDO R&D Project. Efficiency (33.66%) has been confirmed by the AIST. Figure 4 shows a photo of InGaP/GaAs/Si 3-junction solar cell module and I−V characteristics of InGaP/GaAs upper 2-junction solar cell module and Si bottom solar cell module.

(a) Photo

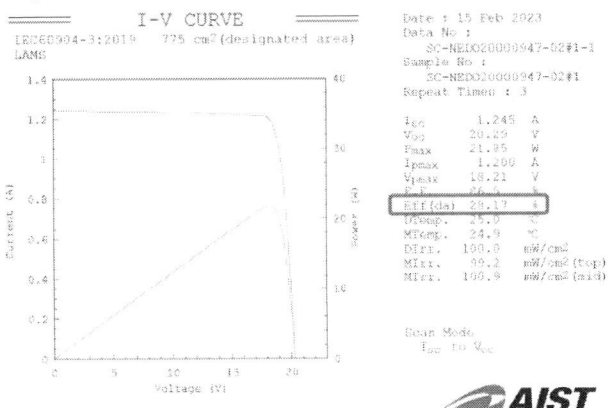
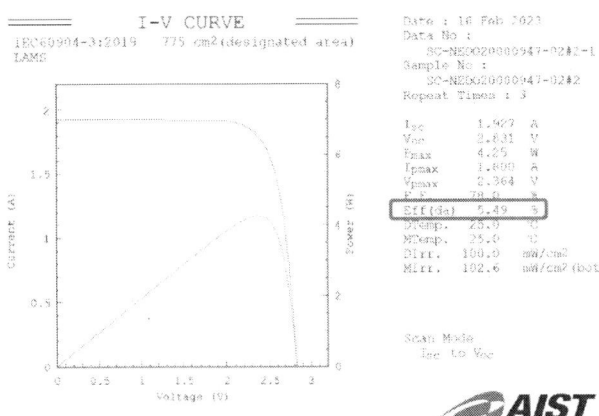

(b) I-V characteristics

Fig. 4. (a) A photo of InGaP/GaAs/Si 3-junction solr cell module and (b) IV charactersitics of InGaP/GaAs upper 2-junction solar cell module and Si bottom solar cell module.

Although our current status is 33.66% for InGaP/Si 4 terminal 3-junction tandem solar cell module with an area of 775 cm^2, development of high-efficiency solar cells with an efficiency of more than 39% is necessary in order to develop high-efficiency solar cell large-area modules with an efficiency of more than 35%,.

4. Comparative studies for Si tandem solar cell modules

Figure 5 shows changes in conversion efficiency of Si, GaAs, CdTe, perovskite single-junction solar cells and modules, and III-V/Si 3-junction and perovskite/Si 2-junction solar cell modules [9] versus area of solar cells and modules including 28.6% efficiency perovskite/Si 2-terminal 2-junction solar cell module with an area of 258.14 cm^2 [9]. Our new record efficiency of 33.66% for InGaP/Si 4 terminal 3-junction tandem solar cell module with an area of 775 cm^2 is also plotted in Fig. 5.

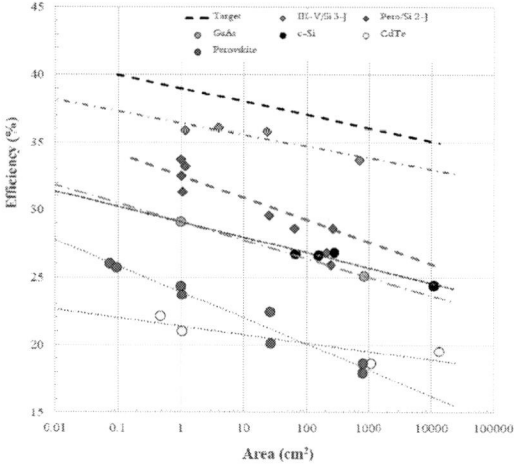

Fig. 5. Changes in conversion efficiency of Si, GaAs, CdTe, perovskite single-junction solar cells and modules, and III-V/Si 3-junction and perovskite/Si 2-junction solar cell modules versus area of solar cells and modules.

Relative decreasing rates of efficiency per one decade of area are sumarrized to be -2.4%/decade for III-V/Si 3-junction and -5.6%/decade for perovskite/Si 2-junction tandem solar cell modules, and -3.1%/decade for CdTe, -4.2%/decade for Si, -5.2%/decade for GaAs, and -9.8%/decade for perovskite single junction solar cell modules. In order to develo high-efficiency solar cell large-area modules with an efficiency of more than 35%, development of high-efficiency solar cells with an efficiency of more than 39% is necessary as shown as the target in Fig. 5. It is clear in Fig. 5 that especially, conversion efficiency of perovskite/Si 2-junction tandem and perovskite single-junction solar cells modules decrease with increase in area of perovskite and organic solar cells and modules. The III-V/Si 3-junction tandem solar cell modules are thought to be very attractive for VIPV-EV appications because of great potential of longer driving distance of more than 30 km/day.

5. Summary

The development of high-performance solar cells offers a promising pathway toward achieving high power per unit cost for many applications. The 2-junction and 3-junction tandem solar cells have great potential of high efficiency of more than 37% and 43%, respectively.

This paper reviewed Japanese R&D activities for tandem solar cells. As the most recent R&D activities, developments of high-efficiency and low cost III-V/Si tandem solar cells and modules were presented in this paper.

Especially, Japanese group has contributed to proposals of double hetero structure tunnel diode for interconnections of sub-cells, wide bandgap AlInP back surface field layer for top cell, and lattice matching of InGaAs middle cell for 3-J applications. New world record efficiency Si tandem solar cell module with an efficiency of 33.66% was also presented in this paper. Comparative studies for Si tandem solar cells and modules were also presented in this paper.

Acknowledgments

The authors express sincere thanks to the NEDO and members of Toyota Tech. Inst. and Sharp for supporting R&D.

References

[1] W. Shockley and H.J. Queisser, J. Appl. Phys. **32,** 510 (1961).
[2] M. Yamaguchi et al., J. Appl. Phys. **129**, 240901 (2021).
[3] M. Yamaguchi et al., J. Phys. D. Appl. Phys. **51**, 133002 (2018).
[4] M. Yamaguchi et al., Proc. 19th IEEE PVSC, (IEEE, New York, 1987) p.1484.
[5] M. Yamaguchi et al., presented at the PVSEC-34, Shenzhen, China, Nov. 5-10, 2023.
[6] NEDO, Interim Report "PV-Powered Vehicle Strategy Committee". ; 2018. http://www.nedo.go.jp/english/index.html.
[7] M. Yamaguchi et al., Prog. Photovolt. **29**, 684 (2021).
[8] Sono Motors, https://sonomotors.com/
[9] M.A. Green et al., Progress in Photovoltaics **31**, 651 (2023).

Space applications for a variety of solar cell technologies

Stephen Taylor

European Space Agency, Keplerlaan 1,
2200AG Noordwijk, the Netherlands

Abstract—**Even if space photovoltaics is not a new subject, technology for both solar cells and satellites has developed rapidly in recent years, in parallel to a dramatic evolution of the market for space applications. This paper aims to explain the status today and the opportunities for solar cell manufacturers in the area of space applications.**

Keywords—component, formatting, style, styling, insert (key words)

I. INTRODUCTION

As for any other component within a system, the best choice of solar cell is typically the one that enables the lowest overall cost to the end customer, while still meeting the requirements that have been defined. For space applications, the overall cost relates to the service or applications provided by the satellite(s), the design of the satellite(s) within this system and the design of the solar array within the satellite. For many years, the most cost-effective approach to solar array design for most applications was to build a relatively smaller solar array using efficient but expensive multi-junction solar cells rather than a bigger solar array using less efficient but cheaper cells. The biggest commercial satellite market was for telecommunications satellites in geostationary orbits with typical requirements for 15 years of operation. Within the bigger trade-off of options at solar array level, the need to function for a long lifetime in a severe particle radiation environment worked in favour of multi-junction III-V cells. More recently, the end user market has shifted towards constellations of satellites in low earth orbit, with shorter lifetime for an individual satellite in a more benign particle environment. This is coupled with a demand for cells that overall has increased with the number of satellites. In the context of this market, there is not a single 'best product' for all situations.

II. RECALL OF SPACE APPLICATIONS

A. The space environment

Space is a vacuum within which the sun spectrum is unfiltered. The presence of ultraviolet light and particle radiation cause degradation to the properties of materials and to the performance of electronic devices, including solar cells. For solar arrays, heat exchange is almost exclusively by radiation and this leads to operating temperatures that are extremely hot when illuminated and extremely cold when in eclipse, with potentially thousands of cycles between these extremes.

Operation of a satellite mission in this environment involves first launching it into space, so additional mass translates to additional cost.

With terrestrial solar cells available at prices below 1$/watt, it might seem that they should always be the most cost effective option also for all space applications. If this is not the case, it's because the cost of the solar cells is only one component of the system cost. To take a trivial example illustrated in Table 1, a solar array made with 15% efficient cells will be twice as big as one made of 30% efficient solar cells. If the cost of 30% efficient solar cells would represent half of the total cost of the array, then use of 15% efficient cells would not reduce the cost of the array, even if they were free of charge, simply because of the array size. On the contrary, the impact on launch cost would be doubled and hence the overall system cost would be much more expensive.

	Cell efficiency	Cost of cells	Other costs	Total cost (launch excluded)
array A	30	X	X	2X
array B	15	0	2X	2X

Table 1: illustration of key contributions to the cost of a satellite solar array. In this example, if an array would be made with 15% cells instead of 30% cells then the array would have to be twice as big and there would be no cost advantage. On the contrary, the mass would be double and the launch cost would be higher. For many years, use of the highest efficiency cells led to the smallest arrays for a given power and the lowest system level cost for many applications. Today, the satellite market is both growing and changing; within it, there are potential applications for different cell technologies.

B. Discussion of mass within a space solar array design

Although the properties of solar cells have a direct impact on the cost of a space solar array, a popular misconception is that production of solar cells with low mass automatically leads also to availability of solar arrays with low mass. In fact, the mass budget of a space solar array is not dominated by the solar cells; the mechanical structure needed to meet mechanical requirements is a major component, as well as the photovoltaic elements needed to meet electrical requirements.

In the future, we can imagine a time when satellites may be assembled in space and requirements associated with the launch environment on assembled sub-systems are no longer relevant. However, the overall mass budget will still have to reflect a set of requirements that will involve transfer of the power generated by wiring in some form and control of the satellite orientation. Hence, even if reduction of solar array mass will always be considered a potential benefit within a satellite system design, the design of a solar array to address all the necessary requirements will imply a mass that remains significantly heavier than the mass only of the solar cells.

III. SOLAR CELL TECHNOLOGIES

A. III-V Multijunction cells , evolution of the market

From the 1990s, availability of expensive, single junction GaAs solar cells on GaAs substrates with efficiencies of 19-20% under the AM0 spectrum became of interest for some applications that benefitted from a performance advantage over silicon cells that had an efficiency of up to 17% and which degraded faster due to particle radiation [1,2].

Although III-V cells were more mechanically fragile than silicon and typically needed by-pass diodes for each cell, the advent of growth on Ge substrates initially reduced cost and also paved the way first to 2, then 3 and 4 junction devices with progressively higher efficiencies. Higher cell efficiencies led to smaller arrays and a cost benefit at system level.

Taking a recent and well documented example [3], use of cells with efficiency of 32% at 25C lead to an array level efficiency of around 20% under operational conditions in space, leading to a 'beginning of life' capability of about 30kW for an array with a mass of 286kg and area of 118m2, hence 105W/kg and 258W/m2.

Even as the performance of III-V cells has continued to improve, a key focus in the last ten years has been within the solar array level mechanical design. Reduction of the volume of a stowed array led to the possibility of deploying more satellites from one launcher. Array level designs based on tensioned, flexible blankets that can be folded or rolled have been demonstrated by several companies and the possibility to approach 200W/kg with current technology has been claimed in some cases.

At the same time, construction of satellite constellations has led to a dramatic growth in the aggregate demand for solar cells for space applications, combined with the possibility of serial production and an associated opportunity to reduce costs.

Assembly of photovoltaic networks at panel level has become much more automated and a market has emerged for standardised solar array designs in parallel to the existing market for space solar arrays, that has historically involved optimisation for individual satellites or families of satellites rather than constellations built in a 'volume production' environment.

In future, the possibility of dramatically increased launch capacity and associated reduction in launch costs may lead to further major changes.

A. Silicon

For the first four decades after the launch of the first satellites into space, silicon solar cells were effectively the only technology available and initially space was the only application for which their use was economical. From this starting point, the worldwide output of silicon solar cells for terrestrial applications is now of the order of hundreds of gigawatts.

When use of silicon cells for large satellites remained prevalent in the 1990s, the choices of cell architecture for dedicated space cells reflected also the techniques available within terrestrial technologies for light trapping and surface passivation (see eg. [4]). Products ranged from relatively simple designs that nevertheless incorporated back surface fields to enhance minority carrier collection, to more complicated but expensive designs that featured the kind of surface texturing featured on 'passivated emitter' structures such as PERL and PERC [5]. A significant difference with terrestrial cell designs was that space cells tended to feature metallisation systems that were evaporated or sputtered, in conjunction with inter-cell connection processes that typically involved welding. This was needed in order to cope with the thermal cycling environment associated with thousands of eclipses when in earth orbit.

Independent of the cell performance, the availability of a worldwide production capacity that is orders of magnitude greater than the market for space solar power is, in itself, an important consideration, given that procurements may be sensitive to delivery schedule or flexibility. The simplified example shown in table 1 does not represent a unique outcome in favour of high efficiency cells and, on the contrary, users have a strong motivation to check if a cost benefit can be derived by finding a way to use cells that are cheap and widely available.

Given that terrestrial silicon solar cell designs have also evolved significantly in the last 20 years, unsurprisingly developers have looked to take advantage of new possibilities also for space applications, with radiation degradation being an area of particular focus. The behaviour of recombination centers in silicon cells can be sensitive to, for example, doping and passivation.

Companies such as mPower and Solestial are now advertising variants of silicon solar cells which exploit annealing phenomena to mitigate the introduction of defects due to particle radiation. It's not the purpose of this paper to comment on the status of the claimed performance either in a positive or negative way. Nevertheless, annealing phenomena have been understood to play a role in the radiation behaviour of most space solar cell materials and it's certainly of great interest to understand whether the performance of silicon cells can be enhanced by deliberately taking advantage of annealing phenomena.

B. CuInGaSe2, CdTe

In very simplified terms, a look at Table 1 explains why cell technologies such as CIGS did not make a big impact on the space market at a time around 2005 when thin film technologies showed a lot of promise, but as yet were not available as products with efficiencies significantly better than half that of multi-junction III-V cells [6]. Even with a promising development perspective, the pathway to maturation for exclusively space related applications was expensive and risky for any manufacturer to follow, even with financial support for R+D from government agencies.

CIGS cells, in particular, were understood to be relatively insensitive to particle radiation due to the annealing effects mentioned above. If development of CIGS cells on a lightweight flexible substrate had been successful, it might have opened new possibilities for space solar array level design. However, the 'thin film' technologies including CIGS and CdTe that have had some success within the terrestrial market are typically deposited on heavy substrates such as glass, in a format that is typically not directly transferable to a space application.

Perhaps the evolution of cell technology available on lightweight substrates, in parallel to that of the growing satellite market, will create a new opportunity.

C. Perovskites

The rapid development of perovskite cell technology is followed with the same interest for space applications as for terrestrial applications, and because the environment in space is more demanding in almost every respect than the terrestrial equivalent, concerns about performance stability are at least as great. Even if space is a vacuum, space solar arrays are nevertheless exposed to humidity for the duration of the lifecycle on ground for manufacturing and testing and the requirements defined for space solar arrays reflect also this.

With many fundamental questions about perovskites still in the process of being answered and the focus of developers understandably on the terrestrial market, it's too early to say whether perovskite based solar cells will become successful products for space applications.

What can be said is that there are multiple potential pathways to space applications for perovskite cells. Firstly, a perovskite cell integrated on a 'conventional' space solar array might offer a future advantage in performance and cost already at cell level, whether as a 'purely perovskite' cell or as a tandem, eg. as a top cell in conjunction with silicon or CIGS. Alternatively, perovskite cells deposited on flexible films may reopen a pathway to lighter solar array blankets.

There have been reports that radiation induced defects may be partially removed by annealing also in perovskite cells.

For other properties of importance to space applications such as temperature coefficient, reverse bias behaviour, stability under extreme temperatures and sensitivity to electronic discharge, there is still a lot of work to do.

[1] See eg. B.E. Anspaugh, GaAs Solar Cell Radiation Handbook, JPL Publication 96-9, 1996

[2] N.Fatemi et al. 'Solar Array Trades Between Very High Efficiency Multi-Junction and Silicon Space Solar Cells', Proc. IEEE PVSC 2000, p.108

[3] C.Zimmerman et al, Proc. European Space Power Confrence 2019

[4] Washio et al. 'Improved Radiation Hardness of Silicon Solar Cells', Proc. IEEE PVSC 2000, p. 1114

[5] G. Strobl et al., &ilicon and Gallium Arsenide solar cells for deep space missions: Proceedings of European Space Power Conf. (Graz, Austria, 1993), pp. 603.

[6] D.M.Murphy et al, Proc. IEEE PVSC 2002

[7] A.Kimani et al, Nature Communications | (2024)1 5:696

41st European Photovoltaic Solar Energy Conference and Exhibition

This presentation was selected by the Sc. Committee of the EU PVSEC 2024 for submission of a full paper to one of the EU PVSEC's collaborating peer-reviewed journals.

IN-DEPTH CHARACTERIZATION METHODOLOGY FOR THE ASSESSMENT OF PASSIVATION IMPACT IN HALIDE PEROVSKITE SOLAR CELLS

Jonathan Parion[1,2,3,4], Santhosh Ramesh[1,4], Sownder Subramaniam[1,4,5], Henk Vrielinck[6], Filip Duerinckx[1,4], Hariharsudan Sivaramakrishnan Radhakrishnan[1,4], Jef Poortmans[1,2,4,5], Johan Lauwaert[3] and Bart Vermang[1,2,4]
[1]imec division IMOMEC, Thorpark 8320, 3600 Genk, Belgium. [2]Hasselt University, Wetenschapspark 1, 3590 Diepenbeek, Belgium. [3]Department of Electronics and Information Systems, Ghent University, Technology Park 126, 9052 Zwijnaarde, Belgium. [4]EnergyVille, Thorpark 8320, 3600 Genk, Belgium. [5]Department of electrical engineering, KULeuven, Kasteelpark Arenberg 10, 3001 Leuven, Belgium. [6]Department of Solid-State Sciences, University of Gent, Krijgslaan 281-S1, 9000 Gent, Belgium.

ABSTRACT: A wholistic characterization approach is introduced, aiming at establishing a link between nano-scale electrical properties and macro-scale device characteristics for the study of interface passivation in perovskite solar cells. IV measurements are combined with admittance spectroscopy (AS) and deep-level transient spectroscopy (DLTS) for the analysis of charge-related performance losses and with Time-of-Flight Secondary Ion Mass Spectrometry (ToF-SIMS) to complete the understanding of ionic accumulation at the perovskite interfaces. An increase of both V_{oc} and FF of approximately 10% are shown for a cell with passivated interfaces compared to a non-passivated reference cell, with an absolute efficiency increase of 4%. The higher FF in the passivated sample could be attributed to a lower influence of some ionic species present in the perovskite. Analysis of the DLTS response yields an activation energy of 0.41 eV, which could be attributed to the presence of either trap states or ionic mobile charges in the perovskite bulk or at one of its interfaces. Finally, preliminary ToF- SIMS results showcase different ionic species and how they are distributed in the device.
Keywords: Perovskite solar cells, Characterization, Surface treatment, Passivation

1 INTRODUCTION

In recent years, passivation has played a key role in pushing state-of-the-art perovskite solar cells (PSCs) close to their silicon counterpart, recently reaching record efficiencies over 25.5% [1]. These high-efficiency PSCs can be combined with silicon bottom cells, to make high-efficiency tandem solar cells reaching efficiencies above 30% [2,3]. In these cells also, passivation plays a key role. A major aspect in defining the best passivation strategy is to clearly identify its impact on the device electrical properties. For this, advanced characterization is required, enabling the extraction of key metrics such as modification of defect properties, improvement of band alignment or reduction in mobile ions density. This important step is however often overlooked, with electrical characterization being limited to the analysis of current-voltage (IV) curves.

In this work, a characterization approach is introduced, paving the way towards more understanding of the impact of passivation on PSCs electrical properties. After the main figures of performance are extracted from IV measurements, admittance spectroscopy (AS) and deep-level transient spectroscopy (DLTS) are performed to extract information about (mobile) defect and passivation-related improvement of the devices. Time-of-Flight Secondary Ion Mass Spectrometry (ToF-SIMS) is then performed to analyze the internal composition of the layer stack and the presence of mobile ions throughout the entire stack. The novelty of this work resides in the characterization approach, in which device-level figures of performance are compared with nanoscale-level defects and displacement of mobile charges. Moreover, this approach is not only applicable to the study of passivation but can also be extended to investigating the stability and degradation mechanisms of PSCs which is also currently a key topic for researchers in the field [4].

2 DEVICE-LEVEL PERFORMANCE ANALYSIS

The cells that are characterized in this work use an inverted p-i-n architecture and FAPbI-based perovskite, with the same layer stack and passivation strategy that was presented by group member Xin Zhang in [1]. The so-called "Reference" cell stack is composed of ITO/NiO/perovskite /LiF/C60/Bathocuproine (BCP)/Ag. The NiO acts as a p-doped hole transport layer (HTL), and the LiF/C60/BCP stack acts as a n-doped electron transport layer (ETL). The "Passivated" cell uses the same architecture as the reference with additional self-assembled monolayers (SAMs) at the perovskite/HTL interface and a surface treatment at both perovskite interfaces. The IV curves and main figures of performance are presented in Figure 1.

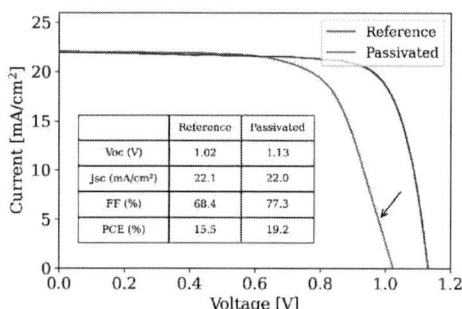

	Reference	Passivated
Voc (V)	1.02	1.13
Jsc (mA/cm²)	22.1	22.0
FF (%)	68.4	77.3
PCE (%)	15.5	19.2

Figure 1: IV measurement of the reference and passivated PSCs, with the main figures of performance given in the table inset. The black arrow shows a slight kink in the reference curve.

As expected, the passivated cell exhibits a higher power conversion efficiency (PCE, + 4%) than the reference. The main contributors are a higher open-circuit voltage (V_{oc}, + 11mV) and a higher fill-factor (FF, + 9%). The short-circuit current (J_{sc}) remains about the same in both cells. This trend is very much in line with what was previously reported in [5] and other works on the subject. A higher Voc is often attributed to a reduced

260

recombination, which could come from a lower density of bulk or interface defects and should be highlighted in the rest of the characterization procedure. Concerning the FF, a lower value can often be attributed to either a higher series resistance or charge-related non-idealities in the device. In the present reference cell, a slight kink in the curve can be observed (black arrow). This phenomenon was previously attributed to interface mismatch between the perovskite and one of its transport layers [6,7]. Again, this hypothesis should be confirmed in the more advanced characterization results.

3 EXTRACTION OF DEFECT PARAMETERS

To get more insight on the reasons behind the V_{oc} and FF improvement in the passivated cell, AS and DLTS measurements are performed and detailed in this section. The capacitance versus frequency (C-f) curve extracted from AS measurements are presented in Figure 2.

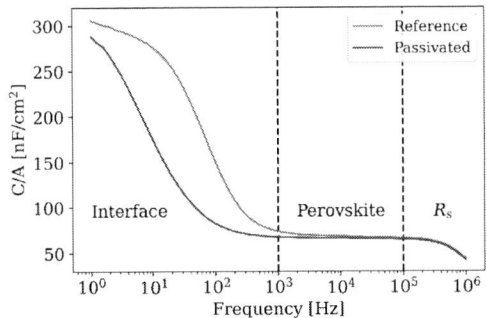

Figure 2: Representation of the capacitance vs frequency (C-f) for the reference and passivated cells extracted at 0V from the admittance spectroscopy data (full lines). The different identified capacitance regions are delimited with dashed lines

The devices exhibit a C-f behavior that is typical of PSCs, with three main frequency regions. The central region, between 1kHz and 100kHz, is the one where the geometric capacitance of the perovskite layer $C_g = \varepsilon_0\varepsilon_{pero}/t_{pero}$ dominates. Since perovskite is an intrinsic material, this value is much greater than the depletion capacitance, often considered as negligible. Having both geometric capacitances at roughly the same value suggests that the perovskite layer itself hasn't been significantly modified by the interface passivation process.

Above 100kHz, the series resistance (R_s) of the devices dominates, resulting in a low-pass filter type of behaviour. The exact frequency at which this transition happens (f_{Rs}) is determined by $1/R_sC_g$. Since both cells have a similar value for C_g and f_{Rs}, it is very likely that their R_s value is also similar. This is interesting, since it excludes the fact that passivation has had a significant impact on the series resistance. It also tends to favour the hypothesis following which the FF in the IV characteristic would have mainly been affected by some interfacial-related phenomena, and not by a large R_s.

The last and most interesting region, below 1kHz, is often attributed to the interface properties of the perovskite with adjacent transport layers. A contribution comes from the transport layers themselves and more specifically, as one of our models inspired from [8] suggests, from the ETL geometric capacitance. Another contribution is the possible accumulation of mobile ions at one of the

perovskite interfaces. This can happen in the presence of an electric field, when sufficient energy is provided. In these circumstances, the ions form a sort of quasi-layer that also contributes to the total capacitance. Unfortunately, it is not possible to discriminate between these two mechanisms based on the AS characterization only, more advanced simulations need to be carried out. What is however visible on the C-f plot is the clear shift in the low-frequency response between the reference and passivated cells. The increase in capacitance happens at frequencies about 1 order of magnitude higher for the reference compared to the passivated sample. Since the ETL properties should be similar in both cases, it is very likely due to a different behaviour of mobile ions. From our basic simulation model, this type of transition could correspond to a higher ion diffusivity in the reference cell, even though it is at this stage not possible to quantify this precisely nor to discriminate on the specific ionic species. Looking back to the IV characteristic, it is very probable that the presence of ionic species at one perovskite interface of the reference cell are linked with the visible "kink" in the IV curve. It was already suspected to be linked to interface mismatch between perovskite and a transport layer, and this seems to be confirmed by the C-f measurements.

To go further on identifying the nature of the charge-related non-idealities at play within the cells, DLTS measurements are performed. These temperature-dependent measurements are usually used to characterize mobile carrier capture and emission by deep-level defects in semiconductors. In particular, it is possible to extract the activation energy (E_A) and pre-exponential factor (k_T, that is linked to the capture cross section σ and the effective mass) of these defects by means of an Arrhenius plot. The basic principle of these DLTS measurements is that a voltage pulse is applied to the device, and the resulting transient in capacitance is recorded within a certain time window. In PSCs, the thermally activated motion of mobile ions can also produce slow capacitance transients in response to voltage pulses, and thus produce DLTS responses. More information on the extraction of an Arrhenius plot based on the measured transients can be found in References [9,10]. The Arrhenius plot that was extracted from the DLTS measurements for the reference and passivated samples is presented in Figure 3.

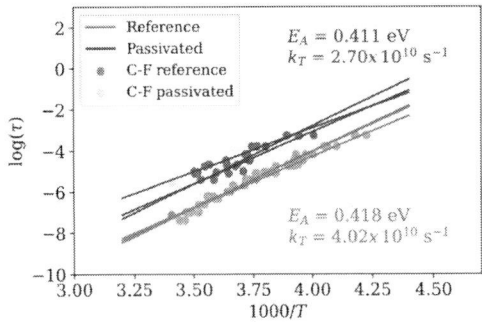

Figure 3: Arrhenius plot as extracted from the DLTS transients, for the reference and passivated cells. The raw data is represented by light-coloured dots, and the linear fits for each measurement with full lines. The activation energy E_A and pre-exponential factor k_T are also given.

For each cell, different voltage pulses are applied, resulting in different Arrhenius regression lines, helping to confirm with more certainty the value of E_A and k_T. These different lines for the reference as well as for the passivated samples are parallel to each other, which logically results in similar E_A for both samples. This suggests that the activation condition of the observed defects doesn't vary much when passivation is introduced in the structure. The value itself of $E_A \approx 0.41$ eV is close to the activation energy for the movement of ionic species, such as methylammonium (MA^+) ions for which $E_A \approx 0.37$ eV was previously reported [10]. Another possible explanation would be the presence of trap states at one of the perovskite interfaces. When introducing the C-f low-frequency transition from Figure 2 for both cells in the Arrhenius plot (one temperature vs frequency coordinate for each sample), it seems to fit perfectly with the extrapolated Arrhenius lines. This indicates that there might be a link between the results of both characterization methods and would tend to confirm that the activation energy obtained in DLTS measurements might be related to an interface-related phenomenon. Another information that is extracted from the DLTS measurements is how k_T decreases by about 65% after passivation. This factor is linked to the ability of the defect to negatively impact recombination, as well as to its effective mass. It can therefore be expected for k_T to be lower in a sample with better V_{oc} and FF.

4 INVESTIGATION OF MOBILE CHARGES

The goal of this final part is to confirm the presence of mobile charges in the material, by non-electrical means. Establishing a link between different kind of characterization techniques is key in ensuring the best possible interpretation and the validity of the results. In ToF-SIMS measurements, a pulsed ion beam is used to gradually remove material from the sample. The removed particles are accelerated and identified by a detector based on their mass. By performing this throughout the entire sample stack, a full profiling of the sample composition can be realized.

A preliminary analysis was performed on a structure similar to the reference cell. Two different samples were considered, one pristine and one that was submitted to heat-induced stress to enhance the stack degradation. In Figure 4, no real difference in halide (I_2^- or Br^-) or MA^+ species are observed between the pristine sample and the stressed sample, showing that the stress has not induced a significant free-out of perovskite mobile charges. However, there seems to be significant diffusion of Ni^- coming from the HTL, in the perovskite layer after degradation. This might be an indication that the layer has deteriorated and that the interface with perovskite might not be optimal. In a similar fashion, it is expected for the reference and passivated samples discussed in this work to show a different quality for the HTL/perovskite interface. This should be confirmed with new ToF SIMS measurements on the (non) passivated structures from previous sections.

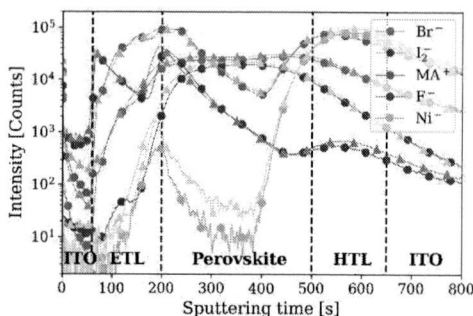

Figure 4: ToF SIMS measurements from the reference architecture. Results before heat stress are shown in dark colour, and results after heat stress are shown in light colour.

5 CONCLUSION

In this work, an in-depth approach was taken to fully characterize and understand the passivation-related improvements in PSCs. IV measurements have shown that the main improvement comes from the V_{oc} and FF. They also exhibit a slight kink in the reference IV curve, that should be related to the perovskite's bad interface properties. C-f measurements tend to confirm this impression, with an improved low-frequency behavior for the passivated sample. E_A and k_T for the supposed interface trap are extracted using DLTS, and either ions or defects are found to be a likely cause for the loss in performance in the reference cell with a reduced impact for passivated devices. Finally, ToF-SIMS preliminary results were showcased, as a non-electrical way of confirming the predominant species accumulated at the perovskite interfaces. Overall, this wholistic approach enables a rather comprehensive understanding of the devices and could also be extended to other scenarios than passivation, including the study of degradation in perovskite solar cells.

6 REFERENCES

[1] Li J. et al., Nature Energy **9**, 308-315 (2024).
[2] Mariotti E. et al., Science **381**, 63-69 (2023).
[3] Aydin E. et al. Nature **623**, 732–738 (2023).
[4] De Wolf S., Aydin E., Science **381**, 30-31 (2023).
[5] Zhang X. et al., ACS Applied Materials & Interfaces **15** (40), 46803-46811 (2023).
[6] Saive, R., IEEE J. Photovoltaics **9**, 1477–1484 (2019).
[7] Ecker B., et al., The Journal of Physical Chemistry C **116** (31), 16333-16337 (2012).
[8] Ravishankar, S. et al., PRX Energy **1**, 013003 (2022).
[9] Hsieh, H.-C. et al., J. Phys. Chem. C **122**, 17601–17611 (2018).
[10] Reichert, S. et al., Nat Commun **11**, 6098 (2020).

Faculty of Engineering
School of Photovoltaic and Renewable Energy Engineering

Pathways for silicon solar cells with molecular singlet fission

Phoebe Pearce, Nicholas Ekins-Daukes

24 September 2024, 41st EU-PVSEC, Vienna, Austria

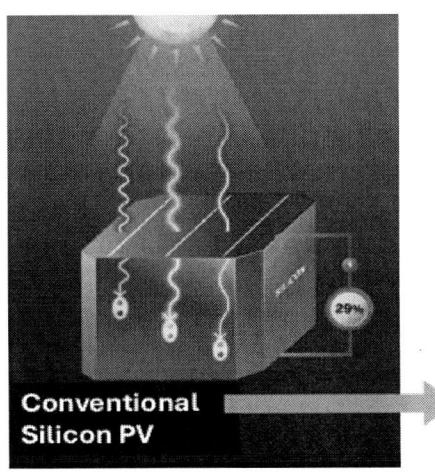

Conventional
Silicon PV

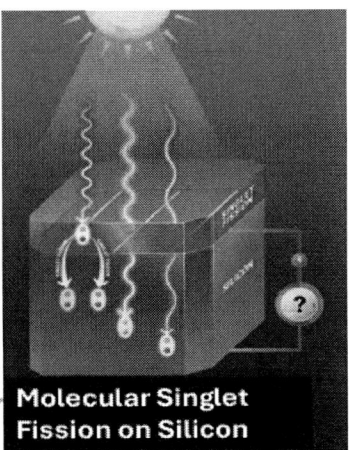

Molecular Singlet
Fission on Silicon

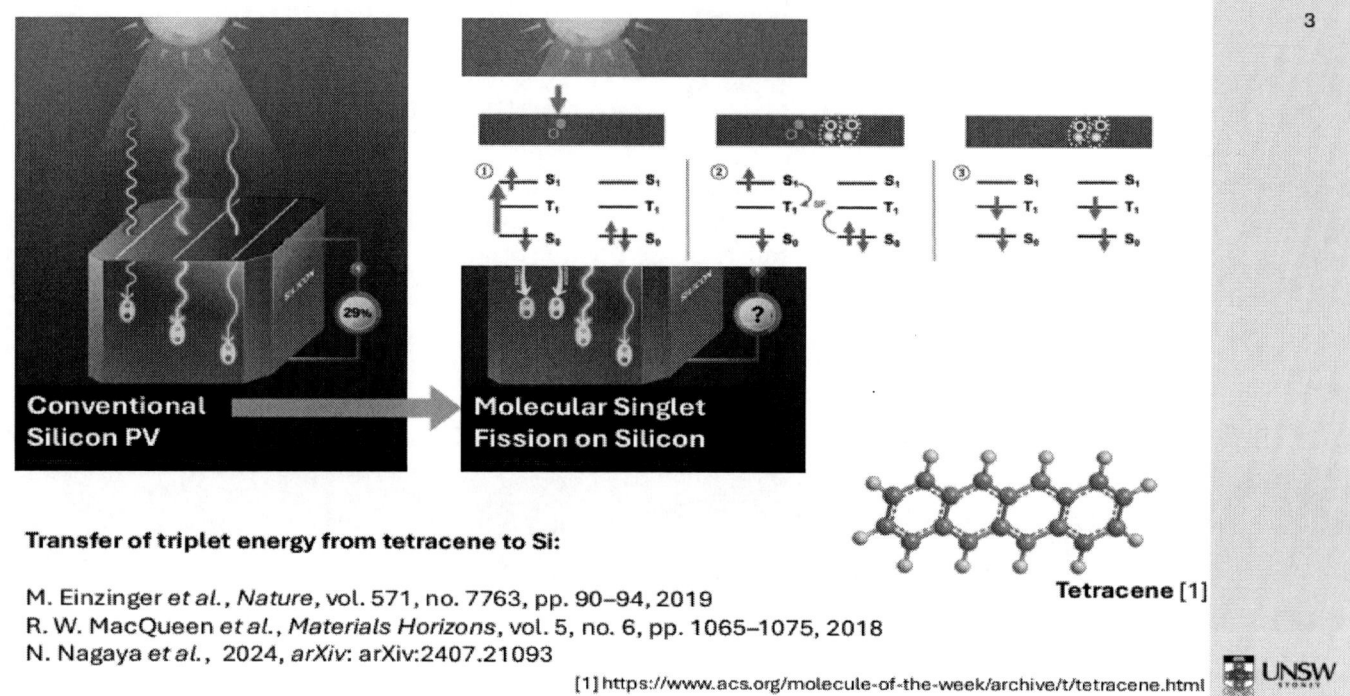

Transfer of triplet energy from tetracene to Si:

M. Einzinger *et al.*, *Nature*, vol. 571, no. 7763, pp. 90–94, 2019
R. W. MacQueen *et al.*, *Materials Horizons*, vol. 5, no. 6, pp. 1065–1075, 2018
N. Nagaya *et al.*, 2024, *arXiv*: arXiv:2407.21093

[1] https://www.acs.org/molecule-of-the-week/archive/t/tetracene.html

---

Observation of IQE > 100%

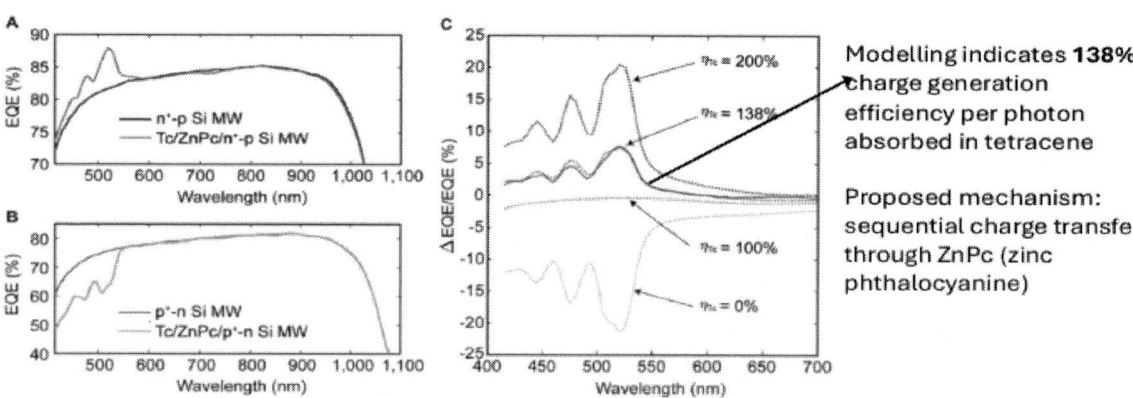

Modelling indicates **138%** charge generation efficiency per photon absorbed in tetracene

Proposed mechanism: sequential charge transfer through ZnPc (zinc phthalocyanine)

Spin & Excitonic Engineering Group, Prof. Marc Baldo, MIT

Figures from N. Nagaya *et al.*, 'Exciton Fission Enhanced Silicon Solar Cell', Jul. 30, 2024, *arXiv*: arXiv:2407.21093. Accessed: Sep. 18, 2024. [Online]. Available: http://arxiv.org/abs/2407.21093

Observation of IQE > 100%

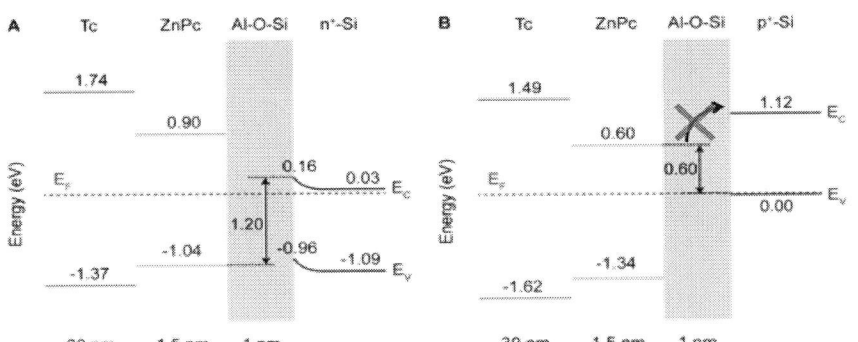

Modelling indicates **138%** charge generation efficiency per photon absorbed in tetracene

Proposed mechanism: sequential charge transfer through ZnPc (zinc phthalocyanine)

Spin & Excitonic Engineering Group, Prof. Marc Baldo, MIT

Figures from N. Nagaya *et al.*, 'Exciton Fission Enhanced Silicon Solar Cell', Jul. 30, 2024, *arXiv*: arXiv:2407.21093. Accessed: Sep. 18, 2024. [Online]. Available: http://arxiv.org/abs/2407.21093

Outline

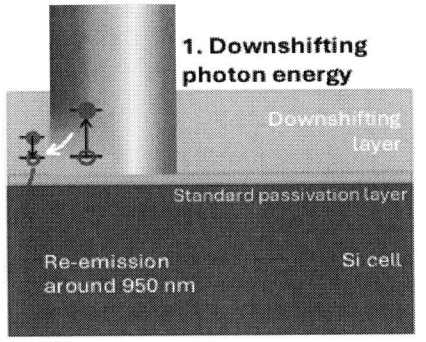

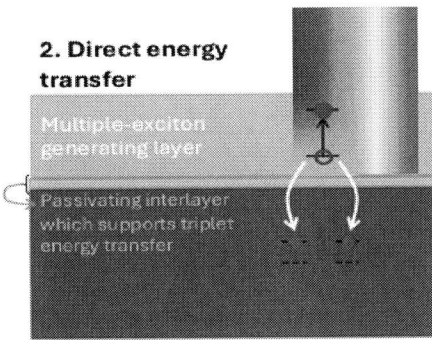

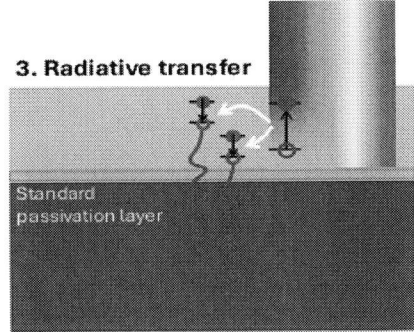

In all cases, consider a layer on top of silicon which absorbs high-energy photons above some threshold energy.

Scenario 1: Downshifting. Layer re-emits photons at a longer wavelength, without carrier multiplication
Scenario 2: Direct energy transfer. High-energy excitations splits into two lower-energy excitations ($E > 1.12$ eV), which are transferred to the silicon
Scenario 3: Radiative transfer. High-energy excitations splits into two lower-energy excitations ($E > 1.12$ eV), which recombine to emit a photon
How well can a known organic singlet fission material do (optically)?

Baseline device model: high-efficiency SHJ cell

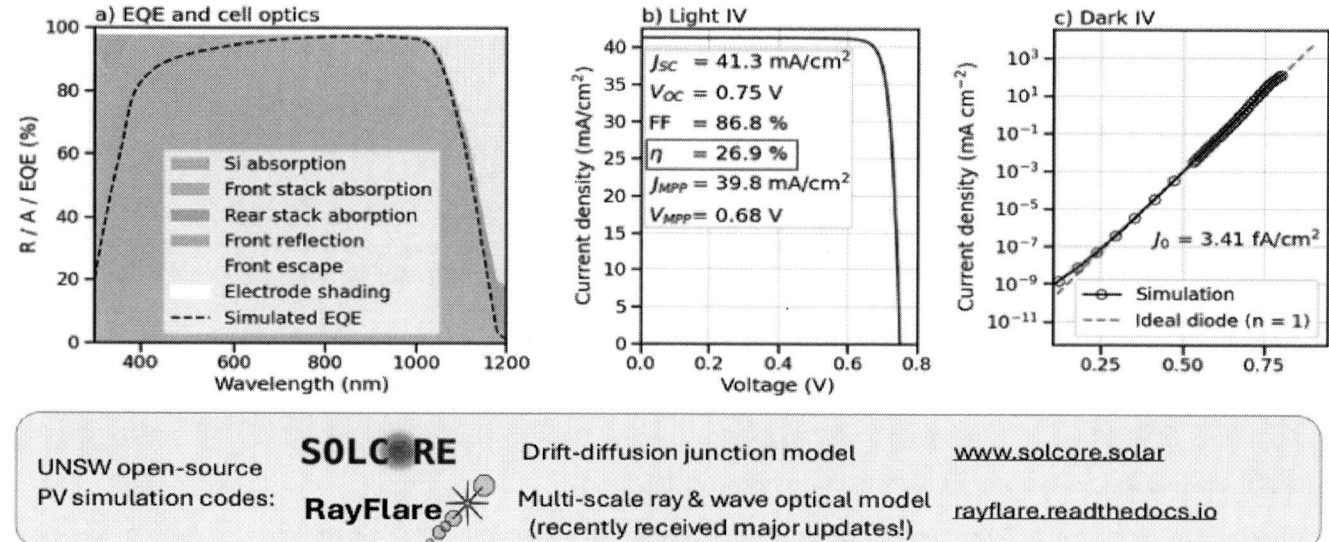

UNSW open-source PV simulation codes:

SOLCORE — Drift-diffusion junction model — www.solcore.solar

RayFlare — Multi-scale ray & wave optical model (recently received major updates!) — rayflare.readthedocs.io

H. Lin *et al.*, 'Silicon heterojunction solar cells with up to 26.81% efficiency achieved by electrically optimized nanocrystalline-silicon hole contact layers', *Nat Energy*, vol. 8, no. 8, pp. 789–799, May 2023, doi: 10.1038/s41560-023-01255-2.

Simple model for downshifting and radiative transfer

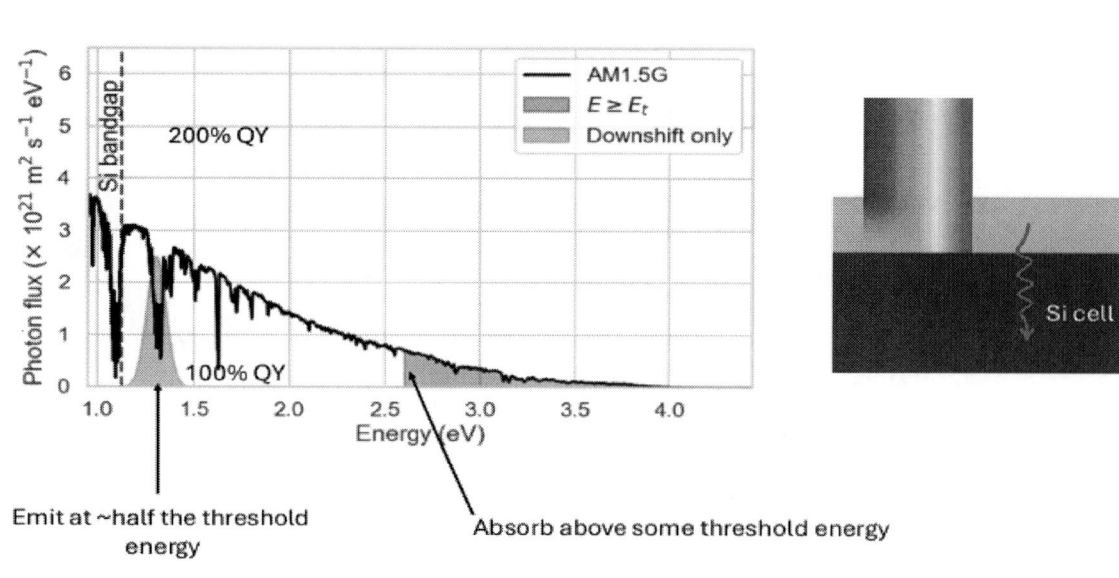

Simple model for downshifting and radiative transfer

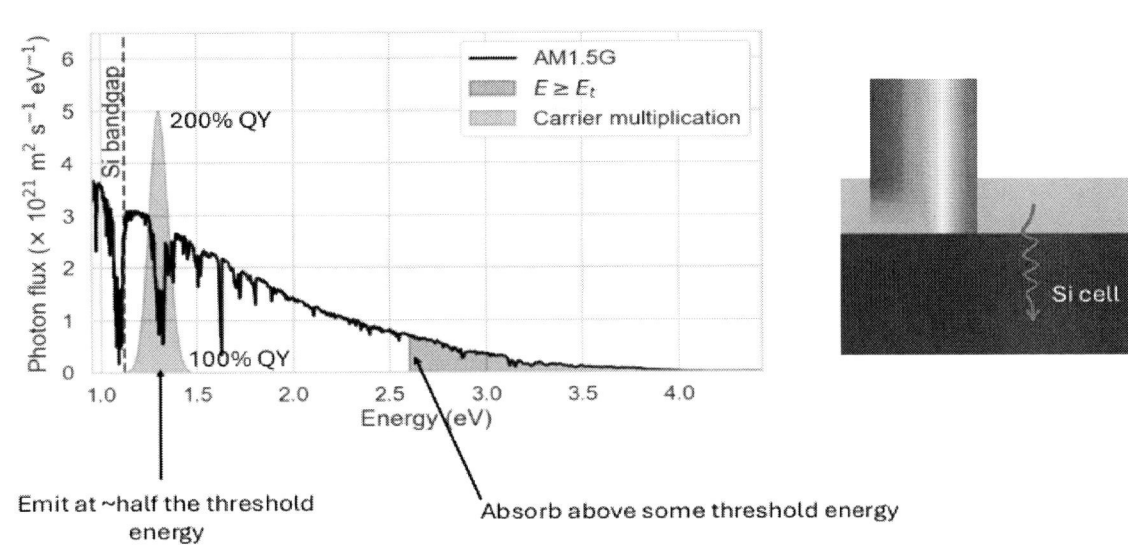

Emit at ~half the threshold energy

Absorb above some threshold energy

Downshifting photon energy

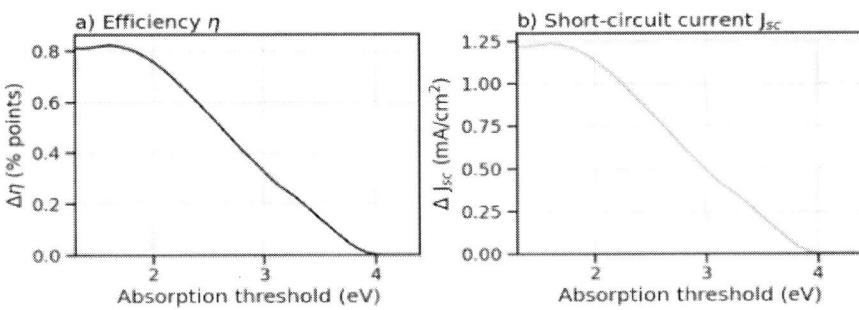

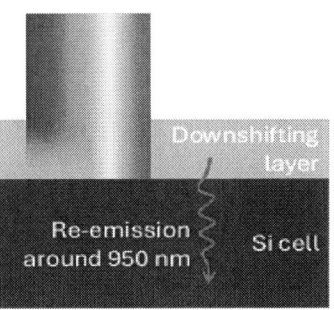

- No carrier multiplication!
- This effect will be most pronounced for cells with low EQE at short wavelengths
- Have neglected losses from re-emitted photons escaping to air (for now)
  - Predicted to be around 12%

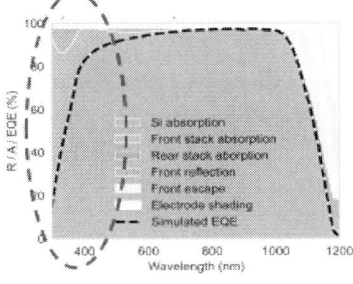

Radiative transfer

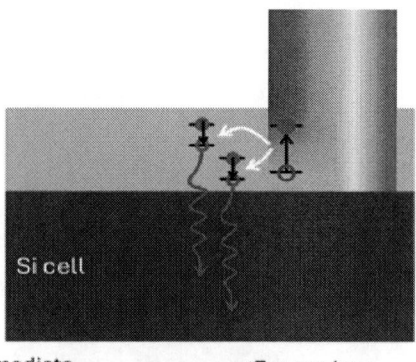

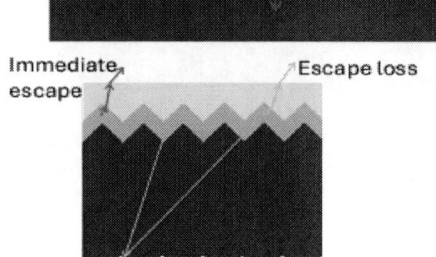

Immediate escape

Escape loss

Si cell

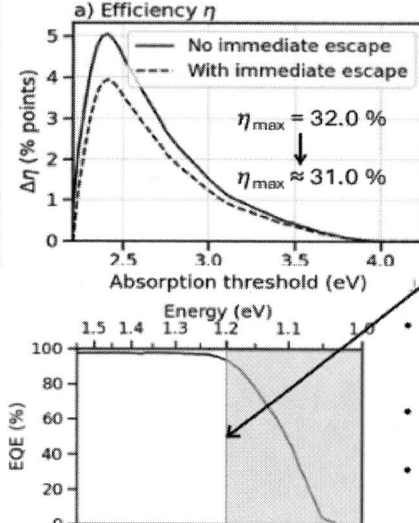

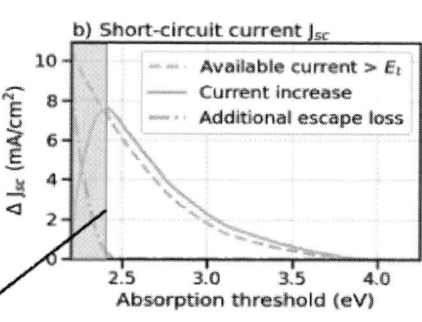

- Assuming emission at half the energy of the absorption threshold: Si becomes transparent from ~ 1050 nm
- High escape loss of downshifted photons
- Will also have an "immediate escape loss" from downshifted emitted photons escaping from front, without ever entering Si (12 - 20%)

Direct energy transfer

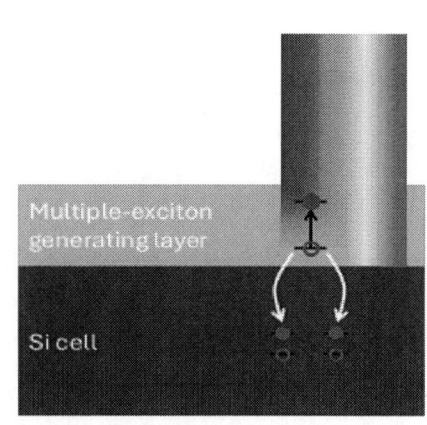

Multiple-exciton generating layer

Si cell

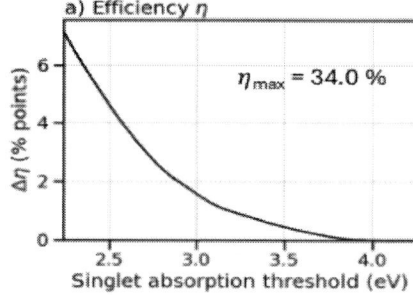

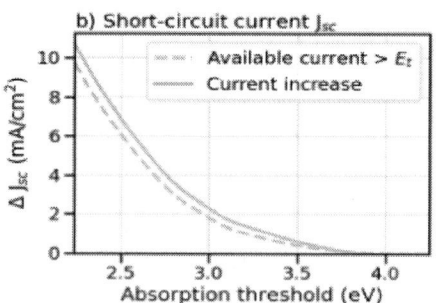

Assuming two electrons can be excited in Si as long as threshold energy is > $2E_{g,\,Si}$ (= 2.24 eV), maximum efficiency is **34%** (baseline: 26.9%)

Direct energy transfer: surface effects

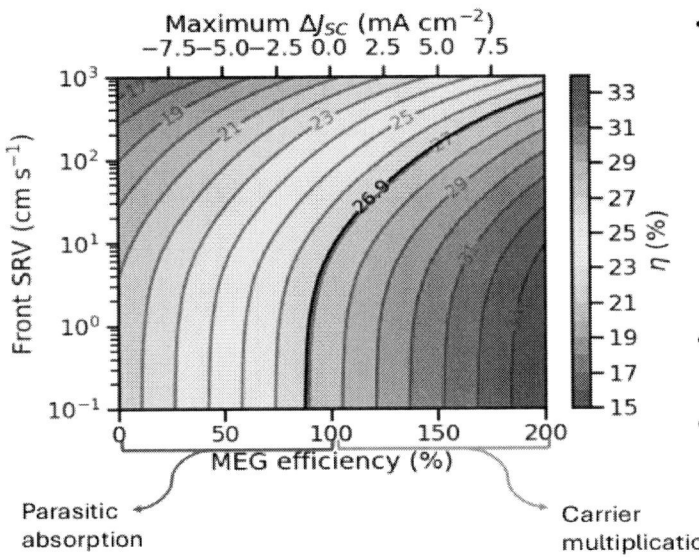

- Actual mechanism for energy transfer not included in this model: extra excited carriers are added near front surface of silicon in the drift-diffusion model:

$$N_{MEG} = \eta_{MEG} \int_{280nm}^{\lambda_t} \Phi_{in}(\lambda)d\lambda$$

$$g_{MEG}(z) = \frac{N_{MEG}}{d} e^{-z/d}$$

- This means high sensitivity to surface recombination (SRV = surface recombination velocity)
- Will need to modify Si cell front surface to enable energy transfer, but **must** maintain low surface recombination

13

Organic layer on silicon: absorption

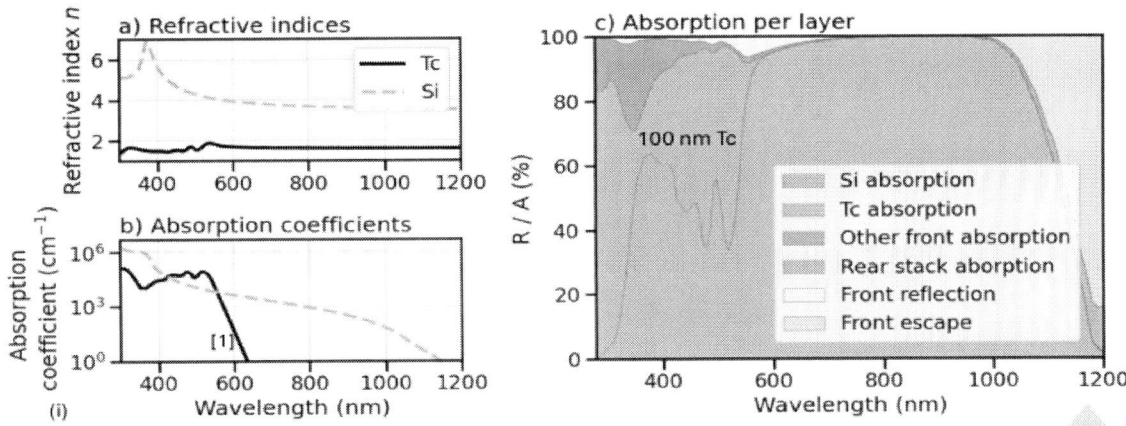

Optical behaviour calculated with ray-tracing + transfer matrix model (TMM) as implemented in RayFlare [2]

Note: neglected longer-wavelength parasitic absorption in Tc.

[1] B. Gompf et al., Eur. Phys. J. E, vol. 27, no. 4, pp. 421–424, Dec. 2008, doi: 10.1140/epje/i2008-10405-5.
[2] P. M. Pearce, Journal of Open Source Software, vol. 6, no. 65, p. 3460, 2021. https://rayflare.readthedocs.io

14

Organic layer on silicon: radiative transfer efficiency

How likely are photons emitted by e.g. tetracene to escape
from the front surface of the device?

Planar case has analytical solution (sum of infinite
geometric series):

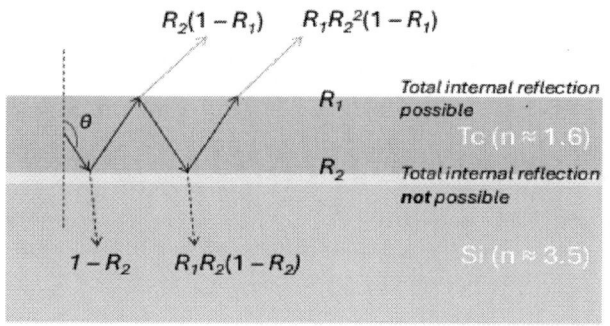

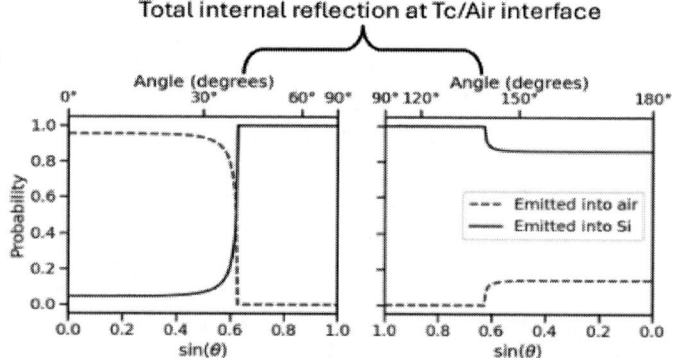

Total internal reflection at Tc/Air interface

Average probability of escape to air: 11.5 %
(isotropic emission: $p(\theta) \propto \sin\theta$ in 3D)

Organic layer on silicon: pyramidal surfaces

- Can no longer write down an analytical solution: Multiple bounces, chance
  of ray exiting from Tc on one pyramid face and re-entering through another;
  this reduces the escape probability
- Use ray-tracing calculations (RayFlare)
- Emission wavelength = 950 nm, no re-absorption of emitted photons

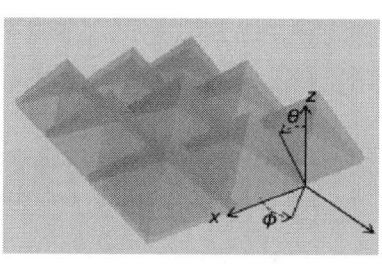

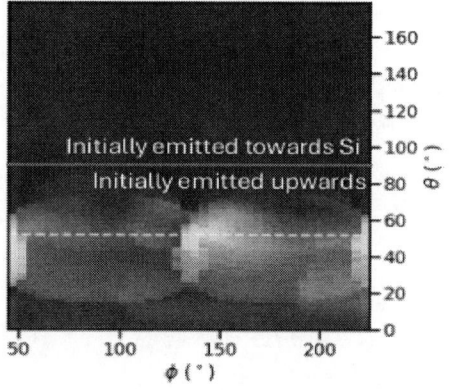
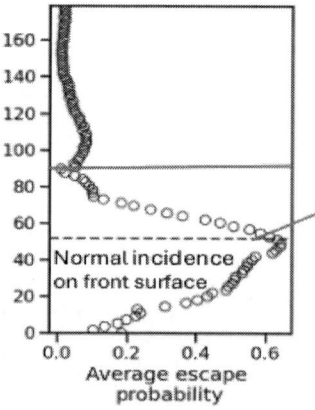

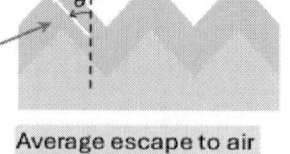

Average escape to air
probability: **20%**

With pyramids + cover
glass/EVA: **13%**

Organic layer on silicon: combining absorption & emission

Absorption of incident spectrum (as a function of Tc thickness)

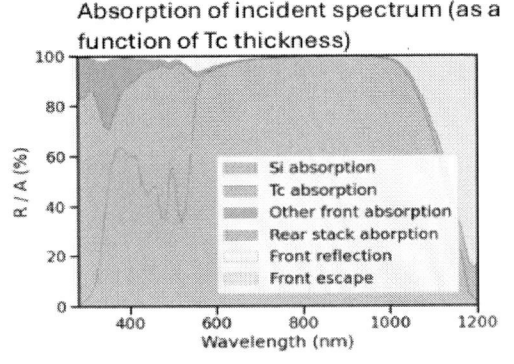

Probability of entering Si* (i.e. not escaping immediately)

	Textured front	Planar front
No cover glass	80 % (i)	87 % (iii)
Cover glass/EVA	87 % (ii)	88 % (iv)

* assuming isotropic emission direction from organic, no further changes to glass or front surface layers to minimize escape

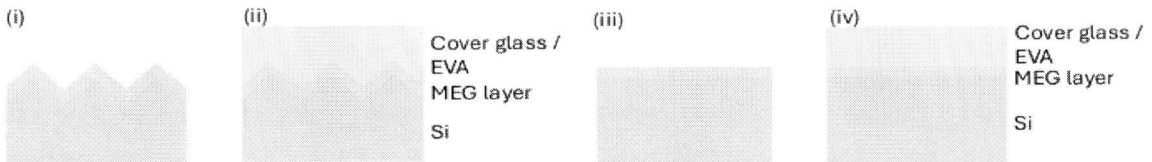

(i)   (ii) Cover glass / EVA / MEG layer / Si   (iii)   (iv) Cover glass / EVA MEG layer / Si

---

Organic layer on silicon: maximum current & efficiency

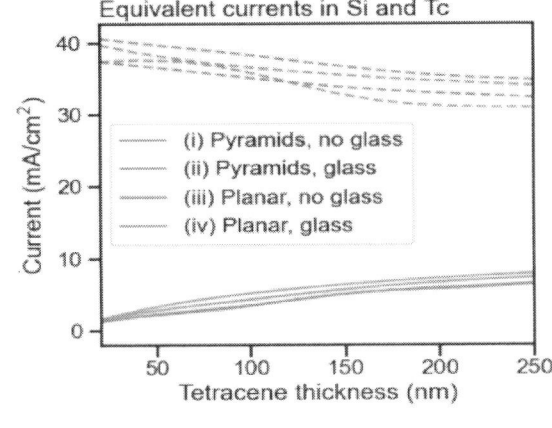

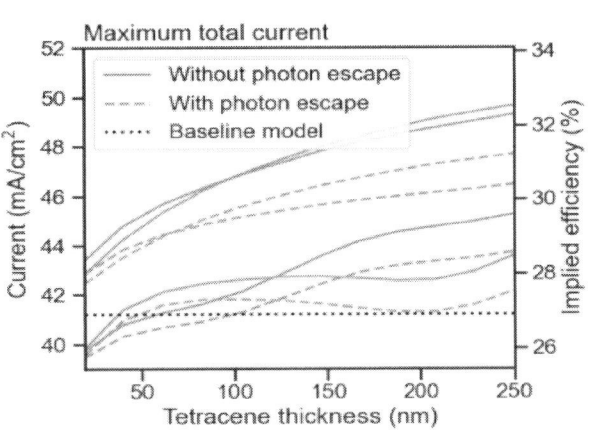

Without photon escape: perfect direct energy transfer from triplets in Tc

With photon escape: 200% yield from an emissive molecule/luminophore with some emitted photons escaping the front surface

Direct energy transfer vs. radiative transfer

Direct energy transfer

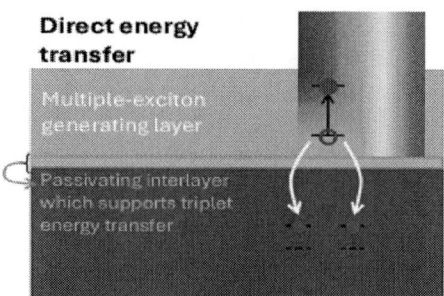

Radiative transfer

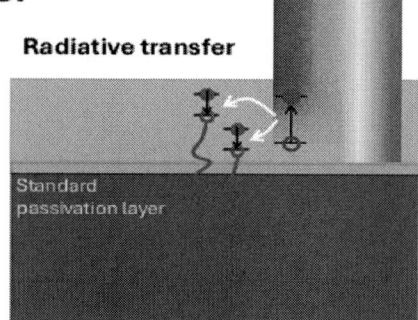

✓ Higher efficiency potential
✓ Well-suited to IBC architecture (contacts on rear only)
▪ Requires :
  ▪ Dedicated singlet fission molecular layer
  ▪ Passivating exciton transport interlayer
    ▪ HfNO$_x$ [1] and ZnPc [2] demonstrated

▪ Lower efficiency potential (still > 30%)
▪ Requires highly emissive lumiphores
✓ Surface passivation and cell structure relatively unchanged
✓ Fast route for efficiency gain?
✓ Compatible with luminescent down shifting films and heterojunction cells

Both: require new, stable singlet fission materials with favourable singlet & triplet energies

[1] M. Einzinger *et al.*, *Nature*, vol. 571, no. 7763, pp. 90–94, 2019
[2] N. Nagaya *et al.*, *arXiv*: arXiv:2407.21093.

Conclusions

- Si cells with singlet fission could be > 30% efficient, with either radiative coupling or direct energy/charge transfer
- Direct energy transfer has a higher efficiency limit but faces additional engineering hurdles
- Surface textures improves absorption but increases chance of emitted photons escaping
- Mixtures/light trapping likely necessary for complete absorption in organic layers
- **Talk 2BO.8.3 (this session), Dr. Shona McNab (UNSW): effect of Si doping on triplet energy transfer**

Download our white paper with these results:

OMEGA Silicon project:
Organic **M**ultiple **E**xciton **G**eneration **A**ugmented **Silicon**

https://www.omegasilicon.solar/about-singlet-fission

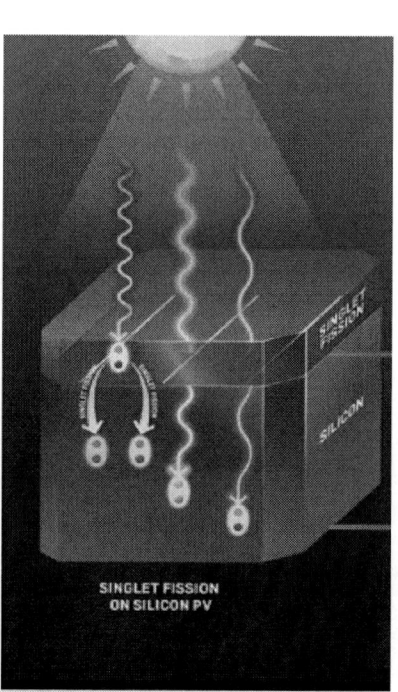

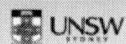

41st European Photovoltaic Solar Energy Conference and Exhibition

CONTROL OF HOT CARRIER THERMALIZATION RATES IN NANOWIRES FOR ADVANCED-CONCEPT PHOTOVOLTAIC SOLAR CELLS

Hamidreza Esmaielpour[a], Nabi Isaev[a], Imam Makhfudz[b], Markus Döblinger[c], Jonathan J. Finley[a], Gregor Koblmüller[a]

[a] Walter Schottky Institut, TUM School of Natural Sciences, Technical University of Munich, 85748 Garching, Germany.
[b] IM2NP, UMR CNRS 7334, Aix-Marseille Université, Marseille 13013, France.
[c] Department of Chemistry, Ludwig-Maximilians-University Munich, Munich, 81377, Germany

ABSTRACT: Hot carrier solar cells are a type of 3rd generation photovoltaic devices. The goal of these cells is to increase the efficiency of solar technology beyond the theoretical limit in single-junction devices. To design an efficient hot carrier solar cell, it is necessary to reduce the rates of hot carrier thermalization in solar cell absorbers. Nanowires (NWs) are potential candidates for hot carrier solar cells due to their one-dimensional geometry and density-of-states. In this study, we are examining the hot carrier effects in core-shell InGaAs/InAlAs nanowires (NWs) with diameters ranging from 110 nm to 200 nm. The results of photoluminescence spectroscopy indicate significant correlations between hot carriers and diameter: as the diameter decreases from 200 nm to 160 nm, the hot carrier effects intensify, but further reduction in diameter (< 160 nm) leads to weaker hot carrier effects. The increase of hot carrier effects by reducing the NW diameter (160 nm < d < 200 nm) is caused by the combined effects of phonon bottleneck and Auger heating. Conversely, in the thin NWs, an increase in microstructure disorder results in higher rates of hot carrier thermalization, leading to weaker hot carrier effects. These findings are consistent with both theoretical and experimental studies, including those using time-resolved photoluminescence spectroscopy and high-resolution transmission electron microscopy.
Keywords: Hot carriers, Nanowires, Thermalization, Auger heating, Shockley-Read-Hall recombination.

1 INTRODUCTION

The latest developments in photovoltaic solar cells have led to devices that have achieved power conversion efficiencies close to the upper theoretical limit for single-junction solar cells (33%). As a result, it is necessary to seek new approaches to reduce some of the main energy loss mechanisms and enhance the efficiency of these devices beyond their upper theoretical limit. One approach is to harness photo-generated hot carriers in solar cells before they dissipate their kinetic energy by interacting with phonons in the system [1]. Nanostructures are promising candidates for hot carrier absorbers because they have demonstrated evidence of hot carriers by confining these particles in space, which leads to slower thermalization rates [2]. In this study, we present our findings on controlling hot carrier thermalization in InGaAs core-shell nanowires for hot carrier solar cell applications.

The integration of III-V nanowires on silicon substrates is an emerging technology with applications in various optoelectronic devices, such as advanced-concept solar cells and light-trapping technology [3]. Nanowires can slow down the relaxation of hot carriers by confining them spatially and provide a condition for the phonon-bottleneck effect [4]. The effects of hot carriers in nanowires have been studied using several techniques, including steady-state photoluminescence (PL) spectroscopy [5], transient absorption pump-probe spectroscopy [6], and electric measurements [3]. In this study, we focus on the influence of the nanowire dimensions on the properties of photo-generated hot carriers under steady-state PL spectroscopy. The goal of this study is to design and characterize nanowire arrays for hot carrier absorber applications in photovoltaic devices.

2 EXPERIMENTAL RESULTS AND DISCUSSIONS

Core-shell $In_{0.2}Ga_{0.8}As/In_{0.2}Al_{0.8}As$ NWs are grown by molecular beam epitaxy using a catalyst-free selective-area growth method. The scanning electron microscopy (SEM) image of the NWs is shown in Figure 1(a). The diameter of the NWs is controlled by creating a template on Si(111)/SiO2 substrates using electron beam lithography. Therefore, it is possible to study the contributions of spatial confinement on hot carrier effects in these 1D nanostructures [5,7]. Figure 1(b) illustrates a scanning transmission electron microscopy (STEM) image of the InGaAs NWs. It is seen that there are stacking faults in the NWs along their length. The STEM results of the InGaAs NWs further evidence that the density of defects increases by reducing the NW diameter [5].

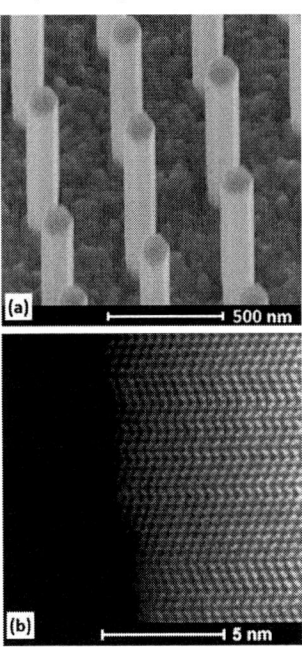

Figure 1(a). SEM image of an array of InGaAs nanowires. (b) STEM image of a typical nanowire showing evidence of stacking faults.

In order to investigate the impact of hot carriers on the InGaAs NWs under steady-state conditions, we performed micro-photoluminescence (PL) spectroscopy at various excitation powers and lattice temperatures. Figure 2(a) displays excitation power-dependent PL experiments of the NWs under 780 nm laser excitation at 10 K. As the excitation power is increased, the slope on the high-energy side of the PL spectra becomes shallower, indicating the presence of hot carrier effects in the NWs.

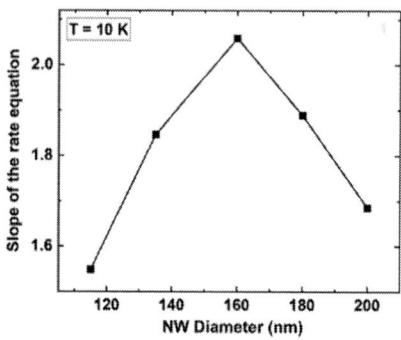

Figure 3. Slope of the rate equation analysis at 10 K as a function of the NW diameter.

Auger heating is another mechanism that can lead to higher temperatures of hot carriers [10]. In this process, when an electron-hole pair recombines, its energy is transferred to another electron or hole, causing it to move to higher energy states. This results in an increase in the temperature of hot carriers, which can improve the efficiency of hot carrier solar cells [11]. The contribution of Auger recombination in the system can be estimated using the rate equation. Figure 3 shows that there is a non-monotonic behavior of the Auger recombination rate versus diameter, with a peak for the NWs with a diameter of 160 nm. This means that when the diameter of the NWs is reduced from 200 nm to 160 nm, the rate of Auger recombination increases, i.e. stronger Auger heating; however, it then decreases for thinner NWs with diameters below 160 nm. The decrease in Auger recombination rates in thin NWs is followed by an increase in the contribution of Shockley-Read-Hall (SRH) recombination due to higher microstructure disorder in the thin NWs [7].

To gain further insights into the behavior of hot carriers and to assess the influence of different recombination mechanisms on these particles, a comprehensive analysis using time-resolved photoluminescence (TRPL) spectroscopy was undertaken. Figure 4(a) illustrates the TRPL results of the InGaAs NWs with a diameter of 160 nm at 10 K under varying excitation powers. It is observed that with an increase in excitation power, the slope of the TRPL signal becomes steeper, indicating a more pronounced Auger recombination within the system [8].

The rate equation analysis is applied to measure the impact of different recombination mechanisms on the TRPL signal by quantifying the time-dependent carrier density. Fitting the entire TRPL signal with the rate equation is demanding due to the numerous variables involved [12]. However, by analyzing the slope of the TRPL signal during the early stages, we can determine the influence of Auger recombination on hot carriers, as Auger recombination occurs rapidly after recombination begins [12]. Figure 4(b) displays the results of the TRPL slope at 650 ps, indicating that the Auger mechanism has the most significant impact on the 160 nm diameter NWs. Furthermore, non-monotonic behavior is observed in this time-resolved spectroscopy, consistent with observations made under steady-state conditions at 10 K [7].

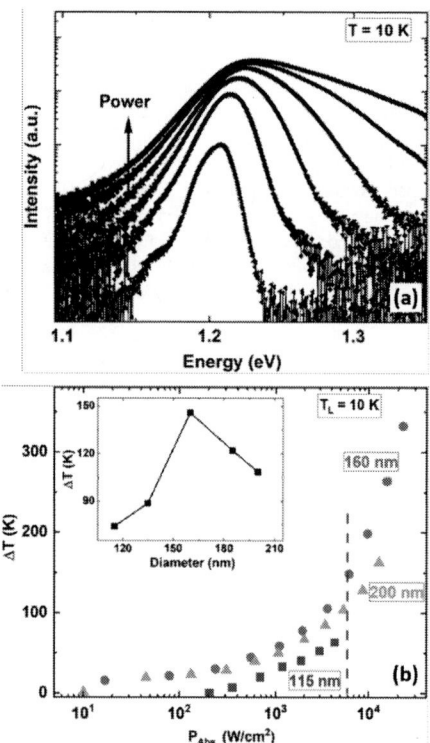

Figure 2. (a) PL spectra of the InGaAs NWs of 160 nm diameter at 10 K under various excitation powers. (b) ΔT of the NWs of various diameters as a function of the excitation power. The inset shows the dependence of ΔT at 6 kW/cm² versus the NW diameter.

To find the temperature of hot carriers, one can analyze the PL spectra using the generalized Planck's radiation law [8]. Figure 2(b) shows the results of the hot carrier temperature (ΔT) versus the excitation power at 10 K for the NWs of different diameters. It is observed that as the excitation power increases, the temperature of hot carriers also increases. The inset of Figure 2(b) demonstrates how the hot carrier temperature changes with the diameter of the NWs at a specific absorbed power density (6 kW/cm²). It is noticeable that as the diameter of the NWs decreases from 200 nm to 160 nm, the hot carrier temperature increases, indicating stronger hot carrier effects. However, when the diameter of the NWs is further reduced (< 160 nm), the effects of hot carriers decrease [5].

According to theoretical studies, an increase in the diameter of the NWs has been found to result in a higher thermalized power density, thereby mitigating the impact of hot carriers [9]. This phenomenon is particularly evident in NWs with diameters exceeding 160 nm, as depicted in the inset of Figure 2(b).

41st European Photovoltaic Solar Energy Conference and Exhibition

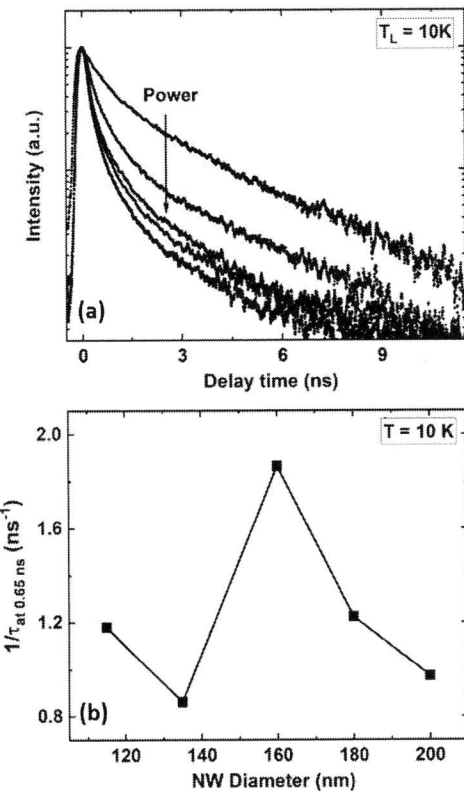

Figure 4. (a) TRPL spectra of the InGaAs NWs of 160 nm diameter at 10 K under various excitation powers. (b) The inverse of the decay lifetime at 650 ps as a function of diameter.

By increasing the lattice temperature, the effects of hot carriers can change due to the temperature-dependent behavior of hot carrier scattering with phonons and ionized impurities [13]. Figure 5(a) indicates the results of the hot carrier temperature for the InGaAs NWs of various diameters versus the lattice temperature from 10 K to 300 K. The figure shows that by increasing the lattice temperature, the effects of hot carriers become weaker (ΔT reduces at elevated lattice temperatures). To study the origin of this effect, the recombination dynamics of hot carriers as a function of the lattice temperature are determined. At lower lattice temperatures, Auger heating has the highest contributions to the recombination mechanism, see Figure 5(b). In contrast, increasing temperature shifts the recombination dynamics towards radiative and SRH recombination [7].

Additionally, the results of TRPL experiments at various lattice temperatures, as illustrated in Figure 5(c), also indicate that the recombination lifetime of carriers through SRH recombination decreases, meaning that its recombination rates increase at higher lattice temperatures. The increased rates of SRH recombination can cause faster hot carrier thermalization, leading to weaker hot carrier effects with increasing temperature [9].

In conclusion, we studied the effects of hot carriers in monolithically grown core-shell InGaAs NWs of various diameters on silicon wafers. The hot carriers in the nanowires exhibit significant dependence on their spatial confinement. This effect is caused by the presence of different recombination mechanisms, especially Auger and SRH recombination, as well as the phonon-bottleneck effect on hot carriers. It has been found that by reducing the density of defects in the NWs and improving their passivation, specifically by lowering the rates of surface recombination velocity, it is possible to enhance the hot carrier effects, making them suitable candidates for photovoltaic solar cell applications.

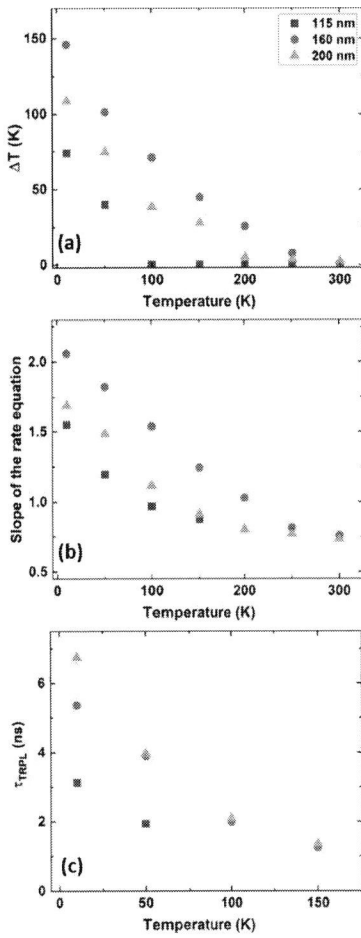

Figure 5. (a) Temperature of hot carriers (ΔT) in the InGaAs NWs of various diameters as a function of the lattice temperature. (b) Results of temperature-dependent rate equation analysis under steady-state conditions. (c) SRH recombination lifetime versus the lattice temperature.

4.3 References

[1] R. T. Ross and A. J. Nozik. "Efficiency of hot-carrier solar energy converters." Journal of Applied Physics 53.5 (1982): 3813-3818.

[2] Y. Zhang, G. Conibeer, S. Liu, J. Zhang, and J-F Guillemoles. "Review of the mechanisms for the phonon bottleneck effect in III–V semiconductors and their application for efficient hot carrier solar cells." Progress in Photovoltaics: Research and Applications 30, no. 6 (2022): 581-596.

[3] J. Fast, U. Aeberhard, S. P. Bremner, and H. Linke. "Hot-carrier optoelectronic devices based on semiconductor nanowires." Applied Physics Reviews 8, no. 2 (2021).

[4] R. Hathwar, Y. Zou, C. Jirauschek, and S. M. Goodnick. "Nonequilibrium electron and phonon dynamics in advanced concept solar cells." Journal of Physics D: Applied Physics 52, no. 9 (2019): 093001.

[5] H. Esmaielpour, N. Isaev, I. Makhfudz, M. Döblinger, J. J. Finley, and G. Koblmüller. "Strong Dimensional and Structural Dependencies of Hot Carrier Effects in InGaAs Nanowires: Implications for Photovoltaic Solar Cells." ACS Applied Nano Materials 7, no. 3 (2024): 2817-2824

[6] D. Sandner, H. Esmaielpour, F. del Giudice, S. Meder, M. Nuber, R. Kienberger, G. Koblmüller, and Hristo Iglev. "Hot Electron Dynamics in InAs–AlAsSb Core–Shell Nanowires." ACS Applied Energy Materials 6.20 (2023): 10467-10474.

[7] H. Esmaielpour, N. Isaev, J. J. Finley, and G. Koblmüller. "Influence of Auger heating and Shockley-Read-Hall recombination on hot-carrier dynamics in InGaAs nanowires." Physical Review B 109, no. 23 (2024): 235303.

[8] H. Esmaielpour, L. Lombez, M. Giteau, J-F Guillemoles, and D. Suchet. "Impact of excitation energy on hot carrier properties in InGaAs multi-quantum well structure." Progress in Photovoltaics: Research and Applications 30, no. 11 (2022): 1354-1362.

[9] I. Makhfudz, H. Esmaielpour, Y. Hajati, G. Koblmüller, and N. Cavassilas, "Interplay of Electron Trapping by Defect Midgap State and Quantum Confinement to Optimize Hot Carrier Effect in a Nanowire Structure" Physical Review B (2024).

[10] M. Achermann, A. P. Bartko, J. A. Hollingsworth, and V. I. Klimov. "The effect of Auger heating on intraband carrier relaxation in semiconductor quantum rods." Nature Physics 2, no. 8 (2006): 557-561.

[11] Y. Takeda, S. Sato, and T. Morikawa. "Effects of impact ionization and Auger recombination on hot-carrier solar cells and hot-carrier photocatalysts." Japanese Journal of Applied Physics 62.SK (2023): SK1003.

[12] L. Krückemeier, B. Krogmeier, Z. Liu, U. Rau, and T. Kirchartz. "Understanding transient photoluminescence in halide perovskite layer stacks and solar cells." Advanced Energy Materials 11, no. 19 (2021): 2003489.

[13] D. K. Ferry, "Hot Carriers in Semiconductors." Bristol, UK: IOP Publishing, 2021.

Design and Prototyping of Spectrum-Split-Type Concentrating Photovoltaic-Thermoelectric Hybrid Power Generator

Kenji Kamide[1], Ryoji Funahashi[1], Tomoyuki Urata[1], Yoko Matsumura[1], Jun Sakuma[2], Hidefumi Akiyama[2], Katsuto Tanahashi[1],

AIST[1], ISSP UTokyo[2], Japan

E-mail: kenji.kamide@aist.go.jp

PVTE：　Our target is

Photovoltaic-Thermoelectric hybrid generation

Heat usually has negative effects

negative temp. coeff.
& shorten the lifetime

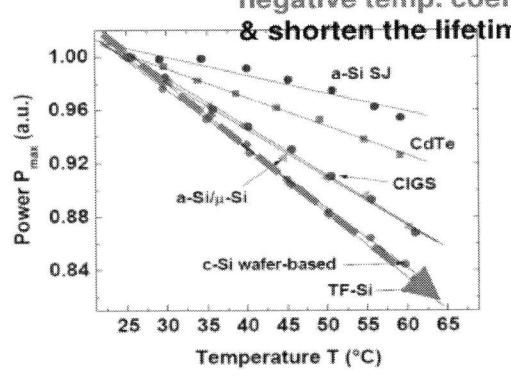

A. Virtuani, D. Pavanello, G. Friesen, 25th EUPVSEC (2010);
doi: 10.4229/25thEUPVSEC2010-4AV.3.83

"active utilization of waste heat from solar cells"

- In solar power generation, most of the light energy eventually becomes waste heat (for PV cells with 20% efficiency, the remaining 80% becomes almost entirely heat).
- Is it possible to create solar cells that utilize heat and turn it into an advantage?

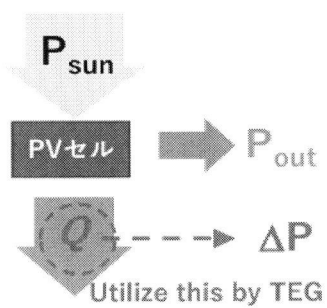

Utilize this by TEG

PVTE (Contact vs Spectrum Split)

Contact type

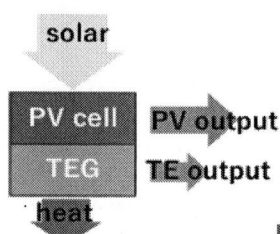

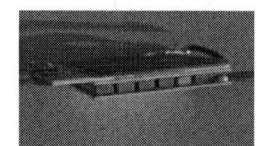

Direct utilization of the heat emitted from the PV cells. The PV and TE are thermally coupled directly.

- **Simple structure**
- **Large portion of heat can be used**

Kenji Kamide, Toshimitsu Mochizuki, Jun Sakuma, Hidefumi Akiyama, Hidetaka Takato, EU PVSEC 2021. DOI: 10.4229/EUPVSEC20212021-1BO.16.4

Spectrum split type

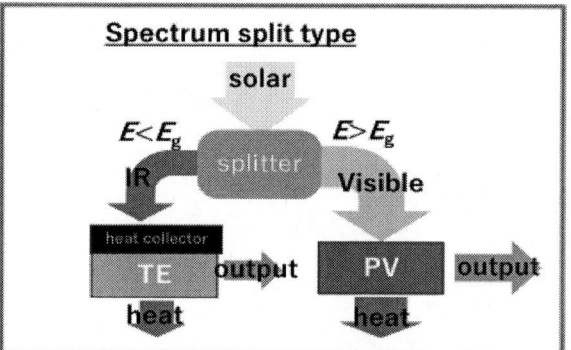

Light is separated into the long and short wavelength sides of the absorber's bandgap, and the long wavelength side is utilized for TEG.

The temperature of the TE element can be increased without raising the temperature of the PV cell, providing a cooling effect on PV cell.

---

cf) Usable heat energy in two types

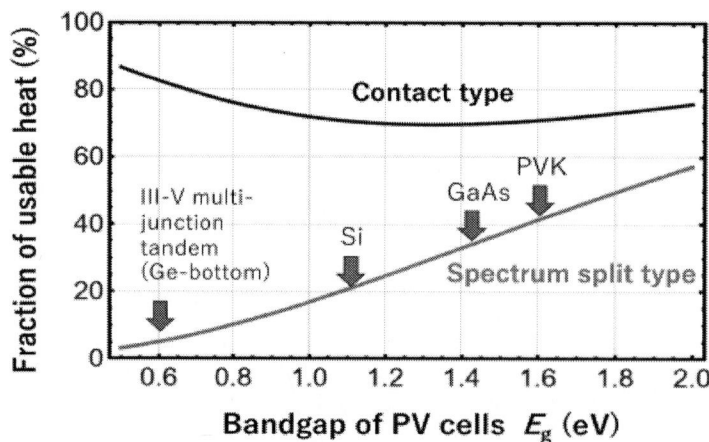

※solar spectrum
=6000K blackbody sun

Advantages in application

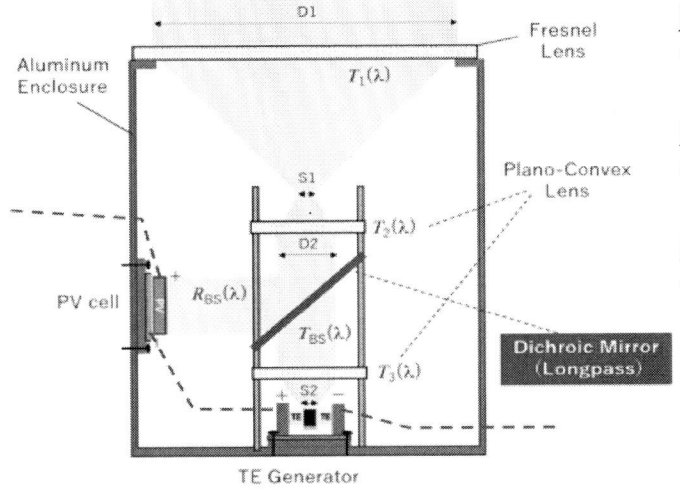

Spatial separation of light and heat
- degree of freedom in heat and light collection design for PV and TE

Concentrator
- small devices make the maintenance easier

Current-matching design
- appropriate TE-pair number avoids mismatch loss for two-terminal connection.

Jun Sakuma, Kenji Kamide, Toshimitsu Mochizuki, Hidetaka Takato and Hidefumi Akiyama, 2023 Appl. Phys. Express 16, 014003. DOI 10.35848/1882-0786/acb12d

---

Modeling & simulation for the device

How much efficiency benefit can ideally be obtained with a spectrum split types?

Thermal Modeling

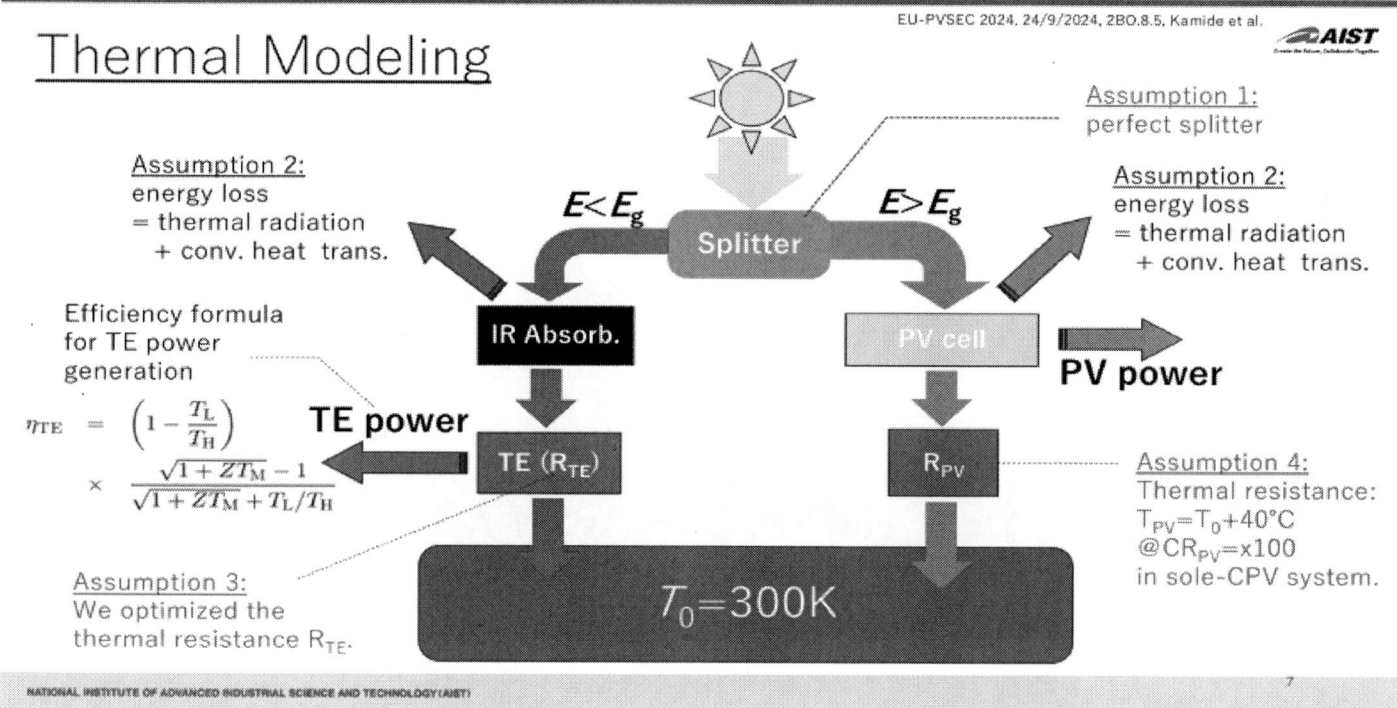

Assumption 1: perfect splitter

Assumption 2:
energy loss
= thermal radiation
+ conv. heat trans.

Assumption 2:
energy loss
= thermal radiation
+ conv. heat trans.

Efficiency formula for TE power generation

$$\eta_{TE} = \left(1 - \frac{T_L}{T_H}\right) \times \frac{\sqrt{1 + ZT_M} - 1}{\sqrt{1 + ZT_M} + T_L/T_H}$$

Assumption 3:
We optimized the thermal resistance R_{TE}.

Assumption 4:
Thermal resistance:
$T_{PV} = T_0 + 40°C$
@$CR_{PV} = ×100$
in sole-CPV system.

$T_0 = 300K$

PV cell modeling

concentrator Si cell
(Amonix's rear-junction back-contact cell)

"Slade A, Garboushian V. 27.6% efficient silicon concentrator cell for mass production. Technical Digest, 15th International Photovoltaic Science and Engineering Conference, Shanghai, 2005."

CR	Eff. [%]	Voc [mV]	FF [%]
7.7	25.0	737	81.2
17.4	26.2	761	82.4
34.0	26.8	780	82.2
65.4	27.2	797	81.9
92.3	27.6	808	82.0
122.4	27.5	815	81.0

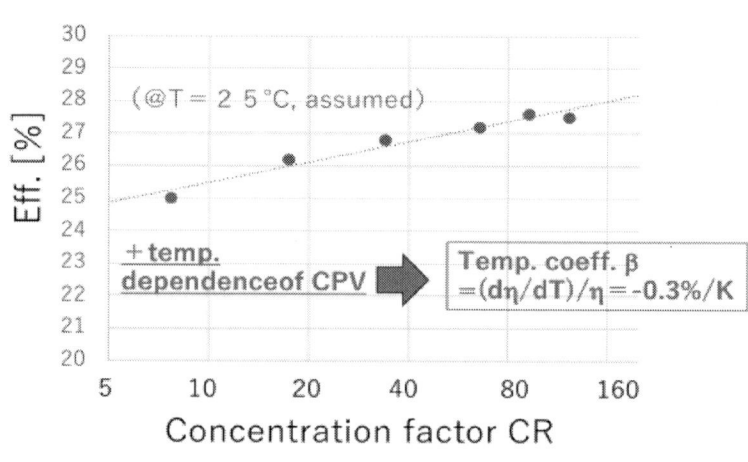

(@T = 25°C, assumed)

+temp. dependence of CPV ➡ Temp. coeff. β =(dη/dT)/η = -0.3%/K

TE device modeling

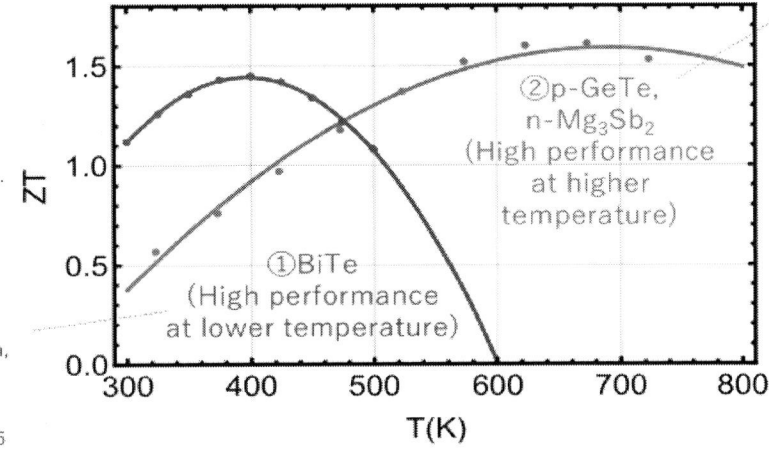

①BiTe (High performance at lower temperature)

②p-GeTe, n-Mg$_3$Sb$_2$ (High performance at higher temperature)

Y.S. Wang, L.L. Huang, D. Li, J. Zhang, X.Y. Qin, Journal of Alloys and Compounds 758, 72 (2018). https://doi.org/10.1016/j.jallcom.2018.05.035.

p-GeTe/n-Mg$_3$Sb$_2$ module

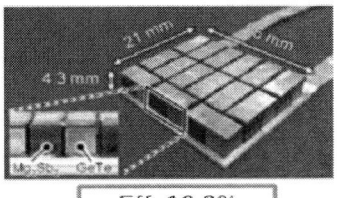

Eff. 10.3% at ΔT=520K.

Fuyuki Ando, Hiromasa Tamaki, Yoko Matsumura, Tomoyuki Urata, Takeshi Kawabe, Ryosuke Yamamura, Yuriko Kaneko, Ryoji Funahashi, Tsutomu Kanno, Materials Today Physics, 36, 101156 (2023). https://doi.org/10.1016/j.mtphys.2023.101156. Reprinted with permission from Elsevier.

Efficiency simulation

※ @ steady state

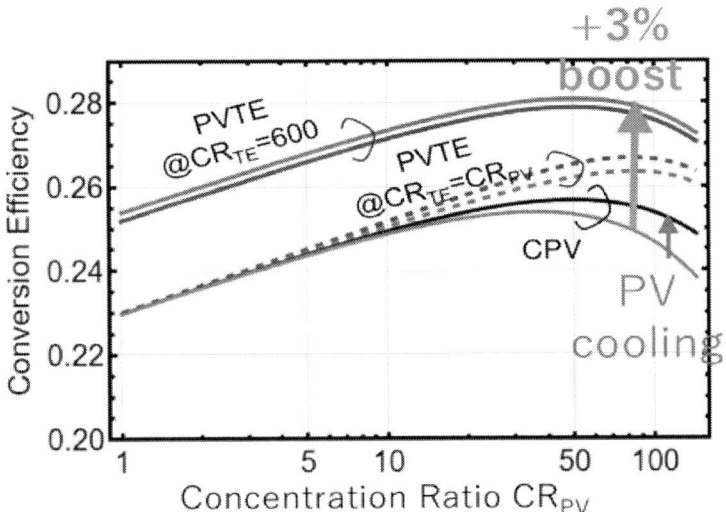

+3% boost

PV cooling

- —— PVTE (BiTe) @ CR$_{TE}$=600
- —— PVTE (Mg$_3$(Sb,Bi)$_2$) @CR$_{TE}$=600
- – – PVTE (BiTe) @CR$_{PV}$=CR$_{TE}$
- – – PVTE (Mg$_3$(Sb,Bi)$_2$) @CR$_{PV}$=CR$_{TE}$
- —— PV in the hybrid
- —— PV only

TE hot-side temperature

	CR$_{TE}$=CR$_{PV}$=100	CR$_{TE}$=600
BiTe	490 K	620 K
Mg$_3$(Sb,Bi)$_2$	520 K	770 K

Design & prototyping

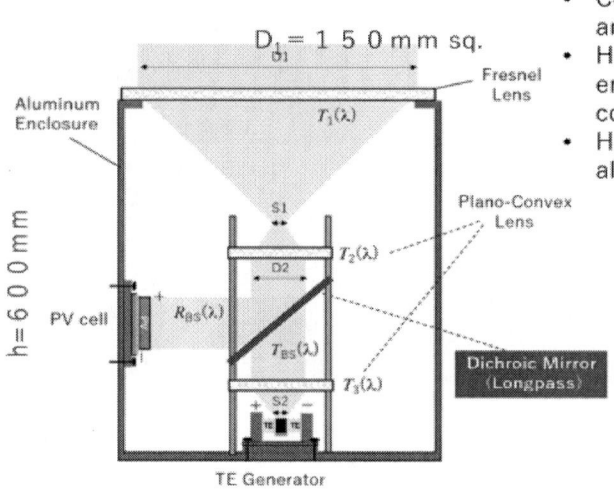

$D_{d1} = 1 5 0$ mm sq.

Fresnel Lens

Aluminum Enclosure

$T_1(\lambda)$

h=600mm

S1

PV cell

$R_{BS}(\lambda)$

$T_2(\lambda)$

D2

Plano-Convex Lens

$T_{BS}(\lambda)$

$T_3(\lambda)$

Dichroic Mirror (Longpass)

TE Generator

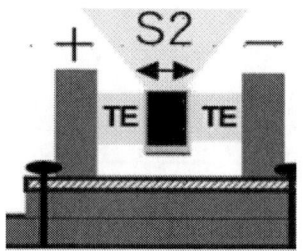

- Concentration factors for PV and TE are freely designed
- Horizontal arrangement enables reducing heat collection area.
- Heat dissipation from the aluminum enclosure

---

transmission / reflection spectra of the splitters A and B

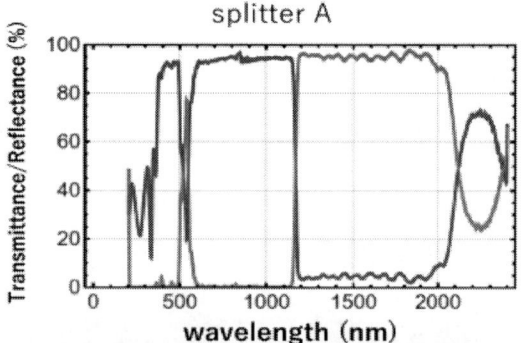

splitter A

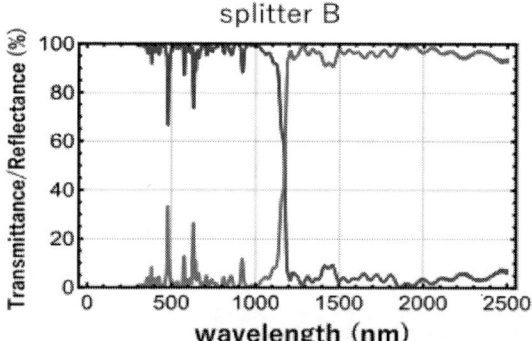

splitter B

splitter A	fraction	expected input
PV current $T_1 * T_2 * R_{BS}$	73.6%	**6.08 A**
heat into TE $T_1 * T_2 * T_{BS} * T_3$	13.95%	**3.14 W**

splitter B	fraction	expected input
PV curret $T_1 * T_2 * R_{BS}$	81.7%	**6.75 A**
heat into TE $T_1 * T_2 * T_{BS} * T_3$	12.3%	**2.77 W**

Initial results

35mm sq. PV cell   25mm sq. heat collector   a pair of TE device (Co349)

TE output and temperature

TE material	splitter	Output (estimate) [mW]	Temp. Diff. (estimate) $\Delta T[K]$
p-Co349/ n-TiNiSn$_{0.98}$Sb$_{0.01}$	A	2.6 (2.9)	86 (84)
BiTe	B	45 (50)	56 (65)

reasonable agreement!

The PV cooling effect of splitter

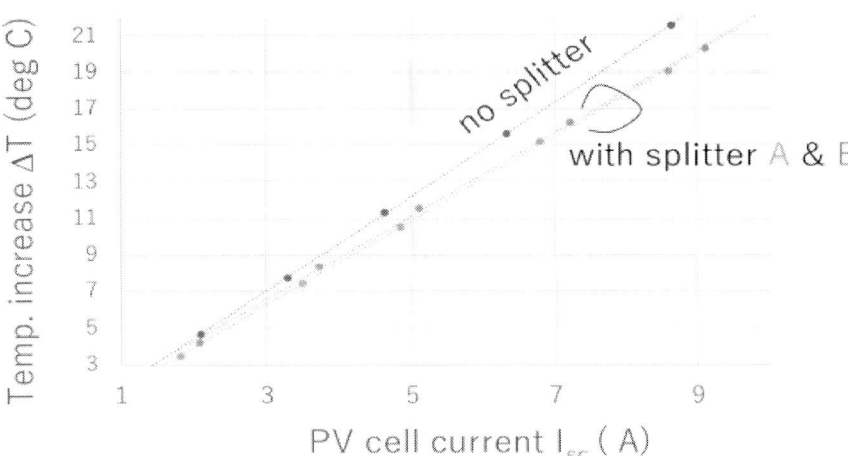

Mitigation of PV cell temperature rise with splitter, hence the cooling effect was confirmed.

EU-PVSEC 2024, 24/9/2024, 2BO.8.5, Kamide et al.

Summary

Efficiency estimation simulation, design, and prototyping of spectrum-split-type concentrator PVTE hybrids.

1. Efficiency boost of +3% over conventional CPV (including cooling effect) by independently optimizing CR_{PV} and CR_{TE}.

2. Characteristics similar in the simulation and cooling effect are observed for the prototype unit.

41st European Photovoltaic Solar Energy Conference and Exhibition

DEVELOPMENT OF AN INTERDIGITATED BACK-CONTACTED SOLAR CELL ARCHITECTURE AS A PLATFORM TO ASSESS EMERGING ABSORBERS AND NEW SELECTIVE CONTACTS

Juan de Dios Castillo[1], Gerard Masmitjà[1], Pau Estarlich[1], Pablo Ortega[1], Cristobal Voz[1], Arnau Torrens[1], Oriol Segura[1], Edgardo Saucedo[1], Massoud Karimipour[2], Sonia Ruiz[2], Mónica Lira-Cantu[2] and Joaquim Puigdollers[1]

[1] Universitat Politècnica de Catalunya, Micro and Nanotechnolologies Group, Barcelona, Spain
[2] Institut Català de Nanociència i Nanotecnologia, Nanostructured Materials for Photovoltaic Energy Group, Barcelona, Spain
e-mail: gerard.masmitja@upc.edu

ABSTRACT: This work shows the development of an interdigitated back-contacted (IBC) platform which could be used to develop both novel-materials as selective contacts and emerging absorbers. The first approach was based on amorphous silicon layer as absorber in combination with vanadium oxide (V_2O_x) and zinc oxide (ZnO) as hole- and electron-selective contacts, respectively. Short circuit current density (J_{sc}) and open circuit voltage (V_{oc}) values of 0.6 mA/cm^2 and 562 mV, respectively, prove the viability of the proof-of-concept platform to manufacture IBC solar cells. New absorbers will be analyzed in the future using this concept to demonstrate its practicability to study and develop solar cells based on an IBC scheme.

Keywords: Interdigitated back-contacted structure; selective contacts; transition metal oxides; thin-film solar cells.

1 INTRODUCTION

A solar cell consists, in essence, of an absorber and two charge-carrier selective contacts. The absorber is the material where the photons of the incident light are converted in electron-hole pairs. These charge-carriers have to be selectively collected by the contacts; one for electrons and one for holes, namely electron (ETL) and hole (HTL) transport layers, respectively [1].

A broad number of emerging absorbers are being studied by the PV community in the last years, such as perovskite, chalcogenide and organic compounds. One goal is to circumvent the limit efficiency that is almost achieved for single junction crystalline silicon (c-Si) solar cell [2] by applying a tandem configuration. In this way, a top absorber with a higher band gap is deposited (monolithic tandem) on top of a bottom absorber (generally based on c-Si) with a lower band gap [3]. Another focus is to use emerging absorbers in thin-film technology to obtain flexible and/or semi-transparent devices suitable for building-integrated photovoltaics (BIPV) [4], indoor PV applications [5] and agrivoltaic systems [6].

Charge-carrier selective contacts must exhibit a strong asymmetric conductivity, depending on the carrier type, moreover, they have to passivate the surface to maximize the effective carrier lifetime and carrier extraction. In addition, they have to avoid some of the limitations of thin-film solar cells to be widely spread in the industry, that is the use of toxic-based contacts (e.g., CdS) or scarce materials (e.g., Indium) often used in transparent conductive oxide (TCO) films that cause high price volatility. In this way, Cd-free ETL and In-free TCO are being extensively investigated [7,8].

The main goal of this work is to develop the contact scheme of an interdigitated back-contacted (IBC) configuration placed on a glass substrate. In this approach, the HTL electrode is defined using a high-work function transition metal oxide, vanadium oxide (V_2O_x), whereas the ETL electrode is based on zinc oxide (ZnO), both contacts deposited by atomic layer deposition (ALD) technique. Therefore, this IBC platform can be used to not only study the behavior of novel selective contacts but also

to assess emerging absorbers. In addition, by using an IBC scheme, the top surface of the absorber is unnecessary for contact, thus it can be dedicated, for instance, to optimize light management and/or long-term stability (i.e., encapsulation), as well as to deposit easily a top absorber of a tandem solar cell using three-terminal (3T) configuration [9].

2 EXPERIMENTAL

2.1 IBC platform integration

The IBC architecture is implemented onto an off-the-shelf glass substrate covered with a film of SnO:F based TCO. Fabrication process is summarized in Figure 1 and it has the next steps: the process starts (*i*) with the deposition of an insulation layer over the conductive TCO/glass substrate, thus preventing any conductive path between HTL and ETL electrodes of the IBC architecture. The dielectric alumina (Al_2O_3) film (~50 nm) was deposited by ALD at 200 °C using trimethyl aluminum (TMA) and deionized water (Di-H_2O) as Al and oxidant precursors, respectively.

Then, a first photolithography stage (*ii*) is done to open a contact window by wet etching, i.e. hydrofluoric acid (HF) at 1%, through the alumina layer to access the bottom TCO layer, which serves as the ETL electrode. The next step (*iii*) is to define the ETL contact, which is patterned by lift-off technique, consisting of a stack of AZO (40 nm) and ZnO (10 nm) films deposited both by ALD at a temperature of 150 °C and 100 °C, respectively. TMA, diethylzinc (DEZ) and Di-H_2O were used as Al, Zn and oxidant precursors, respectively, for those films. Next, a second lithography stage (*iv*) is made to protect the ETL regions and pattern the HTL contact by lift-off technique over the Al_2O_3 film.

The whole HTL electrode consists of an aluminum conductive film with thickness of about 25 nm, followed by a suitable HTL which is a stack of AZO (50 nm) and V_2O_x (15 nm) films deposited both by ALD at a temperature of 150 °C and 125 °C, respectively. Tetrakis(ethylmethylamino)-vanadium (VTIP) and Di-H_2O were used as vanadium and oxidant precursors,

285

respectively. At his point, the IBC platform is ready (*v*) for the next procedure to finish the whole solar cell structure (*vi*), i.e., absorber deposition plus top layer and device definition.

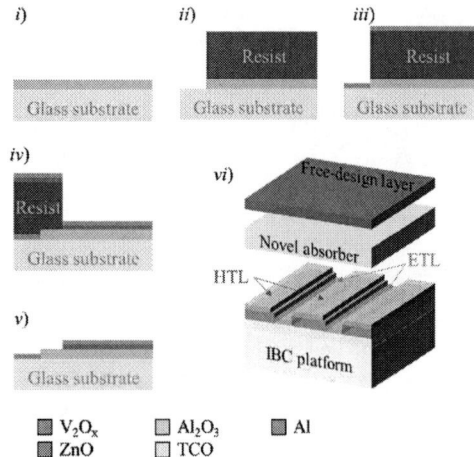

Figure 1: Main fabrication stages of the IBC platform and a 3D sketch of the final device.

2.2 Solar cell fabrication

In order to evaluate the IBC platform, intrinsic hydrogenated amorphous silicon carbide (a-SiC$_x$:H) film was used as absorber material. First, 50 nm a-SiC$_x$:H film was deposited by plasma enhanced chemical vapor deposition (PECVD) technique using a mixture of silane and methane gases at 140 °C and a chamber pressure of 1×10^{-5} mbar. Then the absorber bas covered with a phosphorous doped a-SiC$_x$:H film (~15 nm) and an aluminum doped ZnO (AZO) layer (75 nm) deposited by PECVD and ALD, respectively. This stack serves to facilitate the electrons flow to the ETL electrode, since it is well known the poor lateral conduction of amorphous silicon-based films. Finally, to define the device active area a last photolithography was done in combination with a wet and dry etching done by HF dip and CF$_4$ plasma treatment. This last lithography also serves to open the electrodes, thus enable the contacting. Figure 2 shows a picture of manufactured solar cells using the IBC platform.

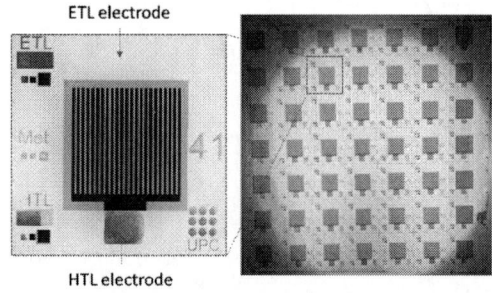

Figure 2: Images of the a-SiC$_x$:H solar cell using the IBC platform.

3 RESULTS AND DISCUSSION

To correctly analyse the solar cell, it is important to understand the flow of photo-generated electron-hole pairs in the absorber of the IBC platform. It is a well-known fact that a-Si based absorbers have a relatively poor transport properties, which mean that the diffusion length (*L$_d$*) and mobilities of charge-carriers are small [10]. The *L$_d$* parameter is well below one micron, thus the lateral charge-carrier transport distance inside the absorber (labeled as *i* in figure 3a) is almost negligible due to the tens of microns of distance between HTL and ETL regions. This inherent property emphasizes the necessity of the use of an alternative path to transport the majority carriers, i.e., electrons generated above the HTL region to the ETL contact. In this regard, an ETL film plus a TCO is placed on top of the absorber, enabling the collection and transport of the electrons (*ii*). Finally, hole- and electron-carriers are selectively collected (*iii*) using proper HTL and ETL electrodes, respectively.

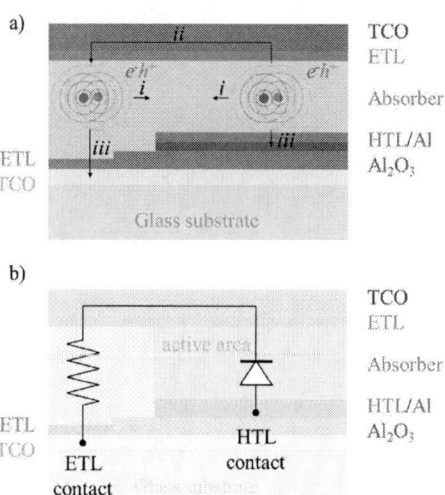

Figure 3: (a) Cross section sketch of the IBC solar cell with the photo-generated electron-hole pairs ($e^- h^+$) and their paths through the absorber to the both selective electrodes, ETL and HTL contacts. (b) Proposed simplified electrical model of the IBC solar cell.

It is worth to mention that the HTL region defines an effective active area of the solar cell, i.e., photo-generated carriers outside this region will not contribute to the short circuit current of the device. Therefore, the HTL region has to be designed larger than the ETL region. Figure 3b represents an electrical model of the above explained charge-carrier collection, where the diode behavior, i.e., active area is defined by the HTL region, whereas the ETL region is characterized as a series resistance.

Before solar cell characterization, the electrical isolation between ETL and HTL electrodes was checked. Figure 4 shows the current-voltage (*I-V*) measurement between both electrodes in which the low current values (~5 nA) corroborate that nonconductive path exists between the FTO layer and the metal (Al) of the HTL electrode. In addition, measurements between two positions of the TCO-electrode were realized to show that it is able to reach the electrode at the end of the fabrication process. Note, the TCO layer is a commercial one, thus its resistivity is well-known.

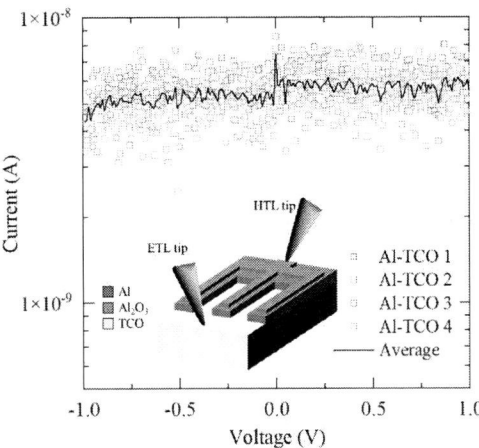

Figure 4: Current-voltage measurement between HTL and ETL electrodes. The inset shows a sketch of the IBC platform before absorber deposition with the contacting tips.

The IBC platform has been used to characterize semi-transparent solar cells based on a-SiC$_x$:H films, which not only serves to demonstrate the usefulness of the IBC structure, but fulfils the needs of reference for emerging absorbers, such as chalcogenide or perovskite compounds. Figure 5 shows dark and light I-V curves of an IBC a-SiC$_x$:H solar cell, showing remarkable values of short circuit current density (J_{sc}) and open circuit voltage (V_{oc}) with 0.6 mA/cm^2 and 562 mV, respectively. Table I summarizes the most relevant electrical results.

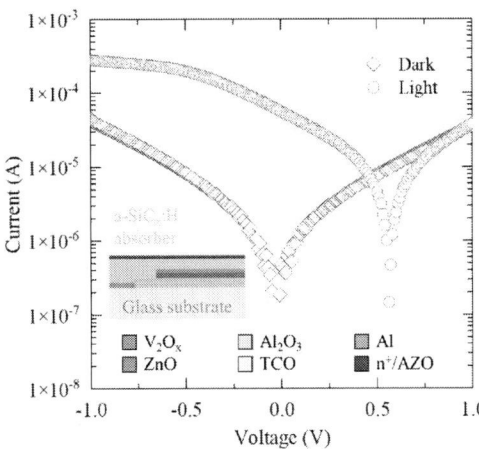

Figure 5: Current-voltage measurement under dark and light conditions of an IBC a-SiC$_x$:H solar cell.

A low shunt resistance (R_{sh}) value can be observed in dark J-V curves in Figure 5, i.e., relatively high current leakage due to shunt resistance. This current leakage limits in part the fill factor (FF) of the device, thus the final power conversion efficiency (η). This phenomenon might be overcome by using a redefined HTL structure. In fact, if a close-up view is made on the edge of the HTL region, the side of the wall-electrode might exhibit a lack of coverage and as a consequence the aluminum directly interacts with the absorber. Or even worse, the front ALD based TCO could contact the Al metal if the absorber layer is too thin. These two assumptions could be the reason for

these low R_{sh} values. Further IBC platform development have to be done to ensure that the V$_2$O$_x$-based HTL covers all the metallic electrode. In addition, the aluminum pad to contact the anode of the solar cell is over-etched, see that the metallic color vanished in the photography of Figure 2, which also contributes to a low FF because of an increase of the series resistance.

Although remarkably good results are reported using the IBC platform with a-Si based absorber, further improvements are required to circumvent technological issues, e.g., the use of alternative metal with a lower HF etch-rate. In addition of starting to use this platform for next generation of absorbers based on selenium or perovskite compounds.

Table I: Electrical characteristics of the best fabricated IBC solar cells using standard test conditions (AM1.5G 1 kW/m^2 solar spectrum an T = 25 °C).

IBC	Area (cm^2)	J_{sc} (mA/cm^2)	V_{oc} (mV)	FF	η (%)
a-SiC$_x$:H	0.1	0.6	562	19.8	0.67

4 CONCLUSIONS

In this work, we report the technological development of an IBC platform, dealing with the integration of the charge-carrier selective contacts on a commercial TCO-coated glass used as a substrate. The platform has been tested using an a-Si based absorber, which enables to demonstrate de viability of the IBC platform to analyze the solar cell behavior, demonstrating V_{oc} and J_{sc} values of 562 mV and 0.6 mA/cm^2, respectively. Taking into account these results, the exploration of alternative absorbers will be addressed in future works to study their performance as absorbers using selective contacts in an IBC configuration.

ACKNOWLEDGMENTS

This work has been supported by the Spanish government under projects PID2022-138434OB-C51 (SCALING) and TED2021-129758B-C32 (TransEl) funded by MCIN/AEI/10.13039/501100011033. TransEl project has been also supported from the European Union "NextGenerationEU"/PRTR program.

REFERENCES

[1] U. Wurfel, A. Cuevas, P. Wurfel, Charge carrier separation in solar cells, IEEE J Photovolt 5 (2015) 461–469. https://doi.org/10.1109/JPHOTOV.2014.2363550.

[2] Z. Sun, X. Chen, Y. He, J. Li, J. Wang, H. Yan, Y. Zhang, Toward Efficiency Limits of Crystalline Silicon Solar Cells: Recent Progress in High-Efficiency Silicon Heterojunction Solar Cells, Adv Energy Mater 12 (2022) 2200015. https://doi.org/10.1002/AENM.202200015.

[3] M. Wright, B. Vicari Stefani, T.W. Jones, B. Hallam, A. Soeriyadi, L. Wang, P. Altermatt, H.J. Snaith, G.J. Wilson, R.S. Bonilla, Design considerations for the bottom cell in

perovskite/silicon tandems: a terawatt scalability perspective, Energy Environ Sci 16 (2023) 4164–4190. https://doi.org/10.1039/D3EE00952A.

[4] P. Kumar, S. You, A. Vomiero, Recent Progress in Materials and Device Design for Semitransparent Photovoltaic Technologies, Adv Energy Mater 13 (2023) 2301555. https://doi.org/10.1002/AENM.202301555.

[5] K.S. Srivishnu, M.N. Rajesh, S. Prasanthkumar, L. Giribabu, Photovoltaics for indoor applications: Progress, challenges and perspectives, Solar Energy 264 (2023) 112057. https://doi.org/10.1016/J.SOLENER.2023.112057.

[6] S. Gorjian, E. Bousi, Ö.E. Özdemir, M. Trommsdorff, N.M. Kumar, A. Anand, K. Kant, S.S. Chopra, Progress and challenges of crop production and electricity generation in agrivoltaic systems using semi-transparent photovoltaic technology, Renewable and Sustainable Energy Reviews 158 (2022) 112126. https://doi.org/10.1016/J.RSER.2022.112126.

[7] A. Wang, J. Huang, J. Cong, X. Yuan, M. He, J. Li, C. Yan, X. Cui, N. Song, S. Zhou, M.A. Green, K. Sun, X. Hao, Cd-Free Pure Sulfide Kesterite Cu2ZnSnS4 Solar Cell with Over 800 mV Open-Circuit Voltage Enabled by Phase Evolution Intervention, Advanced Materials 36 (2024) 2307733. https://doi.org/10.1002/ADMA.202307733.

[8] F. Jay, T. Gageot, G. Pinoit, B. Thiriot, J. Veirman, R. Cabal, S. De Vecchi, W. Favre, M. Sciuto, C. Gerardi, M. Foti, Reduction in Indium Usage for Silicon Heterojunction Solar Cells in a Short-Term Industrial Perspective, Solar RRL 7 (2023) 2200598. https://doi.org/10.1002/SOLR.202200598.

[9] E.L. Warren, W.E. Mcmahon, M. Rienäcker, K.T. Vansant, R.C. Whitehead, R. Peibst, A.C. Tamboli, A Taxonomy for Three-Terminal Tandem Solar Cells, ACS Energy Lett 5 (2020) 1233–1242. https://doi.org/10.1021/ACSENERGYLETT.0C00 068.

[10] T. Tiedje, C.R. Wronski, B. Abeles, J.M. Cebulka, Electron transport in hydrogenated amorphous silicon: drift mobility and junction capacitance, Solar Cells 2 (1980) 301–318. https://doi.org/10.1016/0379-6787(80)90034-4.

41st European Photovoltaic Solar Energy Conference and Exhibition

ANNEALED PHOSPHOROUS-DOPED AMORPHOUS SILICON AS ELECTRON SELECTIVE CONTACT FOR CRYSTALLINE GERMANIUM THERMOPHOTOVOLTAIC CELLS

G. Rivera[1a)], M. Gamel[1], G. López[2], M. Garín[3] and I. Martín[1]

[1]Departament d'Enginyeria Electrònica, Universitat Politècnica de Catalunya, Barcelona, Spain.
[2]Departament d'Enginyeria Gràfica i de Disseny, Universitat Politècnica de Catalunya, Barcelona, Spain.
[3]Department of Engineering, Universitat de Vic—Universitat Central de Catalunya (UVIC-UCC), Vic, Spain
[a)] E-mail for correspondence: gerard.rivera@upc.edu

ABSTRACT: Thermophotovoltaic devices based on crystalline germanium (c-Ge) substrates that avoid the epitaxial growth of III-V compounds are a promising solution for reducing technological costs of such technology. Heterojunction based on n-type silicon layers deposited by PECVD on c-Ge are good candidates. However, deposited layers with an efficient electron extraction have proven to be difficult given the instability of c-Ge interface that leads to significant amount of recombination. Herein, we present a novel strategy consisting of a high-temperature anneal of a phosphorus-rich amorphous silicon layer deposited by PECVD. The annealing improves the conductivity and passivation of the structure by inducing a diffusion process of not only phosphorus but also silicon, as demonstrated by ToF-SIMS. We optimize the thermal process for high conductivity and surface passivation. The morphological changes are observed using Raman spectroscopy and X-Ray Diffraction, indicating a partial recrystallization of the amorphous silicon layer and confirming the formation of a SiGe alloy at the surface. Finally, the optimized structure is implemented in a device and tested under 1-sun illumination, yielding an efficiency of 4.53% and validating the feasibility of the approach.

Keywords: phosphorous diffusion, crystalline germanium, thermophotovoltaics, PECVD.

1 INTRODUCTION

Crystalline germanium (c-Ge), with a bandgap of 0.67 eV, is a suitable absorber material for thermophotovoltaic cells which convert thermal radiation into electricity. With the aim of an efficient carrier extraction, several techniques have been employed to develop electron transport layers (ETL) on p-type c-Ge substrates. The epitaxial growth of III-V compounds such GaInP has already been demonstrated, leading to excellent results, but at the expense of significant costs [1]. Another possible route is PECVD-based silicon heterojunctions such as amorphous (a-Si) [1] or microcrystalline silicon (μc-Si) [2]. Although cost-effective, PECVD-based solutions are very sensitive to the interface quality, requiring critical surface preparations.

A potential solution to this problem involves a high temperature annealing of a phosphorus-doped amorphous silicon film (a-Si(n)) with the objective to diffuse the phosphorus inside the germanium substrate. In this way, a n+/p homojunction is created into the germanium, while the a-Si(n)/c-Ge interface can provide some degree of surface passivation. Some years ago, our research group developed a similar approach using phosphorus-doped amorphous silicon carbide films on p-type crystalline silicon, leading to successful results [3].

With the general objective of making cost-effective c-Ge TPV devices without enquiring into complex surface treatments, our work aims to explore the effect of a high-temperature annealing on a PECVD-based ETL consisting of a ~50 nm phosphorus-doped amorphous silicon film (a-Si(n)) on a p-type c-Ge substrate. Using symmetrical test structures, we first explore different temperature regimes which provide different degrees of conductivity and c-Ge surface passivation. Next, the morphological changes induced by the temperature step are examined in terms of diffusion profiles and changes in silicon layer crystallinity.

Finally, devices are implemented with this technique, with a preliminary assessment of its performance under 1-sun illumination conditions.

2 METHODOLOGY

Our work used commercial (100) p-type c-Ge wafers, with a resistivity of 1.2 Ω.cm and a thickness of 150 μm. A wet cleaning procedure was followed just before introducing the wafers into the PECVD chamber. This procedure consisted of a 1 min HF 1% dip, followed by a 30 s DI water rinse and ended with a 3 min dip in a HCl:H2O 1:1 solution. The a-Si(n) layer was deposited in a 13.56 MHz RF direct plasma PECVD reactor chamber. Hydrogen (H_2) and silane-diluted phosphine ($SiH_4 + PH_3$, 95 % + 5 %) were used as gas precursors, with flow rates set at 200 and 4 sccm, respectively. The annealing took place in a horizontal tube furnace with N_2 atmosphere. Samples were introduced at 400ºC, and temperature increased with a controlled ramp of 15 ºC/min to the explored range of 685ºC up to 730ºC. After maintaining the temperature for annealing times of 30 to 60 minutes, samples were cooled down to 400ºC with a 5 ºC/min ramp.

The surface passivation and layer conductivity were measured using the Sinton WCT-120 lifetime tester tool. The layer conductivity, characterized by its sheet-resistance (R_s), was estimated from the conductance needed for measuring the effective lifetime. For the case of surface passivation, our metric of interest was the saturation current density J_0, expressing the hole recombination current in our annealed ETL structure. In its simplest form, and using a symmetrical structure, the parameter J_0 can be derived from an Auger-corrected effective lifetime (τ_{eff}) vs. a spatially constant excess carrier density (Δn) using the following expression [4]:

$$\frac{1}{\tau_{\text{eff}}} - \frac{1}{\tau_{\text{aug}}} = \frac{2J_0}{q} \cdot \frac{(N_A + \Delta n)}{n_i^2 \cdot w} \qquad (1)$$

where τ_{aug} is the Auger lifetime, q is the fundamental charge, N_A is the bulk doping density, n_i is the intrinsic carrier concentration and w is the wafer thickness.

After its characterization, the annealed a-Si(n) layer was etched in a stirred TMAH solution at 65°C. Then, we repeated the measurements again, which allowed us to compare the effect of the conductivity and passivation offered by the combination of the annealed layer and the diffused region, and that of the sole diffused region, which remained unetched. In Fig. 1a and 1b, we show a cross-section of the test samples with the mentioned two different cases: with and without the annealed a-Si(n) layer.

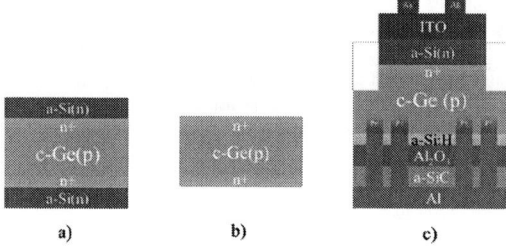

a)   b)   c)

Figure 1: Cross-section of the different structures processed in this work. a) Symmetrical structure with annealed layer and the induced n$^+$ diffusion. b) Symmetrical structure with only an n$^+$ diffusion, after etching the annealed layer. c) Proof of concept PV device structure.

The silicon and phosphorus diffusion profiles were examined by using Time-of-Flight Secondary Ion Mass Spectrometry (ToF-SIMS). Different annealing temperature and times were tested, and the results were expressed as depth profiles. The morphological changes within the structure were examined using Raman spectroscopy and X-Ray diffraction.

Finally, the ETL structure was introduced in 1x1 cm^2 devices (structure also shown in Fig. 1c). The rear surface was passivated by using a dielectric stack composed of a PECVD-deposited ~2 nm intrinsic hydrogenated amorphous silicon carbide (a-SiC$_x$:H(i)), a 50 nm ALD-deposited alumina (Al$_2$O$_3$) layer, and a 45 nm thick, almost stoichiometric amorphous silicon carbide (a-SiC) film deposited by PECVD. This combination has been studied by our group to offer a high-quality passivation while being highly reflective for TPV applications [5]. After depositing an aluminum film, this rear surface was contacted by laser-firing the structure [6-7]. On top of the ETL, 80 nm of indium-tin oxide (ITO) were deposited by sputtering and silver fingers were thermally evaporated. J-V curves under 1-sun illumination were taken to evaluate the performance of the devices.

3 RESULTS

3.1 Conductivity and surface passivation

The conductivity and surface passivation of the annealed ETL films is shown in Fig. 2.

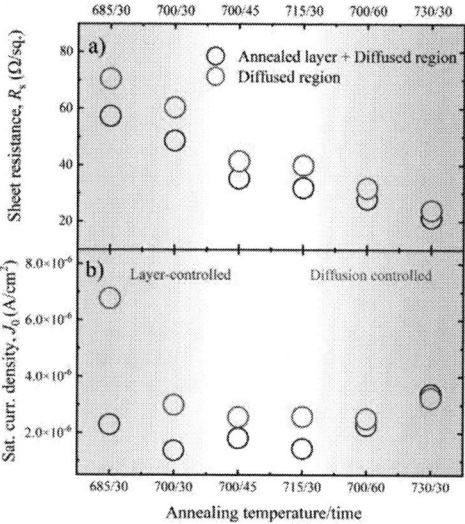

Figure 2: (a) Sheet resistance (R_s) and (b) saturation current density (J_0) evolution as a function of annealing temperature in °C and time in minutes. The measurements after etching the annealed layer are included for direct comparison.

Different trends can be identified in the figure. Firstly, as it can be seen in Fig. 2a, the increase in annealing temperature and time results in an ETL that becomes more conductive (i.e. its sheet resistance decreases) suggesting a diffusion of phosphorus into c-Ge creating an n$^+$ region. The difference in sheet-resistance before and after etching the annealed layer becomes less-apparent as temperature/time increases, indicating a diffusion process in which the conductive region beneath the c-Ge substrate becomes more and more dominant. Secondly, J_0 evolution represented in Fig. 2b shows how the same tendency prevails where the effect of the annealed layers on the passivation vanishes for high temperature/time. A possible explanation for such results is the following: the recombination (J_0) with short temperature/time is strongly dependent on the surface passivation provided by the silicon layer as the phosphorus diffusion is not strong enough to efficiently repeal holes to reach the surface. Thus, the stronger the diffusion the less the saturation current density depends on the surface passivation. As a result, comparing the cases before and after etching the passivation tends to go from a layer-controlled mechanism (increase in J_0 after etching the annealed layer) to a diffusion-controlled mechanism (no significant changes in J_0 after etching the layer). Interestingly, the best passivation seems to be given by a combination of both mechanisms, being the lowest J_0 at the intermediate temperature regime between both mechanisms. For the implementation in devices, we chose a 45 minute anneal at 700°C.

41st European Photovoltaic Solar Energy Conference and Exhibition

3.2 Material characterization

To confirm the diffusion inside c-Ge, we measure the silicon and phosphorus species by ToF-SIMS, whose results are shown in Fig. 3.

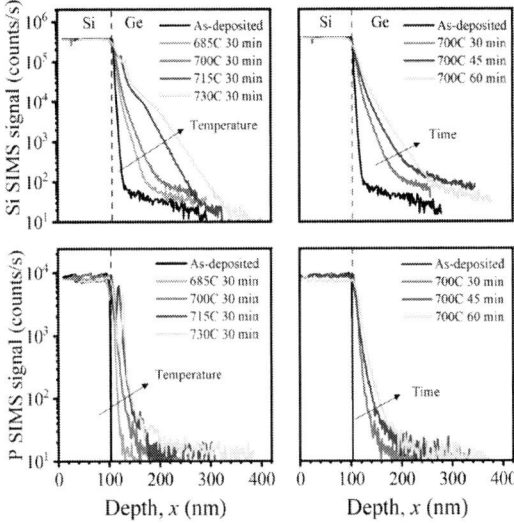

Figure 1: ToF-SIMS diffusion profiles for silicon (top) and phosphorus (bottom) signals. Thermal dependence is shown on the left graphs, while temporal dependence is shown on the right. The dashed line represents the silicon-germanium interface.

As seen from the profiles, there is indeed a diffusion process for both species into c-Ge that increases with temperature and annealing time. It is worth noting that the silicon diffusion is more pronounced than the phosphorous one, despite the higher diffusivity in Ge of the latter. This can be understood upon the fact the phosphorus content into the silicon layer is in the range of about 1 % maximum which limits the availability of phosphorus atoms. Moreover, the phosphorus must diffuse in the silicon layer before going into the germanium substrate. It is well known that the diffusivity of phosphorus in silicon at 700°C is extremely small and much lower than silicon in germanium [8-9]. The evident diffusion of silicon into germanium suggests the possibility of having a SiGe alloy at the surface that may help in the reduction of the saturation current density, i.e. reduction of holes injected into the n^+ region.

To confirm the formation of a SiGe alloy and explore the structural changes induced by the thermal process, Raman and XRD measurements are carried out. As seen in Fig. 4, The Raman spectra of the as-deposited PECVD layer shows two small peaks, one at approximately 300 cm^{-1} corresponding to the Ge-Ge phonon mode and a broad Si-Si band centred around 480 cm^{-1}, characteristic of an amorphous silicon film [10]. After the annealing process, the silicon peak sharpens and shifts to about 520 cm^{-1}, indicating a partial recrystallization process in the silicon layer which is well reported in Raman studies [10]. An interesting feature in the annealed spectra is the appearance of an intermediate peak around 390 cm^{-1} corresponding to a Si-Ge mode [11], thereby confirming an alloy due to the species intermixing. The two minor features between the Si-Ge and Si-Si peaks also reveal the nature of an intermixing

process, being attributed to Si vibrations in different SiGe mixed environments [12]. The XRD spectra shown in Fig. 5 corroborates the crystallization process of the silicon layer as the diffractogram after annealing exhibits a slight appearance of a (400) Si feature, which indicates a preferential Si orientation in the (100) direction.

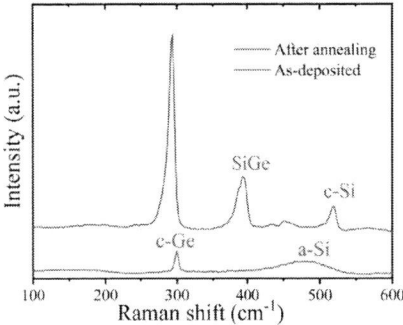

Figure 2: Raman spectra of the ETL structure before and after the thermal process at 700°C for 45min.

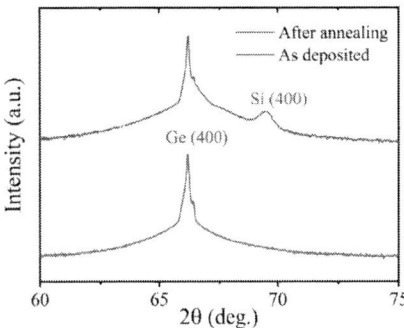

Figure 5: Region of interest in the diffractogram of the ETL structure, before and after the thermal process at 700°C for 45min.

3.3 PV devices tested under 1-sun illumination

Finally, as a previous step to TPV c-Ge devices, we applied the developed ETL into devices designed for solar spectrum. Its performance under 1-sun illumination and AM1.5g spectrum can be appreciated in Fig. 6 where the photovoltaic figures are also indicated.

291

41st European Photovoltaic Solar Energy Conference and Exhibition

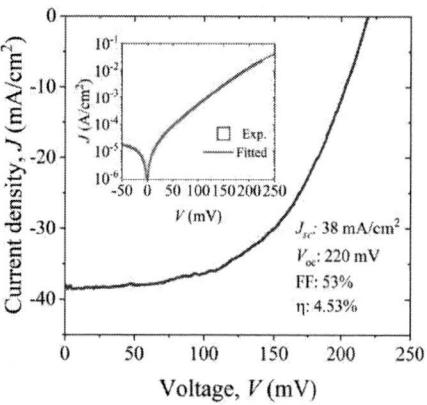

Figure 6: Device performance using the annealed ETL structure under 1-sun illumination. The inset image shows the dark J-V curves, which are fitted using a single-diode model.

Focusing first in the dark J-V curve (shown in the inset), it has been fitted using a single-diode model, with an ideality factor of n=1.17. The fact that this parameter in the exponential behavior of the device closely approaches unity demonstrates the feasibility of the furnace step to create high quality n^+/p junctions.

Under illumination, the device exhibits reasonable open-circuit voltage and fill factor. However, the short-circuit current falls below expectations. This is supported by the external quantum efficiency (EQE) measurements shown in Fig. 7, which exhibit a uniform drop across all absorption wavelengths. A plausible explanation can be found in the observed roughness of the a-Si layer after annealing causing photon scattering at the front surface, but further investigation will be required to identify and assess these optical losses.

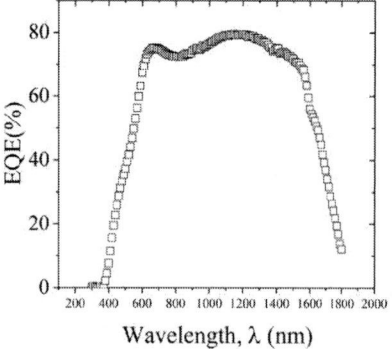

Figure 7: External quantum efficiency (EQE) measurement of the device using the annealed ETL structure.

4 CONCLUSIONS

In this work, we have optimized a structure consisting in the deposition of a phosphorus-rich amorphous silicon layer on a p-type c-Ge substrate, which is annealed at a high temperature. The thermal process induces phosphorus and silicon diffusion into the c-Ge substrate, creating an n^+/p homojunction which relaxes the need for complex c-Ge surface treatments. A high-quality c-Ge surface passivation with a conductive transport layer have been obtained with a 45 minute anneal at 700°C. A silicon-germanium intermixing has been observed, leading to the creation of a SiGe alloy, and a partial recrystallization of the amorphous layer. The device implemented with this structure exhibits promising photovoltaic figures and proves to be a feasible strategy in the development of epitaxial-free c-Ge TPV devices.

5 REFERENCES

[1] E. U. Onyegam *et al.*, "Exfoliated, thin, flexible germanium heterojunction solar cell with record FF=58.1%", *Solar Energy Materials and Solar Cells*, vol. 111, pp. 206–211, Apr. 2013.

[2] B. Hekmatshoar, D. Shahrjerdi, M. Hopstaken, K. Fogel, and D. K. Sadana, "High-efficiency heterojunction solar cells on crystalline germanium substrates", *Appl Phys Lett*, vol. 101, no. 3, Jul. 2012.

[3] L. F. Marsal, I. Martin, J. Pallares, A. Orpella, and R. Alcubilla, "Annealing effects on the conduction mechanisms of p+ -amorphous- $Si_{0.8}C_{0.2}$:H/n-crystalline-Si diodes", *J Appl Phys*, vol. 94, no. 4, pp. 2622–2626, Aug. 2003.

[4] D. E. Kane and R. M. Swanson, "Measurement of the emitter saturation current by a contactless photoconductivity decay method", in *IEEE 18th Photovoltaic Specialist Conference*, Oct. 1985, pp. 578–578.

[5] M. Gamel *et al.*, "Highly reflective and passivated ohmic contacts in p-Ge by laser processing of $aSiC_x$:H(i)/Al_2O_3/aSiC films for thermophotovoltaic applications", *Solar Energy Materials and Solar Cells*, vol. 265, p. 112622, Jan. 2024.

[6] J. van der Heide, N. E. Posthuma, G. Flamand, W. Geens, and J. Poortmans, "Cost-efficient thermophotovoltaic cells based on germanium substrates", *Solar Energy Materials and Solar Cells*, vol. 93, no. 10, pp. 1810–1816, Oct. 2009.

[7] J. Fernández, "Development of Crystalline Germanium for Thermophotovoltaics and High-Efficiency Multi-Junction Solar Cells [Thesis]", Konstanz University, 2010.

[8] A. Chroneos and H. Bracht, "Diffusion of n-type dopants in germanium", *Appl Phys Rev*, vol. 1, no. 1, p. 011301, Jan. 2014.

[9] H. H. Silvestri, H. Bracht, J. Lundsgaard Hansen, A. Nylandsted Larsen, and E. E. Haller, "Diffusion of silicon in crystalline germanium", *Semicond Sci Technol*, vol. 21, no. 6, p. 758, 2006.

[10] T. Deschaines, J. Hodkiewicz, and P. Henson, "Characterization of Amorphous and Microcrystalline Silicon using Raman Spectroscopy", 2009. [Online]. Available: https://api.semanticscholar.org/CorpusID:59380 550

[11] O. Pagès, J. Souhabi, V. J. B. Torres, A. V Postnikov, and K. C. Rustagi, "Re-examination of the SiGe Raman spectra: Percolation/one-dimensional-cluster scheme and ab initio calculations", *Phys Rev B*, vol. 86, no. 4, p. 45201, Jul. 2012.

[12] M. I. Alonso and K. Winer, "Raman spectra of c-$Si_{1-x}-Ge_x$ alloys", *Phys Rev B,* vol. 39, no. 14, 1989.

41st European Photovoltaic Solar Energy Conference and Exhibition

Sensitization of crystalline silicon with organic dye molecules

Lukáš Gdula [1], Branislav Dzurňák [1], Tom Markvart [1, 2]

E-mail: lukas.gdula@fel.cvut.cz, branislav.dzurnak@fel.cvut.cz, T.Markvart@soton.ac.uk

[1] Department of Electrotechnology, Faculty of Electrical Engineering,
Czech Technical University in Prague, Prague, Czech Republic
[2] School of Engineering Sciences, University of Southampton

MOTIVATION

- Reducing silicon thickness used in solar cells
- Charge generation in Si via non radiative (Förster) energy transfer (NRET/FRET)
- Organic dye molecules (donors) deposited on the surface of Si (acceptor)

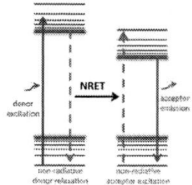

- Both FRET and photon tunnelling ongoing [2]

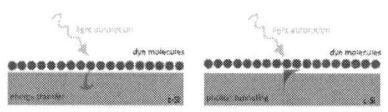

- Demonstrated by photoluminescence lifetime quenching
- Previously observed on quantum dots, Langmuir-Blodgett monolayers and covalently attached molecules on silicon [3, 4, 5, 6]

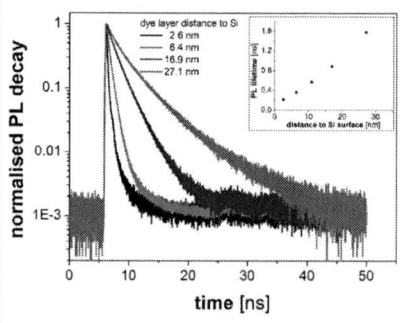

Fig. 4. Normalized photoluminescence decays at different dye-Si distances;
inset: fluorescence lifetime dependence od dye-Si distance

DYE DEPOSITION

- Spin coating method
- Hight quantum yield organic dye Lumogen F R305 – ultrathin layer d ≈ 5Å
- Spacer layer – PMMA variable thickness d = 2.6 – 27.1 nm
- Si, p-type, orientation <100>, resistivity 0.001 – 0.005 Ω.cm, 500 μm thick

Dye R305
↓PMMA
Silicon dioxide
Silicon

TIME-RESOLVED MEASUREMENTS

- Absorption and photoluminescence spectra of dye (Fig. 2)
- Photoluminescence lifetimes measured at spectral maximum (600 nm) by time correlated single photon counting (TCSPC) with 250 ps time resolution

- Liquid dye solution photoluminescence shows single exponential decay 5.8 ns, dye layers show double exponential decays (Fig. 3a, 3b)
- Significant lifetime quenching with decreasing dye layer distance from Si indicates efficient energy transfer (Fig. 4)

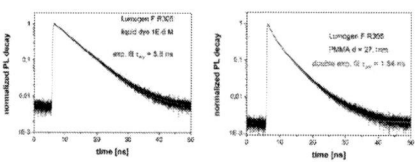

Fig. 3a. Photoluminescence decay of liquid dye in cuvette with (one exp. fitting)

Fig. 3b. Photoluminescence decay of thin dye layer at 27.1 nm distance from Si surface (two exp. fitting)

Average lifetime formula: $\tau_{AVZ} = \dfrac{\sum_i A_i \cdot \tau_i}{\sum_i A_i}$

DYE THICKNESS

- Variable angle spectroscopic ellipsometry (VASE) thickness measurement
- Uniform layers deposition confirmed

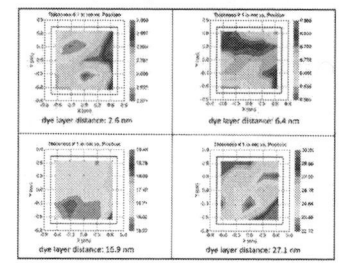

Fig. 1. Thickness of spacer layer (PMMA)

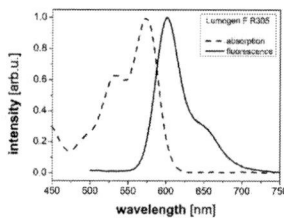

Fig. 2. Absorption and fluorescence spectra of dye solution (concentration 10-5 M)

- Spectral shift of dye photoluminescence at different distances form Si (Fig. 5) suggests possible reabsorption processes

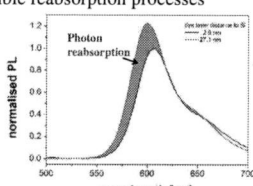

Fig. 5. Normalized photoluminescence spectra

CONCLUSION AND PLANS

- We achieved high precision of depositing ultrathin layers on Si surface
- Observed quenching of PL lifetime dependent on dye-Si distance indicates energy transfer by photon tunneling or FRET
- Photon reabsorption could occur in deposited thin layers
- Removing SiO₂ layer could enhance energy transfer

REFERENCES

[1] Gdula, L.; Dzurňák, B.; Prohaszová, A.; Markvart, T.; Ultrathin organic dye layers for sensitisation of silicon, Journal of Chemical Technology and Metallurgy, 2024, 59(2), 329-334. ISSN 1314-7978.
[2] L. Fang, K. Kiang, N. Alderman, L. Danos, T. Markvart, Photon tunneling into a single-mode planar silicon waveguide, Opt. Express 23, 24, 2015, A1528-A1532
[3] N. Liu, H. Chen, F. Chen, M. Wang, Förster resonance energy transfer from poly (9-vinyl carbazole) to silicon nanoparticles in their composite films, Chemical Physics Letters, 451, 1-3, 2008, 70-74
[4] A. Yeltik, B. Guzelturk, P. L. Hernandez-Martinez, A. O. Govorov, H. V. Demir, Phonon-Assisted Exciton Transfer into Silicon Using Nanoemitters: The Role of Phonons and Temperature Effects in Förster Resonance Energy Transfer, ACS Nano, 7, 12, 2013, 10492-10501
[5] M. I. Sluch, A. G. Vitukhnovsky, M. C. Petty, Anomalous distance dependence of fluorescence lifetime quenched by a semiconductor, Physics Letters A, 200, 1, 1995, 61-64
[6] L. Danos, T. Markvart, Excitation energy transfer rate from Langmuir Blodgett (LB) dye monolayers to silicon: Effect of aggregate formation, Chemical Physics Letters, 490, 4-6, 2010, 194-199

This work was supported by the project "The Energy Conversion and Storage", funded as project No. CZ.02.01.01/00/22_008/0004617 by Programme Johannes Amos Comenius, call Excellent Research.

23-27 Septmber 2024, Vienna (Austria)

  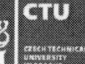

EU PVSEC 41st European Photovoltaic Solar Energy Conference and Exhibition

41st European Photovoltaic Solar Energy Conference and Exhibition

SELF-ORGANIZED FILMS OF CARBAZOLE DERIVATIVES ON STRUCTURED SILICON SUBSTRATES FOR PHOTOVOLTAIC APPLICATIONS

Sergii Mamykin[1], Daria Kuznetsova[1], Nina Roshchina[1], Petro Smertenko[1], Saulius Grigalevicius[2], Gintare Krucaite[2],
Raminta Beresneviciute[2], Simona Sutkuviene[3]
[1]V. Lashkaryov Institute of Semiconductor Physics National Academy of Sciences of Ukraine
41, prospect Nauky, 03028, Kyiv, Ukraine
[2]Department of Polymer Chemistry and Technology, Kaunas University of Technology, Radvilenu plentas 19, LT50254
Kaunas, Lithuania
[3]Department of Biochemistry, Faculty of Medicine, Lithuanian University of Health Sciences, Tilzes str. 18, 47181 Kaunas,
Lithuania
E-mail: mamykin@isp.kiev.ua

ABSTRACT: New organic molecules based on aryl-substituted carbazoles (RB-14, RB-18, RB-22, RB-24, RB-25, RB-27, RB-70, and RB-75) were used to form organic-inorganic hybrids on Si(100) n-type substrates with a resistivity of 2.7 Ohm*cm, structured by standard technology in acid solution. The aforementioned organic materials are commonly used to make organic LEDs. Here, these materials are used for the first time in a hybrid structure for solar energy conversion. Aryl-substituted carbazole-based films were deposited by spin-coating and self-organization in drops of 1% chloroform solution. The deposition process was carried out in a glove box at room temperature under an inert atmosphere. The layers were deposited at 500 rpm for 15 minutes. The morphological features of the deposited and self-organized films were studied by optical microscopy. Current-voltage characteristics were measured using a standard automated tester and then analyzed by the value of dimensionless sensitivity to recognize charge flow and recombination mechanisms. Two dominant forms of morphology were observed: square-like and network-like, which depend on the substituents of carbazole derivatives. Some hybrids have demonstrated a photoelectric effect. The mechanism of charge flow in the obtained hybridsis is bimolecular recombination. Additionally, the current-voltage characteristics of the obtained hybrids and the substitutional structure of organic compounds based on the substituted carbazole derivatives are discussed.
Keywords: aryl-substituted carbazoles; dimensionless sensitivity; organic-inorganic hybrid solar cells

1 AIM AND APPROACH

The aim of this study is to find opportunities to use substituted carbazole derivatives as organic/Si hybrids for photovoltaic applications. A number of carbazole derivatives are widely used in the fabrication of organic light emitting diodes [1]. Earlier [2], aromatic heterocyclic amines were proposed for use in organic on Si hybrids with energy conversion efficiency of about 8.5%. In addition, self-organization of organic molecules has been used for cheaper and simpler organic film formation techniques [2]. Here, we combine these methods for the formation of carbazole derivatives/Si hybrids. New organic molecules based on aryl-substituted carbazoles (RB-14, RB-18, RB-22, RB-24, RB-25, RB-27, RB-70, and RB-75) were used to form organic-inorganic hybrids on structured n-type Si (100) substrates with the resistivity of 2.7 Ohm*cm. The patterned Si substrates were prepared by standard technology in an acid solution. Films of the presented derivatives were deposited by spin-coating and self-organization in drops of 1% chloroform solution. Dark and illuminated (30 mW/cm2) current–voltage characteristics (CVCs) were measured using a standard automated tester with a measurement error of about 0.25% and then analyzed by the value of dimensionless sensitivity (DS) to recognize charge flow and recombination mechanisms [3].

2 SCIENTIFIC INNOVATION AND RELEVANCE

The search for new materials to improve the efficiency of organic/Si hybrids and reduce their cost is expanding to include substituted carbazole derivatives. These organic materials are well known, in particular for the creation of organic LEDs. Using the same materials to convert electrical energy into light energy and vice versa opens up new possibilities in printed electronics, where, for example, a limited number of chemicals can provide a wide range of functions for printing devices. This study attempts to combine the best features of simple methods: fabrication of thin films at room temperature, application of chemical baths and spin-coating, with a new approach to analyze the fine behavior of current-voltage characteristics: the use of dimensionless sensitivity for injection and recombination diagnostics [3].

3 RESULTS

Chemical structures and surface morphology of the carbazole based derivatives films on patterned Si substrate are presented in Table 1. Two morphology images are presented for spin-coated and self-organized from the drop films. Table 1 clearly shows at least two different forms of surface morphology: square-like (samples RB-14, RB-22, RB-24 and RB-27) and network-like (samples RB-18, RB-25, RB70 and RB-75).

The current-voltage characteristics of carbazole derivative hybrids on patterned Si substrates can also be divided into two groups: structures with photovoltaic effect (RB-18, RB-70, RB-75 both for spin-coated and self-assembled films) and structures without a photovoltaic effect. Typical dark and illuminated CVCs in both directions are shown in Fig. 1. The CVCs of the first group of samples are characterized by the following features: (i) the curve of the "-" polarity on Si exhibits predominantly rectifying behavior with DS = 0.5, Fig. 1, a-c; (ii) the DS of the curve increases up to DS = 1.5 in the range from 0.1 V to 1.0 V, which corresponds to the bimolecular recombination, i.e., the presence of a sufficient number of both types of charge carriers in the structure; (iii) the DS of the curves decreases to DS=0.5 at large bias from 1.0 V to 10.0 V; (iv) the curves with "+" polarity on Si show the photovoltaic effect at V_{OC}=(0.08 - 1.0) V, Fig. 1, a; (v) all curves start with DS=1, which corresponds to Ohm's law, Fig. 1, a, d; (vi) dark curves have DS≈1.5 at V=(2-3) V, which corresponds to bimolecular recombination; and (vii) illuminated curves have DS≈1.0 at V=(2-3) V, what corresponds to current restriction by the contact; (viii) DS curve increases at large bias voltages from 3.0 V to 10.0 V, which suggests that the contacts in these structures have a good injection capacity. As for the

CVCs of the second group of samples, they have the following features: (i) all curves have ohmic behavior with DS = 1.0 at low voltages on both polarities, Fig. 2, a-d; (ii) the dark and illuminated curves of the "-" polarity on Si exhibits rectifying behavior with DS = 0.5 at high voltages, Fig. 2, a-c; (iii) for the illuminated curves, as well as for the curves of the first group, the DS of the curves increases to DS=1.5 in the range from 0.1 V to 1.0 V, which corresponds to bimolecular recombination; (iv) the DS of the curve with polarity "+" on Si at high bias is 1.5 in the dark (Fig. 2, d) and 1.0 under illumination (Fig. 2, e).

Table 1: Chemical structure and surface morphology of substituted carbazole derivatives films on patterned Si substrate

Material	Chemical formula	Spin Coater	Drop
1	2	3	4
RB-14			
RB-18			
RB-22			
RB-24			
RB-25			
RB-27			

41st European Photovoltaic Solar Energy Conference and Exhibition

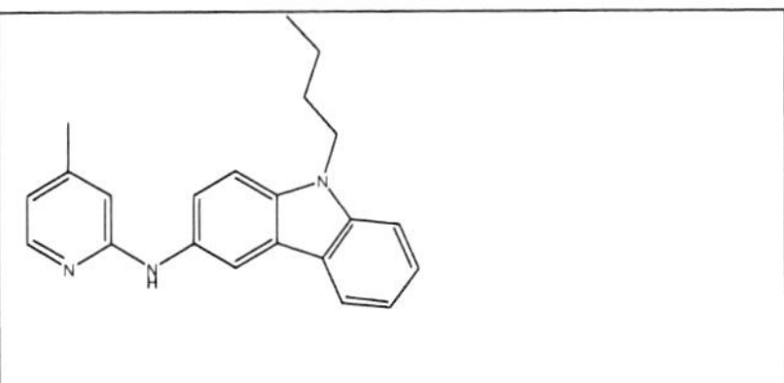

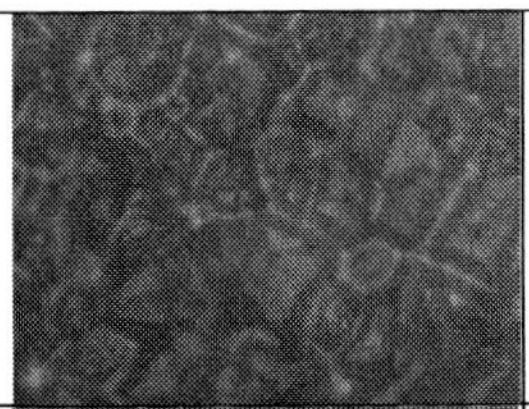

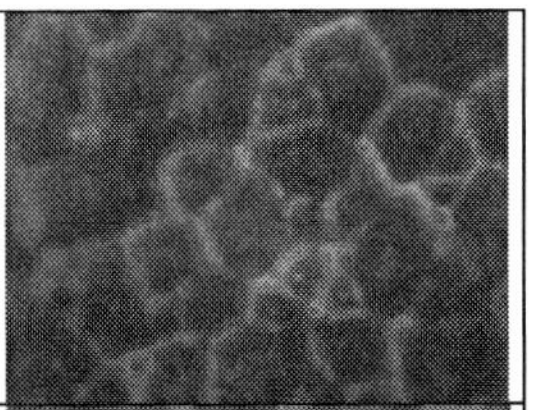

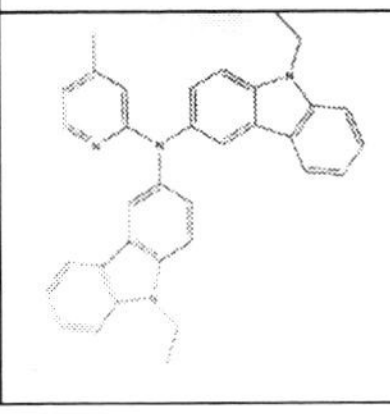

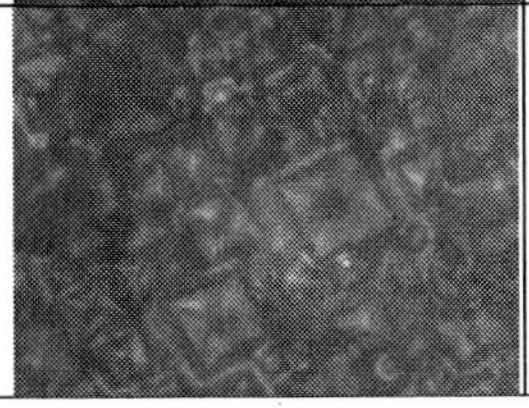

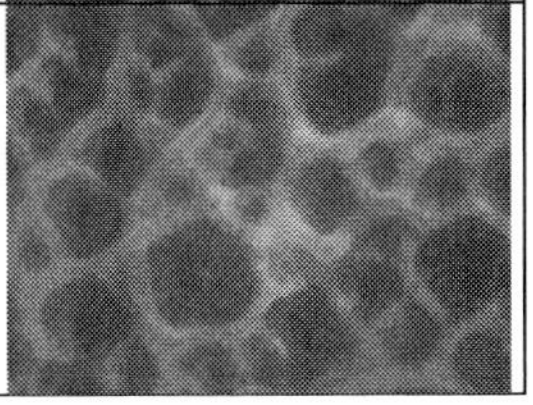

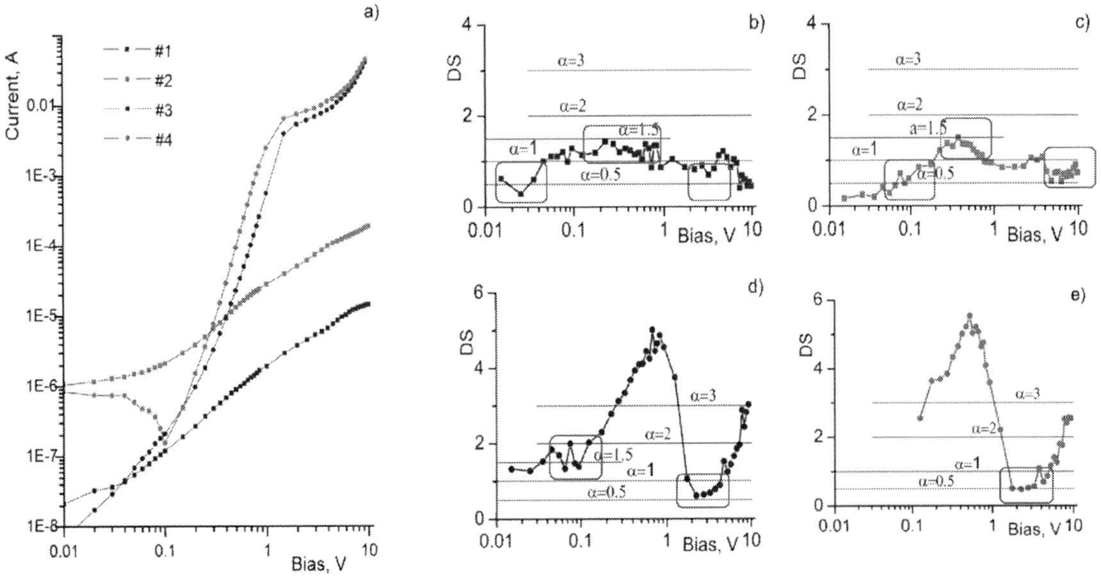

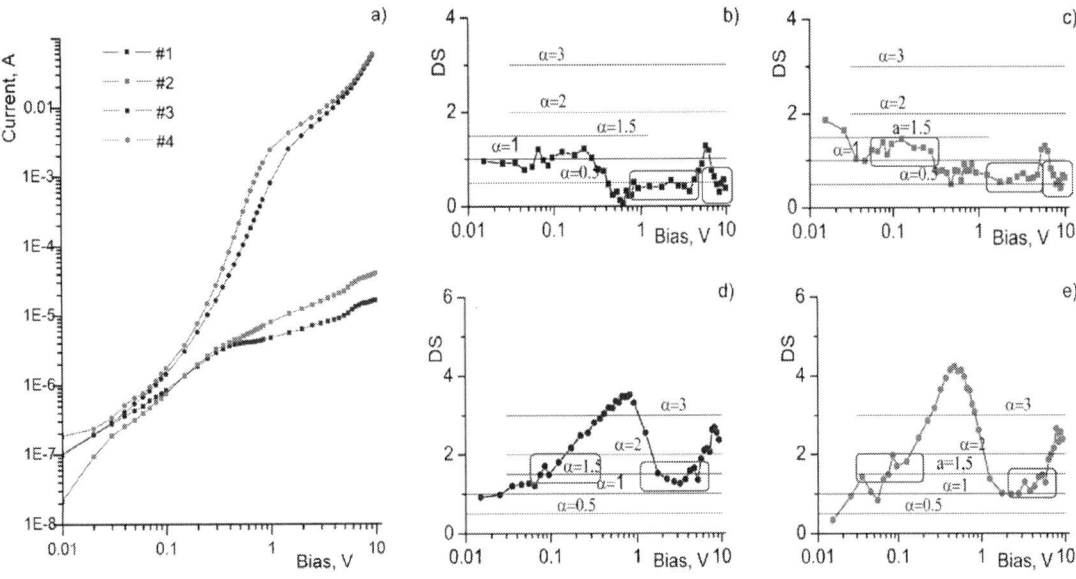

Figure 1: Typical CVCs (a) and their DSs alpha (b-e) for the first type of curves with photovoltaic effect.

Figure 2: Typical CVCs (a) and their DSs alpha (b-e) for the second type of curves without photovoltaic effect: 1, 3 is dark, 2, 4 is light, 1, 2 is "-"silicon, 3, 4 is "+" silicon.

4 CONCLUSION

The morphology of self-organised and spin-coated films of carbazole derivatives showed two different forms of surface morphology: square-like and network-like. The current-voltage characteristics of carbazole derivative hybrids on patterned Si substrates can also be divided into two groups: structures with photovoltaic effect (V_{OC}=(0.08 - 1.0) V) and structures without it. For the first group, the contacts have good injection capability.The CVCs of the second set of samples showed ohmic behavior at low voltages in both directions under dark and light conditions. All carbazole derivative hybrids used in this study have three main behaviors: Ohm's law, rectifying and bimolecular recombination. This makes these hybrids promising for use in solar cells.

FUNDING

This work was supported by Research Council of Lithuania and NATO SPS project G6197 "Plasmonically Enhanced Perovskite Thin-Film Solar Cells".

REFERENCES

[1] S. Grigalevicius, D. Tavgeniene, G. Krucaite, D. Blazevicius, R. Griniene, Yi-Ning Lai, Hao-Hsuan Chiu, Chih-Hao Chang. Optical Materials, 79 (2018) 446.
https://doi.org/10.1016/j.optmat.2018.04.018.

[2] T.Ya. Gorbach, P.S. Smertenko, E.F. Venger, Ukrainian Journal of Physics, 59 (2014) 601. https://doi.org/10.15407/ujpe59.06.0601.

[3] S.V. Mamykin et al. 40th European Photovoltaic Solar Energy Conference and Exhibition (EU PVSEC-40), September 26-30, 2023, Lisbon, Portugal, 2BV.1.12. pp. 020083-001- 020083-005.

41st European Photovoltaic Solar Energy Conference and Exhibition

PLACEMENT ANGLES FOR LUMINESCENT SOLAR CONCENTRATORS: SIMULATING AND EXPERIMENTING WITH BIFACIAL PHOTOVOLTAIC MOSAIC DEVICES

Xitong Zhu[1, *], Frits Reijners[1], Michael Debije[2], and Angèle H.M.E. Reinders[1]
1. Energy Technology Group, Department of Mechanical Engineering, Eindhoven University of Technology, Eindhoven, Netherlands
2. Department of Chemical Engineering and Chemistry, Eindhoven University of Technology, Eindhoven, Netherlands
*e-mail address: x.zhu@tue.nl

ABSTRACT: In this paper, we simulated and measured the performance of luminescent solar concentrator photovoltaic (LSC-PV) devices. Our model and experiments aimed at optimizing the performance of LSC-PV devices of the mosaic series upon variation of their interconnection scheme and angle with respect to irradiance source for various application scenarios. All the simulations and measurements are based on a module we designed earlier, namely the LSC bifacial photovoltaic mosaic device. In the first stage, we used ray-tracing simulation to simulate the optical performance of the LSC cubes and used a computational electrical model to simulate the electrical performance of the entire LSC PV device. In the second stage, we used a solar simulator to measure the performance of the LSC PV devices. In the third stage, the result of simulation and measurement were analyzed to explore the effect on LSC PV device orientation and the limitations of simulation. The results show the simulation mirrors to trends of the measurements and both show the LSC PV device has its highest performance at 40 degrees placement.

Keywords: Luminescent solar concentrator, bifacial cells, ray-tracing simulation, placement angle, experiment measurement

1 INTRODUCTION

Luminescent solar concentrator (LSC) devices have been in development for nearly fifty years. Their primary application is to realize a transparent, reduced cost photovoltaic system, and to improve photovoltaic efficiency under diffuse light [1],[2]. Even though luminescent solar concentrator photovoltaics (LSC-PVs) have the advantage of capturing diffuse light, its placement angle still affects its performance. Therefore, it is necessary to explore the impact of placement angle on the performance of LSC PV.

In this study, we simulated and measured the performance of LSC-PV devices. Our model and experiments aimed at optimizing the performance of LSC-PV devices on the basis of variation of their angle towards an irradiance source in various application scenarios. All the simulations and measurements are based on a module we designed earlier [3], [4], namely the LSC bifacial photovoltaic mosaic device.

2 AIM

The purpose of this research includes the following three parts:
1. Model complex structure LSC PV devices and perform optical simulations.
2. Establish a general electrical model based on optical simulation results and spectral response of PV cells for future simulation research of complex structured LSC PV devices.
3. The impact of placement angle on the performance of LSC PV devices was explored through laboratory measurements and computer simulations.

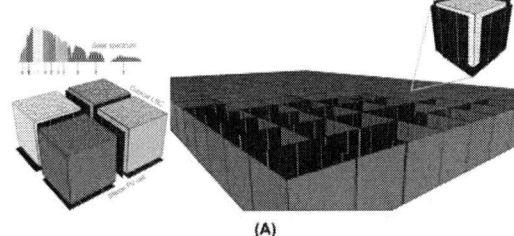

Figure 1: (A) Left: schematic of a mosaic luminescent solar concentrator photovoltaic (LSC-PV) device with edge- and bottom-mounted bifacial and monofacial PV cells. Right: a mosaic module consisting of interconnected cubical LSC PV units. (B) Photographs of (left) a mock-up of a small array of 3 by 5 cubical LSCs and (right) a 10 by 10 device which visually represent mini-PV-modules [3].

Figure 2: 3x3 LSC PV model in LightTools and 3x3 LSC-PV sample.

3 METHOD

In the first stage, we used LightTools ray-tracing simulation to simulate the optical performance of the LSC cubes and used a Matlab electrical model to simulate the electrical performance of the entire LSC PV devices. In the second stage, we used a NeonSee solar simulator to measure the performance of LSC PV devices. Finally, in the third stage, the result of simulation and measurement was analyzed to explore the effect on LSC PV devices of placement angle and the limitations of simulation.

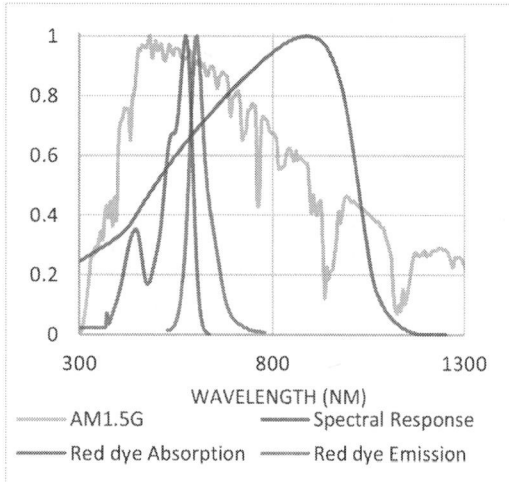

Figure 3: Normalized AM1.5G spectral distribution, spectral response for typical silicon PV cells and the absorption and emission curve of the red dye used in this work.

In the first stage, the optical simulation model was built up following the same method as previous described[5]. For the electric model, as shown in Figure 3, given the absorption and emission characteristics of the dye, the power conversion efficiency (PCE) for LSC-PV device is calculated by the following formula:

$$\eta_{PCE} = \frac{\sum P}{\dot{Q}_{in}}$$

$$= \frac{\sum_{i=1}^{n=total\ PV\ cells} \int_{\lambda=0}^{4500\ nm} S_{PV}^i (\lambda)\ SR\ (\lambda)\ FF\ Voc\ A_{PV}^i\ d\lambda}{\int_0^\infty S_{ap}(\lambda) A_{ap}\ d\lambda}$$

where:
P is the power production of each cell (W),
$\dot{Q}_{in}$ is the radiant flux which falls on the aperture area (W),
S is the spectrally distributed irradiance (W/m²nm),
SR is the spectral response of the spectral distribution (A/W),
FF is the fill factor of the solar cell,
Voc is the open-circuit voltage of the solar cell (V),
A_{PV} is the area of the solar cell (m²),
A_{ap} is the area of the aperture (m²),
S_{ap} is the spectral distributed irradiance on the aperture area (W/m²nm), and
S_{PV} is the spectral distributed irradiance on each solar cell (W/m²nm).

We used the application programming interface (API) to import the optical simulation results from LightTools into Matlab and built the electrical model based on the formula of η_{PCE} to calculate the energy output and energy

conversion efficiency of the LSC PV devices. To compare the results between simulation and measurement, we collected the spectral distribution data from the NeonSee solar cell analysis system (Figure 4) and input the data into the LightTools light source setting.

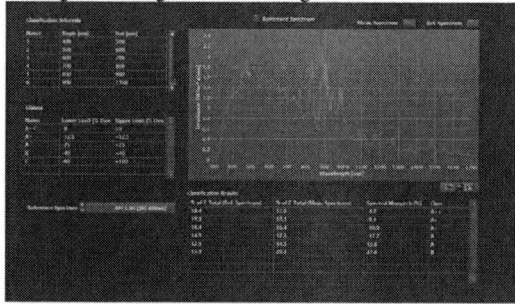

Figure 4: Spectral data used in this research form the output of NeonSee solar cell analysis system[6]

4 RESULTS

Figure 5 below shows the optical power incident on different regions of cells attached to the LSC cube under two different placement angles.

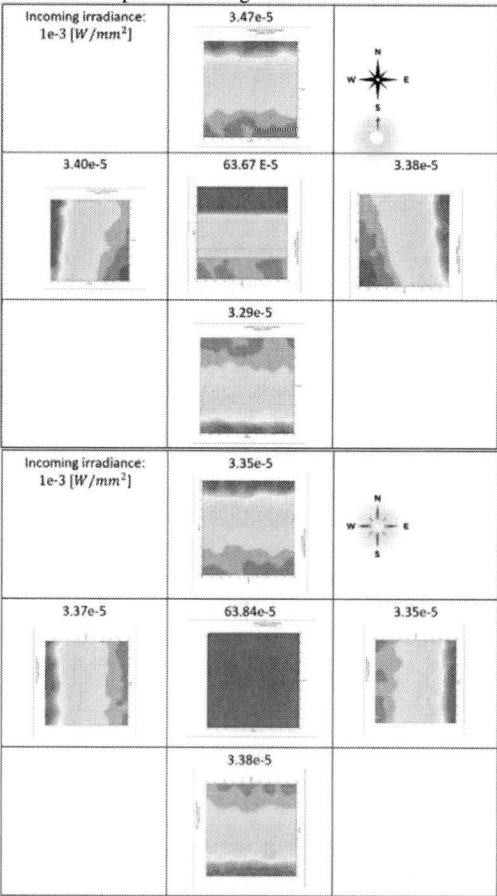

Figure 5: Simulated data from LightTools of cells along central cube, with the centre plot corresponding to the bottom cell. Above: placement angle of 65 degrees. Bottom: sun directly overhead. The value shown above each graph is the average irradiance received by the cell in W/mm².

Figure 6 shows the electrical performance of 3x3 LSC Mosaic PV device with series-parallel wiring under different angles of incidence of the light source. By analyzing the measurement and simulated results, we ascertain the 3x3 cm LSC PV device has the greatest electrical output when placed on a 40-degree plane under direct light.

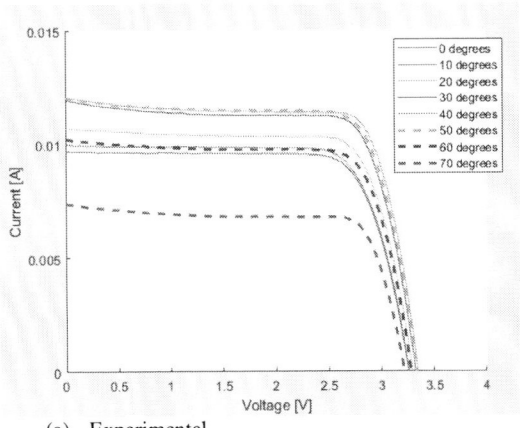

(a)　Experimental

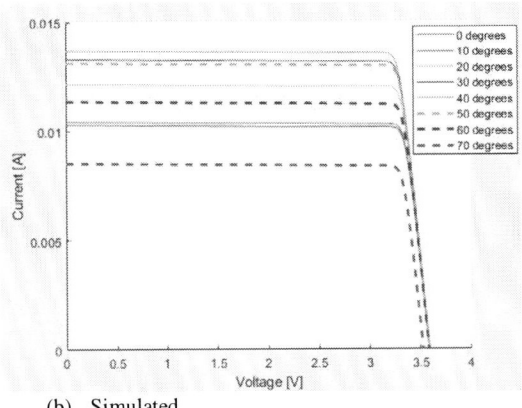

(b)　Simulated

Figure 6: IV curves of LSC PV device with series-parallel wiring under different illumination incidence angles (0 degrees represents the sun being directly overhead)

5　CONCLUSIONS

A more complete LSC-PV device simulation model including optical and electrical components was established. Through simulations and experimental measurements, it was found that the mosaic LSC with bifacial cell PV devices has the greatest electrical output when oriented at an angle of 40 degrees to the horizontal. In the subsequent design and deployment of LSC-PV devices, the placement angle and location of attached cells should be carefully considered to obtain higher electrical conversion efficiency.

REFERENCES

[1] M. G. Debije and P. P. C. C. Verbunt, "Thirty years of luminescent solar concentrator research: Solar energy for the built environment," *Adv. Energy Mater.*, vol. 2, no. 1, pp. 12–35, Jan. 2012, doi: 10.1002/aenm.201100554.

[2] M. Rafiee, S. Chandra, H. Ahmed, and S. J. McCormack, "An overview of various configurations of Luminescent Solar Concentrators for photovoltaic applications," *Opt. Mater. (Amst).*, vol. 91, no. March, pp. 212–227, 2019, doi: 10.1016/j.optmat.2019.01.007.

[3] M. Aghaei, R. Pelosi, W. W. H. Wong, T. Schmidt, M. G. Debije, and A. H. M. E. Reinders, "Measured power conversion efficiencies of bifacial luminescent solar concentrator photovoltaic devices of the mosaic series," *Prog. Photovoltaics Res. Appl.*, vol. 30, no. 7, pp. 726–739, 2022, doi: 10.1002/pip.3546.

[4] M. Aghaei, X. Zhu, M. Debije, W. Wong, T. Schmidt, and A. Reinders, "Simulations of Luminescent Solar Concentrator Bifacial Photovoltaic Mosaic Devices Containing Four Different Organic Luminophores," *IEEE J. Photovoltaics*, vol. 12, no. 3, pp. 771–777, 2022, doi: 10.1109/JPHOTOV.2022.3144962.

[5] X. Zhu, M. G. Debije, and A. H. M. E. Reinders, "Simulation of the Effects of Geometry on Performance of Luminescent Solar Concentrator Photovoltaic Devices," *IEEE J. Photovoltaics*, vol. 14, no. 1, pp. 116–122, 2023, doi: 10.1109/JPHOTOV.2023.3323821.

[6] "neonsee.com." https://www.neonsee.com/en/technology/#irradiance-control.

Low Emissive Molybdenum-Doped ITO for High Vacuum Photovoltaic-Thermal applications

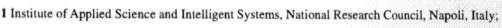

Daniela De Luca[1,2], Umar Farooq[1,3], Paolo Strazzullo[1,4], Eliana Gaudino[1], Antonio Caldarelli[1], Anna Krammer[5], Andreas Schueler[5], Marilena Musto[1,4], Emiliano Di Gennaro[1,3], Roberto Russo[1]

1 Institute of Applied Science and Intelligent Systems, National Research Council, Napoli, Italy;
2 Nanyang Technological University, Energy Research Institute,638798 Nanyang ave, Singapore
3 Physics Department "E. Pancini", University of Napoli "Federico II", Napoli, Italy;
4 Industrial Engineering Department, University of Napoli "Federico II", Napoli, Italy;
5 Swiss Federal Institute of Technology (EPFL),1015 Lausanne, Switzerland

Roberto.Russo@na.isasi.cnr.it

Abstract: The incorporation of thermal insulation through high-vacuum can lead to higher thermal conversion efficiency in PV-T systems [1]. Transparent Conductive Oxide (TCO) coatings, particularly Indium Tin Oxide (ITO) films, have been developed to fully exploit the advantages of high-efficiency vacuum insulation [2]. Molybdenum-doped ITO (ITMO) is a promising candidate for enhancing the performance of ITO in High Vacuum Photovoltaic Thermal (HV PV-T) devices [3]. Notably, ITMO exhibits low reflectivity in the visible and near-infrared regions, maximizing light absorption and energy conversion efficiency

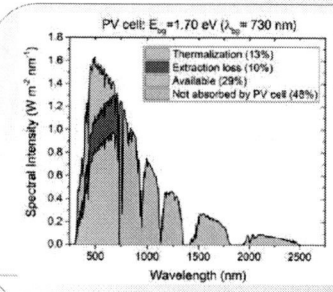

All energy that is not converted or absorbed by the PV cell can be used for thermal energy production

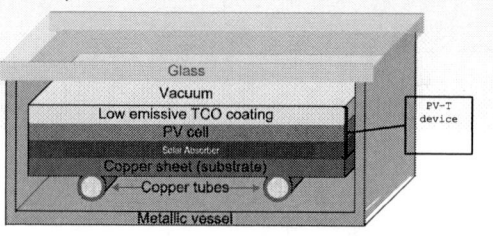

OBJECTIVE: Develop ultra-low emissive TCO films for HV PV-T systems ⟶ Incorporating Mo into the ITO thin films (**ITMO**), may increase the number of carriers and their mobility

We systematically investigate the variations in optical and electrical properties of ITMO thin films by analyzing samples deposited by magnetron sputtering under different doping concentrations, coating conditions, and thermal annealing treatments (250 °C, 300 °C, 350 °C).

Deposition Parameters of ITMO Thin Films Fabricated on Glass (G) and Silicon (S) Substrates

Coating	Deposition Temperature	A_r/O_2 (sccm)	Mo Power (W)
1	20 °C	18/6	35
2	20 °C	18/6	30
3	20 °C	18/6	20
4	20 °C	18/6	15
5	20 °C	18/6	10
6	250°C	18/6	15

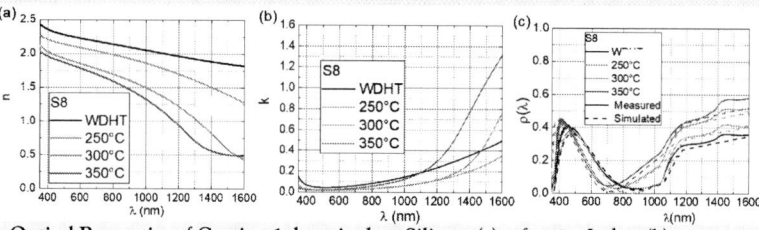

Optical Properties of Coating 1 deposited on Silicon: (a) refractive Index, (b) extinction coefficient, and (c) comparison of measured and simulated reflectivity for samples without thermal treatment (WDHT) and with thermal annealing at 250 °C, 300 °C, and 350 °

The emissivity of deposited samples is attained by using the formula: $\varepsilon(\lambda)=1-\rho(\lambda)-\tau(\lambda)$

Thermal Emittance at temperature T (K): $\varepsilon(T) = \dfrac{\int_0^\infty \varepsilon_\lambda\, E_{bb}(\lambda, T)\, d\lambda}{\sigma T^4}$

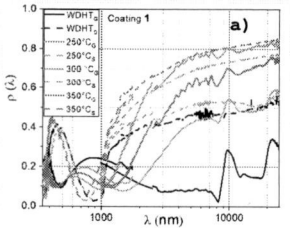

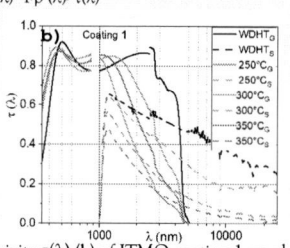

Reflectivity $\rho(\lambda)$ (a) and transmissivity $\tau(\lambda)$ (b) of ITMO coating 1 on glass (G) and silicon (S) samples without and with thermal annealing at various temperatures. Measurements were performed in the visible (VIS), near-infrared (NIR), and mid-infrared (MIR) regions using integrating sphere, spectrophotometer and FTIR.

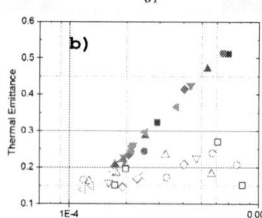

(a) Thermal Emittance vs. thermal annealing temperature for ITMO samples fabricated on glass (full symbols) and silicon (empty symbols).
(b) Thermal Emittance vs. resistivity for samples fabricated on glass (full symbols) and silicon (empty symbols)

Bibliography:
1. A. Mellor et al.,Solar Energy, vol. 174, pp. 386–398, nov. 2018.
2. Alonso-Álvarez et al, Solar Energy, 155, 82-92.
3. M. Micali et al, Solar Energy Materials and Solar Cells, vol. 221, p. 110904, mar. 2021.
4. D. De Luca et al., Applied Energy, vol. 352, sep. 2023.

OPTIMIZATION OF A PLANAR PEROVSKITE SOLAR CELL LAYER THICKNESSES: OPTICAL AND ELECTRICAL EFFECTS

Aleksi Kamppinen, Kati Miettunen
Department of Mechanical and Materials Engineering, University of Turku
Vesilinnantie 5, 20500, Turku, Finland

ABSTRACT: In this paper, the layer thicknesses of a planar perovskite solar cell (PSC) are computationally optimized. Layer thickness optimization is important for all thin film solar cells to maximize photogeneration and charge extraction which typically react in opposite ways to an increasing absorber layer thickness. Moreover, charge transport layers also affect device operation. The thickness of one functional layer may affect the optimal thickness of another. The thicknesses of perovskite (PVK) and hole transport layer (HTL) showed correlated behavior meaning that the optimal PVK thickness changed by altering the HTL thickness. Correlation between the preferred thicknesses of the electron transport layer (ETL) and PVK or HTL was not observed. Instead, the thinnest ETL resulted in the best performance in all cases. The correlation between the optimal PVK and HTL thicknesses mainly arose from device optics, that is, absorption and photogeneration, and the correlation sustained in optoelectronic operation. Charge transport model was, however, required to find the true optimum because PVK thickness is not bounded by the optical operation. Based on the results, it is recommended to optimize rear transport layer and PVK thicknesses simultaneously considering both optical and electrical operation and minimize front transport layer thickness.
Keywords: perovskite solar cells, optoelectronic modeling, layer thickness optimization

1 INTRODUCTION

Solar energy is a promising option meeting the increasing need for renewable and carbon-free energy. Given the diversity of potential applications of solar cells, such as from small to large scale, from outdoors to indoors, and rigid or flexible substrates, there may also be variety in the different types of solar cells meeting the need of a specific application. Perovskite solar cells (PSCs) have attracted attention by their high power conversion efficiency (PCE) [1] as well as low-temperature and solution processable preparation providing potential for low-cost and low energy demanding manufacture [2-4]. PSCs also are an interesting research topic due to their versatility in device design, including for example a number of possible materials and device structures.

Planar structure, which was studied here, is a common PSC device structure where different homogeneous functional layers form a multilayer stack of thin planar films. Planar multilayer structure also describes other thin film solar cells [5]. Thus, the considered structure is common across the different thin film solar cell technologies despite the different materials.

The need for layer thickness optimization is also very general. The optimal absorber thickness is typically found as a compromise between charge generation and extraction. Photogeneration (G_{ph}) is defined by light absorption which benefits from thick layers whereas the charge extraction is typically more efficient for thinner layers due to increasing recombination (R) with the thicker film. Therefore, both optics (absorption) and electrics (charge extraction) need to be considered. From the computational point of view, coupled optoelectronic modeling is needed.

Furthermore, it is not only the absorber thickness that needs to be optimized. Charge transport layers also affect the device performance. Their thicknesses have been reported affecting the optical optimum of perovskite (PVK) layer thickness [6]. While the layer thickness optimization of PSCs is common, it is rare to consider all the different layer thicknesses together instead of, for example, one layer at a time. Optimizing layer thicknesses individually may provide the global optimum if there are no correlations between the thicknesses of different layers and there is only one local optimum, i.e. the global one. However, if the layer thicknesses correlate with each other and especially if there are many local optima, one being the global one, the thicknesses should be optimized considering all the correlated layers at once.

Here, the layer thicknesses of a typical planar PSC are computationally optimized by applying optoelectronic simulations based on transfer matrix and finite element methods. While optimizing the layer thicknesses of a well-known PSC, the focus is to study the correlations between the different layer thicknesses and specify the optical and electrical origins of the preferred layer thickness combination.

2 METHODS

2.1 Simulated device

The modeled planar PSC consisted of SiO$_2$/ ITO/ SnO$_2$/ CH$_3$NH$_3$PbI$_3$ (MAPI)/ Spiro-OMeTAD/ Au layers as shown in Figure 1. The optical properties, i.e. complex refractive indices, were taken from the literature [7-10]. The contacts were modeled as ideal Ohmic contacts and SnO$_2$, MAPI, and Spiro-OMeTAD as semiconductors with the electrical material properties given in Table I: relative permittivity (ε), band gap (E_g), electron affinity (χ), acceptor/donor density (N_A/ N_D), conduction/valence band effective density of states (N_c/N_v), electron/hole mobility (μ_e/μ_h), electron/hole lifetime (τ_e/τ_h), carrier thermal velocity (v_{th}), carrier capture cross-section ($\langle \sigma \rangle$), and bulk/interface trap density ($N_{t,bulk}$/$N_{t,interface}$).

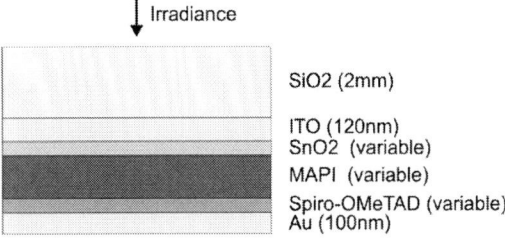

Figure 1: Modeled PSC device. Figure not to scale.

Table I: Applied parameters of the modeled semiconductor materials

	SnO$_2$	MAPI	Spiro-OMeTAD
ε	9 [11]	30	3 [12]
E_g (eV)	3.6 [11]	1.57 [9]	3 [12]
χ (eV)	4.1	3.93 [12]	2.45 [12]
N_A (cm^{-3})	-	10^{13} [12]	2×10^{18} [12]
N_D (cm^{-3})	10^{17} [13]	-	-
N_c (cm^{-3})	4.36×10^{18} [13,14]	2.2×10^{18} [12]	10^{20} [15]
N_v (cm^{-3})	2.52×10^{19} [13,14]	1.8×10^{19} [12]	10^{20} [15]
μ_e (cm^2/Vs)	95 [16]	10 [17]	2
μ_h (cm^2/Vs)	25 [13,18]	10 [17]	10^{-3} [19]
τ_e (ns)	2.83×10^7 [13,20]	*	0.1 [15]
τ_h (ns)	2.83×10^7 [13,20]	*	0.1 [15]
v_{th} (cm/s)	-	10^7* [17]	-
$\langle\sigma\rangle$ (cm^2)	-	10^{-15}* [17]	-
$N_{t,bulk}$ (cm^{-3})	-	5×10^{15}* [17]	-
$N_{t,interface}$ (cm^{-3})	-	5×10^{16}* [17]	-

*For MAPI, local charge carrier lifetime based on local trap density was applied [21].

2.2 Simulation methods

The device model applied transfer matrix method [22,23] for the optics (implemented in MATLAB) and drift-diffusion equations for the charge transport (COMSOL Multiphysics). The transfer matrix method allows, for example, computing reflection and local absorption by wavelength. The local absorption gives the photogeneration as a function of position which was fed into the drift-diffusion model. Steady-state voltage and charge carrier concentrations were solved by the drift-diffusion model providing information, for example, about current density and recombination. More detailed description of the model implementation is given in our recent article [21]. Simulations were conducted at constant 25 °C lattice temperature and perpendicular irradiation was considered.

3 RESULTS

3.1 Hole transport layer

Multiple local PVK thickness optima appeared for every considered hole transport layer (HTL) thickness when maximum achievable photocurrent (MAPC) was optimized, as shown in Figure 2a. MAPC is a direct conversion of G_{ph} to photocurrent, and thus, represents the optical, absorption-based figure of merit. The multiple local optima arise from oscillations in G_{ph} as a function of layer thicknesses. The oscillations are caused by the interference of the forward- and back-propagating waves of optical electric and magnetic fields. Changes in the path length of light in the different layers may affect the local optical electric field intensity in the PVK layer, and consequently, absorption, photogeneration, and MAPC. Therefore, interference gives rise to the local maxima and the presented correlation between the layer thicknesses.

Considering the local PVK thickness optima for each hole transport layer (HTL) thickness separately, the preferred PVK thickness decreased with the increasing HTL thickness (Figure 2a). Thus, the optimal PVK thickness depended on HTL thickness as reported in the literature [6]. However, these are not real local optima in the sense that MAPC continuously increased toward

thicker PVK and thinner HTL layers resulting in all the local maxima, including the global one, lie on the edge of the search space (Figure 2a). Therefore, even though HTL thickness affected the optimal PVK thickness via optics, the purely optical consideration unlikely suffices to find the true optimal thickness combination. That is because higher MAPC values are always expected to occur by increasing the search limit of especially PVK thickness despite the fact that the marginal gain of the thicker layer is likely to decrease with the ever increasing PVK thickness.

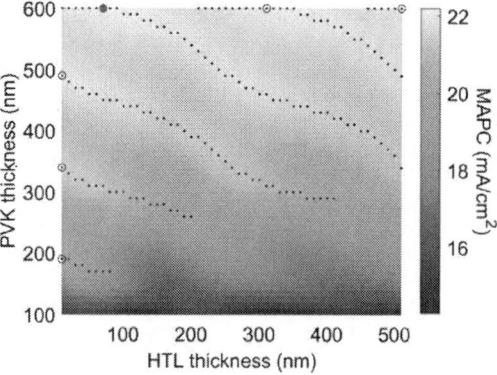

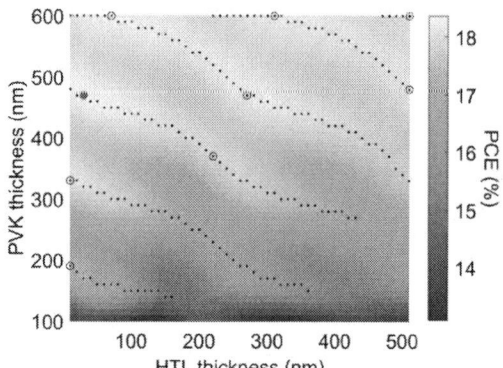

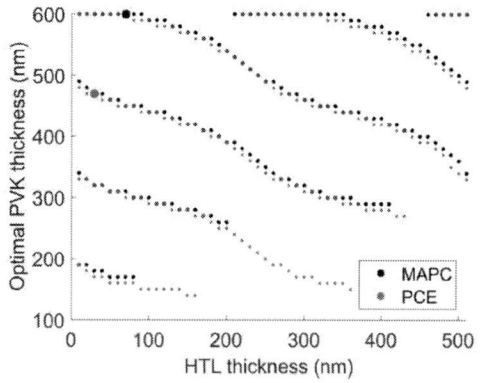

Figure 2: a) MAPC and b) PCE as a function of HTL and PVK thicknesses. Local maxima for each HTL thickness separately (dots) and for both thicknesses (circles) as well as the global maximum within the search space (red filled circle) are noted. c) Optimal PVK thickness as a function of HTL thickness when optimizing MAPC and PCE (small dots refer to the local PVK thickness optima of specific HTL thickness and large dots to the global optimum). ETL thickness of 35 nm was applied for this figure.

The electrical model was coupled to the optical one enabling the optimization of PCE. As shown in Figures 2b and 2c, the different objective function did not change the correlation between PVK and HTL thicknesses. HTL thickness still affected the preferred PVK thickness and the HTL specific PVK optima were very close if not the same as in the case of MAPC (Figure 2c). Thus, the correlation between HTL and PVK thicknesses arising from the optics was also observed for the optoelectronic figure of merit, that is, PCE (Figures 2b and 2c). Unlike in the case of MAPC, a few real local optima of PCE were observed: Three local optima were found within the interior points of the search space in addition to those on the edge (Figure 2b). PCE is naturally bound and starts declining with the increasing PVK thickness beyond the optimal range as should be expected. Because this is not the case for MAPC, electrical operation needs to be considered in the layer thickness optimization in addition to the optical despite the fact that the device optics alone gives very similar values for the local optima.

3.2 Electron transport layer

No correlation between the electron transport layer (ETL) thickness and the optimal PVK thickness was observed, as shown in Figure 3. There were three to four local PVK thickness optima (ca. 140 nm, 280 nm, 420 nm, and 570 nm) for each ETL thickness and these distinct optima were all basically equal independent of ETL thickness. Only very minor, 10 nm (one step size) differences in the local PVK thickness optima occurred for the different ETL thicknesses. The different objective functions (MAPC and PCE) also provided very similar local PVK optima (Figure 3b). For all ETL thicknesses, the difference in the local PVK thickness optima was at most 10 nm between MAPC and PCE.

It should be noted that the HTL thickness applied in the Figure 3 (160 nm) determined the specific optimal PVK thickness values shown in the figure due to the HTL-PVK thickness correlation as shown above (Figure 2). This explains the different global PVK thickness optima in Figures 2 and 3. Instead of the specific optimal value, the focus should be in the uncorrelated nature of optimal PVK and ETL thicknesses. The result is somewhat different to a previous study in the literature [6] where the thickness of TiO_2 front transport layer (FTL) affected the optimal PVK thickness. The difference likely arises from the different refractive indices of the different transport layers. In the previous study, lower refractive index materials did not either show correlation between the optimal FTL and PVK thicknesses which agrees with the present results. The independence of the optimal layer thicknesses allows optimizing these layer thicknesses independently.

In addition to the optimal PVK thickness being independent of ETL thickness, PCE increased monotonically toward thinner ETLs for all PVK thicknesses (Figure 3a). Thus, ETL thickness should be minimized.

3.3 The role of finite charge extraction

As shown in Figures 2c and 3b, the local PVK thickness optima were very similar between MAPC and PCE, but if there was a change, PCE optima occurred at thinner PVK thickness. In the case of PVK-HTL dependence (Figure 3c), global maximum shifted more than the local maxima because totally different local maximum was preferred for PCE compared with MAPC.

The difference between the purely optical (MAPC)

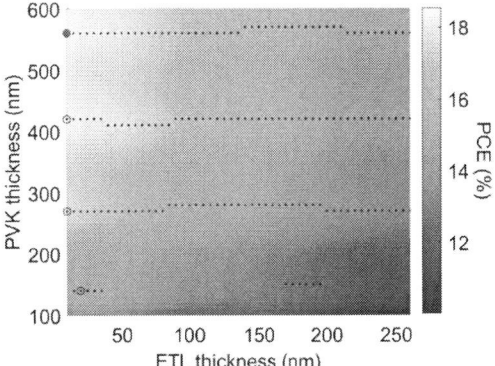

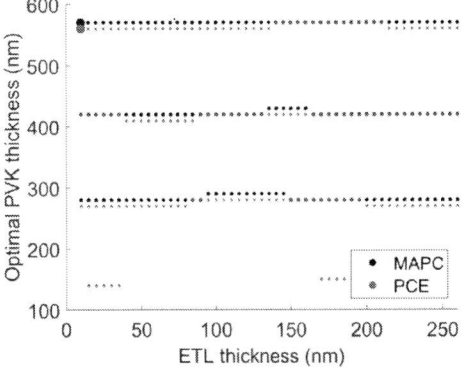

Figure 3: a) PCE as a function of ETL and PVK thicknesses. Local maxima for each ETL thickness separately (dots) and for both thicknesses (circles) as well as the global maximum within the search space (red filled circle) are noted. b) Optimal PVK thickness as a function of ETL thickness when optimizing MAPC and PCE (small dots refer to the local PVK thickness optima of specific ETL thickness and large dots to the global optimum). HTL thickness of 160 nm was applied for this figure.

and optoelectronic (PCE) cases originates from the finite charge collection, that is, carrier lifetime and diffusion length are limited because of recombination which is only included in PCE optimization.

Figure 4 illustrates how the recombination increased with the increasing PVK thickness reducing the open circuit voltage (V_{OC}). The dominating recombination term, non-radiative Shockley-Read-Hall recombination (R_{SRH}) increased almost linearly with the increasing PVK thickness when considering the total R_{SRH} at a constant voltage of 1V ($R_{SRH,1V}$). $R_{SRH,1V}$ was applied to show how recombination limits V_{OC} (and maximum power point voltage V_{MPP}). Recombination at the cell specific V_{OC} or V_{MPP} is not a practical metric for the effect of PVK thickness because R at V_{OC} equals the corresponding G_{ph} and R at V_{MPP} does not increase because of smaller V_{MPP} caused by the increasing recombination itself. Therefore, V_{OC} and $R_{SRH,1V}$ were chosen as metrics to depict the effect of PVK thickness on the recombination, and eventually, voltage. Maximum power point voltage showed similar behavior as V_{OC} as a function of PVK thickness.

HTL or ETL thicknesses did not affect the effect of increasing PVK thickness on V_{OC} and $R_{SRH,1V}$ (Figure 4). However, thicker ETL reduced V_{OC} somewhat probably due to smaller G_{ph} (Figure 4b). Higher light intensity has been reported diminishing charge trapping by defects [24].

Despite the light intensity was not changed here, it is the photogeneration that affects the number of charge carriers so the effect should be the same. Overall, PVK layer affects the recombination most because that is where the charge carriers coexist in the largest concentrations.

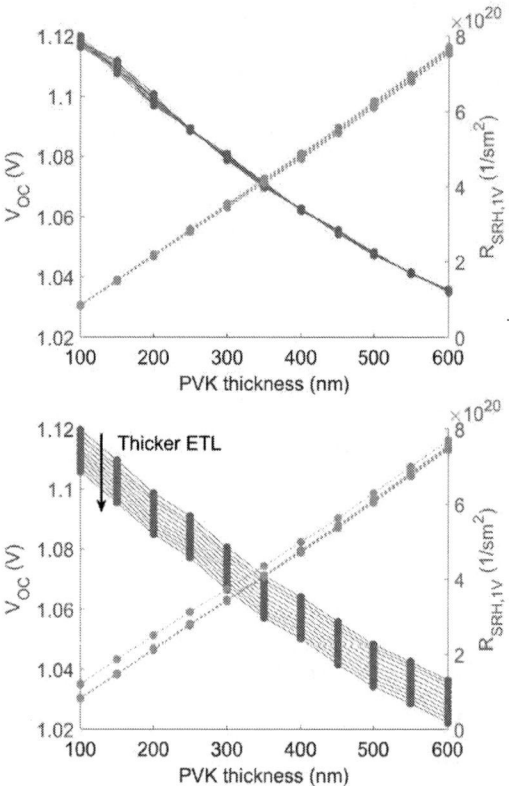

Figure 4: V_{OC} and $R_{SRH,1V}$ as functions of PVK thickness for different (a) HTL and (b) ETL thicknesses.

4 DISCUSSION

4.1 Inverted device and rear contact

The fact that the HTL and PVK thicknesses showed correlated behavior in the present results probably relates more to the position of HTL between PVK and Au rather than which type of carriers it collects from the active material. That is because the reflection from the back contact (Au in this case) is expected to dominate the back-propagating wave and it is the back-propagating wave which causes the interference as discussed above. Thus, ETL thickness would be expected to correlate with the optimal PVK thickness, and HTL not, in an inverted device structure.

Further, if the back contact did not strongly reflect light, the correlation between rear transport layer and PVK thicknesses would probably be weaker or even negligible. Weak reflection could occur, for example, if a transparent or semi-transparent back contact with a relatively similar refractive index compared with the rear transport layer was applied.

4.2 Effect of incidence angle

Because constructive or destructive interference depends on the phase difference between the interfering waves and optical path length determines the phase shift

in a layer, all properties that change the optical path length are expected to affect the interference within the device. Thus, the optimal layer thicknesses of the correlating layers (absorber and the rear transport layer) may depend on the incidence angle because it affects the optical path length in all layers. The potential gain of fine-tuning the layer thicknesses can therefore depend on several factors affecting the incidence angle of irradiation, such as device installation. It also means that optimal layer thicknesses for one incidence angle are likely to be suboptimal for another.

4.3 Film thickness and interface control

The potential benefit from the specific optimal thickness combination depends also on the actual experimental realization of the given thicknesses. The required level of control over different layer thicknesses may be difficult to achieve with solution processes and soft materials. In addition, interfaces should be relatively smooth because scattering from interfaces affects the light path. Evaporation could be an option improving the precise control of layers [25,26].

5 CONCLUSIONS

Based on optoelectronic modeling, it was observed that the optimal thickness of the active layer in a planar thin film solar cell may depend on the thickness of the charge transport layer between the absorber and a reflective back contact. Such dependence arises from the device optics. However, electrical operation needs to be considered to find the true optimum because optical absorption can in principle benefit from ever increasing absorber thickness which is not the case for power conversion efficiency. Further, realistic fluctuations in conditions and device preparation, such as the varying incidence angle and imperfect layers, may hinder the actual gain of interference-based photogeneration improvement. The thickness of the front transport layer did not affect the optimal thickness of the absorber. Furthermore, the thinnest ETL layer resulted in the best performance in all cases.

6 ACKNOWLEDGMENTS

A.K. thanks the Jenny and Antti Wihuri Foundation, University of Turku Graduate School (UTUGS), and the Finnish Foundation for Technology Promotion. K.M. thanks the Research Council of Finland (BioEST, 336577).

7 REFERENCES

[1] M. A. Green et al. Solar cell efficiency tables (Version 64). Prog Photovolt Res Appl. 2024; 32(7): 425-441.

[2] M. Cai et al. Cost-performance analysis of perovskite solar modules. Adv. Sci. 2017, 4, 1600269.

[3] Z. Song et al. A technoeconomic analysis of perovskite solar module manufacturing with low-cost materials and techniques. Energy Environ. Sci., 2017,10, 1297-1305.

[4] A. Martulli et al. Towards market commercialization: Lifecycle economic and environmental evaluation of

scalable perovskite solar cells. Prog Photovolt Res Appl. 2023; 31(2): 180-194.

[5] T. Sinha et al. A review on the improvement in performance of CdTe/CdS thin-film solar cells through optimization of structural parameters. J Mater Sci 2019, 54, 12189–12205.

[6] M. Koç et al. Guideline for Optical Optimization of Planar Perovskite Solar Cells. Adv. Optical Mater. 2019, 7, 1900944.

[7] M. R. Vogt, Development of physical models for the simulation of optical properties of solar cell modules. Ph.D. thesis, Leibniz Universitat Hannover, 2016.

[8] E. Raoult, et al. Iterative method for optical modelling of perovskite-based tandem solar cells. Opt. Express 2022, 30, 9604–9622.

[9] S. Manzoor et al. Optical modeling of wide-bandgap perovskite and perovskite/silicon tandem solar cells using complex refractive indices for arbitrary-bandgap perovskite absorbers. Opt. Express 2018, 26, 27441–27460.

[10] R. L. Olmon et al. Optical dielectric function of gold. Phys. Rev. B 2012, 86, 235147.

[11] J. A. Spencer et al. A review of band structure and material properties of transparent conducting and semiconducting oxides: Ga2O3, Al2O3, In2O3, ZnO, SnO2, CdO, NiO, CuO, and Sc2O3. Appl. Phys. Rev. 2022, 9, 011315

[12] T. Minemoto, M. Murata. Device modeling of perovskite solar cells based on structural similarity with thin film inorganic semiconductor solar cells. J. Appl. Phys. 2014, 116, 054505.

[13] P. Zhao et al. Numerical Simulation of Planar Heterojunction Perovskite Solar Cells Based on SnO2 Electron Transport Layer. ACS Appl. Energy Mater. 2019, 2, 6, 4504–4512.

[14] G. Sanon et al. Band-gap narrowing and band structure in degenerate tin oxide (SnO2) films. Phys. Rev. B 1991, 44, 5672.

[15] Q. Zhou et al. Two-dimensional device modeling of CH3NH3PbI3 based planar heterojunction perovskite solar cells. Solar Energy 2016, 123, 51-56.

[16] M. Feneberg et al. Anisotropy of the electron effective mass in rutile SnO2 determined by infrared ellipsometry. Phys. Status Solidi A 2014, 211: 82-86.

[17] Z. Ni et al. Resolving spatial and energetic distributions of trap states in metal halide perovskite solar cells. Science 2020, 367, 6484, 1352-1358.

[18] A. Kanevce et al. Optimizing CdTe Solar Cell Performance: Impact of Variations in Minority-Carrier Lifetime and Carrier Density Profile. IEEE Journal of Photovoltaics 2011, 1, 1, 99-103.

[19] H. J. Snaith, M. Grätzel. Enhanced charge mobility in a molecular hole transporter via addition of redox inactive ionic dopant: Implication to dye-sensitized solar cells. Appl. Phys. Lett. 2006, 89, 262114.

[20] K. Wijeratne et al. Enhancing the solar cell efficiency through pristine 1-dimentional SnO2 nanostructures: Comparison of charge transport and carrier lifetime of SnO2 particles vs. nanorods. Electrochimica Acta 2012, 72, 192-198.

[21] A. Kamppinen et al. Self-Heating of Planar Perovskite Solar Cells Depending on Active Material Properties. ACS Appl. Energy Mater. 2024, 7, 10, 4324–4334.

[22] R. Santbergen et al. Optical model for multilayer structures with coherent, partly coherent and incoherent layers. Opt. Express 2013, 21, S2, A262-A267.

[23] L. A. A. Pettersson et al. Modeling photocurrent action spectra of photovoltaic devices based on organic thin films. J. Appl. Phys. 1999, 86, 487–496.

[24] S. Ryu et al. Light Intensity-dependent Variation in Defect Contributions to Charge Transport and Recombination in a Planar MAPbI3 Perovskite Solar Cell. Sci Rep 2019, 9, 19846.

[25] H. Li et al., Sequential vacuum-evaporated perovskite solar cells with more than 24% efficiency. Science Advances 2022, 8, 28.

[26] Z. Zhang et al. Semitransparent Perovskite Solar Cells with an Evaporated Ultra-Thin Perovskite Absorber. Adv. Funct. Mater. 2023, 2307471.

41st European Photovoltaic Solar Energy Conference and Exhibition

Photoluminescence imaging of perovskite solar cells in full sunlight

Zhiwen Zheng, Félix Gayot, Juergen W. Weber, Yan Zhu, Ziv Hameiri
The University of New South Wales, Sydney, Australia

Introduction

- Perovskite solar cells (PSCs) have emerged as a promising candidate for next-generation solar cells.
- Field stability is vital for PSC's commercial deployment.
- Photoluminescence (PL) imaging [1] is a powerful inspection method for analysing key electrical parameters in solar cells.
- Due to the challenges of extracting PL under direct sunlight, only a limited number of outdoor PL methods exist, all of which are designed for silicon (Si) solar cells [2]-[3].
- **Aims:** (a) To develop a guideline for selecting an optical bandpass filter (BPF) for outdoor PL imaging, and (b) to apply outdoor PL imaging to PSCs in order to obtain their **world's first field images.**

Methodology

- For outdoor PL imaging, the first image is acquired under open-circuit (OC) conditions, followed by a second image captured under short-circuit (SC) conditions.
- Subtracting these two images allows the extraction of PL images.

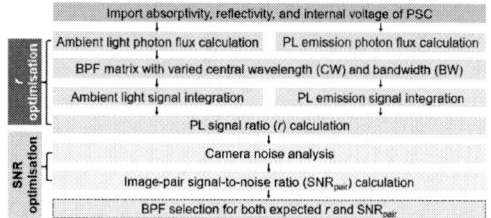

Figure 1: Workflow of the outdoor PL BPF optimisation

- Selecting suitable BPFs is essential for achieving high r and high SNR_{pair} to obtain clear PL images.

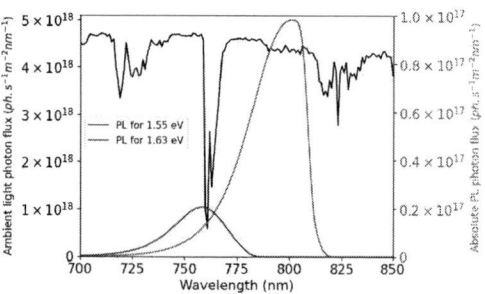

Figure 2: The AM1.5G spectra and PL emission of the investigated PSCs

- BPF optimisation was carried out to investigate PSCs with bandgaps (E_g) of 1.55 eV and 1.63 eV.

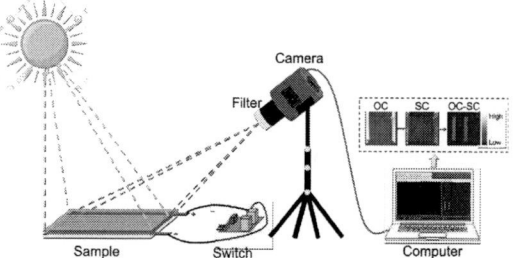

Figure 3: System configuration of outdoor PL imaging

- The BPF with CW of 760 nm and BW of 10 nm was selected due to the existence of a local absorption dip at 763 nm in the AM1.5G.
- The outdoor PL images were captured in Sydney under a clear sky.

Results

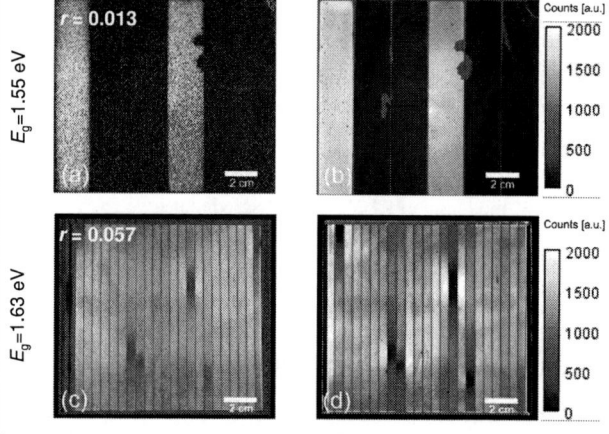

Figure 4: Healmaps under AM1.5G for PSCs: (a) 1.55 eV − r, (b) 1.63 eV − r, (c) 1.55 eV − SNR_{pair}, and (d) 1.63 eV − SNR_{pair}. The triangles mark the selected BPF. Note that different colour scales are used.

- Maximum r and SNR_{pair} align with the ambient spectral absorption dip.
- Similar distribution of r and SNR_{pair} – a strong positive correlation.

Figure 5: World's first outdoor PL images of perovskite modules: (a) 1.55 eV − outdoor, (b) 1.55 eV − indoor, (c) 1.63 eV − outdoor, and (d) 1.63 eV − indoor.

- Similar features are visible in both the indoor and outdoor PL images.
- PSC boundaries, high recombination regions, and other nonuniformities can be easily identified in the outdoor PL images.
- Measured r values closely align with the predicted values.
- The 1.63 eV PSC module exhibits higher contrast due to its higher r.

Conclusions

- We obtained the world's first outdoor PL images of PSCs using the sun as the excitation source.
- We demonstrated that the proposed method can effectively detect various features within perovskite modules.
- The proposed method offers an effective capability for investigating the long-term stability and degradation of PSCs under field conditions.
- The method can be easily extended into tandem devices.

Acknowledgements
This work was supported by the Australian Government through the Australian Renewable Energy Agency (ARENA, Grant 2020/ RND016).

References
[1] T. Trupke et al. Appl. Phys. Lett., vol. 89, no. 4, 2006.
[2] R. Bhoopathy et al. Prog. Photovolt. Res. Appl., vol. 26, no. 1, pp. 69–73, 2017.
[3] G. Rey et al. Prog. Photovolt. Res. Appl., vol. 30, no. 9, pp. 1115–1121, 2022.
[4] P. Würfel et al. J. Lumin., vol. 24–25, pp. 925–928, 1981.

Contact Details
Zhiwen Zheng
zhiwen.zheng@unsw.edu.au
www.acdc-pv-unsw.com

41st European Photovoltaic Solar Energy Conference and Exhibition

ANALYSIS OF COLOR ALTERATION AS A NOVEL DEGRADATION ASSESSMENT METHOD FOR PEROVSKITE SOLAR CELLS

Rustem Nizamov*, Aapo Poskela, Mahboubeh Hadadian, Maryam Esmaeilzadeh, Mikael Nyberg, Kati Miettunen
Department of Mechanical and Materials Engineering, Faculty of Technology, University of Turku, Turku, FI-20014,
Finland
*e-mail address: rustem.nizamov@utu.fi

ABSTRACT: In this study, we present a non-invasive, in situ method of assessing the degradation of perovskite solar cells (PSCs) by analyzing color alterations. We utilized a custom-built photographing chamber with controlled light conditions paired with a high-resolution digital camera. This method offers a cost-effective solution for assessing degradation and could simplify the monitoring process across a range of PSCs. Unencapsulated, hole transport layer-free carbon-based PSCs (C-PSCs) and metal-based PSCs (Au-PSCs) were aged following an ISOS protocol (D-1, dark storage) for 9200 hours at an average temperature of $22 \pm 1°C$ and humidity of $40 \pm 20\%$. Periodic digital photographs and short-circuit current density (J_{SC}) measurements were taken to map electrical performance against color change. A statistically significant correlation ($r = 0.8$ for C-PSCs, $r = -0.7$ for Au-PSCs) was found between color alterations and J_{SC}. These findings suggest that color changes can serve as reliable indicators of PSC degradation.
Keywords: Perovskite solar cell, degradation, color alteration

1 INTRODUCTION

Perovskite solar cells (PSCs) have become a focal point of photovoltaic research due to their high power conversion efficiencies and relatively low fabrication costs [1]. Among the various PSC architectures, those incorporating carbon electrodes (C-PSCs) [2] and metal (Au) electrodes (Au-PSCs) [3] have emerged as promising candidates [4]. C-PSCs are attractive for their stability under ambient conditions, with carbon electrodes being chemically inert, low-cost, and environmentally benign [5]. In contrast, Au-PSCs are known for their excellent electrical conductivity, but they tend to suffer from ion migration, which can lead to faster degradation [6]. Understanding and assessing the degradation patterns of PSCs is critical to optimizing their performance and extending their operational lifetimes.

The current methods for assessing degradation in perovskite solar cells (PSCs) are typically invasive, expensive, and time-consuming [7], [8]. These methods often require complex equipment like scanning electron microscopy (SEM), X-ray diffraction (XRD), among others, which complicates the process of routine degradation monitoring. Notably, degradation in PSCs often manifests visually, particularly in the form of color alteration in the perovskite layer. For instance, C-PSCs, depending on the fabrication recipe, exhibit a dark-to-grey coloration that shifts over time as degradation progresses [5], [9], while Au-PSCs start with a dark hue that fades as the devices age [7], [10]. Interestingly, in dye-sensitized solar cells (DSSCs), a similar phenomenon was observed where the color of the electrolyte correlates with its degradation, providing a pathway for predictive modeling of performance loss [11]. This successful correlation between color changes and performance degradation in DSSCs motivates the exploration of similar non-invasive monitoring methods for PSCs.

In this study, unencapsulated hole transport layer-free C-PSCs and metal-based Au-PSCs were subjected to aging according to ISOS-D-1 [12] tests for approximately 9200 hours, with periodic digital imaging and current-voltage (IV) measurements. Despite minimal color changes that are within the range of just noticeable differences (JND) [13], indications of a correlation between the color alterations and the electrical performance of the cells were observed. This method provides an accessible and economical approach to assess degradation non-invasively.

2 MATERIALS AND METHODS

2.1 Materials

Seven C-PSCs were prepared using commercial carbon monolithic electrodes (CMEs) from Solaronix. The methylammonium lead iodide ($MAPbI_3$) perovskite precursor solution, which includes lead iodide (PbI_2), methylammonium iodide (MAI, CH_3NH_3I), and 5% 5-ammonium valeric acid iodide (5-AVAI) dissolved in gamma-butyrolactone (GBL) and ethanol mixture, was purchased from Solaronix.

For the five of Au-PSCs, the materials include fluorine-doped tin oxide (FTO) glass substrates purchased from Greatcell Solar, zinc powder (Sigma-Aldrich), hydrochloric acid (Sigma-Aldrich), titanium diisopropoxide bis(acetylacetonate) (Sigma Aldrich), acetylacetone (Sigma Aldrich), TiO_2 nanoparticles (Greatcellsolar, 30 NR-D), absolute ethanol (Altia, 99.5%). The perovskite precursors used for this structure includes lead iodide (PbI_2) (TCI, >98%), lead bromide ($PbBr_2$) (TCI, >98%), formamidinium iodide (FAI) (Greatcell Solar, >99.99%), methylammonium bromide (MABr) (Greatcell Solar, >99.99%), and cesium iodide (CsI) (Abcr). Solvents used included dimethylformamide (DMF), dimethyl sulfoxide (DMSO), chlorobenzene (CB), and isopropyl alcohol (IPA) (Sigma-Aldrich). Spiro-MeOTAD (Lumtec) was used as the hole transport material, doped with lithium bis(trifluoromethanesulfonyl)imide (LiTFSI), 4-tert-butylpyridine (tBP) (Sigma Aldrich) and tris(2-(1H-pyrazol-1-yl)-4-tert-butylpyridine) cobalt (III) tris-(bis(tri-fluoromethane) sulfonimide) (FK209, Dyenamo).

2.2 PSC fabrication

The CMEs consist of a stack of mesoporous layers including carbon, ZrO_2, and TiO_2 on a compact TiO_2 layer coated on FTO glass substrate. The electrodes were sintered at 400°C for 30 minutes with a heating ramp of 30 minutes to reach the target temperature. After cooling to room temperature, a polyimide adhesive mask was applied, and the mesoporous layers were infiltrated with 5.74 µL of the $MAPbI_3$ precursor solution under inert nitrogen atmosphere. The substrates were then heated at 50°C for 10 minutes to complete the fabrication process, following the procedure described by Akulenko et al. [9].

For Au-PSCs, FTO substrates were etched using a zinc powder and 4 M hydrochloric acid mixture to define the

309

electrode area. The compact TiO_2 layer was deposited via spray pyrolysis of a precursor solution containing titanium diisopropoxide bis(acetylacetonate) and acetylacetone in absolute ethanol, at 460°C, followed by spin coating a mesoporous TiO_2 layer and sintering at 460°C. For perovskite layer, a triple-cation mixed-halide perovskite with the formula $Cs_{0.05}(FA_{0.83}MA_{0.17})_{0.95}Pb(I_{0.83}Br_{0.17})_3$, was prepared by mixing solutions of formamidinium lead iodide (FAPbI₃), methylammonium lead bromide (MAPbBr₃), and CsI, as detailed in Valdez García *et al.* [14]. The perovskite layer was deposited via spin coating using a two-step program, with chlorobenzene dropped as an antisolvent during the spin process. The films were then annealed at 100°C for 30–40 minutes.

Spiro-OMeTAD was used as the hole transport material (HTM), doped with lithium bis(trifluoromethanesulfonyl)imide (LiTFSI), cobalt additive, and tBP. The spiro-OMeTAD solution was spin-coated onto the substrates at 4000 rpm for 20 seconds. An 80 nm gold electrode was thermally evaporated onto the samples as the top contact. Finally, silver paste was applied on the contact side of the Au-PSCs to ensure better electrical contact.

2.3 Dark storage aging

The unencapsulated PSCs were aged according to the ISOS-D-1 [12] protocol under dark storage conditions. Throughout the aging period, the temperature and relative humidity were continuously monitored using a RuuviTag Bluetooth sensor. The average temperature recorded was 22 ± 1°C, and the average relative humidity was 40 ± 20%. These uncertainties represent the standard deviation of temperature and relative humidity over the aging period.

2.4 IV characterization

For both C-PSCs and Au-PSCs, the IV characteristics were measured using a PalmSens4 potentiostat. A Peccell PEC-L01 class A solar simulator was used to replicate the AM 1.5G solar spectrum at an intensity of 1000 W/m² during testing.

The IV measurements were initiated after a 3-second equilibration phase. For the C-PSCs, an integrated mask [15] was used in the custom holder, fabricated similar to the protocol described by Valdez García *et al.* [14]. A single scan was performed in both forward and reverse directions, with a starting potential of -0.2 V, increasing to 1 V, and returning to -0.2 V. The voltage was incremented in steps of 0.01 V with a scan rate of 100 mV/s. For Au-PSCs, a metal mask with an aperture of 12 mm² was applied to ensure consistent light exposure during the measurement [15]. The IV measurement setup remained similar to that of C-PSCs, with a single scan direction from -0.2 V to 1.2 V. The data from these measurements was analyzed using custom Python code available in [16].

2.5 Digital imaging

A custom-built photography system was used to capture high-resolution images of the PSCs under controlled lighting conditions. The system consisted of a digital camera and a lighting setup designed to ensure consistent illumination throughout the measurement process [17]. The cells were placed on a gray polyvinyl chloride (PVC) plastic background within the photo chamber, and due to their random placement, various patterns may be observed in the transparent regions of the substrates.

The raw images were imported in Adobe RGB (1998) JPEG format. The RGB color space consists of three-color channels—red, green, and blue—with values ranging from (0, 0, 0) to (255, 255, 255). For color extraction, C-PSCs were imaged from the glass side, and color data were collected from five different zones across the device to ensure consistent measurement and minimize variability [18]. For Au-PSCs, eight zones were selected (two per pixel of the solar cell) from the glass side, focusing on the active area. However, RGB does not accurately reflect human visual perception, since it is device-dependent and non-uniform. Small changes in RGB values might result in unequal perceptual changes in color. The extracted color data, with respect to the illuminant, were converted to CIELAB color space [19], which provides a device-independent, perceptually uniform way of quantifying color. Converting the color data ensures that equal changes in CIELAB values correspond to equal perceived changes in color, allowing for a more accurate and objective quantification of color changes over time.

The CIE 1976 $L*a*b*$ color space, known as CIELAB, was introduced by the International Commission on Illumination (CIE) in 1976 to create a perceptually uniform color space [20]. In CIELAB, the lightness component $L*$ ranges from 0 (black) to 100 (white). The $a*$ axis represents the green–red opponent colors: negative values indicate green, while positive values indicate red. The $b*$ axis corresponds to the blue–yellow opponent colors, with negative values for blue and positive values for yellow. While $L*$ is bounded between 0 and 100, the $a*$ and $b*$ axes are theoretically unbounded and can exceed ±150, depending on the reference white, to cover the full range of human color perception.

An alternative representation derived from CIELAB is the CIELCh (also known as CIE HLC) color space, which uses polar coordinates instead of Cartesian coordinates. In this system, $C*$ (chroma) denotes the relative saturation of the color, and $h°$ (hue angle) indicates the specific hue as an angle on the color wheel. This polar coordinate system can provide more intuitive insights into color relationships compared to the $a*$ and $b*$ axes.

3 RESULTS AND DISCUSSIONS

3.1 Visual degradation of PSC.

As presented in Figure 1, the active areas of the C-PSCs initially appeared nearly black and over time shifted slightly towards lighter shades, although the change was not significant enough to be noticed by the human eye (below JND). A yellow pattern could be observed on both sides of the cell around the active area, which has been suggested primarily to be due to the formation of non-photoactive lead iodide (PbI₂) [5].

In contrast, the Au-PSCs exhibited a noticeable color shift in the perovskite layer between the contacts, particularly in the layers sandwiched on top of the gold strips (see Figure 1). Initially, the perovskite layer in Au-PSCs was as black as the carbon electrodes. However, after 1500 hours, a noticeable color change occurred, and by the end of the aging period, the change was undeniable. This significant visual degradation suggests considerable ion migration in the Au-PSCs, leading to decomposition of the perovskite layer [21]. Possible sources of color change in Au-PSCs include the degradation of the spiro-OMeTAD hole transport layer, which can change color upon oxidation [22], and interactions between the gold electrode and the perovskite layer. Moisture ingress can also induce hydration of the perovskite, forming yellowish lead iodide PbI_2 and other compounds [8]. Some pixels were more affected by this phenomenon, resulting in uneven degradation across the device. Due to this uneven degradation, the performance of the pixels was averaged to represent a single device.

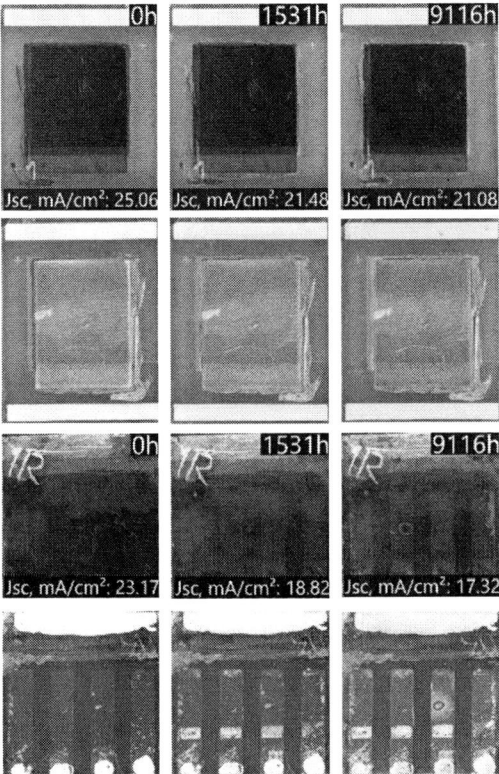

Figure 1: Periodic images of both sides of a C-PSC and Au-PSC, showing visual degradation over 9200 hours of aging (ISOS-D-1).

3.2 Performance and color change

Quantitative measurements of performance and color change were conducted to assess the extent of degradation in both C-PSCs (Figure 2) and Au-PSCs (Figure 3) over the aging period. To evaluate the relationship between color changes and performance degradation, statistical analyses were performed using Pearson correlation coefficients between specific color parameters and electrical performance metrics for each type of PSC. A Pearson correlation coefficient close to 1 or −1 suggests that the variables are linearly related, with the sign indicating the direction of the relationship [23]. Multiple color channels from various color spaces (including RGB, CIELAB, HSL, and HSV) and performance metric such as efficiency (η), open circuit voltage (V_{OC}), short-circuit current density (J_{SC}) and fill factor (FF) were analyzed separately for the C-PSCs and Au-PSCs samples to identify the most suitable indicators of degradation.

For the C-PSCs, the b^* component of the CIELAB color space demonstrated the strongest correlation with J_{SC}. Figure 2 illustrates this relationship, alongside with η, V_{OC}, and FF. During the first 72 hours, significant fluctuations in the b^* values were observed, due to

stabilization processes occurring immediately after fabrication; therefore, these data points were excluded from the analysis. After approximately 300 hours, the fluctuations diminished, and the b^* value began to mirror the trends in J_{SC}, indicating a strong correlation between the active area color changes and electrical performance of the cell.

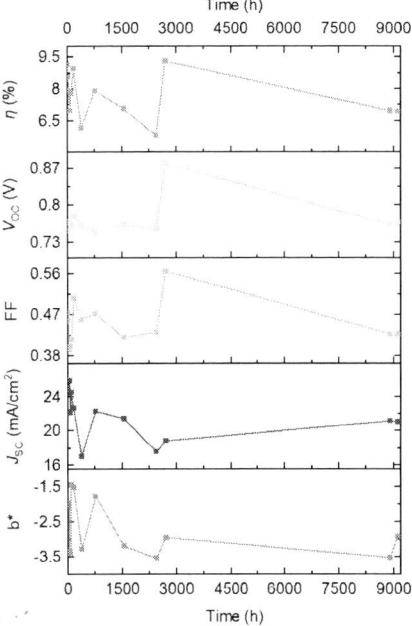

Figure 2: Evolution of η, V_{OC}, FF, J_{SC}, and b^* value (CIELAB color space) for the selected C-PSC over aging time.

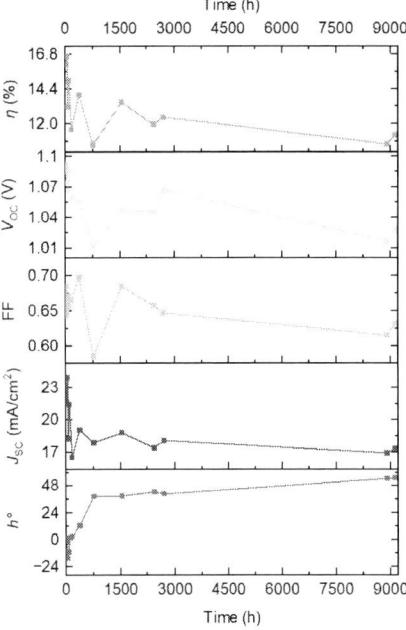

Figure 3: Evolution of η, V_{OC}, FF, J_{SC}, and $h°$ value (CIELCh color space) for the selected Au-PSC over aging time.

The average Pearson correlation coefficient between $b*$ and J_{SC} for C-PSCs was calculated to be 0.76 with a p-value of 0.019, indicating a statistically significant positive linear relationship. The high correlation suggests that as the $b*$ value decreases (indicating a shift towards blue), there is a corresponding decrease in J_{SC}.

For Au-PSCs a strong negative correlation was observed between the $h°$ channel of CIELCh color space and J_{SC} as shown in Figure 3. Initially current exhibited dramatic changes indicating cell stabilization; interestingly the $h°$ values displayed a similar negative pattern during this period. This synchronized behavior between J_{SC}, and color persisted until the end of the experiment.

The calculated average Pearson correlation coefficient between $h°$ and J_{SC} for Au-PSCs was −0.67 with a p-value of 0.018, confirming a statistically significant negative linear relationship. This finding suggests that an increase in the $h°$ value (indicating a shift in hue) corresponds with a decline in J_{SC}. This relationship reflects the observed color shifts, which are linked to degradation mechanisms such as perovskite decomposition and potential oxidation of the spiro-OMeTAD hole transport layer. The linearity implies that the rate of performance degradation is directly proportional to the change in the specific color parameter within the studied range. This suggests that color measurements can be used not only as indicators of degradation but also to quantify the extent of degradation in PSCs.

Furthermore, the Au-PSCs showed a more significant average J_{SC} loss of 28% compared to 17% average for C-PSCs. The higher percentage loss in J_{SC} for Au-PSCs suggests that these devices are more susceptible to degradation mechanisms under the given aging conditions.

The German Society of Color Science and Application (DIN) 99 color difference (DIN99, ΔE_{99}) [24] was used to quantify the visual changes in the device over time. ΔE_{99} provides a metric for quantifying the perceived color difference between two samples, with higher values indicating more significant color changes. The average ΔE_{99} value for Au-PSCs was significantly larger at 22.6, compared to 2.8 for the C-PSCs. This substantial difference in color change indicates that the Au-PSCs underwent more pronounced visual degradation during the aging process, correlating with their greater loss in J_{SC}.

The correlation between the greater ΔE_{99} values and higher J_{SC} losses in Au-PSCs reinforces the relationship between the extent of color change and the degree of performance degradation in PSCs. The smaller color change of the C-PSCs suggests that the carbon electrode provides better stability, protecting the perovskite layer from environmental factors that contribute to degradation. The hydrophobic nature of the carbon electrode may inhibit moisture ingress, reducing the formation of degradation products [5]. However, subtle degradation of the perovskite layer still occurs, affecting performance [6]. In contrast, Au-PSCs showed greater susceptibility to degradation, evidenced by both higher J_{SC} loss and larger ΔE_{99} values [4].

These findings suggest a relationship between the extent of color change and the degree of performance degradation in PSCs. Greater color changes are associated with larger current losses, supporting the hypothesis that color measurements can serve as indicators of device degradation.

Comparing our findings with previous studies on DSSCs, a fundamental difference in the temporal relationship between color alteration and performance degradation was observed. In DSSCs, the degradation of tri-iodide (I_3^-) in the electrolyte reduces the concentration of charge carriers, which initially manifests as a color change due to electrolyte bleaching [11], [25]. This precedes the increase in recombination losses and the subsequent decline in J_{SC}. Therefore, by monitoring the early color changes, it is possible to predict the performance degradation over extended periods, starting from a few hundred hours and forecasting over thousands of hours [11].

In contrast, our study on PSCs shows that both the color alterations and electrical performance degradation occur simultaneously. The synchronous behavior observed in both C-PSCs and Au-PSCs, as evidenced by the strong correlations between the color parameters and J_{SC} over time, leaves limited room for establishing predictive models based on early color changes alone. Since the degradation mechanisms in PSCs, such as perovskite decomposition and interfacial degradation, directly affect both the optical properties (color) and the electrical performance concurrently, the opportunity to predict long-term performance based solely on initial color changes is diminished. Thus, color measurements in PSCs serve more effectively as real-time indicators of device degradation, rather than as tools for predicting long-term performance.

3.3 Influence of aging conditions on degradation

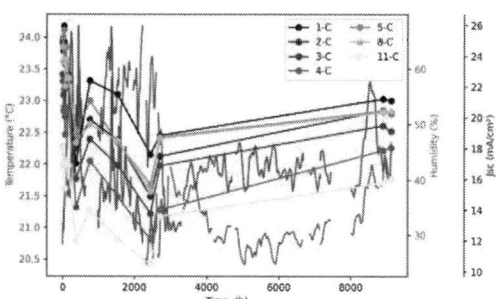

Figure 4: Environmental factors (humidity and temperature) and J_{SC} of the C-PSCs over aging time.

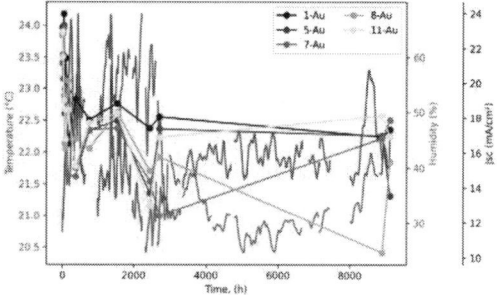

Figure 5: Environmental factors (humidity and temperature) and J_{SC} of the Au-PSCs over aging time.

The hydrophobic nature of the carbon electrode provides better protection against moisture [5], contributing to greater device stability. C-PSCs, even after 9200 hours of ISOS-D-1 aging, still maintained more than 80% of their initial efficiency.

Figure 4 depicts the humidity and temperature alongside J_{SC} of C-PSCs, while Figure 5 shows the same metrics for Au-PSCs. Although PSCs are prone to temperature degradation [7], the temperature in this experiment remained relatively low and stable with small deviations, making it unlikely to be a major factor contributing to cell degradation. Interestingly, during the period between 2400 and 6000 hours, the average humidity was at its lowest. This coincided with a slight stabilization in the performance of both Au-PSCs and C-PSCs, suggesting that humidity may influence the degradation process. However, further investigation is needed to confirm this hypothesis.

The humidity and temperature data were smoothed using a moving average for clearer data representation. The gaps are due to data collection inaccuracies.

4 CONCLUSION

In this study, we developed a non-invasive, in situ method to assess the degradation of PSCs by analyzing color alterations using a custom-built photographing chamber with controlled lighting conditions. Despite minimal color changes that are within the range of JND, we observed significant correlations between specific color parameters and J_{SC}. For C-PSCs, the $b*$ channel of the CIELAB color space showed a strong positive correlation with J_{SC}. For Au-PSCs, the $h°$ component exhibited a strong negative correlation with J_{SC}. Statistical analyses using Pearson correlation coefficients confirmed the significance of these relationships.

Our findings indicate that color alterations, even when subtle, can serve as reliable, non-invasive indicators of performance degradation in PSCs. This method offers an economical solution for assessing degradation and could simplify the monitoring process across a range of PSC devices.

ACKNOWLEDGEMENTS

R.N., A.P., and K.M. acknowledge financial support from the Research Council of Finland (BioEST, project numbers 336577 and 336441, ECOSOL 347275), M.H. thanks SUSMAT profiling funding (Research Council of Finland and University of Turku), and M.E. is grateful for UTUGS doctoral program funding. Strategic Materials Research Infrastructure (MARI) at University of Turku was used for this study.

REFERENCES

[1] S. Chen, X. Dai, S. Xu, H. Jiao, L. Zhao, and J. Huang, "Stabilizing perovskite-substrate interfaces for high-performance perovskite modules," *Science (1979)*, vol. 373, no. 6557, 2021, doi: 10.1126/science.abi6323.

[2] A. Mei *et al.*, "A hole-conductor-free, fully printable mesoscopic perovskite solar cell with high stability," *Science (1979)*, vol. 345, no. 6194, 2014, doi: 10.1126/science.1254763.

[3] M. A. Green, A. Ho-Baillie, and H. J. Snaith, "The emergence of perovskite solar cells," 2014. doi: 10.1038/nphoton.2014.134.

[4] C. Yang *et al.*, "Achievements, challenges, and future prospects for industrialization of perovskite solar cells," *Light Sci Appl*, vol. 13, no. 1, p. 227, 2024, doi: 10.1038/s41377-024-01461-x.

[5] M. Hadadian, J. H. Smått, and J. P. Correa-Baena, "The role of carbon-based materials in enhancing the stability of perovskite solar cells," 2020. doi:

10.1039/c9ee04030g.

[6] S. P. Dunfield *et al.*, "From Defects to Degradation: A Mechanistic Understanding of Degradation in Perovskite Solar Cell Devices and Modules," 2020. doi: 10.1002/aenm.201904054.

[7] R. Wang, M. Mujahid, Y. Duan, Z. K. Wang, J. Xue, and Y. Yang, "A Review of Perovskites Solar Cell Stability," 2019. doi: 10.1002/adfm.201808843.

[8] J. Yang, B. D. Siempelkamp, D. Liu, and T. L. Kelly, "Investigation of CH3NH3PbI3 degradation rates and mechanisms in controlled humidity environments using in situ techniques," *ACS Nano*, vol. 9, no. 2, 2015, doi: 10.1021/nn506864k.

[9] E. S. Akulenko, M. Hadadian, M. Esmaeilzadeh, R. Nizamov, and K. Miettunen, "Bottlenecks in Perovskite Solar Cell Recycling," in *EU PVSEC 2023*, 2023, pp. 001–004. doi: 10.4229/EUPVSEC2023/2BV.2.13.

[10] G. Tumen-Ulzii *et al.*, "Detrimental Effect of Unreacted PbI2 on the Long-Term Stability of Perovskite Solar Cells," *Advanced Materials*, vol. 32, no. 16, 2020, doi: 10.1002/adma.201905035.

[11] A. Poskela, A. Tiihonen, H. Palonen, P. D. Lund, and K. Miettunen, "Predictive Modeling of Dye Solar Cell Degradation," *Solar RRL*, vol. 6, no. 6, 2022, doi: 10.1002/solr.202101004.

[12] M. V. Khenkin *et al.*, "Consensus statement for stability assessment and reporting for perovskite photovoltaics based on ISOS procedures," *Nat Energy*, vol. 5, no. 1, 2020, doi: 10.1038/s41560-019-0529-5.

[13] S. Hecht, "The visual discrimination of intensity and the weber-fechner law," *Journal of General Physiology*, vol. 7, no. 2, 1924, doi: 10.1085/jgp.7.2.235.

[14] J. Valdez García *et al.*, "Simplifying perovskite solar cell fabrication for materials testing: how to use unetched substrates with the aid of a three-dimensionally printed cell holder," *R Soc Open Sci*, vol. 11, no. 9, p. 241012, Sep. 2024, doi: 10.1098/rsos.241012.

[15] D. Kiermasch, L. Gil-Escrig, H. J. Bolink, and K. Tvingstedt, "Effects of Masking on Open-Circuit Voltage and Fill Factor in Solar Cells," 2019. doi: 10.1016/j.joule.2018.10.016.

[16] R. Nizamov, "Solar_cells_measurement_and_plotting." [Online]. Available: https://gitlab.com/mateng-utu/solar_cells_measurement_and_plotting

[17] A. Lawrynowicz, E. Palo, R. Nizamov, and K. Miettunen, "Self-cleaning and UV-blocking cotton – Fabricating effective ZnO structures for photocatalysis," *J Photochem Photobiol A Chem*, vol. 450, p. 115420, 2024, doi: https://doi.org/10.1016/j.jphotochem.2023.1154 20.

[18] R. Nizamov, "RGB_recognition." Accessed: May 31, 2023. [Online]. Available: https://gitlab.com/mateng-utu/RGB_recognition

[19] R. Nizamov, "Color_and_IV_data_plotter." Accessed: Feb. 19, 2024. [Online]. Available: https://gitlab.com/mateng-utu/color_and_iv_data_plotter

[20] International Commission on Illumination. CIE., "CIE 15: Technical Report: Colorimetry. 3rd

edition," 2004.

[21] J. W. Lee, S. G. Kim, J. M. Yang, Y. Yang, and N. G. Park, "Verification and mitigation of ion migration in perovskite solar cells," *APL Mater*, vol. 7, no. 4, 2019, doi: 10.1063/1.5085643.

[22] G. Tumen-Ulzii *et al.*, "Understanding the Degradation of Spiro-OMeTAD-Based Perovskite Solar Cells at High Temperature," *Solar RRL*, vol. 4, no. 10, 2020, doi: 10.1002/solr.202000305.

[23] K. Pearson and A. Lee, "On the Laws of Inheritance in Man: I. Inheritance of Physical Characters," *Biometrika*, vol. 2, no. 4, 1903, doi: 10.2307/2331507.

[24] ASTM, "D2244 − 16: Standard practice for calculation of color tolerances and color differences from instrumentally measured color coordinates," *ASTM International*, 2016.

[25] A. Poskela, K. Miettunen, A. Tiihonen, and P. D. Lund, "Extreme sensitivity of dye solar cells to UV-induced degradation," *Energy Sci Eng*, vol. 9, no. 1, 2021, doi: 10.1002/ese3.810.

41st European Photovoltaic Solar Energy Conference and Exhibition

STATISTICAL MODEL OF OUTDOOR PEROVSKITE PERFORMANCE

P. Manshanden[1], M. Späth[1], M.J. Jansen[1], V. Zardetto[1], A. Aguirre[2,3], V. Depauw[2,3], M. Heydarian[4,5], J. Borchert[4,5]

1] TNO, Westerduinweg 3, 1755LE Petten, The Netherlands / High Tech Campus 21, 5656 AE Eindhoven, the Netherlands

2] Hasselt University, imo-imomec, Martelarenlaan 42, 3500 Hasselt, Belgium

3] imec EnergyVille, imo-imomec, Thor Park 8320, 3600 Genk, Belgium

4] Fraunhofer Institute for Solar Energy Systems, Heidenhofstr. 2, 79110 Freiburg, Germany

5] Department of Sustainable Systems Engineering (INATECH), University of Freiburg, Emmy-Noether-Str. 2, 79110 Freiburg, Germany

ABSTRACT: Perovskite outdoor reliability is at the moment insufficient to allow for commercial use. The first step in improving the reliability is correctly assessing the outdoor performance, which is complicated by varying irradiance, module temperature and spectrum.

In this contribution we propose a statistical model for outdoor performance of perovskites: a multiple linear regression model which can be trained on first month's data. We show that with this model, we can accurately predict the performance of the perovskites under any occurring weather conditions, which can then be compared to the measured outdoor performance. With respect to alternative analysis methods, this offers an improved accuracy both in time and in the offered variable.

Keywords: perovskite, tandem, outdoor performance, performance model

1 INTRODUCTION

At the moment, the barrier preventing market penetration of perovskite into the global solar energy market is no longer its power conversion efficiency which can be equal to crystalline silicon, but its reliability and durability which is still a long way from the required 0.1% degradation per year. Although much progression has been made to the reliability [1,2] and the testing of the reliability [3], as for any technology, the gold standard is and remains outdoor performance.[4]

However the problem with using direct outdoor performance is that the irradiance, angle of incidence module temperature and spectrum vary over the day, and so the power, and indeed power conversion efficiency varies over the day and over the seasons. Assessing a change in module parameters is not trivial for these reasons, unless the change is near catastrophic for the sample. A possibility is to take the samples indoor for measuring, but we have previously shown that invariability of perovskite samples during these actions is not guaranteed.[5]

Plotting multicolored dot graphs with irradiance – power(voltage/current) with time providing the color index will show a degradation in the angle change, but the time of change will remain vague, unless the change occurs in abruptly. Therefore, a prediction of the outdoor performance of perovskite would come in useful.

In this contribution we show a versatile model which can predict the outdoor performance with good accuracy, thereby giving direct

2 EXPERIMENTAL SETUP

2.1 Samples

Four types of outdoor samples were used for the verification.

1. Commercial crystalline silicon module
2. Monofacial perovskite module
3. Bifacial perovskite module
4. Perovskite-silicon tandem cell

Sample type 1, the crystalline silicon module (Elsun [6]) comprised of 60 M2 PERC cells in a lightweight construction.

Sample type 2, the monofacial perovskite module had p-i-n structure, deposited on glass. The module had laser scribed interconnection and had an active area of 10x10 cm. It was encapsulated in inert atmosphere, lacking encapsulant. The rear was covered in black foil to create the monofacial characteristic.

Sample type 3, the bifacial perovskite module had p-i-n structure deposited on glass. The module had laser scribed interconnection and had an active area of 2x2 cm. It was encapsulated using TPO and additional edge sealant.

Sample type 4, the perovskite-silicon tandem cell used a perovskite with p-i-n structure on a polished silicon heterojunction cell. It was encapsulated using TPO and additional edge sealant.

2.2 Outdoor testing facilities

The outdoor test facility used is a small scale facility facing south at an angle of 3 degrees, situated in Petten, the Netherlands within a kilometer of the North Sea. The local climate is a moderate maritime climate, with at the specific location an elevated air salinity.

The devices were monitored with PT100 temperature sensors and pyranometers and filtered silicon reference cells to measure the irradiance. A spectrometer with a range of 300-1700 nm is situated within a couple of meters. A full weather station is situated within 100 m.

The full range adaptable IV tracer is set to measure an IV sweep every 10 minutes in concord with the temperature, irradiance and spectral measurements. The weather data is taken continuously. The samples are at maximum power point between measurements.

The albedo of the background is 10%; in practice rear irradiance when applicable is directly measured with rear irradiance sensors.

3 MULTIPLE LINEAR REGRESSION MODEL

Linear regression is a statistical model which estimates

the relationship between a response and variables. In case of one variable, the correct term is simple linear regression, and can graphically be shown to be a best fit through a line. In case multiple variables are used, it is called a multiple linear regression.

3.1 Linear regression

The method of linear regression is a standard statistical method which can be found in any statistical textbook of basic level. Therefore the description given here only serves as a short reminder, and is not meant as a rigorous introduction.

The goal of linear regression is to fit a linear relation of the variables to a dataset, where the distance of all given training points to the fitted relation is minimized. This is achieved by simultaneously optimizing multiple linear equations. This task is typically iteratively performed by software. The result is an equation in the form of:

$$y = y_0 + ax_1 + bx_2 + \cdots \qquad (1)$$

During the linear regression, the analysis reveals which factors (x_1, x_2 in the equation) are not significant to the model. These can then be eliminated from the equation to create a more optimized form.

3.2 Modifying parameters

Although a linear regression exclusively deals with linear relations, it is not necessary to restrict the variables to linear. If a relationship should physically be e.g. a (co)sine or a logarithm, one can transform the variable before performing the linear regression. In that case, the linear regression is performed of the (co)sine or logarithm of the variable.

3.3 Model parameters for outdoor fitting

Depending on the exact sample used, several parameters are more or less important to achieve a physically meaningful fit to the data.

For samples of all types and makes, the irradiance and sample temperature are important to fit all parameters. It is well known that for a specific sample the sample temperature can be estimated from the ambient temperature, irradiance and windspeed as well. However, as the sample temperature was in this case known we omitted this step.

The irradiance is modified by the angle of incidence (AOI). With an AOI of 0° denoting perpendicular irradiation, the cosine of the AOI is the factor that denotes the change in irradiance. It is obviously possible to incorporate the AOI directly into the irradiance. However, as perovskite samples have a different spectral response than the sensors, the spectral change due to AOI (aka Rayleigh scattering) can be handily intercepted by using this as a separate factor.

Since all samples can be at any time be approximated with a diode, the logarithmic dependence of the voltage on the irradiance should be included for all samples on the voltage and power analysis.

Specific for bifacial samples, we can include the rear and total irradiance as factors.

For tandem samples, the spectral irradiance for each component device is additionally needed.

3.4 Data filtering

The raw data of the outdoor measurements must be filtered in order to be usable. This corresponds to several types of occasions which does not yield useful data.

Intermittent cloudy weather cannot be used when the irradiance changes by more than 5% during a measurement. This is because the IV data then does not have a specific irradiance it belongs to, and the IV curve does not correspond to a diode curve.

At high latitudes sun rises and sets towards the north, meaning the sun can be behind the modules. Although the solar cells do function with only diffuse light, we prefer to keep to more direct irradiance and only use AOI of 10° and more.

3.4 Competing models

Several competing simpler models have been used, mostly for crystalline solar cells. The following optional analysis methods have been compared for the perovskite samples.

The first option is double simple linear regression of temperature and irradiance. This is slightly different from a multiple linear regression, as the temperature is first corrected with known temperature factors, and then the irradiance. For sample types with a long history, e.g. crystalline silicon, these computations are easy to perform, as the temperature factors can be estimated within 5%.,

The second option is irradiance/temperature matrix comparison of indoor and outdoor results. This is properly not a model, but rather a display method which yields insight into the outdoor performance based on the indoor irradiance/temperature behavior. A key assumption of this method is that the sample does not change significantly when placed outdoors.

5 APPLIED MODELS

5.1 Verification: silicon module

Using a multiple linear regression model on silicon modules is overcomplicating the matter as simpler models and analysis methods do the job equally well as possible changes to the modules are well catalogued and can be predicted. However, using an uncomplicated sample is useful for verification.

A silicon module has as variable input for the linear regression model only the basics: angle of incidence, irradiance and temperature, and their modified variations. The computation reveals also non-significant contributors, which can then be eliminated if need be. As for this module the irradiance was sourced from a calibrated silicon module, the angle of incidence yielded no new information and could be eliminated from the list.

The linear regression was performed on power, open circuit voltage and short circuit current, all independently. The model was trained on one month data. One year data is shown below in the response graphs.

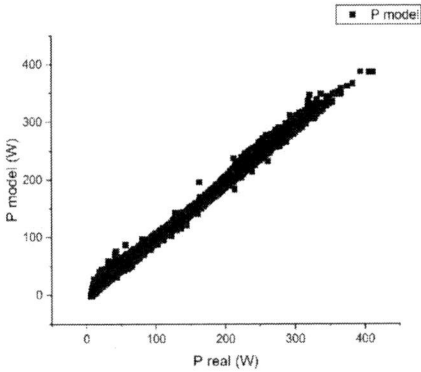

Figure 1:Pmpp model for silicon module, with a mean absolute error of 2%

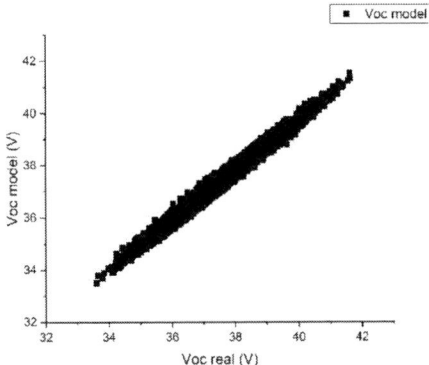

Figure 2:Voc model for silicon module, with a mean absolute error of 0.3%

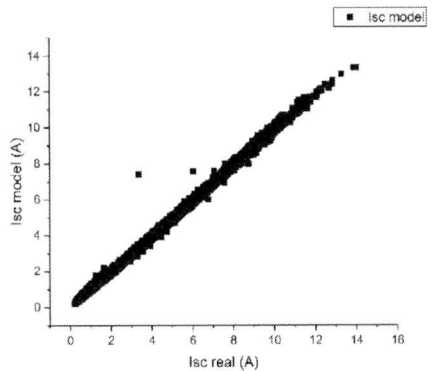

Figure 3:Isc model for silicon module, with a mean absolute error of 1%

The accuracy of the model in training is not significantly different from the accuracy in the first year. Visually the mean absolute error of the Voc would seem to be larger, which is caused by the relatively short value range the voltage can reach.

5.2 Monofacial perovskite module

The easiest perovskite sample to try a linear regression model on would be a monofacial perovskite module, and so that was the second sample to be used as co-verification.

For the monofacial module, the same variables were used as for the silicon module. However, because the module shows slight degradation at later stages, for estimations of the mean absolute error we only consider data before the degradation occurred.

Instead of showing the full plots, for brevity we summarize the model plots with the mean absolute error in a table.

Table 1:mean absolute error for multiple linear regression analysis of bifacial perovskite module

	Mean absolute error
Pmpp	4%
Voc	1%
Isc	2%

The multiple linear regression model gives us an estimate of what the performance should be, given the time of the day, temperature and irradiance. This can be used to compute the expected power, voltage or current. Whereas it is possible to do this exercise for any given point in time, it is useful to suppress the variation further by averaging the values over the week, and comparing this to the measured weekly averages.

As indication of the validity of the method, the time colored dot-graphs of the corresponding irradiance / variable graphs are supplied underneath the comparisons

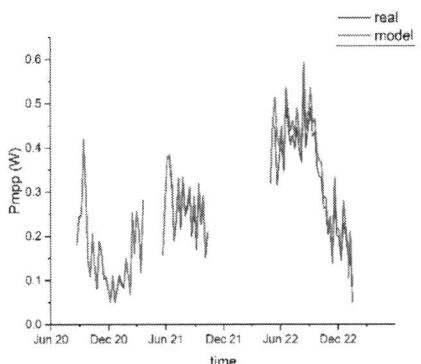

Figure 4:power comparison for perovskite module. After winter 2021, the module is degraded by 10%. Note the deviation in June 2022.

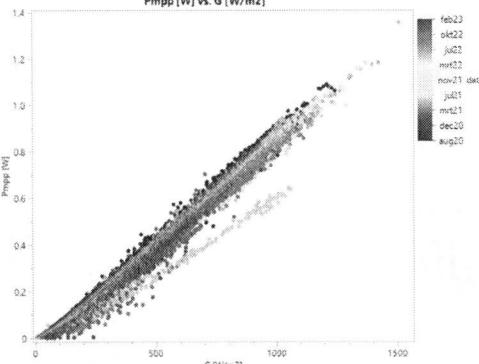

Figure 5: time colored dot-graph of the power versus the irradiance. The deviation in June 2022 can be seen, but the final degradation is harder to distinguish

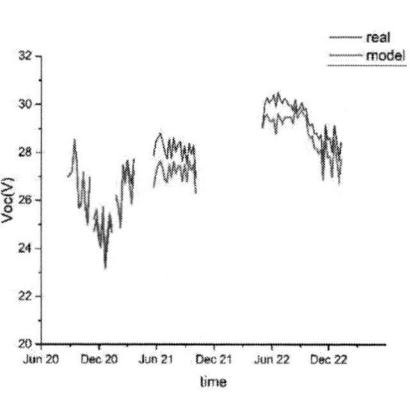

Figure 6:voltage comparison for perovskite module. After winter 2020, the module parameters have changed.

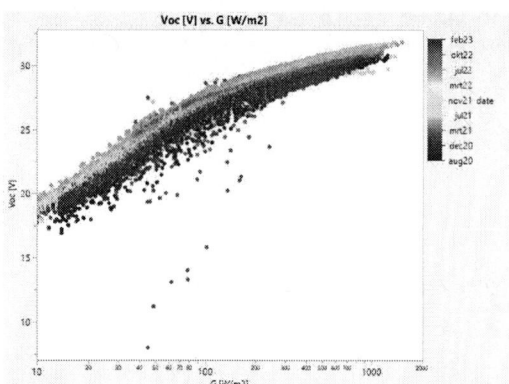

Figure 7: time colored dot-graph of the voltage versus the irradiance. The increase since winter 2020 can be seen. We can also see a high series resistance in general; the log irradiance/voltage plot should ideally be a linear relationship.

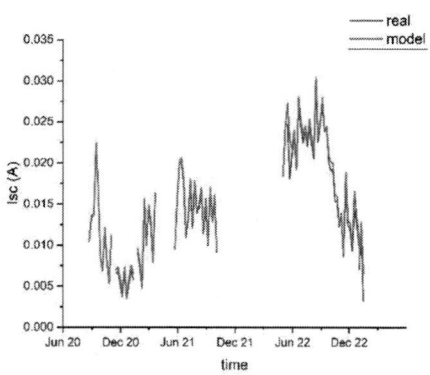

Figure 8:current comparison for perovskite module. A small degradation is seen from November 2022. Note the deviation in June 2022

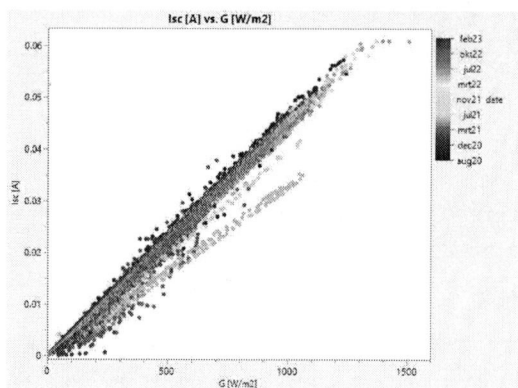

Figure 9: time colored dot-graph of the current versus the irradiance. The deviation in June 2022 is easily visible. The slight degradation from the beginning is just visible.

From the three models we can estimate that the 10% degradation in the module mostly originates from a fill factor problem, as the current decrease is not sufficient to make up the difference. The dot graphs versus irradiance of the three factors show good correspondence. However, the dot graphs are notably worse in pinpointing exact timing, and spotting smaller effects is harder.

The bad performance in June 2022 is a temporary current issue, which was solved on its own, which is consistent with soiling.

This module suffered fatal storm damage, which was an unfortunate end to a good stability run.

4.3 Bifacial perovskite module

A bifacial perovskite module is a natural halfway product to generate a tandem. For a bifacial perovskite module, besides the common variables as seen in the two previous paragraphs, we also need the rear irradiance. This could be estimated from the albedo. However, since it was directly measured we just use the real rear irradiance.

Table 2:mean absolute error for multiple linear regression analysis of bifacial perovskite module

	Mean absolute error
Pmpp	2%
Voc	1%
Isc	2%

The fit for the bifacial perovskite module is slightly better than for the monofacial perovskite module. However, the time range that could be used is less, as the degradation set in at a much earlier stage.

41st European Photovoltaic Solar Energy Conference and Exhibition

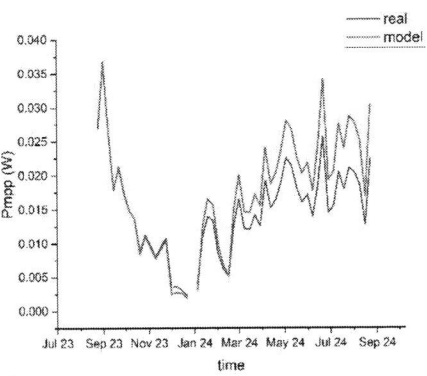

Figure 10: power comparison for bifacial perovskite module. After the interruption, the module starts to degrade

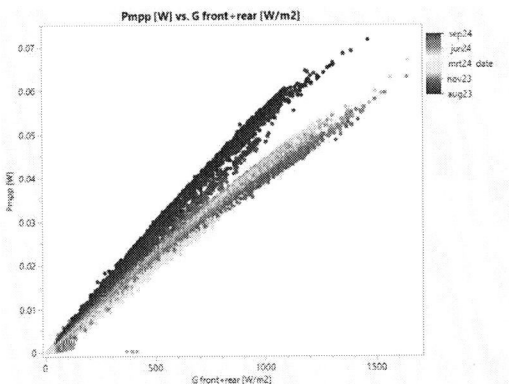

Figure 11: time colored dot-graph of the power versus the irradiance. The degradation point is easy enough to spot, as the magnitude is fairly large.

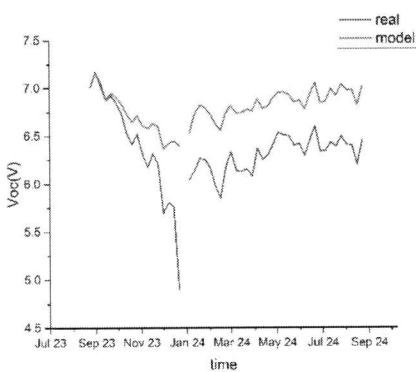

Figure 12:voltage comparison for bifacial perovskite module. The voltage model was initially good, but the deviation from the model started quite early — the low voltage numbers indicate this is likely a degradation and not a fault in the model.

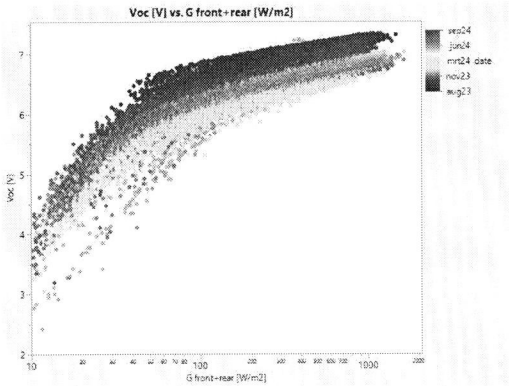

Figure 13: time colore dot-graph of the voltage versus the irradiance. Decrease in voltage can easily be seen from this graph. Again, the non-linerity of the voltage-log irradiance relationship is attributed to series resistance.

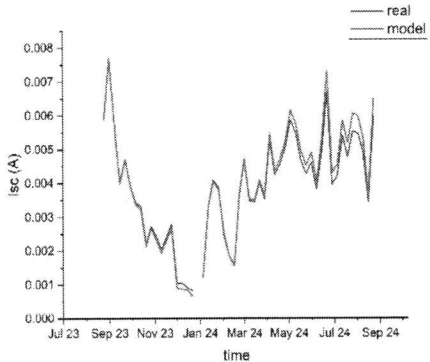

Figure 14:current comparsion for the bifacial perovskite module. The model has a nearly perfect match until April, and is slowly degrading after.

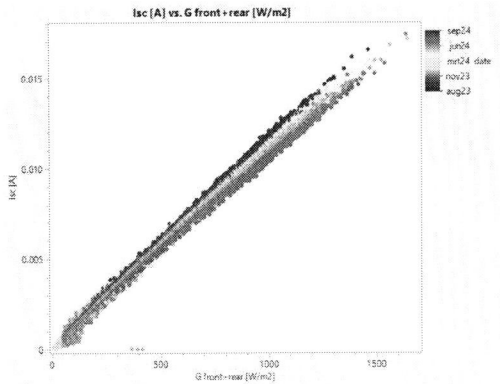

Figure 15: time colored dot-graph of the current versus the irradiance. The current is slowly degrading.

From figure 12+13 we can see that the voltage has changed signficantly already after the first month. The module is at the last measurement point still slowly degrading.

For this module, the model and dotgraphs show good correspondence, where again the model is better at showing timing and smaller effects.

4.4 Perovskite silicon tandem module

41st European Photovoltaic Solar Energy Conference and Exhibition

A perovskite-silicon tandem module module requires the most complicated model. All standard variables need to be applied, but in addition we need the partial irradiance for each component device. This partial irradiance is computed by partial integration over the spectrum as measured at that time by the roof spectrometer.

As for the bifacial perovskite, as verification we only show the mean absolute error. Similarily, we limit the range for which we consider the mean absolute error to the time period before noticable degradation occurs.

Table 3:mean absolute error of the multiple linear regression analysis of the perovskite silicon tandem module

	Mean absolute error
Pmpp	1%
Voc	2%
Isc	2%

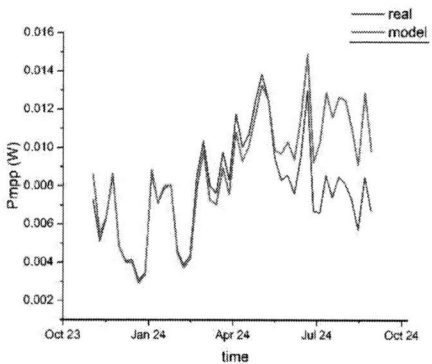

Figure 16:power comparison for the perovskite silicon tandem module. From May the power degrades.

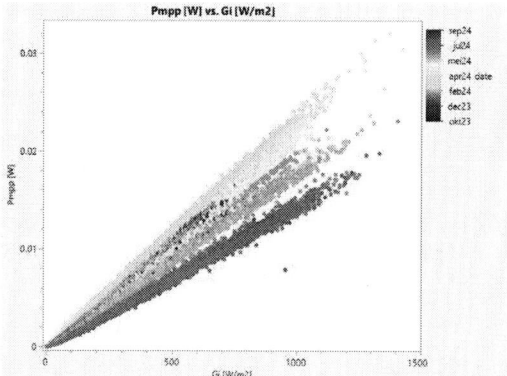

Figure 17: time colored dot-graph of the power versus the irradiance. From May the power degrades in apparently two steps.

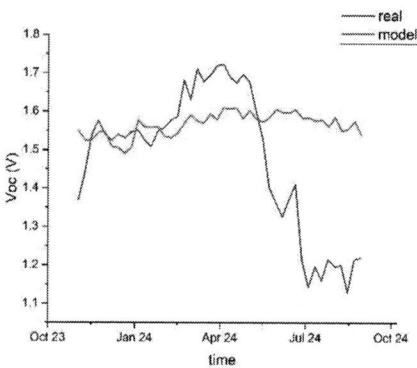

Figure 18: voltage comparison for the perovskite silicon tandem module. The voltage is unstable (also see Mean absolute error in table), and degrades from May.

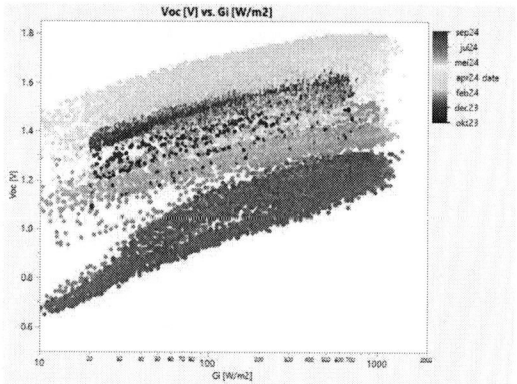

Figure 19: time colored dot-graph of the voltage versus the irradiance.The voltage rises in February-April, and then lowers in two steps from May.

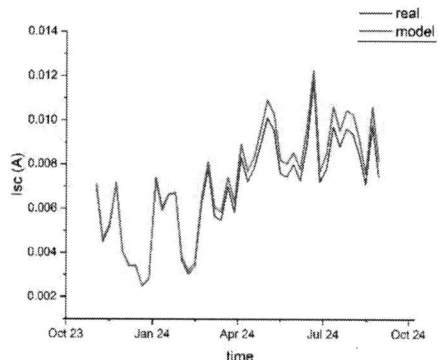

Figure 20:current comparison for the perovskite silicon tandem module. The current rises slightly after May.

41st European Photovoltaic Solar Energy Conference and Exhibition

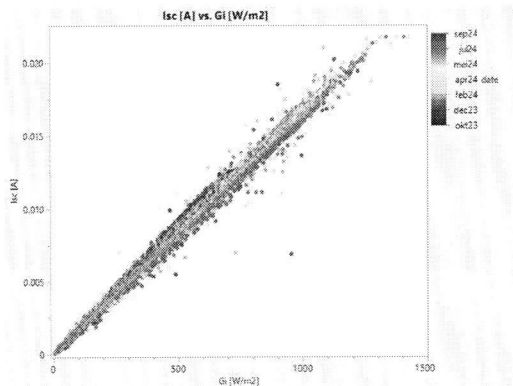

Figure 21:time colored dot-graph of the current versust he irradiance. The current seems higher in winter

The main driver for the degradation is the loss in open circuit voltage, which can also easily be followed from the dot-graphs.

The data in the dot-graphs is not compensated for temperature, which is probably the reason for the seasonal effect in the dot-graph for the current.

5 MODEL COMPARISONS

5.1 Multiple linear regression compared to corrections

For several modules simple linear regression as described in the model section have been compared to multiple linear regression.

For both models the training range is the first month of data, excluding obvious deviations. The compared parameter is the mean absolute error between the model prediction and the outdoor performance. In relevant cases, only data before degradation was noted was taken for the comparison.

Table 4:comparison of the mean absolute error of simple temperature and irradiance corrections and multiple linear regression analysis, only of the maximum power variable

	Mean absolute error (%)	
	Temperature + irradiance correction	Multiple linear regression
Silicon module	2%	2%
Monofacial perovskite module	4%	2%
Perovskite-silicon tandem module	13%	1%

For silicon modules, the correction for temperature and irradiance works quite well, which makes sense because the multiple linear regression does not include anything else but temperature and irradiance. The addition of the logarithmic of the irradiance only has minimal influence on the error in the maximum power – although for the voltage the influence would be larger.

The monofacial perovskite module benefits more of the inclusion of additional parameters – in this case the implied spectral contribution of the angle of incidence.

Perovskite-silicon tandem modules require a multiple linear regression, containing among others the specific irradiance for each component device.

5.2 Temperature irradiance graphs

For silicon modules, outdoor performance can be quite well predicted by the STC performance or the temperature irradiance matrix [6], especially with available real module temperatures. A key assumption is that the samples are only allowed to vary on fairly long timescales, in the order of years.

The indoor graphs are measured fairly precisely at the given temperatures and irradiances. One cannot do that outdoor, so the full outdoor data is filtered with a precision of 1 °C/ 10 W/m². This is a little more stringent than the suggested 2.5 °C/ 25 W/m² from the reference but we had sufficient data available to se the filtering stricter.

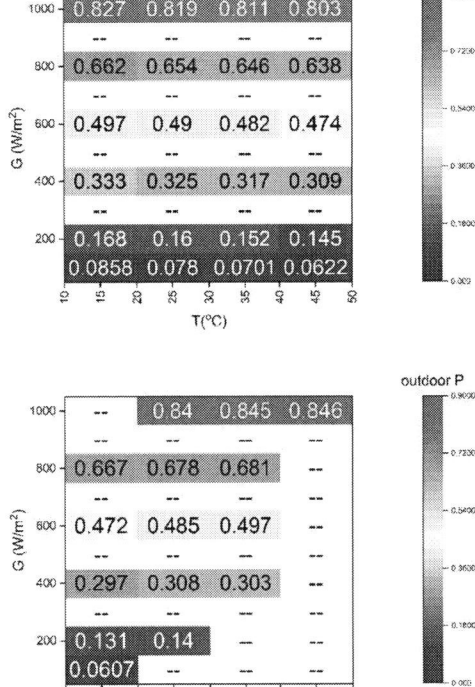

Figure 22: indoor and outdoor maximum power for different temperature and irradiance values. The values are dissimilar, with outdoor higher high irradiance values and lower low irradiance figures.

The graphs are dissimilar, and do not even indicate similar temperature coefficients, for a very simple reason: a key assumption of the method is invalid. Perovskite modules and perovskite-silicon tandem modules do typically change in first couple of days outdoor, and also change each day over the day due to ion migrations. Indoor measurements do not capture those changes accurately as they are typically done on stabilized modules.

321

6 DISCUSSION

A common approach for silicon modules is to measure the temperature and irradiance dependencies indoors and use that to correct outdoor performance. This approach leads to two separate problems. One problem is inherent in using metastable samples: the sample the indoor measurements were performed changed so much that the outdoor results seem unrelated. In addition, while separately correcting for temperature and irradiance is a good approach for silicon modules in general, for perovskite modules this strategy fails to capture some important variables.

Comparing the model results to the raw data in the colored dot graphs, we see that the model shows the same trends as the raw data. However, as the model is temperature corrected and the time resolution is more precise, the data can be analyzed with more precision.

Therefore, training a multiple regression model on the first month of outdoor data is an efficient and useful way of analyzing the long term outdoor performance for perovskite modules.

7 ACKNOWLEGEMENTS

We gratefully acknowledge funding by the EU Horizon grant Triumph 10107525.

8 REFERENCES

[1] S. Baumann *et al*, "Stability and reliability of perovskite containing solar cells and modules: degradation mechanisms and mitigation strategies", Energy Environ. Sci., 2024, Advance Article, https://doi.org/10.1039/D4EE01898B

[2] T.-H. Le *et al*, "Perovskite solar module: promise and challenges in efficiency, meta-stability and operational lifetime", Adv. Electron. Mater., 2023, 9 , 2300093

[3] M.V. Khenkin *et al*. "Consensus statement for stability assessment and reporting for perovskite photovoltaics based on ISOS procedures." Nat Energy 5, 35–49 (2020). https://doi.org/10.1038/s41560-019-0529-5

[4] G.M. Meheretu *et al*. "The recent advancement of outdoor performance of perovskite photovoltaic cells technology", Heliyon 10 (2024), https://doi.org/10.1016/j.heliyon.2024.e36710

[5] P. Manshanden *et al*. "Predicting outdoor performance of perovskite modules with modified reliability testing", EUPVSEC 2023

[6] www.elsun.nl

[7] R. Valckenborg, B v. Aken, "Outdoor performance quantification and understanding of various PV technologies using the IEC 61853 matrix", EUPVSEC2019

41st European Photovoltaic Solar Energy Conference and Exhibition

CHARACTERIZATION AND DEGRADATION OF PEROVSKITE MINI-MODULES

R. Ebner[1], A. Mittal[1], G. Ujvari[1], M. Hadjipanayi[2], V. Paraskeva[2], G. E. Georghiou[2], A. Hadipour[3], A. Aguirre[4,5,6], T. Aernouts[4,5,6], T. Fontanot[7], S. Pechmann[7], S. Christiansen[7,8]

[1]AIT Austrian Institute of Technology, Center for Energy, Giefinggasse 2, 1210 Vienna, Austria, T +43 50550-6628, F +43 50550-6390, rita.ebner@ait.ac.at, www.ait.ac.at
[2]University of Cyprus, 1 Panepistimiou Avenue, 2109 Aglantzia, Nicosia, Cyprus, www.foss.ucy.ac.cy
[3]Kuwait University, College of Science, Department of Physics, Sabah Al Salem University City, Kuwait,
[4]Imec, imo-imomec, Thin Film PV Technology, Thor Park 8320, 3600 Genk, Belgium
[5]EnergyVille, imo-imomec, Thor Park 8320, 3600 Genk, Belgium
[6]Hasselt University, imo-imomec, Martelarenlaan 42, 3500 Hasselt, Belgium
[7]Fraunhofer Institute for Ceramic Technologies and Systems IKTS, Äußere Nürnberger Str. 62, 91301 Forchheim, Germany, www.ikts.fraunhofer.de
[8]Max Planck Institute for the Science of Light, Günther-Scharowsky-Strasse 1, 91058 Erlangen, Germany, mpl.mpg.de

ABSTRACT:
Organic-inorganic hybrid metal halide perovskites are poised to revolutionize the next generation of photovoltaics with their exceptional optoelectronic properties compatibility with low-cost and large-scale fabrication methods. The leap forward in the power conversion efficiency (PCE) enabled by lead halide perovskites is unprecedented, with PCEs emerging from 3.8% in its first study to a current certified value of 25.5% in single-junction and 33.7% in perovskite-silicon tandem devices [1-3]. The main challenge for the successful commercialization of perovskite solar cells is to achieve high stability at the module level. The commercially available solar modules undergo a series of characterization procedures that analyze their properties and ensure their quality. However, these procedures and protocols cannot unambiguously be applied to perovskite solar modules (PSM) due to its unpredictable degradation mechanisms. Therefore more advanced characterization methods are needed to understand the degradation mechanisms in the PSM. In this context, optical and electrical measurement methods are effectively employed in quality control and development support and are essential characterization tools in industry and research.

KEYWORDS: Perovskite, Optical and electrical characterization, Degradation

1 INTRODUCTION

In the proposed work, we employed optical and electrical characterization methods to understand the degradation of perovskite mini-modules. Optical techniques such us electroluminescence (EL), photoluminescence (PL), and dark lock-in thermography (DLIT) are non-destructive measurement techniques and provide high-resolution images showing a two-dimensional distribution of the characteristic features of PV cells and allowing the investigation of cracks, defects, shunts, and stacking faults in the cells [4]. Furthermore, electrical measurements like current-voltage (IV) characteristics and external quantum efficiency (EQE) can provide information on the power output and other device parameters, which could be used to identify the possible degradation route.

2 EXPERIMENTS
2.1 Perovskite mini-modules

A - double cation-double halide perovskite active layer with the composition $Cs_{0.18}FA_{0.82}PbI_{2.82}Br_{0.18}$ was used. To make large-area devices, so-called mini-modules were produced by laser scribing, to generate 7 sub-cells connected in series. To prevent penetration of metallic particles of the top electrode into the soft perovskite layer, ITO (indium-tin-oxide) was used. ITO was also selected as a top electrode to obtain semi-transparent modules. The module stack was as follows: glass/ITO/Hole transport layer (HTL)/560nm 2C Perovskite with bandgap of 1.6 eV/Electron transport layer (ETL)/ITO/glass. Fig.1 depicts the structure of the perovskite sub-cell and Fig. 2 illustrates the cross-section of the mini-module. Fig.3 shows two images of the front and back side of one perovskite mini-module (substrate size: 3cm x 3cm and module size: 2cm x 2cm) and Fig.4 shows the EL-images (front and backside) of one perovskite mini-module.

Four perovskite mini-modules ("S9", "S10", "S11" and "S12") were produced and characterized by applying DLIT-, EL- and PL methods alongside IV and EQE measurements.

IV measurements as well as EL measurements were performed regularly from **July 2021 (07/21) to January 2024 (01/24)** to determine the **aging behavior** of the mini-modules.

The IV measurement setup at AIT was the following:
TRI-SOL Solar Simulator 1-1.6 KW:
Class AAA (IEC 060904-9), Steady State Continuous, Air Mass Filter 1.5 Global, Typical Power Output: 100mW/cm2 (1Sun) ± 20%, vary Intensity from 0 - 1.2Sun, Lamp Type and Power Xe Arc Lamp (Lamp Power:1600W), contacting with crocodile clips.

The results are presented in section 3.1.

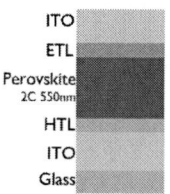

FIG.1 PIN-structure of each sub-cell.

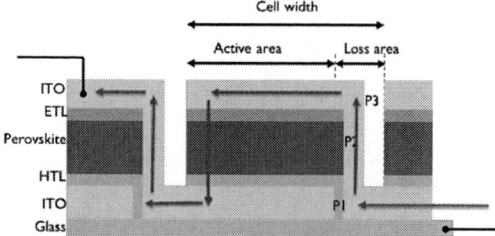

FIG.2 Cross-section of the mini-module.

FIG.3 Perovskite mini-module: Front side (left), Back side (right).

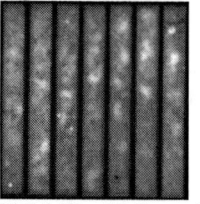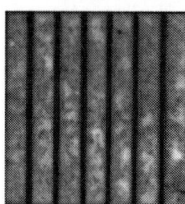

FIG.4 EL of a Perovskite mini-module: Front side (left), Back side (right).

3 RESULTS

3.1 Perovskite mini-modules

The perovskite mini-modules were kept indoors at room temperature and measured every month. The modules were stored in the dark between the measurements. As can be seen in **Fig.7**, the mini-modules behaved in a very stable manner over this period. Only after some measurements (e.g. DLIT, Raman) there is sometimes a clear increase or decrease in power, partly reversible and also irreversible.

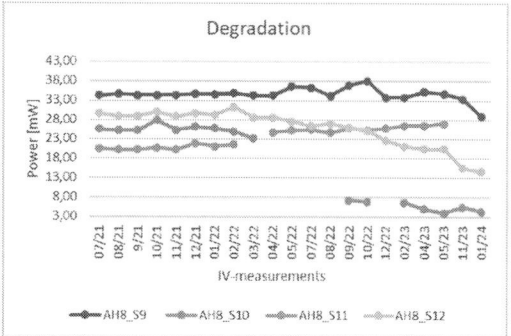

FIG.7 Aging behavior of the perovskite mini-modules

3.1.1 Mini-module "S10"

Fig. 8 shows the EL images of mini-module "S10" with two inactive cells.

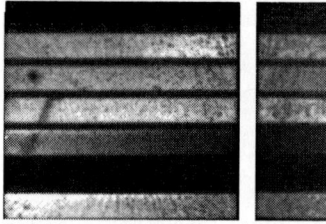

FIG.8 Perovskite mini-module "S10": Front side (left) and back side (right).

On mini-module "S10" DLIT-measurements were carried out.

3.1.1.1 DLIT-measurements

DLIT measurements were carried out on the selected mini modules. However, it soon became evident that these measurements put a great strain on the modules, with the possibility to induce degradation, interrupt electrical contact or even re-establish broken contacts.

In the case of the mini-module "S10", it can be clearly seen in the EL image (see Fig.9), that an interrupted contact was reactivated, after the DLIT measurement performed on 10/2021 was carried out. After DLIT measurement there was only one inactive subcell instead of two.

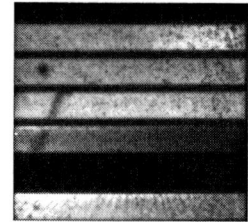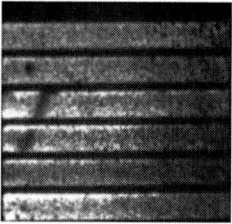

FIG.9 Mini-module "S10" before (left) and after (right) DLIT- measurement

The IV measurement results also showed a significant increase in performance after the DLIT measurement was performed (see Table 1).

Table 1: IV-measurement results of the mini-module "S10", before and after DLIT measurement (10/21).

Mini-Modul	I_{SC} [mA]	V_{OC} [V]	FF [%]	P_{MPP} [mW]	Jsc [mA/cm^2]
S10	10.56	6.106	39.70	25.59	13.48
S10 DLIT	10.39	6.756	39.81	27.94	13.27

However, there was already a drop in the measurement performed on 11/2021 and a subsequent return to the initial values (see Table 2).

Table 2: IV-measurements on "S10"

Mini-module	I_sc [mA]	V_OC [V]	FF [%]	P_MPP [mW]	Jsc [mA/cm²]
S10 (initial value)	10.56	6.106	39.70	25.59	13.48
S10 DLIT Okt. 21)	10.39	6.756	39.81	27.94	13.27
S10 (Nov. 21)	10.48	6.748	38.18	26.99	13.38
S10 (Dez. 21)	10.41	6.737	37.31	26.18	13.30
S10 (Jan. 22)	10.43	6.764	36.95	25.78	13.17
S10 (Feb.22)	10.45	6.843	34.80	24.89	13.35
S10 (March 22)	10.38	6.397	34.73	23.07	13.26

After the performance of some optical measurements in April 2022 the minimodule "S10" was completely degraded (see Fig.7).

3.1.2 Mini-module "S11"

Fig.10 shows the EL recordings of the mini-module "S11" with three inactive cells. The IV measurement results of the mini-module "S11" are presented in Table 3.

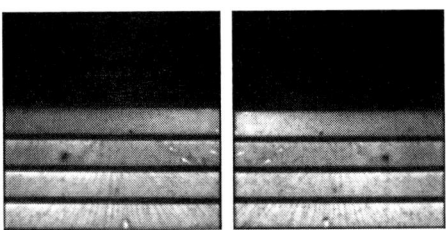

FIG.10 Perovskite mini-module "S11": Front side (left) and back side (right).

Table 3:
IV-measurement results of the mini-module "S11"

I_sc [mA]	V_OC [V]	FF [%]	P_MPP [mW]	Jsc [mA/cm²]
10.37	5.260	37.60	20.50	13.24

In the case of the minimodule "S11", some optical measurements were carried out in March. After the return of the minimodule, a power increase (red arrow) was also determined, which is still present and further increasing (see Fig.11).

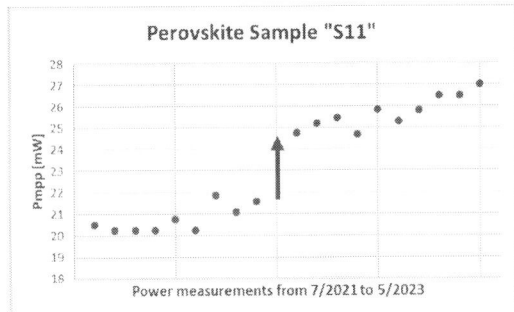

FIG.11 Performance measurements of mini-module "S11"

3.1.3 Mini-module "S12"

Fig.12 shows the EL images of the minimodule "S12"

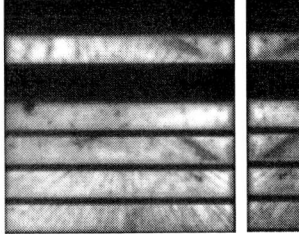

FIG.12 Perovskite mini-module "S12": Front side (left) and back side (right).

The IV measurement results of the mini-module "S12" are listed in Table 4.

Table 4:
IV-measurement results of the mini-module "S12"

V_OC [V]	Jsc [mA/cm²]	P_MPP [mW]	FF [%]
7.789	31.10	34.29	42.91

A significant increase in power was measured for the minimodule "S12" in February 2022. However, there was already a decrease in power in March 2022. The reason for the increase in February 2022 and the decrease in March 2022 is unclear. Since March 2022, a slight drop in performance has been noticeable for the "S12" mini-module.

4 CONCLUSIONS

It should be noted that some measurements (DLIT, Raman) led to an increase or decrease in power, partly reversible but also irreversible. Thus, perovskites are very sensitive to some optical measurements.

For this very long measurement time, both for the perovskite cells and the mini modules, the degradation is much lower than assumed. These are very good results, which predicts a successful future for the use of perovskites.

Further indoor tests and outdoor tests of differently structured perovskite samples and perovskite tandem cells have to be performed in order to identify more defects and thus be able to improve the perovskite cell and module structure. The aging behavior of the perovskite mini-modules will also be further analyzed. In addition, measurement protocols for indoor and outdoor tests of perovskites will be established.

5 REFERENCES

[1] Antonio Urbina 2020 J. Phys. Energy 2 022001
[2] "Oxford PV retakes tandem cell efficiency record",https://www.pv-magazine.com/2020/12/21/oxford-pv-retakes-tandem-cell-efficiency-record/
[3] Enzheng Shi et al., "Two-dimensional halide perovskite lateral epitaxial heterostructures", *Nature*, 2020; 580 (7805): 614
[4] R. Ebner et al., "Non-destructive techniques for quality control of PV modules", 39th Annual

Acknowledgement:

This work was funded through the European Regional Development Fund and the Republic of Cyprus in the framework of the project **"DEGRADATIONLAB"** with grant number INFRASTRUCTURES/1216/0043. http://www.foss.ucy.ac.cy/degradationlab/

"VIPERLAB" has received funding from the European Union's Horizon 2020 research and innovation programme under grant agreement N°101006715

41st European Photovoltaic Solar Energy Conference and Exhibition

Subcell-resolved electroluminescence imaging of monolithic perovskite-silicon tandem solar cell for high throughput characterization

I. J. Djeukeu[1,4], J. Horn[1], M. Meixner[1], E. Wagner[2], S. W. Glunz[3,4] and K. Ramspeck[1]

[1] halm elektronik gmbh, Friesstraße 20, 60388 Frankfurt am Main, Germany,
[2] Frankfurt University of Applied Sciences, Nibelungenpl.1, 60318 Frankfurt am Main, Germany,
[3] Fraunhofer ISE, Heidenhofstr. 2, 79110 Freiburg, Germany,
[4] Albert-Ludwigs-Universität, INATECH, Emmy-Noether-Strasse 2, 79110 Freiburg, Germany

halm.

Motivation

- Examine the resistive network of subcells and interlayer and its impact on electroluminescence (EL) in two terminal (2T) perovskite-silicon (pero-Si) tandem solar cells

- Develop an EL setup suitable for fast and reliable defect recognition and characterization in mass production of pero-Si tandem solar cells

- Determine meaningful EL measurement conditions for metastable pero-Si tandem samples

- Interpret signatures observed in EL images of both subcells

Model: Impact of resistive networks in 2T pero-Si tandem on EL[1]

- Qualitative model proposed to understand the impact of the resistive coupling mechanisms in pero-Si tandem cells on EL emission of the subcells

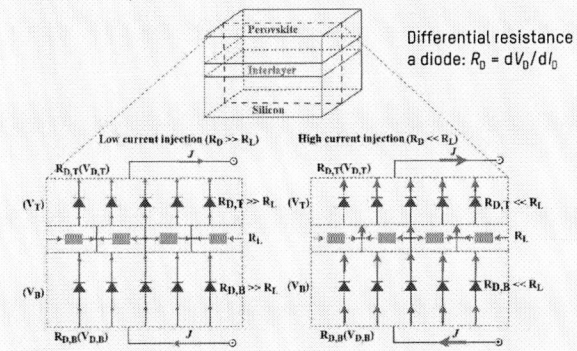

Differential resistance of a diode: $R_D = dV_D/dI_D$

- **Low current injection:**
$R_D \gg R_L$; Equipotential surfaces defined by metal grids, $V_{D,T} + V_{D,B}$ = const., lateral current flow through interlayer possible → high voltage of one subcell forces low voltage in the other subcell

- **High current injection:**
$R_D \ll R_L$ → lateral current flow through interlayer strongly limited → $I_{D,T} = I_{D,B}$ → Defect appearance in EL must comply with local coupling of subcell currents

Industry compatible setup [1]

IV-tester integrable setup with synchronized, triggered EL cameras, for separate and simultaneous EL image acquisition for each subcell within <100 ms

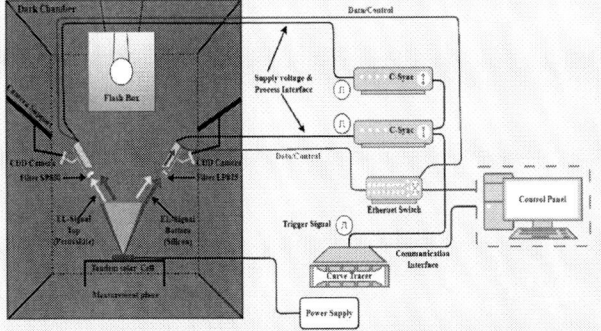

References

[1] I.J. Djeukeu, J. Horn. M. Meixner, E. Wagner, S.W. Glunz, K. Ramspeck, "Subcell-Resolved Electroluminescence Imaging of Monolithic Perovskite/Silicon Tandem Solar Cell for High-Throughput Characterization" Sol. RRL 2024, 2400469

[2] O. Fisher, A. Fell, C. Messmer, R. Efinger, F. Schindler, S. W. Glunz, C. Schubert, "Understanding Contact Nonuniformities at Interfaces in Perovskite Silicon Tandem Solar Cells Using Luminescence Imaging, Lock-In Thermography, and 2D/3D Simulations" Sol. RRL 2023, 7, 2300249

Luminescence experiment

- To check whether relevant defects require special conditions to become visible we capture EL images at various levels of current injection density

- To examine whether relevant defects are continuously detectable, we examine dark forward biasing as possible means for preconditioning while acquiring EL images at regular intervals during the biasing period

- Measurement of J-V characteristics before and after prolonged forward biasing

Luminescence images of pero-Si tandem solar cells at various current injection levels [1]

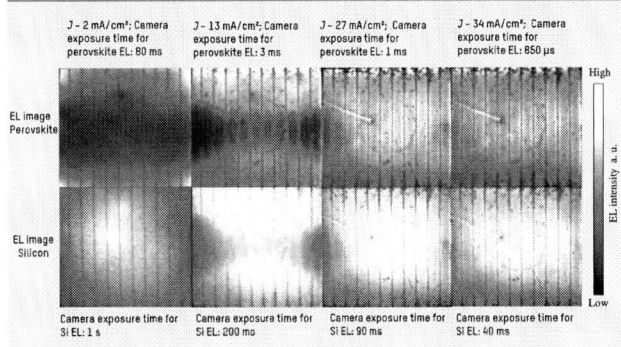

- Defect signatures in EL images vary strongly with current injection level

- Resistive coupling: dark spots (shunts[2]) in pero EL image correlate with bright spots in Si EL image (at $J \sim 2$ mA/cm²)

- Image correlation at $J \gtrsim 13$ mA/cm² reveals that series resistance dominates visible structures

- Signatures from cell processing tools visible (at $J \gtrsim 13$ mA/cm²)

Luminescence images under prolonged forward biasing [1]

- Low current injection → Reversible degradation, complete recovery of the cell during overnight dark storage

- High current injection → Irreversible degradation, incomplete recovery of the cell during overnight dark storage

Conclusion

- Defect correlation: Significant correlation found in defect patterns of 2T tandem solar cells due to resistive coupling mechanisms

- Impact of current injection level: Strong dependence of appearance and correlation of defect signatures on current injection

- Stabilization conditions: Importance of defining clear stabilization conditions for effective integration into mass production, dark-forward biasing found to be detrimental for samples examined in this study

41st European Photovoltaic Solar Energy Conference and Exhibition

工業技術研究院
Industrial Technology
Research Institute

41th EU PVSEC
24 Sept. 2023
2BV.1.35

A Case Study of certainly I-V measurement of the perovskite solar cell under dim light intensity for Solar/Indoor lighting Application

Yean-San Long[1]*, **Min-An Tsai**[1], Hsin-Hsin Hsieh[1] and Fan-Hsuan Yeh[3]

Industrial Technology Research Institute (ITRI)[1] and Taipei First Girls High School[2]

*mickeylong88@itri.org.tw

ABSTRACT

The measured I-V hysteresis is complicated by the vast array of different perovskite solar cell (PSC) device architectures. In our study, we used a dynamic I-V, RTOS method, for I-V then the results showed better accuracy by eliminating in real time the acceptance effect. We also used this method with the testing procedure to compare emerging PV hysteresis behavior under dim light intensity of indoor lighting simulator and solar simulator. Therefore, we will compare the difference between delay-time and RTOS method under dim light intensity, there are shown RTOS method more certainly I-V measurement of PSC under the dim light intensity for Solar/Indoor lighting Application.

The Indoor lighting simulator (Fig. 1a) included D65 (6500 K, Average North USA sky daylight) / TL84 (4100 K, European shop fluorescent) / CWF (4150 K, cool white fluorescent, shop lighting) / U30 (3000 K, shop lighting) / A (2856 K, typical home lighting), meeting CIE Standard illuminants and SEMI PV80. The adjustable light intensity (ranging from 0 to 2500 lx) is supplied with an ultraviolet light source for aging testing, with non-uniformity of less than 2 % (at 15 cm x 15 cm) and temporal instability of less than 2 %. During the I-V measurement in the dark and under Indoor standard lamp system and the scan direction was forward and backward, the sample temperature should be stable at 25 °C with a fluctuation of less than 1 °C, and the irradiance and luminance were determined using a NIST-traceable spectroradiometer (StellarNet) and an NML (National Measurement Laboratory, Taiwan)-traceable lux meter (HIOKI, model FT3424).

I-V measured by RTOS method during the real-time monitoring chart (see Fig. 1b) can be removed the capacitance effect of forward/backward under different level lighting. The first advantage can monitor I-V immediately without capacitive effect, available as Fig. 2(a). The other advantage of forward over to backward in I-V curve is very closely to unity, more accurately and rapidly when evaluating sample performance, Ex. Pmax/Isc/Voc/FF.

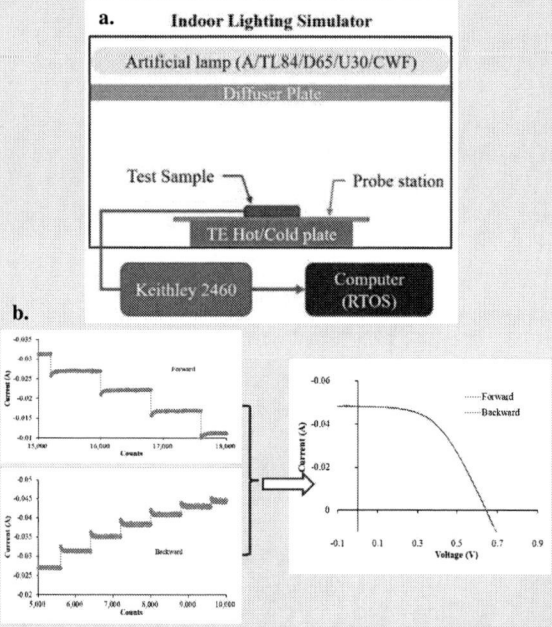

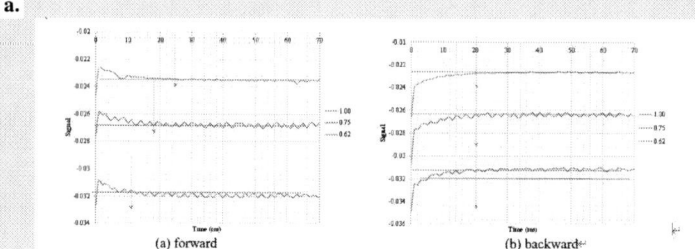

Figure 1: a. Construction of indoor lighting simulator with RTOS measuring method; **b.** Real-Time One-Sweep (RTOS) method, we have developed the special step I-V curves to analyze stability by optimization algorithm under the stepwise figure.

Figure 2: a. Capacitance effect comprised difference between forward and backward at level lighting (1.00/0.75/0,62). **d.** Principle of operation and energy level scheme of the DSC device in forward/backward

Conclusions

Nowadays, the time constant for equilibrium of PSC and DSC can be estimated by the measurement of the transient photocurrent under the application of a stepwise-changed voltage, an overshot current appears immediately after an abrupt increase of applied voltage, and then the current gradually decreases to an equilibrium state. The response time, which relies on lighting level and property of dye, is to decrease while increasing level lighting in the forward and to keep constant in the backward. Because the external biased potential can affect dye difference caused by the conversion step like capacitance effect of forward/backward under different level lighting. While using the way of fixed delay time to get the I-V curve, it will cause different set benchmarks, and not enough to achieve complete removal of measurement errors caused by capacitance effect

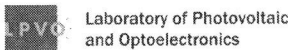

Laboratory of Photovoltaics and Optoelectronics

41st European Photovoltaic Solar Energy Conference and Exhibition

Perovskite Solar Cell Light-Soaking and Relaxation Modelling for Improved Energy Yield Predictions in Indoor Environments

Matija Pirc, Špela Tomšič, Marko Jošt, Marko Topič
University of Ljubljana, Faculty of Electrical Engineering, Tržaška 25, SI-1000 Ljubljana, Slovenia

FE UNIVERSITY OF LJUBLJANA
Faculty of Electrical Engineering

Motivation

- Light-Soaking is at least as important in indoor conditions as in outdoor conditions
  Pirc et al, "Indoor Energy Harvesting With Perovskite Solar Cells for IoT Applications–A Full Year Monitoring Study," ACS Appl. Energy Mater., https://doi.org/10.1021/acsaem.3c02498.

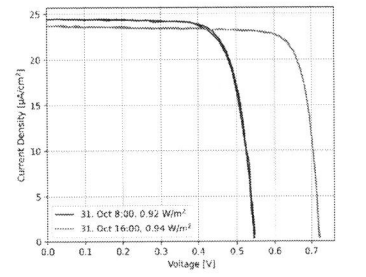
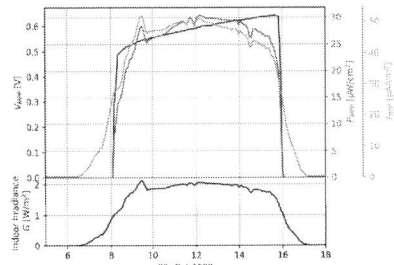

- Can the Light-Soaking effects be predicted based on the history of irradiance G?

Basic SC model

- One-diode model w/o shunt resistance

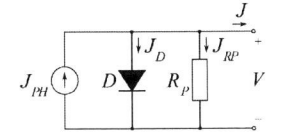

$$J = J_{PH} - J_S \left(\exp\left(\frac{V}{nkT/q} \right) - 1 \right) - \frac{V}{R_P}$$

- Fit based on a month of IV scans (710)

Direct correlations?

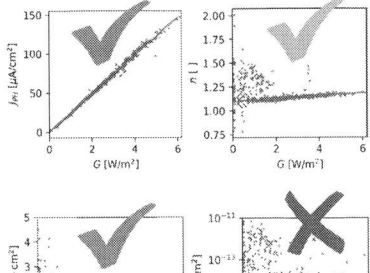

The Light-Soaking model

- J_S – huge range, strongly depends on n → unsuitable to be model output
- Combine n, J_S, R_P → $V_{OC@J_{PH}=const}$ ← make it the Light-soaking model output

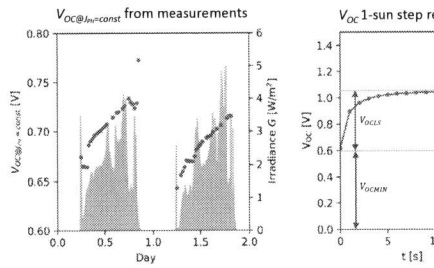
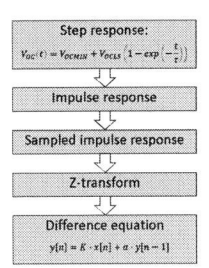

Step response:
$$V_{OC}(t) = V_{OCMIN} + V_{OCLS} \left(1 - \exp\left(-\frac{t}{\tau} \right) \right)$$
↓
Impulse response
↓
Sampled impulse response
↓
Z-transform
↓
Difference equation
$$y[n] = K \cdot x[n] + a \cdot y[n-1]$$

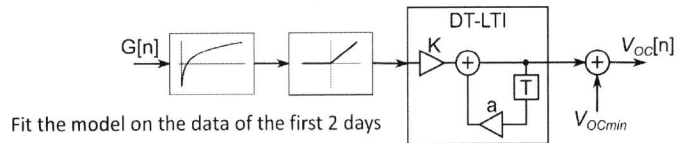

- Fit the model on the data of the first 2 days

Results

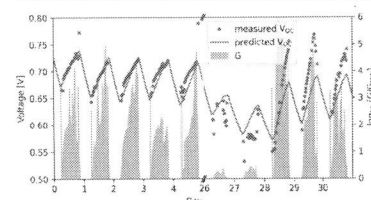
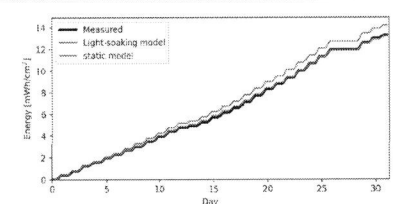

Further details in...

M. Pirc, "Empirical Light-soaking and Relaxation Model of Perovskite Solar Cells in an Indoor Environment," Informacije MIDEM, vol. 54, no. 2, pp. 149–161, Jul. 2024, https://doi.org/10.33180/InfMIDEM2024.207

Acknowledgements

- Žan Ajdič for preparation of the samples and other laboratory colleges, especially Gašper Matič for many fruitful discussions
- Slovenian Research and Innovation Agency: Research Programme P2-0415

41st EU PVSEC 2BV.1.39, 23 - 27 September 2024, Vienna, Austria

matija.pirc@fe.uni-lj.si

Modelling the effects of tandem module circuit configurations

M. Ignacia Devoto[1], Daniel Tune[1*], Ahmer A.B. Baloch[2], Karl Wienands[1], Rüdiger Farneda[1], Bhaskar Parida[2],
Omar Albadwawi[2], Vivian Alberts[2], Andreas Halm[1]
[1] International Solar Energy Research Centre (ISC) Konstanz
[2] DEWA Research & Development Center, P.O. Box: 564 Dubai, United Arab Emirates
*daniel.tune@isc-konstanz.de

ABSTRACT: To study the effects of tandem PV module circuit configurations without the influence of measurement-induced changes to sensitive perovskite layers in the system, a tandem PV simulation module is produced based on specially designed interdigitated back contact (IBC) silicon solar cells with three electrical terminals and such that the module can be reconfigured into any possible terminal circuit configuration (2T, 3T, 4T). Using the measurements obtained from this tandem module as validation, a circuit model of the emulated module is constructed in the MATLAB Simulink environment to study electrical properties. The module and experimentally validated tool are then used to assess the effect of bypass diodes during shading conditions.

1 INTRODUCTION

Progress toward the commercialization of perovskite technology has driven significant research in tandem solar cell technology. To best capture the solar radiation, perovskite-silicon tandem cells are designed with perovskite on top (sunny side) and silicon on the bottom. Over the last decade, much focus has been placed on improving their efficiency. Now that they have surpassed the efficiency of silicon-based solar cells, attention has shifted toward improving their durability, which is primarily driven by the ability of the top perovskite sub-cell to remain stable over longer periods, since the bottom silicon cells' stability has been developed to be so good as to warrant 30+ year performance warrantees on commercial products. However, at this time the durability of perovskite–silicon tandems, especially in large-area modules, is still unacceptable, with a wide variability in reported service life at the laboratory level and relatively few reports on their actual outdoor performance over extended periods [1]. Partial shading and reverse bias degradation have become some of the many challenging topics for tandem technology, where reverse bias is one of the main stressors that cannot be mitigated through module packaging, and must, therefore, be addressed either at the cell or interconnection level. Reverse bias occurs due to unequal current generation in the series-connected cells, which can result from partial shading, local differences in aging, manufacturing defects, and more.

During the partial shading of a module, one shaded cell in a string can either reduce the current through the non-shaded cells, or go into breakdown by causing the non-shaded cells to produce higher voltages than the reverse breakdown voltage of the shaded cell. Bypass diodes have traditionally been used in standard crystalline silicon modules to protect solar cells under reverse bias conditions. Assuming an open-circuit voltage (V_{oc}) of 0.7 V and a bypass diode forward bias (V_{bpd}) of 0.5 V, the maximum reverse bias (V_{rev}) that a shadowed cell could experience in a standard 60-cell module with one bypass diode per 20 cells is approximately -14 V ($V_{rev} = -19 \times V_{oc} - V_{bpd}$). Since Passivated Emitter Rear Contact (PERC) solar cells have a breakdown voltage (V_{bd}) greater than -20 V [2], bypass diodes are activated before any shadowed cell approaches its breakdown bias, thus protecting the cell. Once a bypass diode switches on, the power dissipation of the shaded cell is reduced by limiting its reverse voltage drop, which in turn limits the current passing through it, thereby preventing the development of hot spots.

Perovskite sub-cells in tandem cells are not as robust compared to, for example, PERC cells. Although increasing the breakdown voltage of perovskite cells is an ongoing research effort, and one group has already demonstrated V_{bd} exceeding -15 V [3], most values for perovskite cells range between -1 V and -5 V, depending on processing conditions [4, 5, 6]. This suggests that protecting perovskite cells from reverse bias would require one diode per cell, which is impractical and costly. Therefore, alternative solutions must be sought to resolve this issue.

According to a study by Di Girolamo and Dupré et al., silicon bottom cells are capable of protecting perovskite top sub, but this is limited by their V_{bd} and R_{sh} (shunt resistance) [6]. Similar conclusions were given in a study made by Xu et al. [7]. Assuming a bypass diode architecture for a tandem-based module similar to a standard 60-cell module, the reverse bias of a shaded tandem cell could extend to about -38 V (for $V_{oc} = 2.0$ V). Considering a $V_{bd} = -3$ V and $R_{sh} = 500$ Ω cm² for the perovskite top cell, Di Girolamo and Dupré et al. concluded that a silicon bottom cell would require a $V_{bd} \leq -40$ V to prevent a large reverse bias on the perovskite top cell before the bypass diode activates. Moreover, they demonstrated that silicon cells with a shunt resistance of 5 to 500 kΩ cm² are able to delay perovskite cell reverse bias until the bypass diode activates. However, in the event of multiple shaded cells, higher shunt resistances provide a better protection, as the current allowed through the shaded cells before the bypass activates is greatly reduced. This means that high shunt resistances prevent early breakdown degradation, which is also known to be potentially irreversible in some cases [4].

It is also important to note that for three-terminal tandems, the bottom cell must be a rear-contact technology (with an additional front terminal). This requirement imposes a particular constraint on interdigitated back-contact (IBC) silicon technology, as it is known for its relatively low soft breakdown voltage ($V_{bd} \sim -3.7$ V) [8]. In the case of single-junction modules, cells with such soft breakdown characteristics have a distinct advantage, as partial or complete shadowing of one cell does not necessarily trigger the bypass diode, which increases the energy yield of the photovoltaic system. Additionally, a reverse-biased cell will intrinsically dissipate less power under operating conditions, as its reverse breakdown is uniform across the entire cell.

2 AIM AND RELEVANCE

The aim of this work is to create tools to better understand the real-world effects of different tandem terminal circuit configurations and how bypass diodes can be used to protect tandem cells in different shading scenarios.

3 METHODOLOGY

As the main purpose is to understand diode behavior and requirements in a tandem module similar to standard module, the performance of different tandem terminal circuit configurations (e.g., 2T, 4T, 3T s/r/x, etc.) under STC and a variety of relevant shading scenarios is first compared without the use of any diode. To achieve this without the influence of measurement-induced changes to sensitive perovskite layers in the system, a 22-cell module using IBC technology is manufactured. Accompanying this test module, models for each terminal configuration are develop in MATLAB Simulink and they are validated against experimental data using experimental results from the test module. Several shading conditions with and without bypass diodes are experimentally evaluated and compared to predicted simulations. Finally, models are used to predict other shading conditions for tandem modules with increased number of cells and their response with and without bypass diodes.

4 TEST MODULE

The 22-cell module (see **Figure 2a**) consists only of specially produced IBC silicon solar cells with three electrical terminals (see **Figure 1**). The cells have open circuit voltage of around 0.65 V and two of these cells are used in series to simulate a "perovskite" sub-cell with 1.3 V. By bringing all the contacts out of the rear side of the module, the module circuit can be reconfigured to simulate any current- or voltage-matched (vm) tandem module configuration and thus isolate and evaluate many purely circuit-related effects (see **Figure 2b**).

Due to the voltage matching condition in 3T strings, end losses are intrinsic [9]. As shown in **Figure 3**, under 3T configuration some cells are observed to be partially operating (under reduced voltage) and others are completely disconnected (no current passing through the cell).

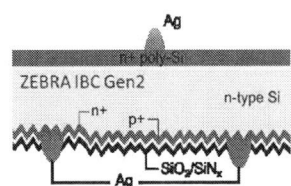

Figure 1. Cross section of a 3T 6BB ZEBRA IBC cell.

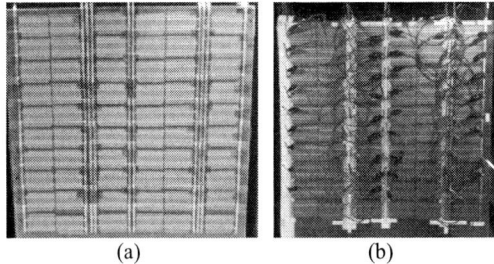

Figure 2. Images of (a) actual test module and (b) test module wiring.

As reported by McMahon *et al.*, end losses in 3T-s configuration are larger than in 3T-r [9]. This is because sub-cells that are completely to the left of GND or completely to the right of V$_{load}$ are disconnected, which only occurs in 3T-s. For 2:1 voltage-matched s-type strings as shown in **Figure 3**, bottom cells Nr. 1 and Nr. 2 and top cell Nr. 20 (from left to right) are not connected (hence dark EL) while top cell Nr. 1 and Nr. 19 are operated at approximately half-voltage (hence reduced EL signal). Similarly, for 2:1 voltage-matched r-type string, bottom cell Nr. 1 is not connected while top cells Nr. 1 and Nr. 20 operate at about half-voltage. In both cases, the end losses shown in the EL images agree well with the McMahon *et al.* study [9].

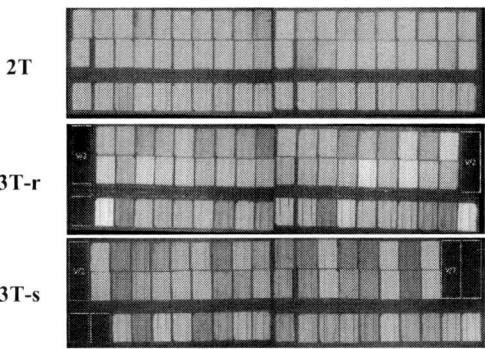

Figure 3. Electroluminescence images of 2T, 3T-r and 3T-s tandem configurations showing end losses.

5 MATLAB Simulink models

To better understand the effects observed and to predict other scenarios that cannot be easily experimentally tested, different circuit models for the test model are constructed in the MATLAB Simulink environment. These models consider a tandem cell as two cells in series with different voltages: one for top (perovskite) cell and one for the bottom (silicon) cell with V$_{oc}$ = 1.3 V and V$_{oc}$ = 0.65 V, respectively. The same input parameters such as incident radiation and diode parameters (R$_s$, R$_{sh}$, V$_{oc}$, I$_{sc}$) are assumed across all solar cells in their respective top and bottom strings. Between top and bottom cells and also between tandem cells several lumped resistors are used to include the resistances of the actual wiring in the test module. Also, initial values for the model are derived from the average I–V parameters (three devices) obtained from STC measurements. The arrangement of the individual top and bottom cells including the wiring follows the schematics shown in **Figure 4**, **Figure 5** and **Figure 6**, which only differ in the terminal configuration.

41st European Photovoltaic Solar Energy Conference and Exhibition

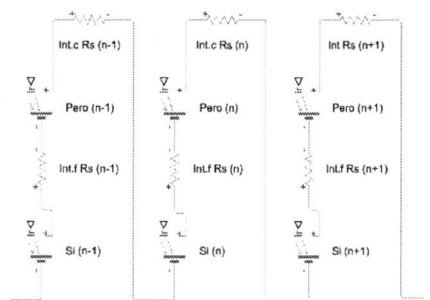

Figure 4. MATLAB Simulink model of a **2T** tandem circuit.

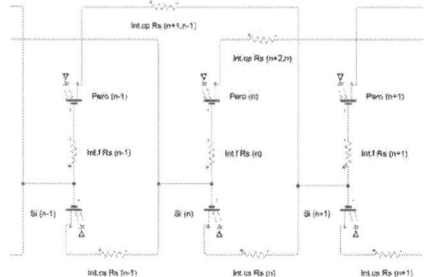

Figure 5. MATLAB Simulink model of a **3T-r** tandem circuit.

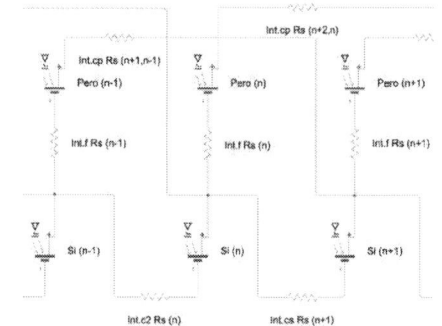

Figure 6. MATLAB Simulink model of a **3T-s** tandem circuit.

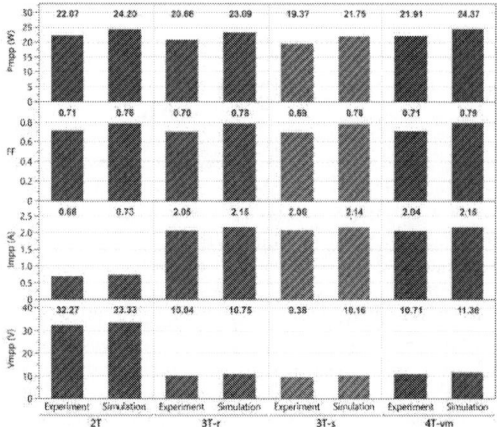

Figure 7. Comparison between module I–V measurements and initial simulation output for the four terminal configurations modelled.

Measurements obtained from the simulation module coupled with reported material and performance characteristics of real perovskite devices are used to validate the initial model (see **Figure 7**) and the model is then improved through successive iterations of increasing circuit complexity. **Figure 7** shows that, depending on the terminal configuration, predictions from simulations have a relative percentage deviation from experimental results of about 9 % to 13 % for maximum power and fill factor and 3 % to 8 % for current and voltage at the maximum power point.

7 SHADING, REVERSE BIAS AND BYPASS DIODES

There is an important question of whether and in what situations diodes can protect several tandem cells connected in series, as perovskite cells have a higher voltage compared to silicon cells. As shown in **Figure 8**, in a 2T tandem configuration, one single diode is able to protect up to 9 cells. Under STC conditions the module generates 45 W at its maximum power point and when a single cell is complete shaded (top and bottom) the output power drops to 22 W. However, by connecting a *single* bypass diode (bpd) between the shaded cell the output power partially recovers (42 W). Moreover, when the bypass diode is connected to a larger number of cells (3 to 9), the power output decreases each time a single cell is included, which means that the diode is able to protect the added cell. Thus, a single bypass diode connected in the same way as that in a 60-cell standard c-Si module can still protect tandems cells but the number of cells protected is limited. This is observed in **Figure 8**, as adding 10 to 14 cells to the bypass diode does not have any effect on the power output (23 W), which is similar to the shaded cell scenario without having any bypass diode in the circuit (22 W). Furthermore, the capacity of the bypass diode to protect tandem cells is dependent on the silicon bottom cell V_{bd}.

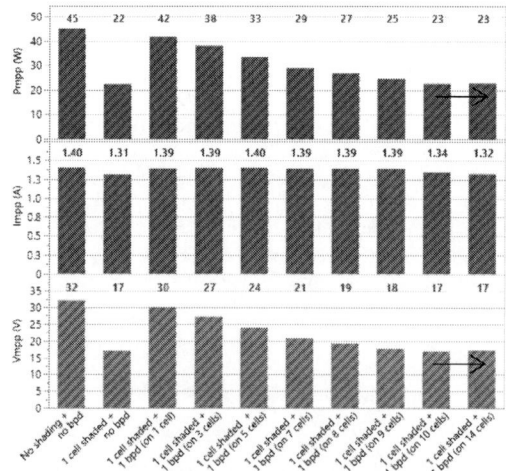

Figure 8. Experimental I–V results for 2T tandem configuration under STC, single shaded cell, shaded cell and single bypass diode (bpd) connected in different ways.

The 2T tandem configuration model was also validated under different shading conditions with and without a single bypass diode connected to a single or

332

several cells. **Table 1** shows a summary of the relative deviation (in %) between experimental and simulated results. Overall, the difference between experimental and simulated values is relatively low, considering that the module performance was affected by differences in intercell connection resistance, whereas in the model all such resistances were equal

Table 1. Relative deviation (in %) between experimental and simulated results for 2T configuration under different shading conditions and number of cells protected by a single bypass diode (bpd).

Scenario	ΔI_{sc}	ΔV_{oc}	ΔI_{mpp}	ΔV_{mpp}	ΔP_{mpp}
No shading & No bpd	−0.40	−0.07	+5.51	+0.66	+6.20
1 cell shaded No bpd	+2.33	−13.77	−9.93	+7.10	−3.54
1 cell shaded **1 bpd on 1 cell**	+0.03	−5.46	+5.25	−7.24	−2.36
1 cell shaded **1 bpd on 3 cells**	−0.35	−3.79	+5.66	−4.28	+1.14
1 cell shaded **1 bpd on 9 cells**	−0.70	−3.23	+5.45	−2.15	+3.19
2 cells shaded 1 bpd on 9 cells	−0.78	−1.49	+5.56	−1.55	+3.92
1 cell shaded 1 bpd on **10 cells**	−0.41	−3.23	+5.70	−3.54	+1.96

As 3T tandem configurations have two interwoven circuits in the strings, there is still the question of if and how diodes can be used; how many and where? Although not shown in this publication, 3T-r experimental results are identical to 2T, i.e., one diode can protect a limited number of cells (up to 9 for the test module in this study). However, one polarity of the bypass diode must be connected to one end of the string, thus limiting the strings/module design layout. No effective implementation of diodes in the 3T-s circuit was found in this work. However, further studies are ongoing to determine if and how bypass diodes could be used in such circuits.

8 CONCLUSIONS

In conclusion, a tandem simulation module consisting of specially fabricated ZEBRA IBC silicon cells with an additional frontside electrode was produced and this was used to validate the output of a Matlab Simulink circuit model of tandem circuits, with relatively low difference in I-V characteristics of the model output compared to the simulation module.

To the question of whether or not the bottom silicon cell can protect the perovskite top cell from reverse bias damage in 3T circuits;

- In principle, yes, if the silicon reverse breakdown voltage V_{BR} is high enough
- However, the cell must be silicon-limited
- There exists a strong interplay between V_{BR} and number of cells under diode

Furthermore, the positive characteristic of soft breakdown in IBC cells in single-junction modules becomes undesirable for bottom cells in tandem configurations, as the low V_{bd} of IBC cells does not protect the perovskite top cell. Given that in IBC technology, p$^+$ and n$^+$ regions are positioned close to each other due to both positive and negative contacts being on the rear side, the IBC soft breakdown characteristic must be re-engineered to be higher for their use in 3T tandem cells [8].

9 ACKNOWLEDGEMENTS

AB, OB, BP, and VA would like to acknowledge the support from DEWA R&D for the granted project on three-terminal perovskite-silicon tandem solar cells.

10 REFERENCES

[1] S. Baumann et al., "Stability and reliability of perovskite containing solar cells and modules: degradation mechanisms and mitigation strategies," *Energy Environ. Sci.*, p. 10.1039.D4EE01898B, 2024, doi: 10.1039/D4EE01898B.

[2] R. Witteck, M. Siebert, S. Blankemeyer, H. Schulte-Huxel, and M. Kontges, "Three Bypass Diodes Architecture at the Limit," *IEEE J. Photovolt.*, vol. 10, no. 6, pp. 1828–1838, Nov. 2020, doi: 10.1109/JPHOTOV.2020.3021348.

[3] F. Jiang et al., "Improved reverse bias stability in p–i–n perovskite solar cells with optimized hole transport materials and less reactive electrodes," *Nat. Energy*, Aug. 2024, doi: 10.1038/s41560-024-01600-z.

[4] R. A. Z. Razera et al., "Instability of p–i–n perovskite solar cells under reverse bias," *J. Mater. Chem. A*, vol. 8, no. 1, pp. 242–250, 2020, doi: 10.1039/C9TA12032G.

[5] C. Wang et al., "Perovskite Solar Cells in the Shadow: Understanding the Mechanism of Reverse-Bias Behavior toward Suppressed Reverse-Bias Breakdown and Reverse-Bias Induced Degradation," *Adv. Energy Mater.*, vol. 13, no. 9, p. 2203596, Mar. 2023, doi: 10.1002/aenm.202203596.

[6] D. Di Girolamo et al., "Silicon / Perovskite Tandem Solar Cells with Reverse Bias Stability down to −40 V. Unveiling the Role of Electrical and Optical Design," *Adv. Sci.*, vol. 11, no. 31, p. 2401175, Aug. 2024, doi: 10.1002/advs.202401175.

[7] Z. Xu et al., "Reverse-bias resilience of monolithic perovskite/silicon tandem solar cells," *Joule*, vol. 7, no. 9, pp. 1992–2002, Sep. 2023, doi: 10.1016/j.joule.2023.07.017.

[8] H. Chu, L. J. Koduvelikulathu, V. D. Mihailetchi, G. Galbiati, A. Halm, and R. Kopecek, "Soft Breakdown Behavior of Interdigitated-back-contact Silicon Solar Cells," *Energy Procedia*, vol. 77, pp. 29–35, Aug. 2015, doi: 10.1016/j.egypro.2015.07.006.

[9] W. McMahon et al., "Homogenous Voltage-Matched Strings Using Three-Terminal Tandem Solar Cells: Fundamentals and End Losses," *IEEE J. Photovolt.*, vol. 11, no. 4, pp. 1078–1086, Jul. 2021, doi: 10.1109/JPHOTOV.2021.3068325.

41st European Photovoltaic Solar Energy Conference and Exhibition

Lise Watrin[1,2], François Silva[2], Cyril Jadaud[2], Pavel Bulkin[2], Jean-Charles Vanel[2], Kassiogé Dembélé[2], Erik V. Johnson[2], Karim Ouaras[2], and Pere Roca i Cabarrocas[1,2]

[1]Institut Photovoltaïque d'Ile-de-France (IPVF), 18 Boulevard Thomas Gobert, 91120 Palaiseau, France
[2]LPICM, CNRS, Ecole Polytechnique, Institut Polytechnique de Paris, route de Saclay, 91120, Palaiseau, France

III-V thin films growth by RP-CVD: *Towards a reduction of industrialization costs*

Context and motivation

- III-V solar cells (SCs) hold the highest efficiency of any SC technology available today: i.e. 47,6% (4J, 665 suns)[1]
- **However**, the cost of III-V SCs is a hundred times higher than that of c-Si SCs[2] and most of this cost difference is due to:
  - i) the expensive growth process (MBE or MOCVD)
  - ii) the expensive III-V or Ge substrate required to epitaxially grow the III-V layers
- **Remote-Plasma Chemical Vapor Deposition (RP-CVD)**, a process compatible with cost reduction:
  - ✓ Low temperature → e.g. GaN: 500°C[3] vs. ~ 1000°C for MOCVD
  - ✓ Low gas consumption thanks to low pressure (e.g. GaAs: 0.5 mbar) → few 10^{-3} vs. 10^2 L/min for MOCVD

Illustration of the III-V RP-CVD reactor

① Growth of homoepitaxial GaAs

TMGa/AsH₃, 500°C, 150 W, V/III = 18, 0.5 mbar.

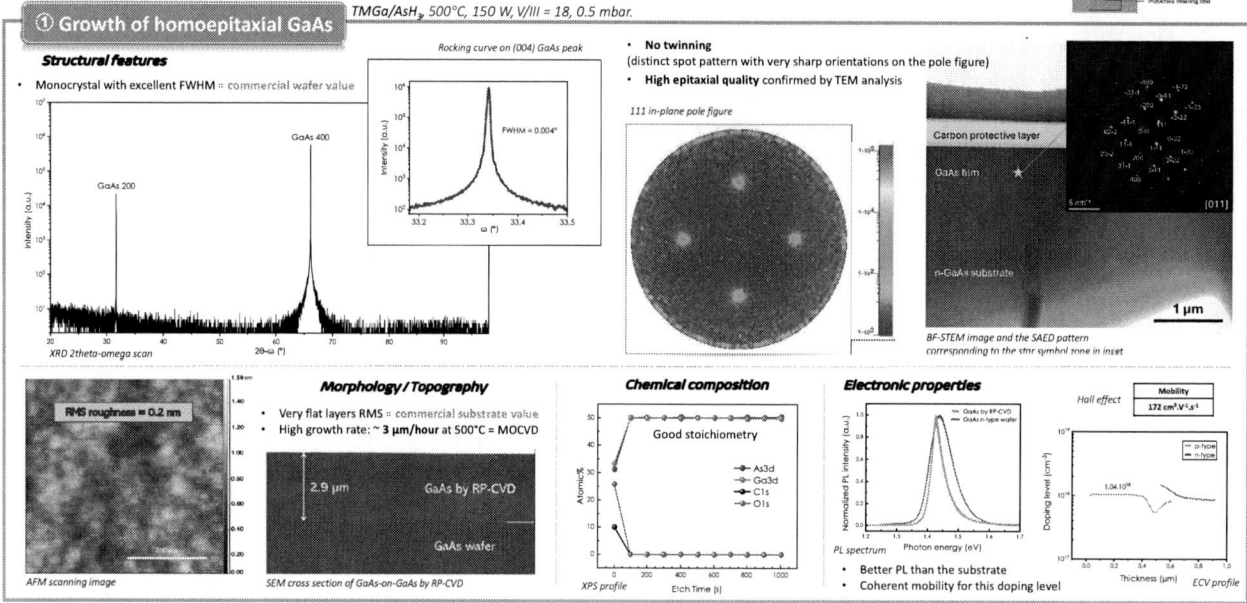

Structural features
- Monocrystal with excellent FWHM ≈ commercial wafer value

Rocking curve on (004) GaAs peak
FWHM = 0.004°

XRD 2theta-omega scan

- **No twinning** (distinct spot pattern with very sharp orientations on the pole figure)
- **High epitaxial quality** confirmed by TEM analysis

111 in-plane pole figure

Carbon protective layer
GaAs film
n-GaAs substrate
BF-STEM image and the SAED pattern corresponding to the star symbol zone in inset

RMS roughness ≈ 0.2 nm
AFM scanning image

Morphology / Topography
- Very flat layers RMS ≈ commercial substrate value
- High growth rate: ~ 3 µm/hour at 500°C ≈ MOCVD

2.9 µm GaAs by RP-CVD
GaAs wafer
SEM cross section of GaAs-on-GaAs by RP-CVD

Chemical composition
Good stoichiometry
As3d, Ga3d, C1s, O1s
XPS profile

Electronic properties
PL spectrum
- Better PL than the substrate
- Coherent mobility for this doping level

Hall effect — Mobility 172 cm²·V⁻¹·s⁻¹
$1.04 \cdot 10^{18}$
ECV profile

② MOCVD versus RP-CVD: the benefit of plasma

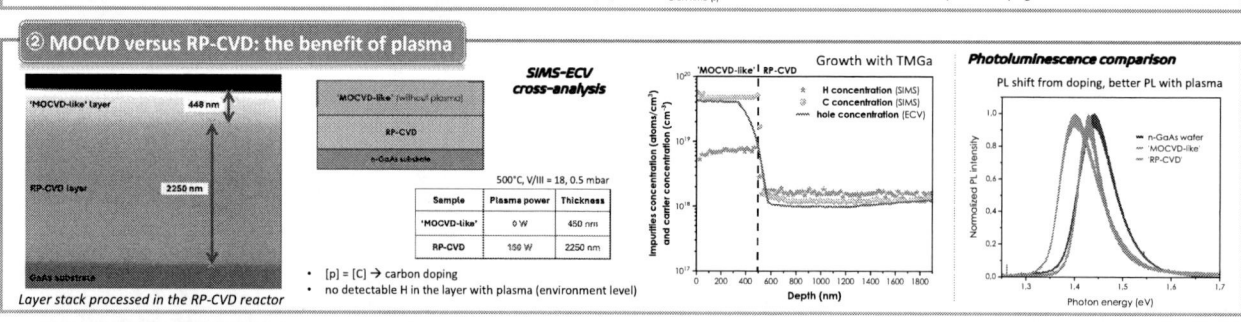

'MOCVD-like' layer 448 nm
RP-CVD layer 2250 nm
GaAs substrate
Layer stack processed in the RP-CVD reactor

SIMS-ECV cross-analysis
'MOCVD-like' (without plasma)
RP-CVD
n-GaAs substrate
500°C, V/III = 18, 0.5 mbar

Sample	Plasma power	Thickness
'MOCVD-like'	0 W	450 nm
RP-CVD	150 W	2250 nm

- [p] = [C] → carbon doping
- no detectable H in the layer with plasma (environment level)

Growth with TMGa
H concentration (SIMS), C concentration (SIMS), hole concentration (ECV)

Photoluminescence comparison
PL shift from doping, better PL with plasma
n-GaAs wafer, 'MOCVD-like', RP-CVD

③ RP-CVD layer integration in solar cell devices

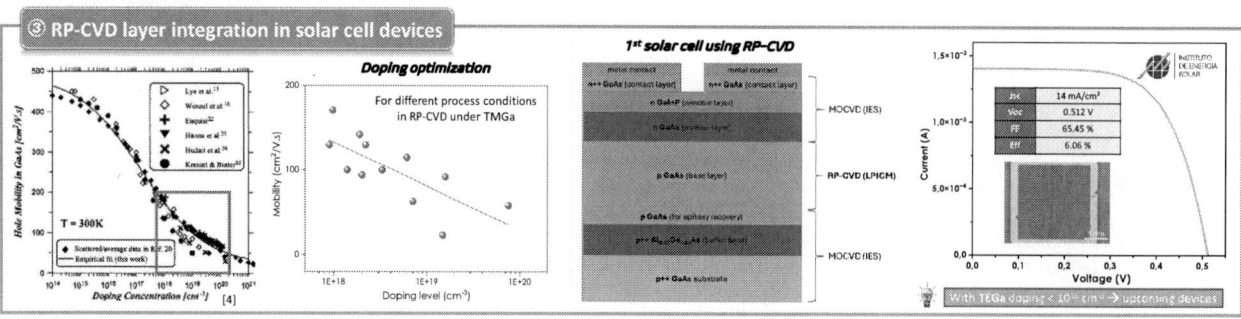

T = 300K
Doping Concentration [cm⁻³] [4]

Doping optimization
For different process conditions in RP-CVD under TMGa
Doping level (cm⁻³)

1ˢᵗ solar cell using RP-CVD
metal contact — n++ GaAs (contact layer)
n GaInP (window layer) — MOCVD (IES)
n GaAs (emitter layer)
p GaAs (base layer) — RP-CVD (LPICM)
p GaAs (for epitaxy recovery)
p++ AlₓGa₁₋ₓAs (buffer layer) — MOCVD (IES)
p++ GaAs substrate

Jsc	14 mA/cm²
Voc	0.512 V
FF	65.45 %
Eff	6.06 %

With TEGa doping < 10¹⁹ cm⁻³ → upcoming devices

Summary

✓ **Optimization of a new reactor** for III-V RP-CVD designed at LPICM

✓ Compatible with **cost reduction**
- Low temperature (~ 500°C)
- Low gas consumption thanks to low pressure (~ 0.5 mbar)

✓ **Homoepitaxial GaAs** growth
- GaAs of the same quality as industrial references
- 1ˢᵗ functional solar cell using GaAs by RP-CVD as absorbing layer

References
[1] Fraunhofer ISE Develops the World's Most Efficient Solar Cell with 47.6 Percent Efficiency
[2] Smith et al., CO: National Renewable Energy Laboratory, 2021
[3] Watrin, L.; Silva, F.; Jadaud, C.; Bulkin, P.; Vanel, J.-C.; Muller, D.; Johnson, E. V.; Ouaras, K.; Cabarrocas, P. R.. J. Phys. D: Appl. Phys. 2024, 57 (31), 315106.
[4] M. Sotoodeh, A. H. Khalid, A. A. Rezazadeh; J. Appl. Phys. 15 March 2000; 87 (6): 2890–2900.

41st European Photovoltaic Solar Energy Conference and Exhibition

MODELING AND MEASUREMENT OF LUMPED SERIES RESISTANCE WITH VARYING ILLUMINATION AND CURRENT CONDITION OF LOW-BANDGAP SOLAR CELLS

Shipei Zhang, Xiawa Wang*
Division of Natural and Applied Sciences, Duke Kunshan University
No. 8 Duke Avenue, Kunshan, Jiangsu Province, China 215316
email address: xiawa.wang@dukekunshan.edu.cn

ABSTRACT: Series resistance (R_S) significantly impacts the fill factor (FF) in photovoltaic devices, especially at high current levels, by causing a voltage drop that reduces the voltage at the maximum power point. Thermophotovoltaic (TPV) cells, which use thermal radiation from high-temperature sources like furnaces, generate large photocurrents due to their proximity to the heat source. Understanding R_S in TPV cells is essential for optimizing their performance and enhancing power conversion efficiency. This study examines the illumination and current dependence of R_S in low-bandgap solar cells, specifically germanium heterojunction cells and commercial GaSb cells from JX Crystal, Inc. The current density-voltage (J-V) curves of these cells are measured under blackbody irradiation, and simulations are conducted using SCAPS-1D software at temperatures ranging from 800°C to 1200°C. R_S values are extracted using three methods: analyzing the area under the J-V curve, determining the slope near the open-circuit voltage (V_{OC}), and applying the double illumination method. The study explores trends in R_S variation with blackbody temperature and finds that R_S values decrease with the rise in blackbody temperature across these three methods.
Keywords: lumped series resistance, thermophotovoltaic cells, low-bandgap solar cells, blackbody irradiation

1 INTRODUCTION

Low-bandgap solar cells are promising candidates for thermophotovoltaic (TPV) systems that have promisng applications in waste heat utilization, military application and space explorations. These cells can absorb a broader spectrum of solar radiation, extending into the near-infrared range, thus effectively converting infrared photons into electricity. This broad absorption capability renders low-bandgap solar cells particularly suitable for TPV applications, where they can utilize high-temperature thermal radiation from sources such as industrial waste heat [1], concentrated solar power [2], and even nuclear reactors [3]. The prospects for developing TPV applications are significant as industries and technologies increasingly move toward sustainable and efficient energy solutions [4]. Continuous advancements in materials science, including the development of new low-bandgap materials with improved thermal and electrical properties, are pushing the performance and viability of TPV systems forward.

In TPV systems, the power density received by the cellis significantly higher than the standard 1-sun illumination intensity. Consequently, the photogenerated current in TPV cells is typically two to three orders of magnitude greater than that in conventional solar cells. This substantial increase in current results in significant power losses due to enhanced series resistance (R_S) [5]. Therefore, accurate analysis and minimization of series resistance are critical for optimizing low-bandgap solar cell performance in TPV applications. The characterization and modeling methods for low-bandgap cells are derived from those used for silicon cells, using standard solar simulators for current density-voltage (J-V) measurement, and can be integrated into a diode model. However, assuming a constant R_S often proves inadequate for accurately predicting the cell's J-V curves under varying illumination conditions. Previous research has investigated this issue and developed models for silicon cells under a variable solar spectrum [6]. The variation in illumination intensity is also based on the solar spectrum with shading to represent real-world application scenarios.

The propose of this project is to bridge this gap by exploring how R_S values change as a solar cell's photocurrents increase by orders of magnitudes. By exploring the two most commonly used low-bandgap solar cells, germanium heterojunction and commercial GaSb homojunction TPV cells [7], under different illumination conditions, we aim to identify the dominant factors that govern these trends and provide insights for future device improvements. We characterized the cells' J-V curves under specific blackbody irradiation conditions and used SCAPS-1D software to simulate J-V curves under different blackbody emitter temperatures and irradiation intensities with the same view factor. The resistances will be summarized and compared using three methods, and we will further investigate their evolution with changing blackbody temperatures.

2 METHODOLOGY

In this study, three methods for determining R_S values are introduced in detail. These methods are based on the analysis of the J-V curve under varying conditions of blackbody temperature and irradiation intensity. For this portion of the study, we have chosen the J-V curve of a germanium heterojunction cell under blackbody calibrator irradiation at 1000°C as a reference to outline the specific steps required to calculate R_S values.

2.1 Compute area bordered by the J-V curve

In 1982, Araujo et al. proposed a novel method to determine the R_S value based on the single-diode model [8]. They began by solving an implicit equation for current density derived from this model. Next, they integrated the solution from zero to the short-circuit current density (J_{SC}). This process yielded the area A under the J-V curve, as shown in Figure 1(a). The simplified equation they described for calculating the R_S value is:

$$R_S = 2 \cdot \left(\frac{V_{OC}}{J_{SC}} - \frac{A}{J_{SC}^2} - \frac{k_B T}{q} \cdot \frac{1}{J_{SC}} \right)$$

where k_B is Boltzmann's constant, T is the temperature, q is the electronic charge, V_{OC} and J_{SC} are open-circuit voltage and short-circuit current density, respectively.

2.2 Calculate inverse slope near the V_{OC}

Figure 1(a) illustrates the R_S value as the inverse of the slope near the V_{OC} within the framework of the single-diode model [9]. The calculation of R_S is performed through a manual method that approximates the value of the ideality factor, applying the subsequent equation:

$$R_S = -\frac{\Delta V_{OC}}{\Delta I_{OC}}$$

where ΔV_{OC} and ΔI_{OC} are the changes in voltage and current density near the V_{OC}, respectively.

2.3 Compare two J-V curves under different illumination conditions

The double illumination method involves comparing the performance of a solar cell under two different illumination levels to determine the R_S value, a method initially introduced by Wolf et al [10]. This approach is based on analyzing shifts in the J-V curves that occur due to varying light intensities. The first shift in the curves reflects a change in current density corresponding to the alteration in incident power, which is directly related to the photo-generated current. The second shift, observed in the voltage, arises from the R_S losses encountered between the two illumination levels and is crucial for determining R_S.

Under lower illumination conditions, both the photocurrent and the consequent voltage drop across R_S are diminished compared to those under higher illumination. To determine the R_S value, the current difference is calculated by selecting the same current value between J_{SC} and an arbitrary current point, as shown in Figure 1(b). This lumped series resistance is calculated by inverting the slope of the line that intersects the corresponding points on the curves. The calculation uses the following equation:

$$R_S = \frac{V_1 - V_2}{J_2 - J_1}$$

where J_1 and V_1 are the current and corresponding voltage points on the J-V curve at 100% intensity, respectively, and J_2 and V_2 are the current and corresponding voltage points on the J-V curve at 90% intensity, respectively.

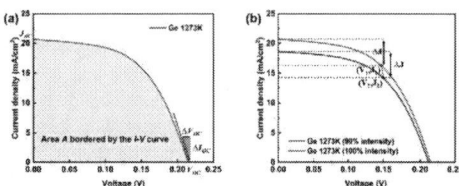

Figure 1: Illustration of the three methods used to determine R_S: (a) calculating the area bordered by the J-V curve, determining the slope near the V_{OC}, and (b) comparing two J-V curves obtained under different illumination intensities.

3 RESULT AND DISCUSSION

The J-V curves of two low-bandgap cells were first characterized using a source meter (Keithley 2636B, Tektronix) under irradiation from a blackbody system (IR-564/301, Infrared Systems Development) set at a temperature of 1000°C with a specific view factor. Subsequently, SCAPS-1D software was used to construct models for the two cells and simulate J-V curves consistent with the measured data, as shown in Figures 2(a) and 2(b). The models were then used to simulate J-V curves over a

temperature range from 800°C to 1200°C. This simulation for different emitter temperatures ensured that the view factor remained constant, eliminating errors caused by varying distances between the measured cells and the blackbody emitter.

Based on the J-V curves of the two low-bandgap cells under different blackbody temperatures, we used simulated J-V curves with different intensities (90% and 100% of the blackbody irradiation intensity) at each emitter temperature to extract the R_S values. By plotting the R_S value as a function of current density, the characteristics of $R_S(J)$ for the two low-bandgap cells are clearly demonstrated, as shown in Figures 2(c) and 2(d). In these two cells, the R_S values at low current densities are unstable and difficult to measure accurately, exhibiting a distinct drop, particularly under lower blackbody temperature conditions. This instability may be attributed to limitations in the stability of the input source [11].

Selecting the R_S values at the maximum power point (MPP) to represent the results of the double illumination method, we found that the R_S values tended to decrease with the rise in blackbody emitter temperature. When the emitter temperature exceeded 1000°C, R_S values tended to stabilize. Comparing the values between the two low-bandgap cells, the germanium cell exhibited lower R_S values than the GaSb cell, with a more pronounced difference at lower emitter temperatures due to the more significant reduction in carrier activation and mobility in GaSb as the temperature decreases.

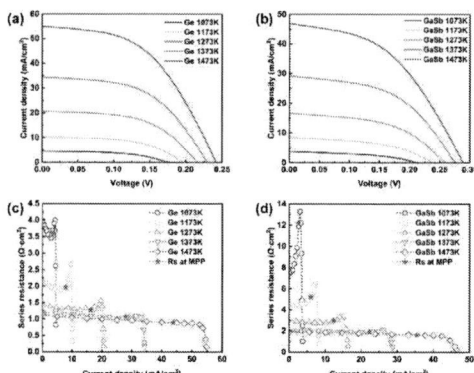

Figure 2: The J-V curves of (a) germanium heterojunction and (b) GaSb homojunction cells under different emitter temperatures. The R_S values against current density for (c) germanium heterojunction and (d) GaSb homojunction cells under different emitter temperatures.

By summarizing the R_S values obtained by the other two methods, namely the area bordered by the J-V curves and the slope near the V_{OC}, we plotted and compared the R_S evolution curve with emitter temperature, as shown in Figures 3(a) and (b). The detailed values are presented in Table I. Both of these low-bandgap cells show a similar trend in R_S values, with a significant decrease as the emitter temperature increases from 800°C to 1000°C. The rate of decrease slows and eventually stabilizes when the emitter temperature reaches 1100°C. Moreover, all three methods show the same difference in R_S values between germanium and GaSb cells: the slope method yields the highest R_S values, while the double illumination method yields the lowest R_S values.

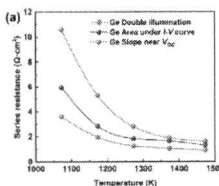 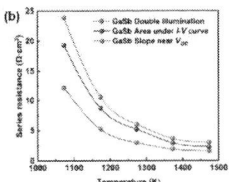

Figure 3: The R_S revolution curves of (a) germanium heterojunction and (b) GaSb homojunction cells as a function of emitter temperature, analyzed by three methods.

Table I: R_S values of germanium and GaSb cells measured using the slope near V_{OC}, the area bordered by the J-V curve, and the double illumination method.

Resistance (ohm·cm^2)	Slope	Area	Double
Ge 1073K	10.58	5.93	3.59
GaSb 1073K	23.89	19.31	12.17
Ge 1173K	5.30	2.85	1.97
GaSb 1173K	10.57	8.73	5.25
Ge 1273K	2.81	1.86	1.27
GaSb 1273K	6.05	5.24	2.99
Ge 1373K	1.89	1.65	1.07
GaSb 1373K	3.66	2.94	1.91
Ge 1473K	1.63	1.30	0.89
GaSb 1473K	3.01	2.35	1.62

4 CONCLUSIONS

In this study, we investigated the impact of R_S on low-bandgap solar cells under varying thermal illumination conditions. By employing three distinct methods—analyzing the area under the J-V curve, determining the slope near the V_{OC}, and using the double illumination method—we were able to gain comprehensive insights into the behavior of R_S. Our results indicate that R_S values decrease with increasing blackbody emitter temperature, finally stabilizing above 1000°C. Additionally, the germanium cell consistently demonstrated lower R_S values compared to the GaSb cell, with a more pronounced difference at lower temperatures due to reduced carrier activation and mobility in GaSb. The study finds that the slope method tends to yield the highest R_S values, while the double illumination method provides the lowest. The observed trends and differences in R_S values enhance our understanding of the factors affecting cell performance and contribute to the ongoing development of more efficient energy conversion technologies.

5 REFERENCES

[1] Z. Yang, W. Peng, T. Liao, Y. Zhao, G. Lin, J. Chen, Energy Conversion and Management 149 (2017) 424.

[2] H. R. Seyf, A. Henry, Energy Environmental Science 9 (2016) 2654.

[3] A. Datas, D. L. Chubb, Photovoltaic for Space (2023) 197.

[4] A. Allouhi, S. Rehman, M. S. Buker, Z. Said, Sustainable Energy Technologies and Assessments 56 (2023) 103026

[5] M. J. Heredia-Rios, L. Hernandez-Martinez, M. Linares-Aranda, M. Moreno-Moreno, J. F. Méndez, Energies 17 (2024) 1520

[6] F. Khan, S. N. Singh, M. Husain, Solar Energy Materials and Solar Cells 94 (2010) 1473

[7] L. Fraas, L. Minkin, J. Avery, H. She, L. Ferguson, F. Dogan, IEEE 42nd Photovoltaic Specialist Conference, (2015) 1

[8] G. L. Araujo, E. Sanchez, IEEE Transactions on Electron Devices 29 (1982) 1511

[9] M. Diantoro, T. Suprayogi, A. Hidayat, A. Taufiq, A. Fuad, R. Suryana, International Journal of Photoenergy 1 (2018) 9214820

[10] M. Wolf, H. Rauschenbach, Advanced Energy Conversion 3 (1963) 455

[11] K. C. Fong, K. R. Mclnrosh, A. W. Blakers, Progress in Photovoltaics: Research and Applications 21 (2013) 490

41st European Photovoltaic Solar Energy Conference and Exhibition

Color Implementation of Cu(In,Ga)Se$_2$ Thin-film Solar Cells with Multilayered Conductive Optical Filters

Yong-Duck Chung[1,2*], Dae-Hyung Cho[1,2], Rina Kim[1], Woo-Jung Lee[1,2], Tae-Ha Hwang[1], Soyoung Lim[1,2], Donghyeop Shin[3,4], Kihwan Kim[3,4], and Mangu Kang[1]

[1]Superintelligence Creative Research Laboratory, Electronics and Telecommunications Research Institute (ETRI), Daejeon, Korea

[2]Department of Semiconductor and Advanced Device Engineering, Korea University of Science and Technology (UST), Daejeon, Korea

[3]Photovoltaics Research Department, Korea Institute of Energy Research (KIER), Daejeon, Korea

[4]Department of Renewable Energy Engineering, Korea University of Science and Technology (UST), Daejeon, Korea

*ydchung@etri.re.kr

ABSTRACT: Traditional photovoltaic (PV) cells and modules exhibit a black color to maximize light absorption and minimize light reflection to achieve high efficiency. However, this monotonous color limits their installation in urban environments where aesthetic harmony is important. Therefore, instead of traditional black-colored PV modules, there is a need for colors that can naturally blend with the surroundings. Multilayer optical filters with interference properties at specific wavelengths are mainly used for color implementation. Multilayer coating is applied on a glass substrate, which is mostly used as PV cover glass. so that the transparent conductive oxide layer itself exhibited both color and excellent electrical conductivity, colorful PV modules can be fabricated without the need for additional components. In this study, the number and thickness of repeated Indium tin oxide (ITO) multilayer films were simulated by the Essential Macleod simulation tool. Next, highly conductive ITO thin films with two different SnO2 compositions were repeatedly deposited to implement red, green, blue, and dark blue color filters. Finally, the conductive optical filters were applied to Cu(In,Ga)Se$_2$ (CIGS) thin-film solar cells, and PV performances were investigated.

Keywords: conductive color filter, indium tin oxide, CIGS solar cell

1. INTRODUCTION

Traditional Photovoltaic (PV) cells and modules are typically black in color to maximize light absorption and minimize reflection to achieve high efficiency. However, this monotonous color limits their application in various fields, such as buildings, vehicles, and mobile devices for energy self-sufficiency [1]. In particular, black PV modules in urban environments are difficult to harmonize with the surroundings, which restricts the possible installation area of the PV modules. Therefore, aesthetic considerations are crucial for PV modules installed in highly visible urban environments. Instead of traditional black-colored PV modules, there is a need for colors that can naturally blend into the surroundings [2].

Multilayer optical filters with interference properties at specific wavelengths are mainly used for color implementation [3-4]. Mainly oxide-based multilayer thin films with high refractive index and low refractive index are alternately prepared to make constructive and destructive interference of a desired specific wavelength to realize reflection or transmission [5]. The color can be implemented by adjusting the refractive index, thickness, and the number of repeating layers in the thin films. The multilayer coating is applied on a glass substrate which is used as the cover glass of PV modules. If the transparent conductive oxide layer itself exhibited both color and excellent electrical conductivity, colorful PV modules could be fabricated without the need for additional components.

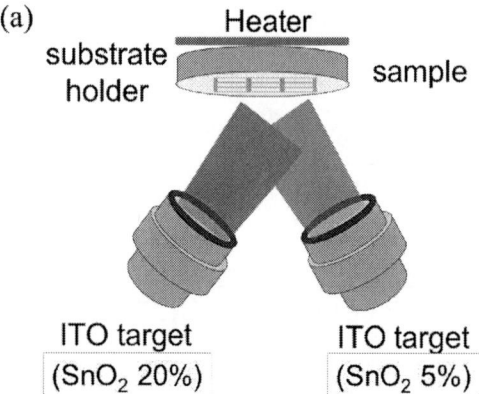
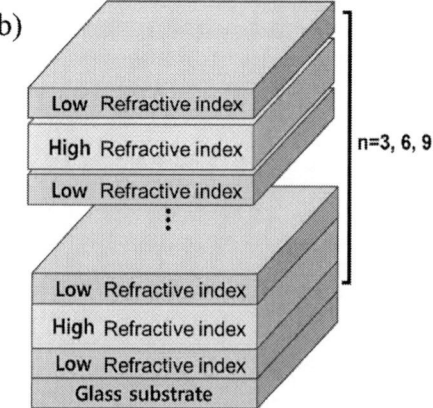

Figure 1. Schematic of (a) RF sputtering process for repeated multilayered optical filter and (b) the color filter fabricated on glass substrate

This study used the RF-sputtering method to fabricate a multilayer structure on a soda-lime glass (SLG) substrate by depositing indium tin oxide (ITO) thin films with two different SnO_2 compositions alternately. The multilayers were adjusted, varying the number of layers and thicknesses to implement specific colors, including red, green, blue, and dark blue color filters. The optical, electrical, and structural properties of the conductive color filters deposited on the SLG substrate were characterized. Red, green, blue, and dark blue color filters were applied to the $Cu(In,Ga)Se_2$ (CIGS) thin-film solar cell.

2. EXPERIMENTAL PROCUDUES

2.1 Deposition of ITO thin film

The ITO thin films were repeatedly deposited using sputter targets with two different SnO_2 compositions of 5 and 20 wt% via radio frequency (RF) magnetron sputtering as shown in Fig. 1. The argon gas flow rate was set to 50 sccm, and a working pressure of 5 mTorr was maintained throughout the process. The set substrate temperature and base pressure were 185°C and under 5.0 × 10^{-6} Torr, respectively.

2.2 Cell fabrication

The CIGS thin-film solar cells were fabricated following a structural configuration of Al/Ni/ITO/i-ZnO/CdS/CIGS/Mo/SLG. The Mo back contact was applied to SLG through DC sputtering. Subsequently, a CIGS absorption layer with a thickness of 2.2 μm was deposited by employing a multi-stage evaporation sequence. The CdS buffer layer was deposited through a chemical bath deposition process. An i-ZnO thin film and an ITO thin film were then deposited by utilizing RF sputtering. Finally, the solar cell areas were defined by mechanical scribing for electrical isolation.

3. RESULTS AND DISCUSSION

The Essential Macleod simulation tool simulated the number and thickness of the repeated multilayers. The optical transmittance and reflectance spectra of the multilayered color filters for red, green, blue, and dark blue colors with different numbers of repeated multilayers were compared with simulation and experiments (Fig. 2). As the number of repetitions increased, the full width at half maximum of the color filters in the target wavelength range became narrower, resulting in more distinct colors.

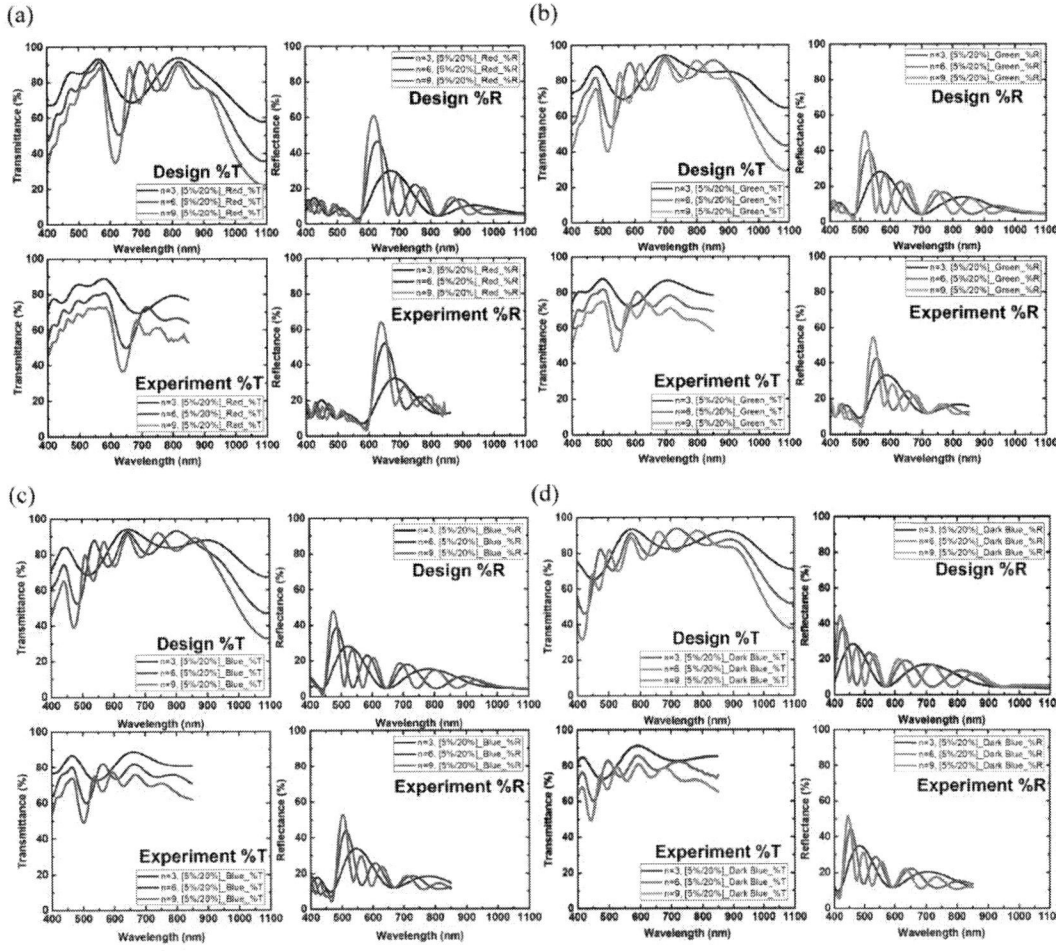

Figure 2. Transmittance and reflectance of multilayered optical filters for (a) red color, (b) green color, (c) blue color, and (d) dark blue color with different number of repeated multilayers.

41st European Photovoltaic Solar Energy Conference and Exhibition

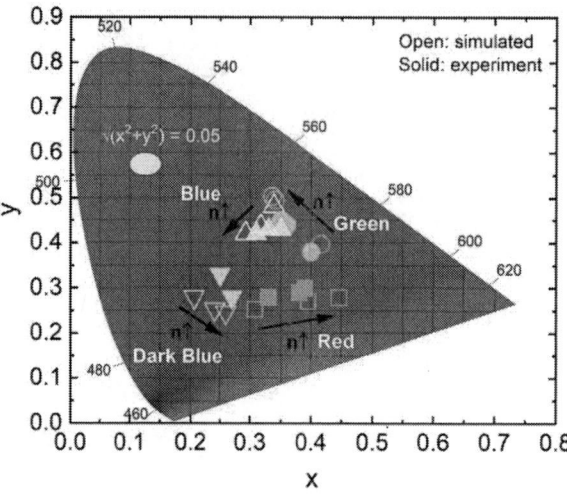

Figure 3. The color indices of the ITO (SnO_2 5 wt%)/ITO (SnO_2 20 wt%) stack structures were deposited on a glass substrate to realize red, green, blue, and dark blue colors.

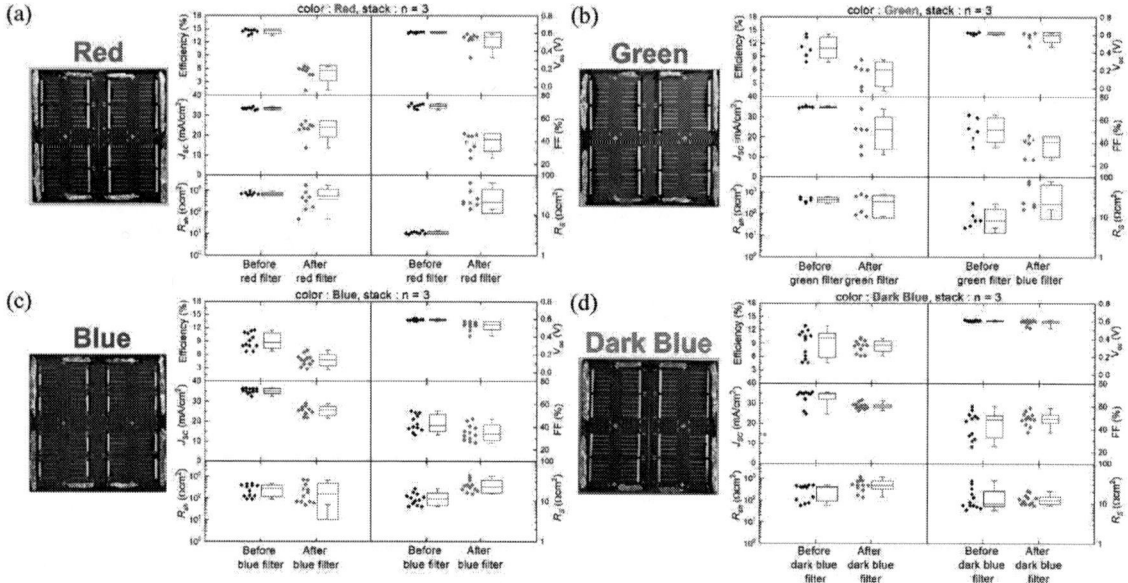

Figure 4. The photovoltaic performances before and after coating with the (a) red color, (b) green color, (c) blue color, and (d) dark blue color of the CIGS thin-film solar cells.

The reflectance spectra in the simulation and experiment were similar, but the peak wavelength shifted towards the long-wave direction due to the slightly increased thickness compared to the simulated values. The color performance of the color filters was analyzed as shown in Figure 3 which showed the simulation and measured values of the xyY color chromatic diagram. As the number of repeated layers increased, the xy color index moved slightly counterclockwise. As a result, more than seven distinct colors were implemented with a distance of $\sqrt{x^2 + y^2} \geqq 0.05$ in the xyY color space.

The photovoltaic performances of the fabricated CIGS solar cells with red, green, blue, and dark blue color filters were investigated (Fig. 4). After the deposition of the color filters, the decrease of the fill factor is observed. The reduction in J_{sc} was caused by optical loss due to the

decreased transmittance of the color filters. Also, there may be a possibility of solar cell degradation due to the substrate temperature during the deposition.

4. CONCLUSIONS

We fabricated multilayered conductive color filters using ITO thin films with two different SnO_2 compositions. The results showed increasing number of stacks narrowed the full width at half maximum of the reflectance peaks in the target wavelength range, resulting in more distinct colors. We evaluated the photovoltaic performances both before and after application of the color filters. As a result, we anticipate that various colors can be produced and applied to PV cells and modules.

5. REFERENCES

[1] H. Lee et al., Wiley Interdiscip. Rev. Energy Environ. **10** (2021) e403.

[2] J. Palm et al., 7th World Conference on Photovoltaic Energy Conversion (2018) 2561.

[3] D. -H. Cho et al., Ceram. Int., **49** (2023) 30586.

[4] D. -H. Cho et al., Proceedings 40th European Photovoltaic Solar Energy Conference, (2023) 020124.

[5] Y. D. Chung et al., Ceramist, **23** (2020) 389.

ACKNOWLEDGMENTS

This work was supported by the program of phased development of carbon·neutral technologies through the National Research Foundation of Korea (NRF) funded by the Ministry of Science, ICT (2022M3J1A1063019) and was also supported by Electronics and Telecommunications Research Institute (ETRI) grant funded by the Korean government [24YB1510].

41st European Photovoltaic Solar Energy Conference and Exhibition

FLEXIBLE THIN-FILM CZTS SOLAR CELL BASED ON AN ELECTROPLATED METALLIC PRECURSOR DEPOSITED ON A MOLYBDENUM/GLASS COATED STAINLESS STEEL FOIL

Io Mizushima[1], Peter Torben Tang[1], Christoph Kammerlander[2] and Andreas Zimmermann[2]
1: IPU P/S, Bredevej 2B, 2830 Virum, Denmark
2: Sunplugged GmbH, Affenhausen 1, 6413 Wildermieming, Austria[2]
io@ipu.dk

ABSTRACT: The thin-film solar cells have recently become economically competitive to crystalline silicon and a much higher fraction of these cells are expected in the future. Especially, CZTS (copper, zinc, tin and sulfur) is a promising absorber for thin-film solar cells. This study is focused on the development of a process for fabrication of flexible CZTS thin-film solar cells based on an electroplated copper/zinc/tin (CuZnSn) precursor deposited on stainless steel foil with molybdenum back contact layer and glass barrier layer. The most challenging part is to develop a reproduceable process for electroplating on the thin molybdenum coating, since the molybdenum requires etching to remove the natural oxide layer and such a treatment is depending on the unknown thickness if the oxide layer. If too little is removed there will be poor adhesion to the subsequently deposited precursor, if too much is removed (i.e. also molybdenum metal is etched away) the electric contact and conductivity of the molybdenum electrode is compromised. The problem has been solved by an anodizing treatment of the foils. In our previous work of fabrication of CuZnSn precursors on molybdenum foil substrate, a two-step electroplating process of tin and CuZn alloy was found to be the most feasible [1]. However, the pure tin (Sn) top layer gives rise to formation of secondary phases during the sulfurization process. By introducing an additional step of copper electroplating at the top, homogenous CZTS formation on the Mo/Ti/glass coated stainless foil substrate has been obtained.

1 INTRODUCTION

Fabrication of CZTS absorbers by electroplating is an effective, fast and economical approach, however, there are few studies on flexible CZTS solar cells - probably since the annealing temperature of the CuZnSn precursor with sulfur, leads to limited options of flexible substrate material. State-of-the-art of flexible solar cells describes mostly absorbers fabricated by vacuum processes [2]. Direct fabrication of CZTS by electroplating onto ITO coated polymer substrate [3] and fabrication by electroplating/annealing using 0.5 mm thick molybdenum foil [4] have been reported.

In this study we have developed a process for fabrication of flexible thin-film CZTS solar cells, investigating the combination of electroplating of the CuZnSn precursor and using a stainless-steel foil substrate with molybdenum back contact layer and a glass layer which is a diffusion barrier for iron atoms, and which also provides electrical insulation. This combination creates flexible solar cells, benefiting from low cost, high flexibility, ductility, and thermal stability.

Molybdenum is a typical electrode for solar cells, however, electroplating on a thin molybdenum layer – deposited by magnetron sputtering – on the dielectric glass material is difficult. The substrate requires etching to remove the natural molybdenum oxide layer on the surface and this treatment can create uneven thickness and thus also conductivity of the substrate. Uneven conductivity reduces the reproductivity of both the chemical composition and the thickness of alloy electrodeposits on the foil. By anodizing the Mo/glass coated stainless steel, formation of molybdenum oxide results in a surprisingly good adhesion to subsequently electroplated CuZn alloy and high reproductivity of the CuZn alloy composition. Furthermore, the top coating of the electrodeposited stack of materials – particularly if it is pure tin - has been found to be critical for formation of secondary phases.

2 EXPERIMENTAL

Electroplating of CuZn and Sn were performed in a 1 L glass beaker, using the flexible foil as cathode (substrate) and either pure Sn or pure Cu as anode. The current density was controlled using a DC power supply being able to accurately control the overall current. The solutions were agitated with a magnetic stirrer and heated (if necessary) with the integrated hotplate. Electroplating of Sn was done at 50 °C using a tin anode and electroplating of CuZn was done at 25 °C using a copper anode. The electrolyte for CuZn electroplating contained 15 g/L copper sulphate, 30 g/L zinc sulphate, 20 g/L glycerol and 120 g/L sodium hydroxide. For Sn electroplating an electrolyte containing 15 g/L tin pyrophosphate and 165 g/L potassium pyrophosphate was used. Copper electroplating was performed in a 50 L tank with an electrolyte containing 90 g/L copper sulphate, 350 g/L pottasium pyrophosphate and 80 g/L dipotassium phosphate at 55 °C.

For fabrication of the CZTS absorber, the electroplated CuZnSn precursors were annealed in a graphite box placed inside a temperature-controlled quartz tube in an inert atmosphere by vacuuming and purging with nitrogen gas. The pressure inside the tube was 175 mbar at the start of each annealing. The temperature was raised at a rate of 20 °C/min. and kept at 585 degrees for 15 minutes. For sulfurization, 50 mg sulfur powder is placed inside the graphite box holding the CuZnSn precursors. After annealing, the phases were identified by Raman Spectroscopy, where the samples were excited by a green (532 nm) continuous wave diode laser.

3 RESULTS

3.1 Treatment of molybdenum/titanium/glass coated stainless steel foil substrate

Substrates of stainless steel foil with 300 nm molybdenum back contact layer, titanium adhesive layer and glass barrier layer are prepared by Sunplugged GmbH. Cross-sections of the substrate are shown in Fig.1.

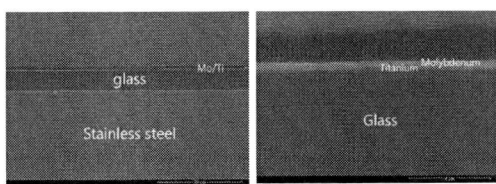

Figure 1: SEM images of a cross-section of molybdenum/titanium/glass coated stainless steel foil substrate.

However, reproductivity of the chemical composition/thickness of CuZn alloy deposits is a critical issue for direct electroplating on such a substrate. Activation in hydrochloric acid of the foil often caused electric contact loss for electroplating. Instead, anodizing treatment in an alkaline electrolyte improved the problem. The treatment changed the color of the surface to black (Fig. 2), which is predicted to be oxide layer, however, the conductivity is high enough and improved adhesion of the subsequent CuZn electrodeposits were observed.

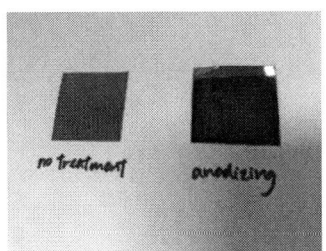

Figure 2: Photo of molybdenum/titanium/glass coated stainless steel foil substrate treated by anodizing.

3.2 Fabrication of CuZnSn precursor

In our previous work, fabrication of CZTS absorbers were obtained by electroplating first CuZn and then Sn followed by annealing of the precursor with sulfur powder [1]. However, the efficiency of these cells never reached more than 0.02%. By following a suggestion from literature [5] of having copper at the top, an electroplated Cu/Sn/CuZn stack was investigated and compared to the previously used Sn/CuZn precursor. Precursors of electroplated Sn/CuZn and Cu/Sn/CuZn on the Mo/glass coated stainless steel foil were prepared, and the samples were annealed with 50 mg sulfur powder in nitrogen gas at 585 °C for 15 minutes (Fig. 3).

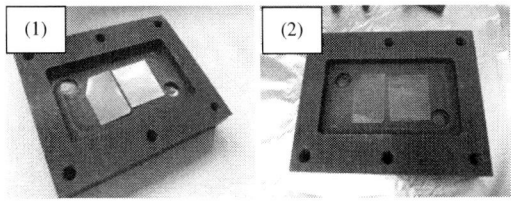

Figure 3: Photos of Sn/CuZn and Cu/Sn/CuZn electrodeposited on molybdenum/titanium/glass coated stainless steel foil substrate before (1) and after annealing at 585 °C (2).

SEM images of CZTS formed from Sn/CuZn and Cu/Sn/CuZn (Fig. 4) show a difference in the presence of secondary phases. The chemical compositions of the

CZTS, as analyzed by EDS, are 49.4%S-13.0%Cu-8.0%Zn-29.6%Sn and 50.8%S-24.4%Cu-12.2%Zn-12.5%Sn respectively. The additional step of copper electroplating on top of the precursor stack resulted in more homogeneous CZTS formation.

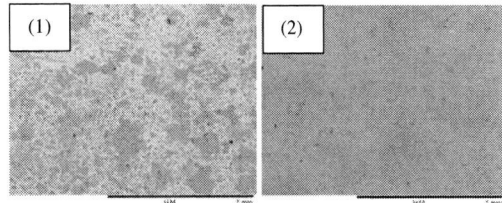

Figure 4: SEM images of CZTS on molybdenum/titanium/glass coated stainless steel foil substrate fabricated by electroplating of Sn/CuZn (1) and Cu/Sn/CuZn (2) precursor and sulfurization at 585 °C.

CZTS layer on the thin molybdenum layer is shown in the SEM image of the cross-section of the sulfurized CuZnSn electrodeposited on Mo/Ti/glass coated stainless steel substrate (Fig.5). The sulphur contents increases towards the substrate, which indicates that molybdenum is sulfurized. CZTS from Cu/Sn/CuZn precursor has a more ideal composition.

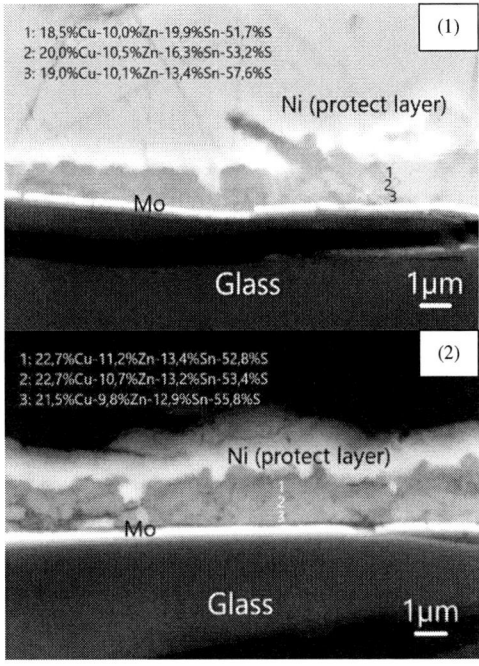

Figure 5: SEM images of cross-sections of CZTS on molybdenum/titanium/glass coated stainless steel foil substrates fabricated by electroplating of Sn/CuZn (1) and Cu/Sn/CuZn (2) precursors and sulfurization at 585 °C. The Ni coating is there to protect the CZTS layer from mechanical damageduring preparation.

3.3 Fabrication of CZTS on flexible foil

Molybdenum/titanium coated stainless steel foil was attempted to be used as substrate for fabrication of CZTS absorber, since the flexible foil with the thin molybdenum coating on glass layer made it difficult to control the chemical composition of the electroplated CuZn due to high resitivity. And the glass coating gives rise to a problem with partial delamination of the molybdenum coating under annealing at 585 degree (as shown in Fig. 6), which could be due to a difference in thermal expansion coefficient between molybdenum and glass.

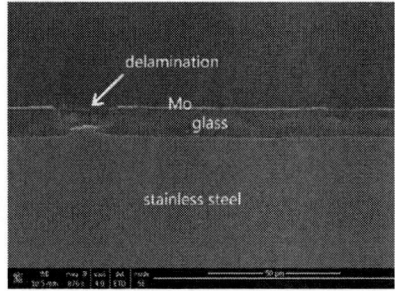

Figure 6: SEM images of cross-sections of molybdenum/titanium/glass coated stainless steel foil annealed at 585 °C.

However, the substrate with no glass coating never allowed CZTS formation during sulfurization. The photos show a difference in the appearance of sulphurized CuZnSn electrodeposited on the two differenet foils with/with no glass coating (Fig. 7).

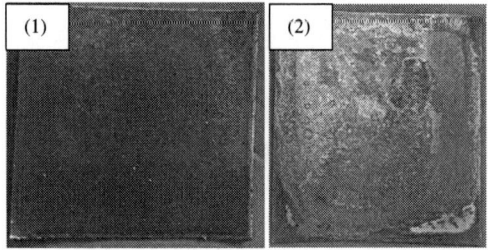

Figure 7: Photos of Cu/Sn/CuZn electrodeposited on molybdenum/titanium/glass coated stainless steel foil substrate (1) and molybdenum/titanium coated stainless steel foil substrate (2) after annealing at 585 °C

The Raman spectrum of the CZTS on no glass coated foil mainly showed molybdenum sulfide phases and no phases containg iron, chromium or titanium (Fig. 8).

3.4 Fabrication of flexible solar cell

The fabrication into solar cells was done by chemical bath deposition of a CdS buffer layer on top of the CZTS, then RF sputtering was used to deposit a thin (≈50 nm) resistive layer of i-ZnO, and finally a thicker (≈300 nm) conductive window layer of AZO (Al-doped ZnO). The fabricated cells were divided into 3 mm by 3 mm squares through manual scribing with a thin needle. The efficiency was measure to only 0.3%. Delamination of the molybdenum coating of the back contact, could be contributing to the the low efficiency.

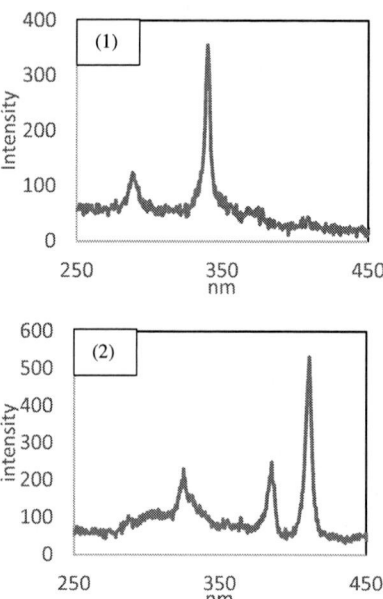

Figure 8: Raman spectra of CZTS absorber layers on molybdenum/titanium/glass coated stainless steel foil substrate (1) and molybdenum/titanium coated stainless steel foil substrate (2) after annealing at 585 °C.

4 CONCLUSIONS

By introducing an additional step of copper electroplating as the top layer, homogenous CZTS formation on the Mo/Ti/glass coated stainless foil substrate has been obtained. However, partial delamination of the molybdenum coating from the glass layer has been seen, and it could be critical for the photovoltaic performance. CZTS fabrication using flexible stainless steel foil without glass coating never succeeded.

5 REFERENCES

[1] I. Mizushima, 40th European Photovoltaic Solar Energy Conference (2023) 020122
[2] X. Li, Material Reports: Energy 1 (2021) 100001
[3] M. Farinella, E-MRS Spring Meeting, Strasbourg, France, 2013
[4] Y. Zhang, Royal Society of Chemistry Advances, 4 (2014) 23666
[5] J. Scragg, Ph.D. thesis, ISSN 2190-5053, Springer (2011)

6 ACKNOWLEDGEMENTS

This work is part of a project called "PlateCell" financed by Danish Ministry of Energy (EUDP) under file number 64020-1088.

41st European Photovoltaic Solar Energy Conference and Exhibition

MANUFACTURING, CHARACTERISATION AND STABILITY TESTS OF PRINTED ORGANIC PHOTOVOLTAIC DEVICES FOR INDOOR APPLICATIONS

I. Ballesteros Garcia[1], A. Khodr[1], D. Fredj[3], C. Ruiz Herrero[2], H. Alkhatib[3], O. Margeat[1], S. Ben Dkhil[3], J. Le Rouzo[2], J. Ackermann[1]

[1] CINaM (CNRS), Aix Marseille Université, CNRS, CINAM UMR 7325, Marseille, France
[2] Aix Marseille Université, CNRS, IM2NP UMR 7334, Marseille, France
[3] Dracula Technologies, Valence, France
Email contact : ignacio.ballesteros-garcia@etu.univ-amu.fr

INTRODUCTION

Organic Photovoltaic (OPV) have seen an increase of interest due to the improvements in performances after the introduction of non-fullerene acceptors (NFA), driving the improvement in efficiencies up to 33%[1] under indoor illumination. Furthermore, the expected internet-of-things devices' (IoT) market increase will drive a huge need of low-power energy supply in indoor environments (iOPV).

The good characteristics of OPV for these applications, including bandgap tuning for selective absorption, mechanical flexibility, light weight, absence of toxic heavy metals and facile module manufacture by high-throughput printing methodologies[2,3] make them an interesting choice as iOPV. Furthermore, their environmental conditions indoor seem to be less critical, creating a favourable context for the use of OPV devices. However, under dim light, improved fill factor and balanced charge transport and extraction characteristics, with suppressed recombination are of outmost importance[4]. The effective utilisation of indoor photons and the establishment of new characterization protocols are crucial for the development of iOPVs field[5]. For that reason, in the IOPV-lab, between CINaM, IM2NP and the company Dracula Technologies, we are working to improve indoor OPV efficiencies and stability that may help us in the optimisation of devices before transferring them to inkjet printing.

METHODOLOGY

Organic PV cells

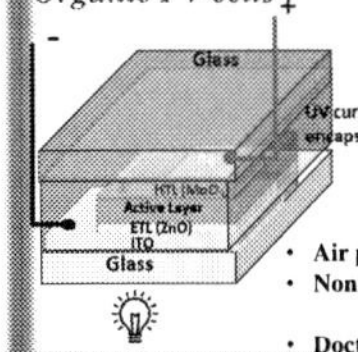

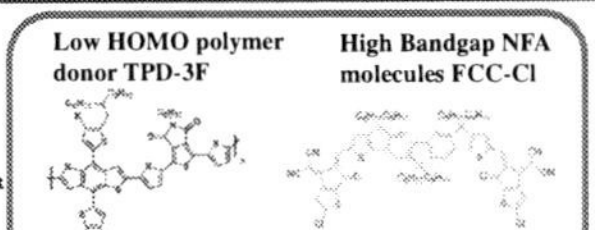

Low HOMO polymer donor TPD-3F
High Bandgap NFA molecules FCC-Cl

- Air processing
- Non-halogeneated solvents/additives: <u>cosolvent: 15% DPE vs. mix of additives: 3,5% DPE+ 5% tetralin</u>
- Doctor blade manufacturing: <u>processed using 1 pass vs. 2PASS samples</u>

Indoor Samples
- Thick active layer samples (>200 nm)
- Absorption in the visible part of the spectra
- Optimisation of ETL/HTL layers done for low illumination

Indoor Measurements
- IV Measurements done at different illumination levels: from 200 lux to 5000lux
- Using a 4000K LED lamp inside the GB

Indoor Ageing test
- Combined indoor ageing conditions in a climatic chamber
- Characterised along the test

AGEING TEST CONDITIONS	
Temperature	40ºC
Relative Humidity	80 %
Illuminance	1000 lux

RESULTS

SAMPLES:
A): cosolvent by single pass — Th ≈ 300 nm
B): mix of additives by single pass — Th ≈ 260 nm
C): cosolvent by 2 passes — Th ≈ 450 nm

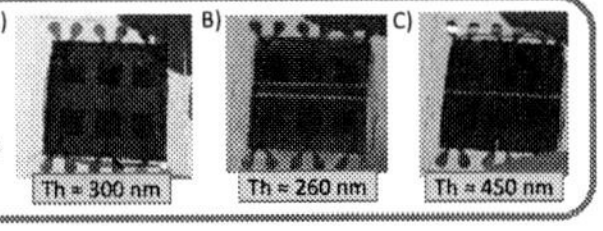

Initial (t0) IV Measurements @ 1000 lux (4000K LED lamp)

	Illuminance	Voc (V)	Jsc(µA/cm²)	FF(%)	PCE (%)	Rs (Ω/cm²)	Rsh (Ω/cm²)
A) Single pass Cosolvent	200 lux	0,88±0,003	25,179±0,5	62,58±3,38	18,49±1,09	370250	2305
	500 lux	0,914±0,003	55,748±1,07	63,83±1,85	18,16±0,62	186841	1032
	1000 lux	0,939±0,002	107,74±2,16	64,23±1,24	18,41±0,51	100019	498
	2000 lux	0,964±0,002	210,71±4,24	64,01±0,96	18,61±0,46	52412	278
	5000 lux	0,996±0,02	512,4±10,4	63,26±0,69	18,68±0,42	21857	116
B) 1 pass Additives	200 lux	0,873±0,001	26,502±0,78	64,59±0,78	19,92±0,6	466174	2368
	500 lux	0,907±0,001	58,719±1,73	64±0,65	19,04±0,59	200000	786
	1000 lux	0,932±0	113,7±3,37	63,59±0,42	19,11±0,63	105520	570
	2000 lux	0,959±0	222,66±6,63	63,14±0,41	19,29±0,7	53956	276
	5000 lux	0,992±0	542,6±16,35	61,91±0,37	19,28±0,7	20968	148
C) 2PASS Cosolvent	200 lux	0,885±0,003	27,339±0,15	66,64±0,22	21,5±0,24	451012	2122
	500 lux	0,916±0,002	60,558±0,33	66,44±0,21	20,6±0,19	206667	907
	1000 lux	0,94±0,002	117,22±0,62	65,95±0,23	20,58±0,17	111330	515
	2000 lux	0,964±0,002	229,46±1,2	65,13±0,28	20,6±0,16	57701	302
	5000 lux	0,993±0,002	558,43±3,13	63,33±0,32	20,33±0,16	24497	167

Charge carrier dynamics in indoor conditions

➤ Trap assisted recombination can be quantified using n parameter. n >1 increases with the trap assisted recombination based on:

$$V_{oc} \propto n \cdot \left(\frac{k \cdot T}{q}\right) \cdot \ln(I_L)$$

➤ Bimolecular recombination quantified from the bimolecular recombination efficiency:

$$\eta_B = \frac{1}{s} - 1, \text{ where } J_{sc} \propto C \cdot I_L^{s}$$

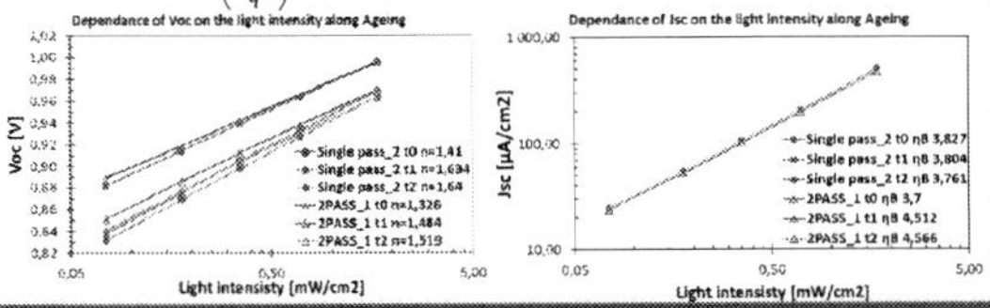

Dependance of Voc on the light intensity along Ageing
Dependance of Jsc on the light intensity along Ageing

AGEING COMPARISON OF SAMPLES USING COSOLVENT VS MIX OF ADDITIVES

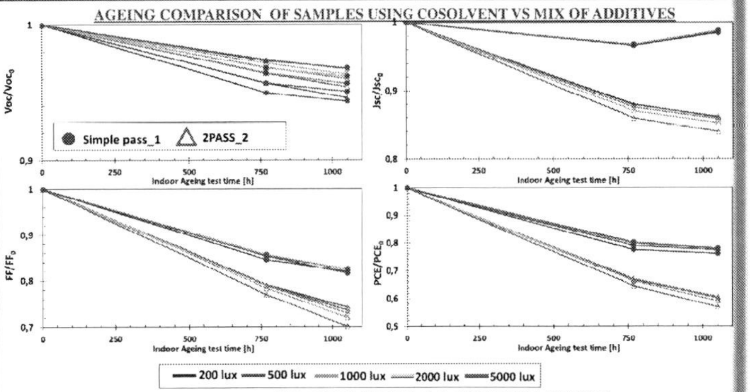

● Simple pass_1 △ 2PASS_2

— 200 lux — 500 lux — 1000 lux — 2000 lux — 5000 lux

AGEING COMPARISON USING SINGLE / 2 PASS DBC PROCESSING

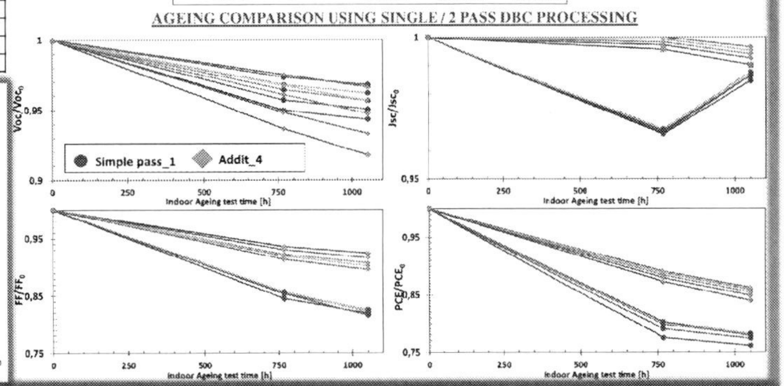

● Simple pass_1 ◆ Addit_4

CONCLUSIONS

➤ Formulations with cosolvents or mix of additives seem to give comparable device performance parameters after encapsulation and comparable thickness (250-300nm).
➤ The double pass processing (2PASS) in DBC gives better IV performance for samples under indoor measurements, improving all the parameters with thicker layers (> 400nm).
➤ Changing the cosolvents (20% loss) for a mix of additives in lower percentage in the formulation of the active layer ink seems to increase slightly the stability of the devices: lower decrease in Jsc, FF and PCE to 15% loss after more than 1000h of indoor ageing test.
➤ Double pass samples with cosolvents have poorer stability (PCE losses higher than 40%) than samples with simple pass (20%) in the indoor ageing test after 1000h.
➤ Along the ageing test, trap assisted recombination increases for both types of processed samples: bimolecular recombination increases more in the 2PASS samples, while keeping almost constant for the single pass samples.
➤ Samples of different cosolvents and additives combinations separately need to be tested to check the influence of different additives separately on the stability of the devices.
➤ Further sensitivity tests for the different parameters (temperature, light stability) could be studied in order to clarify the degradation triggers and stability problems of the devices.
➤ Deeper optical/chemical characterization needs to be done to understand the factors affecting both types of recombination in the devices, relevant for mild illumination and their evolution during ageing.

REFERENCES

1) Kim, et al. - High-efficiency (over 33 %) indoor organic photovoltaics with band-aligned and defect-suppressed interlayers • *Applied Surface Science* 610 (2023): 155558.
2) Liao, Chuang-Yi, et al. - Processing Strategies for an Organic Photovoltaic Module with over 10% Efficiency • *Joule* 4, nº 1 (15 janvier 2020): 189-206.
3) Marrocchi, Assunta., et al. - Poly(3-Hexylthiophene): Synthetic Methodologies and Properties in Bulk Heterojunction Solar Cells • *Energy & Environmental Science* 5, nº 9 (15 août 2012): 8457-74.
4) Saeed, et al. - 2D MXene Additive-Induced Treatment Enabling High-Efficiency Indoor Organic Photovoltaics • *Advanced Optical Materials* 11, nº 1 (2023): 2202135.
5) Lin, Xingting, Shanlei Xu, Bo Tang, et Xin Song. - Indoor Organic Photovoltaics for Low-Power Internet of Things Devices: Recent Advances, Challenges, and Prospects. • *Chemical Engineering Journal* 497 (octobre 2024): 154944.

ACKNOWLEDGEMENTS:
We acknowledge financial support by the ANR for the IOPV-LAB project (Projet-ANR-21-LCV2-0001) between CINaM, IM2NP Laboratories of Aix-Marseille University and the company Dracula Technologies.

41st European Photovoltaic Solar Energy Conference and Exhibition

FABRICATION OF HIGHLY EFFICIENT CDSETE/CDTE THIN FILM SOLAR CELLS WITH EMITTER-LESS CELL STRUCTURE

Yanbo Cai, Hongxu Jiang, Kai Yi, Fei Liu, Guangwei Wang, Deliang Wang*
Hefei National Research Center for Physical Sciences at the Microscale, University of Science and Technology of China,
Hefei 230026, P.R. China
* E-mail: eedewang@ustc.edu.cn

ABSTRACT: With excellent optical and electronic properties, tin oxide (SnO_2) is a promising electron transport layer (ETL) in CdTe thin film solar cell. However, CdTe solar cells with undoped SnO_2 ETLs demonstrate poor efficiencies with relatively low open-circuit voltage (V_{OC}). In this report, Mg-doped SnO_2 (MTO) thin films were fabricated by magnetron co-sputtering technique and employed as the ETLs/window layers in CdTe-based thin film solar cells. The V_{OC} has been significantly improved by doping Mg into the SnO_2 layers, regardless of whether the contacting absorber layer is CdSeTe or CdTe. Upon Mg doping the trap density was reduced from 4.02 to 1.62×10^{13} cm^{-3}, indicating that significant part of the recombination defects at the interface were effectively passivated upon Mg doping in the SnO_2 films, which is responsible for the significant improvement in V_{OC}.
Keywords: CdTe solar cell, magnesim doping, electron transport layer, carrier recombination, defect passivation.

1 INTRODUCTION

Cadmium telluride (CdTe) thin film solar cell is capable of competing with crystalline silicon for commercialization for its advanced growth techniques and relatively low production costs. By incorporating selenium into CdTe to form a CdSeTe alloy and using CdSeTe/CdTe as a double absorber layer, the power conversion efficiency (PCE) of CdTe solar cell has been improved rapidly over the last decade, but it still remains below the Shockley-Queisser limit, primarily due to the relatively low open-circuit voltage (V_{OC}) [1,2].

CdS has traditionally been chosen as the n-type window layer due to its suitable band offset and lattice commensurability with CdTe. In recent years, more transparent oxides have been investigated as potential alternatives for CdS to enhance J_{SC} [3–5], forming a emitter-less cell structure. Over the past decade, Mg-doped ZnO (MgZnO) has emerged as a common window layer, however, brings about some drawbacks including being sensitive to moisture or oxygen and suffering from unfavorable S-kink behavior in the current density–voltage (J–V) curves. In comparison, tin oxide (SnO_2) possesses several advantages, including excellent transparency, high electron mobility, wide bandgap and outstanding chemical stability, becoming a promising candidate for high-performance electronic devices [6,7]. Doping in SnO_2 has proven to be an effective strategy for improving semiconductor device performances [8–11]. High-efficiency perovskite solar cells with Mg-doped SnO_2 (MTO) electron transport layer (ETL) demonstrated suitable energy band alignment and reduced recombinations at the perovskite/ETL interface [12–15].

In this work, we significantly improve the photovoltaic efficiency in SnO_2/CdTe and SnO_2/CdSeTe/CdTe solar cells by using MTO as ETLs, fabricated by magnetron co-sputtering technique. The trap density was significantly reduced, leading to the enhancements in V_{OC} and PCE.

2 EXPERIMENTAL

The SnO_2 thin films with different Mg doping contents were deposited by a radio frequency magnetron co-sputtering system from a pure SnO_2 target and a MgO target. The Mg doping concentration in SnO_2 was controlled by changing the sputtering power of the MgO target. As-deposited Mg-doped SnO_2 thin films were treated at 500 °C in dry air for 30 minutes to improve the optical, electrical and crystallinity properties of the films. The CdTe absorbers with thickness of ~4 μm were deposited on the MTO/FTO or CdSe/MTO/FTO stacks by close-spaced sublimation technique (CSS) at a substrate temperature of 560 °C and source temperature of 660 °C. To promote recrystallization, the stacks were subsequently annealed with $CdCl_2$ at 410 °C in dry air, followed by Au layer deposition as the back contacts. X-ray diffraction patterns (XRD) were obtained using a multifunctional rotating-anode X-ray diffractometer (Rigaku, MiniFlex600). The light current density–voltage (J–V) characteristics of the solar cells were carried out under standard AM 1.5G illumination using a solar simulator (Oriel Sol 3A, USA). Steady-state photoluminescence spectra were obtained using excitation source of a 638 nm laser (Horiba, LabRAM SoLeil Raman spectrometer).

3 RESULTS

The Mg concentrations, named in the form of atomic ratio of Mg/(Mg+Sn), were determined by using X-ray energy dispersive spectrometer (EDS) to be 0, 3.4%, 6.0%, 8.9% and 11.6% for the five MTO thin films fabricated in this work. X-ray diffraction measurements (XRD) reveal the rutile-type tetragonal structure for the MTO films and MgO phase was not detected, meaning that no detectable phase separation in the Mg-doped thin films even when the Mg concentration was 11.6%. As shown in Figure 1 (a), the (101) diffraction peak at ~ 34.0 °, in which the peak of the Mg-doped SnO_2 films showed a slight shift toward higher diffraction angle as compared with the undoped SnO_2, due to the formation of Mg_{Sn} [16]. The optical bandgaps shown in Figure 1 (b) were estimated to be 4.16, 4.13, 4.15, 4.13 and 4.14 eV for these five fims. Relatively large bandgaps of the undoped and Mg-doped SnO_2 thin films reported in this work show the excellent potential for photoelectric devices, specifically for the solar cells. As shown in Figure 1 (c), X-ray photoelectron spectroscopy (XPS) spectra for the undoped and Mg-doped SnO_2 thin films show that the intensity of the Mg 1s peaks was increased with the increasing Mg concentration, confirming that Mg atoms were incorporated into the SnO_2 thin films successfully. Figure 1 (d) shows that the electrical conductivity was decreased rapidly with the increasing Mg doping in the SnO₂ thin films.

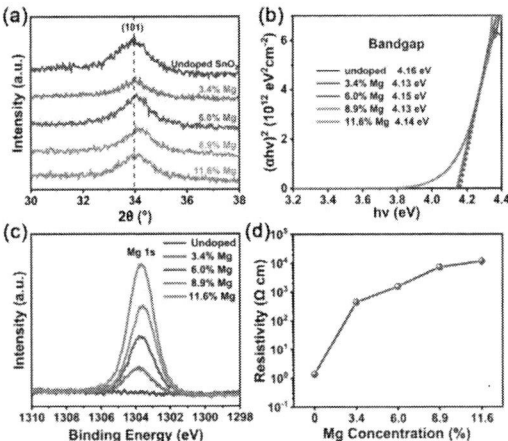

Figure 1. (a) XRD patterns with diffraction angle range of 30 − 38°, (b) Tauc plots, (c) Mg 1s XPS peaks and (d) electric resistivities of the 200-nm-thick Mg-doped SnO₂ thin films with different Mg concentrations.

Electrochemical impedance spectroscopy (EIS) testing was performed on the solar cells under one sun irradiation. Figure 2 (a) shows the Nyquist plots and the inset is the equivalent circuit model used for the fitting [17]. The recombination resistance, related to the resistance of charge recombination at the SnO₂/CdTe interface, were estimated to be 153.8 and 775.8 Ω for the cells with undoped and 6% Mg-doped SnO₂, respectively, indicating that the charge recombination at the SnO₂/CdTe interface was effectively suppressed by Mg doping. Space-charge-limited current (SCLC) measurements were employed to explore the effect of Mg doping in SnO₂ on the trap density (N_t) at the MTO/CdTe interface [18]. The N_t of the undoped SnO₂/CdTe cell was estimated to be 4.02 × 10¹³ cm⁻³, which was reduced to 1.62 × 10¹³ cm⁻³ for the 6.0% Mg-doped cell, as shown in Figure 2 (b). The results demonstrate that a significant part of the trap centers were passivated and the trap-assisted recombination was consequently dramatically reduced.

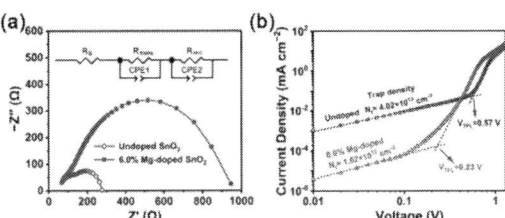

Figure 2. (a) Impedance spectroscopy curves measured for SnO₂/CdTe-based heterostructure devices with and without 6.0% Mg doping in SnO₂. (b) Trap densities measured for the electron-only devices with undoped and 6.0% Mg-doped SnO₂ layer.

Undoped, 3.4%, 6.0%, 8.9% and 11.6% Mg-doped SnO₂ thin films were deposited onto glass/FTO substrates directly, followed by deposition of CdSeTe and/or CdTe absorbers to fabricate CdTe-based solar cells. The current density–voltage (J–V) characteristics of MTO/CdTe and MTO/CdSeTe/CdTe solar cells are shown in Figures 3 (a) and 3 (b). S-kink behavior was not observed in the J–V curves of all the solar cells, although all the MTO films

were thermally annealed at 500 °C in the air and the CdTe absorbers were deposited in a gas mixture of argon and oxygen.

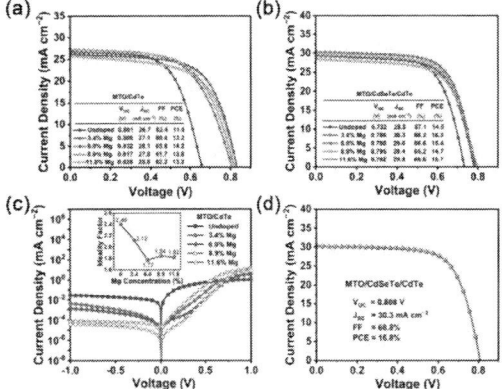

Figure 3. Light J–V curves of the (a) MTO/CdTe and (b) MTO/CdSeTe/CdTe solar cells with different Mg concentrations. (c) Dark J–V curves of the MTO/CdTe solar cells with different Mg concentrations. (d)) J–V curve of the best-performing MTO/CdSeTe/CdTe solar cell in this study.

For the MTO/CdTe solar cells, the values of J_{SC} are relatively high, namely, in the range of 26.1 ~ 27.1 mA cm⁻², which was attributed to the high light transmittance of the SnO₂ window layer with large bandgap of over 4.10 eV. The most remarkable result in the solar cell performance is that the solar cells with Mg-doped SnO₂ films showed substantial increase in V_{OC} and consequently in PCE. The solar cells with an undoped SnO₂ window layer had a V_{OC} of 0.661 V and an inferior PCE of only 11.0%, whereas the V_{OC} was improved to 0.832 V with a PCE of 14.2% for the solar cell with 6.0% Mg-doped SnO₂. With further Mg doping, the V_{OC} of the solar cells maintained at high values. This can be attributed to trap passivation at the SnO₂/CdTe interface by Mg doping in SnO₂.

The dark current density–voltage (Dark J–V) characteristics are shown in Figure 3 (c). The reverse saturation current, and ideality factor were obtained by fitting the dark J–V curves. In the case of the cell with undoped SnO₂, the ideality factor shown in the inset of Figure 3 (c) was estimated to be 2.40, indicating that the undoped SnO₂/CdTe solar cell was made of a diode junction far from an ideal solar cell. Ideality factor with a value of larger than 2 indicated that the carrier transport in the junction with undoped SnO₂ was primarily dominated by severe trap-assisted recombination. After Mg dopoing in SnO₂ the ideality factor implied a significantly reduction in the carrier recombination loss. The highest efficiency MTO/CdSeTe/CdTe solar cell fabricated in this study is 16.8% (without group V doping or anti-reflective coating), as shown in Figure 3 (d).

4 CONCLUSIONS

Mg-doped SnO₂ thin films were prepared by magnetron co-sputtering technique and used as the ETL in CdTe thin film solar cells. The V_{OC} have been remarkably enhanced both for the CdSeTe/CdTe double and CdTe single absorber thin film solar cells. Upon Mg doping in SnO₂, the carrier trapping defects at the SnO₂/CdTe interface are effectively passivated, leading to significantly decreased interface recombination and thereby facilitating electron transport at the front contact. This work demonstrates that passivating the recombination defects at the SnO_2/CdTe interface is an effective route to significantly increase the solar cell V_{OC}, and Mg-doped SnO_2 is a promising ETL for fabricating highly efficient CdTe thin film solar cells. This work provides significant breakthrough towards the in-depth understanding of the role of oxide ETL at the front contact in highly efficient CdTe solar cells.

ACKNOWLEGEMENTS

This work was financially supported by the National Key Research and Development Program of China (No. 2023YFB4202700) and the National Natural Science Foundation of China (No. 61774140).

REFERENCES

[1] W. Shockley, H. J. Queisser, J. Appl. Phys. 1961, 32, 510.
[2] S. Rühle, Sol. Energy 2016, 130, 139.
[3] J. M. Kephart, J. W. McCamy, Z. Ma, A. Ganjoo, F. M. Alamgir, W. S. Sampath, Sol. Energy Mater. Sol. Cells 2016, 157, 266.
[4] J. P. Lemmon, E. Polikarpov, W. D. Bennett, L. Kovarik, Appl. Phys. Lett. 2012, 100, 213908.
[5] M. A. Scarpulla, B. McCandless, A. B. Phillips, Y. Yan, M. J. Heben, C. Wolden, G. Xiong, W. K. Metzger, D. Mao, D. Krasikov, I. Sankin, S. Grover, A. Munshi, W. Sampath, J. R. Sites, A. Bothwell, D. Albin, M. O. Reese, A. Romeo, M. Nardone, R. Klie, J. M. Walls, T. Fiducia, A. Abbas, S. M. Hayes, Sol. Energy Mater. Sol. Cells 2023, 255, 112289.
[6] Q. Jiang, L. Zhang, H. Wang, X. Yang, J. Meng, H. Liu, Z. Yin, J. Wu, X. Zhang, J. You, Nat. Energy 2016, 2, 16177.
[7] H. J. Snaith, C. Ducati, Nano Lett. 2010, 10, 1259.
[8] Y. D. Wang, T. Brezesinski, M. Antonietti, B. Smarsly, Acs Nano 2009, 3, 1373.
[9] A. Andersson, N. Johansson, P. Broms, N. Yu, D. Lupo, W. R. Salaneck, Adv. Mater. 1998, 10, 859.
[10] L. B. Xiong, M. C. Qin, C. Chen, J. Wen, G. Yang, Y. X. Guo, J. J. Ma, Q. Zhang, P. L. Qin, S. Z. Li, G. J. Fang, Adv. Funct. Mater. 2018, 28, 1706276.
[11] M. W. J. Prins, K. O. Grosse-Holz, G. Müller, J. F. M. Cillessen, J. B. Giesbers, R. P. Weening, R. M. Wolf, Appl. Phys. Lett. 1996, 68, 3650.
[12] C. Altinkaya, E. Aydin, E. Ugur, F. H. Isikgor, A. S. Subbiah, M. De Bastiani, J. Liu, A. Babayigit, T. G. Allen, F. Laquai, A. Yildiz, S. De Wolf, Adv. Mater. 2021, 33, 2005504.
[13] L. B. Xiong, Y. X. Guo, J. Wen, H. R. Liu, G. Yang, P. L. Qin, G. J. Fang, Adv. Funct. Mater. 2018, 28, 1802757.
[14] L. B. Xiong, M. C. Qin, G. Yang, Y. X. Guo, H. W. Lei, Q. Liu, W. J. Ke, H. Tao, P. L. Qin, S. Z. Li, H. Q. Yu, G. J. Fang, J. Mater. Chem. A 2016, 4, 8374.
[15] S. Lan, W. T. Zheng, S. Yoon, H. U. Hwang, J. W. Kim, D. W. Kang, J. W. Lee, H. K. Kim, ACS Appl. Energy Mater. 2022, 5, 14901.
[16] P. Wu, B. Z. Zhou, W. Zhou, Appl. Phys. Lett. 2012, 100, 182405.

[17] F. Fabregat-Santiago, J. Bisquert, G. Garcia-Belmonte, G. Boschloo, A. Hagfeldt, Sol. Energy Mater. Sol. Cells 2005, 87, 117.
[18] R. H. Bube, J. Appl. Phys. 1962, 33, 1733.

Characterisation of Degradation Pathways of 3-Terminal Perovskite-Silicon Tandems After Outdoor Monitoring

Miha Kikelj, Laurie-Lou Senaud Florent Sahli, Benjamin Lipovšek, Marko Topič, Christophe Ballif, Quentin Jeangros and Bertrand Paviet-Salomon

FE UNIVERSITY OF LJUBLJANA
Faculty of Electrical Engineering

:: csem

University *of Ljubljana*
Faculty *of Electrical Engineering*

Why a 3-Terminal Tandem (3TT)?

Configuration:	2TT	4TT	3TT	IBC-based 3TT
Design:	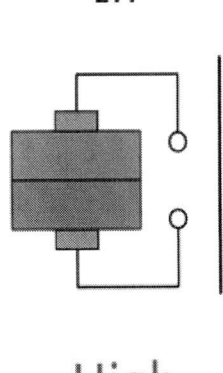	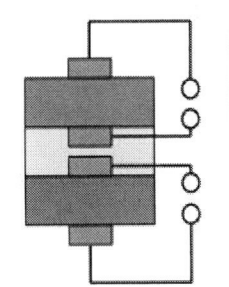	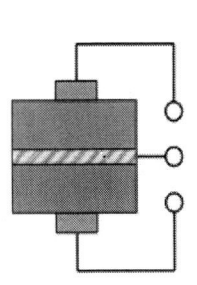	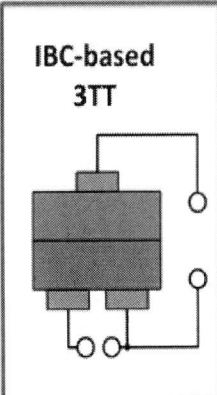
Sensitivity current mismatch:	High	Low	Low	
Energy yield:	++	+++	+++.	
Other costs:	€	€€€	€€	

University *of Ljubljana*
Faculty *of Electrical Engineering*

A Three Terminal Module? – AI solutions?

2T module

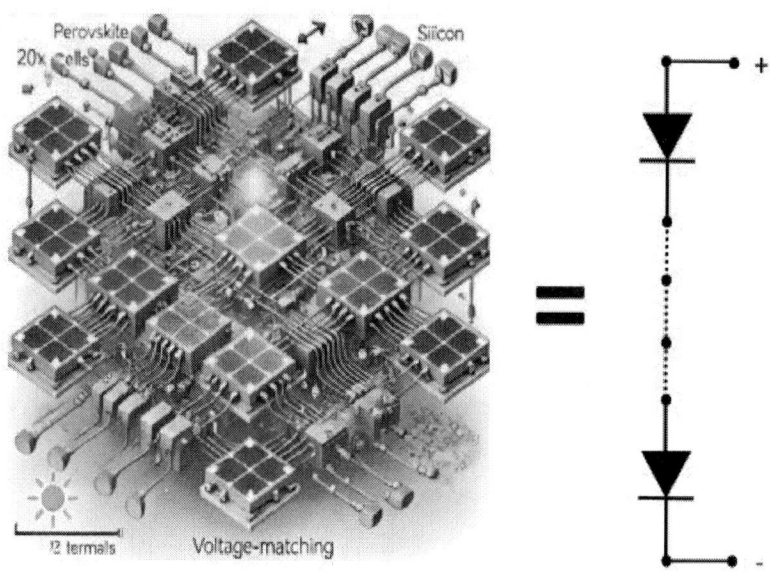

2CO.2.5 - Miha Kikelj 41st EUPVSEC – Vienna 25th of September

---

University *of Ljubljana*
Faculty *of Electrical Engineering*

3TT Voltage Matched (VM) Strings[1] – non-AI solution!

2T module

3TT	**2/1s**	**3/2s**	**m/ns**
S-type	$2 \mathbf{x} V_{bot} = V_{top}$	$3 \mathbf{x} V_{bot} = 2 \mathbf{x} V_{top}$	$m \mathbf{x} V_{bot} = n \mathbf{x} V_{top}$

or   **or**   **=**

[1] W. E. McMahon *et al.*, IEEE JPV, 2021

2CO.2.5 - Miha Kikelj 41st EUPVSEC – Vienna 25th of September

University of Ljubljana
Faculty of Electrical Engineering

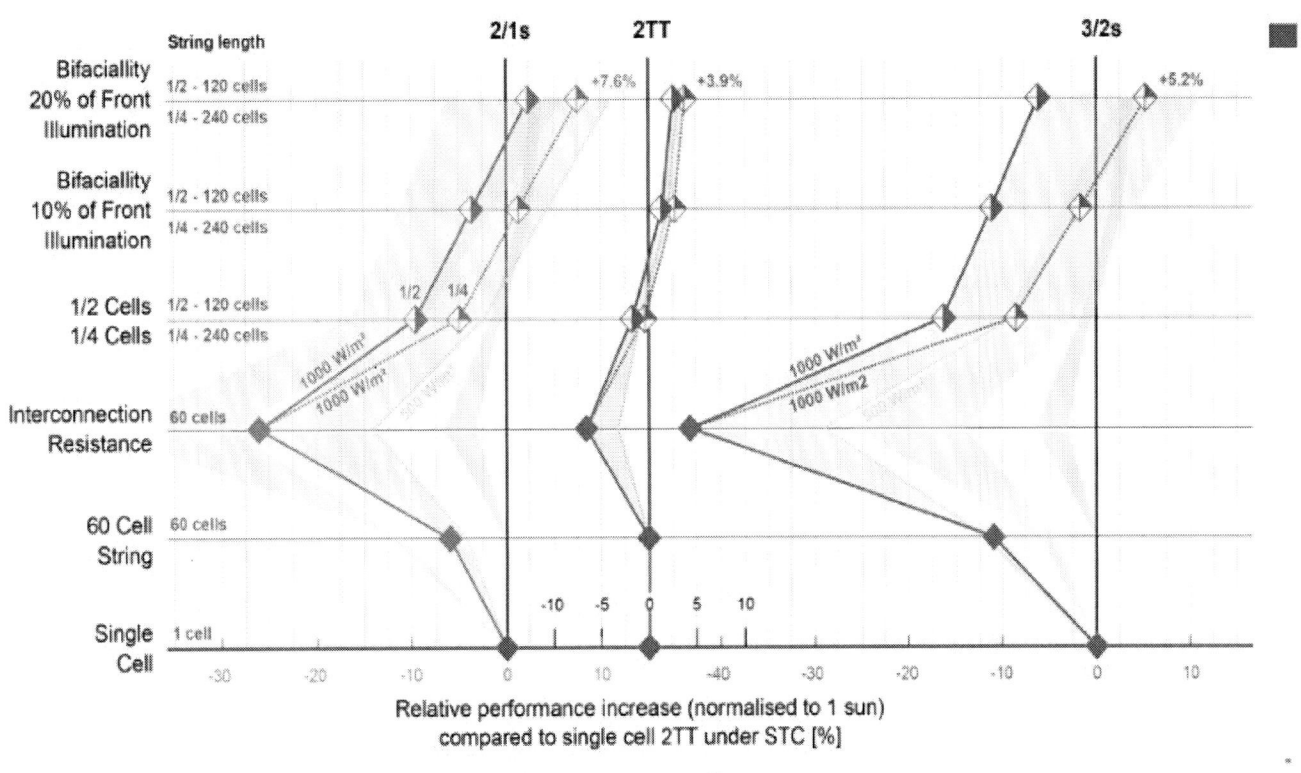

Hypothetical VM Module under Realistic Conditions

University of Ljubljana
Faculty of Electrical Engineering

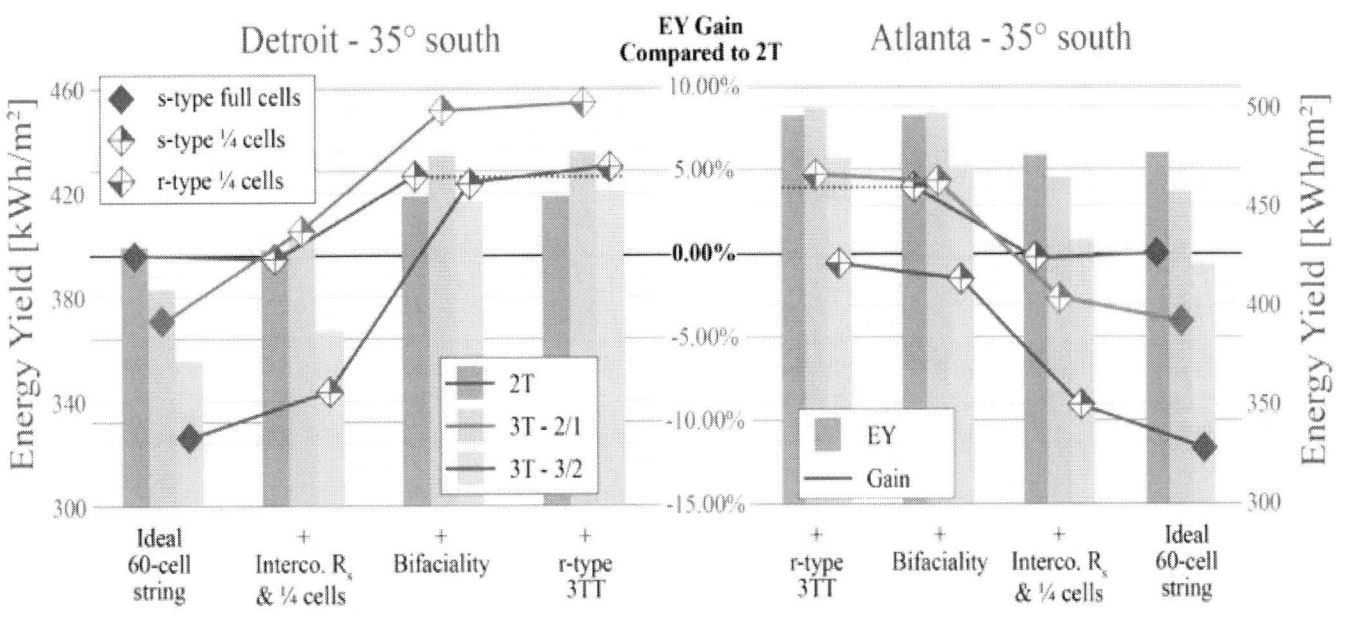

University of Ljubljana
Faculty of Electrical Engineering

Joule

Article

Do all good things really come in threes? The true potential of 3-terminal perovskite-silicon tandem solar cell strings

Miha Kikelj,[1,*] Laurie-Lou Senaud,[2] Jonas Geissbühler,[2] Florent Sahli,[2] Damien Lachenal,[3] Derk Baetzner,[3] Benjamin Lipovšek,[1] Marko Topič,[1] Christophe Ballif,[2] Quentin Jeangros,[2] and Bertrand Paviet-Salomon[2,4,*]

University of Ljubljana
Faculty of Electrical Engineering

A Physical Three-Terminal Perovskite/Silicon Module

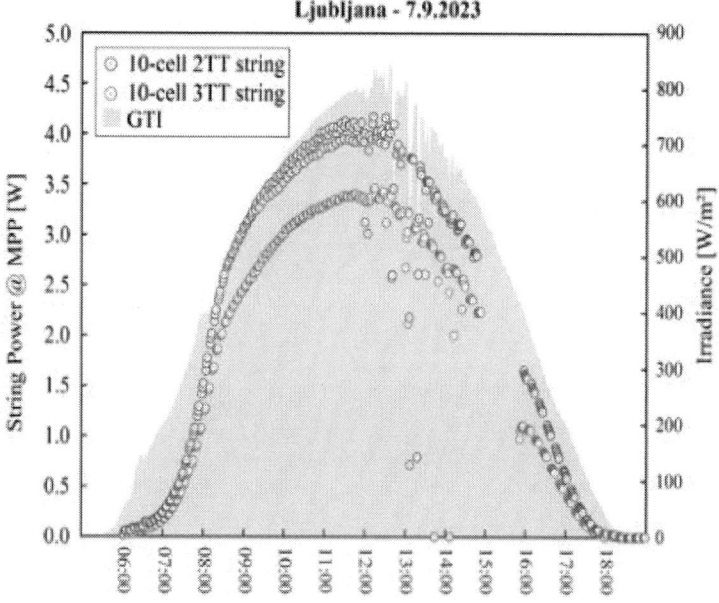

University *of Ljubljana*
Faculty *of Electrical Engineering*

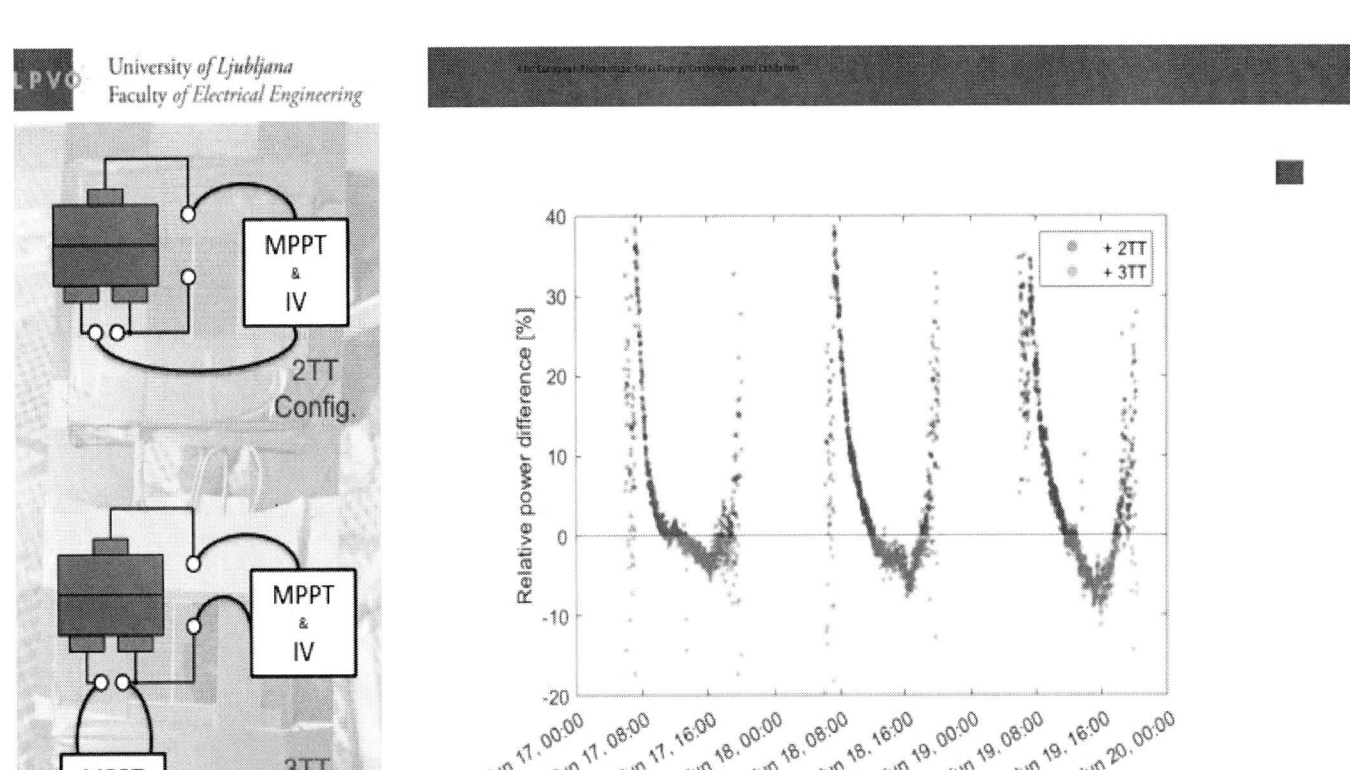

2CO.2.5 - Miha Kikelj   41st EUPVSEC – Vienna 25th of September

---

University *of Ljubljana*
Faculty *of Electrical Engineering*

Outdoor test of 5x5 cm² 2TT/3TT tandems

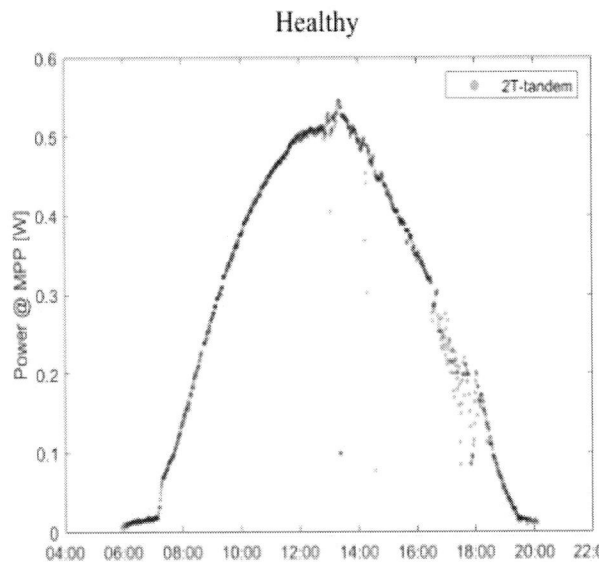

2CO.2.5 - Miha Kikelj   41st EUPVSEC – Vienna 25th of September

University *of Ljubljana*
Faculty *of Electrical Engineering*

Outdoor test of 5x5 cm² 2TT/3TT tandems - degradation

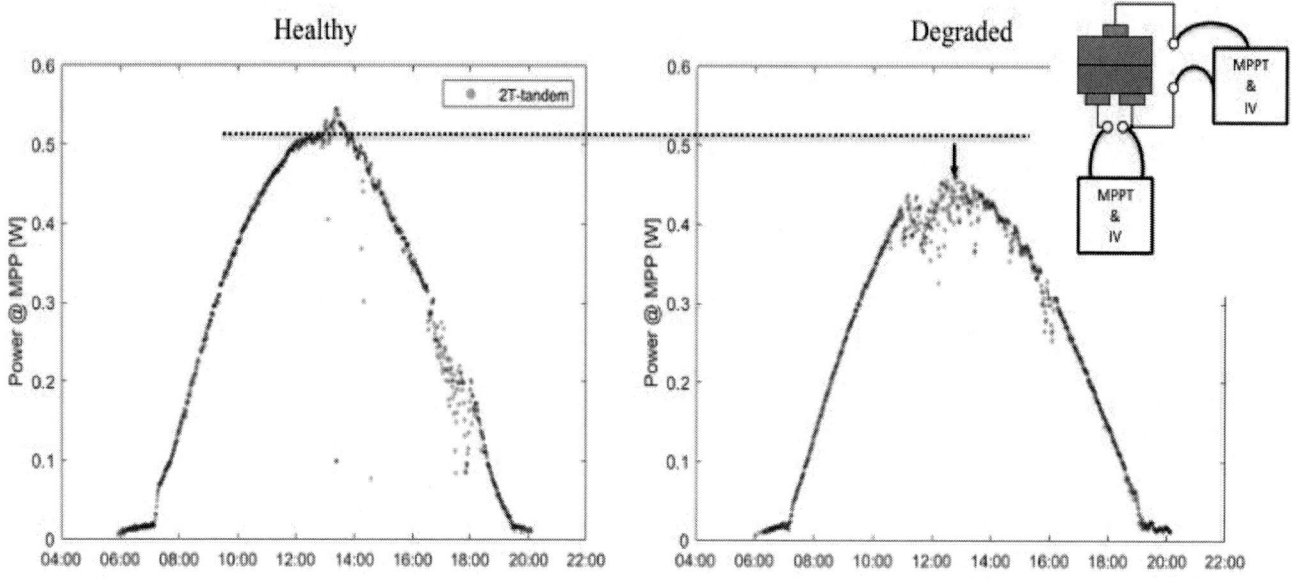

University *of Ljubljana*
Faculty *of Electrical Engineering*

Outdoor test of 5x5 cm² 2TT/3TT tandems - degradation

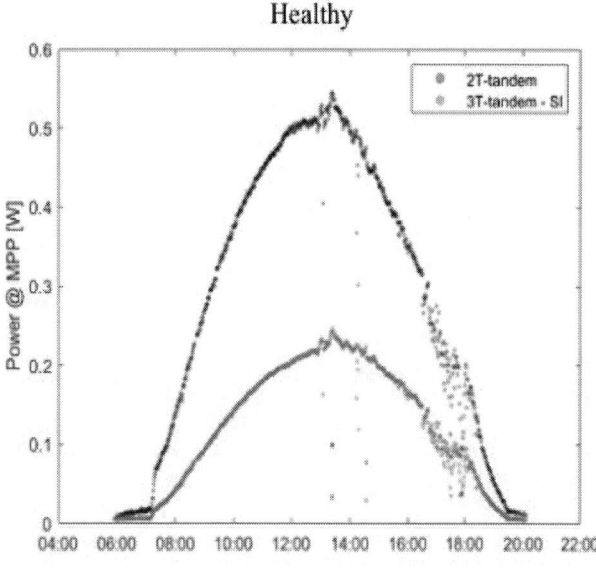

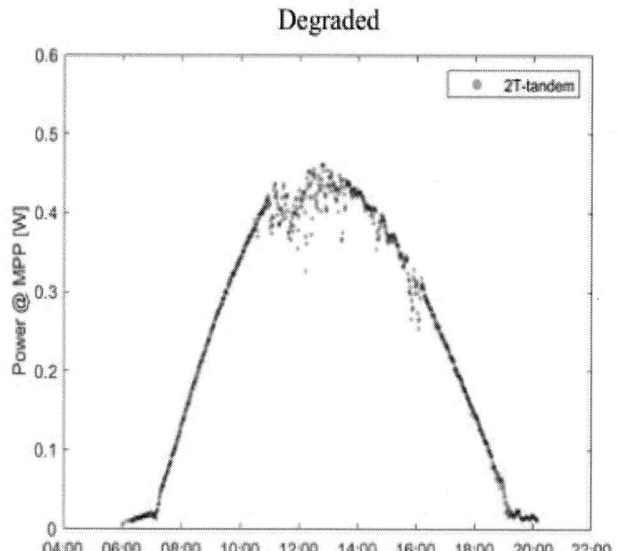

University of Ljubljana
Faculty of Electrical Engineering

Outdoor test of 5x5 cm² 2TT/3TT tandems - degradation

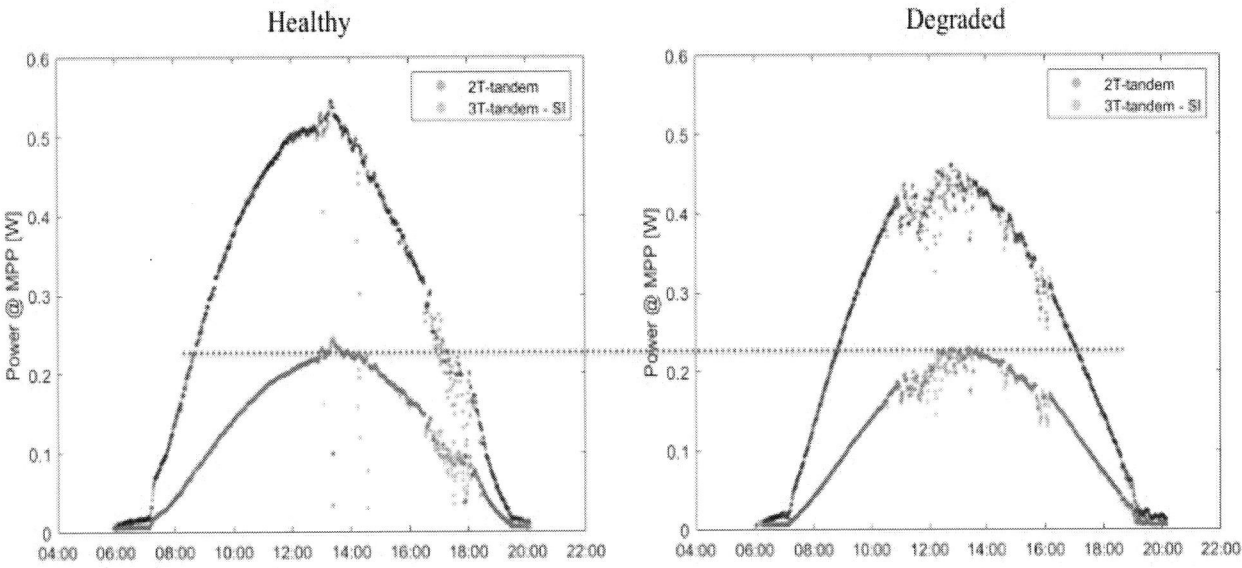

Healthy — Degraded

2CO.2.5 - Miha Kikelj   41st EUPVSEC – Vienna 25th of September

---

University of Ljubljana
Faculty of Electrical Engineering

Outdoor test of 5x5 cm² 2TT/3TT tandems - degradation

Healthy

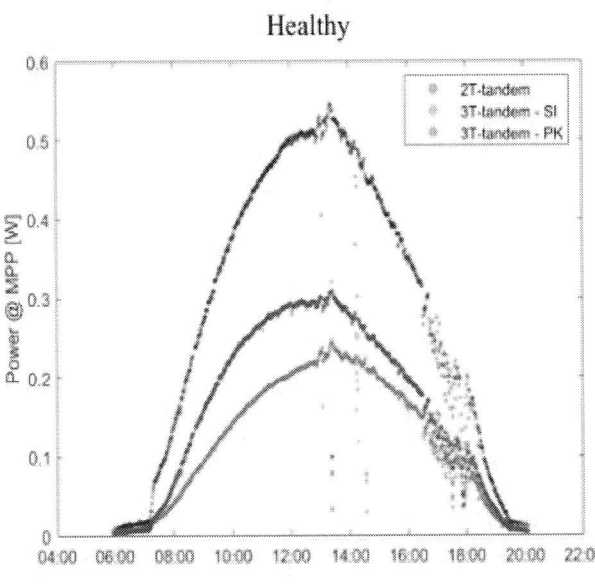

Degraded

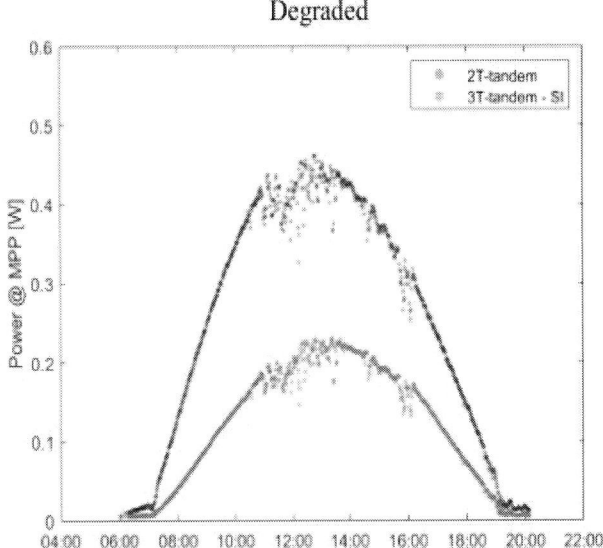

2CO.2.5 - Miha Kikelj   41st EUPVSEC – Vienna 25th of September

Outdoor test of 2TT/3TT tandems - degradation

Healthy   Degraded

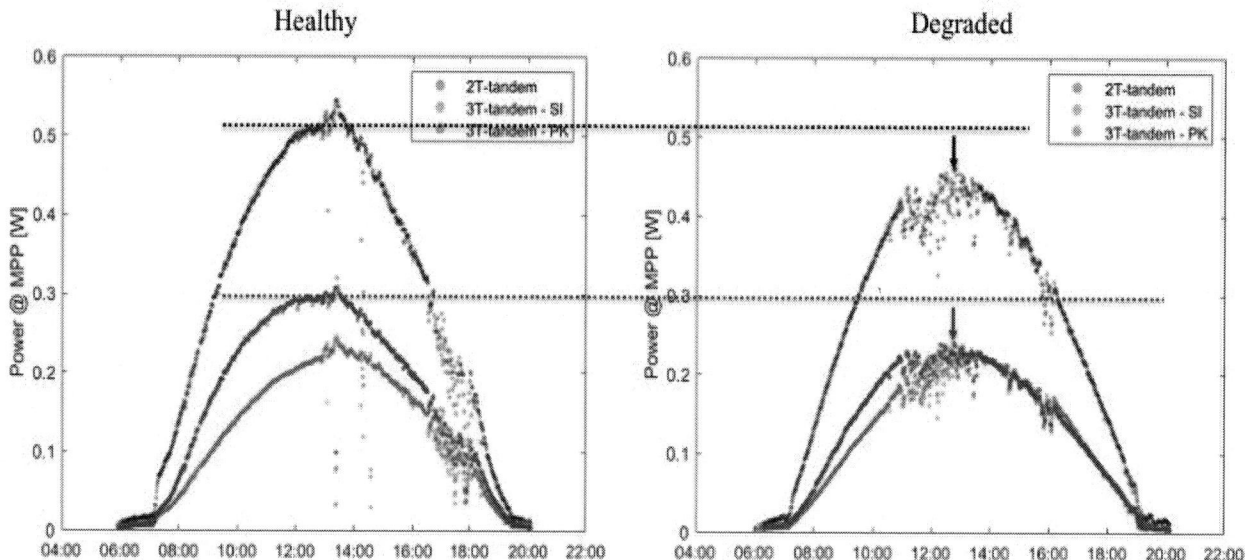

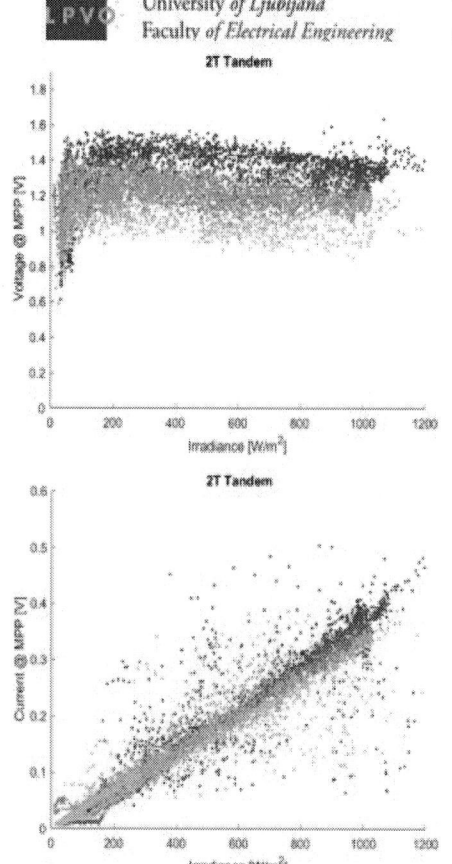

University *of Ljubljana*
Faculty *of Electrical Engineering*

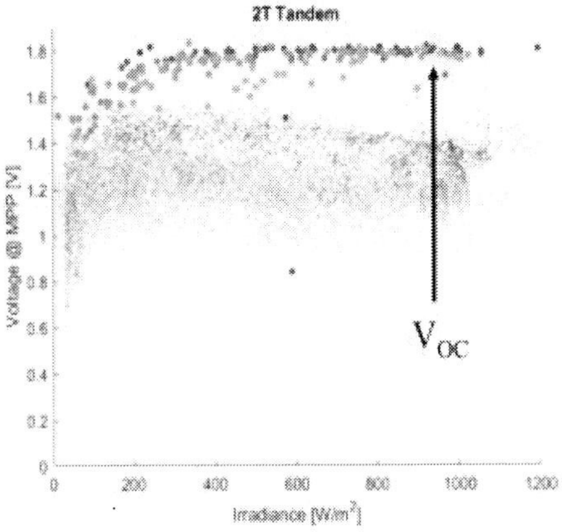

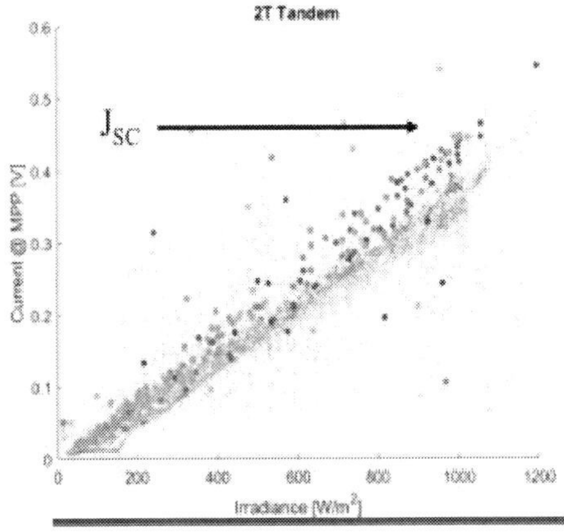

2T Tandem

V_{OC}

2T Tandem

J_{SC}

University *of Ljubljana*
Faculty *of Electrical Engineering*

2T Tandem

V_{OC}

Silicon

Perovskite

2T Tandem

J_{SC}

Silicon

Perovskite

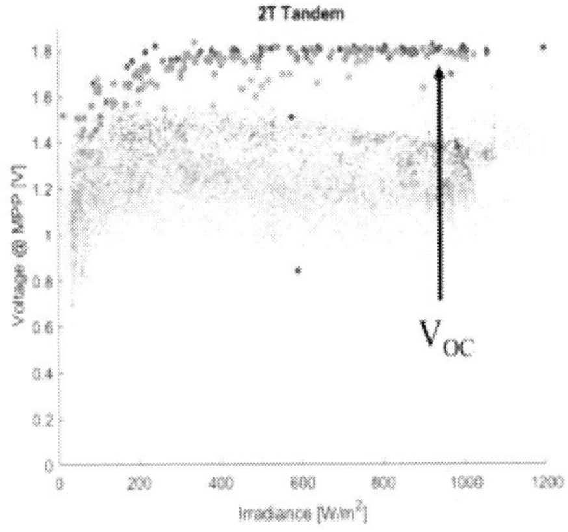

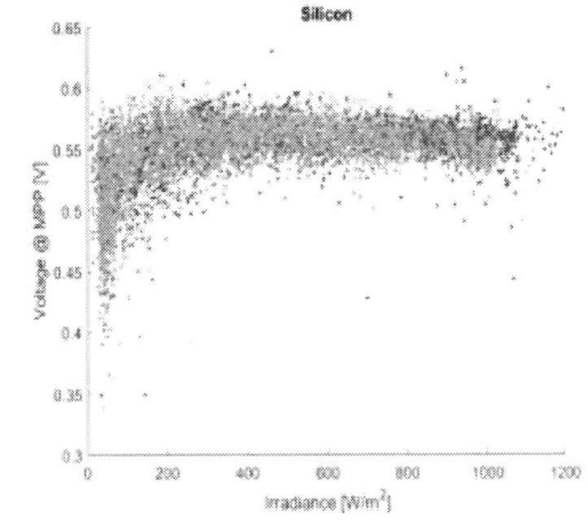

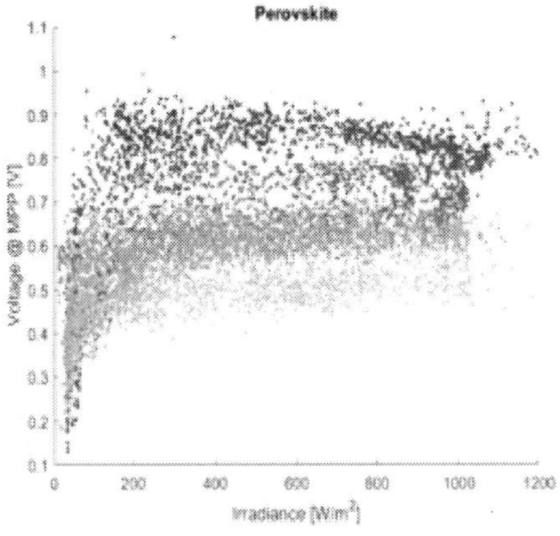

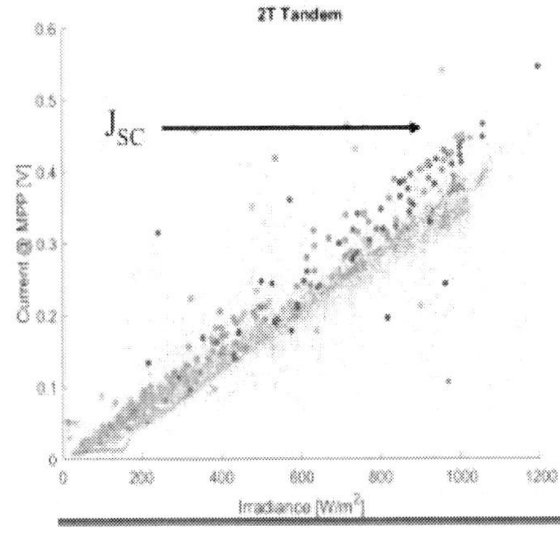

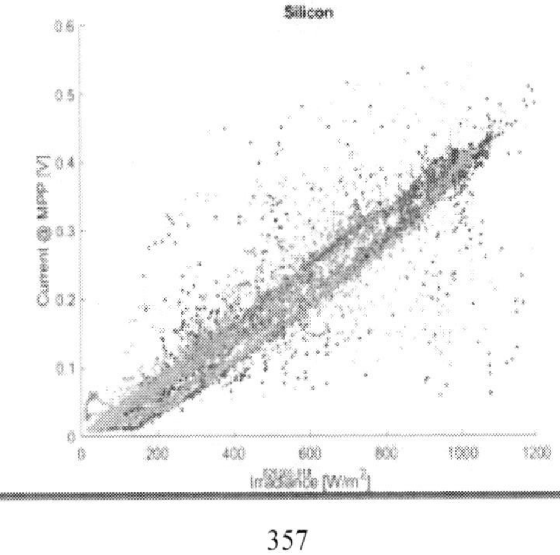

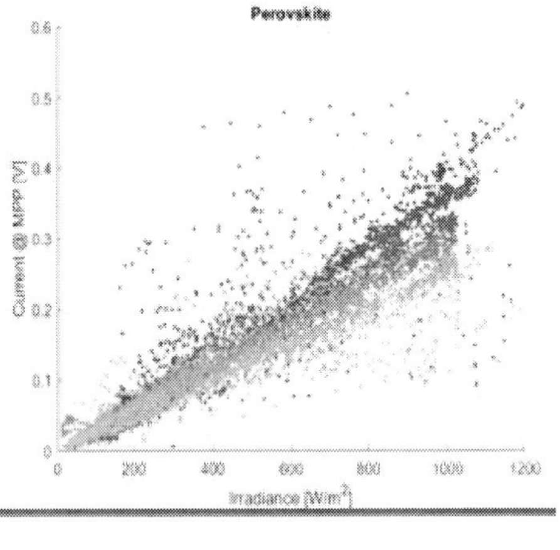

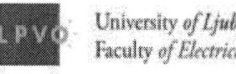

University *of Ljubljana*
Faculty *of Electrical Engineering*

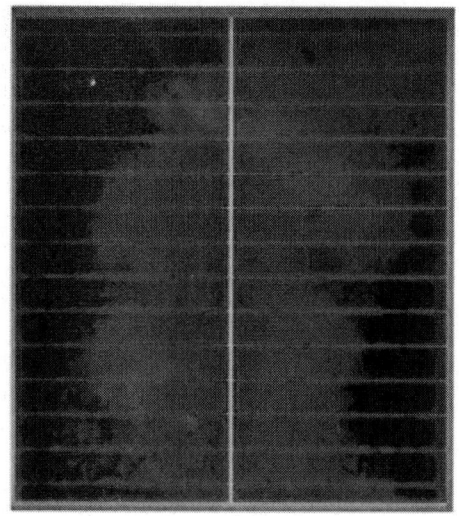

2CO.2.5 - Miha Kikelj   41st EUPVSEC – Vienna 25th of September

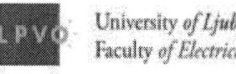

University *of Ljubljana*
Faculty *of Electrical Engineering*

2-terminal @ OC

Perovskite @ OC

Perovskite @ SC

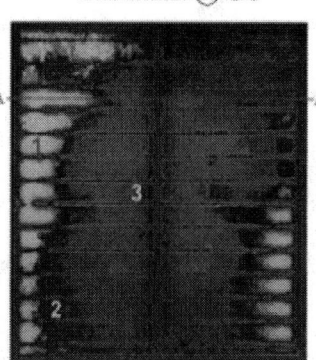

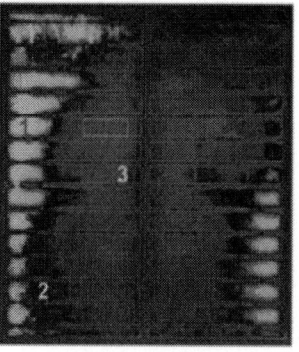

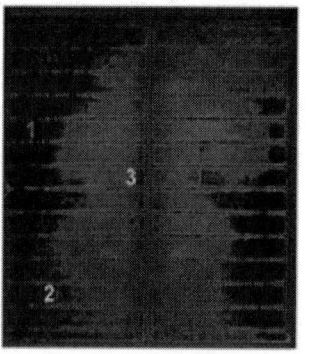

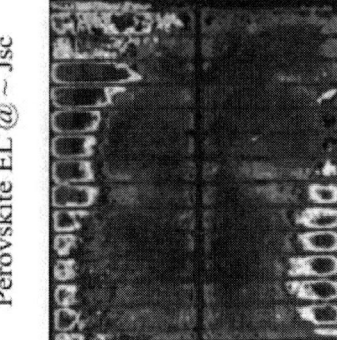

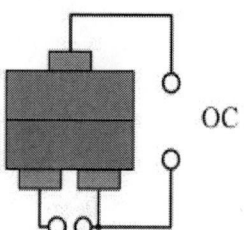

OC

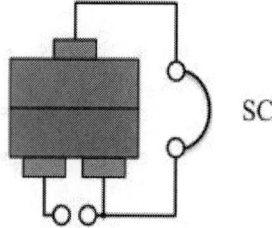

SC

2CO.2.5 - Miha Kikelj   41st EUPVSEC – Vienna 25th of September

University of Ljubljana
Faculty of Electrical Engineering

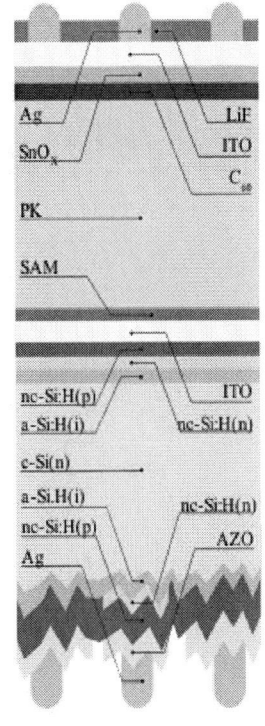

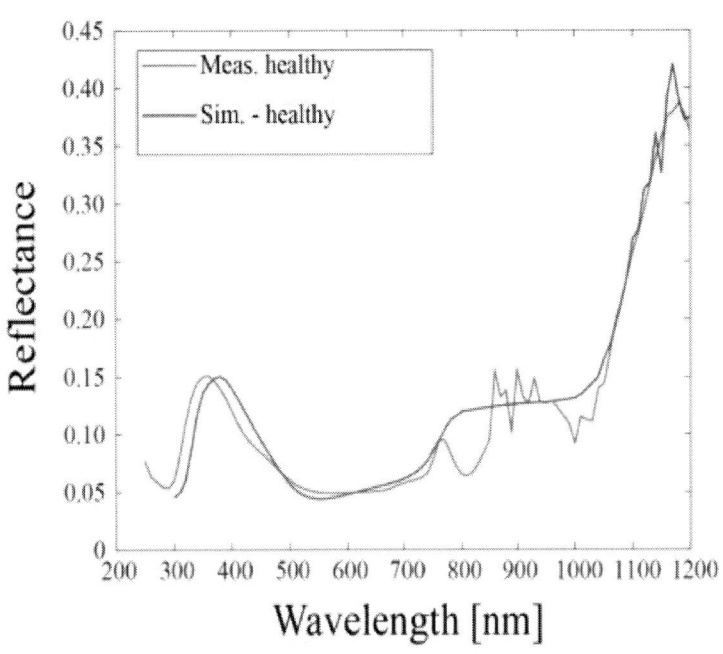

2CO.2.5 - Miha Kikelj   41st EUPVSEC – Vienna 25th of September

University of Ljubljana
Faculty of Electrical Engineering

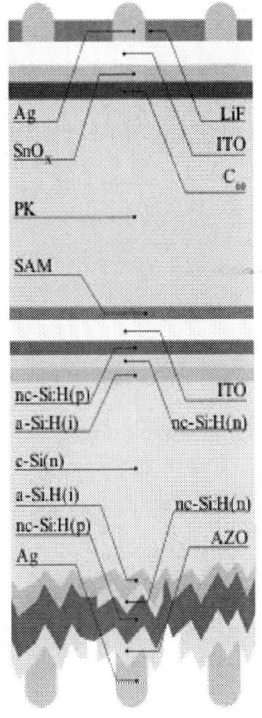

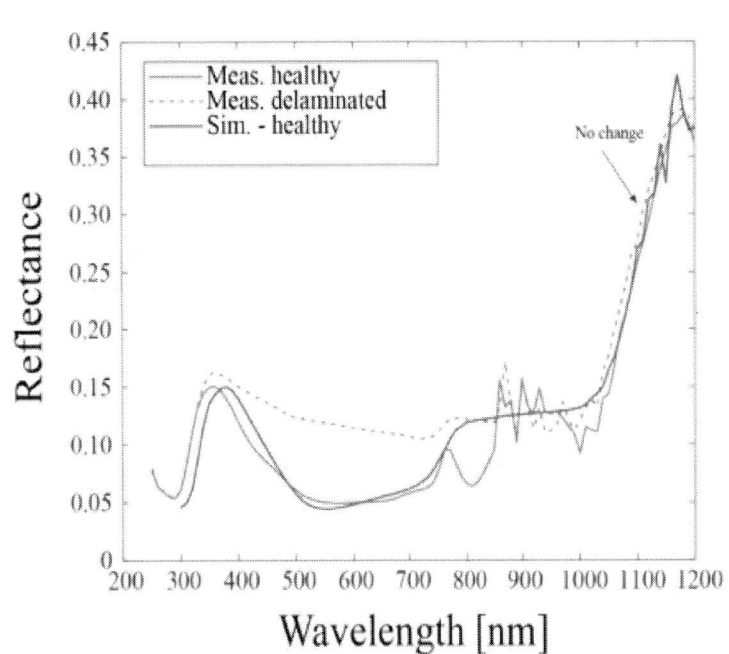

2CO.2.5 - Miha Kikelj   41st EUPVSEC – Vienna 25th of September

University of Ljubljana
Faculty of Electrical Engineering

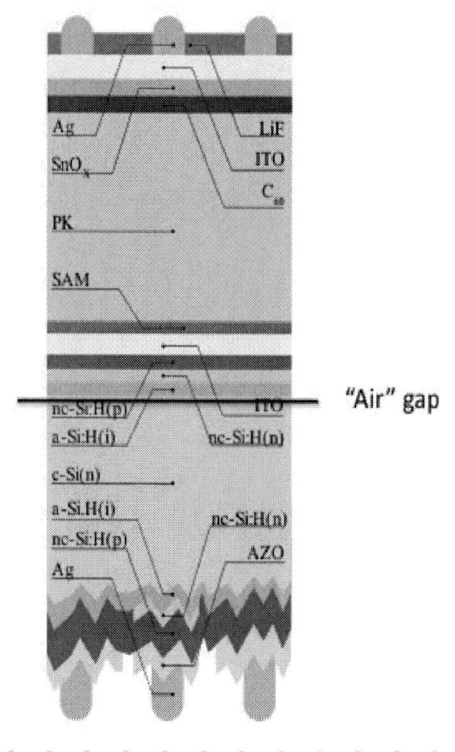

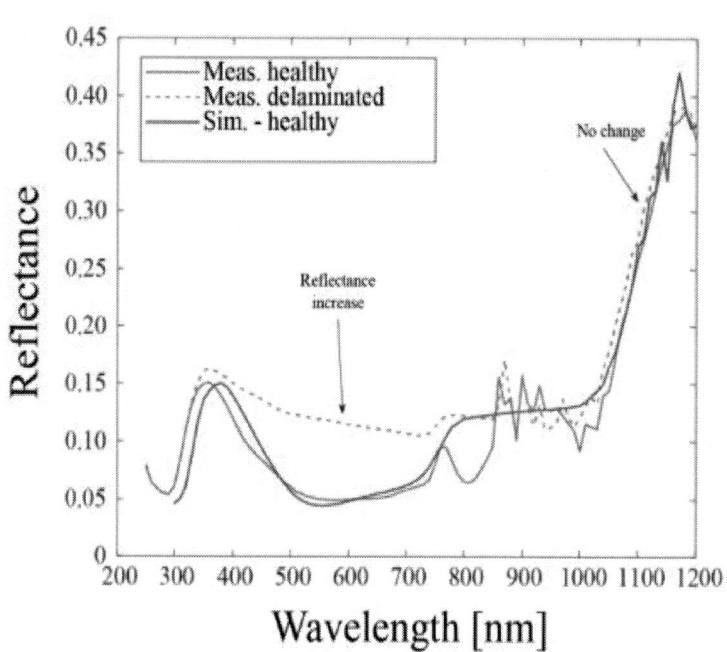

"Air" gap

2CO.2.5 - Miha Kikelj   41st EUPVSEC – Vienna 25th of September

---

University of Ljubljana
Faculty of Electrical Engineering

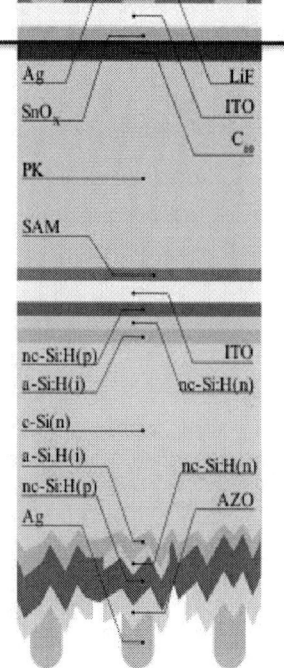

"Air" gap

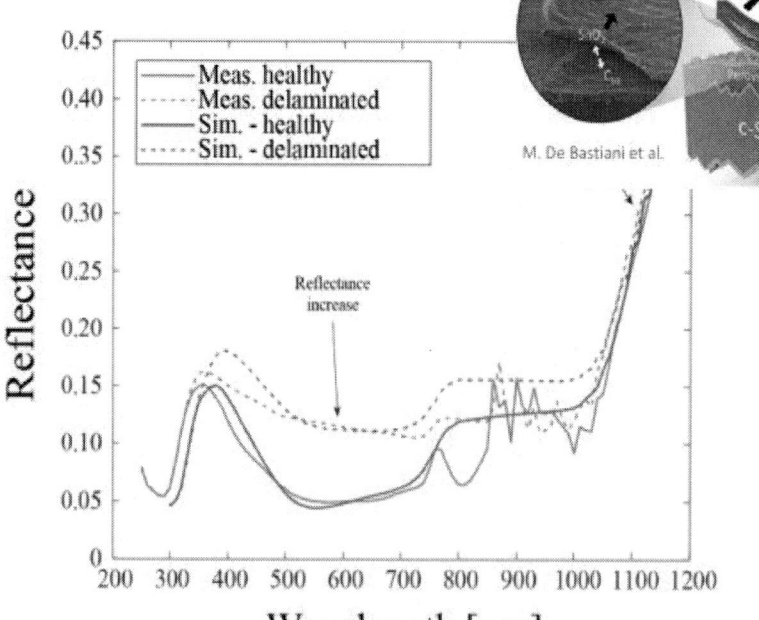

M. De Bastiani et al.

2CO.2.5 - Miha Kikelj   41st EUPVSEC – Vienna 25th of September

University of Ljubljana
Faculty of Electrical Engineering

Conclusions for 3TTs

- 3T - tandems are plausible tandem contenders for the future if appropriate module design considerations are followed
- 3TT's are a great tool for monitoring and evaluating outdoor behaviour of monolithic tandems
- 3TT's allow for in-depth "post-mortem" analysis as they allow access to individual subcells

2CO.2.5 - Miha Kikelj   41st EUPVSEC – Vienna 25th of September

University of Ljubljana
Faculty of Electrical Engineering

Acknowledgment

Slovenian Research Agency under the research programe P2-0415

Slovenian Research and Innovation Agency

Innosuisse AdAstra project

 Schweizerische Eidgenossenschaft
Confédération suisse
Confederazione Svizzera
Confederaziun svizra

Innosuisse – Schweizerische Agentur für Innovationsförderung

:: csem

Jonas Geissbühler,
Christophe Ballif

 MEYER BURGER

Laurie-Lou Senaud · Florent Sahli · Quentin Jeangros · Bertrand Paviet-Salomon

Damien Lachenal & Derk Baetzner

University *of Ljubljana*
Faculty *of Electrical Engineering*

Thank you for your attention!

2CO.2.5 - Miha Kikelj   41st EUPVSEC – Vienna 25th of September

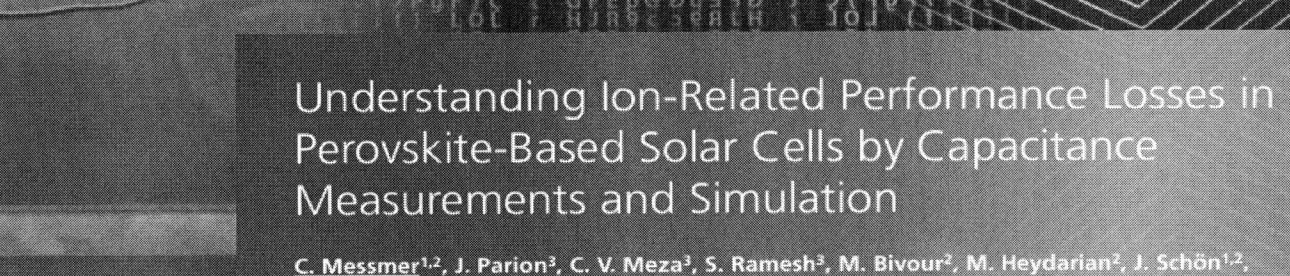

Understanding Ion-Related Performance Losses in Perovskite-Based Solar Cells by Capacitance Measurements and Simulation

C. Messmer[1,2], J. Parion[3], C. V. Meza[3], S. Ramesh[3], M. Bivour[2], M. Heydarian[2], J. Schön[1,2], H. S. Radhakrishnan[3], M. C. Schubert[2], S. W. Glunz[1,2]

[1] INATECH, University of Freiburg, Germany
[2] Fraunhofer ISE, Germany
[3] IMEC, Belgium

EU PVSEC, Vienna, Austria, 25th September 2024

Motivation

Understand Ion-related losses in Perovskite-Based Solar Cells (PSCs)

* **Hysteresis effects** impact the performance of PSCs [1-3]
* Origin: **Ion migration** within the perovskite absorber
* Goal: Understand and mitigate ion-related performance losses
* Approach: Small AC signal analysis [3,4]
  * Investigate frequency-dependent capacitances of PSCs
* **Learn about** device properties, e.g., ion diffusivities, built-in potential, …

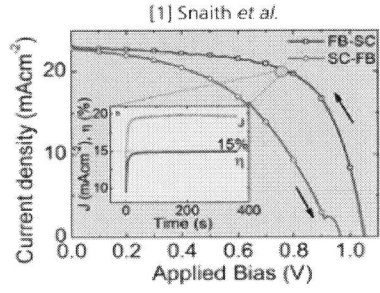

[1] Snaith *et al.*

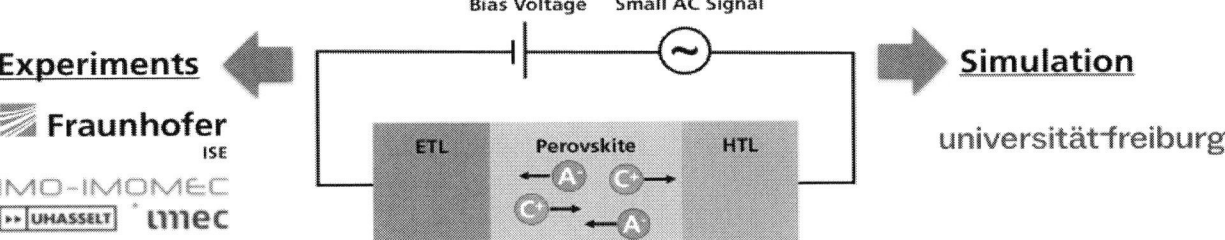

Experiments

Simulation

universität freiburg

[1] Snaith *et al.*, „Anomalous Hysteresis in Perovskite Solar Cells", J. Phys. Chem. Lett. 2014
[2] Le Corre *et al.*, „Quantification of Efficiency Losses Due to Mobile Ions …", Sol. RRL, vol. 6(4), 2022
[3] Ravishankar *et al.*, "Multilayer Capacitances: How Selective Contacts Affect …," PRX Energy, vol. 1(1), 2022
[4] Recart, Cuevas, "Application of junction capacitance measurements…," IEEE Trans. Electron Dev., vol. 53(3), 2006

Modelling Approaches
Small AC signal analysis

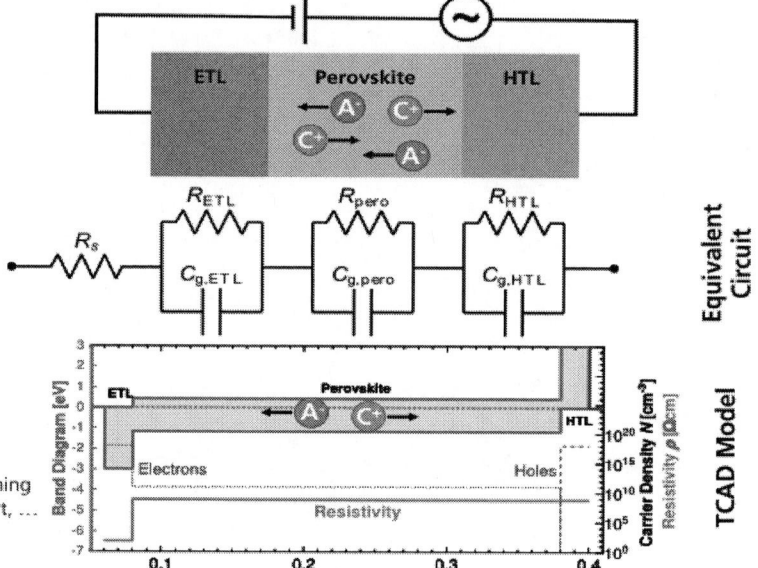

- **Equivalent-Circuit Modelling Approach:**
  - Each layer: RC circuit [1]
  - Geometrical capacitance (plate capacitor):
    $$C_{g,L}/A = \epsilon_0\,\epsilon_r/d_L$$

- **TCAD Modelling (this work):**
  - Detailed device modelling with Sentaurus TCAD [2]
    - Poisson equation and electron/hole drift diffusion (DD)
  - TCAD model was extended to AC signal analysis [3]
    - Compared to Eq.-Circuit Model
  - **Advantages:**
    - DD-Modelling of (several types of) ions incl. preconditioning
    - Study interfaces, full (tandem) stacks, tunneling transport, …
    - Same model for simulation of JV curves, …

[1] Ravishankar et al., „Multilayer Capacitances: …", PRX Energy, vol. 1, no. 1, 2022
[2] Messmer et al., „Toward more reliable measurement procedures…", Prog Photovolt Res Appl. 2024
[3] Messmer et al., „Understanding Ion-Related Performance Losses…", submitted to Solar RRL

universität freiburg   Fraunhofer ISE

Ion-induced Capacitances in Perovskite Solar Cells
TCAD Simulation: Dark State

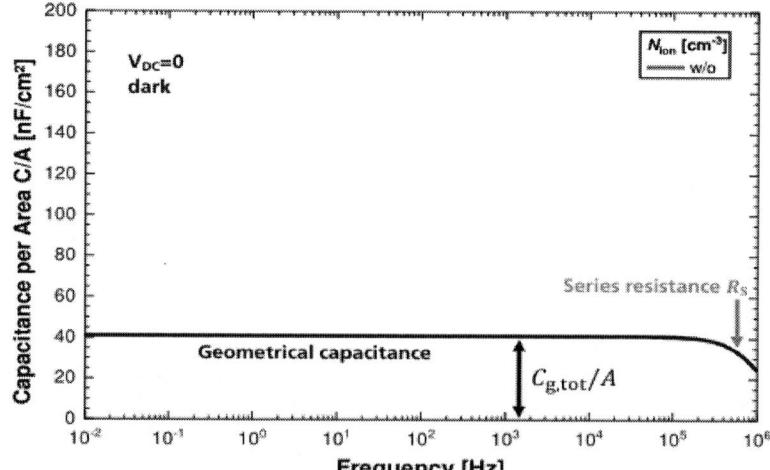

- Dark state, w/o DC bias, w/o ions, varied AC frequency
  - Geometrical capacitances in series connection
  - High-frequency capacitance reduced by series resistance

universität freiburg   Fraunhofer ISE

Ion-induced Capacitances in Perovskite Solar Cells

TCAD Simulation: Dark State

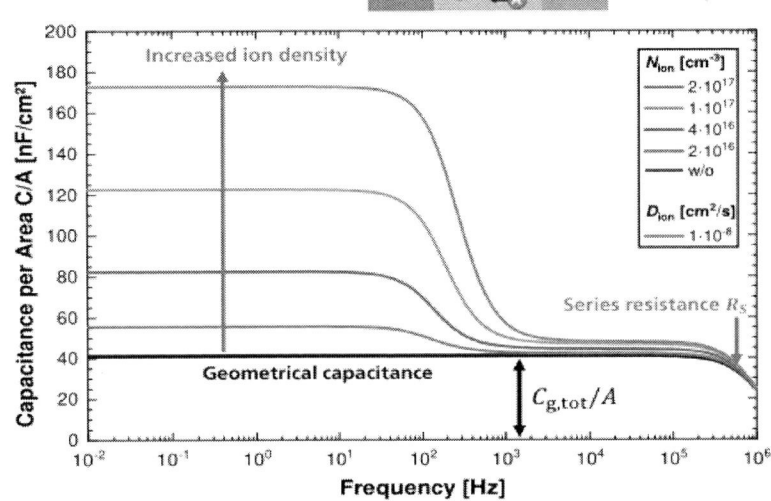

* Dark state, w/o DC bias, w/o ions, varied AC frequency
  * Geometrical capacitances in series connection:
  * High-frequency capacitance reduced by series resistance
* Mobile anions with varied concentration (immobile cations)
  * Increased low-frequency response of capacitance

Ion-induced Capacitances in Perovskite Solar Cells

TCAD Simulation: Dark State

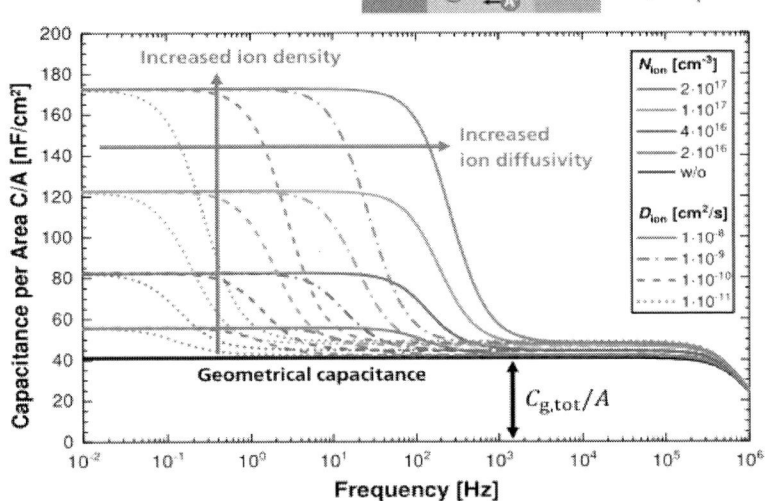

* Dark state, w/o DC bias, w/o ions, varied AC frequency
  * Geometrical capacitances in series connection:
  * High-frequency capacitance reduced by series resistance
* Mobile anions with varied concentration (immobile cations)
  * Increased low-frequency response of capacitance
* Mobile anions with varied diffusivity
  * Characteristic frequency shifts

Ion-induced Capacitances in Perovskite Solar Cells
Experimental Evidence

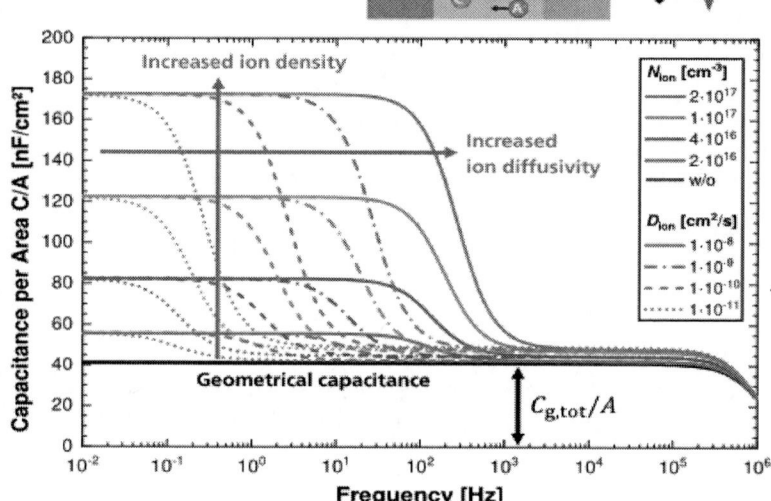

- Perovskite single junction fabricated and measured at IMEC

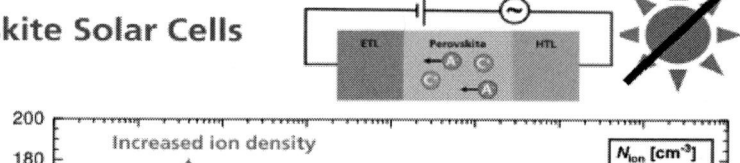

7

universität freiburg   Fraunhofer ISE

Ion-induced Capacitances in Perovskite Solar Cells
Experimental Evidence

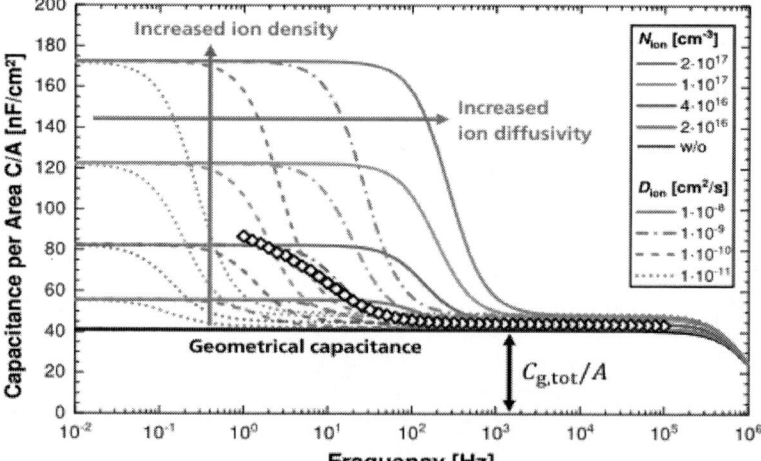

- Perovskite single junction fabricated and measured at IMEC
- What can we deduce from C-f characteristics?

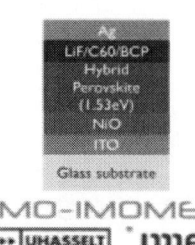

8

universität freiburg   Fraunhofer ISE

Presented at 41ˢᵗ EU PVSEC, 2024, Vienna, Austria

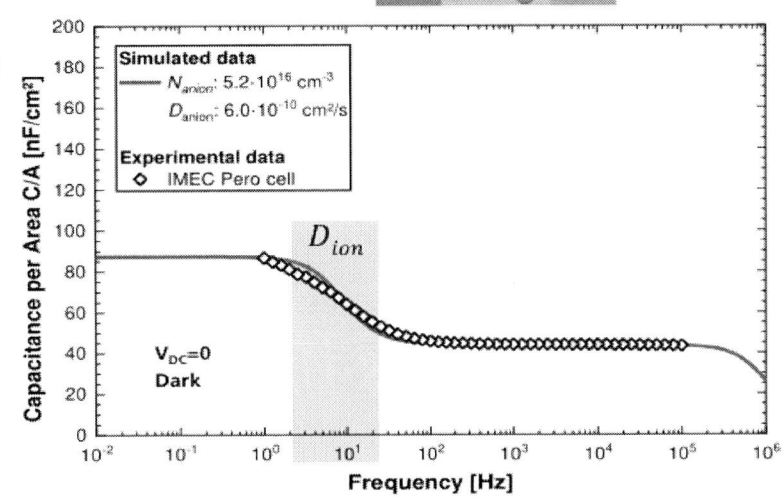

Ion-induced Capacitances in Perovskite Solar Cells
Experimental Evidence

- Perovskite single junction fabricated and measured at IMEC
- What can we deduce from *C-f* characteristics?
  - Characteristic frequency f, here around 10 Hz
    → **Ion diffusivity** of $6 \cdot 10^{-10}$ cm²/s

9   universität·freiburg   Fraunhofer ISE

Presented at 41ˢᵗ EU PVSEC, 2024, Vienna, Austria

Ion-induced Capacitances in Perovskite Solar Cells
Experimental Evidence

- Perovskite single junction fabricated and measured at IMEC
- What can we deduce from *C-f* characteristics?
  - Characteristic frequency f, here around 10 Hz
    → **Ion diffusivity** of $6 \cdot 10^{-10}$ cm²/s
  - High-frequency regime (measured on similar sample)

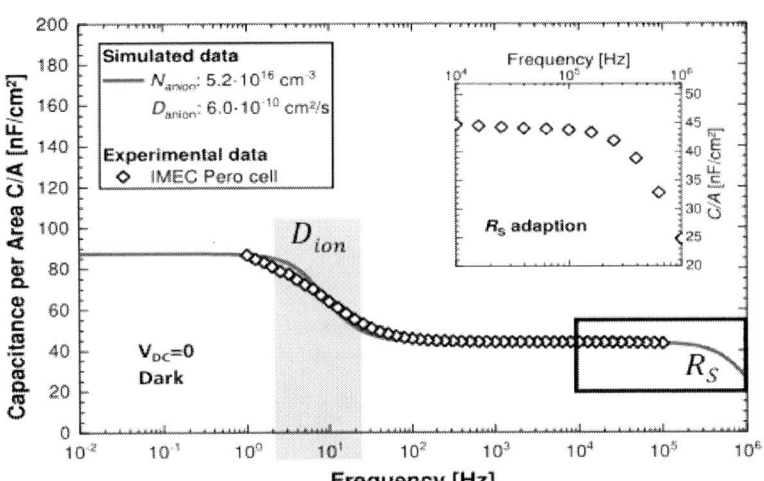

10   universität·freiburg   Fraunhofer ISE

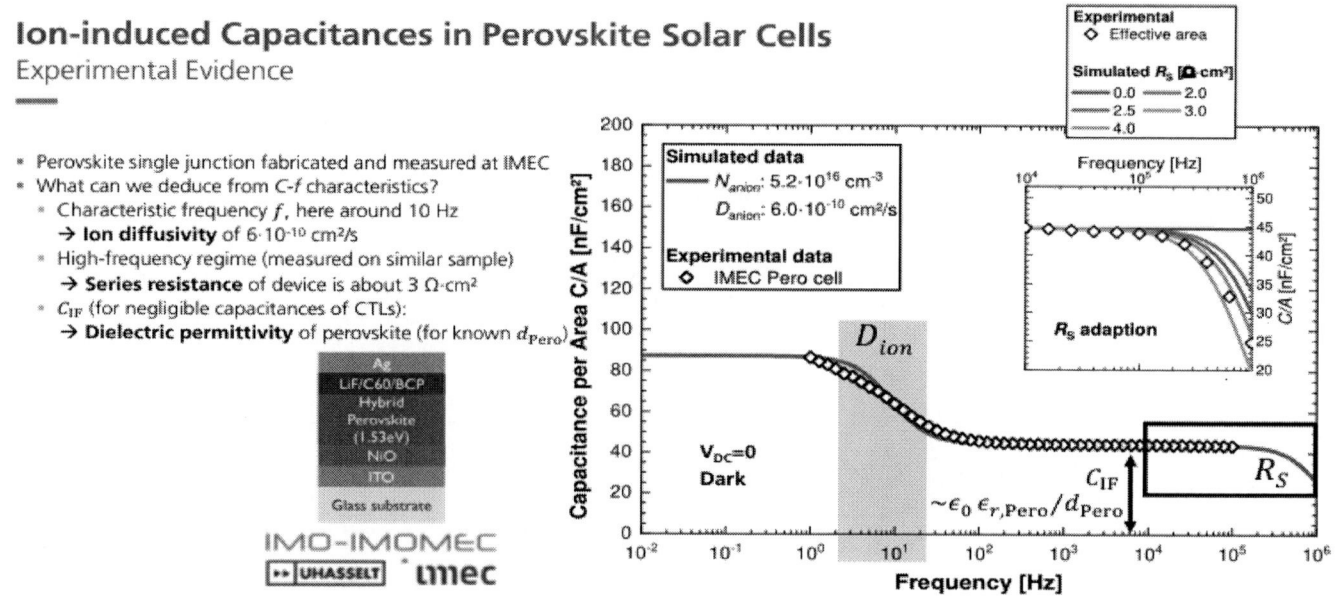

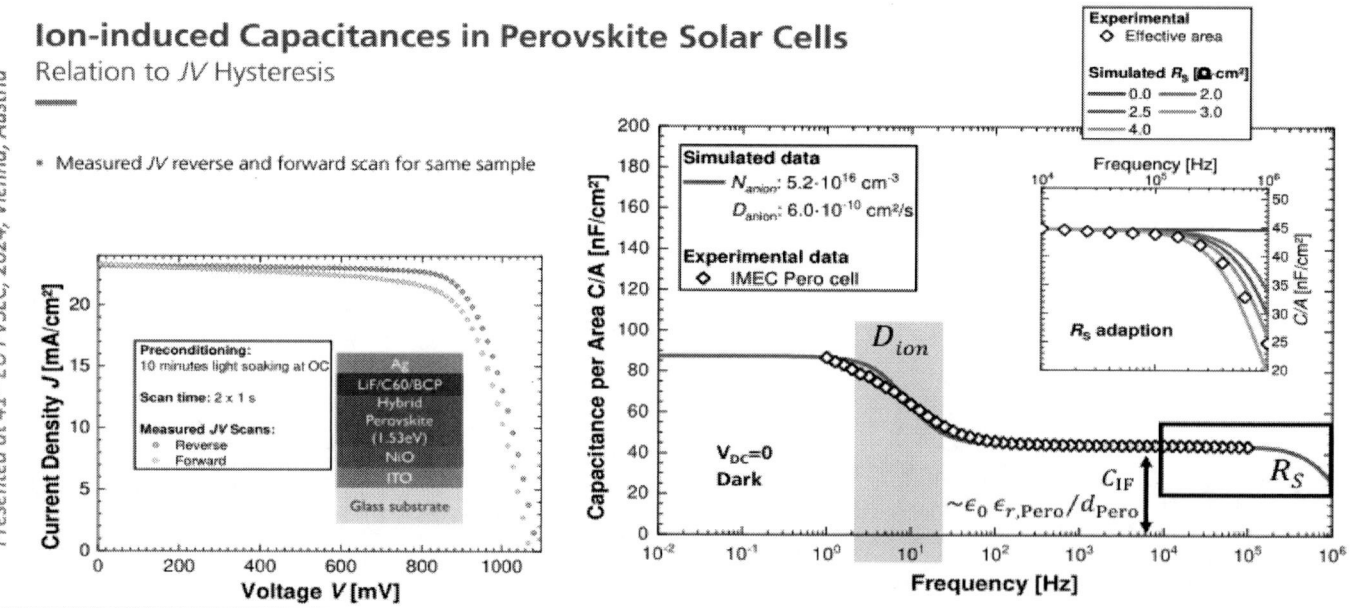

Ion-induced Capacitances in Perovskite Solar Cells
Relation to *JV* Hysteresis

* Measured *JV* reverse and forward scan for same sample
* Simulated *JV* scans with parameters from *C-f* analysis
  * Very good agreement indicating the relation between *C-f* characteristics and *JV* hysteresis

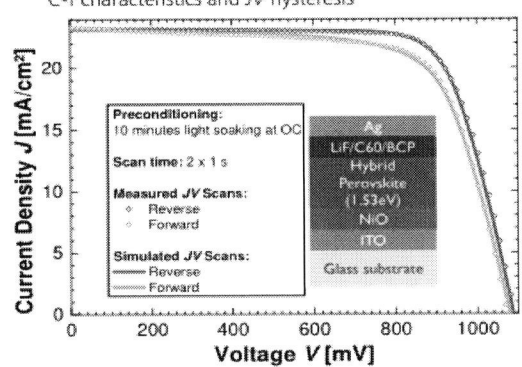

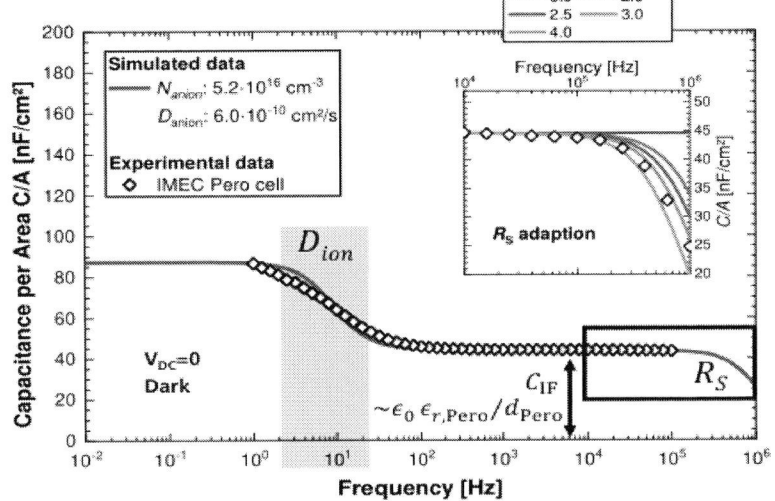

13

universität freiburg   Fraunhofer ISE

Ion-induced Capacitances in Perovskite Solar Cells
Relation to *JV* Hysteresis

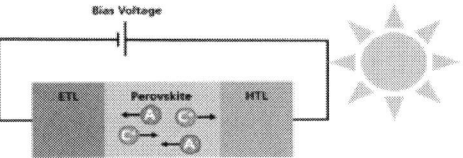

* Measured *JV* reverse and forward scan for same sample
* Simulated *JV* scans with parameters from *C-f* analysis
  * Very good agreement indicating the relation between *C-f* characteristics and *JV* hysteresis

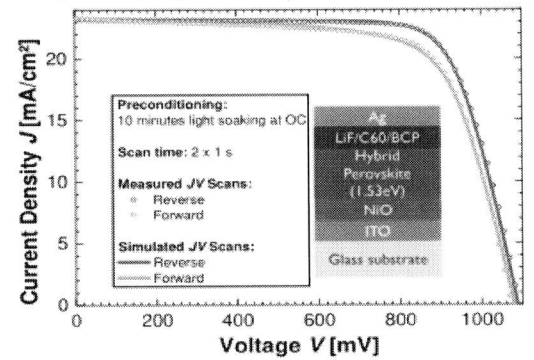

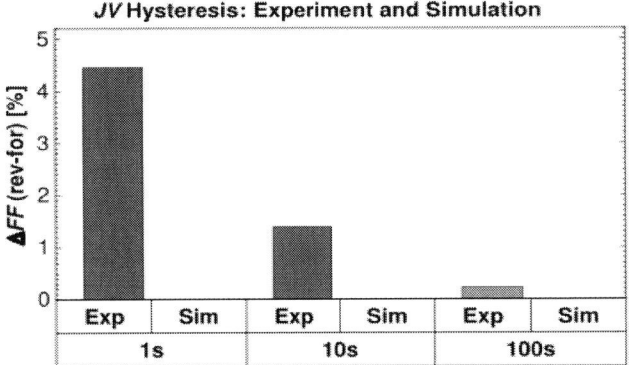

14

universität freiburg   Fraunhofer ISE

Ion-induced Capacitances in Perovskite Solar Cells
Relation to *JV* Hysteresis

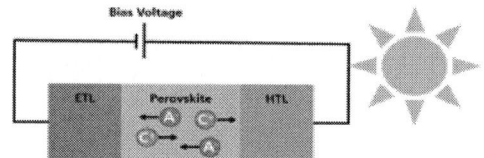

* Measured *JV* reverse and forward scan for same sample
* Simulated *JV* scans with parameters from *C-f* analysis
  * Very good agreement indicating the relation between *C-f* characteristics and *JV* hysteresis for varied scan-times

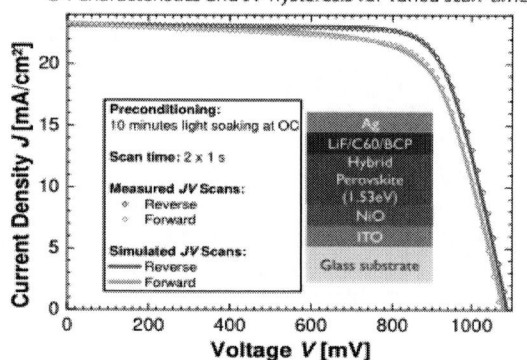

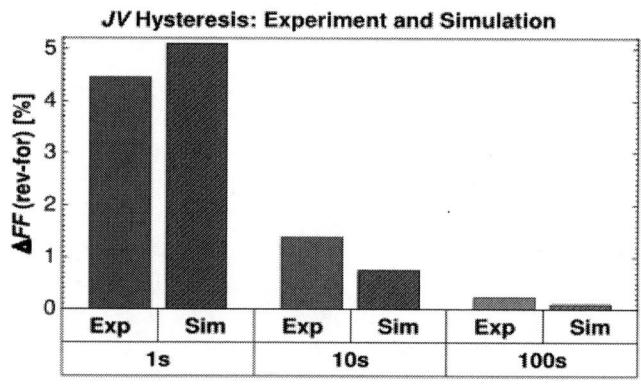

Ion-induced Capacitances in Perovskite Solar Cells
Relation to *JV* Hysteresis

* Measured *JV* reverse and forward scan for same sample
* Simulated *JV* scans with parameters from *C-f* analysis
  * Very good agreement indicating the relation between *C-f* characteristics and *JV* hysteresis for varied scan-times

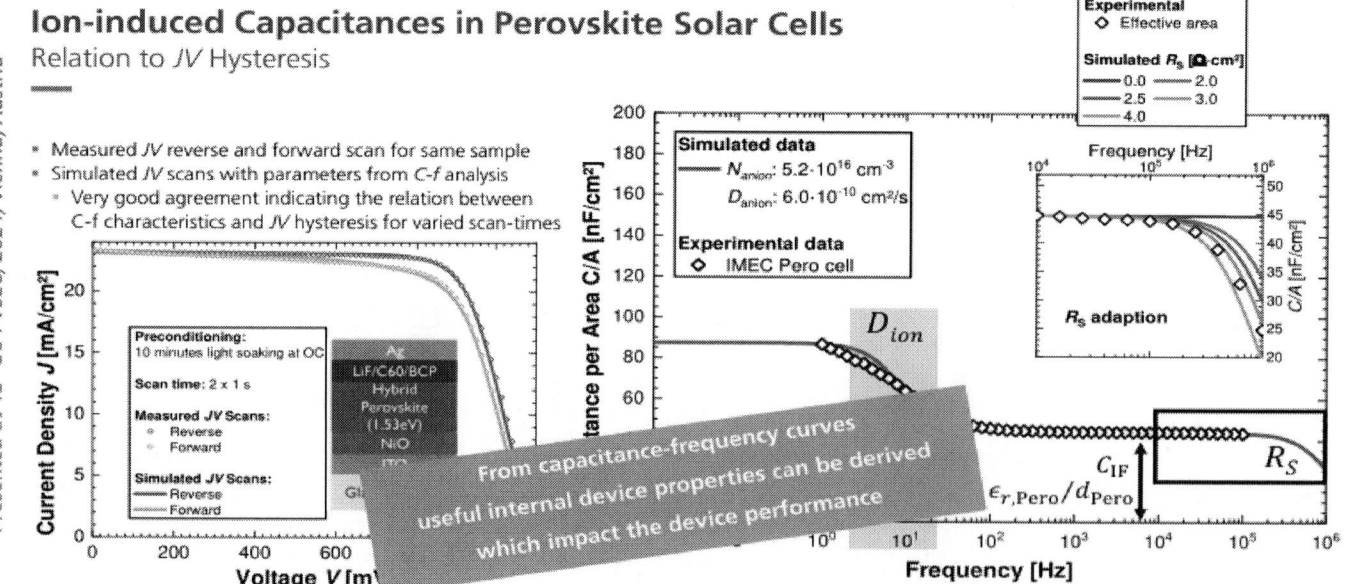

From capacitance-frequency curves useful internal device properties can be derived which impact the device performance

Ion-Induced Capacitances
In-Depth Analysis of the Capacitance Plateau

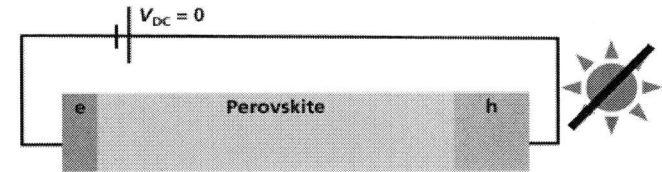

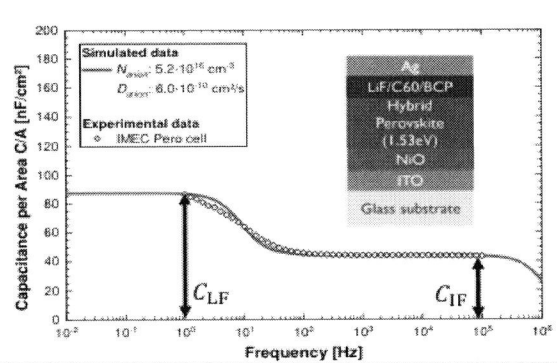

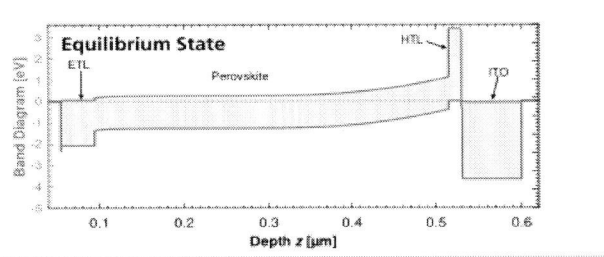

Ion-Induced Capacitances
In-Depth Analysis of the Capacitance Plateau

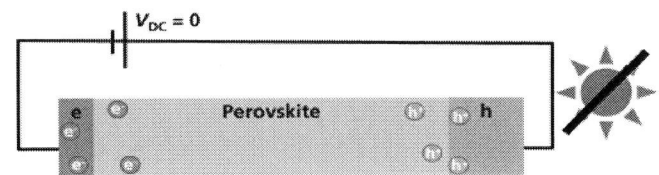

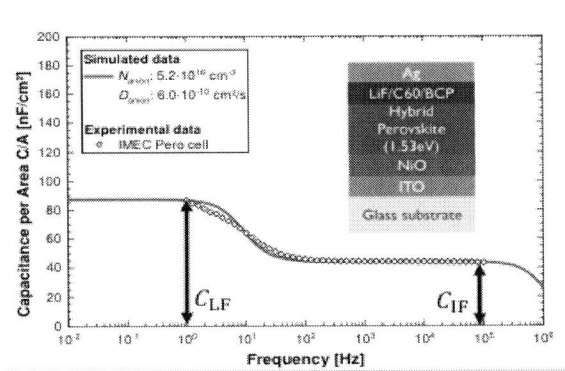

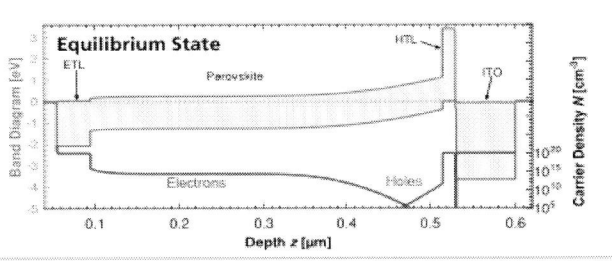

Presented at 41st EU PVSEC, 2024, Vienna, Austria

Ion-Induced Capacitances
In-Depth Analysis of the Capacitance Plateau

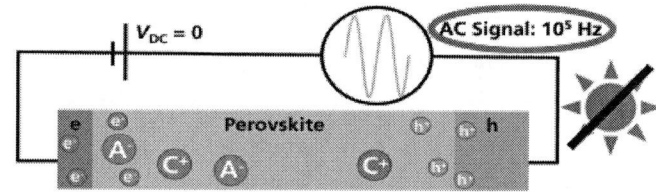

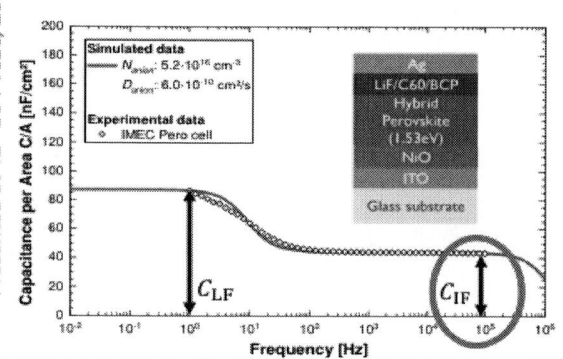

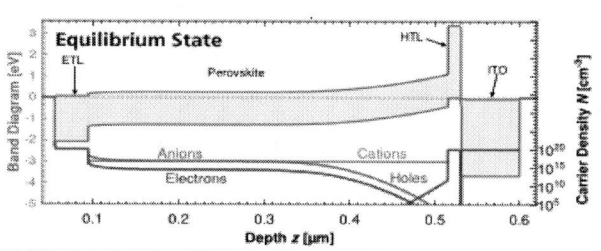

universität freiburg   Fraunhofer ISE

Presented at 41st EU PVSEC, 2024, Vienna, Austria

Ion-Induced Capacitances
In-Depth Analysis of the Capacitance Plateau

* C_{IF} corresponds to charge accumulation at Pero/ETL, Pero/HTL interface

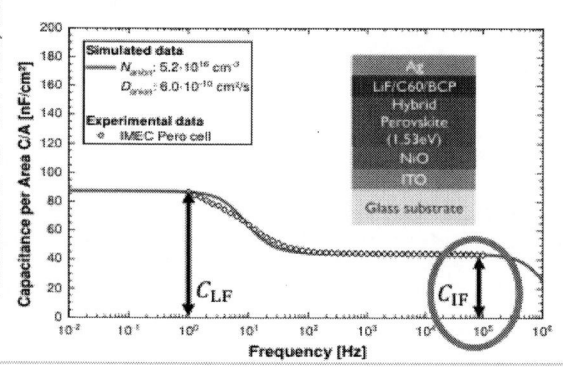

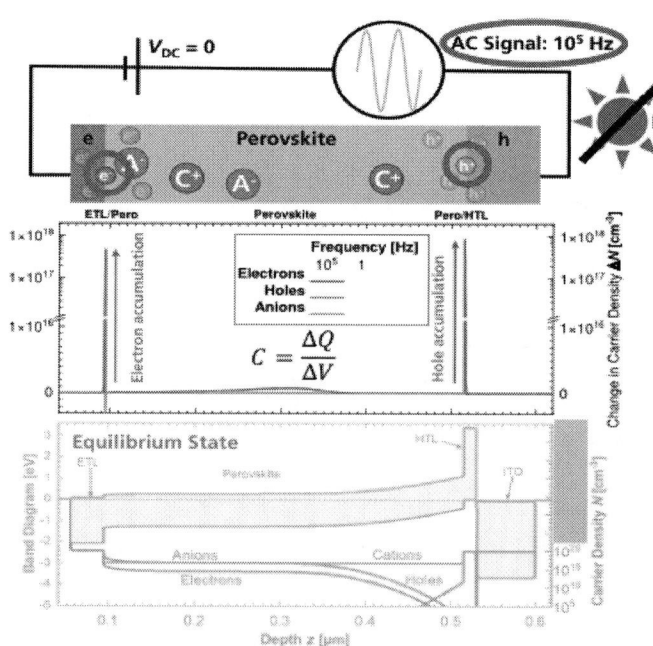

$$C = \frac{\Delta Q}{\Delta V}$$

universität freiburg   Fraunhofer ISE

Presented at 41st EU PVSEC, 2024, Vienna, Austria

Ion-Induced Capacitances
In-Depth Analysis of the Capacitance Plateau

* C_{IF} corresponds to charge accumulation at Pero/ETL, Pero/HTL interface

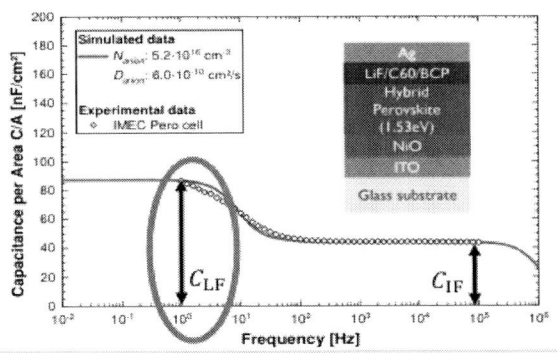

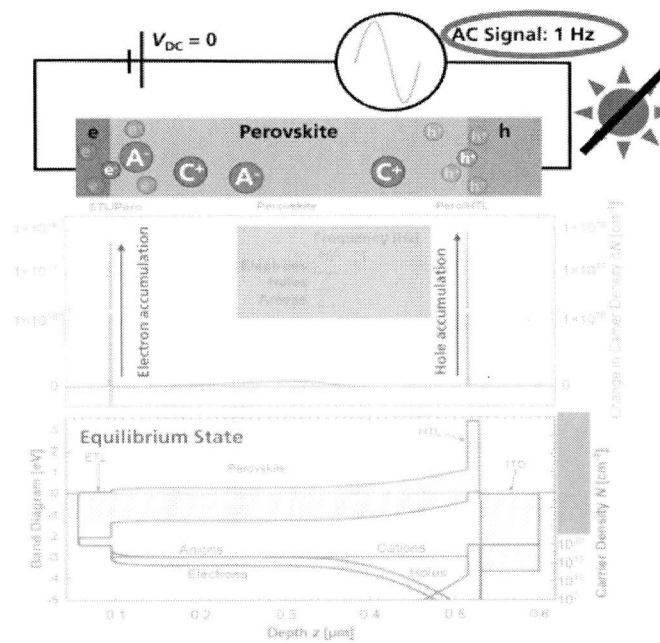

universität freiburg   Fraunhofer ISE

Presented at 41st EU PVSEC, 2024, Vienna, Austria

Ion-Induced Capacitances
In-Depth Analysis of the Capacitance Plateau

* C_{IF} corresponds to charge accumulation at Pero/ETL, Pero/HTL interface

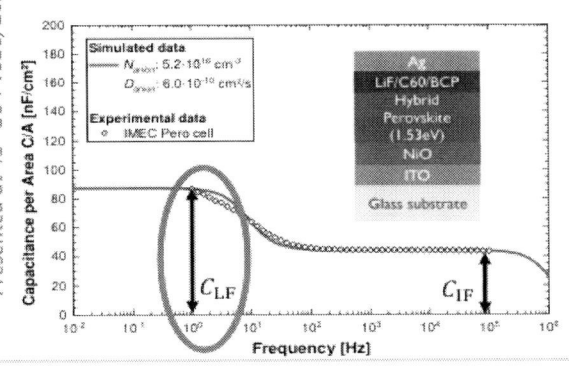

universität freiburg   Fraunhofer ISE

Ion-Induced Capacitances
In-Depth Analysis of the Capacitance Plateau

* C_{IF} corresponds to charge accumulation at Pero/ETL, Pero/HTL interface
* C_{LF}: **Anion displacement** from ETL interface towards bulk leads to further charge carrier accumulation
  * → C_{LF} depends on N_{ion}, **but also on** built-in potential, cations, …

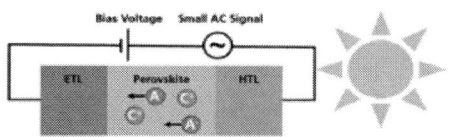

- TCAD modelling helps to interpret C-f plots
- Ionic displacement leads to increased low-f capacitances

Presented at 41st EU PVSEC, 2024, Vienna, Austria

Ion-Induced Capacitances under Illumination
TCAD Simulation

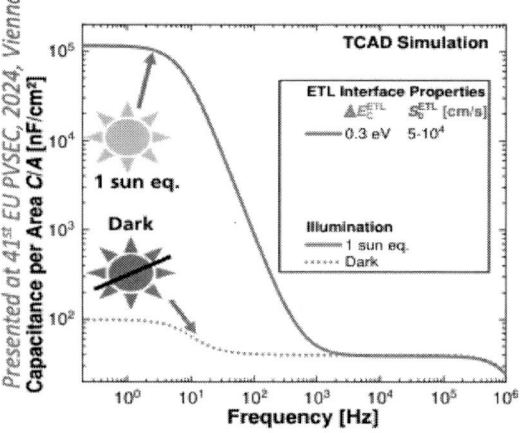

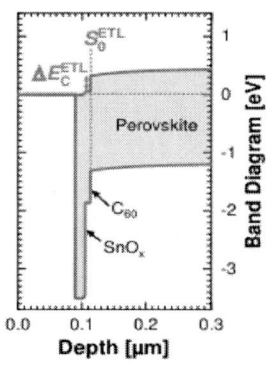

* **Simulated dark and light C-f curves** for different ETL interface properties
  * Capacitance under illumination increases by several orders of magnitude

Presented at 41st EU PVSEC, 2024, Vienna, Austria

Ion-Induced Capacitances under Illumination
TCAD Simulation

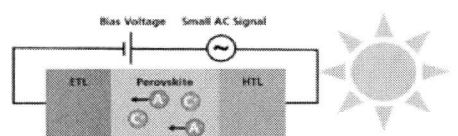

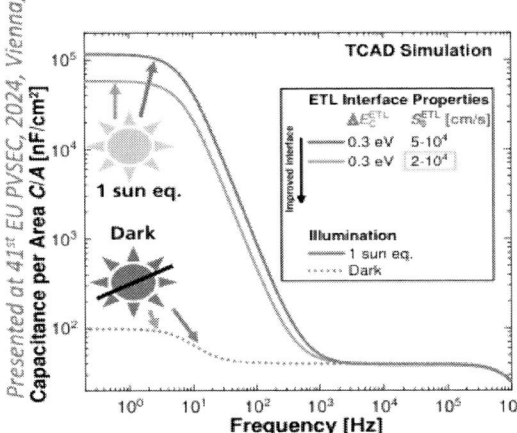

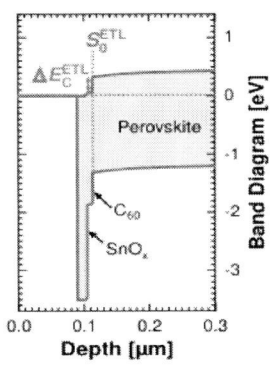

- Simulated dark and light *C-f* curves for different ETL interface properties
  - Capacitance under illumination increases by several orders of magnitude
- For lower surface recombination velocity (SRV) at ETL/Pero interface:
  - Illuminated *C-f* curve shifts down
  - Dark *C-f* curve is unchanged

Ion-Induced Capacitances under Illumination
TCAD Simulation

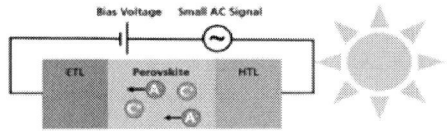

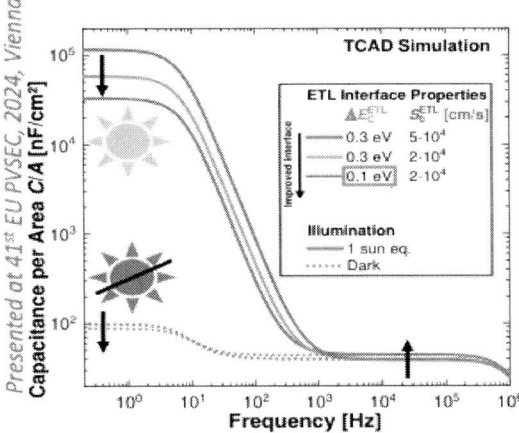

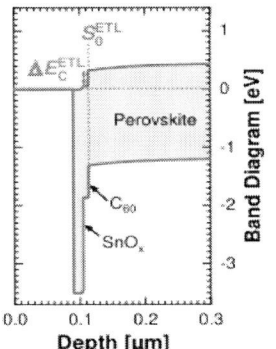

- Simulated dark and light *C-f* curves for different ETL interface properties
  - Capacitance under illumination increases by several orders of magnitude
- For lower surface recombination velocity (SRV) at ETL/Pero interface:
  - Illuminated *C-f* curve shifts down
  - Dark *C-f* curve is unchanged
- For change in band alignment (reduced ETL/Pero conduction band offset):
  - Both illuminated and dark *C-f* curve change
  - **Low frequency plateau**: Capacitance decreases
  - **High frequency plateau**: Capacitance increases

Ion-Induced Capacitances under Illumination
Experimental Comparison

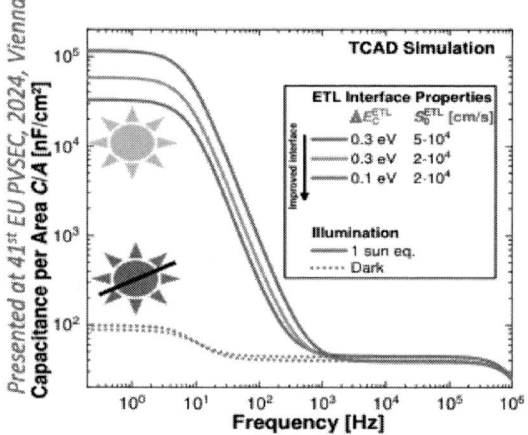

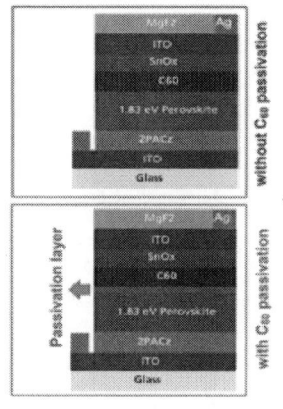

* Perovskite single junctions (used as top cells) fabricated at Fraunhofer ISE and measured at IMEC
  1. **ETL/Pero interface unpassivated**
  2. ETL/Pero interface passivated with PI*

*PI = piperazinium iodide

Ion-Induced Capacitances under Illumination
Experimental Comparison

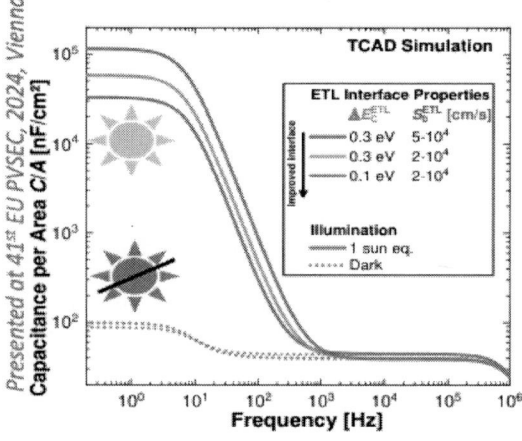

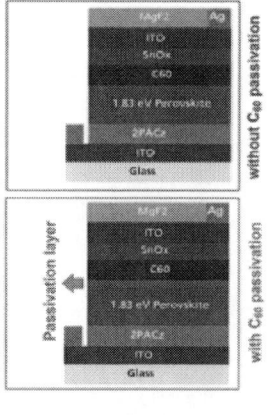

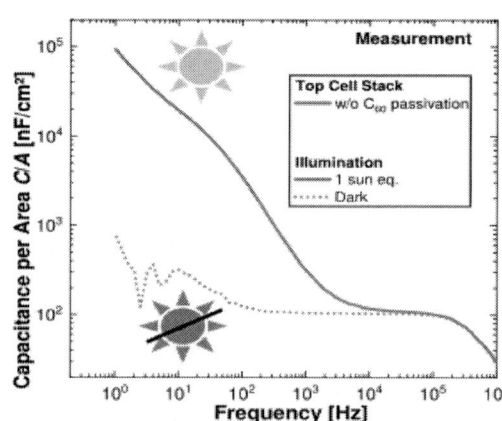

Ion-Induced Capacitances under Illumination

Experimental Comparison

* Good qualitative agreement of TCAD model and measurements

Passivation layer leads to:
* Reduced low-f capacitance
* Increased high-f capacitance

→ Indicates lower recombination
→ Possibly also better ETL/Pero band alignment

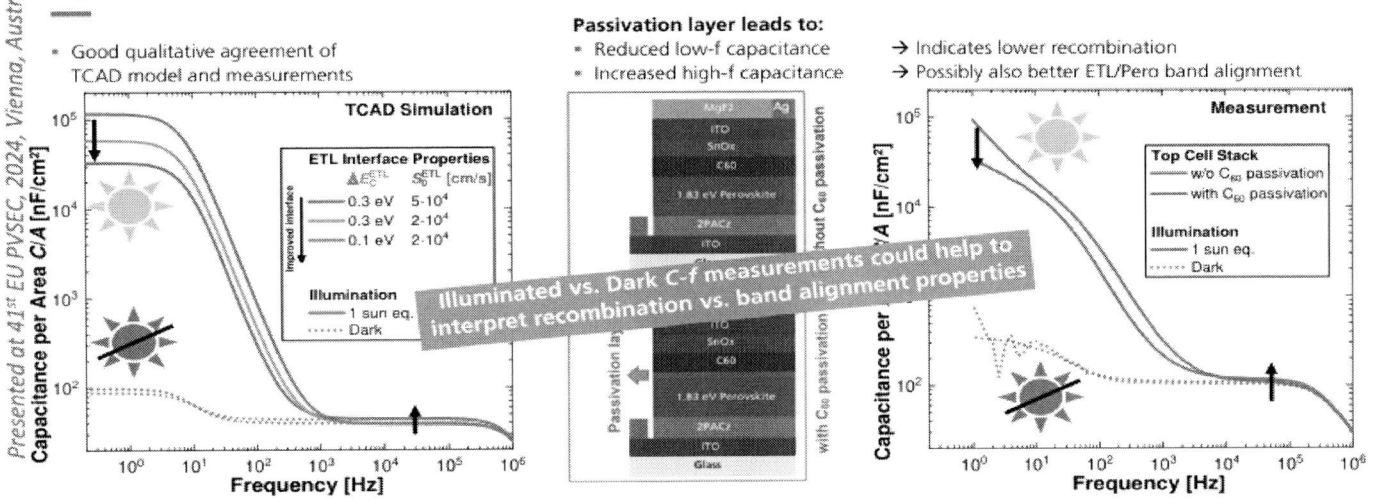

Illuminated vs. Dark C-f measurements could help to interpret recombination vs. band alignment properties

---

Conclusion

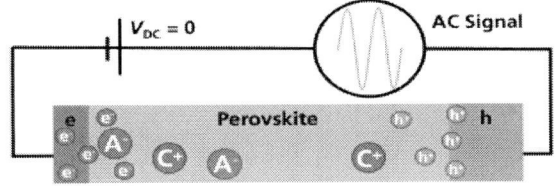

* TCAD Simulation of capacitance response of PSCs:
  * Impact of ionic properties
  * Good agreement with experimental data
  * Insights into physical origin of ionic capacitance

* Link **C-f response** to scan-time dependent **JV hysteresis**

* Investigation of **C-f curves** under illumination
  * Effect of band alignment and interface recombination
  * Experimental evidence

→ Simulation-aided analysis of C-f curves contributes to **enhanced understanding**
→ **Careful interpretation** of data is essential
→ Measurements are **non-destructive** and on device level

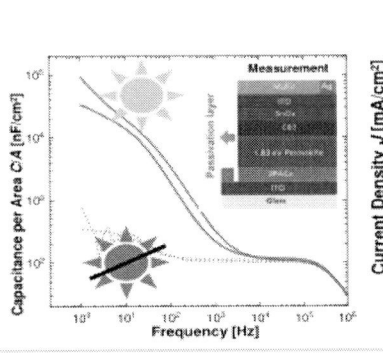

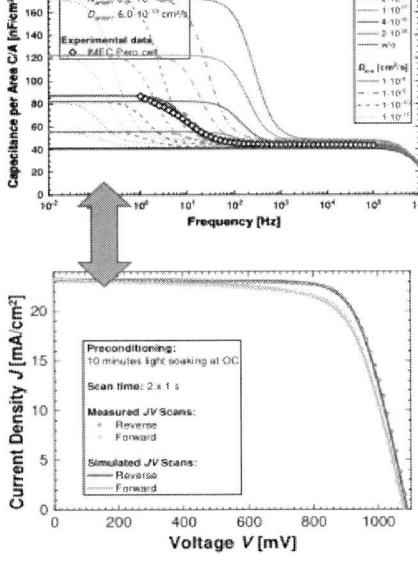

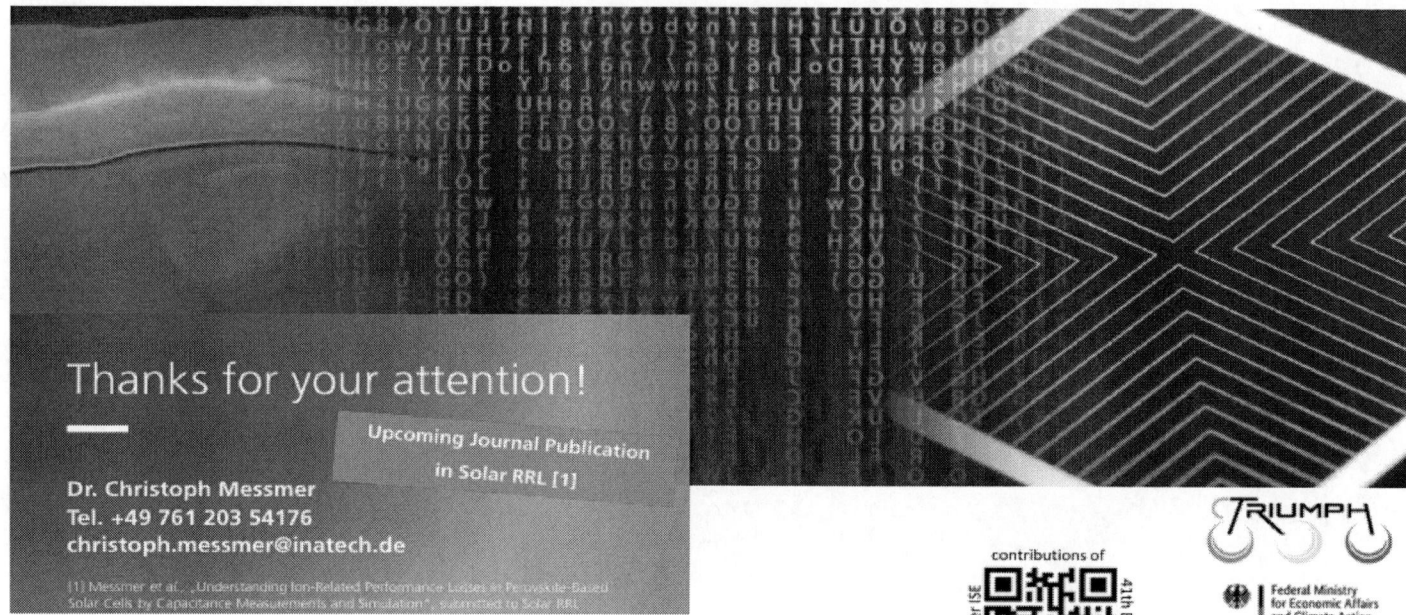

Thanks for your attention!

Upcoming Journal Publication in Solar RRL [1]

Dr. Christoph Messmer
Tel. +49 761 203 54176
christoph.messmer@inatech.de

[1] Messmer et al., „Understanding Ion-Related Performance Losses in Perovskite-Based Solar Cells by Capacitance Measurements and Simulation", submitted to Solar RRL

universität freiburg

 INATECH
INSTITUT FÜR NACHHALTIGE TECHNISCHE SYSTEME

Fraunhofer
ISE

contributions of

Fraunhofer ISE
ise.link/eupvsec2024
41th EU PVSEC

TRIUMPH

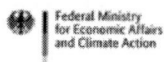 Federal Ministry
for Economic Affairs
and Climate Action

on the basis of a decision
by the German Bundestag

41st European Photovoltaic Solar Energy Conference and Exhibition

This presentation was selected by the Sc. Committee of the EU PVSEC 2024 for submission of a full paper to one of the EU PVSEC's collaborating peer-reviewed journals.

ANALYSIS AND MODELLING OF RECOVERY AND DEGRADATION MECHANISMS IN PEROVSKITE SOLAR CELLS

Guillem Álvarez Pérez[1,2], Arthur Julien[1], Karim Medjoubi[1], Jean Baptiste Puel[1,3], and Jean François Guillemoles[1,2]

[1]Institut Photovoltaïque d'Île-de-France (IPVF), Palaiseau, 91120, France

[2]Institut Photovoltaïque d'Île-de-France (IPVF), UMR 9006, CNRS, École Polytechnique, IP Paris, Chimie Paristech, PSL, Palaiseau, 91120, France

[3]EDF R&D, Palaiseau, 91120, France

ABSTRACT: Perovskite solar cells have reached outstanding results in terms of power conversion efficiency over the last years. However, their stability remains a challenge. Post-degradation studies can be found in the literature, but limited research on the dynamics of recovery and degradation has been conducted. Modelling of experimental behavior provides new insights into the understanding these mechanisms in perovskite solar cells. Drift-diffusion and transfer-matrix methods are employed to simulate optoelectrical characteristics, and the initial performances of solar cells are reproduced by means of a genetic algorithm. The evolution of the correlation among optoelectrical parameters is employed to assess the resemblance of experimental data and simulated mechanisms. This allows to propose feasible mechanisms relevant for recovery and degradation. Simulated mechanisms are computed by varying material parameters such as charge carrier mobility, doping and defect concentration in the absorber and transport layers, as well as in their interfaces. Relevant mechanisms to understand the experimental behavior are proposed, while allowing to discard the unlikely ones. This provides new insights into recovery and degradation of perovskite solar cells and serves already to identify the possible reversibility of such mechanisms, enabling to focus experimental efforts on mitigating irreversible degradation.

Keywords: perovskite solar cells, perovskite stability, modelling

1 INTRODUCTION

Renewable energy will play a crucial role in addressing the escalating global energy demands and reducing emissions from the energy sector, which contributes approximately three-quarters of global greenhouse gases [1]. Among them, solar photovoltaic is called to be an essential part of the energy transition.

The solar photovoltaic (PV) market, predominantly silicon-based, is expected to expand significantly, emphasizing the need for more efficient and cost-effective technologies. Perovskite solar cells have gained attracted the attention of the scientific community because of their remarkable power conversion efficiency improvements over the past decade. While their fabrication costs are competitive, stability remains a major challenge before commercialization.

Although degradation has been already studied [2], [3], [4], many times there is a lack of understanding of its dynamics. Moreover, less work has been done on complementing it with recovery mechanisms, which have been also observed experimentally and remain less understood. This work aims to investigate both phenomena, providing insights into the underlying mechanisms driving these instabilities.

2 METHODOLOGY

2.1 Preprocessing of experimental data

The experimental data studied consists of the temporal evolution of optoelectrical parameters along aging. Preprocessing is necessary for simulations to reduce the noise present in raw data. Smoothing is applied carefully to avoid losing physical information. Achieving the right balance is essential to preserve data integrity.

First, a Savitzky-Golay filter is applied to JV curves recorded in both forward and reverse scan directions. It smooths the JV curves by fitting quadratic polynomials along voltage. Parameters are then extracted, and outliers are removed using low on V_{OC}, FF, and the R_{SH}/R_S ratio as criteria. Time derivatives of the parameters are also checked and data with sudden unphysical variations is discarded. A moving median filter is applied to further smooth JV curves over time, reducing noise and fluctuations.

After processing the forward and reverse scan direction the average is taken to account for the global behavior of the cell. Then, the intervals of recovery and degradation are selected by direct observation of the experimental data. This is done by observing the evolution of electrical parameters simultaneously. With this, it is ensured that all V_{OC}, J_{SC}, and FF follow the same trend, which will allow the analysis of results further on.

In addition, spans are selected based on trends where power changes linearly over time. This approach assumes that within these spans, the device behavior is likely influenced by a simpler combination of physical mechanisms, rather than a complex mix. This simplifies the modeling process and helps in accurately identifying the mechanisms affecting the cells.

These preprocessing steps improve the quality of the data studied, making it possible to proceed with the analysis.

2.2 Modelling recovery and degradation mechanisms

The model used to simulate the devices studied considers a one-dimensional solar cell with uniform layers and parallel interfaces. Perovskite solar cells are modeled with three active layers: the bulk perovskite sandwiched between hole and electron transport layers. These are considered together with a glass front stack with a transparent conducting electrode and a metallic back contact.

The behavior of solar cells is simulated by coupling optoelectrical and drift-diffusion simulations. Optic simulations model light propagation in the cell, and are computed with OptiPV, a code developed at IPVF. It simulates light propagation in thin films using the transfer matrix method. It takes the layer's thicknesses and refractive indices and computes the generation rate. Drift-diffusion simulations are done using SCAPS [5], a one-dimensional numerical solver for thin-film solar cells.

Several physical parameters are required in the model to characterize material and optoelectronic properties. They are adapted from the literature [6], [7], [8], [9]. Some are kept constant, while others vary to simulate recovery or degradation mechanisms proposed, and are presented in Table I. They are computed individually, assuming to act independently from each other.

Table I: Range of physical material parameters explored for simulated recovery and degradation mechanisms. ETL and HTL stand for electron and hole transport layer, respectively.

Parameter	Range
Deep defects in the perovskite (cm^{-3})	$\in [10^{14}, 10^{17}]$
Perovskite/ETL interface defects (cm^{-2})	$\in [10^{14}, 10^{17}]$
Perovskite/HTL interface defects (cm^{-2})	$\in [10^{14}, 10^{17}]$
Electron mobility in perovskite (cm^2 V^{-1} s^{-1})	$\in [5 \ 10^{-2}, 10]$
Hole mobility in perovskite (cm^2 V^{-1} s^{-1})	$\in [5 \ 10^{-2}, 10]$
Doping in the ETL (cm^{-3})	$\in [10^{17}, 10^{19}]$
Doping in the HTL (cm^{-3})	$\in [10^{17}, 10^{19}]$
Electron mobility in ETL (cm^2 V^{-1} s^{-1})	$\in [5 \ 10^{-3}, 5 \ 10^{-1}]$
Hole mobility in HTL (cm^2 V^{-1} s^{-1})	$\in [5 \ 10^{-3}, 5 \ 10^{-1}]$
Series resistance (Ω cm^2)	$\in [10^{-1}, 20]$
Shunt resistance (Ω cm^2)	$\in [2 \ 10^2, 10^3]$

2.2.1 Genetic Algorithm

The genetic algorithm implemented in this work gives access to simulated JV curves that are close to the experimental one, within a given tolerance in V_{OC}, J_{SC}, FF, R_S, R_{SH}. Even when certain material parameters cannot be measured experimentally with precision, they can be treated statistically with this approach. The goal is to generate around 100 parameter sets, which will serve as the starting point for computing recovery or degradation. To better understand how the algorithm works, the steps of the process are summarized as follows.

First, input sets are randomly created within reasonable ranges of the material parameters extracted from the literature. The generation is done in a log-uniform distribution, as material parameters are expected to change on a logarithmic scale. Depending on the sample, 2000-4000 sets are required in the initial step to reach the 100 sets required at the end of the algorithm.

After generation, a selection step filters out sets whose electrical parameters exceed a specified tolerance compared to experimental data in terms of V_{OC}, J_{SC}, FF, R_S, R_{SH}. The 30 closest sets are chosen. Next step comprises the mutation of the selected sets, which is done by multiplying the parameters by a given coefficient selected from a random distribution.

After mutation, selection is performed again, and the mutation-selection process is repeated thrice until the desired group of sets is obtained. In the final step, the threshold for electrical parameters is set to values that vary between 2% - 5% for V_{OC}, J_{SC}, FF, and 5% - 8% for R_S, R_{SH}, depending on the recovery and degradation studied.

The JV curves at the last step of the process are

illustrated in Figure 1, together with the distributions of parameters for generation and mutation steps, showing how simulated results get closer to the experimental values represented by vertical black lines. Moreover, it is also shown how most likely distributions of parameter can be identified after implementing the genetic algorithm. It is interesting to note that some of these quantities are sometimes difficult to be measured precisely experimentally, so the genetic algorithm can provide valuable information also before recovery and degradation are simulated.

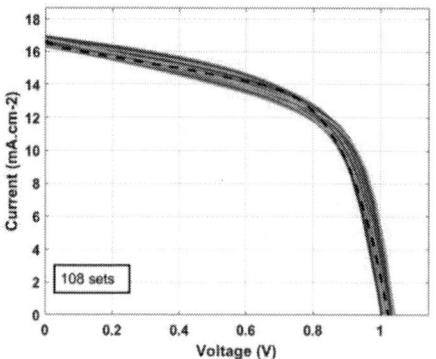

Figure 1: Representation of the experimental JV curve (black cashed lines) and the simulated JV curves after the 8th step of the genetic algorithm for one of the recoveries studied.

2.2.2 Simulating recovery and degradation mechanisms

After determining the starting points of recovery and degradation using the genetic algorithm, the subsequent evolution of these processes can be simulated. This is achieved by modeling the response to variations in a single material parameter for each of the approximately 100 starting sets. Each mechanism is simulated independently. Figure 2 shows an example where recovery is simulated by increasing the hole mobility in the perovskite layer. On the figure it is also shown how, after simulating this mechanism for each starting point, the 108 response sets are averaged, as shown by the solid red line, with a 95% confidence interval shaded based on standard deviation. With approximately 100 sets, we achieve robust averages, filtering out outliers that deviate from the trend. Therefore, the relation between the evolution of the material parameter and its corresponding V_{OC}, J_{SC}, FF and resistances is obtained. This allows us to plot the correlation pathways, used to analyze our results and presented as follows.

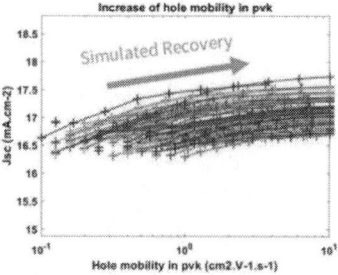

Figure 2: Example of simulated recovery, where the mechanism is the increase of hole mobility in the perovskite. Red line shows the average response of all sets.

2.2.3 Correlation Pathways

Periodic JV measurements are conducted in the lab to track the evolution of electrical parameters during aging. This allows to plot the so-called correlation pathways, which represent the evolution of correlation between electrical parameters as recovery or degradation happen. The experimental correlation pathways can be understood as a footprint of the states that the solar cell undergoes along these phenomena.

Additionally, simulating recovery and degradation mechanisms gives access to the evolution of these correlations computed from the effect of changing material parameters. This enables a direct comparison between experimental data and simulated mechanisms, as depicted in Figure 3. Through this comparison, we can identify which simulated mechanisms align closely with experimental observations and distinguish those that are less likely.

Taking the example shown in Figure 3, it can be seen how the most likely mechanism to be explaining recovery is the healing of deep defects in the perovskite layer. Moreover, the increase electron mobility in the perovskite can be established to not play a role in explaining experimental recovery, as is parts ways when compared to the experimental data, inducing a decrease in Voc.

For each recovery or degradation process studied, this analysis is applied to the correlations between V_{OC}, J_{SC}, FF, as well as pseudo-resistances.

It is important to note that these correlation pathways reflect the device's states during recovery or degradation without accounting for reaction rates or environmental conditions, thus avoiding the need for complex kinetic models.

Additionally, as mentioned earlier, mechanisms are assumed to act independently. While this is a simplification of reality, it enables simulations that provide results for interpreting the experimental observations, which can be tested sequentially afterwards.

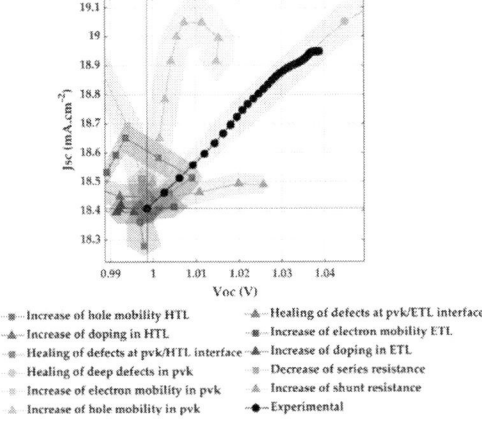

Figure 3: Correlation plot of a recovery studied showing the healing of deep defects in the perovskite to be closely related to experimental recovery.

3 RESULTS

3.1 Description of devices studied

All devices studied have been synthesized and characterized at IPVF, grouped in three batches: Lot 17, Lot 29 and Lot 53. Their absorber layer consists of IPVF's benchmark triple-cation perovskite: $Cs_{0.05}(MA_{0.17}FA_{0.83})_{0.95}Pb(Br_{0.17}I_{0.83})_3$, and all of them are in a nip configuration. Regarding the characterization conditions, JV measurements are performed during aging every 15 minutes, first in the forward and then in the reverse direction with a scan rate of 20mV/s. Specific characteristics on each lot are briefly commented as follows.

3.1.1 Lot 17

This lot examines the impact of a 2D perovskite passivation layer at the perovskite/HTL interface.

Experimental data show initial rapid degradation followed by recovery. The estimated active area of these cells is 0.4 cm², with the structure Glass/FTO/SnO2/triple-cation perovskite/(2D perovskite)/PTAA/Au. Tests were conducted at 32°C under constant AM1.5G illumination (1000 W/m²) in a N_2 atmosphere with high humidity.

3.1.2 Lot 29

For this lot, four different synthesis methods were used for perovskite deposition, each applied to two devices. Preparation method changed the day of preparation of precursor for perovskite solution and the effect of having a preheating of the solution before deposition.

The devices, with a substrate area of 2 cm² and an active area of 0.2 cm², were aged under constant conditions, with periodic JV measurements to track optoelectrical parameters over time. These solar cells have structure Glass/FTO/TiO2/triple-cation perovskite/PTAA/Au, and showed degradation trends alongside two recovery phases, one triggered by accidental light source shutdown and subsequent light soaking.

Aging conditions consisted of ambient temperature in a N_2 environment. Sample temperature being estimated at 32°C, constant illumination AM1.5G, 1000 W/m².

3.1.3 Lot 53

This set of devices was designed to compare cells with and without a perovskite/ETL interlayer. Aging results revealed rapid degradation followed by a slower recovery occurring without light interruption. Each cell has an active area of 0.2 cm² and the structure Glass/TCO/SnO2/(PCBM)/triple-cation perovskite/ PTAA/Au. The tests were conducted at an estimated sample temperature of 32°C under constant AM1.5G illumination (1000 W/m²) in ambient air with high humidity.

3.2 Main learnings

So far, some initial hints are observed across the various devices studied. Nonetheless, more samples are necessary to be able to be more assertive in the conclusions. Despite this, results in Lot 29 show that variations in hole mobility within the hole transport layer and changes in shunt resistance have not shown significant alignment with experimental degradation data, making these mechanisms less likely to explain the observed performance losses.

Conversely, mechanisms involving changes in doping levels within the transport layers and variations in deep defects within the perovskite layer appear to be the most plausible explanations for the degradation trends seen so far. These mechanisms show a closer match to the experimental data and seem to play a role in both the devices' degradation and recovery.

However, results in Lot 17 and Lot 53 do not display a clear pattern in following the simulated mechanisms investigated so far, so further research is needed to provide insight into these devices.

Focusing on Lot 29, it is interesting to note that for some cases, results indicate that the same mechanisms may be contributing to both recovery and degradation, as illustrated in Figure 4. This supposes a first step towards identifying the feasibility of some simulated mechanisms to be acting in a reversible way, affecting both recovery and degradation in perovskite solar cells. This will lead further investigation, serving as a hint to better understand the experimental recovery and degradation of the cells and leading the way to better understanding the causes of these instabilities.

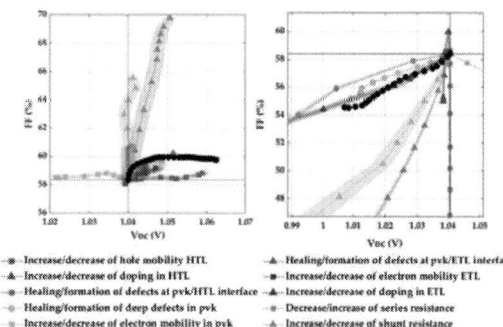

Figure 4: Comparison of correlation plots for recovery and degradation observed in the same device. It can be seen that the healing and creation of defects in the ETL/perovskite interface is the closest to explaining the experimental behavior of both trends.

4   CONCLUSIONS AND OUTLOOK

This work employs novel modeling techniques to explore the recovery and degradation mechanisms in perovskite solar cells, contributing to a deeper understanding of their long-term performance and stability. By focusing on the correlation's evolution of electrical parameters during recovery and degradation, the simulated device's behavior can be directly compared with experimental data. This allows to propose various feasible mechanisms explaining instabilities and discarding those mechanisms that do not align with observed experimental data.

Although many mechanisms may be affecting instabilities simultaneously, the assumption of independent mechanisms allows to isolate some specific mechanisms and possibly identify reversible processes, which are crucial for further research. Reversible mechanisms could play a key role in enhancing the stability of perovskite solar cells. By distinguishing these

from mechanisms that only contribute to degradation, we can focus on identifying those associated with irreversible degradation. This distinction would enable experimental efforts to target and mitigate the irreversible degradation pathways, ultimately leading to significant improvements in device stability.

Future research should also consider modeling additional mechanisms, such as ion migration, which is widely believed to play a significant role in perovskite stability. Including ion migration in the correlation plots could provide new insights into its impact on both recovery and degradation.

Moreover, the ability to propose combinations of mechanisms that align with experimental correlation pathways will be valuable for refining our analysis. As of today, simulating more than one mechanism simultaneously would add up significant complexity to our model. But using machine learning techniques, such as clustering, could streamline this process by predicting the most likely combinations of mechanisms without having to run further simulations.

Finally, expanding the number of devices studied will enhance the statistical reliability of the results and allow for broader generalization. This would help to validate the proposed mechanisms across a wider range of device architectures, contributing to the development of more durable and efficient perovskite solar cells. By continuing to refine models and incorporate new mechanisms, future research has the potential to unlock even greater improvements in the stability and performance of these promising photovoltaic technologies.

REFERENCES

[1] IEA. "Net Zero by 2050: A Roadmap for the Global Energy Sector". In: IEA. License: CC BY 4.0. Paris, France: International Energy Agency, 2021. url: https://www.iea.org/reports/net-zero-by-2050 (visited on 08/09/2023).

[2] Caleb C. Boyd et al. "Understanding Degradation Mechanisms and Improving Stability of Perovskite Photovoltaics". In: Chemical Reviews. Vol. 119. 5. American Chemical Society, 2019, pp. 3418–3451. doi: 10.1021/acs.chemrev.8b00336.

[3] Wanyi Nie et al. "Light-Activated Photocurrent Degradation and Self-Healing in Perovskite Solar Cells". In: Nature Communications. Vol. 7. 1. London, UK: Nature Publishing Group, May 16, 2016, p. 11574. doi: 10.1038/ncomms11574.

[4] Sean P. Dunfield et al. "From Defects to Degradation: A Mechanistic Understanding of Degradation in Perovskite Solar Cell Devices and Modules". In: Advanced Energy Materials. Vol. 10. 26. Wiley-VCH, 2020, p. 1904054. doi: 10.1002/aenm.201904054.

[5] M. Burgelman, P. Nollet, and S. Degrave. "Modelling Polycrystalline Semiconductor Solar Cells". In: Thin Solid Films 361-362 (2000), pp. 527–532. issn: 0040-6090. doi: 10.1016/S0040-6090(99)00825-1.

[6] Abdelhadi Slami, Mama Bouchaour, and Laarej Merad. "Numerical Study of Based Perovskite Solar Cells by SCAPS-1D". In: 2019. url: https://api.semanticscholar.org/CorpusID:221263313.

[7] Farihatun Jannat, Saif Uddin Ahmed, and Mohammad Abdul Alim. "Performance analysis of cesium formamidinium lead mixed halide based perovskite

solar cell with MoOx as hole transport material via SCAPS-1D". In: Optik 228 (2021), p. 166202.

[8] Pankaj Yadav et al. "Exploring the performance limiting parameters of perovskite solar cell through experimental analysis and device simulation". In: Solar Energy 122 (2015), pp. 773–782. issn: 0038-092X. doi: https://doi.org/10.1016/j.solener.2015.09.046.

[9] Jinane Haddad et al. "Analyzing Interface Recombination in Lead-Halide Perovskite Solar Cells with Organic and Inorganic Hole-Transport Layers". In: Advanced Materials Interfaces 7.16 (2020), p. 2000366. doi: https://doi.org/10.1002/admi.202000366.

41st European Photovoltaic Solar Energy Conference and Exhibition

Developments in thermophotovoltaics (TPV)

Esther López and Alejandro Datas

Instituto de Energía Solar - Universidad Politécnica de Madrid, Madrid (SPAIN)

41st European Photovoltaic Solar Energy Conference and Exhibition

Outline

- Introduction to TPV

- Solar-PV vs Thermo-PV

- TPV efficiency: characterization methods and most relevant experimental results

- Electrical power density

- Summary

Introduction to TPV

Solar-PV

Spectrum no controlled
Fixed temperatura: 5500 ºC

Thermo-PV

Spectrum can be controlled
Variable temperatura: 800 - 2000 ºC

Introduction to TPV

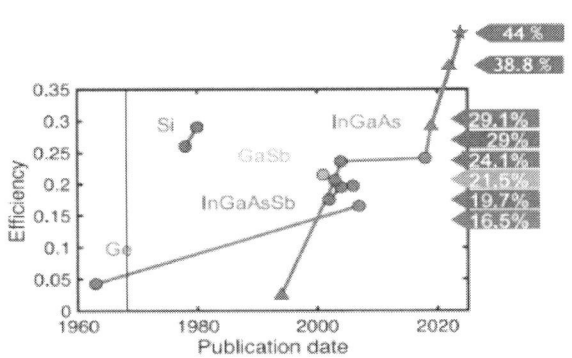

T. Burger. et. al., *Joule*, Vol. 4, Issue 8, 2020

TPV "Batteries"

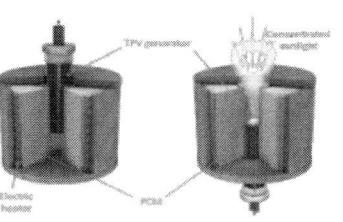

Waste heat recovery

Fuel-fired (portable power)

Solar-thermal power

Solar-PV vs Thermo-PV

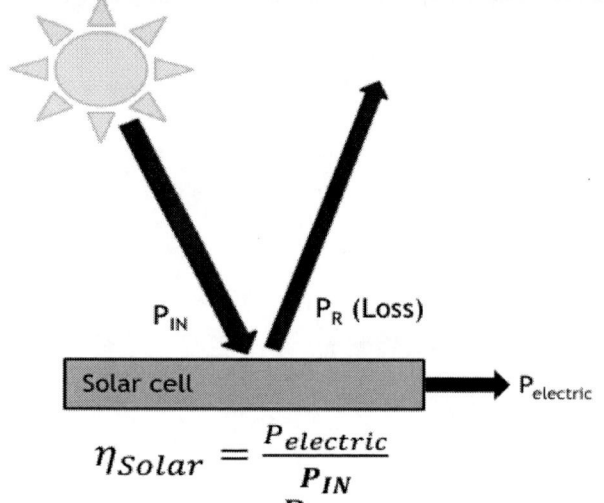

$$\eta_{Solar} = \frac{P_{electric}}{P_{IN}}$$

$$\eta_{Solar,Max} \leftrightarrow P_{electric,Max}$$

$$\boxed{\eta_{TPV,Max} \bowtie P_{electric,Max}}$$

$$\eta_{TPV} = \frac{P_{electric}}{P_{IN} - P_R} = \frac{P_{electric}}{\boxed{P_{abs}}}$$

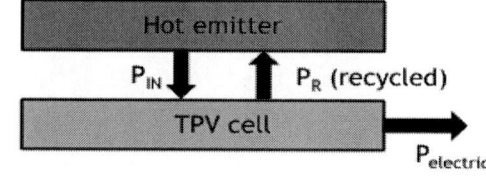

TPV efficiency

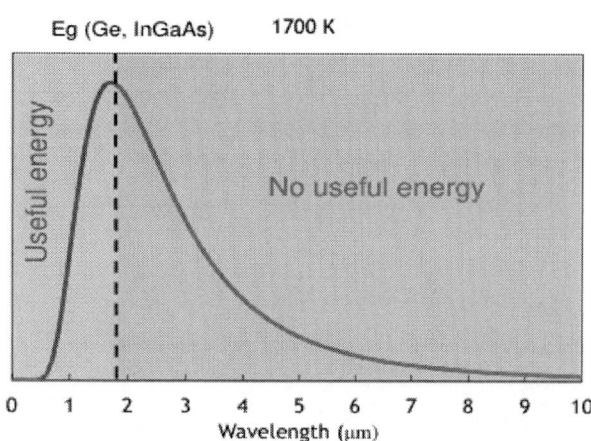

$$\eta_{TPV} = \frac{P_{electric}}{P_{IN} - P_R} = \frac{P_{electric}}{\boxed{P_{abs}}}$$

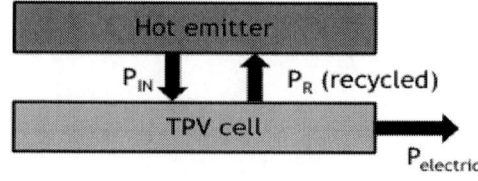

TPV efficiency

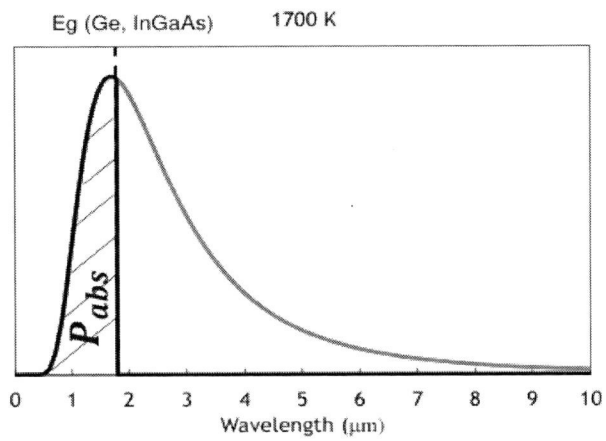

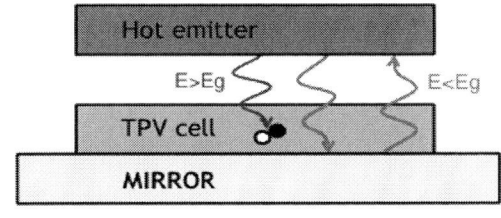

IDEAL MIRROR (100% reflectivity)

TPV efficiency

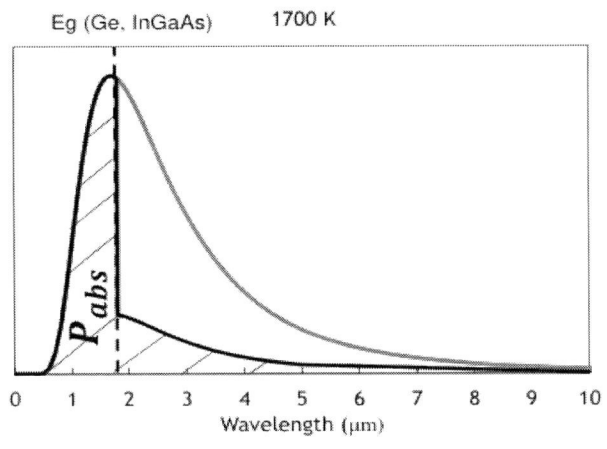

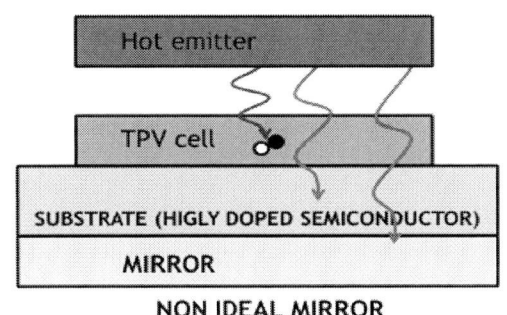

NON IDEAL MIRROR

TPV efficiency: Most relevant experimental results

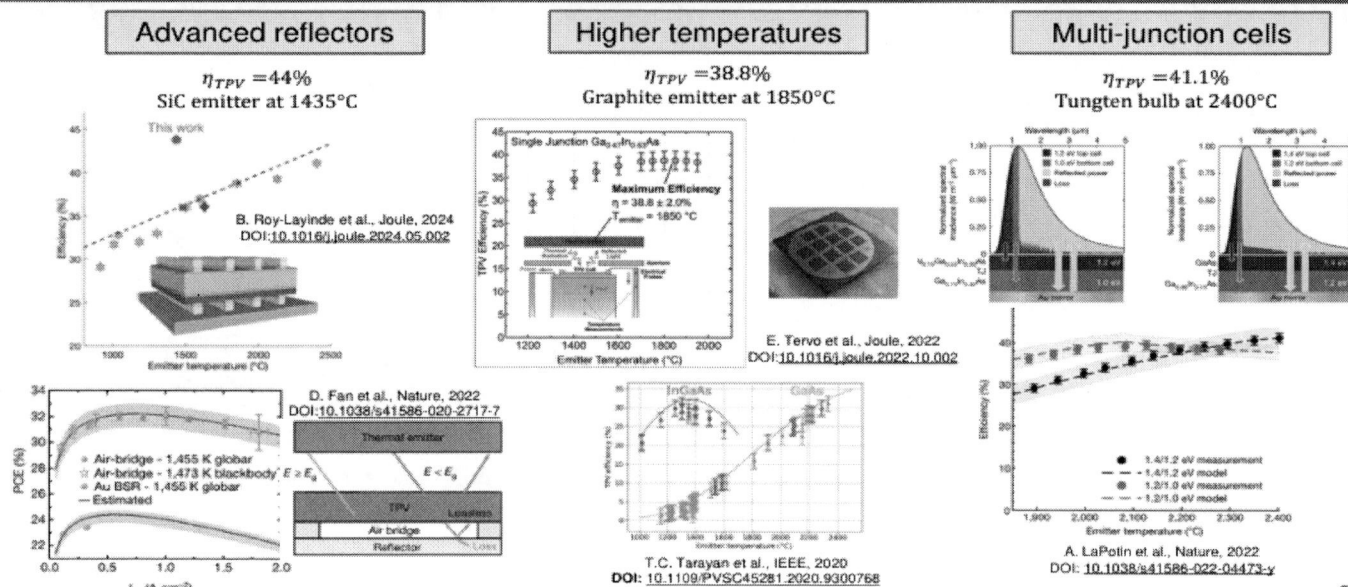

Advanced reflectors

$\eta_{TPV} = 44\%$
SiC emitter at 1435°C

B. Roy-Layinde et al., Joule, 2024
DOI:10.1016/j.joule.2024.05.002

D. Fan et al., Nature, 2022
DOI:10.1038/s41586-020-2717-7

Higher temperatures

$\eta_{TPV} = 38.8\%$
Graphite emitter at 1850°C

E. Tervo et al., Joule, 2022
DOI:10.1016/j.joule.2022.10.002

T.C. Tarayan et al., IEEE, 2020
DOI: 10.1109/PVSC45281.2020.9300768

Multi-junction cells

$\eta_{TPV} = 41.1\%$
Tungten bulb at 2400°C

A. LaPotin et al., Nature, 2022
DOI:10.1038/s41586-022-04473-y

TPV efficiency: Characterization methods

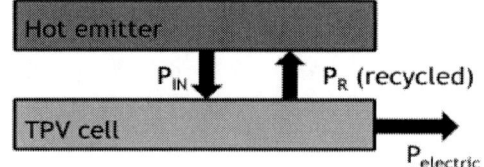

$$\eta_{TPV} = \frac{P_{electric}}{P_{IN} - P_R}$$

TPV efficiency: Characterization methods

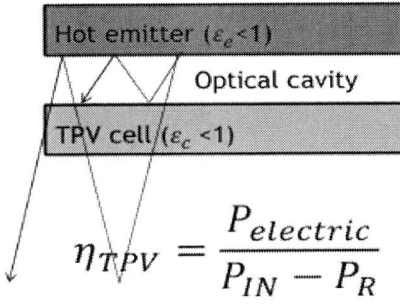

$$\eta_{TPV} = \frac{P_{electric}}{P_{IN} - P_R}$$

$$PCE = \frac{P_{electric}}{P_{IN} - P_R}$$

TPV efficiency: Characterization methods

Indirect optical method

Imaginary optical cavity

$$\varepsilon_{eff} = \frac{\varepsilon_e \varepsilon_c}{\varepsilon_e + \varepsilon_c - \varepsilon_e \varepsilon_c}$$

$$\eta_{TPV}(\varepsilon_{eff}, EQE, VF \dots)$$

Disadvantages:

✖ Assumptions are needed

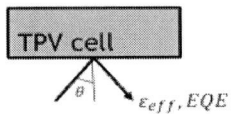

✖ No in-situ measurements

✖ Low power densities are obtained if low VF are used

TPV efficiency: Characterization methods

Direct calorimetry method

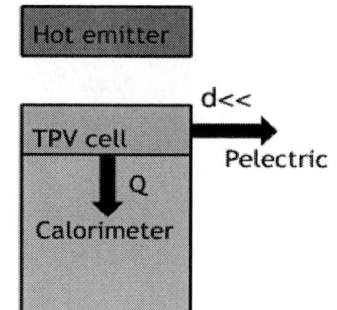

✓ Real optical cavity, like in TPV systems
✓ $PCE \rightarrow \eta_{TPV}$ without assamptions

$$PCE = \frac{P_{electric}}{P_{electric} + Q} = \eta_{TPV}$$

TPV efficiency: Characterization methods

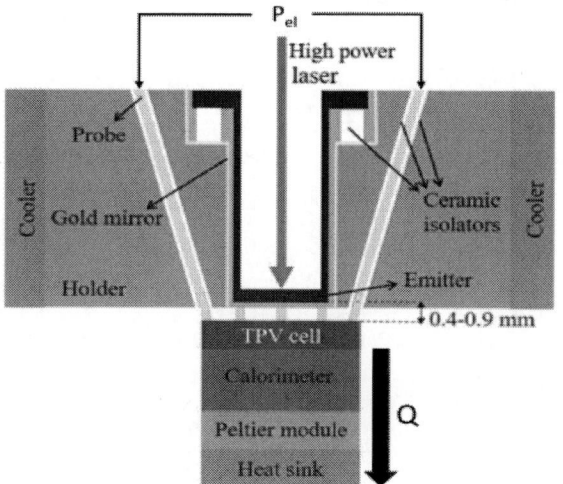

"Thermophotovoltaic conversion efficiency measurement at high view factors"
E López, et al. SOLMAT, 2023

TPV efficiency: Characterization methods

Higher View Factor

Direct calorimetry method:
- Cell electric power
- Heat absorbed

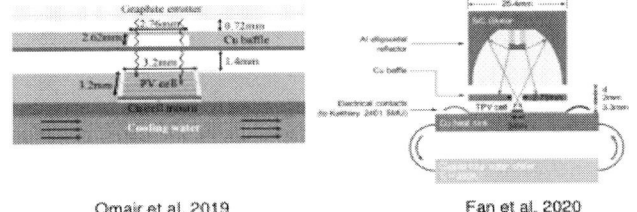

LaPotin et al. 2022
VF=0.01

Tervo et al. 2022
VF=0.77

López et al. 2023
VF=0.91-0.98

Indirect optical method:
- Cell electric power
- Emitter temperature
- Effective emissivity
- View factor
- Cell reflectance

Omair et al. 2019
VF=0.31

Fan et al. 2020
VF=0.472-0.723

Electrical power density

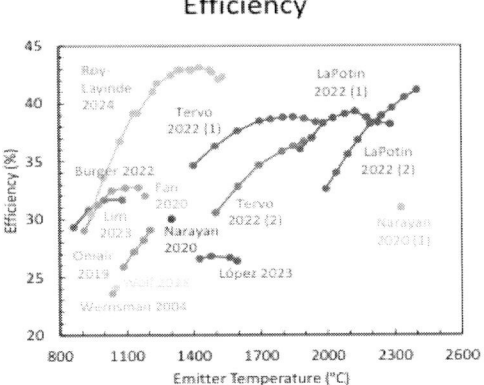

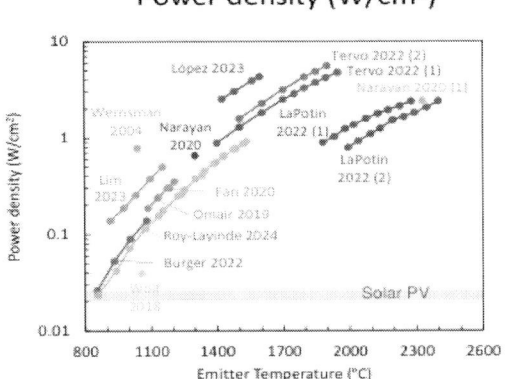

Summary

- TPV efficiency and electrical power density are independent parameters.

- TPV efficiency (pairwise efficiency):
  - There are two methods to obtain the TPV efficiency (no standards yet): indirect optical method and direct calorimetry method.
  - 44% has been obtained at 1435ºC with advanced reflectors (air-bridge), using the indirect optical method.
  - 41.1% has been obtained at 2400ºC with reflectors and multi-junction cells, using direct calorimetry method.

- Electrical power density:
  - Results are highly influenced by the view factors used in the experimental setups.
  - More than 5W/cm^2 has been obtained at 1900ºC.

Thank you for your attention

esther.lopez.estrada@upm.es

Happy to take your questions

We gratefully acknowledge the support of the SUNSON project, funded by the European Union's Horizon Europe Research and Innovation Programme under Grant Agreement No. 101083827

 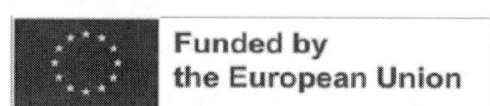

Funded by the European Union. Views and opinions expressed are however those of the author(s) only and do not necessarily reflect those of the European Union or CINEA. Neither the European Union nor the granting authority can be held responsible for them.

41st European Photovoltaic Solar Energy Conference and Exhibition

MONOLITHIC SERIES-INTERCONNECTED TWO-TERMINAL PEROVSKITE-CIGSE TANDEM SOLAR CELLS: VOLTAGE-MATCHED OR CURRENT-MATCHED?

Nicolas Otto[1*], Christof Schultz[1], Guillermo Farias-Basulto[2], Rutger Schlatmann[1,2], Eva Unger[3], Bert Stegemann[1]

[1]University of Applied Sciences – HTW Berlin, Wilhelminenhofstr. 75a, 12459 Berlin, Germany
[2]PVcomB / Helmholtz-Zentrum Berlin für Materialien und Energie, Schwarzschildstr. 3, 12489 Berlin, Germany
[3]Helmholtz-Zentrum Berlin für Materialien und Energie, Department Solution-Processing of Hybrid Materials and Devices, Kekulèstr. 5, 12489 Berlin, Germany
*nicolas.otto@htw-berlin.de

ABSTRACT: Perovskite-CIGSe tandem solar cells represent a promising path toward high-efficiency and low-cost photovoltaics. Typically, tandem solar cells are designed in a two-terminal current-matched (2T-CM) configuration. However, this design suffers from limitations related to spectral variations, which can lead to current mismatch and reduced energy yield. An alternative is the four-terminal (4T) configuration, where top and bottom cells operate independently, but this increases system complexity and cost. The two-terminal voltage-matched (2T-VM) configuration overcomes these challenges by matching the voltages of the top- and bottom cells, making 2T-VM tandems less sensitive to current mismatch while maintaining a simpler system design. This study explores the potential of monolithically interconnected 2T-VM perovskite-CIGSe tandems, focusing on laser patterning techniques for interconnecting the layers. A specific and promising interconnection approach is presented, and the necessary process steps are analyzed to evaluate how they can be adapted from well-established laser patterning techniques. Our findings indicate that the 2T-VM configuration offers a robust and cost-effective solution for achieving high energy yields with minimal complexity, positioning it as a viable alternative for next-generation thin-film solar technologies.
Keywords: Perovskite-CIGSe Tandem Module, Voltage-Matched, Monolithic Interconnection, Laser

1 INTRODUCTION

Perovskite-CIGSe tandem solar cells offer a promising combination of high efficiency and low cost, making them particularly attractive for a wide range of photovoltaic applications. Additionally, these cells are compatible with flexible substrates, which further expands their potential applications, especially in scenarios where lightweight or flexible modules are required, such as in building-integrated photovoltaics (BIPV) or portable solar applications [1-3]. One of the main advantages of this technology is the possibility to monolithically interconnect these thin-film solar cells. This is achieved through the well-established P1, P2, and P3 patterning steps. Here laser patterning is the preferred method due to its high precision, contactless processing, and highly reliable patterning results as it is essential for accurate separation and electrical interconnection of the individual cells [2, 3].

Conventionally, tandem solar cells are designed in a 2-terminal current-matched (2T-CM) configuration. In this setup, the current generated by the individual sub-cells within the tandem must be perfectly matched to ensure optimal electrical performance [4]. However, variations in the solar spectrum - caused by factors such as changing weather conditions, irradiance, angle of the sun or temperature - inevitably cause a mismatch in the generated currents of the top and bottom cells. This mismatch can significantly reduce the overall energy yield of the system, as the current of the tandem is limited by the poorer performing cell [5].

To address this issue, a common alternative is the 4-terminal (4T) tandem configuration, where the top and bottom cells operate independently. While this configuration avoids the limitations related to spectral variations, it introduces greater complexity in system installation and integration, making it a more expensive, heavier, and thus a less practical option in some cases [6, 7].

A promising solution that combines the advantages of both 2T-CM and 4T configurations is the 2-terminal voltage-matched (2T-VM) approach. In this configuration, the interconnection of the cells is realized in such a way that the cells operate at matched voltages, reducing the need for precise current matching. This hybrid approach offers the potential to achieve high energy yields, similar to 4T tandems, without the additional system complexity and installation costs. Consequently, the 2T-VM configuration could provide an optimal solution for maximizing energy yield while maintaining mainly the simplicity of 2T systems [7-9].

Given the proven effectiveness of laser patterning for processing thin-film stacks, this study will not only explore the benefits of the 2T-VM approach but will also investigate how a monolithic realization of the 2T-VM concept can be achieved using laser-based patterning techniques. Moreover, we analyze this promising interconnection concept and discuss how the individual laser patterning steps can be adopted from already established processes and practically implemented in the 2T-VM concept.

2 THIN FILM TANDEM SOLAR CELL CONCEPTS

2.1. Two-terminal current matched (2T-CM) tandems

Functionality. The sketch in Fig. 1 shows a 2T-CM tandem solar cell, where a top cell and a bottom cell are stacked monolithically, sharing two external terminals. The top cell, absorbs high-energy photons while the bottom cell, which has a lower bandgap is designed to absorb lower-energy photons [10].

The cells are interconnected in series, meaning that the current flowing through both cells must be identical (current-matched). Any current mismatch between the top and bottom cells, which can occur due to variations in the solar spectrum, limits the overall performance because the current is restricted by the weaker cell. The substrate provides structural support, while the encapsulation protects the cells from environmental influences (i.e. humidity causing corrosion). The front glass enhances stability. Possible reflection losses can be reduced by an anti-reflection coating. Electrical output is extracted via a junction box (JB), which is connected to the external circuit.

This design enhances efficiency by utilizing a broader range of the solar spectrum compared to single-junction cells (Shockley-Queisser limit) [11], though the performance can be limited by the need to match the current between the two sub-cells.

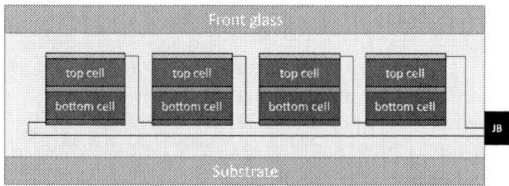

Figure 1: Schematic of a monolithically interconnected 2-terminal current-matched tandem solar cell, showing top and bottom cells connected in series with electrical output via a single junction box (JB).

Limitations. Despite, the enhanced power conversion efficiencies of the 2T-CM approach, it also faces several significant limitations. These systems are typically optimized to perform under Standard Test Conditions (STC), representing ideal laboratory conditions that rarely occur in real-world environments. In practice, varying sunlight intensity, temperature fluctuations, and changes in the solar spectrum cause the cells to deviate from their optimal operating point [4, 12]. This sensitivity to non-ideal conditions can significantly reduce the system's achievable energy yield, as it is unable to maintain peak performance across the full range of daily and seasonal variations. Another challenge is the need for precise tuning of the band gap energies for both, the top and bottom cell, to ensure the generation of the same current density. This could lead to non-optimal material combinations just to fulfill the current-matching criteria [4]. As a result, the choice of materials that can be used is significantly reduced, which in turns limits the flexibility of the design. These dependencies further complicate the fabrication of the tandem cell stack.

2.2 Four terminal (4T) tandems

Functionality. The schematic in Fig. 2 illustrates a 4-terminal (4T) tandem solar cell, where the top cells and bottom cells are electrically independent from each other. Each cell type (top and bottom) is connected to its own set of external terminals, allowing separate current extraction. The separate current extraction enables a higher energy yield, as the performance of one cell does not limit the other.

Nevertheless, the 4T configuration still improves overall efficiency by capturing a broader range of the solar spectrum, but the increased system complexity causes higher efforts and costs.

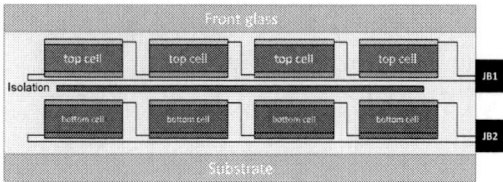

Figure 2: Schematic of a 4-terminal tandem solar cell,

with separate top and bottom cells connected to independent junction boxes (JB1 and JB2).

Limitations. The 4T concept, despite its advantage of independent operation of the top and bottom cells, also faces several significant limitations. Firstly, the manufacturing process is more complex, as each sub-cell requires its own set of contacts and electrical connections. Secondly, the separated wiring increases the overall costs of installation and the system [6, 7]. Another drawback is the introduction of additional layers between the top and bottom cells, necessary to electrically isolate the two cells. These extra layers can cause parasitic absorption, where some of the sunlight is absorbed by the interlayers rather than reaching the bottom cell, reducing overall efficiency [6, 7].

2.3 Two-terminal voltage matched (2T-VM) tandems

Functionality. The schematic in Fig. 3 depicts a 2T-VM tandem solar cell, a configuration designed to overcome the limitations of CM in conventional tandem solar cells. In this design, the top cell is paired with multiple bottom cells (i.e. two bottom cells in the example shown: bottom cell 1 and 2) to ensure that the voltage of the combined cells matches, rather than focusing on current matching. This configuration allows the top and bottom cells to work more efficiently together without the need for precise current alignment. Because the focus is on voltage matching, this design is less affected by spectral changes and can therefore operate more efficiently under varying light conditions [8].

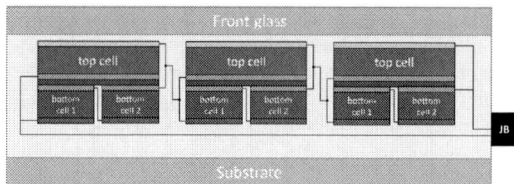

Figure 3: Schematic of a 2-terminal voltage-matched tandem solar cell. The top cell is connected parallel to two bottom cells connected in series, sharing a single junction box (JB).

Benefits. One of the key benefits is that the cells are monolithically interconnected with only two external contacts, which significantly simplifies the system design and reduces the complexity of wiring and integration compared to 4T systems. This makes the 2T-VM approach easier and more cost-effective to implement.

Additionally, this configuration is particularly suited for perovskite-CIGSe combinations, as the differing bandgaps of the sub-cells can be adjusted to complement each other almost ideally for a broad utilization of the incoming spectrum [13, 14]. As a result, an increased robustness against spectral variations is achieved, resulting in energy yields that are comparable to 4T tandem cells, but with the added benefit of lower system and installation costs [8].

In c-Si wafer-based tandems, the 3-terminal (3T) approach combines the advantages of both 2T and 4T configurations. In such a 3T tandem, current is efficiently collected from each cell, while allowing some independence in voltage between the top and bottom cells [15]. This avoids the current-matching issues found in 2T tandems, while simplifying system integration compared to 4T tandems. The 3T design offers greater flexibility in optimizing energy yield and improves efficiency under varying light conditions, as well, providing similar flexibility and robustness against spectral changes. However, the 2T-VM design offers an even simpler configuration while achieving comparable performance by leveraging voltage matching to optimize energy output, without the need for additional complexity found in 3T systems.

Moreover, laser patterning techniques have already been successfully developed for the monolithic interconnection of single junctions [16] as well as for 2T-CM and 4T tandems of perovskite/CIGSe [2, 17]. This potential transfer of the established process windows is a unique advantage of thin film tandems, as these laser patterning techniques are not applied to c-Si wafer-based tandems due to the fundamentally different architectures. The ability to utilize laser scribing for 2T-VM tandems allows, thus, for simplified manufacturing and improved performance, which is not possible with c-Si wafer-based tandems. For this reason, the following section presents an interconnection concept for 2T-VM tandems in which the monolithic interconnection is realized, and implications for laser patterning are discussed.

3 INTERCONNECTION OF 2T-VM TANDEMS

3.1 Monolithic interconnection concept

The illustration in Fig. 4 depicts the layer stack and the monolithic interconnection of a perovskite-CIGSe tandem thin-film solar cell, utilizing six laser patterning steps: P1, P2, P3 for the bottom CIGSe cell, and P1*, P2*, P3* for the top perovskite cell, derived from a proposal by Mantilla-Perez et al. [9].

The layer structure consists of the usual stacking sequences for CIGSe bottom cells (i.e., Mo back contact, CIGSe absorber, CdS buffer, TCO front contact) and perovskite top cells (electron transport layer, perovskite, hole transport layer, TCO top electrode). The isolation layer in between electrically isolates the bottom and top cells to prevent short-circuiting.

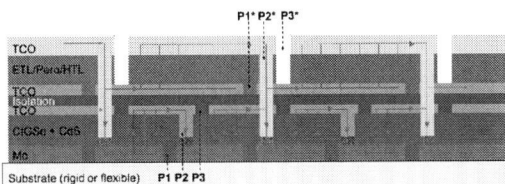

Figure 4: Schematic of a monolithically interconnected 2T-VM perovskite-CIGSe tandem solar cell, showing the layer structure and laser patterning steps (P1, P2, P3 for the CIGSe cell and P1*, P2*, P3* for the perovskite cell).

According to Fig. 4, in total 6 patterning steps, alternating with layer deposition, are necessary to achieve the desired interconnection. In Fig. 4 an example of the interconnection of a perovskite top cell with two CIGSe bottom cells is shown. The three patterning steps P1, P2 and P3 are required for patterning the bottom cell, while the top perovskite cell including the isolation layer is interconnected using three additional laser structuring steps P1*, P2* and P3*, as described below:

- P1: The first laser scribe separates individual cells by cutting through the bottom Mo layer.

- P2: This step removes the CdS + CIGSe absorber layer in a defined line, creating a path for the connection between adjacent cells.
- P3: The final scribe in the bottom cell cuts through the TCO layer to expose the underlying absorber and electrically separates the bottom cells.
- P1*: This step cuts through the top TCO layer, without affecting the underlying isolation layer, creating cell boundaries for the top cell.
- P2*: This laser scribe removes all layers down to the Mo back contact, creating a path for the electric connection
- P3*: Finally, this scribe exposes the back contact of the perovskite top cell isolating it and completing the monolithic series interconnection.

3.2 Implications for laser patterning

Laser-based patterning for monolithic interconnection of CIGSe and perovskite single junction as well as tandem stacks is already known. The procedure for process development is typically as follows: The material properties (thermal, optical, morphological) determine how the laser interacts with the layers. The laser parameters such as pulse length and wavelength, which determine the precision of the material removal, are selected accordingly. To optimize the process, the laser fluence and scanning speed are adjusted to enable precise scribing while minimizing damage to the underlying layers and the surrounding area within the processed layers [3].

In addition, we have analyzed the requirements and state of development of the six aforementioned crucial patterning steps:

- P1: Complete removal of the molybdenum (Mo) back contact without damaging the glass substrate. This step defines the cell widths and ensures effective separation between adjacent cells. This process is well-established for single-junction CIGSe modules, where Mo removal techniques are optimized to ensure precision and avoid damage to the underlying substrate [16].
- P2: The CIGSe absorber layer must be transformed into a Cu-rich phase during this step, with the requirement that no damage is caused to the Mo back contact. The goal is to prepare the bottom cell for electrical interconnection. P2 is an established process for CIGSe solar cells, where this laser scribing technique is reliably used to expose the underlying layers for interconnection [18, 19].
- P3: The TCO layer is ablated without modifying the CIGSe layer below. This step ensures that adjacent CIGSe cells are electrically isolated from one another. Like P1 and P2, this process is established and widely used in the fabrication of monolithic CIGSe solar cells [16].
- P1*: The top transparent conductive oxide (TCO) layer is ablated without causing any damage to the isolation layer between the top and bottom cells. The purpose is to define the boundaries of the perovskite top cells. This step is considered a novel process, as it cannot be directly transferred or adapted from known cell concepts.
- P2*: The perovskite absorber layer is removed, while the CIGSe is mainly transformed into a Cu-rich phase, similar to P2, with no damage to the Mo layer. The goal is to open both the top and bottom cells for

interconnection. This step shares similarities with P2 of the 2T CM approach and is under active development for tandem cell integration [2].

- P3*: Ablation of both the front TCO and perovskite layers is necessary to ensure proper separation, with no damage to the TCO contact of the bottom cell. This is the final step for interconnecting adjacent top cells. This process is similar to P3 in single-junction perovskite cells and is currently being adapted for tandem configurations [20].

Table I: Development status of laser patterning steps for monolithic interconnection of 2T-VM perovskite-CIGSe tandem solar cells

Patterning step	State of development
P1	established (P1 single junction CIGSe)
P2	established (P2 single junction CIGSe)
P3	established (P3 single junction CIGSe)
P1*	novel process
P2*	to be adjusted (P2 2T-CM tandems)
P3*	to be adjusted (P3 single junction perovskite)

While the P1-P3 steps are well-established for CIGSe modules, the corresponding P1*-P3* steps for the perovskite top cell are still in the developmental phase, particularly the P1*. While P2* and P3* show similarities to established processes in 2T and single-junction perovskite cells, which need to be adapted for tandem applications, the P1* has two tasks: (i) Suitable transparent and dielectric materials that can serve as isolation layers to minimize parasitic light absorption, while ensuring their compatibility with the other layers in the stack have to be identified. (ii) Advanced process steps have to be developed to achieve successful ablation of the TCO without damages to the underlying isolation layer.

4   SUMMARY AND OUTLOOK

2T-VM monolithic interconnected tandem cells offer significant potential for enhancing energy yield in thin-film perovskite-CIGSe modules, as they eliminate energy yield losses caused by current mismatches. The 2T-VM concept combines the benefits of both 2T-CM and 4T tandems, without the complexity of adopting a 3T approach. The proposed laser-based 2T-VM interconnection method builds on established or easily adaptable patterning processes, with the only remaining development needed being the novel P1* step. The successful development of this process step is critical for achieving efficient monolithic interconnection in perovskite-CIGSe tandem solar cells.

Future work will focus on the implementation of electrical simulation models to assess and evaluate the energy yield of the 2T-VM approach under real-world outdoor conditions. Additionally, the determination of optimal laser process parameters for fully laser-based patterning of the 2T-VM concept, with particular emphasis on the novel P1* step, will be a key area of investigation. Moreover, an experimental investigation of different isolation materials will be conducted to ensure compatibility with perovskite layers, minimize parasitic light absorption, and guarantee sufficient electrical isolation within the tandem structure.

In summary, the upcoming practical implementation of this approach could pave the way for a cost-effective and high-performance solution in thin-film photovoltaics, utilizing the full potential of tandem technology to achieve higher energy yields while minimizing complexity.

ACKNOWLEDGEMENT

This research has received funding from the European Union Horizon Europe Energy program, project 101122288 — SolMates.

REFERENCES

[1]   F. Fu, S. Nishiwaki, J. Werner, T. Feurer, S. Pisoni, Q. Jeangros, S. Buecheler, C. Ballif, A.N. Tiwari, ArXiv (2019) 1907.10330.

[2]   C. Schultz, G.A. Farias Basulto, N. Otto, J. Dagar, A. Bartelt, R. Schlatmann, E. Unger, B. Stegemann, EPJ Photovolt. 14 (2023) 16.

[3]   B. Stegemann, C. Schultz, in: digital Encyclopedia of Applied Physics (Wiley-VCH Verlag GmbH & Co.). https://doi.org/10.1002/3527600434.eap830, 2019) pp. 1.

[4]   M. Jošt et al., ACS Energy Letters 7 (2022) 1298.

[5]   G.A. Farias-Basulto, M.Á. Sevillano-Bendezú, M. Riedel, M. Khenkin, J.A. Töfflinger, R. Schlatmann, R. Klenk, C. Ulbrich, Solar Energy 266 (2023) 112175.

[6]   K. Alberi et al., Joule 8 (2024) 658.

[7]   H. Liu, C.D. Rodríguez-Gallegos, Z. Liu, T. Buonassisi, T. Reindl, I.M. Peters, Cell Reports Physical Science 1 (2020) 100037.

[8]   M.H. Futscher, B. Ehrler, ACS Energy Letters 1 (2016) 863.

[9]   P. Mantilla-Perez et al., ACS Photonics 4 (2017) 861.

[10]   A. De Vos, Journal of Physics D: Applied Physics 13 (1980) 839.

[11]   W. Shockley, H.J. Queisser, Journal of Applied Physics 32 (1961) 510.

[12]   E. Aydin, T.G. Allen, M. De Bastiani, L. Xu, J. Ávila, M. Salvador, E. Van Kerschaver, S. De Wolf, Nature Energy 5 (2020) 851.

[13]   S. Gholipour, M. Saliba, in: Characterization Techniques for Perovskite Solar Cell Materials, (Eds. M. Pazoki, A. Hagfeldt, T. Edvinsson (Elsevier, 2020) pp. 1.

[14]   P.D. Paulson, R.W. Birkmire, W.N. Shafarman, Journal of Applied Physics 94 (2003) 879.

[15]   P. Wagner, P. Tockhorn, S. Hall, S. Albrecht, L. Korte, Solar RRL 7 (2023) 2200954.

[16]   C. Schultz et al., Materials Today: Proceedings 53 (2022) 299.

[17]   R.K. Kothandaraman et al., Solar RRL 6 (2022) 2200392.

[18]   C. Schultz et al., Solar Energy Materials and Solar Cells 157 (2016) 636.

[19]   C. Schultz, M. Fenske, J. Dagar, A. Zeiser, A. Bartelt, R. Schlatmann, E.L. Unger, B. Stegemann, Solar Energy 198 (2020) 410.

[20]   M. Fenske et al., Energy Technology 9 (2021) 2000969.

41st European Photovoltaic Solar Energy Conference and Exhibition

OPTIMISATION OF MA-FREE LEAD-TIN PEROVSKITE ABSORBER AND INTERFACES IN ALL PEROVSKITE TANDEM SOLAR CELLS

JULES ALLEGRE[a], POLYXENI TSOULKA[a], NOELLA LEMAITRE[a], BAPTISTE BERENGUIER[b], MATHIEU FRÉGNAUX[c], MURIEL BOUTTEMY[c], PHILIP SCHULZ[b], SOLENN BERSON[a]

HTL/PK NBG interface optimizations

Try self assembled monolayers (SAMs) as HTL with Methylammonium (MA) free narrow band gap perovskite, which has higher thermal stability [1, 2].

	Pros	Cons
PEDOT:PSS	• Ease of process • Good energy level alignment	• Poor thermal stability [3], • Parasitic absorption, • Hygroscopic and acid [4, 5]
SAMs	• No absorption • Can promote fast extraction [6]	• Poor surface coverage

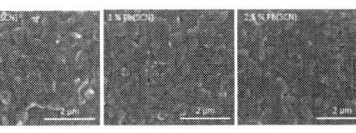

Monolithic 2-T superstrate p-i-n

2PACz
MPA
2PACz/MPA
PEDOT:PSS

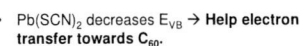

Photovoltaic parameters

- Best parameters with PEDOT:PSS.
- No J_{sc} improvement with SAMs.
- Scans show S-shaped when 2PACz is used → Sign of **potential barrier**.

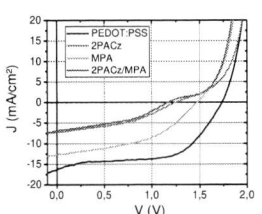

Interface mechanism explanations

□ IPVF
♣ ILV

SAMs have **higher and longer PL signal**
→ **Less non radiative recombination** with PK grown on SAM.

Favourable transfer only with PEDOT:PSS as HTL.
→ Charges are **extracted more efficiently with PEDOT:PSS** as HTL.

Potential barriers at the interface SAMs/PK → **impedes extraction**.
→ Low J_{sc}, V_{oc} and FF!

PEDOT:PSS chosen as the HTL **for PK NBG**.

PK NBG bulk optimizations

- Use lead thiocyanate, $Pb(SCN)_2$ to enlarge grain size and reduced grain boundaries defects [7].
- Add Rbl to passivate bulk and surface defects in MA free perovskite [8].

Material properties

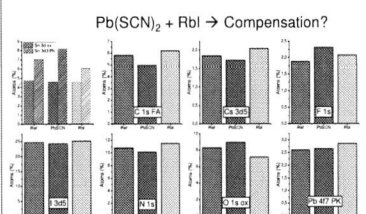

Grain size enlargement with $Pb(SCN)_2$

- $Pb(SCN)_2$ decreases E_{VB} → **Help electron transfer towards C_{60}**.
- Rbl increases E_{VB} → PK **more intrinsic**.

$Pb(SCN)_2$ + Rbl → Compensation?

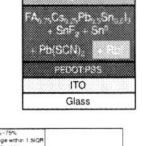

- With $Pb(SCN)_2$ → SnO_xF_y on surface.
- With Rbl → **more stoichiometric**.

Photovoltaic parameters

NBG subcell:
- V_{oc} and FF improvements with additives.
- $Pb(SCN)_2$ **enlarges hysteresis**.
- Adding Rbl **reduces hysteresis**.

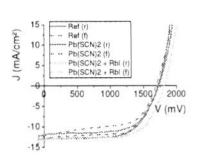

Tandem PK/PK:
- **Improvements of all parameters** with additives.
- Less hysteresis.

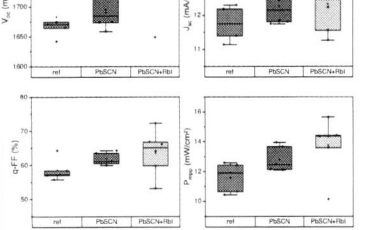

Conclusion & Perspectives

➢ 2PACz and MPA are not suitable HTLs for our PK NBG subcell but promote PK film with less defects.

➢ $Pb(SCN)_2$ + Rbl additives improve PV parameters and reduce hysteresis of solar cell. Underlying mechanisms need futher understanding.

Next steps: Test other SAMs to improve HTL/PK interface and test surface passivation of PK/ETL interfaces.

References

[1] Hernandez et al. *Sustainable Energy & Fuels*, **2022**, 6, n° 20, 4605-13.
[2] Kapil et al. *ACS Energy Letters*, **2022**, 7, n° 3, 966-74.
[3] Chi et al. *Adv. Funct. Mater.*, **2018**, n° 28, 1804603.
[4] Chen et al. *Chemical Engineering Journal*, **2022**, 430, 132701.
[5] Cameron et al. *Materials Horizons*, **2020**, 7, n° 7, 1759-72.
[6] Al-Ashouri et al. *Science*, **2020**, 370, n° 6522, 1300-1309.
[7] Yu et al. *ACS Energy Letters 2*, **2017**, n° 5, 1177-82.
[8] Yang et al. *Advanced Energy Materials*, **2023**, 13, n° 19, 2204339.

AFFILIATIONS :
[a] Univ. Grenoble Alpes, CEA, LITEN, INES, 73375 Le Bourget du Lac, France
[b] Institut Photovoltaïque d'Île de France (IPVF), 91120 Palaiseau, France
[c] Institut Lavoisier de Versailles (ILV), 78000 Versailles, France

ACKNOWLEDGEMENTS :
To CEA and GDR-HPERO for their financial support.

CONTACT :
jules.allegre@cea.fr
solenn.berson@cea.fr

41st European Photovoltaic Solar Energy Conference and Exhibition

 **Laboratory of Photovoltaics and Optoelectronics**

Potential Induced Degradation Free Perovskite-Silicon Tandem Solar Cells

Kristijan Brecl, Matevž Bokalič, Gašper Matič, Marko Topič
University of Ljubljana, Faculty of Electrical Engineering, Ljubljana, Slovenia

Lisa Champault, Quentin Jeangros
CSEM, Neuchâtel, Switzerland

FE UNIVERSITY OF LJUBLJANA
Faculty of Electrical Engineering

:: csem

Abstract ■

Photovoltaic modules are normally connected in series and therefore exposed to a high system voltage between the cells and the grounded frame. A high potential between the encapsulated solar cell in the module and the grounded aluminum frame can cause degradation known as Potential Induced Degradation (PID). PID was discovered and solved in crystalline silicon solar cells more than 10 years ago. More recently, it has also been investigated in PSCs. The results have shown that PSCs can also be susceptible to PID, mainly due to sodium migration between the glass and the perovskite absorber layer. In this contribution, we investigate PID in fully encapsulated tandem perovskite-silicon solar cells. The cells were exposed to a high positive and negative potentials of 1000 V for more than a month and proved to be stable and resistant to PID.

Potential induced degradation ■

Modules in a PV system

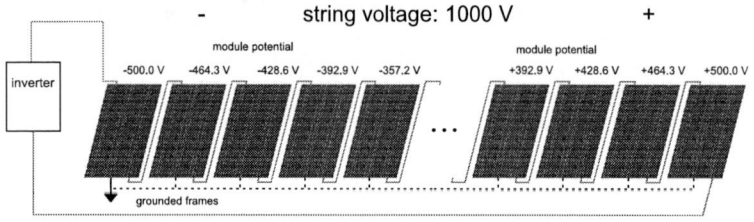

Schematics of module connections in the PV system. Because the modules' frames are grounded and all the modules are connected in series, the individual module bias voltage increases with each additional module. Since the solar cells are »floating« and the typical system voltage is up to 1000 V, the first and the last module may be exposed to the highest voltages of -500 V and +500 V, respectively. (a module voltage of 35.7 V is used in the scheme).

Results ■

PL and EL imaging

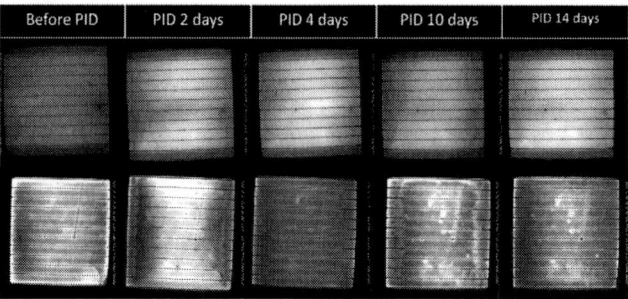

PL (top) and EL (bottom) images of perovskite subcel.

IV measurements

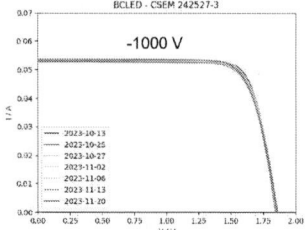

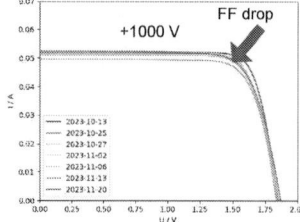

Device under test ■

We have tested fully encapsulated tandem solar cells made of perovskite and silicon with an active area of 4 cm². During the test, the cells were wrapped with a grounded aluminum foil, while the positive and negative contacts were short-circuited and connected to a positive or negative voltage of 1000 V. The cells were periodically characterized by I-V measurements and EL and PL imaging techniques.

Conclusion ■

PID tests of fully encapsulated tandem solar cells made of perovskite and silicon subcells did not show any potential induced degradation. The cells were exposed to the high potential for more than a month, much longer than 96h as suggested by the IEC-62804 standard. The performance of the cells tested did not degrade due to the influence of the high potential.

More information:

K. Brecl et al., "Are Perovskite Solar Cell Potential-Induced Degradation Proof?", SolarRRL, Vol. 6, Iss 2, 2022,
https://doi.org/10.1002/solr.202100815

Acknowledgments ■

- Slovenian Research and Innovation Agency (Research Programme P2-0197).
- This project (PEPPERONI, no. 101084251) is co-funded by the European Union. Views and opinions expressed are however those of the author(s) only and do not necessarily reflect those of the European Union or the European Climate, Infrastructure and Environment Executive Agency (CINEA). Neither the European Union nor the granting authority can be held responsible for them.
- Swiss State Secretariat for Education, Research and Innovation (SERI) under contract number 22.00413.

PEPPERONI

kristijan.brecl@fe.uni-lj.si

41st European Photovoltaic Solar Energy Conference and Exhibition

Modeling of Metastability Behavior in Perovskite-based Solar Cells for Accurate Energy Yield Estimation in Realistic Operating Conditions

Špela Tomšič[1], Marko Remec[1,2], Florian Scheler[2], Mark Khenkin[2], Carolin Ulbrich[2], Rutger Schlatmann[2], Steve Albrecht[2], Marko Jošt[1], Benjamin Lipovšek[1], Marko Topič[1]

[1] University of Ljubljana, Faculty of Electrical Engineering; [2] Helmholtz-Zentrum Berlin für Materialien und Energie
[1]Tržaška 25, 1000 Ljubljana, Slovenia; [2]Schwarzschildstraße 3, 12489 Berlin, Germany

ABSTRACT: In this work, we use an in-house developed energy yield (EY) modeling algorithm as a very effective tool for determining the capabilities of photovoltaic (PV) devices and their production losses associated with different mechanisms of degradation. We first use the EY simulation results to extract and quantify the energy losses attributed to the light-soaking effect (LSE) from the outdoor measurements of the two-terminal (2T) perovskite-silicon (PK-Si) tandem solar cell. The results show that LSE has a non-negligible impact on the device performance under realistic operating conditions, as the energy losses can reach up to 2% on sunny days and even exceed 12% on cloudy days. We then propose and validate a modeling approach for determining the LSE dynamics in PK-based PV devices. Finally, the EY algorithm equipped with the LSE model is used to estimate and analyze the long-term energy losses ascribed to the effect in three different geographical locations (selected from distinct Köppen-Geiger-Photovoltaic climate zones) as well as in two different open-rack configurations of the device. We show that the extent of LSE-induced losses is very different from location to location as well as from season to season. In the case of the vertical orientation of the device, the LSE losses are 3-times higher compared to the location-specific optimal orientation.
Keywords: energy yield algorithm, light-soaking effect modeling approach, two-terminal perovskite-silicon tandem solar cell, realistic operating conditions

1 INTRODUCTION

The successful integration of a perovskite (PK) solar cell together with a silicon (Si) solar cell to form a PK-Si tandem solar cell has shown a tremendously rapid conversion efficiency progress in less than a decade of its development [1]. The progress is mostly driven by the large span of possible improvements in the perovskite solar cell research field, which have already led to the current PK-Si tandem efficiency record of 33.9% [2]. The continuous improvement in power conversion efficiency, stability, and encapsulation techniques brings the PK-Si tandem technology steadily closer to the point of full transition from the laboratories to the rooftops, with numerous outdoor tests already being reported upon.

In contrast to indoor laboratory experiments, however, when studying the PK-Si tandem devices in realistic outdoor operation, their behavior may become significantly influenced by unpredictable and constantly changing environmental conditions. Therefore, to accurately evaluate their performance throughout their lifetime, it is crucial to consider all the physical effects associated with outdoor operation, including the effects of temperature, bandgap variation, as well as sources and mechanisms of degradation.

Specifically, the light-soaking effect (LSE), which refers to the fluctuations of photovoltaic parameters under light exposure, represents a critical factor limiting the accuracy when estimating the power output of PK-based PV devices [3]. Although it is a known effect, it is often neglected during indoor measurements, since it is relatively easy to fully light-soak the cells under continuous one-sun illumination. Under real-world conditions, however, LSE introduces losses that need to be accounted for in order to reliably assess the long-term energy yield (EY) of the investigated devices. According to our best knowledge, the extraction of the LSE still remains a challenge when considering the realistic installation and operation of the device due to the constantly varying environment, be it outdoors or indoors (e.g. in the office for IoT devices).

In this contribution, we combined experimental characterization with detailed EY numerical simulations to investigate the long-term behavior of the two-terminal (2T) PK-Si tandem device operating under real-world conditions. Based on the results of long-term monitoring under outdoor conditions, the EY model first enabled us to extract and quantify energy losses attributed to the LSE. LSE-induced losses would otherwise be impossible to monitor from outdoor measurements alone and also very difficult to observe from indoor laboratory experiments as they are partly characteristic of the dynamic meteorological conditions. From that, we were then able to implement the LSE into the EY modeling algorithm with the ultimate purpose to predict the amount of LSE losses in different environmental and/or installation cases.

2 MODELING

Numerical modeling in this study was employed for two specific purposes. Our first objective was to extract and quantify the losses associated with the light-soaking effect in the 2T PK-Si tandem solar cell directly from the outdoor measurements acquired on the PV monitoring site (Helmholtz-Zentrum Berlin (HZB) Institute, Germany). And our second objective was to implement the LSE mechanism into the energy yield (EY) modeling algorithm in order to estimate, analyze, and compare long-term LSE losses in different meteorological and device installation conditions. For that purpose, we used an in-house developed EY model that has already been used extensively and is described in our previous study [4]. The model is comprised of three modeling approaches (optical, electrical, and thermal), and utilizes realistic structure parameters of the device (t.i. wavelength-dependent refractive indices of the materials, layer thicknesses, and interface textures) as well as measured environmental data (spectrally resolved global tilted irradiance (GTI) and direct normal irradiance (DNI), device operating temperature (Tcell), etc.) obtained on the HZB PV monitoring site. The electrical behavior of the device was determined based on an extensive set of *J-V* measurements attained under well-defined indoor irradiance and temperature conditions, following the methodology provided in [4]. The device was completely light-soaked prior to each *J-V* measurement.

The analysis presented in this study is focused on an encapsulated 2T PK-Si tandem solar cell reported previously in [5] and also presented schematically in Figure 1. For the bottom cell, we used a silicon heterojunction (SHJ) solar cell composed of a 280 μm-thick n-type float zone Si wafer. The rear side of the Si bottom sub-cell was textured to enhance near-infrared absorption, while the front side was polished to facilitate the deposition of spin-coated perovskite. For the top cell, it has been selected the "inverted" design (p-i-n), of which the 500 μm-thick perovskite absorber layer is comprised of a nominal precursor composition $Cs_{0.05}(MA_{0.23}FA_{0.77})_{0.95}Pb(I_{0.77}Br_{0.23})_3$ that exhibits a 1.68 eV wide bandgap. Additionally, the encapsulation stack – glass (2 mm) / polyolefin elastomer-POE (500 μm) – is placed above the described tandem device, as demonstrated in Figure 1.

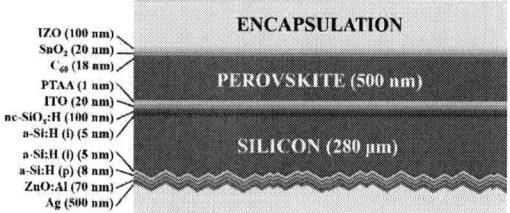

Figure 1: The complete structure of the investigated 2T encapsulated perovskite-silicon solar cell.

2.1 Extraction and quantification of the LSE

With the developed EY model, we were able to accurately calculate the power that would be generated by the investigated PK-Si tandem device under realistic meteorological conditions without any degradation mechanisms taking place. Accordingly, comparing these simulation results with actual outdoor measurements can empower us with the possibility to extract and quantify the amount of LSE losses in realistic operation.

The fact that LSE has a non-negligible impact on the performance of PK-Si tandem devices in real-world conditions can be best demonstrated by comparing the measured and simulated voltage at maximum power point (V_{MPP}). Figure 2 shows the V_{MPP} results of the tandem device over the course of three cloudy and three sunny days in March 2022 (HZB, Germany), together with total irradiance (spectrally-resolved GTI integrated over the wavelength range of 300-1200 nm) and device operating temperature. The discrepancy between simulated (red curve) and measured (blue curve) V_{MPP} values can clearly be observed. On sunny days (March 7th – March 9th), the measurements differ from the simulation results only in the morning hours until the PK sub-cell becomes fully light-soaked and the curves start to overlay. During these three days, the total achievable generated energy of the device is reduced by approximately 2%. On cloudy days (February 25th – February 27th), however, sufficient light absorption is never reached due to the irradiance being too low, and thus the deficit in V_{MPP} persists throughout the day. The extent of LSE-induced losses is much greater in this case, totaling around 12%.

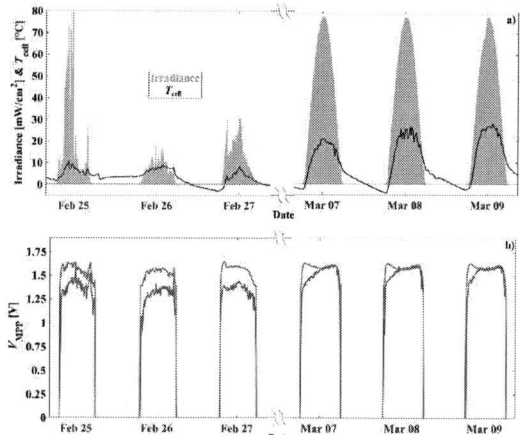

Figure 2: a) Measured total incident irradiance (yellow area) and device operating temperature (black curve) over the course of three consecutive cloudy and three consecutive sunny days. Measured (blue curve) and simulated (red curve) V_{MPP} values of the PK-Si tandem device for six selected days.

The described methodology was then conducted to extract and evaluate the extent of LSE losses over the extended time period of the device operation under real-world conditions. Figure 3a) shows the LSE-induced losses over the period of three weeks (Feb 18th – Mar 12th) at the beginning of the solar cell operation, before any other irreversible degradation mechanisms started to affect the device performance. It can be observed that during these three weeks, the tandem device energy production is reduced by 1.1 kWh/m², which accounts for approx. 6% losses. Furthermore, Figure 3b) represents the daily sum of the incident irradiance (daily irradiation) and the daily average of the device temperature ($\overline{T_{cell}}$). By comparing the results from both figures it can be clearly seen that the daily variations in LSE losses are closely correlated with the device operating conditions. Larger losses can be observed at lower daily irradiation and $\overline{T_{cell}}$ values.

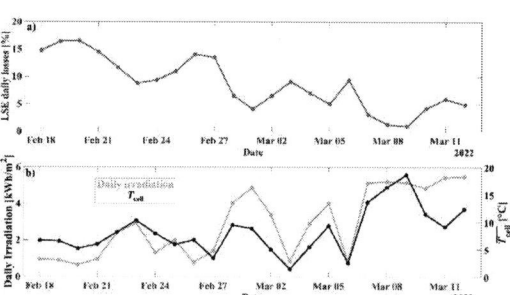

Figure 3: a) Daily energy losses of the tandem device attributed to the LSE effect over the course of three consecutive weeks. b) Daily irradiation (orange curve) and the daily average of device temperature (black curve) over the selected time period.

2.2 LSE modeling

In the preceding section, we showcased the utilization of the *EY* modeling algorithm to extract LSE-induced losses directly from outdoor measurements. However, in order to predict the extent of LSE losses also over longer time periods and in different environmental and installation cases, it becomes necessary to implement the LSE mechanism into the *EY* model. For that purpose, the current section is focused on modeling the light-soaking effect by finding the correlation between its dynamics and realistic operating conditions.

Judging by the results presented in Figure 2, it can be stated with assurance that the difference between the simulated and measured V_{MPP} (ΔV_{MPP}) depends on the total cumulative irradiation that was intercepted by the tandem device during each day from sunrise up to the given time. However, previous reports [6], [7] demonstrated that under standard test conditions (STC) the LSE is affected not only by irradiation but also by the device temperature. This fact that LSE dynamics is indeed significantly influenced by T_{cell} also under realistic operating conditions can again be noticed in Figure 2b) for sunny days. The latter have almost identical irradiance profiles, yet notably different LSE dynamics can be seen. To put it differently, the measured and simulated V_{MPP} results start to align earlier in the day when the device temperatures are higher.

Based on these observations we proposed the following correlation between the LSE-induced voltage deficit and realistic operating conditions:

$$\Delta V_{\text{MPP}} = A e^{-\frac{H_{cum}}{\tau}} \quad (1)$$

H_{cum} in Eq. (1) represents the cumulative irradiation and can be calculated by the following expression (2):

$$H_{\text{cum}}(t) = \sum_{sunrise}^{t} \left(\int_{300\text{ nm}}^{1200\text{ nm}} GTI(\lambda)\, d\lambda \right) \cdot \Delta t \quad (2)$$

where λ stands for wavelength, t is the time of the data point and Δt denotes measurement resolution. Further on, it was assumed that the rate constant τ from Eq. (1) varies with the device temperature according to the Arrhenius equation [], as given in Eq. (3), where k_B represents the Boltzmann constant:

$$\frac{1}{\tau} = B e^{-\frac{E_A}{k_B T_{cell}}} \quad (3)$$

Constants A, B, and E_a were optimized to best fit the ΔV_{MPP} results from sunny days in Figure 2.

The modeled ΔV_{MPP} curve (Eq. (1)) was then used in our EY algorithm to reduce the voltage of the PK-Si tandem device in each data point in accord with H_{cum} and T_{cell}. The comparison between the measured (blue curve) and simulated (green curve) V_{MPP} values is presented in Figure 4, along with ideal simulation results without LSE (red curve), incident irradiance, and device temperature. A very good agreement with the experimental values can be observed, which confirms the validity of the modeling approach. It is important to highlight that for the reason of better visualization, we show in Figure 4 only the results for one cloudy and one sunny day, however, the discrepancy between the simulated and measured generated energy for all six days is less than 0.25%.

(Finally, it should be acknowledged that during extended outdoor operation, LSE dynamics might experience irreversible degradation, leading to potential changes in the parameters from Eq. (1) and (2) over time.)

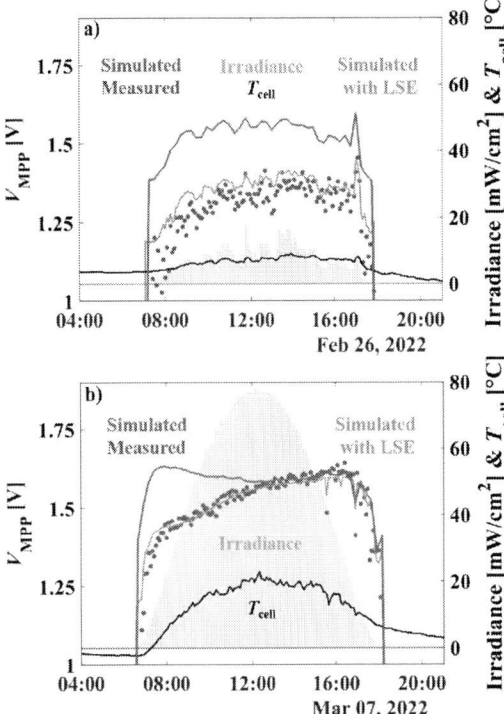

Figure 4: Measured (blue dots) and ideal simulated (red curve) V_{MPP} of the PK-Si tandem device for one specific a) cloudy and b) sunny day in March 2022. Also shown are the modeled voltage values that include light-soaking (green curve) as well as measured device operating temperature (black curve) and incident irradiance (yellow area).

3 ANALYSIS OF ANNUAL LSE LOSSES

In the last stage of this study, the developed EY model upgraded with the light-soaking effect was used to analyze the yearly energy losses ascribed to LSE in different environmental conditions (geographical locations) and in different device installation types. In the LSE modeling approach, it was assumed that the parameters of the decay functions remained constant over time.

The investigated 2T PK-Si tandem device was assumed to be mounted in three North America-based geographical locations from distinct climate zones according to the Köppen-Geiger-Photovoltaic (KGPV) classification [8]. We selected Phoenix, Atlanta, and Columbus since they represent BK, DH, and DM KGPV climate zones, respectively, which have very different environmental conditions and cover a large area of the USA as well as Europe. Meteorological data for each of the selected locations – namely the ambient temperature, wind speed, and spectrally resolved global and diffuse irradiance – were acquired from the National Solar Radiation Database [9] in hourly intervals, given in a total of 8760 data points over the entire typical meteorological year [10]. The operating device temperature was calculated empirically based on the Duffie-Beckman model [11] as a function of the total incident irradiance, ambient temperature, and wind speed. It should be noted that Berlin, from where we obtained the outdoor measurements presented in previous sections, belongs to the DL KGPV climate zone. The latter has similar environmental conditions to the DM zone.

Using the EY model, we calculated the yearly generated energy with and without LSE in order to obtain the yearly production losses attributed to the effect. We analyzed the losses for all three selected locations and for two different open-rack configurations of the device, namely: optimal (location-specific) and South-facing vertical. The results are summarized in Table I. It can be observed for both orientations that the lowest LSE losses are achieved in Phoenix. This was certainly expected since Phoenix represents a location with the highest insolation and ambient temperature, which both influence the LSE dynamics in a favorable way from the device performance viewpoint. In contrast, the device installed at Columbus experiences nearly 3-times higher yearly LSE losses due to the less favorable conditions. Further on, we can notice that in ideal case the generated energy of the tandem cell is reduced by almost 50% when considering vertical device orientation compared to the optimal installation. As a result, the LSE losses are also nearly twice as high in the case of vertical orientation.

Table I: Summary of numerical results for both open-rack configurations of the device and for all three geographical locations.

Orientation	optimal			vertical		
Location	EY ideal [kWh/m²]	EY with LSE [kWh/m²]	LSE losses [%]	EY ideal [kWh/m2]	EY with LSE [kWh/m2]	LSE losses [%]
Phoenix	589.8	583.4	1.1 %	314.5	308.2	2.0 %
Atlanta	477.8	468.7	1.9 %	252.1	243.5	3.4 %
Columbus	420.0	406.9	3.1 %	239.8	228.1	4.9 %

Given the significant impact of LSE dynamics on environmental conditions, it is expected that LSE losses are also notably influenced by seasonal variations. To confirm and further investigate this phenomenon, we distributed the LSE yearly losses over the different months of the year, as demonstrated in Figures 5a) and 5b). The results indicate that the LSE-induced energy losses are indeed two or three times higher in winter months compared to those in summer due to lower insolation and ambient temperatures.

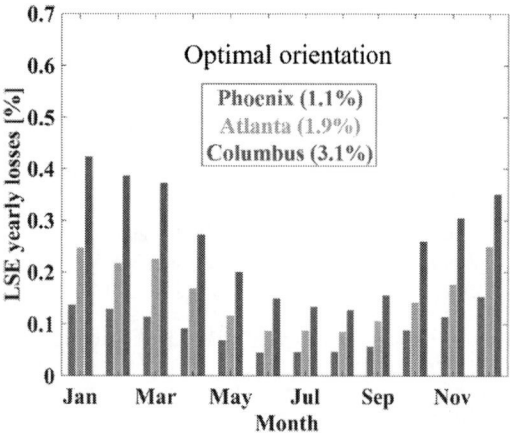

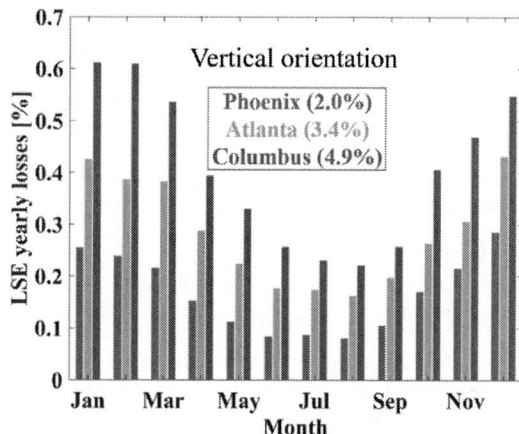

Figure 5: The distribution of the calculated LSE yearly losses over different months for all three selected geographical locations as well as for a) optimal and b) vertical orientation of the tandem device.

4 CONCLUSIONS

We used an in-house developed numerical model for accurate calculation of the electrical energy produced by the two-terminal encapsulated tandem PV device under realistic operating conditions.

The comparison of the simulation results to the outdoor monitoring results first served us to extract and quantify the energy losses induced by the light-soaking effect. We showed that over the course of three consecutive weeks at the beginning of the device operation, the tandem device energy production is reduced by 6% due to the light soaking effect. It was also demonstrated that the LSE losses are closely correlated with the solar irradiation and the device operating temperature.

Based on the observed correlation between the LSE and the operating conditions of the device we proposed an equation to model the voltage deficit ascribed to the effect. After the implementation of the LSE modeling approach in our *EY* algorithm, we obtained a very small discrepancy between the simulated and measured generated energy, amounting to less than 0.25%.

Finally, the developed *EY* algorithm upgraded with the light-soaking effect was used to analyze the long-term LSE energy losses in different geographical locations (selected from distinct KGPV climate zones) as well as in two different device open-rack configurations, namely: optimal and South-facing vertical. The results showed that the highest energy production and the lowest LSE losses are achieved in Phoenix for both configurations. By contrast, the device installed at Columbus experiences nearly 3-times higher yearly LSE losses due to the less favorable conditions.

ACKNOWLEDGEMENTS

The authors acknowledge the financial support from the Slovenian Research and Innovation Agency (research program P2-0415). Š. Tomšič thanks the Slovenian Research and Innovation Agency for her Ph.D. funding. We also thank the Helmholtz Association for funding the project TAPAS (Tandem Perovskite And Silicon solar cells – Advanced optoelectrical characterization, modeling, and stability) within the EU partnering program.

REFERENCES

[1] M. Jošt, L. Kegelmann, L. Korte, and S. Albrecht, "Monolithic Perovskite Tandem Solar Cells: A Review of the Present Status and Advanced Characterization Methods Toward 30% Efficiency," *Adv. Energy Mater.*, vol. 10, no. 26, p. 1904102, Jul. 2020, doi: 10.1002/aenm.201904102.

[2] "Best Research-Cell Efficiency Chart." https://www.nrel.gov/pv/cell-efficiency.html (accessed September 19, 2024).

[3] L. Lin *et al.*, "Light Soaking Effects in Perovskite Solar Cells: Mechanism, Impacts, and Elimination," *ACS Appl. Energy Mater.*, p. acsaem.2c04120, Apr. 2023, doi: 10.1021/acsaem.2c04120.

[4] Š. Tomšič, M. Jošt, K. Brecl, M. Topič, and B. Lipovšek, "Energy Yield Modeling for Optimization and Analysis of Perovskite-Silicon Tandem Solar Cells Under Realistic Outdoor Conditions," *Adv. Theory Simul.*, vol. 6, no. 4, p. 2200931, Apr. 2023, doi: 10.1002/adts.202200931.

[5] A. Al-Ashouri *et al.*, "Monolithic perovskite/silicon tandem solar cell with >29% efficiency by enhanced hole extraction," *Science*, vol. 370, no. 6522, pp. 1300–1309, Dec. 2020, doi: 10.1126/science.abd4016.

[6] B. Li *et al.*, "Revealing the Correlation of Light Soaking Effect with Ion Migration in Perovskite Solar Cells," *Sol. RRL*, vol. 6, no. 7, p. 2200050, Jul. 2022, doi: 10.1002/solr.202200050.

[7] X. Wu *et al.*, "Control over Light Soaking Effect in All-Inorganic Perovskite Solar Cells," *Adv. Funct. Mater.*, vol. 31, no. 28, p. 2101287, Jul. 2021, doi: 10.1002/adfm.202101287.

[8] J. Ascencio-Vásquez, K. Brecl, and M. Topič, "Methodology of Köppen-Geiger-Photovoltaic climate classification and implications to worldwide mapping of PV system performance," *Sol. Energy*, vol. 191, pp. 672–685, Oct. 2019, doi: 10.1016/j.solener.2019.08.072.

[9] M. Sengupta, Y. Xie, A. Lopez, A. Habte, G. Maclaurin, and J. Shelby, "The National Solar Radiation Data Base (NSRDB)," *Renew. Sustain. Energy Rev.*, vol. 89, pp. 51–60, Jun. 2018, doi: 10.1016/j.rser.2018.03.003.

[10] F. E. Vignola, A. C. McMahan, and C. N. Grover, "Bankable Solar-Radiation Datasets," in *Solar Energy Forecasting and Resource Assessment*, Elsevier, 2013, pp. 97–131. doi: 10.1016/B978-0-12-397177-7.00005-X.

[11] J. A. Duffie and W. A. Beckman, *Solar engineering of thermal processes / John A. Duffie, William A. Beckman*, 4th ed. Hoboken: John Wiley, 2013.

41st European Photovoltaic Solar Energy Conference and Exhibition

MICROSTRUCTURAL ANALYSIS ON THE CONFORMITY OF CHEMICAL VAPOUR DEPOSITION (CVD) PEROVSKITE THIN-FILMS ON SILICON FOR TANDEM PV DEVICES

Angela Chen, Emma Holder, Adrian Element, Yong Li, Kenrick F. Anderson,
Tim W. Jones, Benjamin C. Duck, Noel W. Duffy, <u>Gregory J. Wilson</u>
CSIRO Energy, Solar Energy Technologies Newcastle Energy Centre (NSW), Australia
E-mail: <u>Gregory.Wilson@csiro.au</u>

ABSTRACT: Perovskite-based solar cells are a rapidly emerging technology, with lab-scale devices achieving power conversion efficiency (PCE) performance close to commercial scale Silicon solar cells. To enable future commercialisation and large-scale production, CSIRO has developed a chemical vapour deposition (CVD) method that allows for the fabrication of repeatable, highly uniform perovskites layers with a scalability factor of 20× from laboratory scale devices. The sensitivity of perovskites presents handling challenges for both film analysis and production. An extensive study is presented on the conformity of perovskite film via CVD produced sample substrates. In this work we have successfully demonstrated that deposition of ultra-thin (5-50 nm) lead (Pb) precursor layers on industrial silicon wafers and subsequent conversion via a novel CVD technique enables retention of a controlled thickness and conformal perovskite absorber layer on highly variable textured silicon substrates.

Keywords: perovskite solar cells, chemical vapour deposition, film uniformity, textured silicon, lead precursor

1 INTRODUCTION

Photovoltaics (PV) have recently surpassed 1 terawatt (TW) of installed capacity, and multiple models indicate that 75 TW of PV will need to be installed by 2050 to meet growing global energy demands [1]. This requires an installation rate of more than 1 TW per year by 2030. Concurrently, single-junction solar cell efficiencies are rapidly approaching their practical limits, leading to significant investment in tandem solar cell research to achieve higher efficiencies (beyond 40%) and lower their embodied carbon. These advances are crucial for reducing the total system cost by decreasing the system area and, consequently, the balance of system (BOS) costs. Tandem solar cells are projected to reach 2% of the market share by 2030, with lab-scale record efficiencies surpassing 30% for multiple tandem technology pairings [2]. A recent roadmap for tandem PV published in *Joule* outlines the fundamental concepts and status of tandem PV technologies and presents the challenges that must be overcome to translate lab-scale cells to commercial products [3].

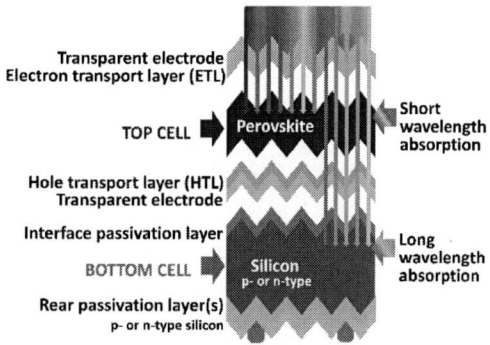

Figure 1: Structure of tandem perovskite/Silicon device.

The shift from silicon to perovskite-based photovoltaics marks a significant advancement in solar technology, driven by the quest for greater efficiency and reduced manufacturing costs. Perovskite solar cells, with their tunable optical and electronic properties, are ideal for coupling with silicon bottom cells in a tandem configuration (see Figure 1 and 2). Perovskite/silicon tandem solar cells can capture a broader spectrum of solar energy, substantially increasing their efficiency beyond what either technology could achieve alone [4]. A reliable and scalable method of perovskite deposition is crucial for constructing efficient wide-bandgap perovskite top cells.

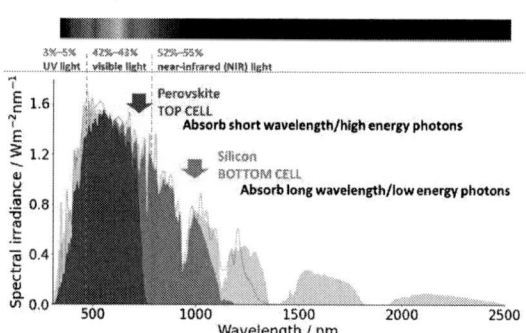

Figure 2: AM1.5 spectrum overlayed with spectral overlap for perovskite and silicon sub cells.

Chemical texturing is employed to introduce a pyramidal morphology on silicon surfaces, significantly enhancing sunlight absorption and thereby improving the photon-to-electron conversion efficiency of solar cells. This texturing process creates a specific morphology of pyramids that plays a crucial role in subsequent processes and overall cell performance. The enhanced light trapping and reduced reflectance due to these pyramidal structures are essential for maximising the efficiency of silicon-based solar cells. Chin *et al.* demonstrated that effective interface passivation, combined with optimised texturing, can lead to efficiencies as high as 31.25% in perovskite/silicon tandem solar cells [5].

However, depositing conformal coatings on silicon pyramidal textures and related morphologies presents a significant challenge. These textures, while beneficial for light absorption, complicate the deposition process, impacting the performance of monocrystalline silicon solar cells. In this work, we address this challenge by optimising an industrial process for top-cells compatible with pyramid features greater than 1 μm. This approach aims to overcome the limitations of solution-processed perovskites, which typically have thicknesses less than 1 μm, ensuring uniform and high-quality coatings on textured surfaces.

Our hybrid deposition approach has been chosen to produce high-quality and reproducible perovskite films. This method combines thermal evaporation and chemical vapour deposition (CVD) techniques to merge the precision and scalability of each process. It is particularly suitable for textured silicon solar cells because it ensures uniform coating. CSIRO has developed a CVD method that allows for the fabrication of repeatable, highly uniform perovskite layers with a scalability factor of 20× from laboratory-scale devices [6]. This method is industrially relevant, fitting in with modern thin-film fabrication techniques. Consistent film thickness and uniformity are crucial for high-performance solar cells, as defects within a film at the sub-micron level can cause short circuits, rendering the cell ineffective [5].

The original concept of the CVD process aimed to produce high-quality perovskite films using the Pb seed layer with a sequence of gas treatments. These treatments included acetic acid (HAc) for lead acetate (PbAc$_2$), hydrogen halides like hydrogen iodide (HI) or hydrogen bromide (HBr) are introduced to transform the lead layer into lead iodide (PbI$_2$) or lead bromide (PbBr$_2$), and *in situ* generated MA gas for generating MAPbI$_3$. However, challenges emerged during this process. Halide acids such as HI are aggressive and caused corrosion issues with the ITO bottom electrode layer, and controlling the MA reaction proved difficult, making the combination of steps unfeasible for achieving high-quality and reproducible perovskite layers. Consequently, alternative methods for converting PbAc$_2$ to PbI$_2$ and PbI$_2$ to perovskite became imperative based on similar observations and reports in the literature [4].

The evolution of our process initially used an atmospheric CVD (ACVD) reactor, where vapours of different reagents had a negative effect and increased operational maintenance due to cross-contamination in the gas line and corrosion was observed in the reactor chamber, chemically hindering the material conversation via CVD. Therefore, engineering of the system was modified towards vacuum-based CVD (VCVD). This approach is advantageous firstly due to the vacuum reducing any residual process gases from being deposited on cold surfaces of the chamber and process lines, and secondly, in practical Si-PVK tandem cell fabrication, the recombination layer (atop the Si bottom cells) and buffer layer (atop the perovskite top cell) are processed using vacuum-based deposition systems. Hence, fabricating the perovskite top cell under vacuum conditions without breaking the vacuum cycle is consistent with long-term applications in tandem solar cells.

2 EXPERIMENTAL METHODS AND MATERIALS

2.1 Materials

Work herein was performed on prepared perovskite thin films deposited on textured silicon. Depositions were performed using our proprietary VCVD, that is described in the more detail in the abstract: 2CV.3.63 Demonstration of industrially scalable Chemical Vapour Deposition (CVD) process for production of high-efficiency perovskite photovoltaics. Silicon substrates were sourced from an industrial partner and were a modified heterojunction stack. For device fabrication NiO$_x$ sourced from Kurt J Lesker was sputtered as the interface layer between the silicon and perovskite sub-cells using a Nanomaster evaporator/sputter system. C$_{60}$ from Sigma-Aldrich was thermally evaporated and capped with sputtered AZO. Ag was evaporated as the top electrode using an Angstrom evaporator.

The surface morphology of the perovskite on Si and ITO substrates was analysed by scanning electron microscopy (SEM) and cross-sectional thin film on silicon images were captured by ThermoScientific Phenom XL and Zeiss Evo MA15. The crystal structure of perovskites was analysed using X-ray diffraction (XRD) measured by PANalytical Aeris Research model X-ray powder diffractometer with Cu Kα radiation ($\lambda = 1.54050$ Å).

2.2 Characterisation techniques – Scanning Electron Microscopy

Scanning Electron microscopy is widely used as a morphological characterisation technique. In this work, this tool played an important role that produced images of the top perovskite thin films that focused on the morphology and uniformity of the material layers that form the basis of a perovskite silicon tandem cell. Further analyses were placed emphasis on the thickness of the layers, the surface coverage and conformity to the bottom silicon cell.

For this technique, an image is formed by the interaction of the specimen and a focused beam of electrons with a voltage between 0.1 and 50 kV. The interaction of the beam and the specimen forms the main signals: secondary electrons (SE), backscattering electrons (BSE), and X-rays. Preparation involves a few straightforward steps of substrate cleaning, sample cutting, stage alignment and sample cross-sectional attachment to stage. To image the surface of the thin films, including the electron and hole selective layers (ESL and HSL, respectively) and the perovskite thin films, no metallisation is typically needed as these layers are usually deposited on conductive substrates, such as indium-doped tin oxide (ITO) and silicon (Si). Therefore, charging, due to lack of electron dissipation, is not usually an issue when using low voltages. In some cases where the SEM tool does not have low-voltage capabilities, metallisation with gold or platinum is needed also if there is the alternative to partially vary the vacuum intensity. Cross-sectional SEM images tend to be difficult to obtain as the cut needs to be near perfect to get a clear picture of thickness and conformity to the substrate in the device. Charging tends to be an issue for cross sections and low-voltage below 10 kV was used. Alternatively, the use of metallisation tends to help issues with sample charging. However, this process can smear the interfaces making it difficult to identify different layers.

SEM is a technique that needs to be used with caution because the top layer cell is lead halide perovskites (LHPs) and they are soft materials that can be easily damaged by the electron beam. This induces artefacts and films, especially under high signal to noise ratio x-ray mapping, tend to degrade under the electron beam causing voids and cracks that make microstructural characterisation difficult. Therefore, a combination of low voltage and low electron dose is needed to obtain desirable results.

3 RESULTS AND DISCUSSION

The work presented demonstrates a CVD method that produces reproducible thin films. As thickness variability is critical to overall performance, we have successfully demonstrated variability of less than 10 nm across a 10 cm² area, free of defects. Most importantly films being produced in this work are at a scale comparable to dimensions of wafers commonly available in industrial supply, thus readily transferable to fabricate tandem devices. In this study we use several methods to critically analyse our thin films, including surface profilometry, x-ray diffraction, scanning electron microscopy and luminescence techniques. Additionally, films are made into complete devices to assess performance against benchmark methods currently used in research. The sensitivity of perovskites presents handling challenges for both film analysis and production. Herein we explore characterisation techniques utilised to ensure the production of highly uniform perovskites layer during the CVD conversion profile. A detailed study is presented on the conformity of perovskite film via CVD processing of sample substrates. The initial physical vapour deposition (PVD) of Pb lead precursor layers on textured silicon substrates was successfully demonstrated to provide precursor film-thicknesses within a targeted range for final film thickness of perovskite absorber layer. Future work, not included in this morphological study, is to stack the perovskite layer on top of silicon substrates to produce high performance tandem cells.

3.1 Peaks & Valleys of Silicon

Figure 3 depicts the initial textured surface used in the CVD process. Pyramid features with heights over 1 μm are clearly observed. Such large and varied features lead to issues when typical spin coating methods are applied.

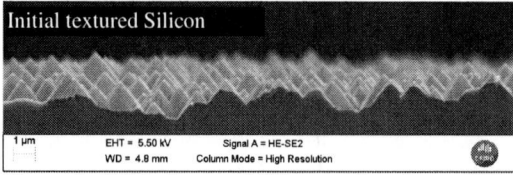

Figure 3: SEM of textured Silicon wafer prior to CVD processing.

The large pyramid heights are incompatible with solution processed layers that have thicknesses typically less than 1 μm. This leads to the type of defects observed in Figure 4a where the tip of many of the silicon sub-cell pyramids is observed to be exposed through the perovskite film. In a complete device these would all represent shunting pathways between the contacts of the perovskite sub-cell.

A second common defect caused by variation in the mean surface height is demonstrated in Figure 4b. Here the film has been made thick enough to eliminate the shunting defect observed in Figure 4a however large gaps can now be observed between the top of the silicon and the bottom perovskite sub-cells. These gaps occur in lower lying regions of the textured surface and result in poor or no contact between sub-cells for these local regions.

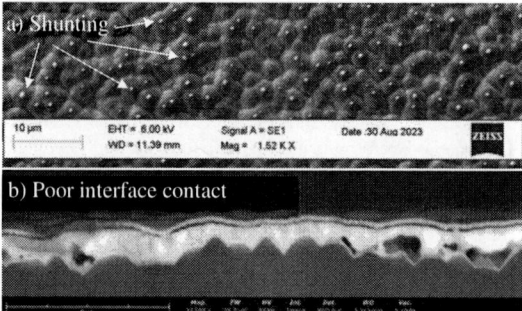

Figure 4: Spin-coating results in interface issues with textured surfaces leading to: a) shunting due to incomplete coverage; and b) poor interface contact due to mean surface height variation.

3.2 Cross-sectional analysis of a Perovskite thin-film on Silicon via VCVD

The VCVD process begins with thermal deposition of a Pb layer onto the textured silicon sub-cell surface. For films to be converted into working tandem devices a blocking layer is sputtered onto this surface prior to the Pb deposition. In both cases a highly conformal Pb coating is achieved as shown in Figure 5. Some granularity in the final Pb layer is observed. The quantity of Pb deposited is a direct control on the thickness of the perovskite layer fabricated. Therefore, the VCVD process allows any arbitrary desired target thickness to be readily achieved.

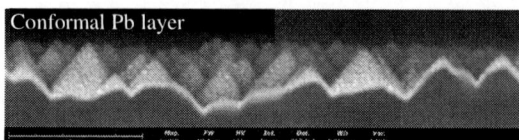

Figure 5: SEM of as deposited thermally evaporated Pb layer onto textured Silicon wafer.

Following the Pb deposition the surface is exposed to an entrained acetic acid vapour stream, resulting in a PbAc₂ intermediate. As depicted in Figure 6 the surface conformity is maintained despite the soft nature of this material.

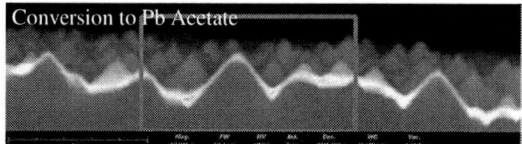

Figure 6: SEM of Pb acetate intermediate in VCVD process.

Full conversion to perovskite is achieved by exposure of the PbAc₂ film to an organic halide precursor in the head of the custom sublimation chamber. The use of multiple sources facilitates control over the composition of the organic halide precursors, which can be varied continuously throughout deposition via both control of the thermalisation temperature of the head and gas flow, allowing for finite control and stoichiometric adjustments in resulting films and semiconductor properties. Figure 7 illustrates a high-resolution SEM of the resulting film which clearly depicts a conformal coating to the underlying silicon is retained. Artefacts observed in the figure are a result of beam damage from the SEM (described briefly in experimental methods) which manifest as crack-like features that appear in the film and grow in size and frequency with continued exposure. For completeness, and to illustrate the intended final device stack, an SEM of a full tandem stack with top blocking layers and metallisation is depicted in Figure 8. The complete conversion to the desired composition is verified by XRD performed as a series of spot analyses across the surface of the large-area coating.

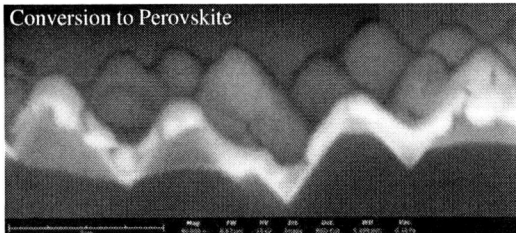

Figure 7: High resolution SEM of VCVD deposited perovskite film. Artefacts (cracks) appear in the perovskite layer due to beam damage from the SEM.

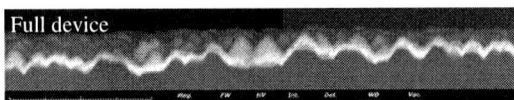

Figure 8: SEM of complete VCVD perovskite/Silicon tandem stack.

4 CONCLUSIONS

Perovskite/Silicon tandem solar cells are a promising path towards higher efficiency photovoltaic devices. Textured silicon sub-cells are required to maintain the highest performance however traditional solution processed techniques for fabricating the perovskite sub-cell layers are incompatible with surfaces that have large pyramids or variations in the mean surface height. A well controlled conformal coating is required to maximise the potential performance of two terminal tandem designs. This means alternatives for solution-based fabrication are required.

Here we have demonstrated the ability of a novel vacuum based chemical vapour deposition (VCVD) technique to create conformal coatings. This technique allows complete control of the fabricated film properties including thickness and composition. Using SEM we have demonstrated the process maintains a high-quality conformal coating onto the textured silicon surface through each of the deposition steps.

4.1 Acknowledgements

This research is supported by the Science and Industry Endowment Fund (SIEF) and the authors acknowledge and appreciate funding from SIEF on behalf of CSIRO.

4.2 References

[1] Haegel, N. M., Atwater, H. A., Barnes, T. M., Breyer, C., Burrell, A. K., Chiang, Y.-M., & van de Lagemaat, J. (2023). Photovoltaics at multi-terawatt scale: Waiting is not an option. *Science*, 380(6649), 1-10. https://doi.org/10.1126/science.abc1234

[2] Wright, M., Vicari Stefani, B., Jones, T. W., Hallam, B., Soeriyadi, A., Wang, L., Altermatt, P., Snaith, H. J., Wilson, G. J., & Bonilla, R. S. (2023). Design considerations for the bottom cell in perovskite/silicon tandems: A terawatt scalability perspective. *Energy and Environmental Science*, 16(10), 4164–4190. https://doi.org/10.1039/d3ee00952a

[3] Alberi, K., Berry, J. J., Cordell, J. J., Friedman, D. J., Geisz, J. F., Kirmani, A. R., Larson, B. W., McMahon, W. E., Mansfield, L. M., Ndione, P. F., Owen-Bellini, M., Palmstrom, A. F., Reese, M. O., Reese, S. B., Steiner, M. A., Tamboli, A. C., Theingi, S., & Warren, E. L. (2024). A roadmap for tandem photovoltaics. *Joule*, 8(4), 1-20. https://doi.org/10.1016/j.joule.2024.03.001

[4] Liu, M., Johnston, M. B., & Snaith, H. J. (2013). Efficient planar heterojunction perovskite solar cells by vapour deposition. *Nature*, 501(7467), 395–398. https://doi.org/10.1038/nature12509

[5] Chin, X. Y., Turkay, D., Steele, J. A., Tabean, S., Eswara, S., Mensi, M., Fiala, P., Wolff, C. M., Paracchino, A., Artuk, K., Jacobs, D., Guesnay, Q., Sahli, F., Andreatta, G., Boccard, M., Jeangros, Q., & Ballif, C. (2023). Interface passivation for 31.25%-efficient perovskite/silicon tandem solar cells. Science, 383(162), 1-10. https://doi.org/10.1126/science.adg0091

[6] Jones, T., Wilson, G., Anderson, K., Hollenkamp, A., Duffy, N., & Firet, N. (2015). Process of forming a photoactive layer of an optoelectronic device. *Google Patents*. https://patents.google.com/patent/WO2016094966A1/en

41st European Photovoltaic Solar Energy Conference and Exhibition

CONTROLLING THE FILM PROPERTIES OF SNO2 IN PEROVSKITE SOLAR CELLS USING SCALABLE SPATIAL ATOMIC LAYER DEPOSITION

H.W. de Vries[1], F.M.M. Souren[1], S.R. Ratnasingham[2], M. Najafi[2]
[1]SALD B.V. Luchthavenweg 10 Eindhoven
[2]TNO

ABSRACT: Nowadays, Atomic Layer Deposition (ALD) is routinely used to deposit tin oxide (SnO2) layers in perovskite solar cells (PSC's). SnO2 is amongst others used as a buffer layer on the C60 layer to prevent it from sputtering damage of the subsequent Transparent Conductive Oxide (TCO) coating. In this contribution we investigate whether the film density of SnO2 layer deposited on the C60 layer by Spatial ALD (s-ALD) can improve the water barrier performance. During the manufacturing of a tandem PSC, tin oxide can protect the underlying perovskite from the water based PEDOT:PSS coating which is required for the narrow band gap perovskite coating on top. First, we systematically explore the s-ALD process parameters, as for example temperature, water dosage and process speed, and analyze their effect on film thickness and refractive index using spectroscopic ellipsometry. Higher refractive index values are generally associated with denser coatings and thus better moisture barriers. Second, 20 nm thick SnO2 layers are deposited directly on the C60 coating using the process parameters explored. Two types of perovskite (PSK) stacks consisting of a dual cation PSK-C60 surface and triple cation PSK-TEACl-C60 surface are investigated. The barrier performance of SnO2 layers are analyzed by mimicking the PEDOT:PSS coating by a water based coating process using a spin coater. The 20 nm thick SnO2 layers deposited on the dual cation PSK-C60-SnO2 cells all survived the water exposure. However, some of the triple cation PSK-TEACl-C60-SnO2 cells clearly degrade upon water exposure
Keywords: spatial atomic layer deposition, tin oxide, electron transport layer, buffer layer, water barrier

1 INTRODUCTION

Atomic-layer-deposited (ALD) SnO2 layers in perovskite solar cells are frequently used as an electron transport layer as well as a buffer layer to prevent damage of the C60 layer from the subsequent TCO sputtering. The objectives of this study are accordingly (i) to explore the SnO2 film properties with spectroscopic ellipsometry, (ii) to investigate the stability of the perovskite device coated with a SnO2 layers upon exposure to water and (iii) identify the factors that control the barrier-ability of SnO2 on the C60 fullerene.

2 EXPERIMENTAL METHODS

Atomic layer deposition of SnO2 as a functional inter-layer in PSCs is frequently applied, because the process allows for growth of conformal, dense and pinhole free layers, even on rough surfaces. Moreover, SnO2 are optically transparent and can thus be used in p-i-n devices.

For the deposition of SnO2 layers we apply the Spatial Atomic Layer Deposition (s-ALD) technique, which could directly be applied to industrial solar cell production. Figure 1 shows a cartoon of the s-ALD principle. The s-ALD reactor consists of a co-reactant step (water vapor) for the oxidation, a precursor step with Tetrakis(dimethylamino)tin (TDMASn) and again the co-reactant step. Precursor and co-reactant are separated by N2 gas curtain to secure that only surface reactions take place and gas phase reaction through mixing of H2O and TDMASn is minimized. The wafer is positioned on a substrate table and moves forward and back below the deposition head. One pass of the wafer through the three zones, corresponds to one ALD cycle of one monolayer of SnO2.

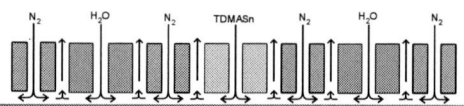

Figure 1: Schematic overview of the s-ALD deposition head, where H2O inlet, Tetrakis(dimethylamino)tin (TDMASn) and H2O inlet are separated by N2 gas curtains.

The film properties were investigated by varying the line speed, the precursor and co-reactant dosage and the process temperature. For the SE measurements the SnO2 layers were deposited on mirror polished silicon wafers with approximately 1.5 nm native silicon oxide layer. The film thickness and refractive index were measured using a Woollam M2000 spectroscopic ellipsometer. The refractive index and film thickness were determined using the general oscillator model for SnO2 (Gaussian model) from the CompleteEase software supplied by Woollam. The refractive index of the films was taken at 1.96 eV. It is well known that to achieve good moisture barriers, besides the density (and thus refractive index) of the barrier layer the underlying surface plays a critical role in the final performance of the barrier film (e.g. due to pinholes). Therefore deposition of SnO2 on the inert C60 surface should be investigated.

Both perovskite (PSK) stacks were deposited on 3x3 cm² glass substrates with an InSnOx (ITO) and PTAA (hole transport layer) coating. The RB-1 consists of a (wide bandgap) 1.8 eV dual cation PSK-C60, the RB-9 consists of a 1.8 eV triple cation PSK-TEACl-C60 coating, see Figure 2. The complete perovskite cells were prepared by TNO except for the SnO2 layer which was processed at SALD B.V..

The barrier performance of the SnO2 layers was assessed by a water exposure test mirrored by a typical PEDOT:PSS spin coating regime. This is the p-type layer that is typically applied as the beginning of the narrow bandgap cell. Typically 300 µl water was applied and spin coated at 3000 rpm. The dwell time was around

408

one minute.

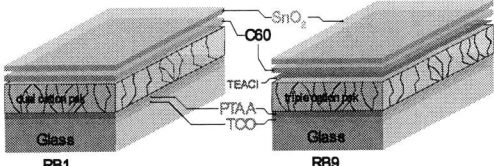

Figure 2: The architectures of a dual and triple cation perovskite device used to assess the barrier properties of a 20 nm SnO_2 coating.

3 RESULTS AND DISCUSSION

3.1 ALD process parameter evaluation

The effect of the process temperature as well as the line speed is depicted in Figure 3. As can be seen the refractive index shows a significant decrease with increasing line speed as well as lower processing temperature. The refractive index differences are significant and suggest that the films become less dense. Surprisingly this is not reflected in the growth per cycle (GPC), although a slight decrease in the GPC is observed at 500 mm/s process speed.

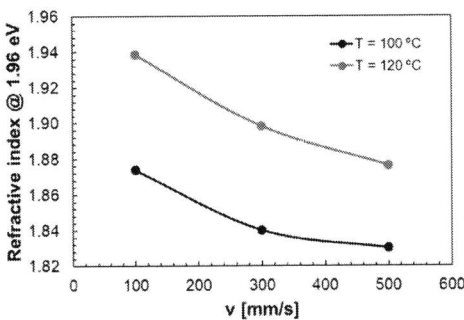

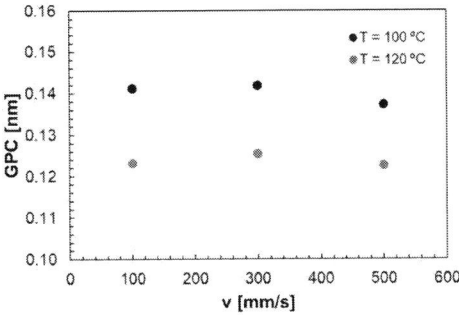

Figure 3: Refractive index (top) and growth per cycle GPC (bottom) as a function of the scan speed and temperature

Subsequently, the effect of the precursor and co-reactant dosage on the refractive index and GPC was investigated for a scan speed of 300 mm/s and 120°C processing temperature. In Figure 4 (left) a very small decrease in the refractive index with increase of precursor dose is observed, whereas in Figure 4 (right) a strong increase of the refractive index with the co-reactant dosage is observed. At the same time from the GPC measurement a clear saturation curve is obtained as a function of the TDMASn flow. The gradual increase in

GPC from 1 to 2 slm is attributed to so-called 'soft saturation effect'. Surprisingly, no sub-saturation is observed for H_2O, in any of the conditions investigated.

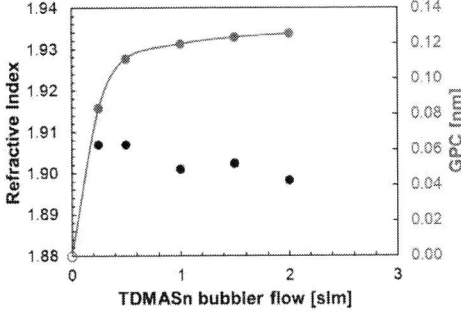

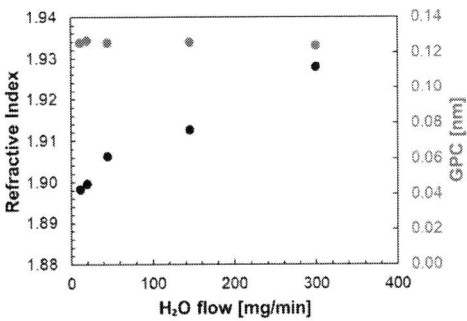

Figure 4: Refractive index and GPC as a function of the precursor (top) and co-reactant (bottom) concentration.

To investigate the influence of the co-reactant in more detail we extended H_2O concentration to much lower values using a bubbler system and we converted the H_2O flow to the partial pressure of H_2O. Despite the much lower H_2O concentration still no sub-saturation for H_2O was observed, however the lower refractive index continuous to decrease suggesting that incomplete surface reaction may lead to less dense films with more remaining TDMASn ligands present in the film.

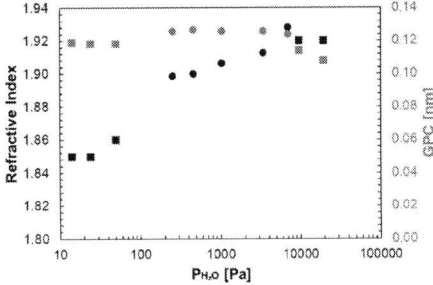

Figure 5: Refractive index and GPC as a function of the H_2O partial pressure. Circles are prepared by the liquid evaporator system the squares by the bubbler system.

3.2 Water barrier assessment

Based on exploration of the process window of the TDMASn/H_2O system the process temperature (100 and 120°C), line speed (50, 100 and 300 mm/s) and the water dosage (40 Pa and 10kPa partial pressure) were selected to explore the H_2O barrier performance. In total 24 samples with 20 nm SnO_2 were processed and during each run one RB1 and one RB9 sample was processed in parallel. The water exposure test samples were visually evaluated and depicted in Figure 5.

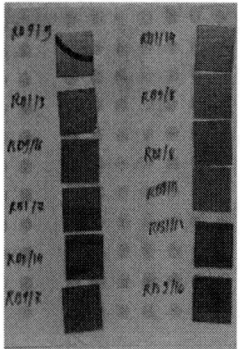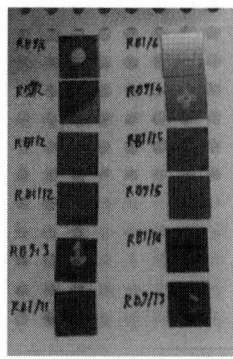

Figure 6: Photograph of the 3x3 cm^2 samples after the H_2O exposure experiment. RB1/6 is an un-coated reference sample.

All the RB1 samples having the dual cation PSK survived the spin coating test except the un-coated sample, see Table 1. The triple cation PSK with the TEACl under-layer appears to be more susceptible to degradation, particularly the samples processed at 300 mm/s. However, the 100°C, 300 mm/s, 10 kPa (level 6) did not degrade which may indicate that the degradation is related to the initial quality of the PSK-TEACl-C60 coatings.

Table 1: Summary of the stability and experimental parameters used for depositing the 20 nm SnO_2 layers on the different substrates, in red are the samples that failed the test, this includes sample RB1/6 which is the un-coated type sample.

RB#/#	T (°C)	v(mm/s)	[H_2O] (Pa)
1/6	-	-	-
9/1	100	50	40
9/2	100	100	40
9/3	100	300	40
9/4	100	50	10,000
9/5	100	100	10,000
9/6	100	300	10,000
9/7	120	50	40
9/8	120	100	40
9/9	120	300	40
9/10	120	50	10,000
9/11	120	100	10,000
9/12	120	300	10,000

Degradation of the RB9 levels 2 and 3 may be attributed to increased porosity. The combination of low H_2O dosage (40 Pa), low temperature 100°C and high line speed significantly decreases the refractive index of SnO_2. However, the failure mechanism of the triple cation PSK is not clear yet. Possibly it can be related to a higher pinhole density, due to a rougher surface and formation of pinholes, etc. The failure at relatively high line speed can also be related to either reduced SnO_2 density (reflected in the lower refractive index) or nucleation delay on the inert C60 layer which can affect the film growth on local or even global scale.

4 CONCLUSIONS

We summarize that no degradation of the dual cation PSK samples (RB1) is observed with a very thin layer of SnO2 of 20 nm, independent of the spatial ALD process parameters investigated. SnO_2 layers deposited by the scalable s-ALD deposition technique can thus enable good barrier layers and provide excellent protection against exposure to H_2O. Thus, thin SnO_2 layer make high throughput Spatial ALD the perfect candidate for high volume PSC production. The stability of the triple cation PSK cannot be pinned down to a single factor like the refractive index and further research is needed to understand the interactions between the C60 and SnO_2 layer.

41st European Photovoltaic Solar Energy Conference and Exhibition

BEYOND THE LAB-SCALE: PEROVSKITE PHOTOVOLTAIC FABRICATION AND INDUSTRIAL ASSESSMENT WITH AUTOMATED SLOT-DIE COATER

Maurizio Stefanelli[1],*, Simon Ternes[1], Luigi Vesce[1], Marco Balucani[2] and Aldo Di Carlo*[1,3]

1 C.H.O.S.E. (Center for Hybrid and Organic Solar Energy), Electronic Engineering Department, University of Rome Tor Vergata, Via del Politecnico 1, 00118, Rome, Italy.
2 RISE Technology, Via Monte Bianco, 18, 35018 San Martino di Lupari, Italy
3 Istituto di Struttura della Materia (CNR-ISM) National Research Council, via del Fosso del Cavaliere 100, 00133, Rome, Italy
emails: maurizio.stefanelli@uniroma2.it*, aldo.dicarlo@uniroma2.it*

ABSTRACT: During the last 15 years, perovskite (PVK) photovoltaics (PVs) attained immense progress in single junction devices and, nowadays, spearhead the development of commercially viable 2-T tandem devices. This achievement became possible due to the exceptional properties of this material class, permitting not only power conversion efficiencies beyond 26%, but also low-cost fabrication from solution by common printing techniques. A wide consensus of researchers and technologist is reached, considering perovskite semiconductors feasible for industrial-scale fabrication alongside the well-establish silicon technology. While laboratory-scale work enabled fast material screening, characterization and process analysis, the transfer of the technology to industry is often not sufficiently treated. Most likely, in situ quality assessment is needed to ensure stable output for failure analysis early in the device fabrication. Here, we present a unique toolkit for simulating industrial conditions by employing a fully automated slot-die coater, sample handling, gas quenching as well as in situ characterization. We succeed in re-creating conditions of high-throughput industrial fabrication in a model box of about 10 m^3 under ambient atmosphere. Using the model box, we can investigate the feasibility of large-scale perovskite fabrication in a very cost-effective manner, identifying possible failure modes and predicting fabrication yield in a close-to-realistic model environment.

Keywords: Narrow band-gap perovskite; perovskite upscaling; slot-die coating.

1 IINTRODUCTION

Perovskite photovoltaics is the hot topic in PV community in the last years thanks to the high efficiency, low-cost production and wide engineering of the material compositions and configurations. [1-3] Upscaling is often a way to prove the possible industrialization of this material in PV context even though the equipment usually used are still at lab-scale level.
The upscaling from laboratory cells to module devices has extreme importance to exploit at market level the PVK PV technology. The PVK deposition is the most critical step process and several techniques (e.g. slot-die, blade coating, bar coating, spray coating etc) in combination with different quenching methods (e.g. antisolvent, gas-quenching, vacuum-quenching) are employed to get homogeneous layer and performance close to the one obtained on small area cell. [4]
The in-situ measurements are more rarely used to understanding crystallization kinetics and nucleation of perovskites in the as-cast solution and the subsequent growth of nanocrystals. [5] The lack of proper probing techniques of the film formation process hinders the construction of a universal design principle that provides a rational guideline to improving the quality of perovskite films.
Here we want to provide a new chapter on perovskite towards industrialization. We used an automatic slot-die coater in a controlled non-inert environment to set up a gas-quenched perovskite deposition till 15x15 cm2 substrate. We consistently change deposition parameters such as height of the slot-die, coating speed and air flow following the process evolution with in-situ reflectance imaging to probe the as-coated film and study crystallization kinetic and growth of PVK crystals in a industrial box environment. [6]
The machine apparatus made by RISE Technology S.R.L. is an automatic slot-die coater equipped with a robot arm

for sample handling, two hotplates, an UV-ozone lamp, a slot die head support and air knife.

The following routine is carried out (numbers in bracket are referred to the machine components involved):

1. Leveling of the slot die head on the sample to achieve a homogenous gap (1,2)
2. Retrieving the sample from the sample holder (0)
3. Uv-ozone treatment (6)
4. Coating of the perovskite ink (1,2)
5. Gas Quenching (3)
6. In situ characterization: Reflectance imaging
7. Modification of Parameters and repetition
8. Annealing (4,5)
9. Storing of Sample (0)
10. Parameters tuning and repetition

In this unique succession of events, we simulate industrial production and test all the steps, which are paramount to achieving high throughput and yield. We believe that, by identifying processes compatible with this setup, we can test all the necessary steps for commercialization of perovskites in a very resource efficient way. Due to the capability of cyclic processing and continuous repetition, we can model the consistency of yield and throughput, identifying possible challenges early on. For example, we found that it is very difficult to implement gas quenching in a way that is repetitive for every sample.

411

41st European Photovoltaic Solar Energy Conference and Exhibition

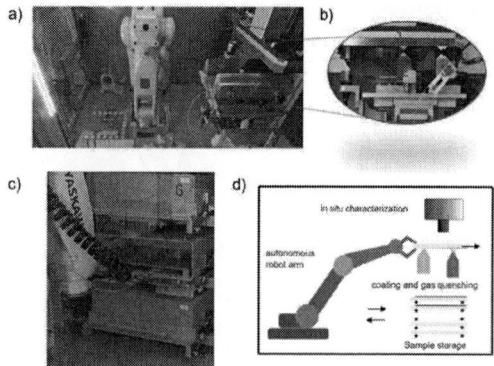

Figure 1: Automatic slot-die coater picture (a,b,c) and in-situ characterization scheme (d)

2 RESULTS AND DISCUSSION

We change deposition parameters such as height of the slot-die, coating speed and air flow according to the in-situ reflectance as simple optical facility. We can the study crystallization kinetics and the growth of PVK crystals in the designed "industrial box" in dry air (RH< 30%) environment.

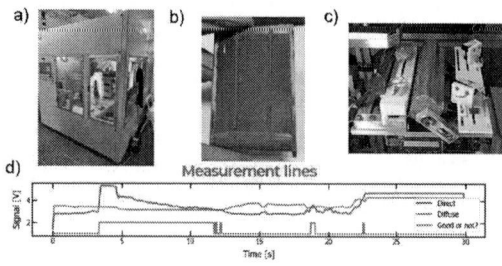

Figure 2: Perovskite reflectance pathway (a) In situ reflectance apparatus (b) m-ONE facility provided by RISE technology (c) plot (V vs s) of direct and diffuse reflectance (d)

The high degree of automation of the machine allows a process loop to be created in which the deposition parameters are evaluated by diffuse and direct in-situ reflectometry. With this setup, diffuse and direct reflections are recorded at the front of the crystallisation during the quenching step. If the diffuse reflectance is greater than the direct reflectance, a false quenching process is taking place due to incorrect crystallisation driven by the breakdown between the nucleation and diffuse regimes of crystal growth. In both cases of imbalance of the crystallization pathways, we have recorded the diffuse and direct footprints of the film, and this fine control and process optimization steps are finally traced in the transition from a greyish perovskite film with high diffuse reflection to a semitransparent brown film with high specular reflection for 7.5x5 and 15by15 cm^2 substrates, respectively.

Moreover, we scale up to 15x15 cm^2 substrate the perovskite film with the optimized process routine carried out from parameters variation and correlation of reflectance.

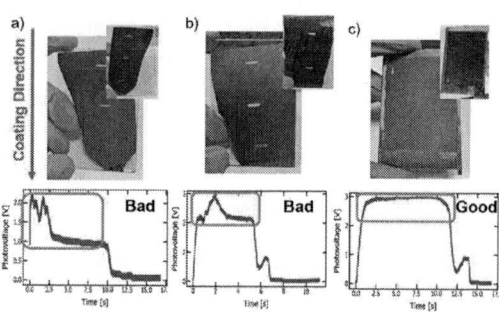

Figure 3: Perovskite film and in situ direct reflectance plot (a,b,c)

3. CONCLUSION

Reflectance measurement can effectively provide information about the crystallization process. A direct high reflectance signal means high mirror-like PVK film. In the bad case, the amorphous phase is present because of solvent trapping, incorrect quenching and poor crystallinity. Tuning coating parameter and quenching time cause high quality perovskite film

This in situ methodology developed in conjunction with the industrial box can effectively provide a tool to understand and probe the gas quenching window for any PVK formulation adopted in principle for this type of process.

4. REFERENCES

[1] Park, J.; Kim, J.; Yun, H.-S.; Paik, M. J.; Noh, E.; Mun, H. J.; Kim, M. G.; Shin, T. J.; Seok, S. Il. Controlled Growth of Perovskite Layers with Volatile Alkylammonium Chlorides. Nature 2023, No. July 2022. https://doi.org/10.1038/s41586-023-05825-y.

[2] Jiang, Q.; Tong, J.; Xian, Y.; Kerner, R. A.; Dunfield, S. P.; Xiao, C.; Scheidt, R. A.; Kuciauskas, D.; Wang, X.; Hautzinger, M. P.; Tirawat, R.; Beard, M. C.; Fenning, D. P.; Berry, J. J.; Larson, B. W.; Yan, Y.; Zhu, K. Surface Reaction for Efficient and Stable Inverted Perovskite Solar Cells. Nature 2022, 611 (7935), 278–283. https://doi.org/10.1038/s41586-022-05268-x.

[3] Green, M. A.; Dunlop, E. D.; Hohl-Ebinger, J.; Yoshita, M.; Kopidakis, N.; Hao, X. Solar Cell Efficiency Tables (Version 58). Prog. Photovoltaics Res. Appl. 2021, 29 (7), 657–667. https://doi.org/10.1002/pip.3444.

[4] Li, D.; Zhang, D.; Lim, K. S.; Hu, Y.; Rong, Y.; Mei, A.; Park, N. G.; Han, H. A Review on Scaling Up Perovskite Solar Cells. Adv. Funct. Mater. 2021, 31 (12), 1–27. https://doi.org/10.1002/adfm.202008621

[5] Hu, H.; Singh, M.; Wan, X.; Tang, J.; Chu, C. W.; Li, G. Nucleation and Crystal Growth Control for Scalable Solution-Processed Organic-Inorganic Hybrid Perovskite Solar Cells. J. Mater. Chem. A 2020, 8 (4), 1578–1603. https://doi.org/10.1039/c9ta11245f.

[6] Ternes, S.; Mohacsi, J.; Lüdtke, N.; Pham, H. M.; Arslan, M.; Scharfer, P.; Schabel, W.; Richards, B. S.; Paetzold, U. W. Drying and Coating of Perovskite

Thin Films: How to Control the Thin Film Morphology in Scalable Dynamic Coating Systems. ACS Appl. Mater. Interfaces 2022, 14 (9), 11300–11312. https://doi.org/10.1021/acsami.1c22363.

5. ACKNOWLEDGEMENT

This research was funded by the European Union's Horizon Europe programme, through a FET Proactive research and innovation action under grant agreement No. 101084124 (DIAMOND).

41st European Photovoltaic Solar Energy Conference and Exhibition

REASONING THE CHANGE IN DEVICE PARAMETERS WITH DEPOSITION POWER OF NIOx FOR LOW-DIMENSIONAL PEROVSKITE SOLAR CELLS

Bhumika Sharma[1][2], Vani Pawar[1][3], Sushobhan Avasthi[1][4]
[1]Centre for Nano Science and Engineering, Indian Institute of Science, Bengaluru, Karnataka, 560012, India
E-mail: [2]bhumikas@iisc.ac.in, [3]vanipawar@iisc.ac.in, [4]savasthi@iisc.ac.in

ABSTRACT: Low-dimensional perovskites have gained attention due to their excellent stability, but still they struggle with achieving competent efficiency. Solar cells' performance depends not only on the absorber layer but also on the transport layer and contacts. Transport layers are responsible for charge separation in perovskite solar cells (PSCs) and hence play a crucial role in the determination of device parameters which collectively give us the device efficiency. Transport layers specific to low-dimensional perovskites have not been rigorously studied and thus remain an open problem to be pondered upon. Here, we demonstrate the effect of modification of transport layer properties on PSCs by linking the properties of NiO_x to the device parameters obtained by current-voltage measurements. The low-dimensional perovskite $(PEA)_2MA_4Pb_5I_{16}$ is used as an absorber for the fabricated PSCs. The deposition power for NiOx during the pulsed DC magnetron sputtering is varied by keeping all the other parameters the same. The power used for deposition is 60W, 75W, and 90W at 200 °C substrate temperature and base pressure of 1e-6 mbar. The champion device is obtained at 75W deposition power for NiO_x, having an efficiency of 10.52%. Improving device parameters by tuning the deposition parameters for NiO_x cumulatively increases the efficiency.
Keywords: NiO_x, low-dimensional perovskite, sputtering, deposition power

1 INTRODUCTION

Perovskite solar cells (PSCs) have an absorber, transport layer, and contacts as components amongst which the transport layer plays a crucial role in charge carrier separation. The properties of transport layers are essential in determining device performance. The hole transport layer (HTL) in the p-i-n architecture of PSCs is required to have high transparency and an optimum band alignment for the devices to be efficient. The deposition procedure of HTL controls its material, optical, and electrical properties which in turn affect the device's performance. NiO_x is being widely used as an HTL for PSCs amongst which 3-dimensional perovskites have been largely reported. Yan et. al. reported a cesium-formamidinium-based 3D PSC with a champion efficiency of 16.23% which was fabricated by using room-temperature sputtered NiO_x as an HTL [1]. Another report was by Seok et. al. which showcased $MAPbI_3$-based PSC using sputtered NiO_x as an HTL demonstrating 15.60% champion efficiency for cells [2]. The transport layers used are often optimized as per 3D perovskites and low-dimensional PSCs are fabricated using the same. It is worth noticing that the properties of low-dimensional perovskites differ from their 3D counterparts, so tuning the recipes is worth thinking. Here, we demonstrate the importance of deposition power for Pulsed DC magnetron sputtered NiOx thin films. The films of equal thickness are deposited at different powers and are used as an HTL for PSCs. The cells are characterized by light and dark current-voltage measurements to understand NiOx-based device performance variations. The impact of transparency, defect density, and other factors is seen in device parameters and an attempt is made to justify device performance by using them.

2 FABRICATION AND CHARACTERIZATION

NiOx thin films are deposited by pulsed DC magnetron sputtering at deposition powers 60W, 75W, and 90W. The base pressure is maintained at 1e-6 mbar, and deposition is done in the presence of Ar and O_2 gas at 200 °C. The films are then annealed at 300 °C for 30s and quenched. $(PEA)_2MA_4Pb_5I_{16}$ (N5) is used as a perovskite absorber which is spin-coated by using a 1.8M solution at

6000 rpm for 45s [3]. The spin-coated films are then annealed at 70 °C for 15 mins. PCBM (20 mg/ml in chlorobenzene), used as an electron transport layer, is spin-coated at 2000 rpm for 60s. BCP (0.5 mg/ml in isopropanol) is then spin-coated at 6000 rpm for 60s followed by 100 nm silver deposition at 1e-6 mbar pressure. The fabricated cells have an active area of 0.045 cm^2.

The thin film morphology is recorded using Carl Zeiss Ultra-55 Field Emission Scanning Electron Microscope (FESEM). The transmission spectrum of NiO_x thin films is recorded using UV-Vis NIR Perkin Elmer LAMBDA 1050 in the range of 250 to 900 nm. Photoluminescence spectra are obtained using LabRAM HR using a 532 nm source laser. Edinburgh instruments-FLS 1000 spectrometer is used to get Time-Resolved Photoluminescence (TRPL) spectrum for carrier lifetime analysis. Light and dark J-V measurements are done using Sol3A Class AAA Solar Simulator.

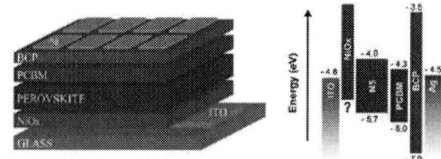

Fig. 1 Device architecture and band alignment diagram

3 RESULTS AND DISCUSSIONS

The topographic images of N5 deposited on NiO_x are taken and shown in Fig. 2. There is no change in the morphologies of perovskite. The low dimensional perovskites do not form a grain-like structure due to which any changes in grain sizes cannot be accounted for. The films have good coverage with a few non-uniformities. The transmission spectrum of NiO_x is one of the most essential parameters for an HTL. The transmission spectrum of NiO_x deposited on three different powers and post-annealed is shown in Fig. 3. It is worth noting that the percentage transmission decreases with increased deposition power even though the thickness of the films is the same. This highlights that with variation in power, the stoichiometry changes, and hence the films transmit differently. [4]

41st European Photovoltaic Solar Energy Conference and Exhibition

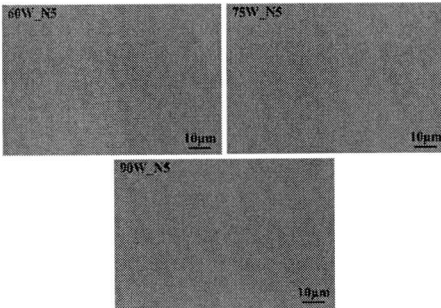

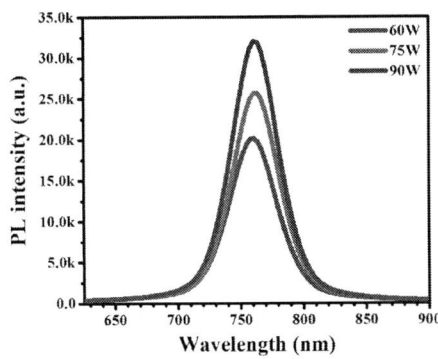

Fig. 2 SEM images of N5 on NiO$_x$

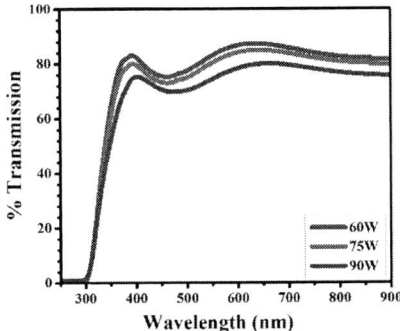

Fig. 3 Transmission spectrum of NiO$_x$ deposited on three different powers

From this data, we expect that the films at 60W should give the highest short-circuit current (J_{sc}). The photoluminescence (PL) spectrum of perovskite deposited on NiO$_x$ is shown in Fig. 4. The PL intensity is highest for perovskite deposited on 90W NiO$_x$ and is lowest for 60W. A decrease in PL intensity can be a signature of either better extraction or trapping of carriers at the interface due to defects. It is difficult to deduce the exact mechanism responsible for the decreased intensity just from PL. Hence, we do TRPL (Time-Resolved Photoluminescence) to get the extractive lifetimes for photogenerated carriers.

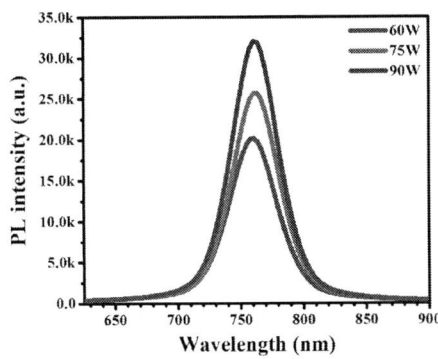

Fig. 4 PL spectrum for perovskite deposited on NiO$_x$

The lifetime decay curves of perovskite on HTL are fitted by a biexponential fit which gives us extractive and SRH lifetimes for carriers. The extractive lifetimes are comparable for all three cases but the SRH lifetime is highest for 75W deposition power. This results in an increase in the effective carrier lifetime for perovskite deposited on 75W NiO$_x$. An increase in SRH lifetime for carriers indicates that the interface is comparatively better

for 75W than 60W and 90W. Thus, we can say that the 75W deposited NiO$_x$ with the highest effective carrier lifetime should give us better efficiency for devices.

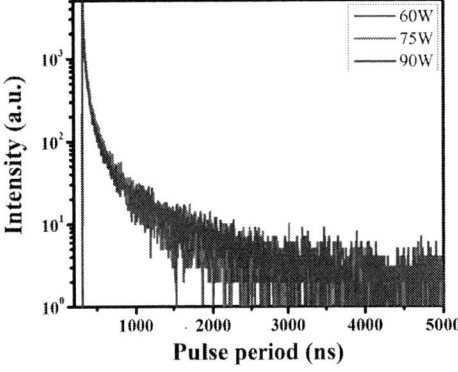

Fig. 5 TRPL lifetime decay curves of perovskite on HTL

Table 1. Carrier lifetime for three different stacks

Stack	τ_1 (ns)	$A_1\%$	τ_2 (ns)	$A_2\%$	τ_{eff} (ns)
60W	32	94	320	6	138±6
75W	38	95	453	5	191±5
90W	32	95	400	5	174±5

The devices are fabricated with NiO$_x$ deposited at three different powers. The dark and light J-V (current density-voltage) characteristics are measured and shown in Fig. 6.

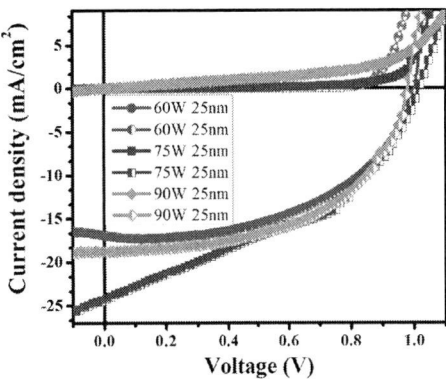

Fig. 6 J-V dark and light characteristics for low-dimensional perovskite solar cells

Table 2. Device parameters for the three device variations

NiO$_x$ deposition power	J_{sc} (mA/cm^2)	V_{oc} (V)	FF (%)	PCE (%)
60W	16.94	1.00	53.5	9.07
75W	23.28	1.01	44.74	10.52
90W	18.78	0.97	54.8	10.06

The short-circuit current (J_{sc}) is the lowest for 60W films contrary to what was expected due to the highest transmission. A possible reason for this could be the lowest effective lifetime (lowest SRH lifetime) due to

415

which carrier might be undergoing non-radiative recombination. The 90W devices have higher J_{sc} but the highest is obtained in 75W films as expected due to TRPL results. The dark characteristics is also better for 75W devices. The V_{oc} (Open-circuit voltage) of 75W devices is marginally better when compared to the other two. The Fill factor (FF) is highest for 90W devices. The FF of 75W devices is lower than 90W for which the reason is yet to be understood. A cumulative of these three device parameters suggests that 75W has the highest efficiency of 10.52%. Close to this is 90W power at 10.06%. These results suggest that higher deposition powers of NiO_x serve well for low-dimensional perovskite solar cells.

4 CONCLUSION

The low-dimensional perovskites have properties differing from their 3D counterparts due to which the optimization of transport layers according to their band alignment is essential. Here we give a comparative of three different deposition powers for NiO_x via pulsed DC magnetron sputtering. The material characterizations of both NiO_x and perovskite deposited on it are analyzed which suggests 75W to be better than 60W and 90W. The change in deposition power changes the ratio of Ni^{2+} and Ni^{3+} which is crucial in determining the defect density and conductivity of NiO_x. This also impacts the transparency of HTL due to which a delicate balance of composition becomes essential in improving device performance. The devices show results in line with this, with the highest efficiency of 10.52%. While 90W devices perform closely to the 75W ones with an efficiency of 10.06%, we can conclude that higher deposition powers work better for low-dimensional perovskite (N5).

5 REFERENCES

[1] M. Yan et al., "Room-temperature Sputtered NiOx for hysteresis-free and stable inverted Cs-FA mixed-cation perovskite solar cells," Mater Sci Semicond Process, vol. 115, Aug. 2020, doi: 10.1016/j.mssp.2020.105129.
[2] H. J. Seok et al., "Transition of the NiOxBuffer Layer from a p-Type Semiconductor to an Insulator for Operation of Perovskite Solar Cells," ACS Appl Energy Mater, vol. 4, no. 6, pp. 5452–5465, Jun. 2021, doi: 10.1021/acsaem.1c00049.
[3]] B. Sharma, V. Pawar, S. Patil, and S. Avasthi, "Effect of Ionic liquids on 2D-3D Lead-based Perovskite Solar Cells," in 2022 IEEE International Conference on Emerging Electronics, ICEE 2022, Institute of Electrical and Electronics Engineers Inc., 2022. doi: 10.1109/ICEE56203.2022.10117645.
[4] P. Salunkhe, M. A. Muhammed, and D. Kekuda, "Structural, spectroscopic and electrical properties of dc magnetron sputtered NiO thin films and an insight into different defect states," Appl Phys A Mater Sci Process, vol. 127, no. 5, May 2021, doi: 10.1007/s00339-021-04501-0.

41st European Photovoltaic Solar Energy Conference and Exhibition

Funded by
the European Union
NextGenerationEU

Analisys of reverse-bias stability of FAPbBr₃ semi-transparent perovskite solar cells

Noah Tormena[1], Alessandro Caria[1], Matteo Buffolo[1], Carlo De Santi[1], Nicola Trivellin[1,2], Andrea Cester[1], Gaudenzio Meneghesso[1], Enrico Zanoni[1], Fabio Matteocci[4,5], Aldo di Carlo[4,6], Matteo Meneghini[1,3].

1. Department of Information Engineering, University of Padova, Padova 35131, Italy.
2. Department of Industrial Engineering, University of Padova, Padova 35131, Italy.
3. Department of Physics and Astronomy, University of Padova, Padova 35131, Italy.
4. Department of Electronic Engineering, University of Rome "Tor Vergata" | UNIROMA2.
5. CHOSE – Centre for Hybrid and Organic Solar Energy.
6. CNR-ISM – Institute for Structure of Matter, National Research Council.

❖ INTRODUCTION: REVERSE-BIAS STABILITY

Stability against reverse-bias operating conditions is crucial in ensuring adequate real-world reliability of current and next-generation PV technology and devices.
Reverse-bias can occur as:
- An unwanted sporadic/systematic condition, when cell shading occurs in a shaded PV module.
- A deliberate systematic condition, when a cell is used in photodetection applications.

❖ TESTED SAMPLES:

FAPbBr₃ SEMI-TRANSPARENT PEROVSKITE SOLAR CELLS

- High EQE (>75%)
- High transmittance (>75%) outside of absorption range (i.e. visible/IR)
  → agrivoltaics / multi-junction cells

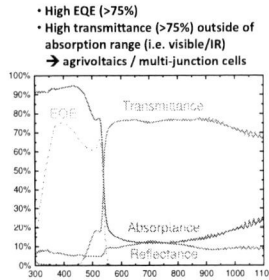

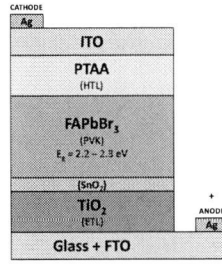

❖ REVERSE-BIAS STEP-STRESS TESTING

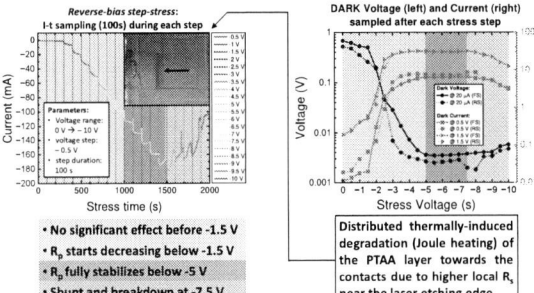

- No significant effect before -1.5 V
- Rₚ starts decreasing below -1.5 V
- Rₚ fully stabilizes below -5 V
- Shunt and breakdown at -7.5 V

Distributed thermally-induced degradation (Joule heating) of the PTAA layer towards the contacts due to higher local Rₛ near the laser etching edge

❖ CONSTANT-BIAS STRESS TESTING AT NEGATIVE VOLTAGE

Stress test parameters:
- 10000 s stress duration, 10 steps
- 10000 s recovery monitoring, 10 steps

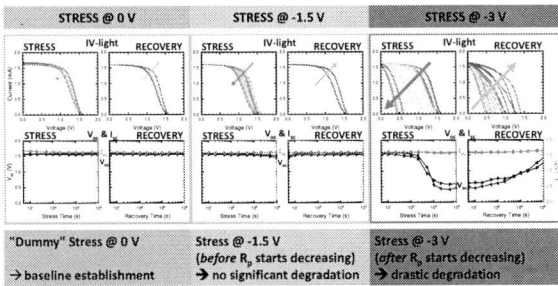

"Dummy" Stress @ 0 V	Stress @ -1.5 V (before Rₚ starts decreasing)	Stress @ -3 V (after Rₚ starts decreasing)
→ baseline establishment	→ no significant degradation	→ drastic degradation

- Characterization measurements themselves affect characteristics and behavior of the cell.
- Degradation involves mainly the open-circuit voltage (affecting all other related parameters).
- The intensity and duration of reverse-bias determine the occurrance and extend of degradation.
- Full recovery was systematically observed, although occurring with a higher time-constant.

❖ PARALLEL RESISTANCE

Parallel resistance (Rₚ) gradual decrease by ~3÷4 orders of magnitude
→ Distributed degradation phenomenon (not localized to a single spot)
→ Not "full"/"regular" shunting (ergo, rectified I-V curve and fully resistive behavior)

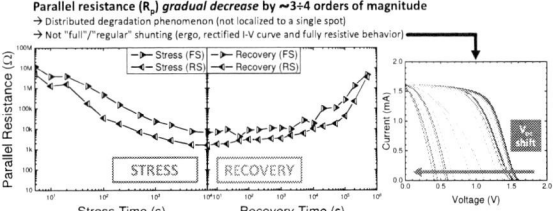

❖ CORRELATING HYSTERESIS WITH Vₒᶜ KINETICS

- Three consecutive tests (stress phase + recovery monitoring phase) on the same cell

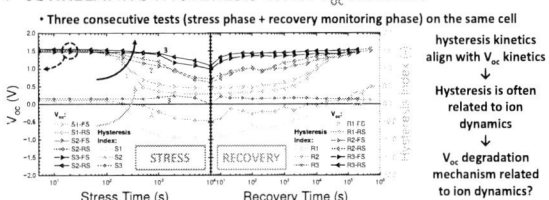

hysteresis kinetics align with Vₒᶜ kinetics
↓
Hysteresis is often related to ion dynamics
↓
Vₒᶜ degradation mechanism related to ion dynamics?

❖ AUXILIARY TESTING: OPEN-CIRCUIT VOLTAGE DECAY

Open-Circuit Voltage Decay

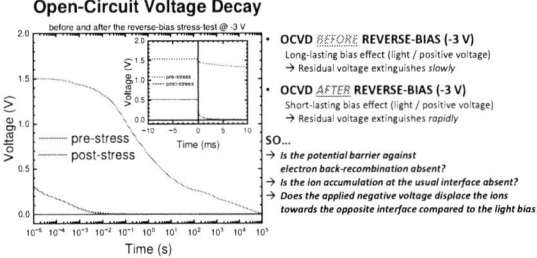

- OCVD *BEFORE* REVERSE-BIAS (-3 V)
  Long-lasting bias effect (light / positive voltage)
  → Residual voltage extinguishes slowly
- OCVD *AFTER* REVERSE-BIAS (-3 V)
  Short-lasting bias effect (light / positive voltage)
  → Residual voltage extinguishes rapidly

SO...
→ Is the potential barrier against electron back-recombination absent?
→ Is the ion accumulation at the usual interface absent?
→ Does the applied negative voltage displace the ions towards the opposite interface compared to the light bias?

❖ INTERPRETATION AND WORKING-THEORY

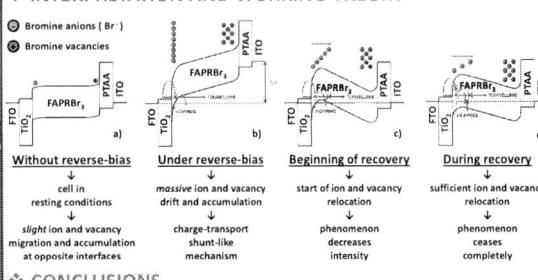

Without reverse-bias	Under reverse-bias	Beginning of recovery	During recovery
↓	↓	↓	↓
cell in resting conditions	massive ion and vacancy drift and accumulation	start of ion and vacancy relocation	sufficient ion and vacancy relocation
↓	↓	↓	↓
slight ion and vacancy migration and accumulation at opposite interfaces	charge-transport shunt-like mechanism	phenomenon decreases intensity	phenomenon ceases completely

❖ CONCLUSIONS

Reverse-bias on PSCs can cause temporary degradation.
- Degradation is caused by ion-related dynamics.
- Degradation mainly affects Vₒᶜ.
- Degradation occurrance and extent depend on reverse-bias intensity and duration.
- Degradation affects the entire surface area of the cell (non-localized).
- Degradation is reversible.

Acknowledgements:
NEST - NETWORK 4 ENERGY SUSTAINABLE TRANSITION Spoke 1: Solar: PV, CSP, CST. This study was developed in the framework of the research activities carried out within the Project "Network 4 Energy Sustainable Transition – NEST", Spoke 1, Project code PE0000021, funded under the National Recovery and Resilience Plan (NRRP), Mission 4, Component 2, Investment 1.3 – Call for tender No. 1561 of 11.10.2022 of Ministero dell'Università e della Ricerca (MUR); funded by the European Union – NextGenerationEU.

UNIVERSITÀ DEGLI STUDI DI PADOVA

DIPARTIMENTO DI INGEGNERIA DELL'INFORMAZIONE

41st European Photovoltaic Solar Energy Conference and Exhibition

ENHANCING EFFICIENCY AND STABILITY OF CSPBI₃ PEROVSKITE QUANTUM DOTS THROUGH CO²⁺ DOPING

Pouriya Naziri[1, 2*], Naeimeh Sadat Peighambardoust[2,3], and Umut Aydemir[2, 3]
1 Graduate School of Sciences and Engineering, Koç University, Istanbul, Turkey
2 Koç University Boron and Advanced Materials Application and Research Center (KUBAM), Istanbul, Turkey
3 Department of Chemistry, Koç University, Istanbul, Turkey
*E-mail of the corresponding author: pnaziri22@ku.edu.tr

ABSTRACT: Inorganic $CsPbI_3$ quantum dots (QDs) have emerged as a promising material for optoelectronic applications, especially in photovoltaics and light-emitting diodes (LEDs), due to their superior photoluminescence quantum yield (PLQY) and high defect tolerance. However, their inherent instability under environmental factors such as light, heat, and moisture remains a critical challenge. To address this, we explored the doping of $CsPbI_3$ QDs with Co^{2+} ions via CoI_2 and $CoCl_2$ sources. The dual role of iodide (I^-) and chloride (Cl^-) ions from the dopants in a mixed doped sample effectively filled iodide vacancies and reduced nonradiative recombination, significantly enhancing the PLQY from 85 % to near 99%. Additionally, the stability of these QDs was improved, maintaining their structural integrity under UV light exposure for over two months. Our research, supported by findings from the doped $CsPbI_3$ QDs, underscores the potential of Co^{2+} doping to improve both the efficiency and longevity of $CsPbI_3$-based devices, positioning them as strong candidates for advanced optoelectronic applications.
Keywords: $CsPbI_3$, perovskite quantum dots, cobalt doping, photovoltaics, LED, stability

1 INTRODUCTION

Inorganic cesium lead halide perovskites, specifically $CsPbX_3$ QDs, have garnered substantial interest in recent years due to their outstanding optoelectronic properties, making them ideal candidates for a wide range of applications such as solar cells, light-emitting diodes (LEDs), displays, and photodetectors [1]. The unique combination of tunable bandgaps, high (PLQY > 90%), and sharp, tunable emission spectra, along with their low-cost synthesis routes, positions them as highly promising materials for next-generation devices [2-4].

Among these materials, $CsPbI_3$ has shown particular promise due to its appropriate bandgap for photovoltaic applications (approximately 1.73 eV in the black α-phase). However, the performance of these QDs in practical applications is hindered by their inherent instability [3,5]. Under environmental stressors such as heat, light, and moisture, $CsPbI_3$ tends to transition from the desirable black cubic α-phase to a yellow, non-perovskite δ-phase, which severely degrades their optoelectronic properties [6-8]. Therefore, improving the stability of $CsPbI_3$ while maintaining its performance characteristics remains a crucial challenge in the development of perovskite-based technologies [3,11].

Various strategies have been explored to enhance the stability of $CsPbI_3$ QDs, such as surface modification, dimensionality reduction, and encapsulation using polymers or inorganic oxides [5,6]. While these approaches have achieved partial success, each comes with certain limitations. For instance, surface modifications may impede carrier mobility, reducing the overall device efficiency. Similarly, dimensionality reduction may negatively affect crystallinity, and encapsulation can present challenges in preserving the functional properties of the QDs [10].

A promising alternative is doping at the Pb-site with metal cations [5]. Metal-ion doping, particularly at the A- or B-site, has been shown to stabilize the phase structure of perovskite QDs without sacrificing their electronic properties [2,9]. Several studies have demonstrated the effectiveness of Pb-site doping using ions such as Sn^{2+},

Cd^{2+}, Zn^{2+}, Sr^{2+}, Mn^{2+}, Ni^{2+}[8], [9], [10], [11], and Ag^+[12], leading to improvements in both stability and performance. These dopants can increase structural integrity, reduce the formation of non-perovskite phases, and optimize optoelectronic properties for practical applications [4,12].

In this work, we investigate the role of Co^{2+} doping in stabilizing $CsPbI_3$ QDs. By incorporating cobalt ions into the perovskite structure, along with iodide (I^-) and chloride (Cl^-) ions for vacancy filling, we aim to address the phase instability that commonly plagues $CsPbI_3$ QDs. Our findings show that Co^{2+}-doped $CsPbI_3$ QDs exhibit a significantly enhanced PLQY, reaching up to approximately 99%, compared to 85% in undoped samples. Furthermore, these QDs demonstrate remarkable stability under UV light for over two months, positioning them as strong candidates for use in PV and LED technologies. This work underscores the potential of synergistic dopants to both stabilize and enhance the performance of metal-halide perovskite nanomaterials, opening new pathways for their application in next-generation optoelectronic devices.

2 EXPERIMENTAL

2.1 Materials

All chemicals were used without further purification. Cesium carbonate (Cs_2CO_3), lead(II) iodide (PbI_2), cobalt chloride ($CoCl_2$), cobalt iodide (CoI_2), oleic acid (OA), oleylamine (OAm), hexane, and 1-octadecene (ODE) were obtained from Sigma-Aldrich..

2.2 Synthesis of Cs-Oleate and Pb-Oleate

To prepare the Cs-oleate precursor, 0.4 g of Cs_2CO_3 was dissolved in 40 mL of ODE with 1.6 mL of OA. The mixture was degassed at 120°C under vacuum for 30 minutes, followed by N_2 purging. This solution was kept at 100°C until needed. In a similar setup, 0.8 g of PbI_2 was dissolved in 40 mL of ODE at 120°C for 1 hour under vacuum. After degassing, OA (4 mL) and OAm (4 mL) were injected. To incorporate Co^{2+} ions into the $CsPbI_3$ quantum dots, the prepared $CoCl_2$ or CoI_2 (in OAm) was injected into the Pb-oleate solution. Once the Cs-oleate solution was prepared, it was rapidly injected into the Pb-oleate solution while the reaction temperature was quickly elevated to 170°C to initiate the formation of CsPbI₃ quantum dots. The sudden increase in temperature, along with the swift injection, ensures controlled nucleation of the quantum dots, a critical step in determining their size and uniformity. The reaction mixture was then maintained at this temperature for a short period to allow for the growth of the nanocrystals.

After the growth phase, the mixture was promptly cooled by immersion in an ice bath, effectively halting further growth of the quantum dots. Rapid cooling is essential to stabilize the nanocrystals, prevent overgrowth, and ensure the desired optical properties and uniform particle size are achieved. This precise control over the temperature profile during both the injection and cooling steps plays a pivotal role in achieving high-quality quantum dots with uniformity and stable optoelectronic properties.

2.3 Purification of QDs

The synthesized CsPbI₃ quantum dots (QDs) were subjected to a thorough purification process to remove unreacted precursors, byproducts, and excess ligands. First, the crude QD solution was washed by adding a nonpolar solvent like hexane, followed by methyl acetate, a polar solvent, to facilitate the separation of QDs from the byproducts.

The mixture was then centrifuged at 8000 rpm for 10 minutes to separate the precipitate, where the QDs settle at the bottom. The supernatant, containing unwanted byproducts and solvents, was carefully discarded. The QD precipitate was then redispersed in hexane, followed by another round of purification, this time centrifuging at 4000 rpm for 5 minutes to ensure a stable colloidal solution free from impurities. This step was repeated several times to achieve highly purified QDs with minimal surface ligands and byproducts, ensuring better optical and electrical properties

3 RESULTS AND DISCUSSION

3.1 Structural and Morphological Analysis

The structural and elemental composition of the Co²⁺-doped CsPbI₃ QDs were examined using Transmission Electron Microscopy (TEM) and Energy Dispersive X-ray Spectroscopy (EDS). As seen in Figure 1b, the TEM image reveals that the synthesized QDs have a uniform and well-ordered cubic structure. The average size of the QDs is approximately 10 nm, consistent with the reported sizes of perovskite quantum dots. Despite Co²⁺ doping, the lattice structure appears undisturbed, indicating that the doping process does not disrupt the crystal morphology and result in a distinctive distribution. Additionally, the EDS spectra in Figure 1a confirm the successful incorporation of cobalt ions along with Cs, Pb, and I into the QD structure, showing distinct peaks corresponding to these elements. This highlights the homogeneity of the doping and its effective integration into the QD matrix, as the elemental distribution maps illustrate a uniform distribution of dopant.

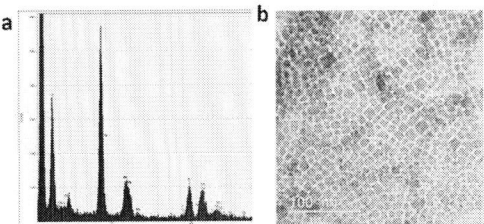

Figure 1: (a) EDS spectra confirming the elemental composition of Co²⁺-doped CsPbI₃ QDs. Peaks corresponding to Cs, Pb, I, Co, and Cl indicate successful doping and the presence of necessary elements.
(b) TEM image showing the crystalline structure of Co²⁺-doped CsPbI₃ QDs with an average size of approximately 10 nm.

The introduction of Cobalt ions plays a crucial role in enhancing the phase stability of the QDs. The structural integrity of these materials is particularly important when considering their application in optoelectronic devices, as maintaining stability at the nanoscopic level can prevent the transition to less favorable phases, such as the yellow δ-phase in CsPbI₃. By stabilizing the crystal structure and minimizing structural defects, Cobalt doping contributes to improved device performance and prolonged operational lifetime.

3.2 Optical Properties and Band Gap Evolution

The optical properties and stability of Co²⁺-doped CsPbI₃ quantum dots (QDs) were carefully monitored using UV-vis absorption spectroscopy. Figures 2a and 2b illustrate the Tauc plots for both the pristine and mixed CoI₂/CoCl₂-doped QDs, showing a slight blue shift in absorption over time. This shift is indicative of a corresponding change in the material's optical bandgap, a critical factor for optoelectronic applications. Initially, the pristine QDs exhibited a bandgap value of approximately 1.820 eV, while the Co²⁺-doped samples showed a slightly higher bandgap of 1.832 eV.

Over the span of 42 days, both samples experienced a gradual decrease in bandgap values. However, the Co²⁺-doped QDs demonstrated significantly better retention of their bandgap properties compared to the pristine QDs. By day 42, the bandgap of the pristine sample had shifted more noticeably toward lower energies, whereas the Co²⁺-doped QDs maintained a more stable bandgap, dropping only slightly over time. The gradual reduction in the bandgap values of the doped QDs suggests that the Co²⁺ dopants play a crucial role in mitigating the degradation processes typically seen in perovskite QDs. This enhanced stability is essential for practical applications, especially in photovoltaic devices, where long-term stability and consistent performance are critical.

The dual CoI₂/CoCl₂ doping strategy offers several benefits for maintaining optical properties over time. By reducing nonradiative recombination and minimizing shifts in the absorption spectra, the doping approach stabilizes the QDs' structure and improves their resistance to environmental factors, such as moisture and light exposure. These findings highlight the potential of Co²⁺-doped CsPbI₃ QDs as viable candidates for stable, high-performance optoelectronic devices.

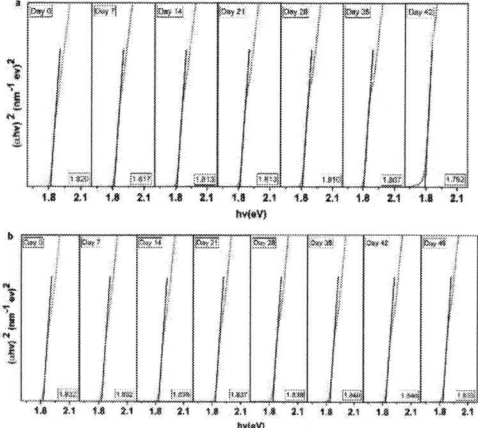

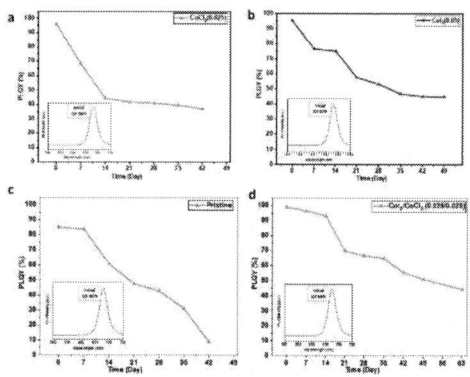

Figure 2: (a) Tauc plots for the pristine CsPbI₃ QDs measured over 42 days, showing a slight blue shift in the absorption edge.
(b) Tauc plots for CoI₂/CoCl₂ doped CsPbI₃ QDs measured over 49 days, exhibiting stable bandgap retention compared to pristine samples.

Figure 3: (a) PLQY stability for CoCl₂-doped CsPbI₃ QDs, with a gradual decrease over time.
(b) PLQY stability for CoI₂-doped CsPbI₃ QDs, showing a slower rate of degradation compared to pristine samples.
(c) PLQY degradation in pristine CsPbI₃ QDs, which drops significantly within 42 days.
(d) PLQY stability for dual CoI₂/CoCl₂-doped QDs, maintaining high PLQY even after 49 days.

3.3 PLQYs and Stability tests:

The stability of the PLQY is a critical metric for evaluating the suitability of perovskite QDs in optoelectronic applications such as PV devices and LEDs. Figures 3a-d track the PLQY of pristine, CoCl₂-doped, CoI₂-doped, and dual-doped QDs over time. Pristine QDs exhibit a rapid drop in PLQY from an initial 85% to less than 30% within 42 days, reflecting their inherent instability. In contrast, the Co²⁺doped samples show significantly improved PLQY stability. Notably, the dual-doped (CoI₂/CoCl₂) QDs (Figure 3d) retain approximately 80% of their initial PLQY after 49 days, demonstrating exceptional long-term stability.

This improvement in PLQY stability is attributed to the ability of Co²⁺ ions to passivate surface defects, which are a major source of nonradiative recombination in perovskite QDs. By filling vacancies at iodide sites and enhancing lattice integrity, Co doping reduces the number of trap states that lead to the loss of photo-generated carriers. The result is a more stable PLQY over time, making these doped QDs highly suitable for use in optoelectronic devices that require both high efficiency and long-term performance.

4 CONCLUSION

This study demonstrated that Co²⁺ doping and I⁻/Cl⁻ ion passivation in a dual doped CoI₂/CoCl₂ sample significantly enhance the stability and optical performance of CsPbI₃ QDs. TEM and EDS confirmed that Co²⁺ doping maintains the QDs' structure, while optical measurements showed long-term stability of the bandgap. PLQY remained high for over 49 days in doped QDs, greatly surpassing the performance of pristine QDs. These results highlight the potential of Cobalt doped CsPbI₃ QDs for use in stable and efficient photovoltaic and LED applications.

5 REFRENCES

[1] K. Galkowski et al., "Determination of the exciton binding energy and effective masses for methylammonium and formamidinium lead tri-halide perovskite semiconductors," Energy Environ. Sci., Vol. 9, 2016, pp. 962-970.
[2] A. Swarnkar et al., "Quantum dot-induced phase stabilization of α-CsPbI3 perovskite for high-efficiency photovoltaics," Science, Vol. 354, No. 6308, 2016, pp. 92-95.
[3] J. Yuan et al., "Non-Fullerene Molecules: Hybrid Perovskite Quantum Dot/Non-Fullerene Molecule Solar Cells with Efficiency Over 15%," Adv. Funct. Mater., Vol. 31, 2021, p. 2170196.
[4] T. Wang et al., "Controllable transient photocurrent in photodetectors based on perovskite nanocrystals via doping and interfacial engineering," J. Phys. Chem. C, Vol. 125, 2021, pp. 5475-5484.
[5] L. Dou et al., "Solution-processed hybrid perovskite photodetectors with high detectivity," Nat. Commun., Vol. 5, 2014, p. 5404.
[6] Y. Wang et al., "Emerging new-generation photodetectors based on low-dimensional halide perovskites," ACS Photonics, Vol. 7, No. 1, 2020, pp. 10-28.
[7] L. Xu et al., "All-inorganic perovskite quantum dots as light-harvesting, interfacial, and light-converting layers toward solar cells," J. Mater. Chem. A, 2021.
[8] W. Van der Stam et al., "Highly emissive divalent-ion-doped colloidal CsPb1-xMxBr3 perovskite nanocrystals through cation exchange," J. Am. Chem. Soc., Vol. 139, No. 11, 2017, pp. 4087-4097.
[9] M. Lu et al., "Simultaneous strontium doping and chlorine surface passivation improve luminescence intensity and stability of CsPbI3 nanocrystals enabling efficient light-emitting devices," Adv. Mater., Vol. 30, No. 50, 2018, p. 1804691.
[10] Z. J. Yong et al., "Doping-enhanced short-range order of perovskite nanocrystals for near-unity violet luminescence quantum yield," J. Am. Chem. Soc., Vol. 140, No. 31, 2018, pp. 9942-9951.
[11] A. De, N. Mondal, and A. Samanta, "Luminescence tuning and exciton dynamics of Mn-doped CsPbCl3

nanocrystals," Nanoscale, Vol. 9, No. 43, 2017, pp. 16722-16727.

[12] D. Ghosh et al., "Heterovalent substitution in mixed halide perovskite quantum dots for improved and stable photovoltaic performance," J. Phys. Chem. C, Vol. 125, No. 10, 2021, pp. 5485-5493.

41st European Photovoltaic Solar Energy Conference and Exhibition

STANDARDIZED TEST ROUTINES FOR THE ASSESSMENT OF POTENTIAL INDUCED DEGRADATION OF PEROVSKITE SOLAR CELLS

Beyza Durusoy[1], David Adner[2], Christian Hagendorf[3], Konrad Wojciechowski[4], Samy Almosni[4], Marko Turek[5]

[1] Middle East Technical University, Ankara, Türkiye
durusoy.beyza@metu.edu.tr
[2] Martin-Luther-Universität Halle-Wittenberg, Germany
[3] HS Anhalt University of Applied Sciences, Saxony-Anhalt, Germany
[4] Saule Technologies, Wroclaw, Poland
[5] Fraunhofer Center for Silicon Photovoltaics CSP, Halle, Germany

ABSTRACT: Potential induced degradation (PID) is a well-studied and understood phenomenon for crystalline silicon solar cells. The most common PID of the shunting type (PID-s) in silicon solar cells is caused by sodium ions' drift and diffusion processes, causing an increased number of local shunts. Based on this knowledge, standard PID tests and equipment have been developed. On the other hand, the physics of perovskite solar cell degradation and the behavior in PSCs under similar stress conditions remains less explored. Therefore, we investigated the applicability of the standard PID tests to three distinct perovskite samples utilizing the commercial PIDcon Bifacial Device. We also checked the effect of the polarity of voltage bias, which helped us to distinguish temperature-induced and voltage-induced effects. The results show that not all PSCs behaved uniformly under PID stress; some displayed improved fill factors, suggesting an enhancement in performance rather than degradation. The varying responses also show the effect of composition on PID resilience. While our findings confirm the promising nature of PSCs' resilience to PID, optical and structural analysis are necessary to pinpoint the exact PID mechanism in PSCs.

Keywords: perovskite tandem, potential induced degradation, PID, perovskite solar cells

1 INTRODUCTION

Perovskite solar cells (PSCs) have shown great promise in the solar energy sector due to their potential to dominate the market. PSCs are distinct from conventional solar cells in several ways. The perovskite material used in PSCs has a wider bandgap than silicon-based solar cells, allowing them to absorb a broader range of light and achieve higher conversion efficiency [1]. They can also be fabricated using cheaper materials and a more straightforward process than silicon-based solar cells. Most importantly, they have tunable optical properties, which offer a wide range of options to reach higher efficiencies, especially making them highly desirable for tandem cell applications [2].

Perovskite/silicon tandem solar cells have recently achieved certified record efficiencies of 34.6%[3], surpassing some established solar cell technologies. However, PSCs face stability challenges primarily due to the nature of the perovskite material [4]. Potential induced degradation is expected to be a challenge for PSCs as well. Previous research has primarily focused on the stability of PSCs under elevated temperature and irradiation, with limited exploration specifically concerning PID [5]. Only a few publications highlight the degradation of PSCs under high-voltage stress. For instance, studies by Carolus et al. reported performance degradation of up to 95% under a 1000V stress [6], while Brecl et al. observed degradation under negative bias but not for positively biased devices [7]. The variation in results can be due to technological disparities and non-standardized PID testing procedures for PSCs. Developing new perovskite tandem solar cells requires testing a wide variety and combination of absorber and charger carrier transport materials. The corresponding development cycles strongly depend on the complexity and duration of tests to be performed for each material combination. Therefore, the implementation of standardized PID testing is crucial in this process.

Our study addresses the susceptibility of three distinct PSC architectures and compositions to accelerated PID testing by employing a commercially available PIDcon test device developed by Freiburg Instruments as a standard tool for PID-testing solar cells. It allows the application of uniform stress conditions: a high voltage bias of 1000V at 60°C for 4 hours. We also investigated the influence of bias polarity on PID effects by analyzing an additional sample set exposed to a -1000V bias. Monitoring PID in perovskite solar cells requires a modified approach as they have lower parallel resistance values than conventional solar cells. Unlike silicon solar cells, where a conductance increase due to shunting indicates PID-s, we had to configure the device to stress the perovskite solar cell and monitor electrical parameters to provide insights into PID effects. Due to the unknown physical nature of potential induced degradation in PSCs, the changes in the shunt resistance might not be a relevant indicator of PID in the device. Therefore, we performed four-wire current-voltage measurements using a solar simulator to analyze the changes in electrical parameters of the samples before and after the PID. The material-specific nature of PID and the need for standardized test routines to assess the PID effect for PSC modules highlight the importance of this research.

2 MATERIALS AND METHODOLOGY

2.1 Sample Details

We investigated two different sample sets from Fraunhofer ISE (Sample Set A) and Saule Technologies (Sample Set B), and we checked whether standardized testing predicts the long-term stability of perovskite modules even though their compositions and structures differ.

422

41st European Photovoltaic Solar Energy Conference and Exhibition

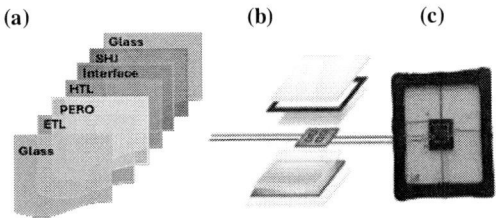

Figure 1. Schematics showing (a) the layers of perovskite silicon tandem modules (from ISE) with the top cell being perovskite and the bottom cell made of silicon heterojunction, (b) the encapsulation structure of the modules, and (c) an actual picture of the module

Sample Set A: These samples are perovskite tandem modules designed with the top cell perovskite and the bottom cell silicon heterojunction (SHJ). Each module has a total substrate area of 2.5x2.5 cm², consisting of 4 cells with an area of 0.25 cm². The layered structure of the module consisted of the following components: Glass, encapsulant sheet, Indium Tin Oxide (ITO) layer, SnOx layer, Electron Transport Layer (ETL), Perovskite absorber, Hole Transport Layer (HTL), interface layers, SHJ, and silver contacts (on both sides) for electrical connection. After the production of the modules, they were exposed to accelerated thermal tests, which resulted in an enhancement of the performance.

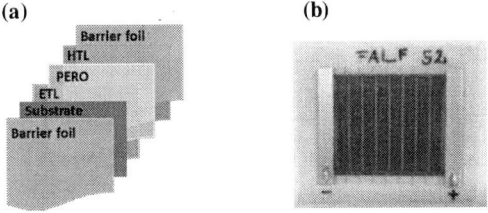

Figure 2. A schematic showing (a) the layers of perovskite modules (from Saule Tech), (b) an actual picture of the module.

Sample Set B: This set includes perovskite solar cells with two different precursor compositions. Each module has a total substrate area of 5.004 cm², consisting of 8 cells in series with an area of 0.556 cm². The layered structure of the module consisted of the following components: barrier layer, substrate, ETL, perovskite absorber, HTL, and the barrier layer.

Both samples arrived in fully laminated conditions, but we attached copper wires to the contacting points for the tests. We stored the samples in a dark nitrogen box between the measurements. After removing them from the nitrogen box, we repeated the I-V measurements to check whether the samples degraded during interim storage.

2.2 Methodology

Our methodology aims at comparing perovskite technologies to crystalline silicon technologies. We aim to employ a standardized test scenario consisting of 1kV, 60°, and 4 hours of stress testing using a widespread, commercially available test tool the PIDCon device developed in Freiburg Instruments. Table 1 shows the technical specifications of the PIDCon bifacial device.

Table 1. Technical specifications of the PIDCon Bifacial Device

Technical specifications	
Dimensions (W x H x D)	310 x 340 x 275 mm³
Electrode (100 x 100 mm)	0.1 to 1.5 kV
Temperature	Up to 150°C
Measurement Time	Minutes to days
Measured Parameters	meas. voltage [mV], meas. current [mA], parallel resistance [Ω], conductance, humidity [%], leakage current [μA], power loss [%]

The IEC has established PID standards for crystalline silicon solar modules. According to these standards, modules that experience less than 5% degradation when subjected to 96 hours at 60°C, 85% relative humidity, and 1kV external voltage are expected to not degrade due to PID during a five-year evaluation in the field [8]. The PIDcon Bifacial Device accelerates the testing and drops the test duration to 4 hours. During the PID testing, the PID software monitors the electrical parameters of the modules. If the conductance increase is greater than 0.2 mS/cm² at the end of the testing, then the module is said to have PID sensitivity; in other words, it failed the PID test. The conductance increase is the determining parameter of PID, as it is directly related to the occurrence of local shunts. If the conductance increases, then the parallel resistance of the cells/modules decreases, indicating that some shunt regions are forming. These local shunts then lead to dead cell areas; hence, the current and power output, and eventually the cell's lifetime, decreases. The mechanism explained here is the PID-shunting mechanism, which is the dominant PID effect seen in conventional silicon solar cells [9].

Unlike the conventional solar cells, we did not take the conductance increase as an indicator for the PID due to the lower resistance value of our perovskite samples. The device could not capture the slight variations in the resistance, and its primary role is to apply uniform test conditions and check the relative performance before and after the PID. Another reason is that the conductance increase implies the occurrence of local shunts, which indicates the PID-shunting effect. However, for the perovskite solar cells, the observed PID effects stated in the literature are not in correlation with the shunting but instead related to more complex and irreversible issues, such as phase segregation [10]. Thus, we used the PIDcon Bifacial Device as a PID stressor rather than a PID observer, and we performed four-wire current-voltage measurements using a solar simulator (SINUS 220 by Wavelabs) to indicate any degradation. Table 2 shows the measurement conditions for both sample sets. We took three runs of J-V measurements and averaged the results to ensure the accuracy and repeatability of the measurements.

Table 2. Measurement conditions for both sample sets.

Sample	J-V measurement	Pre-lamination	PID tests
Sample Set A	1.3V – (-0.2V) sweep range, 1V/s (10mV step, 0.01s or 0.1s delay parameters)	Under 100°C (in ISE, details unknown)	1kV, 60°C, 4 hours -1kV, 60°C, 4 hours
Sample Set B	9V – (-0.2V) sweep range, 1V/s (10mV step, 0.01s or 0.1s delay parameters)	(In Saule Tech, details unknown)	1kV, 60°C, 4 hours

3 RESULTS

We evaluated two sets of perovskite solar cell samples, denoted as sample sets A and B. Sample set A, which consisted of tandem perovskite solar modules, showed interesting fill factor (FF) trends under positive and negative voltage bias conditions.

Positive voltage bias testing (+1kV, 60°C, 4h) shows distinctive FF behavior for the sample set A. Initially, FF values averaged around 63.3% for Cell 4 (a random cell out of four cells in the module). After one hour of high-voltage stress (PID) testing, Cell 4's FF increased to 69.9%. The results are consistent with the thermal tests conducted in Fraunhofer ISE, which also showed performance enhancements. Further, three-hour testing maintained high FF values, indicating stability under positive voltage stress.

(a)

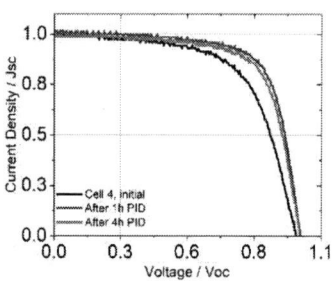

(b)

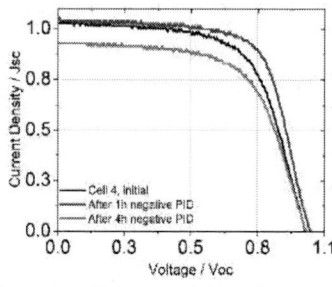

Figure 1: J-V curves of (a) cell_4 on a Pero-Si-Tandem module under positive PID stress (1kV, 60°C), (b) cell_4 on the same module under negative PID stress (-1 kV,

60°C). The samples were stored in a dark nitrogen box between PID tests with varying voltage bias polarity.

Under negative voltage bias conditions (-1kV, 60°C, 4h), the Cell 4 with averaged FF of 63.03% rose to 67.7% after one hour but it stabilized at around 62.5% at the end of 4h testing. Despite minor fluctuations, FF values remained stable, emphasizing resilience to negative voltage stress.

By switching the polarity of the applied voltage, we distinguished temperature-induced and voltage-induced effects. Performance enhancements during positive voltage bias tests can be due to the high temperature rather than the high voltage applied. On the other hand, exposing the sample to negative PID stress for four hours resulted in an FF drop. It increased after the first hour of the testing, showing the effect of temperature; however, at the end of the four-hour testing, the FF dropped by 0.5% compared to the initial value due to the high negative external voltage.

Sample set B, consisting of two types of perovskite solar modules with similar architecture but different precursor compositions ("F" and "O"), demonstrated unique responses. Before PID testing, sample type "F" exhibited an average FF of 63.96%. After PID testing, FF showed a slight decrease of approximately 1.1%, and Voc exhibited a noticeable increase, rising by approximately 2.6%. Although we see a decrease in FF, the PID effect is somewhat different from that of conventional solar cells, in which we often see Isc, Voc, and the FF decrease. The drops in these electrical parameters often indicate the occurrence of local shunts or polarization effects due to the PID.

Conversely, here, the Voc increase for sample type "F" can be due to the changes in bandgap or interfacial properties of the perovskite device. The high temperature and applied voltage can alter the device's internal properties, such as the crystal structure or the composition of the perovskite material. For example, a slight increase in the bandgap of the perovskite could lead to a higher Voc, or changes in the interfacial properties or charge-transport layers could enhance charge carrier collection, resulting in a higher Voc.

(a)

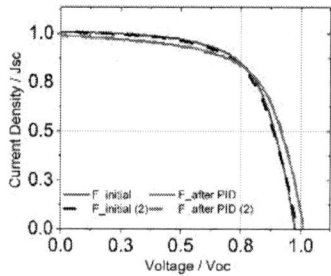

(b)

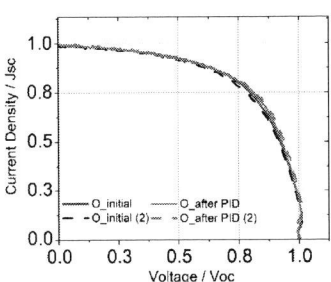

Figure 5. J-V curves of two distinct samples, (a) "F" and (b) "O" (Sample Set B, different precursor composition), before and after PID testing (1 kV, 60°1 kV, 60°,4h)..

Sample type "O" demonstrated an FF increase of around 2.5% after PID testing. While Voc exhibited a minor increase, Isc remained nearly unchanged. The notable increase in FF for sample type "O" indicates that the different precursor composition may have enhanced the device's tolerance to PID.

These results show the resilience of perovskite solar modules to PID and reveal that the observed PID effects are different from those of conventional solar cells. For understanding the underlying mechanisms causing these changes, further characterization studies, such as structural and optical characterization, are necessary.

4 CONCLUSIONS

Applying a standard PID-test protocol using a commercially available PID-test device gives substantial insights into a Perovskite's sensitivity to externally applied high potentials. The change of polarity, i.e., testing with +1kV and -1kV, allows separating voltage-induced effects from temperature-induced effects.

In our investigation, sample set 1 shows similar behavior in either case, implying that the observed changes in the device's fill factor can more certainly be attributed to the temperature than the application of the external voltage. Furthermore, unlike silicon solar cells, we found an improvement in the fill factor instead of degradation due to the stress test. The results can give some insights into process optimization, and they also show that PID test routines must be designed so that one can separate voltage-induced degradation from temperature-induced sample improvements.

Our second set of samples, particularly sample_F, showed a somewhat different behavior with a slight decrease in the fill factor and an increase in the voltage for one sample type. In contrast, sample_O with a different composition did not show these effects. The variation in the results shows the significance of the composition of the perovskite modules; even though their architecture is the same, the composition change made sample_O somewhat resistant to PID, potentially blocking ion diffusion between layers.

The findings of this study about the PID of PSCs have provided insights into the stability, resilience, and potential performance enhancement of these emerging PV devices. However, more exploration into the PID of PSCs is needed than their silicon counterparts. We showed that adapting PID test scenarios into PSCs regarding their structural and compositional differences is possible and can indicate degradation or performance enhancement.

We managed to distinguish temperature-induced and high-voltage effects. Exposing PSCs to high temperatures did not accelerate the degradation but enhanced the performance, especially after 1 hour of PID stress. We found that PSCs experience degradation in a way that alters their structure and composition. However, we must conduct a more detailed structural and optical analysis to pinpoint the exact mechanism. In conclusion, PSCs' response to PID is complex but promising, showing different outcomes than silicon solar cells. For PSCs to reach their full potential, more research and the establishment of standards are necessary.

References

[1] X. Dai, K. Xu, and F. Wei, "Recent progress in perovskite solar cells: the perovskite layer," *Beilstein J. Nanotechnol.*, vol. 11, pp. 51–60, 2020, doi: 10.3762/bjnano.11.5.

[2] A. W. Y. Ho-Baillie, J. Zheng, M. A. Mahmud, F. J. Ma, D. R. McKenzie, and M. A. Green, "Recent progress and future prospects of perovskite tandem solar cells," *Appl. Phys. Rev.*, vol. 8, no. 4, 2021, doi: 10.1063/5.0061483.

[3] "LONGi announces the new world record efficiency of 30.1% for the commercial M6 size wafer-level silicon-perovskite tandem solar cells," 2024. https://www.longi.com/en/news/is-m6-wafer-silicon-perovskite-tandem-cells-new-efficiency-record/ (accessed Sep. 23, 2024).

[4] A. B. Djurišić *et al.*, "Perovskite solar cells - An overview of critical issues," *Prog. Quantum Electron.*, vol. 53, no. June, pp. 1–37, 2017, doi: 10.1016/j.pquantelec.2017.05.002.

[5] H. J. Snaith and P. Hacke, "Enabling reliability assessments of pre-commercial perovskite photovoltaics with lessons learned from industrial standards," *Nat. Energy*, vol. 3, no. 6, pp. 459–465, 2018, doi: 10.1038/s41560-018-0174-4.

[6] J. Carolus *et al.*, "Potential-Induced Degradation and Recovery of Perovskite Solar Cells," *Sol. RRL*, vol. 3, no. 10, pp. 1–3, 2019, doi: 10.1002/solr.201900226.

[7] K. Brecl, M. Jošt, M. Bokalič, J. Ekar, J. Kovač, and M. Topič, "Are Perovskite Solar Cell Potential-Induced Degradation Proof?," *Sol. RRL*, vol. 6, no. 2, pp. 1–10, 2022, doi: 10.1002/solr.202100815.

[8] P. Hacke, "Establishment of a PID Pass/Fail Test for Crystalline Silicon Modules by Examining Field Performance for Five Years", 2017. [Online]. Available: https://www.nrel.gov/docs/fy18osti/70264.pdf

[9] W. Luo *et al.*, "Potential-induced degradation in photovoltaic modules: A critical review," *Energy and Environmental Science*. 2017. doi: 10.1039/c6ee02271e.

[10] Z. Purohit *et al.*, "Impact of Potential-Induced Degradation on Different Architecture-Based Perovskite Solar Cells," *Sol. RRL*, vol. 5, no. 9, pp. 1–7, 2021, doi: 10.1002/solr.202100349.

Acknowledgements

This project has received funding from the European Union's Horizon 2020 research and innovation program under grant agreement N°101006715. We want to especially thank Saule Technologies and Fraunhofer ISE for providing samples and collaboration, Viperlab for research facility support, and Fraunhofer CSP for hosting, alongside the expertise and support of the Diagnostics and Metrology Solar Cells division. B.D thanks the Scientific and Technological Research Council of Turkey (TÜBİTAK) for funding her to be a visiting researcher in Fraunhofer CSP.

EVALUATION OF PEROVSKITE DEVICES UNDER REAL AND EXTREME OPERATING CONDITIONS
A FUNDAMENTAL STEP TOWARD PRACTICAL APPLICATIONS

Marília Braga, Lucas A. Z. Sergio, Anelise M. Pires, Ricardo Rüther
Universidade Federal de Santa Catarina, Fotovoltaica-UFSC, Florianópolis-SC, Brazil

▶ STUDY FRAMEWORK

- **Aim:** Evaluate and optimize performance of PV devices under Brazilian climatic conditions.
- **Location:** Fotovoltaica-UFSC Research Center (www.fotovoltaica.ufsc.br) in Florianópolis-SC (27°S, 48°W) in the south of Brazil (humid subtropical, Cfa). High operating temperature conditions and high irradiance levels with frequent over-irradiance events.
- **Perovskite Devices:** 4 and 12-cell minimodules on flexible substrate with individual cell area > 2 cm².
- **Benchmarking:** Commercially-available devices from various PV technologies and manufacturers are also tested simultaneously.
- **Outdoor Measurements:** Measurements taken every minute, with an additional trigger for over-irradiance events. Full IV curve, MPP, Voc or voltage bias measurements. Additional measurements: device temperature, irradiance, spectral distribution, and ambient variables. 18 small-device measurement channels: 6x 2V/20A for Silicon and tandem cells + 6x 20V/2A for up to 18-cell perovskite minimodules + 12x 4.5V/1A for up to 4-cell perovskite minimodules. Cell-level perovskite measurements also possible.
- **Indoor Lab:** Electroluminescence imaging and a Class A++AA+ solar simulator capable of reaching high irradiance levels (~1.8 suns). The LED-based solar simulator offers a range of spectral distributions and provides continuous illumination for extended testing periods with a test area of 240 x 240 mm.
- **Current Status:** Over 100 perovskite minimodule samples with various configurations tested outdoor.

▶ KEY OUTCOMES

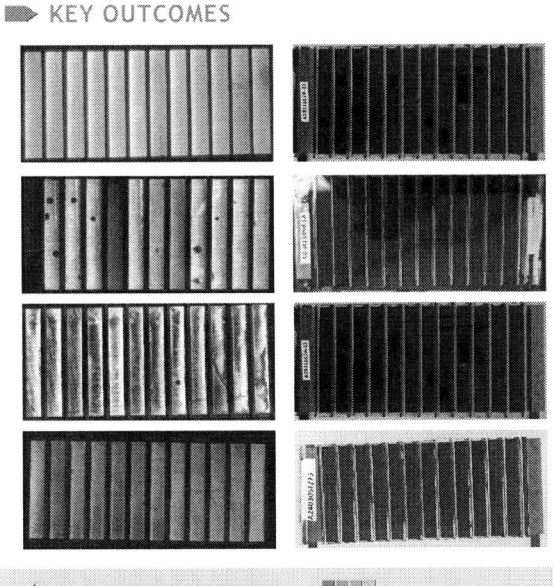

-73%/year

$\left(\begin{array}{c} -0.4\%/year \\ \approx 15k\ days \end{array}\right)$

■ 4-cell minimodules

♦ 12-cell minimodules

100+ SAMPLES

(Graph: Time until 80% of First-Day Efficiency [days] vs First-Day Efficiency [%])

Perovskite's Kryptonites:

 Partners and Acknowledgements:

41st European Photovoltaic Solar Energy Conference and Exhibition

Enhancing Measurement Protocols for Perovskite Photovoltaic Devices: Insights from the VIPERLAB Project

Eugenia Zugasti*[4], Ankit Mittal[2], Lucia V. Mercaldo[5], Javier Diaz[4], Giuseppe Nasti[5], Asier Murillo[4], Natalia Maticiuc[1], Ana Belén Cueli[4], Stephan Abermann[2], Paola Delli Veneri[5], Stephane Cros[3]

[1]Helmholtz-Zentrum Berlin für Materialien und Energie (HZB), Germany; [2]Austrian Institute of Technology (AIT), Austria; [3]Commissariat à l'énergie atomique et aux énergies alternatives (CEA-INES), France; [4]Centro Nacional de Energías Renovables (CENER), Spain; [5] Italian National Agency for New Technologies, Energy and Sustainable Development (ENEA), Italy

*E-mail: ezugasti@cener.com

H2020-VIPERLAB project:

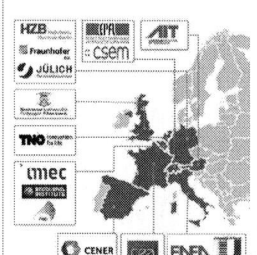

Main goal of the project:
Stimulate, through facilitated and coordinated transnational and virtual access to the best EU perovskite infrastructures and the use of advanced data mining approaches, the starting EU perovskite comunity to work together on the research and development of the perovskite PV technology development in Europe.

Challenge addressed by this work:
The perovskite devices present some measurement challenges that are currently not properly addressed by the existing standard procedures. In order to pave the way for the successful development of this innovative technology, a common approach to measurement protocols should be established

Our objective :

This work describes the outcomes of the three Round Robins performed during the EU project VIPERLAB for the electrical performance assessment of perovskite photovoltaic (PV) technology.

CONFIGURATION

	1st ROUND ROBIN	2nd ROUND ROBIN	3rd ROUND ROBIN
Solar cell Architectures	• Flexible PSK **single junction**, with p-i-n structure, IO + barrier film vacuum lamination • Semitransparent PSK **single junction**, with p-i-n structure, box sealed in glovebox • Rigid PSK **single junction**, with p-i-n structure, PO + barrier film lamination • Rigid PSK **single junction**, with p-i-n structure, cavity glass using epoxy glue lamination • **Tandem** Si/PSK 2T, with p-i-n structure, box sealed in glovebox	• PSK **single junction**, with p-i-n structure, glass/glass encapsulation	• **Tandem** Si/PSK 2T, with p-i-in structure, glass/glass encapsulation
Round Robin configuration	**Binomials**: two laboratories that apply the same protocol within each pair. Different protocols for each binomial: each pair of laboratories uses a different protocol. So, while the two laboratories in each pair use the same protocol, the protocols vary between pairs.	**Star-shaped**: the central laboratory measures the devices and distributes them among the labs which apply the same protocol and bring them back to the central laboratory which performs a second measurement of the samples.	
Measurement	IV Curve	IV Curve and spectral response	

PROCEDURES

SINGLE JUNCTION SOLAR DEVICES

Procedures used for IV measurement: 3 different protocols

PROCEDURE 1	Reverse scan (5 times) Forward scan (3 times) MPP tracking (2 minutes) Reverse and forward scan (1 time)	Voltage range: from (-0.4 to -0,1) V to (1,2 to 1.4) V for single junction From -0,1 V to (1,9 to 2) V for tandem
PROCEDURE 2	Reverse and forward scan (2 - 3 times) MPP fixed voltage Or MPP Tracking (60 s 180 s) Reverse and forward scan (1 times)	Steps: (10 – 50) mV Delay time: 20 ms Scan speed: (17 to 200) mV/s Others: Cell in Voc condition between single measurement points or continuous voltage sweep
PROCEDURE 3	Reverse and forward scan (2 times) MPP Tracking or MPP fixed voltage (180 s) Reverse and forward scan (1 times)	Voltage sweeps ranges from -0.2 V to 1.2 V, Delay time of 20 ms. Scan speed: 50-100 mV/s.

Procedures used for SR/EQE measurement

Wavelength ranges from 300 nm to 850 nm, chopper frequency: 25 Hz, temperature: 25 °C, no preconditioning, no bias light, step 10 nm.

TANDEM SOLAR DEVICES

Procedures used for IV measurement: 3 different protocols

PROCEDURE 4	1 sun illumination: MPP Tracking (60 s) MPP Tracking (120 s) Reverse and forward scan slow (1 time) Reverse and forward scan fast (1 time)	Slow voltage sweeps ranges from -0.2 V to 2.2 V, Swept time of 60 s. Delay time: 10 ms.
	IR illumination: Reverse scan slow (1 time)	Fast voltage sweeps ranges from -1.5 V to 2.2 V, Swept time of 3 s. Delay time: 1 ms.
	Blue illumination: Reverse scan slow (1 time)	
	Dark: Reverse scan slow (1 time)	
PROCEDURE 5	1 sun illumination: Reverse and forward scan slow (1 time) MPP fixed voltage (60 s) Reverse and forward scan slow (1 time) MPP fixed voltage (180 s) Reverse and forward scan slow (1 time) Reverse and forward scan fast (1 time)	Slow voltage sweeps ranges from -0.2 V to 2.2 V, Swept time of 60 s. Delay time: 10 ms. Fast voltage sweeps ranges from -1.5 V to 2.2 V, Swept time of 3 s. Delay time: 1 ms.
	IR illumination: Reverse and forward scan slow (1 time)	
	Blue illumination: Reverse and forward scan slow (1 time)	
	Dark: Reverse and forward scan slow (1 time)	
PROCEDURE 6	1 sun illumination: Reverse and forward scan fast (1 time) Reverse and forward scan fast (1 time) Reverse and forward scan fast (1 time) Reverse and forward scan slow (1 time) Reverse and forward scan slow (1 time) MPP fixed voltage (180 s) Reverse and forward scan slow (1 time) Reverse and forward scan fast (1 time) Reverse and forward scan fast (1 time) Reverse and forward scan fast (1 time)	Slow voltage sweep ranges from -0.2 V to 2.2 V, Swept time of 60 s. Delay time: 50 ms Step: 20 mV. Integration time: 450 ms. Fast voltage sweeps ranges from -0.2 V to 2.2 V, Swept time of 4.8 s. Delay time: 20 ms Step: 20 mV. Integration time: 20 ms.
	IR illumination: Reverse scan slow (1 time)	
	Blue illumination: Reverse scan slow (1 time)	
	Dark: Reverse scan slow (1 time)	

Procedures used for SR/EQE measurement

Wavelength ranges from 300 nm to 1200 nm, chopper frequency: 25 Hz, temperature: 25 °C, no preconditioning,, step 10 nm. Bias light and voltage bias according with IEC 60904-8-1

MAIN FINDINGS

- **Degradation** of the samples affected the measurements, even to the meta-stability effects occurring during preconditioning.
- The spectral distribution of the light source also affected the results, being more critical on tandem devices. For single junction cells this effect can be minimized by using a **spectral mismatch correction** but for tandem cells that is not enough, and a **multi-lamp sun simulator** is needed.
- Positioning of the **measurement mask** is also important to avoid differences. The reflections of the measurement table for semi-transparent solar cells may cause deviations in the determination of J_{sc}.
- Fill Factor is mainly affected by the **connecting system** between the cell terminals and the electronic load.
- Finally, *Voc* is affected by the sample **temperature**, so it is also important to use a temperature probe in order to correct the electrical parameters to STC.
- **Encapsulated** single junction perovskite solar cells delivered with **substrate holders, wires, and masks** reduced the sources of uncertainty in the measurements performed in the different measurement laboratories in these two aspects: aperture area and connection procedure
- The protocol used to determine performance consisting of nine steps (procedure 2, single junction), showed no significant variation or trend in the main electrical parameters at the different steps.
- During the initial measurements, increasing the delay time from 20 ms to 50 ms did not lead to a reduction in hysteresis, with the lowest hysteresis observed in the second reverse and forward scan.
- Discrepancies have been detected in short circuit current that could be related to differences in irradiance among labs: mask positioning, irradiance adjustment, spectral distribution...
- Differences are smaller for those cells which seem to be stable.
- Comparing measurements before and after their shipment to the measuring labs, the short circuit current has been heavily influenced by the measurements. Open circuit voltage has been also slightly influenced. The mean value for the fill factor seems to be stable (although silver paste had to be added to some of the substrates), but there are several samples in which the FF is different between labs. As a result, maximum power is affected by the measurements.

RECOMMENDATIONS

- It is difficult to extract clear information from the Round Robins if the PV devices suffer degradation during measurement or even shipment so the use of the most stable devices possible to prevent variations in the electrical parameters of the samples is advisable. Furthermore, the reduction, as much as possible, of the waiting period in the laboratories is also beneficial.
- Provide a sample holder with samples or encapsulated cells with wires for consistent contacting procedures. To minimize problems with electrical connections, it is recommended to put silver paste on the deposited electrodes to avoid damaging them with the contact tips and to reduce contact resistance.
- Perform also the measurement of the EQE of the samples; otherwise apply a spectral mismatch correction using the data from the manufacturer. In case tandem cells are included in the round robin, ensure that the spectral distribution is appropriate, minimizing the current mismatch between the junctions.
- It would be desirable to exchange a reference cell, ideally a filtered silicon cell, to ensure that all laboratories operate with the correct irradiance settings.
- Temperature control and monitoring of the samples to correct the data to STC is needed.
- The use of larger cells in the Round Robin would decrease the differences due to the positioning of the measurement masks.
- A briefer protocol to determine performance could be sufficient to extract electrical parameters of perovskite solar cells. Further work should be undertaken to check it in different types of perovskite cells.
- Ideally, circular Round Robins would be preferable, where the same cells are measured by all the laboratories

Acknowledgments

H2020 –VIPERLAB is receiving funding from the European Union's Horizon 2020 research and innovation programme under grant agreement N° 101006715. Special thanks to the VIPERLAB project and Associated partners for their continued support in developing technology and promoting knowledge exchange within the Solar Perovskite Community.

41st European Photovoltaic Solar Energy Conference and Exhibition

41st European Photovoltaic Solar Energy Conference and Exhibition
23 – 27 September, 2024, Vienna, Austria

SOLVENT ENGINEERING DRIVEN MORPHOLOGY CONTROL OF PEROVSKITE UNDER AIR AMBIENT DEVICE FABRICATION

Nitin Kumar Bansal, Shivam Porwal, Trilok Singh*

Semiconductor Thin Film and Emerging Photovoltaics (STEP) Laboratory
Department of Energy Science and Engineering, Indian Institute of Technology Delhi, Delhi, 110016, India
esz238389@iitd.ac.in, triloksingh@dese.iitd.ac.in

Perovskite Solar Cells

Mixed cation mixed halide perovskite materials show superior optoelectronic properties -

➤ Tunable bandgap with compositional modification.
➤ Suitable direct bandgap.
➤ High absorption coefficient of $>10^4$ cm^{-1}.
➤ Charge carrier diffusion length exceeds 1 μm.
➤ Low exciton binding energy.
➤ Perovskite solar cells are solution processable and low-cost.

What is the significance of the solvent engineering?

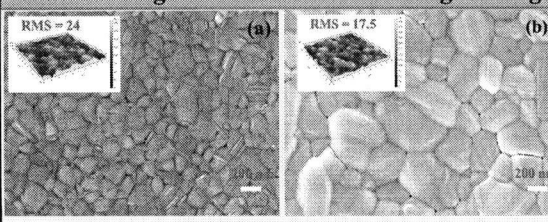

➤ Crucial role in controlling the morphology, crystallinity, and overall quality of perovskite films.
➤ Affects the film formation process, including nucleation, crystal growth, and resulted in reduced defect sites.

Idea of Solvent Engineering and Device Optimization

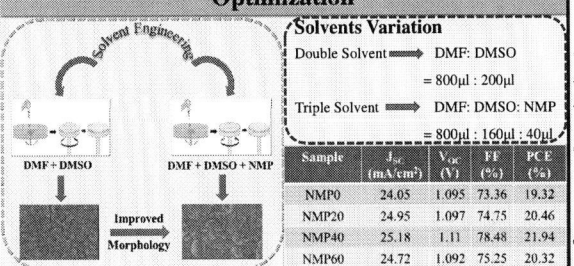

Solvents Variation

Double Solvent ⟶ DMF: DMSO = 800μl : 200μl

Triple Solvent ⟶ DMF: DMSO: NMP = 800μl : 160μl : 40μl

Sample	J_{SC} (mA/cm²)	V_{OC} (V)	FF (%)	PCE (%)
NMP0	24.05	1.095	73.36	19.32
NMP20	24.95	1.097	74.75	20.46
NMP40	25.18	1.11	78.48	21.94
NMP60	24.72	1.092	75.25	20.32

➤ Partially replaced the DMSO with NMP in the solvent
➤ NMP possesses high boiling point and lower vapor pressure compared to DMSO resulting in improved crystallization

Film Crystallinity and Defect Value Analysis

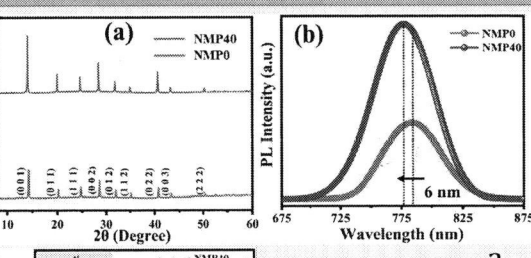

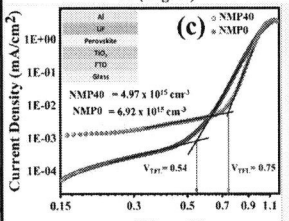
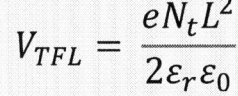

$$V_{TFL} = \frac{eN_t L^2}{2\varepsilon_r \varepsilon_0}$$

Here, ε_r : relative dielectric constant (ε_r =32), e: electron charge, L : perovskite thickness, and ε_0 : vacuum permittivity.

➤ Lower FWHM and high peak intensity of XRD depicted improved film crystallinity.
➤ Enhanced PL peak intensity suggested the reduction in non-radiative defect-assisted recombination losses.
➤ Further, SCLC analysis confirms the NMP-assisted improvement and defect passivation.

Device Performance and Stability Analysis

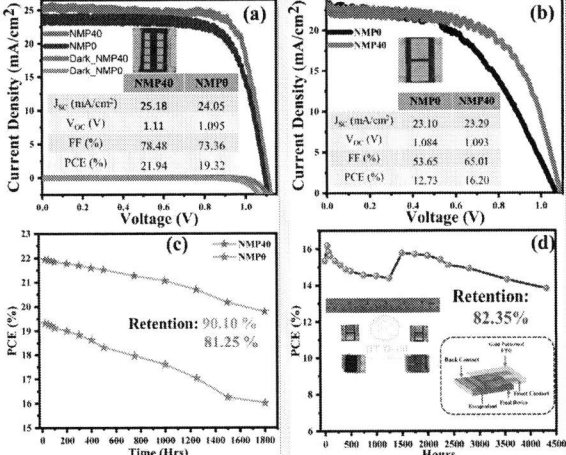

Conclusions

➤ Higher boiling point of NMP helps in controlling crystallization and grain growth of perovskite film.
➤ Morphological improvements confirmed from SEM and AFM.
➤ Increased average grain size after modification resulted in reduced defect sites (lower number of grain boundaries).
➤ Improved crystallinity, lower recombinations, and reduced defect density confirmed from XRD, PL, and SCLC analysis respectively.
➤ Modification strategy showed improved device performance for 0.09 cm² and 1 cm² active area devices.
➤ Both unencapsulated and encapsulated devices depicted significant improvement in device stability.

➤ Improved morphology and reduced defect values ultimately resulted in improved PCE and device stability.

Acknowledgment: N.K.B, S.P acknowledge PMRF, and T.S acknowledge ANRF (CRG/2023/003135) and IIT Delhi for research funding. N.K.B is thankful to SERB for ITS/2024/003653 travel grant.

References:
1. T. Singh, T. Miyasaka, Stabilizing the Efficiency Beyond 20% with a Mixed Cation Perovskite Solar Cell Fabricated in Ambient Air under Controlled Humidity, Advanced Energy Materials 8(3) (2018) 1700677.
2. M.A. Green, A. Ho-Baillie, H.J. Snaith, The emergence of perovskite solar cells, Nature Photonics 8(7) (2014) 506-514.
3. Q. Jiang, L. Zhang, H. Wang, X. Yang, J. Meng, H. Liu, Z. Yin, J. Wu, X. Zhang and J. You, Nature Energy, 2016, 2, 16177.
4. E. Shirzadi, F. Ansari, H. Jinno, S. Tian, O. Ouellette, F. T. Eickemeyer, B. Carlsen, A. Van Muyden, H. Kanda, N. Shibayama, F. F. Tirani, M. Grätzel, A. Hagfeldt, M. K. Nazeeruddin and P. J. Dyson, ACS Energy Letters, 2023, 8, 3955-3961.

Indian Institute of Technology Delhi

41st European Photovoltaic Solar Energy Conference and Exhibition

ROLL-TO-ROLL PRINTED SNO₂ FOR FLEXIBLE N-I-P PEROVSKITE PV

Thomas M. Kraft*, Ville Holappa, and Riikka Suhonen
VTT Technical Research Centre of Finland Ltd., Kaitoväylä 1, Oulu 90590, Finland
*thomas.kraft@vtt.fi

ABSTRACT: For commercial deployment of perovskite photovoltaics (Pk-PV), large area solution based deposition on flexible substrates is recognized as an attractive manufacturing option [1]. Previous studies have developed the deposition process and integration methods, however, for high yield manufacturing process improvements are critical for roll-to-roll (R2R) production [2]. In conjunction, with the multiple film layers in the single junction n-i-p architecture, the transparent electron transport layer (ETL) has a crucial role. Furthermore, for patterning of the layers, gravure printing is used [3] for deposition of the ETL on the R2R wet etched indium tin oxide (ITO) electrode. Herein, the rheology and film characteristics of the tin oxide (SnO_2) layer were investigated with initial Pk-PV devices, on the R2R deposited films, reaching over 11 % PCE. Specifically, two commercially available SnO_2 formulations were tuned for printability and utilized for the ETL in Pk-PVs. The lessons learned from this study can be applied to future developments and independent of sourced materials.

Keywords: R2R printing, ETL, flexible PV, perovskite

1 AIM AND APPROACH

Figure 1: A roll-to-roll printed electronics pilot line at VTT (Oulu, Finland)

Roll-to-roll printed electronics provides a route to large area, flexible device processing at volume (Figure 1). To control the patterning of the layers of the Pk-PVs; printing techniques can be used to avoid laser processing and material waste. Herein, the focus was to investigate the role of the ETL ink formulations, in regards to gravure print quality and PV performance. The study focused on comparing two commercially available SnO_2 inks (Avantama N-31, and Alfa Aesar) and evaluating their use in Pk-PV devices. In both cases, the commercially available metal oxide ink formulations were deposited as water or alcohol based solvent systems. For the R2R printing process gravure printing was used to deposit the patterned ETL film on ITO/PET that was etched in a previous R2R process (Figure 2).

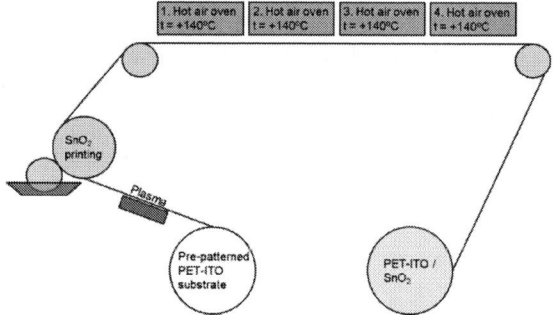

Figure 2: R2R gravure printing of SnO_2 films.

2 RESULTS

2.1 Printed films

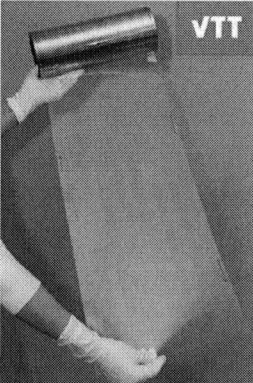

Figure 3: R2R proccessed ETL on etched ITO for scaled manufacturing of Pk-PV.

To evaluate the suitability of metal oxide inks for R2R production extensive rheology, printing, and evaluation must be performed to achieve high quality films (Figure 3) In this study, Avantama N-31, 2,5 wt% of crystalline SnO_2 in a mixture of butanols, was used as received, whereas Alfa Aesar, 15% in H_2O colloidal dispersion, was diluted to 2.5 and higher wt% with water and iso-propanol (IPA). To investigate the role of the solvent system and solute concentration, Alfa Aesar formulation was adjusted and the ink rheology was evaluated prior to printing (Figure 4). The increased SnO_2 wt% resulted in increased viscosity with H_2O:IPA volume ratio of 1:1 whereas with 4:1 ratio, no change in viscosity was observed.

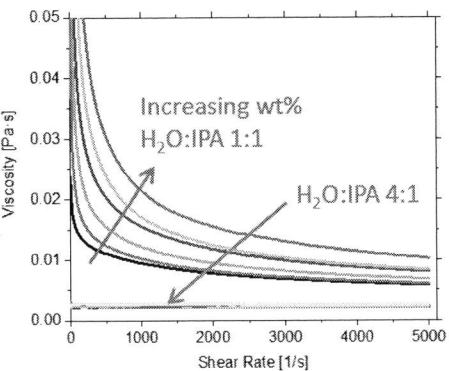

Figure 4: Effect of solvent ratio and metal oxide wt% on ink viscosity.

The printability of the SnO_2 formulations was investigated and found that the Avantama formulation was preferred for gravure printing will homogeneous film quality and minimal edge effects. With the films prepared from the Avantama ink shown in Figure 5 and those prepared from the Alfa Aesar in Figure 6.

The topographical layer quality was investigated with SEM with found that the Avantama film was less densely packed than the Alfa Aesar layer. Based on these results, voids in the Avantama film were found which will have an impact of the successive layer deposition of the Pk-PV n-i-p stack.

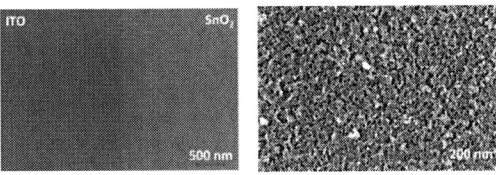

Figure 5: SEM image of the film quality of the R2R gravure printed SnO2 layer using Avantama ink.

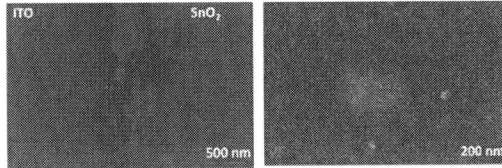

Figure 6: SEM image of the film quality of the R2R gravure printed SnO2 layer using Alfa Aesar ink.

2.2 PV performance

To investigate film quality dependence on PV performance, an n-i-p stack was prepared with PET/ITO/SnO_2/MAPbI$_3$/Spiro-OMeTAD/Au. Initial results indicated that the Alfa Aesar film provided a more appropriate interfacial layer between the MAPbI$_3$ perovskite and ITO (11 % PCE, 55 % FF, 1 V V_{OC}; 20 mA/cm2 J_{SC}). However, these results show promise for

further studies and the steps needed to have both high yield printability and high performing devices.
In general, device PV performance improved with more densely packed SnO_2 nano-particles which enhanced the perovskite metal oxide interface (Figure 7)

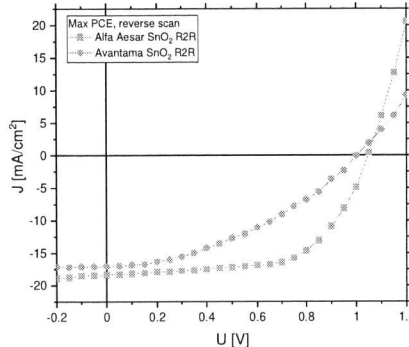

Figure 7: PV performance of the solar cells produced with a R2R printed ETL.

3 CONCLUSIONS

Printability of metal oxide inks is a key variable for R2R processed Pk-PVs, however, printability alone is not the only driving factor for high performance. Furthermore, rheological properties of SnO2 inks can be tuned for high quality, directly patterned, R2R printed thin films.
As the yield and performance of the R2R process is improved with material, ink, and process tuning, so will the opportunities for system integration and exploitation. For example, opportunities for BIPV, Agri-PV, and indoor monitoring all have uses with a flexible Pk-PV product and the role of the development of a high yield manufacturing process will help bring these markets to fruition.

4 AKNOWLEDGEMENTS

Research support provided by the Research Council of Finland (RCF), Printed intelligence infrastructure funding, decision 358621 and the RCF Flagship Programme, Photonics Research and Innovation (PREIN), decision number 346545. Also, we are thankful for funding from the European Union's Horizon 2020 research and innovation program under grant agreement 763977 (PertPV), and the Horizon Europe grant agreements 101122283 (PEARL) and 101082176 (VALHALLA).

We are grateful for funding by the CETPartnership (REFORM, CETP-2022-00348), the European Partnership under Joint Call 2022 for research proposals, co-funded by the European Commission (GA N°101069750) and with funding by Business Finland, decision number 2876/31/2023

5 REFERENCES

[1] Y. Y. Kim et al., "Gravure-Printed Flexible Perovskite Solar Cells: Toward Roll-to-Roll Manufacturing," Adv. Sci., vol. 6, no. 7, p. 1802094, Apr. 2019, doi: 10.1002/advs.201802094.
[2] R. Suhonen, K. L. Väisänen, M. Välimäki, T. Kurkela, M. Ylikunnari, and T. M. Kraft, "Injection Overmolded Flexible Perovskite Solar Modules," p. 487 kB, 4 pages, doi: 10.4229/EUPVSEC2023/2BV.2.36.
[3] M. Ylikunnari et al., "Flexible OPV modules for highly efficient indoor applications," Flex. Print. Electron., vol. 5, no. 1, p. 014008, Feb. 2020, doi: 10.1088/2058-8585/ab6e73.

41st European Photovoltaic Solar Energy Conference and Exhibition

MICRO INVERTED PYRAMID FORMATION IN TITANIUM DIOXIDE LAYER BY PULSED LASER IRRADIATION TO IMPROVE ELECTRON TRANSPORT IN MAPBI₃-BASED PHOTOVOLTAIC DEVICES

L. Ocaña[1,2], C. Montes[1,2], B. González-Díaz[2], S. González-Pérez[3], E. Llarena[1].

[1]Instituto Tecnológico y de Energías Renovables, S. A. (ITER)
Pol. Industrial de Granadilla, s/n, 38600 Granadilla de Abona, Spain.
Ph. +34 922 747 700 / Fax +34 922 747 701 / E-mail lmgonzalez@iter.es
[2]Departamento de Ingeniería Industrial. Universidad de La Laguna.
Camino San Francisco de Paula, s/n, 38206 San Cristóbal de La Laguna. S/C de Tenerife. Spain.
[3]Departamento de Didácticas Específicas. Universidad de La Laguna.
c/ Pedro Zerolo, s/n, Edificio Central Planta 2. Apartado 456, 38200 San Cristóbal de La Laguna. S/C de Tenerife. Spain.

ABSTRACT: This study is aimed to enhance the performance of MAPbI₃-based photovoltaic devices by forming micro inverted pyramid structures using a 1064 nm pulsed laser on the electron transport layer (ETL). These microstructures enhance the hydrophobicity of the ETL, promoting the crystallization of the perovskite layer during the spin coating deposition process. This results in enhanced electron transport and, consequently, improved overall device efficiency. The electrical and hydrophobic properties of the ETL have been investigated in several batches of n-i-p planar perovskite solar cells fabricated using this method. One set of solar cells featuring micro inverted pyramid structures created with varying laser power levels, and another set following the conventional manufacturing process. The samples were analyzed at different deposition stages using a semiconductor characterization system and spectroscopic ellipsometry to evaluate their resistivity and optical characteristics. Additionally, the photovoltaic performance of the devices was simulated using SCAPS-1D software to further assess their electrical parameters and power conversion efficiencies

1 INTRODUCTION

The Solar Cell Laboratory of the Instituto Tecnológico y de Energías Renovables (ITER), in collaboration with La Laguna University (ULL), has been working since 2015 on the development and improvement of photovoltaic devices based on perovskite solar cells (PSCs), based on the use of methylammonium lead triiodide (MAPbI₃), focusing his efforts on two lines of research: the manufacturing of these devices in a cleanroom environment with relative humidity between 40-60% [1] by developing ink-based solutions, in order to provide the means for producing their back contacts as an alternative to thermal evaporation [2], and the development of a new encapsulation method to protect the perovskite cells against environmental agents [3,4].

To improve the efficiency and stability of perovskite-based devices, numerous authors have evaluated the enhancement of the hydrophobic properties of the electron transport layer [5–8]. Harnessing the acquired expertise in this domain has proven significant, particularly within the realm of device manufacturing, specially, the use of the 1064nm pulsed laser, which has been widely employed to perform a selective ablation of the fluorine doped tin oxide (FTO) coating areas, thereby mitigating the occurrence of short-circuits between FTO and the hole transport layer (HTL) [1]. The acquired knowledge enables us to delve into this new line of research, where we are being assessed the formation of micro inverted pyramids in the electron transport layer to create a hydrophobic surface that repels water and attracts non-polar solvents. This surface type, in addition, enhances the crystallization of the perovskite layer during the chlorobenzene-assisted spin coating process [1].

The use of laser microfabrication techniques is on the rise as a means to control the wettability of solid surfaces [9–11]. Several studies assess the application of this tool on TiO₂ surfaces, examining the influence of different machining parameters on the properties of the resulting material, such as superhydrophobicity, underwater superoleophobicity, anisotropic wettability, and smart wettability [9,12–14]. This study researches the enhancement of photovoltaic device performance based on perovskite structures through the formation of micro inverted pyramids using a 1064 nm pulsed laser on the compact TiO₂ layer. The formation of these pyramids aims to improve the hydrophobic properties, thereby promoting the crystallization of the perovskite layer during the spin coating deposition phase. This improvement was expected to enhance electron transport and absorbance range, consequently contributing to the stability and efficiency enhancement of the final device.

2 EXPERIMENTAL

With this objective in mind, we were manufactured n-i-p planar perovskite solar cells using the spin coating technique agents [3]. This process was involved depositing layers on an FTO-coated glass substrate, starting with a layer of TiO₂ as the electron transport layer (ETL), where micro inverted pyramid structures were implemented. For the formation of the micro inverted pyramids, a 1064 nm pulsed laser was used along with a galvanometric scanner, allowing the creation of a pattern with a pulse energy of 100 ns, a speed of 1,800 mm/s, a frequency of 100 kHz, and variable power settings of 5%, 10%, and 20%. This was facilitated the production of micrometric structures with different surfaces and depths. Subsequently, a UV ozone cleaner has been employed to remove any residue. Subsequently, the absorbing layer of MAPbI₃ was deposited to assess its crystallization on the micro inverted pyramids. Based on these results, the final device was simulated, incorporating a hole transport layer (HTL) of Spiro-MeOTAD, gold contacts, and ethylene vinyl acetate (EVA) as encapsulation to seal the device [3,4]. (See figure 1).

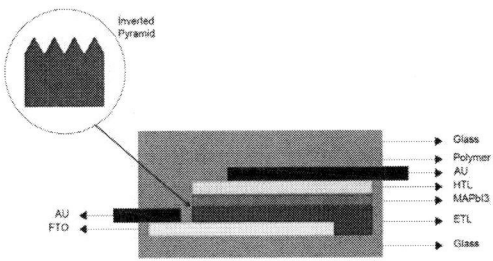

Figure 1: Images of the final device simulated.

3 CHARACTERIZATION

A comprehensive analysis of the ETL layer's behavior was carried out from the perspective of its electrical, optical and hydrophobic properties. For this purpose, two batches of samples were prepared: one with micro inverted pyramids treated at different laser density power (5%, 10% and 20%) and another through a standard manufacturing process. The samples were characterized at three different stages during the process. Characterization was conducted at three stages: first, after depositing the ETL layer; second, following the deposition of the MAPbI3 layer; and third, upon the device's completion. (See table I).

Table I: Instrumentation and measurements conducted during each characterization phase.

Manufacturing phase	Instrumentation/ Measurement	Characterization phase
ETL	SCS: Rsh SE: Transmittance	Phase I: Analysis of the formation of micro inverted pyramids in the ETL layer.
MAPbI3 layer	SE: Absorbance	Phase II: Analysis of the interaction between the ETL layer and the MAPbI3 layer.
PSC	SCAPS-1D: Isc, Voc, FF, PCE	Phase III: Analysis of the final device.

In the first stage, a semiconductor characterization system (SCS) and a spectroscope ellipsometer (SE) were used to evaluate the sheet resistance (Rsh) and the transmittance (T) the analyzed layers, respectively, to analysed the formation of micro inverted pyramids in the ETL layer. In the second stage, a SE was also used to analysed the interaction between the ETL layer and the MAPbI3 layer, through the variation of the absorbance curves. In the third stage, the software simulation SCAPS-1D [3,15] was used to determine the electrical parameters of the complete device, such as short-circuit current (Jsc), open-circuit voltage (Voc), fill factor (FF), and overall power conversion efficiency (PCE). Thus, the influence of the formation of micro inverted pyramids on the hydrophobic properties of the electron transport layer, as well as its implications for the deposition of the absorber layer and the final device performance, were evaluated.

4 RESULTS AND DISCUSSION

4.1 Semiconductor characterization system.

The actual measurements were carried out using a semiconductor characterization system, configured to operate as an ohmmeter.

To identify the optimal laser power for fabricating the micro inverted pyramids, the sheet resistance of the ETL layers, both with and without pyramids, was measured using a semiconductor characterization system configured to operate as an ohmmeter. The results, shown in Table II, clearly demonstrate that sheet resistance increases proportionally with the rise in laser power used for pyramid formation, ranging from 857.1 Ω/sq for samples without pyramids to 3420.1 Ω/sq at 5% power laser ablation, 12,259.6 Ω/sq at 10% power laser ablation, and ultimately becoming non-conductive at 20% power laser ablation. This indicates a reduction in the layer's conductivity, and consequently, a diminished electron transport capability.

Table II: Sheet resistance of the ETL without and with micro inverted pyramids at different laser ablation power levels.

Sample	Rsh
Glass/FTO/ETL (Flat)	857.1 Ω/sq
Glass/FTO/ETL (Power 5%)	3420.1 Ω/sq
Glass/FTO/ETL (Power 10%)	12,259.6 Ω/sq
Glass/FTO/ETL (Power 20%)	-

4.2 Spectroscope ellipsometer

To begin, we conducted an analysis of the transmittance of the ETL without and with micro inverted pyramids at different power levels, in order to assessed the transmittance range of the layer in the spectrum range between 300 and 900 nm. Figure 2 illustrates the results of this analysis.

It is observed that the samples with micro inverted pyramids formed using a 5% exhibit a transmittance similar to that of the flat samples, with no noticeable increase or decrease in any range of the studied spectrum. In contrast, the samples with micro inverted pyramids formed at higher powers show a reduction in transmittance; this reduction is smaller for the samples at 10% laser ablation power and larger for those treated at 20% power laser ablation, ultimately resulting in the complete removal of the layer.

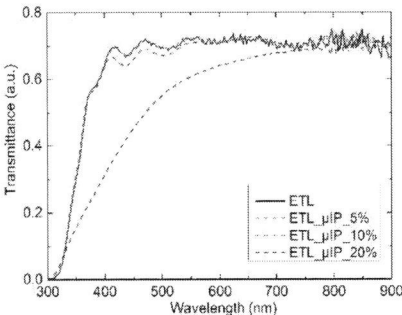

Figure 2. - Transmittance of the ETL without and with micro inverted pyramids at different laser ablation power levels.

To assess the impact of the micro inverted pyramids on the crystallization of the MAPbI₃ layer during the spin coating deposition phase, the absorbance curves of samples were measured, without and with micro inverted pyramids at different laser ablation power levels. As illustrated in Figure 3 and corroborated by the transmittance analysis, the samples were exhibited a variation in their absorbance curve relative to the flat sample. It was observed that the formation of micro inverted pyramids results in an enhancement of the absorption range of the perovskite layer, particularly in its optimal efficiency range between 350 nm and 550 nm. However, there was also a reduction in the absorbance curve observed between 550 nm and 760 nm. This variation is directly proportional to the increase in the energy employed for the formation of the micro inverted pyramids.

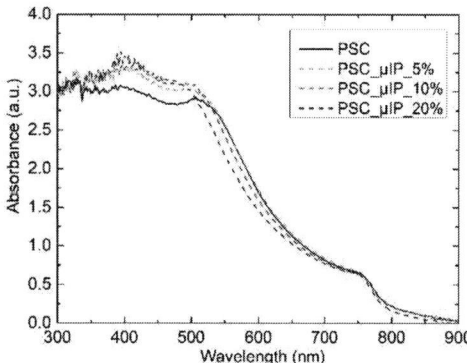

Figure 3. - Absorbance of the MAPbI₃ deposited on ETL without and with micro inverted pyramids at different laser ablation power levels.

4.3 PSC simulations

In order to assess the potential impact of the ETL with micro inverted pyramids in the perovskite layers on complete cells, we conducted simulations using the SCAPS-1D simulation software. Developed by the Department of Electronics and Information Systems (ELIS) at the University of Ghent, Belgium, this software is widely recognized for its accuracy in the study of PSCs [16–21].

Using the sheet resistance, transmittance, band gap, and absorbance curve data for the device layers, we extrapolated the effect of the micro inverted pyramids on the PSCs using SCAPS 1-D software. This enabled us to estimate their potential impact on critical parameters, including short-circuit current (Jsc), open-circuit voltage (Voc), fill factor (FF), and power conversion efficiency (PCE) [3,22].

Figure 4 presents a comparison of the IV curves for the PSCs without and with micro inverted pyramids at different power levels.

As evident from the graph, the PSC with micro inverted pyramids at power level of 5%, show a slight increase in PCE, this was increased from 16.01% to 16.54% following the formation of the micro inverted pyramids. However, the PSCs with micro inverted pyramids formed at higher power settings exhibited a

decline in device efficiency, with those treated at 10% laser ablation power reducing its PCE to 14.35%, while those subjected to 20% laser ablation power to function entirely (as detailed in Table III).

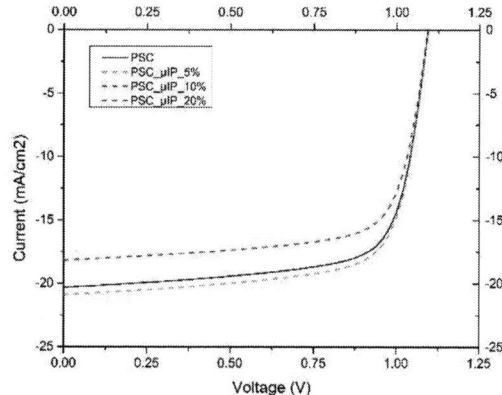

Figure 4. – Curve IV of the MAPbI₃ deposited on ETL without and with micro inverted pyramids at different laser ablation power levels.

Table III: Electrical parameters

Parameters	Flat	5%	10%	20%
V_{oc} (V)	1.09	1,09	1.09	-
J_{sc} (mA/cm²)	20.32	20,90	18.18	-
FF (%)	72.01	72,39	72.14	-
PCE (%)	16.01	16,54	14.35	-
V_{mp} (V)	0.92	0,92	0.92	-
J_{mp} (mA/cm2)	17.32	17,91	15.53	-

5 CONCLUSIONS

The results show an increase in the sheet resistance of the ETL and a rise in the absorbance of the perovskite layer within the 350 nm to 550 nm range of the solar spectrum, where radiation intensity peaks and perovskite-based solar cells achieve maximum efficiency. Additionally, the sheet resistance and absorbance are directly proportional to the laser density power applied during the formation of the micro inverted pyramids. When the data was included into the SCAPS-1D simulation software, it was found that perovskite solar cells with micro inverted pyramids formed at 5% of laser density power exhibited a slight increase in power conversion efficiency. This is likely due to the fact that higher power levels during the fabrication process cause significant degradation of the ETL layer, as indicated by transmittance measurements, thus impairing electron extraction from the device. Therefore, a balance must be achieved when forming the micro inverted pyramids, in order to enhance MAPbI₃'s absorbance without compromising ETL's electron transport capabilities.

This study will be improved by characterizing the samples using scanning electron microscopy (SEM) and atomic force microscopy (AFM) to evaluate their morphology and surface roughness. Furthermore, the formation of the micro inverted pyramids will be analyzed using new laser irradiation sources, such as femtosecond laser techniques.

6 ACKNOWLEDGEMENTS

This work was developed within the MACLAB-PV project framework, which has been co-financed by the INTERREG Madeira-Azores-Canarias Territorial Cooperation Programme (MAC) 2014-2020. 2nd Call. Axis 1. Enhancing research, technological development and innovation.

7 REFERENCES

1. Ocaña, L.; Linares, A.; Llarena, E.; Montes, C.; González, O.; Molina, D.; Pío, A.; Quinto, C.; Friend, M.; Cendagorta, & M. Adaptation of a Crystalline Silicon Solar Cell Laboratory to Produce Perovskite Solar Devices. In Proceedings of the 31st European Photovoltaic Solar Energy Conference and Exhibition; Hamburg, 2015; pp. 1138–1143.
2. 'Montes, C.; Ocaña, L.; González-Díaz, B.; González-Pérez, S.; Llarena, E 'Llarena, E.. Review of a Research Carried out to Produce Conductive Inks and Agglomerates That Make Use of Non-Precious Materials Together with Vehicles Compatible with Thin Layers of Perovskite.; Lisbon, 2023; pp. 020105001–020105005.
3. Ocaña, L.; Montes, C.; González-Pérez, S.; González-Díaz, B.; Llarena, E. Characterization of a New Low Temperature Encapsulation Method with Ethylene-Vinyl Acetate under UV Irradiation for Perovskite Solar Cells. *Appl. Sci.* **2022**, *12*, 5228, doi:10.3390/app12105228.
4. Ocaña, L.; Montes, C.; González-Díaz, B.; González-Pérez, S.; Llarena, E. Evaluation of Ethylene-Vinyl Acetate, Methyl Methacrylate, and Polyvinylidene Fluoride as Encapsulating Materials for Perovskite-Based Solar Cells, Using the Low-Temperature Encapsulation Method in a Cleanroom Environment. *Energies* **2024**, *17*, doi:10.3390/en17010060.
5. Zaky, A.A.; Christopoulos, E.; Gkini, K.; Arfanis, M.K.; Sygellou, L.; Kaltzoglou, A.; Stergiou, A.; Tagmatarchis, N.; Balis, N.; Falaras, P. Enhancing Efficiency and Decreasing Photocatalytic Degradation of Perovskite Solar Cells Using a Hydrophobic Copper-Modified Titania Electron Transport Layer. *Appl. Catal. B Environ.* **2021**, *284*, 119714, doi:https://doi.org/10.1016/j.apcatb.2020.119714.
6. Abbasi, S.; Ruankham, P.; Passatorntaschakorn, W.; Khampa, W.; Musikpan, W.; Bhoomanee, C.; Liu, H.; Wongratanaphisan, D.; Shen, W. A New Single-Step Technique to Fabricate Transparent Hydrophobic Surfaces Utilizable in Perovskite Solar Cells. *Appl. Surf. Sci.* **2023**, *613*, 155969, doi:https://doi.org/10.1016/j.apsusc.2022.155969.
7. Liu, G.-Z.; Du, C.-S.; Wu, J.-Y.; Liu, B.-T.; Wu, T.-M.; Huang, C.-F.; Lee, R.-H. Enhanced Photovoltaic Properties of Perovskite Solar Cells by Employing Bathocuproine/Hydrophobic Polymer Films as Hole-Blocking/Electron-Transporting Interfacial Layers. *Polymers*

(Basel). **2021**, *13*, doi:10.3390/polym13010042.
8. Gong, W.; Guo, H.; Zhang, H.; Yang, J.; Chen, H.; Wang, L.; Hao, F.; Niu, X. Chlorine-Doped SnO2 Hydrophobic Surfaces for Large Grain Perovskite Solar Cells. *J. Mater. Chem. C* **2020**, *8*, 11638–11646, doi:10.1039/D0TC00515K.
9. Yong, J.; Chen, F.; Yang, Q.; Hou, X. Femtosecond Laser Controlled Wettability of Solid Surfaces. *Soft Matter* **2015**, *11*, 8897–8906, doi:10.1039/C5SM02153G.
10. Ijaola, A.O.; Bamidele, E.A.; Akisin, C.J.; Bello, I.T.; Oyatobo, A.T.; Abdulkareem, A.; Farayibi, P.K.; Asmatulu, E. Wettability Transition for Laser Textured Surfaces: A Comprehensive Review. *Surfaces and Interfaces* **2020**, *21*, 100802, doi:https://doi.org/10.1016/j.surfin.2020.100802.
11. Raja, R.S.S.; Selvakumar, P.; Babu, P.D.; Rubasingh, B.J.; Suresh, K. Influence of Laser Parameters on Superhydrophobicity- A Review. *Eng. Res. Express* **2021**, *3*, 22001, doi:10.1088/2631-8695/abf35f.
12. Wang, S.; Jiang, L. Definition of Superhydrophobic States. *Adv. Mater.* **2007**, *19*, 3423–3424, doi:https://doi.org/10.1002/adma.200700934.
13. Cassie, A.B.D.; Baxter, S. Wettability of Porous Surfaces. *Trans. Faraday Soc.* **1944**, *40*, 546–551, doi:10.1039/TF9444000546.
14. S.C., V.; Ambar Kumar, C.; C., T.; Ram Kishor, G.; R.P., G.; R., K.; K.S., B.; John, P. Laser Patterned Titanium Surfaces with Superior Antibiofouling, Superhydrophobicity, Self-Cleaning and Durability: Role of Line Spacing. *Surf. Coatings Technol.* **2021**, *418*, 127257, doi:https://doi.org/10.1016/j.surfcoat.2021.127257.
15. Montes, C.; Dorta-Guerra, R.; González-Díaz, B.; González-Pérez, S.; Ocaña, L.; Llarena, E. Study of the Evolution of the Performance Ratio of Photovoltaic Plants Operating in a Utility-Scale Installation Located at a Subtropical Climate Zone Using Mixed-Effects Linear Modeling. *Appl. Sci.* **2022**, *12*, doi:10.3390/app122111306.
16. Chowdhury, M.S.; Shahahmadi, S.A.; Chelvanathan, P.; Tiong, S.K.; Amin, N.; Techato, K.; Nuthammachot, N.; Chowdhury, T.; Suklueng, M. Effect of Deep-Level Defect Density of the Absorber Layer and n/i Interface in Perovskite Solar Cells by SCAPS-1D. *Results Phys.* 2020, *16*.
17. Slami, A.; Belkaid, A.B. *Numerical Study of Based Perovskite Solar Cells by SCAPS-1D*;
18. Mandadapu, U.; Vedanayakam, S.V.; Thyagarajan, K. Simulation and Analysis of Lead Based Perovskite Solar Cell Using SCAPS-1D. *Indian J. Sci. Technol.* **2017**, *10*, 1–8, doi:10.17485/ijst/2017/v10i11/110721.
19. Lin, L.; Jiang, L.; Li, P.; Fan, B.; Qiu, Y.; Yan, F. Simulation of Optimum Band Structure of HTM-Free Perovskite Solar Cells Based on ZnO Electron Transporting Layer. *Mater. Sci. Semicond. Process.* **2019**, *90*, 1–6, doi:10.1016/j.mssp.2018.10.003.
20. Bansal, S.; Aryal, P. Evaluation of New Materials for Electron and Hole Transport Layers in Perovskite-Based Solar Cells through SCAPS-1D Simulations. In Proceedings of the Conference Record of the IEEE Photovoltaic Specialists

Conference; Institute of Electrical and Electronics Engineers Inc., November 18 2016; Vol. 2016-November, pp. 747–750.

21. Chakraborty, K.; Choudhury, M.G.; Paul, S. Numerical Study of Cs2TiX6 (X = Br−, I−, F− and Cl−) Based Perovskite Solar Cell Using SCAPS-1D Device Simulation. *Sol. Energy* **2019**, *194*, 886–892, doi:10.1016/j.solener.2019.11.005.

22. Ocaña, L.; Montes, C.; González-Pérez, S.; González-Díaz, B.; Llarena, E. Testing Encapsulated Perovskite Solar Cells in a Climatic Chamber by Following the IEC 61215 and IEC 61646 Standards for the Thermal Cycling Test. In Proceedings of the 38th European Photovoltaic Solar Energy Conference and Exhibition; 2021; pp. 406–410.

41st European Photovoltaic Solar Energy Conference and Exhibition

DEMONSTRATION OF INDUSTRIALLY SCALABLE CHEMICAL VAPOUR DEPOSITION (CVD) PROCESS FOR PRODUCTION OF HIGH-EFFICIENCY PEROVSKITE PHOTOVOLTAICS

Emma Holder, Adrian Element, Yong Li, Faiazul Haque, Kenrick F. Anderson,
Tim W. Jones, Benjamin C. Duck, Noel W. Duffy, Gregory J. Wilson
CSIRO Energy, Solar Energy Technologies Newcastle Energy Centre (NSW), Australia
E-mail: Gregory.Wilson@csiro.au

ABSTRACT: Silicon-Perovskite tandem solar cells have the potential to bring 30% conversion efficiency to rooftop and utility scale PV compared to 22% of today. CSIRO has developed a patented novel method for fabricating the perovskite layer on top of any conductive glass or silicon substrate. The method shown here is scalable, producing films of a desired thickness and bandgap to suit the current silicon wafers. The method developed is industrially relevant fitting in with modern thin film fabrication techniques. Consistent film thickness and uniformity are crucial to high performance solar cells. Defects within a film at sub-micron scales are the cause for short circuits rendering the cell ineffective. The work presented here demonstrates how our chemical vapour deposition method produces reproducible thin films. As thickness variability is critical to overall performance, we have successfully demonstrated variability of less than 10nm across a 10cm^2 area, free of defects. Most importantly, films being produced in this work are at scale, compatible to common wafers and ready for translation to silicon substrates and tandem device fabrication. In this study we use several methods to critically analyse our thin films, including surface profilometry, x-ray diffraction, and optical/luminescence techniques. The next step in this work is to stack the perovskite layer on top of silicon substrates to produce high performance tandem cells.

Keywords: Silicon-Perovskite tandem solar cells, chemical vapour deposition, Perovskite film quality and uniformity, Scalable fabrication method, Thin film photovoltaics

1 INTRODUCTION

Silicon-perovskite tandem solar cells have garnered significant interest due to their potential to achieve conversion efficiencies exceeding 30%, a notable improvement over the current 22% efficiency of conventional industrially manufactured silicon solar cells in commercial modules [1]. This efficiency boost is attributed to the complementary absorption spectra of silicon and perovskite materials, which enable more effective utilisation of the solar spectrum [2].

Our group at the Commonwealth Scientific and Industrial Research Organisation (CSIRO) has developed a novel vacuum chemical vapour deposition (VCVD) method for fabricating perovskite layers on planar surfaces such as conductive glass (ITO or FTO) or even textured silicon substrates. This method is inherently scalable and produces films with precise control over thickness and bandgap properties, enabling coupling with existing industrial supply of silicon wafers [3]. Its industrial relevance and compatibility with modern thin-film fabrication techniques make it a promising candidate for large-scale production [4,5].

Uniformity and consistency in film thickness are critical for high-efficiency solar cells. Variations in thickness can lead to sub-micron defects, which are detrimental to cell performance and can cause short circuits [6]. The VCVD method developed by CSIRO addresses these challenges by ensuring a thickness variability of less than 10% across a 10 cm^2 area, free of defects comparable to related processes reported [7]. This precision is crucial for integrating perovskite layers with silicon substrates in tandem solar cell configurations, and particularly if moving to two or more junctions.

Effective interface passivation is essential for achieving high-efficiency tandem solar cells. Chin *et al.* demonstrated that interface passivation can lead to efficiencies as high as 31.25% in perovskite/silicon tandem solar cells [1]. This highlights the potential of

advanced fabrication techniques in enhancing solar cell efficiency.

Recent advancements in chemical vapour deposition (CVD) have shown that high-quality perovskite films with excellent uniformity and minimal defects can be produced [4,6,7]. These films are important for fabricating efficient and stable perovskite solar cells. The scalability of the CVD process developed by CSIRO is a significant advantage over traditional solution-based methods, which often face challenges related to scalability and uniformity [8]. The CVD method enables production of large-area films with consistent properties, making it suitable for industrial applications [3,5,6].

In this study, we employ several analytical techniques, including surface profilometry, x-ray diffraction, and optical/luminescence methods, to evaluate the quality of the perovskite films produced using our CVD method. These techniques provide comprehensive insights into the structural and optical properties of the films, ensuring their suitability for high-performance tandem solar cells. The next stage of this work, not included in this study, involves stacking the perovskite layer on silicon substrates to create high-efficiency tandem cells, translating the laboratory-scale success of our CVD method to commercial applications.

2 EXPERIMENTAL METHODS AND MATERIALS

2.1 Materials

1.1 mm thick 30 mm × 30 mm and 100 mm × 100 mm sputtered ITO/glass substrates were purchased from Kintec, Hong Kong and used as substrates for the CVD perovskite fabrication. The metallic Pb was from Kurt J. Lesker. SnO$_2$ was from Alfa Aesar. Methylammonium Iodide (MAI) and Methylammonium Bromide (MABr), were purchased from Greatcell Solar. All chemicals were used as received without any further purification.

2.2 CVD perovskite depositions

All results presented are from films fabricated in accordance with the device structure shown in Figure 1. The structure was chosen not to be a single-junction cell, yet to represent the optical configuration and materials layers of a top-cell of a tandem device. The purpose of this work was targeting process control for integration to tandem device fabrication (future work). Figure 2 depicts a schematic of the generalised steps involved in the VCVD fabrication process.

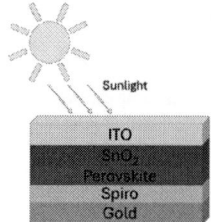

Figure 1: Target perovskite top-cell structure.

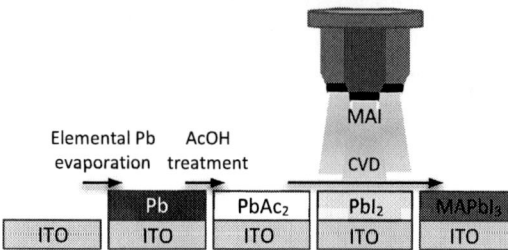

Figure 2: VCVD fabrication process steps.

ITO substrate cleaning: The ITO substrates were cleaned in an ultrasonic bath using a fifteen-minute sequence for each solvent including Hellamanex-III (diluted soapy liquid), Milli-Q deionised water, Acetone and Isopropanol. Then they were dried with N_2 followed by oxygen plasma cleaning for 15 minutes.

ETL Preparation: The SnO_2 ETL solution (~3% diluted) was spin-coated onto the substrate and annealed at 150°C for 30 minutes.

Pb Deposition: After annealing, the films were transferred into an evaporation chamber, where different thicknesses of Pb films were deposited by low-speed physical vapour evaporation.

Pb to lead acetate ($PbAc_2$) Conversion: The Pb films were placed in a custom-built vacuum chamber. The chamber was evacuated to a pressure of ~0.08 MPa then flushed with instrument air for 2 minutes at 1 standard L/min. Acetic acid vapour was introduced into the chamber by turning on the stirring mechanism and a low-flow purge of instrument air. The conversion was visually monitored and once the Pb films became transparent (approximately 2 minutes), they were purged with instrument air for 1 minute at 2 standard L/min. The $PbAc_2$ films were then annealed at 80°C for 5 minutes.

$PbAc_2$ to lead iodide (PbI_2) and Perovskite Conversion: The $PbAc_2$ films were quickly transferred to a clean CVD reactor, where ~1M MAI or MABr precursors were loaded into the sublimator head. The sublimator was held at 160°C and the substrate temperature was held at 90°C via a stainless-steel hotplate. The reaction time in the CVD reactor varied based on the thickness of the original Pb layer.

Characterisation: The UV–vis spectra were measured by a Perkin Elmer model Lambda 950 instrument. XRD measurements were conducted using a Panalytical Aeris XRD at 40 kV and 15 mA. Optical measurements were taken using a custom-built system with a Raspberry Pi HQ camera. Illumination was provided using a diffuse LED backlight. Film thickness measurements were made on a Bruker DektakXT stylus profilometer.

3 RESULTS AND DISCUSSION

3.1 VCVD fabrication process

An internal view of the chamber used for the novel VCVD method is shown in Figure 3. Continuous observation of the reaction progress in the chamber is possible through a porthole window and camera. The chamber is capable of uniform deposition over an area of 120 mm × 120 mm. This enables multiple concurrent experiments to be performed on 30 mm × 30 mm substrates or a single large area deposition. The chamber is equipped with three independently controlled sublimators to perform the organohalide sublimation. This enables control over the precise composition of the film being fabricated. Temperature control of the substrates is achieved with a dedicated hot plate with cold gun for cycling temperature.

Figure 4 depicts images taken at regular 20-minute intervals through a typical conversion run. For each image the reaction has been temporarily halted and a sample removed for ex-situ characterisation. Figure 5 depicts an overlay of XRD data taken at regular time steps. Both indicate a steady conversion of the $PbAc_2$ through to PbI_2 and a simultaneous conversion of the PbI_2 through to the target $MAPbI_3$ perovskite. After 160 minutes the reaction has completed. It should be noted that interrupting the reaction runs for *ex-situ* characterisation increases the overall processing time, not ideal for an industrial or regular processing, yet is necessary to gain mechanistic insight to the chemical reaction in the chamber, a feature not afforded by other rapid conversion process.

All analysis presented here has been performed at multiple spot locations across the large area films to confirm uniformity at scale.

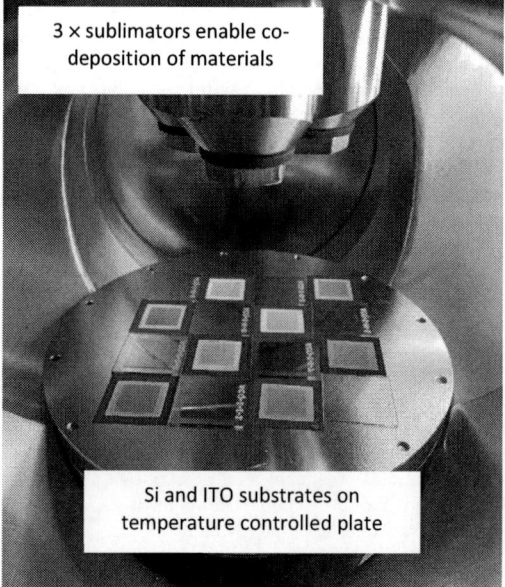

Figure 3: Internal view of CVD reaction chamber (through observation porthole).

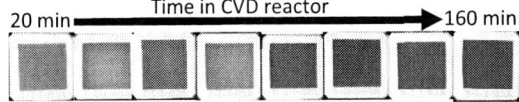

Figure 4: CVD conversion from PbAc$_2$ to MAPbI$_3$ perovskite.

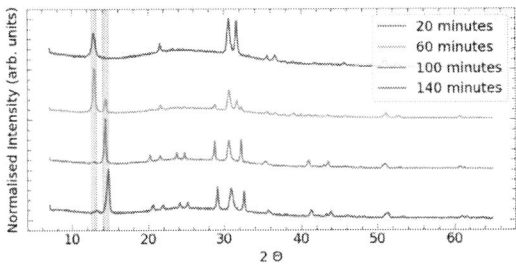

Figure 5: XRD spectra indicates chemical conversion of PbI$_2$ at ~13° 2θ to perovskite at ~15° 2θ with increasing CVD processing time.

3.2 Film thickness

To ensure appropriate optical properties and current balancing within a tandem device stack it is important to target and control a defined thickness. Within the novel CVD process described here this is achieved simply by modifying the thickness of the initial Pb deposition layer. Figure 6 demonstrates the relationship between the final perovskite film thickness and the thickness of the initial Pb film. This simple linear relationship has been demonstrated to be substrate scale invariant.

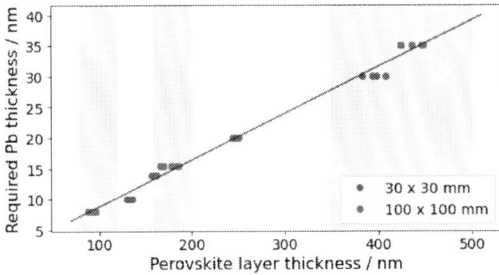

Figure 6: Data showing relationship between final perovskite thickness and the initial deposited Pb thickness.

3.3 Bandgap control

The presence of three independently controlled sublimation heads in the chamber allows control of the composition of the final perovskite. Each head can be loaded with a different material and the deposition tailored to achieve a desired result. An example of formation of a mixed-halide perovskite with varied bandgap has been demonstrated in Figure 7. In Figure 7a a pure MAI sublimation is used while in Figure 7b a combination of MAI and MABr is used to deliver a modified bandgap material.

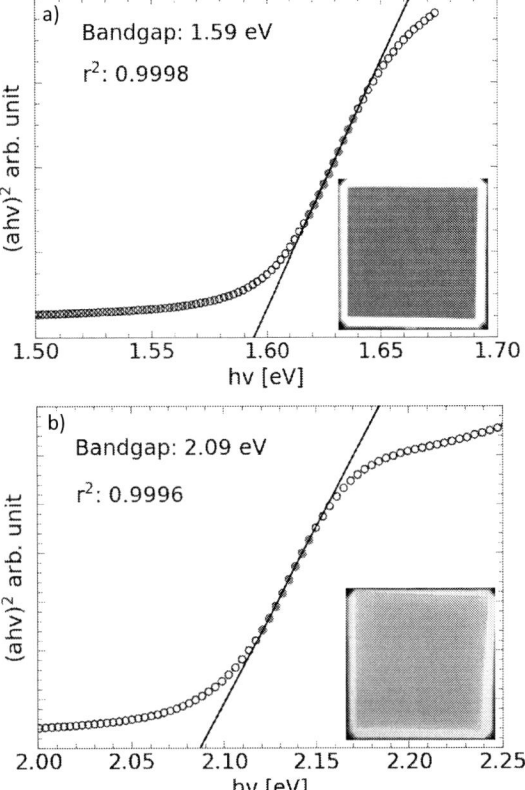

Figure 7: Comparison of UV-Vis spectra of films from CVD runs using different sublimator material compositions: a) MAI used in all three sublimators; and b) MAI and MABr used in sublimators. Inserts show optical images of the films.

4 CONCLUSIONS

The results presented here have demonstrated the realisation of a novel vacuum chemical vapour deposition (VCVD) process that converts a metallic Pb film through to high quality perovskite films. The method used is capable of achieving highly uniform films at scales of up to 100 mm × 100 mm, a limit of the current process chamber configuration. Control of key film parameters including thickness and stoichiometric composition (bandgap) have also been demonstrated. The next step in our investigations is deposition of a perovskite layer to a polycrystalline silicon sub-cell to realise two-terminal (2T) optically coupled tandem devices.

4.1 Acknowledgements

This research is supported by the Science and Industry Endowment Fund (SIEF) and the authors acknowledge and appreciate funding from SIEF on behalf of CSIRO.

4.2 References

[1] Chin, X. Y., Turkay, D., Steele, J. A., Tabean, S., Eswara, S., Mensi, M., Fiala, P., Wolff, C. M., Paracchino, A., Artuk, K., Jacobs, D., Guesnay, Q., Sahli, F., Andreatta, G., Boccard, M., Jeangros, Q., & Ballif, C. (2023). Interface passivation for 31.25%-efficient perovskite/silicon tandem solar cells. *Science*, 383(162), 1-10. https://doi.org/10.1126/science.adg0091

[2] Hossain, M. I., Qarony, W., Ma, S., Zeng, L., Knipp,

D., & Tsang, Y. H. (2019). Perovskite/Silicon Tandem Solar Cells: From Detailed Balance Limit Calculations to Photon Management. *Nano-Micro Letters*, 11(58), 1-20. https://doi.org/10.1007/s40820-019-0287-8

[3] Jones, T., Wilson, G., Anderson, K., Hollenkamp, A., Duffy, N., & Firet, N. (2015). Process of forming a photoactive layer of an optoelectronic device. *Google Patents*. https://patents.google.com/patent/WO2016094966A1/en

[4] Hwang, J.-K., Jeong, S.-H., Kim, D., Lee, H.-S., & Kang, Y. (2023). A Review on Dry Deposition Techniques: Pathways to Enhanced Perovskite Solar Cells. *Energies*, 16(16), 5977. https://doi.org/10.3390/en16165977

[5] Chandler, D. L. (2015). Explained: chemical vapor deposition. *MIT News*. https://news.mit.edu/2015/explained-chemical-vapor-deposition-0619

[6] Leyden, M. R., Jiang, Y., & Qi, Y. (2016). Chemical vapor deposition grown formamidinium perovskite solar modules with high steady state power and thermal stability. *Journal of Materials Chemistry A*, 4(34), 13125-13132. https://doi.org/10.1039/C6TA04267H

[7] Qiu, L., He, S., Jiang, Y., Son, D.-Y., Ono, L. K., Liu, Z., Kim, T., Bouloumis, T., Kazaoui, S., & Qi, Y. (2019). Hybrid chemical vapor deposition enables scalable and stable Cs-FA mixed cation perovskite solar modules. *Journal of Materials Chemistry A*, 7(12), 6920-6930. https://doi.org/10.1039/C9TA00239A

[8] Ma, S., Sansoni, S., Gatti, T., Fino, P., Liu, G., & Lamberti, F. (2023). Research Progress on Homogeneous Fabrication of Large-Area Perovskite Films by Spray Coating. *Crystals*, 13(2), 216. https://doi.org/10.3390/cryst13020216

41st European Photovoltaic Solar Energy Conference and Exhibition

PHOTOLUMINESCENCE AND LIFETIME STABILITY OF PENTACENE AND OXIDE PEROVSKITES NANOPARTICLES FILMS ON NANOTEXTURED SILICON SUBSTRATE

Rémi Ndioukane[1]*, Diouma Kobor[1], Sergio de Armas Rillo[2], Fernando Lahoz Zamarro[2]

[1] Laboratoire de Chimie et de Physique des Matériaux (LCPM), University Assane Seck of Ziguinchor, senegal
[2] Dept. Física, Universidad de La Laguna, Santa Cruz de Tenerife, Spain
Address: Uniersity Assane Seck of Ziguinchor, Néma 2, PoB 523, Ziguinchor, Senegal
Email: remindioukane@gmail.com

ABSTRACT: The present work proposes a new way to obtain stable ferroelectric oxide perovskites with large photoluminescence and very high and stable lifetime beyond that of halogenated perovskites and requiring only ferroelectric nanoparticles, dispersed in phosphoric based gel, thin layers. A protective perovskite layer was deposited on top of the nanoparticle´s film. The typical red-NIR emission band around 650-750 nm associated to the pentacene layer was detected under excitation at 320 nm. In addition to this, a broad visible emission band was observed and became dominant under excitation at around 380 nm, which is attributed to the perovskite nanoparticles. For each pair of doped-undoped, the 1% Mn-doped showed a more intense emission. Consistently, the doping produced an increase of emission intensity for both the pentacene and the perovskites nanoparticles. In the first case with an intensity ratio between 2.27 - 7.70 while, in the second excitation length at 380 nm, the increase of the emission intensity was ranged between 1.6 - 3.14 ration, showing clearly that Mn doping improves the emission intensity. In addition to this, a study of the stability of the luminescence properties was included.
Keywords: Nanoparticle, Perovskite, Pentacene, Thin film

1 INTRODUCTION

Most recently, the inorganic PZN-PT perovskite materials have been investigated for their excellent and stable properties. This ferroelectric material shows other properties such as magnetism when reduced into nanopowder form [1]. PZN-PT single crystals showed properties up to 10 times more interesting than those of the ferroelectric perovskite ceramics currently used as actuators and sensors. However, their fabrication as thin layers form is a challenge due to the presence of a pyrochlore phase around 950°C making difficult to integrate them in electronic or solar cell devices. To integrate them into silicon nanostructures, Ndioukane et al. realized PZN-PT nanoparticles deposition, already synthesized by the so-called solution flux method [2], as thin film on p-type nanoporous silicon substrate.

In this paper we present our results on the photoluminescence (PL) properties of the PZN-PT samples as a function of fabrication parameters, namely HF etching time and doping with Mn atoms. Additionally, a study of the statibily of the PL properties after 5 months is included.

2 MATERIALS AND METHODS

2.1 Samples

The luminescent properties of 6 samples of thin films of nanoperovskites in a gel matrix were determined in January 2023.. In all the cases a pentacene layer was deposited on top of the structures for protective purposes. For the stability study we chose the two samples with the higher intensity luminescence, S49 and S59. Their PL properties were investigated five moths later, in May.

Table I: Presentation of the samples,

Sample	Description	Etching	Doped
S43	Nanotexturized Si + PZN-4-SPT	[HF] = 4.8, 60min	-
S52	Nanotexturized Si + PZN-4-SPT	[HF] = 4.8, 60min	1%Mn
S47	Nanotexturized Si + PZN-4-SPT	[HF] = 9.6, 60min	-
S55	Nanotexturized Si + PZN-4-SPT	[HF] = 9.6, 60min	1%Mn
S49	Nanotexturized Si + PZN-4-SPT	[HF] = 22.8, 30min	-
S59	Nanotexturized Si + PZN-4-SPT	[HF] = 22.8, 30min	1%Mn

2.2 Emission

Fluorescence emission was determined using a FLS-1000 fluorimeter from Edinburgh Instruments, equipped with 450-W xenon arc lamp for excitation.

2.3 Lifetimes

For time-resolved emission spectra an Edimburgh Instruments LifeSpec II Fuorescence spectrometer, equipped with a multichannel plate photomultiplier (single photon counting) for photon detection, was used. The samples were excited with two different Edinburgh Instruments picosecond-pulsed diode lasers, EPL470 and EPL375 for 470 nm and 375 nm, respectively. The laser pulses had a temporal width at half maximum of 80 ps. The instruments were controlled via the Edinburgh Instruments F9000 acquisition software, and the analysis of the decay data was treated with the Edinburgh Instruments FAST software.

3 RESULTS AND DISCUSSION

3.1 Photoluminescence study as a function of fabrication paramenters.

41st European Photovoltaic Solar Energy Conference and Exhibition

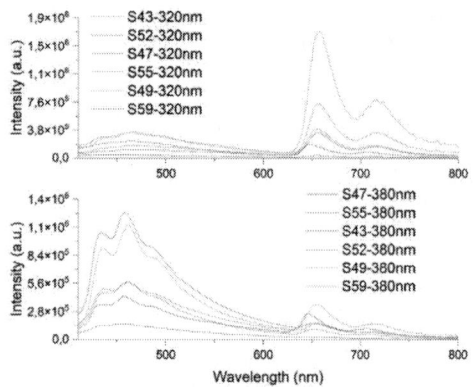

Figure 1. Fluorescence emission of the samples under excitation at 320 nm (top panel) and 380 nm (bottom panel.

The emission band at around 450 nm is due to the nanoperovskites and the band at about 650 nm is assigned to the pentancene protective layer.

The most intense emission from the perovskite nanoparticles is observed in samples S49 and S59.

3.2 Excitation

The excitation was compared to the spectra obtained for S43 at two detection wavelengths: 455 nm, for the nanoperovskites, and 655 nm, for the pentacene. The results can be seen in Fig. 2.

The behaviour remains roughly the same. The most important aspect of the excitation spectra is that the pentacene is excited mainly below 350 nm, as was the case in January. The broad peaks between 360 and 425 nm when detecting at 455nm in May can be due to differences in the etching process of the samples, not to the time that has passed between measurements (notice that the excitation spectra from January comes from a different, less emissive sample).

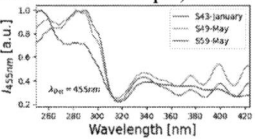

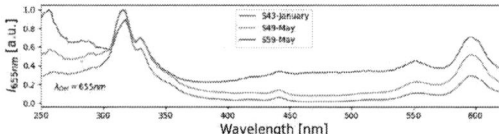

Figure 2: Excitation spectra. Top panel: detection at 455, LP 455 nm. Bottom panel, detection 655 nm, LP645

3.3 Emission

According to the absorption and excitation spectra obtained, two excitation wavelengths were chosen, 320 nm and 380 nm. The first is expected to directly excite the pentacene, which does not absorb much at 380 nm.

There are important differences in the emission spectra when exciting at 320 nm (which is the wavelength that excites the pentacene more directly). In this case, for both S49 and S59, we find a new broad emission band in shorter wavelengths, corresponding to the nanoperovskites. For some reason, the pentacene is not being excited that much, and in turn the nanoperovskites are also absorbing at

shorter wavelengths than in January. It could be that the pentacene layer has degraded while the nanoperovskites in gel have remained stable.

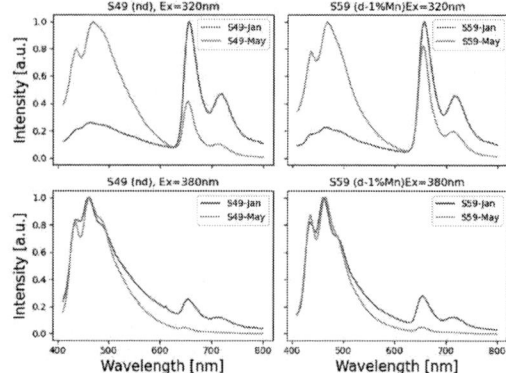

Figure 3: Emission spectra, January vs May. Studies samples: S49 and S59

Pentacene presents an emission peak around 650, while not emitting with significant intensity around 450nm. The peak between 300 and 350 in the excitation spectra when detecting at 650 indicates that the pentacene absorption is important in this range, diminishing above 350 nm, where the fluorophores that emit at 450 nm become dominant.

3.4 Lifetimes

We decided to only check the lifetimes when exciting the nanoperovskites, given that we already stablished in the first measurements that the red emission bands came from the pentacene. However, it could be interesting to interrogate again the pentacene to check if it has in fact degraded or the changes in the emission spectra come from the high inhomogeneity of the sample.

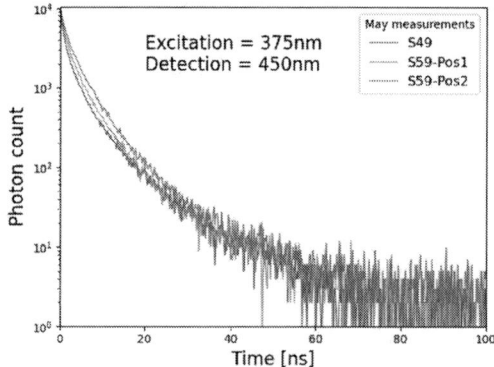

Figure 4: Decay of S49 and S59 in May

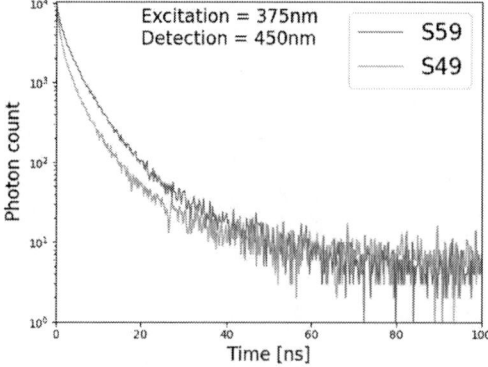

Figure 5: Decay of S49 and S59 in January

The samples exhibit a complex decay, with more than one photon emitting process contributing to the fluorescence. The decays will be fitted to a triple exponential of the form:

$$I(t) = B_1 e^{-t/\tau_1} + B_2 e^{-t/\tau_2} + B_3 e^{-t/\tau_3} \qquad (1)$$

Where τ_1, τ_2 and τ_3 are the decay constants and B_1, B_2 and B_3 represent the pre-exponential factors. Equation [1] is useful to define an average life-time as [3]:

$$<\tau> = \frac{B_1 \tau_1^2 + B_2 \tau_2^2 + B_3 \tau_3^2}{B_1 \tau_1 + B_2 \tau_2 + B_3 \tau_3}$$

The decay curves were fitted to Eq. (1). Table II shows the average life-time obtained from the fitting procedure. Doped samples have longer lifetimes, which is in accordance with the increased emission intensity.

Table II: The average lifetime estimation of the pentacene layer obtained from the intensity decay of the fluorescence detected at 450 nm, under excitation at 375 nm.

Sample	B1	B2	B3	τ1[ns]	τ2[ns]	τ3[ns]	⟨τ⟩[ns]
S49-Jan	8782.1	3500.3	363.17	0.7	2.9	9.1	**3.3**
S49-May	7705,2	3863,2	644,75	0,6	2,8	9,3	**4,1**
S59-Jan	6541.8	4356.2	247.2	1.4	4.3	13.4	**4.5**
S59-May - Pos 1	6866,4	4386,5	560,92	0,8	3,4	9,3	**4,0**
S59-May - Pos 2	6442,3	4764,1	312,73	1,3	4,3	12,7	**4,6**

S59 exhibited 2 very different lifetimes in two separate regions of the sample, which illustrates the inhomogeneity of the sample. The second position had almost exactly the same complex lifetime than 5 months before, while the other position had a much shorter decay time.

S49 also has a longer decay time than before, but it can be explained both by sample degradation or because its inherent inhomogeneity.

3.5 Lifetimes of pentacene emission

In January, the emission lifetime in the pentacene bands (620-650nm) revealed two excitation pathways: first, direct excitation of pentacene (Ex 470nm, ⟨τ⟩ = 7-8 ns) and energy transfer from the nanoperovskites, which exhibited significantly longer lifetimes (Ex 375nm, ⟨τ⟩ = 1.5-2 µs). In order to repeat those measurements, and given that the emission spectra obtained pointed towards degradation of the pentacene, additional emission spectra at those two excitation wavelengths were performed (Fig.6).

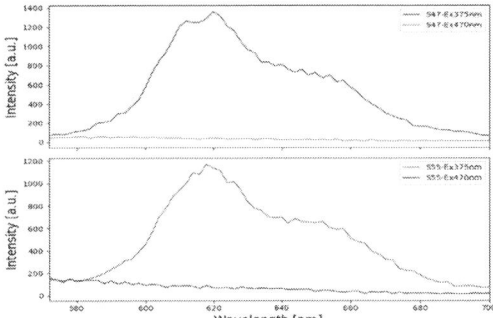

Figure 6: Emission spectra of S47 and S55: pentacene band

Direct excitation of the pentacene (ex 470 nm) produced no emission. Even under long integration times no structure was found, as can be seen in the normalized spectra of fig.7.

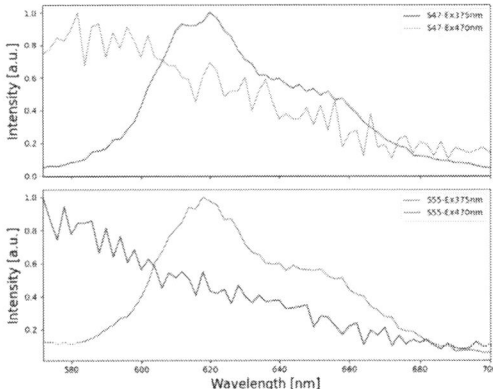

Figure 7: Normalized spectra from figure 5

Consequently, lifetime measurements under 470 nm excitation were impossible, since there was no emission distinguishable from noise. Exciting at shorter wavelengths, however, produced some faint emission (Fig.8).

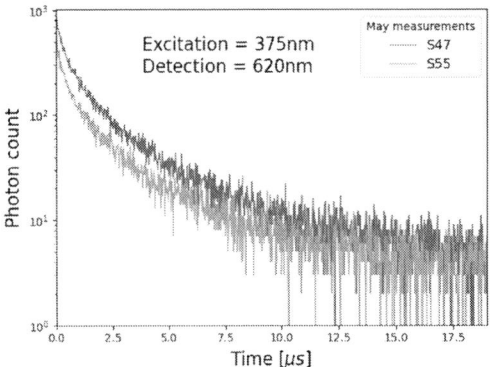

Figure 8: Decay of S47 and S55 in May

Decays that were fitted with the following lifetimes as result in Table III. Which are very similar to the results from January (Table IV).

Table III: The average lifetime estimation of the pentacene layer obtained from the intensity decay of the fluorescence detected at 620 nm, under excitation at 375 nm.

Sample	B1	B2	B3	τ1[ns]	τ2[µs]	τ3[µs]	⟨τ⟩[µs]

S55	1248.8	318.2	125.8	24.3	0.4	2.3	1.6
S47	492.7	339.2	139.7	216.9	0.9	3.0	1.9

Table IV: Comparison of lifespan between January and May

Sample	⟨τ⟩ January [μs]	⟨τ⟩ May[μs]
S55	1.5	1.6
S47	1.8	1.9

These results indicate that the outer pentacene layer has degraded over time. The samples were stored in darkness, but in standard conditions (unregulated temperature and humidity). The nanoperovskites, however, did not change their luminosity characteristics.

4 CONCLUSION

The typical red-NIR emission band around 650-750 nm associated to the pentacene layer was detected under excitation at 320 nm. In addition to this, a broad visible emission band was observed and became dominant under excitation at around 380 nm, which is attributed to the perovskite nanoparticles. Consistent increase of the luminescence properties was observed when doping with Mn atoms. The nanoperovskites did not change their luminosity characteristics. Further testing in the future can determine if this stability is due to protection from the now-degraded pentacene or because inherent stability from the nanoperovskites themselves.

Acknowledgement: This work was carried out with the aid of a grant from UNESCO-TWAS and the Swedish International Development Cooperation Agency (Sida). The views expressed herein do not necessarily represent those of UNESCOTWAS, Sida or its Board of Governors.

References

[1] Ndioukane, R., et al. (2019) EPL, 125, 47004.
[2] Dupret, F. and Van Den Bogaert, N. (1994), Part B. 2.
[3] Lakowicz, J. R. Principles of Fluorescence Spectroscopy **2006** (Berlin: Springer), ch. 04.

41st European Photovoltaic Solar Energy Conference and Exhibition

Compositional Engineering of Double-cation Single-halide Perovskite for Efficient Solar Cell fabrication under Air Ambient Conditions

Mrittika Paul[a,*], Binita Boro[b], Amreesh Chandra[c], Trilok Singh[d,*]

[a] School of Energy Science and Engineering, Indian Institute of Technology Kharagpur, 721302, India

[b] School of Nano Science and Technology, Indian Institute of Technology Kharagpur, 721302, India

[c] Department of Physics, Indian Institute of Technology Kharagpur, 721302, India

[d] Department of Energy Science and Engineering, Indian Institute of Technology Delhi, Hauz Khas, New Delhi 110016, India

Email address(s): *paulmrittika96@gmail.com, *trilok.singh@dese.iitd.ac.in

BACKGROUND

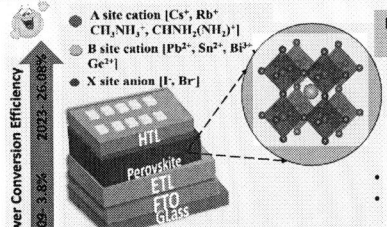

- A site cation [Cs⁺, Rb⁺ CH₃NH₃⁺, CHNH₂(NH₂)⁺]
- B site cation [Pb²⁺, Sn²⁺, Bi³⁺, Ge²⁺]
- X site anion [I⁻, Br⁻]

A typical n-i-p perovskite solar cell structure

Factors affecting Stability of PSCs
Moisture, Oxygen, Light, Heat, Electrical bias

Intrinsic Degradation
- Ion migration
- Metal-perovskite reaction
- Lattice stress

INTRODUCTION

- Compositional tuning of FA-based perovskites by a systematic incorporation of Cs (x= 0.05, 0.10, 0.15, and 0.25), resulting in $FA_{1-x}Cs_xPbI_3$ perovskites under air-ambient conditions (relative humidity (RH) < 30%) is employed.
- Effects of A-site cation composition on the perovskite thin film and overall device performance is studied.
- Precise control of the residual stress management of the perovskite film resulting in enhanced photophysical properties and long-term outdoor stability is achieved.

METHODS

Step 1: FTO substrate cleaning by ultrasonication for 30 mins is each of:
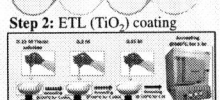

Step 2: ETL (TiO₂) coating
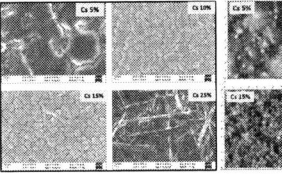

Step 3: Perovskite and HTL coating
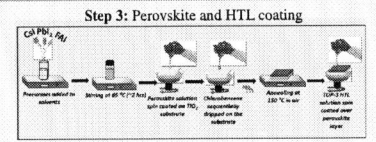
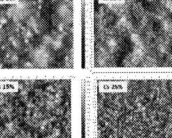
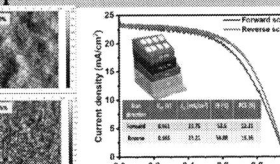

Step 4: Metal contact (Au) deposition via thermal evaporation

RESULTS AND DISCUSSIONS

Effect on the structural and optical properties

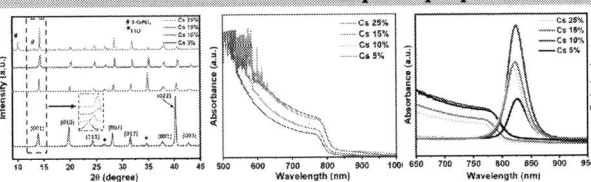

Inferences
- For x= 0.05, 0.10, and 0.15, only black $FAPbI_3$ phase is formed, but x= 0.25 shows δ-$CsPbI_3$ peaks.
- The (001) peak positions shift towards higher angle with increasing x, indicating a decrease in the lattice parameter.
- The decreased FWHMs by adding Cs imply a larger crystallite size. The slight increase in FWHM for Cs 15% might be due to residual stress.
- Absorption onset shows a blueshift when Cs concentration increases beyond 10% (also evident from the magnified normalized PL peaks).

Effect on the interionic bonds of the perovskite structure

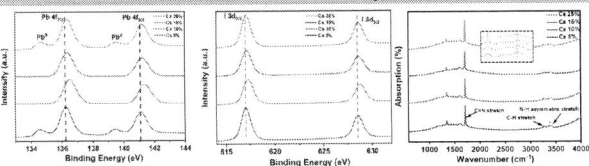

Inferences
- Cs 5% and 25% films show metallic lead peaks (Pb⁰) possibly due to incomplete conversion of PbI₂ to perovskite.
- Pb 4f peaks shift to higher binding energy due to increasing Cs indicates increase in cationic charge of Pb ions, which is responsible for the shrinkage of lattice parameter. This causes Pb-I bond changes.
- FTIR plot shows slight shift of peaks corresponding to C-H and N-H stretch, indicating increased bond strength due to Cs incorporation.

Effect on morphology of perovskite thin film

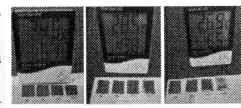

Device performance of $FA_{0.9}Cs_{0.1}PbI_3$ based PSC

Inferences
- Cs 5% film shows inhomogeneous grain growth, unclear grain boundaries.
- Cs 25% film depicts micro sized rods which might be due to presence of δ-$CsPbI_3$
- Cs 10% and 15% films shows compact morphology with homogeneous grains. The grain size is more for Cs 10%.

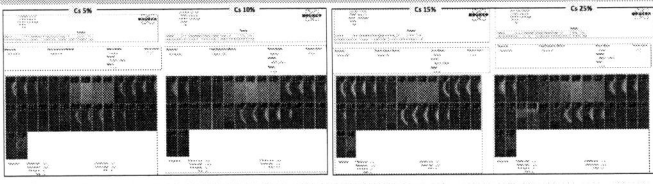

The environmental stability of the as-prepared $FA_{1-x}Cs_xPbI_3$ films with varying Cs concentration (fresh, after 15 days, after 30 days).

Residual stress calculation of perovskite thin films

Inferences
- Nanoindentation analysis done to calculate the reduced moduli (E_r) of the $FA_{1-x}Cs_xPbI_3$ films (x= 0.05, 0.10, 0.15, 0.25).
- The Young's modulus (E) is calculated using the formula:

$$\frac{1}{E_r} = \frac{1-\nu_{sample}^2}{E_{sample}} - \frac{1-\nu_{indenter}^2}{E_{indenter}}$$

- The residual stress of the perovskite films (calculated from XRD) shows positive stress (tensile) for Cs 5% and negative stress (compressive) for Cs 10%, 15%, and 25% due to shrinkage of cubo-octahedral volume of A cation.
- Compressive stress enables better stability of the perovskite structure.

CONCLUSIONS

- Mixed FA-Cs cation results in improved ambient stability of perovskite films.
- Cs 10% film shows the best optoelectronic features in comparison to other concentrations.
- Compressive residual stress leads to better morphology and thus, enhanced stability and low hysteresis in device performance.
- In this study, for HTL, Top-3 is used as a cheaper and more stable alternative of Spiro-OMeTAD.

References: 1. Lee, J. W. et al. Formamidinium and Cesium Hybridization for Photo- and Moisture-Stable Perovskite Solar Cell. Adv. Energy Mater. 5, 1501310 (2015). 2. Binita et al. " Thermal Syability Analysis of Formamidinium-Cesium-based Lead Halide Perovskite Solar Cells Fabricated under Air Ambient Conditions " Energy Technology 2024, 12. 2400034

Acknowledgement: I am grateful to EUPVSEC-2024 for the opportunity to present my research work; School of Energy Science and Engineering, IIT Kharagpur for research facility and MHRD for providing the fellowship.

41st European Photovoltaic Solar Energy Conference and Exhibition

Interface Engineering For Perovskite Solar Cells Using Polymer-based Antisolvent Technique

Lingeswaran Arunagiri, Feng Wang and Feng Gao*
Linköping University, Sweden

Objective

Perovskite solar cells with spiro-OMeTAD as a hole transporting layer (HTL) can attain power conversion efficiencies (PCE) greater than 25%. Despite this astounding performance, there is always a trade-off between efficiency and device stability induced by benchmark dopants of spiro-OMeTAD.

n-i-p perovskite solar cells

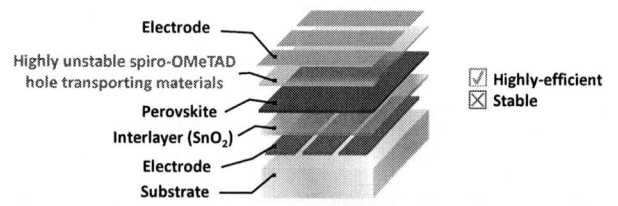

Electrode
Highly unstable spiro-OMeTAD hole transporting materials
Perovskite
Interlayer (SnO_2)
Electrode
Substrate

☑ Highly-efficient
☒ Stable

Development of a stable n-i-p perovskite solar cells without any detrimental additives in the Spiro-OMeTAD HTL.

Method – Antisolvent Strategy

The role of additional additives in spiro-OMeTAD HTL is to fine-tune the energy level alignment to match that of perovskite. We aim to achieve this by tweaking the perovskite surface to modulate the interface between the perovskite and the HTL by utilizing the one-step antisolvent strategy.

Conventional Method

Pure Chlorobenzene → Perovskite

Our Strategy

Mix Organic Polymers in Chlorobenzene → Perovskite

The one-step spin-coating process uses antisolvent pipetting to manage the nucleation and crystallization of perovskite films. Generally, pure solvents such as chlorobenzene (CB), diethyl ether, and toluene have been used as antisolvents. In our case, we employed p-type semiconducting polymer with the antisolvent as a template agent.

Preliminary Data Analysis

≫ How thick is the polymer?
A 12 Å pure polymer layer on top and a ~100 Å mixed region (polymer/perovskite) underneath.

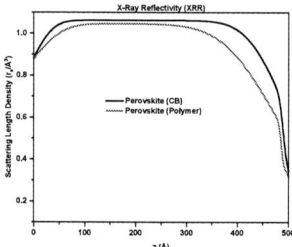

≫ Photoemission electron microscopy (PEEM)
The WF of perovskite (polymer) aligns better with that of spiro-OMeTAD (4.7 eV), allowing for more efficient charge extraction/transportation.

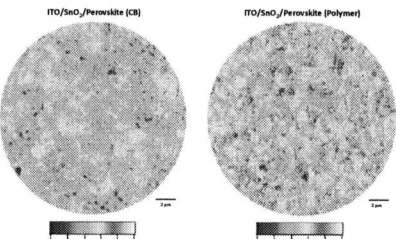

≫ Absorption Spectra
The perovskite absorption increases when polymers are used as a templating agent.

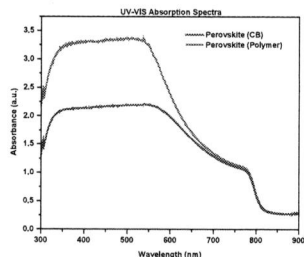

≫ Compared to the control devices (perovskite CB), polymer antisolvent-based perovskite devices had a 0.07 V increase in open-circuit voltage (V_{OC}) and a fill factor (FF) of more than 81%. Time-resolved photoluminescence (TRPL) and the photoluminescence quantum yield (PLQY) studies show that adding polymer significantly reduces non-radiative recombination losses.

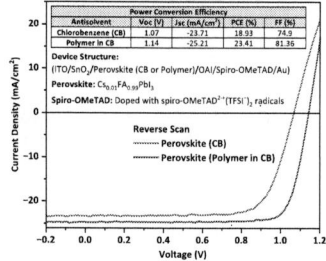

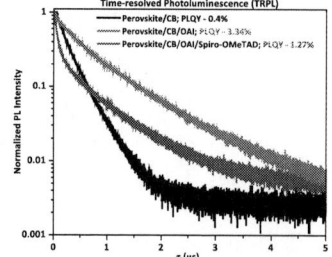

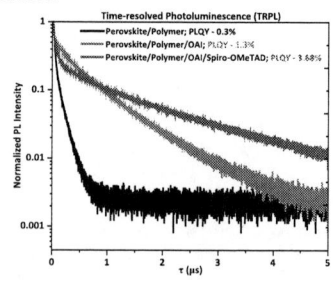

Next step...

Our next step is to answer the following essential research questions:

> Is the polymer affecting the crystallinity of perovskite?

> How do the changes in the WF affect the device performance?

> Can the removal of harmful additives from spiro-OMeTAD increase overall device stability?

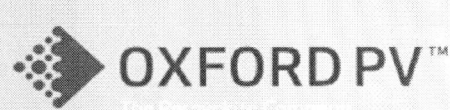

Perovskite record setting silicon tandem modules
Customers expect lower LCOE

41st EU PVSEC
Austria Center Vienna
Vienna, AT

26 September 2024

Christopher Case
Chief Technology Officer

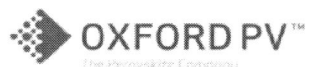

The Oxford PV story

Journey to perovskite solar cell technology

From Professor Henry Snaith's University of Oxford laboratory

2010 Oxford PV established for solid state DSC

2014 Perovskite-on-silicon tandem solar cell development

2016 Thin-film pilot line acquired

2018 Record-breaking cell New research and development campus

2021 Factory build slows with supply chain issues

2024 Commercial modules shipped

2012 OxPV moves to perovskite for BIPV

2015 First perovskite-on-silicon tandem solar cell

2017 First industry-sized perovskite-on-silicon tandem cell produced

2020 World's first perovskite tandem production line under construction

2023 Record-breaking commercial-sized cell production starts

Perovskites have caught attention far beyond the scientific community

"Among the 10 science top breakthrough of 2013."

Science – Dec 2014

"Perovskites are the clean tech material development to watch right now."

The Guardian – Mar 2014

"Perovskite photovoltaic cells have rapidly become one of the hottest areas in energy research over the past few years."

IEEE spectrum – May 2014

" This might be one of the materials that is going to change the game."

NREL – Aug 2014

"Perovskite offers shot at cheaper solar energy."

The Wall Street Journal – Sep 2014

"Perovskites may give silicon cells a run for their money."

The Economist – May 2015

Crystalline silicon technology roadmap

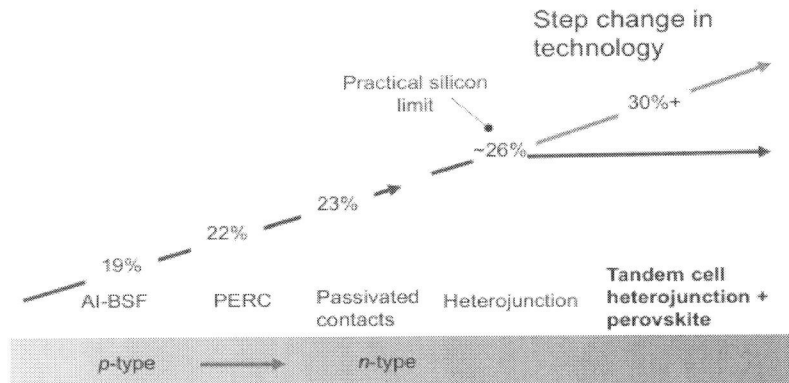

Increased efficiency delivers substantial benefits

Tandem solar modules deliver 20% more power from a panel of the same area

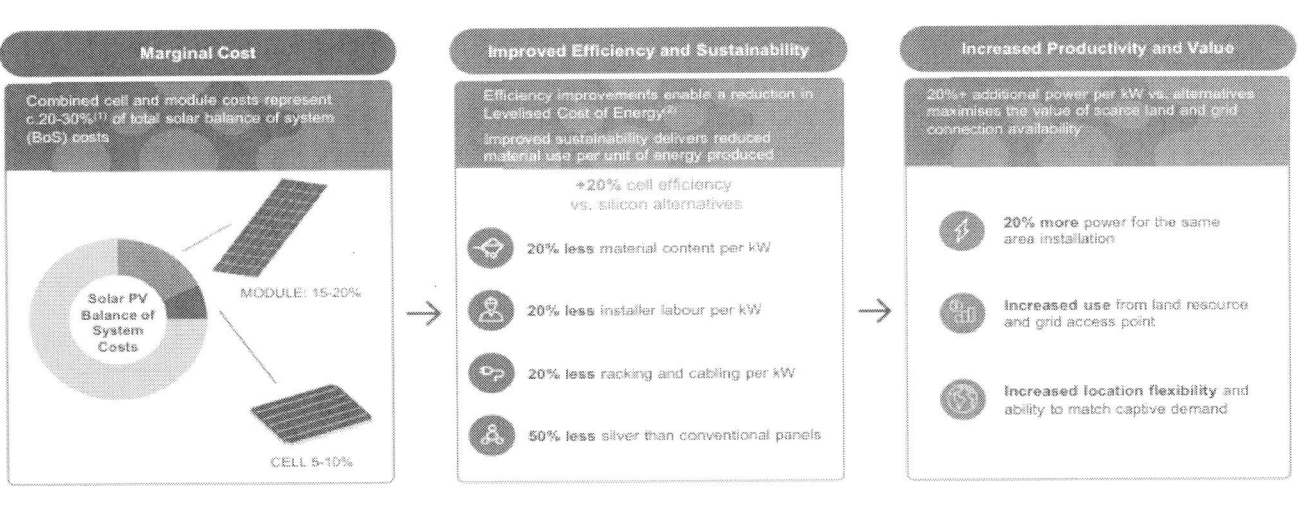

A.W. Blakers *et al.* (1988) PERC cells - 22.8% efficiency

Why did it take so long?

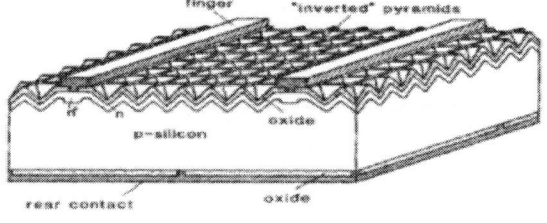

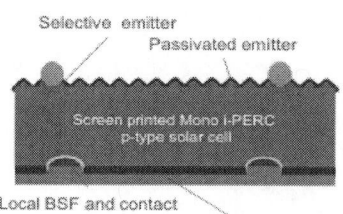

OXFORD PV™

Sidestepping the PV pipeline
Traditional PV vs Oxford PV – managing interdependencies

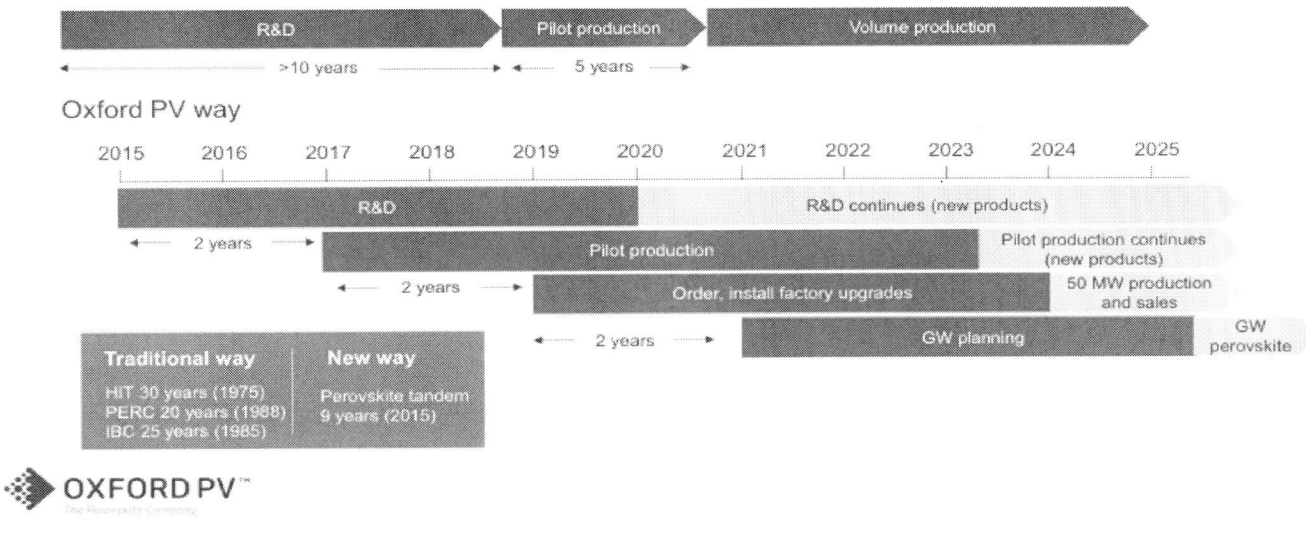

Early devices
Non HJT bottom cells

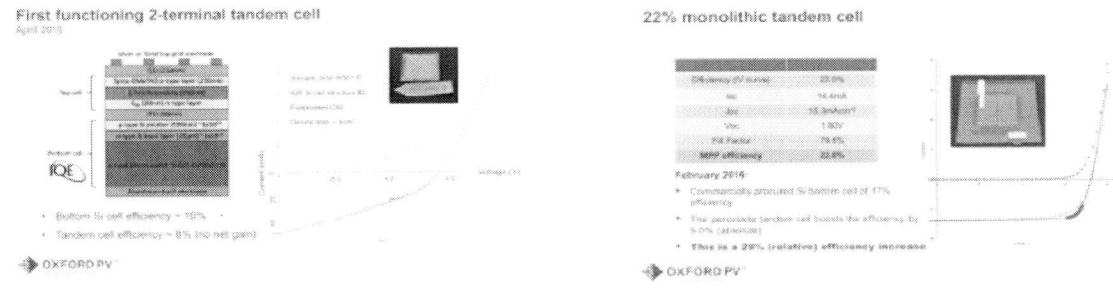

OxfordPV tandem cell development

Concept demonstration large area 21.3% perovskite-Si tandem cell – circa 2015

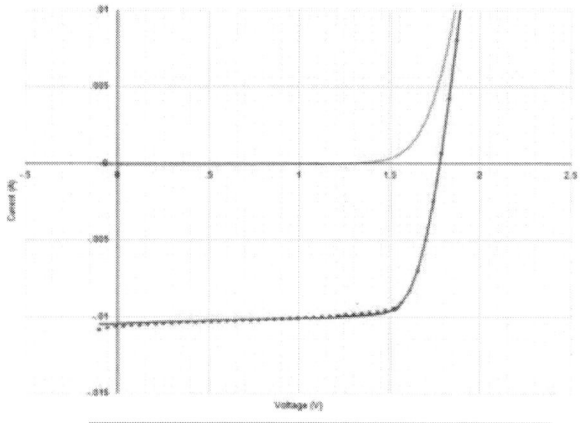

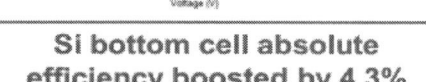

Si bottom cell absolute efficiency boosted by 4.3%

Jsc	16.9mA/cm²
Voc	1.78V
Fill Factor	75.2%
MPP efficiency	22.6%

OXFORD PV™
The Perovskite Company

Selecting the bottom silicon cell

HJT is the best choice

TOPCon vs HJT in tandem

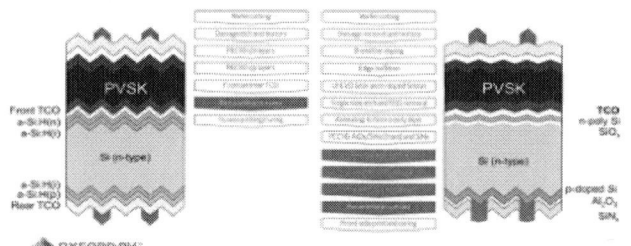

First tandem on commercial HJT bottom cell
April 2016

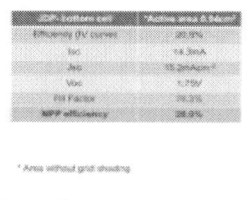

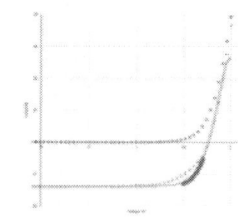

* Area without grid shading

OXFORD PV™

OXFORD PV™
The Perovskite Company

Reliability
Target - multi-decade high-level performance

Challenge
Perovskite photovoltaics have emerged just over ten years ago – how to predict long-term performance on much shorter learning time scale?

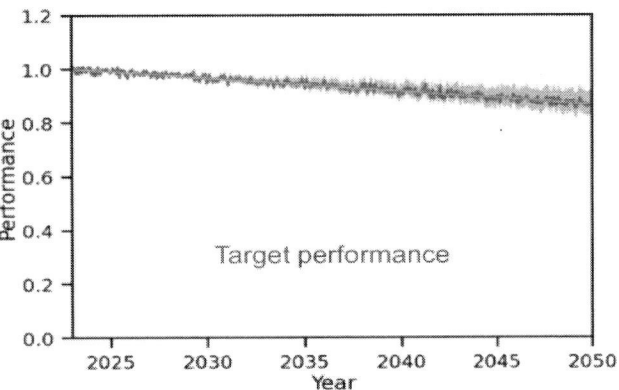

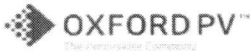

First reliability passes for encapsulated cells
September 2016

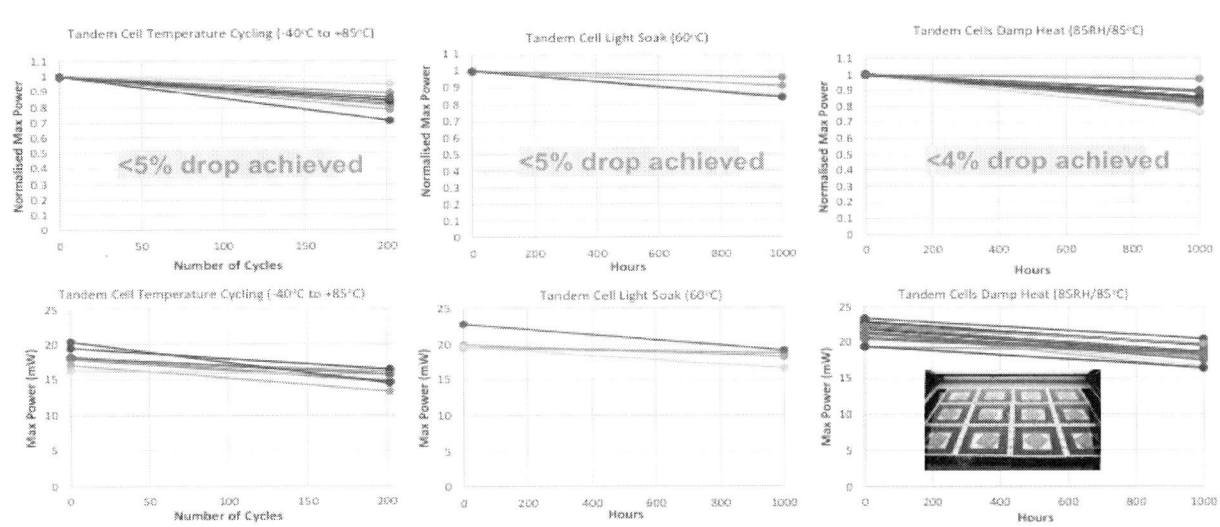

R&D technology development
Critical failure analysis using advanced characterization at all scales

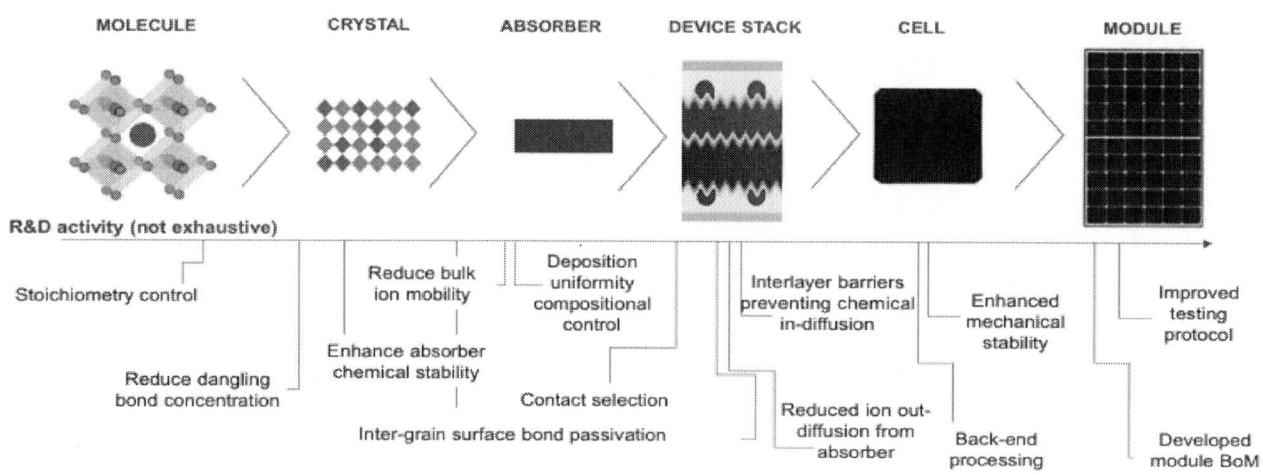

Reliability
Using model insights to iteratively improve stability

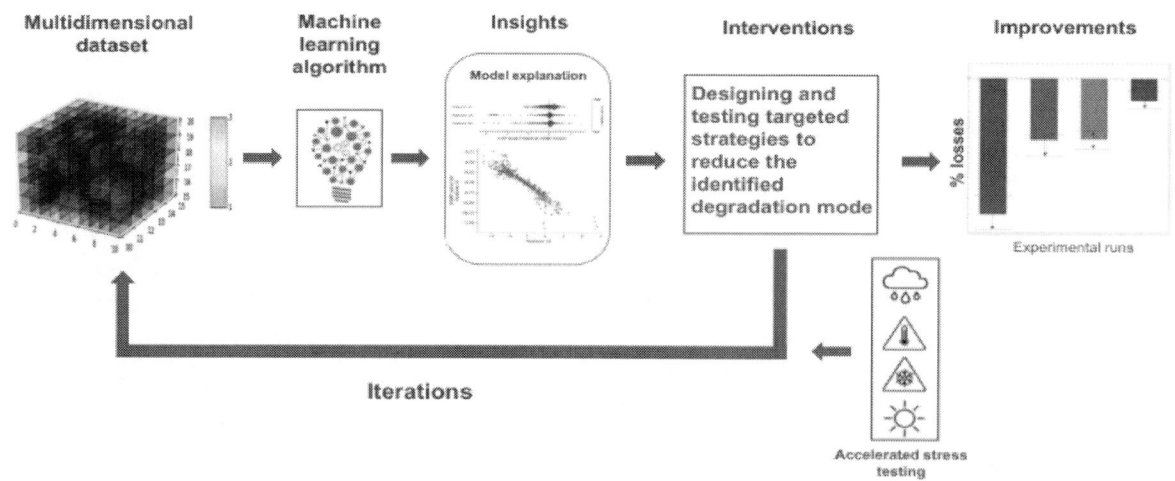

Record large area 28.6% tandem perovskite on silicon
Full area full size measurment certified at Fraunhofer ISE May 2023

Area = (258.14 ± 0.10) cm^2

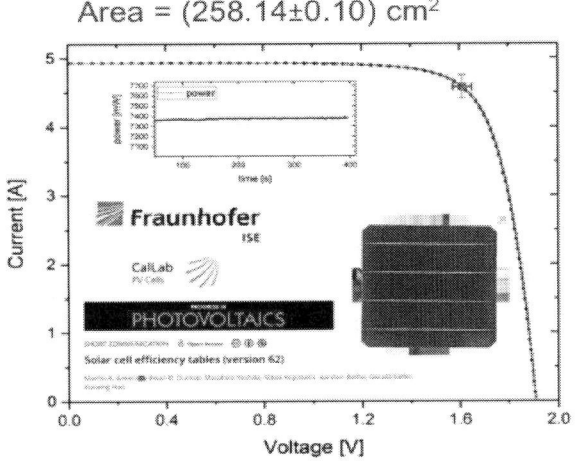

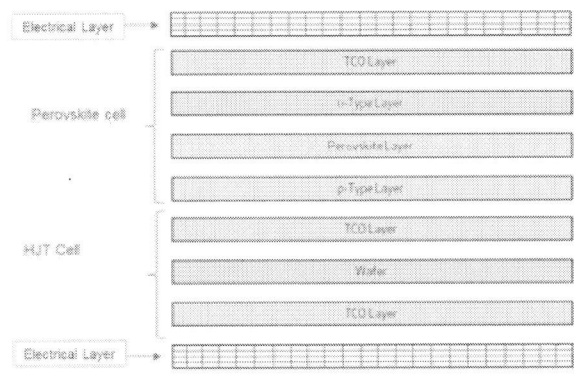

Module approach
Standard materials and assembly

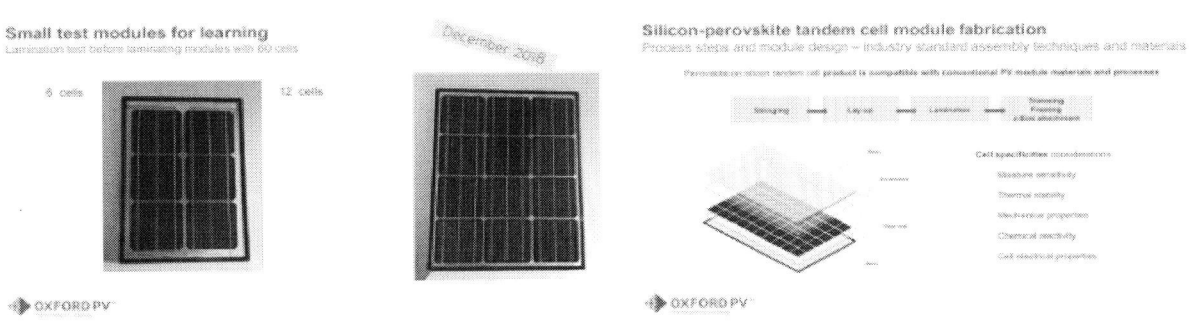

GCL Perovskite

2.0 x 1.0 m perovskite module May 2023

GCL-SI unveils perovskite solar panel with 16.02% efficiency

GCL-SI has launched a new 320 W perovskite solar module. The company guarantees that the 10-year end power output will be at least 90% of the nominal output power, which decreases to 80% after 25 years.

JUNE 26, 2023 EMILIANO BELLINI

Source: www.pvmagazine.com

Microquanta – alpha module

First commercial perovskite module – May 2022

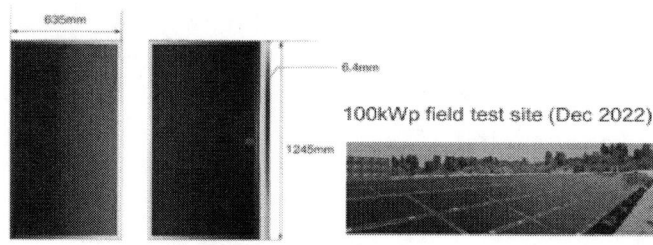

100kWp field test site (Dec 2022)

- 1245 x 635 x 6.4 mm
- Passed IEC 61215 and 61730 (CQC and VDE)

Source: www.pandaily.com/gcl-perovskite-bags-b-round-of-financing-worth-72m

OXFORD PV™
The Perovskite Company

OXFORD PV™
The Perovskite Company

At last:

Production

OXFORD PV™
The Perovskite Company

Major reliability steps

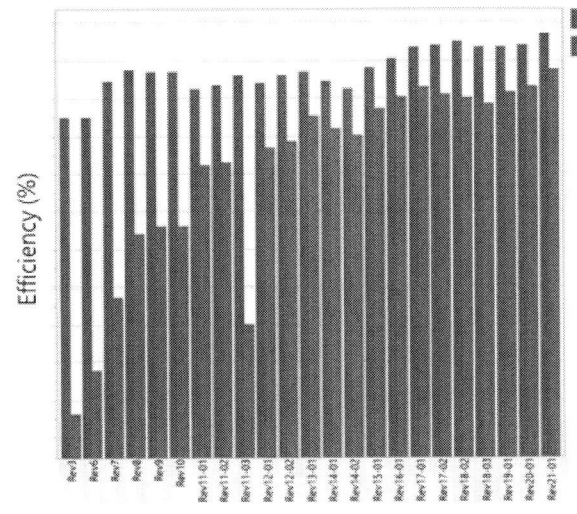

Efficiency
Efficiency after light-and-heat accelerated test

- Continued improvements to Oxford PV baseline build
- Focused development on response to accelerated stress test since 2018

OXFORD PV™
The Perovskite Company

First 25% full area module efficiency
December 2023

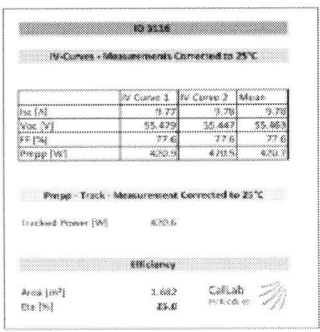

- 25.0% efficiency full size module
- Verified measurement by Fraunhofer
- Record efficiency for full size module*

* Record for full size Si module is 24.8% by Jinko Solar

OXFORD PV™
The Perovskite Company

Record ~27% efficiency module – late news
June 2024

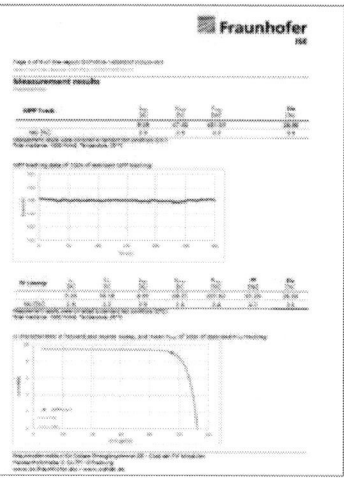

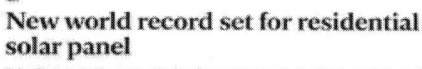

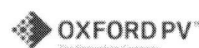

2024 Launch Product – 25-year performance and 15-year product warranty

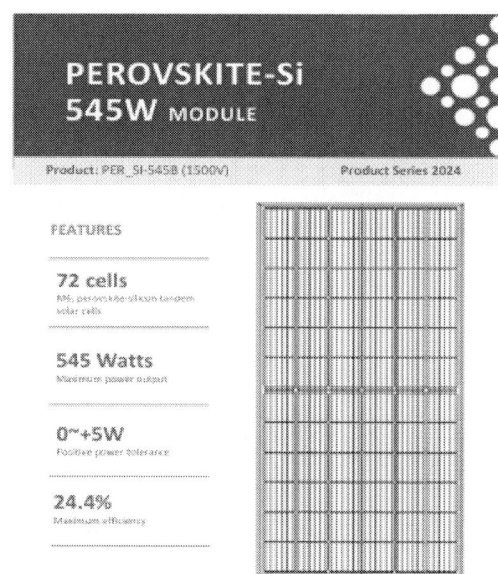

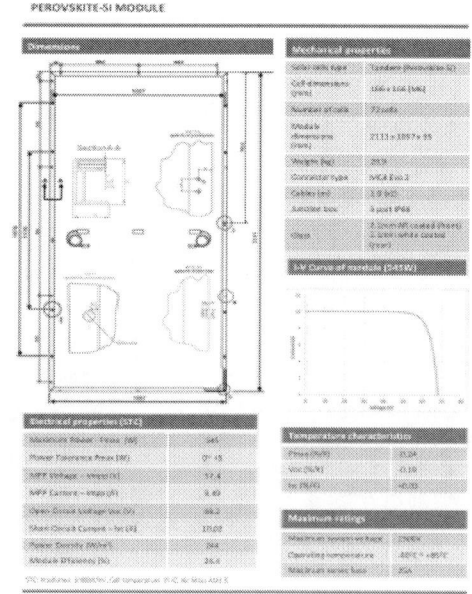

First commercial module product

June 2024

Production module exhibited at Intersolar Munich 2024

Assembled by a EU third party module manufacturer

72 cells "utility" design - up to 545Wp
        50V module and 1500 V string

Up to 24.4% efficiency

Available in limited quantities

Residential versions planned for 2026

Module certification – IEC 61215 protocol

Only a draft certification can be obtained at the moment

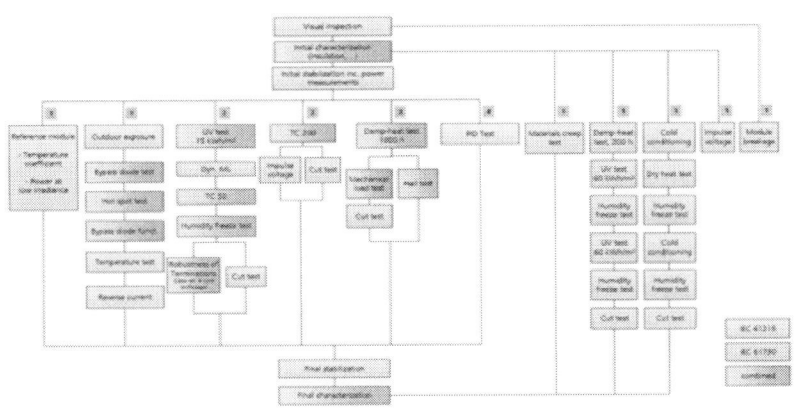

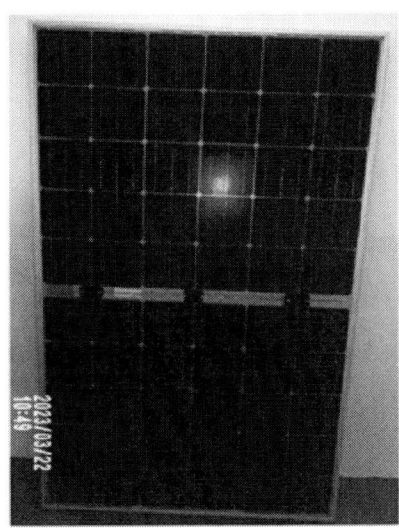

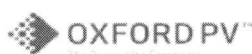

Commercial modules
Post some rel tests

Photos before and after TC200
M05

EL before and after TC200
M05

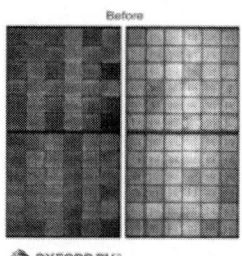
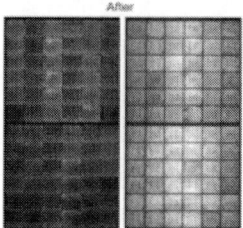

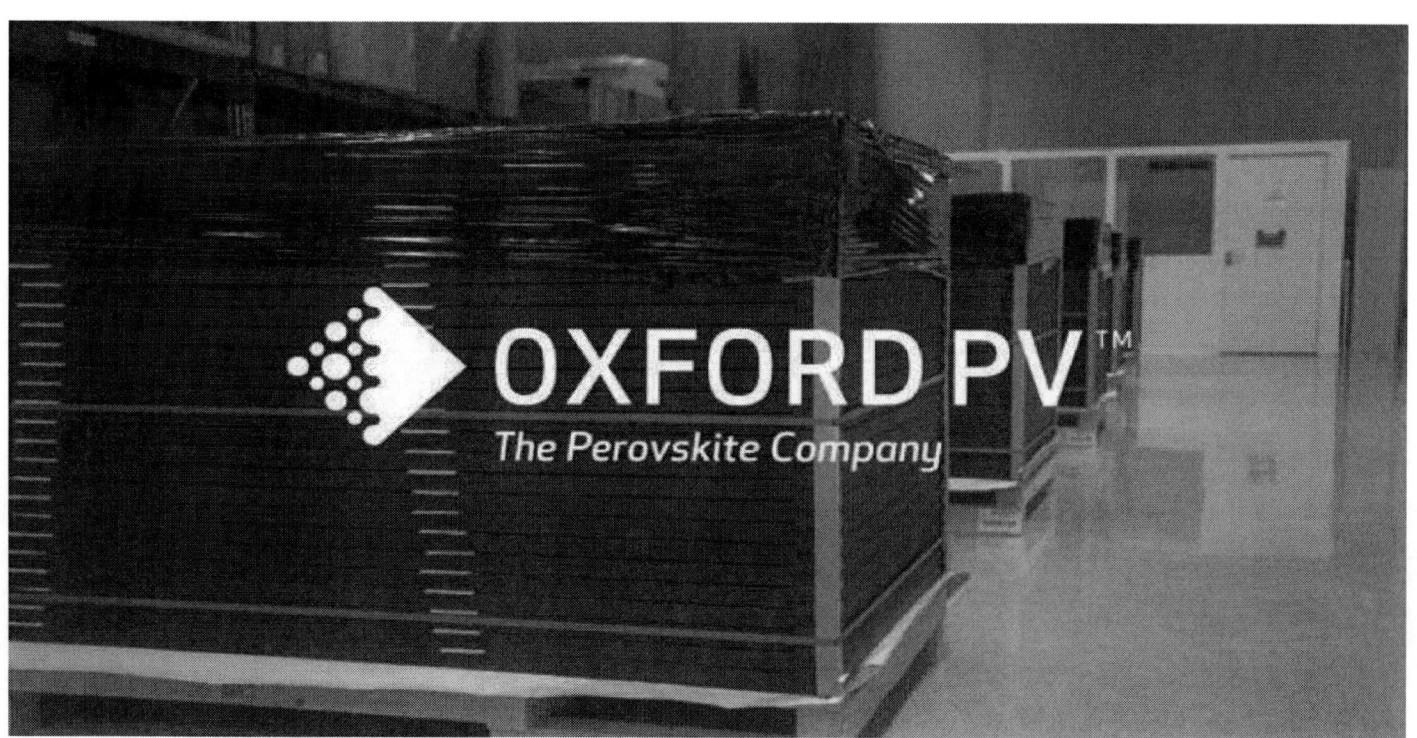

LCOE improves with scaling

Expansion to 1GW plus production levels extends market to utility scale

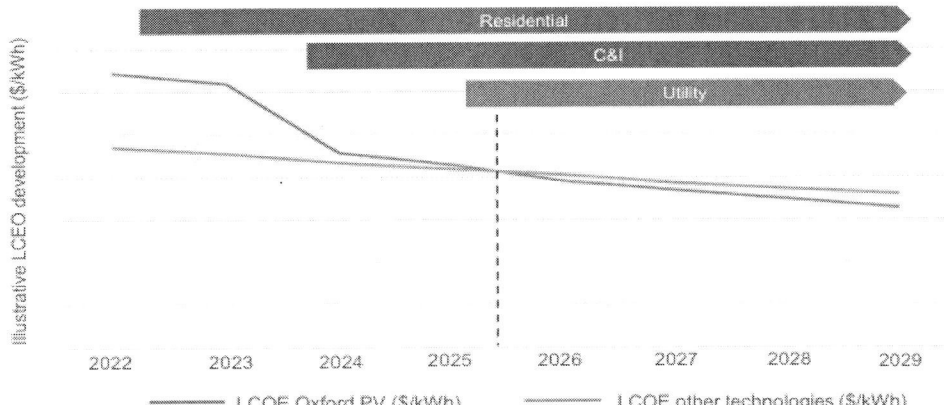

Source: GTM Research, ITRPV 13th Edition, Oxford PV (2023)

Outdoor module testing

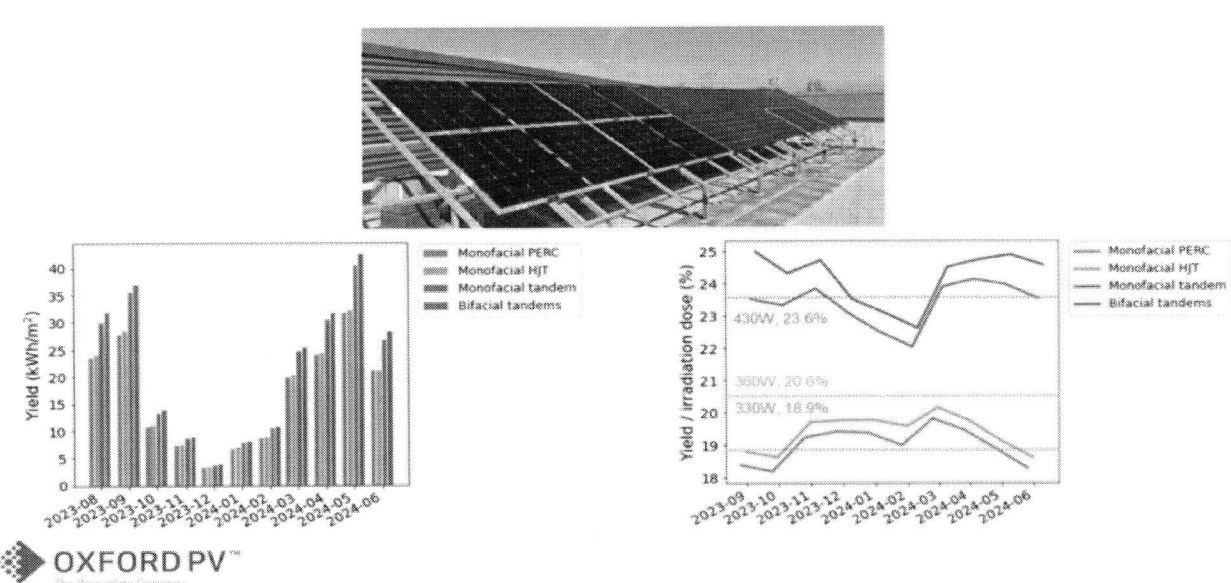

Supersize: towards gigawatt-level manufacturing

Over the rainbow? Or Europe or the US?

OXFORD PV™
The Perovskite Company

Site requirements
Essentials of building and facility

High Volume Tandem Cell Manufacturing		5GWp/a
Land Plot	Land requirement:	Light industrial production
	Size (minimum)	20 ha
	Shape	4:5 or square
Building	Total floor space	60.000 m²
	Hall height	8m
	Mezzanine floor	selected areas
	Basement	none
	Air Conditioning	60% rel.h
	Temperature	21 °C ± 2°C
	Cleanroom	ISO9
	Office area linked to hall	1.000m²
Facility	Power Supply (ave.)	32,5 MW
	Process Cooling Water (ave.)	24 MW
	Compressed Dry Air	7200 m³/h
	Nitrogen (High Grade)	400 m³/h
	City water	10 m³/h
	Argon 5N	Yes
Sepecial Requirements	ESD Floor	main manufacturing only
	Storage for technical gases	Flammable and toxic
	Lighting	500 lm
Accessibility	Truck access	Yes
	Parking	400 lots
	Public transportation	

Oxford PV within the global IP landscape

>570 patents/applications – collaboration and licensing opportunities

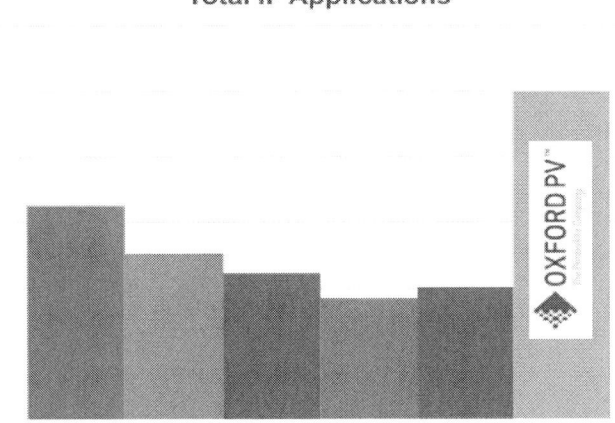

Total IP Applications

5 top companies with large IP portfolio

*Numbers are normalised

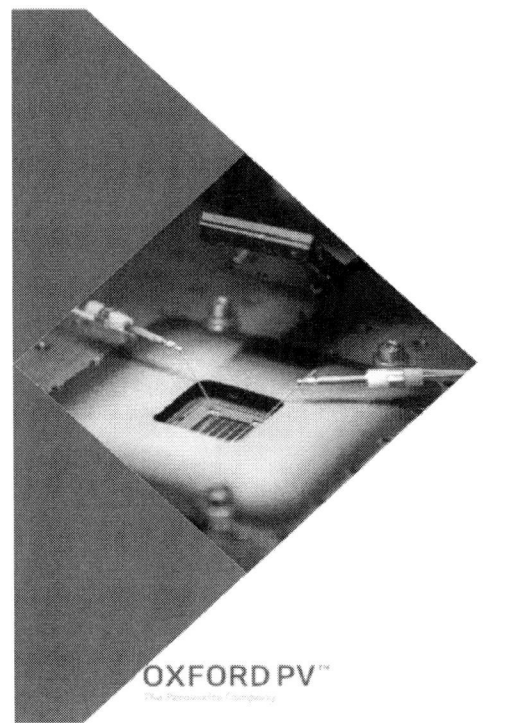

The future

Perovskite in the world
Headquarters locations of **known** companies

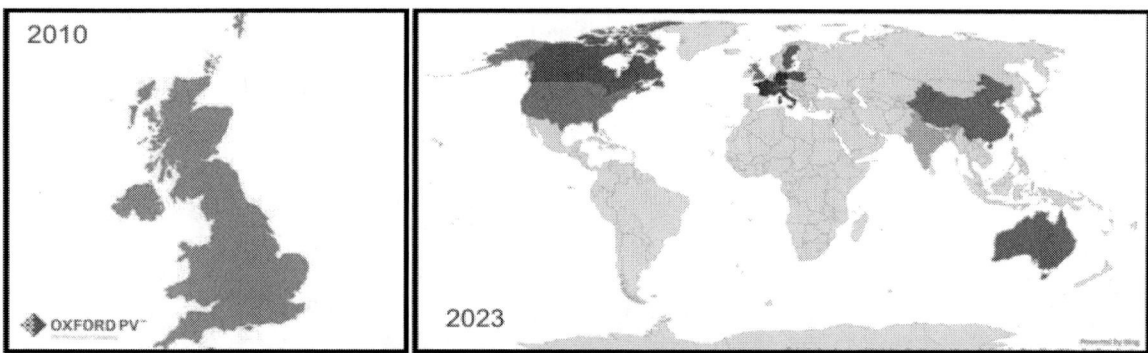

Considers:

- Known perovskite solar manufacturers
- Companies with perovskite PV in their roadmap

Perovskite PV in the world
Known perovskite solar manufacturers or with perovskite PV in their roadmap (not exhaustive)

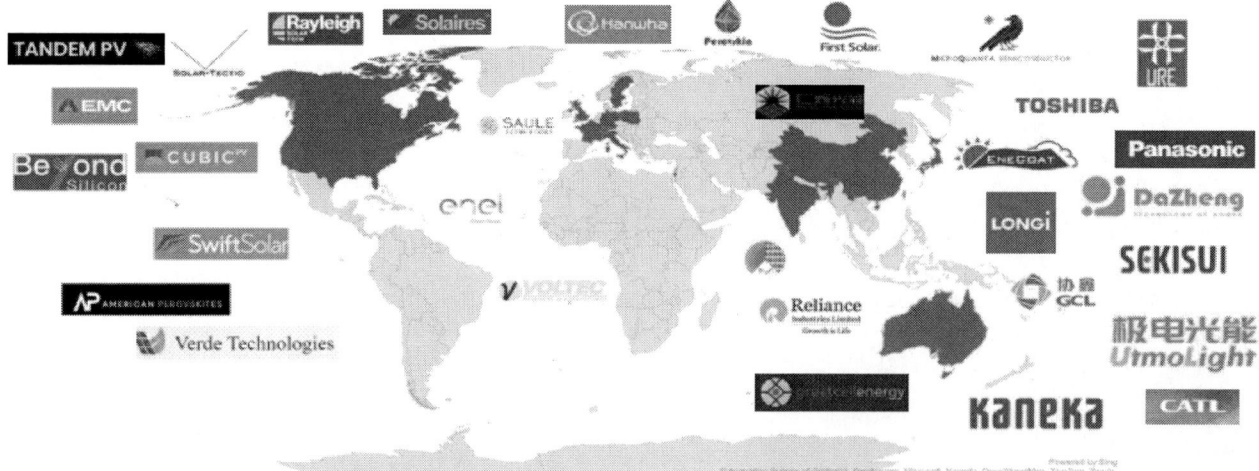

Blue areas correspond to the location of these companies HQ

The future is all-electric

www.oxfordpv.com
chris.case@oxfordpv.com

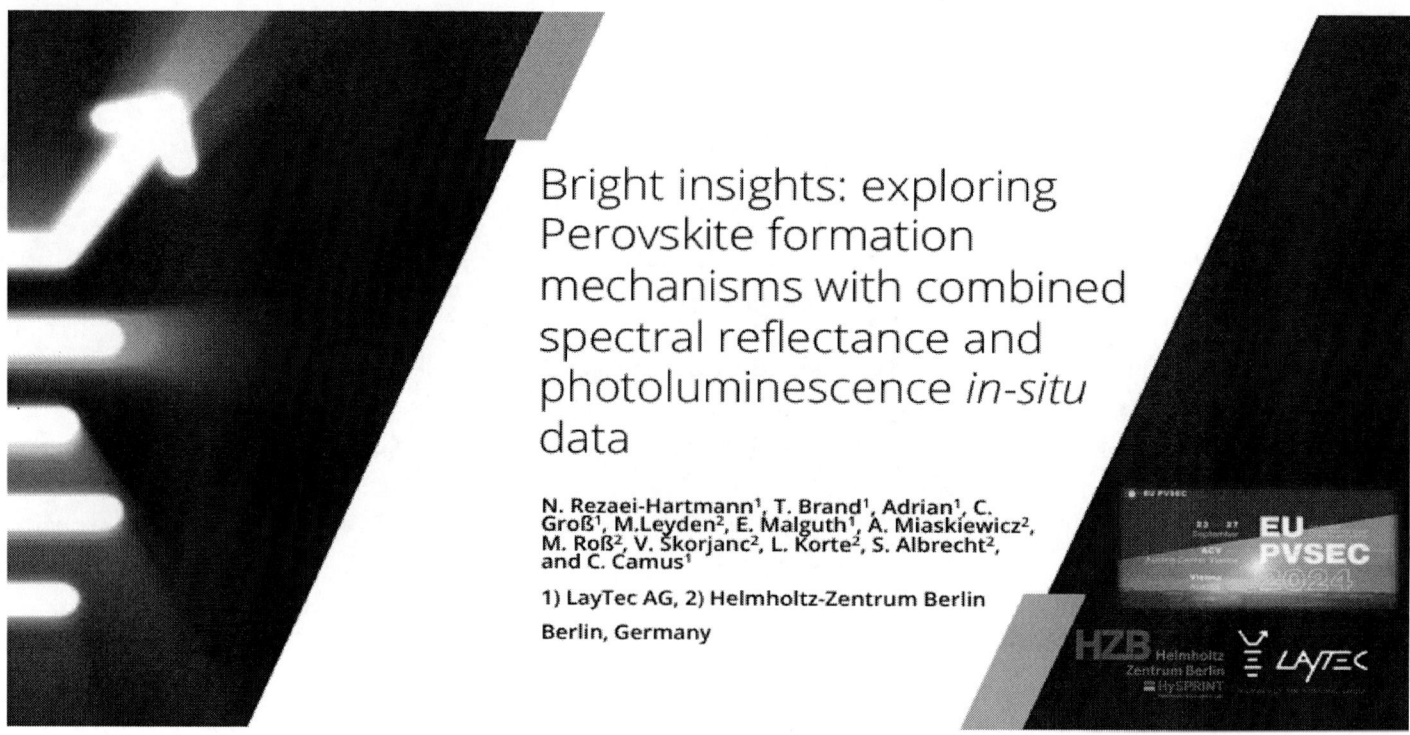

Bright insights: exploring Perovskite formation mechanisms with combined spectral reflectance and photoluminescence *in-situ* data

N. Rezaei-Hartmann[1], T. Brand[1], Adrian[1], C. Groß[1], M.Leyden[2], E. Malguth[1], A. Miaskiewicz[2], M. Roß[2], V. Skorjanc[2], L. Korte[2], S. Albrecht[2], and C. Camus[1]

1) LayTec AG, 2) Helmholtz-Zentrum Berlin

Berlin, Germany

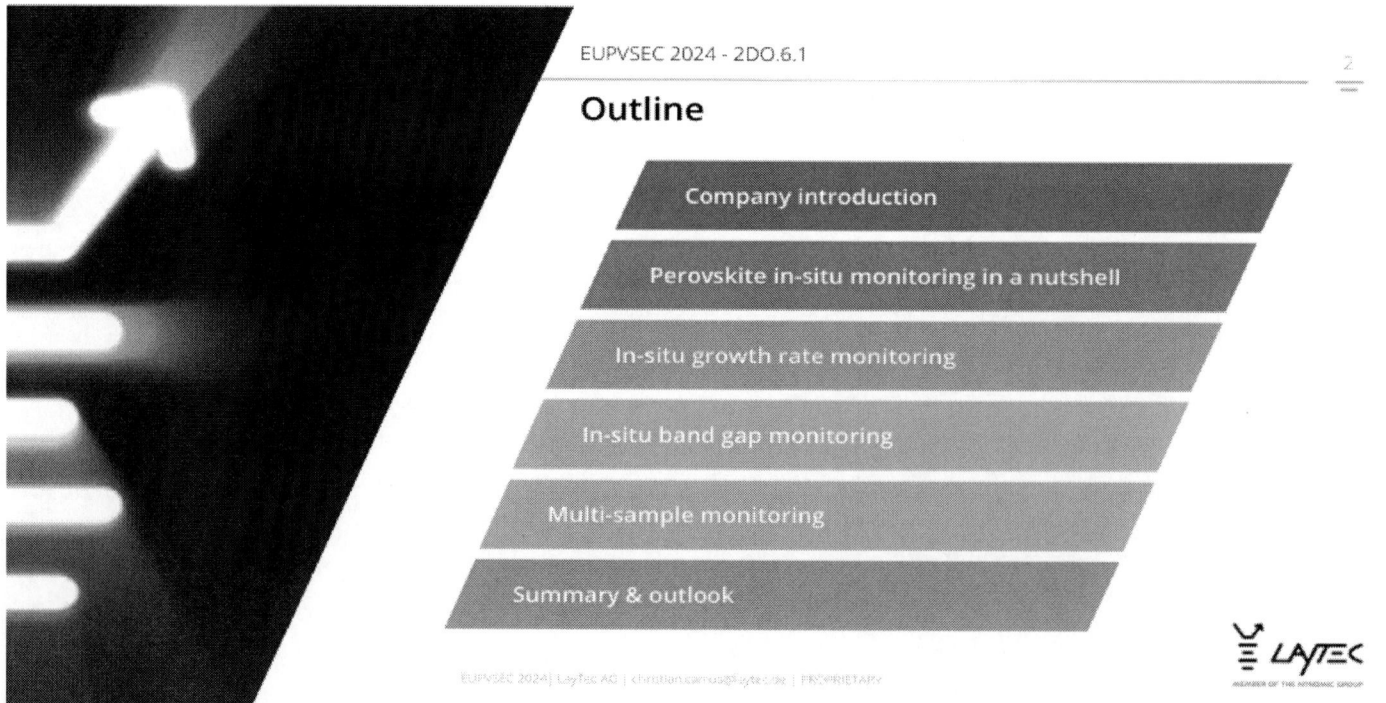

EUPVSEC 2024 - 2DO.6.1

Outline

- Company introduction
- Perovskite in-situ monitoring in a nutshell
- In-situ growth rate monitoring
- In-situ band gap monitoring
- Multi-sample monitoring
- Summary & outlook

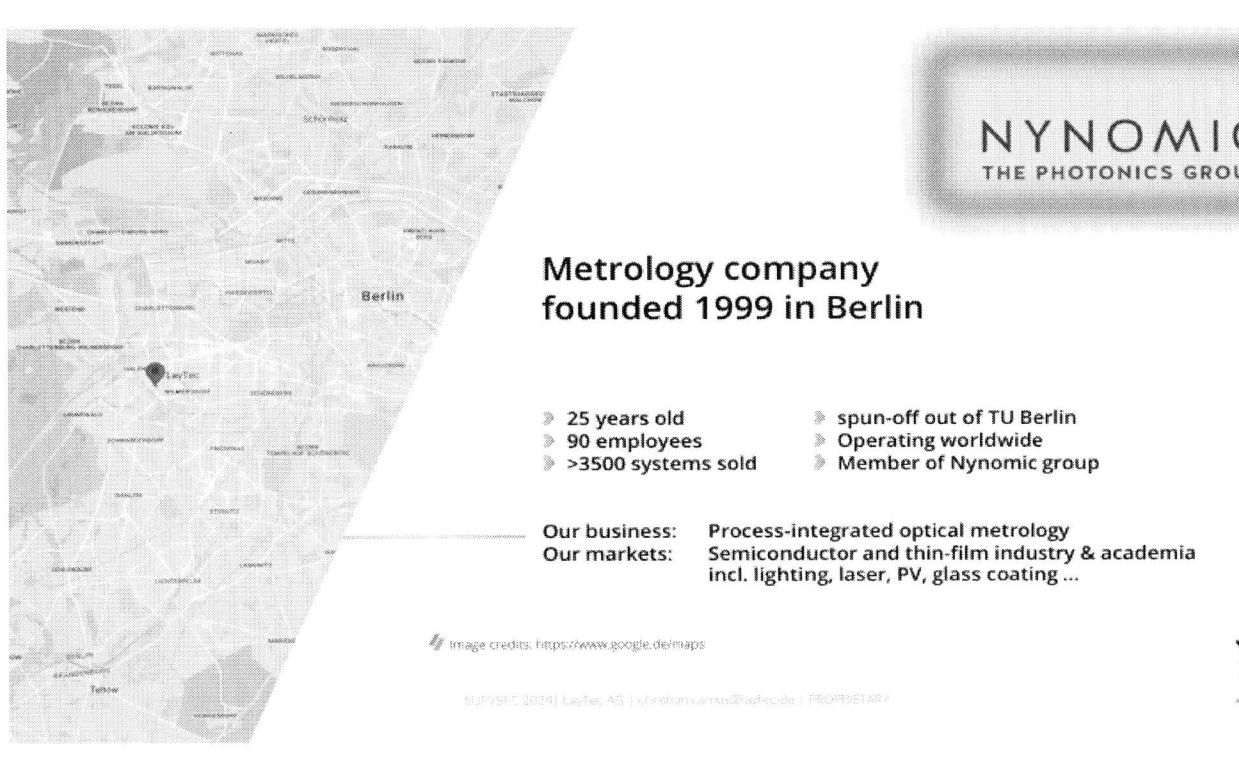

Metrology company founded 1999 in Berlin

- 25 years old
- 90 employees
- >3500 systems sold

- spun-off out of TU Berlin
- Operating worldwide
- Member of Nynomic group

Our business: Process-integrated optical metrology
Our markets: Semiconductor and thin-film industry & academia
incl. lighting, laser, PV, glass coating ...

Company overview

Integrated metrology for various industries and markets

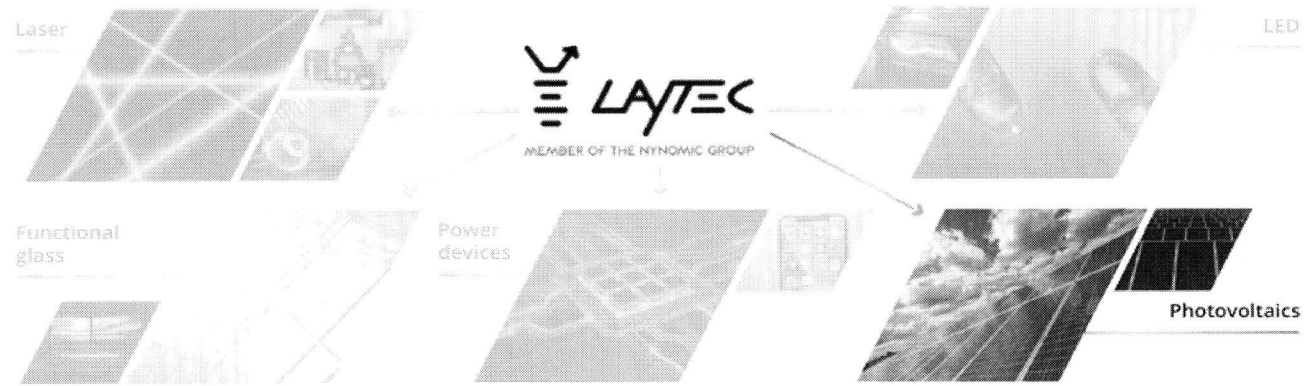

EUPVSEC 2024 - 2DO.6.1

Outline

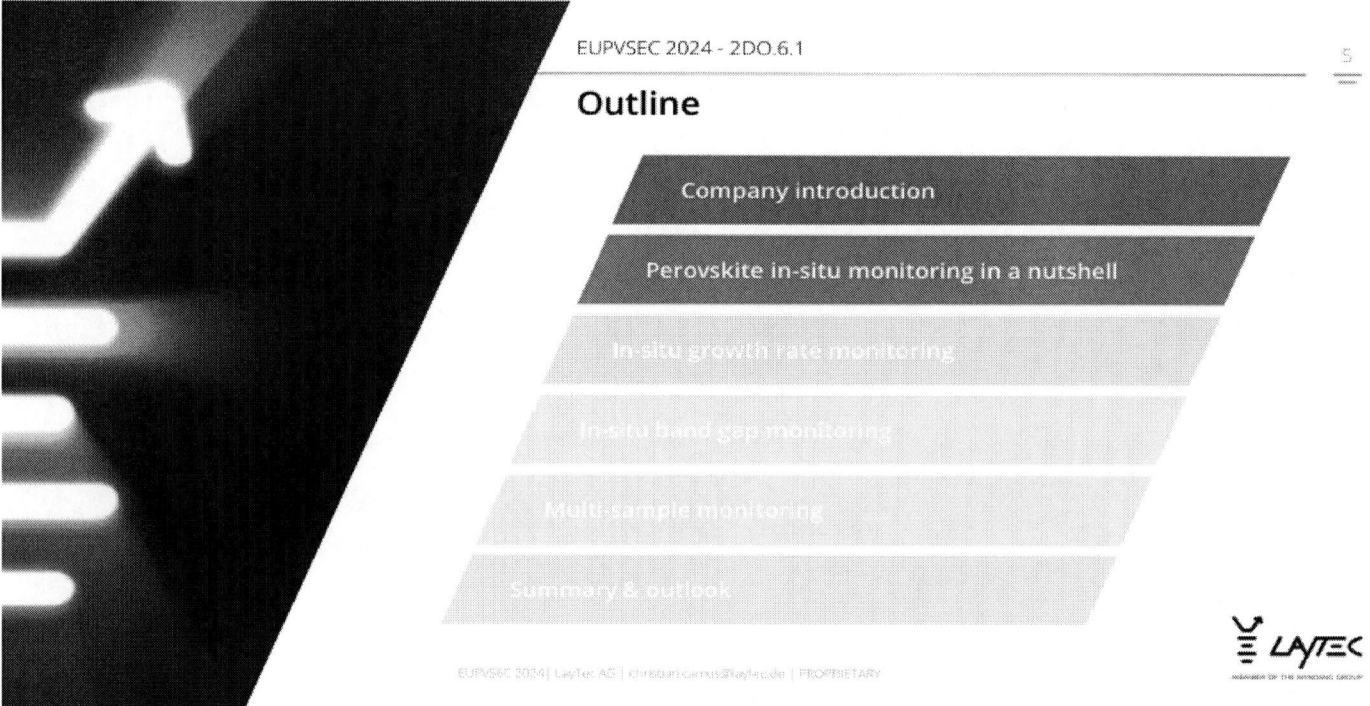

- Company introduction
- Perovskite in-situ monitoring in a nutshell
- In-situ growth rate monitoring
- In-situ band gap monitoring
- Multi-sample monitoring
- Summary & outlook

Perovskite in-situ monitoring in a nutshell

Metrology for perovskite photovoltaics

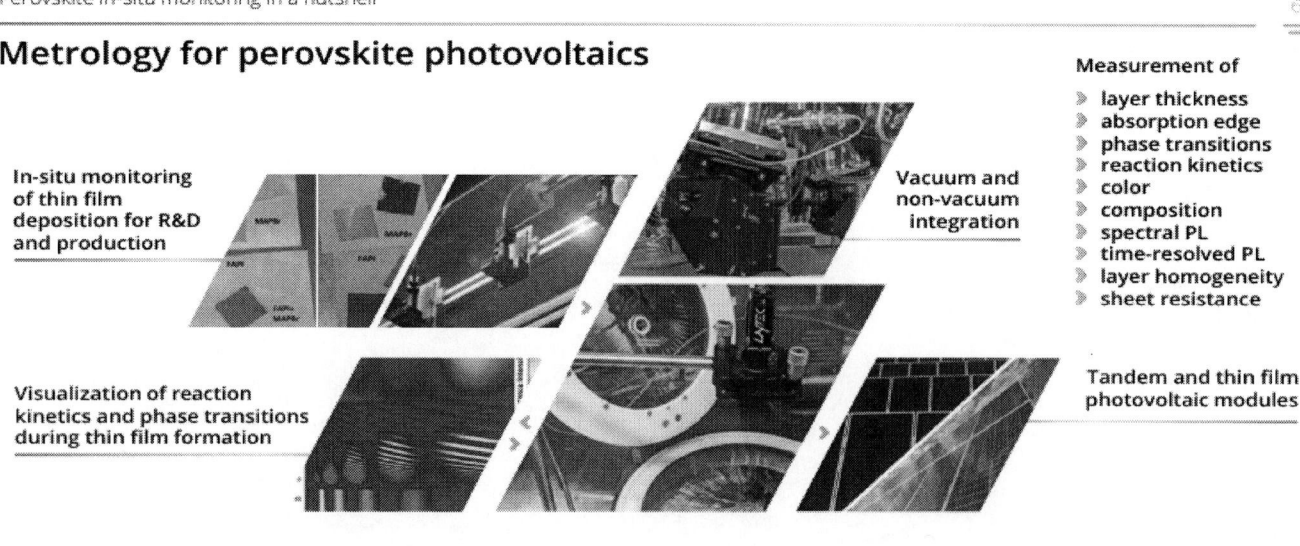

In-situ monitoring of thin film deposition for R&D and production

Vacuum and non-vacuum integration

Visualization of reaction kinetics and phase transitions during thin film formation

Tandem and thin film photovoltaic modules

Measurement of
- layer thickness
- absorption edge
- phase transitions
- reaction kinetics
- color
- composition
- spectral PL
- time-resolved PL
- layer homogeneity
- sheet resistance

Image credits, see last slide

In-situ monitoring of perovskite formation

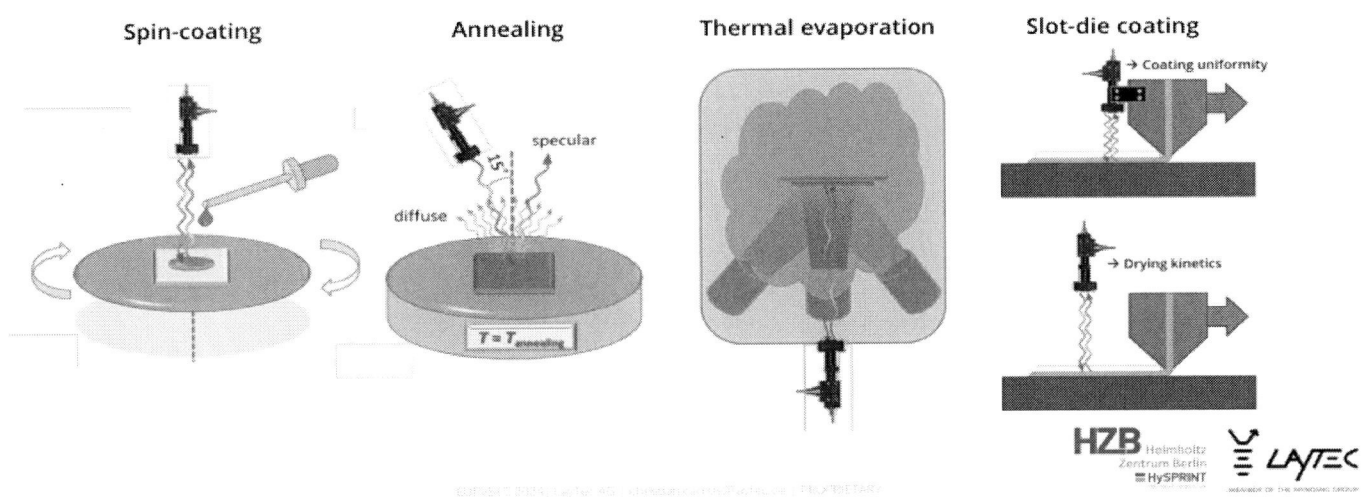

Spin-coating
Annealing
Thermal evaporation
Slot-die coating

In-situ monitoring of perovskite formation

Spin-Coating

Annealing

Thermal evaporation

Slot-die coating

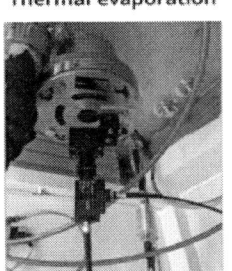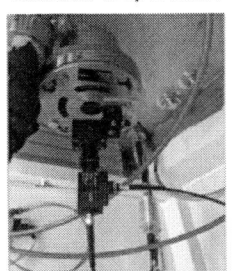

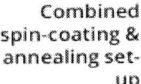

Combined spin-coating & annealing set-up

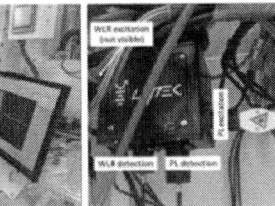

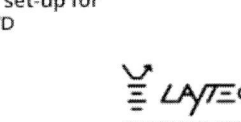

Combined reflectance & PL set-up for PVD

Perovskite in-situ monitoring in a nutshell

In-situ monitoring of perovskite formation by means of reflectance

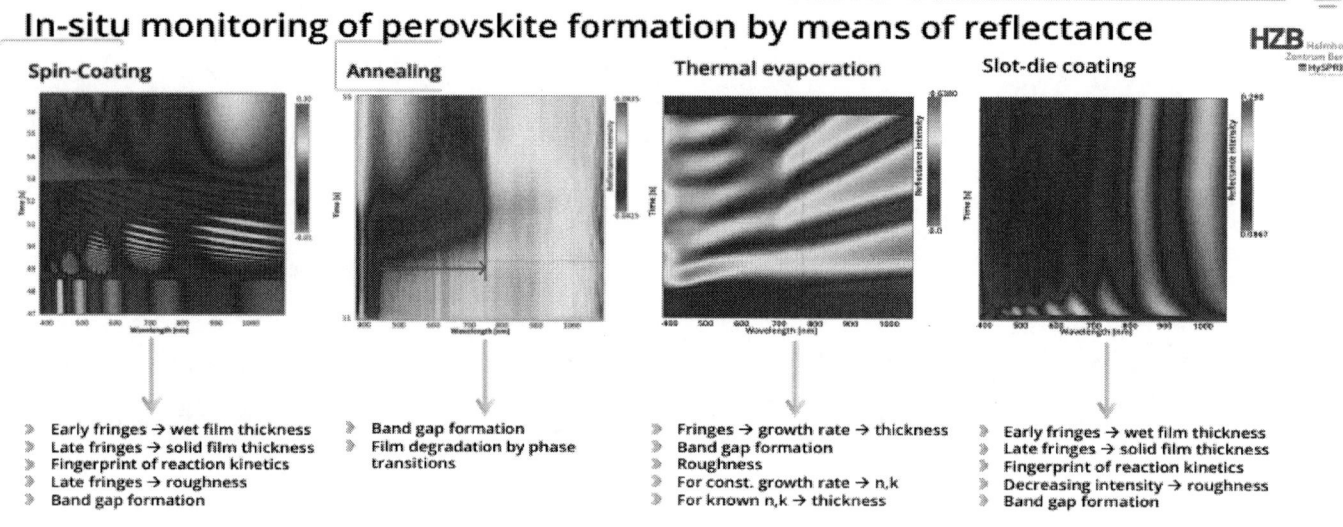

Perovskite in-situ monitoring in a nutshell

In-situ monitoring of perovskite formation by means of reflectance

Experimental & theoretical approach

Theoretical approach

$$R = \left| \frac{r_{10} + r_{12} e^{-2i\beta}}{1 - r_{10} \times r_{12} e^{-2i\beta}} \right|^2$$

with

» Spectral reflectance signal was analyzed based on Snell's law

» Three-layer model (air / solid film / substrate)

$$\beta = 2\pi \frac{d}{\lambda} \hat{n} \cos \varphi$$

d = film thickness
φ = angle of refraction at air / film interface

» Film thickness d,

» Refractive index $n(\lambda, t)$ and

» Extinction coefficient $k(\lambda, t)$...

and complex refractive index

....were determined iteratively from fitting the spectral data

$$(\hat{n}(\lambda, t) = n(\lambda, t) + ik(\lambda, t))$$

EUPVSEC 2024 - 2DO.6.1

Outline

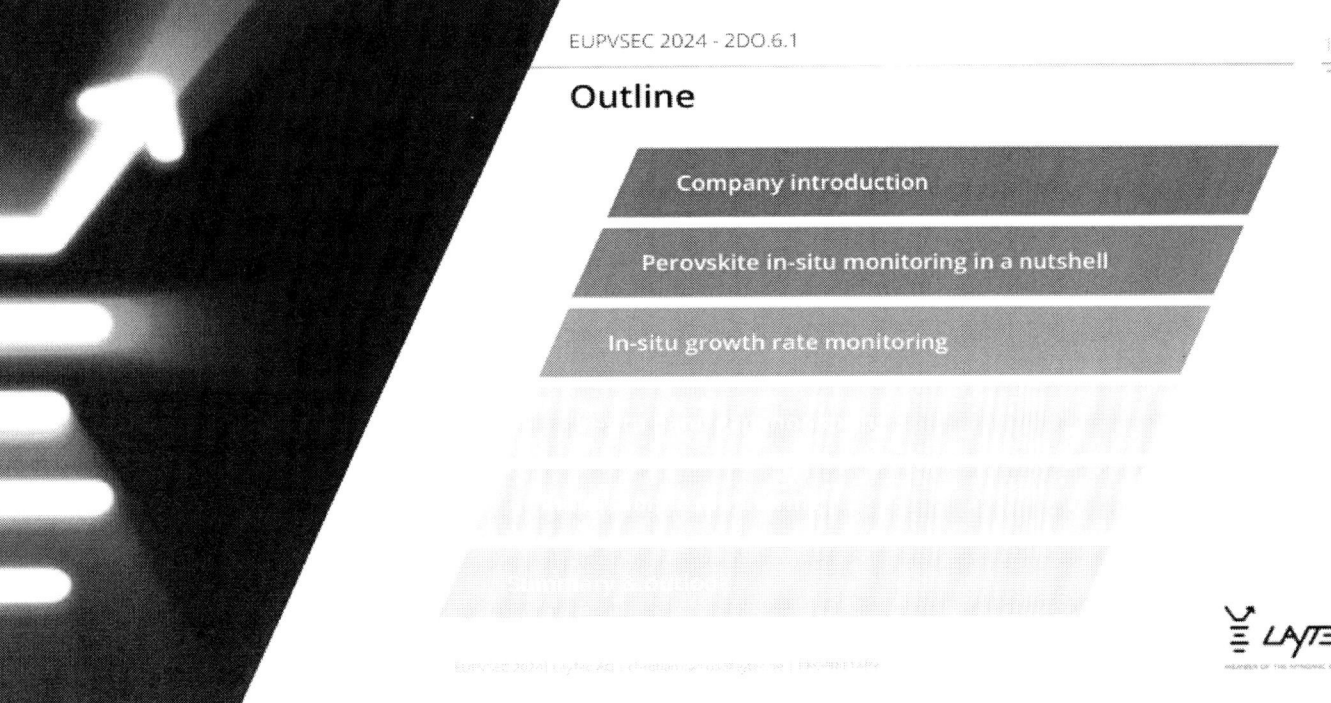

In-situ growth rate monitoring

In-situ monitoring of perovskite formation...
... Quantitative growth analysis...

≫ **...by transient analysis ...**

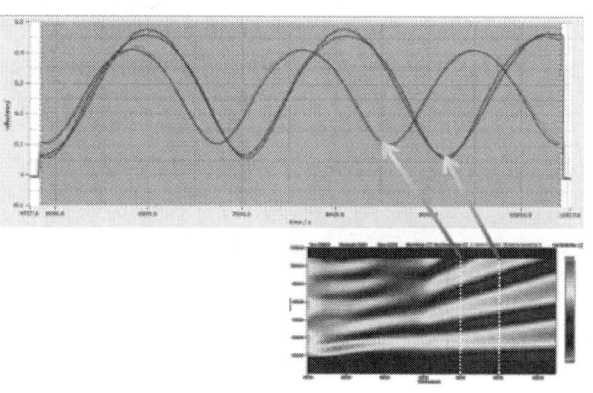

≫ **... for real-time results...**

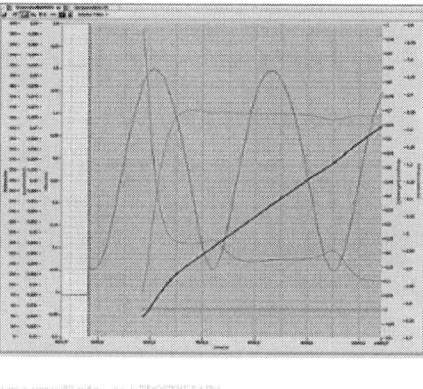

In-situ growth rate monitoring

In-situ monitoring of perovskite formation...
... Quantitative growth analysis...

≫ **...by transient analysis ...**

   ≫ **For constant growth rate:**

      ≫ **Simultaneous measurement of**

         ≫ Reflectance

         ≫ Thickness

         ≫ Growth Rate

         ≫ n & k

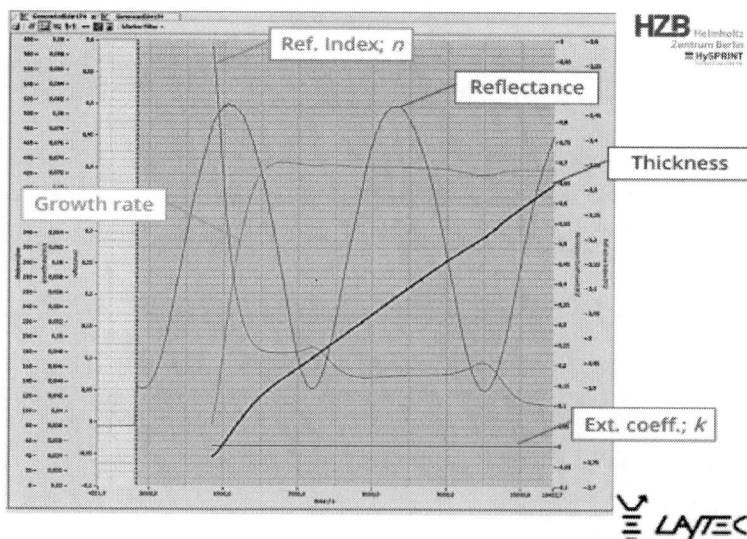

In-situ growth rate monitoring

In-situ monitoring of perovskite formation...
... Quantitative growth analysis...

≫ ...by transient analysis ...

  ≫ For non-constant growth rate:

    ≫ Separation into steps of quasi-constant growth rate

      ≫ Same procedure for each layer:

  ≫ Resulting growth rates for sublayers 1-3

    ≫ 0.14, 0.12 and 0.09 nm/s

    ≫ total thickness of 658 nm for the perovskite film.

    ≫ → growth rate constantly decreases!

≫ Spectral fit also possible for known n, k

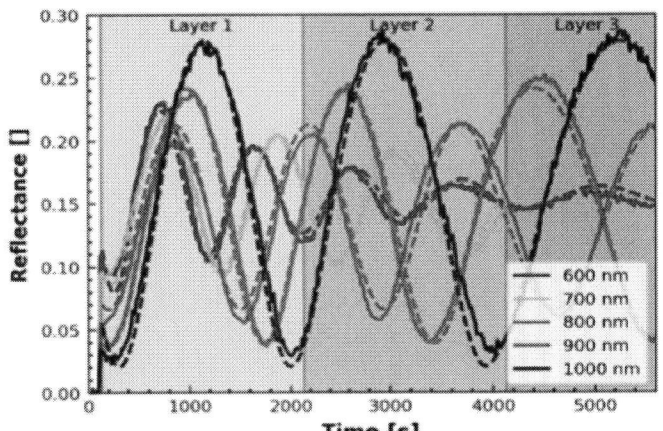

 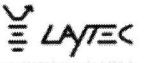

In-situ growth rate monitoring

Active process control based on reflectance data

≫ Distance for adjacent extrema in transient can be used as control parameter

≫ Here, source power was adjusted to keep distance constant (900nm transient)

  ≫ Metrology allowed to keep growth rate constant!

≫ Increasing roughness induced decrease of intensity

≫ Growth rate 0.151 nm/s

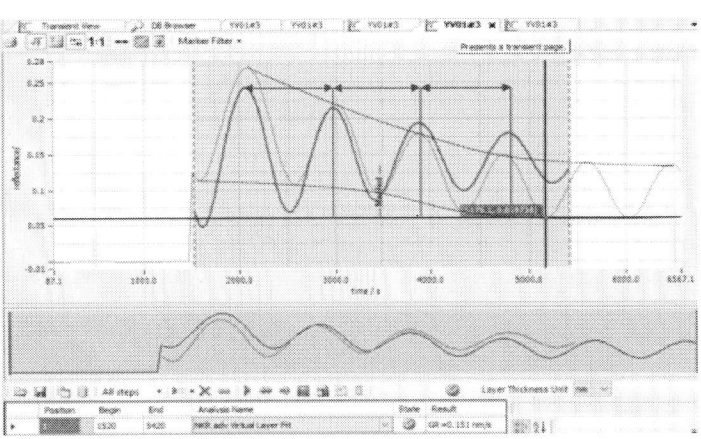

 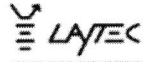

473

In-situ growth rate monitoring

Growth rate monitoring for non-constant rates

- For $r \neq$ const., $n(\lambda)$ and $k(\lambda)$ can be determined by transient fitting for multiple wavelengths

- Thus, effective $n(\lambda)$ and $k(\lambda)$ can be deduced

- Spectral fit with fixed $n(\lambda)$ and $k(\lambda)$ for deducing thickness

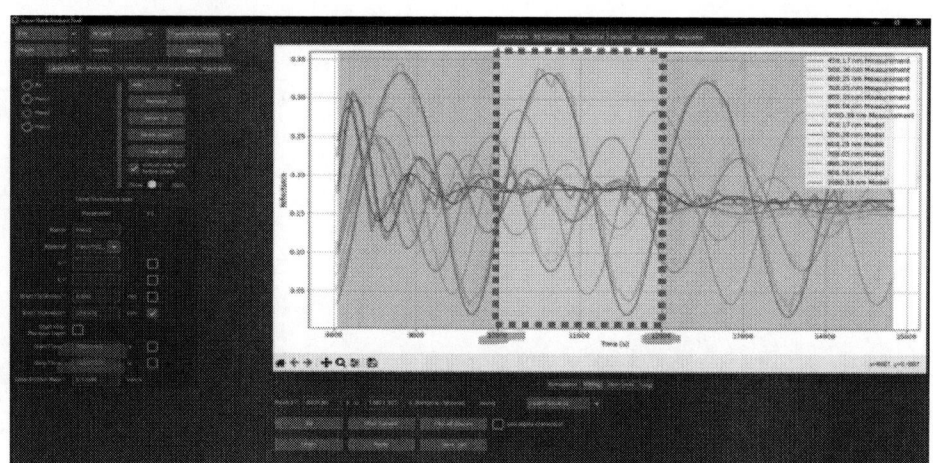

In-situ growth rate monitoring

Growth rate monitoring for non-constant rates

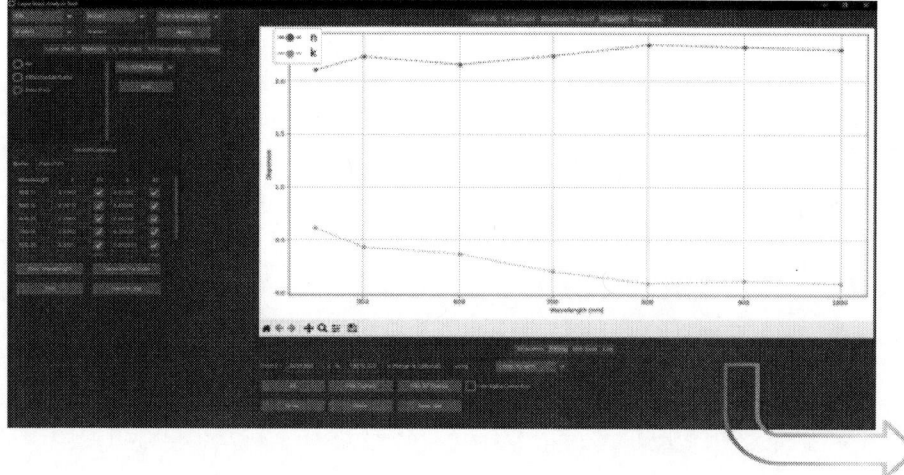

EUPVSEC 2024 - 2DO.6.1

Outline

- Company introduction
- Perovskite in-situ monitoring in a nutshell
- In-situ growth rate monitoring
- In-situ band gap monitoring

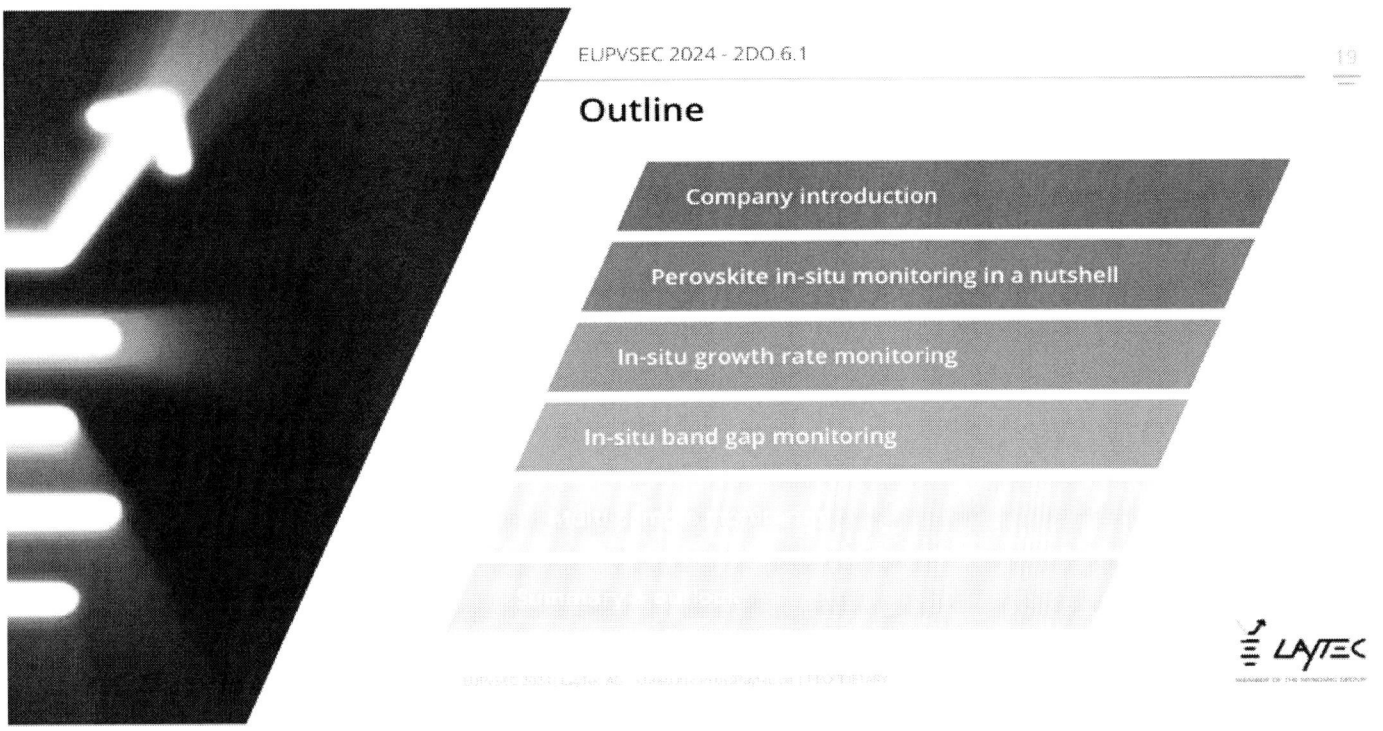

In-situ band gap monitoring

In-situ PL-monitoring of perovskite formation during FaCsPb(IBr)$_3$ PVD

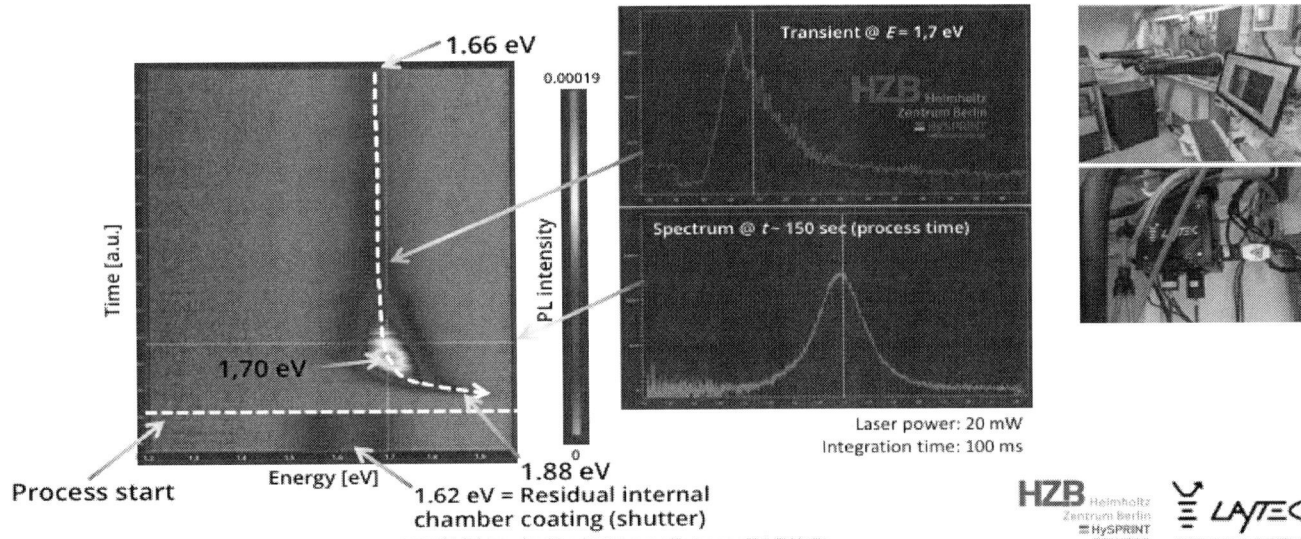

Laser power: 20 mW
Integration time: 100 ms

In-situ band gap monitoring

In-situ PL-monitoring of perovskite formation during FaCsPb(IBr)$_3$ PVD: Comparison with reflectance

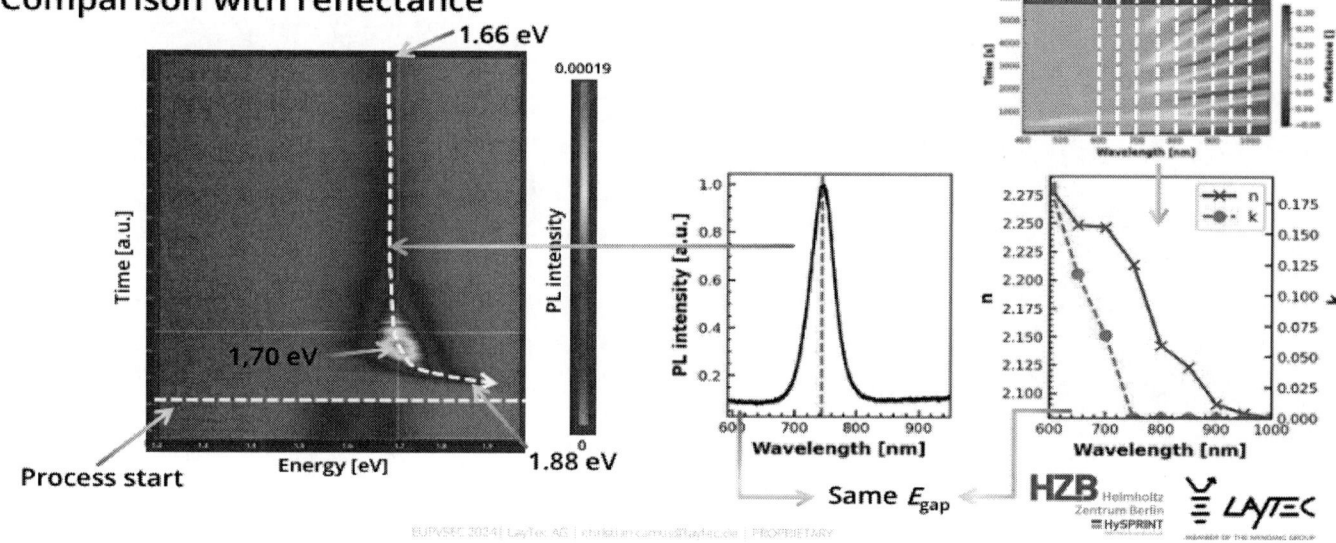

Same E_{gap}

In-situ band gap monitoring

In-situ PL-monitoring of perovskite formation during FaCsPb(IBr)$_3$ PVD: Band gap & compositional shifts

L. Gil-Escrig et al., *ACS Energy Letters* 2021 *6*(2), 827-836
DOI: 10.1021/acsenergylett.0c02445

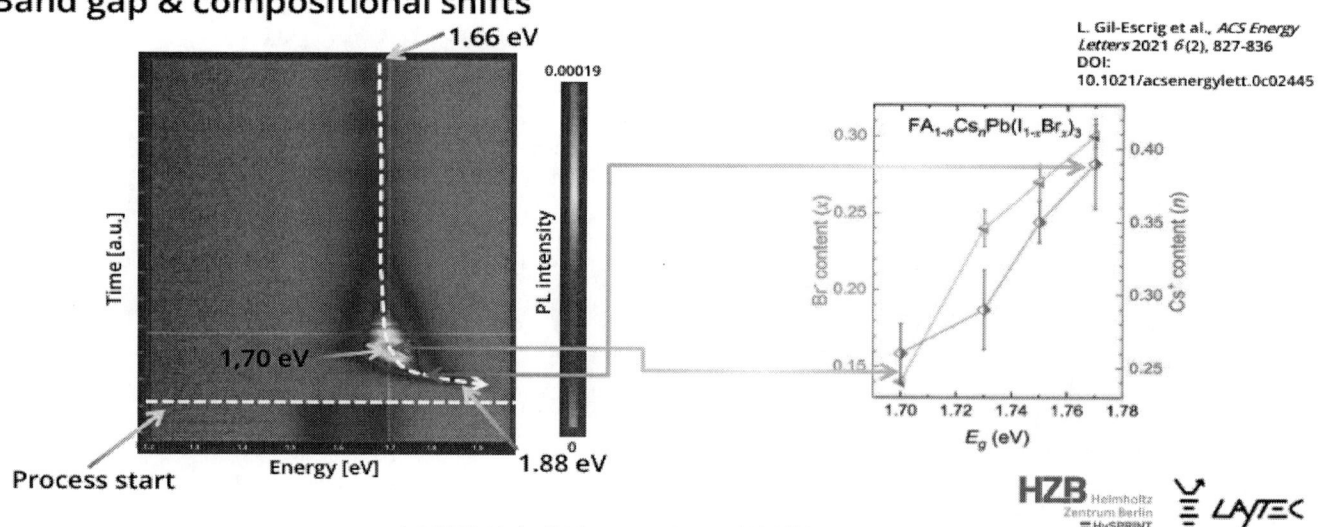

EUPVSEC 2024 - 2DO.6.1

Outline

- Company introduction
- Perovskite in-situ monitoring in a nutshell
- In-situ growth rate monitoring
- In-situ band gap monitoring
- Multi-sample monitoring

Multi-sample monitoring

In-situ monitoring of perovskite formation...
... multi-sample analysis...

» Each sample is analyzed separately

» Comparative studies possible

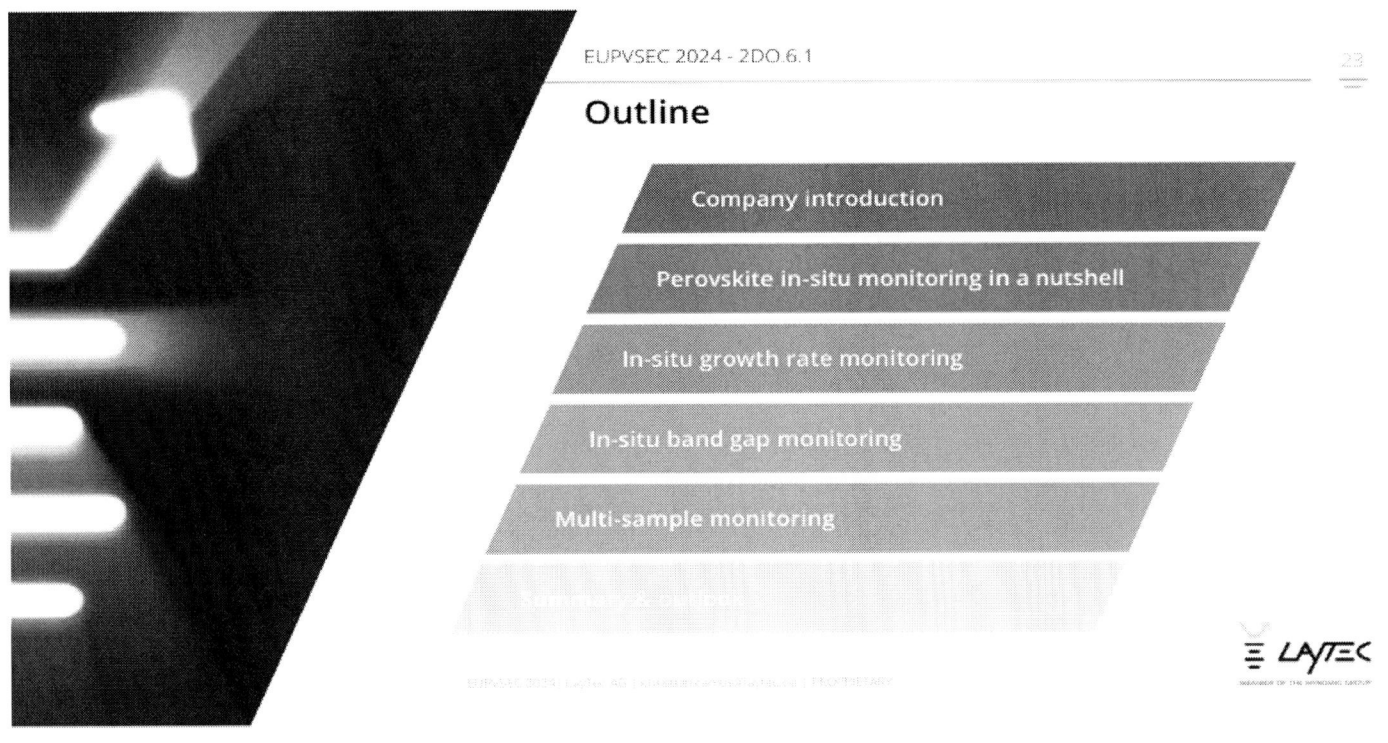

» Reflectance intensity during one rotation

Multi-sample monitoring

In-situ monitoring of perovskite formation...

... multi-sample analysis...

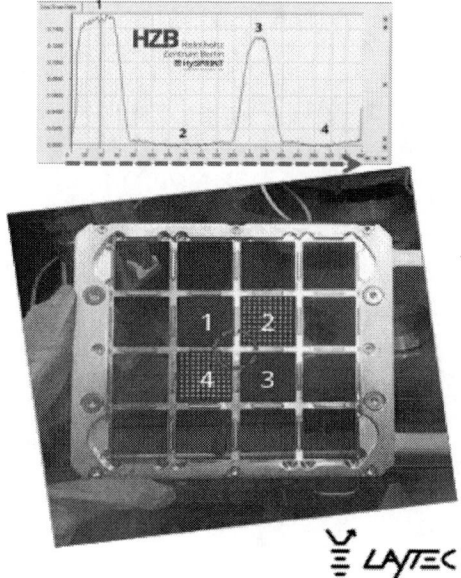

» Indirect monitoring of textures samples by planar control samples

» Almost $R = 0$ on textured samples

» Full signal and "same" process on planar ones

EUPVSEC 2024 - 2DO.6.1

Outline

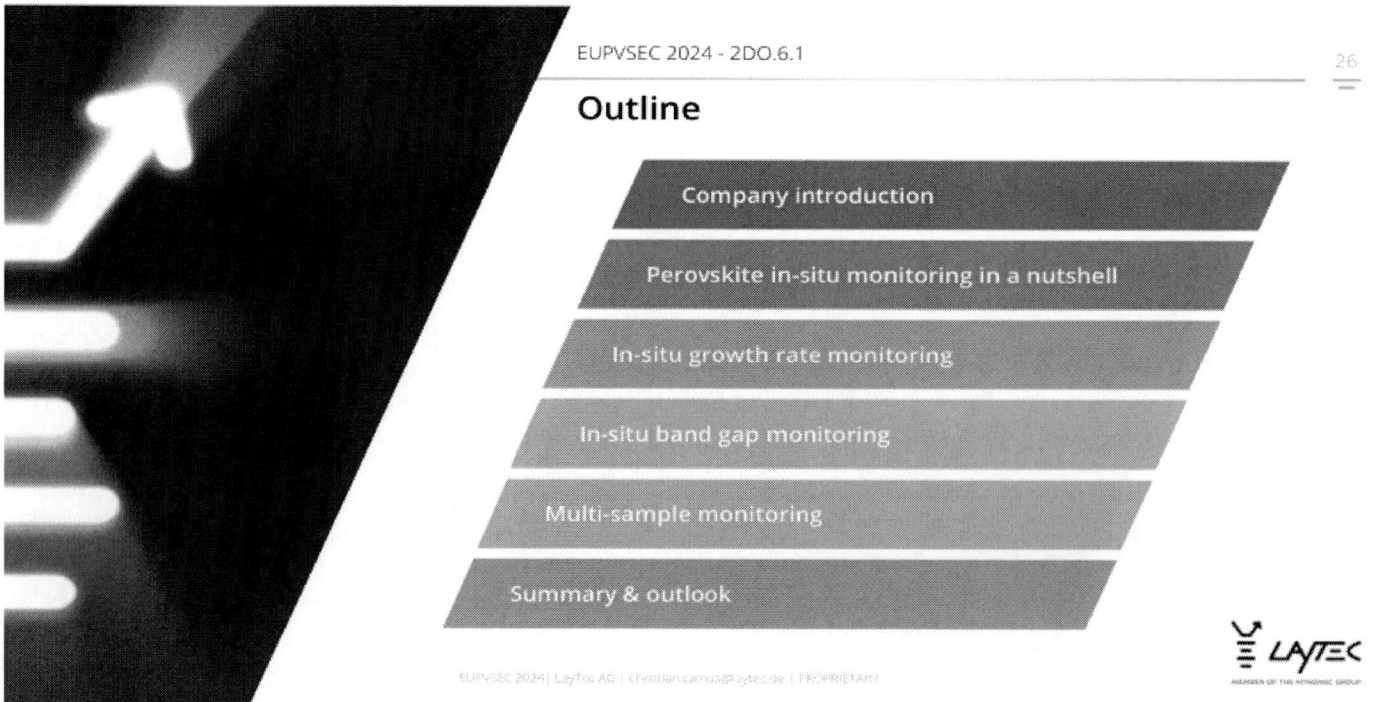

Summary & outlook

- Optical methods are ideally suited for in-situ monitoring of perovskite formation
- For PVD, reflectance spectroscopy allows to determine film thickness as well as optical constants
  - Depending on growth behavior transients or spectra can be used
- Photoluminescence provides further insights about composition in good agreement with reflectance data
- Multi-sample monitoring enables indirect growth control also for textured samples

Knowledge is key

www.laytec.de

Acknowledgements & references

Acknowledgements

» **The SHAPE project has received funding from the Federal Ministry for Economic Affairs and Climate Action under Grant Agreement No 03EE1123D**

» **The Solar TAP project received funding from the Helmholtz Association**

Supported by:

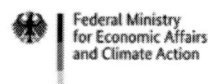

» **HZB groups headed by E. Unger, S. Albrecht and N. H. Nickel**

» **All colleagues at LayTec AG**

on the basis of a decision
by the German Bundestag

41st European Photovoltaic Solar Energy Conference and Exhibition

This presentation was selected by the Sc. Committee of the EU PVSEC 2024 for submission of a full paper to one of the EU PVSEC's collaborating peer-reviewed journals.

ENHANCING CRYSTALLINITY OF PEROVSKITE MATERIALS THROUGH RAPID MICROWAVE ANNEALING

Syed Nazmus Sakib*, David N R Payne, Shujuan Huang, Binesh P Veettil
School of Engineering, Macquarie University, Sydney, Australia
e-mail address: syed-nazmus.sakib@hdr.mq.edu.au*

ABSTRACT: With their unique crystal structure and relatively simple processing, perovskite solar cells have great potential for cost-effective clean energy. This work explores the use of microwave annealing as a fast and efficient processing technique to improve the quality of the perovskite films and overcome some of the main limitations of the traditional hot plate heating method upon scaling up. In this work, MAPbI₃ and CsFA-based triple halide perovskite samples were prepared onto FTO glass substrates and treated under different conditions of microwave power and processing durations. Scanning electron microscopy showed that the microwave-annealed samples illustrated larger grains than hotplate annealing in a shorter time. Such fast growth of larger crystals is favorable for high-quality perovskites since it could reduce recombination pathways and enhance carrier transport. At higher microwave powers, however, partial detachment of the film occurred around the edges of the film due to localized heating because of the deflections of microwaves on the edges of the FTO layer. Again, this was in agreement with COMSOL Multiphysics simulations. Our work demonstrates the opportunities of microwave annealing in fast and efficient preparation of high-quality perovskite films while at the same time pointing out the challenges that have to be met in superior implementation. The study contributes to developing effective methodologies for preparing high-quality perovskite materials for scalable, high-performance solar cell manufacturing.

Keywords: Annealing, Perovskite, Microwave, Crystallinity, Morphology

1 INTRODUCTION

Perovskite solar cells (PSC) have become a critical area of interest for researchers and industrial manufacturers due to the increasing demand for cost-effective and advanced solar cell technologies worldwide [1]. Perovskite solar cells (PSCs) offer many advantages, including easy manufacturing, lower material costs, flexibility in using different substrates, tunable bandgap, potential for combined multijunction architectures, high power conversion efficiency, and a wide range of constituent materials [2-5]. Due to these benefits, there has been a significant increase in the development of new metal-halide perovskite devices in recent times [6-8]. PSCs have made significant progress since their introduction in 2009. Initially, their efficiency was only 3.8%. However, recent advancements have boosted their power conversion efficiencies (PCE) to over 26% [9, 10]. Extensive research efforts have been devoted to enhancing the efficiency and stability of the PSC through improvements in its design and production techniques [11-13]. Several methods have been developed to fabricate perovskite films, including solution depositions, vapor-assisted deposition, thermal vapor deposition, and chemical vapor deposition [14-18]. This diverse range of options allows for flexibility and customization in manufacturing, focusing on creating high-quality perovskite films with specific characteristics such as high crystallinity, homogeneity, low defect density, and excellent material coverage on substrates [19, 20].

Perovskite solar cells can be manufactured inexpensively using various printing methods such as 3D printing, screen printing, roll-to-roll printing, spray coating, knife/blade coating, slot die coating, flexographic printing, gravure printing, and ink-jet printing, enabling large-scale production [21-23]. Perovskite materials need excellent surface morphology and high crystallinity to create high-performing PSCs, which are quality materials for the absorbing layer [24]. Most modern methods for preparing PSCs require the use of a perovskite precursor solution, the addition of an antisolvent such as chlorobenzene to promote nucleation and heating the solution to remove the solvent and facilitate the perovskite crystallization into the desired structure [25].

The thermal annealing process is crucial for the development of perovskite materials, particularly in the production of PSCs, as it has a significant impact on the structural and optoelectronic properties of the films [26]. When perovskite films are exposed to certain temperatures, they crystallize more, leading to larger grain sizes and better film quality. This is helpful because larger grains reduce the number of grain boundaries, which can hinder the movement of charge carriers and lower the efficiency of PSCs [27]. Furthermore, thermal annealing assists in removing residual solvents from the film deposition process, leading to a more uniform and compact film structure. This, in turn, enhances the film's quality and improves its optoelectronic properties by reducing internal defects and increasing carrier lifetimes. As a result, it promotes better light absorption and charge transport, which are crucial for high-efficiency solar cells [28]. Thermal annealing improves the interface between the perovskite layer and other device layers, leading to better charge collection and overall performance. Adjusting thermal annealing parameters fine-tune the perovskite material's crystalline structure, affecting electronic properties such as charge carrier mobility and recombination dynamics. This optimization is crucial for exceptional photovoltaic performance, making thermal annealing a vital step in producing highly efficient solar cells [29].

The conventional hotplate annealing process has several drawbacks. It is unable to reclaim organic solvents, generates a significant amount of waste, and is not suitable for large-scale applications. Furthermore, the perovskite layer needs to be annealed at temperatures exceeding 100°C for over 30 minutes, leading to a substantial increase in production costs. Therefore, challenges still remain in achieving rapid annealing for mass production [27, 30, 31]. Therefore, developing efficient and scalable

481

methods for thermal annealing is a critical step toward increasing the scale of PSC production.

This research aims to explore the viability of microwave processing in preparing perovskite materials for solar cells. Longer annealing time is one of the critical challenges for the mass production of perovskite solar cells [30]. Generally, a selected precursor solution is prepared, spin-coated onto a substrate, and annealed at a certain temperature for a specific time to ensure crystallization. Proper annealing is essential for evaporating the polar solvents and for inducing crystalline transformation of the desired perovskite phase. This process controls the structural and morphological properties of the prepared film [32, 33]. Thermal annealing reduces remaining stresses and internal defects, which results in microstructure adjustments within the specific materials [34-36]. During the crystal growth of perovskites, the annealing duration, temperature, and heating method significantly influence the crystal structure and crystallinity. Hence, various strategies have been applied to find a suitable method for annealing [27, 37]. Microwave annealing has the potential to be an appropriate option for processing perovskite materials. A few studies have shown its potential; however, further exploration is needed to explain the heating mechanism and optimize the process parameters. [38-42].

By investigating the impact of rapid microwave annealing on crystallinity, this research addresses a crucial hurdle of annealing duration and technique that holds back the adoption of these materials in the mass production of perovskite solar cells. The experiment and simulation results can open critical insights into the heating mechanism and the overall crystal growth that could pave the way for a method to enhance the performance of perovskite solar cells.

2 METHODOLOGIES

Two different perovskite compositions were chosen to investigate the crystal growth due to chemical orientation. We have prepared MAPbI3 and CsFA-based triple halide perovskite samples on FTO glass substrates (20 mm x 20 mm x 2.2 mm) using the spin coating method, as shown in Figure 1. After spin coating, one batch of CsFA-based samples was annealed on a hot plate at 100°C for 30 minutes, while another batch of samples was annealed in the microwave oven at different power levels and duration. Similarly, one batch of MAPbI3 samples was prepared on a hot plate at 100°C for 20 minutes and another on a microwave oven at various power levels and times. A SiC block was used to absorb additional microwave energy, as shown in Figure 1.

We developed a microwave oven simulation model using the COMSOL® RF module to understand the heating mechanism involved in annealing the perovskite film. The simulation model involved injecting 1000W of microwave power in TE10 mode through a rectangular port. Transient analysis was conducted for a simulated 60-second duration. The simulation setup only considered the FTO-coated glass substrate condition to simplify and better understand heat generation. The simulation model studies the electric field distribution inside the microwave cavity, the FTO layer's temperature gradient, and other essential parameters necessary for understanding the heating mechanism.

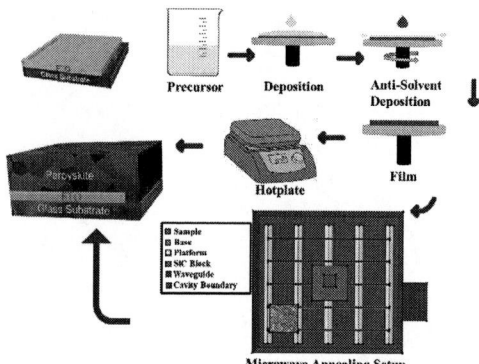

Figure 1 Diagram of the process of the experiment for this work

3 RESULTS AND DISCUSSIONS

A comparative study of two perovskite films annealed on a hotplate and microwave oven has been presented. Various microwave power levels and annealing duration were investigated. The samples with the best results are shown here. For analysis, the samples are named as HA1-CsFA based perovskite annealed on hotplate at 100°C for 30 minutes, MA1- CsFA based perovskite annealed in microwave at 1000W for 60 seconds, HA2- MAPbI3 perovskite annealed on hotplate at 100°C for 20 minutes, MA2- MAPbI3 perovskite in microwave at 700W for 60 seconds. SEM images and the grain size distribution are shown in Figure 2. Based on the analysis, the MA1 sample possesses an average grain size of ~1.83μm whereas the HA1 sample possesses an average grain size of ~1.12μm; the MA2 sample has an average grain size of ~620nm, whereas the HA2 sample possesses an average grain size of ~180nm. Perovskite prepared by microwave annealing exhibits larger grain sizes.

Thermal annealing plays a vital role in forming high-quality crystals for perovskites by removing the solvent and promoting the film crystallization. However, different annealing methods offer distinct functionalities, uniquely impacting the film growth process. Nucleation and growth are the two stages that determine the perovskite films' quality, uniformity, and coverage. It is desirable to have a faster nucleation rate to promote the formation of numerous nucleation sites, leading to a supersaturated state. A critical driving force determines the quality of the interfacial growth to initiate crystal growth. Hence, the crystal growth rate plays a crucial role in uniformity for the desired phase transition. Therefore, duration and temperature govern the nucleation and growth rate [27]. Longer annealing can initiate the decomposition of perovskite structure, resulting in device instability. On the contrary, a shorter annealing time can accelerate the nucleation but may cause rapid volume shrinkage of the solute to diffuse completely. Therefore, the annealing should be done in a particular window to get high-quality crystals. According to classical nucleation theory, rapid heating can increase the nucleation rate exponentially because the supersaturated state can be achieved quickly, leading to a lower free energy barrier for nucleus formation. Additionally, microwave heating for a shorter time can significantly minimize the precursor solution's thermal exposure, lowering the chance of decomposition and resulting in better film quality. So, microwave

41st European Photovoltaic Solar Energy Conference and Exhibition

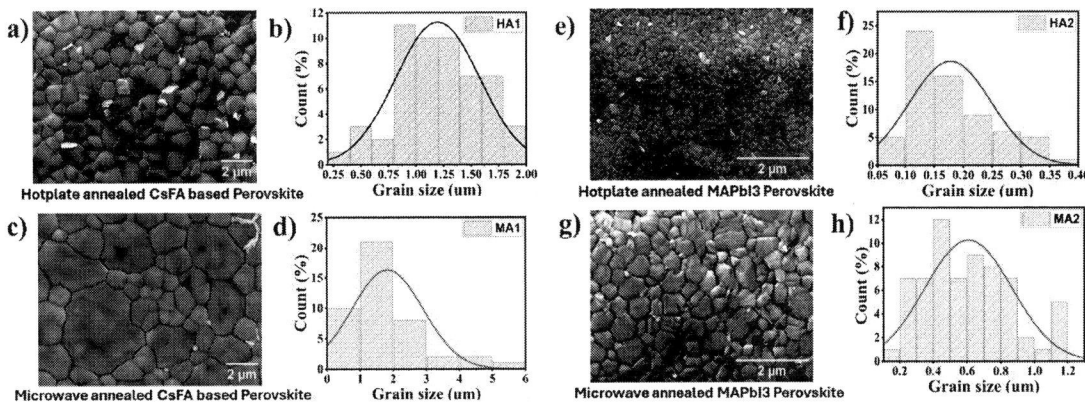

Figure 3 SEM images and grain size distribution of the samples. a), c) SEM images of CsFA-based HA1(Hotplate annealed) and MA1(Microwave annealed) samples. b), d) Grain size distribution for CsFA-based HA1(Hotplate annealed) and MA1(Microwave annealed) samples. e), g) SEM images of MAPbI3 HA2(Hotplate annealed) and MA2(Microwave annealed) samples. f), h) Grain size distribution of MAPbI3-based HA2(Hotplate annealed) and MA2(Microwave annealed) samples respectively.

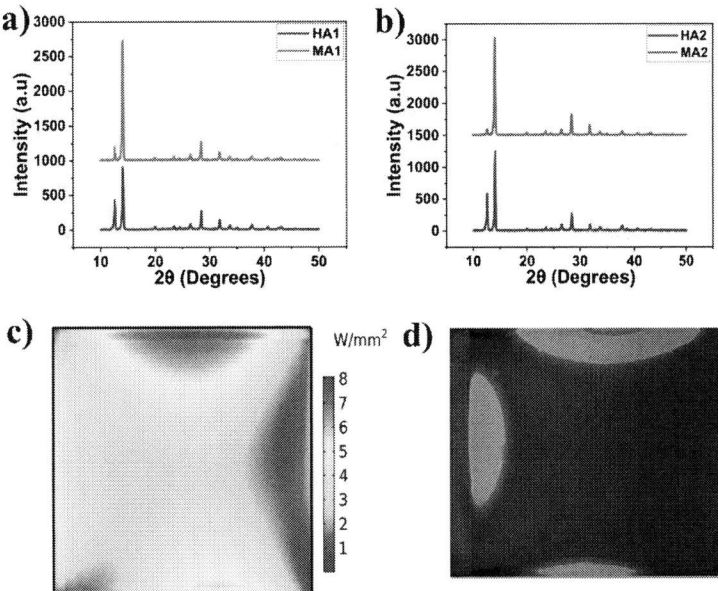

Figure 2 XRD of a) CsFA-based and b) MAPbI3 perovskite samples. c) COMSOL simulation results showing resistive heating on the edges of the substrate. d) Photograph of the microwave annealed sample

annealing can effectively enhance the formation of perovskite films, resulting in larger grain sizes, as seen in Fig 2.

The XRD intensity illustrated in Fig 3 a) and b) supports these findings. The peak at 12.7° refers to the (001) lattice of PbI$_2$, whereas the peaks at 14.2°, 28.37° and 31.8° refer to the perovskite lattices (110), (220) and (310), respectively. For the hotplate-annealed samples, the PbI2 peak is stronger than that of microwave-annealed samples, depicting a more significant presence of PbI$_2$ in the sample, which is detrimental to the device's performance. A few microwave-annealed samples have illustrated edge effects (edges of the substrates became yellow, depicting the depletion of perovskite exposing PbI2) as shown in Fig 3d. To understand the generation of edge effect and get insights into the heating mechanism, an RF simulation model was built in COMSOL and the

FTO substrate was simulated. FTO can absorb a significant amount of microwave radiation through eddy current generation. The free electrons in FTO interact with electromagnetic fields inside the cavity. The finite resistance of FTO (12 Ohm/cm in our samples) results in heating through ohmic losses, as seen in Figure 3 c).

Additionally, the electromagnetic waves diffract at the edges of the thin conducting material inside the multimodal cavity, resulting in a nonuniform current flow along the surface. So, it can be concluded that the FTO generates most of the heat due to resistive losses. Only a tiny amount of heat is generated by the microwave absorption of the polar solvent residue in the film. Consequently, microwave annealing can lead to higher-quality films within a very short time, and the process is scalable for mass production.

483

4 CONCLUSIONS

This work demonstrates the potential of rapid microwave annealing as one of the promising techniques for the fast preparation of high-quality perovskite films of highly efficient solar cells. Comparative analysis of hotplate and microwave annealing reveals that microwave-treated samples exhibit larger grain sizes and improved crystallinity in significantly shorter processing times. Accordingly, in the microwave annealing method, the average grain size for CsFA-based perovskites was 1.83 μm within an extremely short duration of just 60 seconds. In contrast, in the case of hotplate annealing, it was 1.12 μm in 30 minutes. For MAPbI$_3$ perovskites, microwave annealing treatment gave an average grain size of 620 nm after just 60 seconds, which outperformed the hotplate-annealing process that yielded a grain size of 180 nm after 20 minutes. The same observations were further reflected in XRD analysis, where the PbI$_2$ presence was reduced in microwave-annealed samples. COMSOL RF simulations illustrated the heat generation mechanism, showing the main heat generation via resistive losses in the FTO layer. However, even though some of the samples annealed by microwave showed edge effects, this work underlines that microwave power, annealing duration and position of the samples should be optimized to avoid this problem. Fast processing time and better film quality are the two major bottlenecks to upscaling the production of perovskite solar cells. Therefore, the parameters for microwave annealing need optimization in future research, considering the application on different perovskite compositions that may help further understand the implications of device performances and stability. The present work, therefore, provides valuable insight into developing efficient, scalable high-performance perovskite solar cell manufacturing processes.

5 REFERENCES

[1] M. Noman, Z. Khan, S. T. Jan, *RSC Advances* **2024**, 14, 5085-131.

[2] M. Saliba, J. P. Correa-Baena, M. Grätzel, A. Hagfeldt, A. Abate, *Angewandte chemie-international edition* **2018**, 57, 2554-69.

[3] A. Kojima, K. Teshima, Y. Shirai, T. Miyasaka, *Journal of the American Chemical Society* **2009**, 131, 6050-1.

[4] G. E. Eperon, V. M. Burlakov, P. Docampo, A. Goriely, H. J. Snaith, *Adv. Funct. Mater.* **2014**, 24, 151-.

[5] H. Min, M. Kim, S. U. Lee, H. Kim, G. Kim, K. Choi, J. H. Lee, S. I. Seok, *Science* **2019**, 366, 749-53.

[6] Q. Dong, F. Liu, M. Wong, H. Tam, A. Djurisic, *Chemsuschem* **2016**, 9,

[7] H. Dong, C. Ran, W. Gao, M. Li, Y. Xia, W. Huang, *eLight* **2023**, 3, 3.

[8] J. Hao, X. Xiao, *Frontiers in Chemistry* **2022**, 9,

[9] R. X. Lin, J. Xu, M. Y. Wei, Y. R. Wang, Z. Y. Qin, Z. Liu, J. L. Wu, K. Xiao, B. Chen, S. M. Park, G. Chen, H. R. Atapattu, K. R. Graham, J. Xu, J. Zhu, L. D. Li, C. F. Zhang, E. H. Sargent, H. R. Tan, *Nature* **2022**, 603, 73-8.

[10] Z. Liang, Y. Zhang, H. Xu, W. Chen, B. Liu, J. Zhang, H. Zhang, Z. Wang, D.-H. Kang, J. Zeng, X. Gao, Q. Wang, H. Hu, H. Zhou, X. Cai, X. Tian, P. Reiss, B.

Xu, T. Kirchartz, Z. Xiao, S. Dai, N.-G. Park, J. Ye, X. Pan, *Nature* **2023**, 624, 557-63.

[11] R. Sharma, A. Sharma, S. Agarwal, M. S. Dhaka, *Solar Energy* **2022**, 244, 516-35.

[12] M. Sharma, R. F. Hossain, A. B. Kaul 2021 Direct-Band Gap, Solution-Processed 2d Layered Perovskites for Flexible Photodetectors. In: *2021 IEEE International Flexible Electronics Technology Conference, IFETC 2021*: Institute of Electrical and Electronics Engineers Inc.) pp 47-9

[13] A. W. Y. Ho-Baillie, J. Zheng, M. A. Mahmud, F. J. Ma, D. R. McKenzie, M. A. Green 2021 Recent Progress and Future Prospects of Perovskite Tandem Solar Cells. In: *Applied Physics Reviews*: American Institute of Physics Inc.)

[14] J. Burschka, N. Pellet, S. J. Moon, R. Humphry-Baker, P. Gao, M. K. Nazeeruddin, M. Graetzel, *Nature* **2013**, 499, 316-.

[15] Q. Chen, H. Zhou, Z. Hong, S. Luo, H. Duan, H. Wang, Y. Liu, G. Li, Y. Yang, *J. Am. Chem. Soc.* **2014**, 136, 622-.

[16] M. Liu, M. B. Johnston, H. J. Snaith, *Nature* **2013**, 501,

[17] M. R. Leyden, M. V. Lee, S. R. Raga, Y. Qi, *J. Mater. Chem. A* **2015**, 3,

[18] M. R. Leyden, L. K. Ono, S. R. Raga, Y. Kato, S. Wang, Y. Qi, *Journal of Materials Chemistry A* **2014**, 2, 18742-5.

[19] S. Liu, V. P. Biju, Y. Qi, W. Chen, Z. Liu, *NPG Asia Materials* **2023**, 15,27.

[20] C. Liu, Y. B. Cheng, Z. Y. Ge, *Chemical society reviews* **2020**, 49, 1653-87.

[21] M. J. M. Marques, W. Lin, T. Taima, S. Umezu, M. Shahiduzzaman, *Materials Today* **2024**,

[22] B. Parida, A. Singh, A. K. Kalathil Soopy, S. Sangaraju, M. Sundaray, S. Mishra, S. Liu, A. Najar, *Advanced Science* **2022**, 9, 2200308.

[23] S. Valsalakumar, A. Roy, T. K. Mallick, J. Hinshelwood, S. Sundaram 2023 An Overview of Current Printing Technologies for Large-Scale Perovskite Solar Cell Development. In: *Energies,*

[24] P. Kajal, K. Ghosh, S. Powar, *Energy, Environment, and Sustainability* **2018**, 341-64.

[25] N. J. Jeon, J. H. Noh, W. S. Yang, Y. C. Kim, S. Ryu, J. Seo, S. I. Seok, *Nature* **2015**, 517,

[26] M. Luo, C. Wei, Y. Wu, W. Lei, X. Zhang, H. Zeng, *Journal of Materials Chemistry C* **2024**,

[27] L. Wang, G. Liu, X. Xi, G. Yang, L. Hu, B. Zhu, Y. He, Y. Liu, H. Qian, S. Zhang, H. Zai, *Crystals* **2022**, 12,

[28] P. Chen, X. Ma, Z. Wang, N. Yang, J. Luo, K. Chen, P. Liu, W. Xie, Q. Hu, *Physical Chemistry Chemical Physics* **2024**, 26, 14874-82.

[29] S. Wu, C. Li, S. Y. Lien, P. Gao 2024 Temperature Matters: Enhancing Performance and Stability of Perovskite Solar Cells through Advanced Annealing Methods. In: *Chemistry,* pp 207-36

[30] X. Li, H. Yu, H. Liu, J. Huang, X. Ma, Y. Liu, Q. Sun, L. Dai, S. Ahmad, Y. Shen, M. Wang, *Nano-Micro Letters* **2023**, 15, 206.

[31] Z. Ouyang, M. Yang, J. B. Whitaker, D. Li, M. F. A. M. van Hest, *ACS Applied Energy Materials* **2020**, 3, 3714-20.

[32] F. X. Xie, D. Zhang, H. M. Su, X. G. Ren, K. S. Wong, M. Grätzel, W. C. H. Choy, *ACS NANO* **2015**, 9, 639-46.

[33] I. C. Smith, E. T. Hoke, D. Solis-Ibarra, M. D. McGehee, H. I. Karunadasa, *ANGEWANDTE CHEMIE-INTERNATIONAL EDITION* **2014**, 53, 11232-5.

[34] G. Kasperovich, J. Hausmann, *JOURNAL OF MATERIALS PROCESSING TECHNOLOGY* **2015**, 220, 202-14.

[35] L. Wang, M. Q. Huang, X. F. Yu, W. B. You, J. Zhang, X. H. Liu, M. Wang, R. C. Che, *NANO-MICRO LETTERS* **2020**, 12,150.

[36] J. P. Chu, J. S. C. Jang, J. C. Huang, H. S. Chou, Y. Yang, J. C. Ye, Y. C. Wang, J. W. Lee, F. X. Liu, P. K. Liaw, Y. C. Chen, C. M. Lee, C. L. Li, C. Rullyani, *Thin Solid Films* **2012**, 520, 5097-122.

[37] T. B. Song, Q. Chen, H. Zhou, C. Jiang, H. H. Wang, Y. M. Yang, Y. Liu, J. You, Y. Yang, *Journal of Materials Chemistry A* **2015**, 3, 9032-50.

[38] J. Xu, Z. Y. Hu, X. Y. Jia, L. K. Huang, X. K. Huang, L. M. Wang, P. Wang, H. C. Zhang, J. Zhang, J. J. Zhang, Y. J. Zhu, *Organic electronics* **2016**, 34, 84-90.

[39] Q. Chen, T. Ma, F. Wang, Y. Liu, S. Liu, J. Wang, Z. Cheng, Q. Chang, R. Yang, W. Huang, L. Wang, T. Qin, W. Huang, *Advanced Science* **2020**, 7,

[40] M. J. Brites, M. A. Barreiros, V. Corregidor, L. C. Alves, J. V. Pinto, M. J. Mendes, E. Fortunato, R. Martins, J. Mascarenhas, *Acs applied energy materials* **2019**, 2, 1844-53.

[41] M. M. Maitani, D. Iso, J. Kim, S. Tsubaki, Y. Wada, *ELECTROCHEMISTRY* **2017**, 85, 236-40.

[42] Q. P. Cao, S. W. Yang, Q. Q. Gao, L. Lei, Y. Yu, J. Shao, Y. Liu, *Acs applied materials & interfaces* **2016**, 8, 7854-61.

41st European Photovoltaic Solar Energy Conference and Exhibition

This presentation was selected by the Sc. Committee of the EU PVSEC 2024 for submission of a full paper to one of the EU PVSEC's collaborating peer-reviewed journals.

FULLY PRINTED PEROVSKITE SOLAR CELLS AND MODULES

Luigi Vesce[1], Karthikeyan Pandurangan[2], Maurizio Stefanelli[1], Elena Iannibelli[1], Hafez Nikbakht[1], Maria Laura Parisi[2], Adalgisa Sinicropi[2], Aldo Di Carlo[1,3]

[1]CHOSE, Centre for Hybrid and Organic Solar Energy, Department of Electronic Engineering, University of Rome "Tor Vergata", Via del Politecnico 1, 00133 Rome, Italy

[2]Department of Biotechnology, Chemistry and Pharmacy, R2ES Lab, University of Siena, 53100 Siena, Italy

[3]ISM-CNR, Istituto di Struttura della Materia, Consiglio Nazionale delle Ricerche, via del Fosso del Cavaliere 100, 00133 Rome, Italy

vesce@ing.uniroma2.it

ABSTRACT: In few years, the perovskite (PVSK) solar technology reached high efficiency on lab scale cells. The upscaling of PVSK photovoltaic (PV) technology from small area cells (PSCs) to modules, and the related industrial and economical transition, is achievable by scalable and low-cost manufacturing processes/materials and module design/interconnection patterning. Carbon-based PSCs are attracting attention because of low-cost, durability and printability (high throughput), but few works are present about modules. On the other hand, the most efficient and stable FAPI PVSK is mainly deposited by different techniques with a gold counter-electrode. Here, we changed the paradigm by optimizing the materials composition, by blade coating the full stack out of glove-box. The laser ablation strategy in combination with printing techniques were adopted to fabricate a fully printed module (preliminary prototype). We got 20.8%, 15% and 13.2% efficiency on gold- and carbon-based cells (0.5 cm^2), and carbon-based module, respectively.

Keywords: photovoltaic, perovskite, module manufacturing, scaling-up, carbon, printing

1 INTRODUCTION

Owing to their unique physical-chemical properties, [1], [2] organometal halide perovskites permitted to develop a solution process photovoltaic (PV) technology able to deliver power conversion efficiencies higher than any thin film PV and closely resembling the one of silicon. [3] Thanks to a fine tuning of material compositions, device architectures and fabrication processes the efficiency of perovskite (PVSK) solar cells (PSCs) reached 26.1% for the single junction and 33.9% when used in tandem with silicon. [3], [4] The efficiency transfer from laboratory cells to module devices is of paramount importance to exploit at market level the PVSK PV technology. The uniformity of the deposition and the losses induced by front contact and cell interconnections are the main obstacles for scaling-up the cell to a module level. [5] The module design and the interconnection patterning can face and limit these issues. The losses due to the layers' inhomogeneity are related to the difficulties in transferring to module devices deposition processes and material compositions optimized for the small area cells. [6] In this context, the polycrystalline nature of solution-processed PVSK layers induces defects, such as at grain boundaries and vacancies during the fabrication process. [7] A scalable, repeatable and optimized process is the baseline to obtain high efficiency devices and to upscale the photovoltaic technology. [8]–[11] Recently, FAPbI$_3$ has been developed as an effective absorber, with a narrower band gap of 1.47 eV compared to MAPbI$_3$. Therefore, FA is used to replace MA to reduce the bandgap toward a more ideal range. In literature, researchers scaled up to module the FAPbI$_3$ formulation by air-assisted bar-coating, [12] spin-coating, [13] or blade coating techniques, [14] but still adopting unstable HTM (Hole Transporting Materials) and gold counter-electrode. Organic HTMs are considered the ideal p-type semiconductor materials because of the tunable energy level, the good hole mobility and the solution preparation. [15], [16] The drawbacks of some of these HTMs for the commercial applications are

the high-synthetic and doping cost, and the thermal instability. [17] The widely used gold cathode materials for high efficiency cells and large area modules [18]–[20] are corroded by halogen ions, need high energy process during the device fabrication and are not cheap [21], [22]. Carbon-based perovskite solar cell technology can face these issues by replacing the gold counter electrode with the cheap and stable carbon black/graphite layer. [23]–[25]. In the low-temperature architecture, the carbon layer is deposited directly on top of the adsorber layer and the PVSK can have large crystals because it is not constricted in the pore size of the carbon layer. Moreover, the hole transporting material can be easily inserted to improve the hole-extraction and reduce the non-radiative recombination at the perovskite/carbon interface.

In this work, we stabilized the α-FAPbI$_3$ phase by doping with methylammonium chloride (MACl) and achieved a short-circuit current density above 24 mA/cm^2 and efficiency approaching 21%. We performed module designing and related interconnections optimization, exploited interfacial defects passivation to suppress nonradiative recombination losses, to improve charge carrier extraction and photovoltage on module device, and optimized the deposition process to have a homogeneous large area coating. Finally, we report fully low-temperature and fully printed carbon-based FAPI PVSK cells (0.5 cm^2) and modules able to achieve a champion efficiency more than 15%.

2 RESULTS AND DISCUSSION

We adopted the c-TiO$_2$/SnO$_2$ (tin oxide) as ETL, a formamidinium-based lead triiodide (FAPbI$_3$) as PVSK, phenethylammonium iodide (PEAI) as passivation material, Spiro-OMeTAD as HTL (Hole Transporting Layers) and gold as counter-electrode. The full stack, except for the gold, was deposited by the blade-coating technique at low-temperature and out of the glove-box..

We optimized the ETL and the PVSK layer. Although SnO$_2$ ETL fabricated by different methods (spin-coating, chemical bath deposition, atomic layer deposition) could benefit the elimination of the hysteresis, the reason why the results are not reproducible is still unclear. [26] Here, we investigated the SnO_2 colloid precursor from Alfa Aesar by adding NAOH to stabilize the PH value and reduce the hysteresis. Therefore, the material has a wider process window if used out of glove-box by blade-coating technique. Moreover, we worked on the composition to improve the Voc. About the PVSK, we introduced into $FAPbI_3$ formulation CsI and MACl to shrink the lattice and stabilize the PVSK phase. The MACl will be eliminated during the annealing process at 160 °C. The $FAPbI_3$ PVSK deposition process was based on a double step procedure out of glove-box. After the meniscus formation in the first step, the blade-coating machine (Charon, Cicci Research) deposits the PVSK precursor assisted by a heated dry air flow to reach the supersaturated state of PVSK film. A subsequent 2-propanol (IPA-Isopropyl Alcohol) anti-solvent quenching step is performed by the same deposition method to remove the DMSO excess and force crystallization of PVSK. [27] Then the substrate is annealed at 160 °C for 10 min to nullify the yellow phase and to clearly see the PbI_2 phase (**Figure 1**). XRD shows α phase (101), (012), (021), (202), and (211) crystal planes of the cubic PVSK phase (**Figure 1**).

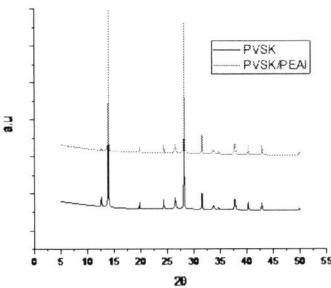

Figure 1: XRD patterns of the PVSK films before and after PEAI deposition.

The synergic effect of the antisolvent and gas quenching causes instant supersaturation of precursor solution and grants fast nucleation and rapid removal of solvents, promoting the homogeneity of PVSK film in the whole substrate. The polycrystalline nature of solution-processed PVSK layers induces defects and trap states, such as at grain boundaries and vacancies during the fabrication process. [7] The PVSK surface passivation by phenethylammonium iodide (PEAI) is one of the most efficient methods to suppress nonradiative recombination losses and to improve charge carrier extraction and photovoltage. The XRD pattern (**Figure 1**) shows a peak at 12.7° (reduced because of the PEAI) related to the PbI_2 excess in the crystalline 3D PVSK film. [28]

The adopted optimization results in an efficiency of 20.8% (Hysteresis Index HI=1.03) on 0.5 cm² active area (**Figure 2**).

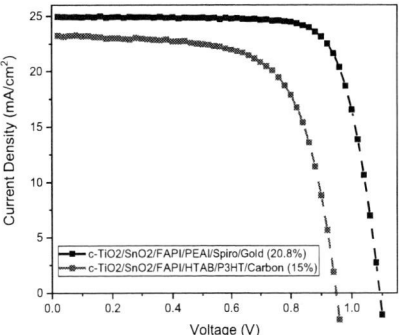

Figure 2: JV curves of the best fabricated 0.5 cm² cells.

Then, we substituted PEAI, Spiro-OMeTAD and the gold counter-electrode with HTAB (hexyl trimethylammonium bromide), P3HT and a low-temperature carbon (Dyenamo) layer, respectively, deposited by blade-coating technique in ambient air to achieve a fully printed perovskite device (**Figure 3**).

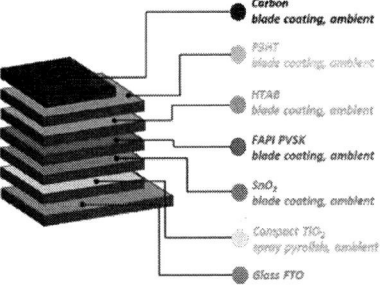

Figure 3: The full stack printed in ambient air.

The efficiency was 15% and the main affected parameters were FF and Voc because of the unoptimized interface between PVSK and carbon (work in progress). The P1 (FTO scribing)-P2 ablation steps by a laser source (here ps-UV laser) are crucial to obtain the module device. The P2 removes the full stack (ETL/PVSK/Passiv./HTL) deposited on FTO from the vertical connection areas to series connect two adjacent cells with the subsequent electrode deposition. The realization of several laser areas with different fluences and number of pulses is useful to evaluate the laser ablation threshold of a material that is a heterogeneous stack. The main constrain is to avoid the TCO damaging to limit the contact resistance between the carbon counter electrode and the TCO. We analyzed the FTO sheet resistance (7 ohm/sq.), the stack thickness (900 nm) and the transfer length measurement (TLM) to minimize possible losses due to the laser process. Then, we deposited by a semi-automatic screen-printing technique the carbon counter-electrode by using an appropriate screen mesh to obtain the electrical insulation between the counter electrodes of adjacent cells. The improved deposition strategy and cells interconnection resulted in a max module efficiency of 13.2% (**Figure 4**).

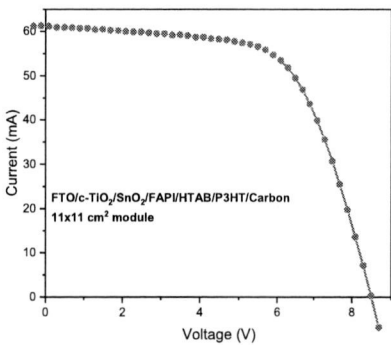

Figure 4: IV curve of the full printed module.

3 CONCLUSION

In conclusion, we presented the preliminary efforts to fabricate low-temperature fully printed FAPI-PVSK cells and modules out of glove-box based on carbon counter-electrode by improving structural design, material chemistry, and fabrication process (coating and laser ablation). The reported simple and reproducible method represent an outstanding baseline to be transferred to a scalable fabrication procedure.

ACKNOWLEDGEMENTS

This research was funded by the European Union's Horizon Europe programme, through a FET Proactive research and innovation action under grant agreement No. 101084124 (DIAMOND).

REFERENCES

[1] A. Kojima, K. Teshima, Y. Shirai, and T. Miyasaka, "Organometal halide perovskites as visible-light sensitizers for photovoltaic cells," *J. Am. Chem. Soc.*, vol. 131, no. 17, pp. 6050–6051, 2009, doi: 10.1021/ja809598r.

[2] H. S. Kim *et al.*, "Lead iodide perovskite sensitized all-solid-state submicron thin film mesoscopic solar cell with efficiency exceeding 9%," *Sci. Rep.*, vol. 2, pp. 1–7, 2012, doi: 10.1038/srep00591.

[3] NREL, "Best research-cell efficiency chart," 2022. https://www.nrel.gov/pv/cell-efficiency.html (accessed May 25, 2022).

[4] M. A. Green *et al.*, "Solar cell efficiency tables (Version 60)," *Prog. Photovoltaics Res. Appl.*, vol. 30, no. 7, pp. 687–701, 2022, doi: 10.1002/pip.3595.

[5] L. Vesce *et al.*, "Ambient Air Blade-Coating Fabrication of Stable Triple-Cation Perovskite Solar Modules by Green Solvent Quenching," *Sol. RRL*, vol. 5, no. 8, pp. 1–11, 2021, doi: 10.1002/solr.202100073.

[6] Z. Yang, C. C. Chueh, F. Zuo, J. H. Kim, P. W. Liang, and A. K. Y. Jen, "High-Performance Fully Printable Perovskite Solar Cells via Blade-Coating Technique under the Ambient Condition," *Adv. Energy Mater.*, vol. 5, no. 13, pp. 1–6, 2015, doi: 10.1002/aenm.201500328.

[7] S. G. Motti *et al.*, "Controlling competing photochemical reactions stabilizes perovskite solar cells," *Nat. Photonics*, vol. 13, no. 8, pp. 532–539, 2019, doi: 10.1038/s41566-019-0435-1.

[8] Z. Li *et al.*, "Scalable fabrication of perovskite," pp. 1–20, 2018, doi: 10.1038/natrevmats.2018.17.

[9] M. Konstantakou, D. Perganti, P. Falaras, and T. Stergiopoulos, "Anti-solvent crystallization strategies for highly efficient perovskite solar cells," *Crystals*, vol. 7, no. 10, pp. 1–21, 2017, doi: 10.3390/cryst7100291.

[10] N. G. Park and K. Zhu, "Scalable fabrication and coating methods for perovskite solar cells and solar modules," *Nat. Rev. Mater.*, 2020, doi: 10.1038/s41578-019-0176-2.

[11] L. Vesce, M. Stefanelli, S. Razza, L. A. Castriotta, F. Di Giacomo, and A. Di Carlo, "Process Engineering and Interfacial Defects Passivation of Large Area Perovskite Solar Modules," *World Conf. Photovolt. Energy Convers.*, 2022, doi: 10.4229/WCPEC-82022-2BO.10.1.

[12] J. W. Yoo *et al.*, "Efficient perovskite solar mini-modules fabricated via bar-coating using 2-methoxyethanol-based formamidinium lead tri-iodide precursor solution," *Joule*, vol. 5, no. 9, pp. 2420–2436, 2021, doi: 10.1016/j.joule.2021.08.005.

[13] M. Jeong *et al.*, "Large-area perovskite solar cells employing spiro-Naph hole transport material," *Nat. Photonics*, vol. 16, no. 2, pp. 119–125, 2022, doi: 10.1038/s41566-021-00931-7.

[14] H. Li *et al.*, "Ink Engineering for Blade Coating FA-Dominated Perovskites in Ambient Air for Efficient Solar Cells and Modules," *ACS Appl. Mater. Interfaces*, vol. 13, no. 16, pp. 18724–18732, 2021, doi: 10.1021/acsami.1c00900.

[15] B. Xu *et al.*, "Tailor-Making Low-Cost Spiro[fluorene-9,90-xanthene] -Based 3D Oligomers for Perovskite Solar Cells Tailor-Making Low-Cost 3D Oligomers for Perovskite Solar Cells," *Chem*, vol. 54, pp. 676–687, 2017, doi: 10.1016/j.chempr.2017.03.011.

[16] S. Calio´, L., Kazim, S., Grä¨tzel, M., and Ahmad, "Hole-transport materials for perovskite solar cells," *Angew. Chem. Int. Ed.*, vol. 55, pp. 14522–14545, 2016.

[17] F. Zhang, X. Yang, M. Cheng, W. Wang, and L. Sun, "Boosting the efficiency and the stability of low cost perovskite solar cells by using CuPc nanorods as hole transport material and carbon as counter electrode," *Nano Energy*, vol. 20, pp. 108–116, 2016, doi: 10.1016/j.nanoen.2015.11.034.

[18] F. Matteocci *et al.*, "Fabrication and Morphological Characterization of High-Efficiency Blade-Coated Perovskite Solar Modules," *ACS Appl. Mater. Interfaces*, vol. 11, no. 28, pp. 25195–25204, 2019, doi: 10.1021/acsami.9b05730.

[19] H. Zhu *et al.*, "Tailored Amphiphilic Molecular Mitigators for Stable Perovskite Solar Cells with 23.5% Efficiency," *Adv. Mater.*, vol. 32, no. 12, pp. 1–8, 2020, doi: 10.1002/adma.201907757.

[20] L. Vesce, M. Stefanelli, and A. Di Carlo, "Low temperature process of homogeneous and pin-hole free Perovskite layers for fully coated photovoltaic devices up to 256 cm 2 area at

ambient condition," *2019 Int. Symp. Adv. Electr. Commun. Technol.*, vol. 764047, no. 764047, pp. 1–5, 2020.

[21] D. Zhou, T. Zhou, Y. Tian, X. Zhu, and Y. Tu, "Perovskite-Based Solar Cells: Materials, Methods, and Future Perspectives," *J. Nanomater.*, vol. 2018, 2018, doi: 10.1155/2018/8148072.

[22] M. G. K. Domanski, J.-P. Correa-Baena, N. Mine, M.K. Nazeeruddin, A. Abate, M. Saliba, W. Tress, A. Hagfeld, "Not all that glitters is gold: metal-migration-in- duced degradation in perovskite solar cells," *ACS Nano*, vol. 10, pp. 6306–6314, 2016.

[23] S. Maniarasu, T. B. Korukonda, V. Manjunath, E. Ramasamy, M. Ramesh, and G. Veerappan, "Recent advancement in metal cathode and hole-conductor-free perovskite solar cells for low-cost and high stability: A route towards commercialization," *Renew. Sustain. Energy Rev.*, vol. 82, no. September 2017, pp. 845–857, 2018, doi: 10.1016/j.rser.2017.09.095.

[24] L. Vesce, M. Stefanelli, and A. Di Carlo, "Carbon-Based Perovskite Solar Cell," 2023, p. 29. doi: 10.3390/iocn2023-14539.

[25] L. Vesce, M. Stefanelli, H. Nikbakht, and A. Di Carlo, "Process Engineering for Low-Temperature Carbon-Based Perovskite Solar Modules †," *Eng. Proc.*, vol. 37, no. 1, pp. 1–5, 2023, doi: 10.3390/ECP2023-14721.

[26] T. Bu *et al.*, "Universal passivation strategy to slot-die printed SnO 2 for hysteresis-free efficient flexible perovskite solar module," *Nat. Commun.*, vol. 9, no. 1, pp. 1–10, 2018, doi: 10.1038/s41467-018-07099-9.

[27] L. Vesce *et al.*, "Perovskite solar cell technology scaling-up: Eco-efficient and industrially compatible sub-module manufacturing by fully ambient air slot-die/blade meniscus coating," *Prog. Photovoltaics Res. Appl.*, vol. 32, no. 2, pp. 115–129, 2024, doi: 10.1002/pip.3741.

[28] P. Chen, Y. Bai, S. Wang, M. Lyu, J. H. Yun, and L. Wang, "In Situ Growth of 2D Perovskite Capping Layer for Stable and Efficient Perovskite Solar Cells," *Adv. Funct. Mater.*, vol. 28, no. 17, pp. 1–10, 2018, doi: 10.1002/adfm.201706923.

41st European Photovoltaic Solar Energy Conference and Exhibition

TÜV RHEINLAND SPECIFICATION ON THE I-V CHARACTERIZATION OF PEROVSKITE-BASED PV MODULES

Giorgio Bardizza[1], Qi Gao[2], Wenhao Xu[2], Yating Zhang[2], Christos Monokroussos[2], Werner Herrmann[1]

[1] TÜV Rheinland (Italy), Via E. Mattei, 3 - 20005, Pogliano Milanese, Italy
[2] TÜV Rheinland (Shanghai) Co., Ltd., No.177, Lane 777, West Guangzhong Road, 200072, Shanghai, P.R. China

Email: giorgio.bardizza@tuv.com

ABSTRACT: Perovskite-based photovoltaics have gained significant attention in recent years due to their potential to surpass traditional silicon-based counterparts in both efficiencies and cost-effectiveness in manufacturing processes. However, challenges such as stability and durability under real-world environmental conditions remain to be addressed. Reliable performance assessment of perovskite-based PV devices is crucial for the further development and commercialization of perovskite PV technology. Several issues are known to affect measurement reliability and must be correctly considered in order to obtain reliable and reproducible results.
At TÜV Rheinland a new internal measurement protocol "2 PfG 2960/12.23 - Requirements for I V Measurement of Perovskite-based Photovoltaic Modules (PSK-PV)" has been developed. It describes the guidelines for adequate electrical stabilization and I-V measurement of perovskite-based PV modules. Together with the IEC 61215 and IEC 61730 standards, which are important references for the performance and safety of PV modules, it forms the basis for the certification of perovskite modules.

Keywords: three to five keywords in order of importance

1 INTRODUCTION

In recent years, there has been a considerable interest in perovskite-based photovoltaics due to their potential to outperform traditional silicon-based counterparts in terms of efficiency and cost-effectiveness during manufacturing. However, challenges related to stability and durability under real-world environmental conditions still need to be addressed.

It is crucial to ensure reliable performance assessment of perovskite-based PV devices to further develop and commercialize perovskite PV technology. Various factors can affect measurement reliability, and it is essential to consider them accurately in order to obtain reliable and reproducible results [1] [2]. Ongoing testing and research aim to overcome these challenges and establish perovskite-based photovoltaics as a viable option for solar energy production.

The testing process involves evaluating the performance of these modules under different conditions, including varying levels of solar irradiation and temperatures. Parameters such as power output, efficiency, and stability are key aspects that are measured.

Additionally, accelerated aging tests are conducted to predict the long-term performance and durability of the modules. These tests play a vital role in the further development and commercialization of perovskite photovoltaic (PV) technology.

At TÜV Rheinland, a new internal measurement protocol called "2 PfG 2960/12.23 - Requirements for I-V Measurement of Perovskite-based Photovoltaic Modules (PSK-PV)" [3] has been developed. This protocol provides guidelines for appropriate electrical stabilization and I-V measurement of perovskite-based PV modules. Together with the IEC 61215 [4] and IEC 61730 [5] standards, which are important references for the performance and safety of PV modules, this protocol forms the basis for certifying perovskite modules.

This work aims to present the new test protocol and demonstrate its applicability to a real case of a large-area commercial perovskite-based PV module. Preliminary results indicate that by following this protocol, measurements under steady-state conditions can be achieved, leading to reliable and reproducible results.

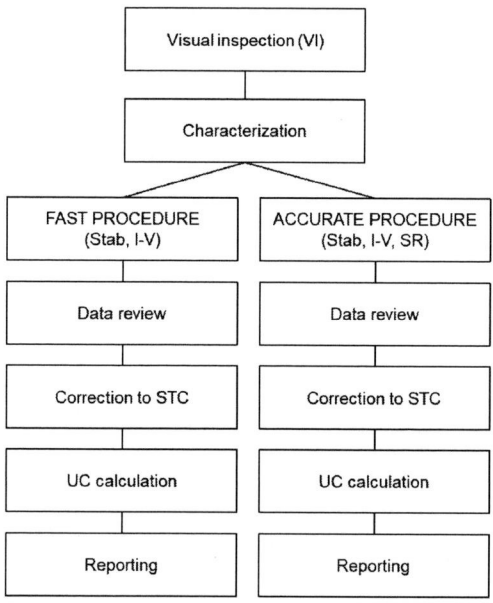

Figure 1 Schematic of the test sequence

2 TEST PROGRAM

2.1 Testing procedures

The test sequence is shown in the schematic of **Figure 1**. In this protocol two different characterization procedures may be selected depending on the duration and precision required:

- FAST procedure: the schematic is shown in **Figure 2** and it is preferable/suitable for comparative measurements or measurements where an uncertainty (UC) on P_{max} up to 3.5%, k=2 is acceptable; it is primarily addressing maximum power determination, which is used to assess a PV module's degradation.

- ACCURATE procedure: the schematic is shown in **Figure** *3* and is more complex and time-consuming, but a low uncertainty on P_{max} in the order of 2%, k=2 can be achieved. The accurate procedure is suitable, when high accuracy is relevant such as creating reference modules and efficiency measurement.

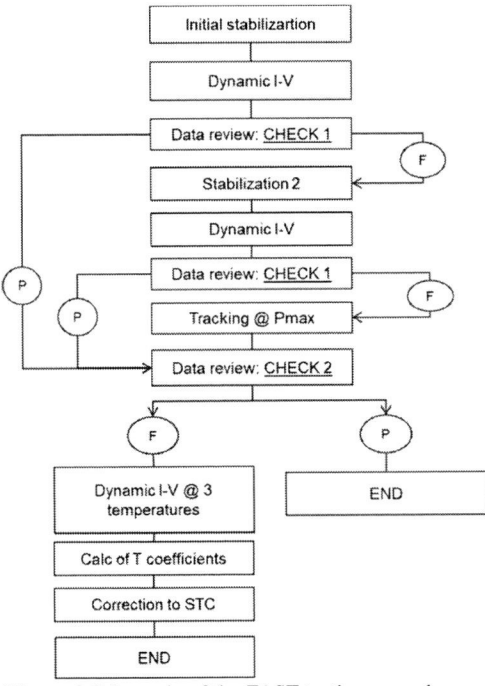

Figure 2 Schematic of the FAST testing procedure

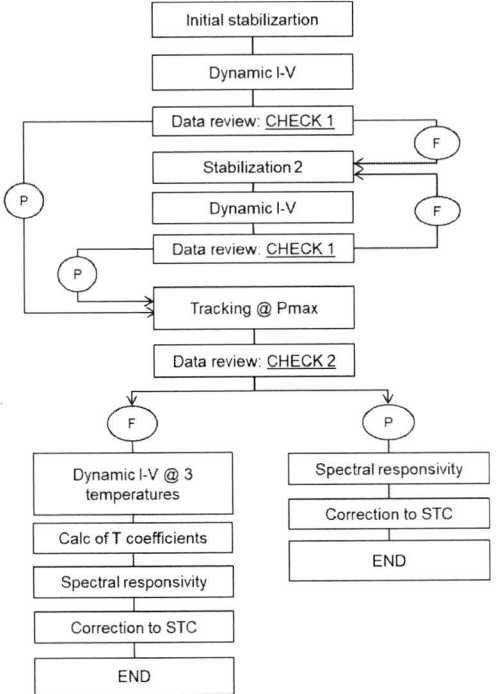

Figure 3 Schematic of the ACCURATE testing procedure

The calculation of the measurement uncertainties is based on a calculation of the expanded uncertainty contribution of the different sources [6].

2.2 Stabilization

The I-V characteristic for the device under test shall be measured such that it reflects, as closely as possible, the performance of the device under steady-state conditions. Stabilisation is requested to be performed before any characterisation (I-V or spectral responsivity measurement). The device under test (DUT) is placed under constant illumination close to standard test condition (STC) and temperature of the device must be controlled (maximum allowed temperature 60 °C, maximum allowed temperature variation once at equilibrium +/- 2 °C). The device is kept at its maximum power point (V_{Pmax}) by applying a voltage bias.
Stabilization must be performed until the P_{max} variation is below ±0.1% in a 5-minute interval.
A second step of stabilization (Stabilization 2) could be necessary after I-V measurements (only the first I-V in case of the FAST PROCEDURE) if conditions of CHECK 1 are not met (see section 2.3).

2.3 Dynamic I-V

The test procedure to perform I-V measurements follows the steps described in the IEC 60904-1 [7] section 7.2. However, the metastability of PSK devices does NOT allow to test them with very fast sweep (< 1s) under flash solar simulator.

A steady state solar simulator is required, and a dynamic I-V technique is the most appropriated to perform I-V measurements of such devices. The dynamic I-V technique is a kind of single step-wise sweep method as defined in IEC 60904-1 section B.4.4.3.1 with customized measurement parameters. After stabilization it is important to keep the DUT under illumination until the sweep starts.

The selection of the appropriate dynamic I-V measurement parameters (number of points and distribution) is fundamental to allow for correct signal processing avoiding digitalization errors. Typically a minimum number of 3 points should be acquired for correct determination of V_{oc} and I_{sc} respectively and at least 5 points for correct determination of P_{max}.

Our recommendation is to acquire I-V curves in both sweeping direction: direct (voltage ramp going from V=0 or slightly negative to V= V_{oc} or slightly higher) and reverse (voltage ramp going from V= V_{oc} or slightly higher to V=0 or slightly negative). This allows afterwards for a correct evaluation of the hysteresis.

2.3 Data review CHECK 1

The first control regards the good match between direct and reverse measured I-V curves. This is done comparing their ΔP_{max} based on the following formula:

$$\Delta P_{max} = \frac{|P_{\max \_direct} - P_{\max \_reverse}|}{P_{\max \_direct} + P_{\max \_reverse}} \times 2$$

The control is passed if $\Delta P_{max} < 1\%$.
We recommend to use polynomial fit to calculate the P_{max_direct} and $P_{max\_reverse}$ from the two I-V curves.

2.4 Data review CHECK 2

The second control regards the temperature of DUT reached during the I-V sweep. The temperature of DUT must be acquired during I-V measurements using standard temperature sensors. The temperature during the I-V sweep must lie in the range (25 ± 4) °C. If this is satisfied the control is passed. Otherwise, temperature coefficients need to be measured and temperature correction according to IEC 60891 [8] must be applied.

For the calculation of the temperature coefficients at least 3 dynamic I-V as described in section 2.2 (reverse direction) at three different temperatures in the range 15 °C, 75 °C must be acquired. The temperatures shall differ of at least 15 °C and cover the range 25 °C, 55 °C. The allowed variation in set-point temperature is ±2 °C. The procedures for the calculation of the temperature coefficients are described fully in the IEC 60891 section 5.

2.5 Measurement tool

A customized integrated hardware and software system (Figure 3) was developed by TÜV Rheinland for the automation control of perovskite PV module pre-conditioning and I-V measurements. The system can perform the DynamicIV measurement, and maximum power point tracking during the pre-conditioning process.

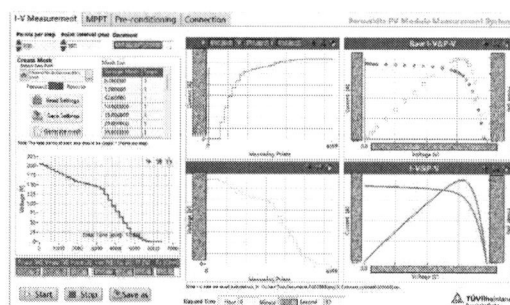

Figure 4 Control software interface for perovskite PV module measurement and pre-conditioning

3 RESULTS

The devices involved in this study consisted of several large-area single-junction PSK PV modules. The modules dimensions are 1 m x 2 m consisting of 190 monolithically interconnected cells in series.

Two I-V curves performed following the fast procedure and executed at two consecutive days are shown in **Figure 5**. The P_{max} values extracted from these I-V curves is shown in **Figure 6**. The testing procedure used to characterize the PSK modules involved in this work allowed to obtain reliable and reproducible results well within the stated UC.

To keep anonymity of the client and product, the data have been normalized. Both I-V curves have been normalized dividing by the same factor in order to keep track of the difference between the data.

A polynomial fit is used to interpolate the data and P_{max} is calculated from the fitting function. Comparing the P_{max} values extracted form the two I-V curves a difference of 0.08 % has been observed. Comparing this P_{max} with the value obtained with P_{max} tracking (5 minutes average) a difference of about 0.32-0.40 % is observed. This confirms that reliable I-V measurements of single junction PSK-PV modules are possible following the procedure described in the 2PfG 2960/12.23 and presented in this work.

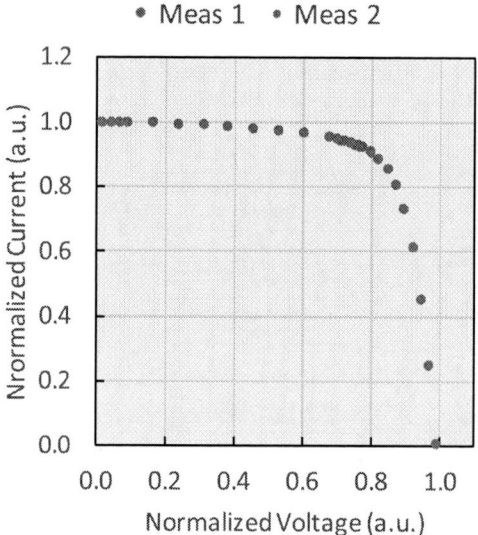

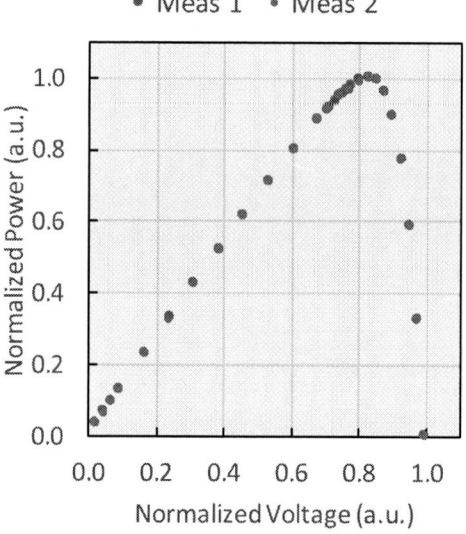

Figure 5 Two I-V curves performed following the fast procedure in two consecutive days. The normalized plot of I-V (top) and P-V (bottom) measured data are shown in the figure.

41st European Photovoltaic Solar Energy Conference and Exhibition

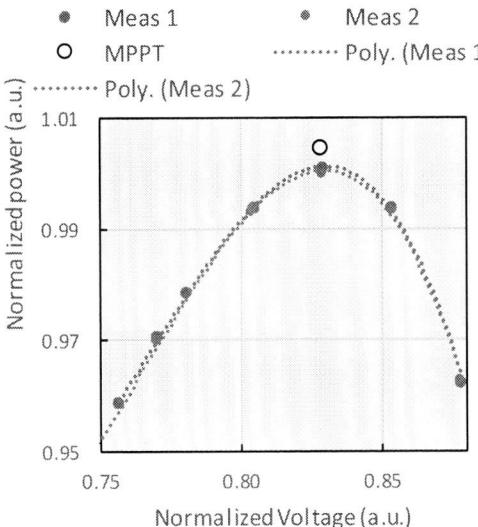

Figure 6 Evaluation of P_{max} from the I-V curves reported in **Figure 5**. Comparison of P_{max} values evaluated from the the fit of the measured data a difference of 0.08 % is observed.

4 CONCLUSIONS

This work introduced a novel measurement protocol called 2 PfG 2960/12.23 specifically designed by TÜV Rheinland experts for characterizing large-scale perovskite PV modules. The protocol provides clear and standardized guidelines for ensuring proper electrical stabilization and measurement of perovskite PV modules, which can vary significantly from traditional silicon-based modules.

Two measurement procedures cater to different needs, ranging from rapid comparisons to high-precision measurement. This flexibility enhances the protocol's applicability across various research and development stages. This makes it directly relevant to industry needs and advancements in commercialization.

In this paper the new measurement procedure is described and applied to a real case of large-scale PSK PV module. First results show that following this procedure it is possible to obtain repeatable and reliable I-V curves assuming an acceptable stability of the device under test. In the example reported two I-V curves result in P_{max} values differing of 0.08%, confirming the very good reproducibility of the measurement. This confirms that reliable I-V measurements of single junction PSK-PV modules are possible following the procedure described in the 2PfG 2960/12.23 and presented in this work. Moreover reproducible I-V measurements of PSK-based PV devices can be achieved in a reasonable time. It is anyway necessary to continuously optimize and develop faster, more accurate and reliable measurement methods to meet the needs of large-scale production.

To the best of our knowledge the proposed test protocol is the first one that has been proposed in the industry for the characterization of large-scale perovskite PV modules, which can accelerate the development and commercialization of this promising technology.

5 AKNOWLEDGMENT

The authors gratefully acknowledge Dr. Fan Bin from Suzhou GCL Nano Co., Ltd for their helpful discussions and sample contribution.

6 REFERENCES

[1] Bardizza, G., Muellejans, H., Pavanello, D. and Dunlop, E., Metastability in performance measurements of perovskite PV devices: a systematic approach, JOURNAL OF PHYSICS-ENERGY, ISSN 2515-7655, 3 (2).

[2] Gao, Q., Lau, J., Lee, E., Monokroussos, C. (2019). Test Method of Current-voltage Characterization of Perovskite PV-module, EU PVSEC.

[3] 2 PfG 2960/12.23 - Requirements for I-V Measurement of Perovskite-based Photovoltaic Modules (PSK-PV).

[4] IEC 61215 Terrestrial photovoltaic (PV) modules - Design qualification and type approval - Part 1: Test requirements

[5] IEC 61730 Photovoltaic (PV) module safety qualification - Part 1: Requirements for construction

[6] Metro-PV deliverable D5 https://www.metro-pv.ptb.de/home/

[7] IEC 60904-1 Photovoltaic devices – Part 1: Measurement of photovoltaic current-voltage characteristics

[8] IEC 60891 Photovoltaic devices – Procedures for temperature and irradiance corrections to measured I-V characteristics.

41st European Photovoltaic Solar Energy Conference and Exhibition

This presentation was selected by the Sc. Committee of the EU PVSEC 2024 for submission of a full paper to one of the EU PVSEC's collaborating peer-reviewed journals.

ONE-YEAR OUTDOOR TESTING OF 4T PEROVSKITE/SI PV MODULES

Matthew Norton[1], Vasiliki Paraskeva[1], Maria Hadjipanayi[1], Elias Peratikos[1], Aranzazu Aguirre[2,3,4],
Anurag Krishna[2,3,4], Santhosh Ramesh[2,3,4], Tom Aernouts[2,3,4], George E. Georghiou[1]
[1]FOSS Research Centre for Sustainable Energy, Department of Electrical and Computer Engineering,
University of Cyprus, 75 Kallipoleos St., Nicosia, 1678, Cyprus
[2]Hasselt University, imo-imomec, Martelarenlaan 42, 3500 Hasselt, Belgium
[3]Imec, imo-imomec, Thin Film PV Technology. Thor Park 8320, 3600 Genk, Belgium
[4]EnergyVille, imo-imomec, Thor Park 8320, 3600 Genk, Belgium

ABSTRACT: Perovskite photovoltaics today demonstrate impressive conversion efficiencies, but more field experience is needed to understand their long-term stabilities. Perovskite-on-silicon tandem modules offer a fast-track route to commercialisation, and as such warrant studies of their field performance. This paper reports on the results of a field measurement campaign involving 5 mini tandem modules installed at the University of Cyprus' outdoor test facility. The modules each incorporate a string of seven perovskite sub-cells layered onto a bifacial TOPCON silicon cell, with 4-terminal connections provided to access the individual layers. To reveal trends under stable conditions, data were filtered to use only clear-sky days, which accounted for over 210 days in 20 months of testing. Over the first 12 months of exposure, the perovskite sub-modules showed a 50% reduction in power conversion efficiency (PCE); however, over the following 8 months, they showed either continuous improvement or stable PCE. Over this latter period, the perovskite PCE was not affected by temperature, yet the samples exhibited a stable negative temperature coefficient and a positive fill factor temperature coefficient. An examination of the current-voltage characteristics showed this behaviour likely results from a change in ion mobility with temperature. Overall, the tandem module arrangement has outperformed a single-junction TOPCON device over the test period.

Keywords: Perovskite, Tandem, Outdoor, Modules

1 BACKGROUND AND MOTIVATION

Perovskite solar cells have demonstrated outstanding performance in recent years, achieving power conversion efficiencies (PCE) as high as 25.5% with theoretical efficiencies calculated up to 31% [1]. Considerable indoor testing has been conducted over recent years at different temperature [2], humidity [3], and bias loading conditions demonstrating the impact of each parameter on the perovskite power conversion efficiency (PCE) and stability. However, while indoor measurements and aging procedures can be used to assess factors impacting stability, they don't fully resemble outdoor operational conditions with regular day-night cycles and continuous changes in irradiance, humidity, and temperature.

Perovskite-on-silicon tandem modules in particular offer advantages as a route to commercialisation of perovskite photovoltaics. Outdoor stability testing of perovskite tandem devices has been reported before [4,5]. It is envisaged that the integration of perovskites into commercially available modules in this manner could involve only a few additional manufacturing steps, and yet would bring all the advantages of high-volume production and supply chains immediately to bear.

Considering the above, the partners of the EU-funded TESTARE project embarked upon a study of the long-term PCE evolution of perovskite/Si tandem miniature modules exposed outdoors for several months.

2 AIM AND APPROACH

The aim of this work is to further the understanding of how tandem solar modules containing a perovskite upper layer on top of a silicon bottom layer perform outdoors over the long term. The central objective of this aim was to collect several years of performance data from a batch of tandem mini modules installed in an outdoor test facility (OTF) with a high annual solar resource. The measurement campaign started on 19th January 2023 and is still ongoing, with the data up until 1st September 2024 presented in this paper.

2.1 Materials

IMEC prepared a batch of five perovskite-on-silicon tandem mini-modules with dimensions 10 x 10 cm for exposure in the field. Each module contained seven series-connected perovskite sub-cells on a single silicon subcell, as shown in Figure 1. These have a total active area of 4 cm^2 with a geometrical fill factor of the perovskite layer of approximately 0.91. The encapsulation was achieved by sandwiching the cells between two layers of outer glass with a butyl edge-sealant. A TOPCON technology, bifacial silicon cell with an approximate conversion efficiency of 20% was used, which also accepted additional rear-side-irradiation of the module through the back glass. This module design has been shown to reliably protect the PV cells from moisture and air, which was necessary to ensure that changes in performance over time were not affected by these factors.

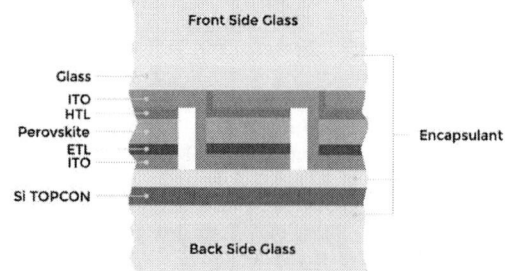

Figure 1: Cross-sectional cutaway of the tandem samples, showing the series-connected perovskite layer stack on top of a bifacial silicon cell. A 4-terminal arrangement allowed access to the perovskite and silicon PV contributions independently.

Additionally, a reference mini module containing just the silicon TOPCON cell was prepared. This sample was kept indoors under dark at room temperature and tested regularly over the campaign, yielding a PCE of 21% including a rear-side boost of 5%. Control tandem devices were also prepared and kept indoors for comparison with the field exposed samples.

2.2 Methodology

The five samples were mounted outdoors at the University of Cyprus' outdoor test facility on a south-facing rack tilted at 35 degrees to the horizontal. The test location has a 'Hot-Summer Mediterranean Climate' (Csa) Köppen-Geiger classification, with an approximate annual solar irradiation resource of 2000 kWhm^{-2} at optimal tilt. The key irradiation and temperature parameters measured over the test period are summarised in Figure 2.

The global irradiance in the plane of the samples was measured using a silicon reference cell. Module temperatures were measured using PT100 sensors pressed on the rear of three of the modules. Electrical performance parameters of the samples were obtained by regularly sweeping the current-voltage (I-V) curves of each sample over the day. Between each I-V measurement, the samples were left open-circuit. Due to the number of experiments running simultaneously at the test site, only one representative silicon sub-cell was measured outdoors over the full measurement period, from the 506T module.

To measure the module I-V curves, the voltage across the samples was swept from -0.1V to 7.8V and immediately back again. The current was measured at a voltage step of 0.1V with a dwell time at each measurement point of 0.2 seconds. Each sample was measured sequentially roughly every 15 minutes.

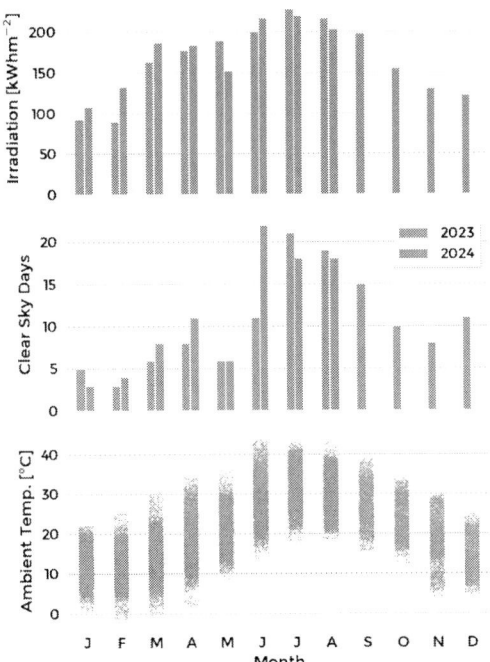

Figure 2: Environmental conditions at the University of Cyprus' outdoor test facility recorded over the test period.

Due to the slow response of the perovskite sub-modules to dynamic field conditions, the analyses presented in this paper are performed on data acquired under steady conditions on clear-sky days. These are defined here as days with uninterrupted sunshine characterised by a peak irradiance of over 750 Wm^{-2} in the plane of the array, a minimum total daily irradiation exceeding 4.5 kWhm^{-2} and a maximum permitted change of irradiance of 200 Wm^{-2} over a 15 minute period. At the test site, 123 such days were identified in 2023 and 90 during the first 8 months of 2024. The distribution of these days over the test period is also presented in Figure 2.

3 RESULTS

3.1 PCE over time

The efficiency over time for all the mini modules are plotted together in Figure 3. These values represent monthly average PCE values based on clear-sky days. Generally, the trends of performance seem to fall within three 6-month stages. In the first 6-month period, the PCE of the perovskite subcells is seen to decrease relatively rapidly, to about two-thirds of initial output. In the second 6-month period, the rate of change of PCE seems to halve, as the PCE drops by a further 20% to roughly 50% of initial. However, in the following 6-month period, the performance of the samples seems to plateau and, in some instances, increases.

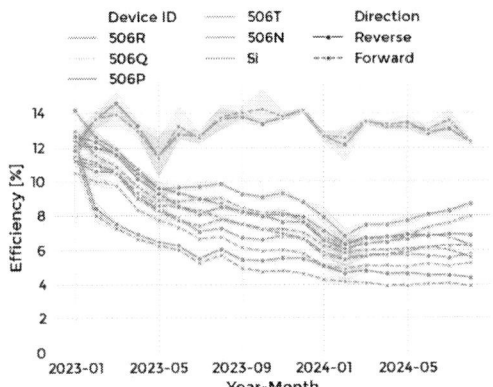

Figure 3: Monthly average PCE evolution over time for the perovskite subcells, and a silicon subcell. PCE is shown from both forward and reverse sweeps.

The larger scatter in the measurement of the silicon subcell is attributed to the lower current and voltage output compared to the perovskite subcells, and the influence of rear-side irradiance. In all perovskite subcell strings, the reverse sweep showed a higher apparent conversion efficiency due to the presence of hysteresis. The magnitude of the hysteresis over the course of the measurements remains approximately stable at around 1% difference in PCE.

Over the duration of the measurement campaign, it was observed that at certain moments changes in performance were common for several of the samples under test, implying that the cause of these were due to the field conditions. For example, most samples showed an inflection in performance in 2023-08 and again in 2024-02, after which PCE either stabilised or increased. Since these two months are directly preceded by the months with maximum and minimum module temperatures respectively, there is a suggestion that this could be temperature dependent, but this hypothesis does not hold across the whole data set. During 2024, the perovskite modules showed impressive stability, and did not seem to be influenced by either ambient temperatures or irradiation.

3.2 Daily variation in PCE

Aside from the long-term trends of PCE, the samples also exhibited daily, or diurnal, variations in efficiency. To analyse the evolution of performance over time, the daily progression of PCE over each day was plotted as a function of the cumulative irradiation absorbed by the samples, to see if days with higher irradiation would produce different effects within the samples. Figure 4 plots the outcome of this analysis for the five perovskite sub-modules over period in question.

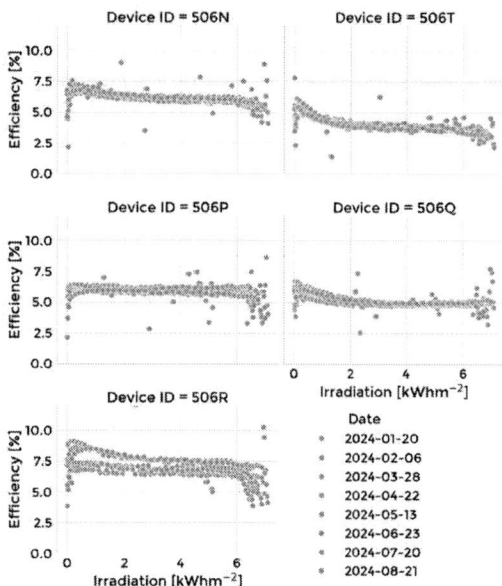

Figure 4: Daily changes in perovskite sub-module efficiency on 8 clear sky days measured over the period January to September 2024. The efficiency is plotted against cumulative irradiation.

What is immediately striking about Figure 4 is that over an 8-month period, the PCE of the modules — with the exception of 506R — remains constant despite changes in environmental conditions. The daily profiles are also very similar for each module; in most cases samples exhibit a slightly higher PCE in the morning which decreases at a higher rate until the samples absorbed approximately 1 $kWhm^{-2}$ of irradiation. After that point, the change in efficiency progresses linearly until the end of the day, where there is a drop-off in apparent efficiency attributed to acceptance angle effects in the modules.

With the exception of sample 506P which is unchanged, the modules show, on average, a relative decrease in efficiency of 25 to 30% over the day, and a corresponding increase overnight. Interestingly, the improvement in efficiency of sample 506R over 2024 seen in Figure 3 is also evident in this plot, and includes relatively higher starting efficiencies at the beginning of the day.

3.3 Temperature coefficients

Whereas prior measurements of single-junction perovskite modules in the field tended to yield confusing temperature coefficients including positive values [6] all five samples here showed a remarkably consistent negative open-circuit voltage (V_{oc}) temperature coefficient over the complete test period. Moreover, as shown in Figure 5 below, the V_{oc} temperature coefficient was consistent across devices. Figure 5 was generated using data taken during the stable period between March and July 2024, but other than a reduced scatter, the average values are obtained consistently over the entire campaign.

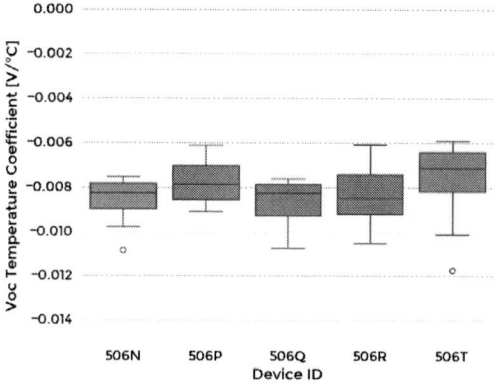

Figure 5: Box and whisker plots of V_{oc} temperature coefficients obtained from the five perovskite sub-modules over the period from March – July 2024. Data were filtered to consider only irradiance levels in the band $950 - 1050$ Wm^{-2}.

The results shown in Figure 5 were also confirmed with indoor tests. Sample 506R was removed from the field for measurement of temperature coefficients under controlled conditions using a solar simulator. The V_{oc} temperature coefficient was determined to be -8.7 mV/°C, which falls within a standard deviation of the outdoor results' mean. The agreement between the indoor and outdoor values is attributed to the level of exposure that the sample had undergone at the point of testing, driving the perovskite cells into a stable condition. Pristine samples kept indoors as controls did not exhibit a stable V_{oc} temperature coefficient when tested under the simulator, and the values obtained were not in agreement with field observations.

Over the first year of exposure, the long-term decrease in PCE of the perovskite modules convoluted the analysis of temperature-dependent behaviour, with different fill factor (FF), maximum power (P_{max}) and short-circuit current (I_{sc}) temperature coefficients determined at different months of the year. However, once the samples had stabilised by the second year of the campaign, it was possible to investigate these with greater confidence.

It was found that the samples exhibited a strong positive FF temperature coefficient in 2024, whereas there was no discernible effect of temperature upon the P_{max} or I_{sc}. This, in combination with the negative V_{oc} temperature coefficient, presented a paradox: normally FF and V_{oc} both reduce with increasing junction temperatures, and reduce power output accordingly.

3.4 I-V curve analyses

To unravel the paradox of why V_{oc} decreases, fill factor increases, and yet PCE remains stable with increased temperature, we examined the I-V curves obtained under steady conditions on clear sky days in 2024. First the dataset was filtered to identify measurements where irradiance was similar to within ± 5 Wm^{-2}, and then from these several curves were selected from different seasons to cover a range of junction temperatures. It should be noted that while the rear-side measurements did not necessarily match the actual junction temperature, the relative changes would have been very similar. The results of this analysis are plotted in Figure 6 for samples 506P (the most stable) and 506R (the most improved). Reverse I-Vs were selected as they are deemed to be more representative than forward traces.

In the case of 506P, we can see that after the first year of exposure, the decreased shunt resistance and increased series resistance combine to shift the maximum power point of the curve above the 'knee' of the I-V curve. At this point, the maximum power point is remains unaffected by changes to the series resistance. Furthermore, the particular shape of the curve now results in a decrease in series resistance of the sample with increasing junction temperature. The cause of this is uncertain, but could be attributed to improved ion mobility at higher temperatures. It is also notable that both the I_{sc} and P_{max} are apparently unaffected by changes to the junction temperature, in agreement with the overall PCE trends.

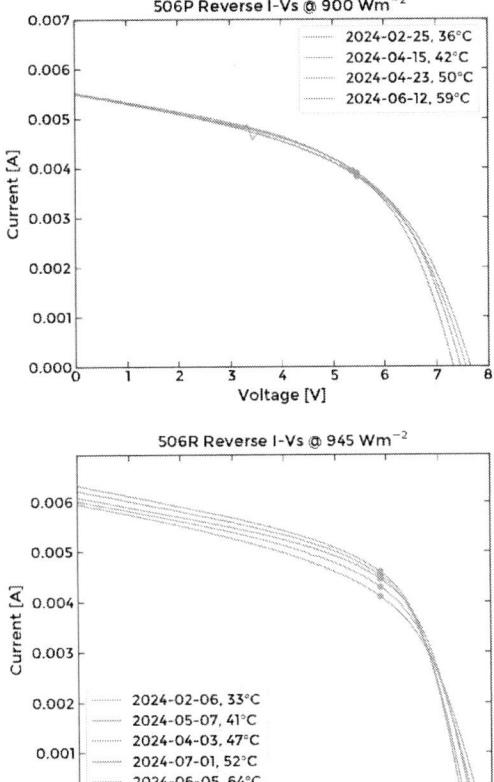

Figure 6: Selected I-V characteristic curves of sample 506P (above) at 900 Wm^{-2} and 506R at 945 Wm^{-2}, taken at different junction temperatures in 2024.

In the case of sample 506R, we see that the improved performance seen in Figure 3 is linked to a recovery in the current generation of the samples over time. This sample also exhibits the same effects of temperature upon the open-circuit voltage, though in this case the performance recovery increases the P_{max} over time. Although it appears as though the device exhibits a positive I_{sc} and P_{max} temperature coefficient, this is an artefact caused by the general increase in performance over time, and is not supported by the wider dataset. It is also noteworthy that the maximum power point voltage does not vary between these measurements.

4 ANALYSIS

4.1 Performance Summary

Indoor measurements of the bare, single-junction bifacial TOPCON cell showed that the 20% efficient cell was reduced to 16% when encapsulated in the module, and that the efficiency was the same when measured from either side. When the module was mounted on the OTF test rack, the reflectance of the mounting structure was found to contribute a PCE boost of around 5% of the front-side irradiance. This resulted in a total apparent PCE of approximately 21%. Indoor measurements also established that the PCE of the front side was reduced to 8% when the perovskite layer was introduced on top of it.

In total, the starting efficiency of the better-performing tandem sample was approximately 26%, which was the combination of the pristine perovskite module and the bifacial silicon subcell. At the end of the 20 months of testing the overall efficiency had returned to 21%, which is the same as the base case of a single silicon subcell. However, over the entire campaign this tandem device delivered an overall energy conversion efficiency of 22%, and at this point the efficiency is still increasing.

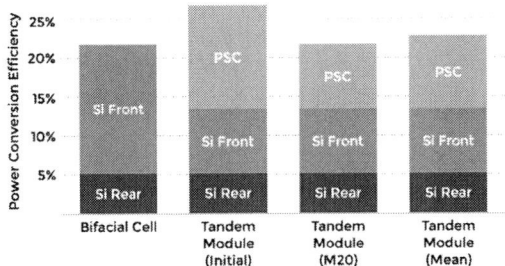

Figure 7: Power conversion efficiency balance for the tandem modules. In the case of the better performing devices, the tandem structure outperforms a base-case bifacial device after 20 months of exposure.

4.2 Performance evolution

The general evolution of the PCE of this batch of mini modules shows an initial decrease in the perovskite performance over time that is attributed to the formation of ionic barriers at the layer interfaces. These act to reduce the current generation over time. However, after a full year of field exposure, it appears that this barrier formation had concluded, and the devices showed stable performance after that point. In the case of two of the modules, particularly 506R, there seems to have been some reversal of this ionic barrier formation that saw an increase in current generation once more, and a corresponding increase in PCE over time.

The change in behaviour of the samples with temperature during the second year of exposure is consistent with the expectation that higher temperatures will reduce the apparent series resistance of samples due to the increase in ion mobility and the higher thermal energy of charge carriers, enabling them to cross over the potential barrier. In this case, the particular shape of the I-V curves showed that it was possible for this reduction to occur in a manner that did not affect the maximum power point, thus explaining the absence of a power temperature coefficient.

8   CONCLUSIONS

The combination of long-term field exposure and filtering for steady measurement conditions has allowed bifacial perovskite-on-silicon tandem mini modules to be studied under stabilised conditions. The outcome of these analyses shows that over the 20 months of outdoor exposure the tandem arrangement has provided a net increase in energy yield compared to a base case of a bifacial silicon photovoltaic module.

Somewhat surprisingly, some of the stabilised modules continue to show an increasing PCE,and have also reached a state where they show no change in efficiency with temperature.

These results provide encouragement that perovskite-on-silicon tandem modules are both technically viable and energetically advantageous,and can offer a compelling route for the incorporation of perovskite photovolatics into mature commercial photovoltaic markets.

[1] M. A. Green *et al.*, "Solar cell efficiency tables (version 62)," *Prog. Photovoltaics Res. Appl.*, vol. 31, no. 7, pp. 651–663, 2023, doi: 10.1002/pip.3726.
[2] H. Zhang, X. Qiao, Y. Shen, and M. Wang, "Effect of temperature on the efficiency of organometallic perovskite solar cells," J. Energy Chem., vol. 24, no. 6, pp. 729–735, 2015, doi: 10.1016/j.jechem.2015.10.007.
[3] A. K. Mishra and R. K. Shukla, "Effect of humidity in the perovskite solar cell," Mater. Today Proc., vol. 29, no. 3, pp. 836–838, 2019, doi: 10.1016/j.matpr.2020.04.872.
[4] M. De Bastiani *et al.*, "Toward Stable Monolithic Perovskite/Silicon Tandem Photovoltaics: A Six-Month Outdoor Performance Study in a Hot and Humid Climate," *ACS Energy Lett.*, vol. 6, pp. 2944–2951, 2021, doi: 10.1021/acsenergylett.1c01018.
[5] J. Liu et al., "28.2%-Efficient, Outdoor-Stable Perovskite/Silicon Tandem Solar Cell," Joule, vol. 5, no. 12, pp. 3169–3186, 2021, doi: 10.1016/j.joule.2021.11.003.
[6] E. Velilla, D. Ramirez, J. I. Uribe, J. F. Montoya, and F. Jaramillo, "Outdoor performance of perovskite solar technology: Silicon comparison and competitive advantages at different irradiances," *Sol. Energy Mater. Sol. Cells*, vol. 191, pp. 15–20, 2019, doi: 10.1016/j.solmat.2018.10.018.

AKNOWLEDGEMENTS

This work has been financed by the European Union through the TESTARE project (Grant ID: 101079488) and by the European Regional Development Fund and the Republic of Cyprus through the DegradationLab project (Grant ID: INFRASTRUCTURES/1216/0043).

41st European Photovoltaic Solar Energy Conference and Exhibition

CHALLENGES FOR SOLDER INTERCONNECTION PUSHED BY HIGH-EFFICIENCY SOLAR CELL DEVELOPMENTS

Benjamin Grübel, Angela De Rose, Achim Kraft
Fraunhofer Institute for Solar Energy Systems ISE
Heidenhofstr. 2, 79110 Freiburg, Germany

ABSTRACT: The solder interconnection to connect solar cells in series is a key process for the manufacturing of reliable high-performance solar modules. The fast development of the cell technologies, interconnector number and the throughput requirements by industrial production lines underlines the necessity of detailed evaluation of the process challenges. Within this work, the developments of the solar cell metallization are addressed by precise evaluation of the metallized area available for the solder joint. The soldering of round wire interconnectors onto small area pads or even directly onto the contact fingers for busbarless designs reveals to become challenging for precise handling and interconnector alignment. Drastic reduction of the size of the solder pad area to realize silver reduction results in poor mechanical and electrical quality of the solder joint. Every cell technology features requirements when it comes to soldering temperatures according to sensitive layers. Therefore, we present the solder alloys available according to their melting temperature. Concerning the criticality of materials potential lead-free alloys are presented. We performed soldering processes with PERC, TOPCon and SHJ solar cells at an industrial stringer machine to evaluate temperature homogeneity of the process. A temperature discrepancy of at least 14 K for SHJ and up to 90 K for PERC solar cells was measured on the cell during interconnection. The challenges and increasing sensitivities arising from developments on solar cell level reveal the need of constant evolution of the soldering tools as well as of the soldering processes.
Keywords: Interconnection, High-efficiency solar cells, Soldering, Stringer

1 INTRODUCTION

In recent years, the PV industry has undertaken significant advancement in the development of new solar cell technologies. While the passivated emitter and rear cell (PERC) technology has dominated the market for years, it is expected to be replaced by technologies with improved performance. Tunnel oxide passivated contact (TOPCon) [1] and silicon heterojunction (SHJ) [2] solar cells are the technologies which are candidates as successor to the PERC technology [3, 4]. Even though the implementation of these new technologies into module application is at its beginning, the effort to develop the subsequent solar cell technology is already intensified. Currently solar cells based on tandem designs are estimated as next step due to their improved behavior in absorbing the light spectrum in a more efficient way [5]. Specifically, tandem solar cells consisting of at least one perovskite absorber are widely investigated in the PV community, which can be deduced by the significant increase in publications on this topic [6].

Apart from the exact solar cell technology, other developments are taking place, influencing module integration. First, optimization of wafer dimensions (size, thickness) [3] affect cell handling, hardware size of tools, industrial throughput and number of busbars/wires. Furthermore, the development of the metallization of the solar cells is driven by to main targets: efficiency and costs. In terms of efficiency, optimization of the metallization focuses on reducing the shaded area while minimizing the resistive losses. A significant drawback here is that state-of-the-art screen-printing metallization for TOPCon and SHJ require a higher amount of silver (Ag) [7] compared to PERC. However, several publications revealed the criticality of Ag facing upscaling targets of a worldwide terawatt PV production [7, 8].

The developments on solar cell level result in more a less sever challenges when implementing into module application to fabricate high-performance long-term stable PV modules. Constantly evolving manufacturing tools and processes are evident to meet this challenge. In this work,

the main challenges faced by interconnection process via soldering are identified and evaluated.

2 DEVELOPMENTS OF SOLAR CELL METALLIZATION AND INTERCONNECTION PROCESSES

2.1 Developments of solar cell metallization

Solar cell technologies are constantly evolving. For the soldering process the metallization of the solar cells is essential for a reliable interconnection. Concerning TOPCon and SHJ solar cells several developments of the metallization have taken place over the last years. Both cell technologies require metallization pastes with high Ag content resulting in an increased Ag consumption compared to PERC solar cells [3, 7]. Consequently, the finger geometry is constantly diminishing while the number of busbars and therefore interconnectors is increasing to compensate the resistive losses of the fingers [3]. Furthermore, busbarless metallization layouts are arising lately to reduce Ag consumption even more requiring adapted interconnection technologies.

Concerning more sensitive solar cell technologies such as SHJ and perovskite-silicon tandem solar cells, featuring temperature sensitive layers, low-temperature metallization pastes are used. These pastes typically show reduced wetting behavior and low mechanical adhesion to the solar cell surface [9, 10].

2.2 Interconnection processes

Lately several geometrical changes of the solar cells have taken place related to format and thickness of the wafers. The increased size of the Si wafers up to 210 mm edge length requires the interconnection of a larger area with more joints to be soldered simultaneously. Therefore, temperature homogeneity during the IR soldering process becomes challenging. The selection of interconnectors for modules has drastically changed regarding the geometry from flat band to round wire with increasingly smaller diameter as the optical properties have shown to be beneficial [11, 12]. These changes require adaptation of stringer tools to be able to process and handle the solar cells as well as the new thin interconnectors. Besides being able to process, other aspects such as precision and alignment combined with high throughput are common challenges.

At this point, PV application remain excluded from the current European RoHS (restriction of hazardous substances) [13]. Consequently, lead-based solder alloys are still preferred especially due to their performance and cost advantages. Nevertheless, it is expected that the exclusion of the RoHS might be withdrawn at some point in the future, which makes it indispensable to anticipate the development of lead-free solder alloys. Alternative lead-free interconnection processes such as Ag-based electrically conductive adhesives (ECAs) are already available, even though this technology is currently quite cost-intensive [14]. Several publications have already shown that lead-free solder alloys can be suitable for PV application but especially the long-term stability plays a significant role [15, 16]. Furthermore, since soldering requires the use of solder fluxes to activate the metal surface, the chemical stability of the solar cell can be questionable. Therefore, the selection of suitable solder fluxes is evident.

3 RESULTS & DISCUSSION

3.1 Metallization area for soldering

So far soldering of solar cells requires metallic surfaces to contact solar cells mechanically and electrically. This means that only the metallization of solar cells is available as interconnection surface. We evaluated the trend of the available metallized area of different solar cell types by analyzing around 100 technical datasheets of solar cells (PERC, TOPCon, SHJ) from 2014 to 2023. In **Figure 1**, the estimated area available for the soldering process is shown in accordance with the number of busbars on the solar cell. As the formats of the solar cells is not identical (M6 – M12) for all datapoints, the area is normalized to short current density J_{sc} of 40 mA/cm². Two clouds of data points can be distinguished between solar cells with continuous full busbars (59 ± 7) mm² and solar cells with busbar consisting of a narrow line with dedicated pads (6.9 ± 1) mm² for the interconnection to reduce the silver consumption of the busbars. It becomes clear that for solar cells with pads, the available area for the solder interconnection is almost a magnitude lower than for solar cells with full busbars. The average pad size is in the range of (0.9 ± 0.1) mm for the width and (0.7 ± 0.2) mm for the length. Furthermore, even though the number of busbars is increasing from 6 to 18, the estimated area remains almost constant in a range between 30 mm² – 90 mm² with a minimum single pad size of 0.5 mm². Another trend is the increasing number of busbars for TOPCon and SHJ compared to PERC. A third data cloud belongs to busbarless solar cells depicted by the violet area. Even though these solar cells technically feature no busbars, they are interconnected with a dedicated number of wires. Therefore, the depicted area shows the estimated area for soldering for 50 fingers on a half cell with a variation of number and diameter of the wires (6 – 20 wires, 0.20 mm – 0.35 mm) and the finger width (10 μm – 40 μm). Independently of the exact configuration, busbarless solar cells have an estimated solderable area an order of magnitude lower than solar cells with busbars.

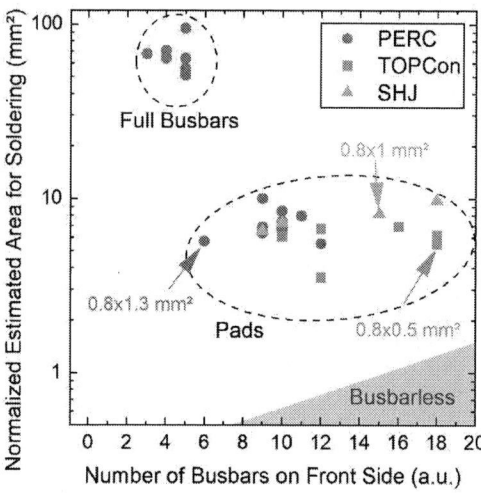

Figure 1: Normalized estimated area available for soldering depending on the number of busbars on the solar cell front side. As the solar cell format of the data points ranges from M6 to M12 the area is normalized to J_{sc} of 40 mA/cm².

The drastic reduction of the metallization area from full busbars to pads or even to busbarless solar cells shows the significant increasing challenge to provide a reliable interconnection in term of mechanical and electrical performance.

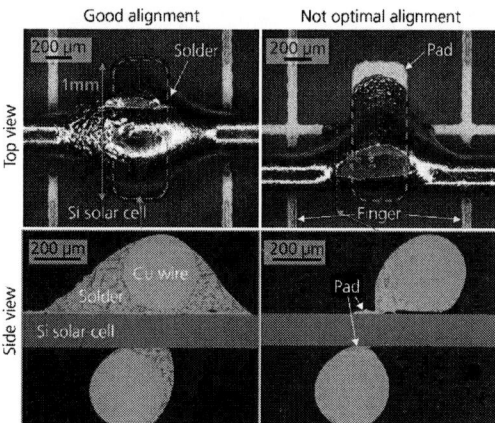

Figure 2: Top view and cross section microscopy images of soldered wires (0.32 mm) onto metal pads (width 1 mm) of a solar cell with good and not optimal alignment.

A second challenge addresses the interconnection process with industrial stringer tools. As shown in **Figure 1**, pads widths are already reduced to 0.8 mm and the wires show a trend to diameters well below 0.3 mm [3]. This requires a high level of alignment precision to solder the wires onto the pads, simultaneously on cell front and rear side. In **Figure 2**, top view and cross section microscopy images of soldered wires onto metal pads are shown. The images on the left show that the wire is perfectly aligned onto the pad, while the right image shows a clear misalignment to one side of the pad. The dimensions of wire (0.32 mm) and pad width (1 mm) shown here are rather conservative so that the precision of the wire deposition is sufficient for this case. However, it is expected that concerning the trends heading to smaller wires and pads due to material savings, such inaccuracies might lead to missing interconnection of some pads and therefore to a decreased fill factor inside the module. Furthermore, the alignment on front and rear side to each other is also important. In the cross-section image of **Figure 2**, a misalignment on the right side shows that in the worst case the wires are fully displaced to each other. This tends the risk of an uneven mechanical stress onto the solar cell which might result in cell breakage at this point.

3.3 Interaction between metallization paste and solder alloy

For a high-quality solder joint, the interaction between metallization and solder alloy has to be considered. In case of an optimized interface, the joint will be mechanically and electrically stable. Characteristics of an ideal interface are:

- Good wetting of metallization pad with liquid solder
- No damage of metallization pad by solder (*e.g.* by Ag leaching)
- Formation of a diffusion zone with limited thickness
- Formation of Ag_3Sn phase with defined thickness within diffusion zone
- Absence of voids, cracks and flux inclusions

These characteristics can be influenced by density, composition and roughness after screen-printing of the metallization paste as well as solder alloy composition, melting point, amount and type of flux and process parameters (temperature, time) during soldering. **Figure 3** shows an example of such an ideal joint.

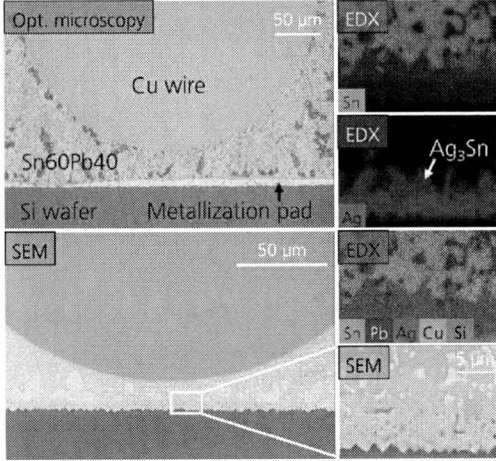

Figure 3: Cross section images of an ideal solder joint on a silicon solar cell. A diffusion zone and Ag_3Sn phase formation is found at the void-free interface between metallization and solder.

With the trend of Ag-reduced metallization, the pads become smaller but also thinner. This could influence the interface quality as well as the resulting solder joint. **Figure 4** shows an example of a joint on a very thin Ag pad. The diffusion of Sn and Ag at the interface leads to partly complete dissolution of the Ag pad in the Sn matrix of the solder, resulting in a direct contact between Sn60Pb40 and Si wafer. This leads to impeded string handling, a reduced mechanical stability < 0.2 N (90° peel

force) and a reduction of the fill factor *FF* of the module due to an increased series resistance R_s.

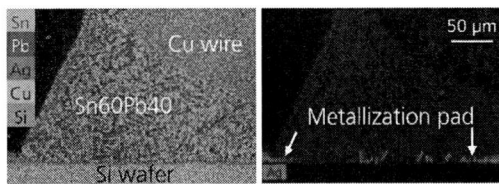

Figure 4: EDX cross section images of a solder joint on a very thin Ag pad on a Si solar cell. During soldering, Ag dissolves in the Sn matrix of the solder, leading to dissolution of the pad.

3.2 Solder alloys

So far, soldering and precisely the use of lead-based solar alloys, represents the state-of-the-art interconnection process [3]. Currently, the exception of PV applications to use lead in accordance with the RoHS is still in place. However, for the case that the exception is withdrawn, lead-free interconnection processes are required.

Figure 5 shows the liquidous temperature T_m of more than 125 different solder alloys according to the base component of each alloy on the *x*-axis. The blue horizontal dotted line represents $T_m = 183\ °C$ for the standard Sn60Pb40 solder alloy as reference. Furthermore, a minimum temperature of 90 °C is suggested according to IEC 61730 - 1 [17] as maximum operating temperature of conventional solar modules [17], even though even higher temperature up to 110 °C are mentioned by Kempe *et al.* [18]. A maximum temperature range up to 300 °C is defined, considering the increasing thermomechanical stress induced by large temperature differences during the cooling phase after soldering. With respect to SHJ and Si tandem solar cells, which feature temperature sensitive layers, the use of Ga-, In- and Bi-based solder alloys allows soldering processes at temperature much lower than for Sn60Pb40. Ga-based solder alloys can be treated as special case since they feature $T_m < 50\ °C$ and might be liquid even at room temperature. In- and Bi- based solder alloys have a broader range of T_m featuring lead-containing and lead-free alloys. The greatest amount of solder alloys is based on Sn. As can be seen, the lead-free Sn-based alloys are located, on average, at higher T_m compared to lead-containing Sn-based alloys. Solder alloys based on Pb or Au have the highest T_m of the shown material combinations.

Figure 5 involves largely all solder alloys available for different applications. However, not all of them are commercially available as research tends to take place at an academic level. For PV application, the solder alloys must fulfill several criteria such as sufficient and long-term reliable mechanical and electrical interconnection within the operation conditions without damaging the solar cells, as well as its availability for scalable purpose while being cost effective. Alloys based on Ga as main component are not suitable apart from its limited availability, due to its very low T_m, which might be beneficial only for very few applications. For In- and Au-based alloys the most critical aspect is presumably the increased material costs. However, for the interconnection of temperature-sensitive solar cells, In would allow low-temperature lead-free solder alloys. Alternatively, Bi-based alloys can be selected. When addressing lead-free soldering for solar cell technologies which tolerate higher soldering temperatures, such as TOPCon, numerous Sn-based Pb-free solder alloys are available. Especially, Pb-free solder alloys used in electronic industry are well known and evaluated due to the RoHS restriction in this industry.

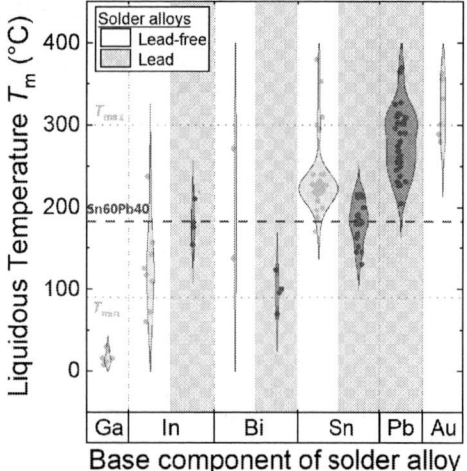

Figure 5: Liquidous temperature T_m of different solder alloys according to the main alloy component and divided between solar alloys containing lead and lead-free alloys.

3.3 Soldering Temperature Homogeneity

To ensure good wetting of the solder during the interconnection process and a stable solder joint, the soldering profile within the stringer has to be adjusted carefully. Both, soldering time and maximum soldering temperature influence the joint formation as well as mechanical stresses induced during cooling due to CTE differences between Cu wire and Si wafer [19]. The development trend towards larger wafer sizes requires homogeneous soldering over a larger area. To realize this, the IR heating lamps have been modified in stringer tools in the past years. A precise temperature control during the process is challenging but necessary due to smaller process windows for the interconnection of high-efficiency solar cells. On the one hand, the soldering temperature has to exceed the liquidus temperature of the solder alloy (183 °C for Sn60Pb40) by about +50 K, on the other hand high temperatures (and long soldering times) might damage layers within SHJ and tandem solar cells [20, 21] and promote the leaching of the metallization.

We performed in-situ temperature measurements on an industrial stringer during IR soldering of different industrial M6 half cells. The temperature is recorded at six different positions on the solar cell front side by thermocouples with a measurement uncertainty of ± 3 K. Each temperature measurement *i.e.* soldering process is repeated for at least six times to determine the statistical error, which is in the range of ± (2-7) K. **Figure 6** shows the deviation of the mean measured peak temperature T_{peak} from the set-temperature for soldering PERC, TOPCon and SHJ solar cells. The set-parameters of the soldering profile are kept the same. **Figure 7** shows the mean value of each measurement position for each solar cell technology with its standard deviation. Under consideration of both systematic and statistical error, it can be clearly seen that the temperature distribution is not homogeneous over the whole solar cell. We determined

temperature differences of $\Delta T_{PERC} = 58$ K, $\Delta T_{TOPCon} = 43$ K and $\Delta T_{SHJ} = 37$ K, which have to be considered respectively to not damage *e.g.* the passivation of the SHJ cell. This difference over the wafer is expected and due to the symmetry of the used hardware (IR lamps, downholder etc.) within the heating zone of the stringer.

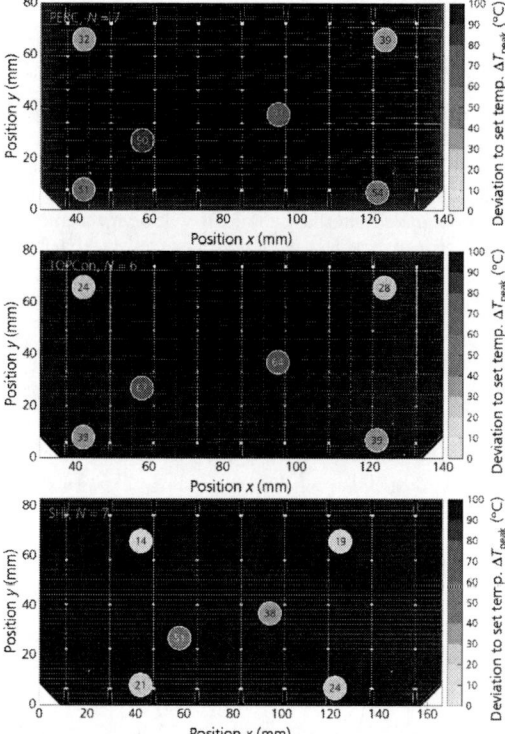

Figure 6: In-situ temperature measurement during IR soldering. Deviation of the measured mean peak temperature T_{peak} to the set temperature of soldering industrial PERC (top), TOPCon (middle) and SHJ (bottom) solar cells.

The measured peak temperatures are directly related to the emissivity of the solar cell. For higher emissivity of the cell (*e.g.* for SHJ), reflections of the IR light are lower (absorption is higher in the relevant wavelength range) and the temperature difference to the set-value is lower [22]. These results of the in-situ temperature measurement are in line with simulations by finite element modelling (FEM), published recently [23]. The FEM model computes the temperature homogeneity during the soldering process and can be used to precisely evaluate an optimal soldering temperature for any type of solar cell.

41st European Photovoltaic Solar Energy Conference and Exhibition

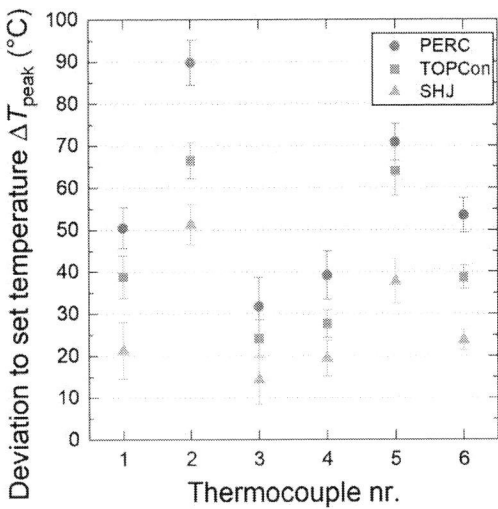

Figure 7: Measured peak temperature of solar cells at six different positions during IR soldering in an industrial stringer. Mean and standard deviation over min. 6 repetitions is given.

4 CONCLUSION & OUTLOOK

In the PV sector, ongoing developments on solar cell and module technology necessitate continuous process optimization to enhance performance while minimizing the costs.

In this work, we focus on the influence of current developments onto the interconnection process using soldering to identify and evaluate process challenges. The transition from PERC to TOPCon and SHJ solar cells has increased the amount of Ag used for the metallization. This results in metal grids with smaller fingers and busbars with an increase of the number of busbars. Larger, thinner wafers and reduced wire sizes demand precise handling and alignment increasing the challenges for the soldering process. Notably, material savings within the metal grid, especially the height of the pads, can negatively affect the quality of the soldering joint resulting in poor mechanical and electrical properties.

While Sn60Pb40 remains a reliable solder alloy for PV application. The shift to lead-free alloys is essential due to possible upcoming of changes to the RoHs directive. Various lead-free solder alloys are suitable for high-temperature applications, whereas low-temperature applications like SHJ or silicon tandem solar cells featuring temperature sensitive layers, have limited options involving Ga, In, or Bi as base component. Especially for these solar cells, precise soldering temperatures are crucial to prevent degradation. However, industrial stringer tools with IR lamps show significant temperature inhomogeneities (14 K to 90 K difference), posing risks to temperature-sensitive solar cells. Therefore, achieving a more uniform temperature distribution during soldering is indispensable.

Given the ongoing evolution on solar cell and module level, our institute will continue to focus on developing interconnection processes and tools.

5 ACKNOWLEDGEMENTS

The author would like to thank all involved colleagues for their support.

The activities have been financially supported by the German Federal Ministry for Economic Affairs and Climate Action withon the research project "Quelle" (FkZ: 03EE1172E)

REFERENCES

[1] F. Feldmann, M. Bivour, C. Reichel, M. Hermle, and S. W. Glunz, "Passivated rear contacts for high-efficiency n-type Si solar cells providing high interface passivation quality and excellent transport characteristics," *Solar Energy Materials and Solar Cells*, vol. 120, pp. 270–274, 2014.

[2] Hao Lin, Miao Yang, Xiaoning Ru, Genshun Wang, Shi Yin, Fuguo Peng, Chengjian Hong, Minghao Qu, Junxiong Lu, Liang Fang, Can Han, Paul Procel, Olindo Isabella, Pingqi Gao, Zhenguo Li, and Xixiang Xu, "Silicon heterojunction solar cells with up to 26.81% efficiency achieved by electrically optimized nanocrystalline-silicon hole contact layers,"

[3] VDMA, "International Technology Roadmap for Photovoltaics," no. 14. Edition, 2023.

[4] China Photovoltaic Industry Association (CPIA), "Photovoltaic Industry Development Roadmap of China," 2023.

[5] Z. Yu, M. Leilaeioun, and Z. Holman, "Selecting tandem partners for silicon solar cells," *Nat Energy*, vol. 1, no. 11, 2016.

[6] M. Dawson, C. Ribeiro, and M. R. Morelli, "A Review of Three-Dimensional Tin Halide Perovskites as Solar Cell Materials," *Mat. Res.*, vol. 25, 2022.

[7] Y. Zhang, M. Kim, L. Wang, P. Verlinden, and B. Hallam, "Design considerations for multi-terawatt scale manufacturing of existing and future photovoltaic technologies: challenges and opportunities related to silver, indium and bismuth consumption," *Energy Environ. Sci.*, vol. 14, no. 11, pp. 5587–5610, 2021.

[8] P. J. Verlinden, "Future challenges for photovoltaic manufacturing at the terawatt level," *Journal of Renewable and Sustainable Energy*, vol. 12, no. 5, 2020.

[9] A. De Rose, D. Erath, T. Geipel, A. Kraft, and U. Eitner, "Low-temperature soldering for the interconnection of silicon heterojunction solar cells," in *Proceedings of the 33rd European Photovoltaic Solar Energy Conference and Exhibition (EU PVSEC)*, Amsterdam, Netherlands, 2017, pp. 710–714.

[10] A. De Rose, T. Geipel, D. Erath, A. Kraft, and U. Eitner, "Challenges for the interconnection of crystalline silicon heterojunction solar cells," *Photovoltaics International*, vol. 40, pp. 78–86, 2018.

[11] A. Protti, A. Welpulwar, J. Shahid, M. Mittag, A. Tummalieh, and C. Reichel, "Analysis of Optical Coupling Gains from Cell Interconnection for the Energy Rating of PV Modules," (en), *40th European Photovoltaic Solar Energy Conference and Exhibition*, 2023.

[12] I. Hädrich, M. Padilla, A. Jötten, M. Mundus, W. Warta, and H. Wirth, "Finger and ribbon optics for increasing module power," in *Proceedings of the 31st European Photovoltaic Solar Energy Conference and Exhibition*, Hamburg, Germany, 2015.

[13] *Restriction of the use of certain hazardous substances, Directive 2011/65/EU: RoHS*, 2011.

[14] T. Geipel and U. Eitner, "Electrically conductive adhesives: An emerging interconnection technology for high-efficiency solar modules," *Photovoltaics International*, vol. 11, no. 21, pp. 27–33, 2013.

[15] T. Geipel, D. Eberlein, and A. Kraft, "Lead-free solders for ribbon interconnection of crystalline silicon PERC solar cells with infrared soldering," in *AIP Conference Proceedings 2156*, Konstanz, 2019, p. 20015.

[16] D. Güldali and A. De Rose, "Material Joint Analysis of Lead-Free Interconnection Technologies for Silicon Photovoltaics," in *45th International Spring Seminar on Electronics Technology (ISSE)*, 2022.

[17] *DIN EN IEC 61730-1 module safety qualification – Part 1: Requirements for construction.*

[18] Michael D. Kempe, Derek Holsapple, Kent Whitfield, and Narendra Shiradkar, "Standards development for modules in high temperature micro-environments,"

[19] L. C. Rendler, A. Kraft, C. Ebert, S. Wiese, and U. Eitner, "Investigation of Thermomechanical Stress in Solar Cells with Multi Busbar Interconnection by Finite Element Modeling," (eng), 2016.

[20] B. A. Korevaar, J. A. Fronheiser, X. Zhang, L. M. Fedor, and T. R. Tolliver, "Influence of annealing on performance for hetero-junction a-Si/c-Si devices," in *Proceedings of the 23rd European Photovoltaic Solar Energy Conference and Exhibition*, Valencia, Spain, 2008, pp. 1859–1862.

[21] J. Haschke, R. Lemerle, B. Aissa, A. A. Abdallah, M. M. Kivambe, M. Boccard, and C. Ballif, "Annealing of Silicon Heterojunction Solar Cells: Interplay of Solar Cell and Indium Tin Oxide Properties," *IEEE J. Photovoltaics*, pp. 1–6, 2019.

[22] A. De Rose, D. Eberlein, O. Parlayan, A. Kraft, and B. Grübel, "Optimized Soldering for the Transition of Industrial Silicon Solar Cells & Modules from PERC to TOPCon or SHJ," Chambéry, Apr. 15 2024.

[23] D. C. Joseph, A. De Rose, D. Eberlein, O. Parlayan, B. Grübel, A. J. Beinert, and D. H. Neuhaus, "Investigation of Temperature Homogeneity During Infrared Soldering of Silicon Solar Cells Using the Finite Element Method," *EPJ Photovolt.*, submitted 2024.

41st European Photovoltaic Solar Energy Conference and Exhibition

OPTIMIZING SUSTIANABILITY: BALANCING ANTIMONY CONTENT FOR ENHANCED OPTICAL PROPERTIES AND ENVIROMENTAL IMPACT IN SOLAR GLASS

Anika Glaubitz[1], Sven Grüttner[1], Selim Yagci[1], Oliver Pfeiffer[1], Ulf Blieske[1]
[1]University of applied Sciences Cologne, Cologne Institute for Renewable Energy (CIRE), Germany
anika.glaubitz@smail.th-koeln.de, sven.gruettner@smail.th-koeln.de, selim.yagci@smail.th-koeln.de,
oliver.pfeiffer@th-koeln.de, ulf.blieske@th-koeln.de

ABSTRACT:
This study investigates the effects of the antimony content in solar glass on its optical properties and the associated environmental factors. Glass samples with high, low and no antimony are analyzed. The spectral transmission and reflectance measurements show that antimony has a significant effect on light transmission, particularly in the near infrared range. Glasses with high and low antimony content show similar transmission and reflection properties, while the glass without antimony shows a significant decrease in transmission beyond 600 nm. The specific absorption values, calculated using spectral and integral measurements, show that reducing the antimony content slightly increases absorption, with glass without antimony absorbing up to 0.44 %/mm more radiation than glass with a high antimony content. Despite minor optical losses, low antimony glass retains comparable transmission properties, doubling the absorption compared to high antimony glass, but still achieving a significant reduction in specific absorption compared to zero antimony.
Taking environmental impacts into account, the analysis suggests that reducing or eliminating antimony in solar glass significantly reduces the risk of toxic emissions during production and recycling, supporting more sustainable and recyclable solar module manufacturing. While integral measurements provide a more accurate assessment of absorption across the spectrum relevant for c-Si solar cells, spectral measurements show that even small amounts of antimony significantly improve the optical performance of solar glass. Overall, the study concludes that sustainable, antimony-reduced solar glass solutions can be developed with minimal impact on energy efficiency, promoting an environmentally friendly approach to photovoltaic production that meets both economic and environmental requirements.
Keywords: solar glass; antimony; glass absorption; recyclability

1 INTRODUCTION

The enormous growth of photovoltaics on the market in Europe and the associated rapid increase in PV scrap at the end of its service life means, that new recycling-related challenges need to be overcome. It is currently estimated that over 200,000 tons of PV modules are thrown away each year, whereas forecasts predict a figure of 400,000 tons by 2030. [1]
The overall objective of this project is to analyze the environmental impact and recyclability of solar glass with different antimony content compared to the positive energetic impact of antimony in solar glass. The results are obtained within the framework of the funded research project "Green Solar Modules" [2], which aims to develop sustainable PV module concepts in a cradle to grave life cycle perspective. Due to its high mass fraction, solar glass is one of the most important optimization goals of this research project. Specially rolled (or cast) glass, which is used for solar glass in several solar glass furnaces around the world, contains antimony components in its material composition. Filtering out antimony during the recycling process is enormously complex and economically inefficient. [1]
It is therefore necessary to investigate the environmental friendliness of solar glass. Environmentally harmful substances such as antimony hinder recyclability, as antimony is toxicologically questionable and poses a health risk to humans if it dissolves in water or air [3]. The leaching of antimony into the environment must be avoided during the recycling process. In the solar glass

itself, antimony is used to improve the optical properties. Iron-2-oxide (FeO) in solar glass absorbs light in the near infrared range and reduces the transmission in this wavelength range accordingly. Added antimony oxide (Sb2O5) reacts with iron-2-oxide and converts it into iron-3-oxide (Fe2O3), which significantly improves the transmission properties of the solar glass. Therefore, the need for optically improved properties must be compared with the environmental impact. [4]
In this study, the optical properties of solar glass samples with different antimony contents are analyzed by measuring transmission and reflection curves with an integrating sphere in order to determine the influence on absorption. The antimony content correlates directly with the absorption of the iron-2-oxide in the glass. In addition to the spectral measurement, integral transmission values are measured via a reference cell as a validation and the values are compared between the solar glasses.
The solar glass samples used are low-antimony solar rolled glass with a thickness of 2 mm and antimony-free solar float glass with a thickness of 3.2 mm. A commercially available solar glass with a high antimony content and a thickness of 3.2 mm serves as the reference glass. The solar glasses are labeled as low antimony, zero antimony and high antimony. The latter glass reflects the industry standard of just a few years ago and will be a recycling problem in the coming years. Zero antimony (float glass) is widely used in production in Europe [1]. In order to keep up with the optical quality of antimony-enriched solar glass from East Asia, attempts are made to use lower

41st European Photovoltaic Solar Energy Conference and Exhibition

concentrations of iron in European production. This results in more valuable material as an input raw material and is an economic competitive disadvantage compared to antimony-enriched solar glass [1]. Low-antimony solar glass is a measure taken by glass manufacturers to reduce the environmental impact without significantly impairing the optical properties. The antimony-free solar glass has an iron concentration of 100 ppm. All solar glasses have a comparable iron concentration. Only the ratio between iron-2-oxide and iron-3-oxide differs. The necessity of the high antimony content in solar glass to reduce the iron content has not yet been scientifically proven on a large scale. This paper examines the optical quality of solar glass with different antimony contents and places them in context with each other in terms of their environmental impact.

2 ENVIRONMENTAL EFFECTS

The use of antimony in solar glass production has a significant environmental impact that must be carefully weighed up. Although antimony improves the optical properties of the glass and thus increases transmission efficiency, it also poses ecological and health risks. Antimony trioxide as an additive is a carcinogenic substance. In the production process, antimony compounds are quite volatile and can produce toxic emissions after melting. The harmful effects of antimony poisoning are comparable to those of arsenic poisoning in that, in addition to respiratory irritation and pneumoconiosis, antimony stains on the skin or other symptoms in the gastrointestinal tract can also occur. [1]

In addition to production, the recycling process for solar glass is also particularly problematic due to antimony. If disposed improperly or during the recycling process, antimony can enter the environment and contaminate both the soil and groundwater. This potential contamination is particularly problematic as antimony is toxicologically questionable and poses serious risks to human health and ecosystems. [5]

The use of antimony in solar glass is expected to cause both carcinogenic and non-carcinogenic risks [5]. The elimination of antimony can greatly reduce the risk of cancer-related health problems and non-carcinogenic toxicity in the life cycle of solar glass. In addition, antimony has a significant impact on resource utilization. The use of antimony-free glass offers the potential to efficiently reduce the use of mineral and metal resources. These effects can be crucial for the development of sustainable solar modules.

While antimony offers economic benefits (through a higher transmission of the solar glass and hence, higher power output of the solar module), the associated environmental risks are significant. A reduction or complete elimination of antimony in solar glass can lead to a significant reduction of negative environmental impacts [1]. Therefore, it is crucial to take a balanced approach that considers both the environmental and economic aspects to ensure a sustainable and environmentally friendly use of solar glass.

3 MEASUREMENT TECHNIQUES

To determine the influence of antimony in solar glasses on their optical properties, glasses with different antimony contents are compared with each other in terms of their transmission behavior. In particular, the optical losses when using low-antimony solar glasses instead of high-antimony solar glasses are investigated and compared with the complete absence of antimony.

Spectral measurements are carried out using an integrating sphere to measure the transmission and reflection curves. The resulting transmission curve and integral transmission value can be validated with the equivalent value of the integral measurement.

3.1 Spectral measurement

To achieve reproducible measurement results, the spectral measurement of the solar glasses is carried out in accordance with DIN EN 62805-2 and in compliance with the specifications regarding the measurement setup and measurement method shown in Figure 1. [6]

To ensure that no extraneous light influences the measurement results, the entire measurement setup is located in a darkroom consisting of two separately connected chambers. For the spectral measurement of the transmission curve over the relevant wavelength range from 280 nm to 1250 nm for silicon solar cells, a water-cooled xenon short-arc lamp is used as the light source, as this is similar to the AM 1.5 spectrum over the required wavelengths and is highly stable. The light source is used together with the filter device to generate the monochromatic light. The filter monochromator used consists of two filter wheels with 25 filters each. By rotating the filter wheels in relation to each other, a total of 40 wavelengths between 338.6 nm and 1188 nm can be set. It should be mentioned here that the measurement results in the wavelength range from 1100 nm to 1188 nm show limitations in the measurement accuracy and are therefore not included in the analysis. The monochromatic light then hits a diaphragm that separates the two dark chambers. In the right-hand darkroom, the radiation hits the glass sample, which is placed in front of the integrating sphere aperture. There is a white standard disk in the integrating sphere, which can be used to set the angle of incidence. The integrating sphere is a measuring component with a diffuse coating on the inside, whereby all the radiation transmitted through the glass sample is diffusely reflected in the integrating sphere. An angle of 9° is set so that the radiation does not reflect back out of the integrating sphere when it hits the white standard disk. The light that is reflected in the integrating sphere is detected by an optometer. For this purpose, a broadband measuring head is connected, which uses a light-sensitive silicon photodiode and is placed in the opening of the integrating sphere. A measurable signal is generated from the monochromatic radiation that is transmitted through the glass sample.

A reference measurement and a sample measurement are carried out to record the transmission. In the reference measurement, the set-up is without a glass sample, whereby air is the reference medium. For the sample measurement, the sample glass is placed in the holder in front of the light inlet opening of the integrating sphere. The generated current measurement signal of the sample measurement is divided by that of the reference measurement to determine the spectral transmittance. To increase the accuracy and reproducibility of the measurement results, the measurement is carried out several times in direct succession. The spectral transmittance curves are averaged over these measurement sequences. This reduces random measurement deviations and statistical fluctuations that can be caused by external influences such as temperature changes, fluctuations in the light source or electronic noise signals. By averaging the measured values, the precision of the spectral transmittance determination is increased, resulting in more reliable and consistent results.

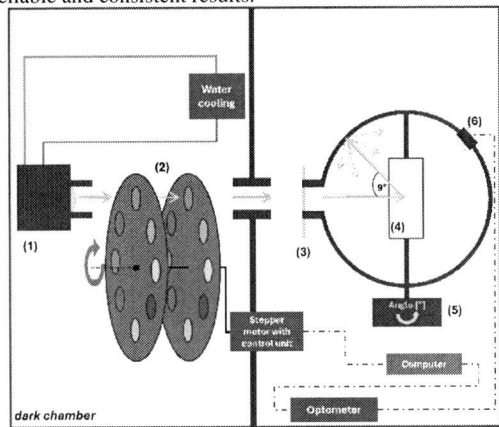

Figure 1: Measurement setup for spectral measurement of the solar glass with (1) xenon short arc light source, (2) double wheel on optical filters as a monochromator, where the solar spectrum is divided into different sub-intervals, (3) the glass sample, (4) white standard disc (for reflection: velvet absorber material), (5) integrating sphere and (6) Si reference cell

In spectral reflectance measurement, the measurement setup is identical, but the glass sample is clamped in front of the white standard disk in the integrating sphere. As a result, the entire radiation at the respective wavelength hits the glass sample and the radiation reflected by the sample reflects evenly inside the integrating sphere. To ensure that only the reflected radiation from the sample is measured, a highly absorbent velvet fabric is placed behind the glass sample and thus serves as a light trap. This prevents radiation that is transmitted through the sample from being reflected back and distorting the measurement result. The optical signal in the integrating sphere can then be recorded as a measurable signal via a crystalline silicon cell connected to an optometer. A reference measurement is also carried out for the reflection measurement, which corresponds to the case of total reflection, as the white standard disk is turned at an angle of 9° to the entrance opening. For the sample measurement, the sample is turned towards the light source at an angel of 9° with the absorbing velvet on the back. The reflectance value at the respective wavelength corresponds to the quotient of the sample and reference measurement.

3.2 Integral measurement

The aim of integral measurement is to determine a transmission value using a reference cell shown in **Fehler! V erweisquelle konnte nicht gefunden werden.**. The entire spectrum of the xenon lamp is directed onto a reference cell via a deflecting mirror. For this purpose, the filter wheels are set to the positions without optical filters. The pinhole diaphragm between the deflecting mirror and the reference cell results in under-illumination, with only a narrow spot of light hitting the solar cell. This minimizes the influence of stray light, which is reflected by the surroundings and can lead to a distortion of the measurement signals. It also ensures that the light spot of the transmitted radiation and therefore the highest radiation intensity hits the solar cell. The radiation leads to a current flow in the solar cell. This can be measured using a measuring device.

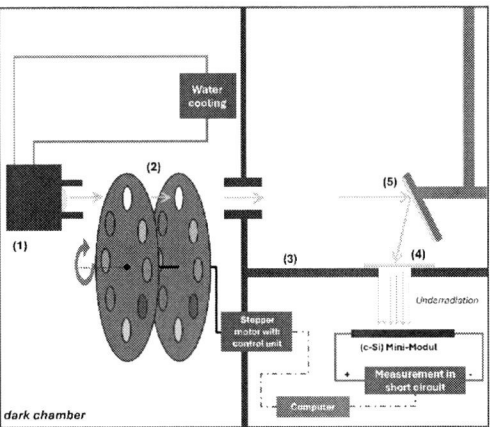

Figure 2: Measurement setup for integral measurement of the solar glass with (1) xenon short arc light source, (2) double wheel on optical filters as a monochromator, where the position without optical filters is selected, (3) pinhole diaphragm, (4) glass sample, (5) mirror for deflecting the light

The transmission value can be calculated using Formula (1) by taking a reference measurement without a glass sample and a sample measurement with the glass sample in front of the pinhole.

$$T_{glass} = \frac{I_{SC,sample}}{I_{SC,reference}} \qquad (1)$$

4   MEASUREMENT RESULTS

The spectral transmission and reflection curves are compared between the analysed glass samples with high antimony, as a reference for the current state of the art for solar glasses, low antimony and no antimony. Then an absorption value for each glass sample is calculated. For validation, the measurement results of the integral measurement are analysed and a further absorption value of the glass samples is calculated. The results of the spectral and integral measurements are then compared and discussed.

4.1 Spectral measurement

The measurement results of the spectral transmission measurement are shown in Figure 3. The measuring points are between 400 and 1120 nm. The transmission value is the quotient of the measurement with and without the glass sample.

It can be seen that in the low wavelength range from 400 nm to 600 nm, the transmittances are in agreement at approx. 0.915. From 600 nm, a clear drop in the transmission curve of the glass sample with no antimony can be seen. It drops to transmission values of up to 0.907 at 900 nm. The transmission curve of the low antimony glass, on the other hand, runs along the transmission curve of the high antimony glass up to 830 nm, rising slightly from 0.915 to 0.917. From 830 nm to 1100 nm, in the near-infrared range of the spectrum, a deviation between low and high antimony can be recognised. The measured values of the low antimony glass are 0.914, while the high antimony glass remains at 0.917. At the measuring points greater than 1100 nm, the transmission of low and high antimony converge again. Based on the measured spectral transmission curves, it can therefore be determined that the use of antimony has an influence on the transmission value of solar glass. However, the reduction in the antimony content only causes slight deviations in the near-infrared range of the spectrum.

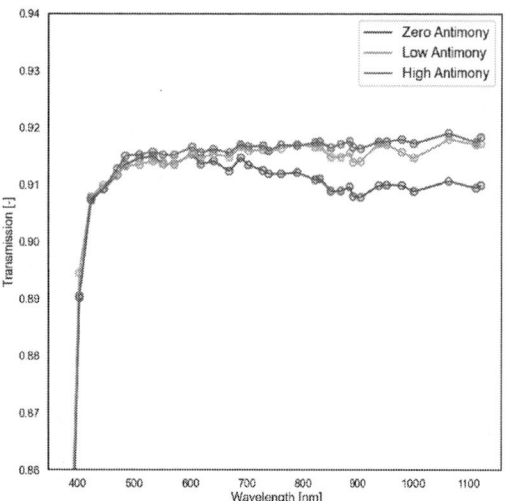

Figure 3: Spectral transmission of the glass samples

The spectral reflectance of the glass samples is shown in Figure 4 The zero antimony glass is a float glass and has a smooth surface structure, while the low and high antimony glasses are rolled glass and have a matt structure due to the manufacturing process. The spectral reflectance curves of low and high are therefore identical, which is why only low antimony is shown. It can be seen in the course of the reflection curves that they only deviate slightly between 400 and 650 nm and 980 to 1050 nm. In the marginal values of the measuring range from 385 to 425 nm, the reflectance values rise from 0 to 0.073. In the low wavelength range from 425 to 500 nm, the values are 0.08, then fall in a parabolic fashion to 0.075 at 650 nm and then rise steadily in a linear fashion to 0.095 at 1050 nm. For the subsequent calculation of an absorbance value for the glass samples, a reflectance value is formed using the integral of the spectral curves. This is 0.075 for the zero antimony glass and 0.078 for the low and high antimony glasses.

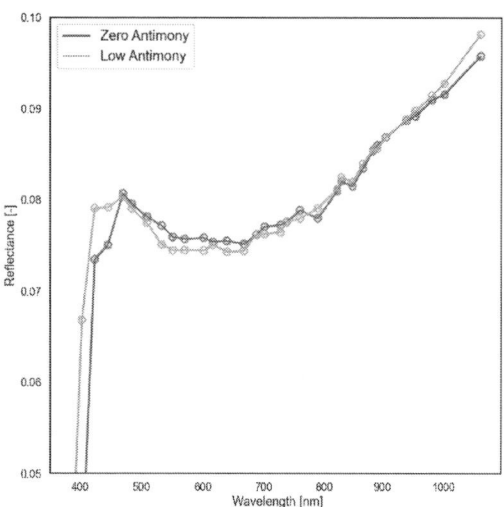

Figure 4: Spectral reflection of the glass samples

In order to be able to make a statement about the effects of the transmission and absorption properties of the glass on the photocurrent density J_{ph} of a solar cell, this is calculated below using Formula (2). For this purpose, the spectral response $S_{cell}(\lambda)$ of an M6 (c-Si) PERC solar cell and the AM 1.5G spectrum of the sun $E_{AM1.5}(\lambda)$ are offset against the transmission values $T_{glass}(\lambda)$ for each measured wavelength. The curves of the spectral response used and the solar spectrum are shown in Figure 5. [7]

$$J_{ph} = \int S_{cell}(\lambda) \times E_{AM1.5}(\lambda) \times T_{glass}(\lambda) d\lambda \qquad (2)$$

Together with the cell area of the M6 cell, this is used to determine the photocurrent. The photocurrent without glass is also calculated for reference. In order to draw conclusions about the absorption values α of the glass, the already calculated reflection value ρ of the glass samples is also taken into account. As the glass samples have different thicknesses, the absorption is specified relative to the glass thickness d and calculated according to Formula (3).

$$\alpha_{spez} = \frac{\left(1 - \frac{I_{ph,sample}}{I_{ph,reference}} - \rho\right)}{d} \qquad (3)$$

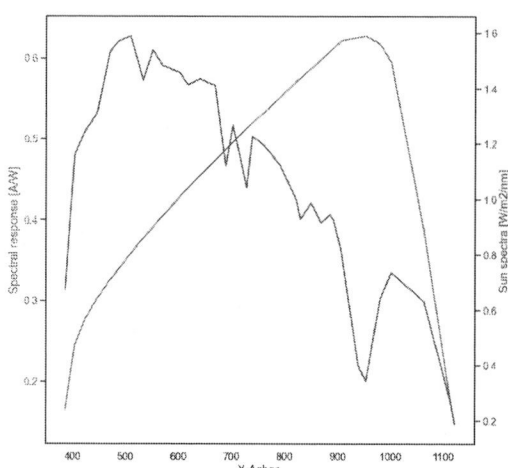

Figure 5: Anomalous photocurrent parameters with (blue) the spectral responsivity of an M6 PERC solar cell and (red) the AM1.5G solar spectrum at the sampling points of the measurement intervals

The results are listed in Table 1. The specific absorption value of the zero antimony glass is 0.34 % per mm glass thickness. For the low antimony glass it is 0.21 %/mm and for high antimony it is 0.10 %/mm. This means that only 0.11 %/mm more is absorbed by reducing antimony in the solar glass. However, if no antimony is used at all, 0.24 %/mm more of the incoming radiation is absorbed.

41st European Photovoltaic Solar Energy Conference and Exhibition

Table 1: Specific absorption of the glass samples by spectral measurement

Glass samples	Photocurrent (uncorrected) [A]	Spez. Absorption [%/mm]
Reference cell without solar glass	10.814	-
Zero antimony Float solar glass (3,2 mm)	9.886	**0.34**
Low antimony Rolled solar glass (2 mm)	9.925	**0.21**
High antimony Rolled solar glass (3,2 mm)	9.936	**0.10**

4.2 Integral measurement

The integral measurement of the transmission is carried out for all three glass samples, as described in section 3.2. This results in a transmission value over the entire spectrum and for the relevant range of a c-Si PERC cell. The results are listed in Table 2. The zero antimony glass achieves a transmission of 90.77 %, the low antimony glass 91.80 % and the high antimony glass 91.87 %. However, it should be noted that the glass samples have different thicknesses, which influences the transmission value. The absorption value is calculated by adding the respective reflection values. For comparability, this is again given relative to the glass thickness. The relative absorption of zero antimony glass is 0.54 %/mm. The low antimony glass is at 0.20 %/mm and the high antimony glass at 0.10 %/mm.

From this it can be concluded that reducing the antimony content doubles the absorption, but only increases it by 0.1 % per mm. If no antimony is used at all, 0.44 % per mm more is absorbed, which is 4 ½ times more.

Table 2: Specific absorption of the glass samples by integral measurement

Glass samples	Uncorrected transmission [%]	Absorp-tion [%]	Spez. Absorption [%/mm]
Zero antimony Float solar glass (3,2 mm)	90.769	1.73	**0.54**
Low antimony Rolled solar glass (2 mm)	91.803	0.40	**0.20**
High antimony Rolled solar glass (3,2 mm)	91.871	0.33	**0.10**

4.3 DISCUSSION OF THE RESULTS

The results from sections 4.1 and 4.2 are compared and discussed below. By calculating the spectral measurements to the photocurrent, it is to be expected that the same or similar absorption results will emerge as from the integral measurement. However, there are slight differences. The specific absorption of the zero antimony

glass is 0.20 %/mm lower in the spectral measurement and is therefore only approx. 2/3 of the integral value. The specific absorption values of the low and high antimony glass are almost identical for both measurement methods. The deviation in the zero antimony glass may have been caused by several factors.

The filters of the measuring stand have the greatest influence. These only go up to 1188 nm, but only measured values up to 1100 nm can be used, as there are too many deviations between the measurements in repeated measurements. However, the cut-off wavelength of a c-Si solar cell extends up to 1200 nm, which means that not all relevant transmission values are recorded in the spectral measurement. This is particularly critical as the influence of antimony is very high in this wavelength range. This is the reason for the deviations in the zero antimony glass. The stronger absorption due to the lack of antimony can be recognised in the integral measurement; in the spectral measurement the influence is less, as decisive measuring points are missing. No relevant deviation can be determined for the high and low antimony glass, as the transmission values remain at a constant level and therefore the extended wavelength range does not cause any changes.

Furthermore, the spectral measurement uses sampling points of the solar spectrum at the measuring points for the conversion to the photocurrent. This reduces the level of detail of this spectrum and thus introduces an error into the results.

In addition, the spectral response used for the conversion is that of a standard cell and not of the cell used for the integral measurement. As a result, deviations from the integral measurement are possible

Moreover, the weighting of the absorption over the spectrum is different for both measurement methods. In the spectral measurement, the absorption is weighted to the solar spectrum when converting to the photocurrent, which decreases in intensity with increasing wavelength. This means that the areas relevant for antimony are less strongly taken into account in the absorption value. In the integral measurement, the transmission value is determined using the xenon spectrum and is not weighted further. This can result in deviations, as the AM1.5G spectrum of the sun and the xenon spectrum do not match 100 percent. However, all measurements are relative to reference measurements without a glass sample, which means that these deviations are classified as minimal. Furthermore, systematic measurement errors that cannot be determined further have an influence on the measurement results.

By capturing the entire spectrum relevant for c-Si solar cells with the integral measurement, the calculated specific absorption values are preferably treated as the final results of the measurement series. It can therefore be concluded that even small amounts of antimony greatly improve the optical properties of solar glass and that a more environmentally friendly solar glass does not have to expect any serious loss of energy yield.

5 CONCLUSION

This study analyses the effects of the antimony content in solar glass on its optical properties and the associated environmental factors. The results show that antimony plays a significant role in improving light transmission, especially in the near infrared range, and thus contributes to optimising the energy efficiency of photovoltaic systems. Although reducing the antimony content leads to a slight loss in transmission, this remains within a range with a moderate effect (about 0.2 % for a solar cover glass of 2 mm thickness). Completely dispensing with antimony significantly increases the absorption of incoming radiation.

From an environmental perspective, the reduction or complete elimination of antimony offers decisive advantages. Antimony-free or low-antimony glass significantly reduces the risk of toxic emissions during manufacturing and the recycling process, helping to conserve natural resources and protect human health and the environment. Given the increasing amounts of photovoltaic waste in the coming decades, the transition to more environmentally friendly solar glasses is a necessary measure to improve the recyclability of solar modules.

Overall, the study shows that the development of sustainable solar glass solutions is possible without significant losses in energy efficiency. This represents an important step towards more environmentally friendly photovoltaic production that takes into account both economic and ecological requirements.

6 ACKNOWLEDGEMENTS

This research was funded by the German Federal Ministry for Economic Affairs and Climate Action (BMWK) under the project 'Green Solar Modules' (grant number: 020E-100550097).

7 REFERENCES

[1] EUROPEAN SOLAR PV INDUSTRY ALLIANCE, "Addressing uncertain antimony content in solar glass for recycling," Oct. 2023. Accessed: Jan. 30, 2024. [Online]. Available: https://solaralliance.eu/wp-content/uploads/2023/10/Recommendation-on-Addressing-uncertain-antimony-content-in-solar-glass-for-recycling.pdf

[2] *"Green solar modules", German project funded by the ministry for economy and environmental protection (BMWK), Funding reference number: 020E-100550097.*

[3] S. Sundar and J. Chakravarty, "Antimony toxicity," *International journal of environmental research and public health*, early access. doi: 10.3390/ijerph7124267.

[4] Thomsen et al., "SOLAR CELL USING LOW IRON HIGH TRANSMISSION GLASS WITH ANTIMONY AND CORRESPONDING METHOD," US 8,802.216 B2. [Online]. Available: https://patentimages.storage.googleapis.com/06/5f/fe/24a390e52fa2d5/US8802216.pdf

[5] P. A. Nishad and A. Bhaskarapillai, "Antimony, a pollutant of emerging concern: A review on industrial sources and remediation technologies," *Chemosphere*, early access. doi: 10.1016/j.chemosphere.2021.130252.

[6] *DIN EN 62805-2 VDE 0126-4-21:2018-06, Verfahren für die Messung von photovoltaischem (PV) Glas: Teil 2: Messung von Transmissionsgrad und Reflexionsgrad*, Sep. 2024. [Online]. Available: https://www.vde-verlag.de/normen/0100469/din-en-62805-2-vde-0126-4-21-2018-06.html

[7] U. Blieske and G. Stollwerck, "Chapter Four - Glass and Other Encapsulation Materials," in *Semiconductors and Semimetals : Advances in Photovoltaics: Part 2*, vol. 89, G. P. Willeke and E. R. Weber, Eds., Elsevier, 2013, pp. 199–258. [Online]. Available: https://www.sciencedirect.com/science/article/pii/B9780123813435000045

41st European Photovoltaic Solar Energy Conference and Exhibition

PHOTOVOLTAIC MODULES COMPRISING III-V CELLS ENCAPSULATED IN COMPOSITE MATERIAL

Francisco J. Cano, Werther Cambarau, Naiara Yurrita, Jon Aizpurua, Juan M. Hernández, Gorka Imbuluzqueta, Eduardo Román Medina, Oihana Zubillaga
TECNALIA-Basque Research and Technology Alliance (BRTA)
Mikeletegi Pasealekua 2, 20009 Donostia-San Sebastian, Spain
paco.cano@tecnalia.com

ABSTRACT: The use of high-efficiency solar cells and lightweight protective encapsulant offers PV modules with high specific power and high-power density values, allowing a sustainable integration of the PV energy in urban and vehicle related applications, in which the available surface and module weight is limited and the energy requirement high. With this aim, modules comprising III-V cells and an innovative fiber reinforced composite encapsulant were studied. The composite consisted of glass fiber reinforcement and epoxy resin matrix, and lab modules were manufactured through resin infusion process. A cell-to-module loss of 6.5% was observed for the composite with standard epoxy matrix, and 5.4% for the one with recyclable epoxy. Regarding the aging performance, the modules encapsulated within the standard composite showed a 3.4% power loss after 1000 hours damp-heat test, and a decrease of 1.6% at the end of the thermal cycling. Concerning the modules containing the recyclable epoxy, a significant degradation was observed after 500 hours damp-heat exposure, being discarded for further study. The work concluded that the lightweight composite encapsulation of III-V cells can be an option to obtain PV modules with high specific power and high-power density, while retaining a suitable PV performance stability.
Keywords: III-V cells, composite encapsulant, integration, lightweight, power density

1 INTRODUCTION

Glass fiber reinforced composite with an acceptable transparency can be used as an encapsulant system leading to lightweight and multifunctional PV modules. The authors studied and reported previously about this system when used as encapsulant for crystalline silicon cells [1,2].

In the present work, the aim was the development of an innovative encapsulation system for PV) modules comprising III-V cells, showing the following features:

- Glass fiber reinforced transparent composite encapsulant leading to monolithic lightweight modules.
- High specific power (W/kg) and high-power density (W/m2) modules, allowing a sustainable integration of the PV energy in urban and vehicle related applications, with restrictions related to weight and available integration surface.
- Composite encapsulant with enhanced recyclability allowing module component recovery [3].
- High performance stability under field temperature, humidity and ultraviolet radiation.

2 EXPERIMENTAL

2.1 Materials and module manufacturing

Lab modules containing one or two cells were manufactured through linear vacuum aided resin infusion process (figure 1). Composites based on standard epoxy and recyclable epoxy system were studied. The recyclable epoxy consisted of a resin containing cleavable functional groups, allowing chemical recycling in mild acidic conditions [3].

Modules with glass-silicone encapsulation were used as the reference baseline encapsulant system, which were manufactured through vacuum lamination.

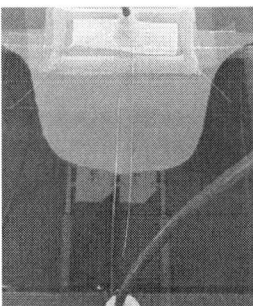

Figure 1: Infusion process during module manufacturing

2.2 Module characterization

The electrical performance of the modules was characterized through electroluminescence (EL) images, current-voltage (IV) curves obtained by solar simulator, and external quantum efficiency (EQE) spectra.

Concerning EQE, spectra were acquired in the 300-1300 nm wavelength range, and short-circuit current density (Jsc) values were further calculated for analysis [1].

Cell-to-module (CTM) losses due to the diverse encapsulation solutions were analyzed.

2.3 Accelerated aging tests

The minimodules were exposed to accelerated indoor tests according to IEC-61215:2021 standard to evaluate their performance stability. Damp-heat (DH) exposure and thermal cycling tests were carried out. The damp-heat test comprised an exposure of 1000 hours to 85°C and 85% relative humidity. The thermal cycling test covered 200 cycles between -40°C and 85°C.

The progress of the weathering was evaluated through current-voltage (IV) curves and external quantum efficiency (EQE) spectra acquired at different exposure intervals.

3 RESULTS AND DISCUSSION

3.1 Module manufacturing and characterization

Monolithic modules where the cells and corresponding connectors were completely embedded in the composite were obtained through linear vacuum resin infusion process (figure 2). No major defects (air entrapment, bubbles, delaminations, etc.) were observed during visual qualitative assessment of the manufactured modules. Further, EL images did not show any damage in the cells (figure 3).

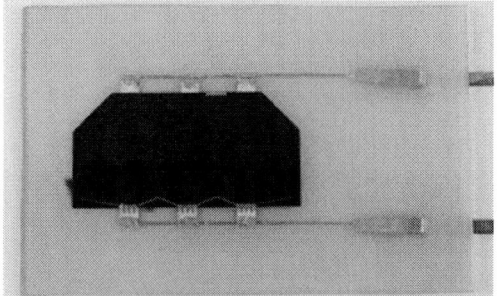

Figure 2: Lab module with one III-V cell encapsulated in standard epoxy composite

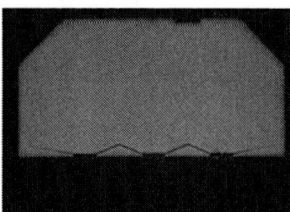

Figure 3: EL image of cell encapsulated in standard epoxy composite

Cell-to-module losses, obtained according to IV curve analysis, are presented in table I. A better CTM performance was obtained for the recyclable system.

Table I: CTM values for studied encapsulants

Encapsulant	CTM (Isc, %)
Standard epoxy composite	6.5
Recyclable epoxy composite	5.4
Glass-silicone laminate (reference)	2.0

3.2 Accelerated aging test results

Concerning the modules encapsulated within the composite containing the standard epoxy, a 3.4% power loss was detected after 1000 hours damp-heat test (figure 4) and a decrease of 1.6% at the end of the thermal cycling (figure 5).

After 500 hours damp-heat exposure, the recyclable composite system showed a significant degradation, with an 8% decrease in short circuit current density (Jsc) according to EQE. Thus, it was decided not to continue with further study of this system presently.

As reference, the glass-silicone modules showed no power loss in thermal cycling, and 3.0% in damp-heat exposure.

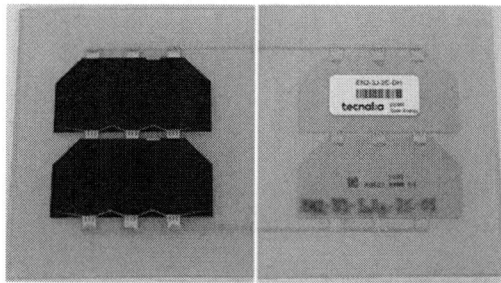

Figure 4: Module after 1000 hours damp-heat exposure

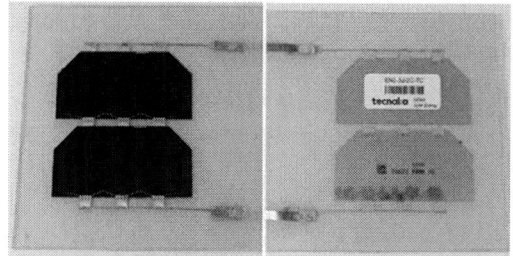

Figure 5: Module after 200 thermal cycles

4 CONCLUSIONS

The study concluded that lightweight composite encapsulation of III-V cells can be an option to obtain PV modules with high specific power and high-density power values, while retaining a suitable PV performance and stability.

More particularly:

- Recyclable epoxy composite showed the lowest CTM, 5.4% loss, but it did not show a good performance stability in damp-heat.
- The standard epoxy composite showed a CTM of 6.5%, and good performance stability in damp-heat (3.4% loss) and thermal cycling (1.6% loss).

5 ACKNOWLEDGEMENTS

This work, performed within the SINFO project, received funding from the Provincial Council of Gipuzkoa under Grant Agreement 2022-CIEN-000043-01.

6 REFERENCES

[1] N. Yurrita, J. Aizpurua, W. Cambarau, G. Imbuluzqueta, J.M. Hernández, F.J. Cano, O. Zubillaga, Photovoltaic modules encapsulated in composite material modified with ultraviolet additives, Solar Energy Materials and Solar Cells 230 (2021) 111250.

[2] N. Yurrita, J. Aizpurua, W. Cambarau, G. Imbuluzqueta, J. M. Hernández, F.J. Cano, I. Huerta, E. Rico, T. del Caño, S. Wölper, F. Haacke, O. Zubillaga, Composite material incorporating protective coatings for photovoltaic cell

encapsulation, Solar Energy Materials and Solar Cells 245 (2022), 111879.

[3] F.J. Cano, G. Imbuluzqueta, N. Yurrita, J. Aizpurua, J. M. Hernández, W. Cambarau, O. Zubillaga, Composite material with enhanced recyclability as encapsulant for photovoltaic modules, Heliyon 9 (2023), e20048.

41st European Photovoltaic Solar Energy Conference and Exhibition

RELIABILITY OF ALUMINIUM-COPPER CONTACT IN PV MODULES

Tobias Messmer[1], Dominik Rudolph[1], Gernot Emanuel[2], Andreas Nägele[2], Andreas Halm[1]
[1]ISC Konstanz e.V., Konstanz, Germany
[2]Fraunhofer Institute for Solar Energy Systems ISE, Freiburg, Germany
Tobias.messmer@isc-konstanz.de, +49 7531 36183 357
Rudolf-Diesel-Str 15, 78467 Konstanz, Germany

ABSTRACT: For the cross connectors of photovoltaic cells interconnected with an aluminium foil, contacts of aluminium and copper are required to contact the copper-based junction boxes in the module. In this study, three concepts for the realization of such a module are presented and mini modules are assembled in which the relevant contacts are investigated with different joining techniques and materials. The mini modules are evaluated regarding their reliability after ageing with damp heat and thermal cycling tests in the climate chamber.
Both the contact between aluminium foil and aluminium bussing as well as the contact between aluminium foil and copper bussing can be realized without significant power loss due to the contact. Furthermore, no significant degradation can be observed after the reliability testing of 3000 hours in damp heat test condition and 600 cycles of thermal cycling test condition.

Keywords: cross connection, silver reduction, IBC, aluminium foil

1 INTRODUCTION

In the POPEI project, the ISC and its partners are developing a p-type IBC cell mainly based on already existing industrial technologies in PERC production lines [1]. A significant part of the costs in the production of PV cells is driven by the costs of metallization pastes. [2] Savings in the consumption of metallization pastes can therefore lead to significant cost savings.

Therefore, in the SPINAT project new aluminium foil based concepts for interconnecting the POPEI-IBC cells are developed, which makes it possible to use significantly less paste for metallization by eliminating the busbars in this approach. In this interconnection approach, aluminium foil is used to interconnect the cells by bonding the foil to the fingers of the cells with a laser process. This aluminium foil is then used instead of ribbons to connect the individual cells of a string eliminating the need for classic ribbons as interconnection [3]. In addition to the cost savings due to reduction of metallization paste, the use of low-cost aluminium foil and laser processes compared to today's standard infrared soldering of copper wires can lead to further cost savings in the bill of material of PV modules.

Since the terminals of available junction boxes are made of copper, an aluminium-copper contact must be implemented within the module to realize this aluminium foil based interconnection approach. This aluminium-copper contact is investigated within this work.

The galvanic corrosion between the metals copper and aluminium can lead to serious degradation of the conductivity and mechanical stability (brittleness) of the connection. The corrosion varies depending on which joining technique was used and how the process parameters were selected. An intermetallic phase is formed which leads to different degrees of corrosion depending on the thickness of the layer of this phase [4]. Because very little has been published on the behavior of this metal compounds in PV modules and possible interactions with module encapsulation materials, this work demonstrates the reliability of such a contact through fundamental experiments in a climate chamber.

2 METHODE

2.1 Concepts of bussing system

Three concepts for the location of the aluminium-copper contact are taken into account. First, contacting the aluminium foil of each string directly to a copper bussing inside the laminate. Second, the aluminium foil is contacted to an aluminium bussing. The aluminium-copper contact is implemented inside the laminate by contacting the aluminium bussing with the copper bussing right before the junction box. Third, in this concept a pure aluminium bussing is used in the laminate but the aluminium-copper contact is implemented inside the junction box.

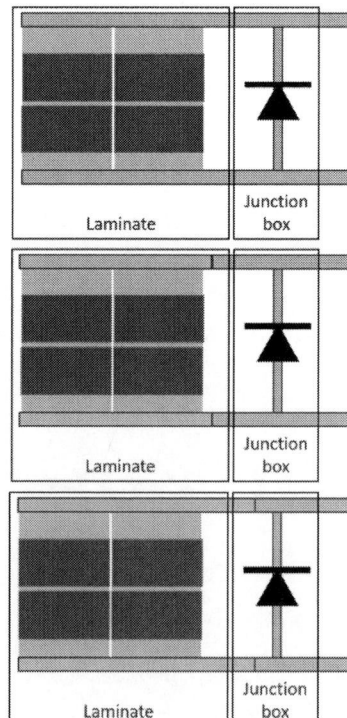

Figure 1: Schematic drawing of the 3 concepts differentiating in the location of the aluminium-copper

contact. TOP: Concept 1. Aluminium foil is connected to a copper bussing. MIDDLE: Concept 2. Aluminium foil is connected to an aluminium bussing. Aluminium bussing is connected to a copper bussing inside the laminate of the module. BOTTOM: Concept 3. Aluminium foil is connected to an aluminium bussing. Aluminium bussing is connected to a copper bussing inside the junction box.

2.2 Bussing material

From the consideration of the three concepts, two materials were identified which are tested for contacting the aluminium foil. The following two combinations were evaluated within this work.
- Aluminium foil on copper bussing
- Aluminium foil on aluminium bussing

2.3 Joining techniques and materials

The soldering of aluminium cannot be carried out using a standard process and materials as they are currently used in PV industry. This is because aluminium forms a non-conductive oxide layer very quickly, which cannot be cracked by standard soldering. As there is no known publication on the reliability of modules with integrated aluminium-copper contacts, various techniques and materials known for the production of aluminium-copper contacts from other industries were compiled from the literature and tested for use in a PV module. In the following the relevant joining techniques are presented.

2.3.1 Soldering with application-specific fluxes

For this joining technique a standard soldering process is used with standard soldering iron and standard soldering alloys but a special flux that is capable to break the oxide layer of the aluminium. The soldering process takes place in the same process step after the flux etched the oxide so that no new oxide layer can form.

2.3.2 Ultrasonic soldering

A special soldering station must be used for ultrasonic (US) soldering. The soldering iron of such a station is capable to heat up the solder alloy as well as send ultrasonic waves to crack the oxide layer of the aluminium. Through the ultrasonic waves the oxide layer is cracked mechanically and similar to the descried process in 2.3.1 the soldering takes place in the same process step after the oxide is broken.

For ultrasonic soldering special soldering alloys (USM) are used that can actively form a bonding to the aluminium.

2.3.3 Aluminium/copper clad

For this joining technique there is no new process introduced but a new material is used. Instead of soldering the aluminium directly to the copper a foil of aluminium and copper is used in between. The foil consists of an aluminium and a copper layer which are rolled to a joined material which is called aluminium/copper clad (Al/Cu clad).

In this work, on the aluminium side of the clad the aluminium foil is soldered using an ultrasonic soldering process as described in 2.3.2. On the copper side of the clad a tin-lead (SnPb) coated copper bussing is soldered using a standard soldering process.

2.3.4 Laser welding

In contrast to the joining techniques described above, no additional solder is applied during laser welding. The bond between the aluminium foil and copper is created by melting the aluminium foil using a laser and bonding it onto the bussing by forming an intermetallic phase.

2.4 Preliminary results and further tested combinations

After the evaluation of preliminary tests of the contacts needed for the three different concepts, it was decided to follow only the concepts 1 and 2. Concept 3 is not discussed further within this paper. Based on the results of the preliminary tests, only the following combinations of joining technique, joining material and bussing material are analysed because they showed the most promising results in the preliminary tests.

Table I: Combinations of joining technique, joining material and bussing material which are investigated in this work.

	Joining technique	Material	Bussing
1	Ultrasonic soldering	USM1	Al
2	Ultrasonic soldering	USM1	Cu
3	Ultrasonic soldering	USM2	Al
4	Ultrasonic soldering	USM2	Cu
5	Al/Cu Clad	USM1	Cu
6	Al/Cu Clad	USM2	Cu
7	Laser welding	-	Al
8	Laser welding	-	Cu

Other combinations were not investigated further because either the adhesion after bonding was too low, the materials corroded very badly in initial tests in the climate chamber or the electrical conductivity of the contact degraded very badly as a result of climate chamber tests. Therefore, other combinations are not discussed here.

2.5 Structure of the samples

To test the reliability of the contact to the bussing, mini-modules (size 220x220 mm²) were produced and exposed to a thermal cycling (TC) and a damp heat (DH) test in the climate chamber. The mini modules consist of a glass on the front side, an encapsulation material of POE, a string of PV cells, a second sheet of POE and a transparent backsheet. The PV strings consist of two PERC shingle cells interconnected with aluminium foil. The interconnection was done with a FoilMet process described in [5]. The aluminium foil that was contacted at the beginning and ends of the string was left very long in order to contact these aluminium foil wings to several different bussings.

It is important for the test to analyse the degradation of the contact between the aluminium foil and the bussing. Therefore, in the samples there is not only the bussings taken out of the modules but the aluminium foil, which is directly connected to the cell. By measuring the module on both the aluminium foil (reference) and the bussing, the part of the degradation caused by the contact to the bussing can be determined very precisely. Both results only differ in the series resistance of the aluminium foil, bussing and the resistance of the contact between aluminium foil and bussing. Since the losses in the aluminium and the bussing are negligible due to the low currents in the sample (Isc approx. 1.3A), the loss due to an approximation of the resistance of the contact between aluminium foil and bussing can be calculated by simply subtracting the IV characterisation values of the reference (aluminium wings) and the IV characterisation values measured at the bussing.

Figure 2: Picture of a string integrated in a mini module. Two shingle cells are interconnected with aluminium foil by a laser process. At both ends of the string the aluminium foil has long wings in order to connect several bussings. The aluminium foil is taken out of the module at both ends of the string to characterise the performance as a reference.

3 RESULTS

In order to be able to implement concept 1, only connections between aluminium foils and copper bussing need to be considered. All other contacts in the module's bussing system would then be comparable to the contacts used in standard modules, which is why these contacts were not investigated further for this work. However, in order to implement the bussing system as described in Concept 2, the connection between the aluminium foil and aluminium bussing must be investigated. In the following, therefore, the results of the reliability tests of aluminium-aluminium contacts and those of aluminium-copper contacts are described.

3.1 Aluminium-aluminium contacts

In Figure 3, there are shown the results of the reliability tests of the aluminium-aluminium contacts tested in this work. For this test aluminium bussings were contacted to the aluminium foils of the strings used in the mini modules with the joining techniques and materials used accordingly to Table I. The diagram shows the power loss related to the contact of the respective joint. The power loss of the related contact is calculated from the difference between the power measured at the aluminium reference contact and the power measured at the respective bussing (more details in 2.5). For all three groups, the top diagram shows the loss due to the contact in the initial state (0 h) as well as the degradation after 1000, 2000 and 3000 hours of damp heat testing. The bottom diagram shows the power losses due to contact in the initial state (0 cycles) as well as the degradation after 200, 400 and 600 cycles of thermal cycling test.

The power loss of the tested aluminium-aluminium contacts are very low. For all samples, the power loss of the contact in the initial state is less than 0.5 %. In addition, the samples show no significant degradation over the

period of the Damp Heat Test (three times IEC). All samples show a power loss due to the contacts of < 0.3 % after DH 3000. For the samples with the combination of ultrasonic soldering and USM2 as well as for laser welding, however, a positive trend can be observed due to the aging in the climate chamber.

Similar results are observed for the samples that were aged in thermal cycling test conditions. Although the power loss of the contacts in the initial state is similar to the results after DH test, the loss of the tested contacts is even lower after TC600 then after DH3000 (both three times IEC). After TC600 the power loss related to the contact is below 0.2 %.

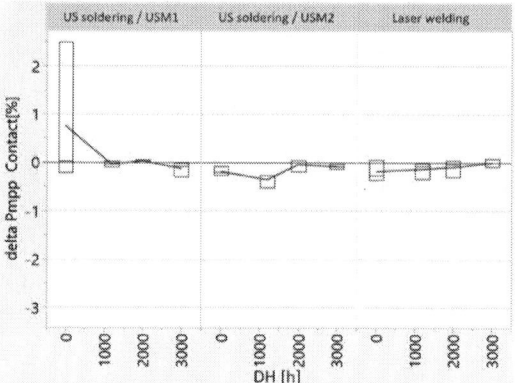

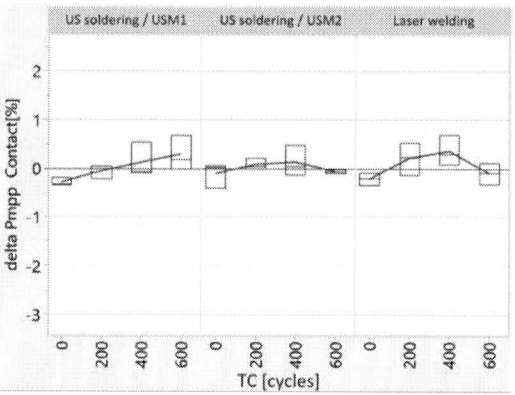

Figure 3: Diagram showing the relative power loss related to the contact between aluminium foil and aluminium bussing over the reliability test duration. TOP: Results of damp heat test in hours. BOTTOM: Results of thermal cycling test conditions in number of cycles.

3.2 Aluminium-copper contacts

In Figure 4, there are shown the results of the reliability tests of the aluminium-copper contacts tested in this work. For this test copper bussings were contacted to the aluminium foils of the strings used in the mini modules with the joining techniques and materials used accordingly to Table I. As described in chapter 3.1 the power loss of the related contact is calculated from the difference between the power measured at aluminium reference contact and the power measured at the respective bussing. The top diagram shows the loss due to the contact in the initial state (0 h) as well as the degradation after 1000, 2000 and 3000 hours of damp heat testing of all tested groups. The bottom diagram shows the power losses due to contact in the initial state (0 cycles) as well as the degradation after 200, 400 and 600 cycles of thermal cycling test.

In damp heat test, the power loss due to the contact in initial state is higher (< 0.7 %) than the results of the aluminium-aluminium contacts in chapter 3.1. In initial state, the power loss of the samples joined with an Al/Cu clad are lower than for the samples joined with ultrasonic soldering or laser welding. For most groups, this diagram shows that no degradation can be observed as a result of the damp heat test. The power loss due to the contact remains constant or even decreases slightly. Only the group for which ultrasonic soldering and USM1 were used shows a degradation. The samples degrade by an average of 1 %. From an initial status of 0.4 % power loss to -0.6 % power loss.

Similar results are observed for the aluminium-copper samples that were aged in thermal cycling test conditions. In thermal cycling test, the power loss due to the contact in initial state is higher (< 0.6 %) than the results of the aluminium-aluminium contacts in chapter 3.1. As with the results in the damp heat test, the samples that were joined with an Al/Cu clad show a lower power loss in the initial state than the samples with ultrasonic soldering or laser welding. This diagram shows that no degradation can be observed as a result of the thermal cycling test. The power loss caused by the contact remains constant or even decreases slightly.

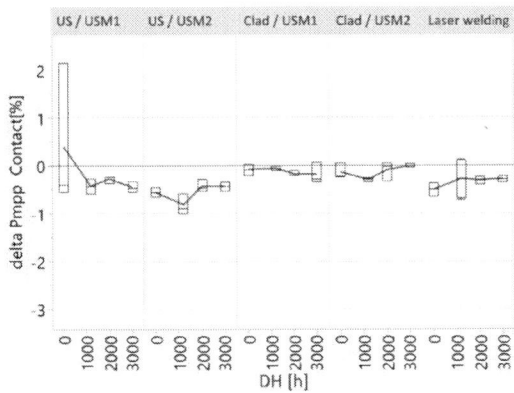

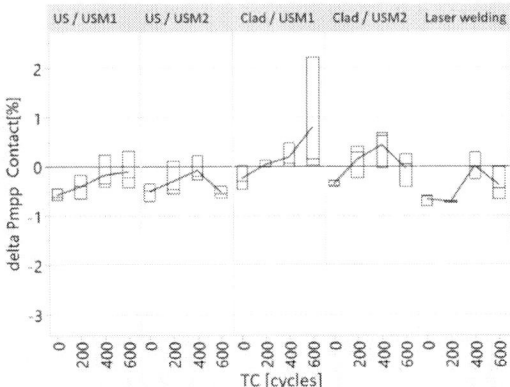

Figure 4: Diagram showing the relative power loss related to the contact between aluminium foil and copper bussing over the reliability test duration. TOP: Results of damp heat test in hours. BOTTOM: Results of thermal cycling test conditions in number of cycles.

4 DISCUSSION

4.1 Aluminium-aluminium contacts

In order to realize concept 2, the reliability of a contact between aluminium foil and aluminium bussing must be investigated. The results of this experimental study show, that there are combinations of joining techniques and materials used to form a reliable contact. The losses are less than 0.3% after DH 3000 and TC 600. With both ultrasonic soldering and laser welding, the contacts do not show any increasing power losses over aging in the climate chamber.

Instead of the expected degradation of the contact, most samples show an improvement in the power losses due to the contact. This could be the case because the aluminium foil is degrading worse than the contact itself. As the degradation of the aluminium foil then has a greater effect on the measurement of the reference than on the other samples, the degradation of the calculated power losses due to the contact has a positive effect on the power loss during aging in the climate chamber.

This would also mean that the method used to calculate the power losses due to the contacts is not entirely correct. The influence of the aluminium foil on the power losses was assessed as neglectable. This would be contradicted by the effect described above. However, the corruption of the result is at a very low level.

4.2 Aluminium-copper contacts

For the realization of concept 1, the reliability of the contact between the aluminium foil and copper bussing was investigated. In this experimental study there are combinations of joining techniques and materials that form a reasonable resistance and show almost no degradation after TC 600 and DH 3000.

In contrast to the results of the aluminium-aluminium contact, which are described in chapter 4.1, no positive trend can be observed after ageing the samples in the DH test. Instead, a constant power loss due to the contact can be observed during the aging of the samples in the climate chamber. The samples do not change much in power loss after ageing compared to the initial measurement. As it can be assumed from the discussion in chapter 4.1 that the aluminium foil might degrade in the DH test, it can be assumed that the aluminium foil also degrades in these samples. This would mean that the actual power loss due to contact after the damp heat test degrades at a very low level.

In the results it was described that the samples joined with ultrasonic soldering and laser welding show higher power losses due to the contact between aluminium foil and copper bussing than the samples joined with a Al/Cu clad. This effect can be observed for TC and DH test condition. The reason for this is most likely not a high resistance contact in this samples but a thinner bussing system used for those samples which results in a higher series resistance. For the samples joined with ultrasonic soldering and laser welding a 1.5 x 0.2 mm² bare copper ribbon was used while for the samples joined with an Al/Cu clad a 5 x 0.3 mm² SnPb coated ribbon was used. Therefore a higher resistance was expected for the samples with thinner ribbon. The power loss due to the loss through the ribbon was expected to be 0.25 % which is very close to the difference of the results shown in chapter 3.2. This indicates that the difference in power loss due to the aluminium-copper contact is very small for the groups shown in chapter 3.2.

When visually inspecting the samples, it can be observed that only the samples with uncoated copper bussing corrode in the DH test. Even if this has hardly any influence on the electrical performance of the bussing system, it must be investigated further to prevent corrosion. The corrosion applies to the samples with Al/Cu clad and the samples with laser welding. In the production of the ultrasonic soldered samples, the bussing system was completely coated with the ultrasonic solder material, which is most likely why these samples do not corrode. By slightly changing the process, the Al/Cu clads could also be coated with solder alloy, so that corrosion would not be expected in these samples either. The laser-welded samples could be produced with tin or silver coated bussing to avoid corrosion at the point in the module. However, this must be tested in further experiments.

4.3 Evaluation of the concepts
In this study, two bussing concepts are investigated, which differ in the location at which the aluminium-copper contact is created. In concept 1, the aluminium foil of the strings is attached directly to the copper bussing. In concept 2, the aluminium foil is first attached to an aluminium bussing and the aluminium-copper contact is implemented in the laminate of the module.

The results of the experiments show that it is possible to attach the aluminium foil to both an aluminium bussing and a copper bussing without any major loss of power due to the contact. As both variants show very similar power losses due to the contact, an interconnection according to concept 1 seems to make more sense, as no further contacts are needed which would lead to hire power losses.

However, the evaluation of the degradation of the contacts has not yet been finalised. As discussed in chapter 4.2 more experiments are needed.

5 CONCLUSION

It can be concluded that very low power losses are possible with an aluminium-aluminium and an aluminium-copper contact in a PV module. Although, it was shown that the method used cannot be applied to precisely evaluate the contact between the aluminium foil and the bussing, the results provide a good estimate of the performance of the contacts. With the values achieved through this experiments, it can be assumed that the power loss due to the contact is about $1 - 3$ W (0.2 - 0.5 %) for a full-size module. Hardly any differences can be observed between the groups shown in this study, which differ in the joining technique, material and bussing material used.

The degradation of the aluminium foil, which is probably the cause of the positively measured trend of degradation of the power loss due to the contact, is also estimated to be very low at module level based on the experience gained in this experiment.

Further new types of samples must be developed in order to investigate the resistance of the contact between the metals more precisely. In addition, the next step is to test the reliability in a full-size module in order to investigate realistic power losses.

6 ACKNOWLEDGMENT

This work was funded by the Federal Ministry for Economic Affairs and Climate Action of Germany as part of the project SPINAT (grant agreement No. 03EE1124C).

7 REFERENCES

[1] J. Lossen et al., Laser structured p-IBC cells interconnected by Al-foil, 11th Back Contact Workshop, Hamelin, 29th Nov. 2023
[2] T. Meßmer et al., Interconnection approach for busbar-less IBC cells based on printed solder paste, 8th WCPEC, 2022.
[3] G. Emanuel et al., Demonstration of an industrial concept for the metallization and interconnection of p-IBC cells using aluminium foil by laser processes, 40th EUPVSEC, Lisbon, 2023.
[4] K. Hofmann et al., Reliable copper and aluminium connections for high power applications in electromobility, 8th International Conference on Photonic Technologies LANE, 2014.
[5] J. Paschen et al, FoilMet®-Interconnect: Busbarless, ECA- and Solder-Free Aluminium Interconnection for Shingled Modules, Progress in Photovoltaics, 30, 2022.

41st European Photovoltaic Solar Energy Conference and Exhibition

This presentation was selected by the Sc. Committee of the EU PVSEC 2024 for submission of a full paper to one of the EU PVSEC's collaborating peer-reviewed journals.

LIGHTWEIGHT PHOTOVOLTAIC MODULES TECHNOLOGIES: RELIABILITY EVALUATION AND MARKET OPPORTUNITY

J. Dupuis[1], C. Abdel Nour[1], J.V. Oliveira Santos[1], P. Lefillastre[2]

[1]EDF R&D, EDF Lab Les Renardières, Avenue des Renardières, Moret Loing et Orvanne, F-77250, France
[2]EDF Renouvelables, 43 Boulevard des Bouvets CS 90310, Nanterre Cedex, F-92741, France

ABSTRACT: This study aims at performing an assessment of lightweight photovoltaic (PV) module's reliability by comparing module's performances and reliability of several manufacturers. Lightweight modules are characterized by a reduced weight compared to classical PV modules with usually less than 10 kg/m² allowing its installation on rooftops with low bearing capacity without the need of reinforcing the roof structure. Even if this PV technology has higher costs than classical modules due to lower capacity production and often specific module material like transparent composite instead of glass, it represents an efficient technical and commercial solution for specific applications with lower transportation, installation, and maintenance costs. Nevertheless, very few information is available on the reliability of these emerging products. The large variety of module components used by the manufacturers do not help into selecting the best modules available on the market.
In this work, after introducing the potential capacity of low bearing rooftop market for France, we detailed a first reliability benchmark of market available lightweight module's products. This benchmark was carried out through indoor experimental work under accelerated aging testing as well as outdoor exposure conditions based on IEC61215 and IEC61730 qualification standards.
The accelerated aging sequence results highlight high power degradation discrepancies between the five technologies investigated with some of them degrading more than 50% after a damp heat − UV − thermal cycling testing sequence. The power generation obtained in outdoor conditions underlines lower energy yield obtained by installing photovoltaic module directly on a flat roof caused by their orientation and tilt (0 to 10°) entailing sometimes higher soiling losses. Based on all these results, LCOE estimations are performed and show an over-cost between 10 and 100% compared to the use of standard module option with roof reinforcement.
Keywords: Photovoltaic module, Lightweight, Reliability, Yield, LCOE

1 INTRODUCTION

With its versatility and low cost, electricity provided by photovoltaic energy has become one of the best options for setting up new power plant. One drawback of this technology is the place needed for the power plant itself which can limit utility scale system in some countries. Putting these photovoltaic systems on rooftop is then an ideal choice, Solar Power Europe has shown that in 2023, 196 GW of the global 447 GW set up in the world was on rooftop [1]. As the cost depends on the size of the system and the ease of the installation, the bigger and the flatter the roof, the lower can be the installation cost.

C&I and Agricultural buildings have great potentials in reason with their high roof surface availability but there is often an issue with the high percentage of roof which can't bear too much weight (low bearing roof). For instance, in France, based on the rooftop data available [2-4] for Agricultural, commercial, and industrial buildings, we have estimated that 26% of the 2 Gm² of the rooftop surface cannot bear more than 15 kg/m², i.e., the weight of a classical module with its roof structure, but it can accept at least 5 kg/m². Considering that only 30% of this surface is available and photovoltaic system efficiency of 20%, the potential will be slightly more than 30 GWp for lightweight module solutions.

Today, structure reinforcement is the main option for deploying photovoltaic on low bearing roofs. The reinforcement option is usually quite expensive and gives often a non-acceptable cost to the PV plant project which is then abandoned.

The continuous decrease of PV costs and technological progress give opportunities for lightweight modules solutions to be used for this kind of roofs. However, lightweight modules are more expensive than classical PV modules (two to four times now), are not always set up with a high tilt to optimize sun light on the plane of array and are fabricated with specific bill of materials that questions their long-term reliability.

The approach studied here consists first, in evaluating the reliability of various lightweight module types from several manufacturers and with different architectures and composition (with or without glass, module for structure fixation or for rooftop direct installation....) using sequential aging tests. Secondly, the qualification of the PV module's yield in outdoor environment in a representative tilt is observed. Lastly, a conclusion about the interest in using these modules in a project is discussed by estimating the levelized cost of electricity of a representative project using the previous results.

2 MATERIAL AND METHODS

The lightweight module relevance is performed by assessing PV modules from five different manufacturers (referred as A, B, C, D and E in Table I) with composite/polymer, often ETFE-based and glassfiber reinforced plastics, and thin glass/polymer materials. The maximum power varies between 200 and 380 Wp.

2.1 Reliability testing

The study was carried out through an original indoor sequential testing procedure including main environmental stresses (Table II) to get global feedback on the module reliability. The sequences of the testing protocol were based on IEC 61215 and IEC 61730 qualification standards. A preconditioning step with current injection at Isc during 3 times 48h and a regeneration step after the damp heat (DH) step of current injection during 48h was performed to avoid any Light Induced Degradation (LID) and Light and elevated Temperature Induced Degradation (LeTID) impacts.

Framed modules were tested normally (B, D, E manufacturers), frameless modules were first glued on a glass (A and C manufacturers).

Table I: Tested lightweight modules technologies and efficiencies from five manufacturers. "Poly." stands for "composite/polymer"; "Glass" for glass/backsheet modules

	A	B	C	D	E
Type	Poly.	Glass	Poly.	Glass	Poly.
Efficiency (%)	16.7	17.1	20.6	18.9	18.3

Table II: Details of the sequential aging testing

	DH	UV	TC
Stress	85°C/85%RH	100W/m²	-40+85°C
Duration	1000h	45kWh	200cycles

Two modules from each manufacturer were evaluated. Electrical characterization (IV) of the PV modules was carried out using a PASAN flash tester SunSim 3b with class AAA equipped with a Dragon Back system. Electroluminescence (EL) measurements were done using a 150 MPixel XT IQ4 Phase One camera with its infrared filter removed and a current injection of Isc in the modules.

2.2 Outdoor testing

An outdoor testing platform was also installed at EDF R&D near Paris in France since September 2023 to evaluate the module's performances over one year operation under real conditions. The modules were installed with two tilt angles: 0° and/or 10° representative of what can be found on flat rooftops PV installations. These configurations were selected to study the energy yield of the module technologies compared to classical PV installations. Each module is connected to an IQ7+ microinverter allowing electrical performances monitoring under real conditions. This installation is also equipped with meteorological sensors to analyze the behavior of the modules under specific weather conditions. A reference classical installation in the same outdoor testing platform south oriented at 30° tilt is used for efficiency comparison.

3 RELIABILITY RESULTS

The Figure 1 and Table III present the IV parameters evolution after each step of the sequential testing. The results shown are the measurement average of the two tested modules per manufacturer.

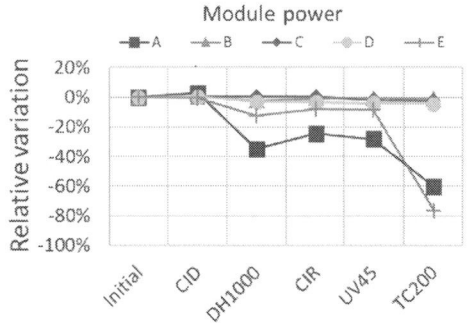

Figure 1: Power losses after each accelerated aging sequence

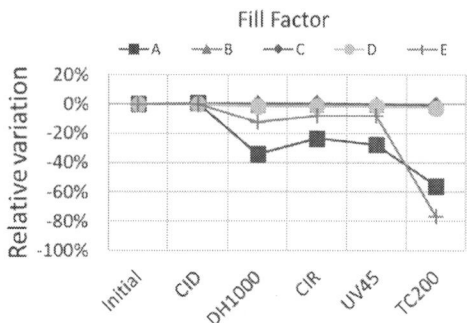

Figure 2: Fill Factor losses after each accelerated aging sequence

As shown in Figure 1, B, C and D modules got excellent results with less than 5% degradation after the whole testing sequence. In contrast, a strong degradation after damp heat testing was visible on A and E modules mainly caused by a high increase in series resistance shown on Figure 2 through the fill factor (FF) evolution and highlighted on the EL images compared to initial state through a global solar cell darkening except around the contacting ribbons as presented in Figure 3. A and E modules did not even pass the requirement of the IEC61215 standard after damp heat with more than 5% degradation.

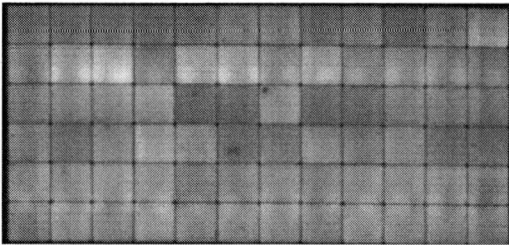

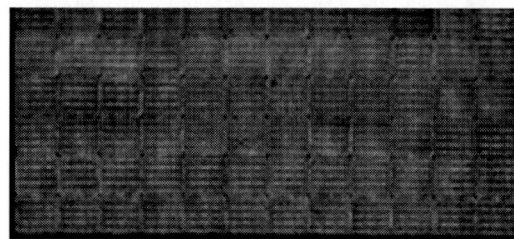

Figure 3: EL images evolution of a module from manufacturer A at initial state (top); after DH 1000h (bottom)

UV exposure did not entail a significant decrease for all modules tested but strong series resistances increase of modules A and E were also observed after thermal cycling sequences with lots of cell cracks.

Figure 4: Overview of a module showing solar cells bending after DH1000 testing sequence.

This high number of cell cracks could be linked to materials degradation and an increase of residual stains in the module from A and E manufacturers. This is illustrated in Figure 4 with a picture of one of the modules after DH1000 showing bended solar cells.

To summarize, for the five manufacturers evaluated, the absence of glass and replacement by composite materials have given mitigate reliability results as only modules from manufacturer C got low degradation rates after the testing sequences. Glass / polymer modules, with a thinner glass compared to classical modules, seem to be a better choice to ensure long term reliability regarding the results.

4 YIELD COMPARISON IN OUTDOOR ENVIRONMENT

As described previously, a performance assessment of the PV module's yield in outdoor environment was carried out as well. One or two modules of each manufacturer were set in outdoor conditions from September 2023 to June 2024. The outdoor platform was composed of 2 modules from manufacturer A, one with 10° and one with 0° tilt, one module from each of manufacturers B, D and E with 10° tilt and C with 0° tilt. No module was cleaned during all the period of the study to analyze the impact of soiling losses. 0° tilted modules are glued on a glass and the glass is fixed on our outdoor platform's concrete surface. The power losses results were computed using PVNOV, a modelling tool developed by EDF to simulate a PV system in a 3D environment and calculate its yield using ray-tracing model [7-9]. The simulation results obtained show power losses varying between 5 and 15% of such configurations compared to classical PV installation with 30° tilt, South oriented with the same global horizontal irradiance.

This behavior was expected since optimal tilt angle for our installation's latitudes maximizing the irradiance in the plane of array is between 30 and 35°. However, we are aware that for rooftops applications we are usually limited by 0 to 10° tilt configuration. In addition, lower tilts can increase the soiling on module's surfaces and makes it more difficult to evacuate dust with the rain.

To better consider module efficiency for various irradiation levels, incidence angle and soiling impacts on the production, an outdoor performance efficiency comparison was carried out over clear sky conditions period in October 2023, May and June 2024 highlighting high efficiencies for C and E modules (Figure.5) compared to their initial efficiencies (see table 1).

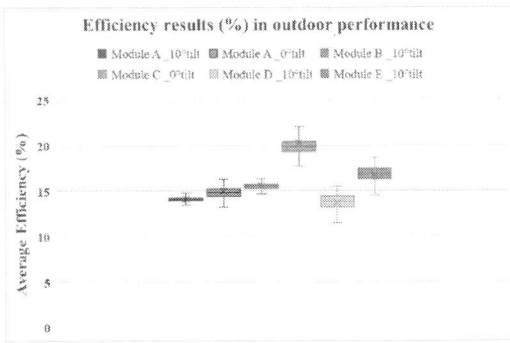

Figure 5: Average efficiency of lightweight modules under real conditions

Additionally, a comparison was established between lightweight modules outdoor platform and a reference PV installation of bifacial technology installed nearby. An assessment of lower tilt angles of lightweight modules adapted for rooftop installations with 0° and 10° modules tilt compared with the reference installation with 30° modules tilt was carried out. For that purpose, the efficiency and energy yields of each studied lightweight module technology were computed and compared to the yield of the reference installation under the same meteorological conditions. The average power losses over the studied period compared to the reference installation were between 8 and 17%.

These losses can reach higher values in some periods as illustrated in figure 6. This figure shows the energy yield losses obtained for a clear sky conditions day during the month of June 2024 compared to the reference installation. The results of module B technology weren't computed for this specific day due to a failure in the micro-inverter connected to this module during the investigated period. The energy yield (kWh/kWp) for modules A, C, D and E were computed with higher losses obtained for Module C reaching 26%. Module C, set up with a 0° tilt presented higher power losses compared to the other technologies with 10° modules tilt. Module A with 0° tilt did not present higher losses than the one with 10°, the efficiency measured for the first one was slightly higher (Figure 5) which can contribute to the yield losses results. The lower energy losses for this day were around 16% for module E with 10° tilt to the horizontal plane. In the month of June, the solar elevation angle is high, thus modules tilt between 0 and 30° should generate similar power with a difference calculated theoretically around ±2%. However, several factors contributed to these high losses:

- With lower modules tilts, increased soiling was observed on the lightweight modules that wasn't evacuated by the rain compared to the reference installation.
- A contribution of the bifaciality factor of the reference installation since it's composed of bifacial modules. According to the albedo measurements collected at the reference installation and simulations performed in our PVNOV modelling tool an average of 5% of bifacial gain is achieved.

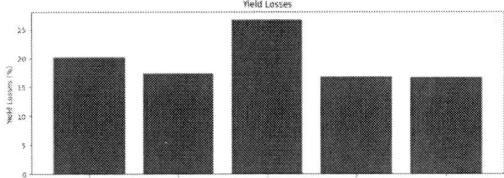

Figure 6: Energy yield losses of each lightweight technology. Two modules were set at 0° tilt (manufacturer A and C) and the other ones at 10° tilt.

Considering that very few days of clear sky conditions data were collected for the results analysis due to the meteorological conditions in the last months of the study, we can conclude that all modules tilted at 10° observed an additional 10% yield losses in June compared to the 30° reference PV plant which could be attributed to an additional soiling.

For 0° tilt, results differ between modules from A and C manufacturers with a 10 to 20% yield losses compared to the reference (excluding bifacial gain). Two elements can explain the losses:

- A higher module temperature for the C module compared to the A module. This would have a negative impact on power generation. As these modules were fixed on the soil, it was not possible to measure the module temperature.
- A difference in rear side rugosity: the aspect of the front side of the modules from the A manufacturer is not uniform and has some round texturing contrary to the surface of modules from manufacturer C which is almost flat. The impact of the front side rugosity on soiling aspect needs a specific study to confirm this trend.

For such PV plant systems, these losses will eventually depend on the location of the installation and the cleaning frequency of the modules. For 0° tilted modules, additional cleaning maintenance should be planned to avoid too many losses.

5 LEVELIZED COST OF ELECTRICITY

Based on reliability, outdoor results and internal data, we can estimate the levelized cost of electricity (LCOE) difference between a classical PV module plant versus a lightweight PV module plant will differ on several aspects:
- Less yield caused by flat surface installation and additional soiling losses
- Higher risk of degradation depending on the chosen lightweight technology
- Higher module cost because of non-conventional materials, reinforced frame if any and a mass production not equivalent to standard modules
- Lower module cost of installation in case of direct module gluing on a rooftop
- Higher cost of O&M if modules are glued to the roof because of additional soiling risk, directly exposed junction box, cables and connectors and difficulty in removing defective modules.
- Weighted Average Cost of Capital (WACC) will also be higher as less return of experience is available on such modules

With these assumptions, we estimate an increase of 10 to 40% for thin glass/backsheet modules and 50 to 100% for polymer-based modules. Even if the use of lightweight solution can be relevant even with its higher cost, the interest in these systems will be strongly linked to the cost of the client's electricity. In addition, several options exist to decrease the cost of these systems: more return of experience will limit the WACC increase, better module reliability will give a lower degradation rate and standardization of the product will lead to a cost decrease of the modules.

6 CONCLUSION

Lightweight modules are a robust solution to address low bearing roof market estimated at several tenth of GW in France only. Despite this potential huge market, we have highlighted some existing issues on today lightweight products present on the market with possible reliability issues, especially for composite/polymer modules. Outdoor performance study put in evidence the risk of having low tilted modules in term of yield but also the impact of soiling that may differ greatly depending on the localization of the PV plant but also on the surface rugosity and,

certainly, module operating temperature in a 0° tilt configuration. Based on all the results obtained, an estimation of the levelized cost of electricity for different hypothesis and configuration explicit an over-cost of 10 to 100% compared to a classical module solution with a roof reinforcement.

Lightweight modules are still new products on the market, as more and more products are coming, manufactured now buy big photovoltaic players, their robustness and cost should improve soon. Nevertheless, reliability testing needs to be carried out and more data from outdoor performance studies are necessary to anticipate the effective yield of such PV plants. Outdoor testing is still undergoing at our laboratory to collect additional data with clear sky and high irradiance levels conditions covering the summer period. The global assessment of the yield comparison in outdoor environment will be finalized after completing the full year data analysis.

7 ACKNOWLEDGEMENTS

The authors are very grateful to the EDF R&D laboratory team for the help provided in performing module testing.

8 BIBLIOGRAPHY

[1] W. Hemetsberger, M. Schmela and S. Dunlo, "Global Market Outlook for Solar Power 2024-2028" (2024)

[2] Bódis, K.; Kougias, I.; Jäger-Waldau, A.; Taylor, N. & Szabó, S., Renewable and Sustainable Energy Reviews 114, 109309 (2019).

[3] « A 100% renewable electricity mix? Analysis and optimization » (2016), Agence de l'Environnement et de la Maitrise de l'Energie (ADEME).

[4] S. Joshi, S. et al., Nature Communications 12, 5738 (2021)

[5] A. Lindsay, M. Chiodetti, P. Dupeyrat, D. Binesti, E. Lutun and K. Radouane, Proceedings of the 31st European photovoltaic solar energy conference, Hamburg, Germany; 2015:1764-1769.

[6] A. Lindsay, M. Chiodetti, D. Binesti, et al., Munich, Germany; 2016:1610-1617.

[7] M. Chiodetti, A. Lindsay, P. Dupeyrat, D. Binesti, E. Lutun, K. Radouane and S. Mousel, Proceedings of the 32nd European photovoltaic solar energy conference, Munich, Germany; 2016:1449-1455.

41st European Photovoltaic Solar Energy Conference and Exhibition

MgO/SIOₓ ADDS HEAT DISSIPATION FUNCTION TO CRYSTALLINE SILICONE SOLAR CELL MODULES

Eiko Shimokata[1], Yasushi Sobajima[1], Keisuke Ohdaira[2], Atsushi Masuda[3]
[1]Gifu Univ., [2]JAIST, [3]Niigata Univ.
*E. Shimokata, shimokata.eiko.w6@s.gifu-u.ac.jp

ABSTRACT: To suppress the temperature increase of crystalline silicon (c-Si) solar cell modules during power generation, a glass film (MgO/SiOₓ) mixed with thermal conductive filler material MgO particles was applied on EVA and back sheets at the bottom of c-Si cells. Vacuum lamination was conducted at a relatively low temperature (90°C) to maintain the shape of MgO/SiOₓ. Increasing the quantity of MgO particles in the film enhanced the formation of heat transfer pathways by the particles in the film, and the temperature elevation during power generation was mitigated. Furthermore, when ethanol was added to the MgO/SiOₓ film formation solution, the temperature rise suppression effect was improved.

KEYWORDS: c-Si cell, module structure, thermally conductive filler, MgO, Heat dissipation function.

1 INTRODUCTION

Crystalline silicon (c-Si) solar cell modules represent the current standard in photovoltaic technology. Photovoltaic modules installed in outdoor environments must demonstrate resilience to the effects of weather and mechanical stress. The c-Si solar cell modules are constructed in a sandwich configuration, with the cells placed between layers of cover glass, an encapsulant, and a backsheet. This configuration provides mechanical strength and inhibits water vapor penetration from the outside, thereby ensuring long-term performance and reliability [1-3].

Since the band gap of c-Si is small, a decrease in power generation efficiency due to a decrease in V_{OC} in response to temperature rise will occur in actual operation [4, 5]. In addition, the c-Si solar cell modules in widespread use are structurally difficult to dissipate heat generated in the cells during light irradiation to the outside, because a material with low thermal conductivity surrounds the c-Si. For the above reasons, operating at the lowest possible temperatures is crucial to ensure the optimal output of c-Si solar cells.

In this study, we investigated using a glass film mixed with MgO particles (MgO/SiOₓ), a thermally conductive filler material, to provide heat dissipation to c-Si solar cell modules to suppress the temperature rise during power generation.

2 EXPERIMENTAL DETAILS

2.1 Preparation of MgO/SiOₓ

The SiOx material is produced by applying liquid glass (Lq. glass) to a substrate, followed by drying. After mixing the required amount of MgO particles in Lq. glass at room temperature, the solution was stirred for 1 hour and used to prepare MgO/SiOx. The solution MgO/SiOx was fabricated on the components using a brush and bar coating technique to obtain the material's heat dissipation properties. Overcoats were applied every hour until the required film thickness was reached, and the film was allowed to dry in air at room temperature for 24 hours. The above solution was applied on the encapsulant (ethylene vinyl acetate (EVA)) or back sheet.

Figure 1 shows an image of the membrane structure. The heat generated in the base material can be dissipated by mixing MgO particles, a thermally conductive filler material, in SiOₓ with low thermal conductivity to form a heat transfer path, as shown in the figure.

This study used 25 to 250 mg of MgO particles as solute and Lq. glass or EtOH-diluted Lq. glass as solvent. The MgO particles used are a commercial product with an average particle diameter of about 5 μm and have a proven track record as a heat-dissipating material for semiconductor devices.

The thickness of the MgO/SiOₓ par coating has been derived from the cross-sectional SEM images of the different coating times.

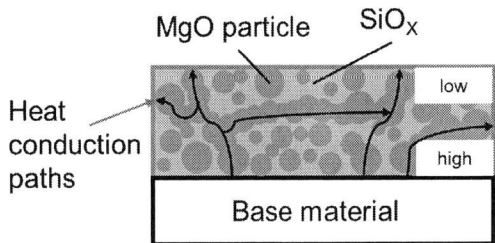

Figure 1: Image showing the formation of heat conduction paths in a MgO/SiOₓ that encapsulates particles of MgO, a filler material with high thermal conductivity

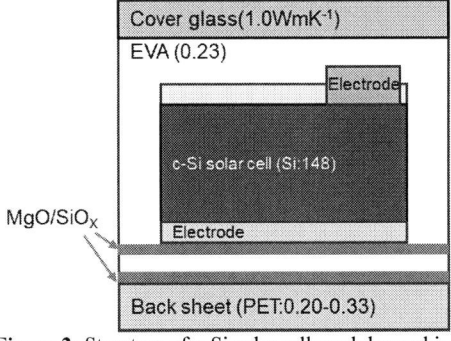

Figure 2: Structure of c-Si solar cell module used in this study and the thermal conductivity of its components, expressed in mWK⁻¹

2.2 Module fabrication by a vacuum lamination process

The structure of the c-Si solar cell module is backsheet/ EVA/ c-Si cell/ EVA/ cover glass (cell area: 15×15 mm², module area: 45×45 mm²). The module structure was formed by stacking and vacuum laminating each component as shown in Figure 2.

523

The module lamination temperature using MgO/SiO$_X$ was determined from the appearance of SiO$_X$ in the pseudo-module structure after lamination with colored Cu/SiO$_X$ instead. Cu/SiO$_X$ was deposited on EVA, and the lamination temperature varied from 85°C to 135°C.

2.3 Continuous light irradiation experiment

The change in *J-V* characteristics of c-Si solar cell modules with MgO/SiO$_X$ was measured as a function of light irradiation (AM 1.5, 100 mW/cm^2) time. As direct measurement of the temperature inside the solar module during continuous operation is difficult, the temperature in the module was estimated based on the temperature characteristics of the c-Si solar cell [5].

This study estimated the module's internal temperature increase due to light irradiation by measuring the change in V_{OC} (ΔV) between immediately after the start and 180 minutes of continuous light irradiation. The ΔV of the conventional structure without MgO/SiO$_X$ (ΔV_S) was used as a reference, and the ΔV of the measured samples (ΔV_F) and temperature characteristics of c-Si solar cells in Ref. [5] were used to calculate the estimated temperature difference (ΔT), an indicator of the heat dissipation effect of MgO/SiO$_X$, using the following formula.

$$\Delta T = \frac{\Delta V_S - \Delta V_F}{|\Delta V_T|}$$

ΔT(°C): Estimated temperature difference
ΔV_S(V): The decreased value of V_{OC} for MgO/SiO$_X$ uncoated module (0 to 180 min of light exposure)
ΔV_F(V): The decreased value of V_{OC} for MgO/SiO$_X$ coated module (0 to 180 min of light exposure)
$|\Delta V_T|$(V/°C): Rate of change of V_{OC} with temperature for c-Si solar cells (calculated from data [5])

The heat dissipation performance was evaluated by the magnitude of ΔT under each condition. Measurements under light irradiation were taken indoors, at room temperature, and no cooling fan was used. Furthermore, materials with low thermal conductivity were used at the module installation points to suppress heat conduction from the bottom of the module to the outside.

3 RESULTS AND DISCUSSION

3.1 Thermally conductive material MgO/SiO$_X$

A thermal imaging camera was used to observe the surface heat dissipation state of the MgO/SiO$_X$-coated (2 layers) glass substrate. Figure 3 shows a surface thermographic image of MgO/SiO$_X$ fabricated on a flat EagleXG glass (20 x 20 mm^2) after one hour of high heat on one side.

A comparison of the two samples shows that with increasing MgO concentration, thermal diffusion occurs at the substrate surface. This suggests that with sufficient MgO particle incorporation, MgO/SiO$_X$ exhibits the characteristics of lateral thermal diffusion.

3.2 Evaluation by SEM-EDX

Cross-sectional SEM evaluation results for samples with different numbers of layers show that the thickness of the MgO/SiO$_X$ film increases monotonically with the number of layers, with a thickness of about 10μm per layer.

Figure 3: Differences in the surface thermal conduction of MgO/SiO$_X$ with different MgO particle contents, (a) 25 mg or (b) 150 mg MgO in 1 ml Lq. glass.

In Figure 4, the solvent in (A) is 1 ml of pure Lq. glass, the solvent in (B) is Lq. glass diluted 1:1 with EtOH (total 1 ml), and the MgO particles in the liquid are constant at 150 mg. A comparison of the EDX Mg mapping results in the figure shows that more Mg was detected in the EtOH dilution (solution (B)). When Lq. glass is diluted with ethanol, the volume of SiO$_X$ decreases because the ethanol evaporates as the Lq. glass solidifies. On the other hand, since the amount of MgO in the solution is constant, the amount of MgO in a unit volume of MgO/SiO$_X$ increases, and the probability of detection in the EDX measurement is considered to have increased.

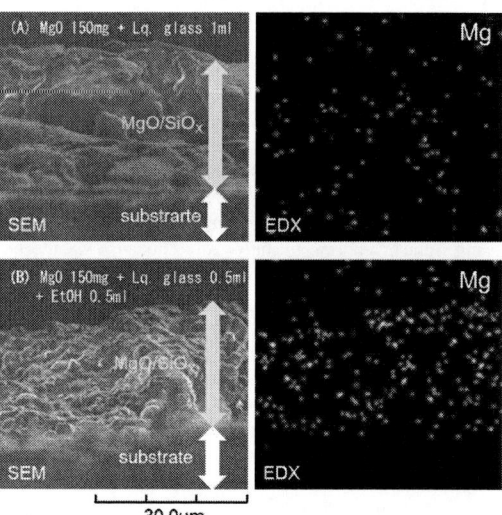

Figure 4: Cross-sectional SEM images of MgO/SiO$_X$ were prepared using solvents without EtOH dilution (top) or EtOH dilution (bottom), and Mg was simultaneously mapped by EDX

3.3 Selection of vacuum lamination temperature

The effect of lamination temperature on SiO$_X$ was observed using a pseudo-module structure in which c-Si was used as the cell without tab wires.

As shown in Figure 5, during vacuum lamination under high-temperature conditions (135 °C), Cu/SiO$_X$ is not in sheet form but is destroyed and diffuses outward from the module. Cu/SiO$_X$ is no longer present around the c-Si cells. This phenomenon has also been observed in MgO/SiO$_X$. To ensure the thermal diffusion path in the lateral direction, SiO$_X$ has to maintain its sheet shape even after the lamination process, which makes it difficult to laminate at high temperatures. In contrast, EVA does not dissolve at low temperatures, making lamination difficult. If lamination is performed below 80 °C, Cu/SiO$_X$ retains its sheet shape. However, sufficient lamination is impossible, and cases of c-Si cell destruction have been observed.

Therefore, to achieve sufficient heat dissipation, vacuum laminating conditions were set at a low temperature (90 °C) where the sheet shape could be maintained and lamination could be performed.

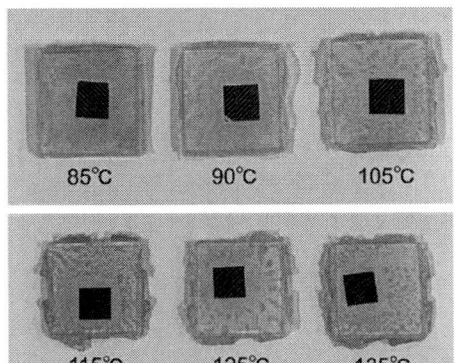

Figure 5: Pseudo structure of a c-Si module with Cu/SiO_X layers deposited on EVA and laminated at different temperature conditions.

3.4 Heat dissipation effect of different solutions

Figure 6 shows the change in V_{OC} of MgO/SiO_X prepared by brush coating plotting against light exposure time. Compared to the uncoated case, the application of MgO/SiO_X suppressed the decrease of V_{OC} during power generation as shown in the figure.

The V_{OC} and conversion efficiency of c-Si solar cells decrease with increasing temperature [5]. As shown in Section 3.1, MgO/SiO_X exhibited lateral thermal conductivity at 150 mg of MgO mixture. Applying MgO/SiO_X suppresses the decrease in V_{OC}, suggesting that MgO/SiO_X provides a heat conduction mechanism at the bottom of the module and the generated heat is expected to be transferred to the outside. The low thermal conductivity of EVA and the backsheet makes it difficult for heat generated in the module to be conducted directly to the back surface. In addition, the SiO_X sheet structure is maintained even after lamination at low temperatures, suggesting that the MgO/SiO_X layer forms a lateral heat conduction path with a heat dissipation effect on the module.

Table 1 shows the ΔT for MgO/SiO_X prepared with different solvents. V_{OC} reduction is effectively suppressed when EtOH-diluted Lq. glass is used as a solvent (B). As shown in section 3.1, the percentage of MgO detected increased with EtOH dilution, indicating increased MgO content in MgO/SiO_X. These results suggest that the ratio of MgO particles in MgO/SiO_X formed using EtOH dilution solution becomes larger, and the formation of heat transfer paths is promoted to obtain sufficient heat dissipation effect.

Table I: The estimated temperature difference (ΔT) between uncoated and MgO/SiO_X coated cases

	ΔT (°C)
(A) Lq. glass 1ml	3.30
(B) Lq. glass 0.5ml + 0.5ml	5.85

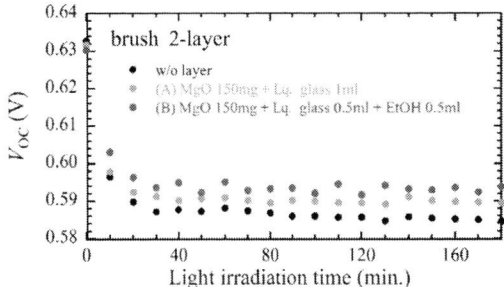

Figure 6: Variation of V_{OC} with light exposure time in c-Si single cell modules with MgO/SiO_X layer prepared in different solvents

3.5 Effect of particulate contamination on heat dissipation effect

The effect of the number of application times on the heat dissipation performance was investigated. Figure 7 shows the change in light irradiation time of V_{OC} due to the different numbers of bar coater applications. Based on the results of the previous section 3.4, Lq. glass diluted 1:1 with ethanol was used as the solvent. The total solvent volume was 1 ml and the amount of MgO particles mixed was 150 mg. Table 2 shows the ΔT for the number of applications. The figure shows that even when a bar coater is used, as in the case of a brush application, MgO/SiO_X inhibits the V_{OC} reduction. However, the bar coater method allows greater film thickness control than the brush method and can be applied to large areas.

The highest V_{OC} reduction was observed when the number of applications was five times. The heat dissipation effect is thought to be enhanced by the formation of stronger heat conduction paths and by the increase in volume, but the results in Figure 5 do not simply correlate with an increase in the number of layers, i.e., an increase in the thickness of SiO_X. When MgO/SiO_X was deposited on glass, the uniformity of SiO_X was lost as the number of layers increased, and at 15 layers or more, there were areas where the film broke. This indicates that the strength of the SiO_X layer structure is low, and even in the case of 10 laminations, the film shape required to ensure a lateral heat transfer path after lamination may be broken after the lamination process. Therefore, increasing the amount of mixed MgO particles and thickness is necessary to achieve sufficient heat dissipation, but it is also important to maintain the MgO/SiO_X sheet shape.

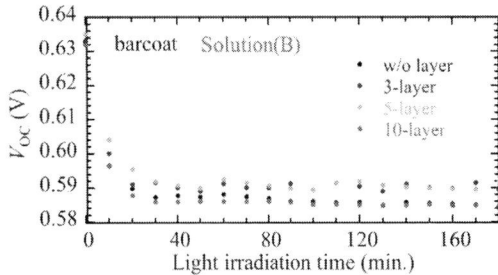

Figure 7: Change in light irradiation time of V_{OC} due to the difference in the number of MgO/SiO_X layer applications

Table II: Estimated temperature difference (ΔT) due to different number of applications

	ΔT (°C)
3-layer	2.20
5-layer	2.55
10-layer	0.44

4 CONCLUSIONS

This study investigated the application of a glass film (MgO/SiO_X) mixed with MgO particles, a thermally conductive filler material, to provide a temperature rise suppression effect during solar cell module operation.

The amount of MgO particles in SiO_X was adjusted to confirm lateral heat conduction. In addition, a low-temperature vacuum lamination process was used to secure the lateral heat conduction path after module formation. Considering the effect of MgO/SiO_X on light incidence, MgO/SiO_X was used at the EVA and backsheet positions directly under the cells to suppress V_{OC} reduction during operation.

MgO/SiO_X was applied to the c-Si module structure in a small area, and a 180-minute continuous light irradiation experiment was conducted for the unapplied c-Si module structure, which showed an estimated heat dissipation effect of about 5°C in terms of V_{OC}.

5 ACKNOWLEDGEMENT

This research was commissioned by NEDO and funded by REFEC.

6 REFERENSE

[1] A.W. Czanderna, F.J Pern, Solar Energy Materials and Solar Cells (1996) 101-181.

[2] Oliveira MCC, Cardosa ASA, Viana MM, Lins VFC, Solar Energy Materials and Solar Cells 81 (2018) 2299-2317.

[3] G. Oreski, G.M. Wallner, Solar Energy 79 (2005) 612-617.

[4] E. Radziemska, E. Klugmann, Energy Conversion and Management 43 (2002) 1889-1900.

[5] E. Radziemska, Renewable Energy 28 (2003) 1-12.

41st European Photovoltaic Solar Energy Conference and Exhibition

This presentation was selected by the Sc. Committee of the EU PVSEC 2024 for submission of a full paper to one of the EU PVSEC's collaborating peer-reviewed journals.

INVESTIGATION OF TEMPERATURE HOMOGENEITY DURING INFRARED SOLDERING OF SILICON SOLAR CELLS USING THE FINITE ELEMENT METHOD

Daniel C. Joseph[1,*], Angela De Rose[1], Dirk Eberlein[1], Onur Parlayan[1], Benjamin Grübel[1],
Andreas J. Beinert[1] and Holger Neuhaus[1]
[1] Fraunhofer Institute for Solar Energy Systems ISE, Heidenhofstrasse 2, 79110 Freiburg, Germany
*Corresponding author: e-mail to: daniel.christopher.joseph@ise.fraunhofer.de

ABSTRACT: In the production of silicon photovoltaic modules, copper ribbons are connected to the electrodes of solar cells using infrared soldering through automated stringer machines. Despite its efficiency and high throughput, infrared soldering results in inhomogeneous heating of the solar cells. Accurately measuring the solar cell temperature during the soldering process within the stringer poses a significant challenge, hindering process optimization to reduce this inhomogeneity. This study presents a novel finite element model to simulate the infrared soldering process, which calculates the solar cell temperature based on factors such as the applied electrical power, the radiation duration from the infrared emitters, and the hotplate temperature. The model is validated using thermocouple measurements at various points on the solar cells during the soldering process, showing a maximum difference of (8 ± 4) K in the peak temperature region between the simulation and experimental data. This FEM model can easily be adapted to new solar cell technologies and various solar cell sizes. Thus, a robust finite element model is developed to accurately determine the solar cell temperature during the infrared soldering process.

Keywords: Finite element method, Infrared soldering, Photovoltaic modules, Radiative heat transfer, Interconnection, Solar cells

1 INTRODUCTION

The soldering process plays a critical role in ensuring both electrical connectivity and mechanical stability in photovoltaic (PV) modules. Precise and homogeneous heating of the solar cells during this process is essential to reduce the thermomechanical stress and ensure the reliability of PV modules [1, 2]. In the photovoltaic industry, automated stringer machines are primarily used to perform industrial soldering processes using infrared (IR) radiation [3] .

Despite its advantages, such as rapid non-contact heating, the IR soldering process results in an inhomogeneous temperature distribution on the solar cells due to the uneven heat distribution and shading effects of the down-holder. This inhomogeneity leads to issues such as badly or non-contacted joints and solar cell overheating. However, measuring the exact solar cell temperature during the soldering process within the stringer presents a complex challenge.

The aim of this study is to develop a finite element method (FEM) model of the industrial IR soldering process used in the PV industry. This model is designed to compute the temperature distribution on industrial silicon solar cells during the IR soldering process. The model aims to enhance the accuracy of temperature predictions during the IR soldering process, thereby contributing to the optimization of the soldering procedure. This model is validated through temperature measurements taken at various positions on passivated emitter and rear cells (PERC) half-cells during the IR soldering process in a stringer. Furthermore, the model is adaptable, enabling temperature calculations for various solar cell technologies, such as TopCon and silicon heterojunction (SHJ), and different solar cell sizes.

2 METHODS

2.1 Infrared soldering

Infrared soldering by industrial stringer utilizes radiative heat transfer [3–5] that operate in the wavelength range of 0.5 to 3 µm [6]. Figure 1 sketches schematically the hotplates and an IR heating zone used in a stringer. The solar cells and the interconnecting wires are held in position by the down-holder during the IR soldering process. The hotplates before the IR heating zones are used to pre-heat the solar cells and the interconnections, while the hotplates after the IR heating zones are used to control the cooling of the soldered solar cells. Each IR zone contains four IR emitters that radiate infrared radiation to heat the solar cells and solder coated copper ribbon.

An IR emitter in the industrial stringer comprises a quartz halogen tube with an internal tungsten filament. The tungsten filament within the emitter is heated by the Joule heating effect [7]. The IR emitters have an efficiency of 90–95 % and exhibit a rapid response to changes in heating [8, 9]. The temperature of the tungsten filament can be regulated using the electric power supplied to the filament (P_{IR}) and the duration of current flow (t_{IR}). These parameters, P_{IR} and t_{IR}, along with the hotplate temperature T_{HP}, are used to control the solar cell temperature and will be referred to as process parameters.

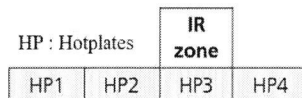

Figure 1 : Hotplates and IR zone position considered for the FEM model.

2.2 FEM Model

The objective of predicting solar cell temperature during the IR soldering process in an industrial stringer is achieved using finite element method (FEM) modeling in COMSOL Multiphysics 6.2. The FEM model consists of the IR emitters, featuring a simplified representation of the tungsten filament, the solar cell with the ribbons and down-holder. The temperature of the tungsten filament (T_F), incorporating process parameters P_{IR} and t_{IR}, is used as the input conditions for this FEM model. The transient temperature of the filament depends on the process parameters P_{IR} and t_{IR}. Temperature-dependent material properties for the tungsten filament, are obtained from [10, 11]. In this FEM model, the temperature values T_F are utilized to calculate the radiation emitted from the IR emitters. Using these radiation values, the solar cell temperature (T_C) is computed through heat transfer in solids and surface-to-surface radiation physics. This process models the radiative heat transfer between the IR emitters and the solar cells with ribbons.

Table I : Thermal properties of the various components used for the simulation.

Material	Thermal conductivity, k [W/m*K]	Specific heat capacity, C_p [J/kg*K]
Silicon solar cell	130[†]	700[†]
Sn60Pb40 solder	50[†]	150[†]
Copper wire	400[‡]	385[‡]
Down-holder	238[†]	900[†]
Tungsten filament	k(T) [10]	C_p(T) [10]
Aluminum reflector	238[‡]	900[‡]
Quartz glass	1.4[‡]	730[‡]

Table II : Optical properties of the various components.

Material	Emissivity, ε [-]	Reflectivity, ρ [-]
Silicon solar cell	0.70 [†]	0.13[†]
Sn60Pb40 solder	0.30 [12]	
Copper wire	0.15 [12]	
Down-holder	0.30[†]	1 - ε
Tungsten filament	ε(T) [11]	
Aluminum reflector	1 - ρ	0.92[‡]
Quartz glass	0.08[†]	0.04[†]

[†] Measured

[‡] Provided by manufacturer

In this work, M6 PERC half solar cell (83 mm × 166 mm) with six busbars (BB) and a down-holder with six metal strips was modelled for the FEM simulation, as the experimental validation was carried using the same configuration. To reduce computational effort, a symmetry plane is applied perpendicular to the longer side of the solar cell. The ribbon comprises a round copper wire attached to the solar cell with Sn60Pb40 solder alloy. The material properties used for the simulation are summarized in Table I and Table II. In this model, the silicon solar cell and the quartz tube are treated as semi-transparent surfaces with transmissivity τ, measured as 0.12 and 0.88 respectively. The reflector is modelled as an opaque surface, and the filament and down-holders are modelled as diffuse surfaces. For opaque and diffuse surfaces, the transmissivity τ is zero.

This model does not consider the pre-heating of the

solar cell at hotplates 1 and 2. Instead, the experimentally measured maximum temperature of the solar cell before the start of IR heating is used as the initial temperature, T_0, for the entire solar cell as it enters the IR heating zone. The solar cell is exposed to two radiation pulses in the IR heating zone. During the first radiation pulse, the solar cell is heated by the radiation from IR emitters 1 and 2 as it enters the IR heating zone. Subsequently, the solar cell moves under IR emitters 2, 3, and 4, where the second radiation pulse heats them. The heating from hotplate 3 is modeled using a convective boundary condition.

3 RESULTS AND DISCUSSION

The FEM model is simulated with process parameter P_{IR}= 60% for all four IR emitters and the temperature of hotplate 3, T_{HP3} = 160 °C. For validation of the simulation with experimental values, the temperature of M6 PERC half-cells was measured using thermocouples attached to the down-holder in the stringer. In this case, the initial solar cell temperature T_0 before IR heating is recorded as 121 °C. From 0 s to 1.2 s, the first radiation pulse is emitted from IR emitters 1 and 2, elevating the solar cell temperature to 176 °C. Over the next 0.8 s, the solar cell transitions to the next position inside the stringer. Between 2 s and 3.2 s, the second radiation pulse from IR emitters 2, 3 and 4 increases the temperature well above the melting liquidus temperature of the Sn60Pb40 solder (190 °C), reaching a maximum of 230 °C. After this pulse, the temperature slightly increases to a maximum of 234 °C at 3.8 s before starting to decrease within the cooling phase. Figure 2 compares the solar cell temperatures obtained from the FEM model with the measured values at position x = 36 mm / y = 71 mm in the main heating zone. The comparison shows that the FEM model accurately predicts the temperature profile during the IR soldering process.

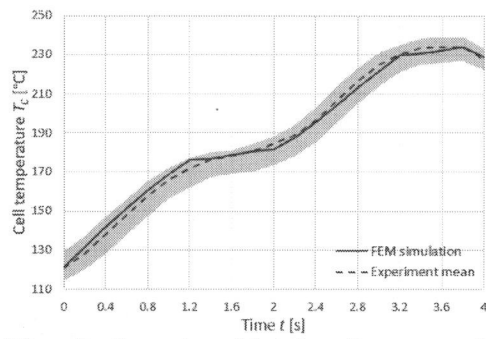

Figure 2 : Comparison of the solar cell temperature from simulation (red) and experiment (gray) during IR soldering of PERC M6 half-cells The gray area represents min./max. statistical error of N = 5 repetitions, while the dashed lines indicate the mean of the measured values.

At 3.8 s, the maximum solar cell temperature is reached. To assess temperature homogeneity across the entire solar cell, the FEM model is compared with measurements from thermocouples placed at three different positions on the solar cells. The results are presented in Figure 3. Similarly, the comparison of the measured temperatures using three thermocouples at positions x1 = 8 mm / y1 = 42 mm, x2 = 36 mm / y2 = 71 mm and x3 = 66 mm / y3 = 42 mm on the solar cells is displayed in Table III.

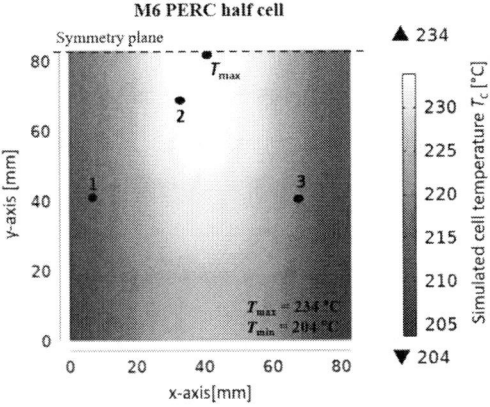

Figure 3 : a) Simulated temperature distribution on the solar cells at t = 3.8 s for different cases of process parameters measured at thermocouples 1,2 and 3.

Table III : Comparison of measured and simulated solar cell temperature at the three thermocouple positions.

Thermo-couple No.	Measured temperature [°C]	Simulated temperature [°C]	ΔT_c [K]
1	203	208	5
2	234	232	2
3	218	210	8

From the table, we can see the FEM model is able to compute the inhomogeneous temperature of the solar cell accurately, with the maximum temperature difference being (8 ± 4) K. The solar cell temperature T_C is around 30 K lower at the edges of the solar cells compared to the middle. The dominant reason for this effect is that the intensity of the radiation reduces radially as the distance from the center of the solar cell increases. Another reason is the shading and reflection of the down-holder design, as it further blocks the incident IR radiation at the edges of the solar cell. This combined effect results in higher temperature at the middle of the solar cells and relatively lower temperature at the edges of the solar cell.

4 CONCLUSION

This research introduces a novel FEM model that determines the inhomogeneous temperature distribution of solar cells during the industrial IR soldering process. The model also accounts for key factors, including the shading from the down-holder and the reflection of radiation from the IR unit reflector. Both of these elements have a significant impact on the maximum temperature reached by the solar cell during IR radiation. The model was validated by measuring the temperature at three different positions on the solar cells during the IR soldering process using thermocouples with PERC half-cells. The maximum temperature differences between the experiment and the FEM model were observed to be less than (8 ± 4) K. This FEM model can be used to determine temperature inhomogeneity during the IR soldering process for other process parameters. The model can also be easily adapted to other solar cell types, such as TopCon and silicon heterojunction (SHJ). Thus, this FEM model emerges as a reliable tool for accurately predicting temperature distribution during any industrial IR soldering process.

5 ACKNOWLEDGEMENT

The authors would like to thank the German Federal Ministry for Economic Affairs and Climate Action for the financial support within the projects "MoQa" (Grant number 03EE1140B) and "Quelle" (Grant number 03EE1172E).

6 REFERENCES

[1] A. J. Beinert, P. Romer, M. Heinrich, J. Aktaa, and H. Neuhaus, "Thermomechanical design rules for photovoltaic modules," *Progress in Photovoltaics*, vol. 31, no. 12, pp. 1181–1193, 2023, doi: 10.1002/pip.3624.

[2] M. Hertl, D. Weidmann, and J.-C. Lecomte, *Microelectronics and Packaging Conference, 2009. EMPC 2009. European*, 2009.

[3] T. Dullweber and L. Tous, *Silicon Solar Cell Metallization and Module Technology*: Institution of Engineering and Technology, 2021.

[4] J. R. Howell, M. P. Menguc, and R. Siegel, *Thermal Radiation Heat Transfer*: CRC Press, 2010.

[5] C. L. Wyatt, "Blackbody Radiation," in *Radiometric Calibration: Theory and Methods*: Elsevier, 1978, pp. 29–37.

[6] B. E. Yoldas and T. O'Keefe, "Deposition of optically transparent IR reflective coatings on glass," *Applied optics*, vol. 23, no. 20, p. 3638, 1984, doi: 10.1364/AO.23.003638.

[7] J. P. Joule, "XXXVIII. On the heat evolved by metallic conductors of electricity, and in the cells of a battery during electrolysis," *The London, Edinburgh, and Dublin Philosophical Magazine and Journal of Science*, vol. 19, no. 124, pp. 260–277, 1841, doi: 10.1080/14786444108650416.

[8] A. Jamnia, *Practical guide to the packaging of electronics: Thermal and mechanical design and analysis,* 2nd ed. Boca Raton: CRC Press, 2009.

[9] R. Strauss, *SMT Soldering Handbook*. Jordan Hill: Elsevier Science & Technology Books, 20.

[10] M. Zhao, Z. Zhou, M. Zhong, J. Tan, Y. Lian, and X. Liu, "Thermal shock behavior of fine grained W–Y 2 O 3 materials fabricated via two different manufacturing technologies," *Journal of Nuclear Materials*, vol. 470, pp. 236–243, 2016, doi: 10.1016/j.jnucmat.2015.12.042.

[11] F. Hu and S. Lucyszyn, "Modelling Miniature Incandescent Light Bulbs for Thermal Infrared 'THz Torch' Applications," *J Infrared Milli Terahz Waves*, vol. 36, no. 4, pp. 350–367, 2015, doi: 10.1007/s10762-014-0130-8.

[12] Y. S. Touloukian and D. P. DeWitt, *Thermophysical Properties of Matter - The TPRC Data Series. Volume 7. Thermal Radiative Properties - Metallic Elements and Alloys*. United States: New York : IFI/Plenum, 1970. Accessed: Jun. 11 2024.

41st European Photovoltaic Solar Energy Conference and Exhibition

IMPACT OF TEXTURED SURFACES AND CLEANING ON SOLAR PANEL GLASS TRANSMITTANCE

Aapo Poskela, Julianna Varjopuro, Tommi Jokikyyny, Aleksi Kamppinen, Heikki Palonen, Kati Miettunen
Department of Mechanical and Materials Engineering, Faculty of Technology, University of Turku
Vesilinnantie 5, 20500, Turku, Finland

ABSTRACT: In outdoor conditions, solar panels face several factors that prevent optimal operation through lowered absorbed light intensity. Luckily, steps can be taken to minimise their effect. In this work we focus on two of these steps: panel cleaning to reduce impact of soiling and textured glass surfaces to improve transmission of high incidence angle light. Removing soiling is relatively simple task, but how do the cleaning chemicals impact the anti-reflection coating on the surface of the solar panel glass? To test this, we apply a series of common cleaning chemicals to the surface of solar panel glass pieces and measure how the transmittance of solar panel glass samples changes. The results suggest that the glass coatings can be safely cleaned using most cleaning liquids, with the exception of dishwashing detergent. The second section of this study investigates how well textured solar panel glass surfaces can improve transmittance on high light incidence angles. Earlier studies suggest that increased haze and light-trapping of a textured glass surface can be used to guide more light to the solar cell when the light source is in a large angle from the panel surface normal, but our results demonstrate that flat glass surface outperforms textured surfaces on all incidence angles.
Keywords: Solar panel glass, Soiling, Cleaning, Textured glass, Transmittance

1 INTRODUCTION

Globally installed photovoltaic (PV) capacity has been constantly increasing in recent years, and is forecast to reach 12 % of overall electricity production by 2028. [1] With such a high share of electricity production dependent on solar energy, it is crucial that the performance of PV is as high and predictable as possible.

The efficiency of photovoltaic devices relies on incident light being absorbed by the active area of the device. Therefore, to reach high performance it is vital that optical losses in the front glass surface of the device are minimized. [2] The most common material for the light-facing surface of solar panels is glass. Glass has a very low absorption, but may have significant optical losses through reflection of incoming light. [2], [3] The optical losses of the glass are generally mitigated with the use of anti-reflective (AR) coatings, which are commonly claimed to cut reflective losses by half. [4], [5] However, this loss reduction is reported for a fresh AR coating, while solar panels installed outdoors are exposed to weather, soiling and, depending on system size and location, periodic cleaning. [6], [7] Therefore, it is important to study how the transmittance of solar panel glass is affected by the soiling and cleaning cycle.

There are several types of AR coatings, but typically solar panel glasses use single layer coatings due to the combination of large area of solar panels and increased manufacturing costs associated with double- and multi-layer coatings. [8] The anti-reflective property of an AR coating utilises destructive interference between light reflected from the air-coating and coating-glass interfaces. Optimal refractive index for a single layer AR coating is between the refractive indices of the surfaces above and below it. [9]

Another technique to alter the optical properties of the front glass is to modify its surface texture. The goal of a textured surface is to trap incident light, especially light with high incidence angle, and guide it to the solar cell. Textured surfaces also offer lowered glare which is beneficial for built environments. [10], [11] The increased transmittance for high incidence light is supported by optical modelling [10], but can these results be achieved in practice?

The goal of this work is to study the impact of cleaning and surface texture on the transmittance of solar panel glass from various incidence angles. Cleaning of AR coated solar panels has several prior studies, but most of these focus on mechanical robustness of the AR coating [5] and some consider the durability of the coating when exposed to acidic liquids. [12], [13] Instead, here we compare the effects of typical cleaning solutions on the AR coating without applying mechanical abrasion. We also test the cleaning methods on soiled samples. In the second part of this study, transmittance of textured solar panel glass samples provided by several glass manufacturers are compared by measuring their transmittance with several incidence angles and comparing them to a flat solar panel glass and a regular window glass.

2 METHODS

2.1 Samples

The solar panel glass samples used for the cleaning experiments were gathered from an unused silicon solar panel. The tempered glass had to be shattered to detach pieces from the ethyl vinyl acetate (EVA) laminated panel. To simplify handling, largest remaining glass shards were chosen for the study. The glass pieces were cleaned by immersing them in a cleaning solution for 20 hours. The used cleaning fluids were: ethanol, acetone, isopropanol, solar panel detergent (OneSystem Solar-Reiniger, Preimess), glass cleaner (Lasi & peili, Mellerud), and dishwashing detergent (Kiilto Asteri, KiiltoClean Oy).

Similar cleaning test is also performed after first soiling the samples by growing algae on them. The algae growth on the samples was cultivated with a mixture of plant nutrient, sugar, and moss which was spread on the glass pieces and left to grow for three days in ambient room temperature.

Latter part of this study focuses on the light transmittance capabilities of textured solar panel glasses. The samples were obtained as free samples from various solar glass manufacturers and they represented a variety of surface patterns. The textured samples were named S1 through S6: S1 was a traditional smooth and flat glass, S2 had a hexagonal honeycomb surface texture, S3 had hexagonally arranged and very pronounced circular pits, S4 was obscured glass with random texture, and S5 and S6 had hexagonally arranged shallow pits (Figure 1). The type of AR coatings on the textured glasses are unknown.

41st European Photovoltaic Solar Energy Conference and Exhibition

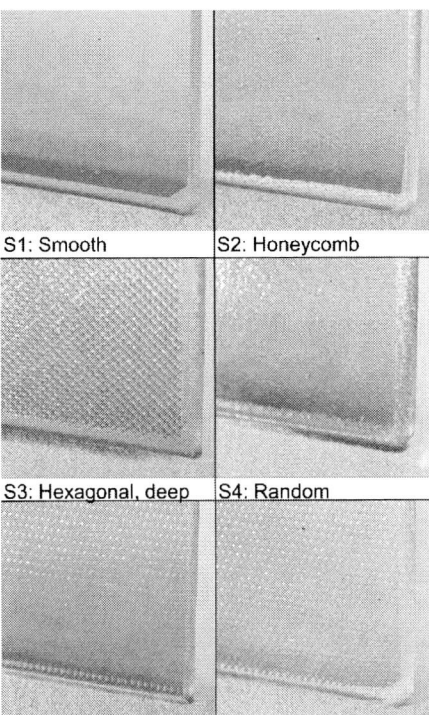

Figure 1 Textured solar panel glass samples. S1 has a flat and smooth surface.

2.2 Measurements

The transmittance of solar panel glass pieces in the cleaning test was measured with a UV-vis Spectrophotometer (Specord 200 plus) with a wavelength range 190 to 1100 nm, 2 nm measurement interval and 0.4 s integration time. Switch from deuterium lamp to halogen lamp occurred at 320 nm.

The angle dependent transmittance of the textured solar panel glass samples was measured with an in-house built measurement device. The device utilizes an integrating sphere and variable angle sample holder to measure transmittance at multiple incidence angles. It is fitted with a CCD, 200-1000nm, Thorlabs CCS200/M spectrometer and a Thorlabs SLS201L/M Stabilized Tungsten-Halogen Light Source. [14] The device was also used to measure the haze of the textured glasses. Transmission haze is used to quantify how much light that passes through an object is scattered. It is obtained as the ratio of scattered transmitted light (i.e. light that is not on the path of incident light beam) and total transmitted light intensities. The haze measurement and calculation procedure followed ISO 14782 standard. [15] Haze is calculated as:

$$\text{haze} = \frac{\tau_4}{\tau_1} - \tau_3\frac{\tau_2}{\tau_1^2}, \quad (1)$$

where τ_1 is the intensity of incident light, τ_2 is the intensity of light transmitted by the sample, τ_3 is the intensity of light scattered by the instrument and τ_4 is the intensity of the light scattered by the instrument and the sample.

3 RESULTS AND DISCUSSION

3.1 Transmittance of cleaned solar panel glass

The first cleaning test was conducted on unsoiled glass samples. Their transmission spectra after they were cleaned, rinsed and dried are shown in Figure 2. With the exception of dishwashing detergent, all of the cleaning chemicals demonstrated an equal increase in the transmittance of the glasses. Notably, solar panel detergent had at best only marginally higher transmittance compared to more generic cleaning solvents. Dishwashing detergent lowered the transmittance by roughly 1 percentage point from the transmission maximum at 520 nm.

The cleaning of soiled samples reached similar results as the cleaning of untreated glass samples. There were differences between algae growth on the samples, which led to significant differences in transmittance between soiled samples (Figure 3). Cleaning removed these differences and recovered the initial transmittance in all cases except dishwashing detergent. The transmittance peak of the sample cleaned with dishwashing detergent was almost 4 percentage points lower than the other cleaned samples.

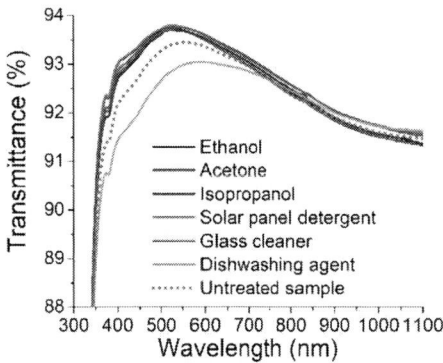

Figure 2 Transmittance of unsoiled solar panel glass samples after cleaning with various chemicals. Modified from [16].

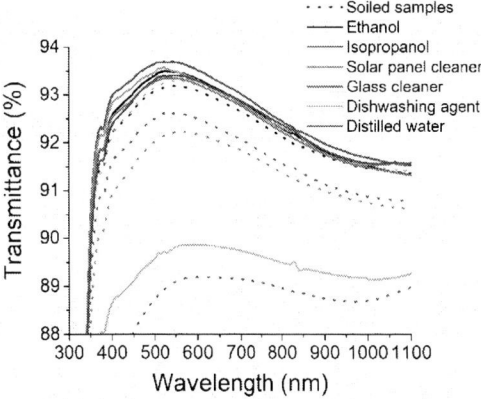

Figure 3 Transmittance of soiled solar panel glass samples before and after cleaning. Dashed lines denote the soiled samples and their colors match with the cleaning chemical that it is later cleaned with. Modified from [16].

The poor performance of the dishwashing detergent is not necessarily caused by damaging of the AR coating. The dishwashing detergent used in this study had a pH of

5, and antireflective coating should generally withstand acidity up to pH of 3. [12] One possibility is that the detergent stained the glass and rinsing with distilled water was insufficient to wash away the residues of the detergent. If that is the case, the AR coating would not be permanently damaged, but based on these results it is still ill-advised to use dishwashing detergent for solar panel cleaning.

The surfaces of the cleaned glass samples were analysed with AFM (Figure 4). The AR coating is still present on the front surface of the glass (Figure 4a), as evidenced by its smoother surface when compared to the rougher bare glass on the back surface (Figure 4b). Additionally, the AFM image also reveals two scratches in the AR coating but they are likely results of earlier rough handling rather than from the cleaning process itself.

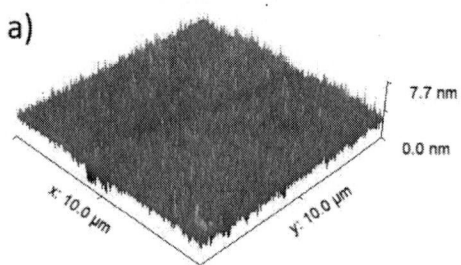

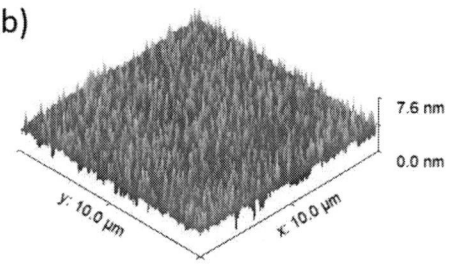

Figure 4 AFM images of the a) AR coated and b) back side of solar panel glass. The valleys crossing a) are scratches in the AR coating.

3.2 Textured solar panel glasses

Transmittance of textured solar panel glasses as function of incidence angle is presented in Figure 5. The results are somewhat surprising since the smooth solar panel glass transmits more light on all incidence angles compared to the textured ones. The initial hypothesis was that the increased haze of textured glasses would translate to improved light transmission on high incidence angles. [10] Best results were achieved with the smooth glass, and close behind were samples S2 and S6 (refer to Figure 1 for their surface texture). Worst performing sample was S3, for which transmittance quickly decreased as incidence angle increased. Generally, plain soda lime glass had lower transmittance than the solar panel glasses, likely due to a lack of AR coating and higher absorption. The excellent performance of the smooth glass on high incidence angles could likewise be explained by it having a higher quality AR coating than the textured glasses, or the texturing would need to be a more optimised structure to outperform the flat glass surface.

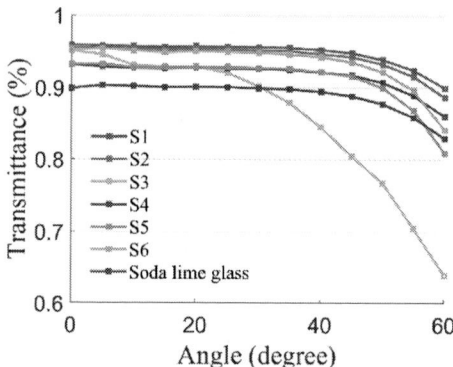

Figure 5 Angle-dependence of solar spectrum weighted average transmittance of textured solar panel glass samples. Soda lime glass sample is ordinary window glass.

Transmission haze of the glass samples can be used to estimate how scattered the light transmitted through a sample is. Haze measurement results show (Figure 6) that sample S3 produces an exceptionally high haze of roughly 70 %, but it fails to transmit light well (Figure 5) on high incidence angles. The hexagonal patterns of S5 and S6 also generate some haze, 10 % and 20 % respectively, while rest of the samples have below 5 %. Expectedly, the smooth glass samples S1 and Soda lime glass had the lowest haze.

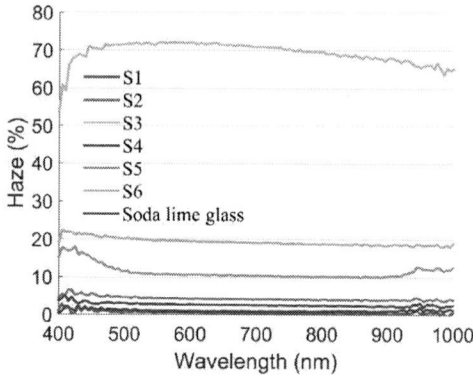

Figure 6 Transmittance haze of the textured glass samples. Soda lime glass is ordinary window glass.

When comparing the haze and high incidence angle transmittance results, it should be noted that there is no significant positive correlation between high haze and high angle transmittance. In fact, the samples which had the highest haze suffer the fastest decrease in their transmittance as incidence angle is increased. This effect might be due to the hexagonal structure of the textured surface, where the elevated parts of the pattern fail to capture incident light while simultaneously shading the valleys.

One aspect to consider is the experimental setup used here: in the transmittance measurements the glass samples were not attached to anything, while in real solar panels the glasses are laminated (usually with EVA) to the solar cells and backsheet. Therefore, the back surface of samples in these measurements were glass-air interfaces instead of glass-EVA interfaces. The refractive index of EVA is very close to the refractive index of glass [17], which leads to reduced reflection especially on higher incidence angles due to a larger critical angle. Since light is more scattered within the textured glass samples, their transmittance benefits more from matching refractive indices.

4 CONCLUSIONS

This work analysed how transmittance of solar panel front glass is impacted by cleaning it with various cleaning liquids, and how surface glass texture affects the transmittance. The cleaning study revealed that typical solar panel AR coating is very robust against most cleaning liquids, with the exception of regular dishwashing detergent. It is possible that the dishwashing detergent merely stained the glass instead of damaging the AR coating, but nevertheless it is advisable to refrain from using it for cleaning solar panels. This avenue of study could be furthered by testing the method used here for different types of soiling, to quantify both potential damage to the AR coating as well as most effective cleaning result.

The comparison of textured glass transmittance revealed that, despite predictions, here flat glass is the best option at all incidence angles. Textured glasses successfully reduce glare, demonstrated by increased haze, but fail to trap light more efficiently at high incidence angles. Since theoretical results suggest that it should be possible to construct a textured surface with improved high incidence angle light trapping, an interesting follow-up study would be to produce a glass surface that imitates the simulated optimal surface geometry as closely as possible.

5 ACKNOWLEDGEMENTS

AP wishes to thank the Research Council of Finland project ECOSOL (project number 347275). A.K. thanks University of Turku Graduate School (UTUGS), KM wishes to thank Research Council of Finland project BioEST (318557, 336577, 320100, 336441)

6 REFERENCES

[1] "Renewables 2023 – Analysis - IEA." Accessed: Jan. 30, 2024. [Online]. Available: https://www.iea.org/reports/renewables-2023

[2] A. S. Sarkın, N. Ekren, and Ş. Sağlam, "A review of anti-reflection and self-cleaning coatings on photovoltaic panels," *Solar Energy*, vol. 199, no. June 2019, pp. 63–73, Mar. 2020, doi: 10.1016/j.solener.2020.01.084.

[3] C. Ji *et al.*, "Recent Applications of Antireflection Coatings in Solar Cells," *Photonics*, vol. 9, no. 12, p. 906, Nov. 2022, doi: 10.3390/photonics9120906.

[4] S. Bashir Khan, H. Wu, C. Pan, and Z. Zhang, "A Mini Review: Antireflective Coatings Processing Techniques, Applications and Future Perspective," *Research & Reviews: Journal of Material Sciences*, vol. 05, no. 06, pp. 36–54, 2017, doi: 10.4172/2321-6212.1000192.

[5] A. M. Law, L. O. Jones, and J. M. Walls, "The performance and durability of Anti-reflection coatings for solar module cover glass – a review,"

Solar Energy, vol. 261, pp. 85–95, Sep. 2023, doi: 10.1016/j.solener.2023.06.009.

[6] T. Sarver, A. Al-Qaraghuli, and L. L. Kazmerski, "A comprehensive review of the impact of dust on the use of solar energy: History, investigations, results, literature, and mitigation approaches," *Renewable and Sustainable Energy Reviews*, vol. 22, pp. 698–733, 2013, doi: 10.1016/j.rser.2012.12.065.

[7] H. P. Garg, "Effect of dirt on transparent covers in flat-plate solar energy collectors," *Solar Energy*, vol. 15, no. 4, pp. 299–302, Apr. 1974, doi: 10.1016/0038-092X(74)90019-X.

[8] C. Ballif, J. Dicker, D. Borchert, and T. Hofmann, "Solar glass with industrial porous SiO2 antireflection coating: measurements of photovoltaic module properties improvement and modelling of yearly energy yield gain," *Solar Energy Materials and Solar Cells*, vol. 82, no. 3, pp. 331–344, May 2004, doi: 10.1016/j.solmat.2003.12.004.

[9] H. K. Raut, V. A. Ganesh, A. S. Nair, and S. Ramakrishna, "Anti-reflective coatings: A critical, in-depth review," *Energy Environ Sci*, vol. 4, no. 10, p. 3779, Sep. 2011, doi: 10.1039/c1ee01297e.

[10] Z. Zhou, Y. Jiang, N. Ekins-Daukes, M. Keevers, and M. A. Green, "Optical and Thermal Emission Benefits of Differently Textured Glass for Photovoltaic Modules," *IEEE J Photovolt*, vol. 11, no. 1, pp. 131–137, Jan. 2021, doi: 10.1109/JPHOTOV.2020.3033390.

[11] "Why to choose textured glass for covering on photovoltaic modules? - Fortemp Technology International Ltd." Accessed: Jan. 29, 2024. [Online]. Available: https://www.fortemp.com/news/why-to-choose-textured-glass-for-covering-on-photovoltaic-modules

[12] X. Wang and J. Shen, "Sol–gel derived durable antireflective coating for solar glass," *J Solgel Sci Technol*, vol. 53, no. 2, pp. 322–327, Feb. 2010, doi: 10.1007/s10971-009-2095-y.

[13] G. Womack, K. Isbilir, F. Lisco, G. Durand, A. Taylor, and J. M. Walls, "The performance and durability of single-layer sol-gel anti-reflection coatings applied to solar module cover glass," *Surf Coat Technol*, vol. 358, pp. 76–83, Jan. 2019, doi: 10.1016/j.surfcoat.2018.11.030.

[14] Tommi Harju, "Automated Transmittance Analyzer," Bachelor's Thesis, University of Turku, Turku, 2023.

[15] International Organisation for Standardization, "Plastics-Determination of haze for transparent materials," 1999. [Online]. Available: https://standards.iteh.ai/catalog/standards/sist/95147130-bf54-4660-a622-

[16] J. Virjonen, "The effect of cleaning on the optical properties of anti-reflection coating," Bachelor's Thesis, University of Turku, Turku, 2023.

[17] M. R. Vogt *et al.*, "Optical Constants of UV Transparent EVA and the Impact on the PV Module Output Power under Realistic Irradiation," *Energy Procedia*, vol. 92, pp. 523–530, Aug. 2016, doi: 10.1016/J.EGYPRO.2016.07.136.

41st European Photovoltaic Solar Energy Conference and Exhibition

Ultra-thin flexible glass as environmental shield for CIGS photovoltaic modules

N. Pervan[1,2], S. Feldbacher[1], M. Harnisch[3], T. Tettenborn[3], A. Zimmermann[3], G. Oreski[1,2]

[1] Polymer Competence Center Leoben GmbH (PCCL), Sauraugasse 1, 8700 Leoben, Austria – nikolina.pervan@pccl.at
[2] Chair of Materials Science and Testing of Polymers, Montanuniversität Leoben, Otto Glöckel-Straße 2/II, 8700 Leoben, Austria
[3] Sunplugged GmbH, Affenhausen 1, 6413 Wildermieming, Austria

INTRODUCTION

➤ With the highest conversion efficiency, amongst the thin-film photovoltaic (PV) materials [1] and a possibility of deposition on flexible substrates, copper indium gallium (di)selenide (CIGS) solar cells are promising candidates for light-weight PV modules in building integrated (BIPV) and vehicle integrated (VIPV) industry.

➤ CIGS solar cells are prone to degradation under environmental conditions, they are especially sensitive to the humidity [2]. Therefore, functional bill of material (BoM), which will provide protection against the environmental stressors, is a key factor to enable long service life-time of the CIGS modules.

➤ Roll-to-roll production enables continuous industrial processes in CIGS module production. Selected BoM should fit in roll to roll process.

➤ Polymers tested within "SOPHOKLES" project have shown insufficient barrier properties towards humidity to protect CIGS. To maintain module's flexibility and thickness while providing insulation from environmental stressors, a new approach is needed compared to the one presented before [3].

➤ Ultra thin glass (UTG) has already been researched as substrate material [4], but in this work UTG will have a role of a frontsheet in the PV module.

➤ Various material combinations were tested through this work, always having as the main component CIGS solar cell deposited on the thin stainless-steel substrate. PV modules were tested for their functionality before and after encapsulation, after storage conditions and during the exposure to the damp heat conditions.

EXPERIMENTAL

MATERIALS
- Ultra Thin Glass (145 µm), Polymer frontsheet
- Casting type: polyurethane (PU), epoxy, polyolefin (POE)
- Foil type: thermoplastic polyolefin (TPO), coloured TPO
- Interconnected CIGS deposited on stainless steel

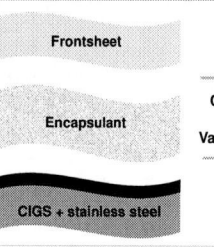

Frontsheet

Encapsulant

CIGS + stainless steel

Casting method + Vacuum lamination

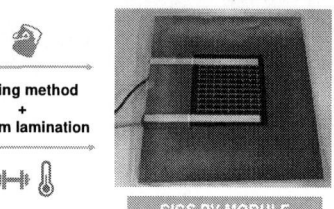

CIGS PV MODULE

RELIABILITY
- IV-curve characterisation (efficiency (η), fill factor (FF))
- Damp heat (DH) exposure @ 85 °C and 55% humidity
- Long-term storing @ 21 °C and 55% humidity

RESULTS and DISCUSSION

PU casting polymer had great adhesion to the steel. Low adhesion to the UTG at 85 °C led to delamination – PV modules lost 14% of their efficiency after 1 h of DH exposure.	Strong shrinkage of the epoxy system during curing/crosslinking led to warpage of stainless steel substrate – failure of PV modules.	Liquid POE system cured in too thin layer. PV modules had voids due to the profile height from the interconnection. Method rejected for safety from glass breaking.	Efficiency of PV module laminates, with commercial polymeric frontsheet and TPO encapsulant, dropped significantly after only a few hours in the DH.	Lamination with TPO encapsulant and UTG gave the best results and performed well under >200 h of DH exposure. Coloured TPO resulted in satisfactory efficiency values.
✗	✗	✗	✗	✓

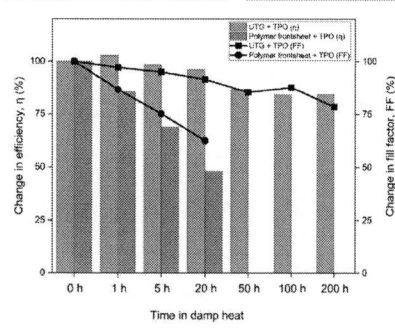

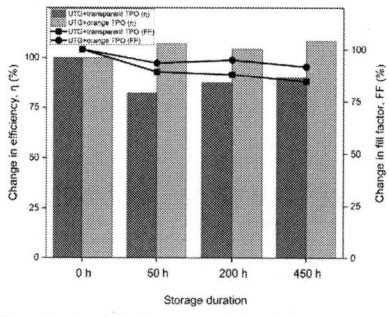

OBSERVATIONS:
➤ UTG fits to the standard lamination process
➤ No glass breakage during lamination
➤ Good adhesion to the encapsulant
➤ No warpage or delamination after lamination

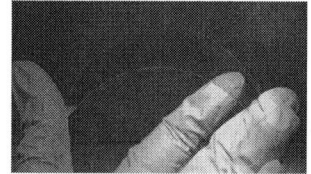

Figure 1. Efficiency and FF values change for TPO PV modules with UTG and polymeric frontsheet during DH exposure.

Figure 2. Change in efficiency and FF values for transparent and coloured TPO PV modules during storing conditions.

Figure 3. Ultra thin glass sheet under curvature.

!!! BE AWARE !!!
➤ Halogen lamp was used for the illumination of modules with intensity of 70 – 75 000 lux, efficiency and fill factor values have not reached full potential, therefore only change in the values during storage conditions and damp heat aging are shown.

CONCLUSIONS and OUTLOOK

Casting method performed poorly – mismatch in material properties.

Polymeric frontsheet showed insufficient barrier to the moisture, after 1h in the DH η dropped by 14%.

Casting method – new materials. Edge sealant addition to extend reliability.

PV module laminated with TPO+UTG combination performed well under damp heat conditions.

Coloured encapsulants gave aesthetic finish, while maintaining good performance of the PV module.

REFERENCES

1. https://www.pv-magazine.com/2022/10/11/swiss-scientists-achieve-22-2-efficiency-for-flexible-cigs-solar-cell/
2. Kessler, F., Rudmann, D., Technological aspects of flexible CIGS solar cells and modules, Solar Energy, 2004, Vol. 77, 685-695
3. Feldbacher, S., Pervan, N., Harnisch, M., Tettenborn, T., Zimmermann, A., Oreski, G., Flexible transparent polymeric front encapsulation as finish layer for CIGS PV cells using additive manufacturing, EU PVSEC, 2023
4. Ramanujam, J., et al., Flexible CIGS, CdTe and a-Si:H based thin film solar cells: A review, Progress in Materials Science, 2020, Vol. 110, 100619

ACKNOWLEDGMENT

This work was conducted as part of the project "SOPHOKLES", which is supported under the umbrella of "Stadt der Zukunft" funded by Austrian Research Promotion Agency (FFG, application number: 43693964).

41st European Photovoltaic Solar Energy Conference and Exhibition

PROCESS DEVELOPMENT AND MATERIAL EVALUATION OF PHOTOVOLTAIC ALUMINUM FACADE ELEMENT FOR BIPV APPLICATION

R. Koepge, M. Pander, S. Großer, B. Jaeckel
Fraunhofer Center for Silicon Photovoltaics CSP
Otto-Eissfeldt-Strasse 12, 06120 Halle, Germany

ringo.koepge@csp.fraunhofer.de

ABSTRACT: The integration of photovoltaic solutions directly into building facades is a key strategy to achieve the European goal of climate neutrality by 2050. The widespread adoption of building-integrated photovoltaics (BIPV) can increase energy security, diversify the energy mix and create new green jobs. By seamlessly integrating renewable energy generation into our cities, BIPV will become a cornerstone of a sustainable future for Europe. It offers aesthetic and functional benefits, such as providing shade, reducing heat gain and improving the overall performance of the building envelope. To achieve these goals, the building and construction sector needs to be supported in integrating photovoltaic technologies into its production and value chains. The initial results of the AluPV project show how PV production processes and material combinations can be adapted to realize PV-integrated aluminum facades as building-integrated photovoltaic elements. A main objective is to adapt the manufacturing process to be suitable for newly designed photovoltaic modules, both in terms of electrical adaptation due to different dimensions and architectural constraints and in terms of mechanical structures due to direct integration into Al facade elements. Therefore, suitable material combinations and process parameters were investigated, focusing on the reduction of mechanical stress and the use of cost-effective standard PV production equipment. Delamination and module bending caused by different thermal expansions during production could be avoided by intelligent pressure distribution. In this article, we will successfully demonstrate our findings using commercially available aluminum facade elements as backside for the integration of a solar cell string. It could be shown that standard industrial laminating equipment is suitable for cost-efficient production and fast implementation of the prototype into the production line. The use of lead-free electrically conductive adhesives (ECA) offered advantages in terms of sustainability and stability for the future use of highly efficient solar cell types. Electroluminescence (EL) images confirm that the solar cell strings have no cell fractures and magnetic field imaging (MFI) proof that the facade module is free of electrical defects. In summary, numerous challenges related to manufacturing and material selection were overcome and the prototype showed promising results. The research results contribute to the further development of building-integrated photovoltaics and show the potential for cost-efficient photovoltaic systems.

Keywords: BIPV, Facade, Magnetic Field Imaging, Electroluminescence, Solar Module, Module Prototype, Manufacturing, ECA, Lamination, Processing

1 INTRODUCTION

The EU has set itself the binding target of achieving climate neutrality by 2050 - emissions are to be reduced by at least 55% by 2030 [1]. Globally, it is important to harmonize development, economic, financial, energy and transport policies with climate protection targets. Energy generation from photovoltaic systems plays a decisive role in achieving these goals. A current draft by the European Commission envisages that 90% of electricity in the EU should be generated from renewable energies - primarily solar and wind energy - by 2040, supplemented by nuclear energy. The BIPV market is still in its infancy and the potential is currently barely being exploited [2]. One challenge is the reliable and form-fitting integration into existing architectural elements. One aspect is the development of BIPV facade elements. The evaluation of suitable material combinations and economic manufacturing process parameters plays an important role in the implementation of prototypes for industrial series production.

As part of the AluPV project, PV solar cell strings were integrated into commercially available aluminum facade elements. Suitable material combinations and encapsulation processes needs to be investigated. In addition, we try to implement an electrically conductive adhesive (ECA) layer due to the growing demand for

sustainable, lead-free products [3,4].

The current investigations deal with four parts of the scientific research. The first part is the manufacturing process. The production of PV-integrated aluminum facades places high demands on the material composite. Due to the different coefficients of thermal expansion, internal stresses are introduced into the facade solar module. This can lead to cracks in the cell connector joints, in the glass and to delamination of the encapsulation. A manufacturing process is presented that demonstrates a defect-free aluminum facade element for BIPV application. Delamination problems as well as module bending caused by different thermal expansions during manufacturing are not detected. Intelligent pressure distribution during manufacturing is the key to defect-free production. The second part of the study aims to quantify material adhesion after lamination. Different aluminum surfaces, such as bare or painted, were investigated. Finally, the third and fourth parts include electroluminescence (EL) and magnetic field imaging (MFI), which provide evidence of properly functioning solar cell strings without cell breakage [4]. In summary, it can be said that the prototype has overcome numerous challenges in the manufacturing process.

2 MATERIALS & METHODS

This Chapter contains the Information about the used samples and methods that were applied in this study. It is subdivided into a material part, a manufacturing part and characterization part.

2.1 Material

The aluminum facade element has a size of 1500 mm x 900 mm and a thickness of 1 mm. It has a structural design with two plateaus for the PV application. The facade is coated with PVDF (see Figure 1) and 15 amm holes have been punched for subsequent connection to the junction box. A foil is covering the surface for protection and is released before lamination. The EVA encapsulation material used corresponds to the current state of the art for encapsulating solar cell strings. Standard PERC solar cells in M6 format with 9bb ECA are connected. Either glass fiber fabric or modified polyester is used for electrical insulation to the aluminum, see Figure 2 and Figure 3.

Figure 1 Aluminum facade before lamination process, a cover polymer foil save the surface for any contamination before material layup.

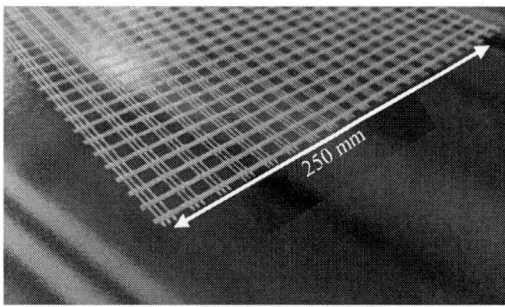

Figure 2 Glass fiber mesh used as insulation layer

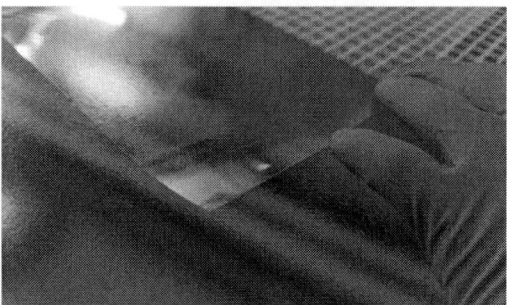

Figure 3 Modified polyester foil used as insulation layer

2.2 Manufacturing

An important part for reliable integration is checking the adhesion between the aluminum facade element and the encapsulant to ensure a reliable connection. The adhesion forces were measured by standard 90° peel tests. The following Figure 4 shows the material stack that was used for peel testing. A standard encapsulation material made of EVA (ethylene-vinyl acetate) and a standard backsheet were laminated onto the sheets to prevent tearing of the EVA layer during the peel test. Five batches of different aluminum materials were produced using standard lamination processes, as applied in the standard solar module manufacturing process. Three batches contain bare aluminum plates (E6EV1, DB703, B40H36), and two plates were varnished (RAL9006, Purple Green).

PET Backsheet	0.27 mm
Encapsulant	0.46 mm
Aluminum plate	1.00 mm

Figure 4 Material stack for peel testing in part one of investigation

The fabrication of lead-free interconnected solar cell strings was performed on a Team Technik Stringer TT Lab i8 ECA. The stringer was set up for M6 full-cell size, with an 18 mm bus bar distance. The cell connection was established using an epoxy-based electrically conductive adhesive. The cross-connection was done manually with a soldering iron at a solder temperature of approximately 300 °C.

The layup of all materials was also done manually. The layer stack consists of the aluminum facade as backsheet material, encapsulant, an additional insulation layer, encapsulant, the solar cell string, encapsulant, and front glass. Figure 5 shows the material stack.

Glass	3.00 mm
Encapsulant	0.46 mm
Cells & EVA Connection	0.20 mm
Encapsulant	0.46 mm
Additional Insulation Layer	0.20 mm
Encapsulant	0.46 mm
Aluminum Facade	1.00 mm

Figure 5 Material stack for process development in part two of investigation

Finally, the material stack was processed by standard fully automated lamination process. The process regime can be seen in the following Table 1.

Table 1 Sample overview

Part	Batch	Quantity
1	Peeling	5 Samples
2	Fabrication	3 Facade-Modules

2.3 Characterization

Peel Testing

In order to determine the adhesion strength between aluminum and the EVA film several strips per specimen were prepared. For each test specimen, ten strips with a width of approximately 15 mm and a length of 100 mm were peeled off. The peeling direction was at 90° to the bonding surface between aluminum and EVA. A testing speed of 51 mm/min was chosen. For comparison of the adhesion strengths, the average peel force was calculated after the initiation of peeling in the range from 10 mm to 100 mm.

Electroluminescence (EL)

For process control, the EL recording was conducted using a CCD camera (Canon EOS 2000D) to ensure manufacturing checks and to verify that the electrical connection to the junction box works properly. The current and voltage were set to short-circuit current and open-circuit voltage. The main goal was to inspect the module, excluding failures in soldering and laying up the insulation layer.

Magnetic Field Imaging (MFI)

In general, the MFI method refers to measuring the magnetic field and its visualization. This method can be applied to both illuminated solar modules or cells, as well as biased modules/cells. Conclusions about the current flow within the module can be drawn; this method can reveal defects that may not be visible using EL. In the current investigation, the background was recorded at 0 A and measured with a current of 10 A. The resolution is around 2.5 mm in the x and y-direction.

3   RESULTS

The following chapters contains the results subdivided into 4 chapters. In the beginning the major important process regime parameter are named. Followed by the quantification of adhesive forces, MFI and EL characterization.

3.1 Lamination Process

The lamination of aluminum facade elements is still challenging. Differences in the thermal expansion coefficient and the geometric structure must be overcome. The expansion differences are a standard material behavior that causes internal shear stresses, resulting in module deformation or bending. The chosen facade structure with its 90-degree wave shape has a high area moment of inertia. In this case, module deformation was not observed, but there are still internal stresses within the laminate. The second issue is related to the pressure distribution during lamination. The waviness of the aluminum facade is the root cause of partial contact between the lamination membrane and the PV part of the facade; air inclusions and cavities result from this and are an additional challenge for the process optimization.

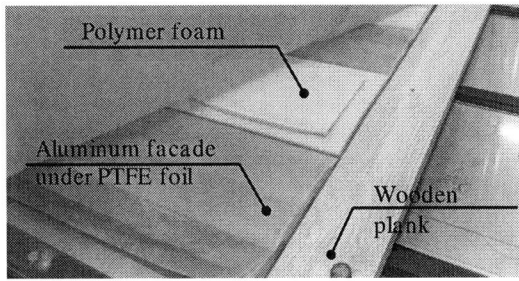

Figure 6 Facade layup for prototype lamination process

The solution was determined by prelamination tests, varying the pressure, melting time, and pressure distribution. Especially the pressure distribution could be

improved by using "filling bodies" such as polymer foam and wooden blanks, as shown in Figure 6. The optimized process regime results in a facade free of cavities, as seen in Figure 7. The microscopic image shows the laminate with a solar cell on the left side and the glass fiber mesh after lamination without any air inclusions.

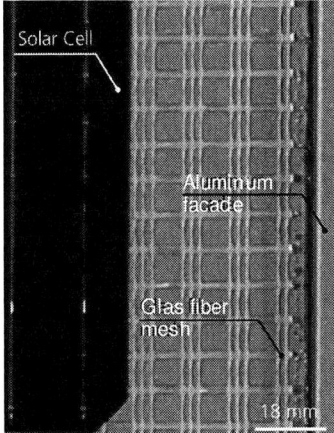

Figure 7: Photovoltaic functionalized aluminum facade after optimized lamination shows no cavities in light microscope image

3.2 Peel Testing Results

The quantification of the adhesion between the encapsulant and the aluminum was performed. An overview of the results is presented in the bar chart in Figure 8. Noteworthy are the coated sheets, RAL 9006 and purple green. Their adhesion strengths are 2 to 3 times higher than those of the three other sheets, which had a bare or anodized surface.

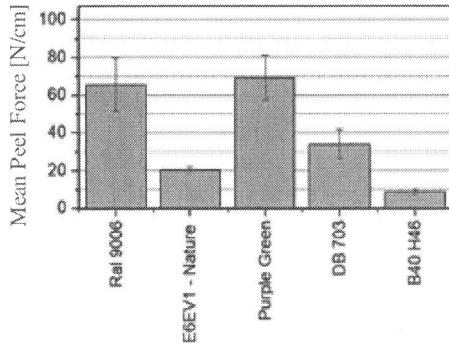

Figure 8: Mean Peel Forces [N/cm] for bare and varnished aluminum samples

In the case of the coated sheets, the backsheet was partially separated from the EVA. This means that the adhesion to the aluminum was better than the adhesion to the backsheet, indicating sufficient bonding between the coated aluminum and the encapsulation material.

3.3 Electroluminescence Imaging

Using a lead-free ECA-based interconnection technique instead of leaded solder could add a further benefit in terms of mechanical and electrical sustainability, which will be part of the reliability investigation in future work. Furthermore, ECA interconnections are suitable for upcoming high-efficiency solar cell types and ensure a stable long-term interconnection due to their potentially higher elastic elongation compared to a rigid soldered interconnection.

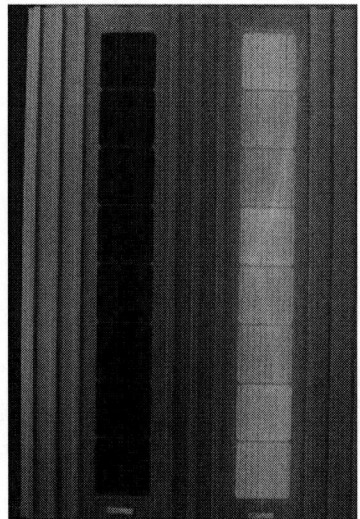

Figure 9: Photovoltaic functionalized aluminum facade prototype electrical function proofing by EL imaging

In this study, the EL image was used to check the manufacturing quality after the lamination process. One sample out of three, as shown in Figure 9, displays a facade module. One of the two strings is biased for EL imaging. Near infra-red emission as well as some additional external visible light (to see the aluminum structure) can be seen. Only the right solar cell string was biased and shows a properly non-defect crack-free cell string. This image is consistent with all the other cell strings that were under investigation. Thus, we concluded that the insulation layer prevents a short circuit that can be caused by the layup and lamination of the facade module. To ensure electrical safety of the façade element the insulation resistance must be tested according to IEC 61730. Comparison of the different insulation layers is part of further research.

3.3 Magnetic Field Imaging (MFI)

The method application on a string inside a aluminum facade was successfully tested on the front and rear side. Figure 10 shows the magnetic field images of the top and rear side of one string position at a facade module. The magnetic flux density B is a vector size and was measured in 3 directions, but only two are shown in the figure. The magnetic flux density in the x direction can be interpreted as the current flow pattern perpendicular to the string direction. These currents should be small for an intact string. For the y direction the magnetic flux density corresponds to the current flow along the cell connectors. The measurement shows the homogeneous current flow through the 9 cell connectors and proofed the intact interconnection after lamination. Due to the sensitivity of the MFI method for cell no. 5 an unexpected defect pattern on the glass-side image was localized, which is planned to be investigated. The method was used for initial characterization and will be applied to modules after accelerated aging tests.

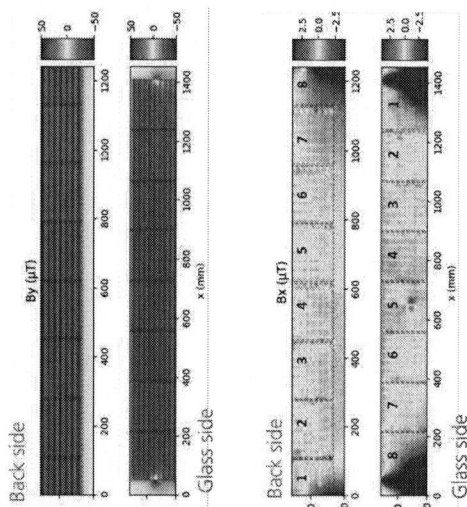

Figure 10 Single facade solar string tested with magnetic field imaging (I = 10 A).

4 SUMMARIZE

The successful direct lamination process of solar cell strings on aluminum facade elements was carried out in the prototype. It showed that there were no air inclusions, no deformations of the facade elements, no delamination, and no glass breakage down to a thickness of 3 mm. Future developments are planned to optimize the weight of the facade elements by reducing the glass thickness and to demonstrate the flexible, reliable connection of ECA cells through IEC testing that also will be characterized by MFI technology.

5 ACKNOWLEDGEMENTS

Financial support by the Federal Ministry for Economic Affairs and Climate Action funded project "AluPV" (FKZ: 03EN1069B) is gratefully acknowledged.

6 REFERENCES

[1] Document 32021R1119, Regulation (EU) 2021/1119 of the European Parliament and of the Council of 30 June 2021 establishing the framework for achieving climate neutrality and amending Regulations (EC) No 401/2009 and (EU) 2018/1999 ('European Climate Law'), http://data.europa.eu/eli/reg/2021/1119/oj

[2] International Technology Roadmap for Photovoltaics (ITRPV), 2023 Results, 15. Edition, May 2024

[3] Stephan Grosser, Matthias Schak, Tudor Timofte, Marko Turek, "Assessment of ECA to Ribbon Interconnection Stability by Current Path and Power Loss Imaging", Proceedings, WCPEC-8, 2022

[4] Dominik Lausch, Marcus Patzold, Maik Rudolph, Chia-Mei Lin, Jens Froebel, Kai Kaufmann, "Magnetic Field Imaging (MFI) of Solar Modules", Proceedings, 35th EU PVSEC, 2018

41st European Photovoltaic Solar Energy Conference and Exhibition

MATERIAL PROPERTIES REQUIREMENTS FOR FRAME SEALANTS AND JUNCTION BOX ADHESIVES

Guy Beaucarne[1], Emmanuel Jadot[1], Dominique Culot[1], Rono Cao[2], Kayla Kenney[3], Suraj Ahuja[4], Valérie Hayez[1]

[1]Dow Silicones Belgium SRL, Rue Jules Bordet Parc Industriel Zone C, B-7180 Seneffe, Belgium
[2]Dow (Shanghai) Holding Co., Ltd., Zhangjiang High-Tec Park, Pudong District, Shanghai, China
[3]Dow Silicones Corporation, 2200 W Salzburg Road, Auburn, MI 48611, USA
[4]Dow Chemicals International, Raheja District II, Navi Mumbai 400705, India

ABSTRACT: We investigate the properties of the most commonly used frame and junction box sealants in the industry and compare them with those of a reference PV sealant with a long track record. Lap shear and peel test studies show that those materials can sometimes show poor adhesion after water immersion aging. With H-bar tensile testing, we also observe variations in material properties depending on the batch, and we show that these sealants are not best suited to withstand complex stress situations. In the last part we discuss the impact of sealant properties on stresses in the glass at the edge of glass-glass laminates.

Keywords: silicone, sealant, adhesion, mechanical stress

1 INTRODUCTION

Frame sealants and junction box adhesives are often considered a less important part of the bill of materials of photovoltaic modules. However, in reality, they are critical for the long-term reliability of photovoltaic modules. Durability and reliability issues in the field are regularly traced back to the use of the wrong sealant or adhesive[1-3]. Unfortunately, such problems are rarely detected during module qualifications because the required IEC standards for qualification, IEC 61215 and IEC 61730, even when extended, are not well-suited to identify long-term durability issues.

In this work we review different aspects of frame sealants and junction box adhesives and indicate how the trends in the industry may come in conflict with trends in the type of sealants and adhesive selection over the last decade or so.

2 DOMINANT TYPES OF SEALANTS TODAY

In the beginning of the PV industry, different types of materials were used to attach and seal module frames and junction boxes. The use of double-sided tape decreased over the years as the industry recognized the superior durability of silicone sealants in the application. The type of silicone has also evolved over the years. One part moisture cure silicones have always been most popular because they are easy to apply and do not require heating or sophisticated mixing systems, as the curing of the material occurs through reaction with moisture in the ambient air and release of a leaving group. While several different chemistries were used in the past, the most common sealants nowadays are oxime sealants, which means that an oxime compound, usually methyl ethyl ketoxime (MEKO) is released upon curing. These sealants usually have a worse toxicology profile than alkoxy sealants, from which methanol or ethanol are released upon curing. The content of filler, which is calcium carbonate particles, has generally increased over the years. If a low-cost calcium carbonate is used, increasing the filler content is a way of reducing sealant manufacturing cost. As we shall see increasing the filler content makes the sealant much stiffer after cure.

3 EXPERIMENTAL

Several batches of the most commonly used frame and junction box sealants in the industry were acquired and tested following standard procedures. The results were compared to those of a reference alkoxy solar sealant, which was tested in an identical way at the same time. The reference alkoxy sealant was DOWSIL™ PV-804 Neutral Sealant, which has been used in the PV industry for a very long time.

The tests included tensile, lap shear, peel, H-bar tests, and also tests that are more specific to the frame sealing application.

3.1 Tensile tests

Tensile testing aims to determine some basic mechanical properties of the material, such as the tensile Young modulus and the tensile strength of the material. Thin sheets (about 2 mm thick) and dumbbell-shaped samples were cut out and then pulled until break in a mechanical test rig (Zwick). The standard followed is ASTM D412.

3.2 Lap shear tests

Lap shear testing measures the adhesive strength of materials by bonding two surfaces and pulling them apart. The shear force moves the substrates in opposite directions, testing the bond's quality. This method is commonly used to evaluate adhesives and sealants' performance, providing insights into their durability and reliability under stress.

The subtrates were anodized aluminum and the standard followed is ASTM D 1002. Some samples were tested after curing was completed, and other samples were additionally aged by water immersion for 7 days. Water immersion is a harsh test that is commonly used in the construction industry. It seems also very relevant for the PV sealant application where sealants can sometimes be exposed to water for a non-negligible period of time. At least 3 pieces were made for each condition, and the reported value is an average of all the test pieces.

When testing the lap shear samples, the maximum force is the not the only property that is looked at, but also the failure mode. If adhesion is good, the sealant will break in its bulk, and one will get 100 % cohesive failure. If the material adheres poorly to the substrate, the whole material joint will detach at one of the substrates' surfaces. In this case one speaks of '100 % adhesive failure', or '0% cohesive failure'. Sometimes there is a mixed failure, in which case the percentage of cohesive failure on the substrate area is reported.

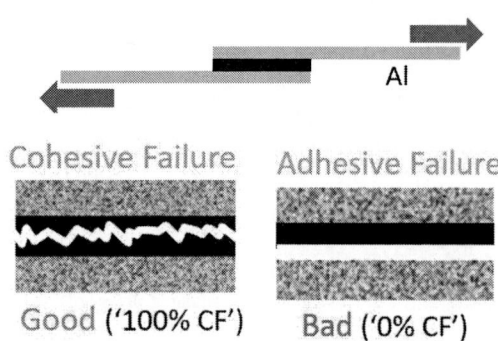

Figure 1: Schematic of lap shear testing, with illustration of failure modes

3.3 Peel tests

In order to assess adhesion on various substrates, peel test structures with mesh were prepared as described in standard ASTM C794 and allowed to cure for 7 days. The substrates were glass, flat pieces of polyphenylene ether (PPE, material used to make junction boxes) and PET backsheet (backsheet with stabilized PET as outer layer). Some of the samples received no further treatment, but some others were then immersed in water at room temperature for 7 days, and some backsheet-sealant samples were exposed to UV light under an irradiance of 50 W/m² with wavelength between 300 and 400 nm for 1000h. All samples were tested by pulling at 180° and the failure mode (% cohesive) was recorded.

3.4 H-bar tests

While tensile tests and lap shear tests test the material in a pure stress mode (tensile stress resp. shear stress), real joints undergo complex mechanical stress situations. To test the ability of the sealants to withstand such complex stresses, H-bar test samples were prepared, which feature thick and wide joints, following the guidelines of the ETAG 02 standard for structural glazing. The substrates were anodized Al and float glass. Once cured the samples are tested in tensile configuration, but as the material is restricted in movement at the substrate-sealant interface, a complex stress structure develops combining tensile and shear stresses, particularly close to the substrates.

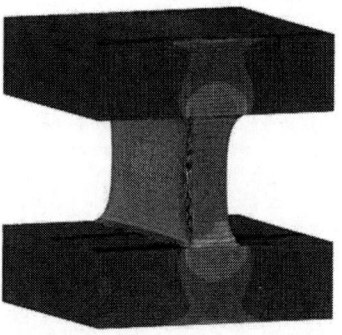

Figure 2: Stress situation in an H-bar test sample as it is pulled for testing

3.5 Frame-glass tests

Test pieces were made that aimed to be closer to the application. Pieces of a PV frame were cut in 5 cm segments. Sealant was dispensed in the U channels of the frame pieces and pieces of 4 mm thick float glass were inserted. The samples were left to fully cure, and then were pulled in a Zwick test rig. The free end of the piece of glass was pulled upwards while the lower part of the frame was fixed to the base of the test rig. The force and displacement was recorded for each sample.

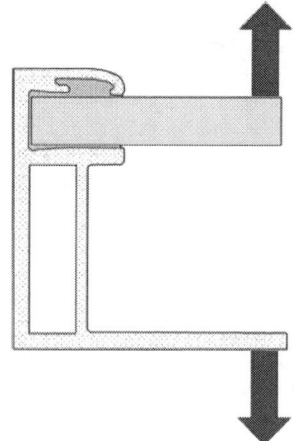

Figure 3: Frame – glass test structure

4 RESULTS

4.1 Tensile test results

Figure 4 shows typical curves obtained during the tensile testing of the samples (in each group, the sample with median tensile strength was selected for curve plotting). As can be seen the oxime PV frame sealants are much stiffer materials, which is related to the high degree of filler loading. The Young modulus estimated by fitting the linear region close to the origin of the curve is 13.4 MPa for the stiffest material, which is very high for a silicone material. The maximum elongation that these materials can take in pure tensile mode before breaking is relatively low. For the material with lowest elongation at break, the value was around half that of the reference alkoxy sealant.

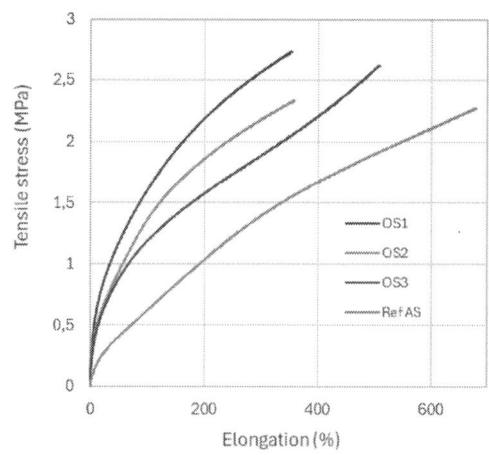

Figure 4: Typical tensile test curves for the different silicone sealants tested

4.2 Lap shear tests

Table 1 gives the lap shear test results. Note that the tests were done in two different campaigns several months apart and that different batches of the sealants were used.

Table 1: Lap shear test results

	As cured		1 week water immersion	
	Shear strength (MPa)	% cohesive failure	Shear strength (MPa)	% cohesive failure
OS1 batch A	1.63	100%	1.77	100%
OS1 batch B	1,8	100%	1.75	100%
OS2 batch A	0,65	83%	0.61	0%
OS2 batch B	1.56	100%	1.42	100%
OS3 batch A	1.44	100%	1.16	40%
OS3 batch B	1.43	100%	1.42	88%
Ref AS Batch A	1.23	100%	1.39	63%
Ref AS Batch B	1.42	100%	1.18	100%

As can be seen, the different sealants perform generally well in the lap shear test. Batch A of OS2 shows low values and complete adhesion failure after water immersion. However batch B behave well in the same test. Batch A of OS3 also shows a fairly high percentage of adhesion failure after water immersion. It should be mentioned that OS2 batch B and OS3 batch B were a couple of months out of shelf life, but that does not seem to have impacted their performance.

4.3 Peel tests

Table 2 shows the results of a preliminary peel test study where a sample of sealant OS1 was compared to the reference sealant Ref AS.

Table 2: Peel test results of initial study (in % cohesive failure)

	As cured			1 week water immersion		
	Back-sheet	PPE	Glass	Back-sheet	PPE	Glass
OS1	100%	100%	100%	30%	100%	100%
Ref AS	100%	100%	100%	100%	100%	100%

While adhesion was good on most combinations substrate-sealant, adhesion on the PET backsheet had degraded for OS1 after water immersion.

A more extensive peel test study was conducted using all the sealants of interest. As can be seen in Table 3, no adhesion issue was observed this time.

Table 3: Peel test results on the various sealants (in % cohesive failure)

	As cured			1 week water immersion			1000h UV
	Back-sheet	PPE	Glass	Back-sheet	PPE	Glass	Back-sheet
OS1	100%	100%	100%	100%	100%	100%	100%
OS2	100%	100%	100%	100%	100%	100%	100%
OS3	100%	100%	100%	100%	100%	100%	100%
Ref AS	100%	100%	100%	100%	100%	100%	100%

4.4 H-bar tests

In Figure 5, the force versus elongation curves for typical H-bar samples (Al substrates) are plotted. In contrast to the tensile testing curve in Figure 4, we see that in case of complex stress situations as occur in H-bar test samples, the oxime PV sealants reach a lower force. The elongation at break is much lower.

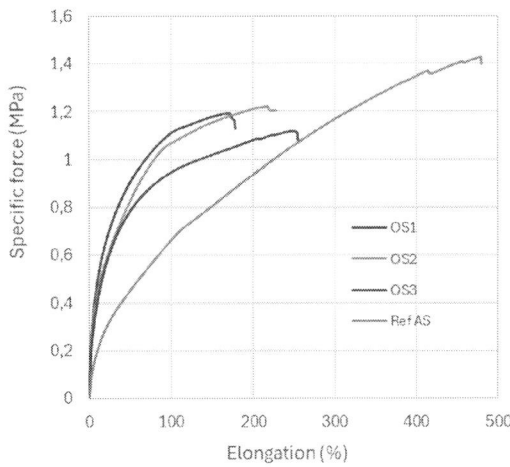

Figure 5: Typical test curves for H-bar sample testing

In table 4, we give the maximum force and the elongation at maximum force for the tested samples (each value an average of 3 test samples). Here as well we have results of two separate testing campaigns several months apart, in which different batches of sealants were used. In the first campaign the H-bars used glass substrate on one side and anodized Al on the other side, while in the second campaign, Al was used on both side. In all cases and for all samples failure was 100% cohesive, and therefore the type of substrate is not expected to impact the measured mechanical parameters.

Table 4: H-bar tensile test results

	First campaign, glass-Al		2nd campaign, Al-Al	
	F_{max} (MPa)	Elongation at F_{max} (%)	F_{max} (MPa)	Elongation at F_{max} (%)
OS1	1.20	72	1.20	157
OS2	0.59	126	1.22	203
OS3	1.18	101	1.14	243
Ref AS	1.3	489	1.45	478

For completeness, we mention here that the OS2 and OS3 sealants used in the 2nd campaign were a couple of months out of shelf life at the time of sample preparation. Here as well, it does not seem to have impacted the results.

We see variable values for elongation, sometimes very low values, for the oxime PV sealants tested. In case of OS2 batch A the force is also very low. The reference alkoxy sealant shows good values of F_{max} and outstanding elongation values, a factor 2 to 4 higher than for the other sealants.

4.5 Frame-glass test structures

We compare the force-displacement curves for OS1 and Ref AS sealants in Figure 6.

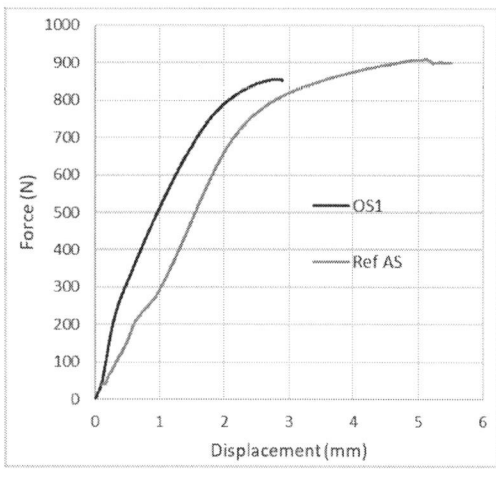

Figure 6: Frame-glass test curves for two sealants

Interestingly, the curve is highly impacted by the type of sealant used. For instance, for a 1 mm displacement during pulling, a force of 500 N is required for OS1 against 300 N for Ref AS.

5 DISCUSSION

5.1 Importance of durable adhesion

As can be seen in previous sections, we often applied water immersion as an aging procedure in our study. This type of test is not included in any photovoltaic standard today, even though one could argue that on long period of rains, it is not uncommon to have water puddling at the lower part of the frame, in direct contact with Al frame, silicone seal and front glass, for extended periods of time. Also, with the emergence of lightweight modules which feature junction boxes at the front of the laminates instead of the back, the frequency of potential occurrences where sealant joints are immersed is increased.

A loss of sealant adhesion has dire consequences for the PV module. For the junction box, it can result in detachment. For the frame, problems will develop as water enters the narrow space between the glass and the frame and finally reach the edge of the laminate. There, water will react with the chemical bonds between the encapsulant and the glass, ultimately resulting in local delamination (Figure 7). A delaminated area at the edge is a place where a conductive path can be created between the cells at high voltage and the frame, and, if detected during visual inspection of a solar installation, results in the module being discarded out of operational and safety reasons.

The results in lap shear and peel tests in sections 4.2 and 4.3 show that some types and batches of sealants lose good adhesion in water immersion. Note that the problem typically would not be detected during incoming inspection because peel and lap shear tests after water immersion is not common practice at module manufacturers.

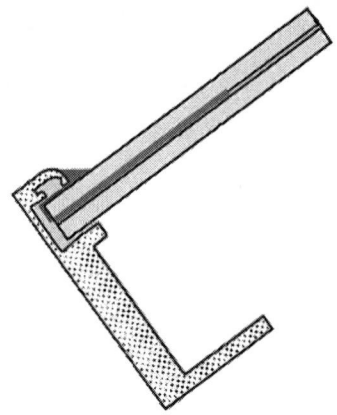

Figure 7: Schematic illustration of delamination caused by loss of adhesion of frame sealant.

5.2 Variability in batches

Not all sealants of a given type show this problematic behavior, as can be seen in Table 1 and Table 2, where the problem was no longer observed for another batch of the material. Also in terms of mechanical properties, there can be variation for different batches, see for instance Table 4.

There is always a certain variability in the properties of the material. This can arise at the stage of sealant manufacturing, for instance because of variability in the incoming material and/or external parameters having some impact such as ambient temperature and relative humidity. Over the shelf life, the properties of the sealant might change because the sealant is aging. This occurs to some extent at a certain rate in standard storage conditions, but can be exacerbated by improper transport and storage.

The sources of natural variability are known and need not be a problem. A proper quality check system at manufacturing should ensure that all material that is released for sales fulfill the minimal requirements. A shelf life study should have been carried to ensure that the material fulfills minimum requirements during its complete shelf life. Some limited variability is therefore expected but controlled.

There is an issue however when variability is extreme and not controlled. This can arise from improper quality check, from excessive material aging during its shelf life, and from improper storage and transport. Long sea freight transport for instance can be a period of excessive degradation. If there is a problem with the material, it might not be detected if incoming checks are insufficient. Introducing water immersion in the inspection of incoming sealants would be good practice in this respect.

5.3 Ability to take load in complex stress situations

The fact that the most used silicone sealants in the industry today are quite stiff materials limits their ability to take on complex stress situations as those that occur in the U-channel of a PV module frame. As can be seen comparing Figure 4 and Figure 5, the value of tensile strength, measured on dumbbell samples and reported in technical data sheets, might give an inaccurate image of the load that the material can take on in practice.

5.4 Stress at glass-glass module edge

Since the recent shift to bifacial glass-glass modules in the industry, the frequency of glass breakage has massively increased. It is regularly observed that the rear glass breaks after a few weeks or months in operation. The larger module size and switch to not-fully-tempered 2 mm glass has definitely played a role, but we suggest here that the situation has been worsened by the use of stiff frame sealants with no allowance of glass rotation and poor ability to sustain complex stress situations. Cracks in the glass originate from the glass edge and the stress situation there is critical.

From basic beam mechanics it is known that a beam that is free to rotate at its ends experiences shear stress at its ends but no tensile bending stress, while a fixed beam under the same load experiences a high bending-related tensile and compressive stress at the beam top and bottom surfaces. The total stress level at the ends is therefore higher, even though the deflection is lower.

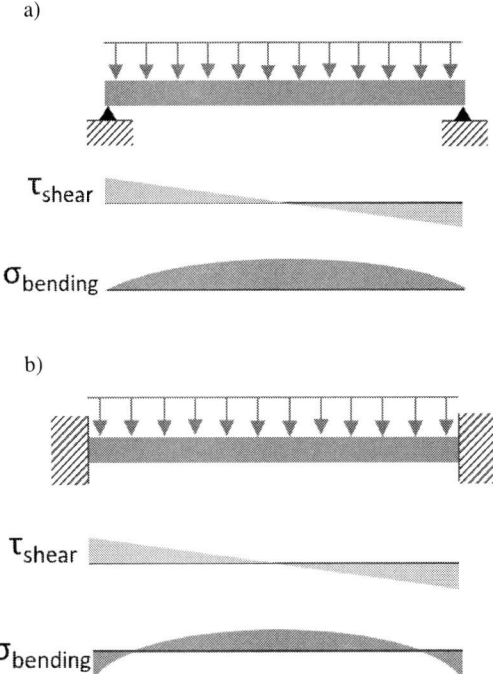

Figure 8: Stress situation from simple beam theory for a constant load in two extreme situations: a) free rotation, b) fixed

These two situations are extremes. Situations in between have a beam fixation with a certain level of elasticity, where the bending moment at the edge is proportional to the rotation $M = k\,\theta$, with M the bending moment and θ the rotation angle at the edge. The fixed beam configuration corresponds to the case where $k=\infty$ while the free rotation case corresponds to $k=0$. The graph in Figure 6 indicates that different sealants lead to different elastic responses, where the stiffer material has a substantial higher k than the lower modulus material. The bending stress at the edge is therefore expected to be higher and therefore the total stress. Quantification of the stress increase will be the topic of future studies.

6 CONCLUSION

The most commonly used sealants in the PV industry industry today are silicones with oxime curing chemistry and a high filler content and a high modulus. Lap shear and peel test studies show that those materials can sometimes show poor adhesion after water immersion aging. With H-bar tensile testing, we also observe variations in material properties depending on batches, and we show that these sealants are not best suited to handle complex stress situations. Finally, based on beam mechanics considerations and results of frame-sealant-glass sample testing, we suggest that stiff sealants worsen the stress situation at the interface of glass-glass laminates and frames.

8 REFERENCES

(1) Mathiak G; Sommer J; Herrmann, W; Bogdanski N; Althaus, J.; Reil F. Impact of Hail and Non-Uniform

Snowload. In *Proceedings 32nd European Photovoltaic Solar Energy Conference.*

(2) Whitfield, K. Durability and Reliability of Polymers and Other Materials in Photovoltaic Modules. **2019**, 235–254.

(3) Oreski G; Barretta C.; Castillon, L.; Christöfl P; Köntges M; Importance of Bill of Material (BOM) Control and IEC 61215 Scope of Application. In *Proceedings 37th European Photovoltaic Solar Energy Conference (presentation)* **2020**,.

DOWSIL™ is a trademark of The Dow Chemical Company ("Dow") or an affiliated company of Dow

41st European Photovoltaic Solar Energy Conference and Exhibition

SOLDER PASTES IN SHINGLED MODULES

Karl Wienands[1], Ignacia Devoto[1], Nils Kopp[2], Carina Hallensleben[2], Rihoko Kizukuri[2], Matthias Helbig[1], Enita Kurtovic,[1] Andreas Halm[1], Daniel Tune[1]

[1] International Solar Energy Research Centre (ISC) Konstanz, Rudolf-Diesel-Str. 15, 78467 Konstanz, Germany

[2] TAMURA-ELSOLD GmbH, Hüttenstraße 1, 38871 Ilsenburg, Germany

ABSTRACT: In this work, we show that solder pastes can be used in the shingled module concept and exhibit excellent durability. We further explore the use of solder pastes for shingled module applications with particular attention focused on reducing the amount of solder paste required per interconnection, the compatibility with relevant industrial module assembly processes, extending the mechanical integrity of connections through the use of non-conductive adhesive, and on the performance and durability of shingled modules made using optimized application of this material.

Keywords: Shingling, low temperature solder paste, conductive adhesive, ECA, non-conductive adhesive, NCA

1 INTRODUCTION

The market prevalence of shingled modules that use electrically conductive adhesive (ECA) interconnects is increasing, but the relatively high cost and silver content of some conductive adhesives creates an impetus for exploring other potential interconnect materials. The use of low temperature solder pastes (LTS) is a mature technology solution that is widely used in microelectronics and some other photovoltaic module concepts and offers a significant cost reduction compared to conductive adhesives on a per weight basis. However, the processing characteristics and material properties of solder pastes are very different to those of conductive adhesives, particularly with respect to the adhesive strength of the solder paste bond and its effect on module performance and reliability, as well as the limits to which the amount of solder paste can be reduced.

The aim of this work is to further explore the use of solder pastes for shingled module applications with particular attention focused on reducing the amount of solder paste required per interconnect, compatibility with relevant industrial shingled module assembly processes, extending the mechanical integrity of connections through the use of non-conductive adhesive, and on the performance and reliability of shingled modules made using optimized application of this material.

2 METHODS AND PROCESS

Solder paste is applied on the busbars of solar cells metalized in a six-busbar shingled layout after laser cutting of the cells into shingle stripes. The application of the solder paste is done via manual stencil printing. For deposition of different amounts of LTS, two stencils with the layout shown in **Figure 1** and with 100 μm and 60 μm thickness were used. The layout consisted of varying patterns of openings in the screens – a continuous line or sequences of pads of different lengths and separations (**Figure 2** and **Figure 3**).

Figure 1. Stencil for solder paste application with varying patterns.

Figure 2. Low temperature solder paste in pads on the busbar of a shingle stripe.

41st European Photovoltaic Solar Energy Conference and Exhibition

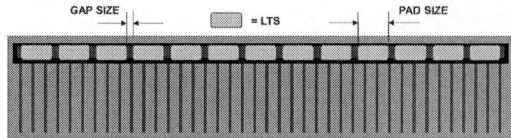

Figure 3 Example of layout for solder paste pads on a shingle stripe.

Five amounts of LTS (labelled Amount 2-6) were selected in a pretest with the varying patterns of the screen and a pick and place robot was used to assemble the shingle stripes into strings before soldering and further processing into one-cell-equivalent minimodules (**Figure 4**). Another variation (Amount 1) was applied via pressure-time dispensing using a needle to achieve the highest amount of the experiment. All amounts used can be seen in **Figure 5**.

Figure 4. One-cell-equivalent shingle minimodule with solder paste interconnects.

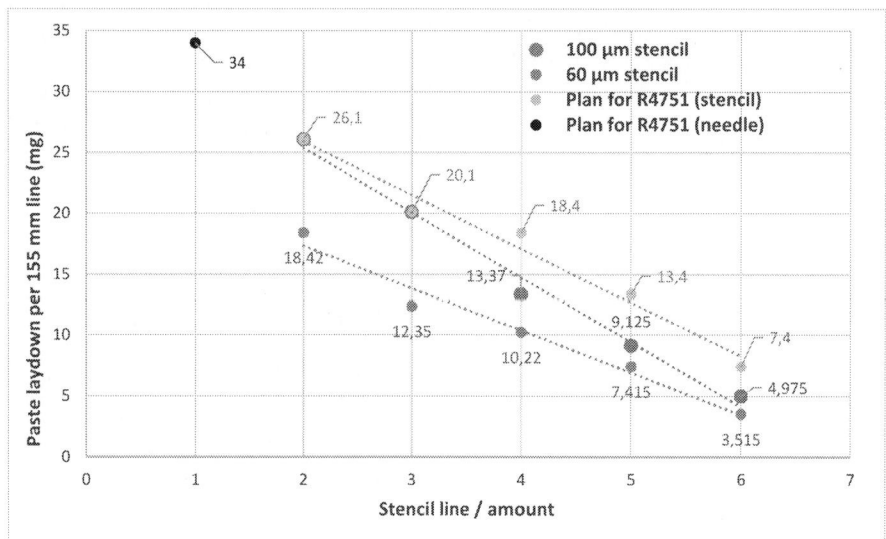

Figure 5. Amounts of LTS as determined in preceding evaluations.

The peel strengths of the various solder paste interconnections were assessed with respect to the physical handling as well as viability for assembly and production of modules. The current–voltage characteristics and electroluminescence images of the minimodules were measured before and after damp–heat and temperature cycling tests as per IEC 61215.

3 RESULTS

Preceding the variation of solder paste amount, an experiment was conducted in which variations in heating time of the solder paste were evaluated. The strong dependence of the peel strength (of Cu ribbons on the shingle busbars) on the soldering time (at 180°C) is shown in **Figure 6** and reveals a narrow process window. Based on this, a standard process was defined as 15 s @ 180°C, giving a median peak peel force of 2.9 N mm^{-1} on the frontside busbars and 1.1 N mm^{-1} on the rear side busbars, with the difference being due to the different metallization pastes used on the front and rear.

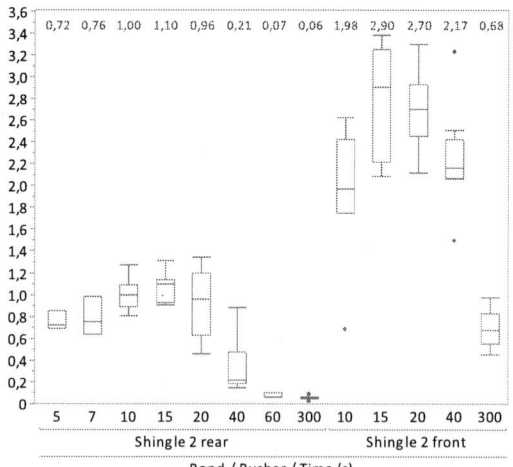

Figure 6. Dependence of peel force on soldering time.

Furthermore, for Amount 1 (pressure time deposition, 2.19 mg cm^{-1}) a difference in degradation for different encapsulation materials (EVA and POE) is seen in EL

(**Figure 7**) and IV (**Figure 8**) measurements, where modules made with the particular EVA used in this study exhibited a significantly higher degradation (26.2 % relative loss of power) than those made with the POE used in this study (5.17 % relative loss of power).

However, up to DH1000 both encapsulants degrade less than 5 % and would thus 'pass' the IEC 61215 pass/fail criteria (see figure 8).

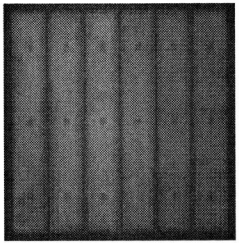

Figure 7. EL images of midimodules encapsulated in either POE (left) or EVA (right) after DH3000.

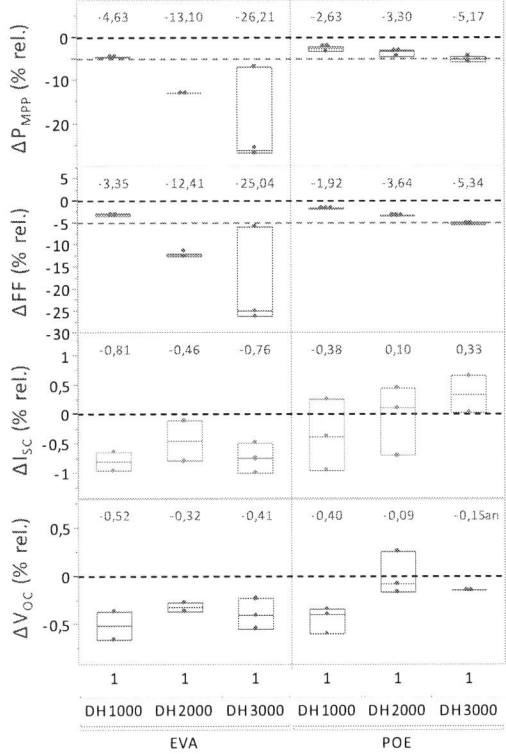

Figure 8. Relative changes in I–V characteristics after DH3000 for minimodules encapsulated using either EVA or POE.

Minimodules with systematic reduction of solder paste amounts were tested in climate chambers. Amount 5 (0.86 mg cm^{-1}, 80.4 mg/cell) was the smallest amount that still providing enough mechanical strength and structural integrity to withstand the subsequent processing steps without failure of the bond and exhibited the best durability with a relative decrease in P$_{MPP}$ of -2.53 %$_{rel.}$ to DH3000 and -2.18 %$_{rel.}$ to TC600 (see **Figure 9**).

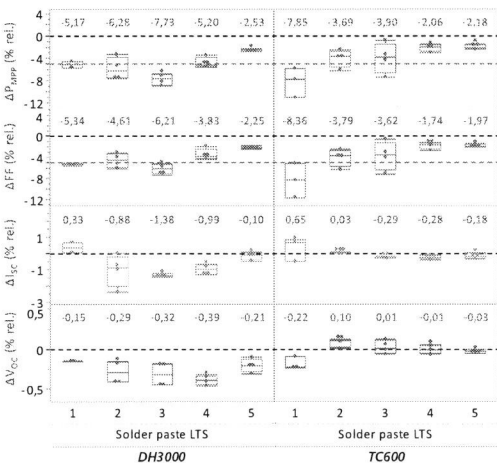

Figure 9. Relative changes in I–V characteristics of minimodules with varying amounts of solder paste after DH3000 and TC600.

To overcome the structural limitations of further reducing the amount of the LTS interconnects, the cell-to-cell bonds were mechanically reinforced by depositing three dots of non-conductive adhesive (NCA) at the centre and ends of the solder paste line as shown in **Figure 10**.

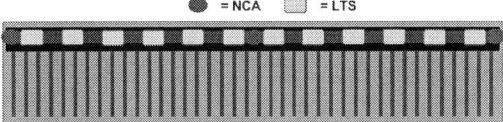

Figure 10. Schematic depiction of NCA and LTS pads on a shingle stripe.

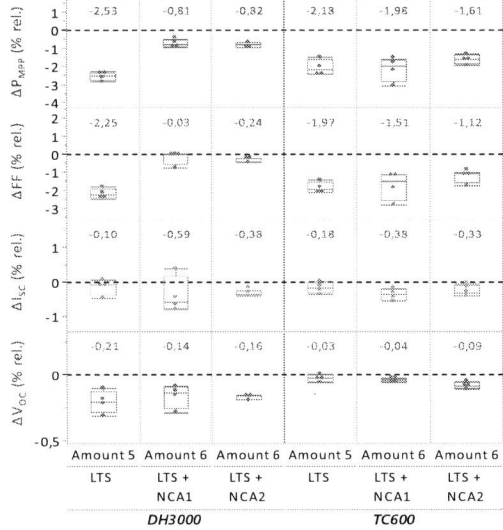

Figure 11 Relative changes in I–V characteristics of minimodules with NCA-reinforced interconnects vs. the lowest amount without NCA after DH3000 and TC600.

This strategy of mechanical reinforcement enabled a further reduction of the LTS amount. Exemplary minimodules with Amount 6 (0.48 mg cm^{-1}, 44.4 mg/cell) strengthened with NCA shows the least degradation after

TC600 (-1.6 %rel.) and DH3000 (-0.82 %rel.) (see **Figure 11**).

4   SUMMARY

LTS was demonstrated to be a viable alternative to ECA for application in shingled modules. The performance and durability of the modules was observed to be excellent for optimized (reduced) amounts. Further reduction of the amount and improvement of the durability was achieved by mechanical reinforcement of the solder bonds with non-conductive adhesive. The lower limit to the amount (i.e., when the amount is reduced to a point where the performance and durability are negatively impacted) was not discovered in this work, suggesting that further reductions are likely possible.

5   ACKNOWLEDGEMENTS

This work was funded by the Federal Ministry for Economic Affairs and Climate Action of Germany (BMWK) as part of Project BIG (grant agreement No. 03EE1116).

41st European Photovoltaic Solar Energy Conference and Exhibition

TiO_2/SiO_X SURFACE COATING ON CRYSTALLINE SILICON-BASED-SOLAR CELL MODULE TO PROVIDE ANTI-SOILING FUNCTIONALITY

Koshiro Iwaki[1] Yasushi Sobajima[1] Keisuke Ohdaira[2] Atsushi Masuda[3]
Japan / Gifu Univ.[1], JAIST[2], Niigata Univ.[3]
E-mail: iwaki.koshiro.x0@s.gifu-u.ac.jp / TEL: +81-58-293-4517

ABSTRACT: This study describes a coating technology of a SiO_X layer mixed with TiO_2 particles (TiO_2/SiO_X), which uses a photocatalytic reaction to reduce the adhesion of organic matter on the cover glass. Low-temperature fabrication conditions were achieved using TiO_2 particles pre-prepared by another method, which is difficult to achieve when depositing the TiO_2 layer directly on the substrate materials. Furthermore, adjusting the number of mixed particles even in SiO_X obtained a layer with photocatalytic reaction and high visible light transmittance.
Keywords: c-Si solar cell module, TiO_2, photocatalytic reaction, surface antifouling coating

1 RESEARCH BACKGROUND AND OBJECTIVES

Solar cells represent a typical renewable energy source, and to be utilized as a power source, they must perform at the requisite level over 30 years [1]. Solar cells operate in a modular structure consisting of a cover glass, an encapsulant, and a backsheet, collectively protecting the cell from the external environment. It is well acknowledged that several factors contribute to the deterioration in the photovoltaic performance of solar cell modules, including damage to the solar cell itself, potential-induced degradation (PID) and electrode failure [2]. The accumulation of matter on the surface reduces the amount of incident light that reaches the solar cell, directly reducing power generation performance. While periodic cleaning is an appropriate method for ensuring optimal functionality, this approach is insufficient in environments where module units are difficult to replace. Enhancing the solar cell module's anti-soiling function is expected to demonstrate stable performance over the long term. The module surface exhibits the adhesion of various objects, including organic matter. The method for decomposing organic matter is the photocatalytic reaction by UV irradiation using TiO_2 [3]. However, TiO_2 exhibits a high refractive index in the visible light region due to its high band gap, and its application in a film structure on the surface may have a suppressive effect on the power generation characteristics [4]. In this study, we investigated using a photocatalytic reaction to use SiO_X film mixed with TiO_2 particles (TiO_2/SiO_X) to impart an anti-soiling function to the top surface of the cover glass.

2 EXPERIMENTAL METHOD

TiO_2 films with high crystallinity be prepared by magnetron sputtering or other methods at high substrate temperatures (over 500 °C) [4, 5]. However, the high-temperature deposition process results in increased costs and restricts the types of substrates that can be employed. Furthermore, a single film of TiO_2 has a large refractive index difference from air, leading to a significant reduction in incident light.

The SiO_X film has a low refractive index and was produced by the solidifying liquid glass (Lq glass). The particle-mixed SiO_X film modifies the refractive index by incorporating a low-refractive SiO_X material with other particles, thereby conferring capabilities not available with SiO_X alone. The addition of TiO_2 particles to Lq. glass

facilitated the decomposition of organic matter through a photocatalytic reaction, thereby imparting an antifouling function to SiO_X. The particle size and the quantity of TiO_2 blended with SiO_X modify the optical properties, affecting the light's quantity reaching the cell. This study employed two types of TiO_2 particles: commercial TiO_2 (c-TiO_2) and nano-sized TiO_2 (nano-TiO_2) particles fabricated under low-temperature conditions in the laboratory. To distinguish between these two types of film structure, the SiO_X structures were designated as either c-TiO_2/SiO_X or nano-TiO_2/SiO_X, respectively. The following section (section 2.1) describes preparing TiO_2 nanoparticles.

2.1 TiO_2 particles preparation

The nano-TiO_2 particles were prepared following the procedure [3]: Firstly, 0.7 ml $TiCl_4$ (99.0%, FUJIFILM) was injected drop by drop into 2.8 ml anhydrous ethanol with stirring to avoid local overheating of ethanol. After the solution cooled down to room temperature, 14 ml of anhydrous benzyl alcohol was added to the previous solution after cooling and stirred for 10 min. The original yellow solution became reddish after the addition of benzyl alcohol. The mixed solution was sealed and stored in an oven at 85 °C for 12 hours without agitation. The nano-TiO_2 particles were precipitated from the solution obtained previously by adding diethyl ether and isolated by centrifugation at 5000 rpm for two minutes. The solid was washed by adding isolated anhydrous ethanol and diethyl ether, followed by a similar centrifugation step (5000 rpm for 2 min). This washing procedure was repeated three times. A pretreatment process using an ultrasonic homogenizer before centrifugation was used to break up clumps of nano-TiO_2 that had adhered and grown huge during fabrication. The ultrasonic homogenizer was set to a high frequency of 40 W with an oscillation frequency of 25 kHz.

2.2 TiO_2/SiO_X films preparation

Lq. glass was used for SiO_X preparation. Lq. glass is a cyclosalane-based material that can dry naturally even at room temperature and has properties such as high light transmittance in the visible light range.

TiO_2 particles are stirred in Lq. glass diluted with EtOH for about 1 hour. In this study, the dilution of EtOH to liquid glass is called x-dilution. The mixed particles varied from 1.5 to 150 mg per 1 ml solution. A bar coater applies the stirred TiO_2/SiO_X to the glass (EagleXG). The substrate is dried at room temperature for 24 hours to form TiO_2/SiO_X.

2.3 Measurement

The crystallite size of the nano-TiO₂ was calculated by applying Scherrer's equation to the X-ray diffraction (XRD) spectra. Moreover, the impact of ultrasonic degradation during the synthesis of nano-TiO₂ was assessed through surface-enhanced scanning electron microscopy (surface SEM).

The film formed on the surface of the module should exhibit high light transmittance. The direct transmittance of the prepared sample was measured.

The presence of a photocatalytic reaction was determined by the degradation of methylene blue deposited on the sample by light irradiation. To check their resistance to UV, the samples were exposed to a high-pressure mercury lamp (150 mW/cm²) for 2000 hours.

Finally, TiO₂/SiOₓ was fabricated on a surface textured cover glass. Actual crystalline Si solar cell modules were fabricated with the above cover glass and power generation characteristics were evaluated. The module structure comprised a backsheet, EVA, commercially monocrystalline cells (p-base 52 x 26 mm²), and cover glass (105 x 70 mm²). The lamination conditions are as follows: vacuum at 135 °C for 5 min, press for 5 min, and pressure hold for 25 min. As in the previous conditions, for TiO₂ nanoparticles, ultrasonic decomposition is used in the fabrication process. The c-Si solar cell module with TiO₂/SiOₓ measured the *J-V* characteristics of light irradiation (AM1.5, 100 mW/cm²) by a solar simulator.

3 EXPERIMENTAL RESULTS

3.1 Evaluation of TiO₂ nanoparticles

XRD measurements of the prepared TiO₂ nanoparticles revealed a particle size of 2.37 nm, calculated using Scherrer's formula. Figure 1 shows the effect of ultrasonic decomposition during nano-TiO₂ fabrication. Although the nano-TiO₂ diameter estimated from the XRD spectra is only a few nm, the apparent particle diameters shown in Figure 1(a) are all several hundred μm or larger. These results indicate that the formed nano-sized particles polymerize to form giant particles. Figure 1(b) shows the image of nano-TiO₂ subjected to ultrasonic decomposition. The figure shows that ultrasonic decomposition can reduce the diameter to about 10 μm on average. The clumping of particles formed during nano-TiO₂ fabrication can be reduced.

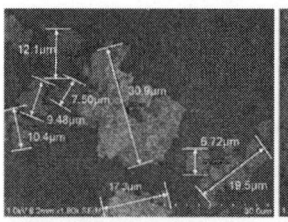 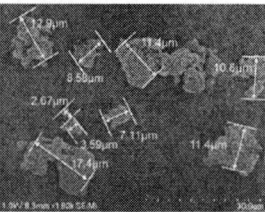

(a) TiO₂ nanoparticles for which ultrasonic decomposition has not been performed.

(b) TiO₂ nanoparticles for which ultrasonic decomposition was performed.

Figure 1: Effect of ultrasonic decomposition during TiO₂ nanoparticle (nano-TiO₂) fabrication

3.2 Permeability Evaluation

Since TiO₂/SiOₓ is intended to be applied to the surface of coverglass, sufficient light transmittance is required. Figure 2 shows the direct transmittance of substrates (TiO₂: 1.5-150 mg, Lq. glass: 0.5 ml, EtOH: 0.5

mL) formed on EagleXG with a bar coater. When TiO particles are mixed at a ratio of about 30 mg in the solution, the diffuse transmittance is high (not shown in this paper), but the direct transmittance shown in Figure 2 decreases, especially in the visible region. This is thought to be due to the scattering and reflection of incident light at the TiO₂/SiOₓ surface due to TiO₂ dispersed in SiOₓ to the order of tens of microns, as shown in sec. 3.1. On the other hand, the direct transmittance in the visible light region has also been shown to be improved by the reduction of TiO₂ content.

Figure 3 compares the direct permeability of nano-TiO₂/SiOₓ and c-TiO₂/SiOₓ. In the case of nano-TiO₂, the permeability is better than that of commercial particles. The effect of the smaller particle size on the permeability was obtained.

The presence of photocatalytic reaction (methylene blue decomposition) by UV irradiation on substrates with small amounts of TiO₂: 1.5 and 3.0 mg has also been confirmed. Figure 4 shows that in the TiO₂/SiOₓ film, the photocatalytic reaction was confirmed with a substrate of c-TiO₂:1.5mg by increasing the EtOH dilution ratio (5x dilution). In the nano-TiO₂/SiOₓ film, the photocatalytic reaction was confirmed on a substrate with 3 mg of nano-TiO₂. Dilution of the solution with EtOH is thought to increase the number of exposed particles on the film surface, thereby enhancing the photocatalytic reaction effect.

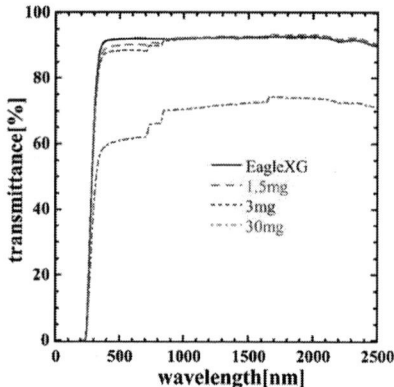

Figure 2: Direct transmittance of the film-forming substrate in bar coater

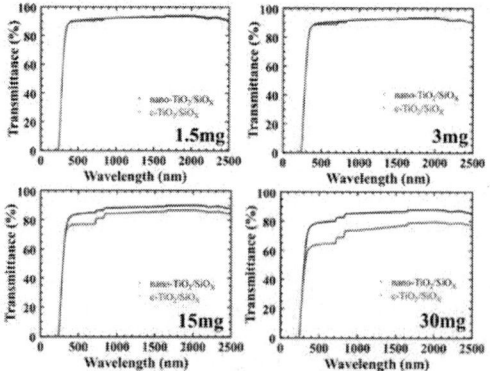

Figure 3: Comparison of direct transmittance between nano-TiO₂ particles and c-TiO₂ particles

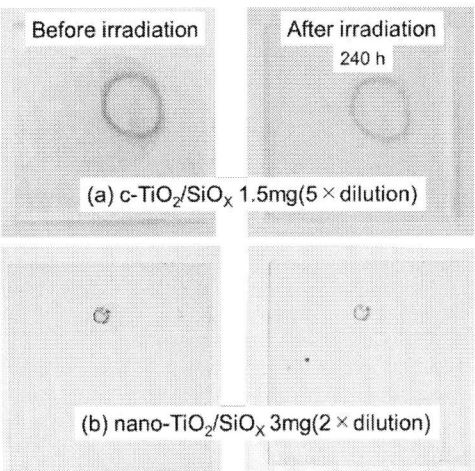

Figure 4: Photocatalytic reaction with a small amount of TiO_2/SiO_X particles in the film

3.3 Evaluation of power generation characteristics in c-Si solar cell modules

Figure 5 shows the modules fabricated with TiO_2/SiO_X films. A film with a large amount of mixed particles, such as 150 mg, would cause the coated glass surface to become cloudy. However, a small amount of film contamination, such as 1.5 mg or 3 mg, will not affect the film's appearance.

Figure 6 shows the J-V characteristics of the module under each condition measured using a solar simulator. The cover glass was common with surface irregularities, and TiO_2/SiO_X was deposited on the surface. J_{SC} did not decrease when TiO_2/SiO_X coated. The power generation efficiencies of w/o, 1.5 mg, and 3 mg at the time of this measurement were 19.39%, 19.40%, and 19.39%, respectively, Photocatalysis on the surface was also confirmed when all TiO_2/SiO_X was in use (not shown in this paper). These results indicate that TiO_2/SiO_X coatings can be applied to c-Si solar cell modules without affecting light transmittance and can impart photocatalytic reaction functions to the surface.

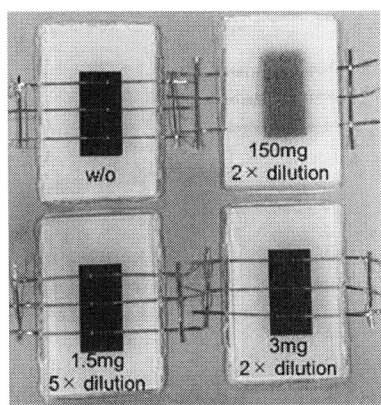

Figure 5: Effects of different amounts of mixed particles on the appearance of nano-TiO_2/SiO_X-surface-coated modules

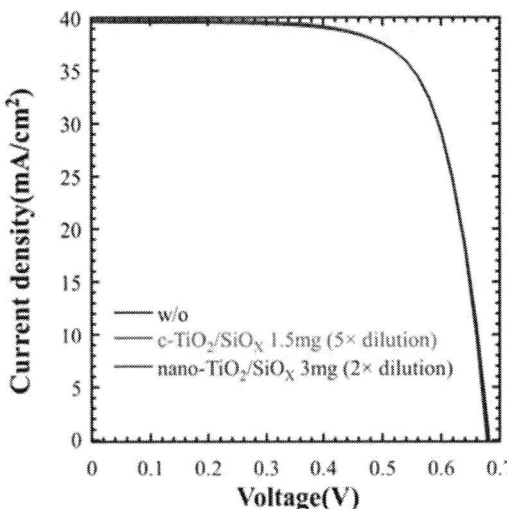

Figure 6: Photo J-V characteristics of c-Si solar cell modules with or without TiO_2/SiO_X coating on the surface

4 CONCLUSION

In this study, TiO_2 / SiO_X was deposited on the module surface to realize the imposition of an anti-fouling function. Similar to monolayer TiO_2, the photocatalytic reaction was manifested by encapsulating TiO_2 in SiO_X as particles. The same functionality applies not only to commercial products but also to nanoparticles fabricated under low-temperature conditions. The light transmittance was improved by reducing the particle content, and the use of EtOH-diluted solvent facilitated the exposure of TiO_2 on the film surface, which enhanced the photocatalytic reaction effect. Highly light-transmissive TiO_2/SiO_X was deposited on the cover glass of c-Si solar cell modules, and both light transmittance and photocatalytic reaction were demonstrated without J_{SC} degradation.

ACKNOWLEDGEMENT

This research was conducted as part of a NEDO project.

REFERENCES

[1] A. Goetzberger, J. Luther, G. Willeke, Sol. Cells 74 (2002) 1.
[2] S. Pingel, O. Frank, M. Winkler, S. Daryan, H. Hoehne, J. Berghold, Proceedings 35th IEEE PVSC, (2010) 2817.
[3] H. Tan, A. Jain, O. Voznyy, X. Lan, F. P. G. de Arquer, J. Z. Fan, R. Quintero-Bermudez, M. Yuan, B. Zhang, Y. Zhao, F. Fan, P. Li, L. N. Quan, Y.Zhao, Z. H. Lu, Z. Yang, S. Hoogland, E. H. Sargent, Science 355 (2017) 722
[4] H. Natsuhara, K. Matsumoto, N. Toshida, S. Nonomura, M. Fukawa, K. Sato, Sol. Eng. Mat. Sol. Cells 90 (2006) 2867.
[5] H. Natsuhara, T. Ohashi, S. Ogawa, N. Yoshida, T. Itoh, S. Nonomura, M. Fukawa, K. Sato, Thin Sol. Films 430 (2003) 253.

41st European Photovoltaic Solar Energy Conference and Exhibition

PERFORMANCE ANALYSIS OF DIFFERENT SHADING-RESISTANT PV MODULE DESIGNS UNDER DIFFERENT PARTIAL SHADING SCENARIOS

Andreas Maixner[1,*], Tales Siquera[1], Matthias Pander[2], Jens Froebel[2], Bengt Jaeckel[2], Hamed Hanifi[1]

[1] AESOLAR, Messerschmittring 54, Koenigsbrunn, Germany

[2] Fraunhofer Center for Silicon Photovoltaics, Otto-Eissfeldt-Strasse 12, 06120 Halle, Germany

*Corresponding Author: a.maixner@ae-solar.com

Climate change and the energy crisis in Europe demand a higher integration of renewable energy into the energy system. To achieve the photovoltaic (PV) installation goals of the European Union for the implementation of 600 GWp of PV installations by 2023, a higher rate of PV installation in both commercial and residential rooftops is expected. Residential rooftop systems suffer from partial shadings in most installations. Even 5% shading of the module area can lead to a total shutdown of the system. In this work, three different shade-resistant PV module designs are simulated, evaluated, and benchmarked against the standard PV module under six different shading scenarios of single-cell, single-row, single-column, multiple-row, multiple-column, and diagonal shading. It is shown that higher shade tolerance can be achieved by the implementation of even fewer bypass diodes compared to the first generation of hot-spot-free and shade-resistant module of AESOLAR with one single bypass diode for each cell. The results show 8%, 10%, and 14% power gain compared to the standard half-cell module over all six shading scenarios. It can be concluded that a promising shade tolerance leading to a higher energy yield, better durability and consequently a lower levelized cost of electricity (LCOE) can be achieved.

Keywords: Shading, Partial shading, Shading resistance, Hotspot, Bypass diodes, Shade-resistant modules, PV, Photovoltaics

1 INTRODUCTION

The climate changing and the current energy crisis in Europe demands more and more renewable energy installation. This is further promoted by the plans from the European Union to double solar power output to 320 GW by 2025 and connect 600 GW to the grid by 2030 [1]. In Germany alone, roofs and façades offer a technical potential in the order of 1000 GWp in terms of PV installations. So far, less than 10% of the roof potential and less than 1% of the façade potential have been utilized, which means there is a high potential for extending such installations in the future [2].

In PV modules, several solar cells are connected in series, which increases the output voltage of the module by each additional cell. The current flow is directly proportional to the size of the illuminated cell area and radiation density and has in unshaded modules an identical value in each solar cell. If a single cell is shaded, this limits the current flow of all other solar cells connected in series. As the shaded cell acts as an electrical resistor, this can lead to a significant reduction of the module power as well as severe heating and even destruction of the shaded cell.

To limit the power losses caused by shading and to prevent damage to the solar cell and the module, bypass diodes are used in common PV modules. These are connected in parallel to the cell strings (a merging of several cells) and can electrically bypass the string in the event of shading of one or more cells inside that string. Typically, commercially available PV modules have three bypass diodes, which means each can bypass a third of the cells in a module. However, in extreme cases, only 5% shading of the module area can lead to a total shutdown of the PV module (in case at least one solar cell per string is shaded) by triggering the bypass diodes [3].

AESOLAR has developed a shade-resistant and hot-spot-free (HSF) photovoltaic module to overcome the challenge of shading and maximize the energy yield on a limited area such as roofs. This module has integrated bypass diodes for each full sized-solar cell, showing up to 80% more energy yield than a similar standard module under partial shading conditions [4]. However, current advances in cell and module technology require a new module design with today's standard half-cut cells. For this reason, several module designs were developed and tested under different partial shading scenarios, using the Spice simulation software. The results of two promising designs are presented in this publication and compared with the performance values of a standard half-cell module and an HSF module under shading conditions.

2 METHODOLOGY

2.1 Defining partial shading

To compare the performance of different module designs, the shading scenarios must be defined first. In total, six shading scenarios of
a) shading of a single solar cell,
b) shading of a single row,
c) shading of a single column,
d) shading of multiple rows (row sweeping),
e) shading of multiple columns (column sweeping) and
f) shading diagonally (diagonal sweep)
were defined and the performance of different module designs was evaluated under these shading conditions. Figure 1 provides an overview of the selected shading scenarios.

For the investigated scenarios a-c, where one cell / one row / one column gets shaded, the total area of the cell was irradiated with an increasing irradiation steps of 200 W/m^2 (20% steps) from 0 to 1000 W/m^2. To keep it uniform for all designs, the shading simulation always starts from the bottom left cell / the bottom row / the left column of the module.

In the shading scenarios d-f, the simulation starts with an unshaded module and shades one row or column after the other with each subsequent simulation step until the module is completely shaded. In the case of diagonal shading, only one cell is completely shaded at the first step. Then, the horizontal and vertical neighboring cells are gradually shaded in the next steps. Also for these shading scenarios, the simulation always starts from the bottom row / the left column / the bottom left cell of the module. The shading intensity for each simulation step is 100% (0 W/m^2).

552

41st European Photovoltaic Solar Energy Conference and Exhibition

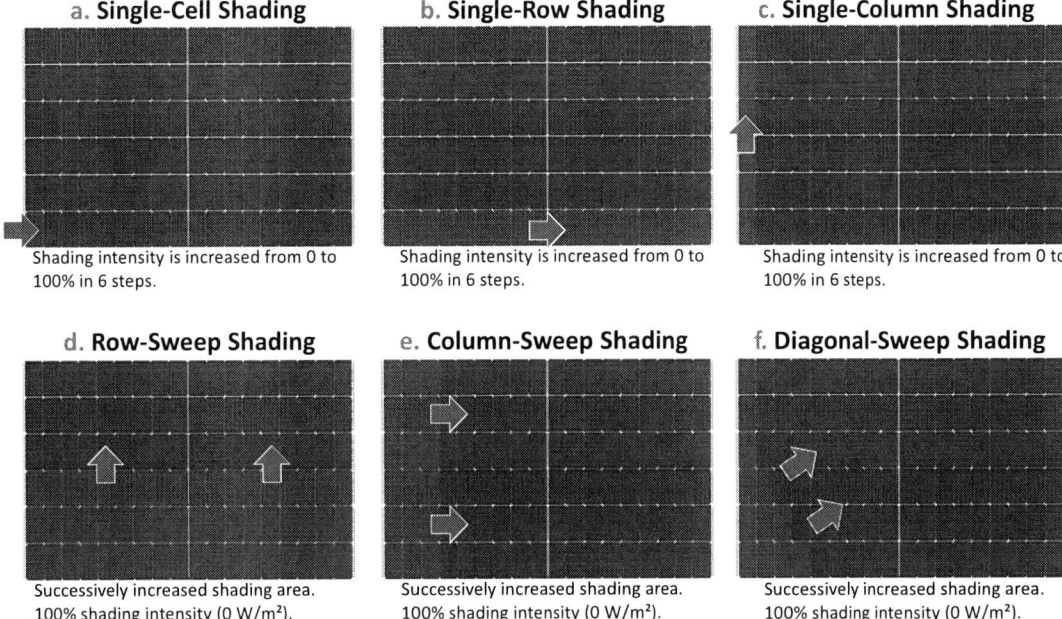

Figure 1: Schematic view of different partial shading scenarios performed in this study. a) Shading of a single solar cell, b) shading of a single row of cells, c) shading of a single column of cells, d) shading of multiple rows (row sweeping), e) shading of multiple columns (column sweeping) and f) diagonal shading sweep from the bottom left corner to the top right corner.

2.2 Investigated PV-module designs

A total of four different module designs are analyzed and evaluated under shading conditions mentioned in 2.1. The interconnection design of each module is shown in Figure 3. A standard butterfly half-cell module with 108 half-cells with three bypass diodes located in the middle of the module is used as the reference module. Furthermore, the first-generation shading resistant "hot-spot-free (HSF)" module of AESOLAR with 60 full-cells is tested and compared with the new designs and the reference module. Each solar cell of the HSF module is protected with an in-laminated bypass diode and there are three extra bypass diodes in the junction box as back-up. In addition, two new designs of shade-resistant module 1 (SR1) and shade-resistant module 2 (SR2) with 108 half-cells are tested and compared with the reference and the first-generation HSF module of AESOLAR.

SR1-module has 18 cell strings, each with a twin parallel connected small sub-string of three solar cells. In addition, there are two extra bypass diodes at the module edges, which are superordinate to the other diodes and can bypass several of the 18 cell strings simultaneously.

SR2 design is similar to the reference module. However, there is an additional bypass diode for every three cells (results in 36 in-laminated bypass diodes) with an extra three diodes located in the junction box in the middle of the module, which works as superordinate bypass diodes to bridge a cluster of subordinated substrings.

The evaluated module designs are structurally uniform. This means that the entire module consists of identical smaller substrings, which contain a bypass diode, the same number of cells and the same electrical interconnection. Therefore, the respective shading scenarios always have the same power-reducing effect on the module, regardless of the position of the shading pattern in the module.

2.3 Shading simulation in LTSpice

The shading simulations are carried out using the LTSpice simulation software. The solar cell model is based on the two-diode model and the module model is based on the SPICE model developed by Hanifi et al. [5]. Each model design is modeled and simulated under the partial shading scenarios shown in Figure 1. Figure 3 shows an example of the SPICE model of the reference module.

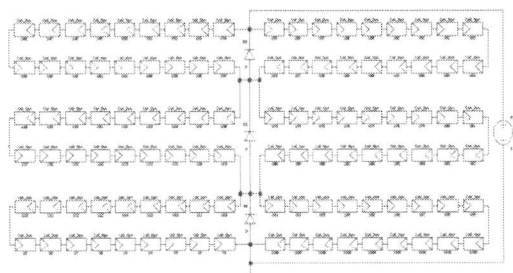

Figure 2: Schematic illustration of the electrical interconnection of the SPICE model using the reference module as an example.

41st European Photovoltaic Solar Energy Conference and Exhibition

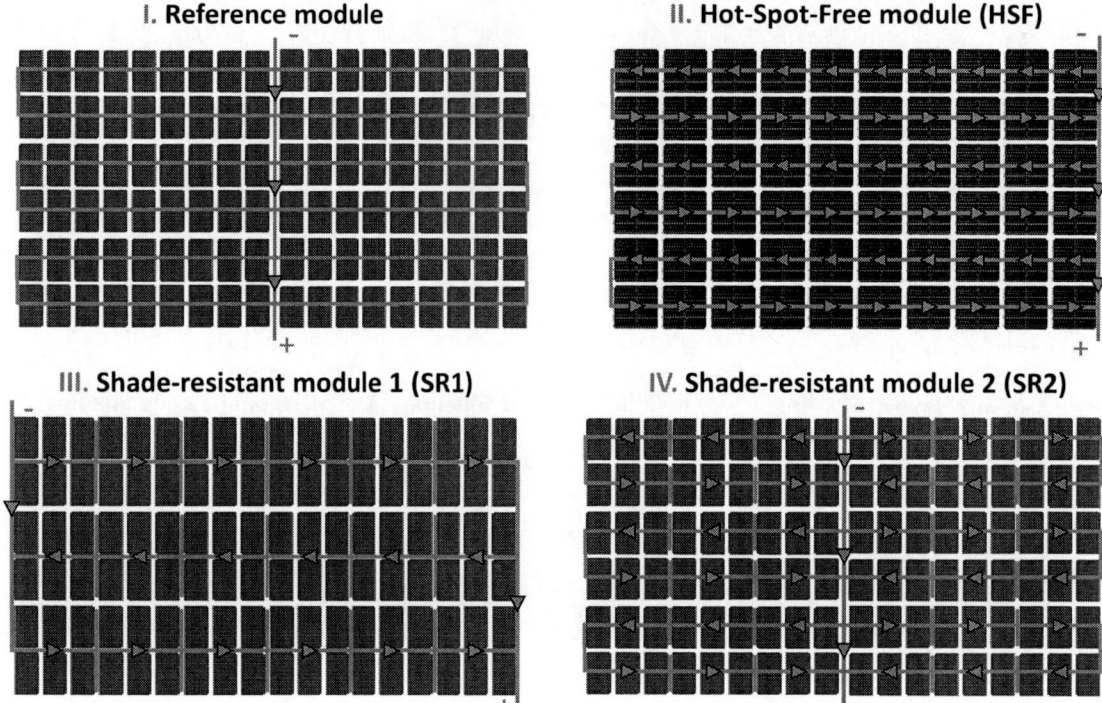

Figure 3: Schematic representation of the four evaluated designs: I. Reference module (standard half-cell module), II. First generation Hot-Spot-Free module of AESOLAR, III. Shade-resistant module 1 (SR1), and IV. Shade-resistant module 2 (SR2).

2.4 Evaluation method

To evaluate the performance of each module design under shading conditions, the average performance of each design is evaluated for each shading scenario. The relative power, i.e. the power of a module when it is shaded compared to its unshaded power, decreases (or remains the same) with each shading step. To obtain a score for each shading step, an integral across all iterations under the relative maximum power point (MPP) of each module is performed. The resulting Area Under the Curve (AUC) is assigned as a score for every design in each shading scenario. Figure 4 illustrates this process graphically.

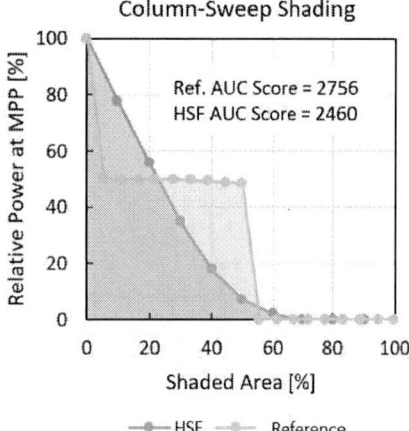

Figure 4: Example of the AUC score concept for the reference and HSF modules in the column-sweep shading scenarios.

Equation (1) defines the AUC as the integral of the relative power of the design over the shading intensity or area, depending on each scenario. Because the relative power function in Equation (2) is composed of discrete values, it is not continuous. As such, the integral can be calculated by partitioning the integration interval by the number of iterations in each shading scenario, as Equation (3) shows.

$$AUC = \int_0^{100} P_{rel}(s)\, ds \qquad (1)$$

$$P_{rel}(s) = \frac{P(s)}{P(0)} \times 100\% \qquad (2)$$

$$AUC = \sum_{k=1}^{n} \frac{[P_{rel}(s_{k-1}) + P_{rel}(s_k)]}{2} \cdot \Delta s \qquad (3)$$

Where:

AUC: Performance score of a design given by the Area Under its relative power Curve

s: Shading intensity or area, depending on the shading scenario [%]

$P(0)$: Module MPP power without shading [W]

$P(s)$: Module MPP power with a given shading intensity or area [W]

$P_{rel}(s)$: Relative module MPP power compared to its MPP power without shading [%]

n: Number of iterations in the shading scenario

Δs: Partitioned integration interval defined by the number of iterations so that $\Delta s = \frac{100}{n}$

The AUC scores of the different designs are then divided by that of the standard half-cell configuration to obtain an average gain (or loss) for each case, according to Equation (4).

$$\bar{\gamma} = \left(\frac{AUC}{AUC_{ref}} - 1 \right) \times 100\% \qquad (4)$$

Where:

$\bar{\gamma}$:   Average performance gain of a design compared to the reference [%]

AUC_{ref}:   Performance score of the reference module

Finally, the individual calculated areas in every scenario are summed together to obtain a total performance score. This final score enables the calculation of the total average gain (or loss) of each design compared to the reference module, also calculated from Equation (4). It conveys a general sense of how well each design performs over all the proposed shading scenarios. It should be noted that the performance of each module in some scenarios can be significantly better or worse compared to the reference module and the average value gives a general indication of how the module is performing.

3 RESULTS

a. Single-cell Shading

The single-cell shading scenario results in the lowest power loss of all shading scenarios analyzed. Figure 5a illustrates the relative power at MPP for this scenario. It can be seen, that the power curve of the modules drops only slightly so that the HSF-module can still deliver 96% of the nominal output power at 100% cell-shading. The SR1 and SR2 designs operate at 93% and 94% of their nominal power at 100% shading respectively. The relative power of the reference module decreases drastically with increasing shading intensity resulting to 35% power loss and 65% of output power relative to the nominal output power when a cell is completely shaded.

The reason why the shading of only one cell has such a strong effect on the power output of the reference module is that each bypass diode protects one-third of solar cells resulting in nearly 35% power loss. Although the string consists of two parallel substrings, the unshaded substring is being bypassed when one cell is completely shaded as the bypass diode will be triggered. This is because the MPP in this shading case is located at full current and two-thirds of the voltage (and not at half current and full voltage), which is the reason why the bypass diode switches through and the power from the unshaded substring can no longer be utilized. The fact that the module power drops to 65 % and is not exactly at two-thirds (66.7 %) of the output power is due to the fact, that the current flow through the bypass diode also causes a small additional power loss, which reduces the power output of the modules.

In contrast to the reference module, the other designs have more bypass diodes, which means that the (sub-) strings contain fewer cells. Therefore, bypassing a string has less of an effect on the performance of the module when a single cell is shaded. This also explains why the HSF module with one diode per cell shows the best result.

b. Single-row shading

As in the previous scenario, the reference module also performs the worst when a single row is shaded (seen in

Figure 5b). The power curve drops sharply up to a shading intensity of 40%, then flattens out slightly and finally ends just under two-thirds of the unshaded output power. The HSF module and the design SR2 show similar behavior, but the drop in power is not as intense as with the reference. The SR1 design, on the other hand, shows a more constant power drop. Compared to the other designs, it has performance advantages up to 50% shading intensity, but from 80% shading intensity the performance is even lower than that of the reference module and ends at 56% of the output power.

c. Single-column shading

In the single-column shading scenario, the reference module shows the lowest resistance to shading (see Figure 5c). The power curve drops almost linearly to 50% of the unshaded output power at full shading, only showing a slight kink at 20% shading intensity. The 50% power drop can be explained by the bypassing of the left half of the module, while the right half of the module delivers full power. As a result, the module supplies half the current but full voltage.

Furthermore Figure 5c shows a clear difference in design performance compared to the single-row shading scenario in Figure 5b. The HSF-module performs best in this scenario and ends with 78% of the unshaded output power, followed by the SR1 design, which ends with 75% of unshaded output power. The SR2 design, which previously showed the best performance with single-row shading, only scores slightly better than the reference and only from shading higher than 80%.

The different performance between single-row and single-column shading scenarios shows, that not only the number of cells affected by the shading is an influencing factor for the power reduction. But also, the power loss depends on the number of affected strings, the number of cells in a string and the quantity of the bypass diodes in operation. The shading of several cells in a string has a smaller influence on the power loss of the PV module than the shading of the same number of cells divided over several strings.

d. Row-sweep shading

In the sweeping scenarios, the performance of all module designs drops to 0% when completely shaded. In the row-sweep shading scenario (Figure 5d), the reference module sees a power decrease in approximately 1/3 increments, as the shading area sequentially obstructs each of its three sub-strings. The small deviation from the 1/3 and 2/3 marks can be attributed to the power dissipated in the bypass diodes. The performance of the HSF module follows a similar pattern. The 1/3 drops in power of this module are due to the three superordinated diodes found on the right edge of the module. It should be noted that the HSF module loses 10 cells as well as power dissipation in 10 bypass diodes leading to a higher power loss. Design SR1 has disadvantages compared to the reference module at low shading percentages, but it shows an advantage when there is high shading. Overall, it shows no advantages compared to the reference module. Design SR2 performs the best in this shading scenario with a 15% performance gain over the reference module.

e. Column-sweep shading

In the column-sweep shading scenario, the reference module immediately loses half of its power output as the first column of cells is shaded (see Figure 5e). This remains unchanged until the shading area reaches the second half of the module when its power output drops to zero.

41st European Photovoltaic Solar Energy Conference and Exhibition

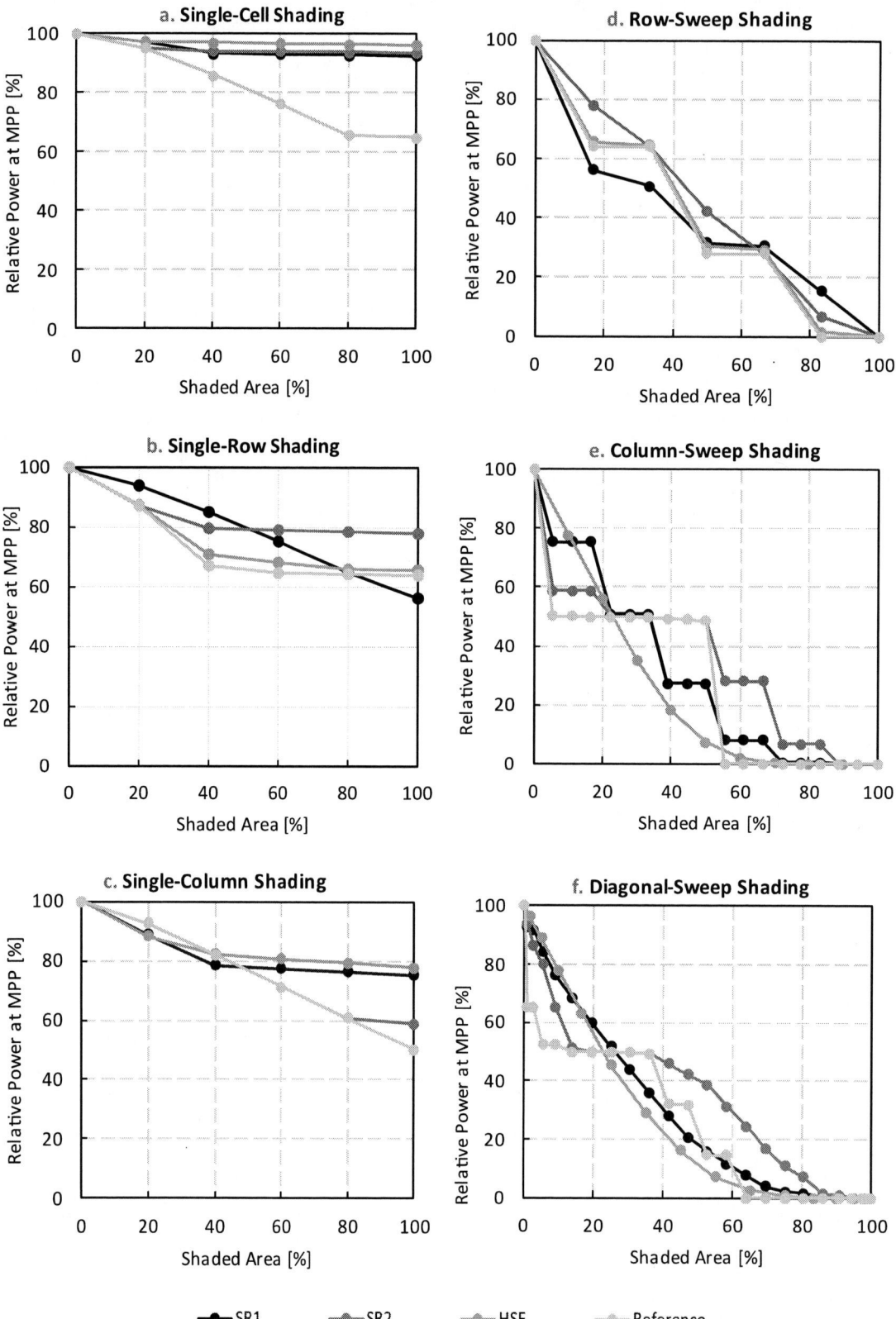

Figure 5: Relative power curves of the analyzed designs at the MPP for the static shading scenarios with a a. single solar cell, b. single row, c. single column, d. row-sweep, e. columns sweep, and f. diagonal sweep.

On the other hand, the SR1 and SR2 modules benefit from an additional power gain of 8% and 26% respectively compared to the reference module.

The HSF-module almost loses power until 60% of the module is shaded. In such high shading conditions, 60% of the module power is lost due to the shading and a significant amount (nearly 35%) is being dissipated in the connected bypass diodes to each solar cell leading to the shutdown of the PV module. The HSF-module has a -11% power loss compared with the reference module due to high dissipation of power in the connected bypass diodes.

f. Diagonal-sweep shading

As the number of shaded cells in each iteration changes, the diagonal-sweep shading scenario is the only non-linear proposed shading scenario. This unique shading

orientation provides interesting results but makes their interpretation more challenging. It somehow combines some of the effects observed in the row and column-sweep shading scenarios. As Figure 5f shows, the output power of the reference module drops rapidly by around 50 % and stops generating electricity at a shaded area of 64 %. All other designs including HSF, SR1 and SR2 show a gain of 2%, 18% and 37% over the reference in this shading scenario respectively. This can be mainly attributed to the larger number of diodes and partitioning of these modules. The 37% gain of SR2 module design is also the highest gain of all the designs and shading scenarios examined.

With all shading scenarios showcased and discussed, Figure 6 summarizes the performance results by showing the average gain of each design at every scenario and in total. Table 1 displays the same data in a numbered format.

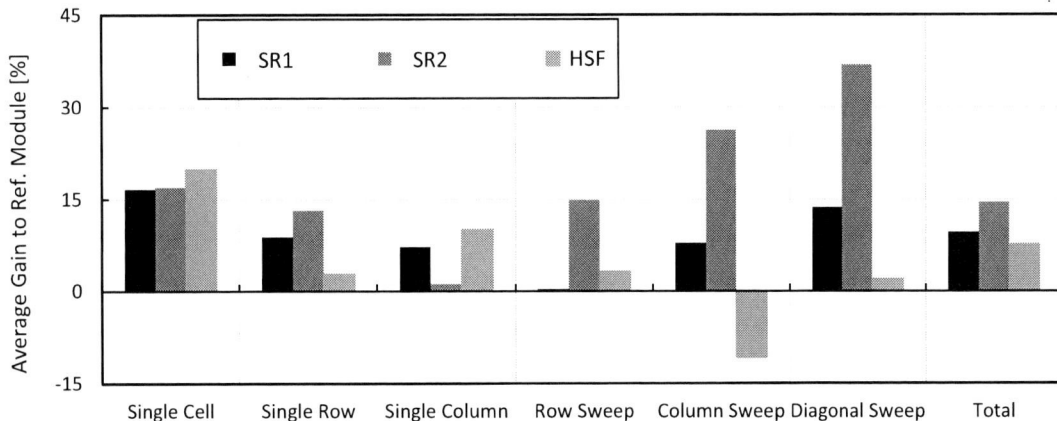

Figure 6: Average gain of the investigated designs compared to the reference module for each shading scenario and in total.

Shading scenario	SR1	SR2	HSF
Single Cell	+ 17	+ 17	+ 20
Single Row	+ 9	+ 13	+ 3
Single Column	+ 7	+ 1	+ 10
Row Sweep	+ 0	+ 15	+ 3
Column Sweep	+ 8	+ 26	- 11
Diagonal Sweep	+ 14	+ 37	+ 2
Total	+ 10	+ 14	+ 8

Table 1: Average gain [%] compared to the reference module.

4 CONCLUSION AND DISCUSSION

When creating the simulation model, care was taken to be able to simulate conditions that are as close to reality as possible. Nevertheless, there are limitations in the simulation model that cause the simulation to deviate from reality. Firstly, the simulation was carried out at a constant temperature of 25°C. Variable temperature conditions such as heating of the module and in particular the solar cells were not taken into account in the simulation model. However, in the absence of a coupled electrical-thermal model for evaluating the cell and module temperature in the simulation (as it was developed by Hanifi et al. [6] to evaluate real cells in real time), the thermal dependence of the series and shunt resistances inside the two-diode model is not considered and their values treated as constant parameters.

The results have shown that not only the number of cells affected by shading has an influence on the shading resistance of the module designs, but also the size and orientation of their internal submodules. The shading of several cells in a string has a lower effect on the module's power loss than the shading of the same number of cells distributed over several strings. For this reason, designs that perform well in one direction of shading tend to show poorer results when the shading pattern is rotated by 90°. Furthermore, the bypass diodes themselves have an influence on the module performance, as a power loss occurs in each diode when it bridges shaded cells. Therefore, extra bypass diodes does not necessarily results in a better performance.

All in all, the results can be seen as a success as it has been possible to develop new module designs with higher shading resistance compared to the standard half-cell module and the first-generation HSF-module from AESOLAR. For the new designs SR1 (10% gain) and SR2 (14% gain), it has also been possible to reduce the number of bypass diodes required compared to the HSF-module, resulting in lower production costs and greater module reliability.

5 ACKNOWLEDGMENT

The authors thank the Federal Ministry for Economic Affairs and Climate Action (BMWK) for funding this work under the SEGMENTPV project with grant number 03EE1180A.

6 REFERENCES

[1] "Energie aus erneuerbaren Quellen | Kurzdarstellungen zur Europäischen Union | Europäisches Parlament." Accessed: Sep. 22, 2024. [Online]. Available: https://www.europarl.europa.eu/factsheets/de/sheet/70/energie-aus-erneuerbaren-quellen

[2] Harry Wirth, "Aktuelle Fakten zur Photovoltaik in Deutschland," 79110 Freiburg, Sep. 2024.

[3] Sebastian Dittmann, Tales Siqueira, Hugo Sánchez Ortiz, Afshin Bakhtiari, Waldemar Meier, Carlos Meza, Ralph Gottschalg, Hamed Hanifi, "Comparison of Indoor Electrical Measurement and Outdoor Energy Yield Evaluation of Shade-Resistant PV Modules under Shading Conditions", EUPVSEC, 2023

[4] H. Hanifi, M. Pander, B. Jaeckel, J. Schneider, A. Bakhtiari, and W. Maier, "A novel electrical approach to protect PV modules under various partial shading situations," *Solar Energy*, vol. 193, pp. 814–819, Nov. 2019

[5] H. Hanifi, J. Schneider, and J. Bagdahn, Redced shading effect on half-cell modules - Measurement and Simulation, EUPVSEC, 2015.

[6] H. Hanifi, S. Regondi, B. Jaeckel, and J. Schneider, "Determination of electrical characteristics and temperature of PV modules by means of a coupled electrical-thermal model," *Journal of Renewable and Sustainable Energy*, vol. 12, p. 023501, Mar. 2020

41st European Photovoltaic Solar Energy Conference and Exhibition

OPTIMAL DESIGN FOR FLEXIBLE SOLAR PANELS
ATTACHED AROUND CYLINDRICAL POLES

Hiroki Sugimoto

PXP Corporation, Nishihashimoto, Sagamihara, Kanagawa, Japan

Hiroki.Sugimoto@pxpco.jp

ABSTRACT: Light-weight flexible solar panels attached on the cylindrical poles standing perpendicular to the ground have been expected as the power source for the high-speed communication infrastructures and the road information infrastructures. The purpose of this paper is to estimate the partial shading effects for the rolled solar panels and to suggest the optimum design for the rolled solar panels. The electrical characteristics of the rolled solar panels under the realistic outdoor environment were investigated by numerical simulation. Two types of panel designs were suggested, Type I had series cell connections to circumferential direction with bypass diodes and Type II had parallel string connections to circumferential direction without bypass diodes. As compare with the flat panel, annual power generation on Type I and Type II panels were about 53% and 67%, respectively. In the case of 12V system, the Type I panel only showed 33% power generation as compare with the flat panel. Type II design also showed the relatively constant output from morning to evening as well as from spring to winter. These results clearly showed that the Type II should be the most suitable design for the solar panels attached on the cylindrical poles.
Keywords: Chalcopyrite, Cu(In, Ga)(Se, S)$_2$, CIGSS, Flexible

1 INTRODUCTION

Recently, light-weight flexible solar cells have taken grate attention for the new field application attached on the light-weight roofs, walls and poles in Japan. Especially the solar panels attached on the cylindrical poles standing perpendicular to the ground have been expected as the power source for the high-speed communication infrastructures and the road information infrastructures. In the case of the solar panels attached on the cylindrical poles, the investigation of the partial shading effects is the most important for estimating the power generation amount for each season in Japan. However, there is few reports on the partial shading effects for the solar panels attached on the cylindrical poles. The purpose of this paper is to estimate the partial shading effects for the rolled solar panels and to suggest the optimum design for the rolled solar panels.

2 EXPERIMENTAL DETAILS

We used numerical simulation for estimating the electrical characteristics of the rolled solar panels under the realistic outdoor environment. The light-weight flexible solar cells were supposed to be the chalcopyrite Cu(In,Ga)(Se,S)$_2$ solar cells fabricated on the thin metal foils with the aperture area size of 156 x 39 mm^2[1]. The substrate of our solar cells is light-weight and flexible Ti foil with the thickness of 50um and the weight of 250g/m^2. Then we estimated two types of models as shown in Fig. 1. Type I had 48 series cell connections to circumferential direction with bypass diodes every 4 cells. Then, the 12 strings were connected in parallel to perpendicular direction. Type II had 48 series cell connections to perpendicular direction without bypass diodes. Then, 12 strings were connected in parallel to circumferential direction. At first, daily changes in the sun light spectrum were calculated by the air mass estimated from the sun altitude on each season in 2014 at Tokyo. Realistic environmental data was obtained from solar-radiation database (METPV-20 by New Energy and Industrial Technology Development Organization, Japan). Then the daily changes in the temperature of the solar panels attached on the cylindrical poles at the same days was estimated from the outside temperature and the sun light intensity on the module surface. Finally, daily changes in the electrical characteristics of the respective solar panels in each season were calculated.

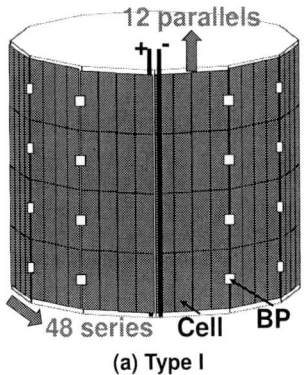
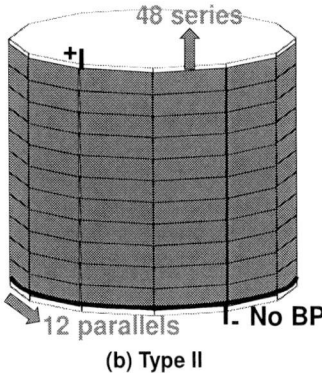

(a) Type I   **(b) Type II**

Figure 1: Schematic images of cell integration for flexible solar panels attached around cylindrical poles. (a) Type I: 48 series cell connections to circumferential direction with bypass diodes every 4 cells. Then, 12 strings were connected in parallel to perpendicular direction. (b) Type II: 48 series cell connections to perpendicular direction without bypass diodes. Then, 12 strings were connected in parallel to circumferential direction.

41st European Photovoltaic Solar Energy Conference and Exhibition

3 RESULTS AND DISCUSSION

Before calculating the daily changes in the electrical characteristics, Current-Voltage (I-V) characteristics of both types of rolled solar panels were calculated under the standard test condition (STC) as shown in Fig. 2. For comparison, I-V characteristics of a flat panel with 48 series and 12 parallel cell connections without bypass diodes was also calculated. Type I panel has multiple optimum points. The notation of OP1 and OP2 in Fig. 2 shows the highest voltage point and the second highest

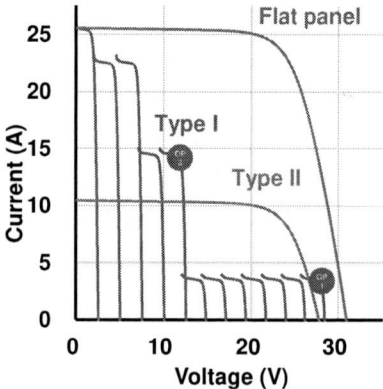

Figure 2: I-V characteristics of both types of rolled solar panels (Blue: Type I, Red: Type II) under STC. For comparison, I-V characteristics of a flat panel with 48 series and 12 parallel cell connections without bypass diodes was shown as gray line. Type I panel has multiple optimum points. OP1 and OP2 shows the highest voltage point and the second highest voltage point, respectively.

voltage point, respectively. In the case of STC condition, the type I panel showed 168W at OP2 and 101W at OP1, while the type II panel showed 211W. The flat panels also showed 551W. Note that the power conditioners for the independent power supply systems cannot trace the optimum point under the system voltage. Therefore, in the 12V system, we can operate only on the OP1.

Then the daily changes in output of the rolled solar panels Type I and Type II in each season were calculated as shown in Fig. 3. For Type I, solid lines show the optimum output on OP2 and dashed lines show the optimum output on OP1 supposed to be used in the 12V systems. For comparison, the output of the flat panel standing perpendicular to the ground and facing to south was calculated as gray lines. The rolled solar panels showed relatively constant output from morning to evening and from spring to winter, while the flat panel showed peaky characteristics. Especially the flat panel standing perpendicular to the ground showed maximum output on winter and minimum output on summer due to the sun altitude and the module temperature. Integrated daily output for respective panels on each season were summarized in Table 1. As compare with the flat panel, annual power generation on Type I and Type II panels were about 53% and 67%, respectively. In the case of 12V system, Type I panels only showed 33% power generation as compare with the flat panels. These results clearly showed that Type II design was the most suitable for the solar panels attached on the cylindrical poles. We also confirmed that relatively constant output with the amount of two-third compared to the flat panels can be obtained by Type II designed panels without bypass diodes.

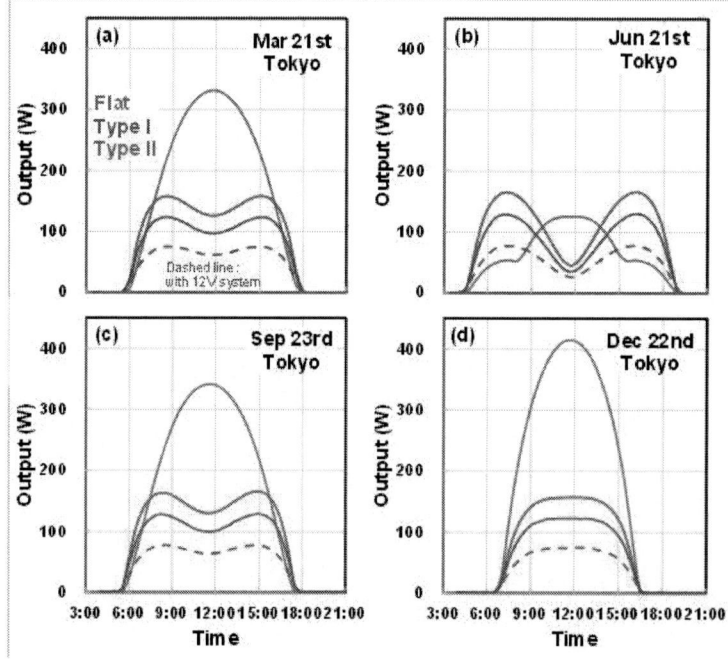

Figure 3: Daily changes in output of the rolled solar panels Type I (Blue) and Type II (Red) in each season at Tokyo, Japan. For Type I, solid lines show the optimum output on OP2 and dashed lines show the optimum output on OP1 supposed to be used in 12V systems. For comparison, the output of flat panel standing perpendicular to ground and facing to south was calculated as gray lines.

560

41st European Photovoltaic Solar Energy Conference and Exhibition

Table 1: Integrated daily output for respective panels on each season.

	Spring	Summer	Autumn	Winter	Ave
Flat	2511 Wh	1034 Wh	2592 Wh	2763 Wh	**2225 Wh**
Type I (OP2)	1206 Wh	1273 Wh	1253 Wh	942 Wh	**1169 Wh**
Type I (OP1)	751 Wh	798 Wh	771 Wh	575 Wh	**724 Wh**
Type II	1546 Wh	1638 Wh	1609 Wh	1201 Wh	**1499 Wh**

SUMMARY

Light-weight flexible solar panels attached on the cylindrical poles standing perpendicular to the ground have been expected as the power source for the high-speed communication infrastructures and the road information infrastructures. The electrical characteristics of the rolled solar panels under the realistic outdoor environment were investigated by numerical simulation. Two types of panel designs were suggested, Type I had series cell connections to circumferential direction with bypass diodes and Type II had parallel string connections to circumferential direction without bypass diodes. As compare with the flat panel, annual power generation on Type I and Type II panels were about 53% and 67%, respectively. In the case of 12V system, the Type I panel only showed 33% power generation as compare with the flat panel. Type II design also showed the relatively constant output from morning to evening as well as from spring to winter. These results clearly showed that the Type II should be the most suitable design for the solar panels attached on the cylindrical poles.

REFERENCES

[1] H. Sugimoto, Y. Hirai, and A. Yamada, Proceedings 34th International Photovoltaic Science and Engineering Conference (2023).

41st European Photovoltaic Solar Energy Conference and Exhibition

Design and Implementation of a CSI Photovoltaic Microinverter Prototype with High Frequency Switching

Francisco Guzman**, Dr. Patricio Valdivia-Lefort*, Dr. Antonio Sanchez**

*Universidad de Santiago de Chile. Av. Libertador Bernardo O'Higgins, 9170022 Estación Central, , Santiago - Chile
**Universidad Técnica Federico Santa María. Av. Vicuña Mackenna 3939, San Joaquín, Santiago – Chile

ABSTRACT

The objective of this work is to demonstrate the viability of reducing the inductance used in a photovoltaic current microinverter that feeds into the electrical grid. In order to reduce the inductance, it was proposed to increase the switching frequency of the semiconductors that compose it. It was proposed to carry out simulations and build a prototype to perform laboratory tests. The simulation included 2 switching frequencies and the quality of the current signals in the DC link as well as in the network was compared. For have a relatively accurate model of a real PV panel, I-V curves were obtained from a module in the laboratory.

The construction of the prototype involved the manufacturing and assembly of a PCB, which was connected and tested where the correct operation of the system was demonstrated; modulation, as well as its steady-state current control capacity. The work concludes with an analysis of the simulations, suggestions to improve designs, and proposals for new research related to investors, current source and high frequency switching.

SETUP

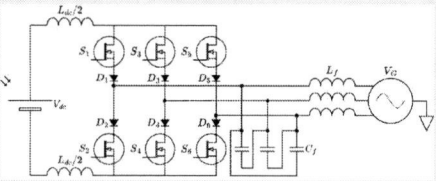

Figure 1 shows the circuit corresponding to the CSI that was used during development of this work. It is observed that an input inductance of total magnitude Ldc is used connected in series to the photovoltaic module, represented by a variable voltage source of magnitude Vdc. The inverter itself is made up of 6 semiconductors (in this case MOSFETs) and 6 rectifier diodes called Si and Di, respectively. Furthermore, it is used a low-pass filter of capacitance Cf and inductance Lf. The network to which this will connect. The whole is represented as a three-phase voltage source VG.

 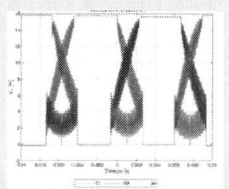

A CSI circuit is used composed of IGBTs available in the software. Although it is not the same to carry out a CSI with IGBT or MOSFET semiconductors, if these are modeled with ideal behavior, both are equivalent. Therefore, an initial approximation of the behavior is obtained of this system using a CSI with ideal semiconductors. The parameters of the simulation are presented in table 1, which are determined from the equipment and components available to build the prototype, for switching frequencies of 50[kHz] and 200[kHz]. A sampling rate of 100[MHz] is used for the microcontroller, and the system sampling rate of the SVM algorithm (from now on it will be called switching frequency at this frequency) is varied to test its performance.

RESULTS

The simulation consists of applying irradiance steps to the photovoltaic module and studying the behavior of the system (particularly the ac and dc currents) in these steps. It starts with an irradiance of 0[W/m2]. Then, every 0.06[s] the irradiance is increased to 200, 500 and 1000 [W/m2] on each step.

From the equations it is determined that for a switching frequency of 50[kHz], a minimum inductance of 6[mH]. The following results are obtained:

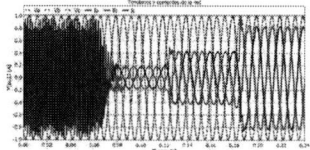 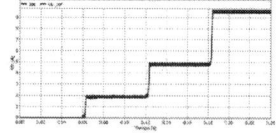

Figure 3: Voltages and currents in the network

Figure 4: Figure 4: Photovoltaic module current

Potencia de entrada [W/m²]	I_{A-RMS} [A]	$I_{A_{out_f}-RMS}$ [A]	THD_{I_A}
200	0.162	0.150	0.424
500	0.312	0.311	0.013
1000	0.594	0.594	0.012

Figure 6: Network power

Table 1: RMS value and THD of the A phase current measured in the network in the simulation at 50[kHz]

A microinverter was implemented to verify the modulation algorithm does not generate open circuits, measurements at the output voltage of the drivers that control the power MOSFETs. The off-state voltage is 0[V], and the on-state voltage is 15[V]. From the graph it is observed that one of the 3 signal outputs is always driving. The same tests are performed with signals S2, S4 and S6. The modulation indices used were M={0.25, 0.5, 0.8, 1}.

CONCLUSIONS

The simulations, being designed to demonstrate the operation of the inverter with the components of those available to build the prototype, show that by increasing the switching frequency at 200[kHz], a higher harmonic content is produced than in the case of switching at 50[kHz]. A possible cause for this effect is due to the lower inductance which is used in the DC link. The 5th and 7th harmonics are the first to appear, and those of greatest magnitude are in the range 29 and 35. These harmonics come from the switching of the CSI inverter. The inductive filter that connects the inverter with the panel allows reducing the magnitudes of these harmonics. As the inductance of the filter decreases, the magnitude increases of these harmonics at the output of the inverter. Since the frequencies are less than the frequency of the AC filter, they pass directly to the network. This must be taken into account as a precaution when seeking to reduce the inductance of the DC filter, since the standards establish maximum values for harmonics that can be injected into the network. A possible solution To keep the inductance value of the DC filter as low as possible is to tune the filter AC at a frequency lower than these harmonics (for example, at 1[kHz]) to reduce their magnitudes and deliver better quality current signals to the network.

ACKNOWLEDGEMENTS

FONDECYT INICIA N°11220697: Dynamic and stationary electrical model for bifacial heterojunction PV modules, PI: Dr. Patricio Valdivia Lefort.

REFERENCES

[1] "Tercer informe bienal de actualización de chile sobre el cambio climático," MMA, 2018, https://mma.gob.cl/wp-content/uploads/2018/12/3rd-BUR-Chile-SPanish.pdf.

[2] "Reporte mensual - sector energético," CNE, 2018, https://www.cne.cl/wp-content/uploads/2022/03/RMensual_v202203.pdf.

[3] Deline, C., "Partially shaded operation of multi-string photovoltaic systems," en 2010 35th IEEE Photovoltaic Specialists Conference, pp. 000394–000399, IEEE, 2010.

[4] Deline, C., Meydbray, J., Donovan, M., y Forrest, J., "Partial shade evaluation of distributed power electronics for photovoltaic systems," en 2012 38th IEEE Photovoltaic Specialists Conference, pp. 001627–001632, IEEE, 2012.

[5] Sahan, B., Notholt-Vergara, A., Engler, A., y Zacharias, P., "Development of a singlestage three-phase pv module integrated converter," en 2007 European Conference on Power Electronics and Applications, pp. 1–11, IEEE, 2007.

[6] Rashid, M. H., "Devices, circuits, and applications," 2007.

[7] Mohr, M. y Fuchs, F. W., "Comparison of three phase current source inverters and voltage source inverters linked with dc to dc boost converters for fuel cell generation systems," en 2005 European Conference on Power Electronics and Applications, pp. 10–pp, IEEE, 2005.

41st European Photovoltaic Solar Energy Conference and Exhibition

This presentation was selected by the Sc. Committee of the EU PVSEC 2024 for submission of a full paper to one of the EU PVSEC's collaborating peer-reviewed journals.

PV MICROINVERTERS: BALCONY POWER PLANTS, LATEST EFFICIENCY RANKINGS, YIELD CALCULATION FOR OVERPOWERED MINI PV SYSTEMS

Stefan Krauter, Jörg Bendfeld
Paderborn University, EET-NEK
Pohlweg 55, D-33098 Paderborn, Germany
E-mail: Stefan.Krauter@upb.de

ABSTRACT: The market for microinverters is growing, especially in Europe. Driven by rising electricity prices and an easing in legislation since 2024, many mini-photovoltaic energy systems are being installed. Since 2014, microinverters have been studied indoors and outdoors at Paderborn University. In the indoor lab, conversion efficiencies as a function of load have been measured with high accuracy and ranked according to Euro and CEC weightings; the latest rankings from 2024 are included in this paper. In the outdoor lab, energy yields have been measured using identical and calibrated crystalline silicon PV modules; until 2020, measurements were carried out using 215 W_p modules. Because of increasing module power, 360 W_p modules were used from 2020 until 2024. In 2024, another upgrade to 410 W_p modules occurred, taking into account the increase from 600 W to 800 W in inverter power suitable for simplified operation permission ("plug-in") in many European countries within a homogenized legislation for such mini-photovoltaic energy systems, or "balcony power plants". This legislation for simplified operation also covers overpowered mini plants, although the maximum AC output remains limited to 800 W. Presently, yield assessments are being carried out in the outdoor lab, which will take at least a year to be valid and comparable. Kits consisting of PV modules, inverters, and mounting systems are also being evaluated. Yield rankings sometimes differ from efficiency rankings due to the use of different MPPT algorithms with different MPP approach speeds and accuracies. To accelerate yield assessment, we developed a simple formula to determine energy yield for any module and inverter configuration, including overpowered systems. This is a linear approach, determined by just two coefficients, a and b, which are given for several inverters. To reduce costs, inverters will be integrated into the module frame or the module terminal box in the future).
Keywords: Microinverter; Efficiency Rating; Energy Yield Rating; Balcony Power Plant; System Performance, AC Modules.

1 INTRODUCTION

Microinverters are inverters that are connected mostly to a single PV module (occasionally to two modules; few are available for four modules, and these are not considered here), so each module–inverter combination acts as an independent power plant. The microinverter consists of a maximum power point tracker (MPPT), the DC-AC inverter, and an islanding protection unit (see, e.g., [1]). For higher power requirements, several module–inverter combinations are interconnected in parallel on the AC output side. This configuration offers various advantages, including easier planning and installation and easy up and downscaling of a plant, including allowing extensions or repairs to be carried out during power plant operation. The logistics are simplified, and the effect of shadowing is very limited. Due to low system voltages, potential induced degradation (PID) does not occur. An excellent overview of the development and advantages of microinverters was compiled by H. Oldenkamp [2]. However, the costs of power plants based on microinverters are about 10–20% higher. Some of the inverters cannot be operated by themselves and require a control unit (often combined with a remote shutdown option and a monitoring system), or a protective device for grid interfacing (depending on national regulations), thus adding extra costs. Additionally, the conversion efficiency may not be as high as for central inverters. Due to smart master–slave concepts, centralized solutions with multiple but relatively large inverters may offer higher yields under weak light conditions. The researchers in [3] provided a performance comparison for systems with microinverters, power-optimizers, and central inverters.

2 METHODOLOGIES FOR MEASUREMENTS

2.1 Measurement of Conversion Efficiency

Due to the reproducible test conditions in the indoor lab, the inverters were examined individually with predefined and controlled input data. While a traditional examination of the electronic circuits inside the inverters was almost impossible due to extensive casting compounds, components with deleted numberplates, and secret control algorithms in the microcontrollers, the investigation followed a "black box" approach, observing the input and output behavior of the device. The input was a PV module simulator, with the data set to correspond to the modules used in the outdoor test. The main output data were the delivered AC power of the inverters which is fed into the public grid. Output is also a function of input voltage. If the input voltage is too low, the inverters stop operating.

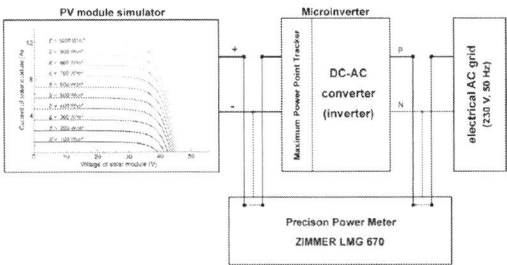

Figure 1: Set up of the indoor power and efficiency measurements using a black box approach via utilizing a precision power meter for DC inputs and AC outputs.

The following examinations were based on the possible range of input data (including voltage) given the specific PV module used for the outdoor investigation

The efficiency measurements were carried out utilizing DC input and AC output measurements obtained from a calibrated precision power meter type ZIMMER LMG 670. The measurement accuracy was $0.025 - 0.1\%$ (depending on the measurement type and range). The output power values used for the inverters (adjusted by controlling the DC input current) were continuously increased in 1024 steps from 0 to maximum. Each step took eight seconds, measurement duration was 500 ms. We tested each inverter for its maximum power, which can sometimes be slightly higher than the rated power. This procedure was only used to determine the general overall course of the efficiency curve. The exact measurements for determining the efficiencies were carried out by directly approaching the relevant numbers for EU and CEC weighted efficiencies. Between 90 and 150 measurements of 500 ms each were carried out, meaning that transient effects were not relevant.

Peak efficiency is often reached close to the maximum load of the inverter. Peak efficiency (often promoted in data sheets) is not a helpful value, since most of the time the inverters operate in the range of 20% to 40% of their rated power under non-arid conditions. Consequently, an adequately weighted efficiency is a more useful value to rate conversion devices. One type of weighted efficiency is the "European Efficiency" η_{Euro}, which is calculated by:

$$\eta_{Euro} = 0.03 \cdot \eta_{5\%} + 0.06 \cdot \eta_{10\%} + 0.13 \cdot \eta_{20\%} \quad (1)$$
$$+ 0.1 \cdot \eta_{30\%} + 0.48 \cdot \eta_{50\%} + 0.2 \cdot \eta_{100\%}$$

The other is the "CEC efficiency" of the California Energy Commission (CEC). The CEC efficiency is computed as an average value of DC–AC conversion efficiencies at six pre-defined relative output values between 10% and 100% of its rated power (with an emphasis on higher irradiance levels), and is determined by:

$$\eta_{CEC} = 0.04 \cdot \eta_{10\%} + 0.05 \cdot \eta_{20\%} + 0.12 \cdot \eta_{30\%} \quad (2)$$
$$+ 0.21 \cdot \eta_{50\%} + 0.53 \cdot \eta_{75\%} + 0.05 \cdot \eta_{100\%}$$

For the "European Efficiency", the weighting factors for high relative power values are lower according to the European irradiance statistics.

2.2. Measurements of electrical energy yield

For the PV outdoor test lab (see Fig. 2, Fig. 3) installed on the roof of Paderborn University ($51.707°$ N, $8.771°$ E), a specific test system was employed. Figure 2 shows the electrical layout of the system, including the PV modules on top with the attached microinverters and the measurement system. Each PV module consisted of 60 solar cells from the same batch of the factory. In the stated plant, these equal and calibrated modules were the input for each microinverter. The goal of the investigation was to analyze the performance of the inverters under real operating conditions, comparing their energy yield simultaneously with the climatic conditions and solar irradiance. The climatic conditions were monitored during the whole test period. The meteorological monitoring equipment consisted of two calibrated pyranometers in the plane of the module (CMP 21 and SP 2 lite by Kipp & Zonen), a 3-D ultrasonic anemometer (by Thies), a

thermo-hydro sensor (by Thies), and a thermo-moisture meter with a wind sensor, WXT 520 (by Vaisala). Each microinverter was directly connected to a calibrated electrical energy meter with an S_0-interface (see Fig. 2). To secure an accurate yield measurement, the calibrated electrical energy meters were replaced on a regular basis with freshly calibrated ones. All S_0-interfaces were connected to a server-based data acquisition system.

215 W_p modules were used until 2020, and from 2020 until 2024, we used ten 360 W_p modules (lower row, from left), as shown in Figure 4. The modules were manufactured by Solarwatt®, and the power output at the STC of each module was measured in the factory in Dresden (Germany). Additionally, one module was sent for a precision measurement to the testing laboratory ISFH in Hameln (Germany). The factory measurements were found to be very accurate (362 W_p vs. 359.34 W_p ±3% at ISFH in July 2021). In 2024, the modules were substituted by modules with a nominal power of 405 W_p, also made by Solarwatt®.

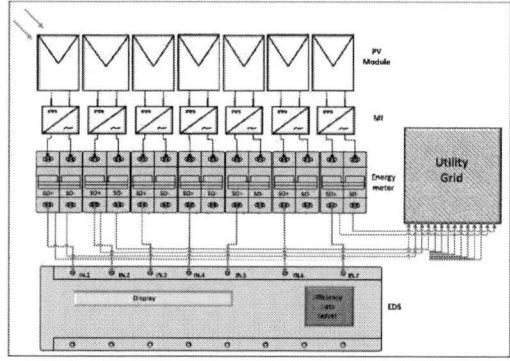

Figure 2: Outdoor measurement setup for the microinverters (MIs).

Figure 3: Configuration of the PV modules of the PV outdoor laboratory in 2023 for the electrical energy yield comparison of microinverters using eight equal, calibrated PV modules (of 360 W_p each) as inputs.

3. RESULTS of MEASUREMENTS

3.1. Conversion efficiency measurements

The measured DC-AC conversion efficiencies over the entire operation range of all inverters are shown in Fig. 4 (with one input for one module) and Fig. 5 (with two inputs for two modules).

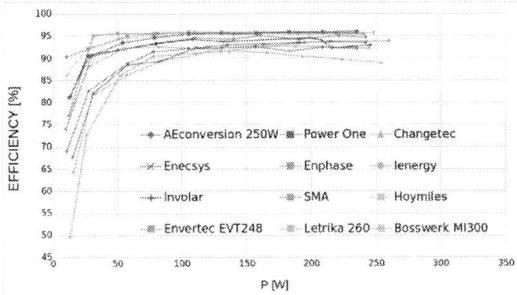

Figure 4: Measured DC-AC conversion efficiencies as a function of power output, for 12 elder microinverters with single PV module inputs.

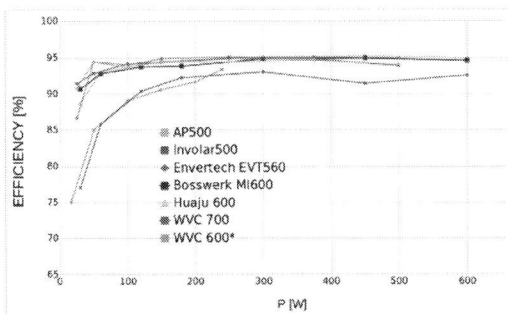

Figure 5: Measured DC-AC conversion efficiencies as a function of power output, for seven microinverters with two PV module inputs.

Based on these measurements, the European (EU) efficiencies and the CEC efficiencies for the microinverters were calculated according to (1) and (2). A total of 11 microinverters were designed for single modules, and 19 inverters had inputs for two PV modules: Anker Solix MI 60; APSystems YC 500, DS3-S; Involar MAC 500; Deye Sun 600 G3; Ecoflow Powerstream 600; Envertech EVT-560; Hoymiles MI 500, 600, 700, 800; Huaju HY 600; NEP BDM 600; Bosswerk Mi 600; Parkside PBKW-300-A1; Technaxx TX 204; Tsun TSOL-MS600; WVC 600, 700.

The ranking, considering the European (EU) conversion efficiency (1), is shown in Table I. Envertech EVT 560 and PowerOne/ABB Micro-0.25-i had the same conversion efficiency, sharing rank number 5. Hoymiles HMS-800W-2T, DeyeSun G3, and Huaju HY600 had the same EU conversion efficiency, therefore sharing rank 7. This applied also to Involar MAC 500 and Bosswerk Mi600, sharing rank 10, as well as AEconversion INV 250-45 and Enecsys SMI-S-240W, sharing rank 25.

WVC 600 stopped operating at a measured power output of 250 W. After a test run at higher temperatures, the inverter consistently failed. Since the functioning of the WVC 600 and WVC 700 inverter has been extremely poor in tests, its rated power has been assumed. For this

reason, WVC 700 is shown first in the table, with a rated power of 600 W, then the assumed 700 W (according to its type name), but finally, the maximum measured power of the WVC 700 inverter was only 600 W.

Table I: Ranking of the tested microinverters by the weighted "European Conversion Efficiency", according to (1). Nominal power (in W) is indicated in the type name.

Rank	Brand/Type	EU conversion efficiency
1	SMA Sunnyboy 240	95.4%
2	Enphase M 215	95.2%
3	Hoymiles MI 500	95.0%
4	Hoymiles MI 600	94.7%
5	Envertech EVT-560	94.6%
5	PowerOne/ABB Micro-0.25-i	94.6%
7	Hoymiles HMS-800W-2T	94.5%
7	Deye Sun 600 G3	94.5%
7	Huaju HY 600	94.5%
10	Involar MAC 500	94.3%
10	Bosswerk Mi 600	94.3%
12	Technaxx TX 204	94.2%
13	APSystems YC 500	94.1%
14	Anker Solix MI 60	93.6%
15	Bosswerk Mi 300	93.5%
16	Envertech EVT-248	93.2%
17	APSystems DS3-S	93.0%
18	Ecoflow Powerstream 600	92.7%
18	Involar MAC 250	92.7%
20	Hoymiles HM 700	92.5%
20	NEP BDM 600	92.5%
22	Tsun TSOL-MS600	92.4%
23	WVC 700 (at 600 W)	91.6%
24	Changetech ELV 300-25	90.9%
25	AEconversion INV 250-45	90.4%
25	Enecsys SMI-S-240W	90.4%
27	Ienergy GT 260	89.9%
28	Parkside PBKW-300-A1	88.9%
29	Letrika 260	88.7%
30	WVC 700 (at 700 W)	73.3%
31	WVC 600 (failed)	0.0%

Table II shows the same type of ranking, but with the CEC efficiency formula (2) applied. Hoymiles HMS-800W-2T 600 and Huaju HY 600 had the same CEC conver-sion efficiency, therefore sharing rank 6. Envertech EVT-560, Involar MAC 500, and Bosswerk Mi600 had the same conversion efficiency (within the accuracy of the measurement), sharing rank 9. The same applied to Bosswerk Mi 300 and Envertech EVT-248, sharing rank 14, while Anker MI 60 and Involar MAC 250 shared rank 16. Ecoflow Powerstream 600 and NEP BDM 600 shared rank 18, and Hoymiles HM 700 and Letrika 260S-C60-P260 shared rank 24.

Table II. Ranking of all microinverters by "CEC Efficiency", according to (2). Nominal power (in W) is indicated in the type name.

Rank	Brand/Type	CEC conversion efficiency
1	Enphase M 215	95.6%
2	PowerOne/ABB 0.25-i	95.5%
3	Hoymiles MI 500	95.4%
4	SMA Sunnyboy 240	95.1%
5	Hoymiles MI 600	95.0%
6	Hoymiles HMS-800W-2T*600*	94.9%
6	Huaju HY 600	94.9%
8	Technaxx TX 204	94.8%
9	Envertech ENV-560	94.6%
9	Involar MAC 500	94.6%
9	Bosswerk Mi 600	94.6%
12	APSystems YC 500	94.5%
13	Deve Sun 600 G3	94.4%
14	Bosswerk Mi 300	94.1%
14	Envertech EVT-248	94.1%
16	Anker Solix MI 60	93.9%
16	Involar MAC 250	93.9%
18	Ecoflow Powerstream 600	92.9%
18	NEP BDM 600	92.9%
20	Tsun TSOL-MS 600	92.8%
21	APSystems DS3-S	92.7%
22	Enecsys SMI-S-240W	92.0%
23	WVC 700 (at 600 W)	91.6%
24	Hoymiles HM 700	91.5%
24	Letrika 260	91.5%
26	Ienergy GT 260	91.4%
27	AEconversion 250	91.2%
28	Changetech ELV 300-25	90.9%
29	Parkside PBKW-300-A1	89.7%
30	WVC 700 (at 700 W)	87.5%
31	WVC 600 (failed)	0.0%

3.2 Influence of operation temperature

During comparative measurements of different PV microinverters, issues arose that are not observed in conventional efficiency measurements, but which may have an impact on the electrical energy yield.

First, either very slow or very nervous maximum power point tracking algorithms were identified, leading to a reduced energy yield. This issue was addressed by the subsequent outdoor yield measurements.

Thermal issues were also identified. As a first explanation for reduced energy yield, it is assumed that conversion efficiency degrades at higher operating temperatures. Therefore, measurements of conversion efficiency and long-term power output at elevated temperatures of 40 °C, 50 °C, and 65 °C were carried out in a heating chamber, specifically a Heratherm Oven (made by Thermo Scientific Inc.), using the same efficiency measurement equipment as mentioned above. The results were published in [4]. A change in efficiency could not be detected at temperatures up to 50 °C, despite high-precision measurements and repeated procedures. These results corresponded to the results presented above. However, it was found that individual inverters temporarily interrupted or completely stopped operation at longer operating times and higher temperatures, and a reduction in maximum power was also observed, which

could result in yield losses. Therefore, attention must be paid to the appropriate selection of inverters for the situation and ambient temperatures. Unfortunately, datasheet information is not always sufficient or reliable, and in the low-cost segment the information is often missing.

3.3 Results of energy yield measurements (outdoor)

Fig. 6 shows an example of the collected data for a single day. Due to the high time resolution, some specific characteristics could be observed, e.g., starting behaviors, MPP tracking (accuracy and speed), dropouts, and performance.

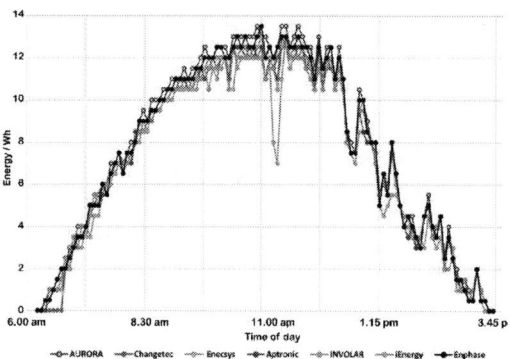

Figure 6: Actual recorded data of AC energy generation (over integrals of 5 min) for seven microinverters during the day on 10/31/2013.

Fig. 7 shows an example of the collected data for a single day during spring of 2015 (including the new SMA microinverter). Some inverters had difficulties following the rapid changes of irradiance levels during that day (e.g., Ienergy).

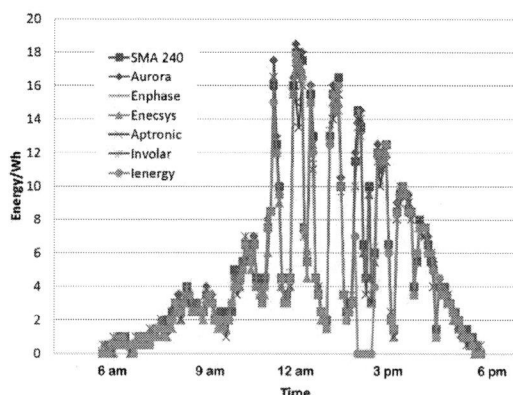

Figure 7: Actual recorded data of AC energy generation of seven microinverters (over integrals of 5 min) during the day on 4/6/2015.

An initial ranking list of the total AC energy yield of the microinverters during the common operation period (winter/spring 2014/15) was published in [3], but the specific results are largely outdated now.

Besides the effects already observed with the 215 W_p modules, such as distinct conversion efficiencies at different irradiance levels, speed, and accuracy of MPPT

41st European Photovoltaic Solar Energy Conference and Exhibition

algorithms, and minimum thresholds for initiating operation, temporal saturation effects were also observed for some inverters with the 360 W_p and 405 W_p modules applied.

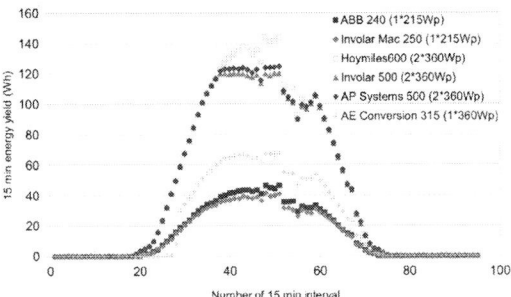

Figure 8: Example of electrical energy yield measurements (during intervals of 15 minutes) of different inverters and two different PV module sizes during a mostly clear day (some clouds in the afternoon).

The resulting electrical energy yields during the course of one day for the different microinverters and module configurations are shown in Figure 8 for a daily course and in [6] for a longer period of time for the 215 W_p modules. To some extent, the above-mentioned effects could be observed. While the different types of effects made it difficult to predict an energy yield for several configurations at some locations, a more consumer-friendly yield-predicting method was elaborated by performing a yield data analysis.

4. UNIVERSAL YIELD ASSESSMENT

To simplify the characterization of a specific combination of PV module and microinverter, a linear equation was applied to well-investigated reference characteristics of a very good inverter, without issues regarding low irradiance, MPPT, and saturation. The inverter chosen as a reference was the Enphase M 215, which was ranked #1 in the CEC efficiency ranking.

Plotting a function of the actual daily average yield (y) over the reference yield (x) gave $y = a\,x + b$ with the trivial coefficients $a = 1$ and $b = 0$ for the reference configuration (Enphase M 215 with the Q-cells 215 W_p module). Figure 9 shows the original configuration with the inverters for single modules and the 215 W_p modules attached. The coefficients of the different inverters for the relative yield equation $y = a\,x + b$ are elaborated in Table III, e.g., it can be observed that for low daily yields, Involar MAC 250 performed a little better than the reference, so b was above 0. For high yields its performance decreased (relative to the reference), so a was above 1. For the Envertech EVT 300 the characteristics were the opposite. The performance at low yields was worse than the reference, so b was negative. The relative performance increased towards high reference yields, so the steepness of the curve was higher, resulting in $a > 1$.

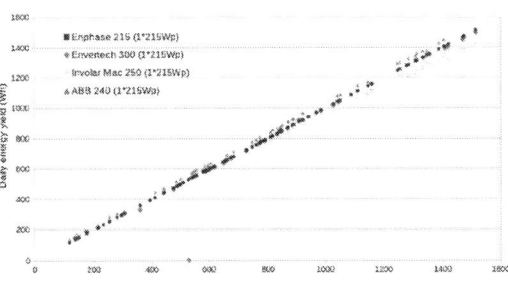

Figure 9: Electrical energy yields of different inverters for single modules with a single 215 W_p module attached. The daily reference yield (x-axis) is the energy yield (AC) achieved by an Enphase M 215 inverter with a single 215 W_p module applied.

Table III: Coefficients for the relative daily yield $y = a\,x + b$ (referenced to Enphase M 215, all with a single 160 W_p, 215 W_p, or 360 W_p module). The yield is given in daily average AC electrical energy, and the order is alphabetical.

Manufacturer	Type (module power)	a	b (Wh)
APSystems	DS-L (1 x 360 W_p)	1.66	− 17
Bosswerk	Mi 300 (1 x 215 W_p)	0.97	+ 5
Enphase	M 215 (1 x 215 W_p)	1.00	± 0
Envertech	EVT 300 (1 x 215 W_p)	1.02	− 33
Involar	MAC 500 (1 x 215 W_p)	0.92	+ 43
Lidl Parkside	PBKW-300-A1 (1 x 160 W_p)	0.67	− 41
Power One /Aurora/ABB	Micro-0.25-i (1 x 215 W_p)	1.01	+ 26

Figure 10 shows the characteristics of different microinverters that can serve two modules, either with two 215 W_p (older measurements) or two 360 W_p modules (latest measurements). Table IV shows the corresponding coefficients a (for "steepness") and b (for "offset") of the relative daily yield curve.

567

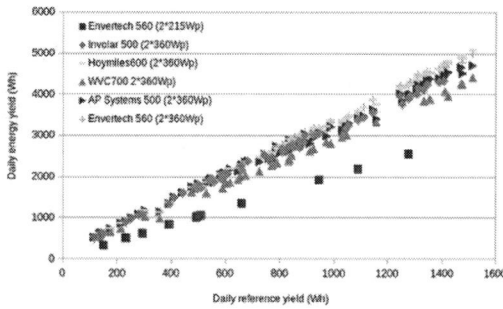

Figure 10: Average daily energy yields (AC) of different inverters for two modules with two 215 W$_p$, 360 W$_p$, or 405 W$_p$ modules attached. The reference yield (x-axis) is the yield achieved by an Enphase M 215 with a single 215 W$_p$ module applied.

Table IV: Coefficients for relative daily yield $y = a\,x + b$ for microinverters serving two modules, either 215 W$_p$, 360 W$_p$, or 405 W$_p$ types (referred to Enphase M 215 with one 215 W$_p$ module). The yield is given in daily average electrical AC energy, and the order is alphabetical.

Manufacturer	Type (module power)	a	b (Wh)
APSystems	DS-S (2 x 360 W$_p$)	3.20	+ 63
APSystems	DS-M (2 x 405 W$_p$)	3.90	−112
APSystems	YC 500 (2 x 360 W$_p$)	2.95	+ 255
Bosswerk	Mi 600 (2 x 360 W$_p$)	3.12	+ 112
Deye	Sun 600 G3 (2 x 215 W$_p$)	1.88	+ 62
Deye	Sun 600 G3 (2 x 360 W$_p$)	3.12	+ 92
Deye	Sun 800 G4 (2 x 405 W$_p$)	3.68	+ 90
Envertech	EVT 560 (2 x 215 W$_p$)	1.98	+ 38
Envertech	EVT 560 (2 x 360 W$_p$)	3.23	+ 110
Hoymiles	MI 600 (2 x 360 W$_p$)	3.19	+ 168
Hoymiles	MI 700 (2 x 360 W$_p$)	3.25	+ 133
Hoymiles	MI 700 (2 x 405 W$_p$)	3.64	+ 296
Hoymiles	HM 800-T2 (2 x 405 W$_p$)	3.81	+ 142
Huaju	HY 600 (2 x 360 W$_p$)	3.14	+ 154
Involar	MAC 500 (2 x 360 W$_p$)	2.89	+ 181
NEP	BDM 600 (2 x 360 W$_p$)	2.70	+ 276
Technaxx	TX 204 (2 x 360 W$_p$)	3.16	+ 190
WVC	WVC 700 (2 x 360 W$_p$)	2.75	+ 172

The coefficients of determination R for all regressions of all measurement values to determine the coefficients a and b were in the vicinity of 0.98.

5   CONCLUSIONS

Efficiency rankings alone do not necessarily reflect the energy yield, so yield measurements are helpful and were carried out in former publications, e.g., [3]. However, due to drastically reduced module prices and relatively stable inverter prices, overpowering has become quite common, which makes specific yield measurements ineffective due to the enormous number of possible configurations of inverter and module sizes.

The use of a reference configuration together with the two coefficients of a linear equation is a simple method to describe the daily yield performance of any microinverter in combination with any PV module, even under or oversized ones. While the prices of PV modules are decreasing at a higher pace than prices for microinverters, we will see more configurations with oversized modules and more saturated microinverters more often in the future. Even the legislation (e.g., in Austria, Germany, Switzerland) is considering overpowered systems and is often limiting power on the AC side only. This underlines the necessity of a method, such as the one described here, to extrapolate the energy yield.

6   OUTLOOK

While the costs of the modules are often now lower than the costs of the inverters, the pressure for cost reduction of the inverters is evident. After a cost investigation of the internal components of a typical microinverter, it was determined that further cost reduction not possible on this side. However, in the external components such as cables, connectors, and casing costs, a reduction may be possible if the inverter can be integrated into the module terminal box or the frame of the module, making most of the external components obsolete. A step in this direction was carried out by the company SolarNative©, which integrated the inverter into a square tube that could be placed inside the module frame. Going further, this would lead to an "AC module" that would use the casing provided by the module frame or by the module terminal box, so the DC wiring would be internal only, making DC connectors obsolete and thus making the installation of the complete PV system easier. Such AC modules have been described in the literature, e.g., [2, 7, 8], however have not yet made it beyond prototype status. A relevant challenge is the stringent tests for PV module certification based on the IEC 61215, which would have to be applied to the integrated inverter as well. These include quick thermal changes between -40 °C and +85 °C and a 1000 h damp heat test at 85 °C and 85% humidity. Usually, electronics fail under such conditions.

7 REFERENCES

[1] W-F. Lai, S-M. Chen, T-J. Liang, K-W. Lee, A. Ioinovici, Proceedings IEEE Energy Conversion Congress and Exposition (2012) 2426.

[2] H. Oldenkamp, I.J. de Jong, Proceedings 24th European Photovoltaic Solar Energy Conference (2009) 3101.

[3] S. Krauter, J. Bendfeld, Proceedings IEEE 42nd Photovoltaic Specialist Conference (2015), 1-4.

[4] D. Stellbogen, P. Lechner, M. Senger. Proceedings 32nd European Photovoltaic Solar Energy Conference (2016) 1654.

[5] S. Krauter, J. Bendfeld, Proceedings 36th European Photovoltaic Solar Energy Conference (2020), 1179.

[6] S. Krauter, J. Bendfeld, Proceedings 37th European Photovoltaic Solar Energy Conference (2020) 935.

[7] R.H. Wills, F.E. Hall, S.J. Strong, J.H. Wohlgemuth, Conference Record 25th IEEE Photovoltaic Specialists Conference (1996) 1231.

[8] D. Leuenberger, J. Biela, IEEE Transactions on Power Electronics, vol. 32, no. 8, (2017) 6105.

41st European Photovoltaic Solar Energy Conference and Exhibition

Aging behavior of polymeric materials used in inverter casings

E. Helfer[1,2], P. Christöfl[1], J. Petro[1], M. Lang[1], V. Reisecker[3], L. Heupl[4], A. Weiermair[4], G. Oreski[1,2]

[1] Polymer Competence Center Leoben GmbH (PCCL), Roseggerstraße 12, 8700 Leoben, Austria – eric.helfer@pccl.at
[2] Institute of Materials Science and Testing of Polymers, Montanuniversität Leoben, Otto Glöckel-Straße 2, 8700 Leoben, Austria
[3] Transfercenter für Kunststofftechnik GmbH (TCKT), Franz Fritsch Strasse 11, 4600 Wels, Austria
[4] Fronius International GmbH, Günter Fronius Straße 1, 4600 Thalheim bei Wels, Austria

INTRODUCTION AND OBJECTIVES

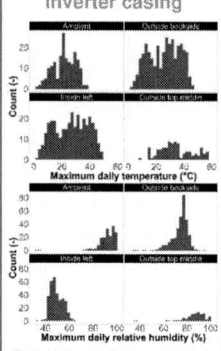

Cover – polymer/metal

Main body – polymer

© Fronius GmbH

Fig. 1: Inverter casing with cover and polymeric main body under environmental stresses.

Objectives
- Assessment of the aging behavior of the polymeric material used in inverter casings.
- Determine which aging process is critical:
  - Thermo-oxidation
  - Hydrolysis
  - Photo-oxidation

EXPERIMENTAL

Polycarbonate/Polybutylene terephthalate (PC/PBT) blend

Inverter casing samples
- Field aged inverters casings from all over the world have been collected.
- Casings have been placed in Arizona and Florida and are collected semiannually.
- Test specimens were cut out of two different sections of the collected inverter casings:
  - on an UV exposed side (sidewall – SW)
  - on a non UV exposed side (backside – BS)

Xenon exposure of plate samples
- Condition A3 of DIN EN ISO 4892-2
- Total of 3000 h with 1000 h steps

Microclimate of the inverter casing
- Placement of temperature/humidity sensors.

Analysis methods
- Attenuated Total Reflectance-Fourier Transform Infrared (FTIR-ATR) Spectroscopy
- Ultraviolet/Visible/Near Infrared (UV/Vis/NIR) Spectroscopy → Change of color (ΔE)
- Differential Scanning Calorimetry (DSC)
- Nanoindentation (NI)

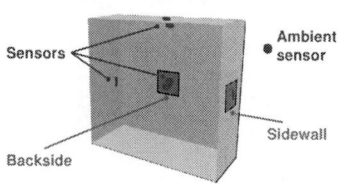

Ambient sensor

Sensors

Backside

Sidewall

Fig. 2: Position of the extracted samples from the inverter casing main body and placement of temperature/humidity sensors.

RESULTS

Micro climate of the inverter casing

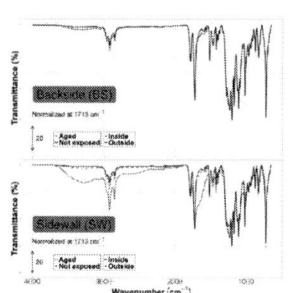

Fig. 3: Maximum daily temperatures (top) and relative humidity (bottom) of the casing.

- Ambient temperatures reach a maximum of close to 40 °C.
- Parts of the casing reach maximum temperatures of up to ~60 °C.
- Backside and top outside show a maximum relative humidity of close to 100%.
- Inside of the inverter casing shows a maximum relative humidity of ~65%.

Inverter casing and test plate samples

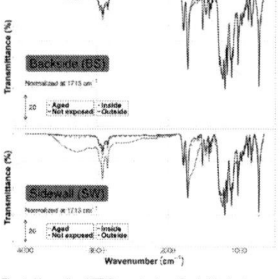

Fig. 4: Normalized FTIR spectra for a Backside (top) and Sidewall (bottom) sample of a field retriever.

- SW sample on the outside (UV exposed) shows a change of chemical structure:
  - e.g. peak in the area 3800-2200 cm⁻¹ (OH)-groups
- BS samples (outside & inside) and SW sample (inside) – non UV exposed – do not show a change in their chemical structure.

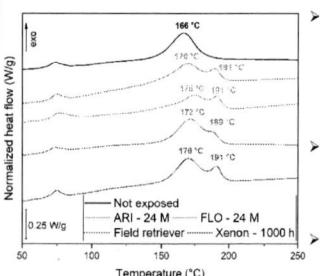

Fig. 6: DSC cooling curves for UV exposed casing (Arizona, Florida, field retriever) and xenon exposed samples in comparison to the not exposed material.

- All UV exposed samples show an increase in crystallization temperature and a double crystallization peak.
- Xenon exposed sample shows a more pronounced double crystallization peak.
- Non UV exposed samples showed no change in their thermal properties.

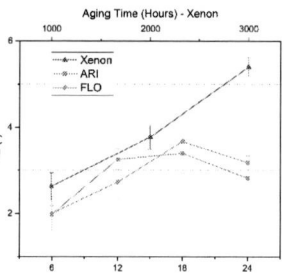

Fig. 5: Change of color of UV exposed casing (Arizona, Florida) and xenon exposed samples.

- Change in color for the xenon exposed samples follows a linear trend.
- After 18 months, change in color for UV exposed samples from Arizona and Florida reaches a plateau.
- Non UV exposed samples showed no change in color.

Fig. 7: Young's modulus obtained by nanoindentation over the cross-section for xenon exposed plate sample.

- Exposed side shows an increase in Young's modulus (~5000 MPa). → Surface embrittlement
- The UV-exposure seems to affect the testing plate up to a depth of ~0.6 mm.
- Non UV exposed side does not show a change in Young's modulus.

CONCLUSION AND OUTLOOK

Thermo-oxidation ✗
Hydrolysis ✗
Photo-oxidation ~

Highly localized degration only at UV exposed surfaces.

Bulk of the material is unaffected.

➡ No failure in the field due to aging has been observed.

Check out:
"Analysis and Material Modeling of Mechanical Property Degradation for Simulation of Weather Exposed Polymers"
3AV.2.20

The research work was performed at the Polymer Competence Center Leoben GmbH (PCCL, Austria) within the framework of the COMET-program of the Federal Ministry for Climate Action, Environment, Energy, Mobility, Innovation and Technology and the Federal Ministry of Science, Research and Economy with contributions by academic and commercial partners. The PCCL is funded by the Austrian Government and the State Governments of Styria, Lower Austria and Upper Austria.

PERFORMANCE OF ARC FAULT CIRCUIT INTERRUPTERS IN PHOTOVOLTAIC INVERTERS CONNECTED TO LONG DC CABLES

Donat Hess, David Joss, Christof Bucher
[1] Bern University of Applied Sciences (BFH), School of Engineering and Computer Science (TI), Institute for Energy and Mobility Research (IEM), Laboratory for Photovoltaic Systems (PV-Lab)
christof.bucher@bfh.ch,

ABSTRACT: Photovoltaic arc fault circuit interrupters (PV AFCI) have been used in the USA for a long time. Today, inverters for the European market are also increasingly incorporating AFCI due to the fact the IEC 63027 standard was launched in 2023, too. This paper examines how well integrated AFCI work with long DC cables, such as large PV systems or PV systems on high-rise buildings whose inverters are installed far away from the PV modules. The AFCIs were tested at different cable lengths, different current-voltage configurations, and various arc positions in the string. It is shown that the different settings influence the performance of the AFCIs, and differences between the individual inverters are also observed. However, all AFCI work well for typical expected cable lengths across all tested manufacturers. Further tests showed that weaknesses only appear with very long cables.

1 AIM AND APPROACH

While arc fault detection is already standard in USA, it is also becoming increasingly important in Europe. Arcs can cause damage to system components due to the high-temperature plasma they generate [1]. Such system failures have led to fires in several instances. Notable examples include the fires in Bakersfield (2009) [2] and Mount (2011) [3]. These fires contributed to the requirement that rooftop PV systems, and later in 2014, all types of PV systems in the USA with a DC operating voltage above 80 V, must include a series arc fault circuit interrupter [4]. Today, inverters for the European market are also increasingly incorporating AFCI. For example, manufacturer like Huawei, SMA, and SolarEdge declare that they have inverters with AFCI function. In 2023 the standard IEC 63027:2023 has been published which is the first with international validity [5].

This study examines the functionality of inverter integrated AFCIs from different manufacturers available on the European market. Despite the increasing number of inverters with integrated AFCIs in Europe, there are still few studies that assess their functionality. The functionality of the AFCIs is tested in approximation on the requirements of the standard IEC 63027:2023. This standard does not specify conditions for cable length. Therefore, this study tests the functionality of AFCIs at different cable lengths and settings. In addition to cable length, the voltage-current configuration and the arc position within the string are also varied, and their impact on the performance of the AFCIs is evaluated.

Inverters of the 10-kW residential class from three different manufacturers were tested. The tested devices are the SUN2000-10KTL-M1 (Huawei), the Sunny Tripower X12-50 (SMA), and the SE10k-RW0TEBEN4 (SolarEdge). Inverters of these brands are selected, as they are among the most prevalent inverter manufacturers and additionally declare that they have inverter with AFCI function. An overview of the devices and the abbreviations used further in the text can be seen in Table I. The devices were properly procured on the market and put into operation in the laboratory with Swiss country settings. With the SE10k-RW0TEBEN4 from SolarEdge Arc detection functionality according to IEC 63027:2023 was only enabled with manufacturer support.

Table I: Overview of devices under test (DUT) and their abbreviations. It is also indicated whether the inverters are operated with or without optimizers.

Abbreviation	Name	Manufacturer	Optimizer
DUT 1	SUN2000-10KTL-M1	Huawei	SUN2000-450W-P (Optional)
DUT 2	Sunny Tripower X12-50	SMA	No
DUT 3	SE10K-RWOTEBEN4	SolarEdge	S440-1GM4MRM

The functionality of the AFCIs is tested under three different current-voltage configurations. The current-voltage configurations are shown in in Table II. Also shown are the separation rate and the maximum gap of the electrodes that generate the light arc during the arc tests. Test A is with low current, low voltage and low separation rate as well as a small gap (test A). Test B is with high current, low voltage, high separation rate and small gap, and test C includes high current, high voltage, high separation rate and opens the electrodes to a large gap. DUT 3 (3-phase) works with power optimizers and a constant bus voltage of 750 V_{DC}. To ensure that its bus current still corresponds to the I_{mpp}, as with the other inverters, its V_{mpp} was set to 750 V for all settings.

Table II: Arcing test conditions. Voltage values in brackets are for DUT 3.

	Test A (low current)	Test B	Test C (high voltage)
I_{mpp} [A]	4	8	8.5
I_{sc} [A]	4.4	8.8	9.4
V_{mpp} [V]	318 (750)	318 (750)	607 (750)
V_{oc} [V]	490 (870)	490 (870)	810 (870)
Sep. rate [m/s]	2.5	5	5
Gap [mm]	0.8	0.8	2.5

Additionally, the AFCIs are tested on arcs located at the start and in the middle of string. For the arc at start of string, the arc generator was positioned next to the inverter at the positive output. For the arc in the middle of string, it was positioned between the 7th and 8th (of 15) PV simulators (see Figure 1). DUT 2 can only be used without optimizers, while DUT 3 must be used with optimizers. DUT 1, however, can be operated with or without optimizers. In these tests, DUT 1 was used without optimizers unless otherwise specified.

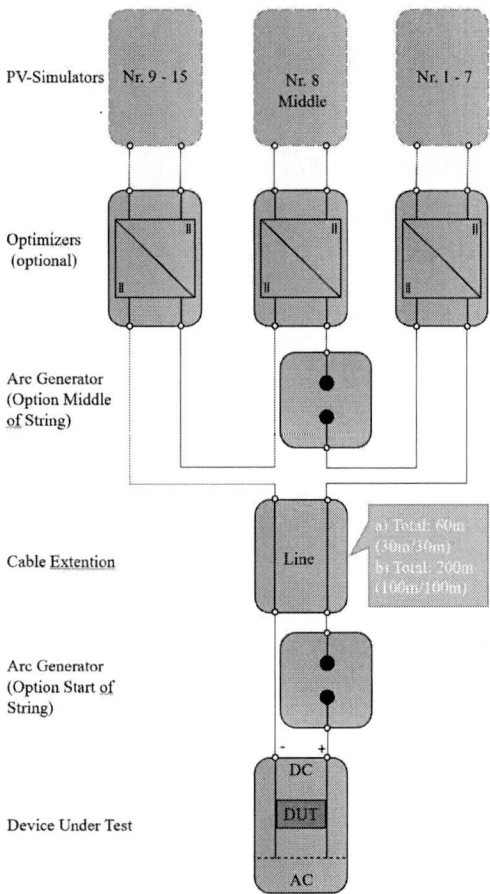

Figure 1: Overview of the test configurations. For the sake of simplicity, only the first, the middle and the last PV simulator are shown. The optimizers were used only with DUT 3 and with DUT 2 when explicitly mentioned. The arc generator was installed either at the "Start of String" or at the "Middle of String" option.

Furthermore, the AFCIs are tested with real PV installation cables of different lengths between the inverter and the PV simulators. One cable length is a short one of 60 m, and the other is a long one of 200 m. The cable length is divided between the positive and negative outputs of the inverter (see Figure 1). The number of

measurements done for each setting is listed in Table III. Due to a configuration failure on the test system side, for DUT 2 only 2 measurements could be used in the end for Test C at the start of the string. Since DUT 3 operates with a constant bus voltage of 750V, Test B and Test C do not differ significantly for this system. Therefore, Test C was omitted for DUT 3. The cables were positioned (DC+ and DC- adjacent to each other) at a distance of 1.5 m from pair to pair on the stone floor in the laboratory.

Table III: Number of measurements done for different inverters, test types (A, B, C), arc positions (start of string, middle of string), and cable lengths (60m, 200m).

	Test A		Test B		Test C	
	S	M	S	M	S	M
DUT 1						
60 m	8	6	5	5	5	4
200 m	12	10	10	10	10	10
DUT 2						
60 m	7	5	5	6	2	5
200 m	5	5	5	5	5	5
DUT 3						
60 m	5	5	5	5	-	-
200 m	5	5	5	5	-	-

Additionally, to explore the system limits, tests were conducted where the inductivity of the 200 m cable was increased. Therefore, the cable was set in series with 2 suppression chokes and with a cable rolled up on a bobbin acting as an air coil. The chokes have an inductivity of 80 µH each, while the inductivity of the cable on the bobbin can be varied by changing the number of windings. In this way, inductivity corresponding to the inductivity of cables with lengths of 478 m ± 45 m, 730 m ± 90 m, 899 m ± 116 m, 1207 m ± 160 m, 1640 m ± 223 m, and 2320 m ± 342 m were generated. For converting inductivity of a system L_S to a corresponding cable length l, it was assumed that the inductivity of a cable increases proportionally with its length. The proportionality factor was determined with the 200 m cable, yielding for the cable length

$$l = L_S \frac{200\,m}{L_{200}},$$

where L_{200} is the inductivity of the 200 m cable. Since the inductivity is frequency-dependent, S was determined for 100 Hz, 1 kHz, 10 kHz and 100 kHz and their mean value was taken. The uncertainty was determined by the standard deviation of the frequency-dependent lengths.

41st European Photovoltaic Solar Energy Conference and Exhibition

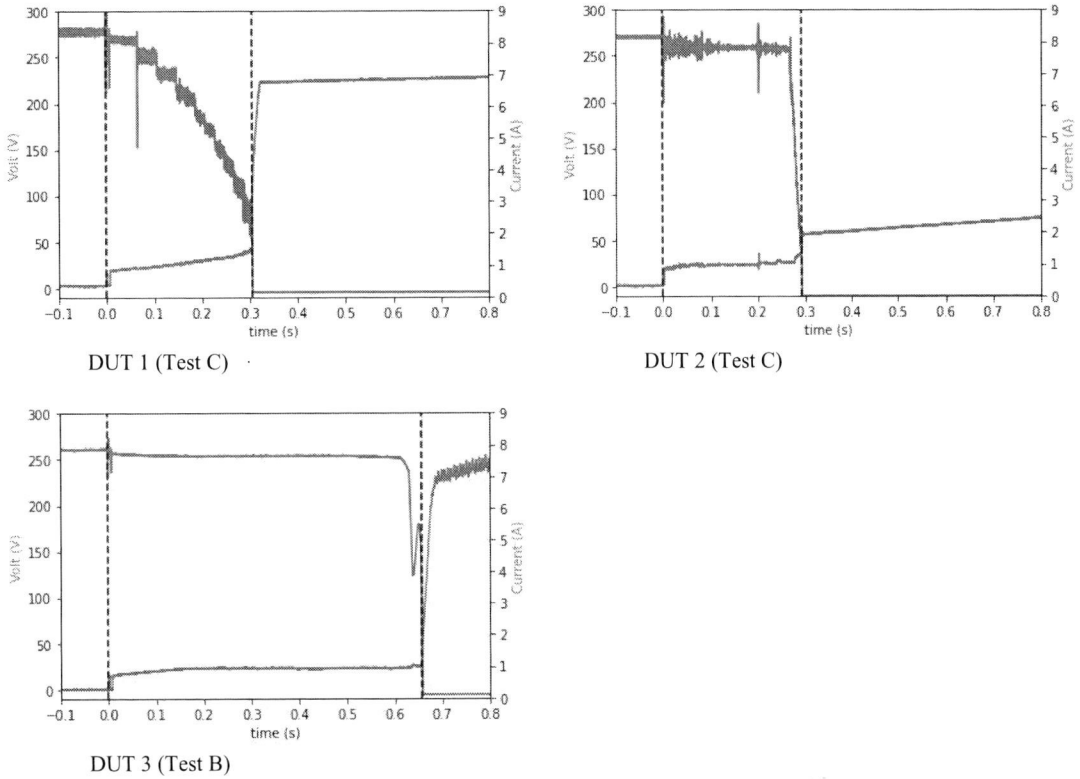

DUT 1 (Test C)

DUT 2 (Test C)

DUT 3 (Test B)

Figure 2: Examples of arc current and arc voltage curves for the different inverters. The first vertical dashed line indicates the start of the arc and the second its end.

To pass the operation criteria of the international standards [5] an arc must be extinguished within 2.5 s and before it has reached 750 J. According to the standards, the arc time starts when the voltage between the two electrodes of the arc generator is greater than 10 V and ends when the arc current is less than 0.25 A. Examples of arc voltage and arc current curves for different inverters can be seen in Figure 2. Additional to the standard, in this study, it was defined that the current must remain below 0.25 A for longer than 10 ms in order to avoid premature termination in the event of unstable arcs. The arc energy E is defined as the energy released during the arc and can be calculated as

$$E = \int_{t_{start}}^{t_{end}} U(t) \cdot I(t)\, dt,$$

where t_{start} and t_{end} are the arc start and end times, U is the arc voltage, and I is the arc current.

An overview of the laboratory set-up can be seen in Figure 3. The PV simulators/ module-level power electronics (MLPE) simulator are divided into 3 racks. The arc generator controller (computer) is used to control the separation rate and the maximum gap between the electrodes of the arc generator. The oscilloscope is used to measure and record the arc voltage and arc current. The DUTs, which are tested one after the other, can be seen in the picture on the right. The arrangement of the cables and

the cable extensions are not shown.

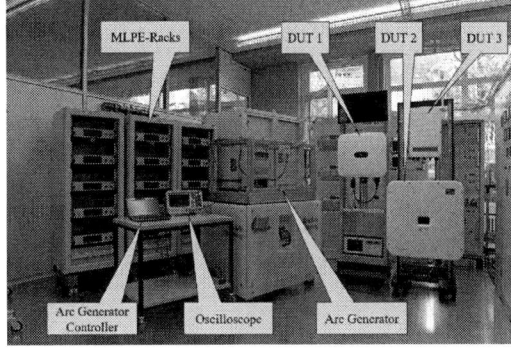

Figure 3: Laboratory set-up during the tests. The cables and cable extensions are not shown.

2 SCIENTIFIC INNOVATION AND RELEVANCE

The standard "IEC 63027: Photovoltaic power systems - DC arc detection and interruption" specifies how arc detectors in PV inverters must be tested. However, the cable length is system-specific and is not tested in the standards. This research project examines whether the AFCI still works with long cables.

41st European Photovoltaic Solar Energy Conference and Exhibition

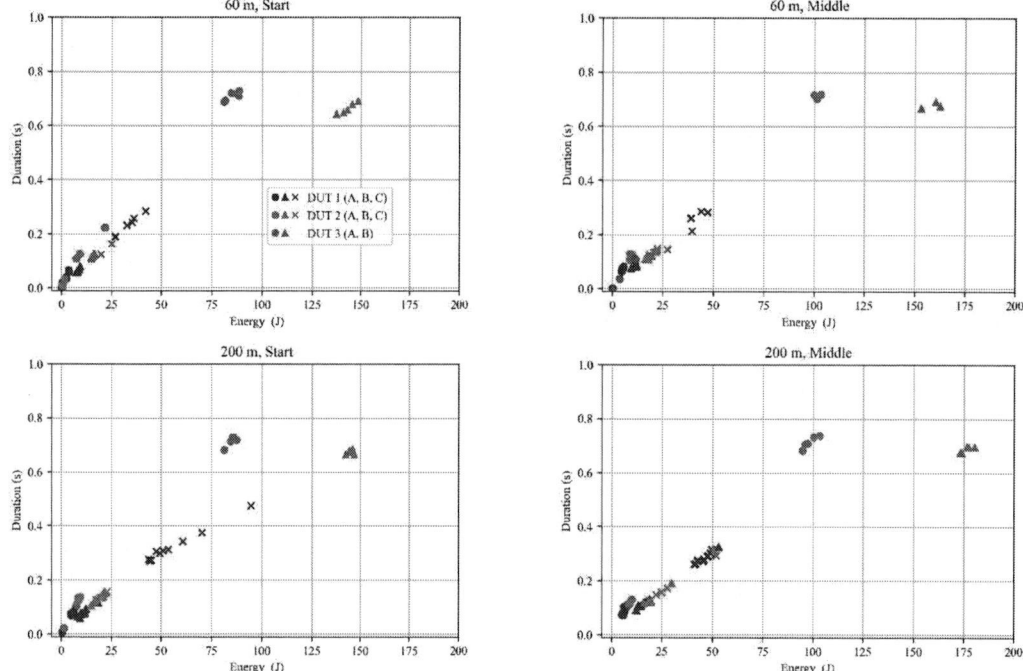

Figure 4: Overview of the arc energy and arc times of all measured arcs in the start (left) and middle of string (right), with cable lengths of 60 m (top) and 200 m (bottom), as well as test settings A, B, and C.

3 RESULTS AND CONCLUSION

3.1 AFCI tests with cable lengths up to 200 m

The arc detectors of all three tested inverters perform well according to the operational criteria of IEC 63027:2023 for typical cable lengths up to 200 m, as well as for both low and high voltage and current scenarios. Figure 4 shows the arc times and arc energies for two cable lengths, two arc positions, three systems, and three (DUT 3: two) current-voltage configurations. Arcs in systems with low bus current (4 A in Test A) and high bus current (8.5 A in Tests B and C), as well as arcs in systems with low voltage (318 V in Tests A and B) and high voltage (607 V, 750 V for DUT 3 in Test C), all passed the test, regardless of the arc position or cable length. No arc time exceeded 0.8 s, and no arc energy was greater than 200 J, while arc times up to 2.5 s and arc energies up to 750 J are allowed.

3.2 Dependencies of arc energy

However, the performance of the AFCI depends on the arc position (start vs. middle of string), as well as on the cable length. An overview of the average arc energies measured under the given conditions can be seen in Figure 5. The standard errors of the mean (SEM), calculated as $SEM = \sigma/\sqrt{n}$ are indicated by the error bars, where σ is the standard deviation and n the number of measurements. A tabular summary of the results in Figure 5 can be seen in Table IV for DUT 1, Table V for DUT 2, and Table VI for DUT 3. In addition to the mean values, the standard errors of the mean (SEM) and for samples also the standard deviations are given. Mean values are also given over the cable lengths or arc position. The uncertainty, after averaging two mean values, was determined using Gaussian error propagation and is based on the SEM.

Since only a few measurements (2-12) are available and therefore the distribution function of the energies is not clearly identifiable, the uncertainties given should be regarded with caution. Nevertheless, they give a sense of the significance of the differences of the mean values.

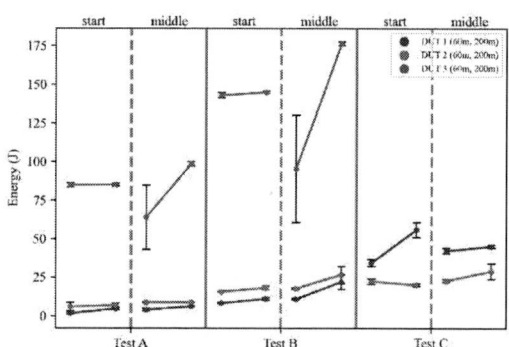

Figure 5: Mean arc energy for different voltage-current configurations (test A, test B and test C), arc positions (middle and start of string), as well as the cable lengths (left 60 m and right 200 m). The error bars indicate the standard errors of the mean.

A small increase in arc energy with increased cable length becomes evident when comparing arc energies in settings with cable lengths of 60 m to settings with cable lengths of 200 m. The average arc energies are slightly lower in most measurement series with cable lengths of 60 m compared to the corresponding series of 200 m (compare the data point at the left end of the lines with the data points at the right end in Figure 5 or see Table IV - Table VI for numeric values). The only exception is DUT

574

2 at the start of the string in test C, for which for the 60 m cable only 2 measurements are given. Averaged over the arc position, however, the arc energy increases for all manufacturers and tests with greater cable length (see mean values in Table IV - Table VI). Although differences in arc energy are observed with different cable lengths, the differences are to small to be critical for passing the test.

In addition to the cable length, the arc position also has a slight influence on the arc energy. In most cases, arcs in the middle of the string have slightly higher arc energies than at the start of string (see Figure 5). Exceptions in the measurement series are DUT 1 at 200 m in test C and DUT 3 at 60 m in test A and B (Figure 5 and Table IV - Table VI for numeric values). Averaged over the two cable lengths, however, the arc energy is larger for arcs in the middle of the string than at the start of string for all manufacturers and tests (See Table IV - Table VI). Even though the arc energy of arcs in the middle of string is slightly higher than for arcs at the start of string, the differences are also small and not relevant for passing the test.

Table IV: Overview of the average arc energies for different measurement settings for the DUT 1. The values are to be read as mean ± SEM (± σ).

Cable Length	Arc Position	Test A [J]	Test B [J]	Test C [J]
60 m	Start	2.29 ± 0.62 (± 1.77)	8.22 ± 0.33 (± 0.74)	34.54 ± 2.21 (± 4.95)
	Middle	4.12 ± 0.75 (± 1.84)	10.86 ± 0.41 (± 0.93)	42.31 ± 1.66 (± 3.32)
	Mean	3.20 ± 0.49	9.54 ± 0.27	38.42 ± 1.38
200 m	Start	4.73 ± 0.60 (± 2.09)	11.08 ± 0.87 (± 2.74)	55.94 ± 4.78 (± 15.12)
	Middle	6.08 ± 0.26 (± 0.81)	21.97 ± 4.72 (± 14.93)	45.21 ± 0.94 (± 2.97)
	Mean	5.40 ± 0.33	16.53 ± 2.40	50.58 ± 2.44
Mean 60m and 200m	Start	3.51 ± 0.43	9.65 ± 0.46	45.24 ± 2.63
	Middle	5.10 ± 0.40	16.42 ± 2.37	43.76 ± 0.95
	Mean	4.30 ± 0.29	13.03 ± 1.21	44.50 ± 1.40

Table V: Overview of the average arc energies for different measurement settings for DUT 2 The values are to be read as mean ± SEM (± σ).

Cable Length	Arc Position	Test A [J]	Test B [J]	Test C [J]
60 m	Start	5.85 ± 2.74 (± 7.26)	15.59 ± 0.28 (± 0.62)	22.33 ± 1.91 (± 2.71)
	Middle	8.80 ± 0.21 (± 0.48)	17.72 ± 0.46 (± 1.11)	22.74 ± 1.06 (± 2.38)
	Mean	7.32 ± 1.38	16.65 ± 0.27	22.54 ± 1.09
200 m	Start	7.02 ± 1.32 (± 2.95)	18.18 ± 1.25 (± 2.80)	20.10 ± 0.78 (± 1.74)
	Middle	8.74 ± 0.29 (± 0.66)	26.67 ± 5.68 (± 12.71)	29.02 ± 5.25 (± 11.73)
	Mean	7.88 ± 0.68	22.42 ± 2.91	24.56 ± 2.65
Mean 60m and 200m	Start	6.43 ± 1.52	16.88 ± 0.64	21.21 ± 1.03
	Middle	8.77 ± 0.18	22.19 ± 2.85	25.88 ± 2.68
	Mean	7.60 ± 0.77	19.54 ± 1.46	23.55 ± 1.43

Table VI: Overview of the average arc energies for different measurement settings for DUT 3. The values are to be read as mean ± SEM (± σ).

Cable Length	Arc Position	Test A [J]	Test B [J]
60 m	Start	84.74 ± 1.42 (± 3.18)	142.92 ± 1.72 (± 3.84)
	Middle	63.89 ± 20.63 (±46.12)	101.99 ± 37.78 (±84.48)
	Mean	74.32 ± 10.34	122.46 ± 18.91
200 m	Start	84.96 ± 0.90 (± 2.01)	145.08 ± 0.69 (± 1.54)
	Middle	98.31 ± 1.39 (± 3.12)	176.27 ± 1.14 (± 2.56)
	Mean	91.63 ± 0.83	160.68 ± 0.67
Mean 60m and 200m	Start	84.85 ± 0.84	143.99 ± 0.92
	Middle	81.10 ± 10.34	139.13 ± 18.90
	Mean	82.97 ± 5.19	141.57 ± 9.46

3.3 AFCI tests with emulated extended cables

Further tests with DUT 1 showed, that only with emulated extended cable lengths any arcs occur that exceed the arc energy and arc time threshold set by IEC 63027:2023. With a simulated cable length of 478 m, investigations using DUT 1 (test C with an arc in the middle of the string) showed that 9 out of 10 arcs were extinguished before reaching the energy value of 750 J. If the cable length is extended to over 2 km, half of the arcs are still detected and extinguished before reaching the energy threshold (see Figure 6). Notably, even if not all tests were passed, those that were passed were very clear. For example, at 2 km, the 5 arcs that were extinguished in time had an arc energy of under 100 J. This suggests that the longer cable length primarily increases the likelihood of non-detection, but only slightly affects the arc time and arc energy in the detected arcs.

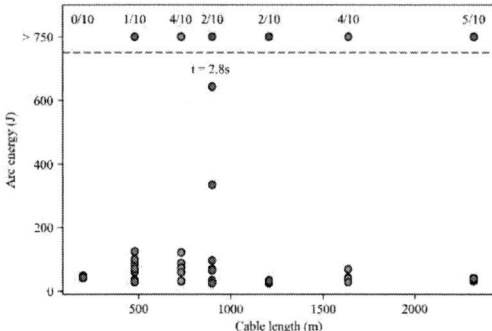

Figure 6: Measured arc energies at emulated extended cable lengths in the setup with DUT 1. The number of arcs that exceeded 750 J in relation to the number of measurements can be seen above the dashed gray line. The arcs that exceeded 750 J were not reported by the inverter. One arc has reached the maximum permissible arcing time of 2.5 s without reaching 750 J.

3.4 Identification of arc fault location

Tests with DUT 1 used with optimizers show that these inverters can approximately identify the position of light arcs. However, in most cases, the exact position of the arcs was missed by one optimizer. If the bus current is interrupted after detecting an arc, the Huawei FusionSolar App can display which optimizers were affected by the disconnection. An example of this display is shown in Figure 7. It indicates an arc between the 1st and 2nd optimizer of the second column (optimizers 6 and 7). In configuration Test B, using 15 PV simulators, arcs were generated successively at 6 different positions. Table VII lists the arc positions and the optimizers affected as reported by the system. In Test 1, the arc occurs at the start of the string (between the inverter and the 1st optimizer), and the system correctly identifies the 1st optimizer but incorrectly also mentions the 2nd or 3rd optimizer. In Test 2, the arc position is correctly identified between the 1st and 2nd optimizer. In test 3 - 6, the arc is consistently localized one optimizer "too early". Thus, in each of the tests conducted, at least one of the two reported optimizers is accurate. Figure 7 shows the displayed arc position between optimizer 6 and 7, while the correct arc position was between optimizer 7 and 8 (Test 5). It remains to be considered that the tests were

conducted with PV simulators with decoupling networks rather than real PV systems and the test setup could potentially influence the results. However, validation measurements were conducted on real PV systems, and thanks to decoupling networks, no significant differences were observed between the simulator characteristicsand the real PV systems.

Table VII: Position of arcs as well as the system-reported position of the arc. Additionally, each test includes the number of measurements conducted.

Test	Arc position	Reported optimizers	Nr. of meas.
1	start of string	1. opt. and 3. opt.	1
		1. opt. and 2. opt.	2
2	1. opt. and 2. opt.	1. opt. and 2. opt.	1
3	3. opt. and 4. opt.	2. opt. and 4. opt.	3
4	4. opt. and 5. opt.	2. opt. and 4. opt.	1
		3. opt. and 4. opt.	2
5	7. opt. and 8. opt.	6. opt. and 7. opt.	3
6	14. opt. and 15. opt.	13. opt. and 14. opt.	5

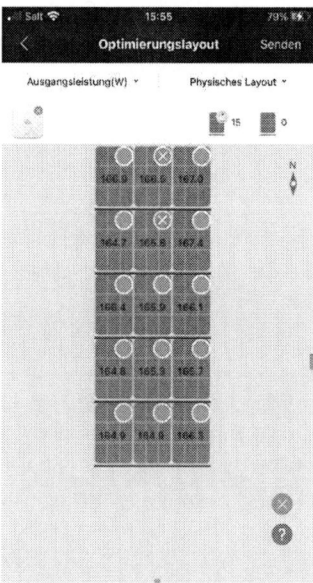

Figure 7: Estimated arc position as displayed in the FusionSolar app. In this example the arc is displayed between optimizer 6 and 7 (first and second), while the correct arc position was between optimizer 7 and 8 (Test 5).

4 CONCLUSION

In conclusion, no AFCI malfunction was detected under ordinary conditions (up to 200 m of cable), however, with extended cable lengths, a fire risk due to undetected arcs remains despite AFCI protection. All tests with cable lengths up to 200 m were passed according to the operational criteria of IEC 63027:2023. The permissible energy and arc time were clearly undercut for each test setting, although the arc energy depended on the arc position (start or middle of the string) as well as the voltage configuration. Nevertheless, it was shown that while the majority of arcs are still detected and extinguished with emulated extended cable lengths, some arcs were not detected. Thus, there is a limited arc protection for systems with long cables.

5 OUTLOOK

The AFCIs need also to be tested for false detection. In this thesis, it was examined whether all arcs are detected and extinguished, thereby minimizing the fire risk. However, for practical applications, it is also important that there are few false detections. Too many false detections due to everyday interference signal lead not only to energy losses but also to increased operational efforts, as the inverter cannot automatically restart after 5 arc detections within 24 hours without manual confirmation (see IEC 63027:2023). The handling of interference signals by AFCIs is therefore another relevant topic that still needs to be investigated.

Furthermore, it remains to be investigated how other inverter manufacturers perform in AFCI tests with emulated extended cables. In this paper, arc detection tests with emulated extended cables were conducted using DUT 1 only. How the AFCIs of DUT 2 or DUT 3 or others perform in these tests still needs to be examined.

6 REFERENCES

[1]   L. Zhu, S. Ji, and Y. Liu, "Generation and Developing Process of Low Voltage Series DC Arc," *IEEE Trans. Plasma Sci.*, vol. 42, no. 10, pp. 2718–2719, 2014, doi: 10.1109/TPS.2014.2330419.

[2]   T. Zgonena, Ji Liang, and Dini David, "Photovoltaic DC Arc-Fault Photovoltaic DC Arc-Fault Circuit Protection and UL Subject 1699B," 2011.

[3]   B. Brook, "Report of the Results of Investigation of Failure of the 1.1135 MW Photovoltaic (PV) Plant at the National Gypsum Facility in Mount Holly," 2011.

[4]   *National Electrical Code (R)*, 70 (R), National Fire Protection Association, Quincy, MA, USA, 2014.

[5]   *Photovoltaic power systems – DC arc detection and interruption*, IEC 63027:2023, Inernational ElectrotechnicalCommission, May. 2023.

41st European Photovoltaic Solar Energy Conference and Exhibition

DESIGN OF THE SUBSTRING MPP TRACKER

Patrick Mader, Sascha Eckerter, Rainer Merz
Karlsruhe University of Applied Sciences
Moltkestraße 30, 76133 Karlsruhe

ABSTRACT: The substring Maximum Power Point tracker increases the output power of a photovoltaic module by operating each substring at the Maximum Power Point. Since the substring Maximum Power Point tracker consists partly of nested DC-DC converters, it is possible to reduce the noise with suitable control. This report therefore derives the control of the DC-DC converters that leads to voltages with as little noise as possible. Measurements prove the theoretically derived relationships.

1 INTRODUCTION

A crystalline photovoltaic (PV) module consists of several solar cells connected in series. To protect the solar cells from thermal overload, bypass diodes split solar modules into substrings and are connected anti-parallel to each substring. If a substring cannot supply the current of the other substrings due to lower irradiation Φ, then the bypass diode conducts and thus limits the power loss

$$P_v = V_F I_s (V_F), \qquad (1)$$

of the shaded substrings with the forward voltage V_F of the bypass diode and the current $I_s (V_F)$ of the substring at the voltage V_F. Bypass diodes thus protect the cells from thermal overload. However, the shaded substring then no longer contributes to the overall power of the module. The so-called substring Maximum Power Point (MPP) tracker [1][2] provides a solution. The substring MPP tracker operates each substring in the MPP and prevents bypass diodes from conducting. It therefore increases the output power $P_{out} = V_{out} \cdot I_{out}$ when the module radiation is inhomogeneous. Figure 1 describes the substring MPP tracker. It is suitable for installation in the junction box and is parallel to the bypass diodes D1, D2, D3 and the substrings S1, S2, S3. The circuit consists of the DC-DC converters 1, 2 and 3. The Pulse width modulated (PWM) signals $v_{PWM1}, v_{PWM2}, v_{PWM3}$ with the period length T and the duty cycles d_1, d_2, d_3 control the DC-DC converters. Since the DC-DC converters work as step-down converters, the transmission behavior of the DC-DC

converters results in the voltages [3]

$$v_1 = d_1 v_0 , \qquad (2)$$

$$v_3 = d_2 v_0 , \qquad (3)$$

$$v_{out} = d_3 v_0 . \qquad (4)$$

This allows to set the three substring voltages v_1, v_3 and $v_2 = v_0 - v_1 - v_3$ by using the duty cycles d_1, d_2, d_3. Figure 2 shows the current-voltage behavior $I_k = f(V_k)$ and the MPP of a substring $k \in \{1,2,3\}$ with different irradiations Φ. In the selected operating range of $400\,\text{Wm}^{-2} < \Phi < 1000\,\text{Wm}^{-2}$ the MPP

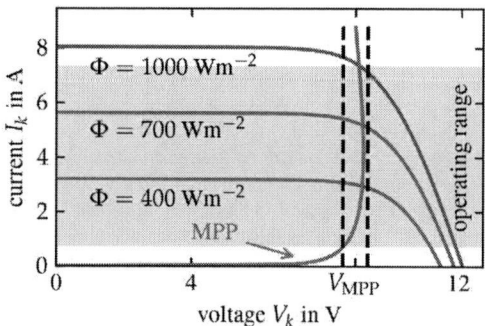

Figure 2: Current-voltage characteristic of a substring with different irradiation values Φ. The MPP shown has an approximately constant voltage V_{MPP} in the operating range.

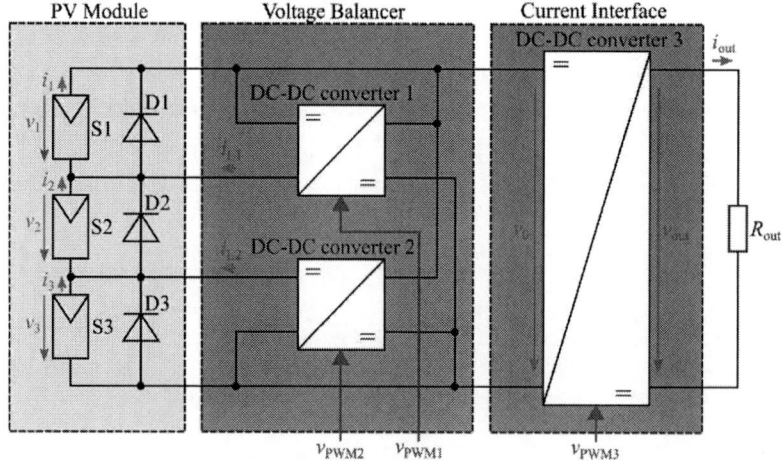

Figure 1: Circuit of the substring MPP tracker consists of three DC-DC converters 1, 2 and 3. Connection directly to the PV module parallel to the three bypass diodes D1, D2 and D3. Ohmic resistance R_{out} models the rest of the module string. PWM signals v_{PWM1}, v_{PWM2} and v_{PWM3} control the DC-DC converters 1, 2 and 3.

voltage V_{MPP} remains approximately constant. Therefore, it is sufficient to balance the substring voltages, so that $v_1 = v_2 = v_3$ applies. The required duty cycles for DC-DC converters 1 and 2 are $d_1 = d_2 = 1/3$. DC-DC converters 1 and 2 thus form a voltage balancer for the substring voltages v_1, v_2 and v_3. DC-DC converter 3 is used to match the module current to the higher string current i_{out} when the PV module is shaded. The resistor R_{out} models the rest of the module string. In order to extract the maximum power P_{out} from the module, a classic MPP algorithm, such as the perturb-and-observe method [4][5], sets the duty cycle d_3. Figure 3 shows the voltage balancer at transistor level. The voltage balancer consists of the two DC-DC converters 1 and 2.

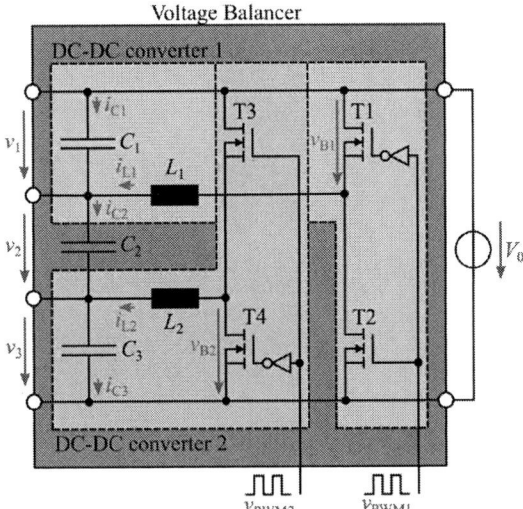

Figure 3: Circuit diagram of the voltage balancer. DC-DC 1 consists of L_1, T1, T2 and DC-DC 2 consists of L_2, T3, T4. Shifting v_{PWM1} and v_{PWM2} changes the current distribution i_{C1}, i_{C2}, i_{C3} through capacitors C_1, C_2 and C_3. This allows to adjust the noise in v_1, v_2 and v_3.

The transistors T1, T2 and the inductance L_1 form the DC-DC converter 1 and the transistors T3, T4 and the inductance L_2 form the DC-DC converter 2. The PWM signal v_{PWM1} with the duty cycle $d_1 := T_{T2,on}/T = 1/3$ and the PWM signal v_{PWM2} with the duty cycle $d_2 := T_{T3,on}/T = 1/3$ control the DC-DC converters 1 and 2. Figure 4 shows the generation of the two PWM signals v_{PWM1} and v_{PWM2} with a microcontroller. The PWM generator consists of the two counters $z_1 \in \mathbb{N}$ and $z_2 \in \mathbb{N}$. Each processor clock T_{clk} increases the respective counter by one. If a counter reaches the first compare register $Z_{11} \in \mathbb{N}$ or $Z_{21} \in \mathbb{N}$, the PWM voltage v_{PWM1} or v_{PWM2} switches from high to low. The first comparison registers thus determine the duty cycle of the PWM signals. If a counter z_1 or z_2 reaches the second compare register $Z_{12} = \{Z_{12} \in \mathbb{N} | Z_{12} \geq Z_{11}\}$ or $Z_{22} = \{Z_{22} \in \mathbb{N} | Z_{22} \geq Z_{21}\}$, then a PWM period T is completed and the counter is reset. The two counters form the carrier signals for the PWM and set the frequency $f = 1/T$. A shift in the two carriers creates a phase shift $\varphi := \angle(v_{PWM1}, v_{PWM2})$ of the PWM voltages v_{PWM1} and v_{PWM2}. The phase shift φ of the two PWM signals to each other determines the superposition of the noise of the voltages v_1, v_2 and v_3. For MPP tracking to function well, it is important that the module voltage v_0 is as low-noise as possible. Unfortunately, there are still no studies on

which phase shift φ leads to a voltage v_0 with the lowest possible noise. This report therefore examines which noise is generated depending on the phase shift φ and when the minimum occurs. For this purpose, this report is divided into a theory part in Chapter 2 and a hardware part with measurements in Chapter 3. The theory part begins in 2.1 with the derivation of an AC equivalent circuit as a basis for the theoretical calculations. The subsequent Section 2.2 calculates the current distribution through the capacitors. Section 2.3 determines the optimal phase shift and Section 2.4 derives dimensioning formulas for the input capacitors. The next Section 3.1 presents the created hardware and Section 3.2 shows the measurement results.

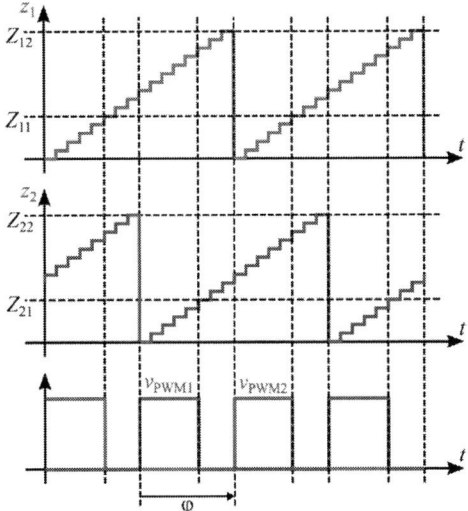

Figure 4: Generation of the PWM signals v_{PWM1} and v_{PWM2} with two counters z_1 and z_2. First compare register Z_{11}/Z_{21} determines the duty cycle. Second compare register Z_{12}/Z_{22} determines the period length. The counters work as carrier signals for the PWM. Phase shift φ indicates the shift of both carrier signals relative to each other.

2 NOISE OF THE SUBSTRING VOLTAGES

2.1 Derivation of the AC equivalent circuit

Figure 5 shows the small-signal equivalent circuit (ESB) of the voltage balancer. The small-signal ESB consists only of AC sources and considers the voltage source V_0 as a short circuit. The ESB also models the transistor T1 as a square-wave voltage source v_{B1} and trapezoidal current source i_{B1} and the transistor T4 as a square-wave voltage source v_{B2} and trapezoidal current source i_{B2}; these correspond to the curves of a half-bridge as derived in [3].

2.2 Current distribution of the substring capacitances

With the same capacitances $C_1 = C_2 = C_3$, the alternating component of the coil currents $i_{L1}(t)$ and $i_{L2}(t)$ leads to the capacitor currents

$$i_{C1}(t) = -\frac{2}{3}i_{L1}(t) - \frac{1}{3}i_{L2}(t), \qquad (5)$$

$$i_{C2}(t) = +\frac{1}{3}i_{L1}(t) + \frac{1}{3}i_{L2}(t), \qquad (6)$$

$$i_{C3}(t) = +\frac{2}{3}i_{L1}(t) + \frac{1}{3}i_{L2}(t). \qquad (7)$$

Equations 5, 6, 7 show that the capacitor currents i_{C1}, i_{C2} and i_{C3} are weighted sums of the coil currents Δi_{L1} and Δi_{L2}. Shifting the coil currents $i_{L1}(t)$ and $i_{L2}(t - \varphi T/2\pi)$ by φ enables the generation of

Voltage Balancer

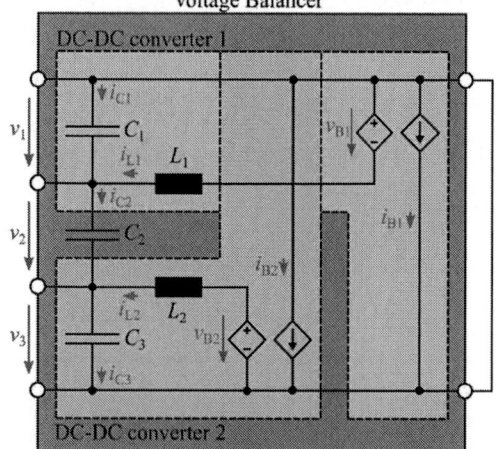

Figure 5: AC equivalent circuit of the substring MPP tracker. Square-wave voltage v_{B1}, v_{B2} and current sources i_{B1}, i_{B2} simulate the half-bridges. Short circuits replace DC voltage sources V_0.

constructive and destructive interference of the currents i_{C1}, i_{C2} and i_{C3} on the capacitors C_1, C_2 and C_3. An optimal shift φ thus leads to minimal noise at v_0. Figure 7.a) to h) show the current and voltage of the voltage balancer in Figure 7.a) to d) for the shifts $\varphi = 0°$ and in Figure 7.e) to h) for $\varphi = 180°$. The graphs Figure 7.a) and e) show the PWM voltages v_{B1} and v_{B2} for the DC-DC converters 1 and 2. Figure 7.b) and f) show the coil currents

$$i_{L1} = \frac{1}{L_1} \int V_1 - v_{B1} \, dt + I_{L1,0} \tag{8}$$

and

$$i_{L2} = \frac{1}{L_2} \int V_1 - v_{B2} \, dt + I_{L2,0} . \tag{9}$$

Figure 7.b) and f) show that a phase shift φ of the PWM signals leads to a shift of $\theta = \varphi - 180°$ of the coil current i_{L2}. Figure 7.c) and g) show the capacitor currents i_{C1}, i_{C2} and i_{C3} resulting from equations 5, 6 and 7. The connected capacitors C_1, C_2 and C_3 integrate the currents i_{C1}, i_{C2} and i_{C3} to the capacitor voltages

$$v_1 = \frac{1}{C_1} \int i_{C1} \, dt + V_{1,0} , \tag{10}$$

$$v_2 = \frac{1}{C_2} \int i_{C2} \, dt + V_{2,0} , \tag{11}$$

$$v_3 = \frac{1}{C_3} \int i_{C3} \, dt + V_{3,0} . \tag{12}$$

Figure 7.d) and h) show the resulting curves of the capacitor voltages v_1, v_2 and v_3. As a result, different shifts φ of the carrier signals lead to different capacitor voltages v_1, v_2 and v_3. The phase shift of $\varphi = 0°$ in Figure 7.d) leads to a larger ripple in the capacitor voltage v_2, but to a smaller ripple in v_1 and v_3. The phase shift of $\varphi = 180°$ in Figure 7.h), on the other hand, leads to larger ripples in v_1 and v_3, but to a smaller one in v_2. Therefore, the following section 2.3 examines which phase shift φ of

the carrier signals leads to the lowest possible noise in v_0.

2.3 Optimal shifting of the carrier signals

For an easier calculation, the following section only considers the fundamental oscillation of the coil currents

$$\underline{I}_{L1}^1 = I_{L1}^1 , \tag{13}$$

with $\underline{I}_{L1}^1 \in \mathbb{C}, I_{L1}^1 \in \mathbb{R}$ and

$$\underline{I}_{L2}^1 = I_{L2}^1 \exp(j(\varphi - 180°)), \tag{14}$$

with $\underline{I}_{L2}^1 \in \mathbb{C}, I_{L2}^1 \in \mathbb{R}$. The fundamental oscillations $\underline{I}_{L1}^1$ and $\underline{I}_{L2}^1$ are distributed between the capacitors C_1, C_2 and C_3 according to equations 5, 6, 7 and result in the capacitor currents $\underline{I}_{C1}^1, \underline{I}_{C2}^1$ and $\underline{I}_{C3}^1$. The capacitor currents in turn cause the effective voltages on the capacitors

$$V_1^1 = \frac{1}{3\omega C} \sqrt{2{I_{L1}^1}^2 + \frac{{I_{L2}^1}^2}{2} + 2I_{L1}^1 I_{L2}^1 \cos(\varphi)} , \tag{15}$$

$$V_2^1 = \frac{1}{3\omega C} \sqrt{\frac{{I_{L1}^1}^2}{2} + \frac{{I_{L2}^1}^2}{2} - I_{L1}^1 I_{L2}^1 \cos(\varphi)} , \tag{16}$$

$$V_3^1 = \frac{1}{3\omega C} \sqrt{\frac{{I_{L1}^1}^2}{2} + 2{I_{L2}^1}^2 + 2I_{L1}^1 I_{L2}^1 \cos(\varphi)} . \tag{17}$$

For a voltage V_0 with as little noise as possible, the effective value

$$V_0^1 = V_1^1 + V_2^1 + V_3^1 \tag{18}$$

should be as small as possible. Figure 6 shows the effective values V_0^1, V_1^1, V_2^1 and V_3^1 over the phase shift φ of the carrier signals. As already stated above, a phase shift of $\varphi = 0°$ leads to a larger ripple in v_2, but to a smaller ripple in v_1 and v_3. A phase shift of $\varphi = 180°$, on the other hand, leads to larger ripples in v_1 and v_3, but to a smaller ripple in v_2. With a shift of $\varphi = 60°$, the three capacitor voltages V_1^1, V_2^1 and V_3^1 are equal. The voltage V_0^1 provides information on how to minimize the noise. A minimum of V_0^1 occurs at $\varphi = 0°$. Therefore, the later circuit does not use a phase shift $\varphi = 0°$ of the carrier signals.

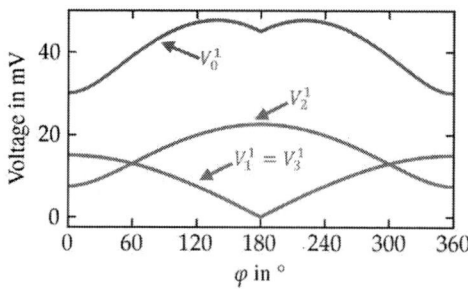

Figure 6: Fundamental effective values of the capacitor voltages V_1^1, V_2^1, V_3^1 and V_0^1 for a shift φ of the carrier signal. The minimum of V_0^1 is at $\varphi = 0°$.

2.4 Dimensioning of the input capacitors

The design of the capacitances C_1, C_2 and C_3 is done by calculating their voltage ripple

41st European Photovoltaic Solar Energy Conference and Exhibition

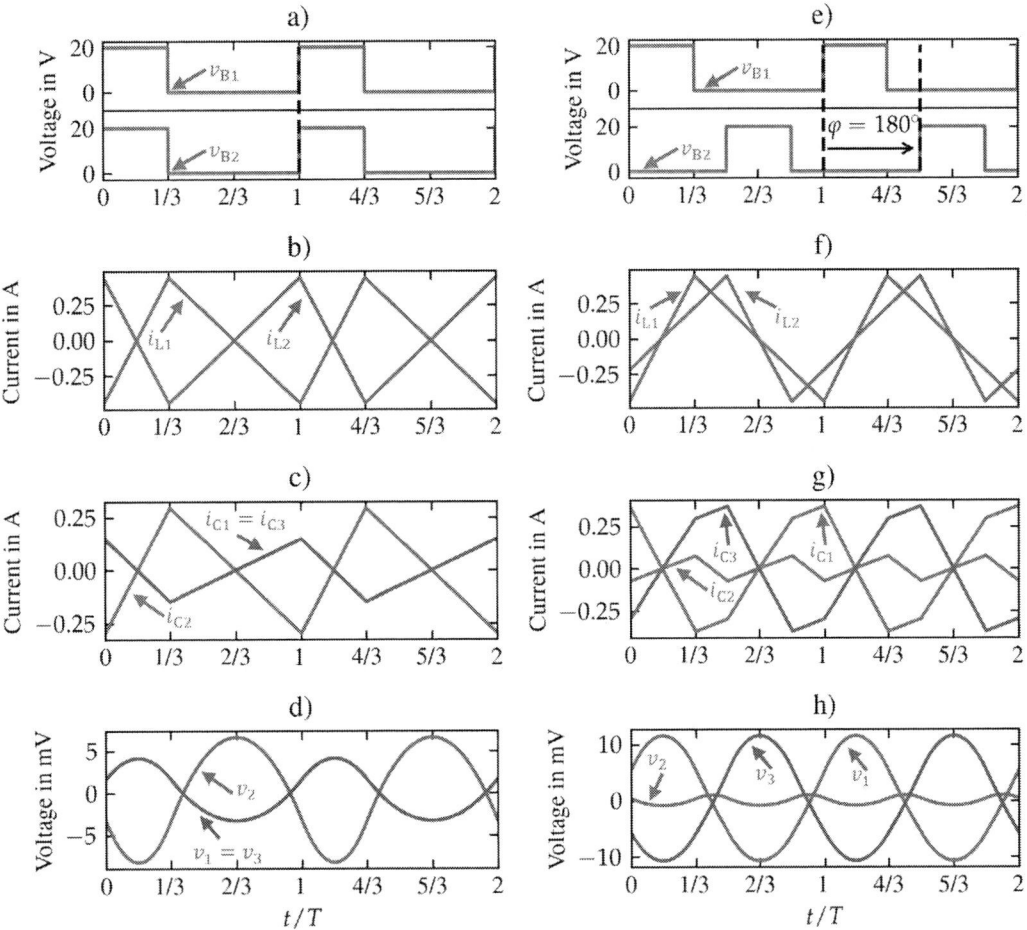

Figure 7: Time characteristics of the balancer for different shifts φ of the carrier signals. a) and e) PWM signal v_{B1}, v_{B2} from DC-DC converter 1 and DC-DC converter 2. b) and f) Coil currents i_{L1} and i_{L2}. c) and g) Capacitor currents i_{C1} to i_{C3}. d) and h) Resulting capacitor voltages v_1, v_2, v_3. Shifting the two carrier signals by φ influences coil currents i_{L1}, i_{L2} and voltages v_1, v_2, v_3.

$$\Delta v_1 = \int_{t_1}^{t_2} \frac{i_{C1}}{C_1}\, dt\, , \{t_1, t_2\} \in (0,T): i_{C1} = 0\, , \quad (19)$$

$$\Delta v_2 = \int_{t_3}^{t_4} \frac{i_{C2}}{C_2}\, dt\, , \{t_3, t_4\} \in (0,T): i_{C2} = 0\, , \quad (20)$$

$$\Delta v_3 = \int_{t_5}^{t_6} \frac{i_{C3}}{C_3}\, dt\, , \{t_5, t_6\} \in (0,T): i_{C3} = 0\, . \quad (21)$$

Solving equations 19, 20 and 21 leads to the capacitors

$$C_1 = C_3 = \frac{V_0}{144 \cdot \Delta v_1 \cdot L \cdot f^2} \quad (22)$$

and

$$C_2 = \frac{V_0}{72 \cdot \Delta v_2 \cdot L \cdot f^2}. \quad (23)$$

3 HARDWARE

Figure 8 shows the designed board of the substring MPP tracker. The half-bridges of the DC-DC converters work with the gallium nitride (GaN) transistors EPC2218. The control is carried out via the microcontroller of the STM32G431 series. The PWM signals have a clock frequency of $f = 1$ MHz. The board is designed for a load current at the output of $I_{out} = 15$ A and an input voltage

$V_0 = 65$ V at an ambient temperature $T = 85°C$ and is therefore suitable for modules up to $P_{out} = V_{out}I_{out} \leq 400$ W.

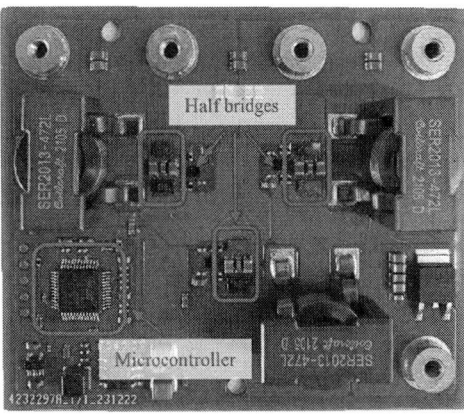

Figure 8: Designed board of Substring MPP Tracker. Half bridges designed with GaN transistors. Microcontroller generates PWM signals.

4 MEASUREMENTS

The measurements are intended to confirm the correctness of the theory derived above. Since a shift of $\varphi = 0°$ leads to the smallest noise in v_0, the measurement is carried out at this shift. Figure 9 shows the measurement of the capacitor voltages v_1, v_2 and v_3. Here Figure 9.a) shows the PWM signals v_{B1} and v_{B2} with the same peak value as in Figure 7.a). Figure 9.b) shows the resulting capacitor voltages v_1, v_2 and v_3. The voltages v_1, v_2, v_3 agree with the theory from Figure 7.d). The voltages v_1 and v_3 are smaller than v_2. The voltages in the measurement have approximately the same amplitude as in the theory.

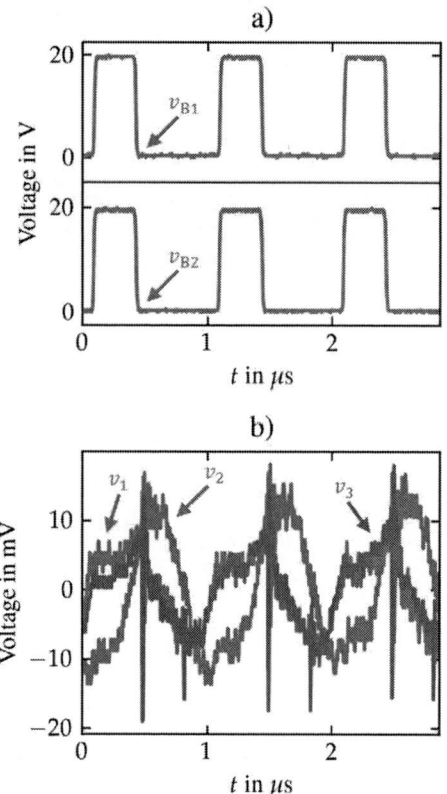

Figure 9: Measurement on the prototype hardware with $\varphi = 0°$. a) Drain-Source voltages v_{B1} and v_{B2} of the transistors T1 and T4 b) Input voltages v_1, v_2 and v_3 have approximately the same course as in theoretical investigation Figure 7.d.).

5 SUMMARY

A typical crystalline PV module consists of three substrings. Bypass diodes protect the substrings by bridging corresponding substrings and thereby limiting the power loss. Bridged substrings then no longer deliver power. The substring MPP tracker prevents bypass diodes from conducting and operates the substrings at the MPP. The circuit of the substring MPP tracker consists of three DC-DC converters. DC-DC converters 1 and 2 are nested DC-DC converters. The input voltages v_0 are noisy due to the switching DC-DC converters. Phase shifting φ of the control signals v_{PWM1} and v_{PWM2} of DC-DC converters 1 and 2 leads to different currents i_{C1}, i_{C2}, i_{C3} and voltages

v_1, v_2, v_3 at the input capacitors C_1, C_2, C_3. This means that noise can be reduced with a suitable phase shift φ. This report selects the phase shift $\varphi = 0°$ for the smallest sum of the effective values of the input voltages v_1, v_2, v_3 and uses this to calculate the capacitance values C_1, C_2, C_3. Measurements of the developed hardware with fixed duty cycles prove the theoretically derived relationships.

5 ACKNOLEGEMENT

This work contributes to the research performed at the University of Applied Science Karlsruhe. The results were generated within the project "Solarpark 2.0" (funding code 03EE1135C) funded by the Federal Ministry for Economic Affairs and Climate Action (BMWK). The authors thank the project management organization Julich (PTJ) and the BMWK.

LITERATUR

[1] R. Merz, S. Eckerter, S. Coenen und P. Mader. „Substring-MPPT steigert Strangleistung bei Teilverschattung". In Tagungsunterlagen 39. PV-Symposium 2024.
[2] R. Brace, A. Neumann, T. Czarnecki, R. Merz "Substring-Mppt For 4-Terminal 3-Substring Modules". In 35th European Photovoltaic Solar Energy Conference and Exhibition.
[3] U. Tietze, C. Schenk and E. Gamm. Halbleiter-Schaltungstechnik, 16. Auflage. Springer Verlag, 2019.
[4] T. Suntio, T. Messo and J. Puukko. Power Electronic Converters: Dynamics and Control in Conventional and Renewable Energy Applications, John Wiley & Sons, 2017
[5] F. Blaabjerg. Control of Power Electronic Converters and Systems, Academic Press, 2018,

41st European Photovoltaic Solar Energy Conference and Exhibition

Testing of Electronic Interface for Diagnostic Functions of Photovoltaic Systems

Edoardo Celi[1], Alessandro Minuto[1], Stefano Rizzi[1], Gianluca Timò[1]

[1] Ricerca sul Sistema Energetico – RSE SpA, Via V. Callegari, 29122 Piacenza, Italy

corresponding author: edoardo.celi@rse-web.it

Motivation. Accurate fault diagnosis is essential for optimizing the performance and maintenance of photovoltaic (PV) systems. Conventional methods based on Supervisory Control and Data Acquisition (SCADA data) often miss faults at the module or inverter level. To overcome these limitations, we present an electronic interface device (EIDPS) capable of measuring key parameters such as short circuit current I_{sc}, open circuit voltage V_{oc}, as well as the voltage V_{wp}, current I_{wp}, and power values of the working point set by the inverter. This device offers a solution to enhance diagnostic functions, allowing for real-time, accurate performance evaluation and early detection of issues such as soiling, partial shadings, module malfunctions. This system can be adapted for single PV modules in a string, isolating degradation factors and eliminating the influence of string-level faults.

Hardware and firmware description

The EIDPS uses a patented circuit/logic previously reported [1]. It is a stand-alone device that could be used with any type of inverter, including older models. Its main components consist of **two solid state relays** (S1 connected in series to the PV string and S2 in parallel), **two sensors** (a **current sensor (I)** in series with the PV string and a **voltage sensor (V)** in parallel) and **a suitably developed control board** to manage all the operations. To minimize the transient rise and oscillatory components during V_{oc} and I_{sc} measurements, a digital IIR filter has been implemented in the microcontroller firmware.

Experimental setup and Preliminary outdoor test results

The EIDPS has been tested with a real PV system and a commercial inverter, confirming its proper functionality. The EIDPS, set for continuous 2-minute interval measurements, did no cause interruptions in the inverter operation. Further testing with 20-second intervals confirmed that frequent measurements do not affect the system normal operation, allowing flexibility in measurement frequency based on diagnostic needs. Accuracy has been validated under clear sky conditions by comparing EIDPS measurements with I-V tracer results taken shortly afterward.

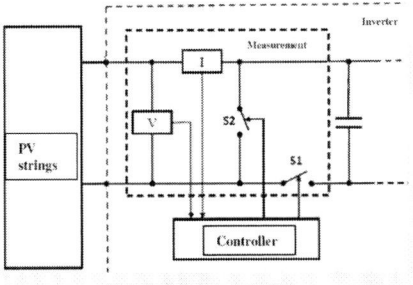

Fig.1 - Conceptual scheme of EIDPS.

Table 1 – Main characteristics of EIDPS.

Prototype data sheet	
Nominal Power P_{nom}	7 kW
Nominal Voltage V_{nom}	780 V
Nominal Current I_{nom}	10 A
Measurement Timing T_m	< 10 ms
Accuracy	< 1 %

Fig.3 – Outdoor EIDPS testing with a real PV system and a commercial inverter.

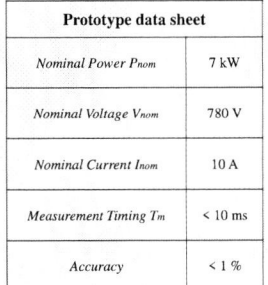

Fig.2 - The graph shows the voltage and current trend both during the V_{oc} and I_{sc} measurement phase (respectively, blue and red areas) and immediately after reconnection to the electronic load (respectively, yellow and green areas). After about 9.5 ms from the start of EIDPS operation, the steady state conditions are restored.

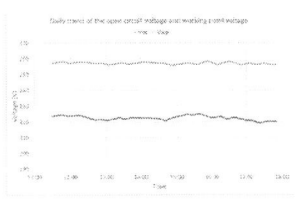

Fig.4 – Daily trends of the V_{oc} and V_{wp} observed during EIDPS testing with a commercial inverter.

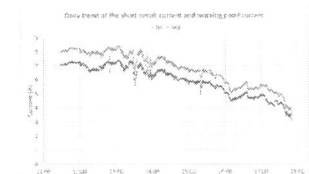

Fig.5 – Daily trends of the I_{sc} and I_{wp} observed during EIDPS testing with a commercial inverter.

Transient analysis capabilities

The device samples voltage and current values at the working point with a sampling time of 40-50 µs. This capability enables the assessment of capacitive effects and internal capacitances, which vary based on PV module technologies and system configurations (series or parallel connections). It also allows for analyzing the PV system response to dynamic working conditions influenced by both environmental factors and the downstream converter.

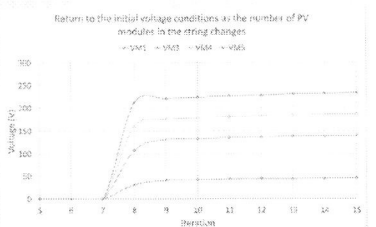

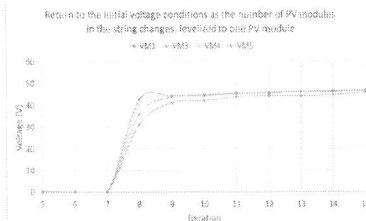

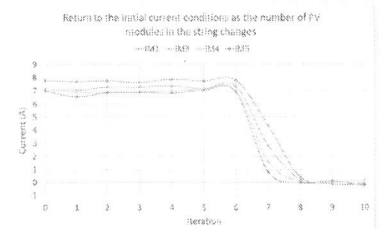

Fig.6 – Starting from the open-circuit condition, the system configuration has been varied from a single module to five modules connected in series. Left graph shows the voltage trends during the transient returning to initial conditions after the measurement of I_{sc} and V_{oc}. The center chart displays the same voltage trends normalized to a single module, while the right graph illustrates current trends. As the number of modules increases, the voltage rise and current drop transients become faster according to the theoretical reduction of the total capacitance. The interval between each iteration is approximately 46 µs.

Conclusions and future steps

A cost-effective and versatile electronic interface for advanced diagnostics of photovoltaic systems (EIDPS) has been tested. This device, measuring some important electrical parameters, allows enhancing the algorithms which identify the more common causes of power losses in real-time, without compromising the PV system energy production. I_{sc} and V_{oc} values can be reliably measured with 1% accuracy in under 10 ms, avoiding any inverter shutdown, as demonstrated during the first tests with a commercial inverter. The next step will be to connect the EIDPS interface to a wider range of commercial inverters, including older generation models still in use in PV systems, to thoroughly validate its compatibility and ensure that it does not cause interference. Future plans include expanding the system hardware and software to measure additional electrical parameters and further optimize energy extraction.

References: [1] European patent N. EP3249492B1.

Acknowledgments: This work has been financed by the Research Fund for the Italian Electrical System under the Three-Year Research Plan 2022-2024 (DM MITE n. 337, 15.09.2022), in compliance with the Decree of April 16th, 2018".

41st European Photovoltaic Solar Energy Conference and Exhibition

Evaluation of degradation and impact of climatic conditions on PV modules exposed to extreme high UV solar radiation

Dr. Patricio Valdivia-Lefort*, Valentina Navarro**

*Universidad de Santiago de Chile. Av. Libertador Bernardo O'Higgins, 9170022 Estación Central, , Santiago - Chile
**Universidad Técnica Federico Santa María. Av. Vicuña Mackenna 3939, San Joaquín, Santiago – Chile

ABSTRACT

This work proposes a methodology for evaluating degradation in the field of PV modules exposed to high radiation climates. It involves IV curves measurements corrected to STC outdoors on a weekly basis and STC measurements in an indoor laboratory using an A+A+A+ solar simulator with a monthly frequency over a period of 16 months. Additionally, degradation is calculated using an empirical model that considers environmental conditions such as UV radiation, relative humidity, and ambient temperature. This allows for obtaining a degradation value based on climatic zones.

The methodology is validated in three photovoltaic plants located in northern Chile. IV curves at STC are obtained for different technologies modules that have been exposed for up to 8 years. To validate the empirical model, a territorial characterization is performed by collecting information from four meteorological stations installed in the field, and degradation values are obtained for each zone.

SETUP

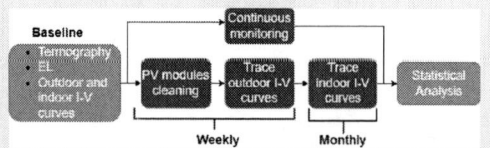

For the methodology development, four photovoltaic plants located in Calama, Antofagasta, Copiapó and Santiago were used. IV curves at STC are obtained for different technologies modules that have been exposed for up to 8 years. Meteorological stations were installed to monitoring environmental variables. In the field, four meteorological stations with UVA, UVB, RH, and ambient temperature sensors were installed.

Annualized degradation rates were calculated based on the type of stress, along with the total degradation rate using data measured at the meteorological stations installed in the field through the model by *Kaaya et al. (2019)*, which are presented in Table 1.

Location	Period	RH	UV_dose [kWh/m^2]	Delta T
Antofagasta	Feb.-Sept. 2023	41%	100.99	71.72
Calama	Feb.-July 2023	31%	114.05	69.46
Copiapó	April-Oct. 2023	42%	100.13	60.62
Santiago	May-Nov. 2023	68%	35.39	49.57

Table 1: Values of relative humidity, incident UV dose on the PV module with E-O tracking, and ambient temperature difference measured in the field for each locality during the specified time periods of the year 2023.

According to the values of the measured meteorological variables, it is possible to perform a classification by zone. In this way, according to the **Köppen-Geiger** criterion, based on the global horizontal irradiance, it is possible generate a classification of 4 categories, low (L), medium (M), high (H) and very high (K), according to the values presented in Table 2.

Categorias	Irradiancia Global Horizontal Anual [kWh/m^2]
L - Bajo	$0 < GHI \leq 1130$
M - Medio	$1130 < GHI \leq 1560$
H - Alto	$1560 < GHI \leq 2070$
K - Muy Alto	$2070 < GHI$

Categorias	Requerimiento	Orden
A - Tropical	$18\,[°C] \leq T_{amm\,min}$	4to
B - Desértico	$Hr \leq 30\%$	2do
C - Estepa	$30\% < Hr \leq 70\%$	3ro
D - Templado	$-3\,[°C] < T_{amm\,min} < 18\,[°C]$	5to
E - Frio	$T_{amm\,min} < -3\,[°C]$	6to
F - Polar	$T_{amm\,max} \leq 10\,[°C]$	1ro

Table 3 shows the **Köppen-Geiger** model adapted by relative humidity, 6 categories can be identified from temperature and relative humidity ranges of the environment, these are tropical, desert, steppe, temperate, cold and polar. Furthermore, there is an order of priority, which is used when a zone meets more than one condition at a time, so it must be choose the higher order option.

RESULTS

Table 4 shows the Annualized degradation rates obtained with the model by *Kaaya et al. (2019)*, using data from variables measured at meteorological stations installed in the field.

Location	k_H (%/year)	k_P (%/year)	k_Tm (%/year)	k_T (%/year)
Antofagasta	0.02	0.05	0.43	0.53
Calama	0.01	0.02	0.43	0.46
Copiapó	0.02	0.05	0.31	0.41
Santiago	0.12	0.15	0.22	0.58

Table 4 presents the annual degradation values obtained with the *Kaaya et al. (2019)* degradation model, using meteorological data collected from field-installed weather stations for the four locations (k_T). Additionally, the annual degradation rate of maximum power obtained from I-V curves measurements at standard test conditions (STC) in an indoor laboratory is presented, taking into account the exposure time of the analyzed modules.

Location	Exposition [years]	k_T [%/year]	DR Pmáx [%/year]
Antofagasta	3	0.53	[0.71, 1.11]
Calama	7	0.46	[0.19, 0.71]
Copiapó	7-8	0.41	[0.39, 1.52]
Santiago	1-3	0.58	[1.37, 3.69]

With the degradation factors and the territorial characterization values obtained, a testing methodology to simulate stress conditions of photovoltaic modules considering the conditions of northern Chile. The proposed methodology is presented in Figure 3

 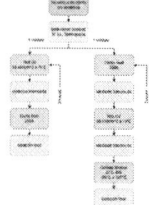

CONCLUSIONS

Degradation calculations were performed in four zones in Chile—Antofagasta, Calama, Copiapó, and Santiago—by comparing the maximum power measured in the laboratory with the manufacturer's specified values for each module. The maximum specified values varied between 0.7% per year and 3.7% per year. Additionally, degradation rates were obtained using an empirical model that considers the environmental conditions of each zone, resulting in values ranging between 0.42% per year and 0.58% per year. It was found that these empirical rates were lower than the maximum degradation rates found with I-V curve measurements at standard test conditions (STC). This difference is attributed to the empirical model not being designed for zones with high UV radiation, such as northern Chile, requiring adjustments to accurately assess the real impact of UV radiation on the PV module's lifespan.

In conclusion, differences in degradation rates were observed among the analyzed modules for similar exposure times. It can be inferred that degradation is a phenomenon dependent on the technology under study, behaving differently in HET, n-PERT, monocrystalline, and polycrystalline modules.

ACKNOWLEDGEMENTS

FONDECYT INICIA N°11220697: Dynamic and stationary electrical model for bifacial heterojunction PV modules, PI: Dr. Patricio Valdivia Lefort.

REFERENCES

1. Escobar RA, Cortés C, Pino A, Pereira EB, Martins FR, Cardemil JM. Solar energy resource assessment in Chile: Satellite estimation and ground station measurements. Renewable Energy. 2014 Nov 1;71:324–32.
2. Cordero RR, Damiani A, Jorquera J, Sepúlveda E, Caballero M, Fernandez S, et al. Ultraviolet radiation in the Atacama Desert. Antonie van Leeuwenhoek. 2018 Aug 31;111(8):1301–13.

PV MODULE BRUSH ABRASION TESTING

Gerhard Mathiak[1,], Nithin. Sha[1], Afra Seenthakath[1], Prashanth Gabbadi[1], Yogesh Kumar[1]
Mark Mirza[2]
[1] DEWA Research & Development Center, P.O.Box: 564 Dubai, United Arab Emirates
[2] Fraunhofer Institute for Silicate Research ISC, Neunerplatz 2, 97082 Würzburg, Germany

Abstract— A test sequence for evaluating the durability of PV modules under cleaning conditions in hot desert environments is introduced, focusing on qualification and benchmarking. This sequence, based on brush abrasion testing in accordance with IEC 62788-7-3, is being assessed using a cleaning robot test stand for full-size PV modules. Changes in surface roughness and glass reflectance are measured. Coaxial microscopy is employed to describe the abrasion of the front glass texture and anti-reflective coating. Additionally, SEM and EDX analyses are conducted to determine the composition of dust found on desert-exposed modules and to evaluate the characteristics of the cleaning brushes.

Keywords—Soiling, robotic cleaning, desert climate, anti-reflective coating, brush abrasion

I. INTRODUCTION

The deployment of photovoltaic (PV) systems in hot, arid desert regions is increasing. These systems face unique challenges, including elevated module temperatures, exposure to ultraviolet radiation, frequent sandstorms, and significant dust accumulation. This study investigates the abrasion effects of cleaning PV modules using robotic systems with bristle-based brushes, comparing the results with modules tested at the Cleaning Test Facility located at the DEWA R&D Center within the 5 GW Mohammed bin Rashid Al Maktoum Solar Park in Dubai. Currently, no standardized method exists for evaluating the abrasion resistance of PV modules and solar glass under cleaning conditions specific to desert environments.

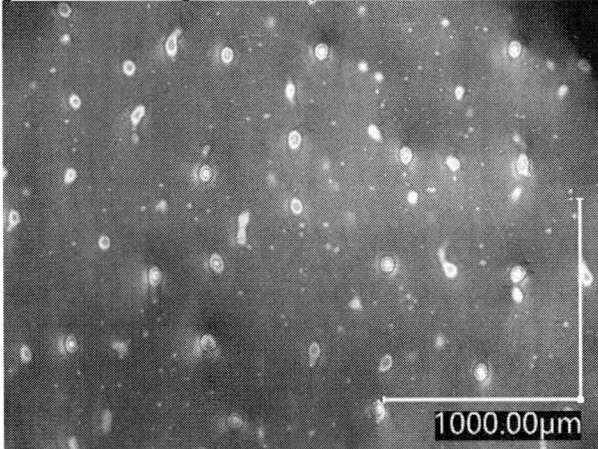

Fig. 1a: Coaxial microscopy of glass surface of PV module

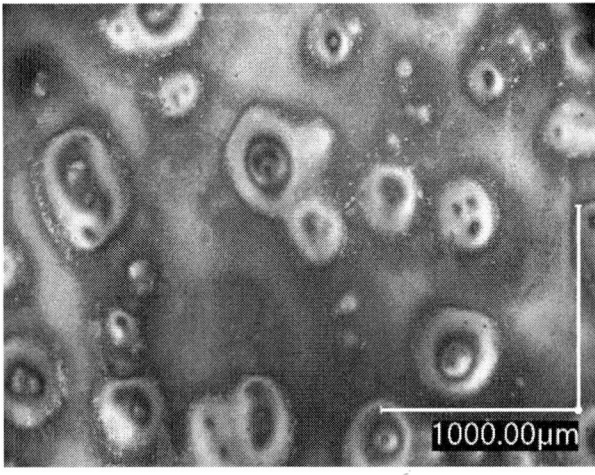

Fig. 1b: Coaxial microscopy of glass surface of PV module

To qualify and benchmark PV modules for hot desert environments, a test sequence for assessing cleaning durability is proposed. This sequence, based on sand and brush abrasion testing as outlined in IEC 62788-7-3 [1], is currently under evaluation using a commercial cleaning robot test stand for full-size PV modules. Coaxial microscopy is employed to analyze the abrasion of the front glass texture and the anti-reflective coating. Additionally, SEM and EDX analyses are conducted to determine the composition of dust found on desert-exposed modules and to assess the characteristics of the cleaning brushes.

The thickness of the anti-reflective coating on most micro-textured solar glass varies between the peaks and valleys due to the roller coating application technique. This variation is observable under a coaxial microscope, where differences in thickness are evident through distinct color patterns. **Figures 1a and 1b** illustrate this phenomenon, showing how the color variations differ due to the micro-texture of the glass and the specific coating processes used by each glass manufacturer.

The observed colors on the surface of the PV module are directly linked to the specific thickness of the anti-reflective coating. This relationship can be further analyzed by comparing the color distribution with the corresponding coating thickness. The color variations result from the interference of white light with the porous silica layer on the surface. The sequence of colors observed as the coating thickness increases, from optically thin to thick, is as follows: 1. brown, 2. dark purple, 3. dark blue, 4. light blue, 5. white, 6. orange, 7. brown, 8. turquoise, 9. yellow, 10. orange, 11. violet, 12. blue, and 13. green, as determined through simulations [2]

This variation in coating thickness may impact the performance and durability of photovoltaic (PV) modules, emphasizing the need for standardized testing and quality control measures to ensure consistent performance across different products.

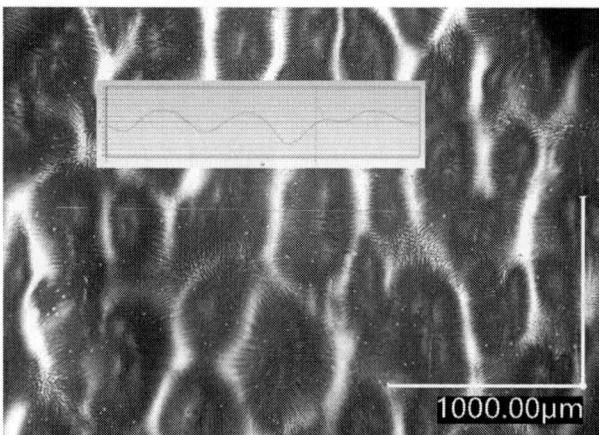

Fig. 2: Profilometry of the PV module's glass surface, embedded in the microscopy image, reveals scratches detected by the profilometer.

Using profilometry (**Fig. 2**) alongside color analysis provides a comprehensive examination of the anti-reflective coatings on PV modules. It is evident that different modules exhibit varying coating characteristics. In **Fig. 2**, the brown areas represent the peaks, while the valleys appear as a whitish-blue hue. The scanned profile ranges from -3 μm to +3 μm over a length of 1.5 mm.

The primary objective of this paper is to investigate the impact of anti-reflective coating degradation on PV modules due to cleaning with bristle-based brushes. The proper selection of anti-reflective coatings and cleaning techniques is crucial, as the coating may degrade more rapidly from mechanical cleaning actions. This paper compares the effects of cleaning with bristles on module performance.

Table 1 outlines the abrasion mechanism of anti-reflective coatings and its impact on reflectance.

Table 1: Anti-reflective coating abrasion mechanism

Mechanism	Cause	Microscopy	Reflectance
Removal of glass (chipping)	Blowing sand	chipping (rainbow pattern)	Increase
Removal of ARC (scratches)	Abrasive brush	White scratches	Increase
Thinning (Chemical Dissolution)	Environment	Color change	Shift to short wavelength
Bristle marks	Dry cleaning	Dark lines	Slight Decrease
Water ingress	Wet Cleaning	Fading of color	Increase

II. METHODOLOGY

Fig. 3a and 3b show the test setup for indoor abrasion testing of PV modules. Commercial PV modules with micro-structured tempered glass and anti-reflective coatings are tested indoors at a tilt angle of 25°. The cleaning robot is operated over the test setup for up to 10, 000 cycles. For dry cleaning abrasion, a helical rotary brush with nylon bristles is used. The bristles are 3.8 cm long, 200 μm thick, and the brush orifice diameter is 4 mm. The robot operates at a cleaning speed of 15 m/min, with the brush rotating at 2.6 cycles/s.

The 10,000 cycles simulate daily cleaning over 25 years. After every 1,000 cycles, the modules are evaluated for changes in power, including short-circuit current, glass surface reflectance, and using coaxial microscopy to assess the impact of cleaning on the glass surface.

Fig. 3a: Indoor Cleaning Test Apparatus

A similar test setup is established at the DEWA R&D Centre

for outdoor testing, designed to evaluate the efficiency of the cleaning robot in desert environments and its impact on PV modules. In this outdoor cleaning test field, modules are subjected to continuous cleaning with the same robot for one year, simulating daily cleaning. The modules are periodically assessed for power output, reflectance, and glass surface degradation using microscopy to monitor the effects of the cleaning process. **Fig. 3c** illustrates the outdoor testing of the robot on PV modules in the Cleaning Test Field at the DEWA R&D Center.

Fig. 3b: Indoor Cleaning Test Apparatus with helical rotatory brush with Nylon bristles

Fig 3c: Outdoor Cleaning Test [3=Mathiak 2023]

III. SAND, DUST AND BRISTLE ANALYSIS

The Cleaning Test Field is located north of the MBR Solar Park, built on a former quartz sand desert. The sand on the ground and the sand collected in dust traps (see **Fig. 4**) exhibit similar grain size distributions and chemical compositions.

Fig. 4: Scanning Electron Microscopy (SEM) of sand and dust collected from dust traps at a height of 1.7 m after a sandstorm reveals quartz sand grains within the range of 100 μm to 200 μm.

Fig. 5a shows sand collected from the top of PV modules in the cleaning test facility. This sand is finer compared to the sand found on the ground at a height of 1.7 m. **Fig. 5b** shows the EDX elemental composition analysis.

Compared to the sand collected at ground level, the dust deposited on the modules contains a higher concentration of calcium, attributed to the finer dust particles.

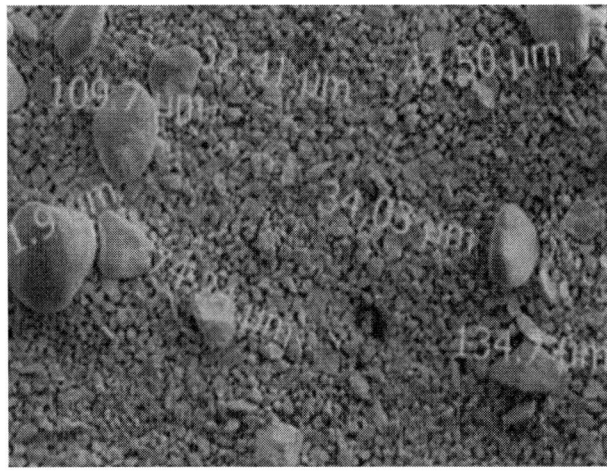

Fig 5a: SEM of Sand and Dust from Module Surface (25° fixed tilt angle)

41st European Photovoltaic Solar Energy Conference and Exhibition

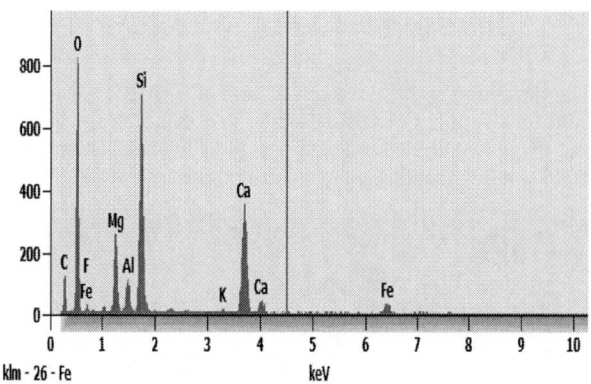

Fig. 5b: EDX of Sand and Dust from Module Surface

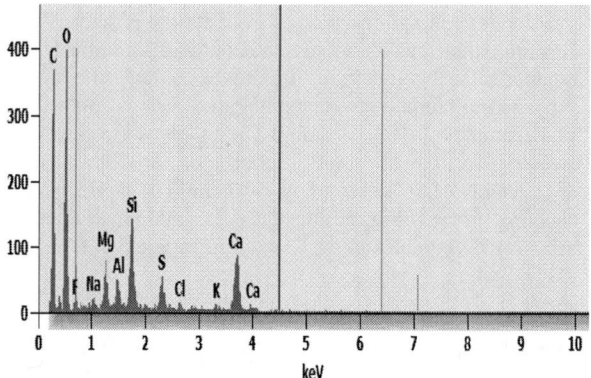

Fig. 6b: EDX of Dust Particle on Bristle (Outdoor)

The bristles are analyzed using scanning electron microscopy (SEM) to assess degradation. Energy-dispersive X-ray spectroscopy (EDX) is performed to determine the overall material composition. The analysis reveals the presence of calcium, which is found in the desert dust. This dust adheres to the bristles, increasing abrasion and accelerating the removal of the anti-reflective coating. **Fig. 6a** shows a microscopic image of the bristles from the SEM, while **Fig. 6b** presents the material composition, highlighting carbon and nitrogen from the nylon bristles, as well as additional sulfur and sodium chloride (NaCl).

IV. Results Indoor Testing

In the test procedure of [5] it is proposed to regularly spread abrasives as the sand of the solar park on the modules. These indoor test runs were performed without additional abrasives. Short-circuit current (Isc) measurements of the module indicate a decrease in Isc with an increasing number of cleaning cycles. Coaxial microscopy reveals visible effects such as scratch marks and color blurring. **Fig. 7a** shows the module before cleaning, while **Fig. 7b** displays the scratch marks on the module's surface after 10,000 cycles of cleaning with the rotary brush.

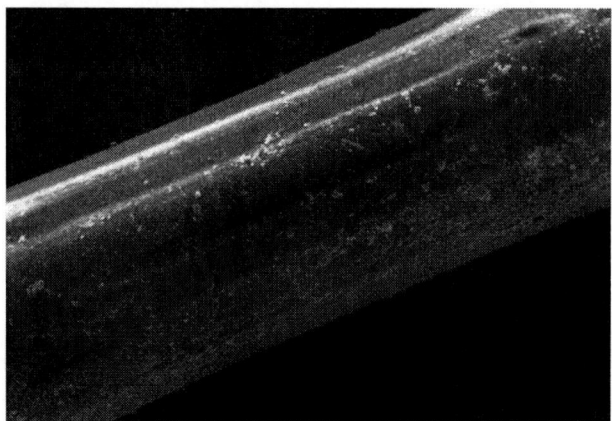

Fig. 6a: SEM of Nylon Bristle (Outdoor)

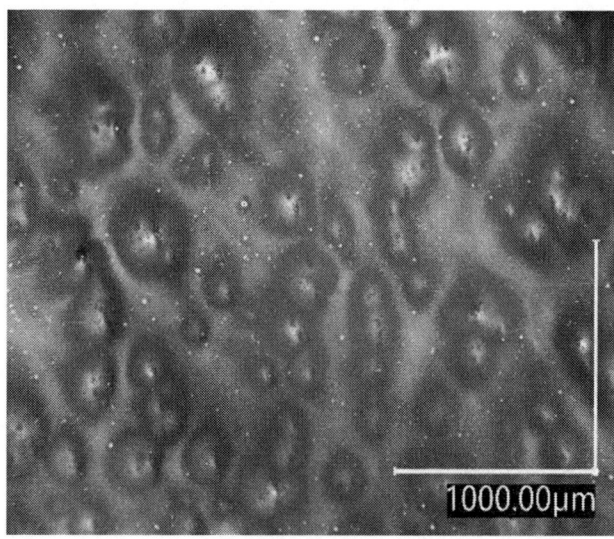

Fig. 7a: Before indoor test

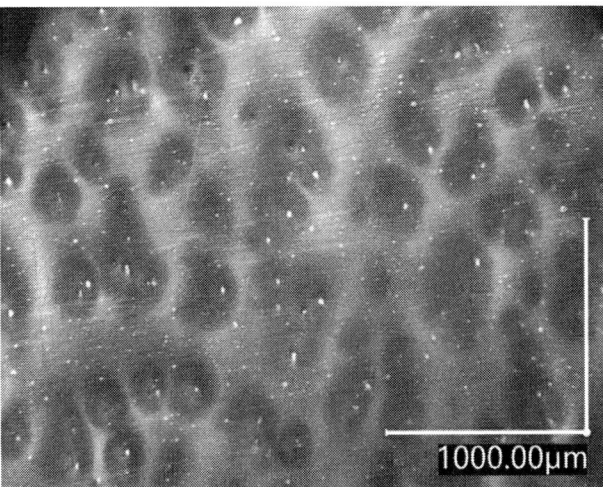

Fig. 7b: Indoor test after 10,000 cycles

The anti-reflective coating withstands the testing without additional abrasives, showing only minor changes in reflectance. Visible effects, such as scratch marks and image blurring, indicate the onset of coating degradation. **Fig. 7c** illustrates the change in reflectance due to indoor testing without abrasives. The shift of the reflectance curve to longer wavelengths can be attributed to the optical thickening of the layer, a phenomenon also observable in the coaxial microscopy images.

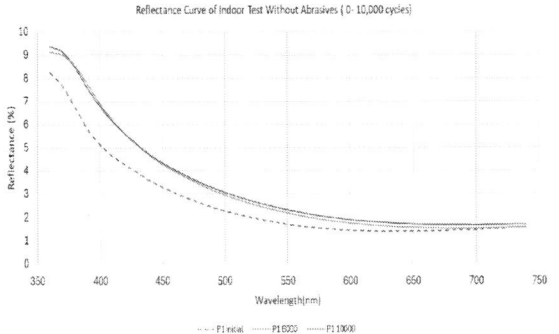

Fig. 7c shows the reflectance curve of the indoor-tested module, which exhibits a shift to longer wavelengths. This shift is attributed to optical thickening of the coating. A shift to shorter wavelengths, in contrast, would indicate thinning of the coating [4=Khan 2020]

V. RESULTS OUTDOOR TESTING

Fig. 8a shows the initial reflectance of the module before field testing, with no coating degradation and an initial reflectance below 4%, indicating the presence of the coating. After one year of field testing (400 cycles), which simulates daily cleaning, significant degradation is observed. **Fig. 8b** shows the

microscopy image of the module, and **Fig. 8c** displays the reflectance curve after field testing. The curve reveals a notable increase in reflectance, which is corroborated by coaxial microscopy showing fading color, indicating significant coating degradation.

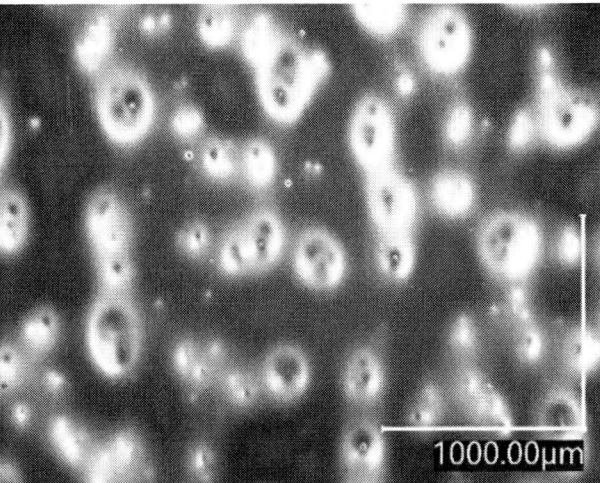

Fig. 8a: Coaxial microscopy of Outdoor tested module before testing

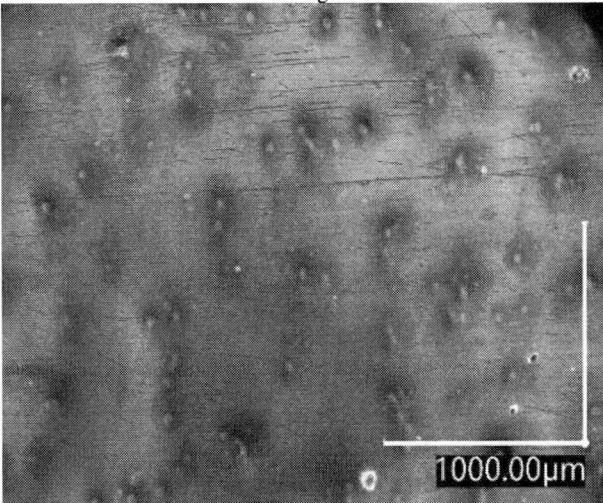

Fig. 8b: Coaxial microscopy of Outdoor tested module (after 400 cycles)

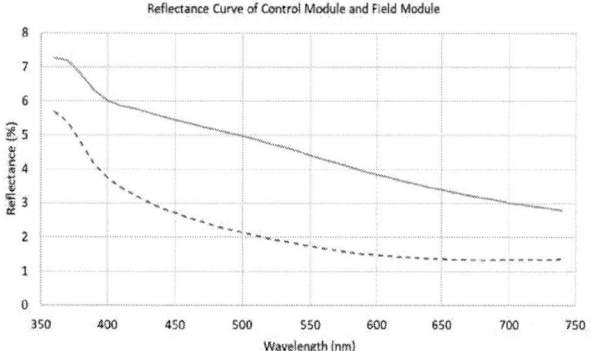

Fig. 8c: Reflectance curve of outdoor tested module before and after testing

VI. CONCLUSIONS

- PV modules exhibit various types of anti-reflective coatings with different thicknesses, observable through coaxial microscopy and color analysis.
- Fine dust on the modules and on the robot's, bristles contain calcium, which can contribute to abrasion through chemical reactions with the solar glass surface.
- Brush abrasion can damage the anti-reflective coating (ARC) on solar panels by scratching or removing it, thereby reducing the panel's ability to effectively absorb light.
- In indoor brush testing without abrasives, a linear increase in reflectance is observed from 0 to 10,000 cycles. The shift of the reflectance curve to longer wavelengths may be attributed to optical thickening effects, such as the ingress of water or bristle particles.
- After one year of daily cleaning in the test field, the anti-reflective coating degrades more significantly compared to indoor tests, primarily due to the presence of abrasives (dust). Microscopic images reveal bristle marks and scratches, indicating partial removal of the coating layer.
- The impact of abrasives is more pronounced in real-world conditions. The anti-reflective coating on the modules experienced significant removal after one year of daily cleaning in the test field.

VII. OUTLOOK

Standardization of cleaning procedures and testing methods is essential. Based on abrasive indoor testing on small samples, as outlined in IEC 62788-7-3, a comprehensive trilateral testing approach should be developed to evaluate full-size modules under real operational conditions [5], [6]. This approach should address:

- Soiling: The types and characteristics of dust.
- PV Modules: The impact on solar glass and anti-reflective coatings.
- Robots: The effectiveness of bristle or microfiber cloth materials and the applied pressure during cleaning.

The goal of this research is to develop efficient and cost-effective cleaning methods for PV modules without causing damage [7]. Future improvements for enhancing PV energy production in hot and dusty regions may include optimizing cleaning schedules, improving water usage efficiency, and exploring alternative soiling mitigation concepts, such as anti-soiling coatings.

VIII. REFERENCES

[1] IEC, 62788 7 3:2022, "Measurement procedures for materials used in photovoltaic modules - Part 7-3: Accelerated stress tests - Methods of abrasion of PV module external surfaces".

[2] Todd Karin, Mason Reed, Jim Rand, Robert Flottemesch, Anubhav Jain: Photovoltaic module antireflection coating degradation survey using color microscopy and spectral reflectance, Progress in Photovoltaics, 30/11, 2022, 1270-1288.

[3] G. Mathiak, A. Seentakath, A. Muthusamy, J. J. John, "PV Module Cleaning Under Hot Desert Conditions: Creating Test Standards for PV Module Cleaning", 2023 Middle East and North Africa Solar Conference (MENA-SC), pp.1-6, 2023.

[4] Khan, M.Z., Pfau, C., Schak, M., Miclea, P.-T., Naumann, V., Debess, A., Hagendorf, C., Ilse, K.: Resilience of industrial pv module glass coatings to cleaning processes. J. Renewable Sustainable Energy 12 (5), 2020, 053504.

[5] N. Ferretti, A. El-Issa and L. Podlowski, "Standardized test procedure for PV module cleaning equipment", *Proc. 36 EUPVSEC*, 2019.

[6] Muhammad Zahid Khan, Ahmed Abuelseoud, Katja Lange, Guido Willers, Mohammed A. Bahattab, Mark Mirza, Hussam Qasem, Volker Naumann, Ralph Gottschalg, Klemens Ilse: Correlation between laboratory and outdoor soiling experiments with anti-soiling coatings, Solar Energy Materials and Solar Cells, 269, 2024, 112751

[7] Klemens Ilse et al.: Techno-Economic Assessment of Soiling Losses and Mitigation Strategies for Solar Power Generation, Joule 3/10, 2019, 2303-2321

41st European Photovoltaic Solar Energy Conference and Exhibition

NUMERICAL SIMULATION FOR COMPARISON OF PV MODULE DESIGNS BASED ON OUTDOOR DATA IN DESERT CLIMATES

Matthias Pander*[1], Bengt Jaeckel[1], Klemens Ilse[1], Amir A. Abdallah[2]
[1]Fraunhofer-Center for Silicon-Photovoltaics (CSP), Halle (Saale), Germany
[2]Qatar Environment and Energy Research Institute (QEERI), Hamad bin Khalifa University (HBKU), Qatar Foundation,
P.O. Box 34110, Doha, Qatar
*Contact: +49 345 5589 5215, matthias.pander@csp.fraunhofer.de

ABSTRACT: Crystalline silicon modules are currently the dominant technology and will remain so in the coming years. The transition to larger wafers in the direction of M10 and M12 goes hand in hand with the optimization of module size in terms of reliability and maximum utilization of freight containers. The reliability of these modules is said to remain unchanged and the performance guarantees by manufacturers are now given for at least 25 years and even lower degradation rates are specified. In moderate climates, this is already questionable and for challenging climates such as deserts, the uncertainty increases further.

An important factor for reliability in desert climates is thermomechanical loads caused by daily high temperature cycles. In a previous study, different cell formats, cell spacing and connector thicknesses were compared and possible design candidates were determined using the simulation of the standard thermal cycling test (TCT). In this work, the cell interconnector loads of design variants for specific desert climate data are determined and compared with the results of the TCT simulation. High-resolution environmental data for Doha (Qatar) for the period from 2018 to 2022 is analyzed, prepared and used for the simulation. First, the module temperatures are determined with the help of models and then the data is reduced to minimize the required calculation points. The determined thermomechanical loads on the cell interconnector are compared for selected designs and contrasted to a TCT. As a result, it can be deduced how many standard temperature cycles correlate with the outdoor loads. Furthermore, differences between individual years can also be assessed.

Keywords: Finite Element Simulation, FEM, Fatigue, Crystalline Silicon

1 INTRODUCTION

Standard crystalline silicon solar modules have demonstrated their reliability in temperate climates. However, to meet the demand for sustainable energy, it is essential to increase both the number of photovoltaic (PV) installations and their application in various climates. Desert regions present an excellent opportunity for solar energy generation due to their high annual solar radiation, which is two to three times higher than that of Central Europe.

Despite this potential, several challenges impact the long-term reliability of the modules and components. In addition to common short-term issues, such as performance losses from soiling, desert environments impose significant thermomechanical stress. This stress arises from consistently high ambient temperatures (averaging around 40 °C) while module temperatures can rise to 70 °C during daytime and subsequent low temperatures during night (15-20 °C). Furthermore, the considerable temperature fluctuations between day and night in desert climates worsen thermomechanical strain on the modules.

In the current edition of ITRPV M6 is currently the smallest cell format. M10 (182 x 182 mm²) and G12 (210 x 210 mm²) formats are dominating the market since 2022. The rectangular formats M10R (182 x 19x mm²) and G12R (182 x 21x mm²) also have certain market share. The larger cell sizes lead to increased module dimensions [1].

In a previous EU PVSEC contribution, the interconnector stress was simulated for diverse cell formats, ribbon geometries, and cell spacing [2]. Setups with high power density due to low cell distance (<2 mm) showed highest ribbon stress which is assumed to result in fatigue-related reliability issues, especially in regions with high daily temperature changes, such as in desert climates. As the size of the solar cells increases as predicted in ITRPV [1], the load on the cell interconnectors increases. Cutting the cells was found to be beneficial for reducing the electrical losses and the load on the cell interconnectors. An increase in the cell spacing can significantly reduce the ribbon stress compared to other geometry changes but reduces module efficiency. The most relevant designs were M6 half-cell, M10 half-cell and M12 half- and third-cell with cell distances between 1 to 3 mm. These designs are further analyzed as they are predicted to be most relevant also in future PV module designs.

In this paper detailed one-minute data from the Outdoor Test Facility (OTF) in Doha (Qatar) is used as the basis for the calculation of module temperatures. The module temperatures are estimated with the SANDIA model [3]. The data must first be filtered to exclude missing or non-physical data sets. Data reduction methods are than applied to minimize the number of simulation steps. The parameterized finite element model from [2] is used with these temperature data sets as input data. As measure of damage the plastic strain energy density is evaluated ([6], [7], [8]) and compared regarding the resulting interconnector stress.

2 DATA BASIS AND PROCESSING

2.1 Outdoor Test Field

Detailed outdoor data (1 min resolution) from the Outdoor Test Facility (OTF) in Doha, Qatar was analyzed for the years 2018 - 2022. The global horizontal solar irradiance, plane of array irradiance, and diffuse horizontal irradiance were measured (Kipp & Zonen CMP21 pyranometers). Direct normal irradiance was measured using a pyroheliometer (Kipp & Zonen CHP1). Albedo measurements were realized using two pyranometers, one facing

591

the ground measuring the reflected solar irradiance and the second pyranometer facing the sky measuring the incident solar irradiance. PV system's electrical parameters are collected by the data acquisition system and modules' temperatures were measured using a thermocouple (SOL. Connect Sensor T, Platinum resistance temperature sensor PT 1000).

The meteorological station was used to measure the ambient air temperature and relative humidity (Thies Clima Hygro-Thermo Transmitter-compact), wind speed and wind direction (Thies Clima wind speed and wind direction Transmitter).

Figure 1: The Outdoor Test Facility (OTF), Doha, Qatar

2.2 Module temperature simulation

Only daytime data is used for further evaluation. In almost all cases the lowest temperature is still covered by this approach, because it occurs in the early morning hours before sunrise. The important variables for the module temperature calculation are irradiance in the plane of array, ambient temperature, and wind speed. The Sandia temperature model is used to calculate the module temperature [3]:

$$T_m = E_{POA} \cdot e^{a+b \cdot WS} + T_a$$

E_{POA} - Solar irradiance incident on the module (POA) (W/m²)
T_a - Ambient air temperature (°C)
WS - Wind speed (m/s)

The parameters a and b depend on the module construction and materials as well as on the mounting configuration of the module. Here $a = -3.47$ and $b = -0.0594$ are used for a glass/cell/glass structure. With this model for each data set a module temperature is calculated.

2.3 Data reduction

The steps for generating the various temperature data sets are briefly presented below and the abbreviations are introduced for later reference. The following steps are performed on the data:

1. Filter for unreasonable, erroneous or missing data sets
2. Determination of the hourly minimum and maximum values from the minutely resolved data [Data set: **"hourlymmax"**]
3. Reduction of data by searching for sign changes (monotonic decrease or increase of temperature) [Data set: **"hourlyamp"**]
   a. Local peak is found when previous and next value is lower compared to actual value
   b. Local valley is found when previous and next value is higher compared to actual value
4. Application of a rainflow counting algorithm to identify significant thermomechanical-load changes and load cycles [4] [Data set: **"rainflow"**]
   a. 4 consecutive temperatures are selected.
   b. Absolute Temperature differences are calculated.
   c. Rainflow counting condition: $S1 > S2 \leq S3$
      $S1 = |T_1 - T_2|$, $S2 = |T_2 - T_3|$, $S3 = |T_3 - T_4|$
   d. When the condition is not fulfilled S1 is significant and added to the data set, else S2 temperatures are discarded and segments S1 and S3 are combined.
5. The maximum reduction is achieved by extraction of the daily maximum and minimum temperatures. This process takes place without the specific time stamp, where the minimum and maximum are always 12 hours apart by definition. [Data set: **"daymmax"**]

The hourly maxima and minima are the basis for further data reduction. In the "hourlyamp" data set only turning points are considered which is the data set with the highest data density in this investigation. The "rainflow" data set is intended to identify only significant cycles by eliminating small cycles during the day. With respect to the daily amplitude the maximum thermomechanical-load cycle per day is given by the "daymmax" data set but excludes all other cycles within the day. Because the minimum temperature of a day can occur, either in the morning or in the evening, the maximum or minimum temperatures of consecutive days can be adjacent when considering the exact time stamp. This sometimes leads to anomalous correlations and fewer load cycles. Therefore, the generic time allocation makes more sense in this case. The data sets are finally exported as a time series (time and module temperature data pair) for use as input parameters for the simulation.

2.4 Simulation of interconnector stress

The parameterized finite element model from [2] is used with the temperature data sets as input data. Material data is taken from literature and internal measurements [6]. The basic module structure is a glass-glass structure with the following geometric parameters:
- cell thickness: 180 µm
- front encapsulant thickness: 0.4 mm
- back encapsulant thickness: 0.4 mm
- glass thickness: 2 mm
- backside glass thickness: 2 mm

Five cells in a row are simulated in the global model which is sufficient to accurately calculate the cell shift and stress on the ribbon. In the present investigation we concentrate on these variants:
- Wafer sizes: M6_HC, M10_HC, M12_HC, M12_TC (HC = half-cell, TC = third cell)
- Cell gaps (distance between cells): 1, 2, 3 mm
- Thickness of ribbon: 0.16 mm, 0.2 mm, 0.24 mm

The simulation process involves two stages. Initially, cell displacements are computed using a global model, followed by the calculation of local stresses at the inter-connectors in a detailed submodel. Plastic strain energy density serves as a damage indicator ([6], [7]). The mesh is adjusted based on the necessary details, with a higher mesh density needed in the detailed model to accurately assess plastic strain from the simulation. Figure 2 illustrates the meshed cell interconnections of the investigated cell gaps.

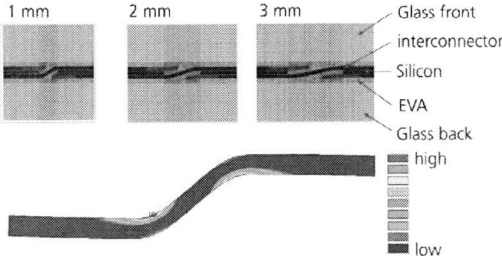

Figure 2: Meshing of the detailed model of the cell gap with interconnector ribbon and exemplary plastic strain energy density plot

For further classification of the calculated damage, the results are related to the results of a standard thermo-cycling test according to IEC 61215 TC200 obtained in [2].

3 RESULTS

3.1 Module temperature data sets

Figure 3 shows an original data set of the module temperature determined by the Sandia temperature model based on data from the OTF in Doha from 2019. In rare cases, there were data gaps in the years 2018 to 2020, which were removed. During the analysis it was found out that the data from 2021 and 2022 had a major data gap due to a hardware malfunction and were therefore not suitable for the intended comparison. It would have been possible to supplement the data with data from another weather station, but this was not done for the demonstration of data processing in this paper.

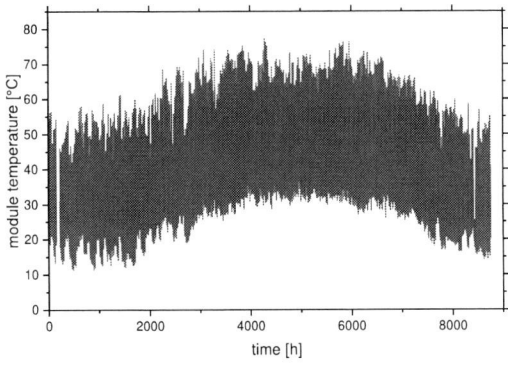

Figure 3: Original 1min resolution module temperature data set for Doha 2019

To visualize the temperatures occurring, Figure 4 shows a histogram of the module temperatures for the 3 years studied. The distribution is similar, although 2020 tends to have higher temperatures.

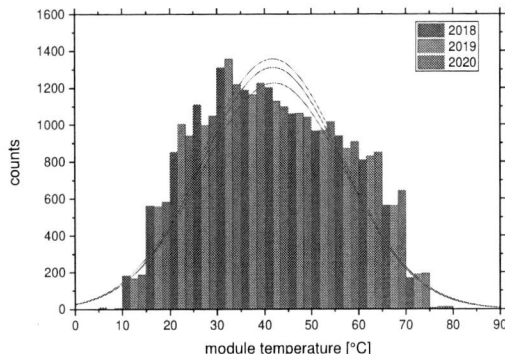

Figure 4: Histogram of module temperatures

In Figure 5 an example for the data reduction is visualized for the year 2019 and the resulting data points are given in Table 1. The initial data set is the 1-minute raw data. The hourlymmax data set is the densest data set with 10,872 data points. Hourlyamp reduces the data points already significantly to 6,505 data points. Especially when the temperature is monotonically falling interim points are discarded (visible especially at the end of the day). The rainflow algorithm discards most small temperature drops and results in now only 934 data points. The daymmax data set results in the smallest data volume with 726 data points (note: two days were excluded because data was missing).

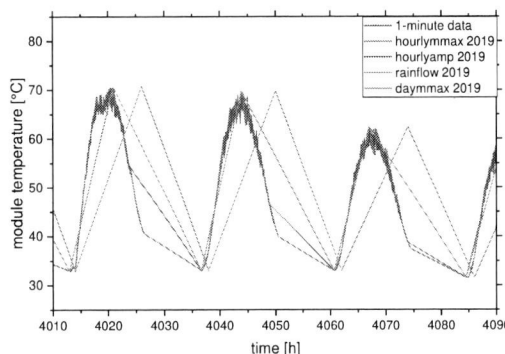

Figure 5: Section of the resulting data sets for module temperature data based on Doha 2019 outdoor data

Table 1: Number of data points of the different data sets (load steps) for the simulation

data set	2018	2019	2020
1-minute	306950	306700	307754
hourlymmax	10936	10872	10980
hourlyamp	6385	6505	6414
rainflow	921	934	887
daymmax	730	726	732

3.2 Comparison of data reduction method and inter-connector stress

To analyze the differences between the individual data reduction methods on the calculated damage during the years under investigation, the M10_HC setup variants were examined in detail. Figure 6 shows an overview of the results.

41st European Photovoltaic Solar Energy Conference and Exhibition

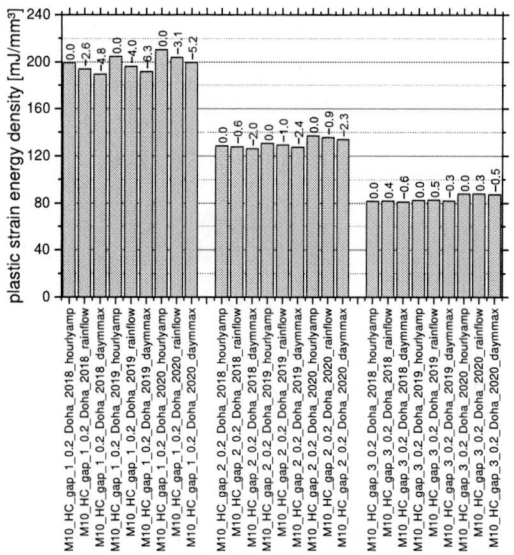

Figure 6: Comparison of simulated plastic strain energy density at the end of each year for M10_HC setups for the different data sets, label on the bars is the relative difference to hourlyamp result of the specific year

The hourlyamp data set with the highest density normally results in the highest interconnector stress. Especially for small cell gaps, the damage per temperature change is higher, and therefore exclusion of smaller cycles leads to differences. The rainflow data set shows differences of up to 4.0 %, and the daymmax data set differences of up to 6.3 % for a 1 mm gap. For a 2 mm gap, the difference in rainflow data is <1.0 %, and for a 3 mm gap, it is negligible. Comparing the years, the highest stress is consistently observed in 2020. This trend is observed for all data sets and is attributed to the higher temperatures as indicated in the histogram (Figure 4).

In terms of simulation time and accuracy, the results for the rainflow data set represent the best compromise for further analysis. Therefore the 2020 rainflow data set is used for further comparison of the other variants.

3.3 Comparison of module variants

The load for the rainflow data set 2020 was determined for the relevant variants. The results are shown in Figure 7. The qualitative results are comparable to the analysis of a TC200 in the previous publication [2]. There is an increase in interconnector stress with smaller cell spacing and larger cells. The lowest stress is determined for the M12_TC (third-cell), while the highest stress is determined for the M12_HC (half-cell). In addition, thicker ribbons also experience higher stress. To classify the results, the calculated stress in the TC200 (data from [2]) is related to the result for the year 2020 (rainflow data set). This is introduced as a stress factor:

$$stressfactor_{TC200} = \frac{pl. \, strain \, energy \, density_{TC200}}{pl. \, strain \, energy \, density_{2020}}$$

The value indicates how many years of the specific year are equivalent to the testing duration of a TC200. A lower value means that fewer years of operation are represented with the standard test. Figure 8 illustrates the result for the stress factor. The value is also not a constant and depends

on the setup. Values between 1.7 and 5.1 are obtained for the variants investigated. For M12_HC with the thick ribbon (0.24 mm) and 1 mm cell gap the worst case appears with a stress factor of 1.7. This means that one TC200 corresponds to only 1.7 years. The opposite is true for M12_TC with a thin ribbon (0.16 mm) and 3 mm cell gap, where the stress factor is 5.05.

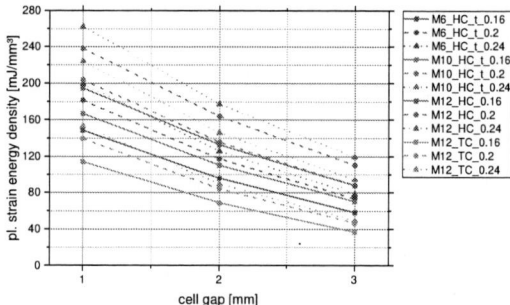

Figure 7: Comparison of simulated plastic strain energy density for the rainflow data set of Doha 2020 for different cell string setups

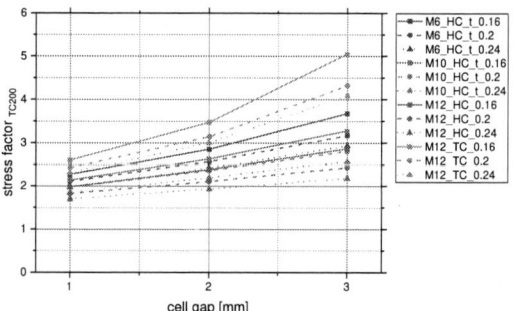

Figure 8: Comparison of stress factor (plastic strain energy density in TC200 related to result of rainflow data set of Doha 2020) for different cell string setups

4 DISCUSSION

The purpose of the data reduction was to optimize the computing time by minimizing necessary simulation steps without losing much accuracy regarding the calculated interconnector stress. Comparably efficient and precise data reduction was achieved by using the rainflow counting method. Since there is already numerical error and inaccuracies in the input data in the estimation anyway, an error of up to 5 % in relation to the final value seems acceptable when reducing the amount of data points by 85 %. The daily amplitude can also be used well for screening data with a small additional error. The strong reduction of data points for the desert climate also indicates that the temperature increases and decreases mostly monotonic during the day. This result may change for other locations and climates.

Higher resolution data is mostly required for configurations where small temperature changes lead to plastic strain which is especially the case for small cell distances (< 1 mm) and thick interconnectors.

The evaluation of the results in comparison to the standard temperature cycle (TC) test showed that for current module structures with a cell spacing of 1-2 mm, the TC200 can only verify 1.7 to 3.5 years in the specific desert climate of Doha. For larger cell spacing, it was at least 5 years. However, it is known that current modules can withstand thermal cycling tests with more than 600 cycles (3x IEC) without significant performance degradation due to interconnector breakage. This seems particularly necessary for long-term use in desert climates based on the presented results. According to the presented results, 2000 to 3000 cycles in the thermal cycling test would be necessary to prove 25 years in desert climates. However, it was assumed here that the accumulation of damage is additive. However, larger load cycles, as in the TC, make a disproportionately large contribution to the damage, so that these large quantities are somewhat too high. Nevertheless, an increased number of cycles is required. To reduce the required number of cycles it may be useful to increase the temperature amplitude like it is proposed in IEC TS 63126 [10].

5   SUMMARY AND OUTLOOK

In this paper detailed one-minute data from the Outdoor Test Facility (OTF) in Doha (Qatar) was used as the basis for the calculation of module temperatures. The module temperatures were estimated with the SANDIA temperature model [3]. Different data reduction methods were used and compared regarding the resulting ribbon stress. Hourly maxima and minima were determined for further data reduction. From these only turning points are considered which is the data set with the highest data density in this investigation. With the rainflow counting algorithm most significant cycles were identified and smaller cycles during the day were eliminated. The maximum load amplitude per day is given by the daily min-max temperatures but excluded all other cycles within the day. This leads to deviations in the overall damage per year of up to 6.3 %. Since this applies for all data sets this method is still feasible for screening and comparison of years and generation of mission profiles [9].

The comparison of the different setups showed that interconnector stress increases with smaller cell spacing, larger cells and thick ribbons. The minimum stress was found for M12_TC (third-cell) with a 3 mm cell gap and a thin ribbon (0.16 mm), while the maximum stress occurred for M12_HC (half-cell) with a 1 mm cell gap and a thick ribbon (0.24 mm). The comparison of the TC200 simulation results with the simulated years showed that there is no simple transfer rule, as the damage that occurs also depends on the module setup. The stress factors determined show that a normal TC200 is not even a minimum requirement for modules to withstand the desert climate for 25 years or more. Significantly more cycles are required or a changed temperature amplitude like it is proposed in IEC TS 63126 [10].

6   REFERENCES

[1]   ITRPV, "International Technology Roadmap for Photovoltaic (ITRPV)", 2024
[2]   M. Pander et al., "Numerical design study for desert climate applications of bifacial PV modules", 40th EU PVSEC, 2023, 10.4229/EUPVSEC2023/3AV.2.18
[3]   D. L. King, W. E. Boyson and J. A. Kratochvil, "Photovoltaic array performance model", Sandia Rep. No. 2004-3535, pp. 1-43, 2004
[4]   L.L. Schluter, H.J. Sutherland, User's Guide for LIFE2's Rainflow Counting Algorithm, Sandia Report SAND90-2259, 1991

[5]   M. Pander et al., "Digital Prototyping – Application of Numerical Methods in Module Development", 36th EU PVSEC, 2019, 10.4229/EUPVSEC20192019-4BO.11.2
[6]   M. Pander et al., "Lifetime estimation for solar cell interconnectors", 28th EU PVSEC, 2013, 10.4229/28thEUPVSEC2013-4CO.10.3
[7]   D. Kujawski, Fatigue Failure Criterion based on strain energy density, Theoretical and Applied Mechanics, Vol. 7, 1, 1989, pp.15-22
[8]   R. Meier et al., "Microstructural optimization approach of solar cell interconnectors fatigue behavior for enhanced module lifetime in extreme climates", Energy Procedia, Vol. 92, 2016, 10.1016/j.egypro.2016.07.020
[9]   B. Jaeckel et al., "Mission profile concept for PV modules: use case – middle east deserts vs temperate European climate", EPJ Photovoltaics 14, 39, 2023 https://doi.org/10.1051/epjpv/2023030
[10] IEC TS 63126 Ed.2, Guidelines for qualifying PV modules, components and materials for operation at high temperatures, to be published 2024

7   ACKNOWLEDGEMENT

This publication was made possible by NPRP grant # NPRP11S-1220-170110 from the Qatar National Research Fund (a member of Qatar Foundation). The authors gratefully acknowledge the financial support of the project "3DProVe" (grant no. 1604/00103) by the State of Saxony-Anhalt and the European Regional Development Fund. The findings herein reflect the work, and are solely the responsibility, of the authors.

41st European Photovoltaic Solar Energy Conference and Exhibition

Impact of Modern Cell Photovoltaic Geometries on Power and Energy Loss due to Cell Cracks

A. Hashem[1], SL. Mortazavifar[1], R. Gottschalg [1,2]
[1]Hochschule Anhalt University of Applied Sciences, Köthen, 06366, Germany
[2] Fraunhofer-Center for Silicon Photovoltaics CSP, Halle (Saale), 06120, Germany
E-Mail: ahmad.hashem@hs-anhalt.de

ABSTRACT: Evaluating microcracks for photovoltaic (PV) modules is challenging as there are varying perspectives among different stakeholders. Evaluation typically occurs on cell level and a certain number of cells is deemed permissible in an entire module. Warranty conditions however, are generally based on power loss of the module. There are many observations of power loss being reported in the literature most focus on 2-3 busbar Al-Back- Surface-Field solar cell technologies. This technology has been replaced by a plethora of different cell types with new cell structures (PERC, TOP-Con, HJT), new wafer sizes (M6, M8, M10, M12), increasing number of busbars (4, 6, 9, 12) and contacting schemes (shingling, smart-wires) as well as wafer-fractions (full cell, half-cell, 1/3rd cell). Despite these advancements, few studies have assessed the impact of these on overall power rating of a module. PV modules can exhibit a variety of crack types. A methodology for evaluating cracks developed in the German research project 'PV-Riss'(PV-crack) will be discussed in this study, the aim is to evaluate the worst possible propagation for the different cell types and geometries mentioned above. Subsequently, the impact cracked cells locations within the module is investigated by carrying out a Monte-Carlo simulation of different crack scenarios.

Keywords: PV Cracks, Cell Geometries, Power Loss

1 Introduction

Solar cells are the fundamental components of PV systems, making them crucial for solar energy conversion. Enhancing the efficiency and longevity of of these cells is essential to optimize energy production and minimize costs. Like any technical system PV systems, might face challenges over time, one significant issue that can arise is cracking within the solar cells. Cracks create inactive or less-active areas that reduce the available carriers for electrical current generation and can block the path of incident light, decreasing the amount of light being converted into electrical energy. This reduction in conversion efficiency of the solar cell affects the entire module and the system. Solar cell technology has evolved tremendously from M0 to M12 over years as shown in **Figure 1** increasing efficiency and reducing costs. The "M" series represents the wafer sizes, with higher numbers indicating larger wafer dimensions Initially, 125 mm solar cells were common in 1995, and M0 wafers were frequently utilized in the early 2010s due to their low cost and efficiency. M2 wafers emerged between 2013 and 2017, and by 2018, the transition to mono Passivated Emitter and Rear Contact (PERC) technology accelerated changes in wafer shapes leading to sizes such as M6, M10,

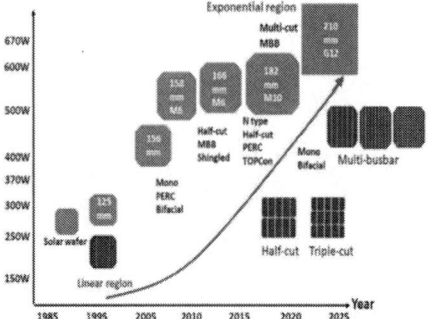

Figure 1: Evolution of PV Cell Technology [1]

and finally M12/G12 [1].

Electroluminescence (EL) imaging has been widely used technique to detect cracks in PV modules as it provides high-resolution images that clearly reveal micro-cracks and other defects, even under low-light conditions. This method is highly effective and non-invasive, ensuring that the solar cells are not subjected to damage during inspection. The current literature primarily explores on simulating the impact of different types of cracks using several techniques mainly in older technologies with few Busbar (BB), with limited coverage on modern technologies. T-test and F-test which are statistical algorithms were applied to identify the significant effect of various cracks types on the output power performance of 2BB poly- crystalline PV modules [2]. The effects of uniform and non-uniform distributions of cracks on open circuit voltage and output power were analyzed on 3BB cell, revealing that non-uniform cracks distribution have higher severity leading localized hotspots which jeoparadizes the output power [3]. Various types of cracks were studied focusing on the percentage of broken (in-active) area in 3BB multicrystalline modules, and an electrical diagram was simulated to connect the active and the in-active area by break resistance (R_b) [4]. Module power is calculated by setting R_b of every cracked cell of all modules to one discrete value in the range of 1 mΩ to 1 kΩ. A direct correlation was observed between the resistance value and the module's output power loss. Few studies tackle the impact of cracks on modern technologies. A comparison was conducted to assess the impact of cracks on Aluminum-Back Surface Field (Al-BSF) and PERC mini-modules [5]. The results indicated that Al-BSF modules and cells exhibited greater variability in recorded values compared to PERC modules, leading to the conclusion that PERC technology offers superior crack resistance and overall reliability. The impact of open cracks on Heterojunction (HJT) cells and the correlation between crack depth and power loss has been explored through the utilization of a 2D finite

596

element model [6]. The direct correlation between crack depth and power loss is uncertain, as a deep crack does not necessarily result in an inactive area, and thus may have minimal impact on power loss.

2 EVOLUTION OF CELL GEOMETRY

Table *1* demonstrates the leading solar cell technologies currently prevalent in the PV market by comparing different cell technologies based on their dimensions and the configuration of busbars , highlighting the evolution from M0 to M12. This data is obtained from ENFSolar [7], a major solar supplier that encompasses over 60,000 solar companies and lists more than 88,900 products. As solar cells dimensions increase, reaching up to 210 mm, there is a corresponding rise in the number of BB from 3 in M0 to 9 in M3, and eventually up to 16 and 15 in M10 and M12, respectively. Additionally, there is a significant decrease in the inter-busbar area and the area between the last BB and the cell corner, thereby reducing the maximum crackable area.

Table 1: Impact ofTechnology on Cell Areas [7]

Cell Technology	Dimensions (mm x mm)	Number of BB	Area Between BB (%)	Area between last BB and corner (%)
M0	156 x 156	3	33.33	15.7
M2	156.75 x 156.75	4	24.88	12.68
		5	19.9	10.19
		6	16.66	8.33
M3	158.75 x 158.75	3	32.75	16.92
		5	19.65	10.14
		9	10,57	7.18
M6	166 x 166	6	16.66	8.23
		9	10.84	6.26
		12	8.07	5.6
M10	182 x 182	10	9.5	6.81
		11	8.9	5.05
		16	5.93	5.13
M12	210 x 210	12	8.3%	4.29
		15	6.49%	4.07

3 SIMULATION RESULTS

This section analyzes the worst-case crackable area within PV cells, a critical factor in determining the impact of cracks on the overall system efficiency. Simulations in this section are utilized using pvlib, an open-source Python library designed for simulating the performance of photovoltaic energy systems. It provides tools for modeling complex environments and PV system operations, initially developed at Sandia National Laboratories [8]. These simulations involve plotting I-V and P-V graphs for various technologies to assess the effect of worst-case crackable area on each technology. The case study focuses on M3 technology, examining different BB configurations. This approach is also applied to other technologies, accounting for variations in their maximum crackable (inactive) areas.

3.1 Maximum Crackable area between BB

Figure 3 illustrates the effect of the worst-case scenario for crackable area between BB. Figure 2a shows the relationship between cell power and cell voltage for different configurations, while Figure 2b presents the

corresponding cell current against cell voltage. The power loss begins at with approximately 32% with 3BB cell, decreases to 20% in 5BB cell, until reaching 10.5% with 9BB cell. There is a clear inverse correlation between the number of BB and the power loss, which contributes to enhanced the cell efficiency.

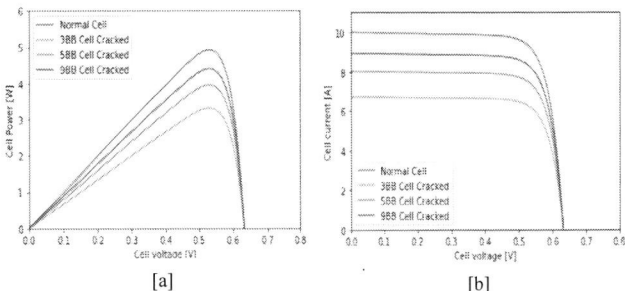

[a]   [b]

Figure 3: I-V & P-V Curves for Max Crackable area between BB

The performance parameters are shown in **Table 2**, the most affected parameter due to cracks is the short circuit current at the maximum power (I_{mp}) however, the open circuit voltage remains nearly unchanged. Consequently, the maximum output power (P_{mp}) of the system is influenced, as it is a product of the voltage and the current.

Table 2: Impact of Max Crackable Area between BB on Perfromance Parameters

Pmp_normal	4.92 W	Imp_normal	9.24 A
Pmp_3BB	3.31 W	Imp_3BB	6.21 A
Pmp_5BB	3.95 W	Imp_5BB	7.42 A
Pmp_9BB	4.40 W	Imp_9BB	8.26 A

3.2 Maximum Crackable area between last BB and Cell Corner

The focus in this section is on the worst-case scenario simulation for the crackable area between the last BB and the cell corner. Figure 3a illustrates the relationship between cell power and cell voltage for different configurations, while Figure 3b presents the corresponding cell current as function of voltage. The crackable area is significantly smaller than the area between BB, resulting in cracks having a reduced impact on the maximum output power.**Error! Reference source not found.** In the worst-case scenario, 3BB cell expriences approximately 18% of power loss, decreases to 10% for 5BB cell and further to around 8% with 9BB cell. Variations in this area can occur among different manufacturers due to differences in design strategies and manufacturing techniques.

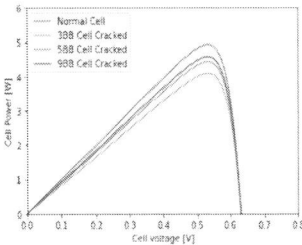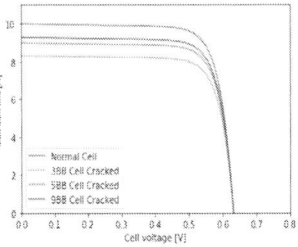

[a]   [b]

Figure 2: I-V & P-V Curves for Max Crackable area between last BB & Cell Corner

Also, during BB fabrication factors such as the number of BB or the choice of materials, can influence how much space is utilized near the cell corners. Resulting in cells have a rectangular or curved shape.

The performance parameters for the second scenarion are shown in **Table 3**. The impact of cracks on the short circuit current and the output power is lower than the first scenario due to the smaller area between the last BB and cell corner.

Table 3: Impact of Max Crackable Area between last BB & Cell Corner on Perfromance Parameters

Pmp_normal	4.92 W	Imp_normal	9.24 A
Pmp_3BB	4.0 W	Imp_3BB	7.67 A
Pmp_5BB	4.42 W	Imp_5BB	8.30 A
Pmp_9BB	4.57 W	Imp_9BB	8.57 A

3.3 Impact of BB number on worst-case power loss

The impact of number of BB across all cell technologies on the average power loss is summarized in **Figure 4** As the number of BB increases, the percentage of power loss significantly decreases, starting from over 30% loss with 3BB cells and dropping to approximately 6% loss with 16BB cells. Thereby enhancing the cell efficiency as well as modules and system level.

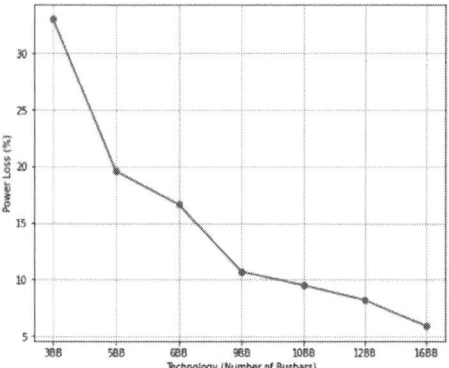

Figure 4: Impact of BB number on Power Loss

3.4 Impact of cracked cell location on module level

Previous studies have primarily focused on the impact of the number of cracked cells within a module on the output power without taking into consideration their position within the module as it's a critical factor influencing the module's performance. A module with cracked cells dispersed across multiple strings impacts the module differently than one with the same number of cracked cells in fewer strings or a single string. This variation occurs because the bypass diodes across the affected strings are activated in all cases, restricting the string's current to that of the cell with the most significant crack. A typical 60-cell module, divided into three strings will be analyzed under three scenarios: same cracked cells in different strings, different cracked cells in different strings, and different cracked cells in the same string. This concept can be also extended to larger modules with more strings.

3.4.1 Same cracked cells in different strings

The impact of identical cracks in different strings of a photovoltaic module is illustrated in **Figure 5**. The the I-V

characteristics of the three strings of the module are shown in Figure 5a, where two strings are affected by the same type of cracks (same in-active area). These string curves are combined to form the module I-V curve, as shown in Figure 5b, which consolidates the impact of these cracks at the module level with the bypass diode.

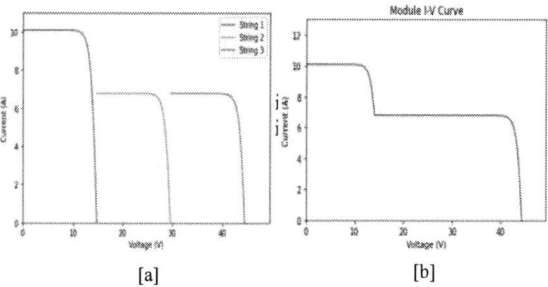

Figure 5: Impact of same cracks in different strings

3.4.2 Different cracked cells in different strings

In the second scenario, cracks of varying severities (different in-active areas) impact two different strings, as illustrated in Figure 6a. This variation influences the module I-V characteristics compared to the first scenario, resulting in multiple step shapes the module I-V curve in Figure 6b. This leads to varying output power values for each type of module based on its structure and number of by-pass diodes.

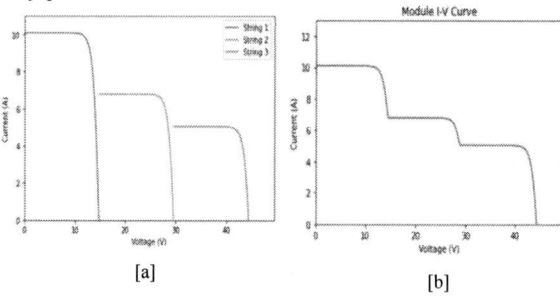

Figure 6: Impact of different cracks in different strings

3.4.3 Different cracked cells in the same string

This scenario, where a multiple cracked cells with different in-active areas are located within the same string, the bypass diode restricts the string current to the cell with the most severe crack. Consequently, only a single step shape appears on the module I-V curve,rather than multiple steps as shown in **Figure 7**. This observation challenges some studies suggesting that the power loss percentage directly correlates the number of cracks. If all the cracked cells are located in one string, the impact will be limited to the most damaged cell.

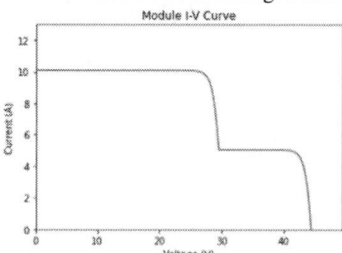

Figure 7: Impact of different cracks in the same string

4 CONCLUSION

This study has shown that modern cell geometries have a considerable influence on power loss due to cracks in PV modules. The results of simulations on various cell types and busbar configurations show that increasing the number of busbars decreases the worst-case crackable area, hence minimizing power loss. Furthermore, it has been demonstrated that the location of cracked cells inside the PV module is an important factor in influencing total module efficiency. These findings contribute to the ongoing development of more robust PV technologies, emphasizing the importance of advanced cell designs in improving module performance and lifetime.

5 REFERENCES

[1] Kurinec, S. K. (2022). Silicon solar photovoltaics: slow ascent to exponential growth. In *Women in Mechanical Engineering: Energy and the Environment* (pp. 221-243). Cham: Springer International Publishing.

[2] Dhimish, M., Holmes, V., Mehrdadi, B., & Dales, M. (2017). The impact of cracks on photovoltaic power performance. *Journal of Science: Advanced Materials and Devices*, *2*(2), 199-209.

[3] Dhimish, M., d'Alessandro, V., & Daliento, S. (2021). Investigating the impact of cracks on solar cells performance: Analysis based on nonuniform and uniform crack distributions. *IEEE Transactions on Industrial Informatics*, *18*(3), 1684-1693.

[4] Morlier, A., Haase, F., & Köntges, M. (2015). Impact of cracks in multicrystalline silicon solar cells on PV module power—A simulation study based on field data. *IEEE Journal of Photovoltaics*, *5*(6), 1735-1741.

[5] Whitaker, C. M., Pierce, B. G., Karimi, A. M., French, R. H., & Braid, J. L. (2020, June). PV cell cracks and impacts on electrical performance. In *2020 47th IEEE Photovoltaic Specialists Conference (PVSC)* (pp. 1417-1422). IEEE.

[6] Ennemrı, A., Mazouz, H., Khouzam, A., & Logeraıs, P. O. (2023). Impact of an open crack on the output characteristics of a heterojunction solar cell. *International Journal of Engineering Science and Application*, *7*(4), 95-104.

[7] ENF Solar. (2024). *ENF Solar – Solar Companies and Products*. Accessed Augus, 2024, from https://www.enfsolar.com/

[8] derson, K., Hansen, C., Holmgren, W., Jensen, A., Mikofski, M., and Driesse, A. "pvlib python: 2023 project update." Journal of Open Source Software, 8(92), 5994, (2023). DOI: 10.21105/joss.05994.

41st European Photovoltaic Solar Energy Conference and Exhibition

Performance Evaluation of the Custom-Made Small PV Modules After Exposure to Saudi Arabia's Climatic Conditions Over 10 Long Years

Amir Al-Ahmed[1]*, Amjad Ali[1], Mohammed A. Alghamdi[2], Osama Asker[2], Ridha Ben Mansour[1]
Firoz Khan[1], Atif S. Alzahrani[1],

[1]Interdisciplinary Research Center for Sustainable Energy Systems (IRC-SES), King Fahd University of Petroleum & Minerals, Dhahran 31261, Saudi Arabia.
[2]Gulf Renewable Energy Lab, A UL GCCLAB Joint Venture, Dammam, Saudi Arabia.
Email: aalahmed@kfupm.edu.sa

Introduction

Seasonal climatic conditions of Saudi Arabia present some unique challenges to photovoltaic (PV) modules. Therefore, it is important to study and understand the environmental effect on different PV modules to draw a characteristic picture for the next generation PV modules. Here, we studied 14 small PV modules fabricated in Fraunhofer ISE, and were deployed in our university campus under a collaborative program. After a decade of field installation and operation, these PV modules were carefully collected and tested in the GCC-UL lab for their IV and EL properties. Key performance parameters, such as, electrical output, efficiency, and degradation rates could be accurately measured and analyzed. The research findings provide critical insights into the behavior of solar PV modules in Saudi Arabia's harsh climate, highlighting the prevalence and consequences of damage or degradation (both internal or external). Analysis reveals that high temperatures, thermal cycling, and mechanical stresses associated with dust deposition leading to slight reduction of electrical performance and some minor internal damages. However, physical appearance (front glass, back sheet, aluminium frame) of all the modules are in good shape [1-3].

Results

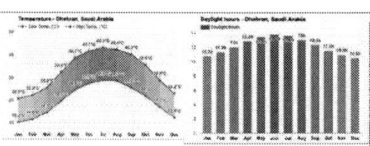

Average yearly temperature and day light hours of Dhahran Saudi Arabia [1]

PV modules glass types

Model name	Number of modules	Glass Manufacturer	Glass type
Module 1-3	3	Saint Gobain	extra-white float, plane
Module 4-6	3	Saint Gobain	extra-white cast, patterned
Module 7-9	3	Saint Gobain	extra-white cast, structured
Module 10-12	3	Interpane	extra-white float, plane, IxAR
Module 13-15	3	CentroSolar	extra-white cast, patterned, IxAR
Module 16-24	9	CentroSolar	extra-white float, plane
Module 25-27	3	CentroSolar	extra-white float, plane, IxAR

PV modules specification (cell-data)

Manufacturer	Fraunhofer ISE
Number of glass types	7
Number of modules	27
Cell type	Mono Crystalline
Number of cells	6
Maximum power rating (Pmax)	4.39 – 4.44 Wp
Open circuit voltage (Voc)	525 mV
Short circuit current (Isc)	8.920 A
Maximum power voltage (Vmp)	525 mV
Maximum power current (Imp)	8.380 A
Cell efficiency	18.22 – 18.43
Normal operating cell temperature (NOCT)	Not available
Dimensions	60x40x5cm³
Weight	n.a
Fill Factor (FF)	78.7 – 78.9 %
Power Tolerance	+/- 1.5 % rel.
Temperature coefficient of current(α)	+0.04 %/K
Temperature coefficient of voltage(β)	-0.33 %/K
Voltage irradiance correction factor(γ)	n.a

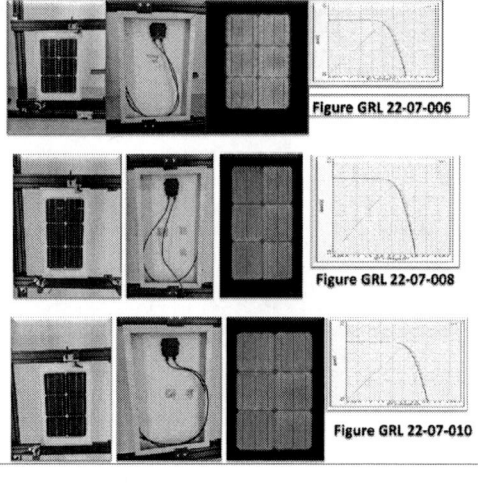

Figure GRL 22-07-006

Figure GRL 22-07-008

Figure GRL 22-07-010

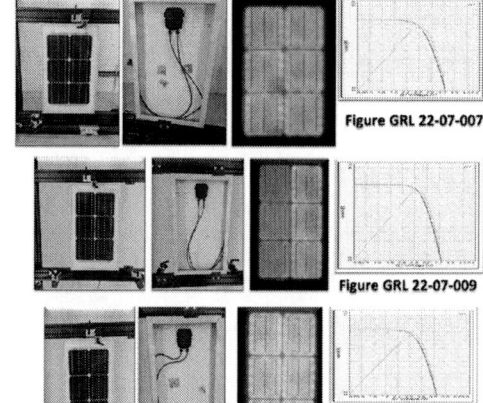

Figure GRL 22-07-007

Figure GRL 22-07-009

Figure GRL 22-07-011

I-V Test results

Module Number	Isc [A]	Voc [V]	Imp [A]	Vmp [V]	Pmp [W]	FF %
GRL 22-07-006	8.59	3.54	7.78	0.368	2.69	66.09
GRL 22-07-007	8.59	3.62	7.98	0.254	2.69	65.71
GRL 22-07-008	8.65	5.56	7.82	0.366	2.69	67.65
GRL 22-07-009	8.61	3.58	7.61	0.261	2.66	67.45
GRL 22-07-010	8.15	3.56	7.38	0.266	1.96	67.78
GRL 22-07-011	8.50	3.60	7.83	0.263	2.05	66.44
GRL 22-07-012	8.52	3.62	7.69	0.264	2.05	66.07
GRL 22-07-013	8.64	3.57	7.57	0.267	2.05	65.18
GRL 22-07-014	8.42	3.60	7.82	0.267	2.08	66.69
GRL 22-07-015	8.60	3.60	6.03	0.264	2.12	67.78
GRL 22-07-016	8.74	3.58	8.02	0.2.54	2.11	67.63
GRL 22-07-018	8.56	3.56	7.46	0.261	2.06	67.50
GRL 22-07-019	8.48	3.63	7.62	0.252	1.91	62.39
GRL 22-07-019	8.03	3.63	7.27	0.245	1.90	65.55

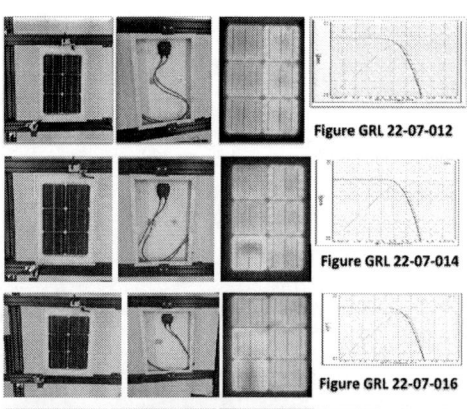

Figure GRL 22-07-012

Figure GRL 22-07-014

Figure GRL 22-07-016

Figure GRL 22-07-018

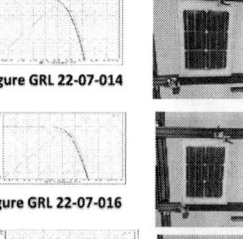

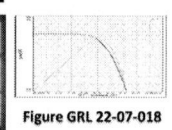

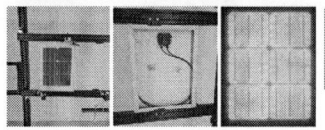

Figure GRL 22-07-013

Figure GRL 22-07-015

Figure GRL 22-07-017

Figure GRL 22-07-019

Summary

- These 14 PV modules were exposed to DHAHRAN climate condition for nearly 13 years.
- IV and EL analysis were carried out using IEC 61215-2:2021 cl. 4.6.3.1, IEC TS 60904-13:2018 test protocol.
- Few modules showed small in the I-V plot, which can be attributed to the internal damage, however, most of the modules showed IV curve without any kink.
- EL image also supports these result.
- There is more than 50% reduction in P_{max} and ~ 15% average reduction in FF.
- Other studies, such as, EL in different applied potential, mathematical modeling are in progress. Which will be presented in the final paper.

References

[1] https://weatherandclimate.com/saudi-arabia/eastern-province/dhahran-january
[2] Solar Energy 173 (2018) 478-486
[3] Renewable Energy 196 (2022) 1170e1186

Acknowledgement

The authors are thankfully acknowledge the funding by IRC-SES, KFUPM, under the project number # INRE2204

41st European Photovoltaic Solar Energy Conference and Exhibition

ANALYZING THE EFFECT OF DAMP HEAT TEST ON VARIOUS PV MODULE TECHNOLOGIES, A COMPARATIVE STUDY.

Ahmad Alheloo (ahmad.ibrahim@dewa.gov.ae), Ali Almheiri, Baloji Adothu, Gerhard Mathiak, Vivian Alberts
DEWA Research and Development Center
MBR Solar Park, Dubai, UAE

ABSTRACT: As solar energy adoption accelerates in desert regions, the durability of photovoltaic (PV) modules under extreme environmental conditions becomes a critical concern. This study examines the effects of Damp Heat (DH) testing on the performance and degradation of three photovoltaic (PV) technologies: Interdigitated Back Contact (IBC), Tunnel Oxide Passivated Contact (TopCon), and Passivated Emitter and Rear Cell (PERC). Conducted at the Mohammed Bin Rashid Al Maktoum Solar Park in Dubai, the study simulated prolonged exposure to extreme heat and humidity over 1000 hours, with plans to extend testing to 2000 hours and beyond. Electrical characteristics, including short-circuit current (Isc), open-circuit voltage (Voc), fill factor (FF), maximum power (Pmax), and series resistance (Rser), were measured before and after the DH test. Electroluminescence (EL) imaging and current-voltage (IV) curve analysis were used to identify early signs of degradation. The results showed minor, but noticeable, degradation, particularly in PERC modules, where increased series resistance and micro-cracks were detected. While IBC and TopCon modules showed better resistance to moisture and heat, reductions in Isc and FF were noted across all technologies. These findings suggest that prolonged DH exposure could lead to further performance declines, underscoring the need for rigorous testing protocols to ensure the long-term durability of PV modules in harsh desert environments.
Keywords: PV degradation, Damp Heat, PERC, TopCon, IBC.

1 INTRODUCTION

In recent years, photovoltaic (PV) technologies have seen rapid advancements, this includes advancements in their ability to withstand harsh environmental conditions. Given the large, and ever expanding, number of PV installations in desert regions like Dubai, where the Mohammed Bin Rashid Al Maktoum Solar Park is situated, assessing their long-term durability is essential. One key factor influencing the longevity and efficiency of these modules is their ability to withstand environmental stressors, such as extreme temperatures and humidity. The Damp Heat (DH) test, a standard in the PV industry, simulates long-term exposure to moisture and heat, aiming to identify potential degradation mechanisms related to large temperature variations and the mechanical and electrical stress they induce on PV modules.

This paper aims to compare the performance of three different PV module technologies: Interdigitated Back Contact (IBC), Tunnel Oxide Passivated Contact (TopCon), and Passivated Emitter and Rear Cell (PERC), under DH conditions to understand how the standard's DH test affects such modules.

2 LITERATURE REVIEW

2.1 Damp Heat testing for desert conditions

While DH is mainly focused on testing a PV modules performance under heat and humidity stress, it is also useful to compare its effect on PV modules to that of desert conditions. Many authors have utilized DH to evaluate outdoor conditions and estimate service life of modules [1]. However, other authors believe that further testing of DH for more hours is unnecessary as it exposes the module to humidity levels and amounts that are unrealistic to what exists in the outdoors, sometimes reaching twice the water exposure, and hence would not cause failures based on what exists in the field [2][3]. Moreover, Kimball et al. highlights the significance of regional climate on the degradation levels caused by DH [4].

2.2 Selected technologies and DH

To further verify and test the technologies and how they perform after DH exposure, it is vital to understand their composition and how it affects their degradation mechanisms. For instance, one study offers a unique take by testing both PERC and TopCon modules through 20 hours of DH after exposing the modules to sodium chloride. The results show that TopCon modules showed a significantly higher degradation when compared to PERC modules, which is mainly caused by corrosion induced by the sodium chloride [5]. Another study concluded that TopCon modules seem to be more vulnerable to humidity than PERC and other technologies, linking the severe degradation they observed to electrochemical reactions involving moisture, soldering flux, and cell metallization [6]. Gebhardt et al. has carried out DH on a large sample of TopCon modules, where results show a median degradation of 1.7% across the modules after 1000 hours [7]. The study also mentions that an additional 1000 hours of DH were carried out and did not result in much added degradation. However, the study reported that most of the degradation in the first 1000 hours occurred in the short circuit current (Isc) while the second 1000 hours showed degradation in the Fill Factor (FF).

3 METHODOLOGY

This study was conducted at the Mohammed Bin Rashid Al Maktoum Solar Park in Dubai, under controlled lab conditions using the DH test. The test conditions were set to 1000 hours, while the study aims to prepare the modules for further phases extending the exposure to 2000 hours and beyond.

The three technologies tested in this study are Bifacial n-type IBC, Bifacial n-type TopCon, and Monofacial p-type mono PERC. These technologies were selected based on having a combination of emerging and developed technologies. One module of each type was randomly selected, making up a total of three modules as the sample used in this study.

These modules underwent a specific testing procedure, shown in Fig. 1. They were tested pre-DH for their electrical characteristics including short-circuit current (Isc), Open-circuit voltage (Voc), Fill factor (FF), Maximum power (Pmax), and Series resistance (Rser), in addition to measuring the current-voltage (IV) curve. Moreover, the modules underwent Electroluminescence (EL) imaging and visual inspection for detecting micro-cracks, shunts, and other defects. Once the 1000-hour DH test was carried out, the modules were tested again using the same techniques, and the results were compared and analyzed with a focus on performance degradation and defect developments.

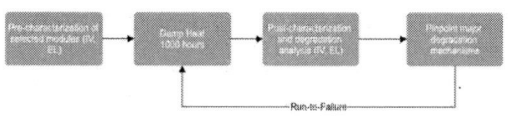

Fig. 1: The testing procedure followed in this study

4 RESULTS AND DISCUSSION

4.1 Electrical performance

After 1000 hours of DH testing, the results for all three PV technologies in terms of electrical characteristics are shown in Table 1. To ensure accuracy and repeatability, the data points in the table represent the average of three consecutive readings for each data point. The percentage difference between the pre- and post-DH measurements was calculated, with positive values indicating a reduction in the respective parameter.

The findings indicate that after 1000 hours of Damp Heat (DH) testing, performance variations across the three PV technologies—bifacial IBC, bifacial TopCon, and monofacial PERC—are mostly within acceptable ranges. However, slight degradation has been observed, particularly in the short-circuit current (Isc) and fill factor (FF), with reductions of around 1-2%. Notably, FF degradation is more prominent in the PERC and IBC modules, while being less significant in TopCon. Moreover, increased series resistance within PERC modules is notable, Although early signs of performance decline, such as increased series resistance in PERC modules, have emerged, the overall degradation remains minimal. These observations suggest that more substantial degradation is likely to occur with extended DH exposure, making further testing essential to fully understand the long-term effects.

4.2 Electroluminescence Imaging (EL) observations

The EL images captured after the DH1000 test showed early signs of degradation, as shown in Fig. 2. For instance, the TopCon modules displayed initial shunt formation, and the PERC modules revealed multiple shunts even before the 1000-hour mark, with some showing up after. These visual indicators, however, do not cause any significant power degradation as of the 1000 hour mark. Therefore, future tests extending to 2000 hours and beyond are expected to possibly reveal more pronounced degradation if these shunts propagate, therefore the shunts were located and will be observed in the 2000 hours mark.

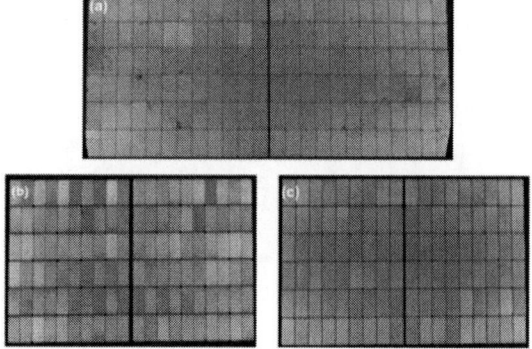

Fig. 2: EL images for the three technologies (a) p-type mono PERC, (b) n-type TOPCon, and (c) n-type IBC.

4.3 Current-voltage (IV) curve analysis

The IV curves of the selected PV modules were measured after 1000 hours of Damp Heat (DH) exposure and compared to their respective pre-DH curves, they are shown in Fig. 3. The analysis revealed only minor variations in the curve, mainly in the current at low voltages. Further DH hours might show promising results.

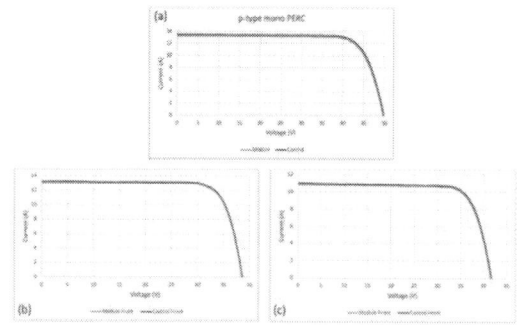

Fig. 3: IV curves for the three technologies (a) p-type mono PERC, (b) n-type TOPCon, and (c) n-type IBC.

Table I: Electrical characteristics of the three technologies

Cell Technology	Module Design		Isc (%)	Voc (%)	Pmax (%)	FF (%)	Rser (%)	EL Observations (After TC200)
n-type IBC	Bifacial	Front	1.3	0.3	0.2	1.9	3.5	N/A
n-type TopCon	Bifacial	Front	1.5	-0.1	-0.8	0.5	6.2	Few initial shunts (before DH1000)
p-type mono PERC	Monofacial	-	1.8	0.1	0.4	2.4	-1.9	Multiple initial shunts, some new shunts

5 CONCLUSION

This comparative study examined the effects of DH testing on three emerging PV technologies: IBC, TopCon, and PERC. After 1000 hours of exposure, the degradation in electrical performance was minimal but noticeable, and particularly noticeable in PERC modules. The IBC and TopCon modules demonstrated better resilience to moisture and heat, with minor reductions in performance. However, early signs of degradation, such as increased series resistance and the appearance of cracks and shunts, were observed, particularly in the PERC modules. The reductions in Isc and FF across all technologies signals the onset of possible significant performance degradation, which is expected to intensify with further DH exposure.

Future testing, by extending to 2000 hours or until failure, will provide deeper insights into the specific degradation mechanisms affecting each technology. This understanding will support the development of more robust PV modules capable of enduring the harsh environmental conditions of desert regions, while highlighting the importance the need for harsher testing requirements for PV modules. By aligning lab-based DH tests with real-world conditions, the study aims to enhance the long-term reliability of PV modules, ultimately promoting broader adoption of solar energy in challenging climates.

6 REFERENCES

[1] Koehl, M., Hoffmann, S., & Wiesmeier, S. (2017). Evaluation of damp-heat testing of photovoltaic modules. Progress in Photovoltaics: Research and Applications, 25, 175 - 183.

[2] Hülsmann, P., & Weiß, K. (2015). Simulation of water ingress into PV-modules: IEC-testing versus outdoor exposure. Solar Energy, 115, 347-353.

[3] Wohlgemuth, J., & Kempe, M. (2013). Equating damp heat testing with field failures of PV modules. *2013 IEEE 39th Photovoltaic Specialists Conference (PVSC)*, 0126-0131.

[4] Kimball, G., Yang, S., & Saproo, A. (2016). Global acceleration factors for damp heat tests of PV modules. 2016 IEEE 43rd Photovoltaic Specialists Conference (PVSC), 0101-0105.

[5]: Sen, C., Wu, X., Wang, H., Khan, M. U., Mao, L., Jiang, F., ... & Hoex, B. (2023). Accelerated damp-heat testing at the cell-level of bifacial silicon HJT, PERC and TOPCon solar cells using sodium chloride. Solar Energy Materials and Solar Cells, 262, 112554.

[6]: Sen, C., Wang, H., Khan, M. U., Fu, J., Wu, X., Wang, X., & Hoex, B. (2024). Buyer aware: Three new failure modes in TOPCon modules absent from PERC technology. Solar Energy Materials and Solar Cells, 272, 112877.

[7]: Gebhardt, P., Kräling, U., Fokuhl, E., Hädrich, I., & Philipp, D. (2024). Reliability of Commercial TOPCon PV Modules–An Extensive Comparative Study. Authorea Preprints.

41st European Photovoltaic Solar Energy Conference and Exhibition

COMPARATIVE DEGRADATION ANALYSIS OF EMERGING PV MODULE TECHNOLOGIES UNDERGOING THERMAL CYCLING

Ali Almheiri (ali.almheiri@dewa.gov.ae), Ahmad Alheloo, Baloji Adothu, Gerhard Mathiak, Vivian Alberts
DEWA Research and Development Center
MBR Solar Park, Dubai, UAE

ABSTRACT: As solar panels increasingly become a main source of energy in harsh desert countries, their ability to withstand extreme temperature fluctuations becomes crucial for long-term reliability. This study investigates the effects of thermal cycling (TC) testing on the durability of three photovoltaic (PV) technologies—Interdigitated Back Contact (IBC), Tunnel Oxide Passivated Contact (TopCon), and Passivated Emitter and Rear Cell (PERC)—to evaluate their resilience in desert climates. The study subjects the modules to 200 cycles according to IEC61215 standards. Electrical parameters, including short-circuit current (Isc), open-circuit voltage (Voc), and fill factor (FF), were measured before and after the test, while Electroluminescence (EL) imaging was used to detect physical damage. After 200 cycles, the degradation in electrical performance across all technologies was minimal, with performance losses below 1%. However, EL imaging detected physical damage, such as microcracks, particularly in IBC and TopCon modules, which could worsen with further testing. While the initial damage had little effect on electrical performance, extended testing is necessary to assess their long-term stability. The results suggest that further TC cycles are needed to reveal more substantial degradation, offering insights into module performance in extreme environments like deserts.
Keywords: Thermal Cycling, PV performance, TopCon, PERC, IBC.

1 INTRODUCTION

Photovoltaic (PV) modules, especially in harsh climates like deserts, are exposed to extreme temperature variations, leading to potential degradation over time. To assess the reliability of emerging PV technologies, this study focuses on subjecting modules of three different technologies to 200 cycles of the IEC61215-based Thermal Cycling (TC) tests as per the IEC61215 standard. The PV technologies are Interdigitated Back Contact (IBC), Tunnel Oxide Passivated Contact (TopCon), and Passivated Emitter and Rear Cell (PERC).

By evaluating degradation through electrical performance metrics and Electroluminescence (EL) imaging, the research, conducted as part of Dubai Electricity and Water Authority's (DEWA) R&D efforts at the Mohammed bin Rashid (MBR) Solar Park, aims to identify early-stage defects and assess the suitability of these technologies for long-term deployment in harsh desert condition environments, providing insights that could guide future PV module design and implementation for such challenging environments.

2 LITERATURE REVIEW

2.1 Thermal Cycling and desert conditions

There is no doubt that Thermal Cycling is an effective test for the identification of PV performance under temperature variations, since various authors have utilized TC to assess PV modules for different reasons. Some authors test the performance of their modules under certain conditions using TC. For instance, Khan et al. have applied TC on their modules installed with concrete slabs for "solar road" applications, showing that attaching the modules to road concrete reduced damage due to TC in EL images as well as IV curves [1]. Moreover, other papers give a glimpse into the expected degradation for the TC test. For instance, Afridi et al. has showcased that 600 hours of TC have led to a degradation of 1.3% in glass-backsheet modules, and various defects in EL images [2]. Meanwhile, Herrmann et al. state that the most significant degradation only occurs after 800 cycles [3].

2.2 Selected technologies in Thermal Cycling

Different technologies react differently to the TC test, and it is crucial to understand how each technology's unique structure and composition fare under the standardized test. For instance, Kasu et al. observed that IBC and TopCon showed weaker temperature dependencies and hence are more resilient to temperature changes when compared to monocrystalline silicon (mono-Si) and multicrystalline silicon (multi-Si) [4]. Moreover, another study tests various types of IBC cells on TC, where they have shown that 200 hours do not show any significant changes, however, 400 hours have begun to show some degradation where two types of IBC cells exhibit notable degradation [5]. Furthermore, Gebhardt et al. has applied 50 cycles of TC on various TopCon modules and reported an increase in EL shunts and cracks even though the performance remained nearly the same, their study also stresses the fact that further TC testing of 200 cycles and beyond is recommended [6].

3 METHODOLOGY

The Mohammed Bin Rashid Al Maktoum Solar Park in Dubai is the location where this study was carried out. The TC test was carried out in a specialized climatic chamber, where the test conditions were set as 200 cycles of TC, with plans for increased cycles based on the analyzed results of the study.

The approach involves the random selection of multiple modules from a pallet for each of the three technologies. The selected samples were tested for current-voltage parameters such as short circuit current (Isc), open circuit voltage (Voc), maximum power (Pmax), fill factor (FF), and PV circuit resistances (Rser, Rshunt). Moreover, Electroluminescence (EL) imaging is also conducted to observe any physical defects like cracks or finger breakages. The modules were then subjected to the IEC61215 Thermal Cycling test, which applies repeated high and low-temperature cycles, and simulates a concentrated version of the temperature differences modules experience in the field.

After 200 cycles, the electrical performance of these selected modules is measured again and is compared to the

41st European Photovoltaic Solar Energy Conference and Exhibition

pre-TC results and analyzed for key alterations and defects. These results serve as a baseline for further cycling and degradation studies, helping to pinpoint when significant performance deterioration begins.

4 RESULTS AND DISCUSSION

4.1 Electrical characteristics

After completing the cycles, several key electrical parameters, including short-circuit current (Isc), open-circuit voltage (Voc), maximum power (Pmax), fill factor (FF), series resistance (Rser), and shunt resistance (Rshunt), were compared, and the resulting percentage difference, the degradation rate, was calculated for each parameter and tabulated in Table 1.

The results reveal a relatively minimal degradation rate in electrical performance for all three module technologies, with degradation rates near or below 1% for most parameters. This level of variation is considered negligible, as it falls within typical measurement accuracy ranges. However, this also suggests that further testing is required to see degradation similar to that seen in desert conditions in relatively short durations. As such, further cycles of TC might lead to more significant degradation.

4.2 Electroluminescence Imaging (EL)

The EL images before and after TC, shown in Fig. 1, provide insight into the physical effects of thermal cycling. For the specific n-type TopCon and IBC modules that were selected, finger breakages were observed, with the IBC module showing a pronounced x-shaped crack that could propagate with further cycling, shown in Fig. 2. Despite these visible defects, the electrical performance of the modules remained largely unaffected after 200 cycles. This suggests that, at this stage of testing, these microstructural changes do not significantly impact the overall module efficiency, however, the propagation of some of the visible defects proves that it can possibly be of significance when exposed to further cycles of TC.

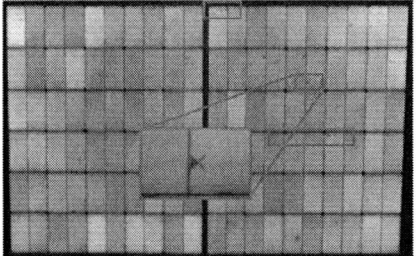

Fig. 2: More detailed representation of the n-type IBC module's EL image showing a crack

The minimal degradation observed across all technologies after 200 thermal cycles suggests that these modules might be suited for deployment in environments with fluctuating temperatures, at least in the short term. However, the observed finger breakages and microcracks in the IBC and TopCon modules could possibly lead to, if exposed to further stress testing, propagation which could in turn result in a reduction in performance, reflecting the harsh conditions in desert environments. It is likely that further cycling beyond TC200 could lead to pronounced degradation, both electrically and structurally, particularly in the n-type modules where internal resistance increases significantly. Monitoring how the observed microcracks and defects evolve with further cycles might assist in understanding their performance in harsh environments.

4.3 Current-voltage (IV) curve analysis

The I-V curve results show that all three PV module technologies—n-type TopCon, n-type IBC, and p-type mono PERC—exhibit minimal degradation after 200 thermal cycles. The p-type mono PERC module demonstrates the highest resilience, with only minor changes in Isc and Voc, and negligible effects on power output. The n-type IBC and TopCon modules show slight reductions in electrical performance, with finger breakages and increased series resistance observed, particularly in TopCon.

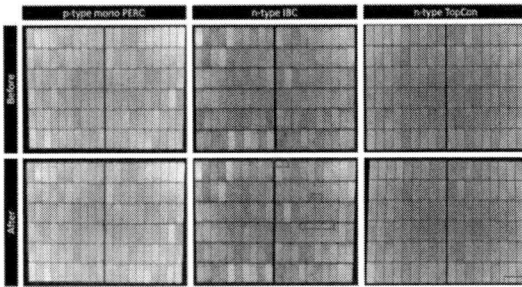

Fig. 1: EL images of each module pre and post TC

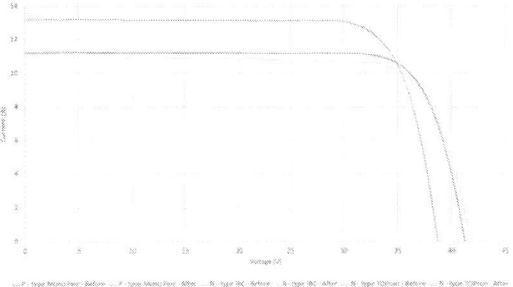

Fig. 3: IV curves for the three technologies

Table I: Electrical characteristics of the three technologies

Cell Technology	Module Design		Isc (%)	Voc (%)	Pmax (%)	FF (%)	Rser (%)	Rshunt (%)	EL Observations (After TC200)
n-type TopCon	Bifacial	Front	-0.5	-0.09	0.4	1.0	-16.8	-2.4	Finger breakage
n-type IBC	Bifacial	Front	-1.2	-0.2	-0.1	1.2	-15.8	-1.8	Multiple finger breakages and x shaped mark propagation
p-type mono PERC	Monofacial	-		-0.6	0.1	-0.5	0.2	-3.0	-0.4	N/A

5 CONCLUSION

The results from this study indicate that all three PV module technologies—n-type TopCon, n-type IBC, and p-type mono PERC—demonstrate some slight degradation but overall resilience after 200 thermal cycles, with some impact on electrical performance. The p-type mono PERC module showed the least degradation, while the bifacial n-type IBC and TopCon modules exhibited slight increases in series resistance and physical defects like finger breakages and micro-cracks appearing and propagating.

Although the current degradation after 200 cycles is minimal, further testing with additional thermal cycles is necessary to fully understand the long-term effects and the potential for more significant performance losses, this is especially true in challenging environments like the desert locale in this study. Hence, the results from this preliminary study form the foundation for more extensive testing, as more cycles (e.g., 400 or 600) are likely needed to fully capture the long-term degradation mechanisms of these emerging PV technologies. While initially, the low degradation of the modules might suggest longevity in such environments, such a result needs further testing and verification.

6 REFERENCES

[1] Khan, F., Rezgui, B. D., & Kim, J. H. (2020). Reliability study of c-Si PV module mounted on a concrete slab by thermal cycling using electroluminescence scanning: Application in future solar roadways. Materials, 13(2), 470.

[2] Afridi, M., Kumar, A., ibne Mahmood, F., & Tamizhmani, G. (2023). Hotspot testing of glass/backsheet and glass/glass PV modules pre-stressed in extended thermal cycling. Solar Energy, 249, 467-475.

[3] Herrmann, W. (2010). How temperature cycling degrades photovoltaic-module performance. Spie Newsroom.

[4] Kasu, M., Abdu, J., Hara, S., Choi, S., Chiba, Y., & Masuda, A. (2018). Temperature dependence measurements and performance analyses of high-efficiency interdigitated back-contact, passivated emitter and rear cell, and silicon heterojunction photovoltaic modules. Japanese Journal of Applied Physics, 57.

[5] Halm, A., Schneider, A., Mihailetchi, V. D., Libal, J., Aulehla, S., Galbiati, G., ... & Peter, K. (2013, June). Evaluation of cell to module losses for n-type IBC solar cells assembled with state of the art consumables and production equipment. In 2013 IEEE 39th Photovoltaic Specialists Conference (PVSC) (pp. 2368-2372). IEEE.

[6] Gebhardt, P., Kräling, U., Fokuhl, E., Hädrich, I., & Philipp, D. (2024). Reliability of Commercial TOPCon PV Modules–An Extensive Comparative Study. Authorea Preprints.

41st European Photovoltaic Solar Energy Conference and Exhibition

ASSESSMENT OF CRITICAL LAMINATE TEMPERATURE INCREASE BY FAST IR-BASED ANALYSIS OF HOT SPOTS ON SOLAR CELLS

Stephan Großer, Matthias Schak, Stefan Eiternick, Bengt Jaeckel, Marko Turek
Fraunhofer-Center for Silicon-Photovoltaics (CSP)
Otto-Eissfeldt-Strasse 12, 06120 Halle (Saale), Germany

ABSTRACT: Defects in solar cell production can lead to thermal heating, posing a risk of damaging and degradation to solar modules in the field. Elevated temperatures can accelerate chemical reactions and material degradation. To address this issue, this contribution demonstrates an experimental approach to approximate the module temperature increase during a full cell shading situation already at solar cells within a measurement time of 100 ms. Fast IR-imaging is applied to determine the time-dependent temperature characteristic. Using a high frame-rate IR camera, single artificially generated hot spots on solar cells are detected and time-dependent temperature profiles extracted. A simplified physical model was used to extract the equilibrium temperature. The determined temperature increase of the encapsulated hot spot was found in a linear correlation to the non-encapsulated extracted value. It could be confirmed, that a 100 ms measurement time is sufficient to predict the module temperature. The contribution demonstrated that fast IR-detection could improve module reliability by enabling an IR-classification of solar cells for quality-driven sorting before module manufacturing.

Keywords: time-resolved thermography, classification, solar cell, module temperature, hot-spot risk, degradation

1 INTRODUCION

Reliability of solar modules in the field is crucial. Types of defects in solar cell production which are causing thermal heating can be a stressor for the solar module in the field. Elevated temperatures accelerate degradation of the module materials due to enhanced chemical reaction velocities. In particular, local heat sources like hot spots result in very high temperatures implying an increased risk of damaging the encapsulation system by thermal decomposition or can lead to glass breakage [1-3].

Hot spots can be detected easily after the metallization of the wafer during solar cell inspection using infra-red (IR) imaging in reverse bias and reverse current measurement. Due to the demand of a high throughput in cell manufacturing and characterization, the feasible time slot for acquiring thermal images should not exceed 100 ms. Furthermore, IR cameras and system integration for solar cell inspection is an additional and cost-pushing equipment. Nevertheless, fast IR-imaging in solar cell production lines can lead to a significant improved quality control if a precise identification of more potential risk-cells can be achieved. A technical feasibility has been demonstrated that hot spots on cells are detectable and assessable by using IR imaging time series approach within 100ms [4].

The study in this contribution was intended to predict from a fast temperature measurement on a solar cell the resulting temperature and extend it by determining the hot spot equilibrium temperature on the module level of the same cell. Initially, artificially generated hot spots on solar cells were implemented to generate a variation of single hot spots. High-framerate IR cameras allow large time-series of infrared images. Datasets are recorded of multiple solar cells for statistics and variation of equilibrium temperature of different hot spots. By interconnection and encapsulation of each cell, mini modules were made allowing to investigate the temperature of the same hot spot on cell as well as on minimodule level. Time-dependent IR images give spatially resolved thermal information of the whole solar cell surface and allow the selection of thermal active spots and their time-dependence. Equilibrium temperatures were measured. With a physical-based model, the approximated

equilibrium temperature (or temperature increase) has been extrapolated. The resulting local equilibrium mini-module temperature was correlated with the equilibrium temperature of the solar cell. From the comparison of solar cell and (mini)module derived temperatures, a linear approximation was confirmed, allowing to approximate the maximum total temperature increase of the encapsulated solar cell caused by a single hot spot within a 100 ms long measurement period.

2 EXPERIMENTAL SETUP

For the investigation industrial-made mono-facial M2 3BB PERC solar cells were used. Standard electrical performance (IV, EL) measurement proofed that no initially defect solar cells were in the group under investigation. To generate a single hot-spot each solar cell was full-area contacted on the rear side contact but only locally on the front side using a single metal pin. The pin was placed for all cells in the same cell area. By applying a short current pulse (of -6 A to -8 A) a single hot spot was generated by the high current density at the local front contact. The local current stress differed and a desired random variation of single hot spot damage intensities were achieved. For time-dependent IR imaging each solar cell has been measured using an InfraTec ImageIR 9400 camera. With a measurement frequency of up to 120 Hz, a time resolution of 8.33 ms could be achieved. The solar cells were placed for IR-imaging in a standard 3BB cell measurement chuck with homogeneous contacting. Initially the solar cell temperature was in thermal equilibrium (room temperature). For simplicity and prevention of burn marks only one reverse current of -1 A has been used for testing of cells and modules. Measured time-series include different large periods like 1 s for solar cells with high time-resolution and around 30 min for mini-modules with low time resolution. Solar cells have been measured from the front side and mini modules from the backside.

Mini module manufacturing has been done by commercially available materials. After interconnection of a single cell with tin-lead coated copper ribbons and cross connectors the cell was encapsulated with EVA in a glass

41st European Photovoltaic Solar Energy Conference and Exhibition

(white) backsheet mini module. Two cross connectors for each polarity where lead through the encapsulation for contacting. The workflow in a chronological manner is shown in Figure 1.

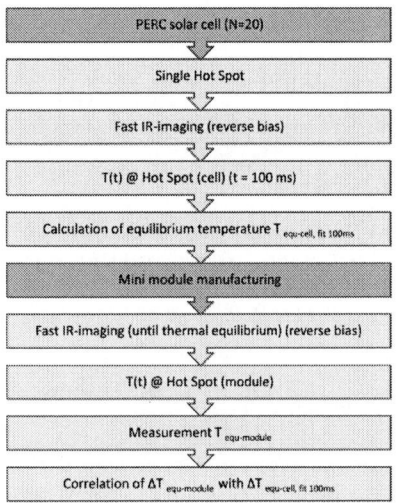

Figure 1: Workflow of the experiment including the manufacturing, the measuring, and the analysis steps.

Comparable temperature images of 3 different solar cells under investigation can be seen in Figure 2 displaying the achieved variation of the hot spots. The measurement condition of all cells is identical with a reverse current of -1 A and a duration of 1 s after the start of excitation. Different temperatures were measured where cell C reach with 64 °C the highest temperature after 1 s of all three cells.

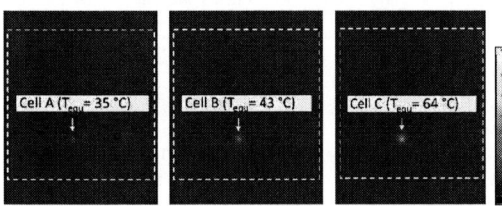

Figure 2: IR-images of 3 different solar cells with same reverse current and heat up time but different equilibrium temperatures of the hot spots (current = -1 A, t = 1000 ms).

3   RESULTS

For each PERC cell with a single hot spot and later the corresponding mini-module time-dependent temperature image series have been measured. By selection of the center of the hot spot center position the temperature at this position was extracted from each image to generate temperature vs. time profiles. Figure 3 show the extracted profile for a time duration of 1 s of current excitation for one solar cell (red dots). Each dot represents one extracted temperature. One can see that the temperature at the hot spot increases fast starting from room temperature. The time duration of 100 ms is indicated by the dashed line. The measurement has been done for the same cell after encapsulation again. The blue dots represent the obtained profile for the module. One can observe for the solar cell that the thermal equilibrium condition was not achieved after 100 ms excitation. Nevertheless, the thermal

equilibrium temperature was reached after 1 s and used to measure the final equilibrium temperature. To approximate the equilibrium temperature before reaching the temperature profile has been fitted (black line in Figure 3) by equation 1 which can be calculated analytically from a heated hot plate.

$$T_{cell}(t) = T_0 + (T_{equ.} - T_0)(1 - e^{-\frac{t}{\tau}}) \qquad (1)$$

This equation describes a homogeneous heated hot plate with heat dissipation. Therefore, it neglects the heat transfer inside the solar cell but allows a simple approximation and fitting of the equilibrium temperature $T_{equ.}$, the ambient temperature T_0, and a time constant τ. In contrast to the cell, the module temperature (in Figure 3) is far from the thermal equilibrium within the measurement time of 1 sec. Much longer measurement period of up to 30 minutes has been used to measure the thermal equilibrium temperature at the hot spot position (on module backsheet).

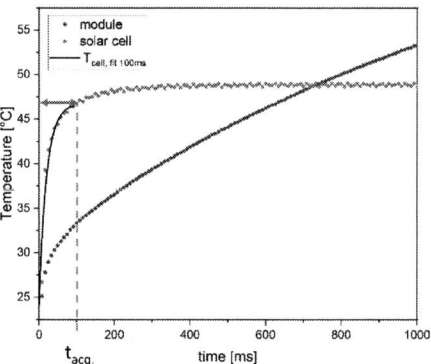

Figure 3: Time-dependent temperatures of the same hot spot on solar cell (red) and (mini)module backsheet (blue). Each point was extracted from the region-of-interest within individual IR-images. Targeted IR-images acquisition time interval (within $t_{acq.}$ = 100ms) is indicated. (I_{rev} = -1 A, $f_{IR-image}$ = 120 Hz)

All solar cells with single hot spots have been measured and analyzed like described before. Figure 4 presents a plot where the fitted thermal equilibrium temperatures increase for each cell are plotted versus the measured thermal equilibrium temperature increase. The difference was calculated by subtracting the final from the initial absolute temperature. Each point represents the result from 1 cell. As shown in Figure 2 before, a large variation between the hot spots were achieved, ranging from temperature differences of around 2.5 °C to 40 °C. At least within the investigated temperature range the correlation between the fitted equilibrium temperature change using the approximation of a hot plate seem to follow a linear behavior. A linear fit has been calculated and plotted within the graph as a red line and describe the data point trend sufficient. Therefore, the equation 1 of a simple approximation for thermal equilibrium fit ($T_{equ.}$) of data within a 100 ms interval is applicable to use it as a parameter to be compared with the determined module temperature difference.

608

41st European Photovoltaic Solar Energy Conference and Exhibition

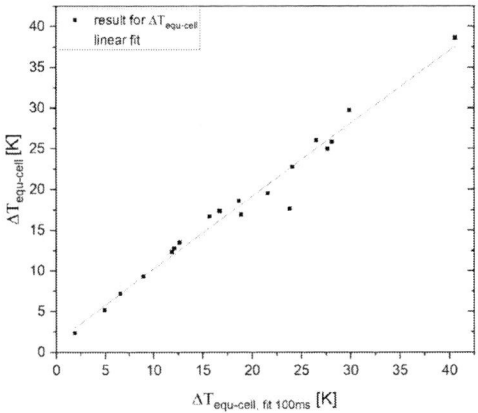

Figure 4: Temperature change of equilibrium temperature (measured) versus the calculated temperature change of the equilibrium from a 100 ms period. Linear fit is plotted in red.

Figure 5 show the graph of the determined temperature changes of the solar cell (from fit of 100 ms) versus the determined temperature changes of the same cell after encapsulation at the hot spot position. A linear trend of the data points can be observed again. In contrast to the solar cell the temperature increase of the mini module is much larger ranging from around 40 °C up to 160 °C. Encapsulation of the solar cell results in a thermal insulation of the solar cell. Glass, EVA and backsheet exhibit in contrast to silicon a low heat conductivity. Heat transfer from the hot spot to the environment is reduced resulting in an increase of the thermal equilibrium compared to the bare cell. Since the trend seems to be linear in the investigated temperature range the data has been fitted by a linear equation as well (red curve).

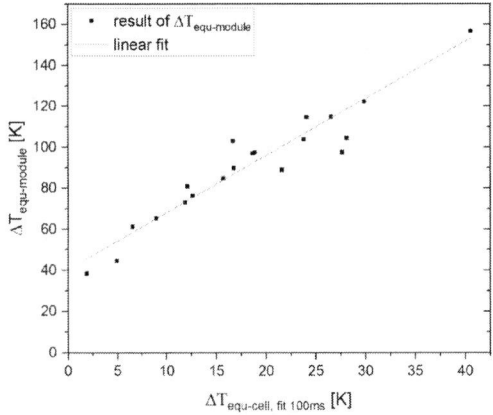

Figure 5: Correlation of hot spot temperature fit and module measurement (equilibrium temperatures).

The determined values of the linear fit of Figure 4 and Figure 5 are listed in Table 1. Parameter B represents the slope of the curve and parameter A the intersection with the y-axis. $\Delta T_{equ-cell}$ in dependence of $\Delta T_{equ-cell, \, fit100ms}$ should be ideally a curve with a slope of 1 and an intersection at the coordinate origin. The extracted slope is around 11 % lower and result in a slight overestimation of the hot spot temperature which is due to the approximation of the function shown in Figure 3. The correlation

coefficient R is close to 1 for both fits, confirming the assumption of a positive linear correlation in the investigated temperature range. $\Delta T_{equ-module}$ in dependence of $\Delta T_{equ-cell, \, fit \, 100ms}$ had shown a larger slope by factor or around 2.8 compared to the cell due to the thermal insulation of the used encapsulation material. Different thermal conditions for solar cell and (mini)module (thermal insulation, heat sink, convection) depend on the individually used manufacturing equipment and materials (cells and module) and must be considered individually.

Table 1: Values extracted from linear fit of the equilibrium temperatures for the solar cells and the mini modules.

	A [K]	B	Pearson R
$\Delta T_{equ-cell}$ ($\Delta T_{equ-cell \, , \, fit100ms}$)	1.29	0.89	0.98
$\Delta T_{equ-module}$ ($\Delta T_{equ-cell, \, fit \, 100ms}$)	40.21	2.79	0.95

4 SUMMARY

In conclusion, the study demonstrated a fast approximation approach for the temperature increase of modules for the used module-encapsulation material and specific testing condition. A solar cell IR-image data acquisition time of 100 ms was confirmed to be sufficient to predict the equilibrium temperature for a module, at least in a worst-case scenario of total shaded cell. While the presented approach applies for a specific equipment and module material, it can directly be transferred to other solar cell measurement setups and module material combinations. The experiment demonstrated the use of fast IR-measurement on laboratory equipment which could be realized also by integration of more cost competitive cameras as concluded by Wasmer et al. [4].

A correlation of realistic (critical) module and solar cell temperatures would allow an improved IR-classification of the solar cell for quality driven sorting before module manufacturing. Throughput of cell manufacturing would hardly been reduced. A further development and a test of implementation in cell tester could enable a classification of temperature on solar cells in example ranked by experience from temperature-accelerated module aging.

ACKNOWLEDGMENT
The financial founding by the BMWK project "Opti-Learn" is gratefully acknowledged (Grant number: 03EE1108A).

REFERENCES

[1] Köntges, M., et al., Review of Failures of Photovoltaic Modules, in IEA PVPS Task 13, Subtask 3.2.(2014) p. 1-140.
[2] Fertig, F., et al. 26th EU PVSEC (2011), doi: 10.4229/26thEUPVSEC2011-2DO.3.1
[3] Ramspeck, K. et al., Energy Procedia 55 (2014) 133 – 140, doi: 10.1016/j.egypro.2014.08.097
[4] Wasmer, S., et al. 33rd EU PVSEC (2017), doi: 10.4229/EUPVSEC20172017-2CV.2.43

41st European Photovoltaic Solar Energy Conference and Exhibition

CORRELATIONAL STUDY ON THE IMPACT OF HARSH ENVIRONMENT STRESS FACTORS ON THE AGEING EFFECTS OF SEVERAL ENCAPSULATION MATERIALS FOR PV MODULES

Dr. Tudor Timofte[1], Maria Ignacia Devoto Acevedo[1],
Dr. Joachim Glatz-Reichenbach[1], Valentina Arias Reyes[2], Andreas Halm[1]
[1] ISC Konstanz e.V., Rudolf-Diesel-Str. 15, 78467 Konstanz, Germany
[2] Federico Santa María Technical University, Avenida España 1680, Valparaíso, Chile
tudor.timofte@isc-konstanz.de

ABSTRACT: The composition of encapsulation materials for photovoltaic (PV) modules was, is, and will be continuously optimized, partly due to longer life time targets, and partly for the necessity of compatibility with novel and future cell technologies. As a consequence, a considerable decrease of outdoor-related failures to this bill of material component and process was noticed in the last decades. Yet, particularly for PV modules installed in harsh environments, such as hot and humid or desert climates, the natural stress factors pose a significant challenge for these encapsulation materials (depending on location, one or a combination of factors such as: high UV irradiation doses, high relative humidity, high temperature, high temperature differences for day-night). Therefore, to understand the ageing mechanism of encapsulation formulations and the interaction with their interfaces under environmental stress, an accelerated stress sequence comprised of damp heat test (DHT), temperature cycling test (TCT) and UV irradiation ageing was defined. Furthermore, two pertinent features were selected to be monitored during ageing at this experiment for an uncomplicated and effective characterization using spectroscopic transmission and adhesion strength (to solar glass and to backsheet foil), which are affected by transmission changes, bond weakening and delamination effects.

Keywords: encapsulation, characterization, accelerated ageing

1 INTRODUCTION

For a fast polymer matrix and additive package optimization loop, the encapsulation material producers need a rapid and solid qualification procedure, to avoid failure such encapsulation discoloration or delamination [1]. From this would also benefit the qualification of different market available encapsulation materials for research activities, as well as for module mass production. The meaning of these environmental accelerated stress tests gain tremendous importance when confronting with the choice of materials for PV modules, which will be installed at locations with extreme climate, e.g. desert [2] or tropical regions. One of the driving ideas was to create a testing procedure, which would require moderate equipment and deployment complexity, while offering fast and reliable results. While similar studies [3] [4] are reported in the literature, this study intends to isolate and quantify the effect of moisture, high temperature and high UV radiation ageing factors on the encapsulation material performance in PV modules. This is done by monitoring and correlating the change of their adhesion [5] at the main interfaces (to glass and to backsheet foil) and the change of the spectroscopic transmission during accelerated ageing compared to initial state. The prepared samples for characterization were aged in climate chambers by TCT, DHT and UV test [6] with the intention to simulate at high pace the natural ageing and hence deteriorate the adhesion of polymers to the contacting surfaces and also to impact their spectral transmission. These characteristics are two of the most important features of encapsulation materials for a long, efficient and reliable operational module life. For the relevance of this study several market available and industrially implemented encapsulation resin types were used: two thermosetting, crosslinking materials: ethylene vinyl acetate (EVA) and crosslinking polyolefin (POE) and one thermoplastic, non-crosslinking material: thermoplastic polyolefin (TPO). The aim of this procedure would be to identify reliably the threshold and degradation mechanism of each tested material, as well as to identify

the most long lasting product for a certain glass-backsheet configuration, if such a test is required. Furthermore, the procedure is considered using a sample design in such a way, that accelerated and a planned natural ageing would be possible, without affecting the comparability.

2 EXPERIMENTAL SETUP

2.1 Materials
- four encapsulation materials with different chemical base polymers and additive packages were selected: a marked available ethylene-vinyl-acetate (EVA1/ G1) with low UV cutoff, two market available cross-linking polyolefins (POE1/ G2, with low UV cutoff, and POE2/ G4, with a regular UV cutoff) and a market available thermoplastic polyolefin encapsulation material (TPO1/ G3) with low UV cutoff.

The additional involved materials were the ones which create together with the encapsulation material the two interfaces toward the exterior of the PV module at both sides:
- solar glass, with a non-geometric, yet recurring microstructure, without anti reflective coating, used as component of produced laminates for peel tests and also used as protective and UV filtering layer for spectroscopic transmission samples;
- backsheet, transparent, multilayer, with the structure: polyethylene terephthalate (PET)/ PET/ primer, whereas the primer is the inner layer which contribute to an enhanced and stable adhesion between backsheet and encapsulation material; the purpose of the backsheet in the experiment is bivalent: one role is to stabilize mechanically (avoid fissure and breakage of encapsulation material) and to reduce the elongation of encapsulation material during peel test, while the other role is to provide the interface for the characterization of peel force between encapsulation material layer and backsheet layer in the laminated stack for the test.

The table I below provides an overview of main relevant features of the involved materials.

610

Table I: Overview of materials used at the reported test and their main features for the test

Material	Type	Thickness (mm)	Size (mm)
EVA1	thermoset	0.5	220x220
POE 1	thermoset	0.5	220x220
POE 2	thermoset	0.5	220x220
TPO 1	thermoplastic	0.5	220x220
Solar Glass	solar glass/ no ARC*	2.0	220x220
Backsheet	PET/PET/ primer	0.3	220x220

*ARC = anti reflective coating

2.2 Sample lamination

Two samples types, which involve laminating the encapsulation films, were produced in this experiment and these will be described in detail in section 2.4 below.

For the lamination of the samples, a laboratory laminator was used. This device is equipped with an electrical heating plate and a silicon membrane lid, capable of applying vacuum and atmospheric pressure during process. All samples were prepared and produced under clean conditions, which imply mainly handling of materials on dry, cleaned surfaces, by wearing protective nitrile gloves and by laminating between cleaned Teflon™ sheets, to avoid contaminations. For each tested encapsulation material the recommendation for lamination parameters from producer were reviewed and considered and at table II are inserted the main lamination recipe features used for each of the polymer resins.

Table II: Lamination parameters used for preparation of samples, assigned to each utilized encapsulation resin

Material	Setup Temp. (°C)	Heating Time (s)	Curing Time (s)	Pressure (mbar)
EVA1	145	360	660	1000
POE 1	155	380	660	1000
POE 2	155	380	660	1000
TPO 1	145	360	460	1000

2.3 Characterization

For the characterization of sample's production process and for quantifying the effect of accelerated ageing on spectroscopic transmission and on the peel force between encapsulation material and glass, as well as between encapsulation material and backsheet, the following devices and methods were used:

- for determining the cross linking degree of encapsulation materials based on copolymers, differential scanning calorimetry (DSC) of cured and uncured samples was measured at a Netzsch 214Polyma unit, by using aluminium crucibles and nitrogen, as protective and purge gas; furthermore, the DSC measurement were complemented by cross linking degree determination through Soxhlet extraction method [7], using toluene as an extraction solvent and 12h extraction time;
- transmission spectroscopy was carried by a UV-VIS-near IR PerkinElmer Lambda 950 spectrometer, fitted with an integrating sphere accessory to evaluate the initial, unaged state and the change in transmission at the considered encapsulation materials during and after the planned accelerated ageing procedures;
- The peel force measurements were made at a ZwickRoell digital force testing unit containing an adapted mounting system for this sample type (see section 2.4)

consisting mainly of a grip with mechanical screw fixation for polymer stripes. For each material and sample, at least three peel stripes for the interface to glass and at least three peel stripes for the interface to backsheet were pulled at a 180° angle between clamped peeled stripe and the laminate. For each measurement, the peel length was at least 100mm and the peel speed was set at 150mm/min.

2.4 Sample types

For the reported experiment two sample types were used, each serving different characterization methods:

- the sample type for measurement of transmission consists of a single laminated encapsulation layer, mounted for mechanical stability during transmission measurements and during ageing between two aluminium frames (50x50mm) and fixed with adhesive tape (Fig. 1A); furthermore, for ageing tests this sample type was introduced between previously mentioned solar glasses and the entire testing body was sealed at all edges by adhesive tape (Fig.1B);

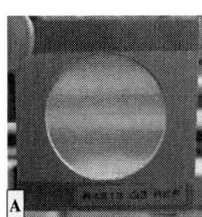 

Fig. 1: Sample for spectroscopy measurements: on left side (A) a single layer cured encapsulation film mounted between two aluminium frames; on right image (B) four samples mounted between solar glasses, sealed at all edges, and mounted in UV chamber

- the sample type used for peel tests is a stack (Fig.2), comprising of a solar glass, on which two layers of encapsulation material with a transparent backsheet on top are added and then laminated, by considering a Teflon sheet separator at one edge (Fig.3); this is required for separating the sample layers after lamination and mounting the cut stripes (after lamination approx. 200mm long and approx. 5.0 mm wide stripes are cut with a cutter knife) at the peel force testing unit;

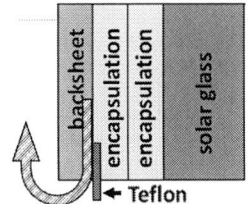 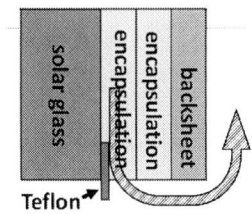

Fig. 2: Cross section of peel sample types *(grey arrows indicate the peel interface and peel direction: at left side for peel of the backsheet; at the right side for peel of encapsulation layer from glass)*

It must be mentioned, that at this study, for the peel test data analysis only the force required to induce adhesive fracture at laminate during the 180° degree peel is processed, without considering the strain component and bending component of the force [5]; these components should be sufficiently low to be neglected, however for accurate measurements these should be introduced in the measurement and data analysis routine.

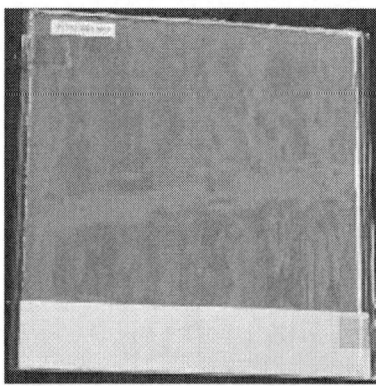

Fig. 3: Sample type used for 180° angle peel tests (the light brown stripe at the lower glass edge is the Teflon™ coated glass fiber mesh separator layer between glass and encapsulation material, as well as between encapsulation material and backsheet)

2.5 Accelerated ageing setup

The above-mentioned samples (one glass-encapsulant-backsheet laminate for each tested encapsulation material) were subjected to degradation by means of three types of accelerated laboratory ageing: DHT, TCT and UV testing. Additionally, for the UV testing only, cured polymer films behind solar glass were aged, for analyzing the effect of this ageing on spectroscopic transmission of the samples.

At damp heat test (DHT) with constant environmental chamber conditions of +85°C setup temperature and 85% relative humidity, the samples (mounted vertically on dedicated sample holder) were aged by three times 1000h testing duration, reaching a total test duration of 3001h (rounded to 3000h) DHT. After each testing session of 1000h, peel test measurements were made by cutting the peel stripes at least 24h after the samples were extracted from DHT climate chamber. Initially and also after each of the ageing steps three stripes for measuring adhesion encapsulation–glass and three stripes for measuring adhesion encapsulation-backsheet were considered.

The temperature cycling test (TCT) was implemented according to the guideline from IEC 61215 norm, by applying to the vertically mounted samples at dedicated sample holder at each testing cycle a temperature variation between -40°C and +85°C (heating and cooling ramps with approx. 1°C/min. and dwell time of 30min. at -40°C, as well as at +85°C); a total of 198 TCT cycles (rounded to 200 cycles) ageing was performed. After 50 and 200 TCT cycles, peel test measurements were made by cutting the peel stripes at least 3h after the samples were extracted from TCT climate chamber. Initially and also after each of the ageing steps three stripes for measuring adhesion encapsulation–glass and three stripes for measuring adhesion encapsulation-backsheet were considered.

The UV chamber testing unit (producer: PI Berlin) was fitted with metal halide low pressure fluorescent UV lamps, testing according to IEC 61215 norm guidelines regarding the homogeneity of UV irradiation, UVA/UVB ratio, UV dose and testing temperature at steady +60°C, without humidity control (testing chamber not isolated air tight from the laboratory environment). Three times 60 kWh/m² UV dose (meaning a dwell of samples in the UV chamber according to calculation of UV dose measured without any filter or object between the sensors and the fluorescent tubes) was applied to the samples, reaching a

total of 180 kWh/m² UV ageing. Samples were measured after each testing sequence: spectroscopy transmission measurements and also peel tests (after each testing session of 60 kWh/m² UV dose, peel test measurements were made by cutting the peel stripes at least 3h after the samples were extracted from UV climate chamber). Initially and also after each of the ageing steps three stripes for measuring adhesion encapsulation –glass and three stripes for measuring adhesion encapsulation-backsheet were considered.

3 RESULTS

For retrieving reliable testing data from samples that contain encapsulation materials, it is important to ensure a suitable processing (lamination) of materials, hence to consider a minimum level of crosslinking reaction [8] at different types of copolymer thermoset blends. Accordingly, the gel content of the produced samples (~ 1g for each sample) with copolymers (EVA1, POE1 and POE2) was verified by Soxhlet extraction, adding 2,6-Di-tert-butyl-4-methylphenol, also simplified named butylated hydroxy toluene (BHT), as an antioxidant agent during initial extraction steps. At least two extraction samples for each thermoset (crosslinking) material, laminated with its specific and above-mentioned parameters and positioned at the center of the lamination unit on a carrier aluminium metal plate (distance between samples of approx. 1m), were considered for the test. Table III below presents for each of the tested thermoset resin the minimal reached gel content (or cross linking) level, which exceeds the recommended minimum level of cross linking specified in literature [8] and at the material producer technical data sheets.

Table III: Minimum recommended and reached cross linking level, determined by Soxhlet extraction with toluene

Material	Min. recommended Cross Linking Level (%)	Min. Measured Cross Linking Level (%)
EVA 1	75	87.2
POE 1	60	71.1
POE 2	70	82.2

Furthermore, the high gel content level reached in the thermoset polymer blends was verified at all thermosets samples by DSC (Fig. 4).

The presented DSC diagrams at Fig. 4 reveal at red lines (first heating segment of cured samples) no measurable residual peak, proving a high gel content level. Same result (no detectable residual DSC peak at first heating of a cured sample) was obtained for POE2 sample at DSC measurement as well. This result highlight the importance of a referencing solvent extraction, e.g. Soxhlet method, since this method is more accurate than currently reported spectroscopy or thermo-analytical (DSC) measuring methods. It is not the sensitivity of the DSC unit responsible for not detecting the residual uncured fraction (Table III) at each material, but the nature of the measurement itself, since during first heating most probably additional curing is induced in the sample, impacting the enthalpy peak during the measurement recording. In contrast, the Soxhlet method keeps a relatively low temperature (< 100°C) in the sample during the entire extraction process and additionally acts against the effect of heating (and eventually undesired triggered crosslinking) at beginning of extracting by the presence of BHT antioxidant.

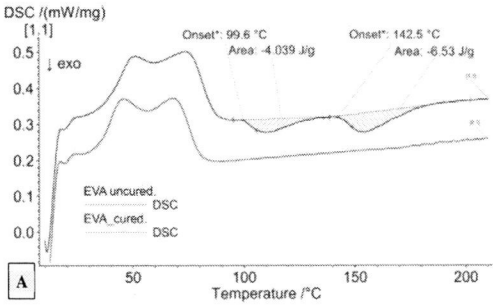

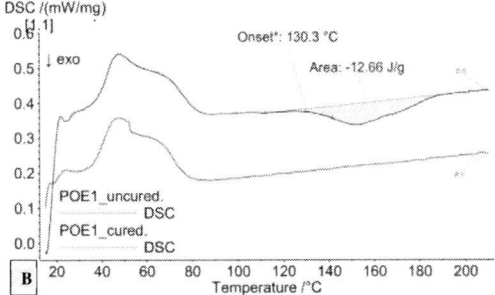

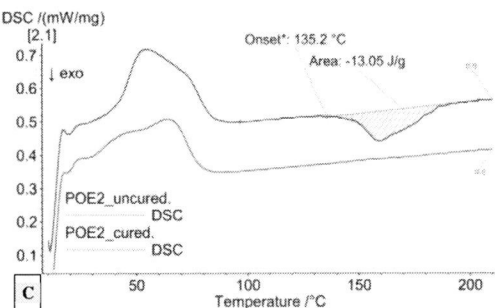

Fig. 4: DSC of EVA1 (**A**), POE1 (**B**) and POE2 (**C**), measured before / after lamination (only first heating is depicted for uncured samples (blue lines) and cured samples (red lines)

Since the (cured or laminated) samples for transmission measurements were positioned behind solar glass during UV ageing, this imply determining how much UV radiation is cut by the solar glass. This investigation was achieved in two ways: first is to measure the spectral transmission of the utilized solar glass (Fig. 5).

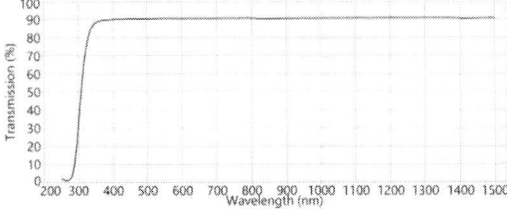

Fig. 5: Transmission spectrum of used 2.0mm thick solar glass with smooth, non-geometric texturing and no ARC

The second method for this purpose measures the UVA and UVB radiation energy in the testing chamber with and without glass in front of measuring head of a radiometer (Opsytec Dr. Groebel RM12), fitted with dedicated UVA

and UVB sensors. The measurement reveal that the glass transmits 13.08% less UVA and 23.30% less UVB radiation energy through, with an overall 13.63% less UV energy reaching the encapsulation layer behind the glass. The transmission spectra provides important details regarding the UV cutoff of the utilized solar glass (6.51% transmission at 285nm; 21.43% at 295nm; and 80% transmission at 330nm).

3.1 Impact of UV ageing on the spectroscopic transmission
The thickness of the samples was measured after lamination with a digital micrometer from company Mitutoyo. The thickness is in range of 0.5mm, a detailed record for each sample was introduced at Table IV below.

Table IV: Thickness of laminated samples for transmission measured with a tolerance of ± 0.020mm

Sample	EVA1	POE1	POE2	TPO1
Thickness (mm)	0.485	0.490	0.530	0.475

At the beginning of test the spectra of all tested encapsulation samples was measured.

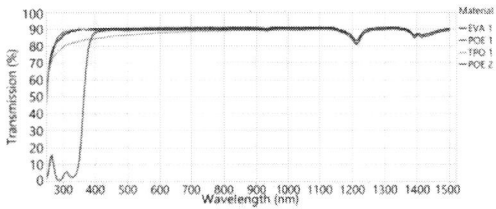

Fig. 6: Spectroscopic transmission for the tested encapsulation materials in wavelength range 250 – 1500nm

Up to POE2, all encapsulation material samples have a low UV cutoff, in the range of 250nm; the POE2 sample cuts UV off at approx. 350nm (Fig.6).
After initial characterization, all samples were mounted between two 220.0x220.0x2.0 mm solar glasses of same type (mentioned and characterized above) and the glass edges were sealed and hold together by Kapton® tape. This testing system was aged three times with 60 kWh/m^2 UV dose (therefore an overall level of UV ageing of 180 kWh/m^2 was achieved) and the transmission was measured after each of the above mentioned ageing steps. The impact of UV ageing on transmission was analyzed at three sections of the light spectra: UV section, therefore in range 250-400nm (Fig.7), visible section, hence in range 400-800nm (Fig.8) and near IR section, with the range 800–1200nm (Fig.9).
At EVA1, POE1 and TPO1 a change in the transmission could be observed: EVA1 and POE1 have approx. 4% less UVB (in range 250-300nm) transmission compared to initial after 180 kWh/m^2 UV ageing. There is no clear change at UVA transmission for these two samples. At TPO1 sample a clearly higher drop of 11% to initial in the UV transmission could be recorded. At this sample the reduction at the UV transmission comprises both, UVA and UVB, whereas UVB range is mostly affected by the change in transmission. This result is interesting, since it was shown previously that the used solar glass cuts most of the UV spectrum from 290-295nm downwards. Therefore, only reduced intensity wavelengths of UVB radiation at their higher wavelength section and UVA radiation affect the encapsulation formulations in such a way, that the transmission is lightly reduced in the UVB

range. At this level of UV ageing it is rather not expected to detect yellowing or change at the transmission spectra in visible spectrum, even with this dose applied by the fluorescent lamp, which should be less severe than a Xenon lamp UV ageing [9]. It must be further investigated, if the observed changes are caused by light coupling effects, meaning more reflection or diffraction of incident light at the sample (e.g. discrete changes of the surface of the sample or within the sample) or if an increase of UV light absorption occurs with ageing at the mentioned UV range. At the POE2 no change at the UV range transmission during UV ageing could be noticed up to the tested level.

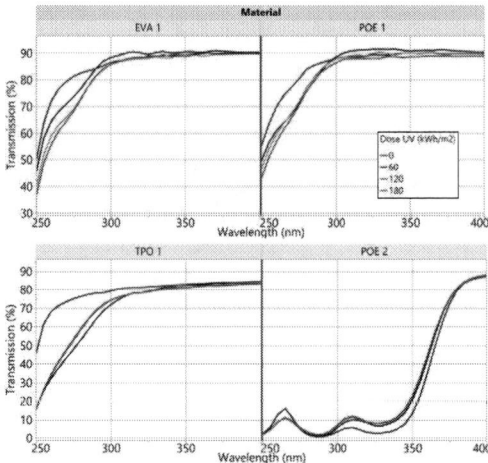

Fig. 7: Change of spectroscopic transmission in range 250-400nm wavelength, after 180 KWh/m^2 UV ageing

Whilst EVA1 proves stability towards UV ageing up to the tested level of only 0.09% change in transmission at visible spectra (400-800nm) to initial transmission, both POE1 and TPO1 reveal a reduction of transmission of 0.7% in the visible part of the spectra (almost constant at wavelength in range 400-800nm for TPO1 and slightly higher loss in range 400-600nm for POE1 compare to its own transmission in range 600-800nm).

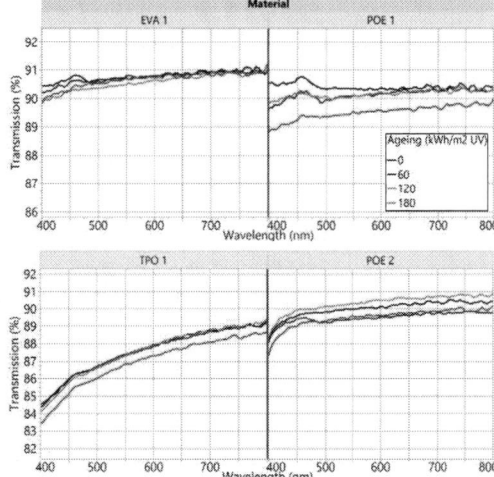

Fig. 8: Spectroscopic transmission change in wavelength range 400 – 800 nm

In the near infrared range of transmission spectra (investigated for the wavelength range 800 – 1200nm) a similar behavior at each of the samples occur, as for the visible range presented in the previous lines.

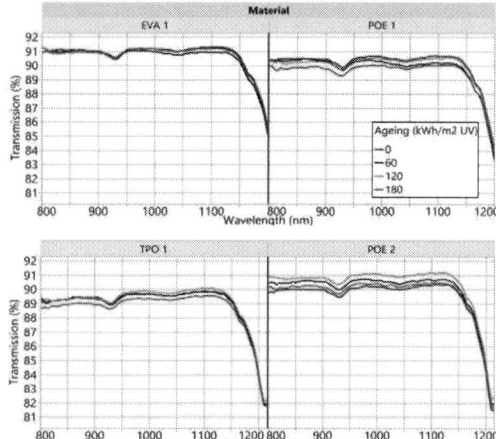

Fig. 9: Spectroscopic transmission change in wavelength range 800 – 1200 nm

At the POE2, again, no change in the transmission at the UV range could be noticed. Also as commented for the results at the UV change of transmission, further investigation is necessary to understand the mechanism behind the change of transmission in visible window of spectra for TPO1 and POE1, however for the visible region, since the change of transmission is similar for wide wavelength range, the assumption would be that the change is related to light coupling effects (higher reflection due to changed surface morphology or change at light coupling within the sample) and rather not caused by light absorption effects within the sample. At TPO1 and POE2 must be further investigated the increase of transmission at visible and IR spectra for the intermediary measurements after 60 kWh/m^2, respectively 120 kWh/m^2 UV ageing. The reported test setup was designed to investigate the conditions and ageing at market available photovoltaic modules with glass-backsheet structure and crystalline silicon solar cells, however the encapsulation material samples were not UV aged and measured under air tight or inert gas conditions, and therefore for deeper investigations the impact of moisture and oxygen should be also considered at future experiments.

3.2 Impact of accelerated ageing on the interlayer adhesion
The focus of the test was to isolate a change of the adhesion force between the encapsulation material and glass, respectively between encapsulation material and backsheet, in form of peel force measurement before and after the defined accelerated ageing. To avoid excess moisture ingress in the laminate or other effects at different climate chamber testing, it was defined to cut the stripes for measurement after each ageing testing sequence not prior, but just after that sequence ends. This increase the risk to induce additional damage when cutting at aged polymers layers, causing micro cracks at the cut edge, which might propagate during testing and bias the result. Up to the current ageing level this procedure seemed to work without causing clear impairment of the testing plan. On the other side the cutting method and its impact on the testing sequence was not part of this study and must be

approached at future experiments related to this subject. At all samples initial measurements, without any ageing, were made (Fig.9). It must be mentioned that already at unaged samples, the peel force at beginning, despite relatively narrow peel stripes, each of just 5.0 (±0.5) mm width, was so high, that at isolated samples, the stripe was detached completely at beginning of test due to cohesive breakage in the stripe at both materials: encapsulation layer and backsheet. The location of the breakage was usually not close to or at the edge of the grip, confirming that the implemented grip was suitable for this experiment: by correct fixation, the stripe didn't slide out of the grip during the test and also the grip do not scissor or damage the sample during fixation and peel procedure.

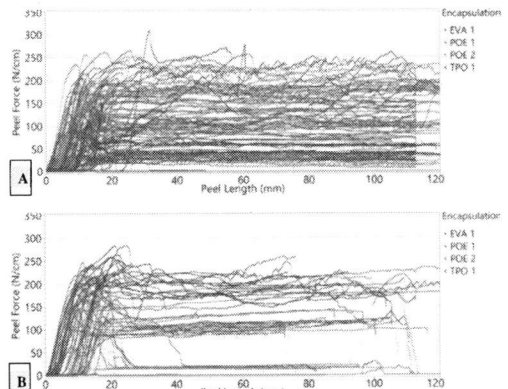

Fig. 10: Complete initial peel diagrams for the mentioned material variations and interfaces

3.2.1 Procedure for peel data filtering and processing

As it can be seen at Fig.10, the displayed bundle of peel diagrams even without additional data after ageing, are difficult to interpret and analyze, therefore a suitable procedure must be identified and used. Furthermore, the initial part of all peel diagrams (containing the initial threshold peel peak) and the last segment (containing the part of test end or eventually a breakage event) of peel stripe, must be removed from data analysis, since these would impair the test analysis.

The break of the peel stripe at beginning of peel measurement is a negative event, since this will cause an exclusion of data set from the evaluation, because only peel measurements, where a peel force could be measured through the peeled length are relevant and were analyzed.

A second reason to remove data from the evaluation was the delamination of peel sample during test at the incorrect interface, which for some samples proved to have weaker adhesion than the intended interface for peel, or weaker than the initial force threshold necessary to initiate and propagate a delamination through the peel stripe at the desired interface.

After cleaning the measurement data from failed measurements and measurement artifacts, the entire remained data package from all of the three peeled stripes available for each encapsulation material, each peel interface and each ageing stage is merged into a boxplot and the peel force is evaluated and analyzed only based on the median value of the boxplot, which is considered to be less affected by outliers and different undesired interfering factors.

The first degradation to be discussed is the damp heat ageing effect on the peel force at both studied interfaces of the encapsulation material (Fig.11).

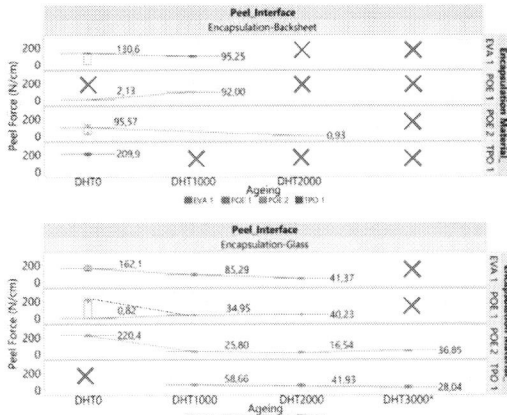

Fig. 11: Change of peel force depicted as median at both interfaces: top diagram for peel force encapsulation-backsheet and bottom diagram for peel force encapsulation-glass, for tested materials up to 3000h DHT ageing (*X marks unavailable data sets due to exclusion of unsuitable data for analysis*)
*3001h were rounded to 3000h in diagram

As clearly visible the upper graph of the diagram at Fig.11, the backsheet-encapsulation interface is strongly affected by the damp heat test, since from DHT2000 onward almost no suitable data sets for evaluation exist. According to observation during the tests, from DHT1000 onward, the backsheet becomes brittle, developing cracks during peel test, most likely due to hydrolysis induced degradation [10], and this impacts the peel at the backsheet-encapsulation material interface, since usually the backsheet material fails at beginning of the peel test. For EVA1 and POE1, the used backsheet retains a fairly high peel force to encapsulation layer up to 1000h DHT, in range of 90N/cm. After this ageing level, at least with the presented testing method and material combination, it seems to be challenging up to impossible to continue the monitoring of interface adhesion. On the other hand, the bottom graph at Fig.11 provide the change to initial of peel force at encapsulation-glass interface, and the tested method provides for the tested materials suitable results for analysis up to 2000h DHT, respectively 3000h DHT for POE2 and TPO1 resins. Since at DHT 3000h analysis data is not available for all polymer resins, at DHT 2000h data set is possible to comment that EVA1 records a drop at adhesion to glass of 74.47%, whereas POE2 at same ageing level and peel interface records a drop in adhesion to glass of 92.49% to initial value. Considering the recorded change in adhesion to glass, all evaluated encapsulation materials on different polymer basis reveal a clear and consistent change of adhesion to solar glass. Whereas EVA1, POE1 and TPO1 have a peel force to glass after DHT2000h in range 40 N/cm, POE2 is definitely closer to a delamination event, since the peel force to glass dropped to only 16.5 N/cm. It is however yet unclear and must be further analyzed, why after further 1000h DHT, hence after total of 3000h DHT, the peel force of POE2 to glass doubles, reaching the range 36.85 N/cm. TPO1 reveal after 3000h DHT the only available dataset, with 28 N/cm. The peel image, according to visual observations, represent adhesive breakage for all materials and tested interfaces.

The second ageing to be reviewed by impact on the peel force of encapsulation material to glass and to backsheet is the TCT testing.

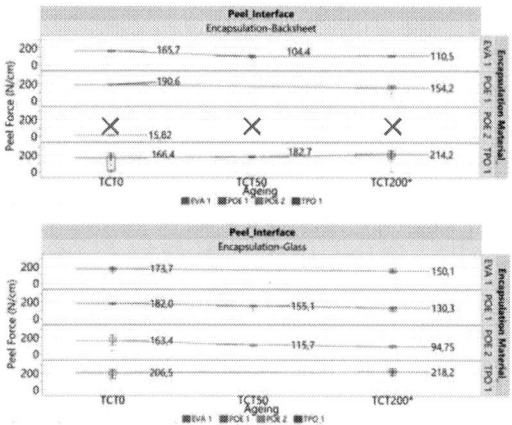

Fig. 12: Change of peel force depicted as median at both encapsulant interfaces: top diagram for peel force to backsheet and bottom diagram for peel force to glass, at ageing up to 200cycles TCT ageing (*X marks unavailable data sets due to exclusion of unsuitable data for analysis*)
*198cycles were rounded to 200cycles in diagram

The peel interface to glass, as well as the peel interface to backsheet are barely affected by the first 200 cycles (Fig.12). EVA1 has from all tested materials the lowest adhesion drop at glass interface (13.5% less to initial) and POE2 has the highest drop at the glass interface (42.0% less to initial). Overall the tested encapsulation materials provide after TCT200 a changed, yet still consistent peel force level. At encapsulation-backsheet interface EVA1 records after TCT200 a lowering of 34.9% peel force to initial, however the absolute median is still at 110.5 N/cm after TCT200. The POE1 reveal after TCT200 a drop to initial at interface to backsheet of 19.09%, while the absolute value is 154.2 N/cm. At TPO1 no degradation at interface to backsheet after TCT200 could be recorded by means of peel force, rather an improvement (increase of peel force). POE2 has no suitable analyzing data at interface to backsheet during TCT. The peel image, according to visual observations, represent adhesive breakage for all materials and tested interfaces.

At UV ageing test, the peel laminates provide a similar, yet not identical degradation behavior like at the damp heat testing for both peel interfaces (Fig.13).

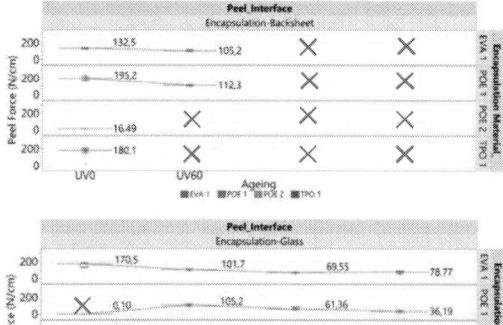

Fig. 13: Change of peel force depicted as median at both

interfaces: top diagram for peel force encapsulation-backsheet and bottom diagram for peel force encapsulation-glass, for tested materials up to 180 kWh/m^2 UV ageing (*X marks unavailable data sets due to exclusion of unsuitable data for analysis*)

The POE2 sample received 15 kWh/m^2 UV dose less, since at the first ageing was mounted by fault with the wrong side towards the UV lamps.
It can be easily seen at Fig.13 at upper graph (peel force at encapsulant-backsheet interface) that for most of the samples already up to, or after 60 kWh/m^2 UV ageing no suitable data for analysis are available anymore. Since the glass cuts a considerable part of the UV radiation, it is interesting to observe that the tested encapsulant-backsheet, respectively the backsheet reveal such a susceptibility towards the applied UV test conditions. EVA1 has a drop of peel force to initial of 20.6% at the backsheet interface, whereas POE1 has a 42.5% peel force drop to initial at the same backsheet interface. Also at the glass-encapsulation material interface all tested groups reveal a considerable lowering of the peel force, however this is of smaller magnitude compared to the damp heat test: 53.6% less at EVA1, 83.3% less at POE2 and 52.9% less at TPO1. Even if the relative value look considerable, the absolute adhesion values are fairly stable after 180 kWh/m^2 UV ageing for EVA1 (78.77 N/cm) and for TPO1 (110.9 N/cm). However POE1 (36.1 N/cm) and POE2 (29.0 N/cm) have quite lower absolute values compared to their initial state and this behavior of adhesion change must be further investigated. The peel image, according to visual observations, represent adhesive breakage for all materials and tested interfaces.

4   SUMMARY

- the applied test provides through relatively simple and widespread analytical equipment procedures a good insight in the ageing behavior of encapsulation materials regarding transmission change and peel force to glass and to backsheet
- care must be taken by the selection of backsheet used at the test, which depending on composition and additivities, as well as on conducted testing scheme, might degrade from a certain level, especially at prolonged ageing sequences and affects the test, mainly by means of cohesive failure and cracks, which would affect adhesion measurements at both interfaces
- transmission at all samples fairly stable with relatively small changes at a narrow wavelength region in UV after UV ageing; yet 0.7% transmission reduction in visible spectra at POE1 and TPO1 after only 180 kWh/m^2 UV ageing could be classified as high and must be further investigated, since light coupling at this test is different to light coupling in a PV module, where the encapsulation material is connected to glass and to each side of the solar cells;
- the tested encapsulation materials reveal a susceptibility at peel force to glass towards DHT and UV ageing, which must be inspected into detail at future tests
- continuation of ageing and further analysis and verifications are necessary before declaring best and worst performance
- one of the next very important for practical field applications and module qualifications, also very challenging approaches, would be to establish a

reliable link between the delamination of layers in the photovoltaic module under real life and long term ageing conditions and a (critical) threshold value for the peel force between encapsulant and glass, respectively between encapsulant and backsheet at unaged or accelerated aged samples

5 ACKNOWLEDGEMENT

This work was supported by the IBC4EU project, which received funding from the European Union's Horizon Europe research and innovation program under grant agreement No.101084259; the authors would like to kindly thank all involved partners and colleagues, for great collaboration and support.

6 REFERENCES

[1] M. Aghaei, et al. – doi: 10.1016/j.rser.2022.112160
[2] R.R. Cordero, et al. – doi: 10.1007/s10482-018-1075-z
[3] C. Barretta, et al. – doi: 10.3390/polym13020271
[4] G.C. Eder, et al. – doi: 10.1002/pip.3090
[5] A.J. Kinloch, et al. – doi: 10.1007/BF00012635
[6] IEC norms 61215-1/-2:2021 and IEC 61730-2:2016
[7] C. Hirschl, et al. – doi: 10.1016/j.solmat.2015.07.043
[8] S. Lust, et.al. – doi: 10.1051/epjpv/2024006
[9] R. Heidrich, et al. – doi: 10.1016/j.solmat.2023.112674
[10] B. Ottersböck, et al. – doi: 10.1002/app.44230

41st European Photovoltaic Solar Energy Conference and Exhibition

Model Calibration of Photovoltaic Modules Photodegradation in High-Radiation Environments Using UV Accelerated Exposure Testing

Patricio Valdivia*, Valentina Arias Reyes**, Dr. Rodrigo Barraza***

*Universidad de Santiago de Chile. Av. Libertador Bernardo O'Higgins, 9170022 Estación Central, , Santiago - Chile
**Universidad Técnica Federico Santa María. Av. Vicuña Mackenna 3939, San Joaquín, Santiago - Chile
*** Universidad Adolfo Ibañéz. Av. Diag. Las Torres 2640, 7941169 Peñalolén, Santiago - Chile

ABSTRACT

There are diverse methodologies based on statistical behavior to predict the lifetime of photovoltaic modules in the field based on their performance during accelerated degradation tests. However, nowadays, those kinds of methodologies do not consider all the Köppen-Geiger climates zones to properly assess the modules' lifetime, causing places like the Atacama desert in Chile, which has the highest radiation in the world, an unreliable option for PV system installations. Build a fully characterized system's response to applied stress over a broad range of levels is the purpose of this study, an accumulation of **300 KWh/m²** is obtained to each module at 60°C and 30 KWh/m² at 70°C.

Keywords: Bifacial module, monofacial module, IV curves, UV, Photodegradation.

SETUP

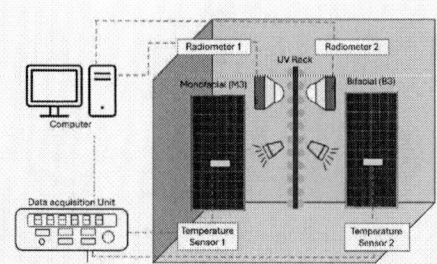

a) The test is carried out in a UV shelf model UV150200 of the EternalSun brand
b) Characterization of electrical measurements under the IEC TS 60904-9:2020 standard with the Temperature Controlled Lab Flasher (TCLF) solar simulator an A+A+A+ Xenon Single Long Pulse

Module	Name	Technology	Pmax [W]	Status
Risen RSM72	B3	PERC+	370	New
Sun Edison SE-F310EzC-4y	M3	Policristaline	310	Used

METHODOLOGY

MQT 10 test according to IEC 61215:2021 was perfomed, exposing frontside of the monofacial module and for the bifacial every 60 kWh/m² was changed for the backside,

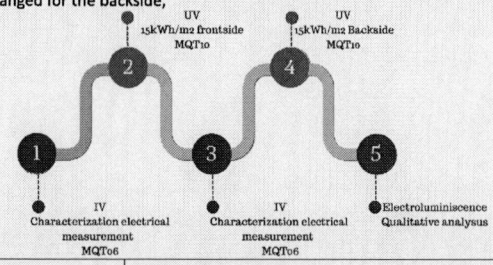

UV Intensity	2 suns UV-A + UV-B
	91,6 W/m2 in range of 280 nm to 400 nm
Spectral Distribution	UV-A Wavelength interval 320 nm - 400 nm: 95%
	UV-B Wavelength interval 280 nm - 320 nm: 5%

Compare kaaya Photodegradation model [4] with results and ajust.

$$k_P = 0.0012 \cdot (UV_{Dose})^{0.63} \cdot (1 + rh^{1.8}) \cdot \exp(-\frac{0.43}{k_B \cdot T_m})$$

RESULTS

Comparing our controlled indoor experiments with SOPHIA results [4]

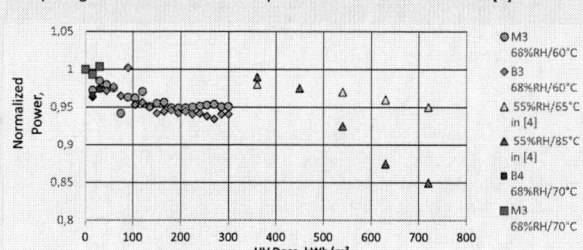

- M3 68%RH/60°C
- B3 68%RH/60°C
- 55%RH/65°C in [4]
- 55%RH/85°C in [4]
- B4 68%RH/70°C
- M3 68%RH/70°C

$$k_P = 0.0012 \cdot (UV_{Dose})^X \cdot (1 + rh^n) \cdot \exp(-\frac{0.43}{k_B \cdot T_m})$$

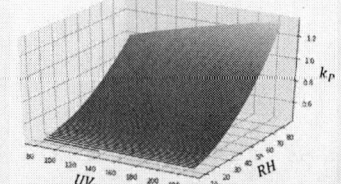

The sensitivity analysis of the Kaaya model highlights that relative humidity exerts the greatest influence on the system's behavior.

The Pearson correlation reveals a statistically significant relationship between the power output and the UV dose, indicating that UV exposure plays a critical role in influencing the performance of the photovoltaic modules.

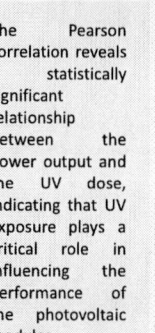

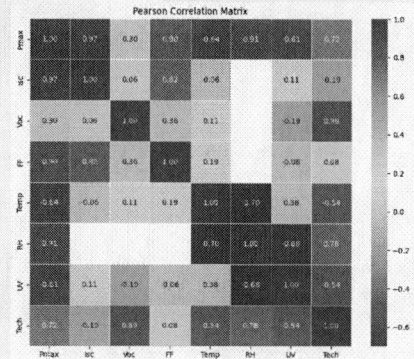

CONCLUSIONS

UV dose has a relevant weight in the photodegradation. Increasing UV weight in the equation by $X = 6.75$ and also adjusting RH exponential with $n = 3.8$ to properly fit with the new measurements.

ACKNOWLEDGEMENTS

FONDEF ID21I10424 : Desarrollo de estándares y certificación de metodologías para componentes fotovoltaicos expuestos a climas de alta radiación. PI: Dr. Patricio Valdivia Lefort

BIBLIOGRAPHY

[1] H. E. Yang, R. French, and L. Bruckman, Durability and Reliability of Polymers and Other Materials in Photovoltaic Modules. William Andrew, 2019.
[2] P. Hacke, M. Owen-Bellini, M. Kempe, D. C. Miller, T. Tanahashi, K. Sakurai, W. J. Gambogi, J. T. Trout, T. C. Felder, K. R. Choudhury, N. H. Philips, M. Koehl, K.-A. Weiss, S. Spataru, C. Monokroussos, and G. Mathiak, "Combined and Sequential Accelerated Stress Testing for Derisking Photovoltaic Modules," inAdvanced Micro- and Nanomaterials for Photovoltaics. Elsevier, 2019, pp. 279–313. [Online]. Available:https://linkinghub.elsevier.com/retrieve/pii/B9780128145012000116
[3] W. J. Gambogi, Y. Heta, J. G. Kopchick, T. Felder, S. W. MacMaster, A. Z. Bradley, B. Hamzavytehrany, B.-L. Yu, K. M. Stika, T. J. Trout, L. Garreau-Iles, O. Fu, and H. Hu, "Assessment of PV module durability using accelerated and outdoor performance analysis and comparisons," in 2014 IEEE40th Photovoltaic Specialist Conference (PVSC). IEEE, jun 2014, pp. 2176–2181. [Online]. Available:http://ieeexplore.ieee.org/document/6925356/
[4] I. Kaaya, M. Koehl, A. P. Mehilli, S. de Cardona Mariano, and K. A. Weiss, "Modeling outdoor servicelifetime prediction of pv modules: effects of combined climatic stressors on pv module power degradation,"IEEE Journal of Photovoltaics, vol. 9, no. 4, pp. 1105–1112, 2019.

ELECTRICAL CHARACTERIZATION OF FRESH AND DEGRADED PHOTOVOLTAIC BACKSHEETS BASED ON TEMPERATURE AND HUMIDITY-DEPENDENT DC CONDUCTIVITY

Anagha E R, S V Kulkarni, Narendra Shiradkar
Indian Institute of Technology Bombay, Mumbai, Maharashtra, India
anaghaer701@gmail.com, svk@ee.iitb.ac.in, naren@ee.iitb.ac.in

ABSTRACT: The operating environment conditions significantly impact the insulation resistance of photovoltaic (PV) modules. The equivalent insulation resistance of the strings connected to an inverter should be above the manufacturer's recommended threshold limit, or else the inverter will trip. The DC conductivity values of the packaging materials decide the module insulation resistance at a given operating condition. Characterizing the backsheet conductivities requires sophisticated facilities, including open electrode fixtures, very low current measurement devices, and environments with controlled temperature and relative humidity (RH). Hence, such studies are not widely reported in the literature. This paper quantitatively determines the temperature and RH-dependent DC conductivities of different types of commonly used backsheets in PV modules. The RH-dependent Arrhenius behavior of DC conductivities can vary from material to material. The objective is also to electrically characterize the PV backsheet insulation properties of fresh and degraded backsheets, which is the preliminary step to understanding the insulation degradation mechanism comprehensively.

Keywords: Backsheet, conductivity, degradation, insulation, RH

1 INTRODUCTION

Continuous and reliable output from PV modules significantly depends on their insulation health. Module insulation resistance/leakage current, a commonly used metric to assess the condition of its insulation, is significantly influenced by the temperature and RH-dependent DC conductivities of the packaging materials [1]. Being the outermost layer, the backsheet is most impacted by temperature and RH in the module package. Hence, it is imperative to study and quantify the DC conductivity of backsheets as functions of the ambient conditions for the reliability assessment of PV modules.

A correlation between the module insulation resistance and the backsheet conductivity is highly dependent on the type of polymer. Currently, there are no reliable tests available to assess the insulation health of field-deployed modules. A test involving passing steam through the backside of the module is adopted in this study to identify modules that can eventually cause insulation-related failures in the field.

The main objective of this study is to quantify the DC conductivity values of commonly used backsheet materials as well as field-degraded backsheets as functions of temperature and RH. It was observed previously that these parameters for a fresh backsheet follow the Meyer-Neldel relation, suggesting a kinetic compensation effect [2].

Moisture ingress into the backsheet can impact the bulk conductivity values. The rates of moisture ingress can vary depending on the type of backsheet in the module [3]. The dependence of DC conductivity on temperature and RH conditions is studied. A comparison of the measured values is made, and suitable conclusions are drawn to see which type of backsheet offers lower conductivity at different temperatures and RH conditions. An investigation like this is a key step toward understanding the insulation degradation of packaging materials and developing methods to estimate their residual life from the insulation strength point of view.

2 EXPERIMENTAL DETAILS

2.1 Detection of Insulation-Related Defects in Modules

The presence of different defects in PV modules can bring down their insulation performance, which can be severe during rain/condensation events. The defects observed in the modules under study include cracking and yellowing of different layers, as shown in Fig. 1.

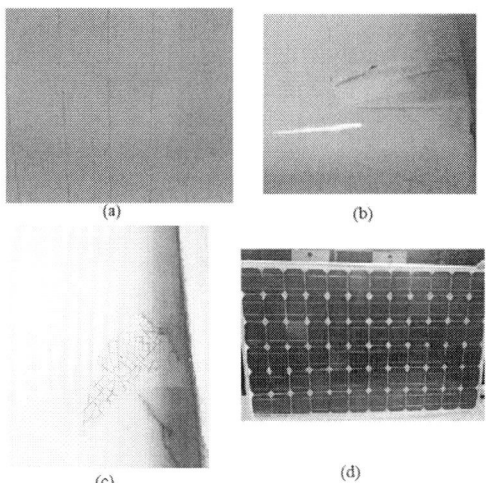

Figure 1: Backsheet defects observed (a) parallel cracks, (b) peeling off of the outer layer, (c) cracking of the inner layer, and (d) yellowing of the backsheet

There needs to be more than conventional dry and wet insulation resistance tests to ensure good insulation health of field-deployed modules. This is evident from the dry insulation resistances measured for the defective modules using an insulation tester, as shown in Fig. 2. Some of these modules showed high values of resistances even during the wet leakage test.

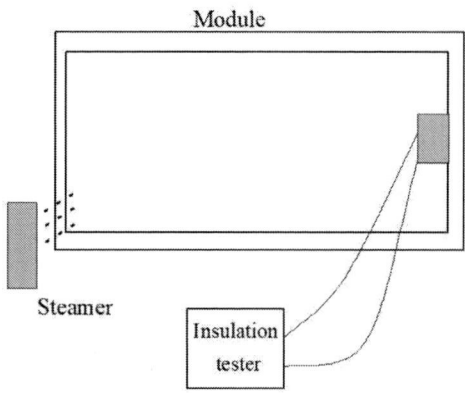

Figure 2: Experimental setup for detection of insulation-related defects in modules

The measured values of dry insulation resistances for the defective modules are given in Table I. Even with visible defects, the insulation resistance values of the tested modules are above the threshold limits set by IEC 61215-2. Modules with polyamide (PA) based backsheets have the lowest insulation resistance values, though well above the threshold. This indicates that the conventional dry and wet insulation resistance tests are sometimes not sufficient to ensure good insulation health of field-deployed modules.

Table I: Measured values of dry insulation resistance of defective modules

Backsheet condition	Outer layer	Insulation Resistance (MΩ)
Fresh	PVDF	-
Parallel cracks in the outer layer	PA	596
Large cracks in the inner layer	FEVE	617
Yellowing of core layer	PA	359
Fine cracks in the inner layer	Unidentified	2100

2.2 DC Conductivity Measurement
The bulk conductivity values of fresh and degraded backsheet materials are measured under different temperatures and RH conditions. An environment chamber in which the temperature ranges from 50°C to 80°C and RH from 10% to 90% is used in the study. A custom three-electrode fixture that allows the backsheet samples to equilibrate with the chamber conditions is employed. The experimental setup is shown in Fig. 3. The backsheet is allowed to equilibrate with the chamber conditions before the resistivity measurements. A Keithley 6517 B electrometer is connected to the fixture and placed outside the chamber to measure the leakage currents through the samples at an applied voltage of 1000 V, from which the resistivity and, hence, the conductivity can be obtained. A fluro-polymer-based fresh backsheet is studied initially, and further experiments are conducted on the backsheets extracted from the degraded modules shown in Fig. 1. As the material composition varies from

backsheet to backsheet, the insulation related electrical behavior under various operating conditions is expected to vary from backsheet to backsheet.

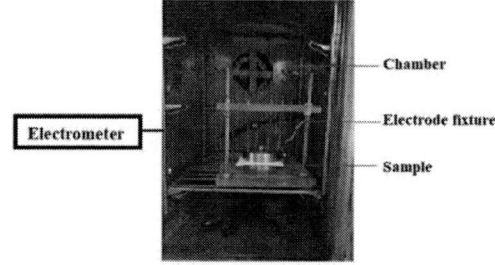

Figure 3: Custom fixture housing backsheet sample inside an environmental chamber

The primary function of backsheets in a PV module is to provide necessary insulation between its active parts. Being in direct contact with the atmosphere and porous in nature, the temperature and RH conditions to which the backsheet is exposed keep changing during the day. This results in changing values of the DC conductivity of the backsheet, as it is a strong function of the module temperature and RH. The result is that the module insulation resistance varies considerably with the ambient conditions, and, in the worst case, this can lead to inverter shutdowns. The degradation of materials can have detrimental effects on the PV plant operation [4]. The present work also aims to compare the insulating behavior of fresh and degraded backsheets as the ambient conditions vary. The module insulation resistance also varies depending on the type of backsheet used [5]. Hence, understanding the insulating behavior of different types of backsheets is a scientifically relevant aspect of PV module reliability.

It is widely known that the leakage current I in a PV module follows the Arrhenius relation with temperature as given in (1),

$$I = I_0 * exp\left(\frac{-E_a}{kT}\right) \qquad (1)$$

where the leakage current pre-exponential factor is denoted as I_0 (A), the activation energy as E_a (eV), the Boltzmann constant as k (eV/K), and the module backside temperature (K) as T. The equation is valid under the assumption that the parameters I_0 and E_a are constants under the operating conditions. However, when another stress, for example, RH, is considered, these parameters become functions of RH. Similar behavior can be derived for the DC conductivity of backsheets as well. As the backsheet in a PV module is continuously exposed to varying temperatures and RH conditions, it is imperative to characterize their conductivities in terms of the same parameters. The facilities for the electrical characterization of PV packaging materials are not common. Hence, such characterization studies are not widely reported in the literature.

A previous work [2] identified that both the Arrhenius parameters of the bulk DC conductivity of fluoro-polymer-based backsheets increase with an increase in RH and that a kinetic compensation effect exists between them. This means that the decrease in the overall conductivity of the backsheet due to an increase in the activation energy is compensated by a stronger increase in the pre-exponential factor. The net effect is that the overall conductivity increases with increasing RH. In this work,

41st European Photovoltaic Solar Energy Conference and Exhibition

the analysis is extended to other types of backsheets to see if the kinetic compensation effect is limited to only fluropolymer-based backsheets or if it is a generic behavior for all types of backsheets. The values of Arrhenius parameters obtained from the conductivity plots and their relation with RH will be of importance in modeling the insulation characteristics of PV modules with respect to the ambient conditions. The variation of conductivity with temperature at 10%, 50%, and 90% RH levels (denoted as 10 RH, 50 RH, and 90 RH) is obtained for various types of backsheets – both fresh and degraded, as shown in Fig. 4.

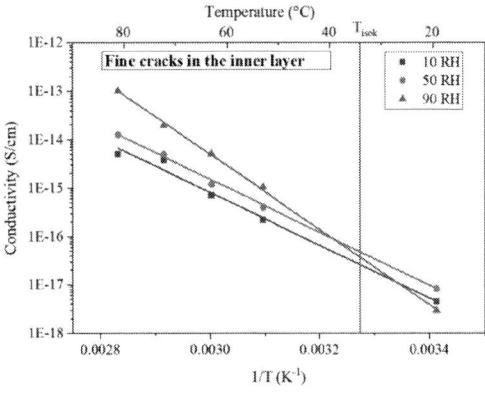

(d)

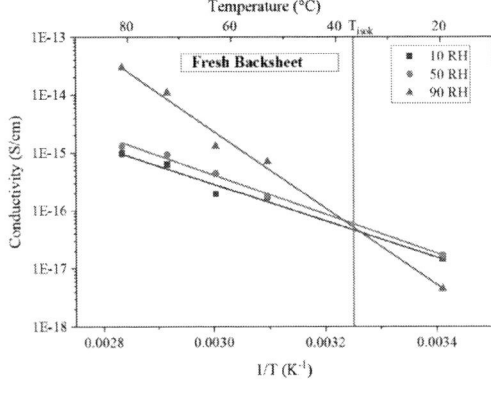

(a)

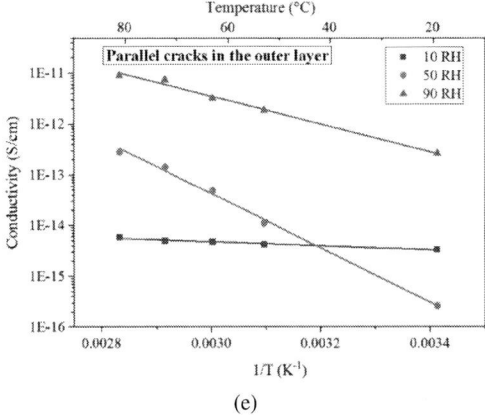

(e)

Figure 4: DC conductivity of backsheets as functions of temperature and RH

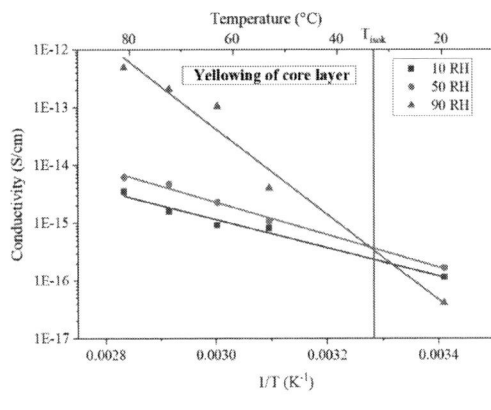

(b)

Among the cases presented in Fig. 4, it can be observed that an isokinetic temperature - the temperature at which the Arrhenius plots tend to coalesce, and the conductivity values become RH-independent, exists for all tested backsheets except the polyamide (PA) based backsheets showing parallel cracks. Even lower levels of RH are found to have an effect on the degraded PA backsheet, as evident from the slopes in Fig. 4(e). The value of isokinetic temperature varies from material to material and defect to defect. It is imperative to consider this temperature, where conductivity values are independent of RH, while comparing the fresh and degraded backsheets to eliminate the effect of RH.

The severity of defects can vary depending on the RH conditions. The conductivity values of degraded backsheets with different defects are compared at 10% and 90% RH, as shown in Fig. 5. It is observed that backsheet defects such as inner layer cracking become more severe at high RH conditions and can cause issues such as inverter trippings.

It is expected that the moisture ingress into the degraded backsheets will be faster as compared to a fresh backsheet. The size of the cracks also determines the extent of moisture ingress and, thus, the conductivity of the backsheet. Hence, the conductivity variation of backsheets with different defects with temperature and RH can be different. Quantifying the conductivity and correlating it with degradation helps in estimating the

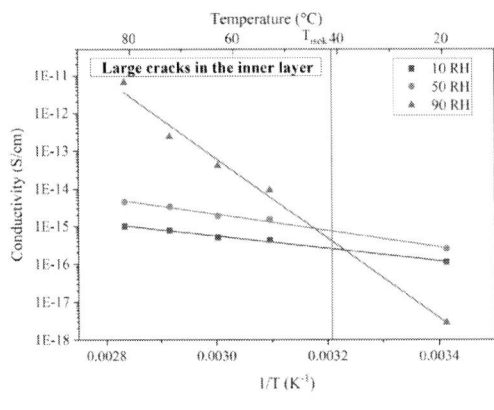

(c)

residual time for the module insulation resistance to fall below the recommended threshold.

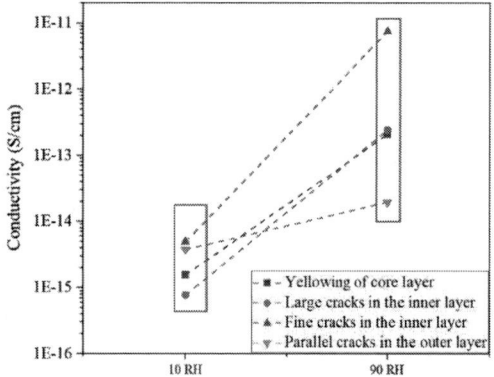

Figure 5: DC conductivity of backsheets with different defects at low and high RH conditions

The characterized values can then serve as references for electrical modeling of module insulation properties. The rate of moisture ingress varies in different types of backsheets because of variations in the water vapor transmission rate. As a result, the dependence of the Arrhenius parameters on RH can vary. Characterization of this kind will also help study the moisture-induced degradation in module packaging materials.

3 CONCLUSION

The dry insulation resistance of modules gives little information about the insulation characteristics of modules with defects. Characterization of conductivities of fresh and degraded backsheets as functions of temperature and RH is done in this work. DC conductivity characteristics of polymer backsheets can be correlated to the insulation health of modules in the field. The Arrhenius behavior of conductivity is RH-dependent and varies from polymer to polymer. The isokinetic temperature has to be considered while comparing the conductivities of fresh and degraded backsheets to eliminate the effect of RH. Certain defects, such as inner layer cracking of backsheets, are more severe at high RH and can lead to inverter tripping. A characterization of this kind is the first step in understanding the insulation degradation mechanisms in the polymer materials used in PV modules.

ACKNOWLEDGMENT

This work is supported by the National Centre for Photovoltaic Research and Education (NCPRE), funded by the Ministry of New and Renewable Energy (MNRE), Government of India, through Project No. 313-21/11/2022-Solar R&D dated 16th March 2023.

References

[1] E. R. Anagha, D. Pratti, S. V. Kulkarni, N. Shiradkar, 2020. Modeling of Leakage Currents in c-Si PV Modules and Investigation of their Arrhenius Behavior. In 47th IEEE Photovoltaic Specialists Conference (PVSC), pp. 1105-1109.

[2] E. R. Anagha, Narasimhan, K. L., Kulkarni, S. V., and Shiradkar, N. S., 2023. Humidity-Induced Electrical Conduction in Fluro-based Photovoltaic Backsheets. Solar Energy, vol. 263, p. 111947.

[3] E.R., Anagha, Kulkarni, S.V., Shiradkar, N., 2021. Moisture Ingress Modeling in c-Si PV Modules using Finite Element Simulations Based on Dual Transport Diffusion. 2021 IEEE 48th Photovoltaic Specialists Conference (PVSC), pp. 2470-2474.

[4] Sitthiphol, N., Sirisamphanwong, C., Ketjoy, N., and Sriprapha, K., 2016. The study of decrement in insulation resistance of PV string and its effects on PV system degradation. Key Engineering Materials (Vol. 675), pp. 734-738. Trans Tech Publications Ltd.

[5] Buerhop-Lutz, C., Pickel, T., Stroyuk, O., Hauch, J., and Peters, I.M., 2022. Insulation resistance in relation to distribution of backsheet types in strings and inverters. Solar Energy Materials and Solar Cells, 246, p.111913.

Investigation of PV Module Degradation on Fixed Structure and Single-axis Tracker in Hot Desert Climate

Baloji Adothu*, Aafra Seentakath, Shahzada Pamir Aly, Gerhard Mathiak, and Vivian Alberts

DEWA Research & Development Center, P.O.Box: 564 Dubai, United Arab Emirates
* corresponding author: phone +971 522 578980 | baloji.adothu@dewa.gov.ae

Abstract— **In the rapidly evolving field of photovoltaic (PV) technology, understanding the impact of environmental conditions on module performance is crucial for optimizing energy production and ensuring longevity. In a year of PV module operation, this paper investigates the early degradation patterns caused by the desert climate. PERC PV modules (monofacial and bifacial) with smart multi-busbar half-cut technology of a tier-1 manufacturer exhibited increased darkness of cells in electroluminescence (EL) images after outdoor exposure. Early discoloration patterns, featuring filled patterns, were observed in fluorescence images. Importantly, these degradation modes were consistent across both single-axis trackers and fixed structures (25-degree tilt angle). High Voc deterioration has been observed in all the modules. The indoor results show that the monofacial fixed tracker module saw a power loss of roughly 2.24%. Whereas 3.23% and 4.55% for bifacial and monofacial modules, respectively on single axis tracker.**

Keywords—desert climate, PV component defects, discoloration patterns, PV module Reliability and Durability.

I. INTRODUCTION

Solar energy is gaining prominence in the Middle East hot desert, offering abundant solar irradiation. ME regions, including Dubai, are increasingly adopting PV modules for residential and utility-scale power plants. Dubai's Clean Energy Strategy aims for 75% clean energy by 2050, featuring initiatives like the Mohammed Bin Rashid Al Maktoum Solar Park. A PV module comprises various components collectively termed the 'bill of materials (BOMs)'. PV module degradation depends on the type of BOM used during the PV module fabrication [1,2]. Depending upon Common defects such as encapsulant discoloration, delamination, corrosion, snail trails, chalking, hot spots, interconnects, finger breakages & discoloration, and cell cracks impact module performance significantly [3]. Different climatic conditions lead to diverse degradation modes and defects in PV modules.

Recent literature suggest that hot desert climates is a good source for solar power generation due to higher irradiance resources [4]. However, the absence of specific guidelines for selecting PV modules for desert climates poses challenges. Desert-induced stressors, including temperature variations,

high UV radiation, and soiling, exacerbate these challenges. The IEC 61215 standard series, designed for moderate climates, falls short in representing desert-induced defects and modes.

The paper's main objective is to show the PV component defects and their identification in PV modules exposed to ME hot desert regions. As is the case with the ME hot desert, which is characterized by harsh weather conditions, it can raise concerns about the PV modules' long-term performance. Collecting harsh weather conditions, and surveys of PV module defects motivate the researcher and PV community to focus on hot desert standard development for representing the desert-induced defects and degradation modes in the PV modules exposed in the ME region.

II. METHODOLOGY

The aim is to address challenges posed by higher module temperatures, ultraviolet radiation, sandstorms, and soiling, which are prevalent in hot desert regions. To assess the performance of PV modules with fixed structures and single-axis trackers, the goal is to compare their indoor results over a first year operational period. This analysis aims to provide insights into the efficiency and effectiveness of these modules under different tracking systems, specifically fixed structures and single-axis trackers.

In this facility, state-of-the-art PERC PV modules, known for their enhanced efficiency, have been installed. The investigation focuses on identifying and demonstrating degradation modes critical in hot deserts. Two monofacial and three bifacial PV modules from a Tier 1 manufacturer were chosen for degradation rate and degradation modes assessment. After a year of operation, comprehensive laboratory measurements were conducted. The results were then compared with the analysis of PV degradation rates and degradation modes, aiming to define effective test procedures that minimize degradation under the harsh climatic conditions of hot deserts. The comprehensive analysis in this scenario involved utilizing I-V measurements, electroluminescence, and UV fluorescence imaging techniques.

III. RESULTS AND DISCUSSIONS

1. Indoor test results of outdoor exposed PV modules

The cutting-edge PERC PV modules with smart multi-busbar half-cut technology installed in outdoor facility. Fig. 1 shows the EL images of the PERC monofacial and bifacial modules installed on the fixed structure (25°) and single-axis tracker respectively. Both bifacial and mono-facial PV modules exhibit fewer dark and bright patterns on solar cells. Although these resemble PID patterns, they may link to light and elevated temperature-induced degradation (LETID) [6,7]. As per published literature and reports, for the module temperature above 60°C, LETID is active in field-operated PV modules [6–8]. It has been noted that in hot climate modules, module temperatures can rise above 100°C [9]. Since the temperature of the PV modules in the desert is nearly always above 60°C, this may be indicative that LETID becomes active in those modules.

The overall electrical parameter degradation of the outdoor PV modules is displayed in Table 1 and the I-V is displayed in Fig. 2. When compared to the other electrical parameters, high Voc deterioration has been seen in all modules (including, both fixed structure and single axis tracker) indicated in Table 1 and Fig. 2. This could be because of increased darkness (LETID) in PV modules. However, further research is necessary to determine the exact cause of the observed dark and bright patterns in PV modules.

The PV modules also show the discoloration pattern in the UV fluorescence (UVF) images. In the case of both monofacial and bifacial modules, filled UVF patterns are distributed over the center of the solar cell (see Fig. 3). These fluorescence patterns are indicative of the early-stage discoloration. One of the known modes of PV module degradation, particularly in hot desert environments, is discoloration. As the discoloration is early stage, it may result in a reduction of the short-circuit current in PV modules, impacting their overall performance.

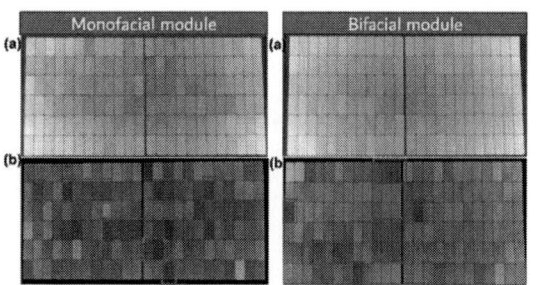

Fig. 1. EL images of PERC mono and bifacial modules installed at fixed tracker (25°) and single axis tracker. The increased darkness cells are represented by red rectangles.

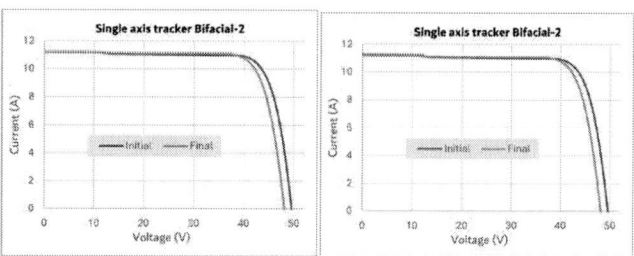

Fig. 2. Initial and final IV curves of the fixed structure monofacial module and the single axis tracker bifacial module. The reduction in the Voc is evident.

The performance of PV modules operating in desert environments is significantly impacted by two degradation mode types i.e. LETID (bright and dark cells) and the early stage of discoloration. The power degradation from single-axis trackers is 3.23% (average of three modules) for bifacial modules and 4.55% for monofacial modules. The power degradation is about 2.24% for monofacial modules from fixed structure. Perhaps because of additional sun exposure, the single axis tracker modules displayed more degradation than the fixed tracker modules.

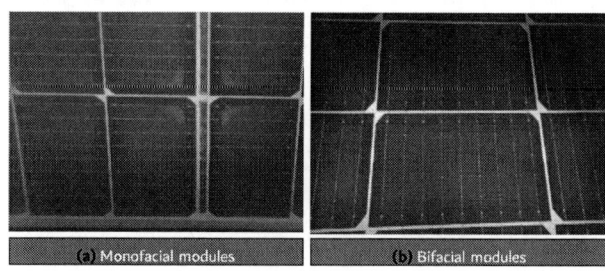

Fig. 3. UVFL images of PERC modules: (a) and (b) Monofacial module and bifacial modules show fluorescence fills over the cell.

Table 1. I-V parameters of monofacial and bifacial PV modules.

Electrical parameters		I_{sc} (A)	V_{oc} (V)	FF (%)	P (W)
Bifacial-1 (single axis tracker)	Initial	11.24	49.68	80.06	447.10
	Final	11.17	47.98	80.30	430.25
	Percentage (loss)	**0.68**	**3.41**	**-0.31**	**3.77**
Bifacial-2 (single axis tracker)	Initial	11.22	49.56	79.89	444.46
	Final	11.17	48.13	80.24	431.21
	Percentage (loss)	**0.53**	**2.88**	**-0.43**	**2.98**
Bifacial-3 (single axis tracker)	Initial	11.21	49.63	79.92	444.63
	Final	11.17	48.17	80.16	431.49
	Percentage (loss)	**0.33**	**2.94**	**-0.31**	**2.96**
Monofacial -1 (Fixed structure)	Initial	11.28	49.51	79.71	445.36
	Final	11.12	47.66	80.19	425.11
	Percentage (loss)	**1.42**	**3.74**	**-0.60**	**4.55**
Monofacial -2 (Single axis tracker)	Initial	11.36	49.82	79.26	448.65
	Final	11.30	48.69	79.74	438.58
	Percentage (loss)	**0.53**	**2.27**	**-0.61**	**2.24**

IV. CONCLUSIONS

In conclusion, the study conducted at the DEWA R&D Center during the first operational year (2023-2024) revealed potential degradation modes in cutting-edge PERC PV modules with smart multi-busbar half-cut technology. Both single-axis and fixed structures (25-degree tilt angle) exhibited consistent degradation patterns, including increased darkness and brightness in EL images (indication of LeTID similar degradation mode). It is suspected that the activity of these degradation modes is accelerated under hot desert climates. When compared to the other electrical parameters, significant Voc deterioration has been seen in all the modules (including, both fixed structure and single axis tracker). The indoor assessment highlighted power losses of 4.55% for monofacial modules and an average of 3.23% for bifacial modules from single-axis tracker. The power degradation is about 2.24% for monofacial modules from fixed structure. The early discoloration mode is also observed in fluorescence images for the same modules in this study.

REFERENCES

[1] B. Adothu, S. Mallick, P. Kartikay, Determination of Crystallinity, Composition, and Thermal stability of Ethylene Vinyl Acetate Encapsulant used for PV Module Lamination, in: Conference Record of the IEEE Photovoltaic Specialists Conference, Institute of Electrical and Electronics Engineers Inc., 2019: pp. 491–494. https://doi.org/10.1109/PVSC40753.2019.8981151.

[2] B. Adothu, R. Pugstaller, M. Tiefenthaler, F. Reny Costa, S. Mallick, G.M. Wallner, Crosslinking Kinetics of Photovoltaic Module Encapsulants – Investigation of Selected EVA and POE Grades, in: 38th Edition of the European Photovoltaic Solar Energy Conference and Exhibition (EU PVSEC-2021), 2021: pp. 779–783. https://doi.org/10.4229/EUPVSEC20212021-4AV.1.35.

[3] B. Adothu, S. Kumar, B. Jäckel, N.L. Muttumthala, Z. Shekason, D. Daßler, K. Chapaneri, P. Gabbadi, Y. Kumar, A. Alheloo, A. Almheiri, J.J. John, G. Mathiak, V. Alberts, R. Gottschalg, Identification and Investigation of Materials Degradation in Photovoltaic Modules from Middle East Hot Desert, in: Submitted EUPVSEC, 2023.

[4] J. Ascencio-Vásquez, K. Brecl, M. Topič, Methodology of Köppen-Geiger-Photovoltaic climate classification and implications to worldwide mapping of PV system performance, Solar Energy 191 (2019) 672–685. https://doi.org/10.1016/j.solener.2019.08.072.

[5] Baloji Adothu, Sagarika Kumar, Bengt Jaeckel, Neha Lyka Muttumthala, Z. Shekason, David Daßler, Kaushal Chapaneri, Prashanth Gabbadi, Yogesh Kumar, Ahmad Alheloo, Ali Almheiri, Jim Joseph John, Gerhard Mathiak, Vivian Alberts, Ralph Gottschalg, Identification and Investigation of Materials Degradation in Photovoltaic Modules from Middle East Hot Desert, in: EU PVSEC 2023, 2023: pp. 001–004. https://doi.org/10.4229/EUPVSEC2023/3AV.2.29.

[6] M. Pander, M. Turek, J. Bauer, BENCHMARKING LIGHT AND ELEVATED TEMPERATURE INDUCED DEGRADATION (LETID), (2018). https://doi.org/10.4229/35thEUPVSEC20182018-5CV.1.46.

[7] Towards a test standard of light and elevated temperature-induced degradation, Technical Briefing (Plant performance), (2020) 53–58. https://solar-media.s3.amazonaws.com/assets/Pubs/PVTP%2023/Toward s%20a%20test%20standard%20of%20light%20and%20elev ated%20temperature-induced%20degradation.pdf (accessed November 2, 2022).

[8] M. Woodhouse, I. Repins, D. Miller, LID and LeTID Impacts to PV Module Performance and System Economics DRAFT Analysis Presentation Outline, in: DuraMAT Webinar, December 14, 20201, 2020: pp. 1–57. https://www.nrel.gov/docs/fy21osti/78629.pdf (accessed May 15, 2023).

[9] M.D. Kempe, D. Holsapple, K. Whitfield, N. Shiradkar, Standards development for modules in high temperature micro-environments, Progress in Photovoltaics: Research and Applications 29 (2021) 445–460. https://doi.org/10.1002/pip.3389.

41st European Photovoltaic Solar Energy Conference and Exhibition

Tackling the fire safety in glass free PV modules

N. Pervan[1,2], S. Feldbacher[1], U. Desai[3], A. Faes[3], C. Ballif[3], G. Oreski[1,2]

[1] Polymer Competence Center Leoben GmbH (PCCL), Sauraugasse 1, 8700 Leoben, Austria – nikolina.pervan@pccl.at
[2] Chair of Materials Science and Testing of Polymers, Montanuniversität Leoben, Otto Glöckel-Straße 2/II, 8700 Leoben, Austria
[3] Ecole Polytechnique Fédérale de Lausanne (EPFL), Maladiere 71b, CH-2002 Neuchâtel, Switzerland

TOWARDS SUSTAINABLE CITIES

DESIGNING CO₂ FREE BUILDINGS WITH ENERGY PRODUCTION

Building integrated and building attached photovoltaic modules (BIPV and BAPV) → Classification by standards for electrical and building-related sectors

PV module designs

Glass / glass
Glass / backsheet
Polymer / polymer

Preferred properties

Light weight
Flexible
Coloured
Safe and reliable

Figure 1. Futuristic building with BIPV and BAPV modules [1].

GLASS-FREE PV MODULES: POLYMER/POLYMER DESIGN

BENEFITS
Complex shapes and Colourful
Flexible and Recyclable
Easier installation and distribution

CHALLENGES
Mechanical and dimensional stability (hail impact, wind and snow loads)
Fire safety

Tackling fire safety

Why are PV modules burning?
Polymeric components are combustible material that ignite in the presence of flame source.

How does it happen?
Ignition: hot spot within the PV module; arching and installation errors. Propagation from external sources.

Figure 2. Burning PV module with flaming melted polymer [2].

PV MODULE'S REACTION TO FIRE

glass → Glass bursting / breakage Detachment

encapsulant → Fire propagation Flaming droplets Smoke

backsheet → Flaming droplets Halogen backsheet: toxic gases and ashes [3]

Figure 3. Standard PV module composition.

2 - 4 min backsheet detachment
7- 10 min glass plate bursting
12 min full disintegration
1.5 - 4 min flaming droplets
6 – 8 min full scale flame

[4]

!!! Polymer/polymer PV module will disintegrate and burn faster !!!

Figure 4. Ignorance on how to deal with a burning PV module [5].

HOW TO MAKE POLYMERS SAFER AND INCREASE PV MODULE'S FIRE RESISTANCE?

Intumescent coatings

➤ When exposed to heat / flame these coatings will swell / char and create an insulating layer for the inner material and slow down (stop) fire propagation.

+ Can be added to the backside of the PV module.

+ Commercially available solutions (aviation industry).

+ Can be implemented on already installed PV modules.

− Not available in transparent version. Activation = loose of transmittance.

− Can be activated during lamination. Aging behaviour – not determined. [6,7]

Flame retardants

➤ These additives when activated via heat / flame will start a chemical reaction and create by-products like water or gasses which act as fire retardants.

+ Suitable for every polymeric component of a PV module.

+ Addition during polymerization process or as additive to a coating.

+ Protects from internal and external sources of fire.

+ Available in transparent version.

− Can be activated during the lamination. Aging and interaction with other components – not determined. [6,7]

Figure 5. Rising Phoenix as representation of PV module's resistance to fire [8].

CONCLUSIONS

BIPV and BAPV can lead towards greener cities. Classification through building and electrical industry standards.

PV modules are combustible product due to the presence of polymeric components.

When exposed to the flame source, PV module will start to decompose within few minutes.

Polymers can be created with higher fire resistance. Either via outer "shield" in form of a coating, or during the polymerization process.

Intumescent coating can be implemented into existing PV modules. Fire retardants fit with every polymeric component in a PV module.

REFERENCES

1. Image created with „COPILOT AI", on 12th of September 2024, 10:34 am 2. Image created with „COPILOT AI", on 12th of September 2024, 10:45 am 3. https://doi.org/10.1080/23311916.2022.2155004
4. Guidline: https://www.dgs.de/news/en-detail/250124-solar-electricity-for-firefighting-forces-and-emergency-services-portal
5. Image created with „COPILOT AI", on 12th of September 2024, 10:51 am
6. https://doi.org/10.1016/j.mser.2020.100604
7. Hull, T. R., et al. Chapter 4 - Environmental Drivers for Replacement of Halogenated Flame Retardants, Polymer Green Flame Retardants, Elsevier, 2014, 119-1798. 8. Image created with „COPILOT AI", on 11th of September 2024, 10:02 am

ACKNOWLEDGMENT

This work was conducted as part of the Solar Era Net Project "DELIGHT", which is supported under the umbrella of SOLAR-ERA.NET Cofund by Austrian Research Promotion Agency (FFG, contract number FO999897443), Swiss Federal Office of Energy (SFOE, contract number SI/502501-01) and Flanders Innovation and Entrepreneurship (VLAIO, contract number HBC.2022.0406). SOLAR-ERA.NET is supported by the European Commission within the EU Framework Programme for Research and Innovation HORIZON 2020 (Cofund ERA-NET Action, N° 691664).

41st European Photovoltaic Solar Energy Conference and Exhibition

ANALYSIS AND MATERIAL MODELING OF MECHANICAL PROPERTY DEGRADATION FOR SIMULATION OF WEATHER EXPOSED POLYMERS

Julia Petro[1,2], Volker Reisecker[3], Eric Helfer[1], Gernot Oreski[1], Thomas Antretter[2], Margit Lang[1]
[1] Polymer Competence Center Leoben GmbH (PCCL), Sauraugasse 1, 8700 Leoben, Austria – julia.petro@pccl.at
[2] Chair of Mechanics, Montanuniversität Leoben, Franz-Josef-Straße 18, 8700 Leoben, Austria
[3] Transfercenter für Kunststofftechnik GmbH, Franz-Fritsch-Straße 11, A-4600 Wels, Austria

ABSTRACT: The photovoltaic (PV) industry is expanding rapidly in order to achieve global climate change goals. The challenges of reducing the cost and increasing lifetime of a PV system are consistently pushing the industry to provide products with increased quality and reliability. PV systems are susceptible to prolonged exposure to various weather conditions as it is an outdoor application installed in different locations all over the world. Like many other industrial applications, PV systems use polymeric components which are known to suffer from aging effects when exposed to harsh weather conditions for long periods of time. Hence, it is rather inevitable to consider potential material degradation effects already in the design phase. This is where virtual design analysis tools such as finite element analysis (FEA) become important for performance prediction. This requires the development of methods to generate suitable long-term mechanical property degradation data that can be modeled in FEA. This work focuses on the mechanical property changes due to accelerated laboratory testing of polymers and presents a method to estimate long exposure performance to be further used in FEA numerical simulations.

Keywords: accelerated weathering; mechanical property degradation; performance prediction; regression analysis; finite element method.

1 INTRODUCTION

The Photovoltaics (PV) industry is required to grow rapidly to achieve the 1.5°C scenario by 2050 [1]. According to the International Renewable Energy Agency (IRENA), PV is expected to increase significantly to account for half of the global electricity generation (~30 TWh in 2023) [2, 3], whereas the current global electricity generation from solar energy is at ~1.4 PWh in 2024 [3]. This poses more challenges on the reliable lifetime performance of the PV systems in the field, and their cost which are key factors that determine the commercial attractiveness of an electricity generation system.

Given the widespread use of PV systems in outdoor settings and their susceptibility to prolonged exposure to various weather conditions, it is crucial to consider the potential degradation of their polymeric components [4, 5] to avoid early failures as reported in numerous past studies [6–8]. This necessitates the development of a methodology for the generation of suitable mechanical property degradation data, which can then be modelled in a format suitable for incorporation into virtual engineering design analysis; e.g., using finite element analysis (FEA). The long-term outdoor weather-affected mechanical property data of polymers are often not readily available, which is essential for successful design analysis of parts made from polymeric materials [7, 9–11]. Modeling approaches in this area of study mainly depend on the specific application of interest and available data. Obtaining detailed information about the degradation mechanism of a polymer has always been a challenge; whereby an accurate description for the evolution of polymer degradation in the form of an ordinary differential equation is rather very difficult [12, 13].

Extensive studies have been conducted on the weathering-induced photooxidative degradation of polymers. Among the changes occurring from photooxidation are discoloration, reduction in refractive index and mechanical properties, and peeling of surface coatings [14, 15]. Several studies have investigated the mechanism of photooxidative degradation for different

polymers. In particular, the reduction in molecular weight and the formation of crosslinks have been identified as the primary causes of observed changes in mechanical properties [7, 9–11]. While many studies have focused on the characterization of polymeric films investigating the kinetics of chemical reactions causing degradation, fewer studies used bulk specimens in their investigations. Bulk specimens were mainly utilized to study the deformation and fracture behavior of the UV exposed polymers [9, 11].

Focusing on the static tensile behavior, thin film specimens have shown significant effects on properties like the Young's modulus, yield strength, and fracture strain. On the other hand, bulk specimens demonstrated consistent behavior up to yielding, showing significant reduction of the fracture strain [16, 17].

In the current study, emphasis has been placed on accelerated laboratory testing (IEC 62788-7-2-A3) to provide data on the long-term durability of the polymer under investigation. The accelerated aging procedure exposes specimens to xenon-arc light in the presence of moisture to reproduce the weathering effects (temperature, humidity and/or wetting) that occur when materials are exposed in actual environments to daylight or to filtered daylight through window glass. Through observation of tensile tests data of UV exposed specimens when compared to unexposed ones, affected properties were determined and hence, a method to estimate long exposure performance has been developed to be used further in FE analysis. The method is then applied to a simple test case of a tensile test simulation and validated against experimental observations.

2 MATERIALS AND METHODS

2.1 Material Preparation

The virgin material used in this study is a flame retardant, injection moldable grade polycarbonate (PC) material. Common applications for this material in the PV industry include inverter casing materials, junction box casing, switches, plugs, and connectors.

Tensile test specimens were injection molded standard

dogbone type (1A) specimens in accordance with ISO 527-2. Tests were performed under standard laboratory conditions of 23 °C and 50 % relative humidity at a loading rate of 1 mm/min for Young's modulus evaluation, followed by a test rate of 50 mm/min on a ZwickRoell Z250 testing machine. Strain measurements were acquired and analyzed using the Digital Image Correlation (DIC) technique. Specimens were aged using an ATLAS Xenontester 440 according to the ISO 4892-2 standard settings summarized in Table 1 for a total of 3000 hours and then results were compared to the unaged material.

Table 1. Testing conditions according to IEC 62788-7-2-A3 standard.

Irradiance	Black-standard temperature	Chamber temperature	Relative humidity
(W/m²nm)	(°C)	(°C)	(%)
0.8	90	65	20

2.2 Tensile test results

The engineering stress-strain behavior of the unaged material as well as the aged specimens at the different ultraviolet (UV) exposure doses in the Xenon tester are shown in Figure 1. Five specimens were tested at each exposure condition.

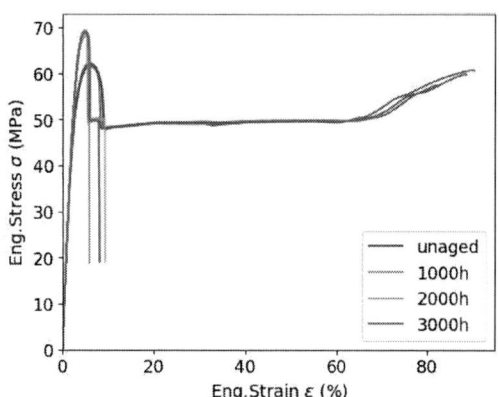

Figure 1. Engineering stress-strain behavior of unaged material compared to its behavior at different UV exposure conditions.

Material properties of interest like the tensile elastic modulus, yield strength and fracture strain have been evaluated for each sample tested and an equivalent value of the respective property has been evaluated to represent the material property measured at a certain exposure condition. Figure 2 shows the results for the elastic modulus variation vs. UV exposure. It shows a random variation of the elastic modulus with no significant increase to indicate embrittlement of the material. On the other hand, a significant drop is observed in the fracture strain upon UV exposure, as shown in Figure 3. The fracture strain, in the case of 1000 hours exposure samples, has dropped by ~90% in comparison to the unaged material. Similar obervations were reported in many UV-exposed polymers [7, 9, 9, 10, 10, 11, 17]. These results are expected, given that UV degradation typically affects a thin surface layer of the specimen leading to the potential

formation of microcracks and consequently, surface embrittlement [18].

Based on these results, the reduction in fracture strain is assumed to be the primary cause of the deterioration in mechanical properties of the material that occurs during the equivalent outdoor weathering conditions. Therefore, the generation of material data for a finite element (FE) analysis would include the reduction of fracture strain in relation to the UV exposure of the material, while maintaining constant values for other mechanical properties.

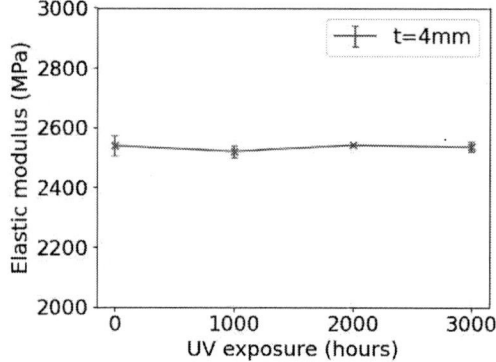

Figure 2. Elastic modulus variation at different UV exposure doses.

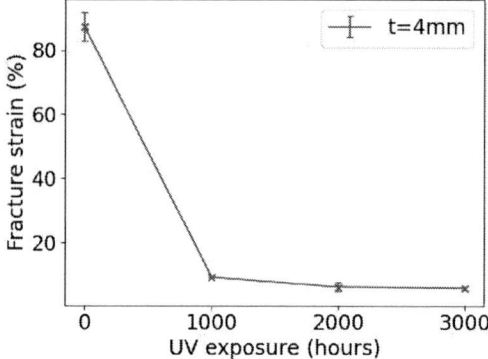

Figure 3. Fracture strain variation at different UV exposure doses.

3   MODELING AND SIMULATION

3.1 Material property degradation modeling

Observing the change in fracture strain values with the different UV exposure times, the characteristic behavior can be expressed by an exponential decaying function, see Figure 3. In this model, the material property decreases rapidly, and converges to some constant value, see Eq. 1.

$$y(x) = ae^{rx} - k \qquad (1)$$

where, y is the normalized fracture strain, x is the UV dosage in hours, a, r & k are fit parameters.

The three fit parameters were determined using least-squares regression analysis. First, the fracture strains normalized by the fracture strain of the unaged material are plotted and then an optimization scheme using genetic algorithm (GA) is implemented to find the least-squares fit function to the available test data, see Figure 4. The quality of the fit has been assessed by evaluating the coefficient of determination or the R-squared (R^2) value. The fit parameters are given in Table 2.

Table 2. Fit parameters for the decaying exponential model for the fracture strain variation.

a	R	k
0.9320	-0.0032	-0.0680

The results from the regression analysis on the fracture strain are then further used to estimate the degraded material stress-strain behavior to simulate the degradation of the mechanical property over exposure time in a FE model. To this end the true stress-strain curve of the unaged material sample and the fracture strain regression model are used to generate the degraded material property data as shown in Figure 5.

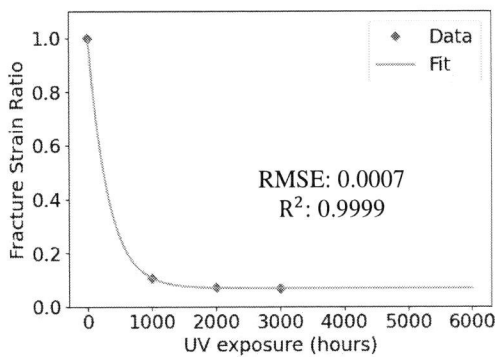

Figure 4. Normalized fracture strain data and regression results for the polymer under study during artificial weathering test.

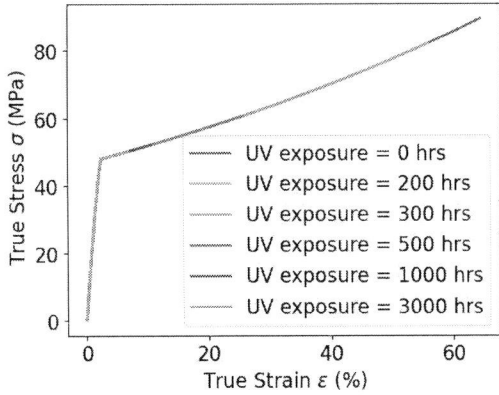

Figure 5. Generated true stress-strain curves for the polymer under study to be used in FE simulations.

3.2 Tensile test case simulation

To demonstrate the implementation of material property degradation in a FE analysis of the weathering process, a time-dependent test case scenario of a simple tensile test simulation was setup in ABAQUS 2019, as shown in Figure 6.

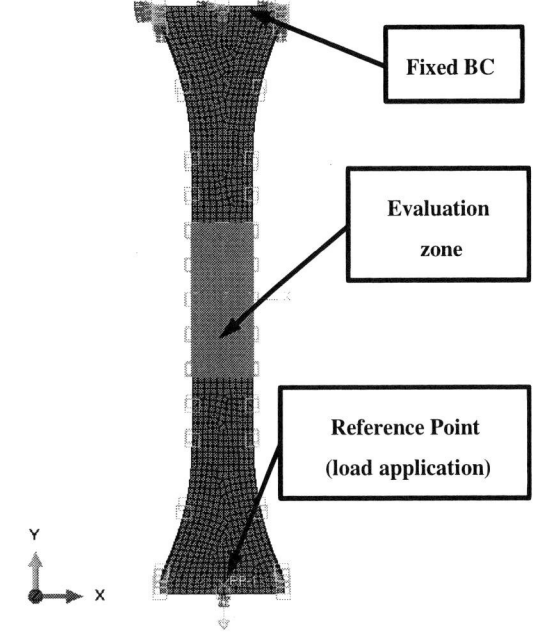

Figure 6. Tensile test simulation setup.

In this simulation, 8-node linear brick elements of type (C3D8R) were used. The initial condition was set at room temperature (~25°C). The load was then applied at a reference point created for one end of the specimen and constrained to move in the loading direction only.

The mechanical behavior of the material was defined in ABAQUS using an incremental elastic-plastic material model, where the total mechanical strain (ε_{total}) is divided into an elastic part and a plastic part, as shown in Eq. 2 and 3.

$$\varepsilon_{total} = \varepsilon_{elastic} + \varepsilon_{plastic} \qquad (2)$$

$$\varepsilon_{plastic} = \ln(1 + e_{nominal}) - \ln(1 + \frac{\sigma_y}{E}) \qquad (3)$$

where, $\varepsilon_{elastic}$, $\varepsilon_{plastic}$ are the true elastic and plastic strains, respectively. $e_{nominal}$ is the nominal true strain, σ_y is the yield strength and E is the Young's modulus of the material.

To incorporate the fracture strain reduction with UV exposure time in the material data, the mechanical properties of the unaged material have been used. In the elastic regime all values from the unaged material have been kept constant (e.g., the elastic modulus). In the plastic regime, the fracture strain of the unaged material is multiplied by the factor (y) calculated from the regression analysis function shown in Eq. 1 to obtain the estimated fracture strain at the desired UV exposure time. The material property degradation datasets generated from the regression analysis model were then input to ABAQUS to define the damage variable.

The simulation results for the normal stresses experienced by the specimen can be seen in Figure 7. The material stress-strain behavior obtained from the simulation shows very good correlation to the experimental observation as depicted in Figure 8.

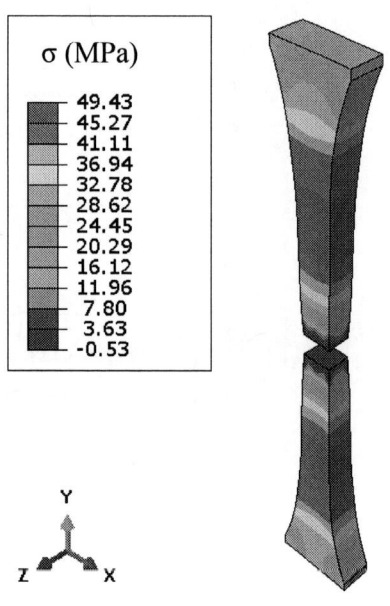

Figure 7. Simulation results for tensile test case scenario.

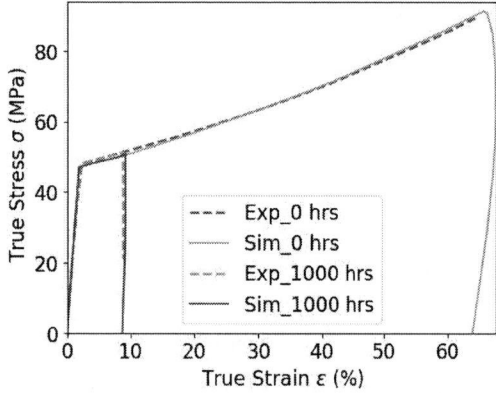

Figure 8. Simulation results validation vs. experimental test result obtained from tensile tests.

4 CONCLUSIONS AND OUTLOOK

This study demonstrated using accelerated weathering test data to provide data on the long-term durability of the polymer under investigation. Affected mechanical properties were identified based on the obtained results from tensile tests conducted on aged specimens in comparison to unaged ones. A method was then introduced to generate material property degradation datasets suitable for implementation in FE codes to predict the performance of polymeric components undergoing degradation of mechanical properties due to weathering.

Future work should validate accelerated weathering results vs. outdoor weathering data and employ material behavior modeling for more complex geometries.

ACKNOWLEDGMENTS

The research work of this paper was performed at the Polymer Competence Center Leoben GmbH (PCCL, Austria) within the framework of the COMET-program of the Federal Ministry for Transport, Innovation and Technology and the Federal Ministry of Digital and Economic Affairs with contributions by Fronius International GmbH. The PCCL is funded by the Austrian Government and the State Governments of Styria, Lower Austria and Upper Austria.

8 REFERENCES

[1] D. Gielen and et al., "World energy transitions outlook: 1.5° C pathway," 2021. [Online]. Available: https://www.h2knowledgecentre.com/content/researchpaper1609

[2] Q. Hassan *et al.,* "The renewable energy role in the global energy Transformations," *Renewable Energy Focus*, vol. 48, p. 100545, 2024, doi: 10.1016/j.ref.2024.100545.

[3] *Statistical Review of World Energy 2024,* 73rd ed.: Energy institute, 2024. [Online]. Available: https://www.energyinst.org/statistical-review/home

[4] S. A. Begum, A. V. Rane, and K. Kanny, "Applications of compatibilized polymer blends in automobile industry," in *Applications of compatibilized polymer blends in automobile industry*, S. A. Begum, A. V. Rane, and K. Kanny, Eds.: Elsevier, 2020, pp. 563–593.

[5] L. Dai, L.-Y. Wang, T.-Q. Yuan, and J. He, "Study on thermal degradation kinetics of cellulose-graft-poly(l-lactic acid) by thermogravimetric analysis," *Polymer Degradation and Stability*, vol. 99, pp. 233–239, 2014, doi: 10.1016/j.polymdegradstab.2013.10.024.

[6] H. Al Mahdi, P. G. Leahy, M. Alghoul, and A. P. Morrison, "A Review of Photovoltaic Module Failure and Degradation Mechanisms: Causes and Detection Techniques," *Solar*, vol. 4, no. 1, pp. 43–82, 2024, doi: 10.3390/solar4010003.

[7] Y. Lv *et al.,* "Outdoor and accelerated laboratory weathering of polypropylene: A comparison and correlation study," *Polymer Degradation and Stability*, vol. 112, pp. 145–159, 2015, doi: 10.1016/j.polymdegradstab.2014.12.023.

[8] M. Aghaei *et al.,* "Review of degradation and failure phenomena in photovoltaic modules," *Renewable and Sustainable Energy Reviews*, vol. 159, p. 112160, 2022, doi: 10.1016/j.rser.2022.112160.

[9] E. S. Sherman, A. Ram, and S. Kenig, "Tensile failure of weathered polycarbonate," *Polymer Engineering & Sci*, vol. 22, no. 8, pp. 457–465, 1982, doi: 10.1002/pen.760220802.

[10] J. Pabiot and J. Verdu, "The change in mechanical behavior of linear polymers during photochemical aging," *Polymer Engineering & Sci*, vol. 21, no. 1, pp. 32–38, 1981, doi: 10.1002/pen.760210106.

[11] J. Tireau *et al.*, "Consequences of thermo- and photo-oxidation on end-use properties of pure PE," in *Consequences of thermo- and photo-oxidation on end-use properties of pure PE*, Ischia (Italy), 2010, pp. 101–103.

[12] A. Plota and A. Masek, "Lifetime Prediction Methods for Degradable Polymeric Materials-A Short Review," *Materials (Basel, Switzerland)*, vol. 13, no. 20, 2020, doi: 10.3390/ma13204507.

[13] T. M. Kruse, O. S. Woo, H.-W. Wong, S. S. Khan, and L. J. Broadbelt, "Mechanistic Modeling of Polymer Degradation: A Comprehensive Study of Polystyrene," *Macromolecules*, vol. 35, no. 20, pp. 7830–7844, 2002, doi: 10.1021/ma020490a.

[14] J. E. Pickett, D. A. Gibson, S. T. Rice, and M. M. Gardner, "Effects of temperature on the weathering of engineering thermoplastics," *Polymer Degradation and Stability*, vol. 93, no. 3, pp. 684–691, 2008, doi: 10.1016/j.polymdegradstab.2007.12.013.

[15] G. Tjandraatmadja, L. Burn, and M. Jollands, "Evaluation of commercial polycarbonate optical properties after QUV-A radiation—the role of humidity in photodegradation," *Polymer Degradation and Stability*, vol. 78, no. 3, pp. 435–448, 2002, doi: 10.1016/S0141-3910(02)00179-9.

[16] H.-C. Hsueh *et al.*, "Micro and macroscopic mechanical behaviors of high-density polyethylene under UV irradiation and temperature," *Polymer Degradation and Stability*, vol. 174, 2020, doi: 10.1016/j.polymdegradstab.2020.109098.

[17] L. Jiang, M. Zhou, Y. Ding, Y. Zhou, and Y. Dan, "Aging induced ductile-brittle-ductile transition in bisphenol A polycarbonate," *J Polym Res*, vol. 25, no. 2, 2018, doi: 10.1007/s10965-018-1443-4.

[18] A. Blaga and R. S. Yamasaki, "Surface microcracking induced by weathering of polycarbonate sheet," *J Mater Sci*, vol. 11, no. 8, pp. 1513–1520, 1976, doi: 10.1007/BF00540886.

RELIABILITY INVESTIGATION OF STRUCTURAL COLOUR INTERLAYERS FOR COLOURED PV MODULES

Markus Babin[1], Roberto Boccardi[1], Aliihsan Bagci[1], Nanna Lysgaard Andersen[1], Peter Behrensdorff Poulsen[1],
Sune Thorsteinsson[1], Karlis Petersons[2], Leif Yde[2], Jan F. Stensborg[2], Catarina G. Ferreira[3,4], Joel D. Cox[3,4,5],
Irina Vyalih[4,6], Jani Lamminaho[4,6], Morten Madsen[4,6]
[1]Technical University of Denmark, Institute of Electrical and Photonics Engineering, 4000 Roskilde, Denmark
[2]Stensborg A/S, 4000 Roskilde, Denmark
[3]University of Southern Denmark, Mads Clausen Institute, Center of Polariton-driven Light Matter Interactions,
5230 Odense M, Denmark
[4]University of Southern Denmark, Climate Cluster, 5230 Odense M, Denmark
[5]University of Southern Denmark, Danish Insititute for Advanced Study, 5230 Odense M. Denmark
[6]University of Southern Denmark, Mads Clausen Institute,
Center for Advanced Photovoltaics and Thin Film Energy Devices, 6400 Sønderborg, Denmark

ABSTRACT: With the increased proliferation of structural colours in PV colouration, concerns are growing regarding the reliability impact of materials used. In the ColorFoil project, structural colour interlayers with integrated three-dimensional diffuser structures are being developed. To determine the degradation behaviour of the different materials within such a laminate, accelerated aging is performed on base polymers coated with resin-based diffuser structures and metal-oxides used in structural colours. To accelerate UV exposure degradation testing, custom LED-based UV light sources have been developed and characterized.
Results show PET foils to be best suited for interlayer substrates as well as good stability for resin-based and SiO_2 coatings after UV exposure. Samples containing TiO_2, however, show significant losses in short-circuit current for both EVA and POE encapsulants. It is suggested to be caused by a photocatalytic effect of the TiO_2 coating, accelerating polymer degradation.
Keywords: reliability, coloured PV, structural colours

1 INTRODUCTION

Coloured PV modules are of high interest for integration into buildings as they allow for a larger variety in architectural design up to colour matching of surrounding materials. This makes it possible to effectively hide the PV system within the building façade or roof. In the past few years there has been significant research on novel colouring solutions as well as development of diverse coloured PV products [1]. While screen- and inkjet-printed ceramic colours have been available on the market for some time, they typically suffer from large optical losses compared to uncoloured modules due to light absorption in the pigments. Recent developments have therefore focused on structural colours, which generate colouration through interference of light reflected from thin-film stacks. The spectrum of reflected light which is responsible for colourful appearance can be tuned through material choice and thickness of the individual layers. While typically resulting in significantly lower optical losses, one drawback of this technology presents itself in form of high angular dependency. As optical path length in the thin-film stack changes with incidence angle, this can lead to changing hue (iridescence). In addition, to achieve colourful appearance in off-specular view angles, satinated glass or other scattering structures are needed to achieve consistent colour saturation.

Traditionally, such thin-film stacks have been deposited directly on the inside of the (satinated) cover glass [2, 3, 4], while newer approaches integrate small particles with structural colours in glass frits [5] or encapsulants [6]. In the ColorFoil project, a solution for structural colour interlayers based on a polymer foil with a resin-based diffuser structure, on which a conformally coated thin-film stack is being developed. The textured surface is expected to lead to a more homogeneous appearance of the coloured PV modules due to reflected light originating from multiple differently oriented surfaces, thus reducing iridescence and increasing off-specular colour saturation.

The main challenges in the development of these interlayer foils are achieving a high base transparency and ensuring high reliability. The latter is especially important in building-integrated PV (BIPV), as construction products are expected to have significantly longer lifetimes than standard PV modules, of up to 50 years. Historically, adding new materials to PV laminates has caused material incompatibilities, resulting in new degradation modes [7]. Thin-film stacks in structural colours typically consist of alternating layers of metal oxide coatings with different refractive indices. Commonly used materials include SiO_2, SiN and TiO_2. While the first two have been used in PV modules in different forms for some time, TiO_2 represents a new addition to PV laminates. This may be especially critical as TiO_2 is known to act as a photocatalyst for several chemical processes, including some responsible for polymer degradation [8].

This work therefore focuses on two areas: Assessing optical losses in material candidates for the base polymer in the ColorFoil interlayer as well as the resin-based diffuser structure and degradation testing of all materials, including metal oxide layers for the structural colour thin-film stacks.

2 METHODOLOGY

2.1 Interlayer requirements and design

In order to create a homogeneous, non-irridescent colour of the PV modules, a base polymer substrate is coated with an acrylate-based, UV-cured resin, textured by nanoimprint lithography. Subsequently a stack of metal-oxide (specifically SiO_2 and TiO_2) thin-films is deposited on this diffuser structure in a conformal manner using physical vapour deposition (PVD) to achieve colouration. A schematic cross-section of the PV module structure is shown in Figure 1.

To maximise module efficiency, both the polymer substrate as well as the resin coating require high optical transmittance. In addition, it must be ensured that UV light and/or damp heat (DH) does not lead to discoloration or delamination of any of these layers.

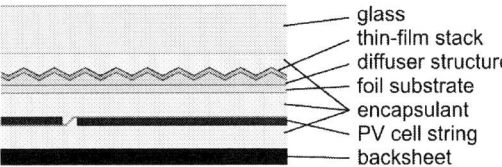

Figure 1: Cross-section of PV module structure for ColorFoil interlayer.

2.2 Degradation testing

To test the effect of UV light on the various materials, a new lab setup for UV aging of single-cell PV modules was designed and built at DTU Electro. It is based on UVA and UVB LEDs in order to achieve the maximum permissible acceleration factor according to IEC 61215-2 MQT 11. To obtain the desired irradiance with good spatial uniformity, 64 UVB and 4 UVA LEDs are mounted on a 100x100 mm PCB surrounded by mirrors. Samples are mounted on a hotplate to allow for similar sample temperatures as when using traditional UV lamps, which have significant visible and thermal emittance. Figure 2 shows a photograph of the system.

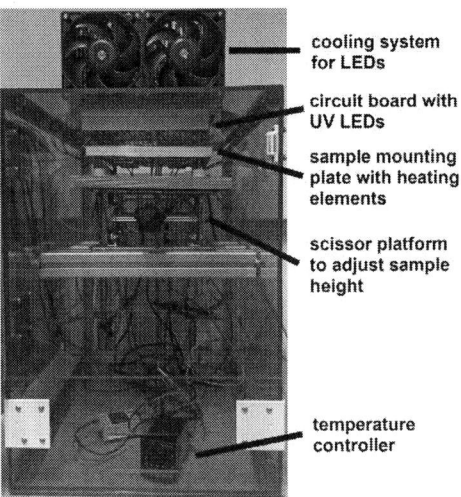

Figure 2: Picture of UV aging setup.

The spatial uniformity and spectral distribution is characterized using a calibrated Ocean Optics QE65000 spectrometer at 64 evenly spaced points in the sample plane. For each measurement point, spectral irradiance within the UVA and UVB regions is integrated and represented as a heat map in Figure 3 and Figure 4.

The average UV irradiance in the sample plane corresponds to approximately 210 W/m², with a maximum deviation of 5.2 %. The UVB irradiance accounts for 6 %, well within the permissible range according to IEC 61215-2 MQT 11. Using this setup, a UV preconditioning dose of 15 kWh/m² can be reached in around 72 hours. Due to the absence of visible and infrared radiation, there is little

parasitic heating in the samples, allowing precise temperature control using the included heating elements.

In addition to UV light, humidity and temperature are also known to accelerate PV degradation. Therefore, a range of samples undergoes damp heat (DH) storage in a climate chamber at 85 °C and 85 %rH.

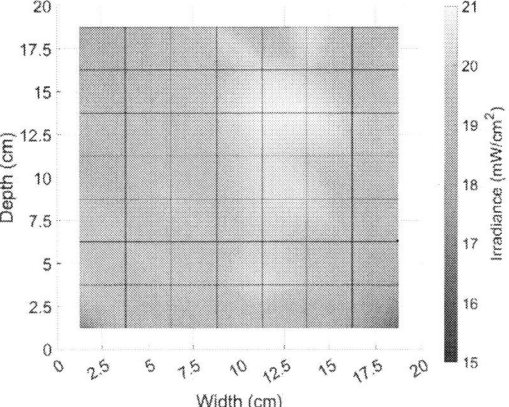

Figure 3: Measured spatial distribution of UVA irradiance in the sample plane.

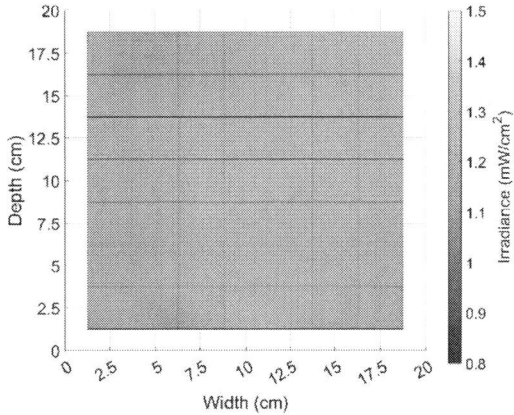

Figure 4: Measured spatial distribution of UVB irradiance in the sample plane.

2.3 Samples and characterization procedures

In a first step, samples of different polymeric foils are tested to determine their optical transmittance and suitability as interlayer base materials: A polyethylene naphthalate (PEN) foil, a polycarbonate (PC) foil and two polyethylene terephthalate (PET) foils. The unencapsulated samples are exposed to UV light for approximately 15 kWh/m² in the UV setup described in section 2.2. Transmittance spectra are measured on the exposed and unexposed foils to determine if there is any material degradation resulting in transmittance reductions following a previously established methodology [9]. An identical analysis is carried out for a PET foil coated by Stensborg A/S with the acrylate-based, nanoimprinted resin diffuser structure to assess any potential increase in optical losses.

In a second step, PV mini-modules are manufactured with two different encapsulants – EVA (3M EVA9110T) and POE (3M PO8110) – using standard low-iron float glass, Al-BSF cSi cells and a black polymeric backsheet. Samples contain PET films coated with either SiO_2 or TiO_2 between two layers of encapsulant between the glass superstrate and the PV cells. Additionally, reference samples containing uncoated PET foils are tested to assess any potential accelerated degradation due to material interactions. The metal oxides were deposited by SDU using PVD in a roll-to-roll process, however achieving a homogeneous layer of TiO_2-coating proved to be difficult due to high sputtering temperatures.

Separate samples undergo either UV degradation (~15 kWh/m^2) or DH storage (1000 h and 2000 h) with IV-measurements before and after to assess performance degradation. Furthermore, EQE measurements are carried out in multiple points of the PV cell to assess any change in the spectral response of the device.

3 RESULTS

3.1 Transmittance of base polymers

Figure 5 shows aging and associated transmission losses in all polymer samples, with highest losses in PC foils. PEN foils also show significant losses as well as an overall poor transmission in the UV region. For both PEN and PC foils, clearly visible yellowing of the polymers could be observed. PET foils show lowest losses in the UV region, with constant overall losses for visible and NIR wavelengths, resulting in no discernible yellowing.

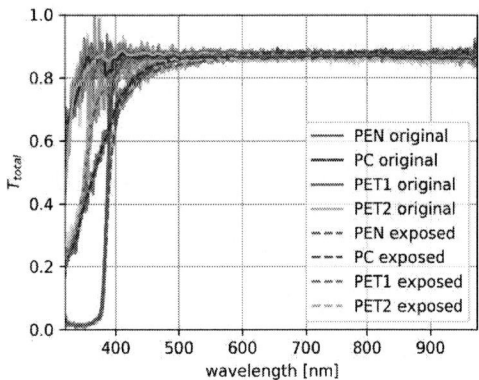

Figure 5: Transmittance before and after UV exposure for base polymer foils.

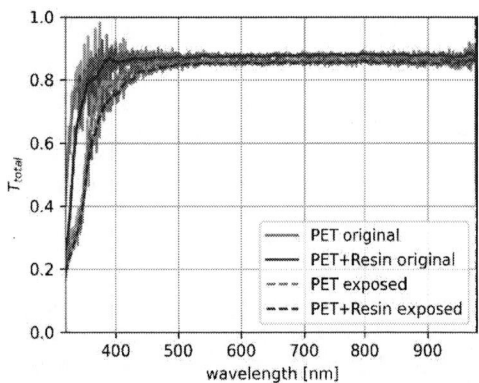

Figure 6: Transmittance before and after UV exposure for resin-coated PET foils.

Subsequently PET foils were coated with different resin formulations and underwent UV exposure as well. The results in Figure 6 show only slight reductions in

overall transmittance – mainly for UV wavelengths – and no significant increase of transmission losses due to degradation for resin-coated samples.

3.2 Reliability of metal-oxide coated polymers

Performance losses in PV mini-modules with metal-oxide coated PET foils after UV exposure are shown in Figure 7. In this case, reference samples as well as samples with SiO_2-coated PET films show 1-2 % losses in both short-circuit current and open-circuit voltage. TiO_2-coated samples, on the other hand, show significantly higher transmission losses, as the Isc decreases by 4-8 %. Losses can be observed for all samples regardless of encapsulant used, with differences between the POE and EVA-containing samples. Due to the previously mentioned inhomogeneous TiO_2 coating, which varies between samples, no final conclusions can be drawn regarding the interaction with different encapsulants, however.

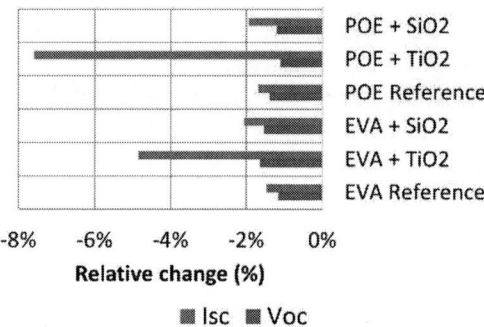

Figure 7: Performance loss of samples with metal oxide-coated PET foils after UV exposure.

After DH storage, performance losses are similar across all samples, as shown in Figure 8. While higher degradation can be observed in POE-containing samples, the relative losses are overall small. No discernible differences exist between measurements after 1000 h and 2000 h of DH storage for most samples.

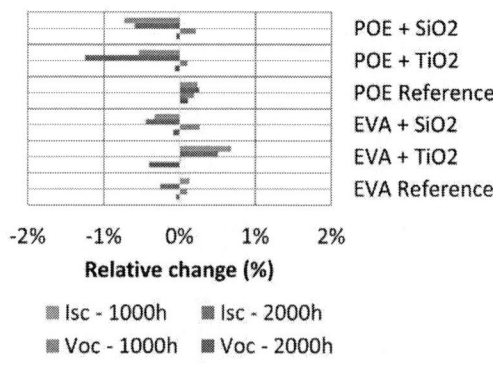

Figure 8: Performance loss of samples with metal oxide-coated PET foils after DH exposure.

The increased transmission losses after UV exposure in the TiO_2 containing samples can tentatively be attributed to photocatalytic degradation of the encapsulant [8], though further research is needed to confirm this. Alternatively, TiO_2 could interact with the PET film, decreasing its transmittance.

Based on EQE measurements, large differences in increased transmission losses were observed for different measurement points. This indicates that differences in coating thickness or surface morphology caused by inhomogeneity in the coating may have an effect on this degradation.

Figure 9 shows the average EQE over 7 measurement points in nearly identical locations on the sample surface before and after UV exposure. It shows some losses between 400 and 500 nm, which is likely due to encapsulant yellowing. In addition, significant efficiency reductions were observed around 1000 nm, possibly indicating a new degradation mode, as typically UV degradation of encapsulants results in even transmission losses over the entire spectral range.

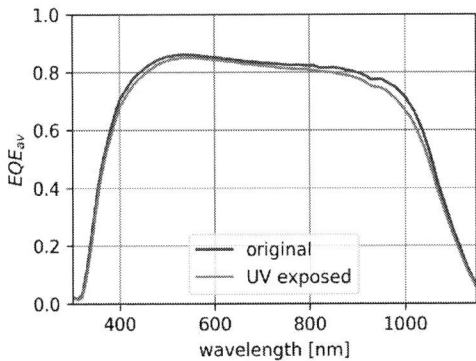

Figure 9: Average measured EQE before and after UV exposure of the POE TiO_2 sample.

Results of degradation testing of different colouration materials including peel testing at the glass-encapsulant interface have been reported in [10].

4   CONCLUSIONS

An investigation of UV stability of materials for use in novel structural colour interlayers was carried out, showing PET as the best candidate for the base carrier material among the tested polymers, both due to high base transmittance and low losses upon UV exposure. Coating the PET foil with a resin-based, nanoimprinted diffuser structure did not lead to significantly increased optical losses or accelerated degradation.

For the metal oxides in the structural colour thin-film stack, PET foils coated with SiO_2 and TiO_2 were tested in combination with EVA and POE encapsulants. UV exposure showed significant transmission losses in samples containing TiO_2, while no significant losses could be observed in DH storage, or either case for samples containing SiO_2. The observed degradation is tentatively attributed to photocatalytic degradation of the encapsulants, though confirmation is required, with a need for further measurements.

Based on these results, alternatives for TiO_2 as thin-film material in the structural colour stack are being investigated. In addition, samples with the full thin-film stack on top of the diffuser structure will be tested to investigate material compatibility within the entire laminate.

This work has also demonstrated a new LED-based UV aging setup with good spectral and spatial adherence to the requirements in the IEC 61215 standard. It allows accelerated degradation testing while minimizing parasitic heating of samples.

Alongside the testing activities, work on upscaling the thin-film deposition is being carried out using SDU CAPE's roll-to-roll (R2R) PVD equipment. Work is underway on matching the small-scale and large-scale results as well as on reducing the losses under upscaling. Furthermore, simulations are being carried out to assess the effect of various diffuser structures on iridescence and to optimize thin-film stacks for a larger variety of colours.

5   ACKNOWLEDGEMENTS

This work was funded by EUDP as part of the "ColorFoil" project under grant 64022-1027.

6   REFERENCES

[1] G. Eder, G. Peharz, R. Trattnig, P. Bonomo, E. Saretta, F. Frontini, C.S. Polo López, H.R. Wilson, J. Eisenlohr, N.M. Chivelet, S. Karlsson, N. Jakica, A. Zanelli, "Coloured BIPV – Market, Research and Development", International Energy Agency, 2019, Report IEA-PVPS T15-07:2019.
[2] Swissinso, "Kromatix", https://www.swissinso.com/technology/ (snapshot 05-July 2022 via https://web.archive.org/)
[3] B. Bläsi, T. Kroyer, O. Höhn, M. Wiese, C. Ferrara, U. Eitner, T.E. Kuhn, "Morpho Butterfly Inspired Coloured BIPV Modules", 33rd EUPVSEC, 2017, doi: 10.4229/EUPVSEC20172017-6BV.3.70.
[4] B. Bläsi, T. Kroyer, T.E. Kuhn, O. Höhn, "The MorphoColor Concept for Colored Photovoltaic Modules", IEEE J.Photovolt., Vol. 11, Nr. 5, 2021, doi: 10.1109/JPHOTOV.2021.3090158.
[5] Ceramic Colors Wolbring GmbH, "ColorQuant", https://colorquant.ceramic-colors.de/en/ (accessed 29-Jan 2024)
[6] Lenzing Plastics, "Colored photovoltaic encapsulation film", https://www.lenzing-plastics.com/en/markets-and-solutions/electronics-and-cable-industry/colored-photovoltaic-encapsulation-film/ (accessed 29-Jan 2024)
[7] M. Aghaei, A. Fairbrother, A. Gok, S. Ahman, S. Kazim, K. Lobato, G. Oreski, A. Reinders, J. Schmitz, M. Theelen, P. Yilmaz, J. Kettle, "Review of degradation and failure phenomena in photovoltaic modules," Renew. Sustain. Energy Rev., vol. 159, May 2022, Art. no. 112160, doi: 10.1016/j.rser.2022.112160.
[8] I. Nabi, A.-U.-R. Bacha, F. Ahmad, and L. Zhang, "Application of titanium dioxide for the photocatalytic degradation of macro- and micro-plastics: A review," J. Environ. Chem. Eng., vol. 9, no. 5, Oct. 2021, art. no. 105964, doi: 10.1016/j.jece.2021.105964.
[9] M. Babin, S. Thorsteinsson, A.A. Santamaria Lancia, P.B. Poulsen, A. Thorseth, C. Dam-Hansen, M.L. Jakobsen, "Dependency of IAM Losses in Colored BIPV Products on the Refractive Index of Colorants", 38th EUPVSEC, 2021, doi: 10.4229/EUPVSEC20212021-4BO.4.2.
[10] R. Boccardi, "Degradation testing of coloured materials for use in PV modules," master thesis, Technical University of Denmark, Roskilde, Denmark, 2024. https://findit.dtu.dk/en/catalog/668496da541df31cdff8147f

41st European Photovoltaic Solar Energy Conference and Exhibition

DIAGNOSING POTENTIAL INDUCED DEGRADATION IN CRYSTALLINE SILICON PHOTOVOLTAIC MODULES

Aysha Mahmood, Rodrigo del Prado Santamaria, Thøger Kari, Peter B. Poulsen and Sergiu V. Spataru.
Department of Electrical and Photonics Engineering, Technical University of Denmark.
Frederiksborgvej 399, 4000, Roskilde, Denmark.

ABSTRACT: In this work, current-voltage (I-V) and electroluminescence (EL) characterization methods are used to diagnose shunting (PID-s) and polarization (PID-p) types of PID in crystalline silicon PV modules. A comparative analysis is carried out on the impact of PID-s and PID-p on the typical diagnostic parameters used in characterizing module degradation. The fraction of the absolute change of the I-V parameters is calculated for PID-degraded PV modules using healthy-state PV modules as references to identify PID sensitive diagnostic parameters. The I-V result for PID-s shows that early-stage degradation can be identified by reduction in I_{MPP} and FF whereas a reduction in V_{MPP} and V_{OC} indicate severe degradation. For PID-p, a reduction in I_{SC} and I_{MPP} can be used to identify early-stage degradation. A deeper insight into the degradation behavior, the cell variability in a PV module, and the current dependency of PID-s and PID-p is achieved by evaluating the mean cell EL intensities from the EL images acquired at multiple current bias levels. From the EL images, PID-s affected PV modules show a higher level of current dependency with increase in cell luminescence mismatch at lower current bias levels as opposed to PV modules that are affected by PID-p.
Keywords: Potential induced degradation (PID), Crystalline silicon (c-Si) photovoltaic (PV) modules, A comparative analysis, Current-Voltage (I-V) characteristics, Electroluminescence (EL) characteristics.

1 INTRODUCTION

Potential induced degradation (PID) can cause significant power loss in photovoltaic (PV) modules that operate in the field with a high system voltage in a relatively short period of time. Power loss up to 53.26 % within one year of operation is documented for PV modules affected by PID in the field [1]. The effect of PID can be partially reversed and its impact on the electrical output of the PV system can be minimized if it is detected in its early stage [1, 2]. However, detecting PID in PV modules with current diagnostic parameters is not always straightforward, as PID exists in different forms and with different degradation signatures depending, among other factors, on the polarity of the voltage stress, cell technology and module bill of materials [1, 2], and can be cofounded in some cases with other degradation modes. Hence, to detect PID of different types, it is necessary to identify the corresponding PID sensitive diagnostic parameters on module level using indoor laboratory based diagnostic methods to minimize variation in the operating condition and eliminate the effect of other degradation modes in the PV system that may affect the electrical performance of the PV module and mask the effect of PID.

PID is caused by the electric field that is generated as a result of a high voltage difference between the operating solar cells and the electrically grounded module frame and surface, causing increased charge recombination losses inside the solar cells [1, 2]. The actual PID mechanism can differ based on the cell technology and its susceptibility to PID [1, 2].

The most widely encountered type of PID, called PID-shunting (PID-s), can affect p-type solar cells in conventional PV modules that operate under negative bias and is well documented in the literature [2]. It is caused by penetration of sodium ions (Na^+) into stacking faults, leading to recombination in the space charge region and shunting of the p-n junction [2]. This mechanism is/may also (be) observed in PV modules based on technologies that have similar front side architecture such as in passivated emitter rear locally diffused (PERL) and passivated emitter rear contact (PERC) PV modules [1].

PID-s is characterized by 1) an increase in second diode saturation current density J_{02} and ideality factor n_2 and 2) a decrease in shunt resistance R_{SH}, fill factor FF, and open circuit voltage V_{OC}, respectively [1, 2]. The latter is mainly used as a diagnostic parameter to identify PID-s in PV modules and is measured at low light condition, a condition where the degradation is more prominent [3]. However, the degradation is not detected by the decrease in the V_{OC} before it becomes severe [3]. On the other hand, extracting the R_{SH} from the single diode model (SDM) is not always possible due to mismatch in the cell performance of the solar cells in a PID affected PV module [3].

Polarization type of PID (PID-p) is/may (be) observed in conventional PV modules under positive bias and in a number of high efficiency (mono or bifacial) PV modules based on technologies such as 1) PERL and PERC: front and rear side degradation when under positive and negative bias, respectively and 2) passivated emitter rear totally diffused (PERT) and tunnel oxide passivated contact (TOPCon): front and rear (PERT only) side degradation when under negative and positive bias, respectively [1]. In the case of PERC rear side degradation, positive charges accumulate within the rear passivation dielectric layer and neutralize it, causing recombination between majority carrier holes and minority carrier electrons [1, 4]. Another proposed mechanism in case of PERC and PERT front side degradation is a change in the net charge of the anti-reflective coating (ARC) layer due to positive or negative charge injection, causing recombination between minority charge carriers and electron or holes [1, 5, 6]. PID-p is characterized by 1) decrease in short circuit current I_{SC} and V_{OC} and 2) increase in first diode saturation current density J_{01} and ideality factor n_1 [1, 11]. This type of PID is not easily detected and/or differentiated from other degradation modes that show similar effect on the electrical parameters [3]. In addition, PID-p does not always show severe degradation at low light conditions [7].

Other identified PID types are: 1) Na^+ penetration type: has similar mechanism as for PID-s but does not cause shunting of the p-n junction and 2) delamination

41st European Photovoltaic Solar Energy Conference and Exhibition

(PID-d) and corrosion (PID-c): are caused by electrochemical reactions/processes within the PV modules in presence of moisture and are characterized by decrease in I_{SC} and increase in series resistance R_S [1].

Electroluminescence (EL) imaging is another effective method used to detect the presence of PID [8]. In the case of PID-s, a qualitative inspection of an EL image acquired at 10 % I_{SC} shows that the PID affected solar cells appear less luminescent than the healthy cells and are located near the module frame [8]. The overall reduction in the EL signal of a cell is due to loss of the injected minority carriers to the shunted areas and decrease in the cell voltage [9]. At a high current bias level (i.e. I_{SC}), only the shunted areas are visibly darker (unless a cell is severely degraded) [9]. For PID-p, the degradation is primarily identified by a homogenous reduction in the EL intensity of PID affected solar cells from EL images captured at I_{SC} [1]. The extent of degradation at lower current bias level is, however, not clearly documented in the literature.

A relatively new approach is to statistically analyze the pixel level EL intensity distributions of the EL images and identify PID signatures on both module and cell level [3].

This paper investigates the impact of shunting and polarization type of PID on the PV module performance with the purpose of identifying PID signatures for diagnosing the presence and type of PID in field degraded PV modules at an early degradation stage. More specifically, it quantitatively examines: 1) the effect of PID-s and PID-p on the electrical parameters of the I-V characteristic curves at high and low light conditions for PV module with different level of degradation; 2) the light dependency of PID-s and PID-p; and 3) the effect of PID-s and PID-p on the EL signal from the PV modules/solar cells at different current bias levels (i.e. the mismatch in cell luminescence and current dependency of PID-s and PID-p).

2 EXPERIMENTAL

2.1 Overview of the data set

The study is carried out on a set of full-size crystalline silicon (c-Si) PV modules of different cell and module technologies, bill of materials (BOM) and number of cells, referred as Type A to M (Table I), obtained from several PID stress testing experiments.

The PV modules are stressed under high voltage to induce PID, using different test conditions, summarized in Table II. The resulting data consist of I-V characteristics curves and EL images measured before (or of a healthy reference module) and after (or of a field degraded module) the PID stress tests.

Table I: Details of the c-Si PV modules. Al-BSF = Aluminium back surface field, G/G = bifacial module with front and rear cover glass, G/BS = mono facial module with front cover glass and backsheet, EVA = ethylene vinyl acetate (encapsulant), POE = polyolefin elastomer (encapsulant), mc = multi crystalline cells.

Type	Cell	Module BOM	Cell type
A, B, C	Al-BSF	G/BS, EVA	p-type
D, E	Al-BSF	G/BS, EVA	p-type, mc-Si
F	PERC	G/BS	p-type
G, H	PERC	G/G	p-type
M	TOPCon	G/G, EVA	n-type

Table II: Details of the test conditions. UNB = under negative bias, RH = relative humidity, h = hours, * = using environmental chamber, ** = using Aluminium (Al) foil method as described in IEC TS 62804-1 2015 [12].

Type	Degradation/test condition
A	Field degraded, UNB
B, C, D, E	60 °C, 85 % RH, -1000 V*
F, G	25 °C, +/- 1500 V, up to 840 h**
H	85 °C, 85 % RH, -1500 V, 1000 h*
M	60 °C, -1000 V, 96 h**

2.2 I-V characterization

The I-V characteristic curves are acquired in accordance with IEC 60904-1 2020 and IEC 60904-1-2 2019 using a triple A sun simulator at standard test condition (AM 1.5, 25 °C, 1000 W/m^2, STC) and at a lower irradiance level (200 W/m^2) [13, 14]. The I-V characteristic curves are corrected using correction procedures from IEC 60891 2021 [15]. The fraction of absolute change of the extracted electrical parameters and the bifaciality factors are determined. A threshold of 5 % of power loss is used to identify the degradation.

A three-parameter extraction method from [16] is applied to determine the SDM parameters: R_S, R_{SH} and n using IV characteristic curve data acquired at STC and 200 W/m^2.

2.3 EL characterization

The EL images are acquired as described in IEC TS 60904-13 using an InGaAs camera at different current bias levels: 0.1, 0.2, 0.3, …, 0.9 I_{SC} and I_{SC} in a temperature controlled dark room [9]. The pixel level EL intensity distributions of the solar cells in a PV module are extracted from the cell segmented EL images [10], statistically analyzed for PID signatures and correlated with power loss.

3 RESULT AND DISCUSSION

3.1 PID-s signatures and diagnostic parameters
3.1.1 Front side degradation under negative bias − mono facial PV module

Fig. 1 shows the impact of PID-s on the I-V characteristic curve of a conventional mono facial p-type mono c-Si Al-BSF PV module that is field degraded under negative bias (Type-A-2). Severe degradation is observed when compared to the I-V characteristics curve of the healthy PV module (Type-A-1, the initial measurements for Type-A-2 were not available). The following observation is made between the healthy (blue curve) and the PID affected (red curve) PV modules: 1) reduction in module maximum power output P_{MAX}, the corresponding I_{MPP} and V_{MPP}, slope near the I_{SC} which indicate reduction in R_{SH}, V_{OC} and FF; and 2) no significant change in I_{SC} and slope near the V_{OC} (i.e. R_S).

The extraction of the SDM parameters confirms the reduction in the R_{SH}. The R_{SH} drops from 2683.4 $\Omega \cdot cm^2$ to 206.4 $\Omega \cdot cm^2$, indicating formation of shunting path(s) across the p-n junction and loss of light generated minority carriers. Moreover, an increase in the ideality factor from 1.039 to 4.638 is determined, revealing recombination process in the space charge region. The change in the R_S is less significant (increases from 1.14 $\Omega \cdot cm^2$ to 1.37 $\Omega \cdot cm^2$).

41st European Photovoltaic Solar Energy Conference and Exhibition

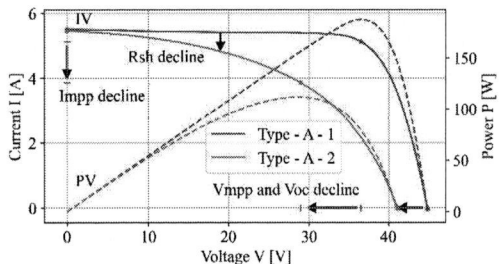

Figure 1: I-V characteristic and PV curves of mono facial p-type mono c-Si Al-BSF PV modules – reference module: Type – A – 1 (blue) and PID degraded module: Type – A – 2 (red) at STC. Showing effects of PID on the I-V and PV curves and the affected I-V parameters.

Table III: Relative change in the I-V parameters of Type - A - 2 (field degraded) PV module using Type - A - 1 (healthy) PV module as a reference, at STC and low light condition (200 W/m²). * = a relative change > 5 % when compared between high and low light condition.

Front	1000 W/m²	200 W/m²
ΔI_{SC} [%]	- 0.73	- 0.2
ΔV_{OC} [%] *	- 8.17	- 33.7
ΔI_{MPP} [%]	- 24.55	- 28.88
ΔV_{MPP} [%] *	- 20.91	- 52.96
ΔP_{MAX} [%] *	- 40.33	- 66.55
ΔFF [%] *	- 34.54	- 49.44

The fraction of the absolute change of the I-V parameters between the degraded (Type-A-2) and the healthy (Type-A-1) PV modules are calculated and shown in Table III. The Type-A-2 PV module shows a power loss of 40.33 % at STC. The main contributor to the power loss is most likely the reduction in the R_{SH} and increase in the ideality factor n as the loss of the light generated minority carriers has a major impact on the I_{MPP}, FF and V_{OC} (the most affected I-V parameters with 24.55 %, 34.54 % and 8.17 % of decrease, respectively). Due to the drop in the latter, the V_{MPP} also decreases by 20.91 %. The I_{SC} is almost unchanged (decrease 0.73 %).

The effect of PID-s is more severe at low light conditions (Table III). While the I_{SC} and I_{MPP} remains almost the same, the relative change of the V_{MPP}, V_{OC}, FF and the resulting P_{MAX} increases negatively (> 5 % of decrease) at 200 W/m² compared to at STC.

The degradation of this PV module is already very prominent at STC due to its severity. However, PV modules that are only slightly degraded or in the early stage of PID, may not show such noticeable visual effects and change in the I-V parameters at STC and thus, would require to be characterized at lower light conditions (e.g. for Type-H-1 PV module, shown later in this section).

Fig. 2 shows the impact of PID-s on the I-V parameters at different degradation levels, for different types of modules. For PV modules with a power loss < 12 %, the indicator for early stage PID-s is reduction in P_{MAX}, FF and I_{MPP} and no significant change in I_{SC}, V_{MPP} and V_{OC}. The V_{MPP} drops for PV modules that are slightly more degraded (power loss > 12 %). PV modules that have a power loss > 40 %, reduction in P_{MAX}, FF, I_{MPP}, V_{MPP} and V_{OC} and no significant reduction in I_{SC} is observed.

The FF and I_{MPP} turns out to be most sensitive towards change in the P_{MAX} and can reveal early stage PID-s when measured at indoor laboratory condition without having

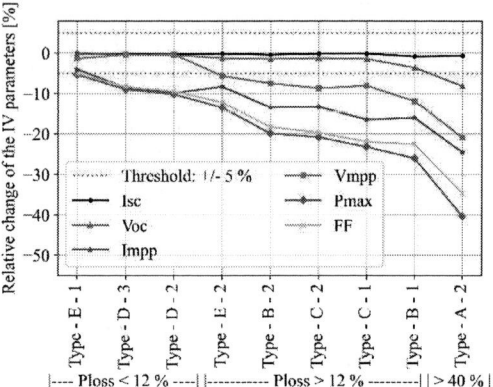

Figure 2: Relative change in the I-V parameters of PID degraded mono facial p-type mono (Type – A, B and C) and multi (Type D and E) c-Si Al-BSF PV modules with different levels of degradation using reference or initial (healthy state) values.

any optical losses as similar effect may be observed when a PV module in the field is shaded and the bypass diodes are activated [3]. Reduction in V_{MPP} and V_{OC}, on the contrary, are mainly an indicator of PID-s severity.

The presence of PID-s is confirmed by the EL measurements. Fig. 3 shows EL images of Type-A PV modules acquired at a high (I_{SC}) and a low (10 % I_{SC}) current bias level. The EL image of Type-A-2 PV module shows homogenous decrease in the EL signal for solar cells that are affected by PID-s. The severely degraded cells appear much darker than the healthy cells already at I_{SC} while other degraded cells become darker at 10 % I_{SC}. Cells near the module frame are most affected by the voltage stress and may have degraded faster than the cells at the center. This type of pattern is similar to what is observed for PID-s in the literature [3, 8, 9]. The reference module (Type-A-1) has cells that are in good condition, except a few that are cracked. The resulting mean EL intensity (shown in Fig. 3) of the PID affected PV module decreases due to reduction in the EL signal from the degraded cells.

To evaluate the cell variability in a PV module and the behavior of PID-s at different current bias levels, the normalized mean cell EL intensities are calculated and plotted against cells that are sorted according to their performance (from best to worst performing) (Fig. 4a, c,

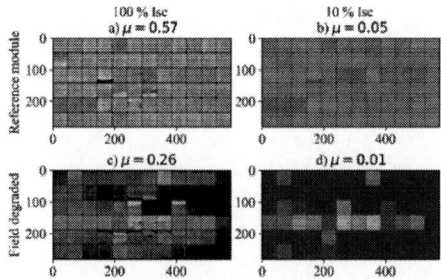

Figure 3: Normalized EL images of reference module Type – A – 1 (top) and PID degraded module Type – A – 2 (bottom) at I_{SC} and 10 % I_{SC} using maximum EL intensity value from EL image acquired at I_{SC}. The corresponding mean EL intensities μ of the PV modules are shown.

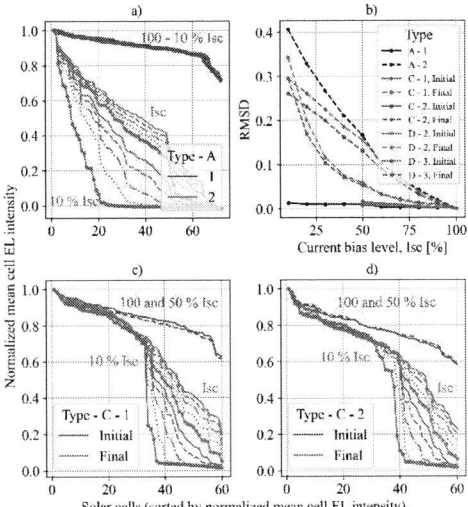

Figure 4: Normalized mean cell EL intensities of solar cells in a) Type − A − 1 (reference module) and Type − A − 2 (PID degraded), c) Type − C − 1 and d) Type − C − 2 at different current bias levels using in-module best functioning cell (fixed at each current bias level). b) RMSD of normalized mean cell EL intensity between I_{SC} and 0.1, 0.2, 0.3, …, 0.9 I_{SC} for mono facial p-type mono (Type − A and C) and multi (Type D) c-Si Al-BSF PV modules.

and d). PV modules that are in a healthy state have cells that emit EL signal of almost the same level. Thus, the mismatch in cell luminescence and the corresponding drop in the normalized mean cell EL intensity curve is only minor (remain above 0.5) (blue curves in Fig. 4a, c and d). Conversely, cells in PID affected PV modules emit different levels of EL signal, showing greater mismatch in cell luminescence and therefore, significant drop (drops below 0.1) (red curves in Fig. 4a, c and d).

The luminescence mismatch increases further at lower current bias levels as more cells emit less/no EL signal compared to the in-module best functioning cell. This current dependency is missing for the healthy state PV modules (blue curves in Fig. 4a, c and d).

Comparing the intensities (red curves, at 10 % I_{SC}) above 0.9 and below 0.1 in Type-A-2 (P_{LOSS} 40.33 %), Type-C-1 (P_{LOSS} 23.25 %) and Type-C-2 (P_{LOSS} 20.85 %) PV modules at the degraded state, show that few severely degraded cells may cause more power loss than more cells that are less affected. The PV modules have 1, 8 and 5 cells nearly identical to the in-module best functioning cell and 54, 24 and 20 cells that are severely degraded, respectively.

To quantify the current dependency, the root mean square deviation (RMSD) is calculated between the normalized mean cell EL intensities at I_{SC} and 0.1, 0.2, 0.3, …, 0.9 I_{SC} as described in [10] for both the healthy and degraded PV modules (Fig. 4b). The RMSD for healthy PV modules is nearly constant at all current bias levels and remains below 0.05, showing no current dependency. For the degraded PV modules, the RMSD increases inversely with the current bias level and gets above 0.1. However, the trend seems to change from an exponential to a more linear increase as the degradation becomes severe (increase in P_{LOSS}). In addition, the RMSD increases with P_{LOSS} of the PV modules at current bias level between 20 % and 50 % I_{SC}.

Table IV: Relative change in the I-V parameters of Type - H - 1 PV module at final/degraded state using initial/healthy state values as reference values, at STC and low light condition (200 W/m²). * = a relative change > 5 % when compared between high and low light condition.

Front	1000 W/m²	200 W/m²
ΔI_{SC} [%]	- 0.56	- 0.05
ΔV_{OC} [%]	- 0.82	- 2.11
ΔI_{MPP} [%]	- 3.97	- 7.53
ΔV_{MPP} [%] *	- 1.7	- 7.97
ΔP_{MAX} [%] *	- 5.6	- 14.9
ΔFF [%] *	- 4.28	- 13.02
Rear		
ΔI_{SC} [%]	1.62	1.8
ΔV_{OC} [%]	- 0.84	- 2.6
ΔI_{MPP} [%]	- 1.3	- 4.34
ΔV_{MPP} [%] *	- 3.5	- 10.35
ΔP_{MAX} [%] *	- 4.76	- 14.24
ΔFF [%] *	- 5.51	- 13.51

3.1.2 Front side degradation under negative bias – bifacial PV module

Table IV shows the impact of PID on the I-V parameters of Type-H-1 PV module, a bifacial p-type mono c-Si PERC PV module that is degraded under negative bias. The front side is slightly degraded with a P_{LOSS} of 5.6 % and is mainly caused by decrease in FF (4.28 %) and I_{MPP} (3.97 %), suggesting shunting type of PID (PID-s). The degradation is more significant at lower light condition and is observed by: 1) decrease in P_{MAX}, I_{MPP}, V_{MPP} and FF from the front side I-V measurement; and 2) decrease in P_{MAX}, V_{MPP}, FF from the rear side I-V measurement.

The P_{MAX} bifaciality factor φ_{Pmax} increases slightly from 0.699 to 0.705, however, we cannot confirm degradation of the front side as the value is within the threshold (70 % +/- 5 %). On the other hand, the I_{SC} bifaciality factor φ_{Isc} does (and should) not change as the I_{SC} is less impacted in the case of PID-s (change from 0.725 to 0.74).

The EL images at the degraded state (Type-H-1) show pattern that is characteristic for PID-s with prominent degradation of cells near the module frame at a lower current bias level and decrease in the mean EL intensity of the PV module (Fig. 5).

Looking at the current dependency plots in Fig. 6c and

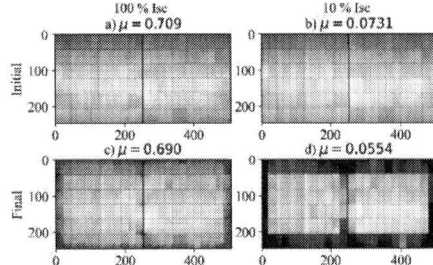

Figure 5: Normalized EL images of Type − H − 1 PV module at initial/healthy state (top) and final/degraded state (bottom) at I_{SC} and 10 % I_{SC} using maximum EL intensity value of EL image at I_{SC}. The corresponding mean EL intensities μ of the PV module are shown.

41st European Photovoltaic Solar Energy Conference and Exhibition

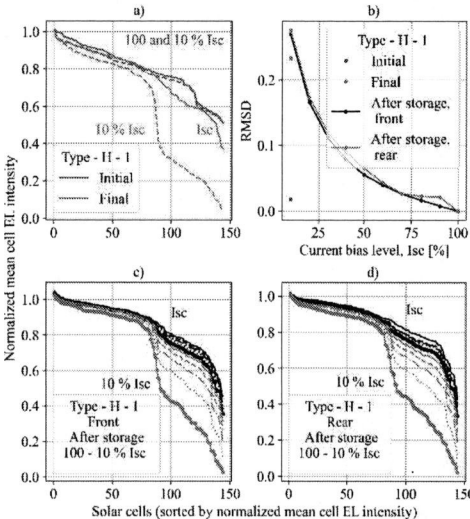

Figure 6: Normalized mean cell EL intensity of solar cells of Type – H – 1 PV module a) at initial and final state and after storage of c) front and d) rear side at different current bias levels using in-module best functioning cell. b) RMSD of normalized mean cell EL intensity between I_{SC} and 0.1, 0.2, 0.3, …, 0.9 I_{SC} front side of Type – H – 1 at initial/healthy state and after storage.

d, no significant difference is observed in the EL intensities from the front and rear side, implying that it is sufficient to only measure the EL signal at the front side of a PV module. Nevertheless, the current dependency trends are consistent with what is observed in the case of PID-s in mono facial PV modules (Fig. 6a and b).

3.2 PID-p signatures and diagnostic parameters
3.2.1 Front side degradation under positive bias

Fig. 7 shows the impact of PID-p on the I-V characteristics curves of a mono facial p-type mono c-Si PERC PV module that is degraded under positive bias by the Al-foil method (Type-F-1). The PV module was slightly field degraded (operated in the field for ~ 2 years) before it was stressed with the high voltage. However, no significant change is observed from the I-V measurements of the PV module before and after the high voltage stress test. A healthy PV module of same cell technology and module construction (never deployed in the field) is used as a reference module to identify PID-p and PID-p sensitive I-V parameters. The following is observed

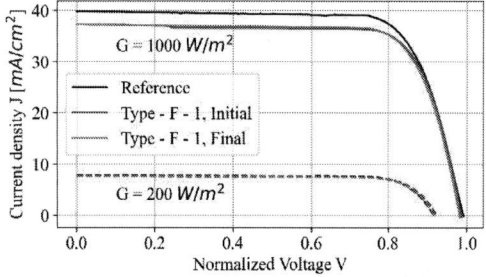

Figure 7: I-V characteristic curves of mono facial p-type mono c-Si PERC PV modules – reference module (black), initial (blue) and final (red) state of Type – F – 1. The voltage is normalized by 40 V. Showing effects of PID on the I-V curves at high (STC) and low light conditions (200 W/m²).

Table V: Relative change in the I-V parameters of Type – F – 1 PV module at final/degraded state using reference module I-V parameters as reference values, at STC and low light condition (200 W/m²).

Front	1000 W/m²	200 W/m²
ΔJ_{SC} [%]	- 6.28	- 3.1
ΔV_{OC} [%]	- 0.98	0.03
ΔJ_{MPP} [%]	- 6.28	- 3.56
ΔV_{MPP} [%]	0.47	0.82
ΔP_{MAX} [%]	- 5.84	- 2.77
ΔFF [%]	1.46	0.31

between reference (black curve) and final state Type-F-1 (red curve) PV module at STC: 1) reduction in P_{MAX}, J_{SC} and J_{MPP}; and 2) no significant change in V_{OC}, V_{MPP}, FF and slopes near J_{SC} and V_{OC} (i.e. R_{SH} and R_S). The J_{SC} and J_{MPP} contribute equally to P_{LOSS} (5.84 %) with 6.28 % of decrease (Table V). There is no clear difference between the two curves and the corresponding I-V parameters at a lower light condition (200 W/m²) (Fig. 7 and Table V). This is, however, different from what is reported in the literature for a PV module that is induced with PID in an environmental chamber (+1500 V, 25 °C, 54 % RH, 168 h): a more severe degradation of P_{MAX}, I_{SC}, V_{OC} and FF are observed at a lower light condition [17].

The shunting type of PID can be excluded as the V_{MPP} and FF are not affected. In addition, the extraction of the SDM parameters of reference and Type-F-1 module shows no significant change in R_{SH} and n. The SDM parameters R_S, R_{SH} and n change from 1.05 $\Omega \cdot cm^2$, 2180.83 $\Omega \cdot cm^2$ and 1.25 to 1.03 $\Omega \cdot cm^2$, 5114.96 $\Omega \cdot cm^2$ and 1.11, respectively (note: the reference and Type-F-1 PV modules have a different datasheet power rating of 295 W and 305 W, respectively).

Hence, the degradation can be attributed to PID-p and may be caused by migration of negative charges into the ARC layer and surface recombination between electron and minority carrier holes [17]. PID-p can be identified with reduction in J_{SC} (or I_{SC}) and J_{MPP} (or I_{MPP}) at STC.

Although the degradation in Type-F-1 PV module is identified using a reference module, the impact of the high positive voltage stress on the PV module is not very clear from the I-V measurements. The EL measurement shows a clear difference in the EL signal before and after the high voltage stress test (Fig. 8). The EL images of Type-F-1 PV module at degraded (final) state show both homogeneous and local decrease (show gray scale gradients) in the EL signal of solar cells that are affected by PID-p at 90 % I_{SC} (Fig. 8c and d). This is similar to what is observed in the literature at I_{SC} [17]. At a lower current bias level (10 % I_{SC}), the EL signal of the cells reduces homogeneously (Fig. 8e and f). Only a few cells, located near the module frame, show severe degradation.

The mismatch in cell luminescence increases in Type-F-1 PV module at degraded (final) state (Fig. 9a). The EL intensity curves at degraded (final) state (red curves) drop below 0.3 whereas the EL intensity curves for reference (black curves) and Type-F-1 PV module at initial state (blue curves) remain above 0.5.

The mismatch in cell luminescence increases at lower current bias levels but is substantially smaller than in the case with PID-s. This may be due to the fact that: 1) the PV module is only slightly degraded and therefore, the performance of cells does not change much compared to the in-module best functioning cell at lower current bias levels; or 2) the absence of shunting paths prevents

640

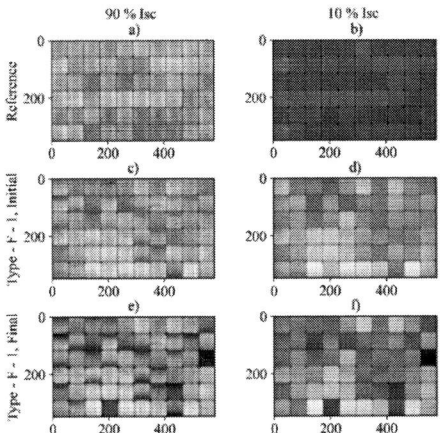

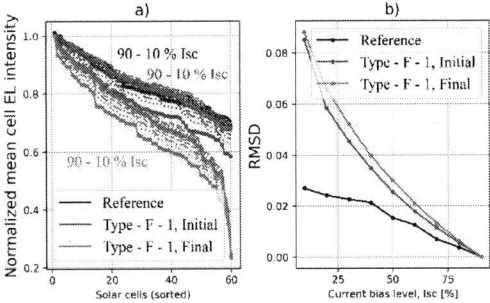

Figure 8: Normalized EL images of Type – F – 1 PV module at initial/healthy (top), intermediate (middle) and final/degraded state (bottom) at 90 % I_{SC} and 10 % I_{SC} using maximum EL intensity value of EL image at 90 % I_{SC} ($I_{SC} = 9.85$ A).

Figure 9: Normalized mean cell EL intensity of solar cells of a) Type – F – 1 at different current bias levels using in-module best functioning cell. The solar cells are sorted by normalized mean cell EL intensity. b) RMSD of normalized mean cell EL intensity between 90 % I_{SC} and 0.1, 0.2, 0.3, ..., 0.8 I_{SC} current bias levels.

additional loss of minority carriers and reduction in cell voltage and EL signal thus, characteristic for PID-p. The RMSD of Type-F-1 PV module at initial and final state gets above 0.05 however, remains below 0.1 (Fig. 9b). The RMSD of reference module is still below 0.05 at all current bias levels as in the case with PID-s. Nevertheless, the relative change in the RMSD curve between the final state and the reference at 10 % I_{SC} is still high.

3.2.2 Rear side degradation under negative bias

Fig. 10 shows the impact of PID-p on the I-V characteristic curves measured at the rear side of Type-G-1 PV module, a bifacial p-type mono c-Si PERC PV module that is degraded under negative bias by the Al-foil method. The degradation is not very significant at the front side (Table VI). The following is observed between the initial (blue curve) and final (red curve) state at the rear side I-V measurement (Fig. 10): 1) reduction in P_{MAX}, I_{SC}, I_{MPP}, and FF; and 2) no significant change in V_{OC}, V_{MPP}, and slopes near I_{SC} and V_{OC} (i.e. R_S and R_{SH}).

The front side of Type-G-1 PV module can be affected by PID-s as seen in section 3.1.2. However, no significant

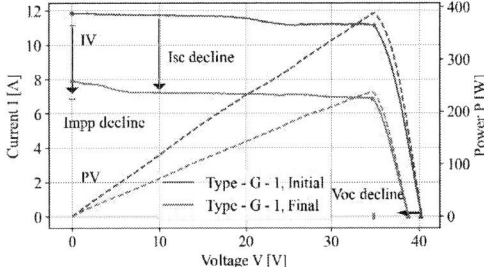

Figure 10: I-V characteristic and PV curves of bifacial p-type mono c-Si PERC PV module – initial (blue) and final (red) of Type – G – 1, rear side. Showing effects of PID on the I-V and PV curves and the affected I-V parameters.

Table VI: Relative change in the I-V parameters of Type - G - 1 PV module at final/degraded state using initial/healthy state values as reference values, at STC and low light condition (200 W/m²). * = a relative change > 5 % when compared between high and low light condition.

Front	1000 W/m²	200 W/m²
ΔI_{SC} [%]	- 2.92	- 3.95
ΔV_{OC} [%]	- 2.63	- 4.08
ΔI_{MPP} [%]	- 2.94	- 4.32
ΔV_{MPP} [%]	- 3.4	- 3.3
ΔP_{MAX} [%]	- 6.24	- 7.48
ΔFF [%]	- 0.81	0.42
Rear		
ΔI_{SC} [%] *	- 33.2	- 43.3
ΔV_{OC} [%]	- 3.62	- 6.44
ΔI_{MPP} [%] *	- 38.54	- 44.4
ΔV_{MPP} [%]	- 0.6	- 4.06
ΔP_{MAX} [%] *	- 38.91	- 46.66
ΔFF [%]	- 5.11	0.55

change is observed in V_{MPP} and FF from the front side measurement (Table VI). The rear side degrades with a P_{LOSS} of 38.91 % and is mainly caused by decrease in I_{SC} (33.2 %) and I_{MPP} (38.54 %), suggesting polarization type of PID. The PID-p may be caused by migration of positive charges into rear AlO$_X$ layer and surface recombination between holes and minority carrier electrons [4].

The rear side degradation is much more severe at a lower light condition and is observed by: 1) decrease in P_{MAX} from the front side I-V measurement; and 2) decrease in P_{MAX}, I_{MPP}, I_{SC} and V_{OC} from the rear side I-V measurement. The rear side degradation is also confirmed by reduction in the bifaciality factors φ_{Pmax} (from 0.69 to 0.45) and φ_{Isc} (from 0.67 to 0.46).

The I_{SC} and I_{MPP} are the most PID-p sensitive I-V parameters in the case of both Type-F-1 and Type-G-1 PV modules. Reduction in V_{OC} may be mainly an indicator for PID-p severity ($P_{LOSS} > 40$ %) as it is only observed for Type-G-1 PV module at lower light condition (P_{LOSS} of 46.66 %).

Similar to Type-F-1, the Type-G-1 PV module is a field-aged module (has operated in the field for ~ 2 years) and was slightly field degraded before it was stressed with the high voltage (Fig. 11a and b). Fig. 11c and d show a clear reduction in the EL signal after the high voltage stress test. The effect of PID-p on the EL signal and the resulting EL pattern is similar to what is observed for Type-F-1 PV modules.

41st European Photovoltaic Solar Energy Conference and Exhibition

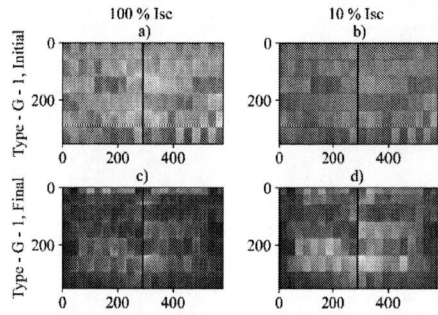

Figure 11: Normalized EL images of Type – G – 1 PV module at initial/healthy state (top) and final/degraded state (bottom) at I_{SC} and 10 % I_{SC} using maximum EL intensity value of EL image at I_{SC}.

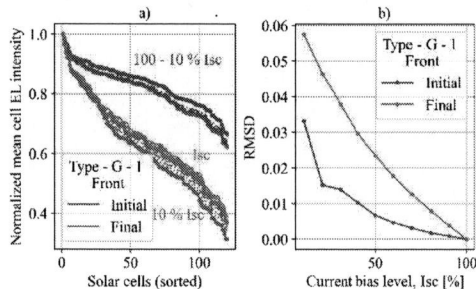

Figure 12: Normalized mean cell EL intensity of solar cells of a) Type – G – 1 at different current bias levels using in-module best functioning cell. The solar cells are sorted by normalized mean cell EL intensity. b) RMSD of normalized mean cell EL intensity between I_{SC} and 0.1, 0.2, 0.3, ..., 0.9 I_{SC}.

The following is observed when comparing the I-V and EL results of Type-H-1 (front side PID-s with P_{LOSS} of 5.6 %) and Type-G-1 (rear side PID-p with P_{LOSS} of 38.91 %), both bifacial p-type mono c-Si PERC PV modules:

1) The impact of PID-s on the EL emission is much more severe. Considering the level of P_{LOSS}, the Type-H-1 PV module has severely degraded cells near the module frame that emit almost no EL signal at 10 % I_{SC} (Fig. 5h);

2) the mismatch in cell luminescence is highly dependent on the severity of the degradation. If all cells are degraded in a PV module, the difference in EL intensities between the in-module best functioning cell (degraded) and other degraded cells will be minor (i.e. low luminescence mismatch with a minor drop in the EL intensity curve). Thus, this alone cannot identify the level of degradation. The magnitude of the drop in the EL intensity curves is nearly the same for Type-H-1 and Type-G-1 PV modules (Fig 6a and 12a).

3) the increase in luminescence mismatch at lower current bias levels is much smaller for PID-p than for PID-s (significant difference in RMSD between I_{SC} and 10 % I_{SC}) (Fig. 6a,b and 12a,b). This may be characteristic of PID-p as discussed previously, and not due to the severity of PID.

3.2.3 Front side degradation under negative bias

Fig. 13 shows the impact of PID-p on the I-V characteristic curve of a bifacial n-type mono c-Si TOPCon PV module that is degraded under negative bias (Type-M-1). The following is observed between initial

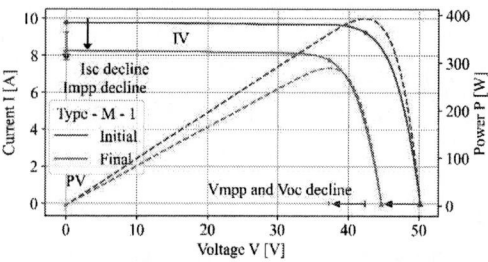

Figure 13: I-V characteristic and PV curves of bifacial n-type mono c-Si TOPCon PV modules – initial (blue) and final (red) of Type – M – 1. Showing effects of PID on the I-V and PV curves and the affected I-V parameters on the front side.

Table VII: Relative change in the I-V parameters of Type - M - 1 PV module at final/degraded state using initial/healthy state values as reference values, at STC and low light condition (200 W/m²). * = a relative change > 5 % when compared between high and low light condition.

Front	1000 W/m²	200 W/m²
ΔI_{SC} [%]	- 15.62	- 12.34
ΔV_{OC} [%]	- 10.9	- 11.67
ΔI_{MPP} [%]	- 16.23	- 17.98
ΔV_{MPP} [%]	- 12.08	- 12.4
ΔP_{MAX} [%]	- 26.35	- 28.15
ΔFF [%] *	- 2.04	- 7.22

(blue curve) and final (red curve) state at front side I-V measurement: 1) reduction in P_{MAX}, I_{MPP}, I_{SC}, V_{MPP} and V_{OC}; and 2) no significant change in FF and slopes near the I_{SC} and V_{OC} (i.e. R_S and R_{SH}).

The reduction in V_{OC} (10.9 %) is one of the main contributors to the P_{LOSS} (26.35 %), unlike in the two previous cases of PID-p where the V_{OC} only reduces for PV module with $P_{LOSS} > 40$ % (Table VII). This suggests a different underlying mechanism and may be caused by migration of positive charges into the ARC layer leading to recombination between holes and minority carrier electrons [1, 5, 6]. The relative change does not increase significantly for the I-V parameters at a lower light condition, except for the FF (7.22 %).

The EL images of Type-M-1 show an overall reduction of the EL signal at the final state (Fig. 14). However, due to the less mismatch in cell luminescence, the corresponding drop in the EL intensity curves at initial and final state are almost the same (Fig. 15). In addition, no current dependency is observed at initial and final state.

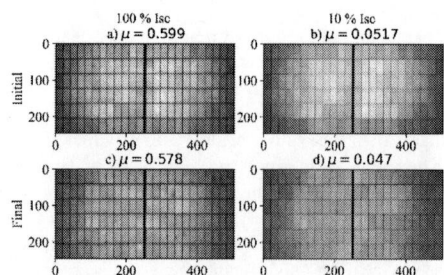

Figure 14: Normalized EL images of Type – M – 1 PV module at initial/healthy state (top) and final/degraded state (bottom) at I_{SC} and 10 % I_{SC} using maximum EL intensity value of EL image at I_{SC}. The corresponding mean EL intensities μ of the PV module are shown.

642

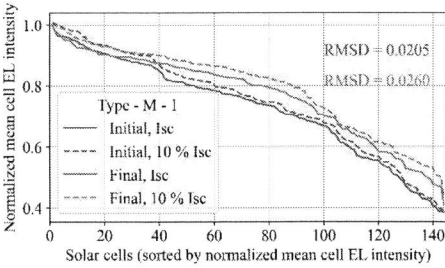

Figure 15: Normalized mean cell EL intensity of solar cells of Type − M − 1 at different current bias levels using in-module best functioning cell. The RMSD of normalized mean cell EL intensity between I_{SC} and 10 % I_{SC} is given in the plot.

4 SUMMARY AND CONCLUSIONS

In this paper, the impact of PID-s and PID-p on the PV module performance is investigated by the I-V and EL characterization methods. The PID sensitive I-V parameters are identified for both PID-s and PID-p at STC and a lower light condition. Table VIII summarizes the diagnostic I-V parameters for identifying the presence of PID-s and PID-p in degraded PV modules.

The impact of PID on the PV module performance is shown to be dependent on the PV technology and polarity of the voltage stress as different degradation behavior is observed on the I-V characteristics curves for Type-A, F, G and M. Thus, a PV module of different technology, bill of materials or construction may have a completely different set of PID sensitive I-V parameters. More data of different PV technologies and with different levels of degradation will be required to validate the impact of PID, including for n-type TOPCon PV modules.

The EL measurement shows that the cell luminescence mismatch increases for a PID affected PV module when compared to a PV module in a healthy state (for PV modules that have at least one healthy cell). Cells near the

Table VIII: Diagnostic parameters for PID-s and PID-p. PID affected (x) and unchanged (-) parameters at different level of PID: early stage (E, P_{LOSS} < 12 %), mild (M, P_{LOSS} > 12 %) and severe (S, P_{LOSS} > 40 %) degradation. * = front side degradation under negative bias in p-type, ⃰ = front side degradation under positive bias in p-type, ⃰ = front side degradation under negative bias in n-type, ⃰ = rear side degradation under negative bias in p-type. Affected parameter: $\Delta_{relative}IV > \pm 5\,\%$, $\Delta_{relative}SDM > \pm 50\,\%$, $\varphi_{I_{SC}}, \varphi_{P_{MAX}} < 65\,\%$ or > 75 %.

	PID-s			PID-p		
	E*	M*	S*	E⃰	M⃰	S⃰
P_{MAX}	x	x	x	x	x	x
I_{SC}	-	-	-	x	x	x
V_{OC}	-	-	x	-	x	x
I_{MPP}	x	x	x	x	x	x
V_{MPP}	-	x	x	-	x	-
FF	x	x	x	-	-	-
R_S			-	-		
R_{SH}			x	-		
n			x	-		
φ_{Pmax}	-					x
φ_{ISC}	-					x

module frame are more affected than the cells located at the center and show homogenous decrease in the EL signal. PID-s shows a greater level of current dependency with prominent degradation at a lower current bias level compared to PID-p. The calculated RMSD value between the EL intensity curves at I_{SC} and 10 % I_{SC} is much higher for PID-s (above 0.1) than for PID-p (below 0.1). However, the EL measurement was able to identify degradation in the case of Type-F-1 PV module, whose impact of PID-p on the I-V measurement was not clear after the high voltage stress test.

5 ACKNOWLEDGEMENTS

We would like to acknowledge contributions from Peter Hacke from NREL, Fraunhofer CSP and European Energy for providing PID data. This research has been carried in the Eurostar's project: Automated Daylight Electroluminescence Inspection of Large Photovoltaic Systems (ADELI).

6 REFERENCES

[1] Molto, C., et al., (2023), Review of Potential-Induced Degradation in Bifacial Photovoltaic Modules. Energy Technol., 11: 2200943.

[2] Luo, W., et al., "Potential-Induced Degradation in Photovoltaic Modules: A Critical Review." Energy & Environmental Science 10, no. 1 (2017): 43–68.

[3] Sergiu Viorel Spătaru, Characterization and Diagnostics for Photovoltaic Modules and Arrays, Phd Thesis, Department of Energy Technology, Aalborg University, 2015.

[4] Luo, W., et al., Elucidating potential-induced degradation in bifacial PERC silicon photovoltaic modules. Prog Photovolt Res Appl. 2018; 26: 859–867.

[5] F. i. Mahmood, et al., "Polarization Type Potential Induced Degradation under Positive Bias in a Commercial PERC Module," 2023 IEEE 50th Photovoltaic Specialists Conference (PVSC), San Juan, PR, USA, 2023, pp. 1-3.

[6] Seira Yamaguchi et al., Polarization-Type Potential-Induced Degradation in Front-Emitter p-Type and n-Type Crystalline Silicon Solar Cells, ACS Omega 2022 7 (41), 36277-36285.

[7] Mahmood, F.I., et al., Susceptibility to polarization type potential induced degradation in commercial bifacial pPERC PV modules. Prog Photovolt Res Appl. 2023; 31(11): 1078-1090.

[8] Koester, et al. Review of photovoltaic module degradation, field inspection techniques and techno-economic assessment. Sept. 2022. DOI: 10.1016/j.rser.2022. 112616.

[9] Photovoltaic devices- Part 13: Electroluminescence of photovoltaic modules, DS/IEC TS 60904-13:2018.

[10] R. del Prado Santamaría, et al., "Evaluating Multi-Bias Modulation for Diagnostics of PV Modules in Daylight Electroluminescence Inspections," 2023 IEEE 50th Photovoltaic Specialists Conference (PVSC), San Juan, PR, USA, 2023, pp. 1-1.

[11] Hacke P, Kumar A, Terwilliger K, et al. Evaluation of bifacial module technologies with combined-accelerated stress testing. Prog Photovolt Res Appl. 2023; 31(12): 1270-1284.

[12] Photovoltaic (PV) modules-Test methods for the detection of potential-induced degradation - Part 1: Crystalline silicon, DS/IEC TS 62804-1:2015, 21.09.2015.

[13] Photovoltaic devices – Part 1: Measurement of photovoltaic current-voltage characteristics, IEC 60904-1:2020.

[14] Photovoltaic devices – Part 1-2: Measurement of current-voltage characteristics of bifacial photovoltaic (PV) devices, IEC TS 60904-1-2:2019.

[15] Photovoltaic devices – Procedures for temperature and irradiance corrections to measured I-V characteristics, IEC 60891:2021.

[16] Bowden, S., et al. (2001). Rapid and Accurate Determination of Series Resistance and Fill Factor Losses in Industrial Silicon Solar Cells.

[17] Mahmood, F. I., et al. (2023). Polarization Type Potential Induced Degradation under Positive Bias in a Commercial PERC Module. In *2023 IEEE 50th Photovoltaic Specialists Conference, PVSC 2023* (Conference Record of the IEEE Photovoltaic Specialists Conference). Institute of Electrical and Electronics Engineers Inc.

41st European Photovoltaic Solar Energy Conference and Exhibition

ON-SITE EVALUATION OF OXYGEN-PLASMA TREATED GLASS SURFACES FOR ANTI-SOILING PROPERTIES

Brahim Aïssa[1*], Ayman Samara[2]

[1]Qatar Environment and Energy Research Institute (QEERI), Hamad bin Khalifa University (HBKU), Qatar Foundation, P.O. Box 34110, Doha, Qatar
[2]HBKU Core Laboratories, Hamad bin Khalifa University (HBKU), Qatar Foundation, P.O. Box 34110, Doha, Qatar
*Corresponding author: baissa@hbku.edu.qa

ABSTRACT: The preparation of anti-soiling surfaces was achieved by subjecting commercial glass substrates to oxygen plasma treatment. Surprisingly, even a brief 60-second exposure to the plasma was enough to drastically alter the glass surface, converting it into a superhydrophilic state. Analysis using atomic force microscopy revealed that increasing the duration of the oxygen plasma treatment not only elevated the surface roughness but also enhanced its superhydrophilic properties. When tested under the harsh outdoor conditions typical of the desert environment at the Outdoor Test Facility in Qatar, the plasma-treated glass surfaces demonstrated a significant anti-soiling effect in comparison to untreated samples. This effect was assessed by measuring dust accumulation on the surface, which was quantified using electron scanning microscopy images. Although the promising potential of plasma treatment for reducing soiling is evident, the complex mechanisms behind particle adhesion and surface interactions remain to be thoroughly explored. A deeper understanding of these interactions is essential for fully unlocking the practical benefits of this treatment in real-world applications.
Keywords: Oxygen Plasma, Anti-soiling, Photovoltaic, hydrophilic.

1 INTRODUCTION

Soiling of photovoltaic (PV) modules refers to the accumulation of dust, dirt, and other contaminants on the surface of the modules, leading to substantial energy losses—often exceeding 1% per day—due to the partial shading of sunlight [1]. The soiling layer absorbs, deflects, and reflects portions of the incoming solar radiation, thereby diminishing the amount of sunlight available for conversion into electricity by the PV cells [2,3]. A range of factors influences the deposition and mitigation of soiling [4-6].

Typically, soiling accumulates during dry periods and can be removed either through natural processes or artificial cleaning. As a result, the loss profile due to soiling is often modeled as a saw-tooth wave, reflecting alternating periods of deposition and cleaning events. Rainfall, being the most common natural cleaning mechanism, can significantly reduce soiling. In standard models, soiling is assumed to accumulate at a steady rate during deposition periods, although extreme events such as sandstorms may result in an exponential increase in soiling.

PV systems installed in desert environments face pronounced soiling challenges due to the arid conditions, though these regions are increasingly favored for solar energy projects due to their high levels of solar irradiation. While frequent cleaning is essential to maintaining efficiency, it is often constrained by economic feasibility, which depends on site-specific cleaning costs and power purchase agreements. Even with an optimized cleaning schedule, annual soiling losses can remain significant, typically within the single-digit percentage range [1]. This underscores the critical need for improved soiling mitigation strategies, both passive and active. Potential solutions include anti-soiling coatings, automated cleaning systems, and other innovative technologies. However, many of these technologies are still in early stages of development, and their widespread application remains limited due to a variety of complex and site-specific factors. These factors include environmental conditions, module and plant characteristics, types of soiling, outdoor durability, and the economic viability of large-scale

implementation. Additionally, the fact that these variables operate across different time scales (e.g., daily or seasonal weather patterns) and size scales (e.g., from plant-level features measured in kilometers to particle-surface interactions at the nanometer scale) adds to the complexity of the soiling issue [2]

Compared to other areas of PV research, such as solar cell design, materials innovation, cell degradation, and yield estimation, the problem of soiling remains underrepresented, despite its potentially massive impact on solar energy production. There is, therefore, an urgent need for more fundamental research on natural soiling processes to serve as a foundation for the development of effective mitigation strategies [3-6].

In this paper, we present an investigation into the optical and morphological properties of glass surfaces modified through oxygen plasma treatments conducted under varying power densities and durations. We systematically studied the total transmission and reflectance of these surfaces. Field emission scanning electron microscopy, atomic force microscopy, and 3D-roughness mapping were used to characterize surface morphology. Additionally, contact angle measurements were performed to evaluate the hydrophilicity of the modified surfaces as a function of plasma treatment duration. The promising results of this study suggest that optimizing commercial glass films using this approach could lead to cost-effective antireflection and self-cleaning coatings, enhancing PV performance in diverse environments, including desert regions.

2 METHODOLOGY

For this study, soda lime glass (SLG) with a thickness of 5 mm and dimensions of 2" × 2" was selected as the substrate material. The SLG samples underwent a thorough cleaning process before any experimental procedures were conducted. This cleaning involved immersion in an ultrasonic bath, sequentially using acetone, isopropanol, and deionized water for 10 minutes each. After the cleaning process, the glass was dried using nitrogen gas to ensure the complete removal of any residual moisture or contaminants.

To analyze the optical properties of the films deposited on the SLG, a combination of UV-Vis spectroscopy and ellipsometry was employed. Specifically, a Perkin Elmer™ Lambda 1050 UV/VIS/NIR spectrometer was used to measure the UV-Vis spectra of the samples, providing insight into their transmittance and reflectance characteristics across the ultraviolet, visible, and near-infrared regions. The ellipsometric data were collected using the Horiba UVISEL 2, allowing precise measurements of film thickness and refractive indices. These measurements are critical for understanding how plasma treatment affects the optical transparency of the silica films.

To investigate the surface morphology of the deposited oxides, field emission scanning electron microscopy (FESEM) was conducted using a JEOL 7610 microscope. This method provided high-resolution images of the surface, revealing any changes in texture or structure due to plasma treatment. Additionally, film thickness and roughness were quantified using a Dektak stylus profilometer. This tool enabled detailed three-dimensional surface measurements, including the average roughness of the treated films. Understanding the changes in surface roughness is essential because it directly influences both the optical and wetting properties of the films.

The topographical characteristics of the surfaces were further examined using Bruker™ atomic force microscopy (AFM) operating in contact mode. AFM allows for nanoscale visualization of surface features, giving deeper insights into how plasma treatment modifies the silica layer's microstructure. The AFM measurements were instrumental in correlating the morphological changes with the treatment duration and plasma power settings.

Wettability, specifically the hydrophilicity or hydrophobicity of the silica films, was assessed through contact angle (CA) measurements. This was done using a Kruss™ tool, which measured the contact angle of water droplets placed on the surface of the treated glass. The contact angle serves as a direct indicator of the surface's affinity for water—lower angles indicate hydrophilicity, while higher angles suggest hydrophobicity. For the optimized silica layer, contact angle measurements were conducted 10 times for each sample to ensure accuracy, with each droplet having a volume of 2 μL. These measurements provided critical data on the plasma treatment's effect on the surface's ability to attract or repel water, which has implications for self-cleaning and anti-soiling applications.

The plasma treatment itself was performed under ambient oxygen conditions using an MTI™ compact plasma reactor. Plasma is created through the ionization of oxygen gas, initiated by the application of high-frequency voltages typically ranging from kilohertz (kHz) to megahertz (MHz). This energetic environment causes significant modifications to the surface, including changes in its roughness and chemical composition. In the case of untreated glass, the root-mean-square (rms) roughness of the surface was measured to be 0.9 nm. However, after just 60 seconds of plasma treatment, the rms roughness increased dramatically to 3.9 nm. This change in roughness plays a crucial role in altering the surface's optical properties, as well as its interaction with environmental particles, potentially enhancing the glass's anti-soiling characteristics.

This study provides a detailed understanding of how oxygen plasma treatment can significantly modify the

surface morphology of soda lime glass, particularly in terms of roughness and wettability. The insights gained from these measurements pave the way for optimizing glass surfaces for various applications, such as improving the performance of photovoltaic systems in harsh environments, including desert regions, by reducing soiling and enhancing light transmission

3 RESULTS AND DISCUSSION

Both hydrophobic and hydrophilic surfaces have been proposed as potential anti-dust coatings in previous studies. As depicted in the image sequence in Fig. 1, there is a clear contrast in wettability between the reference glass and the plasma-treated surfaces. The untreated reference glass exhibited the lowest hydrophilicity, with a contact angle (CA) of approximately 55°. However, as the plasma processing was applied, a noticeable shift in hydrophilic behavior emerged. This is largely due to the well-established relationship between surface roughness and wettability. Specifically, as the plasma treatment duration increased, the surface roughness of the glass was enhanced, leading to a significant decrease in the contact angle of the films [7-11].

In the case of the plasma-treated glass, even after just 60 seconds of treatment, the contact angle was reduced to as low as 12°, reflecting a substantial improvement in hydrophilicity. Furthermore, this trend continued as the duration of oxygen plasma exposure increased, making the surface progressively more hydrophilic. This behavior can be explained through the lens of surface energy theories. According to Young and Wenzel's models, surface topography plays a crucial role in influencing wettability. As surface roughness increases, it creates a higher surface area-to-liquid interface, which facilitates energy minimization between the liquid and the surface, driving the observed increase in hydrophilicity [12-15].

In essence, the plasma treatment process induces morphological changes that enhance the glass's ability to attract and spread water across its surface.

This makes plasma-treated glass surfaces ideal for applications requiring high wettability, such as anti-dust coatings, where the water's ability to spread can assist in removing particles. The results from this study not only underscore the importance of surface roughness in determining wettability but also highlight the potential for fine-tuning glass surface properties through controlled plasma processing to meet specific functional requirements in various environments.

Figure 2 presents representative field emission scanning electron microscopy (FESEM) micrographs of the untreated glass substrate (shown in Fig. 2a, labeled as Pre-Plasma) alongside the micrographs of the glass surface subjected to five minutes of plasma treatment (shown in Fig. 2b, labeled as Post Plasma). To further illustrate the effects of environmental exposure, the corresponding insets display the same samples after they were subjected to dust accumulation over a four-week period at the Outdoor Test Facility (OTF) from January 15 to February 15, 2021.

In both cases, the surface of the glass appears dense, homogeneous, and devoid of any pinholes or cracks, indicating a lack of voids that could otherwise compromise the integrity of the surface. This uniformity is critical for the optoelectronic properties of devices that may utilize these treated glass substrates. A uniform surface ensures consistent light transmission and minimizes scattering, which are essential factors for enhancing the overall efficiency of optoelectronic applications.

Additionally, Figure 3 provides a graphical representation of the average surface coverage by dust particles across glass samples subjected to varying durations of plasma treatment. The coverage was quantitatively assessed using SEM images, with the calculations being averaged over 25 different locations on each surface to ensure a robust statistical representation. This method involved calculating the ratio of the area covered by dust particles to the area that remained uncovered.

The results clearly demonstrate a significant correlation between the wettability of the glass surface and its capacity to repel or attract dust particles. Surfaces with higher hydrophilicity exhibit lower dust accumulation, as water can help to dislodge or wash away particles that adhere to the surface. Conversely, surfaces with lower wettability tend to allow dust particles to adhere more readily, resulting in increased coverage. These findings underscore the importance of optimizing surface properties through plasma treatment, not only to enhance optical performance but also to improve the durability and cleanliness of glass surfaces in practical applications. Overall, this study highlights how modifications in surface morphology and chemistry can play a pivotal role in the management of dust accumulation on optoelectronic devices, ultimately contributing to their performance and longevity.

As the duration of the plasma treatment is extended, there is a corresponding increase in surface roughness, which is attributed to the development of a smaller, grain-like morphology on the glass substrate. This transformation not only enhances the surface roughness but also significantly improves its hydrophilic properties. Consequently, surfaces that exhibit greater hydrophilicity are less prone to retain dust particles, resulting in improved cleanliness and functionality.

In addition to the changes induced by plasma treatment, the concept of surface corrugation has been proposed to further enhance hydrophilicity. By intentionally patterning the surface texture, it is possible to create a more effective interface that promotes water spreading and droplet mobility. This technique leverages the principles of surface energy and wettability, allowing for enhanced management of liquid interactions with the surface.

For a flat surface, the wettability is typically characterized by the contact angle formed between a water droplet and the surface. A lower contact angle indicates better wettability, which is a desired property for applications aimed at reducing dust accumulation. The introduction of corrugation modifies the traditional flat surface profile, influencing how water droplets behave upon contact. This enhanced surface architecture not only increases the effective surface area available for interaction with water but also contributes to energy minimization at the liquid-solid interface, which is crucial for promoting hydrophilicity.

Overall, the relationship between plasma treatment duration, surface morphology, and hydrophilicity is complex and interdependent. Extended plasma processing leads to increased roughness, which in turn enhances hydrophilicity and reduces dust adhesion. Additionally, incorporating surface corrugation strategies could provide further improvements in wettability, offering a promising

avenue for optimizing surfaces in various applications, including optoelectronic devices and self-cleaning technologies.

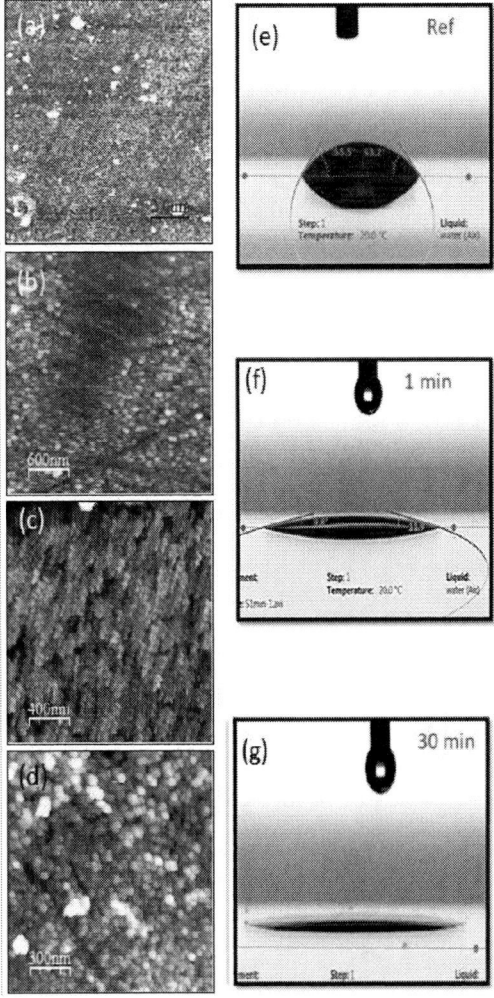

Figure 1: Representative AFM images of (a) the untreated reference glass surface and (b-d) glass surfaces subjected to plasma treatment, illustrating the changes in surface morphology. Panels (e-g) display corresponding contact angle measurements, revealing a progressive increase in superhydrophilic behavior with longer plasma treatment durations.

Figure 2 presents representative field emission scanning electron microscopy (FESEM) micrographs of the untreated glass substrate (shown in Fig. 2a, labeled as Pre-Plasma) alongside the micrographs of the glass surface subjected to five minutes of plasma treatment (shown in Fig. 2b, labeled as Post Plasma). To further illustrate the effects of environmental exposure, the corresponding insets display the same samples after they were subjected to dust accumulation over a four-week period at the Outdoor Test Facility (OTF) from January 15 to February 15, 2021.

In both cases, the surface of the glass appears dense, homogeneous, and devoid of any pinholes or cracks, indicating a lack of voids that could otherwise compromise the integrity of the surface. This uniformity is critical for the optoelectronic properties of devices that may utilize these treated glass substrates. A uniform surface ensures consistent light transmission and minimizes scattering, which are essential factors for enhancing the overall efficiency of optoelectronic applications.

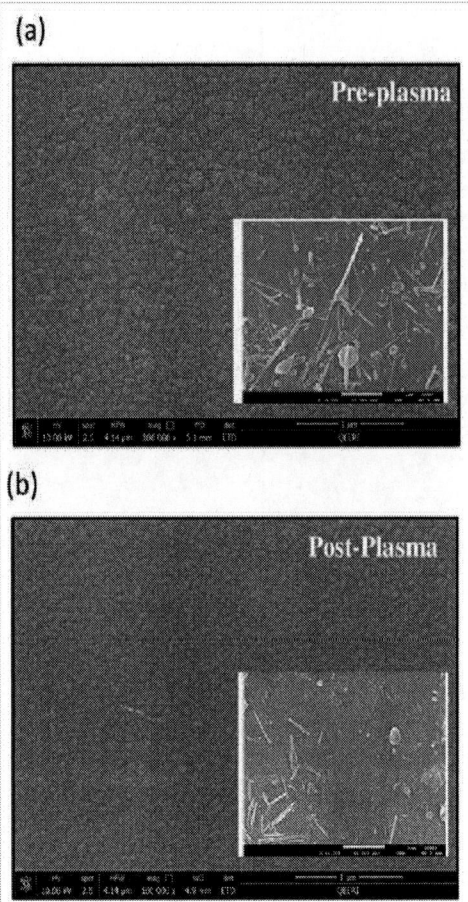

Figure 2: Field emission scanning electron microscopy (FE-SEM) analysis of the glass substrates, illustrating (a) the untreated glass substrate and (b) the glass substrate subjected to a 5-minute plasma treatment. The insets provide a view of the same samples after they were exposed to dust accumulation at the Outdoor Test Facility (OTF).

Additionally, Figure 3 provides a graphical representation of the average surface coverage by dust particles across glass samples subjected to varying durations of plasma treatment. The coverage was quantitatively assessed using SEM images, with the calculations being averaged over 25 different locations on each surface to ensure a robust statistical representation. This method involved calculating the ratio of the area covered by dust particles to the area that remained uncovered.

The results clearly demonstrate a significant correlation between the wettability of the glass surface and its capacity to repel or attract dust particles. Surfaces with higher hydrophilicity exhibit lower dust accumulation, as water can help to dislodge or wash away particles that adhere to the surface. Conversely, surfaces with lower wettability tend to allow dust particles to adhere more

readily, resulting in increased coverage. These findings underscore the importance of optimizing surface properties through plasma treatment, not only to enhance optical performance but also to improve the durability and cleanliness of glass surfaces in practical applications. Overall, this study highlights how modifications in surface morphology and chemistry can play a pivotal role in the management of dust accumulation on optoelectronic devices, ultimately contributing to their performance and longevity.

As the duration of the plasma treatment is extended, there is a corresponding increase in surface roughness, which is attributed to the development of a smaller, grain-like morphology on the glass substrate. This transformation not only enhances the surface roughness but also significantly improves its hydrophilic properties. Consequently, surfaces that exhibit greater hydrophilicity are less prone to retain dust particles, resulting in improved cleanliness and functionality.

In addition to the changes induced by plasma treatment, the concept of surface corrugation has been proposed to further enhance hydrophilicity. By intentionally patterning the surface texture, it is possible to create a more effective interface that promotes water spreading and droplet mobility. This technique leverages the principles of surface energy and wettability, allowing for enhanced management of liquid interactions with the surface.

For a flat surface, the wettability is typically characterized by the contact angle formed between a water droplet and the surface. A lower contact angle indicates better wettability, which is a desired property for applications aimed at reducing dust accumulation. The introduction of corrugation modifies the traditional flat surface profile, influencing how water droplets behave upon contact. This enhanced surface architecture not only increases the effective surface area available for interaction with water but also contributes to energy minimization at the liquid-solid interface, which is crucial for promoting hydrophilicity.

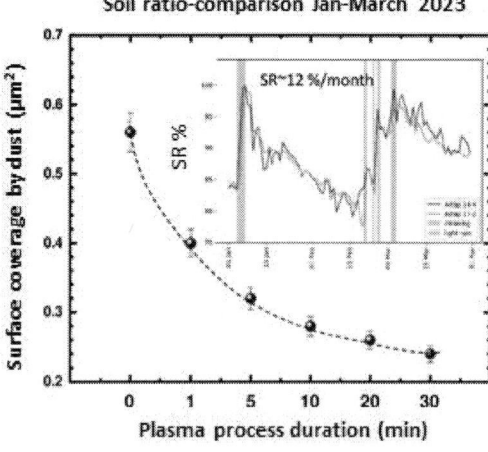

Fig. 3. Dust surface coverage as a function of plasma treatment duration for the glass samples. The inset illustrates the Soiling Ratio (SR %), indicating that the samples experienced an average soiling rate of approximately 12% per month during their exposure to outdoor conditions.

Overall, the relationship between plasma treatment duration, surface morphology, and hydrophilicity is complex and interdependent. Extended plasma processing leads to increased roughness, which in turn enhances hydrophilicity and reduces dust adhesion. Additionally, incorporating surface corrugation strategies could provide further improvements in wettability, offering a promising avenue for optimizing surfaces in various applications, including optoelectronic devices and self-cleaning technologies.

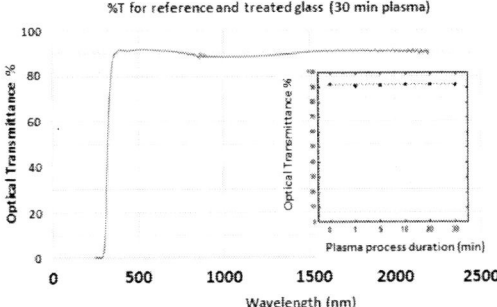

Figure 4: Optical transmittance of both the as-received and plasma-treated glass substrates. The inset highlights the variation in transmittance percentage (T%) at a wavelength of 550 nm in relation to the duration of plasma treatment.

$$\text{Cos } \theta = (\gamma sv - \gamma sl)/\gamma lv \tag{1}$$
$$\gamma lv \times \text{Cos } \theta = (\gamma sv - \gamma sl) \tag{2}$$

With corrugation:

$$\gamma lv \times \text{Cos } \theta = \varphi (\gamma sv - \gamma sl) - (1 - \varphi) \gamma lv \tag{3}$$

Where φ is the fractional area covered by the pattering on surface (i.e. the whole considered surface in the present case), θ is the angle of the droplet, γlv is the liquid to vapor tension, γsv is the tension from surface to vapor, γsl is the surface to liquid tension [16-18].

This mechanism is well-established in the development of potentiometric nano-biosensors, where enhancing sensitivity is achieved by capturing a significant volume of target molecules. In our case, we propose adapting this technique through surface engineering via plasma treatment. This approach has the potential to enhance light management within the anti-reflection layer by introducing controlled surface corrugation, which can significantly affect how light interacts with the surface.

The results from our atomic force microscopy (AFM) analysis, illustrated in Figure 1, strongly support this hypothesis. The data indicate that surface roughness increases dramatically, by more than 400%, transitioning from an initial root-mean-square (rms) roughness of 0.9 nm for untreated glass to approximately 3.9 nm following just 5 minutes of plasma treatment. This substantial increase in surface roughness is expected to facilitate improved light scattering and reflection properties, thereby enhancing the performance of the anti-reflection layer.

Importantly, despite the significant changes in surface morphology, we observed no adverse effects on optical transmittance following the plasma treatment, as demonstrated in Figure 4. This finding is crucial, as it confirms that the enhanced surface roughness achieved through plasma treatment can be integrated into optical applications without compromising the amount of light transmitted through the glass. The preservation of optical performance while improving surface characteristics highlights the effectiveness of plasma treatment as a viable method for optimizing anti-reflective coatings in various applications, including photovoltaic systems and other optoelectronic devices. Overall, these results suggest a promising pathway for utilizing surface engineering techniques to improve light management and efficiency in advanced optical materials.

4 SUMMARY AND CONCLUSIONS

In solar cells, the development of cost-effective transparent layers that function as both anti-reflection and anti-dust coatings is crucial, especially in desert environments where dust accumulation can significantly degrade photovoltaic (PV) performance. This paper proposes a straightforward approach to achieve this by treating PV glass with oxygen plasma. Remarkably, the optical transmission of the treated glass exceeded 90%, demonstrating that the plasma treatment did not adversely affect light transmittance.

To evaluate the wettability of the plasma-treated glass surfaces, we conducted contact angle measurements, which revealed a marked increase in hydrophilicity correlated with the duration of the plasma treatment. This finding aligns with our earlier surface roughness analysis, which was further confirmed through atomic force microscopy (AFM). Notably, reference glass samples treated with oxygen plasma for just 5 minutes exhibited a 400% increase in surface roughness compared to their untreated counterparts. Consequently, these treated surfaces accumulated significantly less dust, as indicated by the measurements of soiling coverage.

These promising results will be elaborated upon in the full version of this paper, which strongly supports the ongoing pursuit of efficient anti-dust and anti-reflection coatings. By leveraging the benefits of oxygen plasma treatment, we can enhance the functionality and durability of transparent layers in solar cell applications, ultimately contributing to improved energy efficiency and longevity in harsh environmental conditions.

References

[1] K. K. Ilse, B. W. Figgis, V. Naumann, C. Hagendorf, and J. Bagdahn, Renewable and Sustainable Energy Reviews," vol. 98, pp. 239–254, 2018.
[2] K. Ilse et al., Joule, 3, 2303, 2019.
[3] T. Sarver, A. Al-Qaraghuli, and L. L. Kazmerski, Renewable and Sustainable Energy Reviews, vol. 22, pp. 698–733, 2013.
[4] A. Sayyah, M. N. Horenstein, and M. K. Mazumder, Solar Energy, vol. 107, pp. 576–604, 2014.
[5] Gujarat Power Corporation Limited, "Aerial view of Gujarat Solar Park." [Online]. Available: https://gpcl.gujarat.gov.in/assets/images/image5.jpg.
[6] Zhou, Y., et al. Journal of Materials Science & Technology, 35(4), 718-726, 2019.
[7] Jiang, Y., et al. Applied Surface Science, 504, 144-150, 2020.
[8] Kumar, S., & Bhat, S. Materials Today: Proceedings, 45, 1046-1051, 2021.

[9] Lehmann, C., et al. Surface and Coatings Technology, 348, 157-162, 2018.
[10] Zhang, Y., et al. Solar Energy Materials and Solar Cells, 161, 93-101, 2017.
[11] Yadav, A., et al. Journal of Vacuum Science & Technology A, 37(4), 041203, 2019.
[12] Sahni, N., et al. Renewable and Sustainable Energy Reviews, 124, 109784, 2020.
[13] Lee, H., et al. Energy Reports, 6, 1961-1970, 2020.
[14] Huang, L., et al. Journal of Photonics for Energy, 11(2), 1-8, 2021.
[15] Mishra, S., et al. Materials Today Communications, 29, 102626, 2022.
[16] Gavi, N. M. H., Ngom, B. D., Beye, A. C., Strydom, A. M., Aissa, B., Srinivasu, V. V., and Chaker, M. Journal of Magnetism and Magnetic Materials, 324(6), 1172-1176, 2012.
[17] El Khakani, M. A., Le Borgne, V., Aïssa, B., Rosei, F., Scilletta, C., Speiser, E., et al. Applied Physics Letters, 95(8), 2009.
[18] Habib, M. A., Barkat, M., Aissa, B., and Denidni, T. Progress In Electromagnetics Research, 88, 135-148, 2018

41st European Photovoltaic Solar Energy Conference and Exhibition

PERFORMANCE, ABRASION RESISTIVITY AND ANTI-SOILING TESTING OF INNOVATIVE, NANOSTRUCTURED ANTI-REFLECTION COATINGS UNDER CONTROLLED AND STANDARDIZED CONDITIONS

C. Pfau[1,2]*, G. Willers[1,2], C. Allagiannis[3], I. Arampatzis[3], and M. Turek[1,2]

[1] Fraunhofer Center for Silicon Photovoltaics CSP, Otto-Eissfeldt-Str. 12, 06120 Halle, Germany
[2] Fraunhofer Institute for Microstructure of Materials and Systems IMWS, Walter-Hülse-Str. 1, 06120 Halle, Germany
[3] NanoPhos S.A., Sci. & Tech. Park of Lavrio, Athens Ave., 19500, Lavrio Greece

*charlotte.pfau@csp.fraunhofer.de

ABSTRACT: Anti-reflection (AR) coatings are an established technology in photovoltaic modules. However, there are still some technological challenges when it comes to combining optimized optical properties with other functionalities like anti-soiling behavior and abrasion resistivity. In our work, we present how implemented test setups and standards including optical characterization, assessment of the susceptibility to dust coverage and abrasion testing can be applied to test new AR coatings. These quick and cost-effective lab test methods are applied to a new type of coating showing its potential of combining high optical performance with reduced dust coverage. Based on an abrasion resistance benchmark test on commercial SiO_2 coatings, we demonstrate that AR coatings can be optimized not only regarding its initial optical performance but also regarding its properties required for an economic and stable operation in the field.

Keywords: anti reflective coatings, PV glass, evaluation, soiling, abrasion, performance and reliability

1 Introduction

Photovoltaic (PV) glass with anti-reflective coating is a major module component with impact on performance and energy yield. Ongoing research and development activities have a focus on developing advanced coatings with improved performance and on reducing manufacturing costs to drive market expansion. Key performance indicators to be evaluated include light transmittance, anti-reflection (AR) effect, durability and the overall impact on the energy yield [1]. Coatings for photovoltaic (PV) module front glasses must not only lead to improved optical performance but must also meet the diverse requirements that arise in the field. For example, in harsh environments with high dust loads and frequent cleaning, high demands are placed on the abrasion resistivity of the coatings. Studies have shown that widely used, first generation nano-porous SiO_2 coatings can largely degrade after just a few years or months, which can be linked to the abrasion caused by dust in combination with brushes in dry or wet cleaning procedures [2]. One approach to reduce coating degradation is using more abrasion-resistant coatings. Another strategy is to improve the anti-soiling behavior of the coatings, which would reduce the amount of cleaning events required. However, currently there are no established anti-reflection coatings on the market that significantly reduce the number of necessary cleaning cycles.

Thus, AR coating developers are still actively working on solar coatings with optimal optical performance, improved abrasion resistivity and, ideally, improved anti-soiling behavior. The assessment of a new coating technology requires complex test scenarios going beyond a single optical measurement to cover all these aspects in a comparable and reproducible manner. Test standards, like the DIN-SPEC 4867 "Cleaning test methods for the abrasion resistance of glass coatings for solar applications" and VDI 3956-1 "Evaluation of the soiling properties of surfaces: Test method for the dust soiling behavior of solar energy systems." have therefore been developed by

research institutes and industry partners. Besides designing appropriate test procedures, the test devices themselves were also developed and implemented into the AR coating test routines, for example a large area brush abrasion test or dust chamber which are adaptable to various soiling and cleaning scenarios, see Figures 1 and 2. In this work, pre-prototype samples of a new generation of innovative, cost-effective non-porous sol-gel coating are investigated regarding their optical performance, abrasion resistance and the anti-soiling behavior following these standardized test procedures. Based on these test procedures, we show that the coating combines improved optical anti-reflection properties with anti-soiling behavior when compared to an uncoated reference. We will also present the results of the abrasion resistivity test.

In this work, we demonstrate a test scheme for the combined assessment of various crucial properties of glass coatings. This test scheme involves several newly established test standards and test devices that have been developed in that aspect. This combined test provides the necessary performance parameters of new glass coating to assess not only its optical properties but also its potential performance in the field. It is applied to an innovative coating technology to further improve the coating properties.

The investigated coating can be applied using standard roller coating technology or by spraying. The coating is intended for initial application by the glass manufacturer as well as for retrofitting in the field on PV modules with, without or partially degraded AR coating. PV park outdoor studies in Greece and China with a previous version of the coatings have shown that the coating leads to a considerable improvement in energy yield gain (5-6%) compared to uncoated reference PV modules [3]. Just like the coating from Ref [3], the investigated coating is an inorganic nano-structed layer containing of a metal oxides mixture mainly of nano-crystalline anatase/rutile titania (TiO_2) anchored with silica (SiO_2) nano-bridges.

To furthermore evaluate and compare the abrasion test results, we performed an abrasion resistance benchmark test consisting of a sample set of commercial SiO_2 coatings including older and newer coating generations of SiO_2 roller sol-gel coatings as well as one magnetron sputter coating.

2 Approach and Methodology

The proposed test sequence includes the determination of optical properties under standard test conditions (STC), the performance measurements using (mini-)modules, soiling and self-cleaning test, and abrasion resistivity testing.

2.1 Optical performance testing

Transmittance measurements on the ARC-glass samples were performed using a double-beam Perkin Elmer Lambda 1050 spectrophotometer with a 150 mm integrating sphere on the basis IEC 62805-2 standard [4]. Furthermore, one-cell mini solar modules were built with these front glasses for an electrical performance characterization. The power performance of the mini modules under standard test condition (STC) was determined by flash testing with a LOANA electrical characterization tool according to IEC 60904-3 [5]. The power performance for low incidence angles was determined with an Abet I-V-characterization tool using a parallel sun simulator light field and different angles of incidence (60° and 0° relative to the surface normal of the mini modules) based on IEC 61853-2 [6] (compare Figure 1, left).

2.2 Soiling and self-cleaning testing

We performed dry treatment as well as light dew treatment soiling test according to VDI 3956/part 1 [7] in a soiling chamber (Figure 1, right) with Middle East Test Dust. The test procedure involves 4 samples à 10 cm x 10 cm per test run in a horizontal sample orientation and humidity at 60% RH. The sequence includes the following steps:

1. dust deposition for 30 min at 22°C,
2. sample exposed to 7 K above dew point for 20 min (in case of dry treatment), or sample exposed to dew point for 20 min (in case of light dew treatment),
3. drying for 10 min at 40 °C
4. measurement of the initial dust coverage (C_{ini}) by microscopy imaging
5. cleaning with wind blow using an air-knife crossing providing about 10 m/s wind speed 1 cm above the sample surface
6. measurement of dust coverage after wind blow (C_{wb}) by microscopy imaging

This sequence of steps results in the initial dust coverage without wind blow and the dust coverage after wind blow.

2.3 Abrasion resistivity testing

Abrasion resistivity testing of the AR coatings was realized using an Amtec Kistler Lawa abrasion testing tool (Figure 2). The test was performed on three replicas per ARC-glass sample type according to DIN SPEC 4867 [8], working mode 1, which is characterized by the following basic test parameters: wet-rotary brush cleaning with

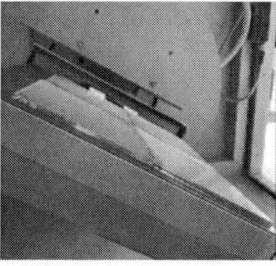

Figure 1: Left: angle dependent electrical performance testing of a mini solar module using a sun simulator with parallel light field and right: comparative soiling and self-cleaning testing in the soiling chamber showing four different ARC-glass samples after wind blow with the air knife (seen in the background).

Figure 2: Brush abrasion testing with "Lawa" tool by Amtec Kistler. The test is applicable for glasses as well as for solar modules (small and full size).

Sunbrush 350x PE brush (brush width = 1.2 m; brush diameter 340 mm to 350 mm, brush rotation speed 120 rpm to 131 rpm, feed rate 5.0 ± 0.2 m/min, brush immersion depth = 20 ± 2 mm and direction of rotation opposite to the horizontal movement of the brush). The test involves a continuous and uniform application of a dust-water suspension (0.9 g/m² of feldspar test dust during each brush pass, spraying of a dust-water suspension with a dust to water mixing ratio of 1.5 g / liter, three spray nozzles before the brush spraying the suspension with an angle of 30° in the direction of the brush). According to the test standard, the test is finished when complete abrasion is detected, i.e. the measured reflectance no longer increases due to reduced ARC coverage. As a result, the number of brush strokes until partial 50% or 90% loss of ARC-performance are reported.

2.4 Samples

Three different ARC-sample types (ARC-A, ARC-B and ARC-C) were produced by applying the coating with a controlled industrial roller coating process on solar grade 3.2 mm thick rolled glass samples. The sample types share the same coating composition while different application process parameters were used. Subsequently, the coated samples and the uncoated reference glass (taken from the same rolled glass pane) were cut into pieces of 20 cm x 20 cm and 10 cm x 10 cm size for mini module production and lab investigation on ARC-glasses, respectively. Finally, the samples were heat-treated at 600°C.

Additionally, for the abrasion resistance benchmark, solar glass samples with two widely used, older generation industrial reference coatings, being one sol-gel roller SiO_2

41st European Photovoltaic Solar Energy Conference and Exhibition

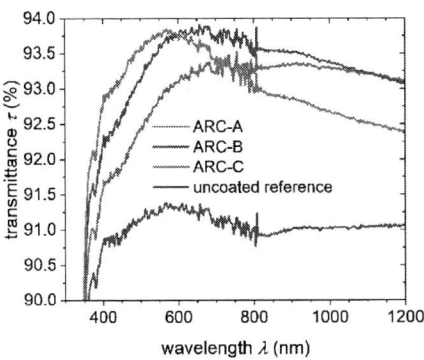

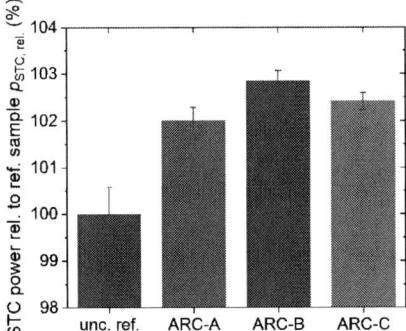

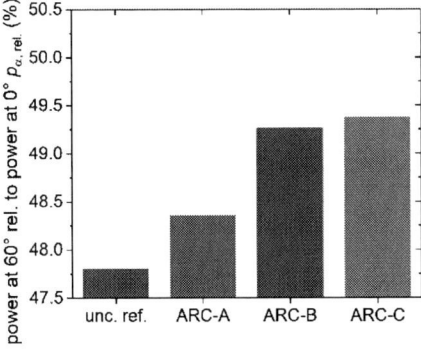

Figure 3: Top: Hemispheric transmittance spectra of the ARC-A, B and C-glasses and of the uncoated reference glass, middle: STC power performance ratio of the respective mini modules relative to the uncoated reference mini module, and bottom: mini module power at a low incident angle ($\alpha = 60°$) relative to the power of the respective sample at vertical incidence ($\alpha = 0°$).

coating ("ind. sol-gel") and one SiO_2 sputter coating ("ind. sputter"), as well as further newer commercial SiO_2 sol-gel ARCs (bm test 1-5) were used.

Three samples / measurements were used per test and the average value is presented; only in the case of angle-dependent flashing just one mini solar module was measured per sample type.

3 Results and discussion

3.1 Optical performance

The transmittance spectra (Figure 3, top) show that the considered coating (especially for sample ARC-B and ARC-C) provides clearly improved anti-reflection performance. Sample ARC-B provides the most suitable spectral distribution with maximum AR effect around the PV design wavelength (about 650 nm) and the highest transmittance gain relative to the uncoated reference, being 2.6% absolute. The transmittance spectra are quite similar to those of usual industrial sol-gel SiO_2 coatings [9], whereas the coating investigated here has about $0.5\%_{abs.}$ lower maximum transmittance than that from Ref. [9]. Compared to usual industrial standard SiO_2 sputter coatings, the considered coating performs much better with about 2% higher maximum transmittance [9].

The transmittance gain translates almost directly into the STC power gain of the respective mini modules, which is shown in Figure 3, middle. A power ratio (module to cell power) increase relative to the uncoated mini module reference is observed for all coated mini module samples with a maximum enhancement for the mini module with ARC-B front glass ($2.4\%_{abs.}$), which can be explained by the highest transmittance gain and the favorable spectral distribution of the ARC-B glass.

The low-angle power performance of the mini modules is presented in Figure 3, bottom. The relative power, which is the ratio of powers at $\alpha = 60°$ and at $\alpha = 0°$ of the same mini module in each case, is for all ARCs higher than for the uncoated reference. ARC-C provides the highest difference to the uncoated sample with $+1.5\%_{abs.}$ directly followed by ARC-B with $+1.3\%_{abs.}$. This is an additional positive effect with respect to the expected energy yield that goes beyond the STC power gain.

3.2 Anti-soiling performance

Figure 4 shows a comparison of the initial dust coverage and the dust coverage after wind blow for the dry treatment soiling test procedure. After cleaning, the coated samples show a significantly lower average coverage of 47%. Sample ARC-A has up to 57% less coverage than the uncoated reference sample. However, the ratios of dust coverage before (C_{ini}) and after the wind blow (C_{wb}) (given in the image) do not differ significantly from each other. Also, in the case of the light dew treatment soiling tests, no significant difference was found (results not shown). This means that there is no significant additionally self-cleaning effect, which is the case for many PV glass ARCs [10]. However, another important effect shows a remarkable result: Despite simultaneous and uniform dusting with identical test parameters, the coated samples show on average 34% less initial coverage than the uncoated reference sample, indicating a clear rebound effect implying an effective anti-soiling property. Sample ARC-C shows an up to 49% lower value.

653

41st European Photovoltaic Solar Energy Conference and Exhibition

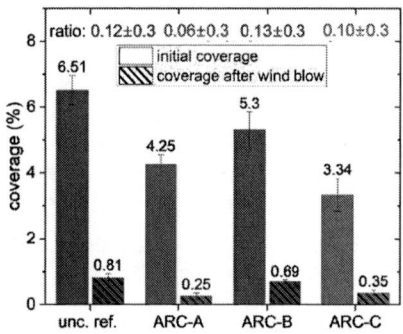

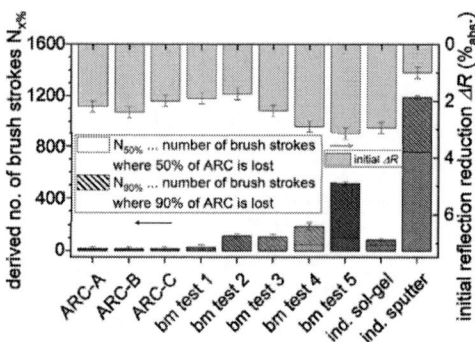

Figure 4: Dust coverages before and after wind blow in dry dust treatment soiling test. The dust coverages were quantified using microscopy imaging analysis according to VDI 3956. The numbers in the top line represent the ratio between coverage after wind blow and initial coverage.

Figure 5: Lower bars (left axis): $N_{90\%}$ and $N_{50\%}$ number of applied brush strokes where 50% and 90%, respectively, of the anti-reflection performance of the coating is gone, derived according to DIN SPEC 4867. Upper bars (right axis): integral, effective reflectance reduction value ΔR of the anti-reflective coatings before abrasion testing, calculated from reflection spectra of AR coated sample and uncoated reference on the basis of equation 1 in DIN SPEC 4867.

3.3 Abrasion resistance benchmark

Figure 5 summarizes the major outcomes of rotating brush cleaning abrasion resistivity test according to DIN SPEC 4867 applied on ARC-A, B and C and a set of commercial ARC samples with the following derived quantities:

o the effective integral reflection reduction value ΔR of the unabraded, initial coating, which is good measure for the AR performance of the coating (calculated on the basis of equation 1, DIN SPEC 4867)
o the $N_{50\%}$ and $N_{90\%}$ number of applied brush abrasion strokes where 50% and 90%, respectively, of the AR effect of the coating is gone (derived by in-situ reflection increase monitoring according to DIN SPEC 4867)

Firstly, for the samples ARC-A, B and C, the clear AR effect of the coating is again evident from the integral reflection reduction value ΔR. On the other hand, the results also show that 90% of coating is abraded after less than 30 brush strokes for all three samples. This abrasion resistivity test result indicates that for harsh environments with high dust loads with and frequent cleaning, a further optimization of the coatings' abrasion resistivity is needed, which is the case for many SiO_2 sol-gel-coatings (see Fig. 5).

The comparison to the other samples in Figure 5 in particular shows that the industrial reference SiO_2 magnetron sputtering coating exhibits much higher abrasion resistivity with $N_{90\%}$ of about 1200 strokes. For the usual industrial sol-gel coating a $N_{90\%}$ of about 100 could be found. On the other hand, the sputtered coating provides a significantly poorer initial antireflection performance, being less than 50% of all tested sol-gel coatings.

For the five tested commercial benchmark samples (bm test 1-5), however, the abrasion resistivity increases with increasing AR effect within this sample series. With 600 strokes needed to abrade 90% of ARC bm test 5 sample provides the highest abrasion resistivity among the tested sol-gels, simultaneously exhibiting the highest AR effect among the tested samples with an effective ΔR of more than 3 %.

4 Conclusion

In summary, we present a combined test scheme to assess not only the transmittance spectra and STC-power performance of a newly developed AR-coating but also its angle-dependent power performance, its anti-soiling behavior and abrasion resistivity. All these key performance indicators are relevant for applying a coating to PV modules. To this end, standardized test procedures like DIN SPEC 4867 and VDI 3956 have been developed and the required test tool designed and integrated into the test procedure.

A new AR coating technology has been assessed showing its anti-reflective behavior and, furthermore, positive effects on the energy yield that go beyond STC power performance. An increased low-angle power-performance and a clear anti-soiling effect, characterized by reduced initial dust coverages, is observed. The results imply a clear potential for energy yield gains especially for operation in low abrasion regions with sparse cleaning events.

The abrasion test indicates that in the current development phase, the pre-prototype SiO_2 sol-gel coating would lose its performance in harsh environments with high dust loads and regular brush cleaning. To evaluate and classify these abrasion test results, we present an abrasion resistance benchmark test on commercial SiO_2 coatings. It shows that an industrial SiO_2 magnetron sputtering coating exhibits a much higher abrasion resistivity than usual sol-gel coatings. However, sputtered coatings are substantially more expensive and provide poorer initial antireflection performance than the SiO_2 sol-gels.

When choosing a coating, a trade-off should be made between various key performance indicators like anti-reflection performance and abrasion resistivity taking into account the costs. In case of the tested sample set of commercial sol-gels shows that improved AR performance is not necessarily accompanied by poorer abrasion resistivity. The coating with the highest abrasion resistivity among the tested sol-gels is also the coating with the highest AR effect.

References

[1] Verified marked research, *Global Photovoltaic Anti Reflection Glass Marked, forecast 2022-2031*, (2021)

[5] IEC 62805-2 (VDE 0126-4-21), *Method for measuring photovoltaic (PV) glass – Part 2: Measurement of transmittance and reflectance* (2017)

[2] A. M. Law et al., *The performance and durability of Anti-reflection coatings for solar module cover glass - a review*, Solar Energy 261, p. 85-95, (2023)

[3] I. Arabatzis et al., *Photocatalytic, self-cleaning, antireflective coating for photovoltaic panels: Characterization and monitoring in real conditions*, Solar Energy 159, p. 251-256, (2018)

[4] IEC 60904-3, *Photovoltaic devices — Part 3: Measurement principles for terrestrial photovoltaic (PV) solar devices with reference spectral irradiance data (IEC 60904-3:2019)*, (2020)

[5] IEC 61215-1-2, *Terrestrial photovoltaic (PV) modules - Design qualification and type approval - Part 2: Test procedures*

[6] IEC 61853-2: *Photovoltaic (PV) module performance testing and energy rating - Part 2: Spectral responsivity, incidence angle and module operating temperature measurements*, (2016)

[7] VDI 3956-1, *Evaluation of the soiling properties of surfaces - Test method for the dust soiling behaviour of solar energy systems*, (2020)

[8] DIN SPEC 4867, *Cleaning test methods for the abrasion resistance of glass coatings for solar applications*, (2022, is currently converted to DIN ISO)

[9] K. Lange et al., *Abrasion testing of anti-reflective coatings under various conditions*, Solar Energy Materials & Solar Cells 240, 111732, (2022)

[10] K. Ilse et al., *Advanced performance testing of anti-soiling coatings – Part I: Sequential laboratory test methodology covering the physics of natural soiling processes*, Solar Energy Materials and Solar Cells 202, 110048 (2019)

41st European Photovoltaic Solar Energy Conference and Exhibition

DEVELOPMENT OF ENCAPSULANT-LESS CRYSTALLINE SILICON PHOTOVOLTAIC MODULES AND THEIR DURABILITY AGAINST POTENTIAL-INDUCED DEGRADATION

Keisuke Ohdaira, Shuntaro Shimpo, Huynh Thi Cam Tu
Japan Advanced Institute of Science and Technology (JAIST)
1-1 Asahidai, Nomi, Ishikawa 923-1292, Japan

ABSTRACT: Conventional crystalline silicon (c-Si) photovoltaic (PV) modules show degradations related to encapsulant, and are difficult to disassemble into individual components. To overcome these issues, we develop a novel crystalline silicon (c-Si) photovoltaic module with openable cover glass. Cells in the modules are not fixed but are just put on hollows in a plastic base. We also investigated the PID tolerance of such encapsulant-less PV modules. In the modules with n-type front-emitter cells, the encapsulant-less modules show significantly high tolerance against polarization-type PID (PID-p). Furthermore, in the modules with p-type Al-BSF cells, shunting-type PID (PID-s) can also be effectively suppressed. We observed that slight PID-s occurs when the cells are in touch with cover glass, and a sufficient gap is necessary between them. An issue for this type of encapsulant-less modules is water ingress, leading to precipitate formation. To suppress water ingress, the development of encapsulant-less modules with O-ring is ongoing.

Keywords: encapsulant-less photovoltaic module, crystalline silicon solar cell,

1 INTRODUCTION

The installations of photovoltaic (PV) modules continue to grow worldwide [1], and a significant amount of PV modules will be discarded in the near future. Currently, conventional crystalline silicon (c-Si) PV modules have a problem of difficulty to be disassembled into individual components, which may result in an increase in landfill waste. This is because the components of PV modules are strongly adhered by encapsulants, such as ethylene–vinyl- acetate copolymer (EVA). In addition to the problem of the difficulty of disassembling, many kinds of degradation modes in the c-Si PV modules are related to the encapsulant, e.g., corrosion of electrodes [2], yellowing [3], and potential-induced degradation (PID) [4,5].

To overcome these issues, some of new PV module structures without encapsulant have been proposed thus far [6–20]. By not using encapsulant, the performance degradation of PV modules related to encapsulant will be suppressed, and more effective reuse and/or recycling of materials is realized since the disassembling of the components becomes easier.

In this study, we have developed a novel c-Si PV module structure with no encapsulant. c-Si PV cells are not tightly fixed but just put on a base made of plastic such as polycarbonate with a cell-size hollow. The schematic structure of the proposed module is shown in Fig. 1 [21]. Unlike in the case of previous encapsulant-less PV modules [6–20], surface cover glass is openable, which enables us to replace failed cells and/or strings. We have also investigated the durability of the novel PV modules with conventional p-type cells and n-type front-emitter cells against PID. Furthermore, we have performed indoor damp heat (DH) tests and outdoor exposure.

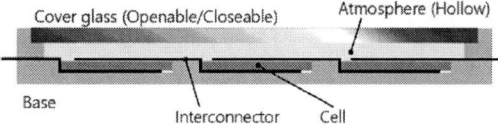

Figure 1: Schematic of an encapsulant-less c-Si PV module [21].

2 EXPERIMENTAL PROCEDURES

Figure 2 shows the structure of the mini-module used in this study [21]. We prepared bases made of plastic such as poly-lactic acid (PLA), poly-tetra-fluoro-ethylene (PTFE) or poly-carbonate (PC) by 3D printing or machining. A p-type Al-BSF cell or n-type front-emitter cell was put in a hollow prepared in the plastic base. Interconnector ribbons were connected with the front-side and rear-side electrodes of the cell. We also prepared mini-modules with standard structure consisting of cover glass/EVA/cell/EVA/backsheet.

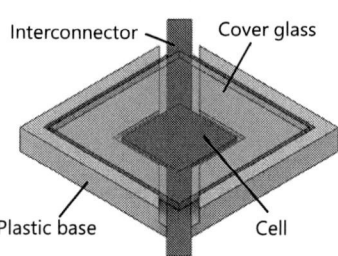

Figure 2: Structure of an encapsulant-less c-Si PV module [21].

We performed PID tests for the mini-sized modules using Al-plate method [22]. A negative voltage of −1000 V was applied to the short-circuited interconnector of the cell with respect to the grounded Al plate fully covering the cover glass. The PID tests were performed at 85 °C in dry air with a relative humidity of <2%. The modules before after the PID tests were characterized by current-voltage–voltage (J–V) measurements in the dark and under 1-sun light illumination. Shunt resistance (R_{sh}) of the cells was evaluated by analyzing the dark J–V characteristics using one-diode model.

Part of the encapsulant-less PV modules received DH test at 85 C and 85%RH. PC base before and after the DH test was characterized by Fourier-transform infrared (FT-IR) spectroscopy.

We also prepared large-sized modules with connected 4 cells connected in series with each size of 156×156 mm². The bases of the modules are made of PC processed by

41st European Photovoltaic Solar Energy Conference and Exhibition

machining. The large-sized modules were installed outdoor for ~6 months and were evaluated with the eye.

3 RESULTS AND DISCUSSION

3.1 PID-of n-type front-emitter c-Si PV modules

Figure 3 shows the J–V characteristics of the conventional and the encapsulant-less n-type front-emitter c-Si PV modules before and after the PID test for 3 days. The conventional module shows significant reductions in short-circuit current density (J_{sc}) and open-circuit voltage (V_{oc}). These are typical characteristics of polarization-type PID (PID-p) [5]. On the other hand, the encapsulant-less module shows much smaller J_{sc} and V_{oc} reductions, indicating high durability against PID-p. Surprisingly, the performance of the encapsulant-less PV module is slightly degraded even in the absence of encapsulant. The usual PID-p is induced by a leakage current flowing through encapsulant. This phenomenon may indicate that there is another carrier flow path, such as a leakage current through the base, as shown in Fig. 4.

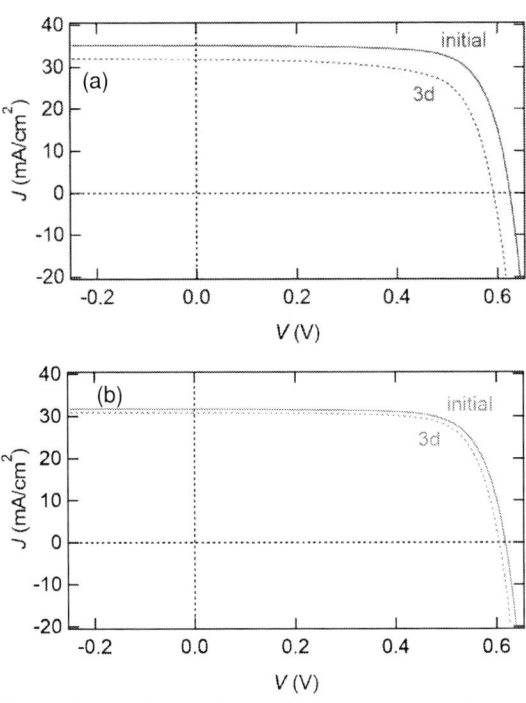

Figure 3: J–V characteristics of (a) the conventional and (b) the encapsulant-less PV modules before and after the PID tests for 3 days.

3.2 PID of p-type Al-BSF c-Si PV modules

Figure 5 shows the J–V characteristics of the conventional and the encapsulant-less PV modules with p-type cells [23]. The conventional module shows clear degradation due to shunting and resulting fill factor (FF) reduction, which is a typical behavior of PID-shunting (PID-s) [24]. We also confirmed the reduction in R_{sh} by the analysis of the dark J–V characteristics [23]. On the contrary, the encapsulant-less module is highly resistive against PID-s. As in the case of n-type modules, we observed a slight degradation in the performance of the encapsulant-less module, indicating the invasion of

sodium ions even in the absence of encapsulant. A possible pathway is the contact between cover glass and cell electrode. We then changed the depth of the hollow so that the cell electrode is not in tough with the cover glass. We consequently confirmed that the slight PID-s can be suppressed, as shown in Fig. 6.

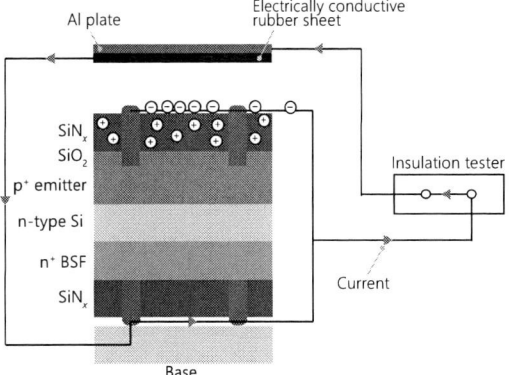

Figure 4: Possible leakage current path in the encapsulant-less PV module.

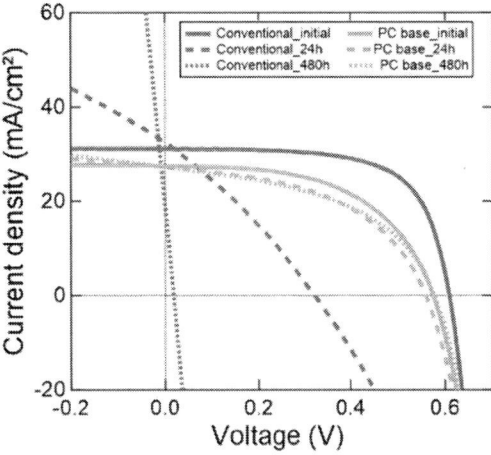

Figure 5: J–V characteristics of the conventional and the encapsulant-less p-type c-Si PV modules before and after the PID test for up to 480 h [23].

3.3 Water ingress to the encapsulant-less PV modules

Figure 7 shows the surfaces of the encapsulant-less c-Si PV modules with the PC base before and after the DH test for 480 h [23]. One can see yellow precipitates in the module after the DH test. According to the FT-IR measurements, the yellow precipitate is bisphenol A, which can be formed by the hydlysis of PC [25].

Figure 8 shows the large-sized encapsulant-less PV modules with the base made of PC. We prepared two types of 4-cell modules and installed in the field, as shown in Fig. 9 [26]. After 6-month outdoor exposure, we observed precipitates at the glass/cell contacts, as shown in Fig. 10.

We thus conclude that water invading into the encapsulant-less module reacts with PC, leading to the formation of precipitates. To avoid the formation of precipitates, we need to prepare a structure for the barrier of water ingress.

41st European Photovoltaic Solar Energy Conference and Exhibition

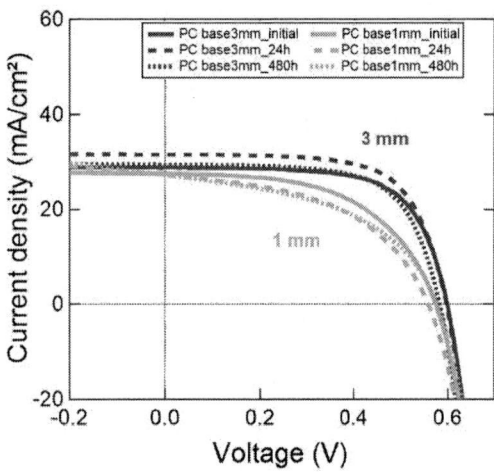

Figure 6: *J–V* characteristics of the encapsulant-less p-type c-Si PV modules with different hollow depths before and after the PID test for up to 480 h. [23].

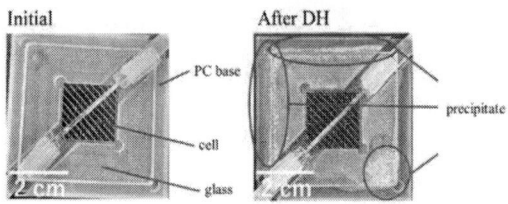

Figure 7: *J–V* characteristics of the encapsulant-less p-type c-Si PV modules with different hollow depths before and after the PID test for up to 480 h. [23].

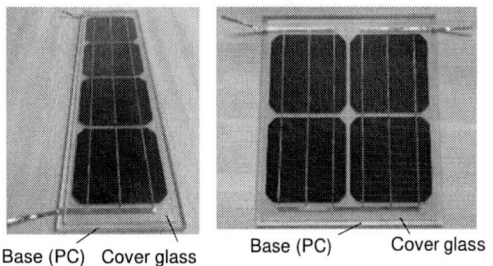

Figure 8: Large-sized encapsulant-less PV modules with 4 cells connected in series.

Figure 9: Large-sized encapsulant-less PV modules installed outdoor [26].

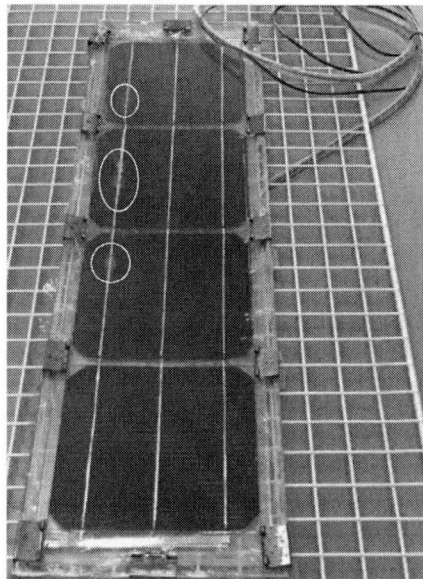

Figure 10: Appearance of the encapsulant-less PV module after the outdoor exposure.

3.4 Module with an O-ring

To overcome the problem of water ingress we are modifying the structure of the encapsulant-less PV modules. Figure 11 shows the schematic structure of the modified encapsulant-less c-Si PV module. An O-ring was put between the PC base and a cover material. We replaced cover glass to transparent PC to realize more light-weight PV modules. The module base was slightly curved with the radius of curvature of ~3 m, aiming at future installation on curved walls and/or roofs. We performed water immersion test for the modules with O-ring packing, and confirmed that the O-ring can suppress liquid water ingress. Figure 12 shows the appearance of the prototype of the modified encapsulant-less PV module.

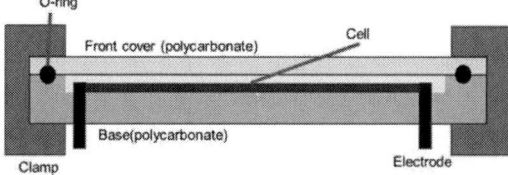

Figure 11: Schematic structure of the modified encapsulant-less c-Si PV module with an O-ring along the edge of the edges.

Figure 12: Schematic structure of the modified encapsulant-less c-Si PV module with an O-ring along the edge of the edges.

4 SUMMARY

We proposed novel encapsulant-less c-Si PV modules and fabricated prototype modules. We confirmed that the encapsulant-less PV modules are highly resistant against PID-p and PID-s. A contact between cover glass and the cell may lead to slight PID-s and sufficient glass-cell gap is required. We observed the formation of precipitates by water ingress. To suppress the water ingress, the development of the modified module structure with an O-ring along the edges is ongoing.

ACKNOWKEDGMENTS

We would like to thank Mr. Kodai Nakamura of JAIST for the fabrication and characterization of n-type c-Si PV modules. This work was supported by NEDO.

References

[1] International Energy Agency, World Energy Outlook (2023).

[2] H. Xiong, C. Gan, X. Yang, Z. Hu, H. Niu, J. Li, J. Si, P. Xing, and X. Luo, Microelectron. Reliab. 70 (2017) 49.

[3] F. J. Pern and A. W. Czanderna, Sol. Energy Mater. Sol. Cells 25 (1992) 3.

[4] W. Luo, Y. S. Khoo, P. Hacke, V. Naumann, D. Lausch, S. P. Harvey, J. P. Singh, J. Chai, Y. Wang, A. G. Aberle, and S. Ramakrishna, Energy Environ. Sci. 10 (2017) 43.

[5] S. Yamaguchi, B. B. Van Aken, A. Masuda, and K. Ohdaira, Sol. RRL 5 (2021) 2100708.

[6] E. Saint-Sernin, R. Einhaus, K. Bamberg, and P. Panno, Proc. 23rd European Photovoltaic Solar Energy Conf. Exhib., 2008, p. 2825.

[7] D. Reinwand et al., Proc. 7th World Conf. Photovoltaic Energy Conversion, 2018, p. 628.

[8] R. Einhaus, K. Bamberg, R. Franclieu, and H. Lauvray, Proc. 19th European Photovoltaic Solar Energy Conf. Exhib., 2004., p. 2371.

[9] P. Leibiger, C. Pönisch, T. Seifert, and D. Kray, Proc. 8th World Conf. Photovoltaic Energy Conversion, 2022, p. 807.

[10] D. Reinwand, B. King, J. Schube, F. Madon, R. Einhaus, and D. Kray, AIP Conf. Proc. 2156, (2019) 020009.

[11] F. Madon, R. Einhaus, J. Degoulange, C. Comparotto, G. Galbiati, and E. Wefringhaus, Energy Procedia 77 (2015) 382.

[12] R. Einhaus, F. Madon, J. Degoulange, K. Wambach, J. Denafas, F. R. Lorenzo, S. C. Abalde, T. D. Garcia, and A. Bollar, Proc. 7th World Conf. Photovoltaic Energy Conversion, 2018, p. 561.

[13] F. Madon, H. Colin, L. Sicot, P. Le fi llastre, J. Degoulange, and R. Einhaus, Proc. 31st European Photovoltaic Solar Energy Conf. Exhib., 2015, p. 2534.

[14] J. Dupuis, E. Saint-Sernin, K. Bamberg, R. Einhaus, E. Pilat, A. Vachez, and D. Bussery, Proc. 25th European Photovoltaic Solar Energy Conf. Exhib., 2010, p. 4148.

[15] J. Dupuis, E. Saint-Sernin, O. Nichiporuk, P. Le fi llastre, D. Bussery, and R. Einhaus, Proc. 38th IEEE Photovoltaic Specialists Conf., 2012, p. 3183.

[16] F. Madon, O. Nichiporuk, R. Einhaus, L. Crampette, B. Semmache, L. Valette, V. Charrier, and B. Damiani, Proc. 28th European Photovoltaic Solar Energy Conf. Exhib., 2013, p. 3149.

[17] R. Couderc, M. Amara, J. Degoulange, F. Madon, and R. Einhaus, Energy Procedia 124 (2017) 470.

[18] M. Mittag, U. Eitner, and T. Neff, Proc. 33rd European Photovoltaic Solar Energy Conf. Exhib., 2017, p. 48.

[19] M. Mittag, I. Haedrich, T. Neff, S. Hoffmann, U. Eitner, and H. Wirth, Proc. 31st European Photovoltaic Solar Energy Conf. Exhib., 2015, p. 93.

[20] C. Pönisch, L. Schanz, J. da C. Fernandes, M. Schmidt, and D. Kray, Proc. 40th European Photovoltaic Solar Energy Conf. Exhib., 2023, p. 020229.

[21] K. Nakamura, H. T. C. Tu, and K. Ohdaira, Abst. 24th Annual Mtg. Jpn. Soc. Appl. Phys. Hokuriku-Shin'etsu Chapter, 2020, p. B09. (in Japanese).

[22] K. Hara, H. Ichinose, T. N. Murakami, and A. Masuda, RSC Adv. 4, 44291 (2014).

[23] S. Shimpo, H. T. C. Tu, and K. Ohdaira, Jpn. J. Appl. Phys. 62 (2023) SK1039.

[24] V. Naumann, D. Lausch, A. Hähnel, J. Bauer, O. Breitenstein, A. Graff, M. Werner, S. Swatek, S. Großer, J. Bagdahn, and C. Hagendorf, Sol. Energy. Mater. Sol. Cells 120 (2014) 383.

[25] N. Shekhawat, A. Sharma, S. Aggarwal, and K. G. M. Nair, Opt. Eng. 50 (2011) 044601.

[26] https://x.com/JAIST16/status/1651007335593304066.

41st European Photovoltaic Solar Energy Conference and Exhibition

EVALUATION OF THE IMPACT OF THE UV EXCITATION INTENSITY ON THE ULTRAVIOLET FLUORESCENCE MEASUREMENT SYSTEM FOR PHOTOVOLTAICS

Zonghan Jiang[1], Carlos Meza[1,2], Hugo Sanchez[1], Ralph Gottschalg[1,2]
[1]Hochschule Anhalt University of Applied Sciences
[2]Fraunhofer Center for Crystalline Silicon Photovoltaics CSP
zonghan.jiang@hs-anhalt.de

ABSTRACT: Ultraviolet fluorescence (UVF) is an optical non-destructive inspection method that is used to evaluate Photovoltaics (PV) modules. By detecting the fluorescence intensity of UV fluorescence images, the degradation of PV module encapsulation can be characterized and visually assessed qualitatively. It would be beneficial to evaluate more quantitatively as there is further information on the state of the module in determining the magnitude and the spectrum of the fluorescence. Quantitative analysis requires an understanding of the correlation of the linearity of the detector, the response (=Fluorescence response) to the signal (=UV excitation), as well as the settings of the camera. Nowadays, the selection of the light intensity of the light source of the UVF measurement device is typically specified to be within a certain range, but the correlation to the fluorescence response as well as the camera settings is normally not given. The camera sensor might saturate to a level where there are pixels bleeding into each other, or it may be below sensitivity. Ideally, the signal (and thus excitation) should be in the linear range of the camera sensor. This paper gives a workflow to link the response of the DUT (device under test), camera sensor and excitation intensity and to arrive at higher quality measurements; the metrological framework will be evaluated in this paper, as well as the importance of sample temperature.
Keywords: Reproducible quantitative UVF, PV, Linearity of the camera, Influence of UV intensity on fluorescence, Influence of temperature on fluorescence

1 INTRODUCTION

UVF is a non-destructive optical inspection method which are widely used to evaluate the defects (Cracks, degradation of the encapsulant, etc) of the PV modules [1]. The principle of the UVF is, through analysis the fluorescence pattern or fluorescence intensity from the encapsulant or backsheet excited by the ultraviolet light, the defects of the PV module can be characterized. In order to gain further information of the degradation of the PV module, the quantitative UVF analysis needs to be developed. In the visual inspection, the impact of the UV excitation intensity has been already founded, furthermore, the absolute homogeneity of the UV light source is difficult to achieve. To achieve a reproducible camera-based UVF quantitative measurement, the relationship between UV excitation intensity and fluorescence result needs to be evaluated and the fluorescence result in every point in the UVF image should be corrected according to this relationship and real UV excitation intensity in the aim of future quantitative UVF measurement.

In the existing UVF studies, the selection of the light intensity of the light source of the UVF measurement device is typically specified to be within a certain range. For instance, in the research of Schlothauer et al., a 375nm UV-diode laser was used as a UV excitation light source [2]. However, the specific ultraviolet light intensity needs for UV excitation were not described in detail. In the experiment of Braden et al., a UVF device using a UV flash as an excitation light source was designed [3]. However, the specific ultraviolet light intensity needed for UV excitation was also not described in detail. In the research of Hobbs et al., A 365nm UV LED array with a total energy of 54W was used as the light source for the UVF device [4]. In the research of Köntges et al., an array of black UV light sources for excitation was installed in the device, and the value of ultraviolet light intensity is mentioned as 100 W/m2 [5]. However, there are no studies showing the selection criteria for the UV intensity of the light source for UVF equipment and the relationship between fluorescence intensity and excitation UV intensity.

In order to quantitatively use the fluorescence intensity or spectral information of the fluorescence to gain more information of the degradation or defects in the encapsulation of photovoltaic modules, in this study, the relationship between the UV intensity used for excitation and the detected fluorescence will be introduced to set a foundation to the future result correction in the UVF quantitative measurement system. Furthermore, the experimental factor (Linearity of the detector) and the environmental factor (temperature) which will influence the fluorescence result will also be evaluated. All these knowledges will provide the basis for the quantitative UVF measurement system.

2 EXPERIMENTAL SETUPS

A measurement system has been developed for the aim of this study. (See Fig. 1-2) The test system consists of two main components. A power supply with controlled output and a test station. The test station has two positions where different wavelengths of light can be installed and an RGB camera. The distance between the different light sources and the center of the camera is 100mm and the distance between the bottom and the top of the test station is 250mm. In the experiment, the settings of the camera should be fixed, Table I showed the settings of camera which should not be changed during the whole experiment. In addition, the aperture of the will be fully opened.

41st European Photovoltaic Solar Energy Conference and Exhibition

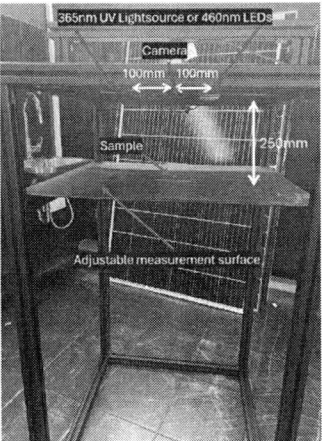

Figure 1: Overview of the measurement station

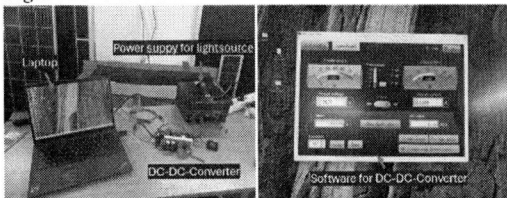

Figure 2: Output-controlled power supply system

Table I: Camera setting

Parameter	Value
Gain of camera sensor	1.0
Gain in blue layer	1.0
Gain in green layer	1.0
Gain in red layer	1.0
Auto exposure time	Off

3 INVESTIGATIONS OF THE LINEARITY OF THE CAMERA RESPONSE TO THE INPUT LIGHT SIGNAL

In order to use camera as detector to analysis the impact of the UV excitation intensity on the fluorescence result, the input light signal must be in the linear range of the detector. In order to evaluate the linearity of the camera response to the input light signal. A measurement system has been developed for that.

In the linearity test of the camera, two 450nm-460nm LEDs were placed at the measurement station above. (see Fig. 1) The reason for choosing this wavelength of the LED to calibrate the camera is that this wavelength is close to the wavelength of the blue fluorescence and is mainly located in the blue layer sensing region of the camera (See Fig. 3). Output-controlled power supply system is used to control the intensity of the LEDs via current change. A gray reference plate with 19 percent reflectivity was placed underneath the in the center of camera range (See Fig. 4). Because the wavelength of the LEDs mainly located in the blue layer, the blue layer will be mainly used to evaluate the linearity of the camera response to the signal.

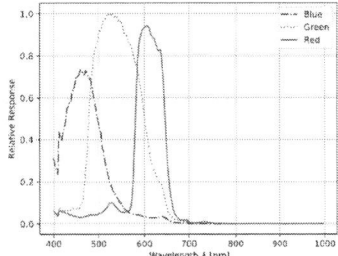

Figure 3: RGB relative response of the basler camera

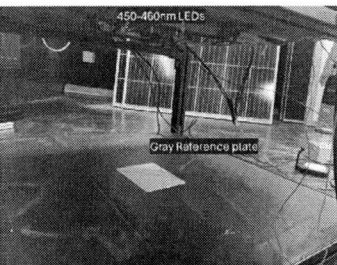

Figure 4: Linearity test of the camera

In the test, the exposure time of the camera will be firstly adjusted to 50000 µs, the current of the LEDs will be adjusted through the output-controlled power supply to achieve different light intensity. The intensity of the light in different current will be measured and calculated through the spectrometer and python (The integration of the spectra). The Figure 5 shows the measurement of the light intensity in the middle point.

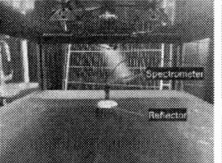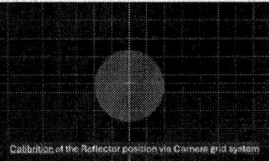

Figure 5: Measurement of the light intensity

After obtaining the light intensity of the center point corresponding to each current value, the pixel values of the center point of the image of the reference plate at different light intensities will be through python calculated and documented. Through the data obtained, the linearity of the camera will be investigated.

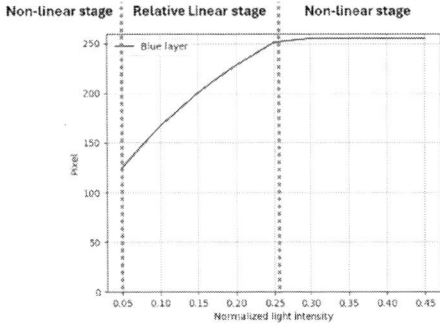

Figure 6: Pixel response to the normalized light intensity

The Figure 6 shows the camera' pixel response to the normalized light intensity. Due to the exposure time and unknown reason, the range from 0 to the first measurement point is non-linear. Furthermore, when the light intensity

661

reached to the threshold, the pixel value starts to saturate and can not be used to evaluate the input light. But there is a range between the 2 threshold which is relative linear. And through the two-lamp method with initial light, (Linearity was verified using the ratio of the pixel value when both lights were on at the same time to the value when both lights were on separately [6].) it was proved that the change of the exposure time (from 2000 μs - 50000 μs) didn't influence the linearity of the linear range.

This section demonstrates that before the UVF measurement, the exposure time of the camera should be adjusted properly so that the maximum measurement signal is under the saturation's threshold. Furthermore, the change from 0 to the first measurement point should be ignored because of the initial shift. In the future UVF quantitative measurement system, the fluorescence pixel value should be correlated with the true input light intensity to achieve a accurate measurement result.

4 CORRELATIONS BETWEEN FLUORESCENCE AND UV INTENSITY

To investigate the fluorescence intensity in relation to the excitation UV intensity, the same measurement station (see Fig. 1) will be installed with two 30W UV LED strip together with the ZWB2 Bandpass filter (See Fig. 7).

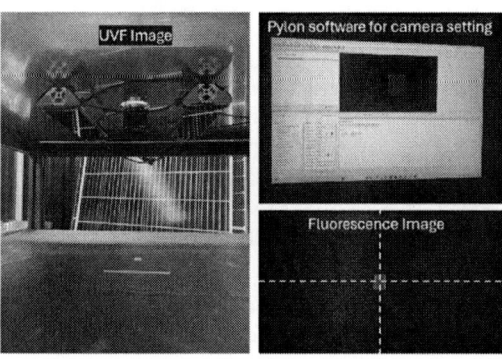

Figure 7: UVF measurement of the sample (left), using software to capture UVF image in the measurement (right top and bottom)

The following steps which is similar as above were carried out.

o Adjustment of the current of the UV LEDs by the controller to realize the change in the excitation UV intensity.
o Calibrate the middle test position of the spectrometer using the grid that comes with the camera software. (See Fig.7)
o The spectra of the middle test points at different currents were collected by the spectrometer and the sum of the light intensity in the UV region of the spectrum was calculated.
o Capture UVF pictures of the sample at different excitation UV intensities and calculate the change in pixel value at the middle test point.

The samples which were used in the test were EVA film, FFC//PET//FFC backsheet, laminated EVA film with FFC//PET//FFC backsheet, laminated TPO film with PPE backsheet, and mini photovoltaic modules encapsulated with EVA. All film-based samples were aged under DH-UV aging conditions for 1250h, while the mini-PV module

was also aged under DH-UV for 250h.

Tests will be performed in a dark room to avoid the influence of other light on the results. The exposure time of the camera will be adjusted to 15000 μs, so that the maximum pixel value is under the saturation's threshold and the range between 0 to the first signal point will be ignored. So that all the measured signal located in the linear range. In addition, each test will be completed in a very short period to avoid the influence of the LED temperature change on the test results.

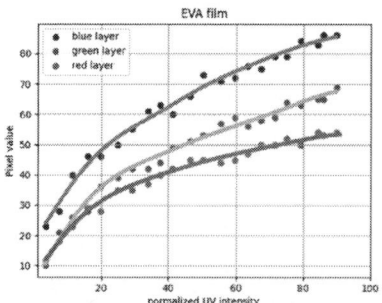

Figure 8: Correlation between excitation UV intensity and Fluorescence pixel value (EVA film)

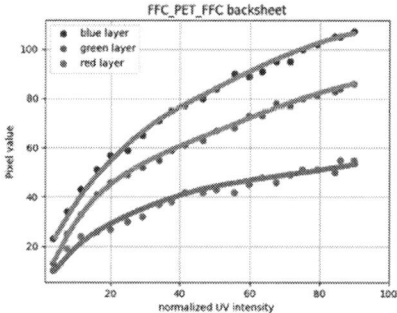

Figure 9: Correlation between excitation UV intensity and Fluorescence pixel value (Backsheet FFC_PET_FFC)

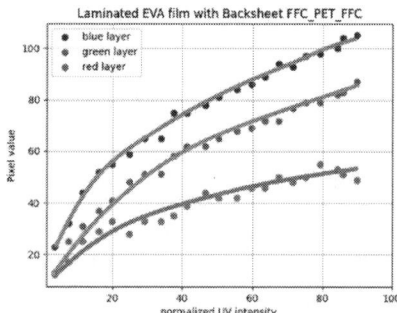

Figure 10: Correlation between excitation UV intensity and Fluorescence pixel value (Laminated EVA film with Backsheet FFC_PET_FFC)

41st European Photovoltaic Solar Energy Conference and Exhibition

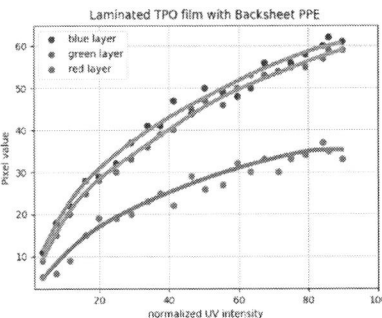

Figure 11: Correlation between excitation UV intensity and Fluorescence pixel value (Laminated TPO film with Backsheet PPE)

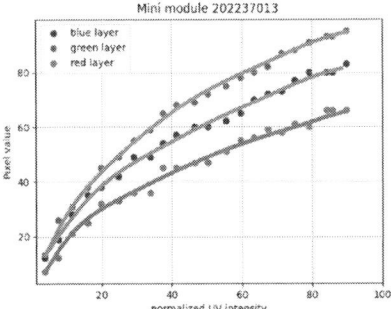

Figure 12: Correlation between excitation UV intensity and Fluorescence pixel value (Mini module after 250h DH-UV aging)

From Fig 8-12, it can be shown that the fluorescence pixels under different excitation UV intensity of all the samples. In all samples, the change in pixels of fluorescence across the layers with growing UV excitation intensity exhibited a parabolic-like growth trend. On the one hand, this increasing trend proves that UV excitation intensity has a great influence on the fluorescence results. It further demonstrates the necessity of excitation intensity control for future quantitative analysis of the degradation of photovoltaic encapsulant by fluorescence intensity. On the other hand, the parabolic type of fluorescence intensity trend illustrates that if the UV excitation intensity is further increased, the intensity of fluorescence will most likely saturate and stop increasing. This phenomenon could not be investigated due to the limitation of the maximum power of the UV source in this experiment (17,035 counts at 368 nm at the test point).

The result above illustrate that the excitation UV intensity affects the intensity of the detected fluorescence,

Through the relationship between the UV excitation intensity and the fluorescence, the fluorescence result in a UVF measurement system with non-uniform UV light source can be corrected according to this relationship.

5 INFLUENCES OF THE SAMPLE TEMPERATURE ON THE FLUORESCENCE

The temperature of the sample is also a factor that affects fluorescence results. To investigate the relationship between them, a mini temperature control device (See Fig. 13) was constructed. The device consists mainly of Peltier elements as well as copper plates. The temperature of the

sample is varied by the good thermal conductivity of the copper and the corresponding temperature change of the two surfaces of the Peltier element in response to the current.

Figure 13: Temperature control device and measurement of the sample temperature

The temperature of the samples (laminated EVA film with FFC//PET//FFC backsheet and laminated TPO film with PPE backsheet) will be raised separately by the temperature control device and measured 5 times by infrarot-thermometer and the average value will be calculated. At the same time, the samples will be excited with a constant UV intensity and 3 UVF images will be captured by the camera with the exposure time 15000 µs. In order to avoid small fluctuations in the camera pixels themselves from affecting the results, the average of the pixels of the test points in multiple pictures will be calculated.

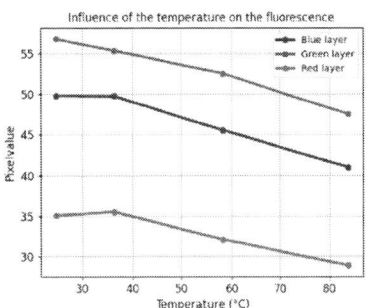

Figure 14: The impact of the sample temperature on the fluorescence pixel value (laminated TPO film with PPE backsheet)

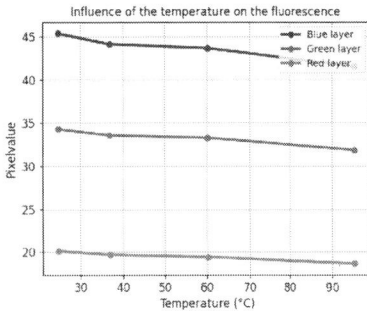

Figure 15: The impact of the sample temperature on the fluorescence pixel value (laminated EVA film with FFC//PET//FFC backsheet)

Fig. 14 and Fig. 15 can show that the fluorescence of different samples clearly has different responses to temperature changes. When the temperature of sample 1 increased, the pixel values of fluorescence showed a clear decreasing trend, but at the same time, the pixel values of fluorescence of sample 2 produced a decreasing trend, but it was weak.

663

The result above demonstrates that the fluctuation of the temperature will lead to the uncertainty of the result. The monitoring of the sample temperature as well as the knowledge of the BOM (Bills of materials) of the measured PV module is essential prerequisite for the future camera-based UVF quantitative analysis.

6 CONCLUSIONS

This paper focuses on the impact of UV excitation intensity as well as external factors on UVF results and describes some necessary control variables and the possibilities of the result correction for future quantitative UVF measurement system.

This investigation showed that the calibration of the detector played an important role in the future quantitative UVF measurement system. The detector should be adjusted or calibrated firstly so that all measured signal located in the linear response range. In the UVF quantitative measurement system, the pixel value should be accurate correlated to the real input light signal.

It was also demonstrated that the relationship between fluorescence intensity and the UV excitation intensity shows a parabolic form. Through this relationship, the uncertainty of the fluorescence result in the UVF image caused by the non-uniform UV light source can be corrected.

Moreover, the fluorescence of the detected samples showed a decreasing trend as the temperature of the samples increased, but the fluorescence response to temperature was different for different samples. The result above demonstrate that the fluctuation of the temperature will lead to the uncertainty of the result, monitoring of the sample temperature and bill of materials of the PV module are essential prerequisite for the camera-based quantitative UVF measurement.

7 ACKNOWLEDGMENTS

This work is supported by the German Federal Ministry of Economics and Climate Protection (BMWK) under the project "EVAplus". Funding code: 03EE1112E

8 REFERENCES

[1] Kontges, Marc; Morlier, Arnaud; Eder, Gabriele; Fleis, Eckhard; Kubicek, Bernhard; Lin, Jay (2020): Review: Ultraviolet Fluorescence as Assessment Tool for Photovoltaic Modules. In IEEE J. Photovoltaics 10 (2), pp. 616–633. DOI: 10.1109/JPHOTOV.2019.2961781.

[2] Schlothauer, Jan; Jungwirth, Sebastian; Köhl, Michael; Röder, Beate (2012): Degradation of the encapsulant polymer in outdoor weathered photovoltaic modules: Spatially resolved inspection of EVA ageing by fluorescence and correlation to electroluminescence. In: Solar Energy Materials and Solar Cells 102, S. 75–85. DOI: 10.1016/j.solmat.2012.03.022.

[3] Gilleland, Braden; Hobbs, William B.; Richardson, Joseph B. (2019): High Throughput Detection of Cracks and Other Faults in Solar PV Modules Using a High-Power Ultraviolet Fluorescence Imaging System. In: 2019 IEEE 46th Photovoltaic Specialists Conference (PVSC). 2019 IEEE 46th Photovoltaic Specialists Conference (PVSC). Chicago, IL, USA, 2019/6/16 - 2019/6/21: IEEE, pp. 2575–2582.

[4] Hobbs, William B.; Johnston, Steve; Gilleland, Braden (2020): Ultraviolet Fluorescence Bleaching Rates for New Cell Cracks. In: 2020 47th IEEE Photovoltaic Specialists Conference (PVSC). 2020 IEEE 47th Photovoltaic Specialists Conference (PVSC). Calgary, AB, Canada, 2020/6/15 - 2020/8/21: IEEE, pp. 2350–2355.

[5] Köntges, M.; Kajari-Schröder, S.; Kunze, I. (2013): Crack Statistic for Wafer-Based Silicon Solar Cell Modules in the Field Measured by UV Fluorescence. In IEEE J. Photovoltaics 3 (1), pp. 95–101. DOI: 10.1109/JPHOTOV.2012.2208941.

[6] Emery, K.; Winter, S.; Pinegar, S.; Nalley, D. (2006): Linearity Testing of Photovoltaic Cells. In: 2006 IEEE 4th World Conference on Photovoltaic Energy Conference. 2006 IEEE 4th World Conference on Photovoltaic Energy Conference. Waikoloa, HI, 2006/5/7 - 2006/5/12: IEEE, pp. 2177–2180.

41st European Photovoltaic Solar Energy Conference and Exhibition

How to Mount PV Modules: the Effect of Different Clamping Configuration on Mechanical Stresses in PV Modules

Pascal Romer, Andreas J. Beinert, Charlotte Hasselblatt, Cornelius Herr
Fraunhofer Institute for Solar Energy Systems ISE, Heidenhofstr. 2, 79110 Freiburg, Germany
Phone +49 761/4588-5044, Pascal.Romer@ise.fraunhofer.de | www.ise.fraunhofer.de/module-fem

- Installation manuals of photovoltaic (PV) modules frequently outlines various mounting configurations
- As per IEC 61215 [1], each mounting configuration must withstand a mechanical load of at least 2400 Pa
- To save cost and time often only most critical configuration is tested
- **Question: which configuration is most critical, and which is most effective in minimizing mechanical stress in PV modules?**
- **To answer: using FEM (finite element method) to examine impact of various mounting configurations and clamp length on stresses in PV modules**

Method

- 3D FEM model of framed 1.8 x 1.1 m² glass-backsheet PV module, containing 144 M6 half-cells
- Simulation of a homogeneous mechanical load of 5400 Pa acting on the frontglass
- Variation of
  1. Clamp length: 50 mm → 100 mm
  2. Clamp position at the long side of the PV module
  3. Clamp position at the short side of the PV module
- Evaluation of the first principal stress in both the solar cells and the front glass
- **Identification of the clamping configuration that results in the lowest stress** in either solar cells or front glass

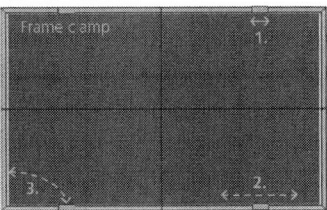

Fig. 1: Module geometry alongside the simulated parameter variations of the module clamping

Results

1. Clamp length

- Reduction of clamp size: 100 mm → 50 mm
  - Deflection increases:
    44.1 mm → 47.2 mm
  - First principal stress in glass increases:
    106 MPa → 122 MPa
  - First principal stress in solar cells increases:
    78 MPa → 84 MPa
- **Larger clamp size results in lower stresses**

2. Clamp position at long side

- Variation of clamp position
  - Minimal deflection using a clamping position of 15 % module length
  - Minimal first principal stress in frontglass at a clamping position of 15 % module length
  - Minimal first principal stress in solar cells clamping at modules corner
- **Clamping at modules corner results in lower stress in solar cells**
- **Optimal clamping position for glass: 15 % of module length**

3. Clamp position at short side

- Variation of clamp position
  - Minimal deflection clamping at modules corner
  - Minimal first principal stress in frontglass clamping at modules corner
  - Minimal first principal stress in solar cells clamping at modules corner
- **Clamping at modules middle results in lower stress in the solar cells**
- **Clamping at modules corner result in lower stress in the frontglass**

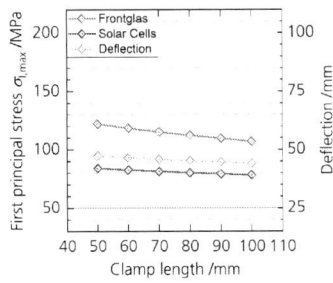

Fig. 2: First principal stress in the frontglass (green), solar cells (blue) alongside the modules deflection (grey) for clamp length between 50 mm and 100 mm.

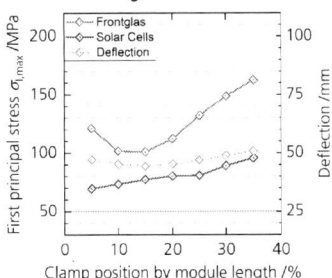

Fig. 3: First principal stress in the frontglass (green), solar cells (blue) alongside the modules deflection (grey) for different clamping positions at the modules long side.

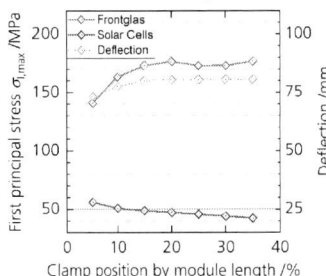

Fig. 4: First principal stress in the frontglass (green), solar cells (blue) alongside the modules deflection (grey) for different clamping positions at the modules short side.

Summary

- Effect of **different clamping configurations** on **mechanical stress** in PV modules is investigated using FEM
- **Longer clamp** reduces both modules deflection as well as stresses in frontglass and solar cells
- **Clamping position on long side at 15 % of module length** minimizes first principal **stress in frontglass**
- **Clamping position on short side** reduces first principal **stress in solar cells**

3AV.2.26 | EUPVSEC

1 International Electrotechnical Commission (IEC), IEC 61215-2:2021 Terrestrial photovoltaic (PV) modules – Design qualification and type approval: Part 2: Test Procedures 2, accessed 21 April 2021.

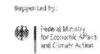

This research was funded by German Federal Ministry for Economic Affairs and Climate Action under the projects SiMilar (Grant number 03EE1131A).

41st European Photovoltaic Solar Energy Conference and Exhibition

FMEA BASED DEGRADATION RATE EVALUATION TO STUDY IMPACT OF DIFFERENT FAILURE MODES AS FUNCTION OF MISSION PROFILES

Bengt Jaeckel*[1], Baloji Adothu[2], Vivian Alberts[2], Matthias Pander[1]
[1] Fraunhofer-Center for Silicon-Photovoltaics (CSP), Halle (Saale), Germany
[2] DEWA Research & Development Center, P.O.Box: 564 Dubai, United Arab Emirates
* Corresponding author: phone +49 345 5589-5135 | Bengt.Jaeckel@csp.fraunhofer.de

ABSTRACT: Crystalline silicon modules were, are and will be the dominant solar cell technology. Associated with that different cell technologies (e.g. Al-BSF, PERC, TOPCon, HTJ, IBC) show different behaviors against outdoor stressors. Furthermore, the PV module encapsulation system (such as front cover, encapsulation material, rear cover, and edge seal) will additionally influence the long-term reliability of the PV system. Therefore, it is important to understand the most relevant factors and clarify expectations. Costs are the main driver to reduce material consumption and increase efficiency in all steps of the value chain, including cell efficiencies. Business case for a particular system is made from system designs (energy yield), assumed annual degradation rates, building cost (CAPEX), and operation costs over the project live time (OPEX). The work laid out in this contribution focuses on the technical risks associated with assumed degradation rates for financial modeling and how to address the risk with proper testing.
A Failure Mode and Effects Analysis (FMEA) study was undertaken to identify, prioritize, and rate different major field observations with the aim to achieve a deeper understanding of the real impact on degradation rates. Different types of pillars are derived to further support testing development with respect to the relevant mission profile (MP) of the PV module/system. As pillars the following categories were used: Design & Quality issues, MP-linked degradation mode (hot desert, temperate climate, etc.), and safety aspects. The reference point for this study is the warranty given "allowed maximal" degradation rates by the manufacturers. If on a contract base other warranty terms are used is outside of the scope of this work but could be adjusted to it.
The major and most relevant topics identified were related to cell metallization issues (e.g. finger interruptions on MBB-design cells), interconnection ribbon fatigue, including a detachment of ribbons from cross connectors, and change in polymeric materials (discolorations).

Keywords: PV Module, outdoor, FMEA, Defects, degradation

1 INTRODUCTION

Historically from the last decade, most modules come with a performance warranty of 25 years. Those are currently increased even more [1][2] to thirty-plus years.
Most known failures today are more infant mortality issues, typically occurring within the first 5 to 10 years of operation [3]-[12]. Current standards are developed for this to screen out engineering failed designs. Additionally, most of the knowledge comes from moderate climates such as middle Europe. The large PV system behavior for more harsh climates like deserts is limited to less than 10 years of operation. The degradation rate and failure modes are different in different climatic zones [13]. However, in recent years the number of reports from those regions increased and several field-relevant observations and failure modes were presented [14]-[16]. The individual findings show different severities and are linked to differences in degradation rate as a function of the mission profile [17]-[21] associated to total in plane irradiance and operating temperatures. This leads to lowest impact on the 90° installation that is most clearly visible in the lowest degradation rate found in the outdoor test facility (OTF) from Dewa Research center and is given in Figure 1.
The aim of conducting the field-relevant degradation mode FMEA was to identify and prioritize the most relevant and critical failure modes for a certain installation location. The focus herein is on separating the most relevant failure modes associated with hot and sunny locations compared to observations that occur everywhere.
Furthermore, it is important to understand the differences between known technologies with yearlong outdoor experiences (e.g. 2-5 bus bar interconnections) compared

to todays and next-gen cells and interconnection technologies such as multi-busbars, low-temperature soldering or the use of ECAs, in conjunction with TOPCon, HJT, IBC and long-term Tandem-cell structures.

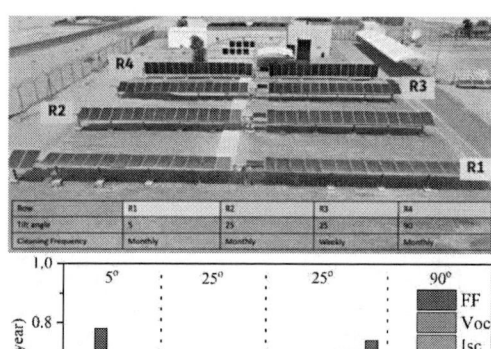

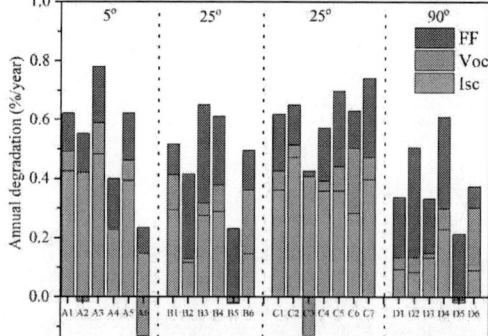

Figure 1: The Outdoor Test Facility (OTF) from DEWA Dubai contains more than 7 different modules of different suppliers manufactured between 2014-2015. Furthermore, the modules are mounted in different tilt angles (5, 25, 25, 90) leading to significant different total in plane irradiance levels.

666

The final goal of the degradation rate-based FMEA is to identify the most critical failure mode for hot climates and to propose field-relevant and failure mode mimicking test sequences to improve PV module testing for such harsh climates.

2 FMEA BASICS

A *Failure Mode and Effects Analysis (FMEA)* study was undertaken to identify, prioritize, and rate field defects with the aim to focus only on desert-specific or at least highly accelerated failure modes in desert applications. Following the methodology of the FMEA approach the Risk Priority Number RPN is the multiplication of the rating values for Severity S, probability of Occurrence O, and probability of Detection D (in the factory). Latter is important as we want to focus on variations of module production processes and not on design or specific climatic / weather events.

Severity, Occurrence, and Detection can have numbers from 1 (no impact, 100% detection probability) to 10 (severe impact, very low probability of detection, safety hazard) resulting in Priority numbers (Risk Priority Numbers RPN) from 1 (best product, no problems → not likely) to 1000 (very high risk of catastrophic failure → can happen for new, fast and not in detail developed

products). Table 1 gives the details for the evaluation of the FMEA criteria.

As known issues and modes of failure/degradation anything within the PV module was evaluated as a potential cause of failure/degradation. Some of them lead to general issues, some to a potential safety hazard and some cause different rates of degradation based on climate. Those are within this paper the most relevant.

Prior to the FMEA possible degradation modes and root causes were identified to better organize the FMEA. The following fundamental modes were identified:

- Cell, grid finger breakage
- Breakage (fatigue) of cell and cross-connector interconnections and ribbons
- Manufacturing issues for cell, ribbon and cross-connector misalignment
- Discolorations at electric circuitry/metal parts
- Discolorations of encapsulant, backsheet, and other polymeric material-based components
- Delamination and bubble formation
- Power loss due to LID, LeTiD, UVID, PID, and other cell effects
- Others such as glass breakage, frame bending

This list is not meant to be complete but covers the most relevant known failures and observations.

Table 1: FMEA evaluation criteria

Probability of occurrence O	Impact / Severity S	probability of Detection D	Points
unlikely	hardly noticeable	The error is **detected in any case by common measures**, so that a very prompt correction is possible	1
Very low	rather insignificant	moderate	2 bis 3
	Aesthetic, very little impact on performance, no impact on safety, does **not fall under warranty**	Procedures are in place to enable **early detection (end of line and/or a few days of lab test)**. However, measurement and test results can be misinterpreted, for example, or incorrect assessments can lead to the defect being overlooked / disregarded. Understood processes (e.g. LID; LeTid, PID) that can be part of QA in cell and module manufacturing.	
low	moderately serious error	low	4 bis 6
	measurable and **clear influence on performance, degradation rate at maximum warranty level**, increased clarification effort with customers to be expected, no influence on safety. Possibilities for compensation or repair are feasible at reasonable expense	**Only detectable with accurate and comprehensive testing** which is rarely performed on small samples - time of detection likely to be after module delivery	
Moderate	serious error	very low	7 bis 8
	Measurable and **clear influence on performance, degradation rate above Warranty**, influence on electrical safety or fire. High liability risk and high follow-up costs to be expected (e.g. module replacement)	**not detectable with common QA measures**. Process monitoring does not provide any clues. Detectable with very elaborate tests, but these are not routinely performed. Time of detection for sure only after module delivery. **Unknown processes or interactions, that take very long to detect in both in Lab and/ or several years in the field.**	
High	extremely serious error	unlikely	9 bis 10
	Can lead to massive injuries or death, severe fire hazard	There are **practically no effective early detection methods**. Defects only **become evident in operation** after (long-term) some time (long-term/ years) in the field	

3 FMEA RESULTS

In the first round of the FMEA process, more than forty different modes were discussed. It turned out that not all of them are specific for a desert climate, however, may be seen as more dominant in such regions. The detailed results are given at the end of the paper in Table 2.

The table provides a short description of the issue, followed by a statement about impact of climate and

mission profile MP. The next columns deal with a possible root cause process during manufacturing, type of module design, and if there is either a power warranty or potential electrical safety hazard. From these columns, the severity S of the mode is defined. To better understand the occurrence a statement is given on how the Occurrence O was determined. Severity and Occurrence combined with give the first prioritization number PN.

The next column defines the possibilities of detection D. Low numbers can be achieved e.g. if a defect can be already identified in the factory so that such modules not even get shipped out. The other end of the scale is 10, where even with extensive testing the failure can't be found because the root cause is not reproduced within the known test protocols.

The sum will lead to the risk priority number RPN where high numbers mean severe or very difficult to find defects and degradation modes. The table is sorted by RPN from high to low.

The last column was added to judge finally on desert relevance. For example, the cracking of a backsheet or discoloration of encapsulation materials are not specific to a desert location. Such findings are well known for most locations where PV is used. However, the time to occurrence is typically shorter in a hot and sunny climate. But depends on the chemical process that induces the change in appearance. Moisture might also be required, and this reactant might not be excessively present in a desert climate. The complexity of such reactions is e.g. documented in [22]-[25]. Typically, not only one reaction runs at the same time and chemical reaction rate constants, which are dependent e.g. on concentration and temperature, mainly drive certain findings in one, but not in another climate.

Finally, it must be acknowledged that the given numbers are not absolute correct and may differ from persons who do the FMEA. But the order of the findings will most likely be the same. And this is the most relevant outcome. Within the group these results were achieved, the highest RPN was 336 for solar cell grid finger breakage, and issues from soldering were solder joints got weak. Based on thermal-mechanical loads the next high RPM come from fatigue of ribbons and other electrical interconnections.

Issues of polymers (encapsulation, backsheet) were also rated high, due to the known impact of UV and temperatures.

Generally low numbers were derived for known effects caused e.g. by cell manufacturing (PID; LeTID) or manufacturing issues because the can be very easy detected within the monitoring of the cell/ module manufacturing and therefore should have quite low occurrence numbers.

Cell breakages were e.g. rated relatively low. Even with a high occurrence the severity is today known not to be to critical [26][27] and by e.g. EL or IR most critical cells/ modules can be sorted out very easily.

4 CONCLUSIONS

Table 2 provides a lengthy list of field observations and obviously is not complete, especially for next-generation cells such as TOPCon and HJT.

The raking provides a good indication to major problems derived from experience and depending on failures will also be relevant for future PV module types.

The last column for "special desert relevance" is a point for discussion. Some failures such as cracking and chalking of backsheets occur everywhere but still have a climate/ time dependence. Where to start and where to end here is difficult. The point here is to further focus on topics that are most relevant for hot and sunny climates but may also include special applications where modules also achieve very high operating temperatures such as in BiPV applications. This thinking also is in alignment with current work within the IEC 63126 project team.

To summarize, the major and most relevant topics were

identified to be:
- Cell metallization issued (e.g. finger interruptions on MBB-design cells)
- Interconnection ribbon fatigue, including detachment of ribbons from cross connectors
- Change in polymeric materials such as discolorations and chalking of backsheets

5 ACKNOWLEDGEMENTS

This work was funded by joint research project between DEWA Research & Development Center and Fraunhofer CSP, and the Federal Ministry for Economic Affairs and Climate Action in the project SegmentPV with grant #03EE1180B.

6 REFERENCES

[1] "International Technology Roadmap for Photovoltaic," 12th Edition Results 2023, 2024.

[2] Solar Power Europe, "Global Market Outlook 2023-2027"

[3] J. C. Coello et al., "Analysis of the Degradation of 735 Commercial Crystalline Silicon Modules after First Operation Year", 25th European Photovoltaic Solar Energy Conference - Valencia, 2010

[4] D. Jordan et al, "Photovoltaic Degradation Rates - an Analytical Review", Progress in Photovoltaics: Research and Applications, 2011, 18

[5] L. G. Garreau-Illes et al., "Backsheet and Module Durability and Performance and Comparison of Accelerated Testing to Long Term Fielded Modules", 28th European Photovoltaic Solar Energy Conference - Paris, 2013

[6] D. Jordan et al., "Compendium of photovoltaic degradation rates", Progress in Photovoltaics: Research and Applications, 2016, 24, 978-989

[7] J. L. Garcia et al., "Degradation analysis of PV module technologies in a moderate subtropical climate", 36th European Photovoltaic Solar Energy Conference - Marseille, 2019

[8] D. Jordan et al., "PV field reliability status - Analysis of 100 000 solar systems", Progress in Photovoltaics: Research and Applications, 2019

[9] K. Kiefer et al., "Degradation in PV Power Plants: Theory and Practice", 36th European Photovoltaic Solar Energy Conference - Marseille, 2019

[10] P. Lechner et al., "Bewertung von Auffälligkeiten und Schäden an der Rückseitenfolie von PV-Modulen", PV Symposium Bad Staffelstein, 2019

[11] M. Bolinger et al, "System-level performance and degradation of 21 GWDC of utility-scale PV plants in the United States", Journal of Renewable and Sustainable Energy, 2020, 12, 043501

[12] N. Bosco et al., "Defining Threshold Values of Encapsulant and Backsheet Adhesion for PV Module Reliability", 45th IEEE Photovoltaic Specialists Conference, 2017

[13] B. Adothu, "Comprehensive review on performance, reliability, and roadmap of c-Si PV modules in desert

climates: A proposal for improved testing standard", Progress in Photovoltiacs, 2024, https://doi.org/10.1002/pip.3827

[14] S. Kumar et al., "Defects and Degradations in Photovoltaic Modules from Hot Middle East Deserts", 40th EU PVSEC, 2023

[15] S. Kumar et al., "Identification and Analysis of Metallization Defects in Desert-Operated Photovoltaic Modules", 40th EU PVSEC, 2023

[16] B. Adothu et al., "Identification and Investigation of Materials Degradation in Photovoltaic Modules from Middle East Hot Desert", 40th EU PVSEC, 2023

[17] B. Jaeckel et al., "Mission profile concept for PV modules: use case – middle east deserts vs temperate European climate", EPJ Photovoltaics 14, 39 (2023), https://doi.org/10.1051/epjpv/2023030

[18] M. Pander et al., "Lifetime estimation for solar cell interconnectors", 28th EU PVSEC, 2013, 10.4229/28thEUPVSEC2013-4CO.10.3

[19] H. Hanifi, "Comparison of optical gains and electrical losses in modules with different designs of partial cells in desert regions", 35th EU-PVSEC, 2017, 10.4229/EUPVSEC20172017-5BV.4.49

[20] M. Pander et al., "Digital Prototyping – Application of Numerical Methods in Module Development", 36th EU PVSEC, 2019, 10.4229/EUPVSEC20192019-4BO.11.2

[21] M. Pander et al., "Numerical design study for desert climate applications of bifacial PV modules", 40th EU PVSEC, 2023

[22] R. Heidrich, "Spatially Resolved Degradation of Solar Modules in Dependence of the Prevailing Microclimate", IEEE Journal of Photovoltaics (Volume: 14, Issue: 5, September 2024), 10.1109/JPHOTOV.2024.3414179

[23] R. Heidrich, "From Performance Measurements to Molecular Level Characterization: Exploring the Differences between Ultraviolet and Damp Heat Weathering of Photovoltaics Modules", Solar RRL, 2024, https://doi.org/10.1002/solr.202400144

[24] R. Heidrich, "UV lamp spectral effects on the aging behavior of encapsulants for photovoltaic modules", Solar Energy Materials and Solar Cells, Volume 266, March 2024, 12674, https://doi.org/10.1016/j.solmat.2023.112674

[25] R. Heidrich, "Diffusion of UV Additives in Ethylene-Vinyl Acetate Copolymer Encapsulants and the Impact on Polymer Reliability", IEEE Journal of Photovoltaics, vol. 14, no. 1, pp. 131-139, Jan. 2024, 10.1109/JPHOTOV.2023.3333198

[26] J. Arp, "Elektrisches Schalten von Mikrocracks: Leistungsverlust: Ja, Nein, Ja," Bad Staffelstein, 2016.

[27] B. Jaeckel, "Long Term Statistics on Micro Cracks and Their Impact on Performance," 31st European Photovoltaic Solar Energy Conference - Hamburg, 2015.

Table 2: Extended table of failure modes covered within this contribution. The Severity runs from 1-hardly noticeable to 10-deadly, Occurance O from 1 – very unlikely to 10 all goods are impacted, and the detection D from 1 easy, always to 10 difficult, not possible to detect

#	Issue, Degradation mode description	Detailed description of issue, impact of Climate and MP	Type of material or process mainly involved to cause issue	Influenced by BOM / module construction? Yes: state influence no: state NOT	Potential degradation rate %/a	Associated risks such as fire, hazardous electrical potentials	Severity S	Description why occurrence is judged to by high or low	Occurance O	PN (S*O)	Description of method of detection. What is the likelihood for detection	Detection D	RPN (S*O*D)	Special relevance for desert applications Yes / No
7	Finger interruptions/breakages	Caused by miss-aligned ribbons during module manufacturing, causing a significant increase in Rs over the long term, more severe an energy yield in desert, early appearance in desert environments	Soldering process	More sever for thin wire	Potentially above warranty	Low	8	Serial defect - all modules will be affected by a single stringer	6	48	Takes time to see in the field, Detection e.g. By EL	7	336	Yes
9	Detachment of ribbon from cell	Due to thermal-mechanical stress the interface between cell (BB) and ribbon (e.g. In ECA or soldering area)	Ribbon/cell interface soldering	No	Potentially above warranty	Low	8	Serial defect - all modules will be affected by a single stringer	6	48	Takes time to see in the field, Detection e.g. By EL	7	336	Yes
8	Ribbon fatigue	Ribbon breaks due to thermal-mechanical fatigue, causing a significant increase in Rs over the long term, more severe an EY in the desert, early appearance in desert environments	Demission of ribbons, cell spacing, type of material	GG/GB both effected	Moderate... Potentially above warranty	Low risk for arcs, medium risk for hot-spot, depending on number of BB	8	For utility-scale used module the likelihood is medium. On special applications this could be very high, also depending on the MP	5	40	Takes time to see in the field, Detection e.g. By EL	7	280	Yes
11	Failure of cross-connection and ribbon	Failure of interconnection between cross connector and ribbon due to corrosion or thermal-mechanical stress; causing significant increase in Rs over the long term, more severe an EY in Desert, early appearance in desert environments	Type of encapsulation materials and composition of metal part	GG/GB both effected	Moderate.... Close to warranty	Low risk for arcs, medium risk for hot-spot, depends on number of BB	8	For utility-scale used module the likelihood is medium. On special applications this could be very high, also depending on MP	5	40	Takes time to see in the field, Detection e.g. By EL if full failure	7	280	Yes
21	Backsheet: cracking	Initially, it is a visual observation that can lead to insulation faults and shut down of power plant	Depends on the material of the backsheet	GB modules	<0.1%/a	High risk of electrical shock	9	Time and location dependent occurrence depends on specific BOM configuration if BOM susceptible all modules of this type will be affected	6	54	Certain materials are known to show this, those should not be used! Depending on MP cracking can occur during expected lifetime, some tests for detection exist (e.g. Solder-bumb test, ext. Stress test)	5	270	No

#	Issue, Degradation mode description	Detailed description of issue, impact of Climate and MP	Type of material or process mainly involved to cause issue	Influenced by BOM / module construction? Yes: state influence no: state NOT	Potential degradation rate %/a	Associated risks such as fire, hazardous electrical potentials	Severity S	Description why occurrence is judged to by high or low	Occurrence O	PN (S*O)	Description of method of detection. What is the likelihood for detection	Detection D	RPN (S*O*D)	Special relevance for desert applications Yes / No
17	Encapsulant: Discoloration	Due to temperature, humidity and uv light parts of the encapsulation material show discolorations general influence on power is actually low, but discolorations are precursors for other reactions that can cause e.g. Finger corrosion and detachment --> increase in Rs	Composition of encapsulation material and moisture ingress	GG/GB both effected	<0.1%/a	Low	4	Lots of modules are showing different color changes. Very strong discolorations are typically a sign of (initial) corrosion time and location dependent. Really depends on MP, including the result of such an optical observation	8	32	Visual inspection in the field, no early detection possible	6	192	Maybe
34	Powerloss: LID / LETID / PID / Temperature / UV	New cell technologies such as HJT, TOPCon etc. N-Type wafers, new dopants (e.g. Ga)	Cell type wafer dopants wafer size	Solar cell technology	Potentially above warranty	Low	6	Statistically not known	6	36	LID and LETID tests exist root cause not known for some observations such as e.g. Voltage drops	5	180	Yes temperature and high irradiance levels are relevant
15	Discoloration: Front of cell Or front of the backsheet	encapsulation change in solar cell coating visually seen from the front side of the module	Encapsulation and backsheet materials, cell technology (ARC)	GG/GB both effected	2% absolute loss of power	Low	3	Time and location dependent occurrence depend on specific BOM configuration if BOM susceptible all modules of this type will be affected	7	21	Visual inspection in the field, not early detection possible for a lot of observations no 100% reproducible indoor test exists	7	147	Maybe
13	Corrosion of internal metallic connections	Corrosion of ribbons and cross connectors, starting with browning, greenish stuff, depends on MP, mainly temperature and humidity	Type of encapsulation materials and composition of a metal part	GG/GB both effected	Moderate.... Close to warranty	Low risk for arcs, medium risk for hot-spot, depending on number of BB	5	Lots of modules are showing different color changes of the ribbons and cross connectors over time. Strong corrosion can lead to fatal failure	5	25	By visual inspection, Severity of corrosion can't be quantified by visual inspection nor EL, MFI might give insights	5	125	No
2	Cell breakage	General handling (Transportation, installation)	Mechanical stress	Depends on design (GG, GF, Frame)	<0.5%/a / warranty	Low risk for arcs, medium risk for hot-spot, depending on number of BB	3	Depends on how personal is trained, and how good the boxing of modules is	8	24	Batch testing, costly, time-consuming, not 100%	5	120	No
4	Cell breakage	Cleaning robots cause vibration during cleaning causing mechanical stress to cells	Mechanical stress	Depends on design (GG, GF, Frame)	Quite small, absolute to be determined	Low risk for arcs, medium risk for hot-spot, depending on number of BB	3	Depends on cleaning protocol and frequency	8	24	Batch testing, costly, time-consuming, not 100%	5	120	Yes

#	Issue, Degradation mode description	Detailed description of issue, impact of Climate and MP	Type of material or process mainly involved to cause issue	Influenced by BOM / module construction? Yes: state influence no: state NOT	Potential degradation rate %/a	Associated risks such as fire, hazardous electrical potentials	Severity S	Description why occurrence is judged to by high or low	Occurrence O	PN (S*O)	Description of method of detection. What is the likelihood for detection	Detection D	RPN (S*O*D)	Special relevance for desert applications Yes / No
31	Discoloration and abrasion of ARC coatings of glass	Discoloration and abrasion of ARC coatings of glass	soiling-induced abrasion, removal of ARC, lead to reduce, transmittance and Isc. Coating thickness reduction.	Compositions of coating, type of hydrophobic/hydrophilic material, and module design	GG/GB both effected 2-3% abs loss in output power	Low	4	Desert climate lead to degradation mode, time and location dependent occurrence depends on specific BOM configuration if BOM susceptible all modules of this type will be affected	6	24	Microscopy, reflectance, UV-Vis-Nir. Spectroscopy, desert-specific, depends on wind speed and dust	4	96	Yes
43	Partial shading and its potential risk for module reliability and safety aspects	Different causes for partial shading are possible (bird drops, chimneys, soiling, building structures etc)	Cell efficiency, module construction, number of cells per bypass diode	GG/GB both effected	Risk for defective insulation system	Low to medium, depends on cells / module design	5	May depend on module installation and tilt-angle of module, stall/night position of tracker system	6	30	Proper cleaning and maintenance, initial check of modules for susceptibility to hot spot heating	3	90	Special soiling accumulation on tracker systems
12	Discoloration: internal metalic parts	Browning, greenish stuff, depends on MP, mainly temperature and humidity	Type of encapsulation materials and composition of a metal part	GG/GB both effected	<0.1%/a	Low	2	Lots of modules are showing different color changes. Very strong discolorations are typically a sign for (initial) corrosion	8	16	Visual inspection in the field, not early detection possible	5	80	No
33	Power loss: PID	Depending on cell and module materials high power losses can occur if modules are operated at high potentials to ground; degradation due to humid conditions, coastal locations, system voltages, and installations	Glass, encapsulation, cell technology	Compositions of glass, type of encapsulant, material and module design, installations	GG/GB both effected potentially very high - up to 100% for parts of the string	Low	5	Humid environment time and location dependent occurrence depends on specific BOM configuration if BOM susceptible all modules of this type will be affected	5	25	EL images, PID test, PV system monitoring	3	75	General for all environments
20	Backsheet Chalking	Outer layer observation! Visual discoloration/decompositi on of outer layer of the backsheet. Typically, this occurs after a few years (3-5 in the field, depending on MP, mainly temperature and humidity	Depends on the material of the backsheet	GB modules	<0.1%/a	Low	2	Time and location-dependent occurrence depends on specific BOM configuration, if BOM susceptible all modules of this type will be affected	7	14	Observation can be reproduced in the lab with a longer testing process Indoor tests can be performed (DH + UV)	5	70	Maybe
19	Backsheet Discoloration	Visual discoloration of both, inner and outer layers of the backsheet. Depends on MP, mainly temperature and humidity	Depends on the material of the backsheet	GB modules	<0.1%/a	Low	3	Time and location dependent occurrence depends on specific BOM configuration if BOM susceptible all modules of this type will be affected	4	12	A clear test to reproduce this kind of observation is not known. It may involve UV, humidity, and temperature, rear-side soiling can also influence the test outcome	5	60	Maybe

41st European Photovoltaic Solar Energy Conference and Exhibition

#	Issue, Degradation mode description	Detailed description of issue, impact of Climate and MP	Type of material or process mainly involved to cause issue	Influenced by BOM / module construction? Yes: state influence no: state NOT	Potential degradation rate %/a	Associated risks such as fire, hazardous electrical potentials	Severity S	Description why occurrence is judged to by high or low	Occurrence O	PN (S*O)	Description of method of detection. What is the likelihood for detection	Detection D	RPN (S*O*D)	Special relevance for desert applications Yes / No
41	Power loss: Glass surface - abrasion	Cleaning robots causing abrasion to ARC coating of front glass, could also be caused by sandstorms	Impact of brushes, sand and dust	Depends mainly on mechanical stability of ARC and type of sand and brush	< 2% abs due to loss of ARC	Low risk for arcs, medium risk for hot-spot, depends on number of BB	5	Depends on cleaning protocol and frequency and used procedure (e.g. Wet vs dry)	6	30	Severity measurable by Flashtest on representative sample (with assumption that other degradations do not exist, and initial measurement exist), all modules will be affected same way. Initial degradation of ARC coating can be e.g. Seen by microscope	2	60	Yes
26	Delamination: Encapsulant – Backsheet Or Delamination: Inner layers of Backsheet Or Glass-Encapsulant	Delamination: Encapsulant – backsheet Or Inner layers of Backsheet Or Glass to encapsulation	Improper lamination, Insulation loss, bubble formation in the field after some T and UV exposure. Desert climate lead mode and MP	Lamination process, depends on material and module design	GG/GB both effected	Low, if far from edges of module, close to active circuitry risk for electric shot (high)	4	Desert climate lead degradation mode, time and location dependent occurrence depends on specific BOM configuration, if BOM susceptible all modules of this type will be affected	3	12	DH, PID, UV and mechanical load tests,	4	48	General for all environments
29	Defective cables	Defective cables, e. G defective insulation	Soil interaction, no proper coverage, climate and MP	Depends on material and connection during the installation	GG/GB both effected	High risk, safety issue	7	Low time and location dependent, depends on MP and climate movement of cables due to wind, birds or other possibilities to damage insulation	3	21	Continuity test visual inspection and proper O&M	2	42	General for all environments
14	Corrosion of external metallic parts (e.g. Frame)	Corrion of metallic external parts due to moisture and soils It can be due to soiling, cementations, humidity. Highly dependent on MP, location more relevant for mounting structure as this is in constant contact with soil and potentially with water	Depends on used metals and their protection.	GG/GB both effected	0	Typically low, except system relies on bonding	3	Depends on location	4	12	Visual inspection, occurrence can be forecasted by understanding soils and climate corrosive testing such as Salty chamber and humidity tests under bias	3	36	No

#	Issue, Degradation mode description	Detailed description of issue, impact of Climate and MP	Type of material or process mainly involved to cause issue	Influenced by BOM/module construction? Yes: state influence no: state NOT	Potential degradation rate %/a	Associated risks such as fire, hazardous electrical potentials	Severity S	Description why occurrence is judged to by high or low	Occurrence O	PN (S*O)	Description of method of detection. What is the likelihood for detection	Detection D	RPN (S*O*D)	Special relevance for desert applications Yes / No
22	Backsheet Burn marks (e.g. From hot-spots)	Visual discoloration of backsheet due to e.g. Partial shading	Depends on material and module design and installation	GB modules burning could also occur in GG modules within the encapsulation material	<0.1%/a	If it occurs it can cause breakdown of insulation	9	Partial shading typically needed, or some severe cracking of cells, only a few modules will typically be affected, less severe for MBB technologies, but may also depend on number of cells in series and cell technology	2	18	EL, Hot-spot endurance tests visual inspection in the field, typically not a manufacturing issue visually seeable burn marks	2	36	No
18	Encapsulation: Bubble formation	Bubble formation in the field after some T and UV exposure. Often caused by improper lamination or soldering flux if bubbles are large, this could lead to cell cracking and power loss in long term, including different light reflections excluding edge seals	Type of flux, encapsulation	GG/GB both effected	<0.1%/a	Low	2	Time and location dependent occurrence depends on specific BOM configuration, if BOM susceptible all modules of this type will be affected	4	8	Visual inspection in the field, even after a few days often caused by bad processes	4	32	No
25	Delamination: Cross-connector	Delamination: Cross-connector	Delamination in front of cross-connector, thermomechanical stress, typically caused by over soldering and improper lamination, cross-connector decompositions or swelling, Desert climate and MP	Lamination process, depends on material and module design	GG/GB both effected	Low, if far from edges of module, close to active circuitry risk for electric shot (high)	4	Time and location dependent occurrence depends on specific BOM configuration, if BOM susceptible all modules of this type will be affected	2	8	DH, PID, UV and mechanical load tests,	4	32	General for all environments
28	J-Box and bypass diodes failures within junction box	J-Box and bypass diodes failures within junction box defective bypass diode (short or open circuit)	Risk for heating of J-box (short) or a general increased hot-spot risk on cells (open)		GG/GB both effected		7	Low time and location dependent, depends on MP and climate	2	14	J-box, Bypass diode failure test	2	28	Do we need a stronger bypass thermal test
28	J-Box and bypass diodes failures within junction box	J-Box and bypass diodes failures within junction box silicon adhesion problems defective housing	Due to water ingress, internal short circuit, arcing, higher heating, detachment	Type of glues used and module design	GG/GB both effected	High risk, safety issue	7	Low time and location dependent, depends on MP and climate	2	14	Pull test of j-box, impact test, check quality of materials as incoming in module factory	2	28	General for all environments
30	Connector failures	Connector failures	Soil interaction, no proper coverage, climate and MP	Depends on material and connection during the installation	GG/GB both effected	High risk, safety issue	7	Low time and location dependent, depends on MP and climate; mechanical damage due to work on system	2	14	Continuity test visual inspection and proper O&M	2	28	General for all environments

#	Issue, Degradation mode description	Detailed description of issue, impact of Climate and MP	Type of material or process mainly involved to cause issue	Influenced by BOM / module construction? Yes: state influence no: state NOT	Potential degradation rate %/a	Associated risks such as fire, hazardous electrical potentials	Severity S	Description why occurrence is judged to by high or low	Occurrence O	PN (S*O)	Description of method of detection. What is the likelihood for detection	Detection D	RPN (S*O*D)	Special relevance for desert applications Yes / No
1	Cell breakage	Broken cells caused by manufacture	Cell and module handling	NOT	<0.5%/a / warranty	Low risk for arcs, medium risk for hot-spot, depending on number of BB	3	Cell cracks always occur, but typically are not of high risk	8	24	Detection at end of line EL-Test	1	24	No
24	Delamination: Encapsulant - Cell/ribbons	Delamination in front of cell at the ribbon, typically caused by flux remaining and improper lamination	Lamination process depends on material and module design, desert climate, and MP	GG/GB both effected	Low	Low, if far from edges of the module, close to active circuitry risk for an electric shot (high)	4	Relevant for desert, time and location dependent occurrence depends on specific BOM configuration if BOM susceptible all modules of this type will be affected	3	12	DH, PID, UV, and mechanical load tests	2	24	General for all environments
6	Finger interruptions/b reakages	Caused by screen printing of the cell (cell fabrication)	Printing	No	Included in cell and module sorting	Low	1	A lot of cells have it, but this is part of the initial flash	8	8	it's part of the cell rating, but difficult to 100% avoid this issue	2	16	No
10	Miss-alignment of cells and ribbons	During lamination internal components of the module can move, movement depends on viscosity of the encapsulant and lamination protocol	Lamination process	GG/GB both effected	Low	Low	2	Happens from time to time	4	8	Visual inspection end of line 100% to keep within safety required limits	2	16	No
3	Glass / Fatal module damage	Cleaning robots damaging modules, replacement of modules needed, increasing O&M costs	Mechanical stress due to "defective" robot behavior	Depends on design (GG, GF, Frame)	100% for brocken module	Low risk for arcs, medium risk for hot-spot, depending on number of BB	7	Very low	2	14	Easy to identify due to glass breakage	1	14	Yes
5	Cell breakage	Due to heavy snow loads	Mechanical stress due to heavy snow	The resilience of module depends on the construction and mounting	100% for broken module	Low risk for arcs, medium risk for hot-spot, depending on number of BB	7	Rare occurrence, not relevant for desert applications	1	7	Mechanical load MQT 16	2	14	No
34	Power loss: LETID	Today known only for PERC cells	Cell type	Solar cell technology	1-2% abs	Low	2	Typical degradation below 1%	2	4	LID and LETID tests exist	2	8	Has a temperature aspect
35	Power loss: LID	For Al-BSF and PERC cells it is known however new cells may have other LID-type of effects - not known, yet	Cell type	Solar cell technology	1-2% abs	Low	2	Typical degradation below 1%	2	4	LID and LETID tests exist	2	8	General for all environments

41st European Photovoltaic Solar Energy Conference and Exhibition

NUMERICAL SIMULATION OF THE BYPASS DIODE FAILURE RESISTANCE AND THOSE POWER CONSUMPTION IN A PHOTOVOLTAIC SOLAR MODULE WITH FAILED BYPASS DIODE

Ibuki Kitamura[1], Toshiyuki Hamada[1], Ikuo Nanno[2], Norio Ishikura[3], Masayuki Fujii[4], Shinichiro Oke[5]
[1]Osaka Electro-communication University / Graduate School of Engineering
[2]Yamaguchi Gakugei University, Yamaguchi College of Arts
[3]National Institute of Technology, Yonago College / Department of Integrated Engineering and its Divisions
[4]National Institute of Technology, Oshima College / Electronic-Mechanical Engineering Department
[5]National Institute of Technology, Tsuyama College / Department of Integrated Science and Technology
hamada@osakac.ac.jp

ABSTRACT: The use of photovoltaic systems (PVSs) has significantly increased. However, the number of cases in which accidents or failures in PVS result in burnout has also witnessed a surge. In particular, there have been cases in which the bypass diode (BPD) in a photovoltaic (PV) module fails because of lightning damage. bypass diode failure is associated with the risk of burning. Previously, we reproduced failure cases in which a single bypass diode fails at various resistance values in a three-cluster PV solar module to identify bypass diode failures and their burnout risks. However, when using simulated power supplies that reproduce the output of PV solar modules and PV solar modules, the verification of the risk of burnout owing to bypass diode failure at the string or array level requires a large-scale PVS. In addition, a PVS has a variety of system output characteristics. However, it is desirable to easily verify the risk of burnout owing to the bypass diode failure of PV solar modules under various PVS output conditions. Hence, in this study, numerical simulations were performed using MATLAB/Simulink to simulate the power consumption of a failed bypass diode in a solar module when the bypass diode fails at various bypass diode failure resistance values (R_F). In addition, an experiment reproducing the same conditions as were used in the numerical simulation was conducted to investigate both the risk of burnout in the case of the failure of the bypass diode of the solar module and confirm the validity of the numerical simulation. In the numerical simulation, when R_F was approximately 4 Ω, the power consumption of R_F was maximized at 14.0 W. Therefore, during the maximum power point tracking operation of a three-cell string PV module with one failed bypass diode, the amount of heat generated was high when R_F was approximately 4 Ω. In an experiment that used a power supply to simulate PV characteristics, the R_F power consumption was also maximized at 14.6 W when R_F was approximately 4 Ω. Therefore, the proposed numerical simulation model successfully reproduces the output characteristics of a PV module with a failed bypass diode. Hence, analyzing bypass diode failures in large-scale PV systems is possible, even though it is difficult to perform such experiments.

Keywords: photovoltaic system (PVS), bypass diode (BPD), open failure

1 INTRODUCTION

The use of photovoltaic systems (PVSs) has experienced a substantial rise. However, there has also been a rise in the number of instances in which accidents or failures in PVSs have resulted in burnout [1]. In particular, there have been cases in which the bypass diode (BPD) in a photovoltaic (PV) module has failed because of lightning damage. The failure of a bypass diode carries the risk of burnout [2-3]. bypass diode failures in solar panels are a global phenomenon with several causes, including lightning, initial diode failure, and aging. When a bypass diode fails, the module string voltage decreases, which results in a reverse current from the string. In several cases, blocking diodes are installed in the string to prevent this; however, blocking diodes often fail when lightning damage occurs [4]. A failed bypass diode loses its rectifying characteristics and acts against a reverse bias. The reverse current flow to a failed bypass diode can cause the failed bypass diode to burn out, depending on the electrical characteristics of the failed bypass diode. Previously, we reproduced failure cases in which a single bypass diode (BPD) failed at various resistance values in a three-cluster PV solar module to identify bypass diode failures and their burnout risks [5-6]. However, when using simulated power supplies that reproduce the outputs of PV solar modules, the verification of the risk of burnout due to bypass diode failure at the string or array level requires a large-scale PV system. In addition, the PVS has a variety of system output characteristics. However, it is desirable to easily verify the risk of burnout owing to the

bypass diode failures of PV solar modules under various PVS output conditions.

In this study, numerical simulations are performed using MATLAB/Simulink to simulate the power consumption of a failed bypass diode in a solar module at various failure resistance values. In addition, the risk of burnout when a bypass diode in a solar module fails is examined.

2 NUMERICAL SIMULATION OF A PV MODULE WITH A FAILED BPD

2.1 Simulation Model

Figure 1 shows the numerical simulation model of a three-cell-string PV solar module with one bypass diode failure. Figures 1 (a)–(b) show this model, proposed by Bishop [1], which is based on the polarity-reversed diode characteristic equation and photocurrents. The solar cell output was set to an open-circuit voltage of 0.64 V, and the short-circuit current I_{sc} was set to 8.95 A. A commercially available PV solar module (Choshu Industry Co. Ltd., CS-236B31) was used as a reference. The current characteristics of the diode are given in Equation (1):

$$I_d = I_{sat}\left\{exp\left(\frac{q_e\,v_r}{n\,k_b\,T}\right) - 1\right\}. \quad (1)$$

The reverse saturation current I_{sat}, breakdown voltage V_r, electron charge q_e, Boltzmann constant k_b, temperature T, and coefficient ideal n were set to 1.1×10^{-7} A, -100 V,

41st European Photovoltaic Solar Energy Conference and Exhibition

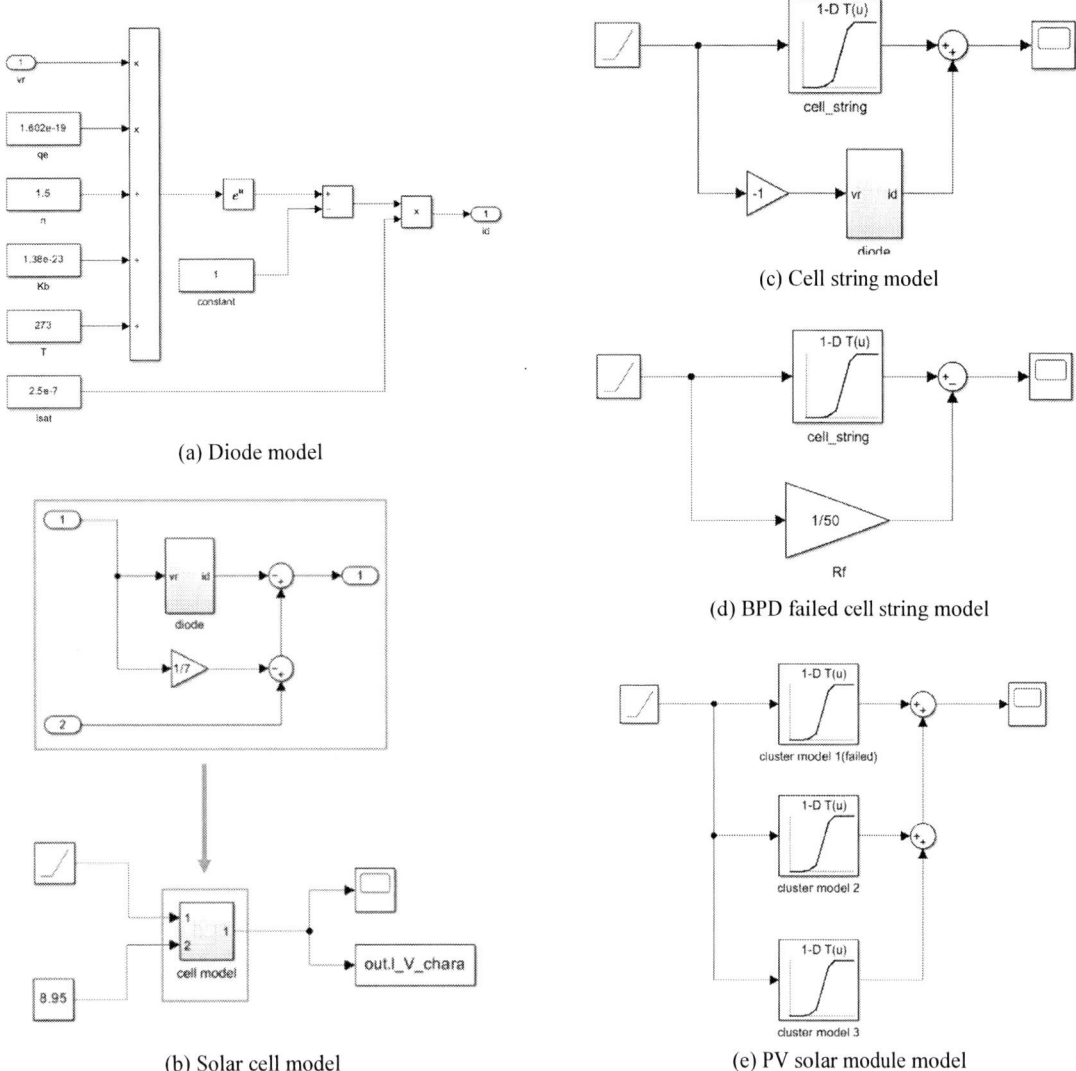

(a) Diode model

(b) Solar cell model

(c) Cell string model

(d) BPD failed cell string model

(e) PV solar module model

Figure 1: Numerical simulation model of a failed bypass diode (BPD) in PV solar module.

1.6×10^{-19} C, 1.38×10^{-23} J/K, 273 K, and 1.5, respectively. Figure 1 (c) shows an overview of the photovoltaic solar module model. The model consists of a cell string of 18 cell models connected in series, with diodes connected in parallel as bypass diodes. Three of these cell strings are connected in series to form a three-cell string PV solar module model. In this simulation, the bypass diode in the three-cell string photovoltaic solar module was replaced by a resistive element (R_F) to reproduce the failure of bypass diode in the module. The simulations measured the current–voltage (*IV*) characteristics of the PV solar module model for different R_F values. The R_F values were simulated in the range of 0.1–50 Ω, and the power consumption of the R_F at the maximum power point of the *IV* characteristics of the PV solar module model was examined. This allowed us to examine the heat generated by the failed bypass diode when a three-cell string PV solar module with one failed bypass diode was operated at its maximum power point.

2.2 Numerical Simulation Results

Figure 2 shows the *IV* characteristics obtained from the PV solar module simulation model for various failed bypass diode resistances (R_F). As R_F decreases, the output voltage and maximum power performance of the photovoltaic solar module decrease because of the lower operating voltage of the cell string to which the R_F was connected. Figure 3 shows the R_F power consumption at the maximum power point of the *IV* curve obtained from the PV solar module model (Fig. 2). Figure 3 shows that the R_F power consumption reached its maximum value of 14.0 W at approximately 4 Ω. Therefore, during the maximum power point tracking (MPPT) operation of a three-cell string PV solar module with one failed bypass diode, the heat generation was found to be higher when the R_F was approximately 4 Ω and the R_F electric power was approximately 14 W.

41st European Photovoltaic Solar Energy Conference and Exhibition

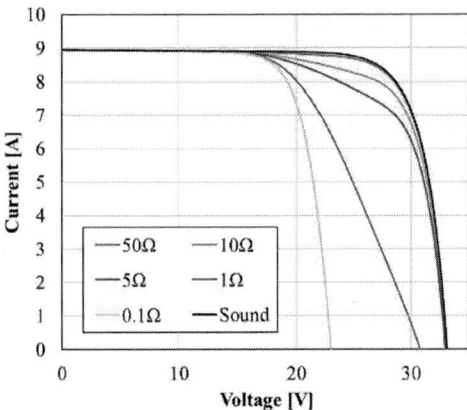

Figure 2: *IV* characteristics of a PV solar module model with a failed bypass diode (BPD).

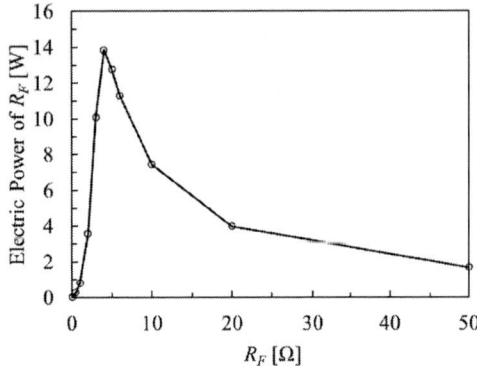

Figure 3: Relationship between bypass diode (BPD) failure resistance and power consumption in a PV solar module.

3 EXPERIMENTAL MEASUREMENTS OF A PV MODULE WITH A FAILED BPD

3.1 Experimental Method

Figure 4 shows an overview of an experiment that simulated bypass diode failure in a PV solar module. In this experiment, we used three multi-module power supply units (Nippon Kernel System Co., Ltd.; MEP12281; open circuit voltage of V_{oc} = 10.9 V; short-circuit current I_{sc} = 8.95 A; fill factor of $F.F.$ = 0.8) that could reproduce the current–voltage characteristics of PV solar modules to simulate a PV solar module with three cell strings. A Schottky barrier diode (FSQ30A045, Kyocera Corp., repetitive peak reverse voltage of 45 V, average rectified current of 30 A) was connected to the power supply units (Units 1 and 2) to simulate a sound bypass diode. An electronic load (RIGOL, DL3031A) or variable resistor (RSSD 25X158, 1 Ω) was connected to power Unit 3 to simulate a faulty bypass diode with various resistance values R_F. Consequently, we reproduced a three-cell string PV solar module with a bypass diode failure. Hereafter, the PV solar module described above is referred to as the "bypass diode failure simulation PV solar module." An electronic load R_L (RIGOL, DL3031) was connected to the bypass diode failure simulation PV solar module, and the

IV characteristics were measured while sweeping in the constant current mode to evaluate the output characteristics of the bypass diode failure simulation PV solar module when the bypass diode failed at each R_F. Simultaneously, we measured the voltage and current between the terminals of the resistor R_F, which simulated the failed bypass diode and examined the heat generated by the electric power of the R_F in the failed bypass diode.

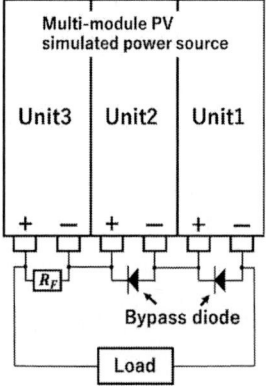

Figure 4: Overview of the reproduction test of the PV solar module with bypass diode (BPD) failure.

3.2 Experimental Results

Figure 5 shows the *IV* and power–voltage characteristics of the bypass diode failure simulation PV solar module at each R_F. When the R_F value of the bypass diode decreased, the open circuit and operating voltages decreased because the operating voltage of the cell string with R_F (Unit 3) decreased.

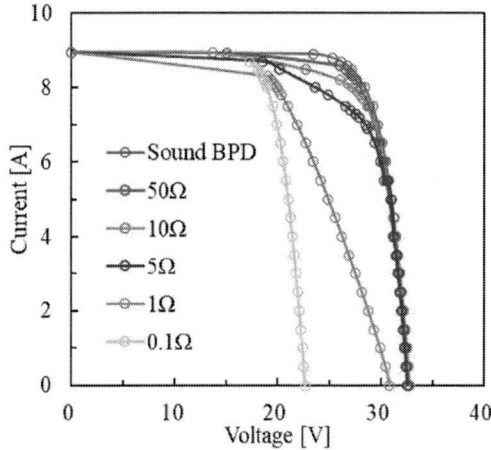

Figure 5: *IV* characteristics of the bypass diode (BPD) failed PV solar module at each R_F.

Figure 6 shows the electric power consumed at each R_F value when the output of the bypass diode failure simulation PV solar module was operated at its maximum power point. The PVS operates under MPPT control using inverters. Therefore, the electric power of the failed bypass diode (R_F) during the PVS solar module operation was evaluated by the electric power of the R_F when the bypass diode failure simulation PV solar module was operated at

678

its maximum power point. Figure 6 shows that the electric power of the R_F in the bypass diode failure simulation PV solar module reached its maximum at approximately 4 Ω, where its R_F electric power was approximately 14.6 W. This result is in close agreement with the numerical simulation results presented in Section 2. Hence, it is believed that bypass diode failure with a resistance value of approximately 1–10 Ω results in the increased heat generation of the faulty bypass diode. Therefore, we propose a numerical simulation model of bypass diode failure in PV solar modules that successfully reproduces an actual PV solar module with a failed bypass diode and thereby indicates that bypass diode failure analysis methods can be developed for large-scale PV systems.

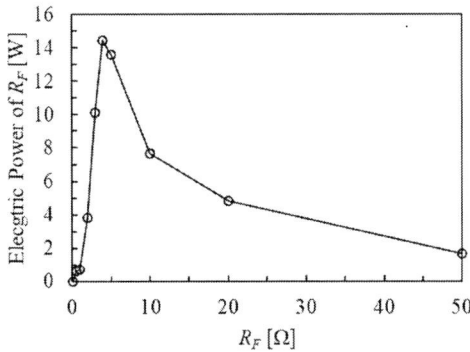

Figure 6: Relationship between the electric power P_{RF} in the faulty bypass diode (BPD) and the fault resistance value R_F of the faulty BPD when the BPD failure simulation PV operates at its maximum power point.

4   CONCLUSIONS

In this study, we numerically simulated the current–voltage characteristics of a PV solar module with a failed bypass diode using MATLAB/Simulink. Our proposed numerical simulation model agrees well with the actual measured output of a PV solar module with a failed bypass diode. Thus, we were able to successfully reproduce the *IV* characteristics of a PV solar module with a failed bypass diode. The proposed numerical simulation model can be applied to analyze the output of a PVS with bypass diode failure and inform failure diagnosis techniques.

Acknowledgements
This study was supported by JSPS KAKENHI (Grant No. JP21H01580).

References
[1] Ministry of Economy, Trade and Industry and Agency for Natural Resources and Energy, "Japan's Energy 10 Questions for Understanding the Current Energy Situation," Tokyo, Japan, 2022.
[2] M. Koentges et al., "Review of Failures of Photovoltaic Modules," Photovoltaic Power Systems Program, Report IEA-PVPS T13-01, 2014.
[3] S. Oke, S. Sakai, H. Tottori, Y. Shimizu, N. Choo, I. Nanno, T. Hamada, N. Ishikura, and M. Fujii, "Proceedings of the 35th EU-PVSEC," 2018, pp. 1996.
[4] N. Ishikura, T. Okamoto, I. Nanno, T. Hamada, S. Oke, and M. Fujii, "Proceedings of the International Conference on Renewable Energy Research and Applications

(ICRERA2018)," 2018.
[5] T. Hamada, I. Nanno, M. Fujii, N. Ishikura, and S. Oke, "Effect of Failure Characteristics of Bypass Diode in Photovoltaic Solar Module on Burnout," J. Inst. Elect. Instal. Engnr. Jpn., vol. 42, pp. 16, 2022. [in Japanese]
[6] T. Hamada, T. Azuma, I. Nanno, N. Ishikura, M. Fujii, and S. Oke, "MDPI Energies," vol. 16, pp. 5879, 2023.

41st European Photovoltaic Solar Energy Conference and Exhibition

IMPACT OF THE MATERIAL COMBINATION ON THE BARRIER PROPERTIES AND THEIR STABILITY IN THE COURSE OF ACCELERATED WEATHERING

Daniel Schüsler[1], Patrick Wessel[1,2], Michael Wendt, Anton Mordvinkin[1*]
1 Fraunhofer Center for Silicon Photovoltaics CSP, Otto-Eissfeldt-Str. 12, D-06120 Halle
2 Anhalt University of Applied Sciences, Bernburger Str. 55, D-06336, Koethen (Anhalt)
anton.mordvinkin@csp.fraunhofer.de, *+49 345 5589-5129

ABSTRACT: Polymer films (encapsulants and backsheets) serve as an important barrier against water ingress into a PV module both in a vapor and liquid state. Once in the module, water firstly initiates degradation of its polymeric components, and thus alters their properties including the barrier properties, leading to various failure modes such as delamination, discoloration, cracking, and corrosion. This results not only in reduced module lifetimes, frequent invertor shutdowns, and associated lower yields, but also poses safety risks. Therefore, the barrier properties, and especially their long-term stability must be thoroughly scrutinized. The long-term stability of polymeric films depends on the polymer type, its additivation, and also interactions between individual layers. In this contribution, 6 different backsheets, based on poly(ethylene terephtalate) (PET), poly(propylene) (PP), and poly(lactic acid) (PLA), 2 different encapsulants (Ethylene-Vinyl acetate Copolymer (EVA) and polyolefin elastomer (POE)) as well as glass-encapsulant-backsheet coupons, based on them, were prepared and exposed to three different accelerated weathering conditions for 2000 h. The weathering tests encompassed a classical damp heat (DH) (85°C 85% relative humidity (RH)), combined DH with a UV irradiation (85°C 60% RH), and UV irradiation in dry conditions (IEC 62788-7-2 A3). The water ingress was monitored non-destructively using near infrared spectroscopy (NIR) every 250 h. The water absorbance data, measured by NIR, were found to positively correlate with the solubility and water vapor transmission rates (WVTR), found by permeation measurements. Further, the crack formation could be directly observed in the NIR signal. The PP-based samples featured the best barrier properties initially and in the course of accelerated weathering, whereas PLA samples exhibited the worst performance. Encapsulants enhanced the barrier properties (with the only exemption being EVA for the PP backsheets) and functioned as a UV filter, protecting backsheets from cracking. However, the compatibility and interactions between the encapsulants and backsheets must be additionally considered, since discoloration could be observed for a number of polymer combinations. The backsheet cracking could be correlated with post-crystallization, assessed by DSC, leading to embrittlement. Additional tests were performed in the liquid water, mimicking the wet-leakage tests, enabling an assessment of isolation on the material level, which can be used as pre-qualification of module's BOMs. Further, an NMR sensor was employed for the first time to study the water ingress into the polymeric coupons over the sample cross-section with a resolution of 50 µm. A water-content gradient could be observed within the polymeric layers.

Keywords: water uptake, encapsulant-backsheet coupons, accelerated weathering, non-destructive characterization

1 INTRODUCTION

The water uptake was recently shown to directly correlate with the deterioration of the module electrical isolation [1]. Also, it was previously shown that barrier properties of PET-based polymeric films can change due to their aging and lead to module performance losses. However, the study [2] was limited only to PET- and AAA-based polymeric films combined with EVA exposed only to DH conditions. Therefore, there is still a gap in understanding how backsheet types of other chemistries (e.g. emerging PP-based backsheets), combined with encapsulant of different chemistries (EVA vs POE) will change their barrier properties in response to various stressors such humidity, UV irradiation, and a combination thereof. This work is filling this gap with a special design of experiment encompassing three different accelerated weathering conditions and sample compositions. The changes in the barrier properties were monitored for individual polymer films (encapsulants or backsheets) and coupons based on them. Thus, interactions between the polymeric layers can be revealed. The changes in the barrier properties were correlated with the changes in the microstructure. Further, NMR spectroscopy was employed for the first time to bring new insights into the water ingress, namely a water content gradient over the polymer sample thickness could be resolved.

2 EXPERIMENTAL PART

2.1 Samples
Two sets of samples were prepared by the vacuum lamination of a stack of several layers, mimicking a typical module composition, as shown in **Figure 1**. To prove the effect of interactions between an encapsulant and a backsheet, a backsheet was either laminated alone or in combination with an encapsulant with a glass piece on top. To avoid edge effects and purely probe barrier properties of the rear side, edges were sealed using an aluminum tape. To facilitate later sample extraction, a release film was placed between the glass and the polymer layer. As backsheets, 4 different poly(ethylene terephthalate) (PET 1-4), 2 poly(propylene) (PP 1-2) and one poly(lactic acid) based backsheets were used. As encapsulants, either ethylene vinyl-acetate copolymer (EVA) or poly(ethylene elastomer) (POE) were utilized.

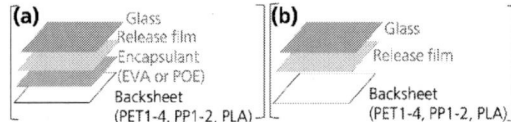

Figure 1 Samples compositions: (a) coupons based on a backsheet and an encapulant combination, (b) coupons based on backsheet.

2.2 Methods
Near-Infrared (NIR) spectroscopy was performed using a mobile handheld spectrometer trinamiX Model SYS-IR-R-P. The spectral region covers the range between 1450–2450 nm with a resolution of 1 % of the

wavelength. In **Figure 2**, a representative NIR spectrum is shown along with its second derivative. The second derivative was used to reduce the baseline effects and improve the spectral resolution. The water peak in the second derivative was then integrated to obtain an estimate for the water content.

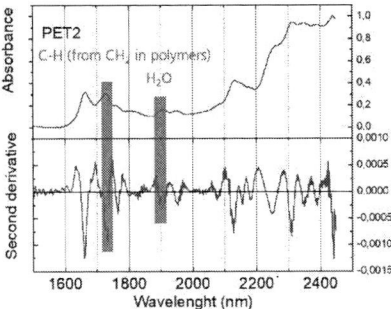

Figure 2 Representative NIR spectrum of a PET2 film and its second derivative. Water and reference polymer signals are marked.

NMR profiling was carried out with an NMR MOUSE PM10 featuring a penetration depth of 2.8 mm and a resolution of 50 µm. 90° and 180° pulses were set to 5.5 µm, 180° pulses had a double power of 90° pulses. The signal was obtained by means of a Carr-Purcell-Meiboom-Gill pulse sequence to overcome the effects of the heterogeneous magnetic field.

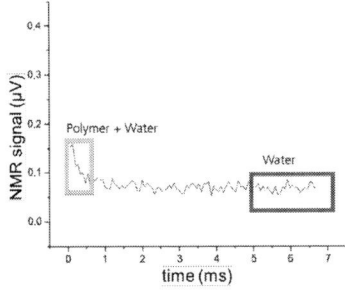

Figure 3 NMR signal of a polymer backsheet exposed to water. Polymer and water contributions to the signal are marked.

2.3 Weathering tests

The samples were exposed to accelerated weathering for 1500 h using conditions specified in **Table 1**. For the tests with UV irradiation, the glass side was always illuminated.

Table 1 Conditions of used accelerated weathering tests

Parameter	Damp Heat (DH)	Combined DH and UV irradiation	IEC 62788-7-2 (dry UV irraditaion)
Temperature, °C	85	85	65
Relative Humidity, %	85	60	20
UV strength (300-400 nm), W/m²	N/A	104	81
Lamp type	N/A	Metall halide	Xenon

3 RESULTS AND DISCUSSION

Figure 4 shows the NIR data for the water absorption recorded in the course of the DH weathering for the backsheets of different chemistries. The data show that the water absorption firstly achieved saturation after 500 h, and then continued to grow after 1000 h of exposure. The NIR data correlate well with the solubility and WVTR values from permeation measurements, as seen in **Figure 5**. Some deviations originate from different experimental conditions and assumption of the applicability of the Henry's law. The PP-based backsheets feature better barrier properties than PET-backsheets.

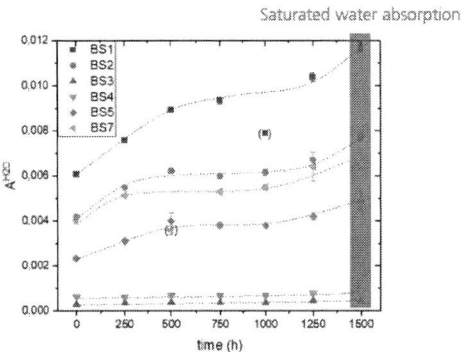

Figure 4 Water absorbance data for studied backsheets, exposed to Damp Heat (DH), obtained by NIR

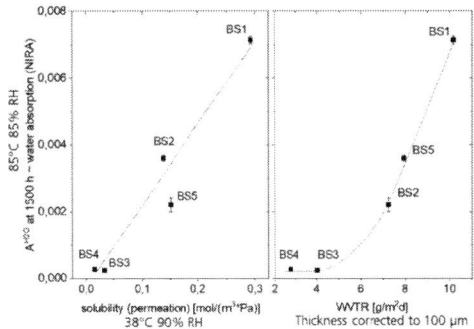

Figure 5 Correlation of the NIR water absorbance measured after 1500 h DH with the solubility and water vapor transmission rate, measured at the permeation setup.

Figure 6 shows the NIR data for the water absorption recorded in the course of the combined damp-heat and UV weathering for the backsheets as well as corresponding coupons using EVA and POE encapsulants. One can see that the introduction of the encapsulant improves the barrier properties, except for the PP backsheets combined with EVA. POE improves the barrier properties to a more extent than EVA. The data after IEC 62788-7-2 are not shown here. The NIR water intensities can be calibrated with the Karl-Fischer titration technique to obtain quantified water content, which will be a matter of the next publication.

Some backsheets (PET2-4 and PP1) showed cracking in the course of the combined damp-heat and UV irradiation. This could be attributed to the post-crystallization

41st European Photovoltaic Solar Energy Conference and Exhibition

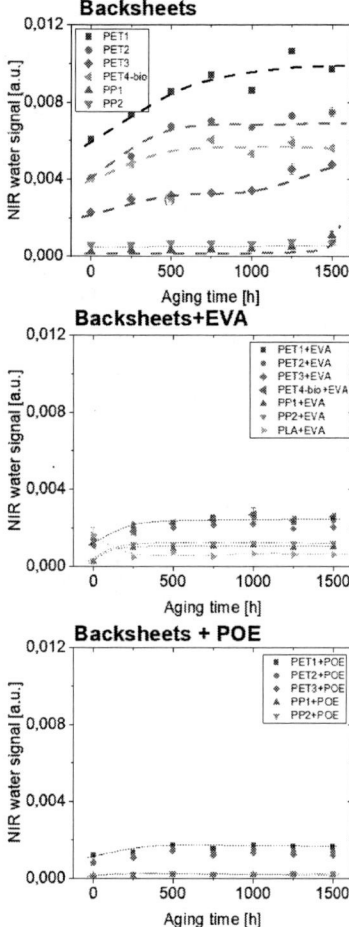

Figure 7 NIR water signal changes in the course of combined damp-heat and UV weathering for backsheets (top), combinations of backsheets with EVA (center) and POE (bottom).

resulting in the embrittlement, subsequently leading to deterioration of barrier properties. No cracking was observed for the coupons, suggesting that encapsulants act as UV-filters contributing to the reliability. However, interactions between backsheets and encapsulants must be additionally considered, as discoloration was observed for some combinations.

An NMR sensor was used to probe the water ingress non-destructively over the samples' cross-section (**Fehler!**

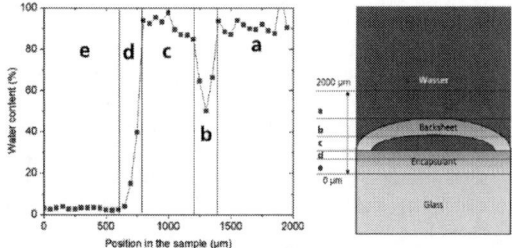

Figure 6 Left: Depth profile of water content in a glass-EVA-PLA coupon after its exposure to liquid water. Letters (a-e) refer to the measured positions schematically shown in the sketch on the right.

Verweisquelle konnte nicht gefunden werden.). In the case of glass-EVA-PLA coupons, water could pass through the backsheet already within the first hour of water exposure and accumulate in the encapsulant within the layer of around 100 µm. Other materials will be likewise investigated by the NMR technique, which will be a matter of the next publication.

4 CONCLUSIONS

To avoid false positive or negative statements, the reliability and durability testing of barrier properties of polymeric materials must be always performed reproducing the module composition, namely combining encapsulants and backsheets. Encapsulants introduce additional interactions which either positively (UV filter effect preventing cracking) or negatively (e.g. increased water uptake of EVA, combined with PP as compared to PP alone) impact the backsheet aging.

Non-destructive spectroscopic methods such as NIR and NMR spectroscopy are powerful tools which can probe the water ingress over time, which is relevant the module lifetime estimations.

5 ACKNWOLEDGEMENTS

This work was funded by the German Federal Ministry for Economic Affairs and Climate Action within the project Folie40, with the reference number of 03EE1173D

REFERENCES

[1] C. Buerhop et al. Prog. Photovolt: Res. Appl. 2022, 30 (8), S. 938–947.

[2] Y. Voronko et al. Prog. Photovolt: Res. Appl. 2015, 23 (11), S. 1501–1515.

41st European Photovoltaic Solar Energy Conference and Exhibition

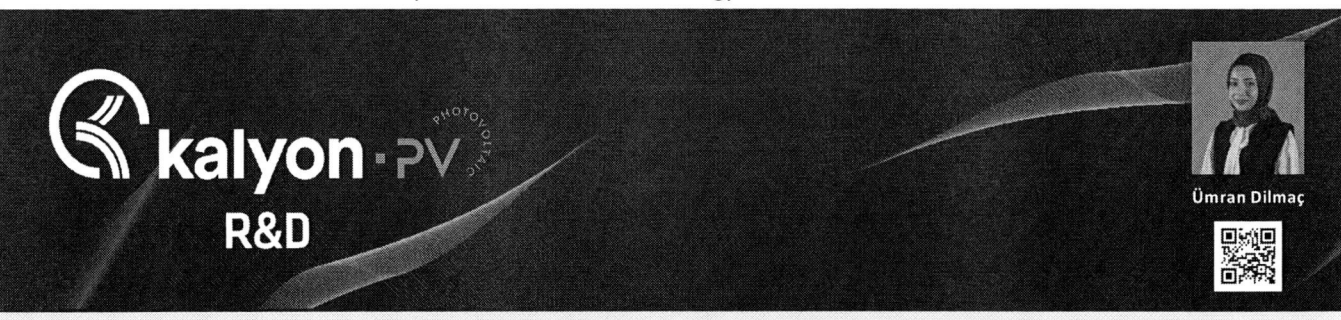

Ümran Dilmaç

INVESTIGATION OF THERMO-MECHANICAL BEHAVIOR OF ENCAPSULATION MATERIALS USED IN SOLAR PANEL PRODUCTION

Ümran Dilmaç[*1], Merve Çorak[1], Meriç Çalışkan Arslan[1], Yıldırım Aydoğdu[2]

[1]KalyonPV Research and Development Center, Kalyon Güneş Teknolojileri Üretim A.Ş., 06909 Ankara, Turkey   e-mail: udilmac@kalyonpv.com
[2]Gazi University, Faculty of Science, Physics, 06560 Ankara, Turkey

INTRODUCTION & MOTIVATION

The encapsulation film is the key material for protecting the solar module, which isolates cells from environment. These solar grade films require tolerance all sorts of environment stress, and have reliable insulation, water resistance and aging resistance. They are crucial for the outdoor reliability, power degradation and lifetime of the PV modules. Since solar panels must be able to adapt to harsh climatic conditions, the thermo-mechanical behavior of encapsulation materials is very important. Conventional encapsulation agents need to undergo chemical cross-linking reaction to ensure strong adhesion during the lamination process of solar PV modules. Cross-linking reactions are chemical processes in which polymer chains are connected to each other, creating 3D network structure. After this reaction, thermoplastic polymers become harder and thermally more stable. One way to prove this is to provide coefficient of thermal expansion (CTE) control. Hence, the final product quality and properties of encapsulant material mostly depend upon the mentioned cross-linking reaction.

Polyolefin elastomer (POE) is a widely used polymeric material for the encapsulation of photovoltaic modules. It melts at elevated temperatures and isolates the module layers via crosslinking reaction. The degree of crosslinking, which is an important parameter for the long-term stability of the photovoltaic module, identified by determining the gel content ratio.

This study aims to comprehensively understand the relationship between CTE, cross-linking, and residual thermal stresses in encapsulation materials for solar panel production. Through an in-depth examination of thermal characteristics of cured and uncured polyolefin elastomer (POE) samples via Thermomechanical Analysis (TMA) and Differential Scanning Calorimetry (DSC) methods, we investigated the effects of cross-linking on the thermal expansion coefficient and residual thermal stresses of the encapsulation film.

EXPERIMENTAL APPROACH

Materials:

Uncured & Cured Polyolefin elastomer—POE

Methods:

- Soxhlet Extraction

to determine degree of cross-linking

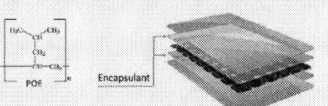

- Thermo-mechanical Analysis (TMA)

to measure the thermal expansion coefficient

- Differential Scanning Calorimetry (DSC)

to examine thermo-mechanical stability

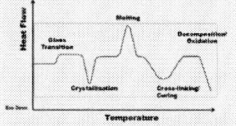

RESULTS

Cross-linking

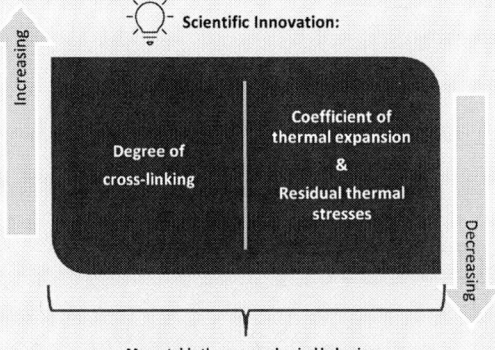

Before cross-linking   After cross-linking

Method for measuring degree of cross-linking:

Figure 1. Soxhlet extractor.

Cured sample's degree of cross-linking:

Table 1. Soxhlet Extraction test results.

M1 (g)	M2 (g)	Degree of cross-linking (%)
0.3833	0.2932	76.49

M1: Initial mass of the sample

M2: Mass of the extraction/insoluble residue

$$\text{Gel Content (\%)} = \frac{M2}{M1} \times 100$$

Relationship:

Lower CTE → Reduced Residual Thermal Stresses: Reducing CTE through cross-linking makes the material less sensitive to temperature changes, minimizing the degree of thermal expansion and contraction. This, in turn, reduces the residual thermal stresses, helping to improve the material's stability and durability.

! **increased long-term stability**

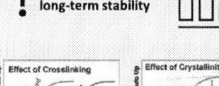

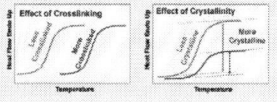

💡 **Scientific Innovation:**

Degree of cross-linking   **Coefficient of thermal expansion & Residual thermal stresses**

Increasing / Decreasing

More stable thermo-mechanical behavior

Decreasing CTE with increasing degree of cross-linking:

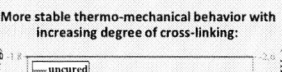

Figure 2. TMA curves of cured and uncured samples.

More stable thermo-mechanical behavior with increasing degree of cross-linking:

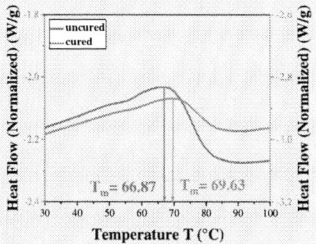

$T_m = 66.87$ $T_m = 69.63$

Figure 3. 2nd heat DSC thermograms of the samples.

CONCLUSIONS

- It was observed that an optimum increase in **cross-linking** directly corresponds to a decrease in the **thermal expansion coefficient** as a result of the studies conducted on cured and uncured samples. Furthermore, this study highlights the critical role of increasing cross-linking in reducing **residual thermal stresses** within the material.

- It has also been observed in the DSC graph that as the degree of cross-linking increases, **melting temperature** and **crystallinity** of the material also increase. The reason for this is that materials with a higher degree of crystallinity have a more ordered and tightly packed molecular structure, which enables them to possess greater **thermal stability.**

- Finally, it was concluded that increasing the degree of cross-linking leads to a more stable thermo-mechanical behavior of the material.

REFERENCES

[1] Omazic, A. (2019). Weathering stability of polymeric materials developed for PV modules operating in harsh climatic conditions. Montanuniversität Leoben. https://pure.unileoben.ac.at/en/publications/weathering-stability-of-polymeric-materials-developed-for-pv-modu
[2] Wu, D. (2015, January 1). Investigation of the reliability of the encapsulation system of photovoltaic modules. Figshare. https://repository.lboro.ac.uk/articles/thesis/Investigation_of_the_reliability_of_the_encapsulation_system_of_photovoltaic_modules/9540926

UV exposure of glass/glass coupons with edge seal and different encapsulants

C. Barretta[1*], L. Meinhart[1], A. Brandstätter[2], D. Geier[3], R. Einhaus[3], A. Gok[4], G. Oreski[1,5]

[1] Polymer Competence Center Leoben GmbH (PCCL), Roseggerstraße 12, 8700 Leoben, Austria – chiara.barretta@pccl.at
[2] Lenzing Plastics GmbH & Co KG, Lenzing, 4860, Austria
[3] Zentrum für Sonnenenergie- und Wasserstoff-Forschung Baden Württemberg (ZSW), 70563, Stuttgart, Germany
[4] Gebze Technical University, 41400, Gebze, Kocaeli, Turkey
[5] Montanuniversität Leoben, Chair of Material Science and Testing of Polymers 8700, Leoben, Austria

INTRODUCTION AND OBJECTIVES

- Photovoltaic (PV) modules based on glass/glass configuration are expected to become dominant on the market within the next 10 years.
- With the further development of bifacial cells, it will be necessary to have a transparent backsheet (such as glass) to increment the electricity that can be produced.
- Additionally, the encapsulant market is shifting from ethylene vinyl acetate (EVA) to alternative polyolefin-based materials.
- The objectives of this work are to:
  - Compare the performances of encapsulants alternative to EVA in a glass/glass based PV module configuration.
  - Assess the reliability of the material combinations exposed to artificial weathering tests and especially UV.

EXPERIMENTAL

- Glass/glass coupons were produced using four different encapsulants: EVA, TPO, POE and an ionomer (ION). The samples were laminated using and edge seal.
- The materials were exposed to UV conditions (IEC TS 62788-7-2-A3) up to 2000 hours. The samples were characterized, before and after the exposure to UV test, by means of:

Visual inspection	UV-Vis spectroscopy	FTIR ATR spectroscopy	DSC	TD-GCMS
Defects identification	Optical properties, additives	Chemical degradation	Thermal properties, morphology	Qualitative additive analysis

RESULTS AND DISCUSSION

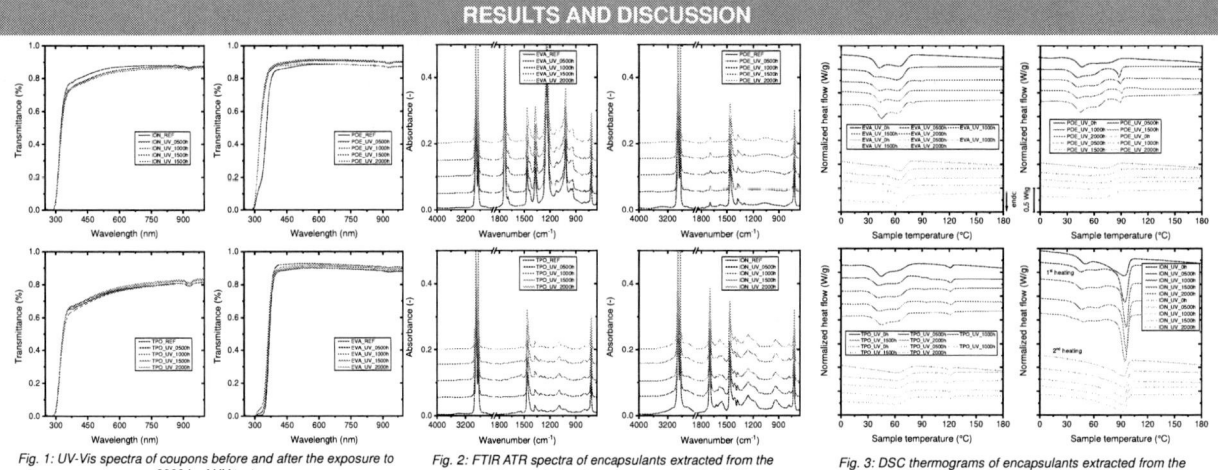

Fig. 1: UV-Vis spectra of coupons before and after the exposure to 2000 h of UV test.

Fig. 2: FTIR ATR spectra of encapsulants extracted from the coupons before and after the exposure to 2000 h of UV test.

Fig. 3: DSC thermograms of encapsulants extracted from the coupons before and after the exposure to 2000 h of UV test.

- Decrease in transmittance for samples containing EVA and ION as encapsulants.
- Changes in transmittance in UV region for EVA encapsulants, possibly due to consumption of stabilizers.
- Shift of UV cut-off for samples with POE encapsulants.

- In particular, the ION is based on ethylene acrylic acid copolymers, the TPO and POE are ethylene α-olefin based copolymers and the EVA is an ethylene vinyl acetate copolymer.
- No significant changes in chemical structure upon UV exposure.

- No significant changes in chemical structure upon UV exposure for all the encapsulants in the second heating step (→ no relevant chemical degradation).
- Changes in the first heating curves due to slight changes in morphology (reversible).

Table 1: Qualitative additive analysis of encapsulants extracted from the coupons before and after the exposure to 2000 h of UV test.

Encapsulant	Antioxidant	HALS	UV Absorber	Encapsulant	Antioxidant	HALS	UV Absorber
EVA	-	yes	yes	TPO	yes	-	-
EVA_UV_1000 h	-	yes	yes	TPO_UV_1000 h	yes	-	-
EVA_UV_2000 h	-	yes	yes	TPO_UV_2000 h	yes	-	-
POE	-	-	yes	ION	-	-	-
POE_UV_1000 h	-	yes	yes	ION_UV_100 h	-	-	-
POE_UV_2000 h	-	yes	yes	ION_UV_2000 h	-	-	-

The encapsulants used within this study showed in general very good stability towards UV.

No significant material degradation could be detected.

CONCLUSIONS AND OUTLOOK

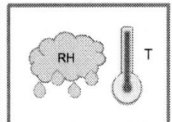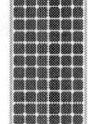

Coupons were laminated with glass/glass, edge seal and four encapsulants.

Exposure of coupons to artificial weathering tests with and without UV did not cause significant chemical degradation of the encapsulants [1, 2, 3].

Presence of the edge seal did not prevent encapsulant degradation for coupons with POE, TPO and ION. Edge seal helped to reduce the effects of EVA degradation on mini-modules and full-scale modules with PERC cells [3].

Interactions with newly developed cell technologies?

REFERENCES
[1] C. Barretta et al., Damp Heat Exposure of Glass/Glass Coupons with Different Encapsulants, 51th IEEE Photovoltaic Specialists Conference (PVSC), 2023
[2] C. Barretta et al., Design and Testing of PV Modules Based on Glass/Glass Configuration to Achieve Extended Lifetime, 40th European Photovoltaic Solar Energy Conference and Exhibition (EU PVSEC), 2023
[3] R. Einhaus et al., Test and Evaluation of Combinations of Encapsulant Materials towards a Long Service Lifetime of PV Modules: PV 40 Plus, 40th European Photovoltaic Solar Energy Conference and Exhibition (EU PVSEC), 2023

ORCID ID

This work was conducted as part of the Austrian "e!MISSION.at – Energy Mission Austria" project "PV40+" (FFG No. 881868) funded by the Austrian Climate and Energy Fund and the Austrian Research Promotion Agency (FFG).

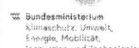

41st European Photovoltaic Solar Energy Conference and Exhibition

Material Screening for the Development of a Photovoltaic Module Using Biodegradable Materials from Renewable Raw Materials

Fraunhofer Center
for Silicon Photovoltaics CSP

Matthias Pander*, Ringo Koepge, Bengt Jaeckel, Anton Mordvinkin

Fraunhofer Center for Silicon Photovoltaics CSP, Otto-Eissfeldt-Strasse 12 | 06120 Halle (Saale) | Germany
*contact: +49 (0) 345 5589-5215 | matthias.pander@csp.fraunhofer.de

3AV.2.50

MOTIVATION

※ Sustainability and resource conservation are key issues of the 21st century

※ EU target: climate neutrality by 2050, reduction of emissions by at least 55% by 2030 [1]

※ PV modules generate no greenhouse gas emissions and significantly contribution to climate protection targets

※ Service life of the modules is finite and raw material cycle and final disposal of the components are increasingly coming into focus [2]

※ Main mass fraction of PV modules (glass, aluminum, copper) can be recycled but polymer fraction is currently based on fossil sources and is mostly thermally recycled [2]

※ Focus of "E-Quadrat" project (2021-2024): Development of PV modules with biodegradable materials from renewable resources

※ **This Work:** Introduction of a material selection and results of qualification tests for a module made of biodegradable and bio-based materials

MATERIAL CANDIDATES, TEST SPECIMEN AND TEST SCHEME

※ A composite material made of polyethylene and wood, which is fully cradle-to-cradle recyclable, is being investigated as frame material substitute [3]

※ After pre-qualification 4 main variants and 2 experimental variants were selected for manufacturing of mini-modules

※ To demonstrate suitability IEC 61215 and IEC 61730 accelerated aging tests are used [4][5]

Layer | Variant	Var1	Var2	Var3	Var4	Var5	Var6	Ref2	Ref
Front side	Glass							
Front Encapsulant	EVA #1 BODG add.	PLA #1 Bio-based BODG	EVA #1 BODG add.	EVA #1 BODG add.	EVA #1 BODG add.	EVA #1 BODG add.	EVA #1 Standard²	EVA #2 Standard³
Cell technology	PERC							
Back Encapsulant	EVA #1 BODG add.	PLA #1 Bio-based BODG	EVA #1 BODG add.	EVA #1 BODG add.	EVA #1 BODG add.	EVA #1 BODG add.	EVA #1 Standard²	EVA #2 Standard³
Back sheet	PET #1 bio-based	PET #1 Standard²	PA11 Bio-based recyclable	PLA #2 Bio-based BODG	PLA #2 Bio-based BODG	2xPLA #3 2xPET #1 Bio-based coating	PET #1 Standard²	PET #1 Standard²
Sample amount	15	15	15	15	3¹	4¹	15	15

Overview of produced Mini-Module variants
Abbreviations: BODG – biodegradable, BODG add. – an additive that promotes biodegradation
¹limited amount of material ²Standard-commercial-material used at Fraunhofer CSP and ³Master batch material

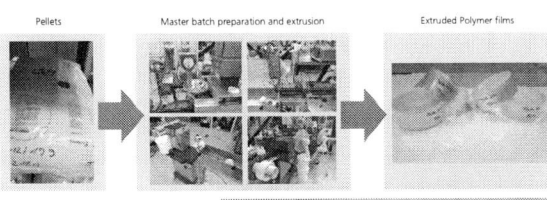

Pellets   Master batch preparation and extrusion   Extruded Polymer films

Schematic overview of the customized in-house extrusion of EVA encapsulants at the Fraunhofer IMWS

Representative Photos of the mini-modules

RESULTS – AGING OF WOODEN FRAME MATERIAL CANDIDATE

※ Humidity is a crucial factor for the mechanical properties

※ Significant drop of flexural modulus after first 500h of DH and remains constant at around 88%

※ Elongation at break drops to 81% and breaking stress to 83% of original level after 2000 h

※ In IEC 61730 Seq B (not shown) flexural modulus drops to ~88% of the original level at HF10-1 and HF10-2, breaking stress and elongation at break only change more significantly in final aging stage (87 % of the original level)

※ Material is stable regarding weathering → suitable candidate for frame, further improvement of material characteristics may be achieved by fiber and additives

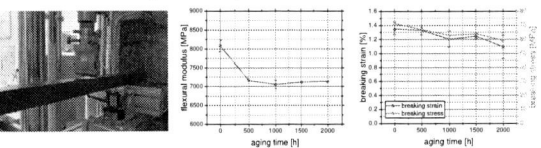

Bending test setup for wooden frame material candidate, Results during DH aging: flexural modulus, breaking strain and breaking stress

RESULTS – MINI-MODULE AGING TESTS

※ IEC 61730 SEQB most challenging test of current certification

※ Moisture ingress and UV irradiation have a much higher effect due to the sample structure

※ Var 4 already completely fails after DH200 as the PLA cracked and delaminates

※ Var 5 and Var 6 outer coating of the films has already peeled off after DH200

※ Var 1 and Var 3 are better or comparable to the reference

※ Some references also show significant performance degradation (4 – 11 %) during this test

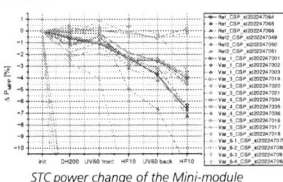

STC power change of the Mini-module samples during IEC 61730 sequence B

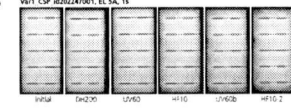

Test	Var1	Var2	Var3	Var4	Var5	Var6	Ref2	Ref
Seq B	Pass	-	Pass	Fail	Fail	Fail	Pass	Pass
Seq C	Pass	-	Pass	Fail	-	-	Pass	Pass
Seq D	Fail	-	Pass	Fail	-	-	Pass	Pass
Seq E	Pass	Fail	Fail	Fail	-	-	Pass	Pass
overall	Suitable	Not suitable	Suitable	Not suitable	Not suitable	Not suitable	Suitable	Suitable

Evaluation of the Mini-module test results   *Degradation at the limit and not all samples affected

CONCLUSION

■ Composite wood-based material
   ■ good resistance in aging tests and manageable disadvantages compared to aluminum
   ■ great potential for further optimization through targeted material improvement to use it as frame material substitute

■ Variants Var 1 (bio-based PET) and Var 3 (bio-based PA11) identified as possible candidates for next stage of module development with sustainable materials

■ Applicability of bio-based materials was demonstrated

■ Truly biodegradable materials considered critical, as biodegradation occurs at low thermal activation and materials have low moisture stability

[1] Document 32021R1119, Regulation (EU) 2021/1119 of the European Parliament and of the Council of 30 June 2021 establishing the framework for achieving climate neutrality and amending Regulations (EC) No 401/2009 and (EU) 2018/1999 ('European Climate Law'), https://eur-lex.europa.eu/eli/reg/2021/1119/oj
[2] Isherwood, P.J.M., Reshaping the Module: The Path to Comprehensive Photovoltaic Panel Recycling. Sustainability 2022, 14, 1676. https://doi.org/10.3390/su14031676
[3] GCC wood material, https://www.greencomposite.de/en/home/greencompositewood/
[4] IEC 61215-2:2021, Terrestrial photovoltaic (PV) modules – Design qualification and type approval – Part 2: Test procedures
[5] IEC 61730-2:2023, Photovoltaic (PV) module safety qualification - Part 2: Requirements for testing

The authors gratefully acknowledge the financial support by the German Federal Ministry for Economic Affairs and Climate Action (BMWK) of the project "E-Quadrat" under grant number 03EE1114

41st European Photovoltaic Solar Energy Conference and Exhibition

FAILURE MODE ANALYSIS OF AUSTRIA'S FIRST ROAD-INTEGRATED PHOTOVOLTAIC SYSTEM

Alexander Erber, Bernhard Grasel

University of Applied Sciences Technikum Wien, Competence Field Renewable Energy Technologies, Vienna, Austria

Introduction

The exploration of traffic areas as a novel photovoltaic (PV) integration opportunity within the traffic sector has been demonstrated in various projects. Limited data and publications on the performance and failure modes of these innovative road-integrated PV (RIPV) modules highlight the need for a comprehensive failure analysis. The PV parking place in Teesdorf, built in 2022, enables such an analysis. Due to the predominant evening use of the RIPV system, both the systems performance and the effect of vehicle loads during nighttime can be analysed. This study focuses on assessing failure modes of RIPV modules installed at Austria's first RIPV system.

Fig. 1. Topview of RIPV system with a marked string (a); inverter box (b); surrounding area (c)

Material and Methods

o Area of the RIPV system: 100 m²
o Power: 16.8 kWp or 12.4 kVA
o Wiring of the PV elements: 42 strings with 18-20 elements wired to 42 microinverters
o RIPV module: commercially available, 2 t wheel load per module
o Applied failure analysis methods: see **Fig. 2.**

Fig. 2. Applied methods and the detectable failure modes

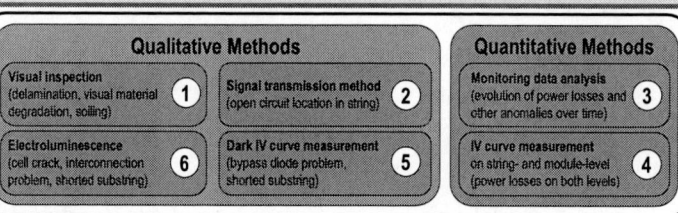

Results and Analysis

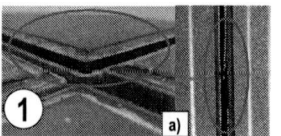

Material degradations:
a) Module detachment and broken module edges: enables water ingress → bypass diode problems (see **Fig. 5.**)
b) Delamination: in combination with a), before and behind the cell encapsulation composite → cell short ciruits (see **Fig. 4.**)

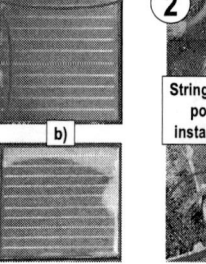

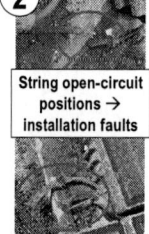

String open-circuit positions → installation faults

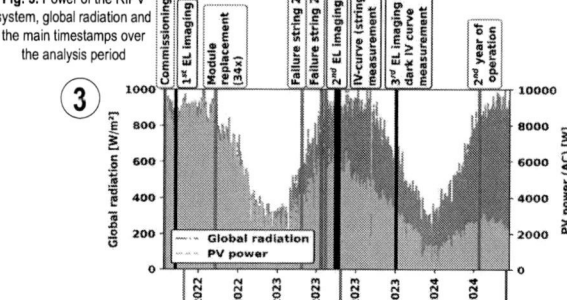

Fig. 3. Power of the RIPV system, global radiation and the main timestamps over the analysis period

Main Findings

o The RIPV system produced 10.2 MWh in its first year of operation, 27.2 % less than the simulated due to a lower module power than specified by the manufacturer.
o Visual inspections reveal material degradations (detachment of the module top layer, delamination, and broken module edges).
o In the analysed monitoring date continuous power losses are observed over the systems operation time. String-level power losses of up to 47.8 % and 77.5 % are calculated for the first and second year of operation.
o Cell cracks are identified as the main cause of the power losses, attributed to vehicle loads, through electroluminescence images.
o At least one open bypass diode path is found for three out of the 16 measured sample strings based on the measurement of the dIV.

Fig. 4. String EL images of String 2.38

Tab. 1. String degradation rates

Calculated from STC corrected monitoring data	First year (2022-05-31 - 2023-05-28)	Second year (2023-05-28 - 2024-06-17)
String 2.38	-30.0%	no power
All strings (mean)	-33.5%	-56.2%
All strings (median)	-34.1%	-56.2%
All strings (max)	-47.8%	-77.5%
All strings (min)	-13.8%	-29.2%

Power values were validated with IV curve measurements.

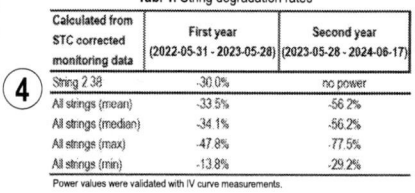

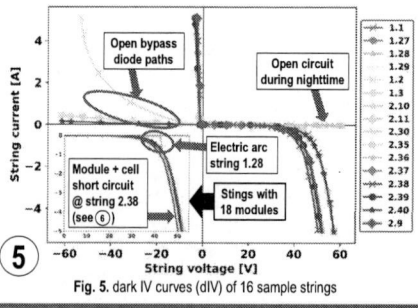

Fig. 5. dark IV curves (dIV) of 16 sample strings

Conclusions

o Material and performance degradation: Failure modes such as delamination, edge breakage, and water ingress were found, resulting in cell short circuits and bypass diode failures. Vehicle-induced cell cracks and fractures, causing the main power losses, were identified. A combination of failure analysis methods is essential to quantify the effects of failure modes and identify their causes.
o Inadequate quality control and necessary design improvements: Improved quality control and adherence to certification standards are recommended for the analysed RIPV module to reduce/prevent the found material failure modes. Changes in module design, such as switching to a glass-glass design or thin-film cell technology, are suggested to enhance durability and mechanical stability. The results show that standardised test procedures are necessary for RIPV modules to ensure durability, safety and performance in real environments.

This research was conducted as part of the project "Accompanying research Solarer Platz Teesdorf" (C177537), which is funded by the Austrian Climate and Energy Fund.

41st European Photovoltaic Solar Energy Conference and Exhibition

EFFECTS OF ENCAPSULANT- BACKSHEET COMBINATIONS ON DURABILITY OF OPTICAL PROPERTIES

Jishnu Ramachandran Nair[1,2] Daniel Schülser[1], Michael Wendt[1], Ralph Gottschalg[1,2] Anton Mordvinkin[1*]

1-Fraunhofer Center for Silicon Photovoltaics (CSP), Otto-Eißfeldt-Str. 12, 06120 Halle, Germany,
2- Anhalt University of Applied Sciences, Bernburger Str. 55, 06366 Köthen (Anhalt)

*Corresponding Author: anton.mordvinkin@csp.fraunhofer.de,

ABSTRACT: The compatibility and interactions between encapsulants and backsheets in photovoltaic (PV) modules have received limited attention in aging studies, despite their critical role in module reliability. This research addresses this gap by investigating the effects of encapsulant-backsheet combinations starting with optical durability, a key factor influencing PV module performance and longevity. To this end, the optical durability of two common encapsulant materials, Ethylene Vinyl Acetate copolymer (EVA) and Polyolefin Elastomer (POE), a transparent Polyethylene terephthalate backsheet (PET) and combinations thereof was investigated. The individual components as well as their combinations were subjected to accelerated weathering under Damp Heat, Damp Heat with UV exposure (DHUV), and the environmental conditions specified in IEC 62788-7-2 A3. The optical durability was subsequently characterized by means of the visual inspection and UV-Vis spectroscopy, whose results were correlated with the Fourier Transform Infrared Spectroscopy (FTIR) data to provide molecular insights.

Results revealed distinct degradation patterns among material combinations. Remarkably, a EVA/PET combination exhibited superior resistance to optical degradation compared to POE/PET, particularly under DHUV conditions. Notably, materials behaved differently in combinations versus when weathered alone, observed by PET's reduced degradation when combined with the encapsulants under DHUV exposure.

This study establishes a foundation for understanding encapsulant-backsheet compatibility in the context of optical durability. The findings highlight the importance of considering material interactions in PV module design and testing protocols. This approach paves the way for developing mechanistic degradation models, potentially enhancing material selection processes, and improving predictions of module lifetime and performance.

Keywords: encapsulants, backsheets, compatibility, optical durability

1 INTRODUCTION

Polymeric materials used in photovoltaic applications are susceptible to various degradation mechanisms that can significantly impact module performance and longevity as seen in Figure 1. Exposure to combined stressors of temperature, humidity, and UV radiation degrades encapsulants and backsheets, resulting in discoloration that reduces transmittance and power output [2].

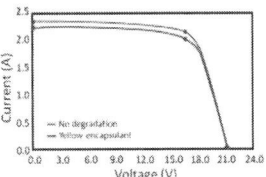

Figure 1. Variation in power output due to discolored encapsulant [1]

Few studies [3] [4] [5] have been done to probe encapsulant-backsheet interactions, either focusing on limited encapsulant or backsheet type or having limited characterization studies relevant to polymeric applications for photovoltaics. Therefore, the research question of which encapsulants are more compatible with which backsheet in context of ageing must be addressed for ensuring module reliability. Further to this, encapsulant degradation studies with respect to optical durability has

been studied [6], however without compatibility studies with backsheets.

2 MATERIALS AND METHODS

2.1 Materials

Commercially available encapsulants (POE and EVA) and backsheet (Transparent PET) were taken for the study and are described in the table below.

Abbreviation	Type	Material
EVA	Encapsulant	Ethyl-Vinylacetate Copolymer
POE	Encapsulant	Polyolefin elastomer
PET trans	Backsheet	Polyethylenterephth alate (transparent)
EVA / PET trans	Combination	EVA and PET trans
POE / PET trans	Combination	POE and PET trans

Table 1 : Material selection and types

2.2 Sample Design

Coupon samples were made for encapsulants and backsheets as a combination and as individual encapsulants or backsheets as shown in below in Figure 2.

41st European Photovoltaic Solar Energy Conference and Exhibition

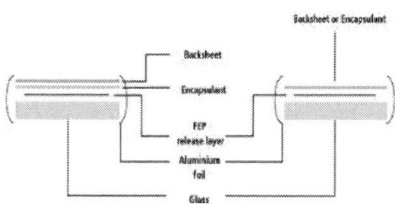

Figure 2. Sample design for materials as a combination(left) and weathered alone(right)

2.3 Methods

The samples were weathered for 2000 hours under the below conditions.

type	conditions	Observation purpose
Damp Heat **(DH)**	85°C and 85% relative humidity	Hydrolysis degradation
Damp Heat +UV**(DHUV)**	85°C and 65% relative humidity. UV intensity 104 W/m²(300-400nm)	Hydrolysis and photooxidative degradation
Damp Heat +UV**(IEC)** IEC 62788-7-2	65°C and 20% relative humidity. UV intensity 81 W/m²(300-400nm)	photooxidative degradation

Table 2: Weathering conditions and purpose

The samples were evaluated for 0h and 2000h after the weathering through macroscopic and microscopic evaluation.

The macroscopic evaluation involved visual inspection and UV – Vis spectroscopy. UV-Vis was done on the Thermofischer scientific Evolution One Plus in the wavelength of 190-1100 nm with steps of 1 nm. The Reduced representative solar weighted transmittance (τrsw) was calculated as shown below where $\tau[\lambda]$ is the transmittance, and $Ep\lambda$ being the photon flux in the wavelength range (280-1100 nm).

$$\tau\,rsw = \frac{\int \tau[\lambda]Ep\lambda\,[\lambda]d\lambda}{\int Ep\lambda\,[\lambda]d\lambda} \qquad (1)$$

Furthermore, the transmittance at 450nm wavelength was also evaluated. The microscopic evaluation was done through Fourier Transform infrared spectroscopy (FTIR) using Thermofischer Nicolet iS20. The carbonyl index was calculated using the ratio of the integral of the C=O carbonyl peak at around 1700 cm-1 to the area of the CH₂-rocking vibration peak at 720 cm-1 taken as normalization peak.

3 RESULTS AND DISCUSSION

EVA encapsulant weathered alone for 2000h of DH is taken as an example to show the analysis procedure.

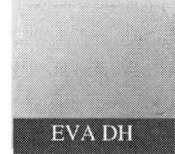

Figure 3. Discolored EVA DH sample

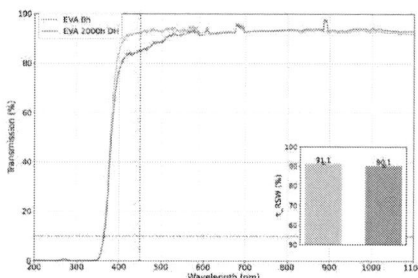

Figure 4. Reduction in transmittance and τrsw for EVA DH sample weathered alone.

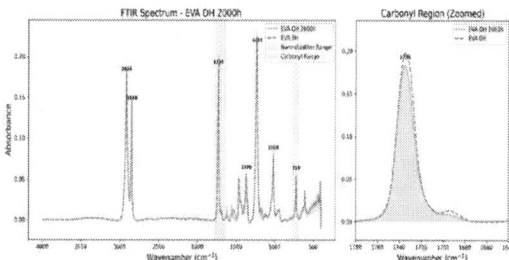

Figure 5. FTIR evaluations showing decrease in the carbonyl peak for EVA DH

The discoloration is reflected in the reduced transmittance values between 400 and 600 nm (Figure 4), suggesting chromophore formation. It is supported by the FTIR data (Figure 5), the decrease in the carbonyl peak is attributed

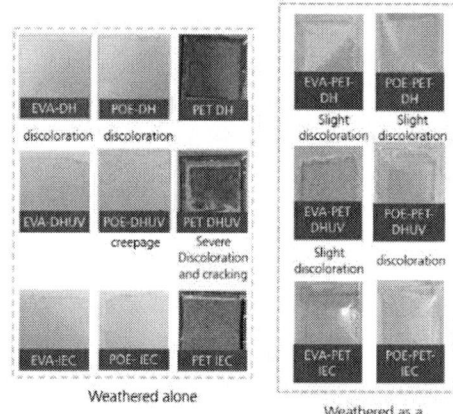

Figure 6. Visual inspection for all samples

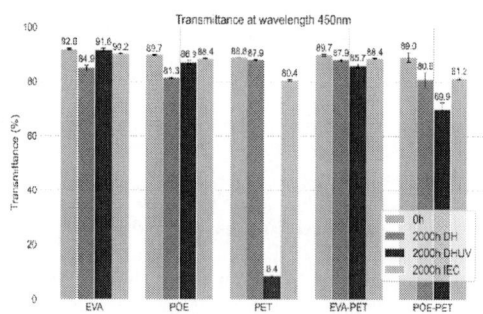

Figure 7. Transmittance at 450nm for all samples

to the degradation due to acetic acid from hydrolysis

688

reactions [7] leaving chromophoric double bonds behind that cause discoloration and reduced transmittance.

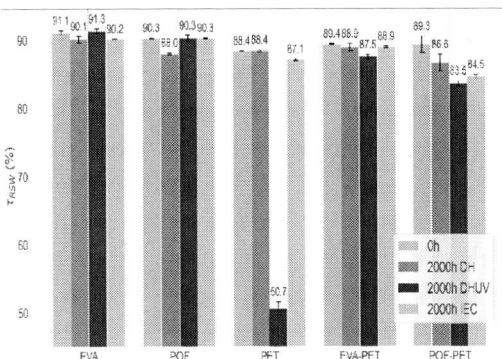

Figure 8. $\tau\,rsw$ for all samples

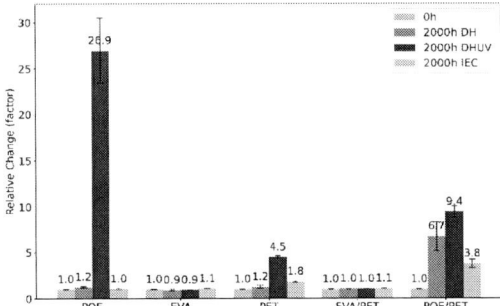

Figure 9. Relative changes(as factor) in carbonyl index

3.1 Individual vs. Combination

The intense degradation of PET after 2000 h of the DHUV exposure sample as seen in visual inspection in Figure 6 as compared to the slight discoloration in EVA/PET and POE/PET combination, shows the intense difference between materials weathered alone and as combinations. This difference is also observed in the transmittance at 450 nm in Figure 7 and $\tau\,rsw$ in Figure 8. This indicates that EVA and POE were acting as a UV protective filter to prevent it from intense photodegradation.

POE exhibited a decrease in reduced solar-weighted transmittance values after 2000 h DH+UV as shown in Figure 8. However, the change in the carbonyl index was significant, with a factor of nearly 27 as seen in Figure 9. When combined with PET, the relative change was only a factor of 9.4. Oxidative reactions can introduce carbonyl defects into ethylene units, triggering Norrish Type I reactions where photochemically induced homolytic cleavage generates free radical intermediates, or Norrish Type II reactions that produce ketones and vinylidenes [8]. This undesired formation in POE due to UV exposure and hydrolysis was reduced by using the POE/PET combination, where PET's higher barrier properties to oxygen prevented the intense degradation.

3.2 Combinations

While the POE/PET combination exhibited noticeable discoloration after 2000h of the DHUV exposure, the EVA/PET combination demonstrated superior stability. This enhanced optical durability is evident in the stable transmittance values at 450 nm and minimal changes in the carbonyl index observed for

EVA/PET across all weathering conditions. This suggests that EVA/PET offers greater resistance to chemical degradation, specifically hydrolysis and photooxidative degradation, which is remarkable since EVA typically is considered to be a less chemically stable material. Further investigation into the interactions between additives within the EVA/PET and POE/PET combinations using Pyrolysis Gas Chromatography – Mass spectroscopy is planned for the future research.

4 CONCLUSIONS

This study highlights the crucial role of polymeric interactions in the performance of backsheets and encapsulants for photovoltaic applications. Notably, the EVA/PET combination demonstrated superior optical durability and degradation resistance compared to the POE/PET combination. This finding, supported by both macroscopic and microscopic analyses, underscores the importance of considering material combinations rather than individual components in isolation. Future research should prioritize investigating a wider range of commercially available and developing materials for photovoltaic applications, employing a combined macro-micro analysis approach to elucidate the underlying degradation mechanisms. This comprehensive understanding will ultimately enable a realistic performance assessment and subsequently the development of accurate lifetime prediction metrics based on mechanisms for these critical components.

REFERENCES

[1] S. &. M. F. &. M. R. &. E. A. Diallo, „Understanding Photovoltaic Module Degradation: An Overview of Critical Factors, Models, and Reliability Enhancement Methods,“ in *E3S Web of Conferences,* 2023.

[2] C. Kim, M. Jeong, J. Ko, M. Ko, M. Kang und H. Song, „Inhomogeneous rear reflector induced hot-spot risk and power loss in building-integrated bifacial c-Si photovoltaic modules,“ Nr. 163, 2021.

[3] G. T. Farrukh ibne Mahmood, „Impact of different backsheets and encapsulant types on potential induced degradation (PID) of silicon PV modules,“ *Solar Energy,* Bd. 252, 2023.

[4] A. O. G. E. M. E. G. H. C. P. G. E. M. Omazic, „Increased reliability of modified polyolefin backsheet over commonly used polyester backsheets for crystalline PV modules,“ *Appl Polym Sci,* Bd. 137, Nr. 30, 2020.

[5] O. I. E. G. e. a. Brune B, „Quantifying the influence of encapsulant and backsheet composition on PV-power and electrical degradation,“ *Prog Photovolt Res Appl,* pp. 716-728, 2023.

[6] L. S. C. A. G. F. M. P. Valeria Fiandra, „New PV encapsulants: assessment of change in optical and thermal properties and chemical degradation after UV aging,“ 2024.

[7] H. RiedlG, „Environmental fatigue crack growth of PV glass/EVA laminates in the melting range.,“ Bd. 32(9), 2024.

[8] R. G. W. Norrish und C. H. Bamford, „Photodecomposition of aldehydes and ketones,“ 1937.

41st European Photovoltaic Solar Energy Conference and Exhibition

PID OUTDOOR MEASUREMENTS,
A NEW TEST SETUP

Jörg Kirchhof
Fraunhofer IEE
Joseph-Beuys-Str. 8, 34117 Kassel, Germany, Joerg.kirchhof@iee.fraunhofer.de

ABSTRACT: Potential induced degradation (PID) still occurs on PV modules in the field. Different approaches have been developed to prevent PID, e.g., at cell- or module-level or by avoiding operation conditions which promote PID at the PV-generator-level. In cases where these measures are not possible, or where a replacement of PID sensitive modules would lead to excessive costs, additional apparatuses have been designed, which offer periodic voltage reversal at night to "heal" PID damages from the beginning. The effect was previously demonstrated under ideal laboratory conditions, but outdoor long-time tests have seldom been done [1]. During the research project PID-Recovery, long-term investigations concerning the PID influence on PV-modules have been carried out. The focus of this paper are the measurement setup and test results.

1 WHAT DID WE DO?

At the outdoor test-field of Fraunhofer IEE, seven crystalline silicon PV modules have been installed, on which the polarization voltage between PV-cells and module frame follows software-controlled test profiles. Simultaneously, the leakage current of the module frames and the IV-curves of each PV module were periodically recorded over several months. Realistic voltage profiles, but also reverse-polarization during the night for regeneration purposes can be applied on these modules. The polarity of the voltage profile can be inversed as well. On one of these PV-modules, a surface-voltage-measurement on the glass surface of the PV module has been added, which gives deeper understanding onto the degradation conditions for each cell. As a reference, six additional PV modules have been operated under the same conditions but without polarization voltage. Here, the IV-curves have been recorded as well. During the test, individual modules have been exchanged between the polarized set-up and the polarization-free set-up to investigate self-healing of the PID-effect.

Figure 1: PV-module with surface electrodes and contact wires stretching to the breakout boxes.

2 HOW TO MEASURE SURFACE VOLTAGES?

Every voltage measurement needs to draw current from the voltage source. In the case of surface voltages, the glass acts as resistor, which is composed of the glass impedance between the solar cells and the glass surface and the surface resistance which is lower due to dirt and moisture. Therefore, the "voltage probe" needs to have an extremely high impedance and on the other hand, it needs to follow a huge voltage range which can almost reach the polarization voltage of the cells against ground. To reduce this glass resistance, an electrode has been attached to the

glass surface which covers the cell-grid of individual solar cells on the module. This increases the current collected by the electrode and it also avoids shading of the cells (see Figure 1).

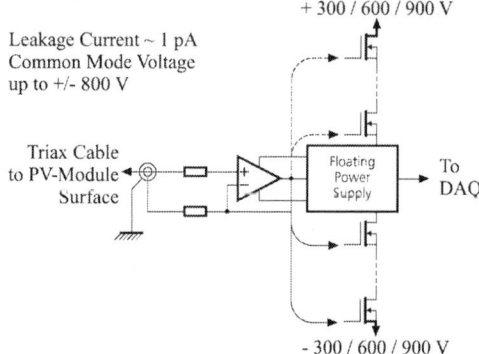

Figure 2: Simplified circuit diagram

16 ultra high impedance amplifiers have been built, which can follow the input voltage thanks to their floating power supply. A series of MOSFETs deliver the operating current for the floating power supply and make the output voltage of the amplifier equal to the input voltage (see Figure 2).

When handling of extreme low currents is needed, impedances come into play which are usually neglected. An important impedance is the cable isolation resistance, which would be parallel to the input resistance of the amplifier. To avoid any current between internal cable Conductor and cable isolation, an active guard shielding has been applied, where the internal shield of the cable is connected to the output of the amplifier. Therefore, there is no voltage difference between shield and inner cable conductor and no current can flow through the cable isolation. A triaxial cable has been used where a second shield is grounded to avoid EMC or oscillation problems.

3 REAL TIME DEGRADATION MEASUREMENTS

The degradation of the six PID-stressed modules over 350 days under outdoor conditions is shown in Figure 3. As indicator, the fill factor is used. Due to shunting in the PID-S degradation [2], the cell is bypassed, which creates a degradation of the I-V-Curve. This leads to the fill factor reduction shown in Figure 3. The modules are differently affected by PID. Module 2 shows the fastest and strongest PID-effect. Especially module 6 shows phases of fast degradation and phases of slow degradation. At the end, module 6 still has the highest fill factor.

After 434 days, the test program for modules 4 and 5 have been changed according to the voltage profile shown in Figure 5. For module 4, a regeneration voltage of + 1000 V at night has been applied while the voltage profile during the day still gives to degradation stress. On the other hand, module 5 is exposed to positive voltages only. Figure 4 shows that within 50 days a regeneration occurs, but that only a fill factor of 60 % has been achieved while the original fill factor before degradation was about 80 %. Regeneration happens on both modules, but thanks to the pure positive bias at module 5, the regeneration happens faster.

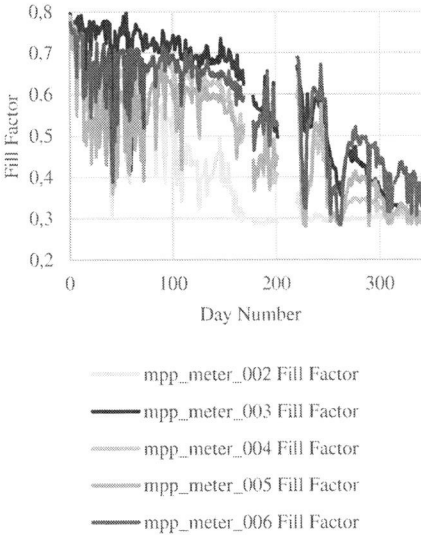

Figure 3: Degradation (Fill-Factor)

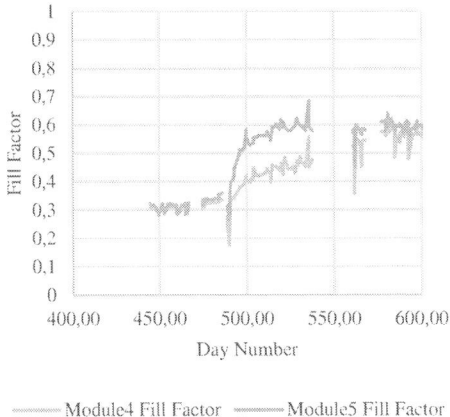

Figure 4: Regeneration of module 4 and 5.

The effect of degradation is clearly visible in the I-V-curves of the modules. In Figure 6 the I-V-curves of modules 2-6 show no sign of degradation, but after 200

days of PID stress, the I-V-curves of these modules indicate a loss of fill factor (see Figure 7). This degradation influences the whole I-V-curve and is not limited to series or parallel resistance of the module. Especially module 2 has changed to a pure resistive behavior which is an effect of PID-S degradation.

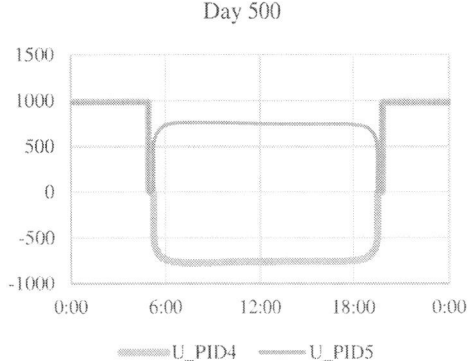

Figure 5: Voltage profile during regeneration of module 4 and 5

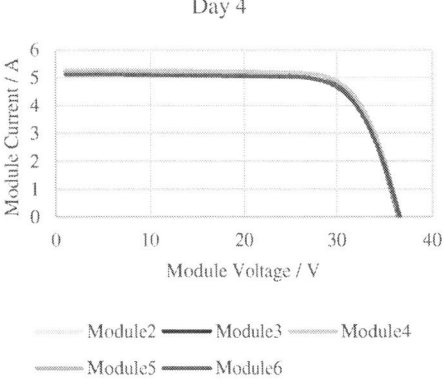

Figure 6: I-V-curves of PV-modules 2 to 6 at day 4. No degradation is visible.

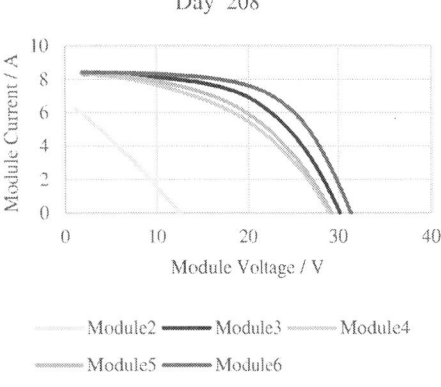

Figure 7: I-V-curves of module 2 to 6 after 208 days of degradation.

41st European Photovoltaic Solar Energy Conference and Exhibition

Figure 8: The position of each surface electrode.

In Figure 8 the electrodes have been assigned to four different groups. The yellow group (electrode 1, 4 and 13) consists of cells which are directly at the border, which means that two sides of the cell are near to the grounded module frame. The green group of electrodes (3, 5, 8, 9, 14, 15, 16) includes cells which only have one side in proximity to the module frame. In contrast to this, the purple and orange groups have larger distance to the frame.

The surface voltages pictured in Figure 9 are showing the voltage stress for each of the observed cells. This stress is the highest for the electrodes with the lowest voltage against ground, because in this case, the voltage difference between cell and glass surface reaches its maximum value. Due to the weather conditions during this day, humidity had a huge impact onto the conductivity of the glass surface. E.g. at 16:00 h, all electrodes reached zero voltage against the frame which gives maximum PID-stress to all cells, no matter if they are close to the frame, or not, when humidity drops, a separation between the different electrode groups is visible. The purple curves are close to the cell voltage while the green curves cover a larger voltage area between cell-voltage and the level of the yellow curves. Unfortunately, the measurement for cell 11 has failed. The red line shows the applied PID-voltage at the cell and the blue curve shows humidity at the test site.

A comparison between environmental temperature and leakage current (see Figure 10) of the module frames show that the current amplitude is mainly influenced by temperature.

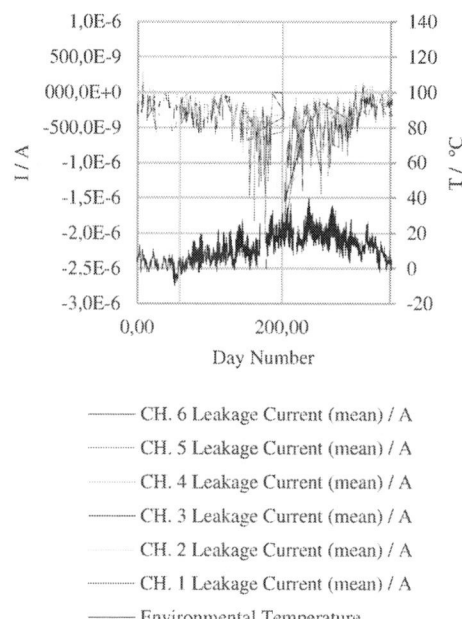

——— CH. 6 Leakage Current (mean) / A
——— CH. 5 Leakage Current (mean) / A
——— CH. 4 Leakage Current (mean) / A
——— CH. 3 Leakage Current (mean) / A
········· CH. 2 Leakage Current (mean) / A
········· CH. 1 Leakage Current (mean) / A
——— Environmental Temperature

Figure 10: Leakage current vs. temperature

An integration of the leakage currents of the module frames over time is pictured in Figure 11. Surprisingly the curve of module 2 shows the smallest amplitude. Another interesting feature is the behavior of the curve of module 6 vs. the curve of module 4. At the beginning, module 6 shows less amplitude than module 4 but after 200 days the behavior is inverted and curve 4 shows a smaller amplitude. It seems that PID sensitive PV-modules show smaller leakage currents, or smaller leakage currents lead to larger PID effects.

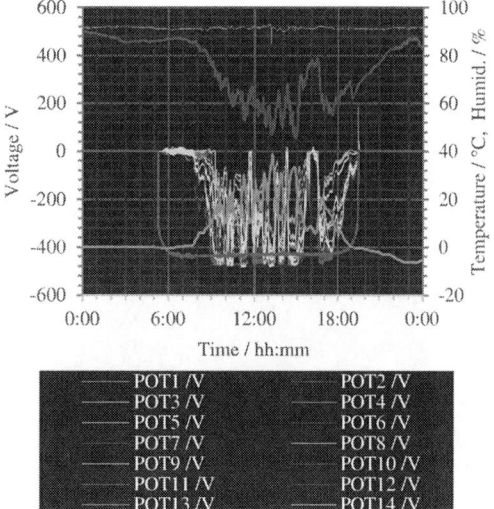

Figure 9: Surface voltages at the PV module

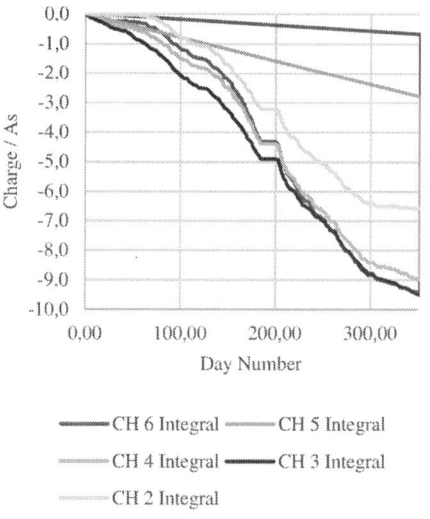

——— CH 6 Integral ——— CH 5 Integral
········· CH 4 Integral ——— CH 3 Integral
········· CH 2 Integral

Figure 11: Integration of leakage current over time

692

4 LESSONS LEARNED

The experiments show that PID takes about one year until the degradation process is finished. A later added nightly recovery does not bring back the original performance. The measurement of surface voltages demonstrates that the PID stress is the largest on PV cells close to the grounded module frame. This stress highly depends on humidity and moisture.

For the surface voltage measurement, high technical effort was necessary to avoid unwanted loads and to minimize influences on the measured values.

Leakage currents depend on temperature and there seems to be a correlation between PID sensitivity and leakage current.

5 ACKNOWLEDGMENTS

The investigations have been carried out within the project PID-Recovery (FKZ 0324184A), which has been funded through the German Federal Ministry of Economic Affairs and Climate Action. Only the authors are responsible for the content of this publication.

Thanks to Peter Funtan for support and to Dr. Volker Naumann of Fraunhofer CSP for a fruitful collaboration within the project.

6 REFERENCES

[1] Kirchhof, Tan: „Neue Untersuchungsergebnisse zur Potenzial Induzierten Degradation – PID", 28. Symposium Photovoltaische Solarenergie, Bad Staffelstein, 2013

[2] Rumiantcev, Schick, Erichsen, Jaeckel, Hagendorf, Naumann: „Prediction of PID-S on the basis of accelerated outdoor module testing and weather data", 37th PV-SEC, Lisbon, 2020

41st European Photovoltaic Solar Energy Conference and Exhibition

Coatings or tapes?
Imaging methods to show the successful repair of backsheets cracks

Raffael Schifferegger [1,2], Yuliya Voronko [1], Anika Gassner [1,3], Gabriele C. Eder [1], Eric Tilly [4]

[1] OFI, Austrian Research Institute for Chemistry and Technology, Vienna, Austria; [2] TU Wien, Institute of Applied Physics, Vienna, Austria;
[3] TU Wien, Institute of Production Engineering and Photonic Technologies, Vienna, Austria; [4] ENcome Energy Performance GmbH, Klagenfurt, Austria
Corresponding authors: raffael.schifferegger@ofi.at, gabriele.eder@ofi.at

Motivation & Status Quo

The PV industry faces the critical challenge of extending the lifespan of PV modules. While new PV technologies with enhanced reliability are one avenue, another effective approach is to **prolong the operational life** of existing modules through **reuse, repair, or refurbishment**. Repair solutions for damaged backsheets (BS) help to ensure that PV modules can meet their expected lifespan of 20+ years without (i) significant energy loss, (ii) safety issues due to insulation breakdown in wet conditions (R_{iso}) or (iii) progressive material degradation.

The ReNewPV research project focuses on creating effective repair solutions for cracked or defective BS, primarily by **restoring R_{iso}** of these modules. Reliable repairs not only extend the service life of PV modules, offering **cost benefits** to solar farm operators and owners, but they also maintain stable electricity yields and restore operational safety by **applying coatings** to damaged BS films. By extending the service life, **PV waste is reduced, resources are saved, logistics costs are reduced** and CO_2 emissions are reduced by the possibility of on-site repairs.

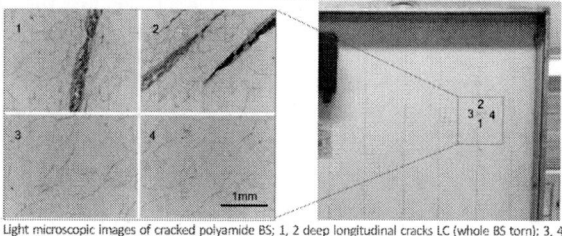

Light microscopic images of cracked polyamide BS; 1, 2 deep longitudinal cracks LC (whole BS torn); 3, 4 micro cracks MC (only outer BS-layer affected)

Repair Method & Evaluation of Crack Filling

After coating / taping of the cracked BS, a wet leakage test was performed to verify successful restoration of the insulation resistance (R_{iso}).

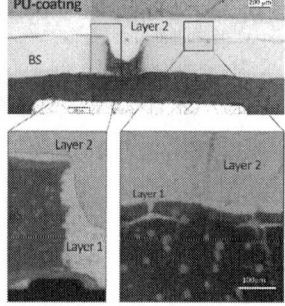
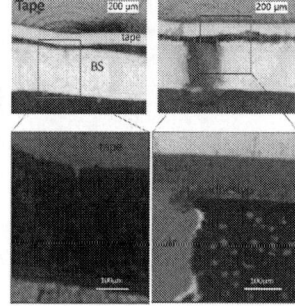

The **developed repair method** focuses on restoring the functionality of cracked backsheets through the application of coatings or adhesive tapes or foils.

Key is the ability of the repair material (1) to **fill the cracks** and (2) to **build a protective barrier layer** on the surface. The table summarizes the characteristics and evaluation of crack filling of **3 repair options**:

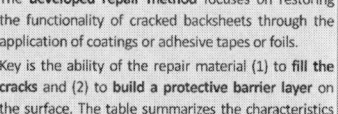

Material	Compo-nents	Contains solvent	Coating layers	Crack filling (ATR)	
				MC	LC
Silicone-coating	1	no	1	yes	yes
PU-coating	2	yes	2	yes	yes
PVC-tape	1	no	1	yes	no

Reliability Testing

Accelerated aging tests IEC 61215-2 MQT 11 temperature cycling (-40 -> +85°C) were conducted to evaluate the long-term stability of the repaired modules. Two intermediate evaluations were conducted during the process to assess the effect of the stress-impact of the TCs on (i) the material stability, (ii) R_{iso} and (iii) adhesion quality and (iv) electrical performance (P_{MPP})

Characterization

The effectiveness of each repair option was assessed through several characterization methods:
Non-destructive methods:
- **Ultrasonic microscopy (USM):** visualization of various interfaces within the PV module stack with the aim to identify which layers within the BS are already cracked -> lateral
- **Electrical performance measurement (P_{MPP}):** evaluation the overall performance pre- and post-repair of the PV-module and after accelerated aging
- **Electroluminescence imaging (EL):** detection of defects/cracks of the solar cells; allows for a pre- and post-repair comparison and the aging effect
- **Isolation Resistance measurement (R_{iso}):** verification of the insulation properties of the module under wet conditions (IEC 61215-2, MQT 15, wet leakage current test;
Destructive methods on module cross sections:
- **Light microscopy (LM):** examination of the physical structure and integrity of repaired sections, revealing the effectiveness of crack filling and material restoration
- **ATR-FTIR imaging:** visualization and spectroscopic identification of materials, facilitating the evaluation of crack filling quality; analysis of chemical stability of the coatings and adhesives

Results

IEC 61215-2 MQT 11 temperature cycling -40 → +85°C

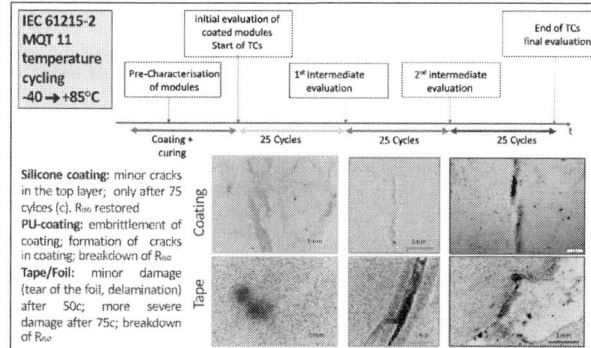

Silicone coating: minor cracks in the top layer; only after 75 cylces (c). R_{iso} restored
PU-coating: embrittlement of coating; formation of cracks in coating; breakdown of R_{iso}
Tape/Foil: minor damage (tear of the foil, delamination) after 50c; more severe damage after 75c; breakdown of R_{iso}

Results of the final evaluation after the accelerated aging tests

Repair/ Material	performance loss (ΔP_{MPP})	Physical stability of repair layer	Adhesion to BS (test tape TESA 4651)	R_{iso} restoration	Chemical stability of repair layer
Silicone-coating	-12.7 %	yes	good	yes	yes
PU-coating	-10.3%	stiffening after aging -> brittle	good	no	yes
PVC-tape	-18.7%	shrinking	good, detachment at the rims	no	yes

Results of the characterisation

→ After repair, the R_{iso} of all repaired modules was restored.
→ **Crack filling** was best characterized on cross sections with light microscopy and ATR-imaging; **USM** is a useful method to analyse the depth of LCs; the lateral resolution of this method, however, is not sufficient to visualize MC or crack filling of the coatings
→ **Repair tape solutions** effectively fill the MCs (with adhesive) and prevent further crack growth; however, they are ineffective in repairing deep BS cracks as they do not fill the crack voids and therefore do not restore full BS functionality.
→ **Coatings (PU and silicone)** can fill MCs and deep LCs (thus replacing the backsheet material) and restore the insulating and protective properties of the backsheets

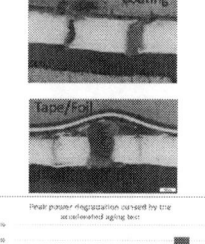
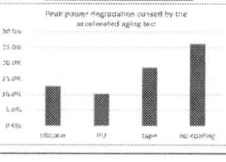

Conclusions & Outlook

After repair, the R_{iso} of all repaired modules was restored. After accelerated aging tests the **silicone coating performed the best**, as the silicone coated modules were the only passing the R_{iso} test. Furthermore, silicone is a **cost effective** and **environmentally friendly** material as it does not contain any solvent or fluor. As a 1-component system, it is easy and **quick to apply even in the field**. The coating process requires only one single layer and **hardens quickly**. The PU coating is effective in crack filling and restoring R_{iso}, but suffered from embrittlement upon temperature cycling. It is currently in the process of being optimized, we can confirm a reduction in brittleness in the second generation. First long-term reliability data with silicone and PU-coating are already available for a test-installation with repaired backsheets (operative since 08.2021). Based on the results of the PVC tape, the use of an additional protective BS foil glued on top of the cracked BS will be tested next.

The work was performed within the project ReNewPV "Beschichtung zur Erhöhung der Lebensdauer von PV Modulen mit beschädigten Rückseitenfolien" under the Kreislaufwirtschaft - Energie- und Umwelttechnologie Ausschreibung 2023; ProjectNr. FO999912440

41st European Photovoltaic Solar Energy Conference and Exhibition

EFFECT OF WEIGHT PERCENT GRAPHENE ON BARRIER PROPERTIES OF ETHELYNE VINYL ACETATE (EVA) FOR IMPROVED PHOTOVOLTAIC MODULE PACKAGING RELIABILITY

Amalu, Emeka H; Fabunmi, Oluwagbemiga A; Hughes, David J; Pang, Yongxin; Short, Michael
Department of Engineering, School of Computing, Engineering and Digital Technologies, Teesside University,
Middlesbrough, Tees Valley TS1 3BX, UK
E.Amalu@tees.ac.uk; O.fabunmi@tees.ac.uk; D.J.Hughes@tees.ac.uk; y.pang@tees.ac.uk; M.Short@Tees.ac.uk
Phone: +44(0)1642342450

ABSTRACT: As the use of appropriate weight percent (wt%) filler in composites is critical in achieving optimal barrier properties, this research develops graphical model for predicting the right wt% graphene (GNP) filler in Ethelyne Vinyl Acetate (EVA) matrix to achieve its optimal water vapour transmission rate (WVTR). Laboratory test vehicles are four 0.16 mm thick films of EVA-GNP composites of composition 0wt%GNP, 4wt%GNP, 8wt%GNP and 12wt%GNP in EVA matrix. ASTM E96 wet-cup method is used to study composites' WVTR. FEA tool is COMSOL Multiphysics software. Modelled geometry is Glass-xwt%GNP-Glass perovskite cell of dimensions 26 x 26 x 0.1 mm thick. The amount of wt%GNP in EVA-GNP composite and ambient temperature are found to influence its barrier properties. Increasing temperature increases composites' diffusivity, permeability, porosity and WVTR, but increase in wt%GNP causes the diffusivity, permeability, porosity and WVTR to decrease to a minimum at 8wt%GNP and increase afterwards. Conversely, increasing wt%GNP and temperature decrease composite's solubility. Although 8wt%GNP composite has the lowest steady state WVTR, the 12wt%GNP composite accumulates the lowest moisture concentration of magnitude 6.03 kg/m^3. It is concluded that increasing wt%GNP in EVA-GNP composite above critical 8wt%GNP results in decrease in moisture accumulation but not further decrease in WVTR.
Keywords: Ethelyne Vinyl Acetate (EVA), Graphene (GNP) nanofillers, Encapsulant barrier properties, Water vapour transmission rate (WVTR), Moisture ingress into EVA.

1 INTRODUCTION

Polymer composites with nanofillers have demonstrated improved barrier properties over their respective matrixes. However, the use of appropriate weight percent (wt%) filler in composite is critical in achieving optimal composite barrier properties. Ethelyne Vinyl Acetate (EVA) has demonstrated improved barrier properties when graphene (GNP) is used as filler in the matrix. Determining the right wt% GNP in EVA matrix will broaden its application scope.

EVA, an acetate copolymer, is used in several applications. It is used in shoe sole where it provides cushion and anti-slip resistance. It is used to insulate cable to prevent short circuits and provide safety. Recently, EVA is widely being used to encapsulate photovoltaic (PV) cells and modules in the PV manufacturing industry owing to its excellent encapsulation properties. These include great flexibility, superb weather resistance, outstanding adhesion/bonding to glass, and cost effectiveness. Fig. 1 shows two architectures of solar PV module and cell. Fig. 1(a) shows a 3-D view of crystalline silicon (c-Si) PV module with critical components laminated in EVA. Fig. 1(b) depicts the critical components of the c-Si PV module with EVA as the encapsulation material. In Fig. 1(c), polyisobutylene (PIB) is used as an edge sealant to protect perovskite cells owing to EVA limitations. In c-Si PV module, EVA encapsulation provides adequate protection to the module against environmental degradation factors which include moisture and dust. It also provides mechanical support to the module against handing, shock, fatigue and impacts loads.

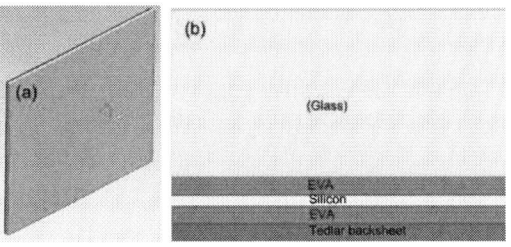

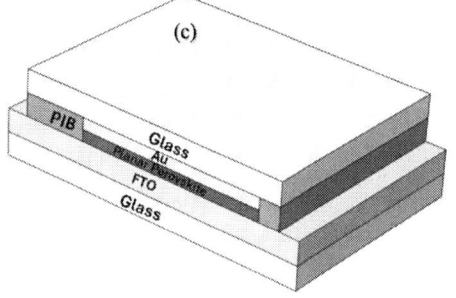

Figure 1: Architecture of solar photovoltaic module/cell showing (a) 3-D view of crystalline silicon PV module with EVA encapsulation; (b) Key components of crystalline silicon PV module with EVA encapsulation; (c)Glass-to-glass PIB edge sealant of pperovskite solar cell.

Despite the many advantages EVA has demonstrated as an excellent encapsulation material, its application in PV technology is limited. EVA degrades substantially over time on prolonged exposure to UV radiation and high temperature. Under this condition, EVA reacts to produce acetic acid which causes its colour to turn yellow. Another key limitation is its barrier against water vapour. Over time, EVA allows moisture to ingress into PV encapsulated with it. Significant permeation of moisture into EVA encapsulants reaching c-Si PV cells has caused

695

delamination and corrosion in many modules operating in the fields. This limitation has hindered its application in encapsulation of perovskite solar cells.

Numerous studies have researched barrier properties of polymer-nanocomposites. The focus has been to optimize the properties of the matrix through addition of nano fillers of various compositions for different applications which include the PV technology. Research on polymer-filler nanocomposite which targets improvement in moisture barrier properties for application in PV technology is found limited. Researchers [1] investigated reinforcement of suryln films with micro-nanofribrillated celluloses (MFC) for application in organic photovoltaic (OPV) encapsulation. The contents of the MFC studied are 0.6%, 0.8%, 1.0%. It is found that the water vapour transmission rate (WVTR) of the composites decreases as the contents of the MFC increases. In piping technology, [2] studies polymer organic-inorganic hybrids of PVC nanocomposites for commercial piping. Fillers of modified type of montmorillonite (MMT) 30B, 25A, 93A and natural type of MMTNa+ were used in the inorganic phase. The materials WVTR diminish with increasing filler concentration. Researchers [3] investigated N-Dimethylhexadecyclamine (DMHDA+) montmorillonite (Mt) random copolymer polypropylene (CPP-R) nanocomposites (AmK10CPP-R) for subsea pipelines using three weight loading of 0.3%, 3% and 4%. The moisture uptake and diffusivity of the composites were compared. The composite with 4% filler recorded the least moisture uptake and lowest diffusion coefficient. Furthermore, [4] studied one single polymer, cellulose cinnamate (CCi) and CCi nanoparticles (CCI-NPs) with three degrees of substitution (DS) of 0.6, 1.0 and 2.0. The water vapor permeability (WVP) reduced gradually from 0.6 to 2 - confirming improvement in barrier properties with use of single-polymer composite. In yet another related work, [5] measured the WVP of polylactide (PLA)-based sustainable composite films with two weight percentages of GNP 2wt% for PLA and 10wt% for polyethylene glycol (PEG). Impregnation with GNP reduced the water permeability from 1.45×10^{-14} to 1.43×10^{-14} [kg m/(m^2 s Pa)]. In similar research, [6] studied PLA using montmorillonite layered silicate (nano clay) as filler. The wt% of the filler studied is from 1 to 6. The test temperature is 38 °C at 90% relative humidity. The results indicate that WVTR decreases with increase in wt% of the nanoclay.

Beyond synthetic polymers and fillers, research work has been conducted on moisture barrier properties of biopolymers. Using biofillers, [7] researched the moisture barrier improvements in biodegradable films based on starch-PVA-nanoclay. Three wt percentages of 0.5%, 1%, and 1.5% are investigated. Results show that nanoclay concentrations higher than 1.0 w/v resulted in worse moisture barrier performance. The best WVTR of 331.366 g.mm/(m^2.day) was at 0.5 w/v and the worst of 884.44 g.mm/(m^2.day) was recorded at 1.5 w/v.

Only a few studies on improving the moisture barrier properties of EVA using fillers are available. Researchers [8] worked on using acid functionalized Graphene Nanoplatelets (GNP) to boost the barrier properties of EVA. Using four GNP of 0.001%, 0.01%, 0.1%, 1 % parts per hundred rubbers (phr), the WVTR of the four compositions were tested and compared with EVA. Measured WVTR for EVA was about 56.35 [g/ (m^{2}day)]. After mixing with GNP, the WVTR of EVA composite

films decreased to the range of 14.54 - 18.37 [g/(m^2 day)]. In similar research, [9] investigated the transport behaviour of aromatic solvents such as benzene, toluene and xylene in EVA-Clay nanocomposites utilizing three wt percentages of 3, 5, and 7. The findings seem to indicate that solvent uptake was minimum for composites with 3wt% filler and increased with increase in filler content. It is proposed that aggregation of clay filler at higher loading is responsible for the observation.

As limited research targeting improvement of barrier properties of EVA – especially for PV improved packaging reliability, the need for this current research is necessitated. The main aim of this investigation is to quantify the effect of weight percent graphene (wt%GNP) on ethylene vinyl acetate (EVA) barrier properties for improved photovoltaic module packaging reliability. The special focus of the aim is to determine the effective wt%GNP in EVA matrix that yields the lowest water vapour transmission rate (WVTR) and concentration of moisture in the composite for improved encapsulation of photovoltaic module. The key objectives include to study the effect of (i) temperature and wt%GNP on mass transport properties of EVA-GNP composite, (ii) temperature and wt%GNP on water vapour transmission rate (WVTR) of EVA-GNP composite, (iii) wt%GNP on magnitude of water vapour accumulated in EVA-GNP composite under extended damp heat test (DHT). The investigation is carried out at three different temperatures with four critical wt% composites.

2 MATERIALS AND METHODS

The materials and three methods used in this investigation are discussed in this section.

2.1 Experimental method

Tables I and II present the materials used in this investigation. In Teesside University materials' laboratory, original sample of Graphene powder is dissolved in xylene solution to form Graphene masterbatch. Similarly, the original sample of EVA pellets is dissolved in xylene solution to form EVA masterbatch. The two masterbatches are used to create four weight percentages of Graphene in EVA matrix. These are 0wt%GNP, 4wt%GNP, 8wt%GNP and 12wt%GNP in EVA matrix. Fig. 2 presents the materials used in this investigation. Fig. 2(a) shows the original samples of Graphene container, Fig. 2(b) is the Graphene, Fig. 2(c) displays EVA material, while Fig. 2(d) depicts the original sample of EVA material.

To create 4wt%GNP in EVA matrix, 1.4 g of Graphene masterbatch is mixed with 33.6 g of EVA masterbatch. The mixture is properly stirred using an 8000 M mixer/mill. Fig. 3 depicts the mixer equipment, and its vial and balls component used in composite mixing. Similar procedure is followed to create the other compositions. All the compositions are in paste form. The four solution blends are poured into four different molds and dried in an oven at 50° centigrade temperature to form identical 0.16 mm thick films. Fig. 4 presents the mold inside the oven, while Fig. 5 displays the four test vehicles of 0wt%GNP, 4wt%GNP, 8wt%GNP and 12wt%GNP in EVA matrix. Three test vehicles are produced for each EVA-xwt%GNP composition.

The ASTM E96 wet-cup method is used to determine the WVTR of the test vehicles. The method comprises a cup, desiccator (calcium carbonate), de-ionised water, temperature and humidity sensor, and scale and a chamber. The process involved filling a cup with water to a level that leaves a small air gap of about 0.25 cm from the cover. Fig. 6 shows the cup, and the cover seal used in this work. A test vehicle film is used to enclose the cup and held in position. The cup is weighed and then placed in a test chamber. Fig. 7 depicts the set-up of the experiment. It consists of a cup, desiccator (calcium carbonate), de-ionised water, temperature and humidity sensor, and a scale. For experiments at temperatures different from ambient temperature, the chamber was put in an oven in the required test temperature. The humidity in the chamber was about 3% while the humidity inside the cup was about 98%. The difference in humidities drives moisture inside the cup through the test film into the desiccator. The weight of the cup is measured every 10 mins for seven hours and subsequently once a day for 5 days. The experiment is repeated three times for each test vehicle in the test temperatures of 19 °C, 36 °C and 50 °C. Employing Eq. (1), the WVTR is computed.

$$WVTR = \left(\frac{G}{tA}\right) = \left(\frac{G}{t}\right)/A \qquad (1)$$

Where: G is weight change in grams, t is time in days, A is area of the film surface. The units of WVTR is g/(day·m²) or g·mm/(day.m²).

Table I: EVA material properties

Material	Manufacturer	Vinyl acetate Percentage (wt %)	Product type	Product form
EVA	Exxon Mobil Chemical	27.6	Escorene ™ Ultra LD 755 Series	Pellets

Table II: Graphene material properties

Material	Manufacturer	Particle Size	Specific Surface area	Product form
Graphene	ThermoFisher Scientific	2 to 10 nm (Thickness)	20 to 40 m^2/g	Powder

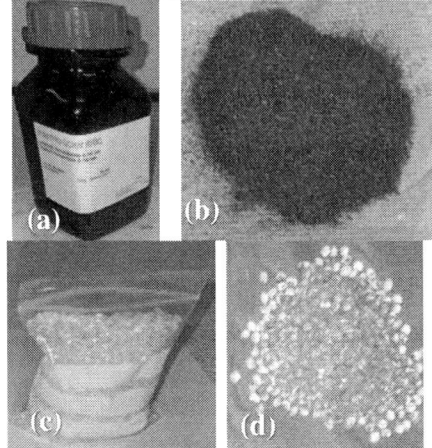

Figure 2: Investigation materials showing (a) Graphene container, (b) Graphene (c) EVA container (d) EVA

Figure 3: 8000 M mixer/mill showing (a) The machine (b) Vial and balls machine component.

Figure 4: Mold inside the oven

Figure 5: Four test vehicles of 0wt%GNP, 4wt%GNP, 8wt%GNP and 12wt%GNP in EVA matrix.

Figure 6: A cup and its cover seal

Figure 7: Experiment set up comprising a cup, desiccator (calcium carbonate), de-ionised water, temperature and humidity sensor, and scale.

2.2 Analytical method

Analytical technique is used to calculate the input parameters required in finite element simulation of water ingress into the EVA-xwt%GNP composites. Fick's law of diffusion, shown in Eq. (2), is used to calculate moisture diffusion through the films; Arrhenius mass transport equation presented in Eq. (3) is applied to determine diffusivity, permeability and solubility variation with temperature. The equations are presented thus:

$$\frac{\partial c_i}{\partial t} = D\left(\frac{\partial^2 c_i}{\partial x^2} + \frac{\partial^2 c_i}{\partial y^2}\right) \tag{2}$$

$$D = D_O\left(e^{\frac{-E_D}{RT}}\right) \tag{3a}$$

$$P = P_O\left(e^{\frac{-E_P}{RT}}\right) \tag{3b}$$

$$S = S_O\left(e^{\frac{-E_S}{RT}}\right) \tag{3c}$$

Where: D represents the diffusion coefficient; D_O is pre-exponential factor; E_D is activation energy of diffusivity; R and T are absolute gas constant and temperature, respectively. Similarly, P represents the permeability; P_O is pre-exponential factor of permeability; E_P is activation energy of permeability. For other paramaters, S is solubility, S_O is pre-exponential factor of solubility, E_S, is activation energy of solubility.

Other analytical equations and theories employed in the computation of input parameters include Kozeny-Carmen equation depicted in Eq. (4). This equation is deployed to determine the variation of porosity with permeability; Henry's law, Eq. (5), and Goff-Gratch formula stated in Eq. (6) are exploited to compute inlet moisture concentration based on ambient temperature and vapour pressure. The equations are represented thus:

$$k = \phi^2{}_s \frac{\epsilon^3 D_p{}^2}{180(1-\epsilon)^2} \tag{4}$$

Where: ϵ is porosity of the encapsulant, D_p is average diameter of sand grains, K is absolute permeability of the encapsulant and $\phi^2{}_s$ is sphericity of the particles that make of the encapsulant (1 for spherical particles).
Also,

$$P = DS \tag{5a}$$

Where all the paramaters retain their definitions. Using Henry's law, the applied concentration $C_{sat(i)}$, can be expressed as:

$$C_{sat(i)} = Sp_{actual(i)} \tag{5b}$$

Where: p_{actual} is actual vapour pressure in the surrounding air and S is the solubility coefficient. The p_{actual} can be determined using Eq. (5c):

$$p_{actual} = \frac{Relative\ Humidity\ P_{Saturate}}{100} \tag{5c}$$

While the $P_{Saturate}$ can be calculated using the Goff-Gratch formula at any given temperature stated in Eq. (6):

$$Log_{10}P_{saturate} = -7.90298\left[\frac{373.16}{T-42.3916}\right] +$$
$$5.02808 \times Log_{10}\left[\frac{T}{373.16}\right] -$$
$$1.3816x10^{-7} \times \left(10^{11.344\left(\frac{1-T}{373.16}\right)}\right) + 8.1328x10^{-3}$$
$$x\left(10^{-3.49149\,x\,\left(\frac{T}{373.16-1}\right)}\right) + Log_{10}(1013.246) \tag{6}$$

The properties of the EVA-xwt%GNP composites at 19 °C, 36 °C and 50 °C, generated from the analytical method are presented in Table III. These are used as the input parameters in the finite element modeing of moisture ingress in the EVA-xwt%GNP composites.

Table III: Materials and their properties at each temperature.

Material	Diffusivity (m²/s)	Permeability (g/m.s.atm)	Porosity (1)	Solubility (g/m³.atm)
19 Degrees Celsius				
EVA	3.82E-12	1.97E-07	0.0091	51675.9
EVA-4%GNP	1.12E-12	1.28E-07	0.0065	49495.2
EVA-8%GNP	8.68E-13	2.38E-08	0.00072	47432.9
EVA-12%GNP	4.74E-12	1.95E-07	0.0202	45370.6

Material	Diffusivity (m²/s)	Permeability (g/m.s.atm)	Porosity (1)	Solubility (g/m³.atm)
36 Degrees Celsius				
EVA	8.46E-12	3.01E-07	0.0138	35463.8
EVA-4%GNP	2.32E-12	2.00E-07	0.0107	34144.9
EVA-8%GNP	1.59E-12	1.15E-07	0.0021	32722.2
EVA-12%GNP	7.90E-12	2.43E-07	0.0242	31299.5

Material	Diffusivity (m²/s)	Permeability (g/m.s.atm)	Porosity (1)	Solubility (g/m³.atm)
50 Degrees Celsius				
EVA	1.54E-11	4.01E-07	0.0189	26344.5
EVA-4%GNP	7.40E-12	3.08E-07	0.0156	24988.2
EVA-8%GNP	5.92E-12	1.62E-07	0.0048	23947.1
EVA-12%GNP	1.33E-11	3.51E-07	0.0276	22905.9

2.3 Finite element analysis (FEA)

Finite element analysis (FEA) employing COMSOL Multiphysics 6.2 is deployed to model moisture ingress into the test vehicles (EVA-xwt%GNP composites) used as encapsulants in Glass-xwt%GNP-Glass perovskite cell assemblies. The dimensions of a cell in length, breadth and thickness are 26 mm x 26 mm x 0.1 mm. The finite element (FE) model of the assemblies is created in COMSOL Multiphysics. Fig. 8 shows a Glass-xwt%GNP-Glass perovskite cell assembly. Fig 8(a) reveal details of the dimension while Fig. 8(b) discloses the materials bonded together. Transport of diluted species solver is deployed to model moisture ingress into the xwt%GNP composites. Modelling set-up included meshing. A fine mesh with an element size of 0.00014 mm was used to mesh the model. The mesh had tetrahedral elements. Fig. 9 presents the FE model of an assembly with adequate mesh. Load application and boundary condition specification are the last processes before the simulation set-up is run. The concentration of moisture in the ambience is applied at the four sides of the cell assembly. The model has fixed support at the bottom and simulation is run.

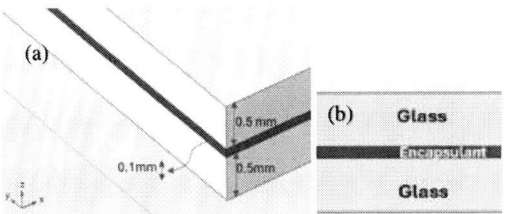

Figure 8: Glass-xwt%GNP-Glass perovskite cell assemble showing (a) Dimensions (b) Assembled materials

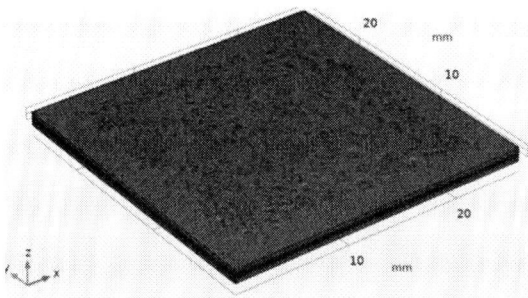

Figure 9: Glass-xwt%GNP-Glass perovskite cell with adequate mesh

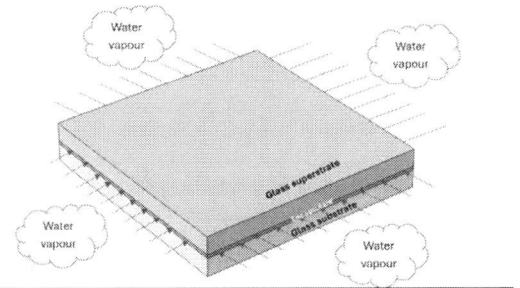

Figure 10: Application of moisture at the four sides of a Glass-xwt%GNP-Glass perovskite cell assembly.

3 RESULTS AND DISCUSSION

The results and discussion are done in three sub-sections. Sub-section 3.1 discusses the effect of temperature and wt%GNP on mass transport properties of EVA-GNP composite. In sub-section 3.2, the eeffect of temperature and wt%GNP on water vapour transmission rate (WVTR) of EVA-GNP composite is treated while the effect of wt%GNP on magnitude of water vapour accumulated in EVA-GNP composite under extended damp heat test (DHT) is presented in sub-section 3.3.

Detailed discussion on these is presented thus:

3.1 Effect of temperature and wt%GNP on mass transport properties of EVA-GNP composite

The mass transport properties of material considered are diffusivity, permeability, porosity and solubility. The results of study on the effect of temperature and wt%GNP on these properties of EVA-GNP composite is presented in Figs. 11 to 14. Figs. 11, 12 and 13 present bar chart plots of Diffusivity, Permeability and Porosity against EVA-GNP composites at 19 °C, 36° C and 50° C temperatures, respectively. It is observed in the figures that diffusivity, Permeability and Porosity of EVA-xwt%GNP composites increases with increase in temperature. Also observed is that as the xwt%GNP added increases from zero in EVA to 4wt%GNP, the properties decrease and reach a minimum at 8wt%GNP and increase afterwards in 12wt%GNP. It is deduced that EVA-8wt%GNP composite is a critical mix. Furthermore, Fig. 14 depicts the plot of solubility against EVA-GNP composites at 19 °C, 36° C and 50° C temperatures. It is detected from the plot that solubility of EVA-GNP composites decreases steadily with increase in temperature. The figure also shows that the solubility of EVA-GNP composites monotonically decreases with increase in xwt%GNP added in the composite. In this investigation, 12wt%GNP composite produces the lowest amount of solubility coefficients and 0wt%GNP yielded the highest amount of solubility coefficients, at all temperatures.

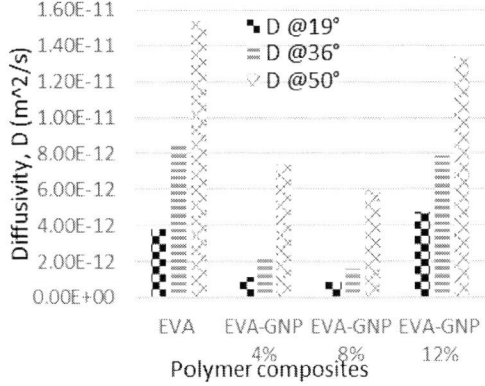

Figure 11: Plot of Diffusivity of x wt%GNP in EVA-GNP composites at 19 °C, 36 °C and 50 °C temperatures

41st European Photovoltaic Solar Energy Conference and Exhibition

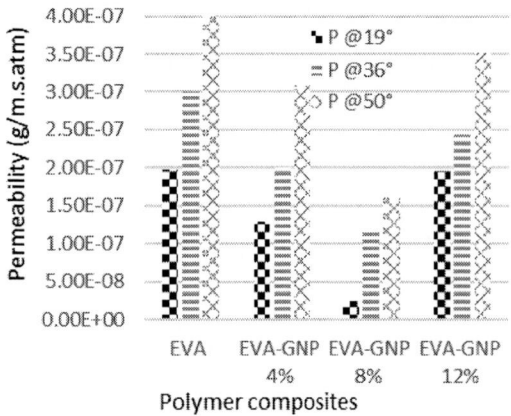

Figure 12: Plot of Permeability of x wt%GNP in EVA-GNP composites at 19 °C, 36 °C and 50 °C temperatures

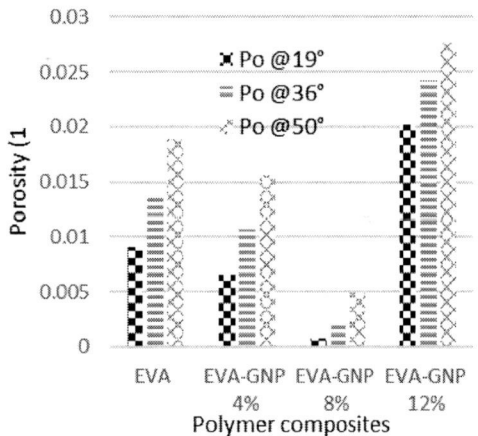

Figure 13: Plot of Porosity of x wt%GNP in EVA-GNP composites at 19° C, 36 °C and 50 °C temperatures.

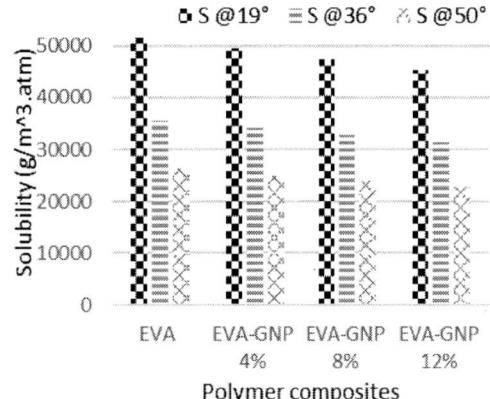

Figure 14: Plot of Solubility of wt%GNP in EVA-GNP composites at 19 °C, 36 °C and 50 °C temperatures

3.2 Effect of temperature and wt%GNP on water vapour transmission rate (WVTR) of EVA-GNP composite

Generated steady state WVTRs of xwt%GNP in EVA-GNP composites is depicted in Fig. 15. The amount of wt%GNP in EVA-GNP composites and the operating temperature are found to influence the mass transport properties of the composites. In Fig. 15, WVTR of EVA-GNP composite is observed to decrease with addition of xwt%GNP in the EVA matrix. It decreased from the highest at 0wt%GNP to reach minimum at 8wt%GNP and increased afterwards in 12wt%GNP. It is noted in the figure that the composites have lowest and highest WVTRs at 19° C and 50° C, respectively. Utilising Fig 15, Fig. 16 is created to be used as model to predict the WVTR of xwt%GNP in the EVA matrix at weight percent graphene $0 \leq wt\%GNP \leq 12$ and temperatures $0\ °C \leq t \leq 50\ °C$ temperatures.

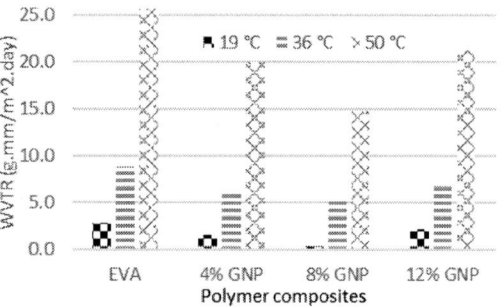

Figure 15: Plot of steady state water vapour transmission rate (WVTR) of xwt%GNP in EVA-GNP composites at 19 °C, 36 °C and 50 °C temperatures

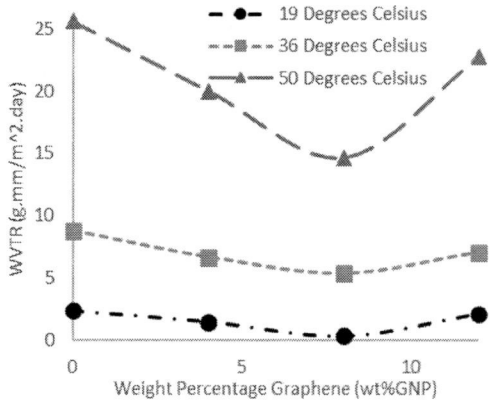

Figure 16: Plot of water vapour transmission rates (WVTR) of xwt%GNP in EVA-GNP composites at 19 °C, 36 °C and 50 °C temperatures.

3.3 Effect of wt%GNP on magnitude of water vapour accumulated in EVA-GNP composite under extended damp heat test (DHT)

The results of the effect of wt%GNP on magnitude of water vapour accumulated in EVA-GNP composite under extended damp heat test (DHT) is plotted in Fig 17. EVA, with 0wt%GNP, is found to accumulate the highest amount of water vapour. The amount of moisture accumulated is observed to decrease as the quantity of the wt%GNP increases. Thus, 12wt%GNP composite accumulates the lowest amount of moisture. The figure shows that the rate of moisture accumulation decreases with increase in wt%GNP in the EVA-GNP composite with EVA highest and EVA-12wt%GNP lowest. Conversely, time to saturation of the composite is inversely proportional to xwt%GNP added in the EVA matrix. EVA is the fastest to saturate and EVA-12wt%GNP composite is the least to saturate with moisture. This observation is because as the weight of graphene added in the EVA matrix increases, the volume of the matrix which can be occupied with moisture decreases owing to graphene not being soluble in water.

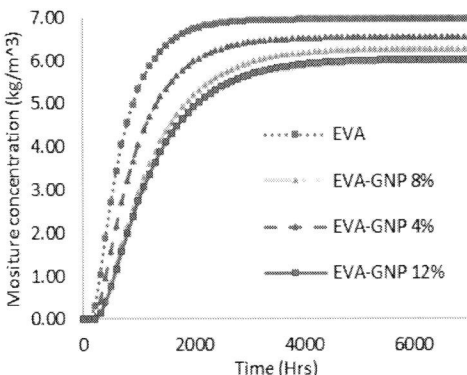

Figure 17: Plot of water vapour concentration against time for the x wt%GNP in EVA-GNP composites.

4   CONCLUSIONS

This investigation develops graphical model to estimate the right weight percent (wt%) graphene (GNP) filler in Ethelyne Vinyl Acetate (EVA) matrix to achieve its optimal water vapour transmission rate (WVTR) at 19 °C, 36 °C and 50 °C temperatures. The model is suitable for weight percent graphene $0 \leq wt\%GNP \leq 12$ and temperatures $0\ °C \leq t \leq 50\ °C$ temperatures.

From the study of Effect of temperature and wt%GNP on mass transport properties of EVA-GNP composites it is concluded that diffusivity, permeability and porosity of EVA-GNP composites increases with increasing temperature while the solubility decreases with increasing temperature. Addition of xwt%GNP in EVA matrix affects its diffusivity, permeability and porosity properties with 8wt%GNP being a critical mix that produces the lowest value of the properties. The solubility properties of EVA-GNP composites do not have a critical mix since the solubility continuously decreases with increasing addition of graphene in the EVA matrix.

From the investigation on effect of temperature and wt%GNP on water vapour transmission rate (WVTR) of EVA-GNP composite 8wt%GNP is the optimal composition because it yielded the lowest magnitude of WVTR. It is also concluded that increase in ambient temperature causes EVA-GNP composites to increase transmission of water vapour.

The research on effect of wt%GNP on magnitude of water vapour accumulated in EVA-GNP composite under extended damp heat test (DHT) lead to the conclusion that increase in addition of wt%GNP in EVA-GNP composite results in the composite absorbing and accumulating less moisture. It also decreases the rate at which it accumulates moisture, increasing the time to achieve saturation by moisture.

However, the realization that increase in wt%GNP in EVA-GNP composite above critical 8wt%GNP results in decreases in moisture accumulation but not further decrease in WVTR in the composite indicates that a trade-off among composites WVTR, moisture accumulation and mechanical properties must be made in selecting the best material for improve photovoltaic module packaging reliability.

Based on the results of this investigation, it is concluded that addition of xwt%GNP in EVA matrix impacts on EVA barrier properties. A critical 8wt%GNP in EVA-GNP composite produces the optimal WVTR. Increase in wt%GNP above this critical weight results in decreases in moisture accumulation in the composite but not any further decrease in WVTR.

REFERENCES
[1] Lertngim, A., et al., Preparation of Surlyn films reinforced with cellulose nanofibres and feasibility of applying the transparent composite films for organic photovoltaic encapsulation. Royal Society Open Science, 2017. 4(10): p. 170792.
[2] Kalendova, A., et al., Polymer/clay nanocomposites and their gas barrier properties. Polymer composites, 2013. 34(9): p. 1418-1424.
[3] Mahdavi, E., M. Haghighi-Yazdi, and M.M. Mashhadi, Coupled thermal stress and moisture absorption in modified a montmorillonite/copolymer nanocomposite: Experimental study. Polymer Testing, 2023. 124: p. 108092.
[4] Wang, J., et al., Self-compounded nanocomposites: toward multifunctional membranes with superior mechanical, gas/oil barrier, UV-shielding, and photothermal conversion properties. ACS Applied Materials & Interfaces, 2021. 13(24): p. 28668-28678.
[5] Ahmed, J., et al., Polylactide/graphene nanoplatelets composite films: Impact of high-pressure on topography, barrier, thermal, and mechanical properties. Polymer Composites, 2021. 42(6): p. 2898-2909.
[6] Duan, Z., N. Thomas, and W. Huang, Water vapour permeability of poly (lactic acid) nanocomposites. Journal of membrane Science, 2013. 445: p. 112-118.
[7] Calambas, H.L., et al., Physical-mechanical behavior and water-barrier properties of biopolymers-clay nanocomposites. Molecules, 2021. 26(21): p. 6734.
[8] Yuwawech, K. and J. Wootthikanokkhan, EVA film reinforced with acid functionalized graphene nanoplatelets as a transparent barrier layer to enhance the durability of solar cells. International Journal of Automotive and Mechanical Engineering, 2019. 16(1): p. 6301-6318.
[9] Wilson, R., et al., Clay intercalation and its influence on the morphology and transport properties of EVA/clay nanocomposites. The Journal of Physical Chemistry C, 2012. 116(37): p. 20002-20014.

41st European Photovoltaic Solar Energy Conference and Exhibition

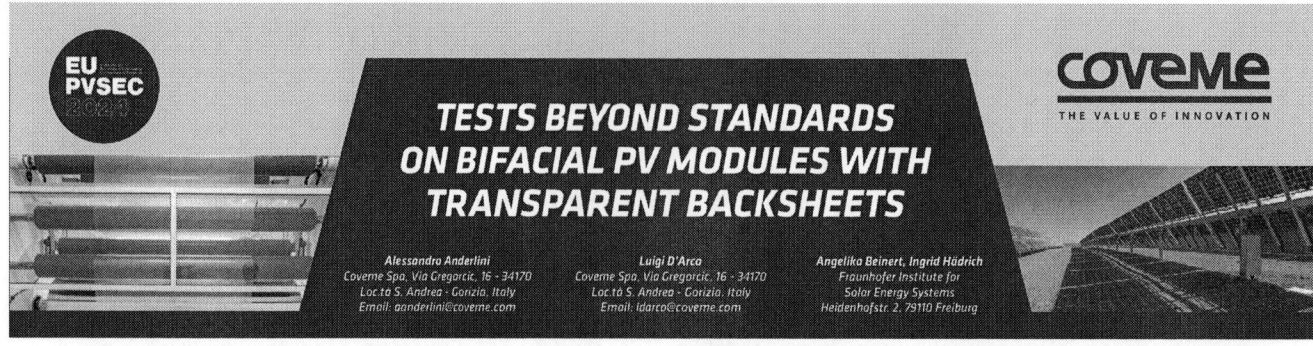

TESTS BEYOND STANDARDS ON BIFACIAL PV MODULES WITH TRANSPARENT BACKSHEETS

Alessandro Anderlini
Coveme Spa, Via Gregorcic, 16 - 34170
Loc.tà S. Andrea - Gonizia, Italy
Email: aanderlini@coveme.com

Luigi D'Arco
Coveme Spa, Via Gregorcic, 16 - 34170
Loc.tà S. Andrea - Gonizia, Italy
Email: ldarco@coveme.com

Angelika Beinert, Ingrid Hädrich
Fraunhofer Institute for
Solar Energy Systems
Heidenhofstr. 2, 79110 Freiburg

INTRODUCTION

Given the rising market share of bifacial cell technologies, there is a need for transparent backsheets. On the one hand these backsheets need to have a high resistance towards water ingress, on the other hand, experience with premature field failures of modules which were equipped with certain backsheet types [1] suggests to investigate the reliability of backsheets towards environmental stresses with tests beyond the existing standards

SEQUENTIAL TEST

In order to test for the general backsheet durability, sequential tests based on a sequence including damp heat exposure, UV exposure and thermal cycling have proven to reveal potential weaknesses such as backsheet cracking. One such test is the so called module accelerated stress test (MAST) proposed by DuPont [2]. A variation of this test was conducted on two full size modules with a transparent PET-based backsheet. The test consisted of a total of 1000 h damp heat exposure, 360 kWh/m² front side exposure UV and 600 thermal cycles. The difference between the modules was the encapsulant with one containing EVA and one using POE. The aim was to understand differences that might arise from these different material combinations. Both modules lost 3.1 % in PMPP after this test sequence. The EVA module exhibited yellowing, whereas the POE module did not show this phenomenon. The transparent backsheet remained on both modules clear and free of cracks and maintained its insulation properties.

DAMP HEAT TEST

With respect to stability against moisture an extended damp heat test (3000 h) sequence has been applied to a module with a water resistant transparent backsheet. The module lost 0,75 % in PMPP. Again, the backsheet remained clear and free of cracks and maintained its insulation properties.

AIM AND APPROACH

The major aim of his work is to understand more about the general quality of transparent PET based backsheets. Therefore, in order to test the general backsheet quality, two full sized glass-backsheet modules (one with EVA (M05), one with POE (M06)) were exposed to a test sequence based on damp heat, UV and thermal cycling exposure (see Figure 1).

In order to evaluate the water resistance of a transparent PET based backsheet a 3000 h damp heat was conducted on a full sized PV module.

TABLE 01 - LIST OF TESTED MODULES

MODULE ID	SIZE/CELLS	FRONT SIDE	FRONT ENCAPSULANT	REAR ENCAPSULANT	BACKSHEET	STRESS TEST
M03	60 cells/ PERC	glass	POE-UV transparent	POE-UV cut	dyMat® CLR HOPVE†	DH 3000
M05	60 cells/ PERC	glass	EVA-UV transparent	EVA-UV cut	dyMat® CLR HOPVE†	DH-UV-TC sequence
M06	60 cells/ PERC	glass	POE-UV transparent	POE-UV cut	dyMat® CLR HOPVE†	DH-UV-TC sequence

The tested modules were all characterized by power measurements, electroluminescence images and isolation tests. Since Boron doped PERC cells were employed a BO-LID repair procedure was employed before the final power measurements in both tests. The procedure consisted of an Isc current soak for 48 h at 80 °C.

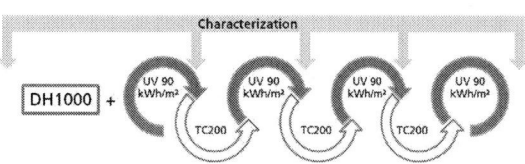

FIGURE 1 - CONDUCTED SEQUENTIAL STRESS TEST ON M05 AND M06

Depicts the applied sequential stress test. It is based on the module accelerated stress test (MAST) defined by DuPont. The major difference is that the UV dose is applied from the front to test the whole module. Additionally, the UV exposure is 360 instead of 270 kWh/m², with a spectrum that is relatively higher in the lower wavelength region.

Considering the spectral distribution, 360 kWh/m² correspond to about 6-7 years outdoor operation, assuming an annual dose of 50-60 kWh/m² in central Europe. In the short-wavelength range it corresponds to about twice as much respectively.

Nonetheless, due to temperature and other effects, a one by one transfer to real conditions cannot be derived based on this. The albedo within the UV chamber is around 10 % which means around 36KWh/m2 of the irradiance reaches the module form the backside.

SCIENTIFIC INNOVATION AND RELEVANCE

To our knowledge, there is no data how a transparent backsheet performs in combination with EVA and POE in a full size module within a sequential test including 1000 h damp heat, 360 kWh/m² UV and 600 thermal cycles or a 3000 h damp heat test.

In particular the sequential test represents the combination of potential field conditions and therefore gives an indication about the long-term durability of the tested bill of materials.

In addition the 3000 h of damp heat exposure gives an indication about the likelihood of water ingress. Furthermore, the relevance of the application of BO-LID repair could successfully be demonstrated.

TEST RESULTS

SEQUENTIAL TEST

Other than the POE module that did not exhibit any significant visual changes, the EVA module developed encapsulant yellowing on the front side of the modules. The yellowing is clearly visible after the second TC200 test (see Figure 2).

The yellowing is strongest around the cell edges. This most likely results from the diffusion characteristics of oxygen and outgasing species, which are specific for EVA. The backsheet remained clear and without cracks for the duration of the whole test sequence.

FIGURE 2 - YELLOWING IN M05 (EVA MODULE) AFTER THE FULLY CONDUCTED TEST SEQUENCE

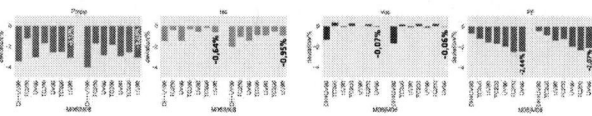

FIGURE 3 - DEVIATION FROM INITIAL POWER VALUES DURING THE COURSE OF THE CONDUCTED MAST TEST SEQUENCES.

Figure 3 shows the deviation of power output associated values from the initial measurement. M05 refers to the module with EVA, M06 contains POE. Both modules exhibit a similar power loss pattern within the sequential test. The first two TC200 tests lead to a recovery of the power loss caused by DH1000+UV90 and UV90 respectively.

This behavior is also reflected in Isc and Voc. It is likely related to a BOLID repair that occurs during current soak at elevated temperatures, which is given during thermal cycling. The remaining TC test does not lead to a significant recovery since most likely the power degradation is mainly resulting from interconnection issues as reflected in electroluminescence images (not presented here).

Both modules show a power loss of 3,1% in PMPP, which is considerably low for such and extensive test. The insulation values were maintained above the device test capacity (5000 MOhm) throughout the test sequence.

DAMP HEAT TEST

During the initial 1000 h of damp heat exposure the module exhibited a high power loss, which could be almost fully recovered by a 48 h Isc current soak at 80 °C. This so called BO-LID repair was repeated at the end of the test.

Throughout the 3000 h of damp heat exposure the backsheet remained clear and without cracks (see Figure 4). The measured change in color (yellowness index and dE based on L,a,b values) was below 2. This is below what is visible with the human eye.

The module remained with a power loss of 0.75 % in PMPP, which is mainly attributed to fill factor loss. This is a very low degradation rate for a glass-backsheet module.
The insulation values were maintained above the device test capacity (5000 MOhm) throughout the test sequence.

FIGURE 4 - VIEW THROUGH MODULE AFTER 3000H OF DAMP HEAT EXPOSURE

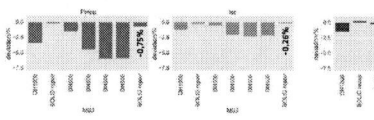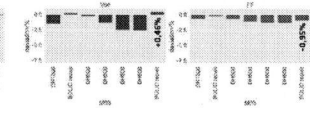

FIGURE 5 - DEVIATION FROM INITIAL VALUES FOR M03.

CONCLUSIONS

In conclusion, this work showed a very good performance of modules with a transparent backsheet in an extensive sequential test independent of the employed encapsulant type as well as a high water resistivity of a transparent backsheet in combination with POE in a 3000 h Damp Heat Test. The low power loss values and high insulation property maintenance are indicators for Glass-Backsheet Bifacial PV module structures as valid alternative to Glass-Glass modules.

REFERENCES

[1] J. Markert, S. Kotterer, D.E. Mansour, D. Philipp, P. Gebhardt. Advanced analysis of backsheet failures from 26 power plants, EPJ Photovolt. 12 (2021) 7
[2] W. Gambogi, T. Felder, S. MacMaster, K. Rey-Choudhury, B.-L. Yu, K. Stika, H. Hu, N. Phillips, T.J. Trout. Sequential Stress Testing to Predict Photovoltaic Module Durability, in: 2018 IEEE 7th World Conference on Photovoltaic Energy Conversion (WCPEC) (A Joint Conference of 45th IEEE PVSC, 28th PVSEC & 34th EU PVSEC), IEEE, 2018, pp. 1593–1596.

41st European Photovoltaic Solar Energy Conference and Exhibition

ENHANCED PERFORMANCE OF PV MODULES USING HIERARCHICALLY STRUCTURED GLASS IN DIFFERENT CLIMATIC CONDITIONS

Cristina L. Pinto and Jaione Bengoechea
National Renewable Energy Centre (CENER),
Ciudad de la Innovación 7, Sarriguren, 31621 Navarra (Spain)
cpinto@cener.com

ABSTRACT: PV modules laminated with hierarchically structured glasses exhibit enhanced performance due to their multifunctional properties. Studies have shown that random nanocones enhance PV module energy production through various mechanisms: they create a broadband omnidirectional anti-reflective surface, reduce soiling rates by over 50%, and offer extreme abrasion resistance. Furthermore, microcylinders on the glass surface increase glass emissivity, thereby reducing the solar cell's operating temperature through passive radiative cooling. Adding these features to the glass, PV modules can sustain higher power output over time. In this study, all properties associated with PV modules laminated with structured glasses are integrated into PVSyst to assess the increase in energy yield compared to commercial PV modules. Three different locations, representative of various environmental conditions (irradiance, angle of incidence, soiling rate, temperature), are selected for the study: Spain with a moderate climate and low soiling rates, Norway with a colder climate and high angles of incidence, and Saudi Arabia with elevated temperatures, higher irradiance and soiling rates.
Keywords: PVSyst, Subwavelength Structures, IAM factor, PV, Soiling losses, temperature losses

1 INTRODUCTION

The performance of photovoltaic (PV) modules is strongly influenced by the materials' properties, particularly the front cover glass. In recent years, significant research has focused on enhancing PV efficiency through advanced anti-reflective and anti-soiling technologies. One promising approach is the use of nano/micro-textured glass surfaces, which offer a range of benefits, including broadband omnidirectional anti-reflectance (AR), anti-soiling (AS), self-cleaning capabilities, and passive radiative cooling (RC), along with improved abrasion resistance [1]–[3].

While laboratory testing is essential for characterizing these properties, it is equally important to evaluate their real-world performance under various environmental conditions. Simulations using tools such as PVsyst allow for detailed modelling of how these nano/micro-structured glasses impact energy production in PV systems across different geographical locations and climates. This simulation-based approach provides a broader understanding of the global advantages these enhanced glasses offer compared to standard commercial modules.

This study evaluates the energy gain of PV installations using modules laminated with nano/micro-structured glass. Despite the demonstrated improvements in energy production due to the multiple functionalities of these glasses, their impact across different climates and locations has yet to be fully analysed. The performance of these structured glasses is simulated using PVsyst, incorporating key enhancements such as broadband omnidirectional AR, AS, RC, and high abrasion resistance. The results are compared with commercial PV modules across three distinct locations—Kjeller (Norway), Noain (Spain), and Rumah (Saudi Arabia)—selected to represent a wide range of irradiance levels, incidence angles, temperatures, and soiling rates.

2 METHODOLOGY

In this work, all simulations were conducted using PVsyst, a widely used software for the simulation and analysis of PV systems. It allows detailed modelling of PV installations by considering various system parameters, environmental conditions, and performance factors. The program is typically employed to evaluate energy production, optimise system design, and assess the impact of different technologies under diverse climatic conditions. Its ability to simulate site-specific data makes it a valuable tool for comparing the performance of PV modules in various geographical locations.

However, PVsyst has limitations when it comes to modelling certain advanced optical properties of materials. For example, it does not allow the direct inclusion of glass reflectance. Instead, it permits the specification of the refractive index for an AR coating. This limitation becomes particularly relevant when dealing with AR layers that create an effective graded refractive index such as subwavelength structures, as these cannot be fully integrated into the software's current modelling framework.

2.1 Effect of Anti-Reflective Property

As said, in the PVsyst program, the module reflectance is a parameter that cannot be directly inputted, thus any improvement achieved in reflectance will not be visibly reflected due to this limitation. However, the Incident Angle Modifier (IAM) factor is a parameter that can be manually included in the program. Therefore, in this study, in order to incorporate reflectance improvements both at normal and at different angles of incidence, the IAM factor will be redefined. To achieve this, the short-circuit current generated at zero angle of incidence will be taken as a reference, assuming that the glass reflectance is zero and all incident light enters the cell. Consequently, commercial modules will not yield a value of 1 at normal incidence, but rather a slightly lower value due to their inherent reflectance, whereas it will be higher for the structured glass. This approach enables the incorporation of reflectance advantages and yields their corresponding improvements in electrical production.

$$IAM'\,factor = \frac{I_{sc}(\theta)}{I'_{sc}(\theta = 0) \times \cos\theta}$$

where $I_{sc}(\theta)$ is the short circuit current measured at θ

41st European Photovoltaic Solar Energy Conference and Exhibition

angle of incidence, and $I'_{sc}(\theta = 0)$ is the short circuit current at normal incidence assuming zero reflection at the air-glass surface. For the I_{sc} theorical calculations, it can be obtained from J_{sc} through next equation:

$$Isc = A \times J_{sc} = A \times \sum_{\lambda < Band\ Gap} SR(\lambda) \cdot E_0(\lambda) \cdot T(\lambda)$$

where A is the area, $SR(\lambda)$ is the spectral response of the solar cell, $E_0(\lambda)$ is the solar irradiance and $T(\lambda)$ is the transmittance. For the case I'_{sc}, the transmittance will be 1 for each wavelength.

At normal incidence, with the transmission values for structured glass and commercial glass from 300 nm to 1100 nm, the $E_0(\lambda)$ and the $SR(\lambda)$ of a PERC solar cell, the J_{sc} can be computed for a PV module with commercial glass, structured glass, and ideal glass. Hence, from these calculations, the structured glass presents a decrease of 2.41% in the I_{sc}, while the commercial one shows a decrease of 5.14% concerning an ideal glass (zero reflectance). Thus, the IAM factor changes accordingly, and it is shown in **Figure 1**:

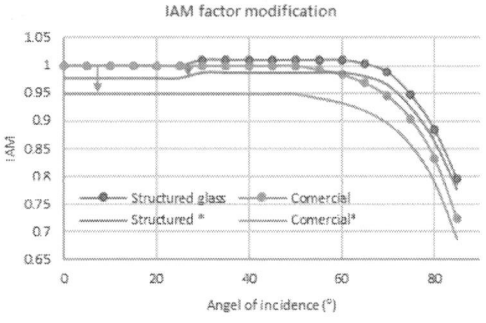

Figure 1: Measured IAM factor of a module with structured glass and a commercial module, with the rescale of the IAM factor to include the effect of reflectance at normal incidence.

Once the new IAM factor is defined, the improvement in the reflectance of the structured glasses will be considered. The other parameters are easily introduced into the PVsyst program (soiling rate and thermal loss coefficient).

2.2 Effect of the Radiative Cooling

PVsyst includes a section for thermal loss parameters, which is crucial for accurately modelling the thermal behaviour of PV modules. One of the key parameters in this section is U_0, which represents the thermal loss coefficient under natural convection conditions. In our case, the PV modules are equipped with a high emission glass, leading to increased thermal radiation losses.

Given that the structured glass emits 14.7% more radiation than a flat glass [2], [3], the thermal losses due to radiation will also be higher. However, the adjustment has been made empirically rather than applying a direct proportional increase to the U₀ value. This correction is based on observed temperature differences between a conventional module and one with structured glass. In clear-sky conditions with high irradiance, the module with the textured glass has been found to operate at a temperature 2.5°C lower than the conventional module. Taking this empirical data into account, the U₀ value has

been adjusted to $U_0 = 31.6\ W/m^2K$ to accurately reflect the improved cooling performance of the structured glass, leading to more precise simulations of the module's thermal behaviour.

2.3 Solar PV farm characteristics

For the present study, a grid-connected solar PV farm of 2.2 MWp constituted by 5,000 modules as 200 modules per string in 25 series has been defined in PVsyst. The modules are installed in a fixed plane at the optimum tilt angle for the location with no shadings. The commercial modules are the generic Mono 440 Wp Twin 144 half-cells and the structured glass' PV modules are the same as the commercial ones but with the IAM factor modified. The generic inverter of 2 MWac central inverter has been selected.

3 RESULTS AND DISCUSSION

The results presented in this study are derived from simulations conducted for three distinct locations: Kjeller in Norway, Noain in Spain, and Rumah in Saudi Arabia. These locations were selected to represent a diverse range of climatic conditions, allowing for a comprehensive evaluation of the performance of PV modules with structured glass under varying environmental scenarios.

Each site offers unique characteristics:

- Kjeller, Norway (Latitude: 59.9750° N, Longitude: 11.0532° E): Located in a humid continental climate region, is characterized by lower irradiance levels, predominantly diffuse radiation, and cooler temperatures (Dfv–Koppen–Geiger classification).
- Noain, Spain (Latitude: 42.7589° N, Longitude: -1.6364° W): With an oceanic climate, exhibits moderate irradiance with a mix of direct and diffuse radiation, and moderate temperatures (Cfb–Koppen–Geiger classification).
- Rumah, Saudi Arabia (Latitude: 25.5594° N, Longitude: 47.1335° E): With a desert climate, is known for high irradiance levels, predominantly direct radiation, high temperatures, and significant soiling (BWh–Koppen–Geiger classification).

To evaluate the impact of the structured glass properties, several comparisons were conducted:

- Anti-reflective effect: The first comparison analysed the impact of the AR property alone, without introducing soiling rates or modifying the thermal loss coefficient U₀.
- Anti-soiling property: In the second comparison, the AS property was included for locations with significant soiling rates, assessing its effect on performance.
- Radiative cooling effect: Finally, the RC property due to enhanced thermal emission was incorporated to evaluate its impact on module performance across the different locations.

3.1 Norway (Kjeller)

This Norwegian city is located in the south of the country. For the simulations, the meteorological data from Meteonorm 8.1 has been used. The optimal tilt angle for maximizing energy production throughout the entire year at this location is 49°. Due to the local climate, soiling is minimal, and thus no significant accumulation of dirt on the modules is expected. While snow is considered a form of "soiling," there is currently no available data on how the nano-structured glass interacts with snow. As a result, no anti-soiling analysis was performed in this location. The focus of the simulations in Kjeller has been on the effects of anti-reflectance and passive cooling. The loss diagrams obtained from the simulations of a commercial PV farm, structured PV modules considering only the AR property and finally the structured PV modules considering the AR and RC properties are presented in **Figure 2**.

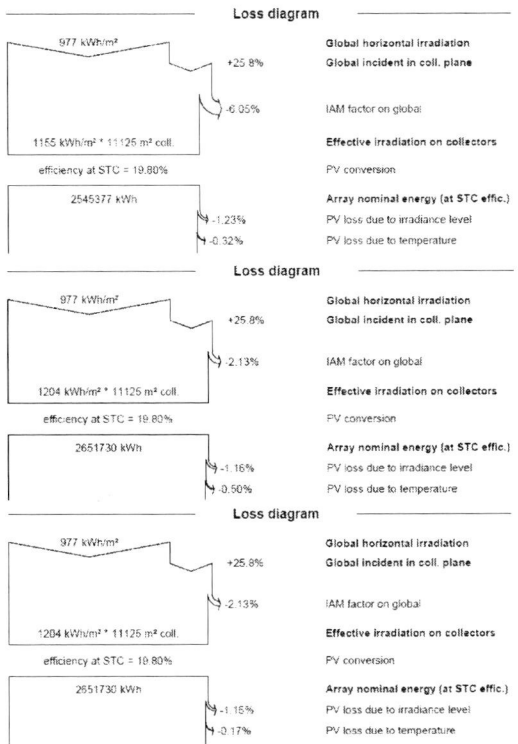

Figure 2. PV loss diagrams from Norway: Up) Commercial PV modules, Middle) Structured PV modules considering only the AR property and, Bottom) Structured PV modules considering both AR and RC properties.

3.1.1 Anti-reflective property

The first comparison considers only the AR property of the structured PV module with no refrigeration functionalities. For that, the AR property has been assessed through the IAM factor modification. The effective annual energy generated by the commercial PV array is 2,441.24 MWh, while the array composed of structured PV modules produces 2,539.09 MWh, representing a relative increase of 4.0% (Table I). In northern countries like Norway, the main component of global irradiance is diffuse, meaning that a significant portion of the light reaches the module from all directions. Therefore, the omnidirectional characteristic of the AR surface becomes particularly important. Any improvement in the IAM factor directly impacts energy generation, making the AR surface's performance crucial in these conditions. From the loss diagram, for the commercial PV modules, the losses regarding the IAM factor correspond to 6.1%, whereas, for the structured glass' PV, this loss is

reduced up to 2.1%. It is also noteworthy to mention that the temperature loss increases from 0.3% to 0.5%. This is due to as more light enters the cells, they will heat up.

3.1.2 Radiative cooling

The structured glasses emit 14.7% more energy than commercial glasses, so the Uc value has been modified, from 29 to 31.6 W/m^2K. Norway is a cold country, and as a result, ambient temperatures are typically low. Consequently, thermal losses due to cell temperature are minimal. In this scenario, with a higher thermal loss coefficient, the annual energy production for a PV farm using structured glass modules is 2,547.49 MWh, representing a 4.4% increase compared to commercial modules (Table I).

From the loss diagrams, the PV loss due to temperature for commercial modules represents 0.3% (a low value due to the cold climate) and for the structured glass, the loss is 0.2%, a reduction of 0.1%.

Table I: Summary of effective energy at the output of the array in Norway.

	AR	+RC
Commercial (MWh)	2,441.240	2,441.240
Structured (MWh)	2,539.086	2,547.490
Relative gain	4.0%	4.4%

3.2 Spain (Noain)

Noain is located in the north of Spain. The meteorological data from PVGIS and a tilt angle of 35° have been selected for the simulations. For this location, the light/energy loss diagrams and the evaluation of the energy produced from the commercial PV arrays and structured PV modules have been computed for each analysed case.

The loss diagrams of the five simulations are presented in **Figure 3 Figure 4**: commercial PV modules, commercial modules with soiling losses, structured PV modules considering only the AR property, considering AR and AS properties and finally, considering all the structured PV characteristics.

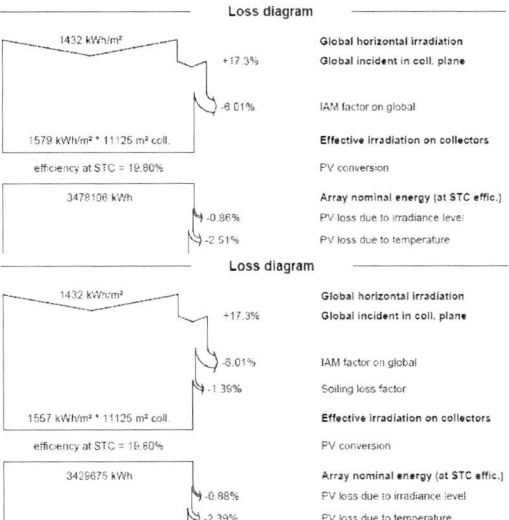

Figure 3. Commercial PV loss diagrams from Spain. Up) with no soiling rate, Bottom) with a monthly soiling rate.

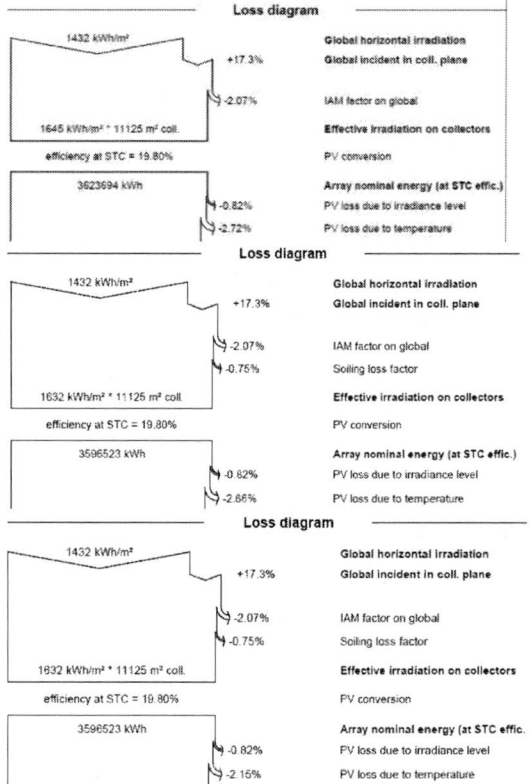

Figure 4. Structured PV loss diagram from Spain. Up) considering only the AR property. Middle) Considering the AR property, a monthly soiling rate and the AS property. Bottom) Considering a monthly soiling rate, the AR, AS and RC properties.

3.2.1 Anti-Reflective property

The first comparison considers only the AR property of the structured PV module with no anti-soling nor refrigeration functionalities. As shown in **Figure 3** up), the losses by IAM factor represent 6.0% for the commercial module, while for the structured module represents only 2.1% (**Figure 4** up). On the other hand, as the structured module absorbs more light, the losses due to temperature increase, from 2.5% for the commercial module to 2.7% for the structured module. With lower loss flows, the energy production increases, from 3,268.893 MWh to 3,397.202 MWh, corresponding to a relative increase of 3.9% (**Table III**).

3.2.2 Anti-soiling property

The second scenario considers the soiling losses, defined for each module as in **Table II**. The soiling values have been modified from [4], taking into account local conditions in Noain, where it rains frequently from autumn to spring, resulting in minimal soiling of the modules during these months. However, in summer, when rainfall decreases and the harvest season begins, dust and suspended particles in the air increase, leading to higher soiling rates. Note that the structured module offers an anti-soiling property, decreasing the glass soiling rate by 50%. In this scenario, the AR and AS properties have been considered. The results are shown in **Figure 3** bottom) and **Figure 4** Middle) where the losses due to soiling are depicted: for commercial modules, it represents 1.4% whereas for structured modules the soiling accounts for

only 0.8%. This reduction in soiling rate increases the energy production by 4.6% (Table III). Note that the losses due to temperature are lower as less irradiance reaches the solar cell due to soiling.

Table II: Monthly soling rate in Spain (Noain)

Jan.	Feb.	Mar.	Apr.	May	Jun.
0.0%	0.0%	0.0%	1.0%	2.0%	3.0%
Jul.	**Aug.**	**Sep.**	**Oct.**	**Nov.**	**Dec.**
3.0%	3.0%	1.0%	0.0%	0.0%	0.0%

3.2.3 Radiative Cooling

Finally, the effect of temperature reduction due to the radiative cooling effect has been added to the simulation for the third scenario. Through the radiative cooling effect, the temperature losses have been reduced from 2.7% (previous case) to 2.2% increasing the energy production by a total of 5.1%.

Table III: Summary of effective energy at the output of the array in Spain.

	AR	+AS	+RC
Commercial (MWh)	3,268.893	3,227.239	3,227.239
Structured (MWh)	3,397.202	3,374.199	3,391.233
Relative gain	3.9%	4.6%	5.1%

3.3 Saudi Arabia (Rumah)

Rumah is a village located in Saudi Arabia, characterized by a hot and arid climate. For the simulations, the meteorological data from Meteonorm 8.1 has been used. The optimal tilt angle for maximizing energy production throughout the entire year at this location is 25°. Although the solar irradiance in this area is very high, a significant portion of the irradiance is diffuse due to the suspended dust in the atmosphere. Additionally, high levels of soiling are expected, which can significantly impact the PV performance.

The loss diagrams from all the simulations performed are shown in **Figure 5Figure 6**, which represents the results obtained from the commercial solar farm without soiling losses, the commercial solar farm with soling losses, and the structured PV solar farm with no soiling, with soiling and considering the effects of radiative cooling effect.

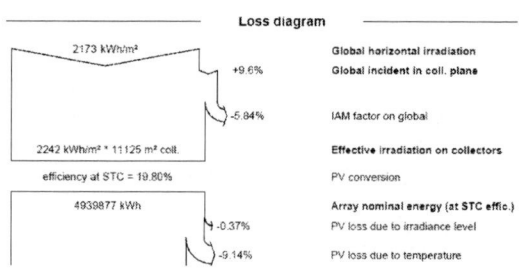

41st European Photovoltaic Solar Energy Conference and Exhibition

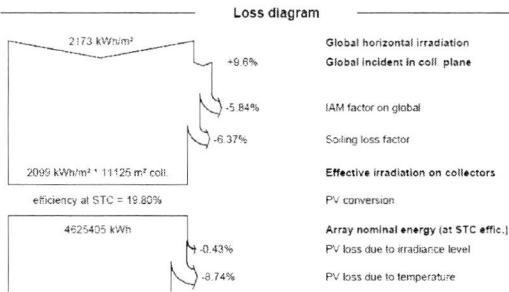

Figure 5. Commercial PV loss diagrams from Saudi Arabia. Up) with no soiling rate, Bottom) with a monthly soiling rate.

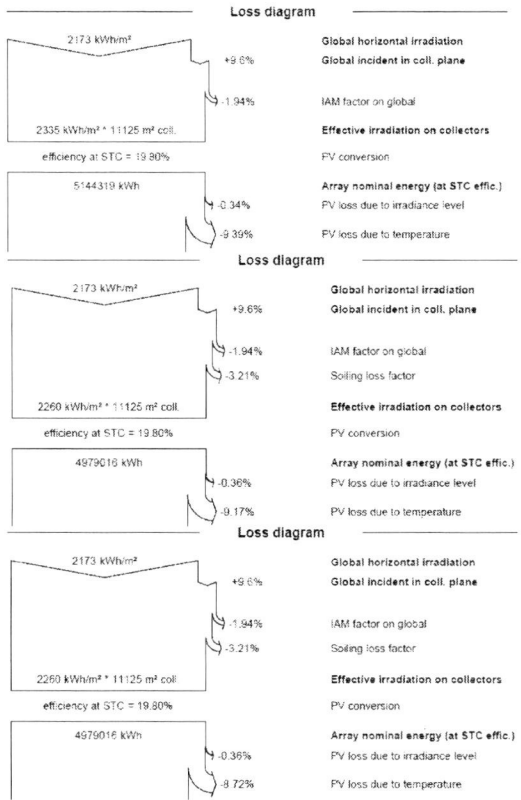

Figure 6. Structured PV loss diagram from Saudi Arabia. Up) considering only the AR property. Middle) Considering the AR property, a monthly soiling rate and the AS property. Bottom) Considering a monthly soiling rate, the AR, AS and RC properties.

3.3.1 Anti-Reflective property

Due to the high diffuse component of solar irradiation, which constitutes approximately 35% of the total global irradiation, the omnidirectional property of the AR surface becomes increasingly important. In this scenario, considering only the improvement in the IAM factor, the optical losses have been reduced from 5.8% to 1.9%, resulting in a relative energy gain of 3.8%.

3.3.2 Anti-Soiling property

The second scenario considers the soiling losses, defined for each month as shown in **Table IV** (values have been obtained from [5]). In arid regions like Saudi Arabia, soiling is a significant challenge due to the presence of sand and the frequent occurrence of sandstorms. These storms

can cover the modules entirely, sometimes completely blocking light from reaching the cells of the module and drastically reducing the energy performance. In this scenario, the structured module offers an AS property, reducing the soiling rate by 50%. The results, shown in **Fig. 7**, demonstrate the impact of soiling: for conventional modules, soiling losses account for 6.4%, while for structured modules, the losses are reduced to 3.2%. With these properties (AR and AS), the energy gain rises to 7.1% (**Table V**). Regarding temperature losses, the soiling reduces the temperature of the module, as less light reaches it.

Table IV: Monthly soling rate in Central Saudi Arabia [5]

Jan.	Feb.	Mar.	Apr.	May	Jun.
5.8%	5.8%	4.9%	1.0%	2.0%	3.0%
Jul.	**Aug.**	**Sep.**	**Oct.**	**Nov.**	**Dec.**
3.0%	3.0%	1.0%	0.0%	0.0%	0.0%

3.3.3 Radiative Cooling

In hot climates, high module temperatures are a significant concern, as elevated temperatures reduce the efficiency of PV cells and result in energy losses. This issue is particularly pronounced in regions with high ambient temperatures, such as Saudi Arabia, where the intense solar irradiance causes substantial heating of the modules. Any improvement in passive cooling can have a considerable impact on energy production, as lowering the operating temperature of the modules leads to reduced thermal losses and increased efficiency. In this study, the results show that temperature-related losses slightly decreased from 8.74% for conventional modules to 8.72% for textured modules with enhanced passive cooling. This improvement translates to a relative increase in annual energy production by 7.7% considering all the structured glass properties.

Table V: Summary of effective energy at the output of the array in Saudi Arabia.

	AR	+AS	+RC
Commercial (MWh)	4,340.887	4,083.326	4,083.326
Structured (MWh)	4,507.348	4,373.797	4,395.791
Relative gain	3.8%	7.1%	7.7%

4 CONCLUSIONS

PVsyst has been used to simulate the energy production of enhanced PV modules at three different locations with varying climates: Norway, Spain, and Saudi Arabia. The PV modules are laminated with hierarchically structured glasses that offer several key functionalities: enhanced broadband omnidirectional anti-reflective properties, a 50% reduction in soiling rate due to their anti-soiling characteristics, and a decrease in operating temperature via passive radiative cooling, attributed to their higher emissivity, along with improved abrasion resistance. Simulations were conducted and compared with standard commercial modules to assess the performance improvements.

The results suggest that, when considering only the anti-reflective property, the relative energy gain is 4.0% in Norway, 3.9% in Spain, and 3.8% in Saudi Arabia. When the anti-soiling property is also factored in, energy losses due to soiling are significantly reduced (by 4.6% in Spain), particularly in regions with high dust levels, such as Saudi

Arabia, where a relative improvement of 7.1% is observed. Finally, the enhanced cooling provided by the structured glass leads to a modest yet valuable reduction in temperature-related losses, especially in hotter climates, resulting in a relative gain of 4.4% in Norway, 5.1% in Spain, and an impressive 7.7% in Saudi Arabia. Overall, the combination of these properties demonstrates the potential for structured glass to improve PV module performance across a wide range of climatic conditions, with the greatest impact seen in regions with higher temperatures and soiling rates.

5   REFERENCES

[1] C. L. Pinto et al., "Random subwavelength structures on glass to improve photovoltaic module performance," *Sol. Energy Mater. Sol. Cells*, vol. 246, no. August, p. 111935, 2022, doi: 10.1016/j.solmat.2022.111935.

[2] C. L. Pinto et al., "Outdoor thermal performance of photovoltaic devices with enhanced daytime radiative cooling glass."

[3] Á. Andueza et al., "Enhanced thermal performance of photovoltaic panels based on glass surface texturization," *Opt. Mater. (Amst).*, vol. 121, no. 111511, 2021, doi: 10.1016/j.optmat.2021.111511.

[4] M. García et al., "Soiling and other optical losses in solar-tracking PV plants in Navarra." pp. 211–217, 2011. doi: 10.1002/pip.1004.

[5] R. K. Jones et al., "Optimized Cleaning Cost and Schedule Based on Observed Soiling Conditions for Photovoltaic Plants in Central Saudi Arabia," *IEEE J. Photovoltaics*, vol. 6, no. 3, pp. 730–738, 2016, doi: 10.1109/JPHOTOV.2016.2535308.

41st European Photovoltaic Solar Energy Conference and Exhibition

A DATA-DRIVEN CALIBRATION OF THE FEM TEMPERATURE MODEL WITH WIND DIRECTION INPUT

Anastasios Kladas, Bert Herteleer, Jan Cappelle
KU Leuven Research Group ELECTA Ghent
Gebroeders De Smetstraat 1, 9000 Ghent, Belgium

ABSTRACT: This paper proposes an enhanced method for photovoltaic (PV) temperature (T_{PV}) modeling using a data-driven approach. By modifying the methodology of Herteleer et al. [1], we aim to make the FEM temperature model easier to calibrate and more accurate. The calibration of Faiman's model [2] coefficients is optimized using modern computational techniques, minimizing the error between measured and estimated temperatures. A novel component is introduced: the inclusion of wind direction as a variable influencing the cooling effect on PV systems. Our results demonstrate an improvement in model accuracy, reducing the mean absolute error (MAE) by 0.29°C compared to previous methods. This advancement provides a more refined tool for estimating T_{PV} in real-world conditions.
Keywords: PV temperature, wind cooling, PV modelling

1 Introduction

The efficiency of PV systems is closely tied to their operating temperature. Elevated temperatures can lead to reduced power output and accelerated degradation of PV modules. Operating temperature is influenced by several environmental factors, including ambient temperature (T_{amb}), irradiance (G_{PoA}), wind speed (WS), and wind direction (WD). Existing models, such as those by [3] et al. and Barry et al. [4], have utilized techniques like weighted moving averages for temperature estimation. However, these approaches are often computationally intensive and difficult to calibrate.

Faiman's model [2] provides a simplified approach for estimating PV module temperature but does not account for dynamic environmental variables like wind direction. The work presented in this paper builds upon the FEM methodology [1] to create a dynamic T_{PV} model that incorporates wind direction—an often-overlooked factor in temperature models. By calibrating Faiman's model coefficients using a data-driven approach, we improve the accuracy and ease of implementation for PV temperature prediction.

2 Methodology

The primary objective of this research is to streamline the implementation and calibration of Faiman's model, enhancing it to accommodate dynamic environmental factors. Building upon the FEM methodology introduced by Herteleer et al. [1], we propose a data-driven approach to optimize the smoothing parameter (α_{EWM}) for the exponential weighted moving average (EWM) filter and the coefficients (U0 and U1) of Faiman's model.

Model Enhancement: We modify Faiman's original equation to incorporate wind direction (WD) as a weighting factor for wind speed (WS), resulting in the Faiman-WDF model (1):

$$T_{PV} = T_{amb} + \frac{G_{PoA,EWM}}{U0_c + U1_c * WS_{EWM} * f(WD)} + k_i \quad (1)$$

Where:

- $G_{PoA,EWM}$ is the plane of array irradiance with an applied EWM filter.
- WS_{EWM} is the wind speed with an applied EWM filter.
- $U0_c$ and $U1_c$ are the calibrated Faiman coefficients.
- f(WD) is the wind direction factor.
- K_i is the bias correction element.

Calibration Process:

1. **Data Filtering:** Exclude data points with $G_{PoA}<100$ W/m² and over-temperature (OT) OT<4°C (OT = T_{PV}-T_{amb}). Further remove outliers and faulty measurements.

2. **Optimization:** Utilize the SciPy library's minimize function with the Powell method to optimize α_{EWM} and the initial coefficients U0 and U1 by minimizing the root mean square error (RMSE) between measured and estimated over-temperature as shown in eq. (2).

$$Z(a_{EWM}, U0m\, U1) = \sqrt{\frac{OT - \frac{G_{PoA,EWM}}{U0 + U1 * WS_{EWM}}}{N}} \quad (2)$$

Where N is the number of samples.

3. **Bias Correction:** After the initial estimation of coefficients, a linear regression analysis is applied between the results of the OT model from the preceding step and the measured OT, as shown in Figure 1. Subsequently, the updated over-temperature model is defined in eq. (3).

$$OT = k_s * \frac{G_{PoA,EWM}}{U0 + U1 * WS_{EWM}} + k_i$$
$$= \frac{G_{PoA,EWM}}{U0_c + U1_c * WS_{EWM}} + k_i \quad (3)$$

4. **Wind Direction Modelling:** If the WD influences the cooling effect on the photovoltaic system, it must be expressed as a weight to WS, acknowledging that winds from different directions may result in varied cooling effects on the PV system. Consequently, to visualize these weights, equation (1) is solved for f(WD), yielding eq. (4).

The outcomes of the weight calculations against the wind direction can be observed in Figure 2. Each direction is assigned its unique weight, and to construct the model, all wind directions are rounded to their nearest integers. Notably, there is a discernible weight pattern when ample data is available, but for directions where data is limited, uncertainty prevails.

41st European Photovoltaic Solar Energy Conference and Exhibition

$$f(WD) = \frac{\frac{G_{PoA,EWM}}{OT - k_i} - U0_C}{WS_{EWM} * U1_C} \qquad (4)$$

To ensure the reliability of coefficients, data is filtered based on the following criteria:

- Standard Deviation (STD): Discard directions with high variance (greater than 0.6, an empirical value).
- Count: Exclude data with an insufficient amount (less than 10% of the median count of all directions).

The final weights are determined as the median value for each direction. The weights for discarded directions are set equal to 1, to maintain the integrity of the overall model. For the examined dataset, the estimated temperature using the Faiman-WDF model under GPoA=1000W/m², Tamb=30°C and WS =5m/s, if the wind direction is 30° will be equal 45°C while if the wind direction is 300° it will be equal to 50°C

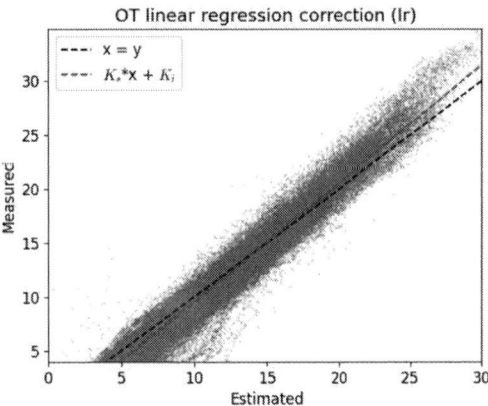

Figure 1 Linear regression between the results of the OT estimations via eq (2) using the coefficients of the first optimization and the measurements. The data used originate from the KUL PV system described in the data used chapter.

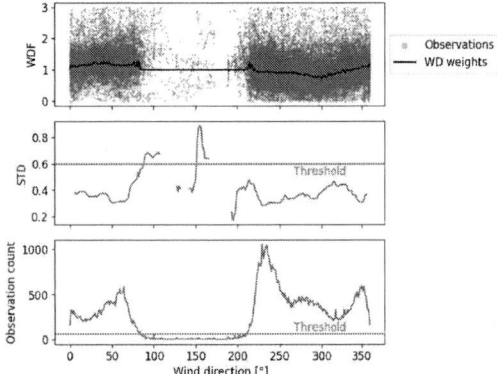

Figure 2 Wind direction weights estimation. The upper graph shows the observed f(WD) and the calculated weights for each direction while the middle and the lower graph show the filtering based on the standard deviation of the weights on every angle as well as the filter based on the observation counts. These two filters are combined.

3 Data Used

The dataset employed in this study was sourced from the KU Leuven rooftop PV system, as detailed in Herteleer et al. [7]. The data spans from May 2015 to March 2016 and has been resampled to one-minute intervals to facilitate both model development and testing.

4 Results

The performance of the proposed Faiman-WDF model was evaluated against traditional Faiman's model with EWM using standard coefficients. The evaluation metrics employed were Mean Absolute Error (MAE) and Root Mean Squared Error (RMSE).

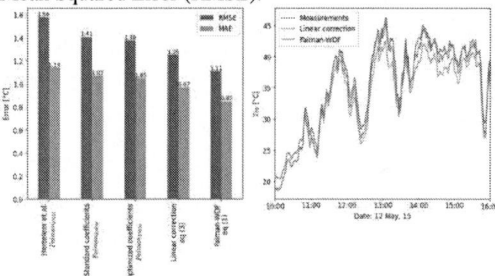

Figure 3 Left bar-plot: A comparison of accuracy, expressed in MAE and RMSE, benchmarked against both the Faiman model (Applied EWM using standard coefficients) and the various development steps of the modified version. Right line-plot: A comparison illustrating the impact of wind direction on model accuracy, contrasting the linear correction step with the final version.

The incorporation of wind direction as a weighting factor resulted in a significant reduction in MAE by 0.29°C (25%) compared to the traditional FEM approach. Figure 3 illustrates the comparative accuracy between the standard and optimized models, highlighting the effectiveness of the data-driven calibration and wind direction integration.

These results underscore the critical influence of wind direction on PV module cooling and the enhanced accuracy achieved through the Faiman-WDF model.

5 Conclusions

This study successfully developed a data-driven calibration approach for Faiman's T_{PV} model, incorporating EWM filters and wind direction as a critical weighting factor. The optimized Faiman-WDF model demonstrated a substantial improvement in temperature estimation accuracy, reducing MAE by 25% compared to traditional FEM approaches. The integration of wind direction addresses a previously overlooked parameter, providing a more nuanced and reliable T_{PV} estimation framework.

Future work will involve expanding the methodology to various datasets to validate its adaptability and performance across different PV systems and environmental conditions. Additionally, a comprehensive stability analysis of the temperature coefficients over extended periods will be conducted to assess the temporal consistency and reliability of the model

6 References

[1] B. Herteleer, A. Kladas, G. Chowdhury, F. Catthoor, and J. Cappelle, "Investigating methods to improve photovoltaic thermal models at second-to-minute timescales," *Solar Energy*, vol. 263, p. 111889, Oct. 2023, doi: 10.1016/J.SOLENER.2023.111889.

[2] D. Faiman, "Assessing the outdoor operating temperature of photovoltaic modules," *Progress in Photovoltaics: Research and Applications*, vol. 16, no. 4, pp. 307–315, Jun. 2008, doi: 10.1002/pip.813.

[3] M. Prilliman, J. S. Stein, D. Riley, and G. Tamizhmani, "Transient Weighted Moving-Average Model of Photovoltaic Module Back-Surface Temperature," *IEEE J Photovolt*, vol. 10, no. 4, pp. 1053–1060, Jul. 2020, doi: 10.1109/JPHOTOV.2020.2992351.

[4] J. Barry *et al.*, "Dynamic model of photovoltaic module temperature as a function of atmospheric conditions," *Advances in Science and Research*, vol. 17, pp. 165–173, Jul. 2020, doi: 10.5194/asr-17-165-2020.

AUTHOR INDEX

Abanda, Amaia 1335
Abdallah, Amir A. 591, 878, 1156
Abdelrahim, Mohamed 1156
Abdin, Zain U. 1391, 1957
Abermann, Stephan 428
Abrahão, Raphael 1528
Acevedo, Maria I. D. 610
Ackermann, Jörg 345
Adam, Zoltan 209
Adier, Marie 25
Adner, David 422
Adothu, Baloji 601, 604, 623, 666, 772, 878, 988, 1168
Adrian, Adrian 466
Aernouts, Tom 323, 494
Aghaei, Mohammadreza 1293, 1297
Aguilera, David M. 1391
Aguirre, Arantxa 315
Aguirre, Aranzazu 323, 494
Aguirre, Miguel 960
Ahnood, Arman 169
Ahuja, Suraj 539
Aiello, Andreas 1782
Aïssa, Brahim 139, 645, 1050, 1156, 1245
Aizpurua, Jon 511
Akiyama, Hidefumi 277
Al-Ahmed, Amir 600
Alam, Muhammad A. 878
Alamy, Philippe 1396
Albadwawi, Omar 330
Al-Bajjali, Saif 2040
Alberts, Vivian 330, 601, 604, 623, 666, 878, 1168
Albinius, Niklas 1348
Albrecht, Steve 399, 466
Albuquerque, Daniel 1304, 1675
Alcocer, Kilian 397
Alet, Pierre-Jean 1548
Algaidy, Sari 163
Algergawy, Alsayed 1085
Alghamdi, Mohammed A. 600
Alheloo, Ahmad 601, 604
Ali, Amjad 600
Alkhatib, Hasan 345
Allagiannis, Christos 651
Allebé, Christophe 182
Allegre, Jules 397
Almeida, José C. 1752, 2021
Almeida, Marcelo P. 1752, 1763, 2021
Almheiri, Ali 601, 604

Almosni, Samy 422
Alonso, Victor 758, 974
Alonso-García, Carmen 1651
Alskaif, Tarek 1354
Altin, Müjde 1439
Alujevic, Neven 1691
Alvarez, Jose M. 1301
Alvarez-Brito, Eduardo 135
Álvarez-Pérez, Guillem 379
Aly, Shahzada P. 623, 988, 1168
Alzahrani, Atif S. 600
Alzate, Juan 2068
Amalu, Emeka H. 695
Anagha, E. R. 619
Anamiati, Gaetana 1124, 1141
Anaya, Julián 974
Anchorena, Oscar 1369
Anderlini, Alessandro 702
Andersen, Nanna L. 632, 1199, 1385
Anderson, Kenrick F. 404, 437
Anderson, Kevin 1089
Andersson, Robin 1799
Andrade, Nathianne M. 1556
Aninat, Remi 943
Antretter, Thomas 627
Antwis, Luke 189
Apostoleris, Harry 772
Appel, Tjade 1476
Arampatzis, Ioannis 651
Ariolli, Daniela 2200
Arslan, Meriç Ç. 225
Arslan, Meric C. 683
Arunagiri, Lingeswaran 446
Asa'A, Shu-Ngwa 1709
Asaa, Shu-Ngwa 1596
Ascencio-Vásquez, Julián 1089, 1885
Asiri, Abdullah M. 800
Asker, Osama 600
Assaid, El M. 776
Aßmann, Nicole 1
Assoa, Ya-Brigitte 1434
Auer, Johann 2033
Avasthi, Sushobhan 414
Axisa, Matthew 1040
Aydemir, Umut 418
Aydogdu, Yildirim 683
Azizi, Ferozan 1828, 2077
Azkona, Nekane 201

Azzopardi, Brian	743, 1305, 1309, 2134, 2145
Azzopardi, Carmel	2134
Babin, Markus	632, 780, 966, 1199, 1385
Baccar, Dorra	125
Bachour, Dunia	1524, 1664
Bagci, Aliihsan	632
Bai, Xueqi	7
Bakovasilis, Apostolos	929, 1519
Balafoutis, Athanasios T.	1304
Balchada, Henrique	1770
Baldacci, Jacopo	1769
Ballif, Christophe	182, 349, 626, 862, 1862
Baloch, Ahmer A. B.	330
Balucani, Marco	411
Bamberger, Evelyn	1234
Bangsund, Audun	1995
Bansal, Nitin K.	429
Baptista, Fátima	1651
Barchi, Grazia	1812, 1921
Bardizza, Giorgio	490, 739, 995, 1046, 1065
Barraza, Rodrigo	562, 584, 618
Barretta, Chiara	684, 809, 892
Barrio, Rocío	163
Barrionuevo, Bruno	1304
Barrou, Alexis	1862
Barth, Vincent	953
Bartholomäus, Martin	1287
Bartolo, Brian	1305, 1309, 2134
Bartsch, Jonas	226
Basler, Felix	1544, 1620
Battisti, Kurt	1782
Baumann, Linus	1079
Baumann, Ulrike	106
Baumgartner, Franz P.	1079, 1501
Bayo, Araceli H.	2145
Bazkir, Özcan	1046
Beaucarne, Guy	539
Behrendt, Julian	63
Beinert, Andreas	527, 665, 1620
Beinert, Angelika	702
Bejat, Timea	940
Belawadi, Aditya G.	812
Belkilani, Kaouther	1776
Bellacicco, Sophie	1496
Bellenda, Giovanni	995
Benatto, Gisele A. D. R.	1241, 1317
Bendfeld, Jörg	563
Bengoechea, Jaione	703, 960, 2052
Benick, Jan	40
Benítez-Fernandez, Rafael	163
Benito, Veronica B.	1156, 1245
Berenguier, Baptiste	397

Beresneviciute, Raminta	295
Bermudez-Benito, Veronica	1664
Berrah, Lamia	1418
Berrian, Djaber	1719
Berson, Solenn	397
Berthet-Rayne, Quentin	1245
Berwind, Matthew	1501
Betak, Juraj	1451
Bhattacharjee, Ankur	2005
Bhoraskar, Akshay	1616
Biezemans, Anne	943
Binani, Ashish	1616, 1729
Bivour, Martin	40, 363
Bizzini, Olivier	1434
Blanc, Philippe	2057
Blankemeyer, Susanne	1433
Bleicher, Friedrich	1704, 1828
Blieske, Ulf	505, 2046, 2139
Blum, Niklas	1466
Boccardi, Roberto	632
Böck, Leonhard	924
Boddaert, Simon	1396
Bodeux, Romain	792
Bogdanov, Dmitrii	1790
Bokalic, Matevž	398, 738, 2138
Bolding, Jons	193
Bolink, Henk J.	50
Bonilla, Ruy S.	110
Bonomo, Pierluigi	1187
Borchert, Juliane	315
Borgers, Tom	101, 929, 1647
Boro, Binita	445
Borowski, Peter	1412, 1534
Borrello, Cosimo	1590, 2027
Borz, Giovanni	1596, 1663
Bosch, Elina	1584, 1682, 1715, 1903
Bosco, Giacomo	1596
Bosman, Johan	943
Bosone, Martina	2129
Bothe, Karsten	35
Boudellioua, Abdelaziz	155
Bouguerra, Sara	854, 1638
Boutov, Dmitri	1705
Bouttemy, Muriel	397
Braga, M.	427, 1029, 1293, 1297, 2009
Brahim, Sarra B.	1085
Braid, Jennifer L.	886
Brand, Andreas	135, 226
Brand, Thorsten	466
Brandstätter, Andreas	684
Brastel, Alexis	940
Braun, Christian	1544

Brecl, Kristijan 398, 738
Breitenbücher, Marian 101
Brendel, Rolf 13, 35
Breniaux, Edouard 2205
Brenneisen, Stephan 1401
Bressy, Vincent 1434
Breyer, Christian 1790
Bridel-Bertomeu, Agnes 1740
Brivio, Elisabetta 2056
Bruckner, Helmut..1957
Bruggeman, M. 113
Bruhwyler, Roxane 1606
Bua, Letizia 2205
Buceta, Alicia.. 960, 2052
Bucher, Christof..................... 571, 721, 1435, 1501
Buchholz, Florian 30, 101, 175
Buddgård, Jonas...751
Bueno, Bruno 1172
Buffolo, Matteo...417
Buijsch, Frans O. 892
Bulkin, Pavel 334
Bunge, Lisa 1658, 1675
Bunme, Pawita 1626
Burgers, Antonius R......................................1729
Burkhardt, Daniel 57
Burri, Matthias 721
Busto, Chiara 2205
Busuttil, Daniel 2145
Byford, Brandon..886
Caballero, David 1663
Cabarrocas, Pere R. I. 334
Caccivio, Mauro 721, 862, 1009, 1286, 1704
Cai, Hanmin 1390
Cai, Yalun 110
Cai, Yanbo 346
Cal, Silvia 960
Caldarelli, Antonio.......................................302
Cambarau, Werther 511
Camus, Christian...466
Candan, Mücahid 1118
Candelise, Chiara 2205
Caneva, Silvia 2205
Cano, Francisco J..511
Cantisano, Jose 1301
Cao, Fangfang 800
Cao, Rono 539
Cappelle, Jan.. 709, 1514
Carbone, Rosario 1590, 2027
Cardoso, Andressa D. S. 1610
Caria, Alessandro 417
Carpintero, Luis A. 758
Carr, Anna J. 1616

Carrillo, Rafael E. 1548
Carroy, Perrine 50
Carvalho, Paulo C. M.....................................1354
Case, Christopher 447
Castellazzi, Luca 1183
Castillo, Juan D. D. 285
Catipovic, Ivan 1691
Cattaneo, Gianluca 1862
Catthoor, Francky 1519
Caudevilla, D. 163
Cavaco, Afonso 2138
Cavalcante, Danielle B. 1556
Cebria, Maria 1687
Celi, Edoardo 583
Celik, Duygu 2205
Centazzo, Massimo 106
Cereceda, Eneko 201
Cermák, Jan 1476
Cesar, Kay 1729
Cesenia, Eduardo M. 2145
Cester, Andrea 417
Champault, Lisa 398
Chan, Catherine 834
Chandra, Amreesh 445
Chang, Han-Chen 148, 222
Chapaneri, Kaushal 988, 1055, 1168
Chapon, Julien 1245
Chasparis, Georgios 2015
Chatzipanagi, Anatoli 1894
Chemnitzer, Rene 25
Chen, Angela 404
Chen, Jin-Cheng 222
Chen, Kexun 189
Chen, Ran 834
Chen, Sung-Yu 148
Cheng, Cheng-Liang 148
Chhapia, Gaurang 1719
Chianese, Domenico 1286, 1331
Chiesa, Samuele 1331
Cho, Dae-Hyung 338
Chopard, Jérôme 1496
Choulat, Patrick 1647
Chowdhury, Gofran 1516, 1663, 1988
Chozas, Sofia 50
Christiansen, Silke 323
Christöfl, Petra 570, 892
Chueh, Wei-Lo 144
Chung, Yong-Duck.......................................338
Ciesla, Alison 834
Clement, Florian 89, 155, 164, 226
Clochard, Laurent 110
Clyncke, Jan 1849

Colberts, Fallon	1638
Coletti, Gianluca	834
Colin, Hervè	1369
Collares-Pereira, Manuel	1687
Collave, Claudia G.	2040
Colvin, Dylan J.	826
Comak, Mertcan	175
Congouleris, Nicolas	1304
Cooper, Emma	886
Çorak, Merve	122, 683
Cordeiro, Diogo	1675
Cornago, Iñaki	1400
Cornaro, Cristina	1020
Cornella, Alessia	2205
Corre, Pierre-Yves	25
Correia, Joana	1696
Corti, Paolo	1187
Coskun, Özlem	35, 101, 213
Cosme, Damien	1245
Costa, Francis	1647
Couderc, Romain	792, 995
Cox, Joel D.	632, 1385
Crawley, Dru B.	2149
Creon, Laura	25
Cristóbal, Ana B.	2138
Cros, Stephane	428
Crozier, Nicole M.	2062
Cuaresma, Jesús	1279
Çubukçu, Mete	1118
Cueli, Ana B.	428, 960
Culot, Dominique	539
Curtis, Taylor L.	1841
Cusenza, Maria A.	2044, 2056
D'Agostino, Delia	1183, 2149
D'Arco, Luigi	702
Daenen, Michaël	854, 1638, 1709
Dahle, Arne	35, 101
Dalibor, Thomas	1412, 1534
Danelli, Andrea	2044, 2056
Danovitch, David	250
Das, Gourab	149
Daschinger, Thomas	1433
Daßler, David	1132
Datas, Alejandro	384
Daume, Darwin	773, 1314
De Biasio, Martin	1825
De Brabandere, Karel	1089
De Castro, C.	758, 974, 983
De Cook, Nicolas	1606
De Jong, Richard	1565, 1576, 1638, 1709
De L'Epine, Mélodie	101, 1305, 1309, 1715, 1906, 2134
De Luca, Daniela	302

De Oliveira, Aline K. V.	1029, 2009
De Rose, Angela	499, 527
De Santi, Carlo	417
De Seoane, Jose M. V.	1682
De Sousa, Joyce A. O.	1418
De Vries, Hindrik	193, 408
Debije, Michael	299
Deckers, Elke	1638
Deckers, Martijn	1443
Deckx, Julien	1089
Defrenne, Nicolas	2068
Del Prado, Alvaro	163
Del Ser, Javier	1335
Delbeke, Oscar	1634
Demant, Matthias	57, 63
Dembélé, Kassiogé	334
Demicoli, Marija	1040
Demir, Melisa	225
Demofonti, Giuseppe	1596
Denafas, Julius	101
Deniz, Esref	1806
Dennenmoser, Martin	1085
Depauw, Valerie	315
Dernis, Michel	1396
Desai, Umang	626
Despeisse, Matthieu	1862
Desrues, Thibaut	50
Devoto, Ignacia	545
Devoto, M. Ignacia	330
Dewallef, Stefan	1647
Di Carlo, Aldo	411, 417, 486
Di Gennaro, Emiliano	302
Di Giusto, Fabio	1638
Di Napoli, Annalisa	2072
Di Sabatino, Marisa	1669
Diarce, Gonzalo	1400
Diaz, Javier	428
Diestel, Christian	57
Dietrich, Andreas	1132
Dijksterhuis, Jakob J.	834
Dilmac, Umran	683
Dippell, Torsten	13
Dirubio, Christopher	826
Dittmann, Sebastian	1606
Dittmar, Hanna	2205
Djeukeu, Ivanol J.	327
Dkhil, Sadok B.	345
Döblinger, Markus	273
Donoso, Jose	1906
Dörn, Markus	1782
Dorn, Silke	35
Dos Santos, Jeremias	1675

Dou, Qizheng 1647
Dozio, Gian C. 1331
Dreisiebner, Andreas 1401, 1736
Driesen, Johan 1443, 1634
Driesse, Anton 1214, 1976
Droudakis, Alexandros I. 2124
Du, Keming 135
Duan, Lian 2099
Duarte, Dorivaldo 1687
Duarte, Sebastian 163
Duck, Benjamin C. 404, 437
Duerinckx, Filip 101, 260, 1647
Duffy, Noel W. 404, 437
Dullweber, Thorsten 35, 101, 106
Duman, Hatice 213, 225
Dunlop, Ewan 999
Dupon, Olivier 1565, 1709
Dupuis, Julien 519, 792
Durand, Salomé 1733
Durusoy, Beyza 422
Dutykh, Denys 1552
Dyson, Paul J. 800
Dzurnák, Branislav 294
Eberlein, Dirk 527
Ebert, Matthias 1132
Ebner, Rita 323, 1305, 1309
Echeverria, Iván G. 618
Eckert, Jonas 155
Eckerter, Sascha 578, 1758
Ecoffey, Serge 250
Eder, Gabriele 694, 966, 1704, 1825, 1828, 2077, 2200
Eggers, Jan-Bleicke 1172
Ehsan, Ali 2145
Eikelboom, Erik 101
Einhaus, Roland 684
Eiternick, Stefan 607
Ekins-Daukes, Nicholas 263
El Ainaoui, Khadija 776
El Mrabet, Yasmine 776
Elamri, Yassin 1496
Element, Adrian 404, 437
Elhamaoui, Said 776
Eliassi, Mojtaba 1988
Ellmann, Martin H. 1178
Emanuel, Gernot 514
Ensslen, Frank 812, 1172
Erber, Alexander 686
Eriksen, Erling W. 1480
Eroglu, Sertaç 122
Eskandari, Aref 1293, 1297
Esmaeilzadeh, Maryam 309
Esmaielpour, Hamidreza 273

Essbai, Soha 743
Estarlich, Pau 285
Esteras, Miguel 1335
Estola, Pirjo 1799
Estrada, Esther L. 384
Ezquer, Mikel 1400
Fabel, Yann 1466
Fabris, Francesca 101
Fabunmi, Oluwagbemiga A. 695
Faes, Antonin 626, 738
Fano, Vanesa 201
Faramarzi, Seyed M. S. 712
Farias-Basulto, Guillermo 393
Farmakis, Filippos V. 2124
Farneda, Rüdiger 330
Farooq, Umar 302
Fath, Moritz 1853
Fath, Peter 76, 1853, 2152
Fava, Luís 1989
Fedrizzi, Maria C. 1763
Feichtner, Markus 1704, 1782
Feldbacher, Sonja 534, 626, 1828, 2077
Feldhof, Anne-Maren 2139
Felipe, Inmaculada C. 898
Fell, Andreas 63
Fernandes, Cláudia 1675
Fernandez, Ana M. 1894
Ferreira, Catarina G. 632, 1385
Fialho, Luís 1651, 1658, 1675, 1687, 1696, 1989, 2138
Fidalgo, Ignacio 898
Figgis, Benjamin W. 1156, 1245
Figueiredo, Gilberto 1752, 2021
Filho, Eduardo S. 1118
Finley, Jonathan 273
Fischer, Marie 1885, 2192
Fki, Rania 2205
Fledderus, Henri 943
Flouchi, Imane 776
Fokuhl, Esther 786
Foles, Ana 1989
Fonseca, Luiz 1528
Fontanot, Thommaso 323
Fooladgar, Ehsan 1799
Formiga, João 1304
Franchi, Norman 1491
Fredj, Donia 345
Frégnaux, Mathieu 397
Friesen, Gabi 721, 862, 995, 1009, 1046, 1286, 1704
Froebel, Jens 552
Frontini, Francesco 1187
Frossard, Pascal 1548
Fuchs, Ida 1273, 1995

Fujii, Masayuki	676
Fumey, Damien	1496
Funahashi, Ryoji	277
Gabbadi, Prashanth	585, 1055
Gagliano, Antonio	1341
Gagnaire, Dimitri	1733
Gaiddon, Bruno	1733
Gaisberger, Lukas	2015
Galán, María I. R.	1610
Galarza, Alejandra	2068, 2099
Gallego-Castillo, Cristobal	2183
Gallmetzer, Sandra	1095, 1225
Gamel, Mansur	289
Gao, Feng	446
Gao, Qi	490, 739
Garabetian, Thomas	2205
García, Fernando	1279
Garcia, Ignacio B.	345
Garcia, José C.	1033
Garcia, Juan L.	878
Garcia, Kévin	1160
García-Hemme, Eric	163
García-Hernansanz, Rodgar	163
Garcin, Jean	1496
Gardeski, Matthew	826
Garín, Moisés	289
Gassner, Anika	694, 1704, 1825, 1828, 2077, 2200
Gatin, Inno	1691
Gaudino, Eliana	302
Gaulding, Ashley	1841
Gayot, Felix	308
Gazbour, Nouha	2205
Gdula, Lukáš	294
Gebhardt, Paul	786, 911
Geerligs, L. J.	113
Geier, Dieter	684
Gelibert, Stephane	1434
Geml, Fabian	1, 7
Genoe, Jan	712
Georghiou, George E.	323, 494, 1209, 1264
Georgilakis, Pavlos	1519
Gerber, Alexander	2138
Gerber, Andreas	1347, 1444, 1543
Gevaerts, Veronique	943
Ghennioui, Abdellatif	776
Ghidesi, Giancarlo	2084
Gioia, Ferdinando	1590, 2027
Girardi, Pierpaolo	2044, 2056
Glarner, Roger	1401
Glatz-Reichenbach, Joachim	610
Glaubitz, Anika	505
Glunz, Stefan W.	327, 363
Glunz, Stefan	40, 176
Göbel, Alexander	158
Godoy-Perez, G.	163
Gohil, Hardik	76, 768
Gok, Abdulkerim	684
Golab, Antonia	2033
Gölboylu, Selin C.	225
Golroodbari, Sara	1501
Golubev, Timofey	1322
Gombás, Z.	209
Gomes, Amanda M. F.	2009
Gomes, João	943
Gonzalez, Alejandra C.	2057
González, Miguel Á.	974, 983
González-Díaz, Benjamín	432
González-Francés, Diego	758, 974, 983
González-Pérez, Sara	432
Goraya, Baljeet S.	2152
Gorchs, Gil	1596
Gordon, Ivan	2205
Gordon, Michael	1543
Gostein, Michael	1245
Gottschalg, Ralph	596, 660, 687, 878, 1005, 1576, 1606
Gou, Yangyang	800
Gouabault, Anaïs	2068
Govaerts, Jonathan	929, 1647
Goverts, Martina	1835
Gracia-Amillo, Ana	2134
Graeber, Dietmar	1776
Graeber, Robin	2046
Grand, Pierre-Philippe	2068
Grasel, Bernhard	686
Gréau, David	1733
Gregory, Geoffrey	106
Greulich, Johannes	63, 164, 176
Grigalevicius, Saulius	295
Groen, Niels	1596, 1663
Grommes, Eva-Maria	2139
Groß, Claudine	466
Großer, Stephan	535, 933
Grosser, Stephan	607
Grübel, Benjamin	499, 527
Grünsteidl, Stefan	1534
Grüttner, Sven	505
Gry, Johannes	40
Guastella, Salvatore	1108
Guerra, Gerardo	1124, 1141
Guerra, Walter	1596
Gueymard, Christian A.	1596
Guidetti, Giulia	2084
Guillemoles, Jean F.	379
Guillevin, Nicolas	101

Guillon, Sebastien 1462
Guštin, Matej 2138
Gutjahr, Astrid 113, 834
Guzman, Francisco 562
Ha, Duy-Long 1220
Haberstroh, René 135, 155, 164
Hacke, Peter 826
Hadadian, Mahboubeh 309
Hadipour, Afshin 323
Hadjipanayi, Maria 323, 494
Hädrich, Ingrid 702, 812
Haedrich, Ingrid 786
Hagendorf, Christian... 422
Hahn, Giso 1, 7
Hallensleben, Carina 545
Halm, Andreas 330, 514, 545, 610, 898
Hamada, Toshiyuki 676
Hamam, Zeina 1434
Hameed, Mohammed A. 1576
Hameiri, Ziv.............................. 308, 761, 980, 1104
Hamon, Gwenaelle 250
Hanifi, Hamed 552
Hansen, Per-Anders 132, 172
Hanser, Mario 40
Haque, Faiazul 437
Harder, Nils-Peter 1033
Harnisch, Martina 534
Harrison, Samuel 101
Hashem, Ahmad.. 596, 1005
Hasselblatt, Charlotte.. 665
Hategan, Sergiu M. 717
Hauer, Martin... 1782
Haug, Franz-Josef 182
Haunschild, Jonas 57, 214
Havasi, Gergely 209
Hayez, Valérie 539
Heer, Philipp 1390
Heim, Manuel 1313
Heinonen, Aleksi.. 1361
Heinrich, Martin 1620
Heinzle, Nino.. 1264
Heitmann, Johannes 205
Helbig, Matthias 545
Helfer, Eric 570, 627
Hennig, Carsten 1132
Hensel, Andreas 1966
Herath, Kristian... 1412
Herceg, Sina.. 1885, 2192
Herguth, Axel... 1, 7
Hermle, Martin 40
Hernandez, Guillermo O. 2200
Hernández, Juan M. 511

Herr, Cornelius 665, 1620
Herrero, Carmen M. R.. 345
Herrmann, Werner 490, 739, 995, 1046
Herteleer, Bert 709, 1089, 1514
Herz, Magnus 1065, 1089
Herzberg, Wiebke.......................... 1456, 1476
Hess, Donat 571, 721
Hessler-Wyser, Aïcha 182
Heupl, L. 570
Heydarian, Maryamsadat 363
Heydarian, Mina.. 315
Higueruela, Francisco R. F. 1610
Hillmann, Martin 1412
Hiltebrand, Roger 1736, 1747
Hitchcock, Will 1770
Hitte, Vincent 1496
Hoex, Bram 878
Hoffman, Hannah 2095
Hoffmann, Erik 106
Hofmann, Marc 158
Hofmann, Rene 1961
Holappa, Ville 430
Holder, Emma 404, 437
Holland, Nicolas... 1544
Hoppe, Georg 135
Horn, Jonas 327
Horta, Pedro 1658, 1675, 1687, 1696, 1989
Hoß, Jan.. 30, 175
Hsieh, Hsin-Hsin 328
Hu, Guang 1392
Huang, Chih-Jeng.. 148
Huang, Meixian 86
Huang, Shujuan 481
Huang, Ying-Yuan.............................. 148, 219, 222
Huerta, Hugo...................... 1361, 1969, 2001, 2102
Hughes, David J. 695
Hurni, Julien 182
Hut, Anouk 1988
Hüttl, Bernd.. 773, 1509
Huyeng, Jonas.............................. 89, 158, 164, 226
Hwang, Tae-Ha 338
Iannibelli, Elena 486
Ibrahim, Nabeel 1055
Idlbi, Basem 1369, 1776
Ilse, Klemens... 591
Imbuluzqueta, Gorka.. 511
Infante, Paulo 1658
Irulegi, Olatz 1400
Isabella, Olindo 1391, 1417, 1957
Isaev, Nabi 273
Isasi, Telmo 201
Ishikura, Norio 676

Iwaki, Koshiro 549
Jääskeläinen, Jaakko 2087
Jachmann, Joseph 1509
Jäckel, Bengt...878
Jadaud, Cyril...334
Jadot, Emmanuel...539
Jaeckel, Bengt...............535, 552, 591, 607, 666, 685, 755, 772
Jaeger-Waldau, Arnulf 1894
Jäger, Philip 106
Jager, Wander...2205
Jahani, Babak 1476
Jahn, Mike 19
Jahn, Ulrike...1501, 1909
Jain, Sachin 1664
Jang, Juhee 1269, 1328
Jänkälä, Matti 1375
Jansen, Mark 315
Jasielec, Jerzy J. 1969
Jason, Daniel...1365
Jay, Frédéric 792
Jeangros, Quentin 349, 398, 892
Jeon, Joonyoung 716
Ji, Jingjia 86
Jiang, Hongxu 346
Jiang, Yongjie 800
Jiang, Zonghan 660
Jiménez-Castillo, Gabino 1341
Jimeno, Juan C. 201
Jo, Sangmin 2160
Job, Enzo 812
John, Jim J. 878, 988, 1168
Johnson, Erik V. 334
Jokikyyny, Tommi..530
Jolivet, Raphaël...1859
Jones, Tim W. 404, 437
Jooß, Wolfgang 76, 149, 2152
Jooss, Wolfgang 768, 1853
Joseph, Christopher D......................................1620
Joseph, Daniel C. 527
Joss, David....................571, 1435
Jošt, Marko 329, 399
Jost, Norman...886
Jouanneau, Corentin 250
Jouttijärvi, Sami...................1361, 1375, 1969, 2102, 2119
Juana, Luis..1654
Juillion, Perrine 1496
Julien, Arthur 379
Jurado, Juan M...1610
Juso, Hiroyuki...254
Kaaya, Ismail............854, 1565, 1576, 1596, 1638, 1709, 1885
Kadota, Naoki 1194
Kaiser, Martin 1544

Kaizuka, Izumi 1906
Kakoulaki, Georgia 1894, 2108
Kalaghichi, Saman S. 30, 175
Kalliojärvi, Heidi 1258
Kalms, Alicia...1400
Kamide, Kenji 277
Kammerlander, Christoph 342
Kamphues, Joshua...7
Kamppinen, Aleksi 303, 530
Kanawala, D. N. 1391
Kang, Mangu...338
Kankanamge, Dilshika H. 2087
Kapeller, Rudolf..2205
Karalus, Steffen.................................1456, 1476
Karatepe, Engin 1806
Karhu, Juha A..1361
Kari, Thøger 636, 1241, 1317
Karimipour, Massoud 285
Karrenbrock, Anne 2139
Karttunen, Lauri 1969, 2102
Kathan, Johannes 2033
Katsikogiannis, Alexandros.............................1596
Kaufmann, Kai 1132
Kazacos, Duarte 1118
Kazantzidis, Andreas.......................................1466
Kazem, Hussein A..878
Keding, Roman 226
Keiner, Dominik 1790
Kenney, Kayla 539
Kenny, Robert 1894
Kentsch, Ulrich 189
Kerekes, Krisztián 1758
Kern, Jonas...205
Kern, Melanie..2205
Kester, Josco 101
Khan, Firoz 600
Khan, Muhammad..106
Khan, Nabeel..106
Khenkin, Mark 399
Khodr, A. 345
Kikelj, Miha 349
Kim, Changki 2160
Kim, Chongmin.................................1269, 1328
Kim, Jinyoung 2160
Kim, Ju-Hee 716
Kim, Kihwan 338
Kim, Moonyong 834
Kim, Rina 338
Kim, Sedong 1775
Kim, Yong H. 716
Kim, Yongil...2160
Kirch, Jochen 755

Kirchhof, Jörg	690
Kirkil, Gökhan	2205
Kishore, Ravi	1647
Kitamura, Ibuki	676
Kizukuri, Rihoko	545
Kladas, Anastasios	709, 1514
Klaus, Daniel	911
Kleinhans, Alexander	1544
Klengel, Robert	1132
Klenk, Markus	1079, 1401, 1736, 1747
Kluska, Sven	19, 135, 155, 164, 226
Klute, Carola	1132
Kobayashi, Nobusato	1194
Koblmüller, Gregor	273
Kobor, Diouma	441
Koduvelikulathu, Lejo	149, 175
Koepge, Ringo	535, 685, 933
Koester, Lukas	1225
Kohn, Norbert	158
Kohno, Tohru	1253, 1953, 1982
Kojima, Nobuaki	254
Kolås, Tore	1381
Könen, Stefanie	2139
Kopp, Nils	545
Korevaar, Marc A. N.	1249
Korkmaz, Güven	213
Korsós, Ferenc	209
Korte, Lars	466
Kossen, Eric J.	113
Kouame, Konan	250
Kousounadis-Knousen, Markos	1519
Kraft, Achim	499
Kraft, Leonard	1132
Kraft, Thomas M.	430
Krähmer, Sabrina	1776
Kräling, Ulli	786
Krammer, Anna	302
Krauter, Stefan	563, 1790, 1932
Krc, Janez	1146
Krieg, Katrin	19, 164
Krishna, Anurag	494
Krisztián, David	209
Kroon, Jan	101
Krucaite, Gintare	295
Kuan, Ta-Ming	144
Kubicek, Bernhard	1305, 1309, 1347, 2134, 2169
Kuhn, Tilmann E.	1172
Kühne, Marcel	892
Kühnert, Jan	1476
Kulhavy, Lukas	125
Kulkarni, Shrikrishna V.	619
Kumano, Kengo	1953, 1982
Kumar, Avinash	76
Kumar, Saravana	57, 214
Kumar, Shubham	1357
Kumar, Sudarshan	1559, 1724
Kumar, Yogesh	585
Kunze, Philipp	63
Kuo, Cheng-Wen	144
Kuraoka, Akihiro	1194
Kurtovic, Enita	545
Kuruganti, Vaibhav	30
Kurumundayil, Leslie L.	63
Kuypers, Ando	943
Kuznetsova, Daria	295
Kyranaki, Nikoleta	854, 1638
Lachowicz, Agata	182
Lacombe, Marie	2068
Lagast, Karel	1514
Laget, Hannes	1647
Lamminaho, Jani	632, 1385
Lampa, Josefin	1799
Landa, Margot	940
Landberg, Lars	1124, 1141
Landes, Dieter	773, 1509
Lang, Margit	570, 627
Lang, Xiting	800
Lange, Gerrit	35
Lanzetta, Ciro	1769
Lappalainen, Kari	1258, 1937, 1977
Larionova, Yevgeniya	35
Lauwaert, Johan	260
Lawrie, Linda K.	2149
Le Rouzo, Judikaël	345
Le, Philip	1638
Lebeau, Frederic	1606
Lee, Chun-Wei	144
Lee, Woo-Jung	338
Lefillastre, Paul	519
Leimgruber, Fabian	2169
Leiva, Amanda M.	50
Leloux, Jonathan	1369, 1596
Lemaitre, Noëlla	397
Leow, Shin W.	995
Leyden, M.	466
Li, Fang	826
Li, Minghui	800
Li, Qiuxian	1390
Li, Yong	404, 437
Li, You-An	219
Li, Yung-Chih	144
Li, Zhuofeng	25
Liang, Tian S.	1187
Lim, Soyoung	338

Lin, Chun-Ping 148, 219, 222
Lin, Shih-Chieh 144
Lin, Yi-Ping 222
Linares, Ana 960
Lindahl, Johan 1903
Linder, Johannes 1719
Lindfors, Anders V. 1361
Lindh, Mattias 1799
Lindig, Sascha 1089
Linke, Jonathan 30, 101, 175
Linsenmeyer, Aswin 755
Lipovšek, Benjamin 349, 399
Lira-Cantu, Mónica 285
Liu, Anyao 25
Liu, Fei 346
Liu, Xirui 800
Liu, Zhipeng 86
Livera, Andreas 1209, 1264
Lizana, Fernando F. 2092
Lizin, Sebastien 2205
Llarena, Elena 432, 960
Lohmüller, Elmar 158, 176
Lohmüller, Sabrina 176
Lomeri, Hamed J. 854, 929
Long, Yean-San 328
Loonen, Roel C. G. M. 1392
López, Gema 289
Lopez-Garcia, Juan 1156, 1245, 1664
Lopez-Velasco, Gerardo 1496
Lorenz, Andreas 226
Lorenz, Elke 1456, 1476, 1544
Lossen, Jan 30, 175
Louwen, Atse 1095, 1225, 1812, 1873, 2205
Lu, Yibo 86
Lübke, Maximilian 1491
Lukinskas, Povilas 101
Luo, Bin 1647
Lyubenova, Teodora S. 999
Maarouf, F. 226
Maaroufi, Hamza 1065
Macarulla, Marcel 1596
Macdonald, Daniel 25
Macdonald, James 1596
Macé, Philippe 101, 1584, 1682, 1715
Mack, Sebastian 164
Mader, Patrick 578, 1758
Madsen, Morten 632, 1385
Maduta, Carmen 1183
Maeda, Kengo 1194
Mahmood, Aysha 636, 1241, 1317
Mahmood, Farrukh I. 826
Maixner, Andreas 552

Makhfudz, Imam 273
Makrides, George 1209, 1264
Malcorps, Philippe 1516
Malguth, Enno 466
Malik, Stephanie 1132
Mamykin, Sergii 295
Mandorlo, Fabien 953
Manito, Alex R. A. 1752, 1763, 2021
Manshanden, Petra 315, 834
Mansour, Djamel E. 911
Mansour, Ridha B. 600
Mansouri, Mathieu 1733
Manzolini, Giampaolo 1921
Marangis, D. 1209, 1264
Marchand, Mathilde 1859
Marcotte, Médérick 250
Marechal, Philippe 1220
Margeat, O. 345
Maria, Enrico D. 1663
Markert, Jochen 786, 812
Markvart, Tom 294
Marrero, Asier M. 428
Marstein, Erik S. 1089
Marteau, Batiste 50
Martín, Isidro 289
Martínez, Mario 746, 1301, 1365
Mártinez, Oscar 758, 974, 983
Martulli, Alessandro 2205
Masmitjà, Gerard 285
Masson, Gaëtan 1584, 1682, 1715, 1903, 1906
Masuda, Atsushi 523, 549
Mateos, Yeray 201
Matheron, Muriel 50
Mathiak, G. 585, 601, 604, 623, 772, 878, 988, 1055, 1168
Matic, Gašper 398
Maticiuc, Natalia 428
Matos, Pedro 1989
Matsumura, Yoko 277
Matteocci, Fabio 417
Maugeri, Giosué 1108
Mazzucchelli, Paola 2205
McCleland, Jacqueline L. C. 2062
Meddahi, Amar 1462
Medina, Eduardo R. 511
Medina, Ismael 999
Medjoubi, Karim 379
Meereboer, Martijn 101
Mehler, Melanie 1
Meinhart, Lisa 684
Meixner, Michael 327
Melgar, David 1118, 1369
Mellone, Celeste 2084

Meneghesso, Gaudenzio 417
Meneghini, Matteo...417
Mercade-Ruiz, Pau 1124
Mercaldo, Lucia V..428
Mermoud, André...1740
Mertens, Verena 35, 101, 106
Merz, Rainer 578, 1758
Messmer, Christoph 363
Messmer, Marius 164
Messmer, Tobias 101, 514
Meuret, Youri 1647
Meusel, Manuel 53
Meyer, Fabian 135
Meyer, Imke...1770
Meyer, Kevin 1433
Meyer, Lukas 1435
Meza, Carlos................... 660, 1305, 1309, 1606, 2134
Meza, Cristian V. 363
Miaskiewicz, Aleksandra 466
Midtgård, Ole-Morten...1273
Miech, Juri 7
Mielich, Niko..89
Miettunen, Kati....303, 309, 530, 1361, 1375, 1969, 2102, 2119
Mignonac, Alexandre 1305, 1309
Mihailetchi, Valentin................................... 30, 101
Miklic, Žiga 1146
Mikulic, Antonio..1691
Milimonfared, Jafar 1293, 1297
Min, Byungsul 13
Minuto, Alessandro...583
Mirza, Mark 585
Misfeld, Heidrun 1476
Mittag, Max 920, 2095
Mittal, Ankit............................... ....323, 428, 743
Mizuno, Hidenori 1626
Mizushima, Io 342
Mjøs, Øyvind 132
Mo, Alvin...834
Mockeviciute-Azzopardi, Austeja.........................2134
Mofakhami, Eeva 940
Moine, Gérard...1733
Möller, Marius C...1932
Mollier, Stéphane 1369
Molto, Cecile 826
Monokroussos, Christos................... 490, 739, 1005
Montes, Carlos 432
Montes-Romero, Jesus..1264
Moradpoor, Iraj 2109
Mordvinkin, Anton 680, 685, 687, 933
Moreda, Guillermo 1651
Moreda, G.-P..1654
Moretón, Rodrigo 1160, 1369

Morisset, Audrey 182
Morlier, Arnaud....................854, 1565, 1638, 1709
Morozova, Olga 189
Mortazavifar, S. L. 596, 1005
Moschner, Jens 854, 1634, 1647
Mosel, Frank 125
Moser, David......... 1020, 1095, 1225, 1565, 1596, 1663, 1812, 1873, 1909, 1921, 2200
Motiwala, Saurabh 1559, 1724
Mühlich, Mona..1172
Muka, Eni...197
Mukherjee, Srijani 1552
Müller, Björn...1118
Müller, Matthias 205
Muñoz, Delfina 50, 1909
Muñoz-García, Miguel-Ángel..............................1651, 1654
Muñoz-Rodriguez, Francisco J............................. 1341
Muntwyler, Urs 1409, 1679
Murillo, Asier 2052
Musto, Marilena 302, 2072
Muthusamy, Arumugham 1055
Mütter, Gerhard... 1966
Myhre, Stine F. 2078
Naas, Tyke 132
Nagel, Henning 40
Nägele, Andreas 514
Nair, Jishnu R. 687
Najafi, Mehrdad 408
Najah, Mohamed 250
Nakajima, Akihiko 1194
Nakamura, Kyotaro 254
Nanno, Ikuo... 676
Naspolini, Helena 1029
Nasser, Hisham 19, 197
Nasti, Giuseppe 428
Navarro, Valentina.. 584
Nazeeruddin, Mohammad K. 800
Naziri, Pouriya 418
Ndioukane, Rémi 441
Nedaei, Amir 1293, 1297
Nekarda, Jan...135, 226
Nemitz, Wolfgang 780
Neuhaus, Dirk H. 920, 2095
Neuhaus, Holger.. 527
Neumaier, Lukas 1825
Ney, Mylana 250
Nguyen, Hieu T. 25
Nguyen, Nathalie 50
Niederhofer, Stefan 1961
Nieto, María B.. 1651
Nigl, Thomas.. 2077
Nikam, Maitheli 1663

Nikbakht, Hafez	486
Nikitina, Veronika	924
Niskanen, Johannes	1969
Nitzel, Damon	1249
Nizamov, Rustem	309
Noack, Philipp	13
Noels, Serge	1849
Nold, Sebastian	2152
Nordseth, Ørnulf	1381
Norton, Matthew	494
Nour, Christine A.	519
Nouri, Bijan	1466
Ntsala, Palisa G.	2062
Nussbaumer, Hartmut	1401, 1736, 1747
Nyberg, Mikael	309
Nygård, Magnus M.	1480
Oberbeck, Lars	1859, 2068, 2099
Ocaña, Luis	432
Ochoa, Lluvia	1245
Oh, Jaewon	826
Oh, Sujeong	1269, 1328
Ohdaira, Keisuke	523, 549, 656
Ohshita, Yoshio	254
Oke, Shinichiro	676
Olea, J.	163
Oliosi, Michele	1740
Oliveira, Helena	1658
Olofsson, Arvid	1799
Oozeki, Takashi	846, 1626
Opatovsky, Martin	1451
Oreski, G.	534, 570, 626, 627, 684, 809, 892, 966, 2077, 2200
Ortega, Eneko	201
Ortega, Pablo	285
Ortiz, Hugo S.	1606
Osama, Amr	1341
Otaegi, Alona	201
Otoo, Edward	1392
Otto, Nicolas	393
Ouaras, Karim	334
Ourinson, Daniel	164, 226
Oyarzun, Aritz L.	2134
Öz, Aksel Kaan	911
Ozaki, Ryo	254
Özar, Nilsah	1439
Özkalay, Ebrar	862, 1009, 1046, 1704
Paesa, Marta C.	1638
Paez, Pablo S. E.	1095
Palitzsch, Wolfram	101
Palonen, Heikki	530
Pampin, Janire	201
Panchabikesan, Karthik	2149
Pander, Matthias	535, 552, 591, 666, 685, 755, 933
Pandurangan, Karthikeyan	486
Panduri, Fabio	721
Pang, Yongxin	695
Panhuysen, Markus	773, 1314
Paraskeva, Vasiliki	323, 494
Parida, Bhaskar	330
Parion, Jonathan	260, 363
Parisi, Maria L.	486
Parker, Danny S.	2149
Parlayan, Onur	527
Parra, Vicente	1279
Parvin, P.	1293, 1297
Passaro, Marcello	2205
Pastor, D.	163
Patel, Dharm	1132
Patha, Andreas	2033
Patton, Daniel J. C.	1279
Paul, Mrittika	445
Paulescu, Marius	717
Pauli, Eva	1476
Paviet-Salomon, Bertrand	182, 349, 1862
Pawar, Vani	414
Payne, David N. R.	481
Pearce, Phoebe	263
Peche, Rene	2045
Pechmann, Sabrina	323
Peibst, Robby	13
Peighambardoust, Naeimeh S.	418
Penas, André	1682, 1715
Peng, Meilin	86
Peratikos, Elias	494
Pereira, Sara	1687
Perelman, Antoine	953
Peres, Paula	25
Perez, Alba	1663
Perez, Inaki	1770
Perez-Astudillo, Daniel	1524, 1664
Perez-Lopez, Paula	1859, 2057
Perrin, Marielle	1733
Persello, Severine	1496
Pervan, Nikolina	534, 626
Peter, Christoph	30
Petersons, Karlis	632, 1385
Petersson, Anna M.	1799
Petro, Julia	570, 627
Petzschmann, Jonas	1313
Pfau, Charlotte	651
Pfeiffer, Oliver	505, 2046
Pfyffer, Selina	1401, 1736, 1747
Phang, Sieu P.	25
Philipp, Daniel	786, 812
Piazzi, Antonio	1769

Pierce, Benjamin G. 886
Pierro, Marco 1020
Pieters, Bart E. 1347, 1444, 1543
Pignatelli, Angelo 1596
Pilat, Eric 1160, 1369
Pillai, Dhanup 1245, 1664
Piluso, Pierre 940
Pingel, Sebastian 57, 226
Pinto, Cristina 703, 2052
Pirc, Matija 329
Pirelli, Barbara 2129
Pires, Anelise M. 427
Pirot-Berson, Lucie 792
Pittalis, Marco 2108
Plaza, C. 1584, 1682, 1715
Plessing, Lukas 780
Plissonnier, Alexandre 1434
Polacchi, Cristina 1873
Polo, Jesus 1214
Polzin, Jana 19, 40
Ponomarenko, Anna 2040
Poormohammadi, Fereshteh 1443
Poortmans, Jef 260, 712, 929, 1647, 1709
Popplow, Laura 2139
Porter, Jennifer 1663, 1770
Porwal, Shivam 429
Poskela, Aapo 309, 530
Pospischil, Maximilian 101
Poulsen, Peter B. 632, 636, 1287, 1317, 1385
Prasad, Manjunath 101
Pravettoni, Mauro 995
Preis, Pirmin 149, 175
Preisig, Janis 1747
Preu, Ralf 158, 176, 226, 2152
Protti, Alexander 920
Puel, Jean B. 379
Puglisi, Lisandro 1304
Puigdollers, Joaquim 285
Purohit, Ishan 1559, 1724
Puthiyapurayil, Aafra S. 623
Qin, Yusen 86
Qiu, Zhiheng 800
Queiroz, Isadora M. 1029
Queiroz, Rodrigo S. 1556
Radfar, Behrad 189
Radhakrishnan, Hariharsudan S. 260, 363, 929, 1647, 1709
Radzevicius, Aurimas 101
Rahdan, Parisa 2183
Rajan, S. Prithivi 1596
Ramesh, Santhosh 260, 363, 494
Ramos-Fuentes, Isaac A. 1496
Ramspeck, Klaus 327

Randle-Boggis, Richard J. 1669
Ransome, Steve 1264
Ranta, Samuli 1361, 1969, 2001, 2102
Ratnasingham, S. R. 408
Rau, Björn 1348
Raval, Mehul 76, 149, 768, 1853
Rebohle, Lars 163
Rebollo, Míguel Á. G. 758
Reekmans, Bart 929, 1647
Rehman, Abdul 250
Reichel, Christian 920, 2095
Reichel, Rene 1412
Reichle, Julian 76, 2152
Reijners, Frits 299
Rein, Stefan 57, 63, 214
Reinders, Angèle H. M. E. 299, 1392
Reise, Christian 1072
Reisecker, Volker 570, 627
Reiser, Elisabeth 966
Remec, Marko 399
Remund, Jan 1435
Ren, Jinlei 25
Rende, Fedele 1782
Rennhofer, Marcus 743, 1305, 1309, 1961
Rentsch, Jochen 2152
Reyal, Jean-Pierre 1396
Reyes, Valentina A. 610, 618
Rezaei-Hartmann, Nasim 466
Riaño, Sandra 1335
Richter, Armin 40
Riechelmann, Stefan 765, 995, 1046
Riehle, Tim 924
Rigaud, Eric 2057
Riise, Heine N. 1480, 2078
Rillo, Sergio D. A. 441
Rinio, Markus 751
Ripke, M. 35
Rist, Tobias 812
Rivas, Jose 746, 1301, 1365
Rivera, Gerard 289
Rivera, Mariella 1072
Rizzi, Stefano 583
Robledo, Jesus 1596
Röder, Julian 1412
Rodríguez, Osbel A. 2205
Rodríguez-Conde, Sofía 746, 1279, 1301, 1365
Rodríguez-Lucas, Delia 1654
Roessler, Florian 135
Roig, Irma 1596
Romer, Pascal 665, 812, 1620
Rosen, Isaac 101
Rosenberg, Eva 2078

Roshchina, Nina	295	
Rosina, Konstantin	1451	
Roß, Marcel	466	
Rossetti, Andrea	1108	
Rößler, Torsten	924	
Rougieux, Fiacre	110	
Röver, Ingo	101	
Røyset, Arne	1381	
Rozanov, Konstantin	2040	
Rudolph, Dominik	514	
Ruiz, Alfonso L.	1610	
Ruiz, Pau M.	1141	
Ruiz, Sonia	285	
Rummelhoff, Stian	1995	
Russo, Roberto	302, 2072	
Rüther, Ricardo	427, 1029, 2009	
Saegebarth, Kai	1085	
Sahli, Florent	349	
Saidi-Chalopin, Elika	1733	
Saint-Cast, Pierre	176	
Sakakibara, Reyu	182	
Sakib, Syed N.	481	
Sakuma, Jun	277	
Salari, Majid	751	
Salis, Fabio	2084	
Salperwyck, Christophe	1369	
Sals, Sem	892	
Salvador, Michael	878	
Samara, Ayman	645	
San Andrés, E.	163	
Sanchez, Antonio	562	
Sanchez, Hugo	660, 1005	
Sánchez-Calvo, Raúl	1654	
Sánchez-Friera, Paula	1909	
Santamaria, Rodrigo D. P.	636, 1241, 1317	
Santbergen, Rudi	1391, 1417, 1957	
Santos, J. V. Oliveira	519	
Santos, Jose D.	1335	
Santos, Leticia D. O.	1354	
Sapkota, Subarna	1412	
Saucedo, Edgardo	285	
Saugues, François	1733	
Savin, Hele	189	
Savisalo, Tuukka	101	
Saviuc, Iolanda	2108	
Saw, Min H.	995	
Scerri, Kenneth	1305, 1309	
Schak, Matthias	607	
Schebek, Liselotte	1885, 2192	
Scheer, Roland	1576	
Scheler, Florian	399	
Schenck, Catherina	2062	

Schenk, Paul	755	
Scherret, Jaqueline	1782	
Schifferegger, Raffael	694	
Schill, Christian	1118, 1369, 1544	
Schimanke, Sabrina	35	
Schlatmann, Rutger	393, 399, 1348	
Schmid, Alexandra	1046	
Schmidt, Jan	35, 193	
Schmiga, Christian	155	
Schmitt, Emmanuel	1434	
Schmitz, Jurriaan	214	
Schnaus, Dominik	1466	
Schneider, Astrid	1782	
Schneider, Jale	135	
Schneider, Simon	1957	
Schneiders, Thorsten	2139	
Schön, Jonas	363	
Schönau, Elisabeth	773	
Schönau, Maximilian	773, 1314	
Schram, Wouter L.	1917	
Schranz, Christian	1782	
Schube, J.	19, 226	
Schubert, Maik	1412	
Schubert, Martin C.	363	
Schubnel, Baptiste	1548	
Schuesler, Daniel	687	
Schüler, Andreas	302	
Schüler, Marc A.	1620	
Schulte-Huxel, Henning	13, 1433	
Schultz, Christof	393	
Schulz, Philip	397	
Schulze, Achim	773, 1314, 1509	
Schulz-Ruhtenberg, Malte	89	
Schüpbach, Eva	1409, 1679	
Schüsler, Daniel	680	
Schutt, Thomas	1412	
Schwabl, Daniel	2077	
Schwarz, Andreas	2046	
Schweigstill, Tadeo	89	
Sciullo, Alessandro	2205	
Scognamiglio, Alessandra	2084	
Scroppo, Sofia	2139	
Seentakath, Afra	585, 1055	
Segura, Oriol	285	
Seick, Cinja	1596	
Seigneur, Hubert	826	
Sen, Nesrin T.	35	
Senaud, Laurie-Lou	349	
Seppälä, Simeon	2109	
Sergio, Lucas A. Z.	427	
Serra, Filipe	1675	
Serra, João M.	1705	

Sgouridis, Sgouris	772
Sha, Nithin	585, 1055
Shakiba, Ali	761, 1104
Sharma, Ashish K.	1559, 1724
Sharma, Bhumika	414
Sharma, Deepak	169
Sharma, Rama	761, 980
Sharma, Ruchi K.	169
Shimokata, Eiko	523
Shimpo, Shuntaro	656
Shin, Donghyeop	338
Shiradkar, Narendra S.	878
Shiradkar, Narendra	619
Shirazi, Elham	1917
Shochet, Ofer	101
Short, Michael	695
Silva, François	334
Silva, José	1658, 1675, 1687, 1696, 1989
Silva, Lucas T.	1556
Simeunovic, Jelena	1548
Singh, Ojas	886
Singh, Trilok	429, 445
Sinicropi, Adalgisa	486
Sinopoli, Alessandro	139
Siquera, Tales	552
Sivaramakrishnan, Hariharsudan	101
Škorjanc, Viktor	466
Slooff-Hoek, Lenneke	1616
Smertenko, Petro	295
Smith, Ligia	1841
Smith, Ryan	826
Sobajima, Yasushi	523, 549
Soeiro, André	1675
Sohani, Ali	1020
Solofra, Nate	1249
Solórzano, Jorge	1160
Søndenå, Rune	132, 172
Sondoqah, Mousa	1095, 1225, 1812
Sorbet, Patxi	1400
Sourd, Francis	1496
Souren, Floor	193, 408
Souza, Francisco A. A.	1354
Sovetkin, Evgenii	1347, 1444, 1543
Spagnolo, Sofia	2056
Spataru, Sergiu V.	636, 1241, 1287, 1317
Späth, Martin	315
Spätlich, Sarah	106
Speer, Volker	1412
Spera, Fabian	2046
Sraisth	768, 911
Srivastava, Sanjay K.	169
Stagno, Luciano M.	1040
Stalmans, Lieven	1647
Stannowski, Bernd	892
Starke, A.	53
Stefanelli, Maurizio	411, 486
Stegemann, Bert	393
Steinlechner, Sebastian	2033
Stellbogen, Dirk	1313
Stensborg, Jan F.	632
Stenzig, Laura	765, 1046
Stepec, Murielle	1220
Stieldorf, Karin	1782
Stierstorfer, J.	101, 2205
Stoicescu, Liviu	1241
Stokkan, Gaute	1669
Stölzel, Marko	1412
Sträter, Hendrik	1046
Straub, Nils	1456
Strazzullo, Paolo	302, 2072
Strömberg, Rich	1849
Stuckelberger, Josua	25
Stueve, William	1245
Suarez, Sergio	746, 1301, 1365
Subbarao, P. M. V.	1357
Subeh, Mosab	1050, 1156
Subramaniam, Sownder	260
Suemitsu, Issei	1953, 1982
Sugimoto, Hiroki	559
Suhonen, Riikka	430
Sulca, Kabir P.	758, 974, 983
Suri, Marcel	1451
Sutariya, Mahesh	1528
Sutkuviene, Simona	295
Sutterlueti, Juergen	1264
Suvarn, Shashank	1055
Svedjeholm, Maria	1799
Syri, Sanna	2087, 2102, 2109, 2119
Szabó, Sandor	1894
Taheri, Nabi	1769
Takamoto, Tatsuya	254
Takashima, Takumi	1626
Talvi, Micke	1977
Tamizhmani, Govindasamy	826
Tanahashi, Katsuto	277
Tanahashi, Tadanori	846
Tang, Peter T.	342
Taylor, Nigel	1894
Taylor, Stephen	257
Ternes, Simon	411
Terrados, Cristian	758, 974, 983
Tervo, Seela	2102, 2119
Tessmann, Christopher	164
Tettenborn, Tuuli	534

Teymouri, Arastoo 834
Thalheimer, Martin 1596
Thawanyavitchajit, Chisanupong.................... 1770
Thebault, Martin 1418
Theelen, Mirjam 943, 1835, 1873
Theristis, Marios............................ 1089
Thomassen, Bent............................. 132
Thony, Philippe.............................. 1434
Thorsteinsson, Sune.................. 632, 1199, 1385
Tierney, Paul 110
Tilly, Eric.................................. 694
Timò, Gianluca 583
Timofte, Tudor.............................. 610
Tina, Giuseppe M. 1341
Tomšic, Špela................................ 329, 399
Topic, Marko.............329, 349, 398, 399, 738, 1146, 2138
Tormena, Noah 417
Torrens, Arnau 285
Torres, Ignacio 163
Torres, Pedro 1752, 2021
Tous, Loic 1647
Trattnig, Roman 780
Traunmüller, Wolfgang 2015
Treberspurg, Christoph 1782
Treberspurg, Martin 1782
Trivellin, Nicola............................. 417
Truong, Thein N............................. 25
Tsai, Min-An................................ 328
Tsanakas, Ioannis................. 1160, 1552, 2200
Tsanakas, John A. 1220
Tsemekidi-Tzeiranaki, Sofia 1183
Tsoulka, Polyxeni 397
Tsunoda, Jun 1253, 1953, 1982
Tu, Huynh T. C. 656
Tucci, Mauro............................... 1769
Tulinski, Lona.............................. 1747
Tune, Daniel 101, 330, 545
Tuomiranta, Arttu............................ 1462
Turala, Artur 250
Turan, Rasit................................ 19, 197
Turek, Marko 53, 422, 607, 651
Turri, Evelyn............................... 1225
Tzinoglou, George I........................... 2124
Übermasser, Stefan 2169
Ugranli, Faruk.............................. 1806
Újvári, Gusztáv 323, 743
Ul-Abdin, Zain.............................. 1417
Ulbikas, Juras............................... 101
Ulbrich, Carolin 399, 1348
Umeda, Kazuhiko 1194
Unger, Eva 393
Urata, Tomoyuki 277

Urban, Harald............................... 1782
Urbina, Antonio............................. 2052
Uygun, Berkay 19
Vähänissi, Ville 189
Vaidya, Haresh 1528
Valckenborg, Roland 1178
Valdivia, Patricio............................ 618
Valdivia-Lefort, Patricio............562, 584, 2092
Valencia, Felipe 1501
Valle, Benoît 1496
Valoti, Flavio 995
Van Aken, Bas.........................834, 1729
Van De Water, Oscar 1616
Van Den Storme, Guy 929
Van Den Storme, Manuel 929
Van Der Heide, Arvid............ 712, 854, 995, 1709, 1849, 2200
Van Der Ploeg, Bas 1943
Van Der Vleuten, Maarten..................... 943
Van Dyck, Rik 101, 929
Van Dyk, Ernest E. 2062
Van Gijlswijk, René 1616
Van Overstraeten, Julien...................... 1715
Van Rossum, Aron 1391
Van Sark, Wilfried....................1943, 1976
Vandamme, Nicolas.......................... 2068
Vanel, Jean-Charles 334
Vanhanen, Tuomas 101
Varjopuro, Julianna 530
Varney, Valérie 2139
Vasquez, Pia 50
Vavilkin, Tatjana 1647
Veettil, Binesh P. 481
Veihelmann, Tobias 1491
Veneri, Paola D. 428
Ventosinos, Federico 50
Verdeil, Olivier 1733
Verkou, Maarten 1957
Vermang, Bart 260
Veronese, Elisa 1921
Vesce, Luigi............................411, 486
Vespermann, Merle 1476
Viani, Lucas 1279
Vicente, Diogo 1705
Victoria, Marta........................2138, 2183
Vidal, Nerea 1687
Vieira, Bruno J. 1763
Vilela, Jonathan 1301
Villa, Simona.........................1178, 1835
Villoslada, Daniel......................746, 1365
Vincent, Robin 1740
Viriyaroj, Bergpob........................... 1375
Virtuani, Alessandro......................862, 1862

Vitale, Simone	1369
Vito, Domenico	2129
Vogt, Aaron	89
Voirol, Alexandre	1234
Völler, Steve	1669
Voronko, Yuliya	694, 966
Voroshazi, Eszter	953
Voz, Cristobal	285
Vrielinck, Henk	260
Vuillon, Laurent	1552
Vulic, Natasa	1390
Vyalih, Irina	632, 1385
Wagenmann, Dirk	158
Wagner, Enno	327
Wagner-Mohnsen, Hannes	205
Waldau, Arnulf J.	1906
Wambach, Karsten	2045, 2200
Wang, Deliang	346
Wang, Feng	446
Wang, Guangwei	346
Wang, Jiali	25
Wang, Li	834
Wang, Lu	86, 1859
Wang, Shuo	1249, 1361, 2001
Wang, Xiawa	335
Wang, Yichun	7
Waschl, Alfred	1782
Watrin, Lise	334
Wattenberg, Bianca	13
Weber, Anne-Kathrin	1435
Weber, Juergen W.	308
Weber, Thomas	743
Wehnert, Danny	1132
Weiermair, A.	570
Weinert, Nicolas	1
Weinrich, Frank	765, 995
Weiß, Karl-Anders	809, 1885, 2192
Wellens, Christine	911
Wendt, Michael	680, 687
Wernke, Luka	1348
Wessel, Patrick	680
Westerberg, Amelia O.	1903, 1906
Widler, Adrian	1079
Wienands, Karl	330, 545
Wienberg, Robin	214
Wilbert, Stefan	1466
Willers, Guido	651
Wilson, Gregory J.	404, 437
Winkelmann, Jan	1966
Winter, Michael	35, 193
Winter, Stefan	765, 995, 1046
Wirtz, Wiebke	1433
Wiss, Olivier	1434
Wittmer, Bruno	1740
Woehler, Wilkin	63
Woernhoer, Alexandra	63
Wöhrle, Nico	57
Wojciechowski, Konrad	422
Wolf, Andreas	164
Wong, George	2099
Woodhouse, Michael	2152
Wörnhör, Alexandra	57
Wright, Brendan	761, 980, 1104
Wright, James	110
Wu, Bang-Hao	148
Wu, Haodong	800
Wu, Li-Guo	144
Wu, Yu	113
Xiao, Chuanxiao	800
Xu, Frank	1005
Xu, Wenhao	490, 739
Yagci, Selim	505
Yamaguchi, Akira	1194
Yamaguchi, Masafumi	254
Yde, Leif	632
Ye, Jichun	800
Yeh, Fan-Hsuan	328
Yi, Kai	346
Yildirim, Nurhayat	122
Yordanov, Georgi H.	712, 1647
You, Chang C.	1381
Young, Jørgen	1430
Yu, Cheng-Yeh	144
Yu, Mingzhe	110
Yun, Changyeol	2160
Yurrita, Naiara	511
Zaimi, Mhammed	776
Zamarro, Fernando L.	441
Zanoni, Enrico	417
Zardetto, Valerio	315
Zaror, Yasmin	101
Zarzalejo, Luis F.	1466
Zaversky, Fritz	1400
Zech, Tobias	1476
Zehndorfer, Jakob	780
Zekri, Atef	1050
Zeman, Miro	1957
Zenteno, Franciso J. P.	163
Zerafa, Steve	1305, 1309, 2134
Zeyen, Elisabeth	2183
Zhang, Junchuan	800
Zhang, Shipei	335
Zhang, Yating	490, 739
Zhang, Yi	800

Zheng, Zhiwen 308
Zhou, Guohua 86
Zhu, Junjie 172, 1430
Zhu, Xitong 299
Zhu, Yan 308
Zilles, Roberto 1752, 1763, 2021
Zimmermann, Andreas 342, 534
Zubillaga, Oihana 511
Zugasti, Eugenia 428, 1305, 1309, 2052
Züll, Laura 2139
Zult, Michiel 1616
Zwahlen, Theo 1435